LIVING IN A CHEMICAL WORLD
Framing the Future in Light of the Past

ANNALS OF THE NEW YORK ACADEMY OF SCIENCES
Volume 1076

LIVING IN A CHEMICAL WORLD
Framing the Future in Light of the Past

*Edited by Myron A. Mehlman, Morando Soffritti,
Philip Landrigan, Eula Bingham, and Fiorella Belpoggi*

*Published by Blackwell Publishing on behalf of the New York Academy of Sciences
Boston, Massachusetts
2006*

Library of Congress Cataloging-in-Publication Data

Living in a chemical world : framing the future in light of the past / edited by Philip Landrigan ... [et al.].
    p. cm.
    ISBN-13: 978-1-57331-653-8 (alk. paper)
    ISBN-10: 1-57331-653-9 (alk. paper)
    1. Cancer–Environmental aspects–Congresses. I. Landrigan, Philip J.

RC268.25.L58 2006
616.99'4071–dc22

2006019682

The *Annals of the New York Academy of Sciences* (ISSN: 0077-8923 [print]; ISSN: 1749-6632 [online]) is published 28 times a year on behalf of the New York Academy of Sciences by Blackwell Publishing, with offices located at 350 Main Street, Malden, Massachusetts 02148 USA, PO Box 1354, Garsington Road, Oxford OX4 2DQ UK, and PO Box 378 Carlton South, 3053 Victoria Australia.

Information for subscribers: Subscription prices for 2006 are: Premium Institutional: $3850.00 (US) and £2139.00 (Europe and Rest of World).
Customers in the UK should add VAT at 5%. Customers in the EU should also add VAT at 5% or provide a VAT registration number or evidence of entitlement to exemption. Customers in Canada should add 7% GST or provide evidence of entitlement to exemption. The Premium Institutional price also includes online access to full-text articles from 1997 to present, where available. For other pricing options or more information about online access to Blackwell Publishing journals, including access information and terms and conditions, please visit www.blackwellpublishing.com/nyas.

Membership information: Members may order copies of the *Annals* volumes directly from the Academy by visiting www.nyas.org/annals, emailing membership@nyas.org, faxing 212-888-2894, or calling 800-843-6927 (US only), or +1 212 838 0230, ext. 345 (International). For more information on becoming a member of the New York Academy of Sciences, please visit www.nyas.org/membership.

Journal Customer Services: For ordering information, claims, and any inquiry concerning your institutional subscription, please contact your nearest office:
UK: Email: customerservices@blackwellpublishing.com; Tel: +44 (0) 1865 778315; Fax +44 (0) 1865 471775
US: Email: customerservices@blackwellpublishing.com; Tel: +1 781 388 8599 or 1 800 835 6770 (Toll free in the USA); Fax: +1 781 388 8232
Asia: Email: customerservices@blackwellpublishing.com; Tel: +65 6511 8000; Fax: +61 3 8359 1120
Members: Claims and inquiries on member orders should be directed to the Academy at email: membership@nyas.org or Tel: +1 212 838 0230 (International) or 800-843-6927 (US only).

Printed in the USA.
Printed on acid-free paper.

Part II. Identifying and Preventing Hazards in the Environment and at Work

Benzene and Other Gasoline Constituents/Additives

The Health Watch Case–Control Study of Leukemia and Benzene: The Story So Far. *By* DEBORAH C. GLASS, CHRISTOPHER N. GRAY, DAMIEN J. JOLLEY, CARL GIBBONS, AND MALCOLM R. SIM 80

Benzene Exposure and Multiple Myeloma: A Detailed Meta-analysis of Benzene Cohort Studies. *By* PETER F. INFANTE 90

Dangerous and Cancer-Causing Properties of Products and Chemicals in the Oil Refining and Petrochemical Industries. Part XXX: Causal Relationship between Chronic Myelogenous Leukemia and Benzene-Containing Solvents. *By* MYRON A. MEHLMAN 110

Causal Relationship between Non-Hodgkin's Lymphoma and Exposure to Benzene and Benzene-Containing Solvents. *By* MYRON A. MEHLMAN ... 120

Genetic Polymorphism of Toxicant-Metabolizing Enzymes and Prognosis of Chinese Workers with Chronic Benzene Poisoning. *By* J-X. WAN, Z-B. ZHANG, J-R. GUAN, D-Z. CAO, R. YE, X-P. JIN, AND Z-L. XIA 129

Use and Misuse of Mechanistic Data in Risk Assessment

Gene–Environment Interactions in Cancer: Do They Exist? *By* KARI HEMMINKI, ASTA FÖRSTI, AND JUSTO LORENZO BERMEJO 137

Manufactured Uncertainty: Protecting Public Health in the Age of Contested Science and Product Defense. *By* DAVID MICHAELS 149

Misuse of Genetic Data in Environmental Epidemiology. *By* PAOLO VINEIS 163

Collision of Basic and Applied Approaches to Risk Assessment of Thyroid Toxicants. *By* R. THOMAS ZOELLER 168

Chlorinated Solvents

Science and Policy in Risk Assessments of Chlorinated Ethenes. *By* CHRISTINA RUDÉN 191

Potential Health Effects of Occupational Chlorinated Solvent Exposure. *By* AVIMA M. RUDER 207

Endocrine Disruptors

Endocrine Disruptors: Challenges for Environmental Research in the 21st Century. *By* MARY S. WOLFF 228

Risk Assessment of Endocrine Disrupters: The Role of Toxicological Studies. *By* ALBERTO MANTOVANI 239

Air Pollution and Nonfibrous Particles

Diesel Exhaust and Coal Mine Dust: Lung Cancer Risk in Occupational Settings. *By* BARBARA HOFFMANN AND KARL-HEINZ JÖCKEL 253

Lung Tumor Risk Estimates from Rat Studies with Not Specifically Toxic Granular Dusts. *By* MARKUS ROLLER AND FRIEDRICH POTT 266

ANNALS OF THE NEW YORK ACADEMY OF SCIENCES

Volume 1076
September 2006

LIVING IN A CHEMICAL WORLD
Framing the Future in Light of the Past

Editors
MYRON A. MEHLMAN, MORANDO SOFFRITTI,
PHILIP LANDRIGAN, EULA BINGHAM, AND FIORELLA BELPOGGI

This volume is the result of a conference entitled **Framing the Future in Light of the Past: Living in a Chemical World**, held by the Collegium Ramazzini in cooperation with the Comuni di Carpi, the European Ramazzini Foundation, and the National Ramazzini Institute, and held on September 18-21, 2005, in Bologna, Italy.

CONTENTS

Dedication. *By* MORANDO SOFFRITTI xv

Introduction. *By* MYRON A. MEHLMAN, MORANDO SOFFRITTI, PHILIP LANDRIGAN, EULA BINGHAM, AND FIORELLA BELPOGGI xvii

Part I. Keynote Addresses

Identification of Carcinogenic Agents and Primary Prevention of Cancer. *By* LORENZO TOMATIS 1

Children's Environmental Health Research—Highlights from the Columbia Center for Children's Environmental Health. *By* FREDERICA PERERA, SHEILA VISWANATHAN, ROBIN WHYATT, DELIANG TANG, RACHEL L. MILLER, AND VIRGINIA RAUH 15

Interpreting Cancer Trends. *By* ELIZABETH M. WARD, MICHAEL J. THUN, LINDSAY M. HANNAN, AND AHMEDIN JEMAL 29

The Anatomy of the Exposures That Occurred around the World Trade Center Site: 9/11 and Beyond. *By* PAUL J. LIOY AND PANOS GEORGOPOULOS 54

Mailing: The *Annals of the New York Academy of Sciences* are mailed Standard Rate.
Postmaster: Send all address changes to *Annals of the New York Academy of Sciences*, Blackwell Publishing, Inc., Journals Subscription Department, 350 Main Street, Malden, MA 01248-5020. Mailing to rest of world by DHL Smart and Global Mail.

Copyright and Photocopying
© 2006 The New York Academy of Sciences. All rights reserved. No part of this publication may be reproduced, stored, or transmitted in any form or by any means without the prior permission in writing from the copyright holder. Authorization to photocopy items for internal and personal use is granted by the copyright holder for libraries and other users registered with their local Reproduction Rights Organization (RRO), e.g. Copyright Clearance Center (CCC), 222 Rosewood Drive, Danvers, MA 01923, USA (www.copyright.com), provided the appropriate fee is paid directly to the RRO. This consent does not extend to other kinds of copying such as copying for general distribution, for advertising or promotional purposes, for creating new collective works, or for resale. Special requests should be addressed to Blackwell Publishing at journalsrights@oxon.blackwellpublishing.com.

Disclaimer: The Publisher, the New York Academy of Sciences, and the Editors cannot be held responsible for errors or any consequences arising from the use of information contained in this publication; the views and opinions expressed do not necessarily reflect those of the Publisher, the New York Academy of Sciences, or the Editors.

Annals are available to subscribers online at the New York Academy of Sciences and also at Blackwell Synergy. Visit www.annalsnyas.org or www.blackwell-synergy.com to search the articles and register for table of contents e-mail alerts. Access to full text and PDF downloads of *Annals* articles are available to nonmembers and subscribers on a pay-per-view basis at www.annalsnyas.org.

The paper used in this publication meets the minimum requirements of the National Standard for Information Sciences Permanence of Paper for Printed Library Materials, ANSI Z39.48-1984.

ISSN: 0077-8923 (print); 1749-6632 (online)
ISBN-10: 1-57331-653-9 (paper); ISBN-13: 978-1-57331-653-8 (paper)

A catalogue record for this title is available from the British Library.

Digitization of the *Annals of the New York Academy of Sciences*

An agreement has recently been reached between Blackwell Publishing and the New York Academy of Sciences to digitize the entire run of the *Annals of the New York Academy of Sciences* back to volume one, issue one.

The back files, which have been defined as all of those issues published before 1997, will be sold to libraries as part of Blackwell Publishing's Legacy Sales Program and hosted on the Blackwell Synergy website.

Copyright of all material will remain with the rights holder. Contributors: Please contact Blackwell Publishing if you do not wish an article or picture from the *Annals of the New York Academy of Sciences* to be included in this digitization project.

Asbestos and Man-Made Mineral Fibers

Measurements of Asbestos Burden in Tissues. *By* RONALD F. DODSON AND
  MARK A.L. ATKINSON 281

Asbestos Ban in India: Challenges Ahead. *By* TUSHAR KANT JOSHI,
  UTTPAL B. BHUVA, AND PRIYANKA KATOCH 292

Radiation

Ionizing Radiation and Cardiovascular Disease. *By* DAVID G. HOEL 309

Power-Frequency Electric and Magnetic Fields in the Light of Draper *et al.*
  2005. *By* JOHN SWANSON, TIM VINCENT, MARY KROLL, AND GERALD
  DRAPER 318

Nanotechnology and Emerging Industries

Getting It Right the First Time: Developing Nanotechnology while Protecting
  Workers, Public Health, and the Environment. *By* JOHN M. BALBUS,
  KAREN FLORINI, RICHARD A. DENISON, AND SCOTT A. WALSH 331

Part III. Reducing and Managing Risk in High-Hazard Sectors
Agriculture

Pesticides and Adult Respiratory Outcomes in the Agricultural Health Study. *By*
  JANE A. HOPPIN, DAVID M. UMBACH, STEPHANIE J. LONDON,
  CHARLES F. LYNCH, MICHAEL C.R. ALAVANJA, AND DALE P. SANDLER 343

The Assessment of Occupational Exposure to Diazinon in Nicaraguan
  Plantation Workers Using Saliva Biomonitoring. *By* CHENSHENG LU,
  TERESA RODRIIGUEZ, AURA FUNEZ, RENE S. IRISH,
  AND RICHARD A. FENSKE 355

Cancer and Pesticides: An Overview and Some Results of the Italian
  Multicenter Case–Control Study on Hematolymphopoietic
  Malignancies. *By* LUCIA MILIGI, ADELE SENIORI COSTANTINI,
  ANGELA VERALDI, ALESSANDRA BENVENUTI, WILL, AND PAOLO VINEIS ... 366

Geographic Model and Biomarker-Derived Measures of Pesticide Exposure and
  Parkinson's Disease. *By* BEATE RITZ AND SADIE COSTELLO 378

Construction

The Construction Industry. *By* KNUT RINGEN AND ANDERS ENGLUND 388

Frequency and Quality of Radiation Monitoring of Construction Workers at Two
  Gaseous Diffusion Plants. *By* EULA BINGHAM, KNUT RINGEN, JOHN DEMENT,
  WILFRID CAMERON, WILLIAM MCGOWAN, LAURA WELCH,
  AND PATRICIA QUINN 394

Evaluation of Exposure to PAHs in Asphalt Workers by Environmental and
  Biological Monitoring. *By* LAURA CAMPO, MARINA BURATTI,
  SILVIA FUSTINONI, PIERO E. CIRLA, IRENE MARTINOTTI, OMAR LONGHI,
  DOMENICO CAVALLO, AND VITO FOÀ 405

Carcinogens in the Construction Industry. *By* BENGT JÄRVHOLM 421

Epoxy Resins in the Construction Industry. *By* TON SPEE,
    COR VAN DUIVENBOODEN, AND JEROEN TERWOERT   429

Hazardous Waste

Health Effects of Hazardous Waste. *By* STEVE M. DEARWENT, M. MOIZ MUMTAZ,
    GAIL GODFREY, THOMAS SINKS, AND HENRY FALK   439

Cancer Mortality in an Area of Campania (Italy) Characterized by Multiple
    Toxic Dumping Sites. *By* PIETRO COMBA, FABRIZIO BIANCHI, LUCIA FAZZO,
    LUCIA MARTINA, MASSIMO MENEGOZZO, FABRIZIO MINICHILLI,
    FRANCESCO MITIS, LOREDANA MUSMECI, RENATO PIZZUTI,
    MICHELE SANTORO, STEFANIA TRINCA, MARCO MARTUZZI,
    AND "HEALTH IMPACT OF WASTE MANAGEMENT CAMPANIA" WORKING
    GROUP   449

Hazardous Waste: Recognition of the Problem and Response. *By* HENRIK
    HARJULA   462

New Directions in Managing Hazardous Wastes from an Industry
    Perspective. *By* LYNN D. JOHNSON   478

Developments in Management and Technology of Waste Reduction and
    Disposal. *By* PHILIP RUSHBROOK   486

Evaluating Energy Technology Risks

Systematic Approach to Evaluating Trade-Offs among Fuel Options: The
    Lessons of MTBE. *By* J. MICHAEL DAVIS AND VALERIE M. THOMAS   498

Environmental Impacts and Costs of Energy. *By* ARI RABL
    AND JOSEPH V. SPADARO   516

Weapon Destruction

Worldwide Governmental Efforts to Locate and Destroy Chemical Weapons
    and Weapons Materials: Minimizing Risk in Transport and Destruction. *By*
    RALF TRAPP   527

Options for the Destruction of Chemical Weapons and Management of the
    Associated Risks. *By* RON G. MANLEY   540

Health and Environmental Threats Associated with the Destruction of Chemical
    Weapons. *By* JIŘÍ MATOUŠEK   549

Part IV. Tools and Strategies to Reduce Risk: Applying Science to Achieve Prevention
Long-Term Carcinogenesis Bioassays and Other Testing Strategies

Results of Long-Term Carcinogenicity Bioassay on Sprague-Dawley Rats
    Exposed to Aspartame Administered in Feed. *By* FIORELLA BELPOGGI,
    MORANDO SOFFRITTI, MICHELA PADOVANI, DAVIDE DEGLI ESPOSTI,
    MICHELINA LAURIOLA, AND FRANCO MINARDI   559

Results of a Long-Term Carcinogenicity Bioassay on Sprague-Dawley Rats Exposed to Sodium Arsenite Administered in Drinking Water. *By* MORANDO SOFFRITTI, FIORELLA BELPOGGI, DAVIDE DEGLI ESPOSTI, AND LUCA LAMBERTINI 578

Use of Carcinogenicity Bioassays in the *IARC Monographs*. *By* VINCENT JAMES COGLIANO 592

Part V. Provision of Health Care
Health Care

Chemical Hazards in Health Care: High Hazard, High Risk, but Low Protection. *By* MELISSA A. MCDIARMID 601

National and International Response to Occupational Hazards in the Healthcare Sector. *By* BRIGITTE FRONEBERG 607

Hazardous Anticancer Drugs in Health Care: Environmental Exposure Assessment. *By* THOMAS H. CONNOR 615

Beyond Managing Healthcare Risks: The Health-Promoting Hospital Initiative in Mexico. *By* CARLOS SANTOS-BURGOA 624

Handling Anticancer Drugs: From Hazard Identification to Risk Management? *By* MARJA SORSA, MERVI HÄMEILÄ, AND EIJA JÄRVILUOMA 628

Chemical Safety and Health Conditions among Hungarian Hospital Nurses. *By* ANNA TOMPA, MÁTYÁS JAKAB, ANNA BIRÓ, BALÁZS MAGYAR, ZOLTÁN FODOR, TIBOR KLUPP, AND JENÖ MAJOR 635

Residual Hazard Assessment Related to Handling of Antineoplastic Drugs: Safety System Evolution and Quality Assurance of Analytical Measurement. *By* ROBERTA TURCI AND CLAUDIO MINOIA 649

Child Health

Framing the Future in Light of the Past: Living in a Chemical World. *By* PHILIP J. LANDRIGAN, JENNY PRONCZUCK DE GARBINO, AND BROOKE NEWMAN 657

Children's Environment and Health in Latin America: The Ecuadorian Case. *By* RAUL HARARI AND HOMERO HARARI 660

Environmental Impacts on Children's Health in Southeast Asia: Genotoxic Compounds in Urban Air. *By* MATHUROS RUCHIRAWAT, PANIDA NAVASUMRIT, DAAM SETTACHAN, AND HERMAN AUTRUP 678

New Developments in Children's Environmental Health in Europe. *By* GIORGIO TAMBURLINI 691

Genetic Environment and Effect Modulators

Toxicogenomics—A New Systems Toxicology Approach to Understanding of Gene–Environment Interactions. *By* KENNETH OLDEN 703

Toxicoproteomics in Liver Injury and Inflammation. *By* B. ALEX MERRICK 707

Gene Expression Alterations in Immune System Pathways following Exposure to Immunosuppressive Chemicals. *By* RACHEL M. PATTERSON AND DORI R. GERMOLEC 718

Transcriptional Profiling and Functional Genomics Reveal a Role for AHR Transcription Factor in Nephrogenesis. *By* KENNETH S. RAMOS 728

Part VI. Open Communications

Results of Long-Term Carcinogenicity Bioassays on Coca-Cola Administered to Sprague-Dawley Rats. *By* FIORELLA BELPOGGI, MORANDO SOFFRITTI, EVA TIBALDI, LAURA FALCIONI, LUCIANO BUA, AND FRANCESCA TRABUCCO 736

Occupational Kidney Cancer: Exposure to Industrial Solvents. *By* NACHMAN BRAUTBAR, MICHAEL P. WU, ELION GABEL, AND LEE REGEV ... 753

Occupation and Breast Cancer: A Canadian Case–Control Study. *By* JAMES T. BROPHY, MARGARET M. KEITH, KEVIN M. GOREY, ISAAC LUGINAAH, ETHAN LAUKKANEN, DEBORAH HELLYER, ABRAHAM REINHARTZ, ANDREW WATTERSON, HAKAM ABU-ZAHRA, ELEANOR MATICKA-TYNDALE, KENNETH SCHNEIDER, MATTHIAS BECK, AND MICHAEL GILBERTSON 765

Adverse Health Effects of Fluoro-edenitic Fibers: Epidemiological Evidence and Public Health Priorities. *By* CATERINA BRUNO, PIETRO COMBA, AND AMERIGO ZONA 778

Glycol Ethers: A Ubiquitous Family of Toxic Chemicals: A Plea for REACH Regulation. *By* ANDRÉ CICOLELLA 784

Controlling Exposure to Chemicals: A Simple Guide. *By* ALASTAIR HAY 790

Progress of Epidemiological and Molecular Epidemiological Studies on Benzene in China. *By* GUILAN LI AND SONGNIAN YIN 800

Carcinogen Exposure and Epigenetic Silencing in Bladder Cancer. *By* CARMEN J. MARSIT, MARGARET R. KARAGAS, ALAN SCHNED, AND KARL T. KELSEY 810

Causal Relationship from Exposure to Chemicals in Oil Refining and Chemical Industries and Malignant Melanoma. *By* MYRON A. MEHLMAN 822

A Regional Approach to Assess the Impact of Living in a Chemical World. *By* CHRISTOPHER T. DE ROSA, HERALINE E. HICKS, ANNETTE E. ASHIZAWA, HANA R. POHL, AND M. MOIZ MUMTAZ 829

Low-Dose Risk, Hormesis, Analogical and Logical Thinking. *By* GIOVANNI A. ZAPPONI AND IDA MARCELLO 839

Part VII. Roundtable on Social and Economic Impact of Occupational and Environmental Diseases
Tools and Strategies to Reduce Risk: Applying Science to Achieve Prevention

Occupational Injury and Illness Meet the Labor Market: Lessons from Labor Economics about Lost Earnings. *By* LESLIE I. BODEN 858

The Economic Costs of Health Service Treatments for Asbestos-Related Mesothelioma Deaths. *By* ANDREW WATTERSON, TOMMY GORMAN, CARI MALCOLM, MAVIS ROBINSON, AND MATTHIAS BECK 871

Valuing the Adult Health Effects of Air Pollution in Chinese Cities. *By* ROBERT W. MEAD AND VICTOR BRAJER 882

Chinese Workers and Labor Conditions from State Industry to Globalized Factories: How to Stop the Race to the Bottom. *By* MARINA THORBORG .. 893

Applying Cost Analyses to Drive Policy That Protects Children: Mercury as a Case Study. *By* LEONARDO TRASANDE, CLYDE SCHECHTER, KARLA A. HAYNES, AND PHILIP J. LANDRIGAN 911

Abstracts of Poster Presentations 925

Index of Contributors 943

Financial assistance was received from:

Major Institutional Sponsors
- National Institute for Environmental Health Sciences, USA
- Centers for Disease Control, USA
- National Cancer Institute, USA
- European Environment Agency, Denmark
- Environmental Protection Agency, USA
- ARPA Emilia Romagna, Italy
- American Cancer Society, USA

Italian Contributors
- Assindustria Bologna
- Banca di Bologna
- Banca Popolare dellEmilia Romagna
- Centergross
- Comitato Regionale ANCPL
- Comune di Argelato
- Comune di Castenaso
- Comune di San Giorgio di Piano
- Conservatorio di Bologna
- Consorzio Cooperative Costruzioni
- Collegio Costruttori Edili ed Imprenditori Affini della Provincia di Bologna
- Coop Costruzioni
- CoInd
- Cooperativa Edificatrice Ansaloni
- Cooperativa Editrice Consumatori
- Cooperativa Trasporti Alimentari
- Cooperativa Trasporto Latte

- FactorCoop
- Finanziaria Bologna Metropolitana
- Fondazione Cassa di Risparmio in Bologna
- Fondazione per la Ricerca sul Cancro Fernanda e Gaudenzio Renzi
- Fondazione Cassa di Risparmio di Carpi
- Fondazione Isabella Seràgnoli
- Fruttagel
- Granarolo Felsinea
- Hera
- Hera Bologna
- Interporto Bologna
- Legacoop Bologna
- Manutencoop
- Protezioni Elaborazioni Industriali
- Provincia di Bologna
- Regione Emilia Romagna
- Unilog Group

This Conference Was Held Under the Auspices of:
- The President of the Italian Republic
- Ministero Istruzione Università e Ricerca
- Ministero della Salute
- Regione Emilia Romagna
- Alma Mater Studiorum Università di Bologna
- Università degli Studi di Modena e Reggio Emilia
- Università degli Studi di Padova
- Provincia di Bologna
- Comune di Bologna
- Lega Italiana per la Lotta Contro i Tumori, sede centrale

The New York Academy of Sciences believes it has a responsibility to provide an open forum for discussion of scientific questions. The positions taken by the participants in the reported conferences are their own and not necessarily those of the Academy. The Academy has no intent to influence legislation by providing such forums.

Dedication

On September 20, 2005, the Cancer Research Center of the European Foundation of Oncology and Environmental Sciences "B. Ramazzini," located in the Castle of Bentivoglio in the province of Bologna, Italy, was dedicated in the name of Professor Cesare Maltoni, its founder and director until his untimely death in 2001. The dedication ceremony was held in occasion of the international conference "Framing the Future in Light of the Past: Living in a Chemical World," organized by the Collegium Ramazzini, of which Professor Maltoni was a founder and for which the Castle of Bentivoglio also houses the General Secretariat headquarters.

Construction of the Castle of Bentivoglio was completed in circa 1480 by Giovanni II Bentivoglio, member of Bologna's ruling elite, to be used as the family's *domus jucunditatis,* a countryside retreat and location for marvelous celebrations. The restoration of the 10,000 square meter compound was commissioned in the late 1800s by then owner marquis Carlo Alberto Pizzardi and carried out by architect Alfonso Rubbiani. Upon his death in 1922, Pizzardi bequeathed his entire patrimony, including the Castle of Bentivoglio, to the hospitals of Bologna. During the second World War, the castle was severely damaged and left in a state abandon until the end of the 1960s.

In the early 1970s, Professor Cesare Maltoni was the young director of the "F. Addarii" Institute of Oncology of the University of Bologna. Thanks to the forward thinking vision of Senator Luigi Orlandi, then President of the Hospital Administration of the city, Maltoni obtained permission to use the immense halls, stables, and warehouses of the Castle of Bentivoglio to create what would over the coming decades, become one of the most important research centers in the world for experimental studies in industrial and environmental carcinogenesis. The collocation of high-tech laboratories, archives, offices, and library now contained in the spacious castle represent a unique union of Renaissance architecture and modern science.

Professor Cesare Maltoni was born in Faenza, Italy on November 17, 1930 and died in his home in Bologna on January 21, 2001 at the age of 70. A world-recognized leader in the research of dangerous industrial carcinogens in the general environment and in the workplace, Professor Maltoni was the first to demonstrate that vinyl chloride is a carcinogen that causes angiosarcomas of the liver. He was also the first to prove that benzene is a powerful, multipotential carcinogen. Other studies conducted by Maltoni identified the carcinogenic potential of vinylidene chloride, trichloroethylene, and many other chemicals. Maltoni pioneered the use of life-span mega-experiments in rodents in order to more closely resemble tumorigenesis in humans where cancers tend to appear in the later years of life. In so doing, he increased the predictive power of carcinogenesis bioassays.

Professor Maltoni was dedicated to the philosophy of Bernardino Ramazzini, an 18th century physician and scholar who discovered the link between occupational exposure to dangerous chemicals and conditions and their adverse effects on workers' health. Together with close friends Irving J. Selikoff, Myron Mehlman, and other eminent scientists, Maltoni founded and served as the first General Secretary of the Collegium Ramazzini, an academy of 180 internationally renowned experts in epidemiology, toxicology, and environmental and occupational health from over 30 countries.

The great achievements of Professor Maltoni were made possible thanks to the strong support of Italian friends and colleagues such as Professor Emilio Bartalini and Professor Francesco Corrado; to the collaboration of Professor Elio Garzillo and Architect Sabina Ferrari, who both served as Superintendent of Cultural and Architectural Heritage of the Emilia Romagna Region; to the talents of Professor Pier Luigi Cervellati, who oversaw the restoration of the castle, updating the structure to meet the technological needs of the Cancer Research Center; to the strong link with the Emilia Romagna Region, the Province of Bologna, and the local municipalities; and to the trust and confidence demonstrated by the President of the European Ramazzini Foundation Marco Vacchi and Vice President Franco Lazzari, as well as that of Isabella Seràgnoli, President of the Isabella Seràgnoli Foundation.

The dedication of the Cesare Maltoni Cancer Research Center of the European Ramazzini Foundation will help to keep alive the memory of Professor Maltoni's work and most importantly, as he did while living, inspire young scientists who seek to protect public heath and defend the quality of our environment.

MORANDO SOFFRITTI
General Secretary
Collegium Ramazzini

Introduction

The Collegium Ramazzini is an independent, international academy founded in 1982 by Irving J. Selikoff, Cesare Maltoni, Myron Mehlman, and other eminent scientists, and comprises 180 internationally renowned experts in the fields of occupational and environmental health. The Collegium offers a bridge between the world of scientific discovery and the social and political centers that must act on these discoveries to protect public health.

The Collegium Ramazzini held its first international conference "Living in a Chemical World" in Bologna in 1985. The second international conference was in Washington, DC in 1995 and focused on "Preventive Strategies for Living in a Chemical World." Twenty years after the first event, "Living" returned to Bologna with the title "Framing the Future in Light of the Past: Living in a Chemical World." The aims of this year's conference were to:

- update the state of knowledge on carcinogenic risks present in the workplace and general environment in order to define the scientific bases for more stringent regulations and more effective prevention;
- present the results of studies on hazardous agents and situations brought about by recent technological innovations and new lifestyles in order to suggest the application of new strategies for prevention;
- evaluate data on the diffusion of industrial hazards from developed to developing nations;
- review the scientific tools currently available for prevention and, if necessary, re-direct the strategies.

Over 300 participants traveled to Bologna from five continents and 35 countries to take part in this unique 3-day conference. Participants included Fellows of the Collegium Ramazzini, leaders of international agencies, scholars from the world's foremost universities, and representatives from various other interested groups.

As in 1985 and 1995, the conference proceedings have been published in the Annals of the New York Academy of Sciences and provide a source of new and useful information for experts in this sector.

We are grateful to all who worked to plan and organize this conference, to all who submitted abstracts and presented their work, for supporters who made the conference possible, and for the planning and organization of the European Ramazzini Foundation. We are especially grateful to Kathryn Knowles and her staff who were always there and always helpful during the Conference and, along with Karyl Norcross Mehlman, M.D., Ph.D., worked long hours for the 7 months following the conference helping the Editors and the New York Academy of Sciences, especially Steven Bohall, to organize and publish the Proceedings in this prestigious volume.

MYRON A. MEHLMAN
MORANDO SOFFRITTI
PHILIP LANDRIGAN
EULA BINGHAM
FIORELLA BELPOGGI
Editors

Identification of Carcinogenic Agents and Primary Prevention of Cancer

LORENZO TOMATIS

International Society of Doctors for the Environment, 52100 Arezzo, Italy

ABSTRACT: *During the annual Ramazzini Days, the Mayor of Carpi confers the Ramazzini Award on scientists deemed by the Collegium Ramazzini to have made outstanding contributions to furthering the aims of Bernardino Ramazzini in safeguarding public health. Dr. Lorenzo Tomatis was the Ramazzini Award recipient in 2005, and the presentation of the award was a highlight of the Symposium. The Ramazzini Lecture given by Dr. Tomatis follows.*

INTRODUCTION

In his introduction to *De morbis artificum diatriba*, Bernardino Ramazzini states modestly that his book was not inspired by a desire for glory but by a sense of duty; he had no pretensions to write a great work of art but wrote it for the good of the community and workers. Ramazzini exemplifies how science, legal justice, and social equity can harmoniously and efficiently coexist in a competent, sensible, committed physician. In our society, these three qualities rarely converge. Social equity is the most consistently maltreated of the three, while science is generally considered, by definition, to be above criticism while deliberately ignoring the possibility that its objectivity is often blurred by conflicts of interests.

One of the main merits of Bernardino Ramazzini is that he made physicians aware of questions other than those raised traditionally, that is about the nature of the work one is doing (quam artem exerceat), and of an area of medical concern that Hippocrates had neglected and scientific medicine did not consider part of its duties, which is the health of workers.[1] Even though Ramazzini's descriptions of working conditions and recommendations for their improvement necessarily refer to the preindustrial period, they are still largely valid today as is his emphasis on primary prevention.

Prevention, and specifically primary prevention, is the subject of my presentation. It might appear unnecessary to recall the distinction between primary and secondary prevention, but a look at the current scientific literature indicates some sort of oblivion, such that secondary prevention appears to be considered

---

Address for correspondence: Lorenzo Tomatis, Cave 25/r, Aurisina TS 34011, Italy. Voice: 0039-040-200284; fax: 0039-040-200284.
    e-mail: ltomatis@hotmail.com

the only means of prevention and chemoprevention in an ill-defined territory between the two. Considerable advances in our understanding of the mechanisms of progression of early cancerous lesions have moved the emphasis from a combined etiopathogenetic approach to the study of disease to a merely pathogenetic one. For instance, the program of an International Conference on Frontiers in Cancer Prevention to be held in October of 2005 and advertised as the "world's most comprehensive, transdisciplinary cancer prevention meeting," shows an excellent scientific level devoted mainly to early diagnosis, screening, genetic predisposition, and chemoprevention, while research on the etiology of cancer, directly related to primary prevention, represents only a minor part of the program. The sort of aristocratic tendency that between the time of Hippocrates and that of Bernardino Ramazzini led to dismissal of the occupational diseases of the working class by scientific medicine, or at least by a large sector of the biomedical establishment, is continuing today with the priority given to intellectually stimulating research, which usually has potential economic outcomes as its implicit but rarely declared goal. In this context, the pharmaceutical industry plays both a direct and an indirect role by the conditioning effect of its conspicuous financial support.

DIFFICULTIES OF PRIMARY PREVENTION OF CANCER

For a long time, research on primary prevention and the promotion of sanitary and social equity was hindered not only by a relative scarcity of funds but also by the difficulty, if not the impossibility, of obtaining financial support for certain projects. Today, financial support to several areas of research is more abundant; thus, the excuse that certain projects cannot be supported because of lack of funds might be questioned. A different, indirect but efficient system has been used for some time to block research that is considered not to be in the interests of the dominant economic power and consists of generously financing preselected areas of research, thus attracting scientists to objectives other than the protection of public health. The major flow of funds is toward large clinical trials and investigations on mechanisms of action or genetic predisposition, the results of which are widely advertised and guarantee access to important scientific journals. Some of the projects are useful and innovative, while others can be implemented only with a huge economic and organizational effort made possible by the availability of funds. This not to say that such research should not be done but that it should not stifle research projects that are uncongenial to the economic power.

Blockage by lack of funds has thus been replaced by blockage via a plethora of funds. To resist the attraction of abundant, secure funding, the publication of results in journals with high impact factors, and an opening to a brilliant career, much courage, determination, and a spirit of sacrifice are needed as they were needed years ago to start and persist in research on prevention with scarce or inadequate funding. That there are still scientists who have such courage and determination is one of the few reassuring signs in the present greedy, ruthless era.

Primary prevention, with the aim of preventing the occurrence of disease, and with indiscriminate universality as one of its main characteristics, should always have high priority. The primary prevention of infectious diseases did not encounter serious resistance, except for the reluctance of certain groups to accept systematic immunization, and was therefore solidly built on international cooperation. If primary prevention of infectious diseases has not been implemented with the same care and efficiency throughout the world, it is not because of doubts about the etiological agents of the diseases, which, once identified, nobody denied their being equally pathogenic at all latitudes, but because of the perverse combination of extreme poverty in certain countries, the irreducible selfishness of the rich countries, and the greed of multinational corporations.

Primary prevention of cancer of occupational and environmental origin, instead, has often stumbled on an obstacle course, and the identification of a chemical or physical agent as carcinogenic has too often met with skepticism, if not open hostility. Some chemical compounds were recognized as carcinogens in some countries and not in others, and even where they have been recognized as carcinogenic, the permitted or accepted concentrations varied considerably from country to country, as if their carcinogenicity could disappear or change at certain borders.[2,3]

The very long delay between the identification of a carcinogenic agent and adoption of adequate measures of prevention cannot be explained by a lack of advanced, specific medical procedures, as was the case in the early fight against some infectious diseases. The measures taken have generally been late and incomplete, coming only after the damage had spread and even then rarely providing total protection. The prevailing assumption, also used as an improper justification, was that the production of certain goods is necessary and vital, even when it was only aimed at increasing consumption of inessential goods, and that the risks involved in their production are an unavoidable price that society must pay. This attitude steadily disregarded the evidence that the highest price is paid by a particular sector of the population, in which morbidity and mortality are considerably higher than those in the rest of the population.

Ionizing radiation is a good example of how the appearance of evidence of the carcinogenicity of an environmental agent is not necessarily followed by adoption of measures of prevention and often not by even elementary prudence. Seven years after the discovery of X rays by Roentgen in 1895,[4] two reports were published describing their induction of malignant tumors of the skin[5,6]; this was an exceptionally short lag between the introduction of an agent into the environment and demonstration of its dangers. Neither radiologists nor health authorities nor the public, however, appeared to pay much attention to the possible risks associated with the use of X rays. Given their immense usefulness for diagnostic and therapeutic purposes, it was perhaps luckily so; nevertheless, we must also regret that their use increased in an almost total absence of caution. The case of ionizing radiation also shows the difficulties that are met in gaining acceptance of the dangers of small doses, both at work and in the general environment. It took 40 more years before the carcinogenicity of natural radiation was recognized when it was finally acknowledged as the cause of tumors in Schneeberg miners.[7] It took several more decades before it was officially accepted that the general population was at risk from exposure to natural radiation at levels much lower than those found in the environment of the mines.[8] An excess risk for cancer was recently shown[9] at doses and dose rates of radiation that for decades have been proclaimed as safe, in a climate of recent evidence that exposure to background radiation has slightly increased over the past few years.[10] As for other environmental agents, economic and political interests have interfered substantially with priorities in the defense of public health.

On the occupational front, benzene, for which evidence of carcinogenicity goes back to the 1920s, is one of the most important examples. The concentration of 100 ppm officially accepted in 1946 was sharply reduced to 10 ppm in 1978, even though the knowledge about its carcinogenicity was substantially the same at those two dates. Nor was existing knowledge much more advanced in the 1990s, when the maximum acceptable concentration was lowered to 1 ppm and a concentration of 0.3 ppm was proposed. The evolution of acceptable concentrations was not driven by progress in understanding of the mechanisms underlying the carcinogenicity of benzene or by an increased attention to occupational risks by the industries concerned or the health authorities, but was instead the result of the struggle for health by workers, unions, and concerned physicians and scientists against formidable economic interests.[11] In spite of the fact that the hemotoxicity of concentrations less than 1 ppm was recently further confirmed,[12] powerful industrial interests are still trying to undermine recognition of the risk of low concentrations.

Asbestos is probably the most dramatic[13,14] of the examples that provide evidence of a discrepancy between scientific evidence of an adverse effect and its translation into adequate preventive measures. Because of the determination of powerful economic interests to maintain the level of their profits at all costs, no international agreement yet exists to ban the production and use of asbestos worldwide, and more than 2 million tons are still produced annually. While progress is slowly being made in the right direction, some rich countries continue to exploit the permissive or absent occupational legislation in poor countries and send them old ships, stuffed with asbestos, to be demolished.[15]

BIRTH OF INTERNATIONAL AGENCY FOR RESEARCH ON CANCER (IARC)

Implementation of primary prevention is associated, to a considerable degree, with attribution of risks. This clearly depends on the availability of data on a number of risk factors so that plans for intervention can be formulated on the basis of urgency, feasibility, and priorities, depending on the relevance of the risks. The best known and most reliable sources of information on human carcinogens are those based on evaluations made by IARC[16] and the National Toxicology Program (NTP).[17] As the IARC started its program of evaluations several years earlier than the NTP, it has data on many more agents.

Retrospectively, the period when the IARC was created can be looked upon as a time of widespread enthusiasm and hope in cancer research. It was also a time when the idea that primary prevention had to be a high priority in the fight against cancer seems to have reached some of the greatest world authorities and when the stated intentions of politicians were implemented with an astonishing rapidity, rarely if ever seen in international affairs. On November 8, 1963, 12 eminent French personalities from widely differing backgrounds, including the oncologist Antoine Lacassagne, the biologist Jean Rostand, the writer Francois Mauriac, and the architect Charles Le Corbusier, approached the President of France, Charles de Gaulle, and invited him to take up a universal strategy of research to fight one of the greatest threats to human kind: cancer. The emotion that President de Gaulle had recently experienced in visiting two persons to whom he was particularly attached, both dying of cancer, may have played a role in the swiftness of his favorable response. General de Gaulle asked the ministers of foreign affairs of wealthy countries that had contributed substantially to cancer research and control worldwide (besides France, the United Kingdom, the United States, and the USSR) and the Director General of the World Health Organization (WHO) to meet in Paris to discuss how to implement a common initiative. The foreign ministers of two additional countries, namely the Federal Republic of Germany and Italy, were later also asked to attend the meeting that took place in December 1963.[18]

Thanks to the strength of its basic ideals, the initiative generously and vigorously proposed by France managed to survive, in spite of some hostility from the International Union against Cancer (the oldest and at the time the most important international organization for cancer control, better known under the French abbreviation UICC), which, while adhering in theory to an increased boost to cancer research, did not hide its lack of enthusiasm about the creation of an institution that would escape its control. There was also some resistance from the heads of national cancer centers, who supported the initiative in principle but feared that a new international organization would divert funds and competent scientists from their own institutes.[18] The initiative kept its original public health orientation, but it was not equally successful in stimulating the financial generosity of the states.

IDEALISM AND REALISM OF FUNDING

The original proposal of General de Gaulle was that the new international institution be endowed with 0.5% of the military expenditure of the participating states. On the basis of estimates of the 1965 defense budgets of the first six states that adhered to the French initiative, 0.5% would have meant about US$ 400 million annually (0.5% of the estimated defense budget of the United States alone would have been about US$265 million). Funding of the new institution at the level originally proposed would have meant a considerable boost for cancer research. The first six member states finally agreed on a much more modest annual contribution of US$150,000 each, so that the initial annual budget of IARC was US$900,000—400 times less than it would have been if the original French proposal had been accepted. The budget did increase over the years as more states joined the agency, and a statutory budget of US$37 million was granted to the agency by its governing council for the biennium 2004–2005; 0.5% of the estimated amount spent for "defense" by the 16 states presently participating to the agency in just 1 of those 2 years would have represented about US$2 billion.

In accordance with the statutes approved by the World Health Assembly on May 20, 1965 in Geneva and with advice given by the governing and scientific councils, it was agreed that the agency, in line with its international role, should establish a program of permanent activities that included: (*a*) collection and dissemination of information on the epidemiology of cancer and cancer research in both developed and developing countries; (*b*) identification of the causes of cancer; and (*c*) promotion of international collaboration in cancer prevention worldwide.[19,20] Lyon, France was chosen as the site for IARC, which officially started its activities in May 1967.

The coexistence under the same roof of offices and laboratories, together with an efficient administration, favored multidisciplinarity, which was a strong characteristic of several IARC programs from the beginning. The fact that the agency could offer laboratory facilities actively involved in planning and conducting research made it possible to attract competent scientists from various backgrounds, including pathologists, biochemists, chemists, toxicologists, and virologists, as well as epidemiologists and biostatisticians. In this way, IARC was able to build up scientific teams that could interact and collaborate efficiently on an equal basis with scientists in any other research institute in the world.

IARC MONOGRAPH PROGRAM

Among the first activities of the agency were the collection and processing of data on morbidity and mortality from cancer, an educational program and evaluation of carcinogenic risks to humans. For the purpose of this presentation, I shall focus on the last of these basic activities, although the other two and several other projects were very successful and deserve full recognition. When, in 1968, the agency was requested to provide a list of human carcinogens, two reliable, albeit incomplete lists of human carcinogens were available: one proposed by Hueper and Conway[21] comprising 17 agents or groups of agents considered definitely carcinogenic to humans and one by WHO that consisted of 16 agents.[22]

The IARC Monograph Program was initiated in 1969; in 1972, the first volume of the *IARC Monographs* was published in a series that became known worldwide as the "orange books." To date, 85 volumes of *Monographs* have been published and four additional volumes are in press, covering 900 agents (chemicals, groups of chemicals, complex mixtures, occupational exposures, biological agents, cultural habits, physical agents).

Individual agents and complex exposures are assigned to different groups, according to the level of evidence for their carcinogenicity: group 1, human carcinogens, presently consists of 95 entries; group 2A, probable human carcinogens, has 65 entries; group 2B, possible human carcinogens, has 240; group 3, not classifiable for carcinogenicity in humans, has 608; and group 4, probably not carcinogenic to humans, has 1. Leaving out for the moment group 3, but keeping in mind that the limitation and inadequacy of the evidence of carcinogenicity for the agents included in this group may not be necessarily related only to the characteristic of their biological interactions but also to the quality and quantity of available data, there are 403 agents for which there is evidence of a causal association with cancer in humans with decreasing strength from group 1 to group 2B.

To my knowledge, in spite of a number of attempts, no one has yet succeeded in dislodging an agent from IARC group 1 and dragging it toward incertitude. Nevertheless, although it has been impossible to deny the evidence for agents in group 1, the carcinogenicity of certain compounds has been limited to the induction of particular tumor types, with doubt cast on or a denial of causal associations with other types of tumors. For instance, while no one dares to deny that vinylchloride causes liver angiosarcoma, a conspicuously vocal fraction of the scientific establishment maintains that there is no causal association between vinylchloride and liver tumors of other histological types or with tumors in other organs.

Another example is formaldehyde, reevaluated by IARC in 2004[23] and transferred to group 1 from group 2A, to which was assigned in 1987 and again in 1995. *A posteriori*, one might wonder whether the hesitancy manifested in 1987 and in 1995 was justified, but at least in 2004 it was finally recognized as a human carcinogen. The acceptance of evidence for its carcinogenicity is, however, limited to the induction of rhino pharyngeal carcinoma, while the evidence for an association with myeloid leukemia is reported to be strong but not sufficient, and only limited evidence is available for an association with sino nasal cancer. One may ask whether the hesitation present even in this much more advanced last evaluation serves public health well. There is, for instance, a real possibility of domestic exposure to formaldehyde indoors, which mainly concerns children. It is not easy to establish the extent to which exposure to low concentrations of a carcinogen represents a risk for the rest of one's life when the exposure occurs at an age of particular fragility in certain aspects and which, by definition, allows the longest possible time for a long-term effect to manifest. The difficulty of designing an adequate epidemiological study on such effects is an expression of the limitations of our methods of investigation. Nevertheless, awareness of these limitations should encourage a greater commitment to primary prevention instead of leading toward an *a priori* denial or even ignoring the possibility of a long-term risk.

The Predominant Role of Epidemiology

It is interesting that the inception of the *IARC Monographs* program in the late 1960s coincided with the rise of epidemiology as the fundamental discipline in the assessment of risk for noncommunicable diseases, including cancer. This was a considerable change from the attitude that had prevailed since 1922, when Passey[24] succeeded in inducing malignant skin tumors in mice with soot extracts; his results were taken as definitive confirmation of the observations of Percival Pott, implying that clinical and epidemiological observations had to be confirmed experimentally in order to be accepted. After it was agreed in the early 1970s that epidemiological results could by themselves prove causation, the view that only the epidemiological approach could provide acceptable evidence for a causal relationship between an exposure and human cancer began to prevail. A first consequence was that experimental results, in particular those of long-term bioassays, were considered of secondary importance. A second consequence was that epidemiologists began to attempt quantification of risks attributable to certain causes and to calculate, admittedly roughly, the proportions of cases that could be avoided by efficient prevention. The best-known attempts were those of Wynder and Gori[25] and Higginson and Muir[26] and, most elaborately and in greatest detail, by Doll and Peto.[27] All three studies attributed the great majority of tumors to environmental causes and agreed that the most relevant risk factors were related to lifestyle, in particular tobacco use and dietary habits, which were reported to be responsible for 60–70% of cancers, with greatly lower estimates for those due to occupational and other environmental exposures.

Tobacco undeniably plays an essential role in increasing the human cancer burden, and there is little doubt that a better social and health-oriented education could help individuals in being more conscious and responsible in the choice of their life habits. However, the emphasis given to lifestyle factors, to the detriment of information on the role of chemical pollutants, favored the uninterrupted production of agents with negative effects on health that remain hidden or secret or are deliberately underestimated. Furthermore, attributing most cancer cases to lifestyle, which is interpreted as being related to free personal choice, unduly amplifies the individual's responsibility, diverts attention from the lack of commitment of health authorities, and obscures the etiological role of other risk factors.[28]

An Incoherent Attribution of Risks

The way in which attributable risks are considered is incoherent, as unequal degrees of evidence for the carcinogenicity of various factors are treated equally.[28] A necessary requirement for declaring an environmental chemical carcinogenic to humans is that conclusive epidemiological studies support a causal relationship, and particularly robust evidence for an association between an occupational exposure and human cancer is required before a causal association is accepted. The evidence for a contribution of dietary factors to the cancer burden is usually circumstantial and, in some instances, rather weak. Punctilious precision is used in calculating occupational and environmental risks, while wide latitude is allowed for risks related to diet, ranging between 10% and 70%. It was recognized, however, that the occupational carcinogens identified so far "tend to be those which increase the risk of some particular type(s) of cancer very substantially," and that other occupational carcinogens might not have been detected simply because they have not been investigated or because the exposure concerns a small number of individuals, and no suspicion was raised.[27]

AN ELEGANT AMBIGUITY

On the one hand, it was recognized that certain industrial products present in the general environment (for instance, pesticides) can increase the frequency of tumors; on the other hand, their identification as risk factors was made to depend on finding that they are carcinogenic in situations in which the exposure is very high, such as during occupational exposure. Such findings, however, depend in turn on an obligatory series of circumstances, such as a preexisting suspicion of carcinogenicity, a sufficient number of exposed individuals to ensure the statistical validity of the observations, and a sufficiently long duration of exposure and of follow-up. Magnifying the difficulties in providing convincing evidence of an increased cancer risk with endless discussions on the statistical credibility of the data, may have also contributed to focus the attention mainly, if not exclusively, on cancer, and in this way divert the attention from other adverse health effects. It was further recognized that some substances but, as carefully specified, certainly not all, for which there is experimental evidence of carcinogenicity, even if obtained at doses much higher than those to which humans are generally exposed, could have the same effect in humans.

This way of presenting and interpreting data casts light on a particularly elegant ambiguity that has allowed a coupling of certainty about risks proclaimed as ascertained and convincing with wide areas of shadow. These shadowy areas, to which consistent components of the etiology of cancer were relegated, have attracted insufficient attention, largely because research on mechanisms, often related to new therapeutic approaches and to the genetic component of risk, has greatly expanded to the detriment of studies on etiology and primary prevention, the latter being focused almost exclusively on the role of lifestyle.[29]

The assignment of a compound to one of the groups proposed by IARC can have important consequences for the determination of attributable risks and translation into measures for primary prevention. Most agents assigned to group 1 and several of those assigned to group 2A are dealt with as human carcinogens and are subjected to strict legislation. Group 2B, however, represents a large parking lot in which 240 agents have been stored because of the relative inadequacy of the experimental and epidemiological evidence of carcinogenicity. The systematic undermining of the significance of long-term carcinogenicity tests and the extreme caution with which some epidemiologists assess evidence for risk for fear of being accused of creating false-positive results emphasize these inadequacies. The relative inadequacy or the absence of epidemiological data, however, cannot be considered equivalent to negative findings, nor can it be considered more relevant for public health than positive experimental findings.

The probability that additional epidemiological data will become available in the near future on compounds assigned to group 2B, to avoid indefinite extension of their storage in this parking lot, is rather remote. Given the objective difficulties of designing adequate studies capable of credibly demonstrating risks of low or medium level and the access of the results of such studies only to journals with low impact factors, agents assigned to group 2B have not raised the interests of epidemiologists. Similarly, there are limited chances that they will be submitted to additional long-term tests, given the dramatic reduction in the number of independent laboratories interested in carrying out long-term bioassays, which, with a few conspicuous exceptions, are at present almost exclusively in the hands of commercial laboratories or of laboratories internal to industries.

NO EASY SOLUTIONS FOR SITUATIONS TYPE 2B

There are no easy solutions for situations that can be defined as type 2B, when the experimental and epidemiological data are relatively limited and do not reach the level of evidence that is defined as sufficient. Group 2B includes agents that are quite disparate in terms of public health and economic relevance as well as in the level of evidence for their toxicity. Some should undergo indepth investigation without delay, such as acetaldehyde, acrylonitrile, gasoline, bitumen, chloroprene, carbon tetrachloride, 1,2-dichloroethane, hexachlorobenzene, and a few agents that were recently downgraded from group 2B to group 3 that include atrazine, phthalates, rock wool, and glass wool.

If the validity of the precautionary principle is not accepted, type 2B situations will create an impasse of which the only outlet is the official perpetuation of risk conditions with possible ominous consequences on health. In practical terms, such situations are as difficult to deal with as those related to exposures to very low doses of recognized human carcinogens. However, admitting such difficulty does not mean that we can deny *a priori* their possible etiological role. The actual role of a long series of risk factors in increasing the cancer burden is still poorly known, and the noxious effects of low or very low concentrations of environmental pollutants have only begun to be elucidated. The adverse effects of extremely low doses of agents that act as endocrine disruptors have raised the greatest attention, such that even a newspaper like *The Wall Street Journal* expressed some concern about exposure to bisphenol A, phthalates, and atrazine.[30]

ROLE OF LOW CONCENTRATIONS

Cadmium, which was recognized as a human carcinogen after relatively high occupational exposures, represents a different spectrum of adverse effects. Considered to be a nongenotoxic carcinogen, at extremely low doses was shown to be mutagenic without causing direct DNA damage but by interfering with the mismatch repair system of DNA replication errors.[31,32] At concentrations that can be found in the general environment, cadmium can thus induce genomic instability, which is not sufficient *per se* to cause neoplastic transformation but is sufficient to increase cellular susceptibility to other exogenous and endogenous agents, possibly contributing in this way to increasing the risk for cancer.

Other examples of the long-term adverse effects of exposure to low concentrations of environmental agents include the possible prenatal origin of certain childhood leukemias. Translocations typical of myeloid leukemia, probably due to maternal exposure to some toxic compound, were shown to be present at birth in children who developed the disease years later. While not sufficient *per se* to cause the disease, they might increase the risk for leukemia by inducing genomic instability.[33,34] Furthermore, an association has been reported between maternal exposure during pregnancy and paternal exposure before conception to a series of chemicals and *Ras* proto-oncogene mutations in children who later develop lymphocytic leukemia,[35] while prenatal and early postnatal exposure to atmospheric pollutants has also been reported to be associated with an increased risk for childhood cancer.[36,37]

OUR RESPONSIBILITY

One of the priorities of research today is the unraveling of the complexity of gene–environment interactions in modulating the susceptibility to chronic diseases. We may expect that substantial progress will be made when it is possible to reliably measure both environmental exposures and genetic variations.[38] Up to now, much more attention has been paid to the study of the individual's genome than to the measurement of individual's environmental exposures; the methodology developed for genotyping is presently far more advanced and accurate than the methodology employed in the measurement of environmental exposures.[39] To compensate for this disparity, an increased effort to upgrade exposure assessment procedures is essential. At the same time, however, we should never forget that uphill to the measurement of exposures, a key role in the protection of public health will be played by an action aimed at banning or sharply decreasing the presence of noxious chemical in our environment.

If we really want to draw up a credible table of attributable risks and, above all, if we want to implement efficient primary prevention, conscious of the responsibility we have toward the present but also future generations, we should seriously consider all the various components of risk that have until now been unjustifiably underestimated or ignored.

REFERENCES

1. CARNEVALE, F. 2002. Introduzione a: Bernardino Ramazzini: Le Malattie dei Lavoratori (De Morbis Artificum Diatriba). Libreria Chiari. Firenze.
2. TOMATIS, L. 2005. Primary prevention of cancer in relation to science, sociocultural trends and economic pressures. Scand. J. Work Environ. Health. **31:** 227–232.
3. CARNEVALE, F., R. MONTESANO, C. PARTENSKY & L. TOMATIS. 1987. Comparisons of regulations on occupational carcinogens in several industrial countries. Am. J. Ind. Med. **12:** 453–473.
4. ROENTGEN, W.C. 1895. Ueber eine neue Art von Strahlen Sitzunggsber. Phys. Med. Gesellsch. Wurtzb. 132–141.
5. FRIEBEN, A. 1902. Demonstration eines Cancroid der rechten Handruckens, das sich nach langdauernder Einwirkung von Roentgenstrahlen entwickelt hat. Fortschr, Roentgenst. **6:** 106–111.
6. SICK, H. 1902. Karzinom der Haut das auf dem Boden eines Roentgenulcus entstanden ist. Muench Med Wochenschr. **50:** 1445.
7. SCHUTTMAN, W. 1993. Schneeberg lung disease and uranium mining in the Saxon ore mountain (Erzgebirge). Am. J. Ind. Med. **23:** 355–368.
8. INTERNATIONAL AGENCY FOR RESEARCH ON CANCER. 1988. Man-made mineral fibers and radon. IARC Monographs on the Evaluation of Carcinogenic Risks to Humans, Vol. 43. IARC, Lyon.
9. CARDIS, E., M. VRIJHEID, M. BLETTNER, *et al.* 2005. Risk of cancer after low doses of ionizing radiation: retrospective cohort study in 15 countries. BMJ **331:** 77–83.
10. EATON, L. 2005. UK agency reports slight increase in radiation exposure. BMJ **330:** 1229.
11. INFANTE, P. & M.V. DISTASIO. 1988. Occupational benzene exposure: preventable deaths. Lancet. **i:** 1399–1400.
12. LAN, Q., L. ZHANG & G. LIG. 2004. Hematotoxicity in workers exposed to low levels of benzene. Science **306:** 1774–1776.
13. CASTLEMAN, B. 1983. Asbestos: Medical and Legal Aspects. Law and Business, Clifton, NJ.

14. LADOU, J. 2004. The asbestos cancer epidemics. Environ. Health Perspect. **112:** 265–290.
15. HARRIS, L.V. & I.A. KAHVA. 2003. Asbestos: old foe in 21st century developing countries. Sci. Total Environ. **307:** 1197–1199.
16. INTERNATIONAL AGENCY FOR RESEARCH ON CANCER. 1972–2005. IARC Monographs on the Evaluation of Carcinogenic Risks to Humans, Vols. 1–85. IARC Press, Lyon.
17. DEPARTMENT OF HEALTH AND HUMAN SERVICES. 2004. 11th Report on Carcinogens, National Toxicology Program National Institute of Environmental Health Sciences. Washington, DC.
18. SOHIER, R. & A.G.B. SUTHERLAND. 1990. La genese du Centre International de Recherche sur le Cancer. IARC technical report No. 6, Lyon.
19. IARC Annual Report, 1969, IARC 1970, Lyon.
20. IARC Annual Report, 1972–1973, IARC 1973, Lyon.
21. HUEPER, W.C. & W.D. CONWAY. 1964. Chemical Carcinogenesis and Cancers. Charles C. Thomas. Springfield, IL.
22. WORLD HEALTH ORGANIZATION. 1964. Prevention of cancer. WHO Technical Report No. 276, Geneva, 1964.
23. INTERNATIONAL AGENCY FOR RESEARCH ON CANCER. IARC Monographs on the Evaluation of Carcinogenic Risks to Humans, Vol. 87, IARC, Lyon, *In press*.
24. PASSEY, R.D. 1922. Experimental soot cancer. Br. Med. J. **11:** 1112–1113.
25. WYNDER, E.L. & G.B. GORI. 1977. Contribution of the environment to cancer incidence: an epidemiologic exercise. J. Natl. Cancer Inst. **58:** 825–832.
26. HIGGINSON, J. & C.S. MUIR.1979. Environmental carcinogenesis: misconceptions and limitations to cancer control. J. Natl. Cancer Inst. **63:** 1291–1298.
27. DOLL, R. & R. PETO. 1981.The causes of cancer: quantitative estimates of avoidable risks of cancer in the United States today. J. Natl. Cancer Inst. **66:** 1191–1308.
28. TOMATIS, L., J. HUFF, I. HERTZ-PICCIOTTO, *et al.* 1997. Avoided and avoidable risks of cancer. Carcinogenesis **18:** 97–105.
29. TOMATIS, L., R. MELNICK, J. HASEMAN, *et al.* 2001. Alleged 'misconceptions' distort perceptions of environmental cancer risks. FASEB J. **15:** 195–201.
30. WALDMAN, P. 2005. Common industrial chemicals in tiny doses raise health issue. Wall Street Journal A1.
31. JIN, Y.H., A.B. CLARK, R.J. SLEBOS, *et al.* 2003. Cadmium is a mutagen that acts by inhibiting mismatch repair. Nat. Genet. **34:**326–329.
32. CLARK, A.B. & T.A. KUNKEL. 2004. Cadmium inhibits the functions of eukaryotic muts complexes. J. Biol. Chem. **279:** 53903–53906.
33. WIEMELS, J.L., Z. XIAO, P.A. BUFFLER, *et al.* 2002. In utero origin of t(8;21) AML1-ETO translocations in acute myeloid leukemias. Blood **99:** 3801–3805.
34. MCHALE, C.M. & M.T. SMITH, 2004. Prenatal origin of chromosomal translocations in acute childhood leukemias: implications and future directions. Am. J. Hematol. **75:** 254–257.
35. SHU, X.O., J. P. PERENTESIS, W. WEN, *et al.* 2004. Parental exposure to medications and hydrocarbons and ras mutations in children with acute lymphoblastic leukemia: a report from the Children's Oncology Group. Cancer Epidemiol. Biomarkers Prev. **13:** 1230–1235.
36. KNOX, E.G. 2005. Oil combustion and childhood cancers. J. Epidemiol. Commun. **59:** 755–760.
37. KNOX, E.G. 2005. Childhood cancers and atmospheric carcinogens. J. Epidemiol. Commun. Health **59:** 101–105.

38. WILD, C.P. 2005. Complementing the genome with an "exposome": the outstanding challenge of environmental exposure measurement in molecular epidemiology. Cancer Epidemiol. Biomarkers Prev. **14:** 1847–1850.
39. VINEIS, P. 2004. A self-fulfilling prophecy: are we underestimating the role of the environment in gene-environment interaction research? Int. J. Epidemiol. **33:** 945–946.

Children's Environmental Health Research—Highlights from the Columbia Center for Children's Environmental Health

FREDERICA PERERA,[a,b] SHEILA VISWANATHAN,[a,b] ROBIN WHYATT,[a,b] DELIANG TANG,[a,b] RACHEL L. MILLER,[b,c] AND VIRGINIA RAUH[b,d]

[a]*Department of Environmental Health Sciences, Mailman School of Public Health of Columbia University, New York, New York 10032, USA*

[b]*Columbia Center for Children's Environmental Health, Columbia University, New York, New York 10032, USA*

[c]*Division of Pulmonary, Allergy, Critical Care Medicine, College of Physicians and Surgeons, Department of Medicine, Columbia University, New York, New York 10032, USA*

[d]*Heilbrum Center for Population and Family Health, Columbia University, New York, New York 10032, USA*

ABSTRACT: A growing body of evidence has been generated indicating that the fetus, infant, and young child are especially susceptible to environmental toxicants as diverse as polycyclic aromatic hydrocarbons (PAHs), pesticides, lead, mercury, polychlorinated biphenyls (PCBs), and environmental tobacco smoke (ETS). Exposures to these toxicants may be related to the increases in recent decades in childhood asthma, cancer, and developmental disability. The Columbia Center for Children's Environmental Health (CCCEH), located in New York City, has developed four cohorts around the world to elucidate the relationships between these exposures and childhood illness. This article summarizes the recent findings from the Center's projects in the context of current research in children's environmental health.

KEYWORDS: children; environmental health; research; review; PAH; pesticides; ETS; asthma; cancer; developmental disability

Address for correspondence: Frederica P. Perera, DrPH, Department of Environmental Health Sciences, Mailman School of Public Health of Columbia University, and Columbia Center for Children's Environmental Health, 100 Haven Avenue, Tower III, 25F, New York, NY 10032. Voice: 212-304-7280; fax: 212-544-1943.
  e-mail: fpp1@columbia.edu

INTRODUCTION

Today's children are growing up in an environment quite different from that of their parents or grandparents. Exponential growth in technology, consumption, manufacturing, and population has defined the past decade. Fueling this growth is the development of new products and chemicals. Over the past five decades, over 80,000 synthetic chemical compounds have been created; 2000–3000 new chemicals are submitted for review by the Environmental Protection Agency (EPA) on a yearly basis.[1] Of high-volume chemicals currently in circulation, only 43% have been tested for potential human toxicity, and only 7% have been studied for possible effects on development.[2–4] In addition to the concerns about exposure to an increasingly diverse array of chemicals, there is mounting evidence that the fetus and infant are significantly more sensitive to a variety of environmental toxicants than adults because of differential exposure, physiologic immaturity, and a longer lifetime over which disease initiated in early life can develop.[5]

The incidences of several childhood diseases and disorders have been increasing in recent decades. For example, there has been more than a 50% increase in asthma-associated school absences between 1980 and 1996 as well as an increase in the estimated annual number of physician office visits and hospital outpatient visits among children aged 5–14 years.[6] The incidence of cancer in children younger than 15 years of age increased from 124.3 per million in 1975–1979 to 139.9 per million in 1990–1995.[7] Developmental disabilities, the name given to a broad group of conditions caused by learning or physical impairments, affect an estimated 17% of U.S. children under the age of 18 years.[8] The high rates of these childhood disorders have significant social impacts and medical costs for individual families and the country as a whole. Therefore, there is a need to understand the role of environmental factors in childhood disease and neurodevelopmental disorders and to identify the primary environmental toxins affecting children's health so that preventive measures can be taken. Here we define environmental factors as toxic exposures due to lifestyle (smoking and diet) and pollutants in the workplace, ambient air, and water and food supply.

Differential Exposure and Susceptibility

It is important to distinguish children from adults when assessing environmental health impacts because of their unique behaviors and biological characteristics that can increase vulnerability to certain toxicants. For example, experimental and human data indicate that the fetus and young child are especially vulnerable to the toxic effects of environmental tobacco smoke (ETS), polycyclic aromatic hydrocarbons (PAHs), particulate matter (PM), nitrosamines, pesticides, polychlorinated biphenyls (PCBs), metals, and radiation.[5] The differential susceptibility of the fetus and newborn in part is due to increased exposure and altered biological susceptibility; nutritional deficits, genetic predispositions, and social stressors also contribute.

Increased Exposure

Behaviors common in childhood, which are not observed in adults, can have a major effect on the biological availability of toxicants in children. For example, young children breathe air closer to the ground, exposing them to particles and vapors present in carpets and soil. While playing and crawling around on the floor, children can inhale or dermally absorb toxicants, which are subsequently absorbed more efficiently in children than in adults.[9] Compounding the effects of this behavior is the fact that infants have twice the breathing rate of the average adult. Also, most young children display hand-to-mouth behavior and thumb-sucking habits that can increase exposure.

Dietary habits of children may also cause increased exposure to some toxicants. Children under 5 years of age eat three to four times more food per unit of body weight than the average adult American; and the average one-year old drinks 10–20 times more juice than the average adult.[10] Dermal exposures may also be higher, as a typical newborn has more than double the surface area of skin per unit of body weight than an adult.[11]

Biological Susceptibility

Human infants and children differ from adults not only in their size and potential for exposure but also in their ability to metabolize environmental toxins. The *in utero* and childhood periods are characterized by rapid physical and mental growth and gradual maturation of major organ systems. In fact, typical newborns double their weight within 6 months of birth and integral parts of the nervous and immune systems are formed during the first 6 years of life.[12] Additionally, sex organ development, myelination, and alveoli formation begin late in pregnancy and continue until adolescence.[9] As cells are proliferating rapidly and organ systems are immature, they are sensitive to the potentially harmful effects of environmental toxins.

Absorption, metabolism, and excretion pathways in infants and children differ from those in adults. These pathways dictate the amount of a toxicant, in its various forms, that is present in the body. Epidemiological studies with biomarkers have demonstrated placental transfer of toxicants, and in some cases slower fetal clearance of chemicals such as PAHs, PCBs, and mercury.[13–15] An infant's kidney filtration rate is lower than an adult's, thus increasing potential susceptibility.[9]

Finally, infants and children have more years of future life than most adults. Thus, there is more time for early exposures to trigger diseases that have long latency periods. For example, early exposure to carcinogens will be more likely to lead to cancer than the same exposure experienced later in life.

Modifying Factors

Increased susceptibility to certain environmental exposures in childhood is not only due to biological and behavioral traits associated with early growth and development. Nutritional factors, genetics, and psychosocial stressors can also modify the effect of these exposures.

Nutrition

Most of the research on micronutrients has focused on their main effects, but there is some evidence of interactions with environmental exposures. Certain micronutrient deficiencies have been associated with childhood asthma, adverse birth outcomes, child development, and childhood cancer. For example, essential fatty acids are associated with low birth weight reduced head circumference, and cognitive and motor function.[16–18] Nutritional status modulates inflammatory response to air pollution and its effects on childhood asthma. Evidence for this includes the presence of antioxidants in the airway surface liquid of the lung, which reduces oxidative stress and prevents airway inflammation, thereby limiting asthma-like symptoms.[19–21] By removing free radicals and oxidant intermediates, antioxidants protect DNA from the genotoxic, procarcinogenic effects of chemicals that bind to the DNA.[22,23] Nutritional deficiencies, in some cases, may be closely related to lower socioeconomic status, although variations exist within each socioeconomic bracket.

Genetics

Genetic susceptibility can take the form of common polymorphisms or haplotypes that modulate the individual response to a toxic exposure. For example, two genes have been identified that can increase an individual's vulnerability to organophosphates (OPs), such as chlorpyrifos, by reducing the reservoir of functioning protective enzymes.[24] The first gene has a prevalence of 4% and results in a poorly functioning form of the enzyme acetylcholinesterase; the second gene results in a relatively inactive form of the enzyme paraoxonase (prevalence of 30–38%), an enzyme that detoxifies chlorpyrifos before the toxin can inhibit acetylcholinesterase.[25] Evidence of an interaction between the paraoxonase 1 (PON1) genotype and OP pesticides includes the finding that the effect of chlorpyrifos on head circumference at birth was significant only among women with low PON1 activity.[26] Other examples of gene–environment interactions involve the gene coding for the d Alanine (d-ALA) enzyme that affects lead metabolism and storage,[25] and genetic polymorphism in the dopamine transporter that is associated with increased behavioral problems in children prenatally exposed to tobacco smoke.[27] Other research has found that the P450 and glutathione-S-transferase gene families play a role in the activation and detoxification of various xenobiotics. For example, the P450 and glutathione-S-transferase genes are involved in the metabolism of PAHs and can influence the level of PAH–DNA damage. PAH–DNA adduct levels in human placenta were significantly higher in infants with the cytochrome-P450 1A1 (CYP1A1) MspI restriction site, a genetic marker associated with lung cancer risk, than in infants without the restriction site.[28] Glutathione S-transferase T1 (GSTT1) null was also shown to be a marker for lower birth weight and preterm birth among babies of pregnant women who actively used tobacco.[23] Children carrying the glutathione S-transferase M1 (GSTM1) null gene who were exposed *in utero* to tobacco smoke had increased risk for persistent asthma and wheezing.[29] GSTM1null in asthmatic children also is associated with increased susceptibility to the harmful effects of ozone, such as reduced forced expiratory flow.[30]

Individual- and Community-Level Psychosocial Stressors

The notion that individual- and community-level conditions can produce profound effects on host susceptibility to disease is derived from the long-standing existence of strong social class gradients in health.[31] Recent studies have shown that women who live in violent, crime-ridden, physically decayed neighborhoods are more likely to experience pregnancy complications and adverse birth outcomes, after adjusting for a range of individual-level sociodemographic attributes and health behaviors.[32,33] Other studies have suggested that the stresses of racism and community segregation are associated with lower birth weight.[34] Several have shown that the effects of individual poverty on birth outcomes are exacerbated by residence in a disadvantaged neighborhood.[35] In one of the few studies that has measured interactions between physical toxicants and individual psychosocial stressors, an analysis of Northern Manhattan mothers and toddlers done at the Columbia Center for Children's Environmental Health (CCCEH cohort) found that the risk of developmental delay among children exposed prenatally to maternal ETS was significantly greater among those whose mothers experienced material hardship during pregnancy.[36]

Children's Disorders Related to Environmental Factors

Cancer Risk/Genetic Damage

Environmental exposures are quickly gaining recognition as potential risk factors for childhood cancer.[37] Several studies suggest that the fetus clears carcinogens less efficiently than the adult and thus may be more vulnerable to genetic damage and the resultant risk of cancer. For example, carcinogen-DNA adducts are a marker of increased cancer risk.[38] Experimental evidence shows that the amount of PAHs crossing the placenta and reaching the fetus is less than one-tenth of the dose to the mother,[39,40] yet the levels of PAH–DNA adducts measured in rodent fetal tissue are higher than expected.[41] Similarly, research in mothers and newborns has consistently shown that PAH–DNA adduct levels in the white blood cells (WBC) of newborns were similar to or exceeded those in paired maternal samples, despite the estimated 10-fold lower dose of the parent compound to the fetus.[42–44] In addition, fetal plasma cotinine levels were higher than in paired maternal samples, suggesting reduced ability of the fetus to clear carcinogenic cigarette smoke constituents.[42,43] This research indicates that the differential effect of exposure to PAHs in the fetus is not limited to a particular ethnic or geographic group.[43] Increased adducts in the fetus relative to the adult could result from lower levels of phase II (detoxification) enzymes and decreased DNA repair efficiency in the fetus.[42,45–47]

Chromosomal aberrations have been associated with increased risk of cancer in multiple studies[48,49] and are another biomarker used to detect the preclinical effects of cancer-causing environmental toxicants. In New York City newborns, maternal exposure to airborne PAHs during pregnancy has been associated with increased frequency of chromosomal aberrations in WBC, suggesting that risk of cancer can be increased by exposure in the womb.[50] Studies have also linked maternal tobacco smoking to increased chromosomal aberrations in WBC of newborns.[51] Other research has shown an approximately 10-fold higher risk of infant acute myeloid leukemia with increasing maternal consumption of DNA topo 2 inhibitor-containing foods, raising concerns about benzene, a topo 2 inhibitor.[52]

Respiratory Disease/Asthma

An estimated 9 million (12.5%) children aged <18 years in the United States have had asthma diagnosed at some time in their lives.[53] A recent study found that over 25% of elementary school children in Harlem had asthma.[54] Environmental factors are known or suspected to contribute to these high rates of disease. There are critical windows in both prenatal and postnatal development during which exposure to irritants and other toxicants can modify the formation and maturation of the lung. The complete development of the human lung occurs through the sixth to eighth years of life.[55] But there is recent evidence from the CCCEH cohort study that important adaptive immune responses may begin *in utero*. These include findings that cord blood T cell proliferation in response to specific allergens can occur independently of maternal sensitization.[56] Moreover, prenatal exposure to PAHs with ETS was associated with an increased risk of respiratory symptoms at the age of 2 years.[57]

The association of diminished air quality with the increased prevalence of respiratory symptoms has been documented internationally, including studies in Holland and Indonesia, indicating that the prevalence of respiratory symptoms increases as air quality decreases.[58,59] Pesticide exposure has also been associated with respiratory disease and multiple chronic respiratory symptoms in children.[60] Finally, it recently has been demonstrated that multiple toxicants (in this case, prenatal PAHs and postnatal ETS) can act synergistically to increase risk of respiratory symptoms in children at 12 months and probable asthma at 24 months.[57]

Neurobehavioral Disorders

The exquisitely sensitive process of the development of the human central nervous system involves the production of 100 billion nerve cells and 1 trillion glial cells, which then must follow a precise stepwise choreography involving migration, synaptogenesis, selective cell loss, and myelination.[61] A mistake at any one step can have permanent consequences. Experimental studies of prenatal and neonatal exposure to the OP chlorpyrifos have demonstrated neurochemical and behavioral effects as well as selected brain cell loss.[62,63] The behavioral and morphologic effects of developmental toxicants are highly dependent on the timing as well as on the dose and duration of exposure. This is illustrated by both rodent and human studies showing that the effect of irradiation on brain malformation is heightened during the window of susceptibility of fetal development.[61] Adverse neurological development, including lowered intelligence, diminished school performance, and increased rates of behavioral problems have been associated with exposure to low levels of a number of environmental toxicants and pollutants. Cohort studies have demonstrated that low-level exposure to lead (even below 10 μg/dL in blood) during early childhood is inversely associated with neuropsychological development through the first 10 years of life.[64–67] There is evidence that even blood lead levels below 10 μg/dL may be associated with reduced cognitive functioning.[68] Prenatal exposure to PCBs and methylmercury, predominantly from maternal seafood consumption, has been associated with neurocognitive deficits.[69]

Selected Research Findings from the CCCEH

PAHs

PAHs are commonly found in ambient air[70] and are listed among the 189 hazardous air pollutants covered under the Clean Air Act. Incomplete combustion of organic material (gasoline and diesel fuels, coal, oil, and tobacco products) is the major source of PAHs.[71,72] Common sources of PAHs in the urban environment include traffic, particularly diesel trucks and buses, heating fuels, and cigarette smoke. A number of PAHs are known human carcinogens, and laboratory studies indicate that the PAH fraction of respirable PM may cause developmental deficits in infants.[22,73] Studies also suggest that some of the PAH fraction of diesel exhaust particulates (DEP) may potentiate allergic responses.[74]

A CCCEH cohort study evaluated the effects of prenatal exposure to airborne PAHs (monitored during pregnancy by personal air sampling of the mother) on birth outcomes, after controlling for the effects of known determinants of fetal growth.[75] Researchers found mean birth weight and head circumference to be lower among newborns born to African American mothers with higher PAH exposures.[75] There was also a significant interaction between PAH adducts and ETS such that the combined exposure to high ETS and high adducts had a significant multiplicative effect on birth weight and head circumference among both African American and Dominican newborns.[76] The observed associations between prenatal PAHs and reduced fetal growth are of concern because several studies have reported that reduced birth weight or head circumference at birth or during the first year of life correlates with poorer cognitive functioning and school performance in childhood.[77] The finding was consistent with the prior observation that the levels of PAH–DNA adducts in cord blood of Caucasian newborns in Poland were significantly associated with lower birth weight, reduced length, and head circumference.[22] More recently, follow-up of children in the CCCEH had shown that elevated prenatal exposure to airborne PAHs (measured by personal monitoring during pregnancy) was associated with a significant increase in risk of developmental delay at age 3 (odds ratio 2.9).[78]

Pesticides

Insecticides are a class of chemicals that are generally designed to disrupt neurological pathways in animals and are thus effective for their original purpose of eliminating bothersome insects and other pests. Although insects are more susceptible to the effects of certain pesticides, including pyrethroids, because of lower levels of detoxifying enzymes, human exposure to pesticides may be of concern because of the widespread application of the toxicants. Approximately 80–90% of American households use pesticides.[79] Children are exposed to pesticides from multiple sources as the substances are sprayed on food, grass, and in homes and schools.

Organochlorines like dichlorodiphenyltrichloroethane (DDT) have been phased out and replaced with less persistent pesticides like OPs, carbamates, and pyrethroids. Chlorpyrifos, diazinon (OPs), *cis*-permethrin, *trans*-permethrin (pyrethroids), propoxur, and bediocarb (carbamates) have been the pesticides most commonly detected in house dust and indoor air.[80–82] Despite their lower persistence, OPs and carbamates can inhibit acetylcholinesterase in humans. The mechanisms of action of pyrethroids, pyrethrins, and organochlorines is to interfere with nerve cell function.[25] The neurotoxic effects of these pesticides may place burdens on early neurodevelopment of the fetus and may possibly have long-term effects on children, although experimental and epidemiological studies are quite limited. Animal studies have shown that exposure to relatively small doses of chlorpyrifos at key developmental moments can cause permanent changes in brain function.[62,83]

A CCCEH cohort study found that, prior to the EPA phase-out of the pesticide chlorpyrifos in 2001, 71% of newborns had detectable levels of the pesticide in cord blood.[84] The mean cord blood chlorpyrifos level was 6.9 pg/g, compared to 1.3 pg/g after the ban.[85] The researchers found that among babies born prior to the EPA phase-out, prenatal chlorpyrifos and diazinon exposures measured by cord blood concentrations were significantly associated with impaired fetal growth.[86] Lower air and cord blood levels of the pesticides as well as lack of association with birth weight and birth length were observed after the 2001 phase-out,[86] indicating the efficacy of the regulatory action.

Future Steps in Children's Environmental Health

In conclusion, much has been learned about certain types of risk factors for environmental health-related diseases in children. Rising rates of asthma, certain cancers, and developmental disability and the growing evidence that risk of certain adult disease is associated with *in utero* and childhood exposures indicate that maintaining an "early focus" can have a significant impact on the overall burden of disease. When preventive measures have been enacted based on this knowledge, children's health has benefited. Incorporating strategic principles to translate existing and future data into public health policy will ensure future benefits in children's environmental health.

REFERENCES

1. U. S. EPA. 1996. Proposed Guidelines for Carcinogen Risk Assessment. Office of Research and Development. Washington, DC.
2. NATIONAL ACADEMY OF SCIENCES. 1984. Toxicity Testing: Needs and Priorities. National Academy Press. Washington, DC.
3. GOLDMAN, L.R. & S.H. KODURU. 2000. Chemicals in the environment and developmental toxicity to children: a public health and policy perspective. Environ. Health Perspect. **108**(Suppl. 3):443–448.
4. LANDRIGAN, P.J., B. SONAWANE D. MATTISON, *et al.* 2002. Chemical contaminants in breast milk and their impacts on children's health: an overview. Environ. Health Perspect. **110**: A313–A315.
5. PERERA, F.P., S.M. ILLMAN, P.L. KINNEY, *et al.* 2002. The challenge of preventing environmentally related disease in young children: community-based research in New York City. Environ. Health Perspect. **110**: 197–204.

6. MANNINO, D.M., D.M. HOMA, L.J. AKINBAMI, *et al.* 2002. Surveillance for asthma—United States, 1980–1999. MMWR Surveill. Summ. **51:** 1–13.
7. RIES, L.A.G., M.P. EISNER, C.L. KOSARY, *et al.* 2002. SEER Cancer Statistics Review, 1973–1999. http://seer.cancer.gov/csr/1973_1999/. National Cancer Institute. Bethesda, MD. Accessed August 12, 2003.
8. BOYLE, C.A., P. DECOUFLE & M. YEARGIN-ALLSOPP. 1994. Prevalence and health impact of developmental disabilities in US children. Pediatrics **93:** 399–403.
9. BEARER, C.F. 1995. Environmental health hazards: how children are different from adults. Future Child. **5:** 11–26.
10. WILES, R. & C. CAMPBELL. 1993. Pesticides in Children's Food. Washington, DC: Environmental Working Group, 1993.
11. INTERNATIONAL PROGRAMME ON CHEMICAL SAFETY. 1986. Principles for Evaluating Health Risks From Chemicals During Infancy and Early Childhood: The Need for a Special Approach. Environmental Health Criteria, 59 ed. World Health Organization. Geneva, Switzerland.
12. SONAWANE, B. & R. BELILES. 1997. The susceptibility of children to immunotoxic and neurotoxic agents. Children's Environmental Health Network. 1st National Research Conference on Children's Environmental Health. Washington, DC.
13. PERERA, F.P. 1996. Molecular epidemiology: insights into cancer susceptibility, risk assessment, and prevention. J. Natl. Cancer Inst. **88:** 496–509.
14. NATIONAL RESEARCH COUNCIL. 2000. Toxicological Effects of Methylmercury. National Academy Press. Washington, DC.
15. RAMIREZ, G.B., M.C. CRUZ, O. PAGULAYAN, *et al.* 2000. The Tagum study I: analysis and clinical correlates of mercury in maternal and cord blood, breast milk, meconium, and infants' hair. Pediatrics **106:** 774–781.
16. CRAWFORD, M.A., W. DOYLE, P. DRURY, *et al.* 1989. n-6 and n-3 fatty acids during early human development. J. Intern. Med. **225**(Suppl. 1): 159–169.
17. VOIGT, R.G., C.L. JENSEN, J.K. FRALEY, *et al.* 2002. Relationship between omega3 long-chain polyunsaturated fatty acid status during early infancy and neurodevelopmental status at 1 year of age. J. Hum. Nutr. Diet. **15:** 111–120.
18. MAKRIDES, M., M. NEUMANN, K. SIMMER, *et al.* 1995. Are long-chain polyunsaturated fatty acids essential nutrients in infancy? Lancet **345:** 1463–1468.
19. HATCH, G.E. 1995. Asthma, inhaled oxidants, and dietary antioxidants. Am. J. Clin. Nutr. **61**(Suppl.): 625S–630S.
20. GREENE, L.S. 1995. Asthma and oxidant stress: nutritional, environmental, and genetic risk factors. J. Am. Coll. Nutr. **14:** 317–324.
21. PEAT, J.K., W.J. BRITON, C.M. SALOME, *et al.* 1987. Bronchial hyperresponsiveness in two populations of Australian school children. III. Effect of exposure to environmental allergens. Clin. Allergy **17:** 297–300.
22. PERERA, F.P., R.M. WHYATT, W. JEDRYCHOWSKI, *et al.* 1998. Recent developments in molecular epidemiology: a study of the effects of environmental polycylic aromatic hydrocarbons on birth outcomes in Poland. Am. J. Epidemiol. **147:** 309–314.
23. WANG, X., B. ZUCKERMAN, C. PEARSON, *et al.* 2002. Maternal cigarette smoking, metabolic gene polymorphism, and infant birth weight. JAMA 2002. **287:** 195–202.
24. MUTCH, E., P.G. BLAIN & F.M. WILLIAMS. 1992. Interindividual variations in enzymes controlling organophosphate toxicity in man. Hum. Exp. Toxicol. **11:** 109–116.

25. GREATER BOSTON PHYSICIANS FOR SOCIAL RESPONSIBILITY. 2000. In Harm's Way: Toxic Threats to Child Development. Greater Boston Physicians for Social Responsibility. Cambridge, MA.
26. BERKOWITZ, G.S., J.B. WETMUR, E. BIRMAN-DEYCH, et al. 2004. In utero pesticide exposure, maternal paraoxonase activity, and head circumference. Environ. Health Perspect. **112:** 388–391.
27. KAHN, R.S., J. KHOURY, W.C. NICHOLS & B.P. LANPHEAR. 2003. Role of dopamine transporter genotype and maternal prenatal smoking in childhood hyperactive-impulsive, inattentive, and oppositional behaviors. J. Pediatr. **143:** 104–110.
28. WHYATT, R.M., D.A. BELL, R.M. SANTELLA, et al. 1998. Polycyclic aromatic hydrocarbon-DNA adducts in human placenta and modulation by CYP1A1 induction and genotype. Carcinogenesis **19:** 1389–1392.
29. GILLILAND, F.D., Y.F. LI, L. DUBEAU, et al. 2002. Effects of glutathione S-transferase M1, maternal smoking during pregnancy, and environmental tobacco smoke on asthma and wheezing in children. Am. J. Respir. Crit. Care Med. **166:** 457–463.
30. ROMIEU, I., J.J. SIENRA-MONGE, M. RAMIREZ-AGUILAR, et al. 2004. Genetic polymorphism of GSTM1 and antioxidant supplementation influence lung function in relation to ozone exposure in asthmatic children in Mexico City. Thorax **59:** 8–10.
31. CASSEL, J. 1976. The contribution of the social environment to host resistance: the fourth Wade Hampton Frost Lecture. Am. J. Epidemiol. **104:** 107–123.
32. ZAPATA, B.C., A. REBOLLEDO, E. ATALAH, et al. 1992. The influence of social and political violence on the risk of pregnancy complications. Am. J. Public Health **82:** 685–690.
33. KLIEGMAN, R. 1992. Perpetual poverty: child health and the underclass. Pediatrics **89:** 710–713.
34. DAVID, R.J. & J.W. COLLINS. 1997. Differing birth weight among infants of US-born blacks, African-born blacks, and US-born whites. N. Engl. J. Med. **337:** 1209–1214.
35. WISE, P.H. 1993. Confronting racial disparatities in infant mortality: reconciling science and politics. Am. J. Prev. Med. **31:** 7–16.
36. RAUH, V.A., R.M. WHYATT, R. GARFINKEL, et al. 2004. Developmental effects of exposure to environmental tobacco smoke and material hardship among inner-city children. J. Neurotoxicol. Teratol. **26:** 373–385.
37. VAN LAREBEKE, N.A., L.S. BIRNBAUM, M.A. BOOGAERTS, et al. 2005. Unrecognized or potential risk factors for childhood cancer. Int. J. Occup. Environ. Health. **11:** 199–201.
38. PERERA, F.P. 2000. Molecular epidemiology: on the path to prevention? J. Natl. Cancer Inst. **92:** 602–612.
39. SRIVASTAVA, V.K., S.S. CHAUHAN, P.K. SRIVASTAVA, et al. 1986. Fetal translocation and metabolism of PAH obtained from coal fly ash given intratracheally to pregnant rats. J. Toxicol. Environ. Health **18:** 459–469.
40. NEUBERT, D. & S. TAPKEN. 1988. Transfer of benzo(a)pyrene into mouse embryos and fetuses. Arch. Toxicol. **62:** 236–239.
41. LU, L.J., R.M. DISHER, M.V. REDDY, & K. RANDERATH. 1986. 32P-postlabeling assay in mice of transplacental DNA damage induced by the environmental carcinogens safrole, 4-aminobiphenyl and benzo(a)pyrene. Cancer Res. **46:** 3046–3054.

42. WHYATT, R.M., W. JEDRYCHOWSKI, K. HEMMINKI, *et al.* 2001. Biomarkers of polycyclic aromatic hydrocarbon-DNA damage and cigarette smoke exposures in paired maternal and newborn blood samples as a measure of differential susceptibility. Cancer Epidemiol. Biomarker Prev. **10:** 581–588.
43. PERERA, F.P., D. TANG, W. JEDRYCHOWSKI, *et al.* 2004. Biomarkers in maternal and newborn blood indicate heightened fetal susceptibility to procarcinogenic DNA damage. Environ. Health Perspect. **112:** 1133–1136.
44. PERERA, F.P., D. TANG, R.M. WHYATT, *et al.* 2004. Comparison of PAH-DNA adducts in four populations of mothers and newborns in the U.S., Poland and China. 94th Annual Meeting. American Association for Cancer Research. Washington, DC.
45. CALABRESE, E.J. 1986. Age and Susceptibility to Toxic Substances. John Wiley and Sons. New York, NY.
46. NATIONAL ACADEMY OF SCIENCES. 1993. Pesticides in the Diets of Infants and Children. National Academy Press. Washington, DC.
47. LAIB, R.J., K.P. KLEIN & H.M. BOLT. 1985. The rat liver foci bioassay: age dependence of induction by vinyl chloride of ATP-deficient foci. Carcinogenesis **6:** 65–68.
48. HAGMAR, L., A. BROGGER, I.L. HANSTEEN, *et al.* 1994. Cancer risk in humans predicted by increased levels of chromosomal aberrations in lymphocytes: Nordic study group on the health risk of chromosome damage. Cancer Res. **54:** 2919–2922.
49. BONASSI, S., L. HAGMAR, U. STROMBERG, *et al.* 2000. Chromosomal aberrations in lymphocytes predict human cancer independently of exposure to carcinogens. European study group on cytogenetic biomarkers and health. Cancer Res. **60:** 1619–1625.
50. BOCSKAY, K.A., D. TANG, M.A. ORJUELA, *et al.* 2005. Chromosomal aberrations in cord blood are associated with prenatal exposure to carcinogenic polycyclyic aromatic hydrocarbons. Cancer Epidemiol. Biomarker Prev. **14:** 506–511.
51. PLUTH, J.M., M.J. RAMSEY & J.D. TUCKER. 2000. Role of maternal exposures and newborn genotypes on newborn chromosome aberration frequencies. Mutat. Res. **465:** 101–111.
52. ROSS, J.A. 1998. Maternal diet and infant leukemia: a role for DNA topoisomerase II inhibitors? Int. J. Cancer Suppl. **11:** 26–28.
53. DEY, A.N. & B. BLOOM. 2005. Summary health statistics for U.S. children: National Health Interview Survey, 2003. Vital Health Stat. **10:** 223.
54. NICHOLAS, S.W., B. JEAN-LOUIS, B. ORTIZ, *et al.* 2005. Addressing the childhood asthma crisis in Harlem: the Harlem children's zone asthma initiative. Am. J. Public. Health **95:** 245–249.
55. PLOPPER, C.G. & M.V. FANUCCI. 2000. Do urban environmental pollutants exacerbate childhood lung disease? Environ. Health Perspect. **108:** A252–A253.
56. MILLER, R.L., G. CHEW, C.A. BELL, *et al.* 2001. Prenatal exposure, maternal sensitization, and sensitization *in utero* to indoor allergens in an inner-city cohort. Am. J. Respir. Crit. Care. Med. **164:** 995–1001.
57. MILLER, R.L., R. GARFINKEL, M. HORTON, *et al.* 2004. Polycyclic aromatic hydrocarbons, environmental tobacco smoke, and respiratory symptoms in an inner-city birth cohort. Chest **126:** 1071–1078.
58. HONG, C.Y., S.E. CHIA, D. WIDJAJA, *et al.* 2004. Prevalence of respiratory symptoms in children and air quality by village in rural Indonesia. J. Occup. Environ. Med. **46:** 1174–1179.

59. JANSSEN, N.A., B. BRUNEKREEF, P. VAN VLIET, et al. 2003. The relationship between air pollution from heavy traffic and allergic sensitization, bronchial hyperresponsiveness, and respiratory symptoms in Dutch schoolchildren. Environ. Health Perspect. **111:** 1512–1518.
60. SALAMEH, P.R., I. BALDI, P. BROCHARD, et al. 2003. Respiratory symptoms in children and exposure to pesticides. Eur. Respir. J. **22:** 507–512.
61. FAUSTMAN, E.M. 2000. Mechanisms Underlying Children's Susceptibility to Environmental Toxicants. Environ. Health Perspect. **108:** (Suppl. 1): 13–21.
62. CHANDA, S.M. & C.N. POPE. 1996. Neurochemical and neurobehavioral effects of repeated gestational exposure to chlorpyrifos in maternal and developing rats. Pharmacol. Biochem. Behav. **53:** 771–776.
63. CAMPBELL, C.G., F.J. SEIDLER, & T.A. SLOTKIN. 1997. Chlorpyrifos interferes with cell development in rat brain regions. Brain Res. Bull. **43:** 179–189.
64. BAGHURST, P.A., A.J. MCMICHAEL, N.R. WIGG, et al. 1992. Environmental exposure to lead and children's intelligence at the age of seven years. The Port Pirie cohort study. N. Engl. J. Med. **327:** 1279–1284.
65. BELLINGER, D.C., K.M. STILES, & H.L. NEEDLEMAN. 1992. Low level lead exposure, intelligence and academic achievement: a long-term follow-up study. Pediatrics **90:** 855–861.
66. NEEDLEMAN, H. & C. GATSONIS. 1990. Low-level lead exposure and the IQ of children: A meta-analysis of modern studies. JAMA **263:** 673–678.
67. CANFIELD, R.L., C.R. HENDERSON, D.A. CORY-SLECHTA, et al. 2003. Intellectual impairment in children with blood lead concentrations below 10 μg per deciliter. N. Engl. J. Med. **348:** 1517–1521.
68. LANPHEAR, B.P., K. DIETRICH, P. AUINGER & C. COX. 2000. Cognitive deficits associated with blood lead concentrations <10 microg/dL in US children and adolescents. Public Health Rep. **115:** 521–529.
69. GRANDJEAN, P., P. WEIHE, V.W. BURSE, et al. 2001. Neurobehavioral deficits associated with PCB in 7-year-old children prenatally exposed to seafood neurotoxicants. Neurotoxicol. Teratol. **23:** 305–317.
70. INTERNATIONAL AGENCY FOR RESEARCH ON CANCER. 1983. Polynuclear Aromatic Compounds. Part 1. Chemical, Environmental, and Experimental Data. IARC Monographs on the Evaluation of the Carcinogenic Risk of Chemicals to Humans. International Agency for Research on Cancer, Lyon, France.
71. CHUANG, J.C., G.A. MACK, M.R. KUHLMAN & N.K. WILSON. 1991. Polycyclic aromatic hydrocarbons and their derivatives in indoor and outdoor air in an eight-home study. Atmos Environ. **25B:** 369–380.
72. LEWTAS, J. 1994. Human exposure to complex mixtures of air pollutants. Toxicol. Lett. **72:** 163–169.
73. COHEN, A.J. & C.A. POPE III. 1995. Lung cancer and air pollution. Environ. Health Perspect. **103**(Suppl. 8): 219–224.
74. TAKENAKA, H., K. ZHANG, D. DIAZ-SANCHEZ, et al. 1995. Enhanced human IgE production results from exposure to the aromatic hydrocarbons from diesel exhaust: direct effects on B-cell IgE production. J. Allergy Clin. Immunol. **95:** 103–115.
75. PERERA, F., V. RAUH, W.Y. TSAI, et al. 2003. Effects of transplacental exposure to environmental pollutants on birth outcomes in a multi-ethnic population. Environ Health Perspect. **111:** 201–205.

76. PERERA, F.P., V. RAUH, R.M. WHYATT, *et al.* 2004. Molecular evidence of an interaction between prenatal environmental exposures on birth outcomes in a multiethnic population. Environ. Health Perspect. **112:** 662–630.
77. HACK, M., N. BRESLAU, B. WEISSMAN, *et al.* 1991. Effect of very low birth weight and subnormal head size on cognitive ability at school age. N. Engl. J. Med. **325:** 231–237.
78. PARERA, F.P., V. RAUH, R.W. WHYATT, *et al.* Effect of prenatal exposure to airborne polycyclic aromatic hydrocarbons on neurodevelopment in the first three years of life among inner-city children. Environ. Health Perspect. doi:10.1289/ehp.9084. [Online 24 April 2006].
79. LANDRIGAN, P.J., L. CLAUDIO, S.B. MARKOWITZ, *et al.* 1999. Pesticides and inner-city children: exposures, risks, and prevention. Environ. Health Perspect. **107**(Suppl. 3): 431–437.
80. WHITMORE, R., F.W. IMMERMAN, D.E. CAMANN, *et al.* 1994. Non-occupational exposures to pesticides for residents of two U.S. cities. Arch. Environ. Contam. Toxicol. **26:** 47–59.
81. LEWIS, R.G., R.C. FORTMANN & D.E. CAMANN. 1994. Evaluation of methods for monitoring potential exposure of small children to pesticides in the residential environment. Arch. Environ. Contam. Toxicol. **26:** 37–46.
82. CAMANN, D.E., J.S. COLT, S.I. TEITELBAUM, *et al.* 2000. Pesticide and PAH distributions in house dust from seven areas of USA. SETAC Conference, Abstract 570. Society of Environmental Toxicology and Chemistry.
83. SONG, X., F.J. SEIDLER, J.L. SALEH, *et al.* 1997. Cellular mechanisms for developmental toxicity of chlorpyrifos: targeting the adenylyl cyclase signaling cascade. Toxicol. Appl. Pharmacol. **145:** 158–174.
84. WHYATT, R.M., D.B. BARR, D.E. CAMANN, *et al.* 2003. Contemporary-use pesticides in personal air samples during pregnancy and blood samples at delivery among urban minority mothers and newborns. Environ. Health Perspect. **111:** 749–756.
85. WHYATT, R.M., D. CAMANN, F.P. PERERA, *et al.* 2005. Biomarkers in assessing residential insecticide exposures during pregnancy and effects on fetal growth. Toxicol. Applied Pharmacol. **206:** 246–254.
86. WHYATT, R.M., D.E. CAMANN, Y. COSME, *et al.* 2004. Persistence of chlorpyrifos and diazinon in the indoor environment following U.S. regulatory action to ban residential use. International Society for Exposure Analysis, Abstract.

Interpreting Cancer Trends

ELIZABETH M. WARD, MICHAEL J. THUN, LINDSAY M. HANNAN, AND AHMEDIN JEMAL

Department of Epidemiology and Surveillance Research, American Cancer Society, National Home Office, Atlanta, Georgia 30329-4251, USA

ABSTRACT: The interpretation of cancer incidence trends is complicated by short-term random variation, artifactual fluctuations introduced by screening, changes in diagnosis or disease classification, completeness of reporting, and by the multiplicity of factors that may affect risk for specific cancer sites. We analyzed trends in 56 different cancer sites and subsites in the U.S. SEER registries in the period 1975–2002 using joinpoint analysis. The increase in cancer incidence for all sites combined that became evident with the inception of the SEER registries in the mid-1970s has abated since the early 1990s. Among the 15 most common cancer sites in men, sites with increasing incidence rates during the most recent time period include melanoma of the skin and cancers of the prostate, kidney and renal pelvis (kidney), and esophagus. Among women, incidence rates are increasing for leukemia, non-Hodgkin's lymphoma, melanoma, and cancers of the breast, thyroid, urinary bladder, and kidney. Incidence rates for all childhood cancers combined increased 0.6% per year from 1975 to 2002. Cancer mortality rates have decreased in the United States since 1991 in both men and in women; site-specific death rates have decreased in the most recent time period for 12 of the top 15 cancer sites in men and 9 of the top 15 cancer sites in women. Similar trends in cancer incidence and mortality have been reported in other industrialized countries. Possible reasons for these trends are discussed.

KEYWORDS: cancer incidence trends; joinpoint analysis; breast cancer; prostate cancer; non-Hodgkin's lymphoma; melanoma; kidney cancer; esophageal cancer; thyroid cancer; testicular cancer; Hodgkin's lymphoma; multiple myeloma; leukemia; childhood cancer

INTRODUCTION

Temporal trends in the incidence of particular types of cancer may reflect changes in exposure to underlying etiologic factors, changes in classification,

---

Address for correspondence: Elizabeth Ward, Department of Epidemiology and Surveillance Research, American Cancer Society, National Home Office, 1599 Clifton Road, NE, Atlanta, GA 30329-4251. Voice: 001 404-327-6552; fax: 001-327-6450.
  e-mail: elizabeth.ward@cancer.org

or the introduction of new screening or diagnostic tests. Inference about the extent to which environmental pollution may contribute to these trends must be interpreted against a backdrop of the large and changing impact of tobacco use, diet, and physical activity; changes in the prevalence of certain infectious agents; and artifacts of changes in classification and diagnosis. This article will provide an overview of cancer trends in the United States and discuss factors thought to be influencing trends for selected cancer sites. We discuss cancers with recent increasing trends because they are a focus of public concern, and provide opportunities for etiologic research and prevention.

METHODS

We present incidence trends for the leading 15 cancer sites in men and women on the basis of the data from the U.S. SEER Registries, 1975–2002, as previously published.[1] We also examine recent trends for 41 additional cancer sites as well as selected anatomic and histological subtypes, classified by ICD-O-3 site and histology codes using analytic software provided in SEER*Stat.[2-4] Incidence rates are based on people of all ages and are age-adjusted according to the U.S. 2000 standard population. Trends are assessed using joinpoint regression analysis by fitting a series of joined straight lines on a log scale to the trends in the age-adjusted rates. A maximum of three joinpoints (four line segments) were allowed. Where possible, trends are adjusted for reporting delay, which accounts for late registration of cases and other corrections that occur after the standard reporting period.

Trends are examined for men and women separately; race-specific trends are presented for selected cancer sites. Trends are described as increasing or decreasing when the annual percent change (APC) for the most recent time period is statistically significant ($P < 0.05$); otherwise, the trend is described as stable (flat). Incidence rates for childhood cancer are grouped by primary site and are presented both for ages 0–14 and ages 0–19 for males and females combined. Joinpoint analyses are presented for all childhood cancers and the most common types of cancer in children, all leukemias combined, acute lymphocytic leukemia (ALL), and brain cancers. In addition, for the nine most common cancer sites in children, we report APCs for the total time interval 1975–2002, as well as two equal segments (1975–1988 and 1989–2002), as previously published by SEER.[4]

RESULTS

TABLE 1 summarizes long-term cancer incidence trends for the top 15 cancer sites in men and women. In men, incidence rates (per 100,000) for all cancer sites combined rose from 466.5 in 1975 to 657.3 in 1992 and then fell to

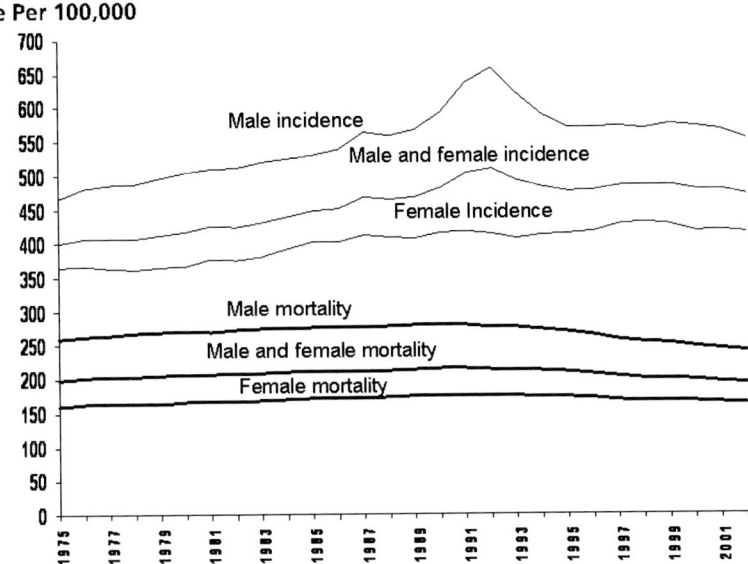

FIGURE 1. Annual age-adjusted cancer incidence and death rates for all sites, by sex, US, 1975–2002.

553.2 in 2002. The most recent trend from the delay-adjusted joinpoint model for all cancer sites combined in men was flat from 1995 to 2002. In women, cancer incidence rose from 365.7 in 1975 to a peak of 431.5 in 1998, and then decreased to 415.1 in 2002. The most recent trend from the delay-adjusted joinpoint model for women encompassed the interval 1987–2002 and increased slightly (by 0.3% per year). In contrast to cancer incidence rates, cancer death rates decreased since 1993 in men (APC = −1.5) and since 1992 in women (APC = −0.8) (FIG. 1). The recent decrease in cancer mortality is largely attributable to declines in death rates from lung, prostate, and colorectal cancer in men and declines in death rates for breast and colorectal cancer in women.

Among the 15 most common cancer sites in men, sites with increasing incidence rates during the most recent period include melanoma of the skin and cancers of the prostate, kidney and renal pelvis (kidney), and esophagus. Decreasing trends among men are observed for cancers of the lung and bronchus, colon and rectum, oral cavity and pharynx, stomach, and larynx. Incidence rates for the remaining 15 cancer sites are stable.[1] A large increase in incidence rates for liver cancer in men (APC = 4.5) occurred from 1984 to 1999, but the incidence rates for liver cancer now appear to be stabilizing in men (TABLE 1). A similar pattern is observed for women (not shown in TABLE 1),

TABLE 1. SEER incidence rate trends with joinpoint[a] analyses for 1975 through 2002 for the top 15 cancers, for races, adjusted for delay in reporting[b]

	Joinpoint analyses (1975–2002)[d]							
	Trend 1		Trend 2		Trend 3		Trend 4	
	Years	APC[e]	Years	APC[e]	Years	APC[e]	Years	APC[e]
All sites[f]								
Both sexes	1975–1983	0.9[g]	1983–1992	1.8[g]	1992–1995	–1.7	1995–2002	0.3
Male	1975–1989	1.3[g]	1989–1992	5.2[g]	1992–1995	–4.7[g]	1995–2002	0.2
Female	1975–1979	–0.2	1979–1987	1.5[g]	1987–2002	0.3		
Top 15 for males[c]								
Prostate	1975–1988	2.6[g]	1988–1992	16.5[g]	1992–1995	–11.2[g]	1995–2002	1.7[g]
Lung and bronchus	1975–1982	1.5[g]	1982–1991	–0.4	1991–2002	–1.8[g]		
Colon and rectum	1975–1986	1.1[g]	1986–1995	–2.1[g]	1995–1998	1.0	1998–2002	–2.5[g]
Urinary bladder	1975–1987	1.0[g]	1987–1996	–0.5	1996–2000	1.6	2000–2002	–2.6
Non-Hodgkin's lymphoma	1975–1991	4.3[g]	1991–2002	0.2				
Melanoma of the skin[h]	1975–1985	5.8[g]	1985–2002	3.8[g]				
Leukemia	1975–2002	0.1						
Oral cavity and pharynx	1975–2002	–1.2[g]						
Kidney and renal pelvis	1975–2002	1.8[g]						
Stomach	1975–1988	–1.2[g]	1988–2002	–2.0[g]				
Pancreas	1975–1981	–1.8[g]	1981–1985	1.1	1985–1990	–2.1	1990–2002	0.1
Liver and intrahepatic bile duct	1975–1984	1.7	1984–1999	4.5[g]	1999–2002	–0.7		
Brain and other nervous system	1975–1989	1.2[g]	1989–2002	–0.3				
Esophagus	1975–2002	0.8[g]						
Larynx	1975–1988	–0.3	1988–2002	–2.8[g]				

Continued.

TABLE 1. Continued

	Joinpoint analyses (1975–2002)[d]							
	Trend 1		Trend 2		Trend 3		Trend 4	
	Years	APC[e]	Years	APC[e]	Years	APC[e]	Years	APC[e]
Top 15 for females[c]								
Breast	1975–1980	−0.4	1980–1987	3.7[g]	1987–2002	0.4[g]		
Lung and bronchus	1975–1982	5.5[g]	1982–1990	3.5[g]	1990–1998	1.0[g]	1998–2002	−0.5
Colon and rectum	1975–1985	0.3[g]	1985–1995	−1.8[g]	1995–1998	1.5[g]	1998–2002	−1.5[g]
Corpus and uterus, NOS	1975–1979	−6.0[g]	1979–1988	−1.7[g]	1988–1997	0.7[g]	1997–2002	−0.6
Non-Hodgkin's lymphoma	1975–1990	2.9[g]	1990–2002	1.2[g]				
Ovary	1975–1985	0.2	1985–2002	−0.7[g]				
Melanoma of the skin[h]	1975–1981	6.1[g]	1981–1993	2.2[g]	1993–2002	4.1[g]		
Pancreas	1975–1984	1.2[g]	1984–2002	−0.2				
Thyroid	1975–1981	−1.2	1981–1993	2.0[g]	1993–2002	5.3[g]		
Cervix uteri	1975–1981	−4.6[g]	1981–1997	−1.1[g]	1997–2002	−4.5[g]		
Leukemia	1975–2002	0.2[g]						
Urinary bladder	1975–2002	0.2[g]						
Kidney and renal pelvis	1975–1990	2.8[g]	1990–2002	1.6[g]				
Oral cavity and pharynx	1975–1980	2.5	1980–2002	−0.9[g]				
Stomach	1975–2002	−1.7[g]						

[a] Joinpoint (JP) Regression Program, Version 3.0. April 2004, National Cancer Institute.
[b] Sources of data are SEER 9 areas (Connecticut, Hawaii, Iowa, Utah, and New Mexico, and the metropolitan areas of San Francisco, Detroit, Atlanta, and Seattle-Puget Sound).
[c] The top 15 cancers were selected based on the sex-specific age-adjusted rate for 1992–2002 for all races combined.
[d] Joinpoint analyses with up to three joinpoints.
[e] APC based on rates that were age-adjusted to the 2000 U.S. Std Population.
[f] All sites excludes myelodysplastic syndromes and borderline tumors; ovary excludes borderline tumors.
[g] APC is statistically significantly different from zero (two-sided $P < 0.05$).
[h] Age-adjusted rates for melanoma of the skin are calculated using white patients only.

TABLE 2. Summary of most recent SEER incidence trends with joinpoint[a] analyses for 1975–2002 for the 56 sites and 13 histological subtypes examined[b]

Trend (APC)[c]	Sites	
	Males (49 sites)[d]	Females (51 sites)[d]
Flat	14 (28.6%)	19 (37.3%)
Increasing	16 (32.7%)	16 (31.4%)
Decreasing	19 (38.8%)	16 (31.4%)

Trend (APC)[c]	Histological subtypes	
	Males (10 subtypes)[d]	Females (13 subtypes)[d]
Flat	2 (20%)	1 (7.7%)
Increasing	4 (40%)	8 (61.5%)
Decreasing	4 (40%)	4 (30.8%)

[a]Joinpoint (JP) Regression Program, Version 3.0. April 2004, National Cancer Institute.
[b]Sources of data are SEER 9 areas (Connecticut, Hawaii, Iowa, Utah, and New Mexico, and the metropolitan areas of San Francisco, Detroit, Atlanta, and Seattle-Puget Sound).
[c]The APC for the trend based on the rates age-adjusted to the 2000.
U.S. standard population was deemed increasing or decreasing when statistically significant (<0.05) and flat otherwise.
[d]Percent of sites/subsites examined.

with an increase of 4.3% per year from 1983 to 1999, but a stable trend for 2000–2002 (data not shown).

In women, incidence rates for the top 15 cancer sites increased in the most recent time period for leukemia, non-Hodgkin's lymphoma (NHL), melanoma, and cancers of the breast, thyroid, urinary bladder, and kidney. Incidence rates decreased for cancers of the colon and rectum, ovary, cervix uteri, oral cavity, and stomach.[1]

Considering all 56 cancer sites and 13 histological subtypes examined, recent trends were approximately evenly divided between those for which rates increased, decreased, and were stable. In contrast, more of the trends for histological or anatomic subsites were increasing, as was expected since these subsites were examined on the basis of a priori expectation of a trend. (TABLE 2)

Cancers with Recent Increasing Trends

Breast Cancer

Breast cancer incidence increased most rapidly among white women between 1980 and 1988, during the period when mammography screening was widely disseminated (FIG. 2). Subsequently, breast cancer incidence rates among white

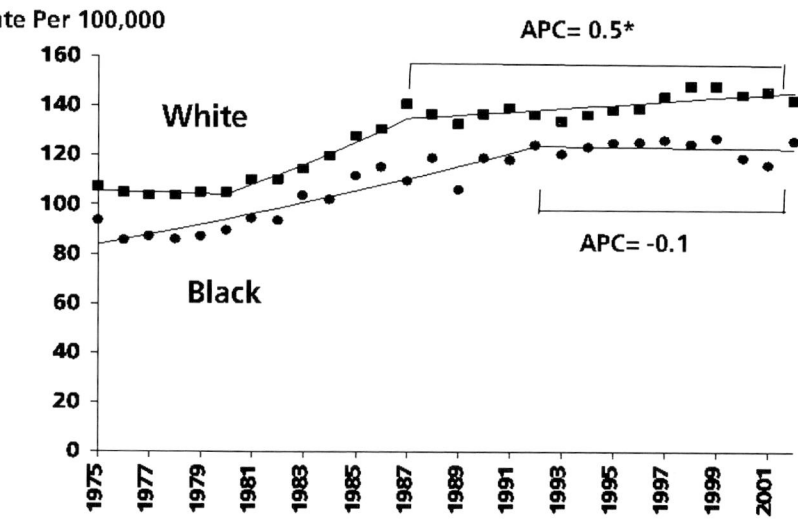

FIGURE 2. Trends in age-standardized incidence of female breast cancer by race, adjusted for delay of reporting, 1975–2002.

women continued to increase, but more gradually. Breast cancer incidence rates among black women increased from 1975 through 1992, then stabilized (FIG. 2). The continuing increase in breast cancer incidence since the late 1980s has been confined to white women over age 50 (data not shown). In contrast, among women under age 50, breast cancer incidence has declined in the most recent time period among both white (APC = −0.2) and black (APC = −1.0) women.

Long-term trends for breast cancer are influenced by historical changes in the prevalence of known risk factors, such as age at menarche and reproductive history, as well as by screening.[5–7] The increase among white women primarily involves small (≤2 cm) and localized-stage tumors, although a small increase in the incidence of regional stage tumors and those larger than 5 cm has occurred since the early 1990s.[8] Likely reasons for the continuing increase in incidence for white women over age 50 include the increasing prevalence, frequency, and sensitivity of mammography; use of combined estrogen-progesterone therapy by postmenopausal women; and the increasing prevalence of obesity.[8]

Although it is well known that breast cancer is associated with cumulative exposure to estrogen from endogenous and medical sources, the potential role

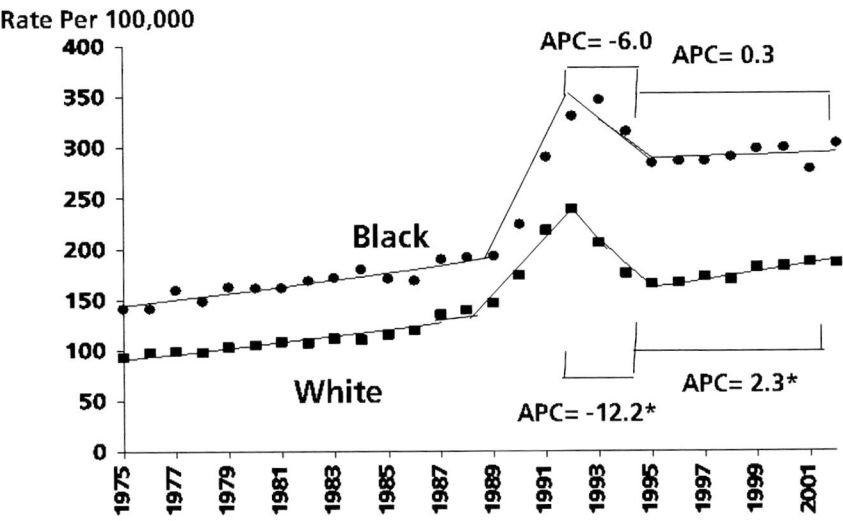

FIGURE 3. Trends in age-standardized incidence of prostate cancer by race, adjusted for delay of reporting, 1975–2002.

of low-level environmental contaminants is unclear. Research priorities include better understanding the biology of breast development, the natural history of breast cancer, and the role of exposures during susceptible periods such as puberty.[9]

Prostate Cancer

Prostate cancer incidence among U.S. men increased slowly from 1975 until the late 1980s, rose sharply with the introduction of prostate-specific antigen (PSA) screening until 1992, and then declined sharply until 1995. Subsequently, rates stabilized in black men and resumed a gradual increase among white men (FIG. 3). The large peak in prostate cancer incidence resulted from the introduction of PSA screening into a population with high prevalence of previously unrecognized disease.[10] Autopsy studies demonstrate a high prevalence of subclinical prostate cancer in older men, therefore, the observed incidence rate is highly susceptible to perturbation by diagnostic changes. Widespread use of transurethral resection of the prostate for benign disease is thought to have contributed to rising incidence rates even before use of PSA.[10,11]

Etiologic factors and reason(s) for the substantially higher rates of prostate cancer in black than white males are not well understood. Known risk factors include age, race, and family history. Diets high in lycopenes have been associated with lower risk in several studies.[12] International ecological studies suggest that diets high in saturated fat may be associated with increased risk, but results of within-country studies of dietary factors have not substantiated this.[13] Human prostate carcinomas are usually androgen sensitive, suggesting that steroid hormones, particularly androgens, play a role in human prostate carcinogenesis.[14]

Non-Hodgkin's Lymphoma

The incidence of NHL increased rapidly among men from 1975 through 1991, then stabilized through 2002 (TABLE 1). In women, a rapid increase in incidence from 1975 through 1990 was followed by a more gradual increase of 1.2% per year. Some of the increase in NHL rates between 1975 and 1990 resulted from an increase in AIDS-related histological subtypes (primary central nervous system [CNS] lymphomas, high-grade immunoblastic and Burkitt lymphoma, and, to a lesser extent, intermediate grade large-cell diffuse lymphomas). After 1990, the incidence of most AIDS-related subtypes decreased with the introduction of highly active antiretroviral therapy (HAART).[15] However, the changing rates of AIDS-related lymphomas do not completely explain the trends in NHL incidence rates. The impact of improved diagnosis and classification changes is thought to be small.[16]

Established risk factors for NHL include congenital immunodeficiency, immunosuppressive drugs, autoimmune disorders, and infections (HIV, HTLV-1, EBV).[17] Occupational risk factors that are associated with NHL include exposure to pesticides and herbicides, especially 2,4-dichlorophenoxyacetic acid; and farming.[17] There are conflicting data on the association between solar ultraviolet light exposure and increased risk of NHL.[18]

Melanoma

The incidence of melanoma of the skin has been rising among white men and women in the United States since 1975 (TABLE 1). Similar increases have been observed among light-skinned individuals in many industrialized countries. These trends are thought to be largely attributable to historical trends in behaviors that affect sun exposure,[19] although increased awareness and detection may have a smaller influence. Skin pigmentation and the presence of melanocytic nevi and freckles are associated with risk.[20] Research questions concern the extent of protection offered by sunscreens[21] and the possible contributing role of ozone depletion.

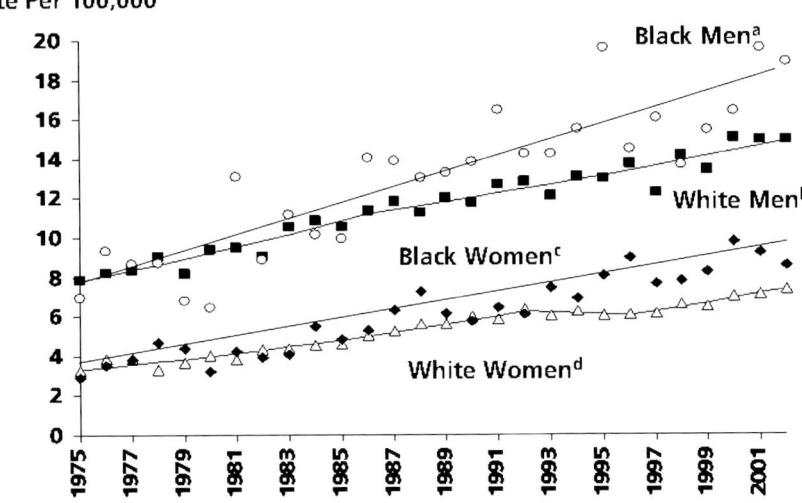

FIGURE 4. Trends in age-standardized incidence of kidney cancer (adenoma) by race and sex, 1975–2002.

Kidney Cancer

Incidence rates for cancer of the kidney (classification includes tumors of the renal pelvis) are rising in both men and women in the United States (TABLE 1). Much or all of this increase involves adenocarcinoma of the kidney (renal cell cancer), which is increasing in all race/gender groups (FIG. 4). Similar increases in renal cell cancer have been observed in many other countries and regions.[22] Increased detection of presymptomatic tumors by imaging procedures, such as ultrasonography, computed tomography, and magnetic resonance imaging contributes to, but does not fully explain the trend.[23] Tobacco smoking is an important risk factor for both adenocarcinomas of the renal parenchyema and transitional call cancers of the renal pelvis. Obesity is a strong risk factor for cancers of the renal parachyma but not the renal pelvis. Hypertension and antihypertensive therapies are also associated with increased risk of renal adenocarcinoma. Past use of phenacetin, no longer available in the United States, was associated with increased risk of cancer of the renal pelvis. Increased risk of renal cell cancer has been reported in a number of occupational groups. Male rats exposed to vapors of unleaded gasoline have an excess of renal cancers;

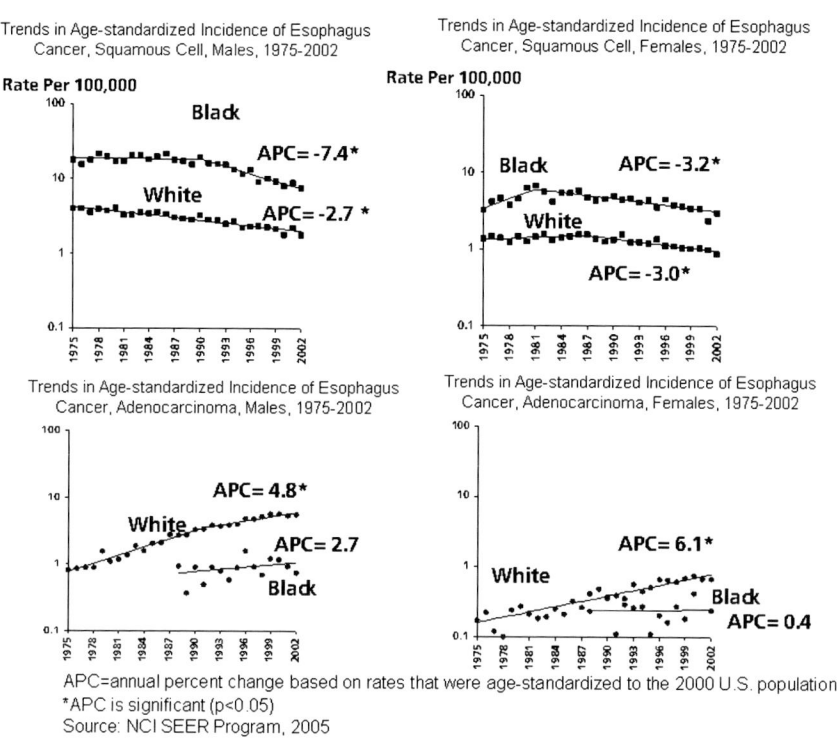

FIGURE 5. Trends in age-standardized incidence of esophageal adenocarcinoma and esophageal squamous cell carcinoma by race and sex, 1975–2002.

kidney cancer elevations associated with gasoline-exposed populations have been observed, but not consistently.[24]

Esophageal Cancer

The incidence of esophageal cancer (all cell types combined) has been rising slowly in men 1975–2002 (APC = 0.5) and is stable in women (APC = −0.3). These site-specific trends mask large changes by histological type (FIG. 5). The incidence of squamous cell carcinoma of the esophagus, historically the most common histological type, decreased among white men and women from 1975 through 2002, among black men after 1990 and among black women after 1981. In contrast, the incidence of adenocarcinoma of the esophagus rose steeply among whites from 1975 to 2002, while remaining stable among black men and women. Similar increases in adenocarcinoma of the esophagus and decreases in squamous cell carcinoma of the esophagus have been observed in other industrialized countries.[25]

Tobacco and alcohol consumption are the most important causes of squamous cell carcinoma that occurs in the upper third of the esophagus. Risk factors for adenocarcinoma of the esophagus, generally occurring in the lower third of the esophagus, include obesity and gastroesophageal reflux disease (GERD), and Barrett's esophagus.[26] Barrett's esophagus is a premalignant condition that involves chronic inflammation and dysplasia in the lower esophagus. One puzzling aspect of the trend is why adenocarcinoma of the esophagus has increased so sharply in white, but not black, men and women. It is hypothesized that the difference may result from a higher prevalence of *Helicobacter pylori* infection in black men and women.[27] Chronic infection with *H. pylori* is a risk factor for stomach cancer but may protect against adenocarcinoma of the esophagus.[28]

Thyroid Cancer

The incidence of thyroid cancer (all histological types combined) has increased since 1980 in both men and women; the most recent trends are sharply increasing in men (APC = 11.6%, 2000–2002) and women (APC 5.3%, 1993–2002). Analysis of this trend suggests that much of this increase is due to a diagnostic artifact. Nearly all of the increase involves small, well-differentiated papillary tumors, by far the most common histological subtype. In contrast, incidence rates for the second most common type of thyroid cancer (differentiated follicular cell) have decreased since 1975. Factors thought to account for the overall increase in the diagnosis of thyroid cancer are the change in coding rules to include benign-appearing papillary lesions (formerly termed *papillary adenomas*) and diagnostic improvements such as the introduction of ultrasound and thin needle aspiration.[29] Diagnostic advances have a large impact on incidence rates because the prevalence of occult (i.e., <1.5 cm) thyroid carcinoma at autopsy is high, ranging from 6% to 36% in various studies.

Incidence rates for thyroid cancer are almost three times higher in women than in men. This female predominance may also reflect greater diagnostic scrutiny in women. Small papillary tumors of the thyroid are often detected through ultrasound and fine-needle aspiration for benign thyroid conditions such as Graves' disease, Hashimoto's thyroiditis and nonendemic goiter, the prevalence of which are severalfold higher in women than men.

Etiologic factors for thyroid cancer include radiation exposure; some of the increase in incidence can be attributed to from past use of radiotherapy for benign conditions and diagnostic X ray exposure. Exposure to I^{131} contamination from the Chernobyl nuclear power plant accident increased risk of thyroid cancer in children a dose-related fashion; both iodine deficiency and iodine supplementation modified this risk.[30,31] Although nuclear weapons tests conducted in Nevada during 1951–1962 may have resulted in a modest increase in thyroid cancer risk among people who were exposed as children, there are considerable uncertainties in estimates of I^{131} dose and effect.[31] Increased thyroid cancer incidence resulting from I^{131} exposure among children born in the 1950s does not appear to be a factor in the increase in thyroid cancer incidence at the national level, because birth cohort analyses for all SEER areas combined show similar increases in incidence among children born in the 1950s as occurred in preceding and subsequent birth cohorts (data not shown). The relationship between thyroid cancer and administration of radioactive iodine for diagnostic and therapeutic purposes is unclear.

Thyroid cancer incidence is increased in geographical areas where a low-iodine diet results in a high prevalence of endemic goiter.[29] Exposure to substances that affect thyroid function may affect thyroid cancer risk in highly exposed subgroups of the population. For example, ethylene bisthiocarbamate (EBDC), a widely used fungicide, is metabolized to ethylene thiourea (ETU), which causes decreased thyroid hormone, increased thyroid stimulating hormone (TSH), and thyroid tumors in rats. A small study of EBDC applicators in Mexico found significantly higher TSH levels in applicators than in controls, suggesting that increased TSH compensated for a thyroid hormone decrease (altered homeostasis).[32] No studies have directly examined the implications of this for thyroid cancer, either in highly exposed workers or at the much lower exposure levels in the general population.

Testicular Cancer

The incidence of testicular cancer increased substantially from 1975 to 2002 in both white (APC = 1.7%) and black (APC = 2.8%) men. Incidence rates in the United States are about three times higher in white than black men. Testicular cancer rates have been rising in white populations of European ancestry worldwide.[33] Increased detection is unlikely to play a major role because most testicular cancers can be palpated at diagnosis.[33]

The worldwide increase in testicular cancer incidence is thought to represent a real increase, although the factors contributing to the change are not well understood. Cryptorchidism is the major identifiable risk factor associated with the development of testicular cancer, with an odds ratio between 2.5 and 14 in case–control studies.[34] Although an increase in cryptorchidism has been reported in England and Wales, the data on temporal trends in cryptorchidism in other populations are limited.[34,35] Use of exogenous estrogens during pregnancy is associated with an increased risk of testicular cancer in male offspring. However, because exogenous estrogens are not widely used in the United States and are not used in other countries where testicular cancer incidence is rising this is an unlikely explanation for the increasing incidence.[34] Ecologic analyses have given rise to the hypothesis that testicular cancer trends may be related to trends in maternal smoking during pregnancy, but this is not supported by most case–control studies to date.[33] Low birth weight and low gestational age have been reported in some studies to increase testicular cancer risk; an increased rate of testicular cancer among survivors of premature birth may contribute to rising incidence.[33]

It has been hypothesized that testicular dysgenesis syndrome may contribute to the rising incidence of testicular cancer and to low and possibly declining semen quality. The antiandrogenic effect of phthalates has been studied as a potential disruptor of embryonal programming and gonadal development during fetal life.[36] Prenatal exposure to phthalates in male rodents impairs testicular function and shortens anogenital distance.[37] A recent study of anogenital distance in 136 boys 2–36 months of age found that anogenital distance was inversely related to concentrations of four phthalate metabolites in maternal urine during pregnancy.[38] The possible relationship of phthalates to testicular cancer is an active area of research.

Hodgkin's Lymphoma, Multiple Myeloma, Leukemia

Incidence rates for Hodgkin's lymphoma were stable from 1975 to 2002 have decreased slightly among men and increased slightly among women (TABLE 3). Rates for multiple myeloma increased steadily for all races (TABLE 3) and for black and white men and women (FIG. 6). The incidence of multiple myeloma is about twice as high in black compared to white men and women. Reasons for the increasing incidence and greater risk among black men and women are unknown. Several studies suggest that multiple myeloma is associated with obesity,[39] but not with cigarette or alcohol consumption.[40,41] The risk of multiple myeloma is significantly elevated in people with a first-degree relative who developed multiple myeloma.[42] An inverse association between risk of multiple myeloma and occupation-based socioeconomic status, education, and income.[43] Other studies provide some support for a relationship between multiple myeloma and ionizing radiation exposure and evidence for an increased risk of multiple myeloma among agricultural workers.[44] Benzene has been associated with multiple myeloma in some studies but not others.[44] Exposure to immunologic challenges and variability in HLA subtypes have been associated with multiple myeloma risk in some studies.[45,46]

The incidence of all leukemia from 1975 to 2002 was stable in men and increased slightly in women (TABLE 3). Incidence rates are declining or stable in both men and women for all of the major subtypes of leukemia except for ALL in women and acute myeloid leukemia (AML) in both men and women for which rates are increasing (TABLE 3). Reasons for increases in ALL in women and AML in men and women and decreases in chronic lymphocytic leukemia and chronic myeloid leukemia in both men and women are unknown. Cigarette smoking is an established cause of AML.[47] Several studies have reported an increased risk of AML associated with overweight and obesity.[48] Occupational exposure to benzene and radiation are associated with increased risk of AML, as are a number of other occupational exposures.[49]

TABLE 3. SEER incidence rate trends with joinpoint[a] analyses for selected hematopoetic cancers, 1975–2002 for selected cancers, all races[b]

	Joinpoint analyses (1975–2002)[c]					
	Trend 1		Trend 2		Trend 3	
	Years	APC[e]	Years	APC[e]	Years	APC[e]
Hodgkin's lymphoma						
Males	1975–2002	−0.7[f]				
Females	1975–2002	0.4[f]				
Non-Hodgkin's lymphoma						
Males	1975–1991	4.3[f]	1991–2002	0.2		
Females	1975–1990	2.9[f]	1990–2002	1.2[f]		
Multiple myeloma						
Males	1975–2002	0.9[f]				
Females	1975–2002	0.7[f]				
Leukemia						
Males	1975–2002	0.1				
Females	1975–2002	0.2[f]				
Acute lymphocytic leukemia[d]						
Males	1975–1988	2.9[f]	1988–2002	−0.6		
Females	1975–2002	1.2[f]				
Chronic lymphocytic leukemia[d]						
Males	1975–1988	0.6[f]	1988–2002	−1.3[f]		
Females	1975–1994	0.1[f]	1994–2002	−2.9[f]		
Acute myeloid leukemia[d]						
Males	1975–1990	−0.7[f]	1990–2002	1.8[f]		
Females	1975–1987	−1.2	1987–2002	1.8[f]		
Chronic myeloid leukemia[d]						
Males	1975–1977	−13.2	1977–1997	0.7	1997–2002	−4.9
Females	1975–2002	−0.4[f]				

[a]Joinpoint (JP) Regression Program, Version 3.0. April 2004, National Cancer Institute.
[b]Sources of data are SEER 9 areas (Connecticut, Hawaii, Iowa, Utah, and New Mexico, and the metropolitan areas of San Francisco, Detroit, Atlanta, and Seattle-Puget Sound).
[c]Joinpoint analyses with up to three joinpoints.
[d]Not adjusted for delay in reporting.
[e]APC based on rates that were age-adjusted to the 2000 U.S. Std Population.
[f]APC is statistically significantly different from zero ($P < 0.05$).

Cancer in Childhood

Trends in childhood cancer incidence were similar in age group 0–14 years and 0–19 years (TABLES 4 and 5); therefore, for simplicity, we limit our discussion to trends in age 0–19. Incidence rates for all childhood cancers increased by 0.6% per year from 1975 to 2002, based on joinpoint analysis (FIG. 6). Contributing to this trend were rising incidence rates of leukemia, which increased 0.7% per year from 1975 to 2002, and of brain cancer, which increased by

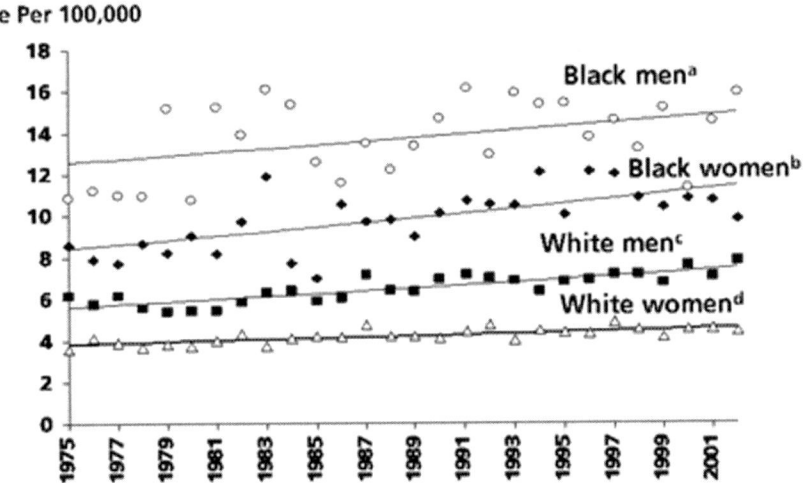

APC= Annual percent change based on rates that were age-standardized to the 2000 U.S. population
[a] APC=0.6*, 1975–2002
[b] APC=1.1*, 1975–2002
[c] APC=1.0*, 1975–2002
[d] APC=0.6*, 1975–2002
*APC is significant (p<0.05)
Source: NCI SEER Program, 2005

FIGURE 6. Trends in age-standardized incidence of myeloma by sex and race, adjusted for delay of reporting, 1975–2002.

11.2% per year from 1983 to 1986 and has since been stable. Because of the rarity of childhood cancers, even the age-standardized rates from SEER are imprecise (FIGS. 7 and 8). Past analyses of childhood cancer incidence rates 1975–1995 suggested that incidence trends for both brain cancer and leukemia might best be represented by a step function, with a lower rate prior to 1984, followed by an abrupt rise and a constant higher rate afterwards.[50] With 7 additional years of data, this pattern is still apparent for brain cancer but not for leukemia. It has been hypothesized that the abrupt increase in brain cancer incidence from 1983 to 1986 was attributable to improved diagnosis because of the introduction of magnetic resonance imaging and stereotactic biopsy.[50] No analogous change in diagnostic sensitivity or classification has been proposed to account for the rising leukemia incidence, although increased use of bone marrow biopsy in the late 1970s and early 1980s may have resulted in improved detection and survival for the 10% of children with acute leukemia who have normal peripheral blood smears at diagnosis. Increased use of selective chemotherapy agents and immunophenotyping in the late 1970s may have improved specification by cell type.[50] Similar increasing trends in childhood cancer, leukemia, and brain cancer incidence have been reported from

TABLE 4. SEER incidence rate trends with joinpoint[a] analyses for 1975 through 2002 for childhood cancers (not adjusted for delay in reporting)[b]

	Joinpoint analyses (1975–2002)[c]															
	Ages 0–14								Ages 0–19							
	Trend 1		Trend 2		Trend 3		Trend 4		Trend 1		Trend 2		Trend 3		Trend 4	
Cancer site or type	Years	APC[d]	Years	APC[d]	Years	APC[d]	Years	APC[d]	Years	APC[d]	Years	APC[d]	Years	APC[d]	Years	APC[d]
All cancers	1975–2002	0.7[e]							1975–2002	0.6[e]						
Brain cancer and other CNS[f]	1975–1977	15.8	1977–1983	−3.4	1983–1986	11.2	1986–2002	0.2	1975–1977	14.6	1977–1983	−3.4	1983–1986	11.2	1986–2002	−0.05
Leukemia	1975–2002	0.7[e]							1975–2002	0.7[e]						
All	1975–2002	0.9[e]							1975–2002	0.9[e]						

[a] Joinpoint (JP) Regression Program, Version 3.0. April 2004. National Cancer Institute.
[b] Sources of data are SEER 9 areas (Connecticut, Hawaii, Iowa, Utah, and New Mexico, and the metropolitan areas of San Francisco, Detroit, Atlanta, and Seattle-Puget Sound).
[c] Joinpoint analyses with up to three joinpoints.
[d] APC based on rates that were age-adjusted to the 2000 U.S. Standard Population using joinpoint regression analysis.
[e] The APC is significantly different from zero ($P < 0.05$).
[f] CNS = central nervous system.

TABLE 5. SEER incidence rate trends for childhood cancer, 1975–2002, 1975–1988, 1989–2002 (not adjusted for delay in reporting)[a]

	Ages 0–14				Ages 0–19			
	Change[b] 1975–2002	APC[c] 1975–2002	APC[c] 1975–1988	APC[c] 1989–2002	Change[b] 1975–2002	APC[c] 1975–2002	APC[c] 1975–1988	APC[c] 1989–2002
All childhood cancers	27.1	0.7[d]	1.1[d]	0.3[f]	28.7	0.6[d]	1.1[d]	0.2[e]
Leukemia	37.1	0.7[d]	1.3[d]	0.2	40.5	0.7[d]	1.0[d]	0.4
Lymphomas and reticuloendothelial neoplasms	−21.0	−0.5[d]	−0.7	0.3	−9.3	−0.2	0.2	−0.6
Central nervous system	60.6	1.2[d]	2.3[d]	0.3[f]	52.8	1.0[d]	2.1[d]	0.1[e]
Sympathetic nervous system tumors	17.9	0.5	0.4	0.9	21.2	0.5	0.6	0.9
Retinoblastoma	−35.1	0.6	1.0	−0.6	−35.1	0.6	1.0	−0.8
Renal tumors	42.2	0.4	0.5	−0.4	39.5	0.5	0.5	−0.5
Hepatic tumors	84.0	2.0[d]	3.1	2.1	84.0	2.0[d]	2.9	2.2
Malignant bone tumors	46.6	0.3	2.2	−0.6	40.8	0.5	2.1[d]	0.0
Soft tissue sarcoma	24.9	1.0[d]	0.6	−0.3	29.9	0.9[d]	1.2	0.5

[a]Sources of data are SEER 9 areas (Connecticut, Hawaii, Iowa, Utah, and New Mexico, and the metropolitan areas of San Francisco, Detroit, Atlanta, and Seattle-Puget Sound).
[b]Change = the difference between the 2002 and 1975 rate expressed as a percentage of the 1975 rate.
[c]APC based on rates that were age-adjusted to the 2000 U.S. Std Population.
[d]The APC is significantly different from zero ($P < 0.05$).
[e]The APC for 1989–2002 is significantly different from the APC for 1975–1988 ($P < 0.5$).
[f]The APC for 1989–2002 is significantly different from the APC for 1975–1988 ($P < 0.10$).

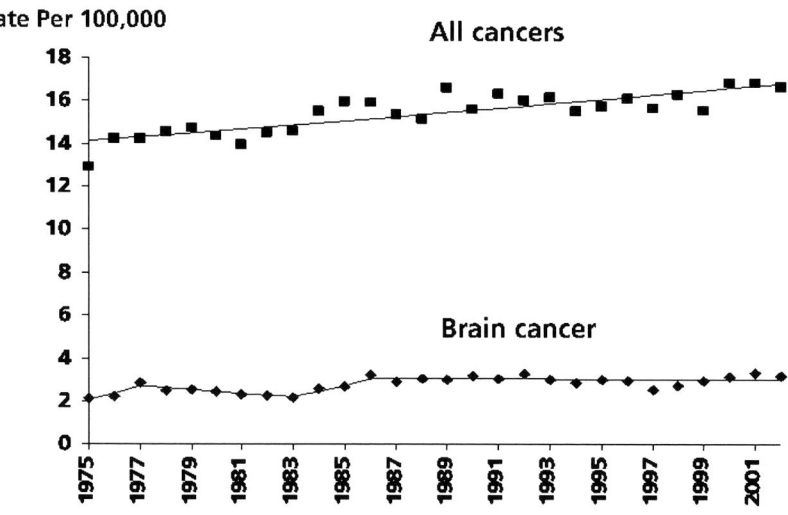

Source: NCI SEER Program, 2005

FIGURE 7. Trends in incidence of childhood cancer, ages 0–19, 1975–2002.

other industrialized countries with longstanding population-based cancer registries, such as Sweden and Northern England, and from a European database of childhood and adolescent cancer cases.[51–53]

Regarding the less common sites of childhood cancer (TABLE 5), the incidence of retinoblastoma has declined and the incidence of tumors of the sympathetic nervous system and renal tumors was stable from 1975 to 2002 (TABLE 5). Increasing incidence was observed for childhood hepatic tumors and soft tissue sarcoma when the entire time interval was considered; the incidence of malignant bone tumors increased during the 1975–1988 time period only (TABLE 5).

Although there have been many studies to identify the etiology of leukemia and brain cancer in children, the causes are largely unknown. Several studies have associated increased risk of childhood leukemia and brain cancer with increasing parental age, which may contribute to the trends in industrialized countries.[54] The increase in childhood leukemia primarily reflects an increase in ALL, the most common type of leukemia in childhood. The only known environmental risk factor for ALL is ionizing radiation, as demonstrated both in populations exposed *in utero* and in children exposed to radiotherapy.[55] The possibility that certain infections may influence risk of ALL has continuing interest, but no clear resolution.[55] Of particular interest are hypotheses about the timing of exposure to infectious agents, either *in utero*, at the time of birth, or at critical stages of development.[56] A systematic investigation of spatial clustering of childhood leukemia cases in Europe found some evidence

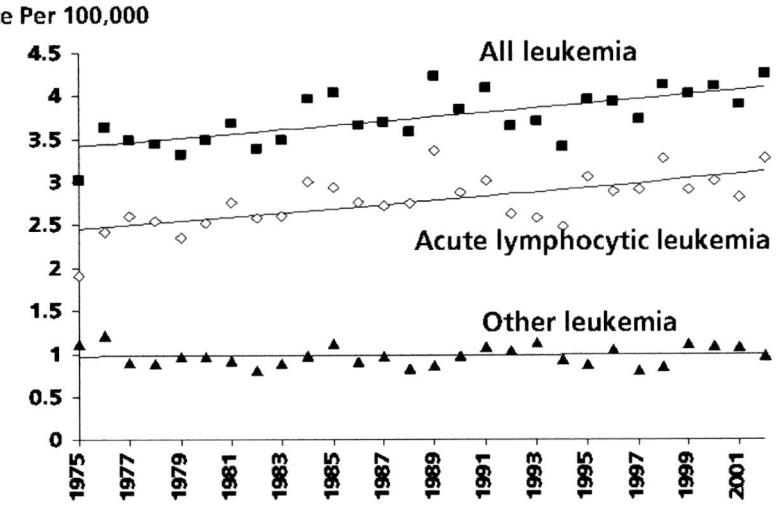

Source: NCI SEER Program, 2005

FIGURE 8. Trends in incidence of childhood leukemia, ages 0–19, 1975–2002.

of spatial clustering, the magnitude of which was small.[57] A number of studies report higher childhood leukemia incidence in areas with increased levels of population mixing, particularly in previously isolated populations.[56] In addition to infection, other exposures such as electromagnetic fields, parental occupation, and medication use during pregnancy have been studied, with inconsistent results.[55] Exposure to benzene from motor vehicle exhaust is also inconsistently associated with risk.[58] Numerous studies have examined the association between childhood leukemia and extremely low-frequency (ELF) magnetic fields. The International Agency for Research on Cancer (IARC) has characterized this evidence as "limited."[59]

The only known environmental cause of childhood brain tumors is ionizing radiation. Children treated with radiation for tinea capitus or leukemia subsequently have increased risk of brain tumors.[55] Other studies have examined the association of childhood brain cancer with maternal consumption of cured meats, paternal smoking, pesticides, farm animals, infection with a polyomavirus, and parental occupation with inconsistent results.[55]

SUMMARY

Recent trends in cancer incidence rates represent a substantial improvement over the large increase that occurred from the inception of the SEER registries in 1973 to the early 1990s. That increase was largely attributable to the epidemic of cancers caused by tobacco and the increased detection of prostate and breast cancer due to the introduction of population screening. The most recent trend has been flat in men from 1995 to 2002 and has increased slightly (0.3% per year) in women from 1987 to 2002. The overall trend represents a mixture of sites for which incidence is increasing and others for which it is decreasing

Several of the cancer sites with increasing trends are plausibly explained by changes in known risk factors or improvements in diagnosis. For example, trends in melanoma can be accounted for by generational trends in sun exposure and the increase in thyroid cancer is largely attributable to diagnostic artifact. For several other cancers with increasing trends, the explanations are less clear. These include multiple myeloma, renal adenocarcinoma, testicular cancer, and non-Hodgkin's lymphoma.

In many respects, trends in cancer mortality provide a more meaningful indication of the success of primary, secondary, and tertiary cancer control measures than do trends in cancer incidence. Death rates are less susceptible to artifactual fluctuations due to introduction of new screening tests, although they too are influenced by changes in disease classification or diagnosis. Death rates from all cancers combined have decreased by 10% in the United States from 1991 to 2002; site-specific death rates have decreased in the most recent time period for 12 of the top 15 cancer sites in men and 9 of the top 15 cancer sites in women. Similar trends have been observed in other industrialized countries.

The fact that the incidence rate of a particular cancer is changing over time is not, in itself, evidence that any specific environmental factor is responsible for that change. Multiple changes occurred in the United States and other industrialized countries during the 20th century. These include large changes in diet, physical activity, tobacco use, medical diagnostic and therapeutic practices, as well as rapid increase in synthetic chemicals. It is both essential and challenging to view environmental factors related to cancer in this broad context. The decrease in incidence and death rates from many cancers in industrialized countries has been achieved by applying knowledge about cancer prevention, early detection, and treatment at the population and individual level. Much progress has been made, but even greater challenges remain. Among the most pressing needs is to apply these lessons in developing countries, where the globalization of tobacco smoking and obesity as well as the uncontrolled exposure to industrial pollutants and agricultural pollutants are compounding environmental risks related to air and water quality, infectious diseases, and inadequate nutrition.

REFERENCES

1. EDWARDS, B.K., M.L. BROWN, P.A. WINGO, et al. 2005. Annual report to the nation on the status of cancer, 1975-2002, featuring population-based trends in cancer treatment. J. Natl. Cancer Inst. **97:** 1407–1427.

2. Surveillance Research Program, National Cancer Institute SEER*Stat software (www.seer.cancer.gov/seerstat) version 6.1.4.
3. Surveillance, Epidemiology, and End Results (SEER) Program (www.seer.cancer.gov) SEER*Stat Database: Incidence - SEER 9 Regs Public-Use, Nov 2004 Sub (1973-2002), National Cancer Institute, DCCPS, Surveillance Research Program, Cancer Statistics Branch, released April 2005, based on the November 2004 submission.
4. RIES, L.A.G., M.P. EISNER, C.L. KOSARY, et al. Eds. 2005. SEER Cancer Statistics Review, 1975-2002. National Cancer Institute. Bethesda, MD.
5. ARMSTRONG, B. 1976. Recent trends in breast-cancer incidence and mortality in relation to changes in possible risk factors. Int. J. Cancer **17:** 204–211.
6. CHU, K.C., R.E. TARONE, L.G. KESSLER, et al. 1996. Recent trends in U.S. breast cancer incidence, survival, and mortality rates. J. Natl. Cancer Inst. **88:** 1571–1579.
7. MILLER, B.A., E.J. FEUER & B.F. HANKEY. 1993. Recent incidence trends for breast cancer in women and the relevance of early detection: an update. CA Cancer J. Clin. **43:** 27–41.
8. GHAFOOR, A., A. JEMAL, E. WARD, et al. 2003. Trends in breast cancer by race and ethnicity. CA Cancer J. Clin. **53:** 342–355.
9. BRODY, J.G., J. TICKNER & R.A. RUDEL. 2005. Community-initiated breast cancer and environment studies and the precautionary principle. Environ. Health Perspect. **113:** 920–925.
10. POTOSKY, A.L., B.A. MILLER, P.C. ALBERTSEN & B.S. KRAMER. 1995. The role of increasing detection in the rising incidence of prostate cancer. JAMA **273:** 548–552.
11. MERRILL, R.M., E.J. FEUER, J.L. WARREN, et al. 1999. Role of transurethral resection of the prostate in population-based prostate cancer incidence rates. Am. J. Epidemiol. **150:** 848–860.
12. ETMINAN, M., B. TAKKOUCHE & F. CAAMANO-ISORNA. 2004. The role of tomato products and lycopene in the prevention of prostate cancer: a meta-analysis of observational studies. Cancer Epidemiol. Biomarkers Prev. **13:** 340–345.
13. BOSTWICK, D.G., H.B. BURKE, D. DJAKIEW, et al. 2004. Human prostate cancer risk factors. Cancer **101**(Suppl. 10): 2371–2490.
14. BOSLAND, M.C. 2000. The role of steroid hormones in prostate carcinogenesis. J. Natl. Cancer Inst. Monogr. **27:** 39–66.
15. ELTOM, M.A., A. JEMAL, S.M. MBULAITEYE, et al. 2002. Trends in Kaposi's sarcoma and non-Hodgkin's lymphoma incidence in the United States from 1973 through 1998. J. Natl. Cancer Inst. **94:** 1204–1210.
16. HARTGE, P. & S.S. DEVESA. 1992. Quantification of the impact of known risk factors on time trends in non-Hodgkin's lymphoma incidence. Cancer Res. **52**(Suppl. 19): 5566s–5569s.
17. MULLER, A.M., G. IHORST, R. MERTELSMANN & M. ENGELHARDT. 2005. Epidemiology of non-Hodgkin's lymphoma (NHL): trends, geographic distribution, and etiology. Ann. Hematol. **84:** 1–12.
18. MCMICHAEL, A.J. & G.G. GILES. 1996. Have increases in solar ultraviolet exposure contributed to the rise in incidence of non-Hodgkin's lymphoma? Br. J. Cancer **73:** 945–950.
19. JEMAL, A., S.S. DEVESA, P. HARTGE & M.A. TUCKER. 2001. Recent trends in cutaneous melanoma incidence among whites in the United States. J. Natl. Cancer Inst. **93:** 678–683.

20. RODENAS, J.M., M. DELGADO-RODRIGUEZ, M.T. HERRANZ, et al. 1996. Sun exposure, pigmentary traits, and risk of cutaneous malignant melanoma: a case-control study in a Mediterranean population. Cancer Causes Control. **7:** 275–283.
21. WOODHEAD, A.D., R.B. SETLOW & M. TANAKA. 1999. Environmental factors in nonmelanoma and melanoma skin cancer. J. Epidemiol. **9**(Suppl. 6): S102–S114.
22. MATHEW, A., S.S. DEVESA, J.F. FRAUMENI, JR. & W.H. CHOW. 2002. Global increases in kidney cancer incidence, 1973-1992. Eur. J. Cancer Prev. **11:** 171–178.
23. CHOW, W.H., S.S. DEVESA, J.L. WARREN & J.F. FRAUMENI, JR. 1999. Rising incidence of renal cell cancer in the United States. JAMA **281:** 1628–1631.
24. WARD, E. 2005. Kidney cancer. In Preventing Occupational Disease and Injury, 2nd ed. B. Levy, G. Wagner, K. Rest & J. Weeks, Eds.: 294–296. American Public Health Association. Washington, D.C.
25. VIZCANIO, A., V. MORENO, R. LAMBERT & D. PARKIN. 2002. Time trends incidence of both major histologic types of esophageal carcinomas in selected countries, 1973-1985. Int. J. Cancer **99:** 860–868.
26. BROWN, L.M. & S.S. DEVESA. 2002. Epidemiologic trends in esophageal and gastric cancer in the United States. Surg. Oncol. Clin. North Am. **11:** 235–256.
27. GRAHAM, D.Y., H.M. MALATY, D.G. EVANS, et al. 1991. Epidemiology of Helicobacter pylori in an asymptomatic population in the United States. Effect of age, race, and socioeconomic status. Gastroenterology **100:** 1495–1501.
28. YE, W., M. HELD, J. LAGERGREN, et al. 2004. Helicobacter pylori infection and gastric atrophy: risk of adenocarcinoma and squamous-cell carcinoma of the esophagus and adenocarcinoma of the gastric cardia. J. Natl. Cancer Inst. **96:** 388–396.
29. FRANCESCHI, S., P. BOYLE, P. MAISONNEUVE, et al. 1993. The epidemiology of thyroid carcinoma. Crit. Rev. Oncog. **4:** 25–52.
30. CARDIS, E., A. KESMINIENE, V. IVANOV, et al. 2005. Risk of thyroid cancer after exposure to 131I in childhood. J. Natl. Cancer Inst. **97:** 724–732.
31. COMMITTEE ON THYROID SCREENING RELATED TO I-131 EXPOSURE IOM, AND COMMITTEE ON EXPOSURE OF THE AMERICAN PEOPLE TO I-131 FROM THE NEVADA ATOMIC BOMB TESTS, NATIONAL RESEARCH COUNCIL. 1999. Exposure of the American People to Iodine-131 from Nevada Buclear-Bomb Tests: Review of the National Cancer Institute Report and Public Health Implications. National Academy Press. Washington, D.C.
32. STEENLAND, K. 2003. Carcinogenicity of EBDCs. Environ. Health Perspect. **111:** A266; author reply A266–A267.
33. PURDUE, M.P., S.S. DEVESA, A.J. SIGURDSON & K.A. MCGLYNN 2005. International patterns and trends in testis cancer incidence. Int. J. Cancer **115:** 822–827.
34. HANNA, N., R. TIMMERMAN, R. FOSTER, et al. 2003. Testis cancer. In Cancer Medicine, Vol. 6. J. Holland, E. Frei, Eds.: 1747–1768. BC Decker Inc. Hamilton, London.
35. TOPPARI, J., M. KALEVA & H.E. VIRTANEN. 2001. Trends in the incidence of cryptorchidism and hypospadias, and methodological limitations of registry-based data. Hum. Reprod. Update **7:** 282–286.
36. SKAKKEBAEK, N.E., E. RAJPERT-DE MEYTS & K.M. MAIN. 2001. Testicular dysgenesis syndrome: an increasingly common developmental disorder with environmental aspects. Hum. Reprod. **16:** 972–978.
37. BARLOW, N.J., B.S. MCINTYRE & P.M. FOSTER. 2004. Male reproductive tract lesions at 6, 12, and 18 months of age following in utero exposure to di(n-butyl) phthalate. Toxicol. Pathol. **32:** 79–90.

38. SWAN, S.H., K.M. MAIN, F. LIU, et al. 2005. Decrease in anogenital distance among male infants with prenatal phthalate exposure. Environ. Health Perspect. **113:** 1056–1061.
39. CALLE, E.E., C. RODRIGUEZ, K. WALKER-THURMOND & M.J. THUN. 2003. Overweight, obesity, and mortality from cancer in a prospectively studied cohort of U.S. adults. N. Engl. J. Med. **348:** 1625–1638.
40. BROWN, L.M., L.M. POTTERN, D.T. SILVERMAN, et al. 1997. Multiple myeloma among blacks and whites in the United States: role of cigarettes and alcoholic beverages. Cancer Causes Control **8:** 610–614.
41. BROWN, L.M., G. GRIDLEY, L.M. POTTERN, et al. 2001. Diet and nutrition as risk factors for multiple myeloma among blacks and whites in the United States. Cancer Causes Control **12:** 117–125.
42. BROWN, L.M., M.S. LINET, R.S. GREENBERG, et al. 1999. Multiple myeloma and family history of cancer among blacks and whites in the U.S. Cancer **85:** 2385–2390.
43. BARIS, D., L.M. BROWN, D.T. SILVERMAN, et al. 2000. Socioeconomic status and multiple myeloma among US blacks and whites. Am. J. Public Health **90:** 1277–1281.
44. MORGAN, G.J., F.E. DAVIES & M. LINET. 2002. Myeloma aetiology and epidemiology. Biomed. Pharmacother. **56:** 223–234.
45. POTTERN, L.M., J.J. GART, J.M. NAM, et al. 1992. HLA and multiple myeloma among black and white men: evidence of a genetic association. Cancer Epidemiol. Biomarkers Prev. **1:** 177–182.
46. LEWIS, D.R., L.M. POTTERN, L.M. BROWN, et al. 1994. Multiple myeloma among blacks and whites in the United States: the role of chronic antigenic stimulation. Cancer Causes Control **5:** 529–539.
47. ANONYMOUS. 2004. IARC Monographs on the Evaluation of Carcinogenic Risks to Humans. Volume 83. Tobacco Smoke and Involuntary Smoking. International Agency for Research on Cancer. Lyon, France.
48. ROSS, J.A., E. PARKER, C.K. BLAIR, et al. 2004. Body mass index and risk of leukemia in older women. Cancer Epidemiol. Biomarkers Prev. **13:** 1810–1813.
49. LINET, M.S. & R.A. CARTWRIGHT. The leukemias. In Schottenfeld, Ed.
50. LINET, M.S., L.A. RIES, M.A. SMITH, et al. 1999. Cancer surveillance series: recent trends in childhood cancer incidence and mortality in the United States. J. Natl. Cancer Inst. **91:** 1051–1058.
51. DREIFALDT, A.C., M. CARLBERG & L. HARDELL. 2004. Increasing incidence rates of childhood malignant diseases in Sweden during the period 1960–1998. Eur. J. Cancer **40:** 1351–1360.
52. COTTERILL, S.J., L. PARKER, A.J. MALCOLM, et al. 2000. Incidence and survival for cancer in children and young adults in the North of England, 1968-1995: a report from the Northern Region Young Persons' Malignant Disease Registry. Br. J. Cancer **83:** 397–403.
53. STELIAROVA-FOUCHER, E., C. STILLER, P. KAATSCH, et al. 2004. Geographical patterns and time trends of cancer incidence and survival among children and adolescents in Europe since the 1970s (the ACCIS project): an epidemiological study. Lancet **364:** 2097–2105.
54. DOCKERTY, J.D., G. DRAPER, T. VINCENT, et al. 2001. Case-control study of parental age, parity and socioeconomic level in relation to childhood cancers. Int. J. Epidemiol. **30:** 1428–1437.

55. BUNIN, G.R. 2004. Nongenetic causes of childhood cancers: evidence from international variation, time trends, and risk factor studies. Toxicol. Appl. Pharmacol. **199:** 91–103.
56. MCNALLY, R.J. & T.O. EDEN. 2004. An infectious aetiology for childhood acute leukaemia: a review of the evidence. Br. J. Haematol. **127:** 243–263.
57. ALEXANDER, F.E., P. BOYLE, P.M. CARLI, et al. 1998. Spatial clustering of childhood leukaemia: summary results from the EUROCLUS project. Br. J. Cancer **77:** 818–824.
58. BUFFLER, P.A., M.L. KWAN, P. REYNOLDS & K.Y. URAYAMA. 2005. Environmental and genetic risk factors for childhood leukemia: appraising the evidence. Cancer Invest. **23:** 60–75.
59. IARC Working Group on the Evaluation of Carcinogenic Risks to Humans. 2002. Non-ionizing radiation, Part 1: static and extremely low-frequency (ELF) electric and magnetic fields. IARC Monogr. Eval. Carcinog. Risks Hum. **80:** 1–395.

The Anatomy of the Exposures That Occurred around the World Trade Center Site

9/11 and Beyond

PAUL J. LIOY AND PANOS GEORGOPOULOS

Division of Exposure Science, Environmental and Occupational Health Sciences Institute, Robert Wood Johnson Medical School – UMDNJ, Piscataway, New Jersey 08854, USA

ABSTRACT: The attack on the World Trade Center (WTC) resulted in a new era of awareness on terrorism in the United States and the issues surrounding the potential for acute and/or long-term health outcomes caused by personal exposures to toxicants released during a terrorist event or an accident. The aftermath of the collapse yielded a situation usually not encountered in environmental health science: a large population's exposure to a previously uncharacterized complex mixture of airborne gases and particles, and re-suspendable particles (>2.5 μm in diameter). This led to a series of rapidly changing potential and actual exposure categories, both in space and time that were associated with the complex mixture of heterogeneous composition and character; e.g., very large particles mixed with much smaller amounts of fine particles, and gases released by uncontrolled combustion. The four categories of outdoor exposure that were encountered will be discussed over the period from September 11 until the fires ended on December 20, 2001. Further, the complex issue of indoor exposure to deposited dust will be highlighted from the beginning through the residual exposure issues being examined today (Category 5 period). The strength of the information on the initial WTC dust and smoke, and the smoke plumes from the fires and the continuing (permanent) gaps in our knowledge within the exposure sciences will be discussed, as well as our attempt to reconstruct exposure for various segments of the population in southern Manhattan and the surrounding areas. This all will be tied to lessons that must be considered in response to future events, natural or otherwise.

KEYWORDS: settled dust/smoke; exposure science; WTC aftermath; indoor/outdoor

Address for correspondence: Paul J. Lioy, Ph.D., Division of Exposure Science, Environmental and Occupational Health Sciences Institute, Robert Wood Johnson Medical School – UMDNJ, 170 Frelinghuysen Road, Piscataway, NJ 08854. Voice: 732-445-0155; fax: 732-445-0116.
e-mail: plioy@eohsi.rutgers.edu

INTRODUCTION

Exposures occur as the result of events that can be short or long in duration and they could be caused by a continuous or a single instantaneous release.[1,2] Such events can affect individuals or a population. A main purpose of exposure science is to understand the relationship between how the individual or population comes into contact with a contaminant that could be physical, chemical, or biological in nature and the relevance of that contact with a health outcome.[3] Exposures can be studied indirectly or directly. For example, one can make direct measurements in the human body of the level of toxicants or metabolites, etc., or exposure can be estimated indirectly by measuring the levels of materials in media or exposure pathways that can eventually lead to contact with a contaminant or be modeled from source to receptor.[4] These contacts and events come under the view of exposure science, which has been applied to address environmental issues over the past 25 years.

THE TERRORIST ATTACK ON THE WTC

On September 11, 2001, the United States suffered a terrorist event that resulted in a series of complex exposures to dust and/or smoke that could or could not be analyzed quantitatively by current direct measurement or indirect measurement techniques. The fundamental reason was that the initial impact of the events of 9/11 was almost instantaneous or very short in duration. When the planes struck the building, they produced a series of explosions that led to fires that engulfed both buildings and ultimately, about an hour an a half later, both towers collapsed on the 16-acre site.[5] This collapse was unprecedented in that the structures and their contents were literally pulverized to dust as they fell.[6,7] The settling dust and smoke from the intense fires were distributed in Southern Manhattan and Brooklyn. After the collapse, which affected all of southern Manhattan, a large fraction of the material continued to be distributed in the air to the east and southeast of the WTC site because of the west to northwesterly winds present that morning.[7]

The initially exposed population encountered very large quantities of suspended dust and smoke; in fact, one could estimate that upward of 100,000 μm per cubic meter of particles were probably present in the air during the initial minutes after the collapse of each structure. However, no local ambient air measurements were immediately available, nor could be expected to be available, to quantify the initial airborne particles and gases exposures. The intensity of the atmospheric concentrations/exposures were illustrated qualitatively in the graphic pictures of individuals walking through southern Manhattan blanketed in the dust and smoke that had been initially released and settled on them during the collapse. However, the event also yielded longer-term exposures due to the conditions at Ground Zero and in the surrounding outdoor and indoor locations during subsequent months.[8,9]

THE CATEGORIES OF POTENTIAL HUMAN EXPOSURE

The exposure periods associated with the aftermath of 9/11 were not just confined to the initial minutes to hours post event. There were five specific post 9/11 environmental/occupational exposure categories.[6] Four of them were defined as outdoor exposure categories, and one was the indoor exposure category. The outdoor exposure categories are illustrated in FIGURE 1 with the fraction of the original source strength estimated by the y-axis.[8] This conceptual model of post-event contamination indicates how rapidly the plume emissions and dust re-suspensions were reduced after the first day. In fact, Category 1 exposures cover the period associated with the initial 5–6 h post collapse of the Twin Towers. That period of time included the initial release and distribution of the pulverized building materials, and the most intense period of the WTC plume emissions caused by the jet fuel fires. The Category 2 exposures covered the period of time when: (*a*) the fires burned throughout the 16 acres that comprised Ground Zero, and (*b*) the highest levels re-suspendable particulate mass could be mobilized from surfaces. These direct gaseous and fine particle plume emissions, and the re-suspendable total particle emissions occurred from 9/11 through 9/14. The period ended because of the first post 9/11 rain that reduced the intensity of the fires and washed away the settled re-suspendable materials into the Hudson and East Rivers. As a note, one must remember, however, that there is a degree of uncertainty associated with the durations of Categories 1, 2, and 3 outdoor exposure since there were no systematic measurements taken in the ambient air during this period of time for materials other than asbestos.[10] The exposure Category 3 was associated with lower concentrations of both re-suspendable dust and direct emissions of

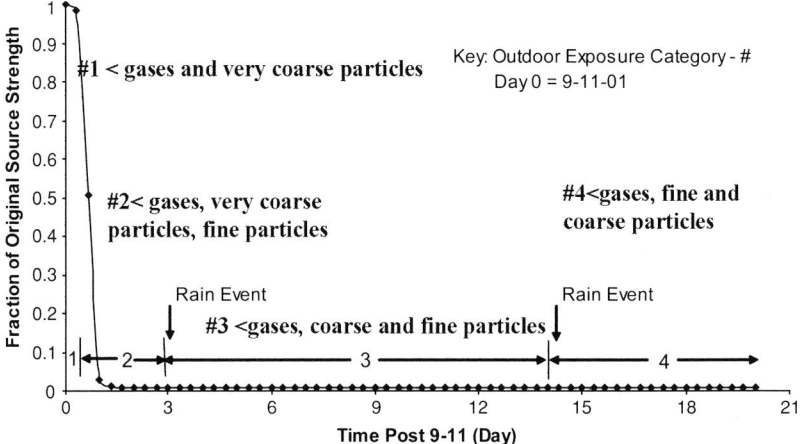

FIGURE 1. Conceptual WTC outdoor plume impacts and exposure categories: decay model and time post event (adapted from Lioy *et al.*[8]).

smoke and gases from the fires into the atmosphere. During this period there were intermittent releases of high concentrations of gaseous substances and particulate matter into the local atmosphere within Ground Zero caused by fires uncovered by the rescue workers or the movement of debris. The period of time associated with Category 3 exposures covered day 4 through day 13/14 post 9/11. It ended when the second rain event occurred on September 25, 2005.[7,8]

Exposure Category 4 includes the period of time when the strength of the emissions released by the fires diminished and the plume affected the local air quality only on particular days (e.g., October 3 through 6, 2001). However, most of the time the ambient air contained levels of particulate matter that were usually close to the daily New York City (NYC) average concentrations.

TABLE 1 provides information on the Category 5 exposure period, which was the indoor environment. This has been a very difficult exposure category to define or characterize because many people were affected by very different types of contacts with settled WTC dust/smoke. The reason was that the conditions of individual residences and buildings were not uniform, since each was affected differently by the material released during the collapse. Thus, the level of contact would vary from location to location, structure to structure, and from time to time. It is clear that the indoor air or total indoor environment contamination issues were caused primarily by the initial settled dust and smoke that were distributed in homes, apartments, and commercial and government buildings. The dust could be from a millimeter in depth up to several inches in depth depending upon how close the structures were to the collapse.[7] Further, other variables affecting the levels of indoor contamination included (*a*) whether or not the integrity of the building had been broken, (*b*) whether or not windows were open during the time of the collapse and, (*c*) the condition and integrity of ventilation systems. Ventilation systems in buildings with central air conditioning were affected more if the central air conditioners were not turned off by workers just after the collapse. The indoor

TABLE 1. Category 5 exposure period for environmental and occupational exposures after the attack on the WTC

Category	Descriptor	Pollutants	Duration	Material
5	Indoor environment	1. Initial dust and smoke (very coarse particles) 2. Smoldering fires 3. Re-suspended dust and smoke 4. Diesel emissions infiltration	9-1* through?	1. Settled dust/smoke 2. Diesel emissions

*A second round of indoor testing by EPA is anticipated in 2006, some indoor cleanups and building demolition continue.

environment issue was primarily associated with the period from September 9, 2001 to the end of 2003. However, there has been a continuing concern about residual dust/smoke in buildings and ventilation systems, which led to Environmental Protection Agency (EPA) and the Council of Environmental Quality creating a committee called the World Trade Center Expert Panel. The panel has assisted EPA and the local community in the design of a new indoor testing program to determine if any residual WTC dust still exists indoors primarily in southern Manhattan and western Brooklyn, New York, and are the levels of the residual settled dust high enough to warrant a cleanup.[11]

Characterization of Exposure Categories 1 and 2

During the period from 9/11 until 9/13 the WTC plume had a large amount of buoyancy (see FIG. 2) which was caused by the intensity of the fires (temperature $>1000°F$).[6,7] Thus, instead of depositing additional large quantities of smoke during the first 2 days in downtown Manhattan, Brooklyn, and New Jersey, most of the particles and gases were advected (rose vertically) to above 1000 feet in the lower troposphere. Then, because of the westerly winds present on 9/11, the plume moved downwind above southern Manhattan and Brooklyn and out to sea.[7] On 9/12 and 9/13 the plume intensity was much weaker (FIG. 2); however, late on the September 12 and during September 13 higher atmospheric concentrations occurred locally, and there were periods

FIGURE 2. The structure of the WTC plume: 9/11 through AM on 9/13; Category 1 and 2 exposure periods (courtesy of NESCAUM).

when the WTC plume impacted New Jersey and Upper Manhattan. On the evening of September 12 and into the early morning of September 13 the weakened WTC plume did affect a number of local areas because a nocturnal inversion (see last panel of FIG. 2). However, the ambient air exposures would still be much lower than experienced by individuals on 9/11.

The downwind movement of the plume on 9/11, illustrated in FIGURE 3 (a NASA satellite image), was above Brooklyn and Southern New York City, New York Bay, the edge of Sandy Hook, NJ, and the Atlantic Ocean. This was a wide plume; however, because of the buoyancy discussed above, it did not affect the people below as severely as it might have if the winds were higher or there were more subsidence in the atmosphere. The satellite picture and computer image in FIGURE 4 shows the plume from the standpoint of the intensity of emissions released and how it moved downwind. One can see that on 9/11 southern Manhattan was affected most by the plume, although there were hot spots (red) in Western Brooklyn. The impact estimated by the computer simulation was consistent with the satellite imagery (see figure) that was taken during the same period of time.[12] The issue of outdoor settled dust and smoke re-suspension persisted for about 3 days until a major rain event occurred on September 14, 2001.[6] From the images shown on television and in videos it was apparent that there were short periods of time when the concentrations

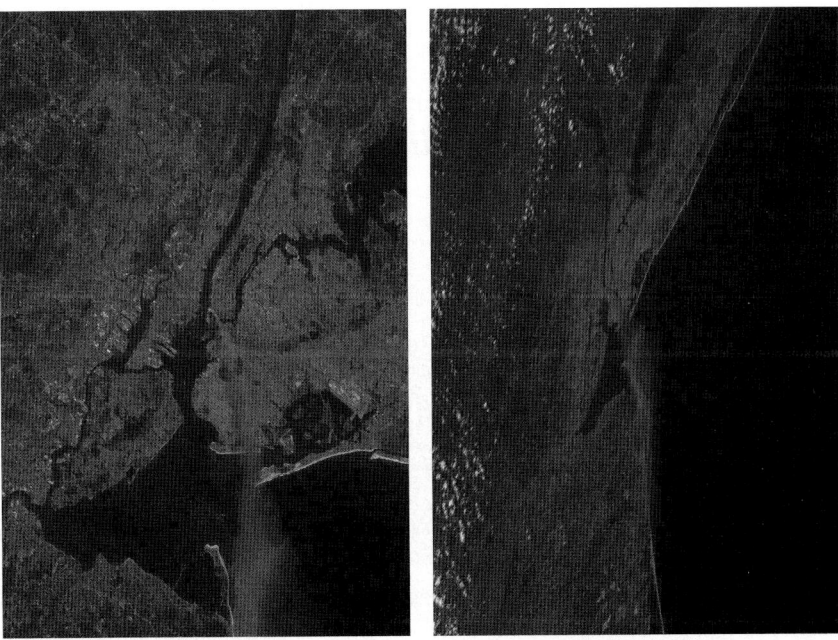

9/11/2001 12:10 PM   9/11/2001 12:30 PM

FIGURE 3. NASA space station images of the WTC plume on 9/11.

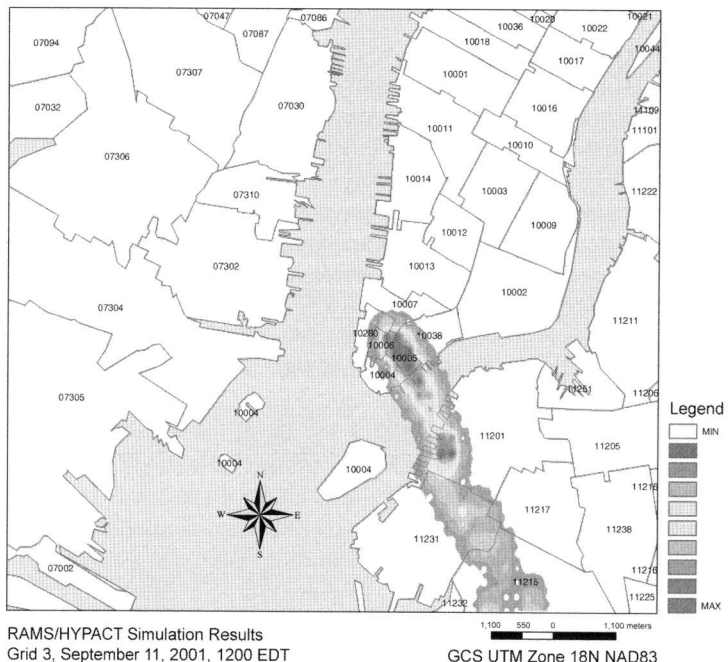

RAMS/HYPACT Simulation Results
Grid 3, September 11, 2001, 1200 EDT
GCS UTM Zone 18N NAD83

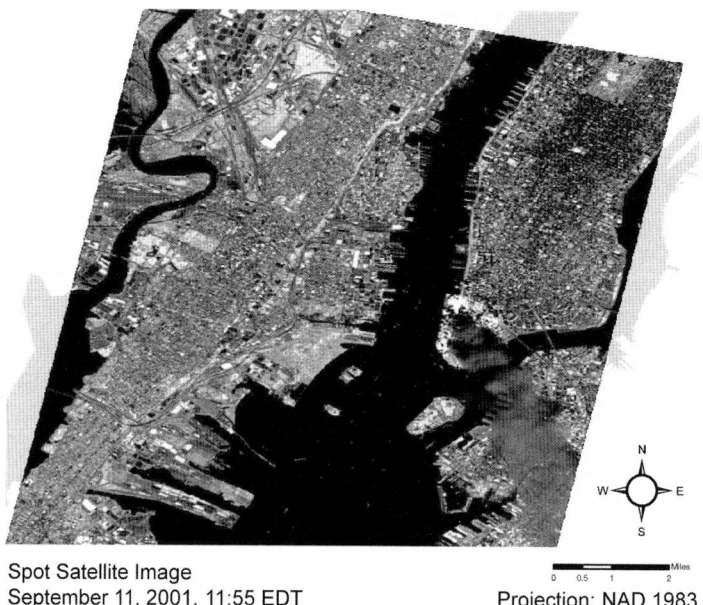

Spot Satellite Image
September 11, 2001, 11:55 EDT
Projection: NAD 1983

FIGURE 4. WTC plume dispersion modeling concentration estimates and performance test comparisons with satellite images for 9/11.

of total re-suspended particles in the atmospheres were probably in excess of 1000 μm per cubic meter. This is not a quantitative estimate since there was still no sampling. The releases were periodic and caused the upwelling of materials by trucks, cars, and police vehicles that drove over the settled material on the streets during rescue attempts at Ground Zero.

The Category 2 exposures did include periods of time when the plume, caused by the continuing fires, could be seen over the Ground Zero area as it drifted vertically from the 16-acre site. There were high concentrations at Ground Zero, but because the plume had vertical buoyancy it did not spread horizontally within Southern Manhattan around Ground Zero. The highest levels of particles and gases would have affected those rescue workers and emergency personnel located at Ground Zero, and people in buildings when the plume passed an elevated intake.

The individuals and groups potentially exposed to initial dust/smoke during the different post collapse periods are represented in Table 2.[6,7] The primary groups of people who were affected by Category 1 exposure, were the survivors of the initial collapse, and local residents, rescue workers, commuters, and shop owners, etc., who were within the area immediately surrounding Ground Zero. Category 2 exposures, included people who were residents and workers trying to get back to their homes or finding it very difficult to leave Manhattan. At this point in time the groups exposed expanded to the professional and volunteer rescue workers, who began to assist the Police Department, Fire Department, and other emergency rescue people in NYC who were on the scene during and immediately post collapse. Also included in the Category 2 exposure period were the press and public officials who immediately went to Ground Zero to observe or assess the magnitude of damage. In most cases, individuals, including the professional and volunteer rescue workers, were not wearing respirators.

TABLE 2. Who could have been or could be exposed to the materials present in the *initial dust/smoke*?

- Survivors of the initial collapse of the WTC
    - Local and downwind residents
    - Rescue workers
    - Commuters
    - Shop/business owners, operators, and customers
- Re-suspension of the dust/smoke during the following week
    - Professional and volunteer rescue workers (not wearing respiratory protection)
    - Outdoor and indoor cleanup workers
    - Residents and workers on Wall Street area downwind
- Re-suspension of dust/smoke during the next weeks/months
    - Workers not wearing respiratory protection at WTC site
    - Indoor cleanup workers not wearing respiratory protection
    - Residents and workers returning to poorly cleaned buildings
- Although not measured, gases would be associated with many of these exposures

Category 3 Exposures

During the first 3 days post collapse cleanup was limited. The start of outdoor and indoor cleanup was associated with Category 3 exposure, and was most notable in the Wall Street area, which was being cleaned prior to re-opening on Monday September 17, 2001. The groups that could come into contact with WTC dust/smoke, the particles and gases from the diminishing fires would be outdoor workers and indoor workers cleaning up buildings and residences. Exposure would continue among volunteer and rescue workers, especially those who were still not wearing respiratory protection at the beginning of Category 3 exposure period.

Category 4 Exposures

In addition to workers associated with the recovery efforts (e.g., post rescue period) Category 4 exposures primarily affected residents and workers returning to buildings and participating in outdoor activities. The buildings might have been fully cleaned, partially cleaned, or were exposed to combustion products within the fire plumes that were still being released (at much lower levels) in a variety of places within Ground Zero. The Category 4 exposure period lasted the longest, about 3–4 months. It was marked by periodic releases into the ambient air of combustion emissions from smoldering fires and the diesel emissions from construction vehicles used during the recovery period post collapse.

Category 5 Exposures

The Category 5 exposure period was most significant in terms of the level of uncertainty about the intensity and types of contacts that occurred with the dust/smoke. This level of uncertainty would start with poorly cleaned buildings that, in some cases, had to be re-cleaned because these units were cleaned before such efforts were completed on the entire building. This was a significant problem because individuals failed to recognize that there would be re-suspension and recontamination within an entire building and residences if all residences and inside spaces were not cleaned thoroughly at the same time.[9,13] Some buildings still remain closed to date, with a number that are expected to be demolished, or are in the process of being demolished, and sites will be prepared for reconstruction. One important point to note is that most of the materials associated in the indoor exposure issue contamination were components of the settled dust and smoke particles not gas species.

COMPOSITION OF SETTLED DUST/SMOKE AND OTHER AIR POLLUTANTS

The gaseous species emitted by the burning jet fuel, buildings, etc., would have dissipated quickly after release and not settled indoors or outdoors. However, no systematic measurements were obtained of gases released during Categories 1, 2, and 3 exposure periods and there will never be such measurement data. In addition, there were few measurements of the concentrations of total particulate matter that were in the air during each of these exposure periods.[6] Thus, what are we left with in terms of data for use in exposures analyses for Categories 1, 2, 3, and 5 are derived primarily from analyses of the dust and smoke that had settled to the ground. FIGURE 5 shows the settled dust and a micrograph of the individual particles that settled throughout lower Manhattan, etc. Pie charts shown in FIGURE 6 provide the general characteristics of the total settled dust obtained from analyses completed by Lioy et al. and the United States Geological Survey (USGS).[7,14] Most of the materials released was associated with three major component classes and include gypsum, concrete, and man-made fibers. To a lesser extent the contaminants quartz, chrysotile, zinc, iron, and lead were found in the particles. Furthermore, particle size analyses showed that in contrast to the particle size distribution normally found in ambient air, over 99% of the mass was associated with particles and fibers that were greater than 2.5 μm in diameter. After inhalation, particle deposition in the lung would occur primarily in the upper airways. What is also important to note as a major concern about the potential for acute health affects was that the pH of these samples ranged from 9 to 11.[6,7,9] Thus, the samples were basic and would also cause irritation at the locations where the particles were deposited in the respiratory tract.

Although the diameter of the particles was large compared to what is normally associated with air pollution, the materials within the settled dust/smoke included long glass fibers released during the collapse of the building as 110 stories of windows on the buildings disintegrated. These would penetrate farther into the lung because of their smaller aerodynamic diameter, and, as noted by Lioy et al., these fibers had much smaller particles coagulated along the surface that could easily be dislodged and move deeper into the respiratory system.[7] This is explained in detail in Lioy et al.[7] and was probably a contributing factor along with the pH of the dust/smoke for the onset of the "WTC cough" that was detected initially among firefighters by Prezant et al.[15] This health outcome occurred mainly in workers who came to Ground Zero on day 1 and stayed for a number of days. Those who came to Ground Zero after the Category 2 exposure period did not show many of these same symptoms. It should be noted again that we are missing the gaseous materials that were present during the first 3 days. Thus, any irritation detected in the lung, whether it was initial irritation or persistent irritation, would have to be considered as a combination of the large particles and fibers deposited in the lung and

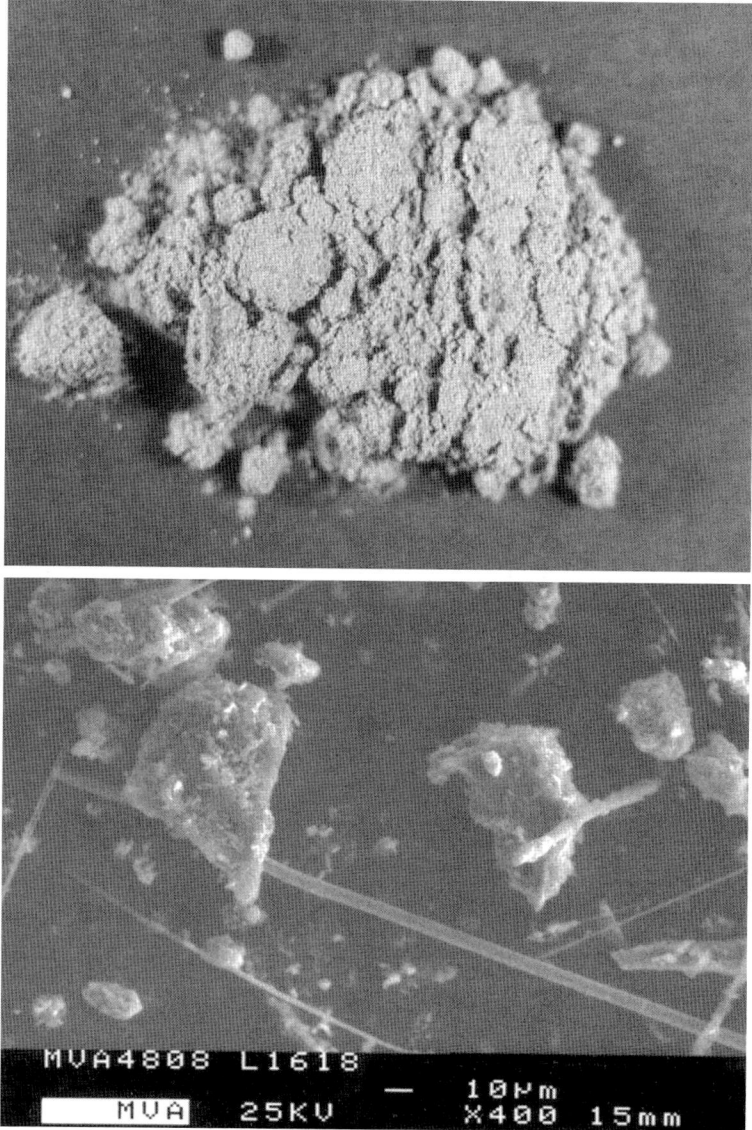

FIGURE 5. The general appearance of the bulk dust (outdoor sample) (P. Lioy and J. Millette pictures) taken during September 2001.

co-occurring gaseous combustion species of unknown composition at the time of inhalation.

Information was obtained by Lioy et al. about the semi-volatile fraction of the materials found in the settled dust/smoke.[7] Included were polycyclic

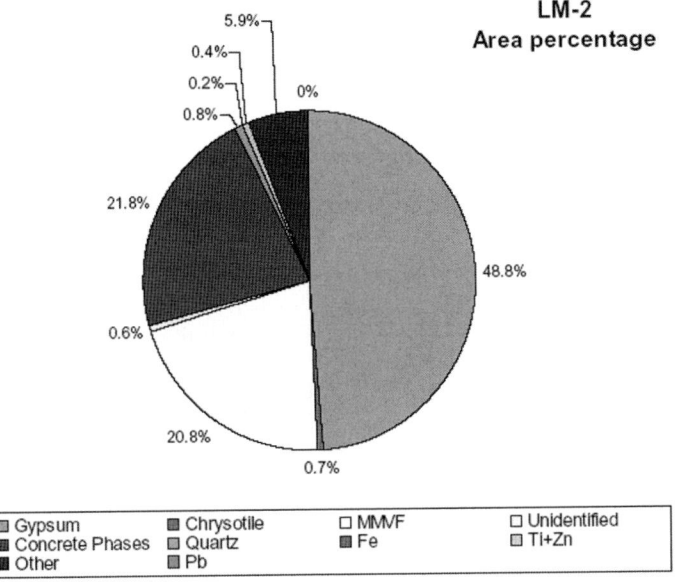

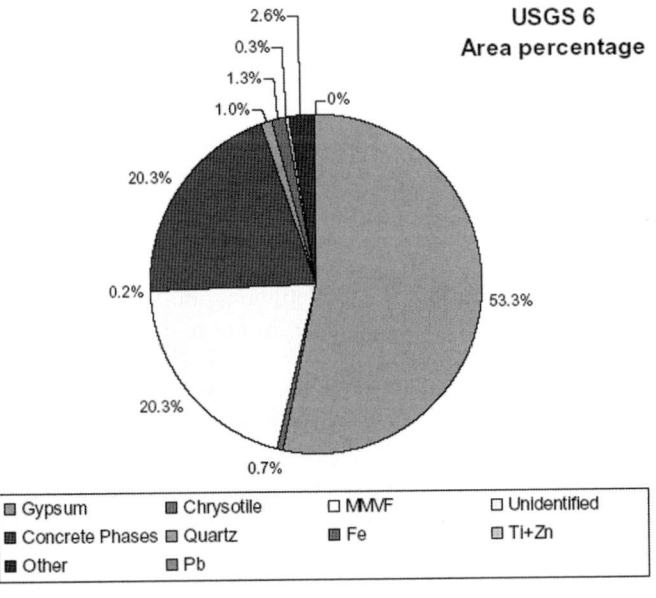

FIGURE 6. General characteristics of settled dust/smoke of representative samples taken by Lioy et al.[7], or the USGS[14].

hydrocarbons, phthalates, and other products of incomplete combustion and unburned jet fuel. Much lower concentrations of polychlorinated chlorinated biphenols (PCBs) and polybrominated diphenol ethers (PBDEs), and even lower concentrations of dioxins and furans were measured in the settled dust/smoke. The largest fraction of semi-volatile materials was the polycyclic aromatic hydrocarbons (PAHs) that comprised above 0.1% of the settled dust/smoke mass.

There actually was only one known fine particulate matter air sample, <3.0 um in diameter, taken during the Categories 1 and 2 exposure periods, which was directly affected by the WTC plume. That sample was taken by NYU Medical Center at 25th St, north and east of Ground Zero (L.C. Chen and Z. Fan, personal communication, 2005). The air sampler did clog because of the high levels of smoke in the air, but based upon the satellite imagery and plume simulations done by Environmental and Occupational Health Sciences Institute (EOHSI) we were able to determine that the monitor was within the plume for approximately one-half hour on the evening of September 12, 2001. The sample was analyzed for PAHs and, as would be anticipated, significant quantities of PAHs were found in the sample. Since the plume stayed over the area for a period of one-half hour, the ambient fine particles mass concentration was estimated to be approximately 500 μm per cubic meter (L.C. Chen and 2 Fan, personal communication, 2005). This estimated value was at least a factor of 5 times over what was found in samples of that duration on a day that the Canadian forest fires of 2002 affected the air quality in New York City and New Jersey.[16]

Since the Category 5 exposure period was dominated primarily by initial dust and smoke released on 9/11, it is natural to discuss the characteristics of these materials next. The major difference between components and materials that found their way indoors, and those setting outdoors was the particle size distribution.[9,13] The particles that settled on surfaces within buildings in southern Manhattan and the impact they had on individual buildings was determined by whether or not: (*a*) there was a breach in the building or the ventilation system caused of the explosion (as shown in FIG. 7); (*b*) the particles were pushed around windows, doors etc., into ventilation ducts or window ACs; or (*c*) the particles entered through openings in the building (e.g., open doors). The levels of materials that deposited across indoor surfaces looked more like the level of contamination (millimeters to inches thick) one would have seen after a mud slide or flood, except for one variable—the dust, which was dry, light, and not wet.

The materials deposited were associated with the initial plume of particles that ranged from around below 1 μm in diameter to well over 100 μm in diameter. As shown in TABLE 3 the particles present in the outdoor settled dust and smoke samples were primarily greater than 53 μm in diameter.[7,13] For the indoor samples the particles detected were distributed primarily between the 10 and 53 μm size range; although, in some samples the 53 μm and larger size

FIGURE 7. Example of buildings breached by the collapse on 9/11 (P. Lioy pictures taken on September 17, 2001).

range was still dominant.[13] The reason for the differences had to do with the nature of the breaches that occurred in a building, or the pathways by which the particles entered the building. For example, if the building remained intact (no broken walls) the particles would enter around the doors, door jams, window panes, and window frames. The largest particles would settle in and around the window or door frame reducing the range in particle sizes that would penetrate into the building. The deposited and re-suspendable particles would be more likely to stay in the air longer if they were distributed by friction or mechanical or natural ventilation.

Indoor PAH profiles previously published by Offenberg *et al.* were similar to the outdoor profiles.[17,18] The PAH concentrations found in the indoor dust were still above 0.1% of the mass and included a wide range of semi-volatile material.

TABLE 3. Range of indoor and outdoor mass (%) by aerodynamic particle size for all analyzed initial dust/smoke and re-suspendable settled dust/smoke samples

Location	<2.5 μm	2.5–10 μm	10–53 μm	>53 μm
Outdoors	0.88–1.33	0.30–0.40	34.6–46.6	52.2–63.6
Indoors	0.40–0.80	0.20–2.30	20.1–78.5	19.1–79.1

Source: Yiin et al.[13]

Composition of Particles and Gases During the Categories 3 and 4 Exposure Periods

Categories 3 and 4 periods of exposure were associated with emissions that would occur after the first and second rain event, respectively. The first rain event occurred on September 14, 2001 and the second rain event occurred on September 25th. The results shown in FIGURE 8 were previously published by Landrigan et al.,[6] and obtained from NYU and NYC fine particulate matter samplers, particles less than 2.5 um in diameter, during the NYU Medical Center's air sampling program set up post-9/11. As discussed by Landrigan et al., the fine particulate matter in the atmosphere from about 9/14 to 9/26, 2001 was slightly elevated above the normal levels seen in NYC. The levels measured at the NYU Hospital, which was the site closest to the WTC were about a factor of 2 to 5 times higher than the measurements taken at other locations. This was due to the periods of time when the WTC plume passed by the NYU Hospital.[6] When the daily average concentrations are put into perspective with the daily national ambient air quality standard for $PM_{2.5}$, the levels measured at NYU were not above the daily standard and the atmospheric concentrations were only slightly above the daily National Ambient Air Quality Standard (NAAQS) of 65 μm per cubic meter for 24 h on 9/26 at Pace College. The conclusion was that although there were increases in $PM_{2.5}$ in the locations around Ground Zero during the Category 3 exposure period, the air quality was not severely impacted on a continuing basis. This is in contrast to what one might have anticipated based upon the fact that fires were still burning at the WTC site.[6]

The most detailed analyses of the post-9/11 Category 4 exposure were completed by the United States Environmental Protection Agency (USEPA).[19] The EPA started monitoring at the beginning of the Category 4 exposure period with, as seen in FIGURE 9, most of the sites operated by the EPA located in and around the WTC site. By the end of October, the daily measured concentrations of $PM_{2.5}$ were significantly reduced. In fact, at the EPA sites that were adjacent to Ground Zero there were only a few days at the beginning of the sampling period when the daily $PM_{2.5}$ concentration exceeded 65 μm per cubic meter and those were limited to a period of time that ended around the middle of October. Thus, these results would be similar to the data obtained by the NYU site and the sites operated by the City of New York, indicating that $PM_{2.5}$ concentrations were above the NAAQS for $PM_{2.5}$ for a limited number

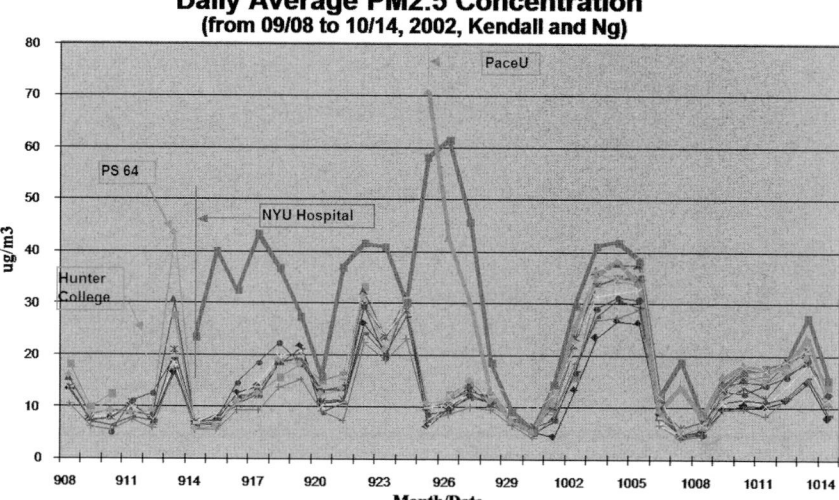

FIGURE 8. Fine particle levels in lower Manhattan post September 11, 2001 (from Landrigan et al.[6]).

of days in areas during Category 4 exposure period immediately adjacent to Ground Zero. These measurements were obtained approximately one to two blocks from Ground Zero with sites A, C, and K receiving the highest concentrations. From mid-October through the rest of the year and the beginning of 2002 the concentrations measured by the EPA at all sites were well below the NAAQS for $PM_{2.5}$.

A few interesting observations during this period of time (Category 4 exposure period) were associated with the first comprehensive gaseous measurements in areas surrounding Ground Zero. The benzene concentrations measured by the EPA post-9/11 in the Category 4 period did exceed 25 ppb.[19] However, these daily average concentrations were not significantly different from levels that would be found the kitchen of a home with an attached garage. If one takes these levels as an indicator of the significance of benzene emissions prior to this date, there were probably concentrations released during the combustion of all materials at the site well in excess of 100 ppb near Ground Zero. This concentration estimate is speculative, but some spot measurements collected on the pile itself were in excess of 1 ppm.[19] Similar to $PM_{2.5}$ levels measured during the Category 4 period, however, air pollution was generally returning to typical urban concentrations.[19]

Lead levels in the atmosphere never exceeded the City benchmark of 1.5 μm per cubic meter for a 30-day average during the post-9/11 period. The average concentrations measured throughout the Category 4 exposure period were well below 1 μm per cubic meter.[19]

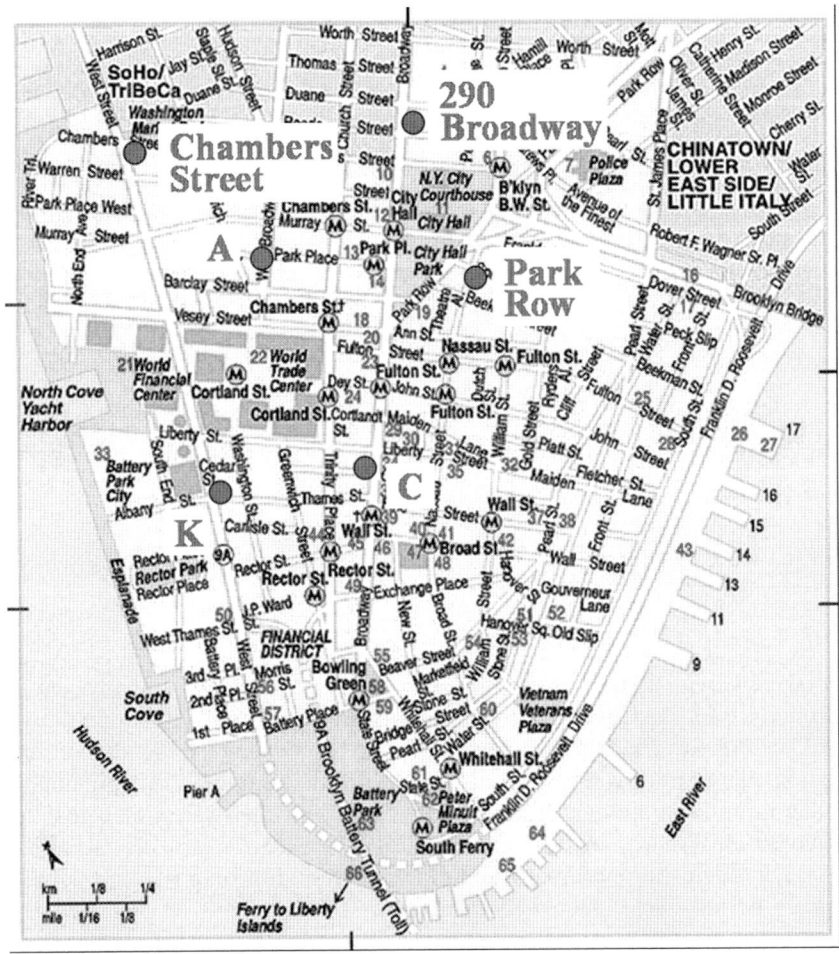

FIGURE 9. EPA WTC air monitoring sites (EPA website).

Work conducted by Cahill et al. focused on ultra fine particles in the atmosphere.[20] They found during their period of sampling, which commenced in Category 4 exposure period, that concentrations of ultra fine particles would increase and those increases were probably due to the WTC fires. They determined that the chlorine increases were associated with the combustion of plastics, and that the plume levels of ultra fine chlorine were significantly above background. These levels were observed during the beginning of October and as with many other pollutants measured during the Category 4 exposure period these were progressively reduced by the end of October. As previously noted in Landrigan et al. the levels increased during the period when there was a nocturnal inversion during the evening while at Ground Zero there were still smoldering fires and periodic bursts of flames during the recovery operations.[6]

CONTACTS OF VARIOUS MEMBERS OF THE AFFECTED POPULATION WITH WTC EMISSIONS

In addition to the different types of materials that were in the air or on surfaces during each exposure period, there were large uncertainties associated with defining actual types of contacts that occurred during Categories 1, 2, and 5. These uncertainties exist for different reasons. For Category 1, the potential exposure of individuals was caused by extremely high concentrations (unmeasured) plus the presence of people "in harm's way" but in contact with the material for a relatively short period of time (<1–2 h).[15,21] For individuals who were in contact with WTC emissions during Category 2 exposure the concentrations would be somewhat lower, but the contacts were probably longer and of different types because of the fact that individuals were completing and participating in activities that were very near to the WTC. Some individuals were being exposed to re-suspended settled dust/smoke particles while others were close to direct emissions at Ground Zero without a respirator. The ambient air contacts during the Categories 3 and 4 exposure periods were consistent with the concentrations found in the ambient air in an urban environment. There were relatively low concentrations and occasional periodic increases due to other WTC-related activity and sources. The contacts occurred while outdoors in Southern Manhattan, and when the wind shifted among the different downwind directions.[6,10,12]

Personal exposure estimations were developed by Wolff *et al.* These were published for women who were pregnant and who at the time of the collapse were at the WTC site or were participating in activities around Ground Zero.[22] These estimates were derived from diaries kept by each woman during the month post-9/11, and were matched for time and location with computational simulation made of the magnitude, extent, and direction of the WTC plume during 1 month post-9/11. A representation of these results is shown in FIGURE 10, and clearly identifies the differences between Categories 1, 2, 3, and 4 exposures, and the influence of Category 5 (indoor) exposures during that period. The exposure indices that were developed for each of 187 women indicate that the highest WTC exposures (indoors and outdoors) occurred during first few days post collapse (Categories 1 and 2 exposure periods). Since these exposures represent individuals who were active at Ground Zero during the event, or just after the event, they provide a basis for estimating the exposures of others who were most affected by contact during or after the event. The analysis show that for at least for the first 10 days, which includes periods Categories 1, 2, and 3, the highest exposures were clearly dominated

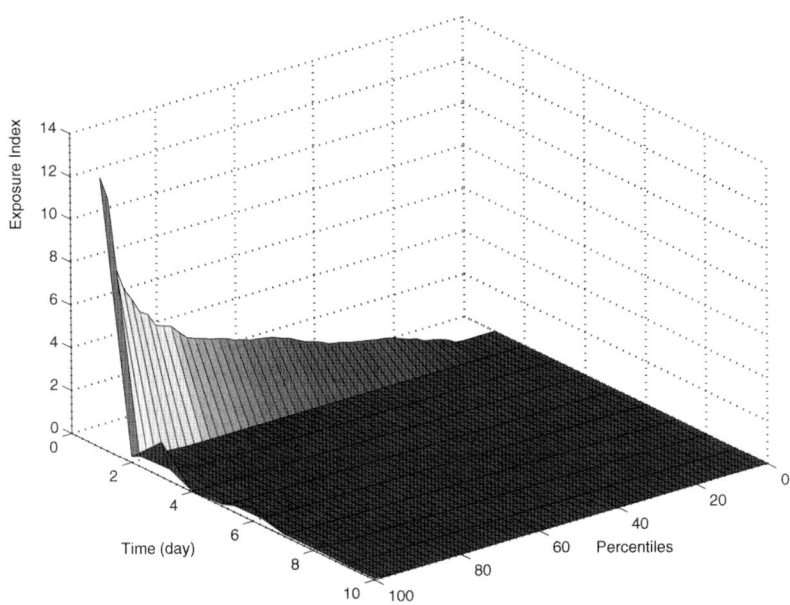

FIGURE 10. Estimated exposure profiles of pregnant women at the WTC on September 11, 2001 and for the first 30 days post 9/11 (adapted from research of Wolff et al.,[23]).

by those exposures that occurred during the Category 1 and somewhat less during the Category 2 exposure period.

Long-Term Category 5 Exposures—Indoors

Unfortunately, most exposure characterizations cannot deal directly with Category 5 contacts that occurred throughout the post-9/11 period. Many of the Category 5 exposures were individual in nature, and were associated with very individual life experiences of the people who were exposed to the settled dust found in their work place or residence, etc. Category 5 exposures will be associated with indoor activities that include cleaning indoors before people re-enter, re-entry of people into homes or businesses that were not totally cleaned. This occurred and the recontamination of homes that were cleaned but were still in buildings that were not totally cleaned. This occurred before a unit was cleaned up by a resident or owner.[23,24] There is no way to truly document the magnitude and extent of these exposures. However, it can be said that many businesses and homes, etc., were cleaned up eventually, and at the present time the EPA is examining the issue of whether or not there are still any residual exposures that are occurring indoors.[11] This will be addressed by a second round of sampling indoors by the EPA.[11] The EPA's first volunteer cleanup program was reviewed by the Inspector General (IG).[23] The one flaw detected by the IG was that the Agency did not completely evaluate HVAC systems prior to a cleanup.[23] The major finding during the first EPA cleanup was, however, that relatively few of the volunteer locations were still highly contaminated before cleanup. The homes that volunteered and found to still have contamination were offered a cleanup, and after no more than three or four attempts all chemicals of potential concern were reduced to background.[24] However, it should be remembered that there are still buildings that have not reopened, and others that are in the process of being demolished instead of being cleaned up, and then are scheduled for reconstruction.

DISCUSSION

The one result of the application of exposure science to the WTC terrorist event was to define, with noted limitations, the categories of exposure that occurred post-9/11. Clearly, there was an attempt to reconstruct "contact" that occurred instantaneously for Categories 1 and 2, since under the best of circumstances it was nearly impossible to monitor or evaluate the actual intensity of exposure to many individuals and groups. The exposure estimations associated with each category have been developed qualitatively or quantitatively by a number of investigators and used in health evaluations, risk assessments, and cleanup.[25-35] With this in mind the issue becomes how do we move forward? The is no simple answer to the question because the nature, complexity, and suddenness of catastrophic terrorist or natural events, but there are some obvious issues.[21]

Monitoring

Under the best of circumstances it would have been very difficult to monitor and evaluate what was in the air during the collapse and the period of time immediately after the event. Still, there needs to be a set of systems developed and available that can provide an opportunity to monitor the air during an acute exposure event using a portable monitoring system. This, however, is not simple; in fact, it is a daunting task, since the system must be capable of handling high concentrations of relatively unknown materials over a short period of time. The best one could hope for is the development of simple instrumentation that can continuously detect indicators that may give you an idea of the concentrations of materials there, but not necessarily the actual concentrations of individual chemicals. For example, you obtain an indication of the concentrations of all particles in the atmosphere during the event but not the composition. For an acute event this is not adequate since you do not know what is present that can lead to harm. You can also obtain the concentrations of total volatile organics, or some important individual toxic volatile organics over a relatively short period of time with continuous monitoring or short-term spot samples. The downside for the gaseous materials is again the same as would be experienced for particles: an indication of chemicals present but not necessarily the levels of toxic materials that can cause acute health effects including death. We must develop continuous monitors for key toxicants that one might expect to have in "harm's way" during a major event. If it is a combustion event, it is relatively easy to obtain spot measurements of benzene and other highly toxic materials that may be present in the plume. However, one also must be sure that a device(s) can obtain accurate measurements of high concentrations. There is also the possibility of measuring PAHs. Although they are not acute toxins, the PAHs could be measured for longer sampling times than one uses for the measurement of benzene or other highly toxic materials (e.g., chlorine, or sarin) during an event. For particles, the task is still very difficult and one important variable to measure simultaneously is particle size. A major problem with the WTC aftermath was that even during the post-9/11 periods described as Categories 3 and 4 there were no accurate measurements of coarse particles (>10 μm in diameter). The reason is that we did not and still do not have the appropriate monitors. Usually, $PM_{2.5}$ is measured in the atmosphere. This has been established as a NAAQS to prevent long-term health effects and asthma attacks and acute health outcomes among the elderly or the highly infirm population.[36] In dealing with 9/11 we needed samplers to give us an idea of the *total mass* burden in the atmosphere so that one could determine whether or not the amount of material in the air was at a sufficiently high enough concentration to recommend that people should have been wearing surgical masks during the initial days post event. However, we could not complete such measurements since the country's focus before 9/11 was primarily preventing long-term health risks among sensitive individuals. The conditions presented by 9/11 resulted in acute exposures to total particulate matter among members of the entire downtown Manhattan population.

Standards for Particles and Other Materials

Prior to 9/11 the EPA developed criteria for air pollutant standards (NAAQS) for preventing both long-term exposures and relatively short-term health risks among sensitive subgroups. The EPA and other agencies also developed emission standards to lower levels of toxicants in the air and to minimize long-term health risks (e.g., cancer). What was not recognized prior to the terrorist attack was the need to be using guidelines like AEGLS (Acute Exposure Guidelines) that are being developed by the EPA. They can be used to establish whether or not one could safely go into an area after an event. There are a limited number of AEGLS, but each has a defined set of criteria for implementation.[37] There is an AEGL 1: it includes exposures for very short periods and is designed to identify risks that yield acute reversible effects. There is an AEGL 2: to protect against nonreversible but not lethal effects. Finally there is an AEGL 3: it is designed to indicate that if exceeded during an exposure period there will be lethal effects. It is important that the EPA and others aggressively pursue the development of the AEGLS for a variety of materials (e.g., coarse particles that can be emitted in any collapse or fire) and materials that can be released in a plume; for example, chlorine (industrial chemical), sarin, or a biological warfare weapon. Such guidelines would clearly define the significance of an acute exposure event to the health of a local population. What was troubling during 9/11, and we think was very difficult for people to understand, was that in some cases asbestos was being used as an indicator for acute exposure. This was not reasonable because asbestos does not typically yield acute health effects; it produces asbestosis and possibly mesothelioma after 30 years of exposure.[6] These responses do not usually result from one acute exposure that is what occurred in post collapse Category 1 exposure period. There were better indicators (e.g., pH and man-made vitreous fibers) that could have helped focus on the main issue: what materials could have led to effects like the WTC cough.[15] In the future it would be important to make sure guidelines are in place to prepare for the kind of exposures that occur in a hazardous event, a natural disaster, or terrorist event.

Sampling Strategy

The sampling strategies available post-9/11 were primarily those developed for long-term air sampling programs in areas where there might in fact be an individual source. These kinds of studies take a long time to set up, deploy, and yield accurate data. They usually require electricity and other types of equipment and a reasonable monitoring location, which were not available immediately post-9/11. The weakness was not having a strategy available that could be implemented quickly and efficiently in the area affected by the releases into the atmosphere. Such a strategy is essential to get the data necessary to evaluate whether or not there is potential for an acute exposure over a relatively short period of time, and then apply modeling analyses to interpret the results for making decisions to manage acute risks during rescue or recovery operations. A "new" sampling strategy is needed that can be tested systematically within current programs or training exercises to determine whether or not a new sampling strategy can actually work. We have not seen a national program implemented yet that demonstrates how we would respond in an acute event with the sampling strategies that have been defined or identified for use by the EPA Red Team. This would be a very important application or simulation that is needed define the most efficient and "safe" way to enter "harm's way" and obtain accurate information. Further, it will be very important to see how much of a turn-around time would be necessary to provide the kind of data needed to define accurately what in fact is in harm's way.

Clearly the most important need during an exposure event similar to the WTC or other terrorist attack would be respirators suitable for rescue and cleanup. Emergency responders need respirators and other personal protection devices that are commensurate to protecting their health, but still have the flexibility to rescue victims. The type of respirators required must allow the rescue workers to move quickly, and have a very good internal communications system.

The kind of strategies needed to be developed for use in acute exposure events involve two steps: (*a*) determining what the concentration or what the material may be in "harm's way" so that people who are doing emergency medical service and other types of rescue efforts are properly protected to reduce exposures, and (*b*) provide information about what rescue workers are facing when they reach harm's way. Finally, strategies need to be developed in such a way that you can get materials and devices to the site that can be operated with relatively little or no electricity in strategic locations to determine there may be exposures that may have a consequence to health.

CONCLUSIONS

The five categories of exposure from post-9/11 indicate that the most important periods of times for exposure were probably during Category 1 and possible Category 2 of the event. Category 3 and 4 exposure periods indicate that the exposures were lower and the levels were reducing to ambient background. The Category 5 exposures will never be able to be assessed completely since we do not have documentation of how the individuals spent their time and where they went and how they went back into their homes and businesses post-9/11. Included are the level of cleanup done in their businesses, time they spent in their business post-event before the fires burnt out in December, and the level of recontamination in their building due to the fact that other locations were not cleaned up and air drifted into the building or throughout the building.

There are lessons to be learned from this event that can improve our ability to respond effectively and maintain protection of rescue workers and others as they approach the area where an event occurred or would continue to occur. These involve sampler development, acute exposure guidelines, sampling strategies, and respiratory protection devices.

ACKNOWLEDGMENTS

Collaborators on Outdoor Samples: C. Weisel (1,6), J. Millette (3), S. Eisenreich (1,5), D. Vallero (4), J. Offenberg (5), B. Buckley (1), R. Hale (7), B. Turpin (1,5), M. Zhong (2), M.D. Cohen (2), C. Prophete (2), I. Yang (1), R. Stiles (1), G. Chee (2), W. Johnson (1), S. Alimokhtari (1), C. Weschler (1,6), Z. Fan (1), and L.C. Chen (2).

Additional Collaborators on Indoor Samples:
V. Ilacqua (1), L.M. Yiin (1), M. Gallo (1,6) A. Vette (4), M. Kendall (2), J. Gorczynski (2), J. Xiong (2), X. Lu (2), C. Quan (2), and G. Thurston (2).

Collaborators on WTC Plume:
The EOHSI Chemical Chemodynamics Laboratory N. Lahoti, S. Isukapalli, S.W. Wang, E. Jayjock (1,6), S. Perry (4), G. Foley (4), and G. Stenchikov (5).

Discussions with many members of the NIEHS Centers Consortium, and other organizations including M. Gochfeld (1), H. Kipen (1), M. Wolff (8) and P. Landrigan (8):

1. Environmental and Occupational Health Sciences Institute of NJ (UMDNJ - R.W. Johnson Medical School & Rutgers University)
2. Nelson Institute of Environmental Medicine (NYU School of Medicine)
3. MVA, Norcross, GA
4. National Exposure Research Laboratory, US EPA, RTP, NC
5. Department of Environmental Sciences, Rutgers University
6. Department of Environmental and Community Medicine, UMDNJ – R.W. Johnson Medical School
7. William and Mary College, VA
8. Mount Sinai School of Medicine, NY

Initially voluntary activities, and subsequently funded by NIEHS Center Supplements NIEHS P30–ES05022-1551, and EPA University Partnership Supplements to CERM–CR827033, CR-883162501.

REFERENCES

1. LIOY, P.J. 1999. The 1998 ISEA Wesolowski Award lecture. Exposure analysis: reflections on its growth and aspirations for its future. International Society of Exposure Analysis. J. Expo. Anal. Environ. Epidemiol. **9:** 273–281.
2. NATIONAL RESEARCH COUNCIL. 1991. Human Exposure to Airborne Pollutants: Advances and Opportunities. National Academy Press. Washington, DC.
3. LIOY, P.J. 1990. The analysis of total human exposure for exposure assessment: a multidiscipline science for examining human contact with contaminants. Environ. Sci. Technol. **24:** 938–945.
4. LIOY, P. 1995. Measurement methods for human exposure analysis. Environ. Health Perspect. **103:** 35–43.
5. CLAUDIO, L. 2001. Environmental aftermath. Environ. Health Perspect. **109:** A528–A537.
6. LANDRIGAN, P.J. et al. 2004. Health and environmental consequences of the World Trade Center disaster. Environ. Health Perspect. **112:** 731–739.
7. LIOY, P.J. et al. 2002. Characterization of the dust/smoke aerosol that settled east of the World Trade Center (WTC) in lower Manhattan after the collapse of the WTC 11 September 2001. Environ. Health Perspect. **110:** 703–714.

8. LIOY, P., C. WEISEL & P. GEORGOPOULOS. 2005. An overview of the environmental conditions and human exposures that occurred post 9–11. *In* ACS Symposium Book. J.S. GAFFNEY & N. MARLEY, Eds. Oxford Publishers. Oxford. pp. 23–28.
9. NEW YORK CITY DEPARTMENT OF HEALTH AND MENTAL HYGIENE. 2002. Final Technical Report of the Public Health Investigation to Assess Potential Exposures to Airborne and Settled Dust in Residential Areas of Lower Manhattan as part of the World Trade Center Environmental Assessment Group.
10. LORBER, M. *et al.* 2004. Assessment of inhalation exposures and potential health risks that resulted from the collapse of the World Trade Center Towers. Environ. Manage. 2004: February 27–29.
11. EASTERN RESEARCH GROUP., I. 2004. Summary report of the first meeting of the World Trade Center Technical. Review Panel.
12. HUBER, A. *et al.* 2004. Modeling air pollution from the collapse of the World Trade Center and potential impact on human exposures. Environ. Manage. **2004:** February 35–40.
13. YIIN, L.M. *et al.* 2004. Comparisons of the dust/smoke particulate that settled inside the surrounding buildings and outside on the streets of southern New York City after the collapse of the World Trade Center, September 11, 2001. J. Air Waste Manage. Assoc. **54:** 515–528.
14. MEEKER, G. *et al.* Open File Report 2005 - 1031. Determination of a diagnostic signature for World TradeCenter dust using scanning electron microscopy point counting techniques. US Geological Survey (USGS) Denver, CO.
15. PREZANT, D.J. *et al.* 2002. Cough and bronchial responsiveness in firefighters at the World Trade Center site. N. Engl. J. Med. **347:** 806–815.
16. NJDEP. 2003. Annual Report for Air Quality, state of NJ, Trenton, NJ.
17. OFFENBERG, J.H. *et al.* 2003. Persistent organic pollutants in the dusts that settled across lower Manhattan after September 11, 2001. Environ. Sci. Technol. **37:** 502–508.
18. OFFENBERG, J.H. *et al.* 2004. Persistent organic pollutants in dusts that settled indoors in lower Manhattan after September 11, 2001. J. Expo. Anal. Environ. Epidemiol. **14:** 164–172.
19. USEPA. 2004. Response to September 11. (USEPA, WTC website).
20. CAHILL, T. *et al.* 2004. Analysis of aerosols from the World Trade Center collapse site, New York, October 2 to 30, 2001. Am. Chem. Soc. Meet. 2003. **38:** 165–183.
21. LIOY, P.J. & M. GOCHFELD. 2002. Lessons learned on environmental, occupational, and residential exposures from the attack on the World Trade Center. Am. J. Ind. Med. **42:** 560–565.
22. WOLFF, M.S. *et al.* 2005. Exposures among pregnant women near the World Trade Center site on 11 September 2001. Environ. Health Perspect. **113:** 739–748.
23. O.I.G. 2003. EPA Response to the World Trade Report: Challenges. Successes and Areas for Improvement, Report 2003-p-00012, Washington, DC April 2003.
24. USEPA. 2003. Residential Confirmation Study. Interim Final. New York Response and Recovery Operations, EPA Region 2, NYC, NY.
25. BANAUCH, G.I. *et al.* 2003. Persistent hyperreactivity and reactive airway dysfunction in firefighters at the World Trade Center. Am. J. Respir. Crit. Care Med. **168:** 54–62.
26. BECKETT, W.S. 2002. A New York City firefighter: overwhelmed by World Trade Center dust. Am. J. Respir. Crit. Care Med. **166:** 785–786.
27. BERKOWITZ, G.S. *et al.* 2003. The World Trade Center disaster and intrauterine growth restriction. JAMA **290:** 595–596.

28. BOSCARINO, J. et al. 2002. Utilization of mental health services following the September 11th terrorist attacks in Manhattan, New York City. Emerg. Ment. Health **4:** 143–155.
29. FIREMAN, E.M. et al. 2004. Induced sputum assessment in New York City firefighters exposed to World Trade Center dust. Environ. Health Perspect. **112:** 1564–1569.
30. GALEA, S. et al. 2002. Psychological sequelae of the September 11 terrorist attacks in New York City. N. Engl. J. Med. **346:** 982–987.
31. GALEA, S. et al. 2002. Posttraumatic stress disorder in Manhattan, New York City, after the September 11th terrorist attacks. J. Urban Health **79:** 340–353.
32. GAVETT, S.H. et al. 2003. World Trade Center fine particulate matter causes respiratory tract hyperresponsiveness in mice. Environ. Health Perspect. **111:** 981–891.
33. LEVIN, S. et al. 2002. Health effects of World Trade Center site workers. Am. J. Ind. Med. **42:** 545–547.
34. REIBMAN, J. et al. 2003. Respiratory health of residents near the former World Trade Center: the WTC residents respiratory health survey [abstract]. Am. J. Respir. Crit. Care Med. **167:** A335.
35. ROM, W.N. et al. 2002. Acute eosinophilic pneumonia in a New York City firefighter exposed to World Trade Center dust. Am. J. Respir. Crit. Care Med. **166:** 797–800.
36. US EPA. 2005. Review of the National Ambient Air Quality Standards of Particulate Matter: Policy Assessment of Scientific and Technical Information 511... 514. QAQPS Staff paper, EPA-452/R-05-005a, December, 2005.
37. THE NATIONAL ACADEMIES. 2000. Acute Exposure Guideline Levels for Selected Airborne Chemicals. National Academy Press. Washington, DC.

The Health Watch Case–Control Study of Leukemia and Benzene

The Story So Far

DEBORAH C. GLASS,[a] CHRISTOPHER N. GRAY,[b] DAMIEN J. JOLLEY,[b] CARL GIBBONS,[b] AND MALCOLM R. SIM[a]

[a]*Monash University, Victoria 3004, Australia*

[b]*Deakin University, Victoria 3217, Australia*

ABSTRACT: A case–control study nested in the Health Watch cohort of petroleum industry workers, investigated whether the excess of lympho-hematopoetic cancers, identified among male members of the Health Watch cohort, was associated with benzene exposure. Cases of non-Hodgkin's lymphoma ($n = 31$), multiple myeloma ($n = 15$), and leukemia ($n = 33$) were identified between 1981 and 1999. Cases were age-matched to five controls. Exposure was retrospectively estimated for each occupational history using an algorithm in a relational database. Benzene exposure measurements, supplied by Australian petroleum companies, were used to estimate exposure for specific tasks. The tasks carried out within the job, the products handled, and the technology used, were identified from interviews with contemporary colleagues. More than half of the subjects started work after 1965 and had an average exposure period of 20 years. Exposure was low, 85% of the cumulative exposure estimates were <10 ppm years. Matched analyses showed that non-Hodgkin's lymphoma and multiple myeloma were not associated with benzene exposure. Leukemia risk, however, was significantly increased for the subjects with greater than 16 ppm years cumulative exposure, odds ratio (OR) 51.9 (5.6–477) or with greater than 0.8 ppm intensity of highest exposed job. Cumulative exposures were similar to those found in comparable studies. The inclusion of occasional high exposures, for example, as a result of spillages, reduced the ORs, when the exposure was treated as either a continuous or a categorical variable. Our data demonstrate a strong association between leukemia and modest benzene exposure. The choice of cut-point and reference group has a marked effect on the ORs, but does not change the overall conclusions.

KEYWORDS: case–control study; exposure; benzene; leukemia; non-Hodgkin's lymphoma; multiple myeloma; petroleum industry

Address for correspondence: Deborah C. Glass, Monash University, Department of Epidemiology and Preventive Medicine, Central and Eastern Clinical School, Alfred Hospital, 89 Commercial Road, Melbourne, Victoria 3004 Australia. Voice: 61-3-9903-0554; fax: 61-3-9903-0556.

e-mail: deborah.glass@med.monash.edu.au

HEALTH WATCH COHORT

Health Watch is a prospective cohort study of employees in the Australian petroleum industry who have worked for more than 5 years. It compares the mortality and cancer incidence of the cohort with that of the Australian population. The cohort includes employees from offices, upstream extraction and processing sites, refineries, terminals, and airports all over Australia. The cohort includes about 18,000 employees of which some 1300 are women.

Health Watch commenced in 1981 at Melbourne University, with a face-to-face survey and this survey was repeated in 1986, 1991, 1996, and 1999. At the surveys, subjects provided demographic details, health status information, smoking, and alcohol data and details of their work history. In 1999 the cohort was closed to further recruitment and transferred to Adelaide University. In 2005 it was moved to Monash University.

About 95% of eligible employees have participated.[1] Subjects were actively followed and matched with the Australian national death and cancer registries. The standardized mortality ratio (SMR) for men has remained stable over 25 years at about 0.7.[1] However, an excess incidence of lymphohematopoietic cancers (LH cancers) was identified among the male members of the cohort in 1997.[2] The LH cancers included leukemia, non-Hodgkin's lymphoma (NHL), and multiple myeloma (MM). The observed excess of these cancers has reduced over the years (TABLE 1), but a nested case–control study was set up in 1995 to investigate whether the excess was related to occupational benzene exposure.

Benzene has been designated a class 1 carcinogen by International Agency for Research or Cancer (IARC) because of its leukemogenic properties.[3] It is present as a small percentage of crude oil, gasoline, and at many stages in the refining process.[4] Historically much higher percentages of benzene were handled at some petroleum industry sites in Australia. One or two refineries used BTX (benzene/toluene/xylene mixture) a byproduct of coke operations, which contains 70% benzene. Some refineries and terminals also used benzene as a feedstock or drummed benzene itself. At most sites, the consequent exposures were brief, intermittent, applied to relatively few employees, and ceased around 1975. One refinery used benzene as a feedstock until relatively recently, taking delivery of benzene by tanker a few times a year.

TABLE 1. Cancer incidence for males reported over time in the Health Watch cohort

Male SIR	7th Report 1987[2]	8th Report 1988–89[21]	9th Report 1992[22]	10th Report 1998[23]	11th Report 2000[24]	12th Report 2005[1]
Leukemia	2.0(0.6–4.6)	3.4(1.7–5.9)	2.8(1.7–4.5)	2.0(1.3–2.9)	1.5(1.0–2.2)	1.1(0.7–1.5)
NHL	1.1(−)	1.7(0.8–3.1)	1.3(0.7–2.2)	0.9(0.6–1.3)	1.0(0.7–1.4)	0.9(0.7–1.2)
MM	1.1(−)	2.2(0.6–5.6)	1.8(0.7–3.9)	1.9(1.0–3.3)	1.7(0.9–2.8)	1.1(0.6–1.8)

Standardized by age, sex, and calendar period of follow-up. SIR: standardized incidence ratio.

CASE–CONTROL STUDY METHODS

A nested case–control study was carried out in which each case was individually matched on age and sex to five controls from the cohort, drawn with replacement.[5] Cases were defined as male members of the Health Watch cohort, who had:

- first diagnosis of LH cancer after entering the Health Watch cohort; and
- diagnosis confirmed by pathology report, cancer registration, letter from medical practitioner, or death certificate; and
- had reported LH cancer to Health Watch either by self or by family, unless they were lost to contact by Health Watch, or were deceased.

There were 31 NHL, 15 MM, and 33 leukemia cases meeting these criteria identified in the cohort between 1981 and 1999.

Exposure to benzene was estimated for individual subjects based on their job history, to which an algorithm was applied, which used Australian exposure data provided by participating oil companies.[5–7] A job history was prepared for each subject, based on data that had been collected, largely prospectively, from the four cohort interviews that took place between 1980 and 1999.[5] The job histories were complete for 98% of subjects ($n = 494$), and were then checked by the employing company against their records. We used each subject's job history to identify which of several standard structured job-specific questionnaires was appropriate for each job. A contemporary colleague or colleagues were then interviewed about each job using the appropriate job-specific questionnaire, identifying the time spent on each of the tasks carried out, the technology used, and the products handled, for example, petrol, diesel, benzene. The interviews were carried out by one of the two experienced occupational hygienists, blind as to case status.

The basic tasks in refineries and terminals have not changed greatly over time, although the technology used and the frequency with which they are carried out, have changed. We collected information about the task frequency and technology change at the site visits and allowed for this in the exposure assessment. We did this by, for example, allocating a longer time for tank farm operators doing dipping in the 1950s because of more frequent excise checks.

We collated Australian personal benzene exposure monitoring data provided by participating companies and calculated the arithmetic mean exposure for each task or job represented. This included more than 3870 individual benzene measurements. If more than one technology could have been used, for example, top and bottom loading for road tankers, separate base estimates (BEs) were calculated for each technology. The BE was then allocated to individuals carrying out that task in a relational database. Few or no local data were available for some short-term tasks and data for these were sought from the literature. The BEs were validated against relevant exposure data identified from the literature.[8]

Cumulative exposure was estimated using a task-based algorithm in a relational database. The algorithm was similar to that used in other petroleum industry studies,[9,10] it took into account: the time spent on different tasks with a job, the proportion of time spent handling each product within the task, the percentage of benzene in the product handled, the technology used for the task, and the years spent on each job.[6] If the BE was applied to a task with different technology, a multiplier was used to scale the exposure, for example, the BE for top loading was attributed to a driver who used top splash loading and a factor of three was applied to increase the exposure estimate. Some of the multipliers were derived from the literature; others were agreed by a panel of local industry occupational hygienists. These expert-derived multipliers were used for fewer than 160 tasks from a total of 3457 tasks assessed.

These data were used to estimate for each subject, the cumulative exposure to benzene in ppm years, and to identify the intensity of the most highly exposed job in ppm (cumulative exposure divided by years spent on that job). The subjects were categorized by cumulative exposure in ppm years and intensity of exposure in ppm.

We also identified jobs where there may have been infrequent high exposure events (HEEs) as a result of spillages or work practices that are no longer carried out and so would not be represented in the exposure data used for the BEs. We estimated the exposure that might have resulted from these HEEs and added them to the mean exposure for that subject.[7] Odds ratios (ORs) were then recalculated.

ORs for leukemia, NHL, and MM were calculated from conditional logistic regression using Stata.

CASE–CONTROL STUDY FINDINGS

The cases and controls were well matched. Subjects worked in the industry for an average of 20 years (range 4 years to 42 years), the majority started employment after 1965.[5] Exposure was low, the estimated lifetime cumulative benzene exposures ranged from 0.005 to 57.3 ppm years, with a mean of 4.9 ppm years. Eighty-five percent of the cumulative exposure estimates were <10 ppm years. For cases the mean exposure was 10.6 ppm years, for controls it was 3.86 ppm years.[5] The addition of occasional high exposures, increased exposure for 25% of subjects but for most, the increase was less than 5% of total exposure (FIG. 1).[7]

In matched analyses, NHL and MM were not associated with benzene exposure.[5] However, the OR for leukemia increased with cumulative exposure when exposure was treated as a continuous variable in a matched analysis, OR 1.10 (95% CI: 1.04–1.16) per ppm year. When the HEEs were added to the cumulative exposure, the continuous variable OR was 1.03 (1.01–1.05).[7]

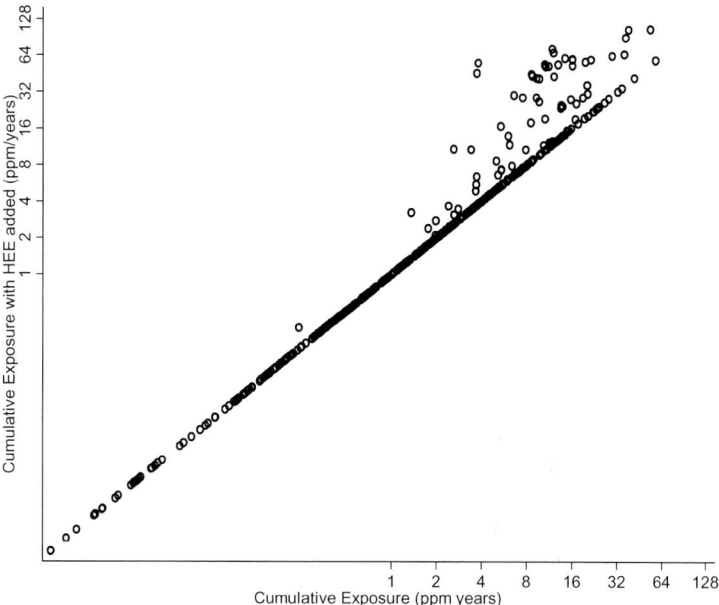

FIGURE 1. Comparison of benzene exposures with and without HEEs in a case–control study of petroleum workers.

The OR also increased with increasing cumulative exposure when exposure was treated as a categorical variable. The OR for the most highly exposed group, including the 7 cases with >16 ppm years cumulative exposure, was 98 (8.8–1090) when compared to those with less than 0.5 ppm years exposure to benzene.[5] A later reanalysis was carried out because it was recognized that the original reference group was small and because the precision of the exposure estimation process made it difficult to justify separating those with less than 0.5 ppm years from those with 0.5–1 ppm years exposure. Combining the two lowest exposure groups made a larger and more stable reference group containing the nine lowest exposed case sets. When the highest exposed group was compared to this new grouping the OR was 51.9 (5.6–477)[11] without HEEs and OR 7.79 (2.34–25.89) with HEEs included.[7] The reduction in ORs when the HEEs are added is probably because the leukemia risk is associated with higher exposures and hence the risk per ppm year was reduced.

We also examined risk by intensity and found that the risk was increased for subjects whose highest intensity job was greater than 0.8 ppm (exposure group >0.8–1.6 ppm, OR 6.34 [1.54–26.18]; exposure group >1.6–3.2 ppm, OR 5.58 [1.00–31.04]). However, exposure intensity and cumulative exposure are highly correlated and neither goodness-of-fit statistics nor the stepwise conditional logistic regression algorithm allowed us to distinguish which was the better predictor of risk.[5]

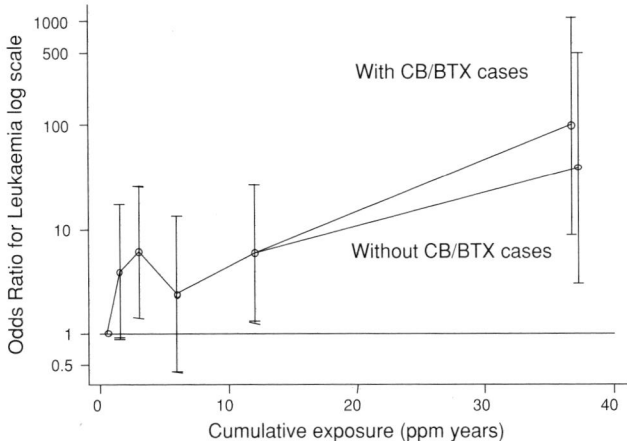

FIGURE 2. Conditional (fixed-effects) logistic regression, leukemia by cumulative exposure to benzene (ppm-years) with and without cases exposed to CB/BTX.

Risk of leukemia was strongly associated with exposure to 100% benzene (concentrated benzene or CB) or to BTX (FIG. 2). There was very little difference in the estimated cumulative exposures between the non-CB/BTX exposed group (mean 32.67 ppm years, median 29.85 ppm years) and the CB/BTX exposed group (mean 32.25 ppm years, median 28.08 ppm years). In fact, the mean and median exposure of the CB/BTX exposed group was slightly lower than that of the non-CB/BTX exposed group. The difference in OR for the highest exposure group with and without the CB/BTX exposed subjects is not of itself statistically significant. However, it is a large difference which, if real, cannot be explained in terms of differential exposures as estimated.

We showed that the risk of leukemia is associated with exposure within 15 years of diagnosis, the association with exposure prior to this period is weak.[12] There was no association between leukemia (with or without adjustment for cumulative benzene exposure) and date of starting work in industry, period of employment, or duration of work. After adjusting for cumulative benzene exposure, there was no association between leukemia and industry sector. We found no association between tobacco or alcohol consumption and leukemia. We consider it unlikely that there were other confounding exposures, for example, radiation or retroviruses.[5]

ORs were calculated for the leukemia subtypes acute nonlymphocytic leukemia (ANLL), chronic lymphocytic leukemia (CLL), and chronic myeloid leukemia (CML). Because there were relatively few cases of the leukemia subtypes, it was necessary to combine the three lowest exposure groups and the two highest exposure groups. The OR in the exposure group >8 ppm years were significantly raised only for ANLL OR 7.17 (1.27–40.44).[5] An association between ANLL and higher benzene exposures has been reported in several papers previously.[13] The OR for CLL was high 4.52 (0.89–22.90) and this has not been reported previously. It should be noted that this is an incidence study, however, and of the 11 CLL cases in the study only 1 has died.[1] Most previous ORs in this field have been calculated on cancer deaths and an OR based on deaths might well have a different result for this disease.

CRITICISMS OF THE STUDY

Three main criticisms that have been made of the study are:

- that there was recall bias in the exposure assessments[14];
- that the exposure estimates were too low; and
- that there is diagnostic bias, more highly exposed subjects had their blood screened and thus more cases were identified in this group.[15]

With respect to the first criticism, our exposure estimates were very detailed, based on verified job histories and very largely on exposure data collected by the companies occupational hygienists. The job history was checked against company records and the interviews were case blind and structured, so the potential for recall bias is limited.[5] In addition, the lack of association between benzene exposure and NHL or MM, for whom the exposure data were collected in exactly the same way, suggests that recall bias is unlikely to explain association with leukemia.

As regards the extent of exposure, most of the exposures in this study are similar to those estimated in previous comparable studies although they do not range as high. However, the cohort is more recent than the Canadian[16] and UK[17] studies, which have subjects whose job histories go back to before 1920. Thus the use by subjects of older technologies, resulting in higher exposure, will be less frequent in our study. There is, for example, only one top splash loader in this study. Most of the BE values were derived from local exposure monitoring data and have been validated.[8] One of the most highly exposed jobs is that of drum filler, however, drum fillers in Australia have always been required to have open-sided rather than fully enclosed drum sheds as needed in Canada. The increased ventilation is likely to have resulted in lower exposure and hence it is expected that an Australian study would have a lower drum filling BE than the Canadian study. The association between blood changes and very low exposure to benzene (in terms of cumulative and intensity of exposure) has been confirmed by a recent study from China.[18]

With respect to diagnostic bias, no cases in the study were identified from routine blood screening carried out by petroleum companies. We understand in fact, that no cases of leukemia have ever been diagnosed on the basis of such routine blood screening in Australia.

THE FUTURE

There were too few cases for much detailed analysis by leukemia subtype or to establish clearly what time window is pertinent to exposure assessment for benzene with respect to increased risk of leukemia. In 2002, we drew up a proposal with colleagues who had conducted similar case–control studies in the UK and Canada to combine the data in the three studies to provide a more powerful analysis of these questions. We may also be able to analyze the data using the new WHO or REAL[19] criteria for grouping leukemias and NHL.

In 2004, the Australian, Canadian, and UK petroleum industry case–control studies were audited by a team from two external research institutions. They considered that "all of the studies had been well performed, there were no issues of subject selection, methods, or general data quality that were likely to have distorted their internal comparisons." They also concluded that "the evidence of an increased risk at higher exposures in Australia was convincing."[20] The audit team saw no obstacles to the data pooling and recommended that this should happen.

CONCLUSIONS

Our data demonstrate a strong association between leukemia and modest benzene exposure measured as cumulative exposure in ppm years or as intensity of exposure of the highest exposed job in ppm. The study was carried out to high standards and is robust to external audit.

Combining the data with that of the UK and Canadian studies would provide more power to establish the nature and extent of the association between benzene exposure, leukemia, and its subtypes and to derive more precision on time frames associated with this increased risk.

REFERENCES

1. GUN, R.T., P. RYAN, D. RODER, et al 2005. Health Watch twelfth report 2005. Department of Public Health, Adelaide University. Adelaide, Australia.
2. CHRISTIE, D.K., I. ROBINSON, I. GORDON, et al. 1988. Health Watch seventh annual report 1987. The University of Melbourne, Department of Community Medicine. Melbourne, Australia.
3. IARC. 1982. IARC Monograph on the evaluation of the carcinogenic risk of chemicals: some industrial chemicals and dyestuffs. IARC WHO. Lyon.
4. WALLACE, A. 1996. Environmental exposure to benzene: an update. Environ. Health Perspect. **104:** 1129–1136.
5. GLASS, D.C., C.N. GRAY, D.J. JOLLEY, et al. 2003. Leukemia risk associated with low level benzene exposure. Epidemiology **15:** 569–577.

6. GLASS, D.C., G.G. ADAMS, R.W. MANUELL, *et al.* 2000. Retrospective exposure assessment for benzene in the Australian Petroleum Industry. Ann. Occup. Hyg. **44:** 301–320.
7. GLASS, D.C., C.N. GRAY, D.J. JOLLEY, *et al.* 2005. Health Watch exposure estimates: do they underestimate benzene exposure? Chem. Biol. Interact. **154/155:** 23–32.
8. GLASS, D.C., C.N. GRAY, G.G. ADAMS, *et al.* 2001. Validation of exposure estimation for benzene in the Australian petroleum industry. Toxicol. Ind. Health **17:** 113–127.
9. ARMSTRONG, T.W., E.D. PEARLMAN, R.A. SCHNATTER, *et al.* 1996. Retrospective benzene and total hydrocarbon exposure assessment for a petroleum marketing and distribution worker epidemiology study. Am. Ind. Hyg. Assoc. J. **57:** 333–343.
10. LEWIS, S.J., G.M. BELL, N. CORDINGLEY, *et al.* 1997. Retrospective estimation of exposure to benzene in a leukaemia case-control study of petroleum marketing and distribution workers in the United Kingdom. Occup. Environ. Med. **54:** 167–175.
11. GLASS, D., C. GRAY, D. JOLLEY, *et al.* 2002. The Health Watch case control study. *In* Proceedings of a Conference Institute of Petroleum 19th October 2002. London.
12. GLASS, D., M.R. SIM, L. FRITSCHI, *et al.* 2004. Leukaemia risk and reevant benzene exposure period-Re: follow-up time an risk estimates. Am. J. Ind. Med. **47:** 481–489, 2002 [comment]. Am. J. Ind. Med. **45(2):** 222–???.
13. BERGSAGEL, D.E., O. WONG, P.L. BERGSAGEL, *et al.* 1999. Benzene and multiple myeloma: appraisal of the scientific evidence. Blood **94:** 1174–1182.
14. SCHNATTER, R.A. 2003. Benzene exposure and leukemia. Epidemiology **15:** 509.
15. GOLDSTEIN, B.D. 2003. Benzene exposure and leukemia. Epidemiology **15:** 509–510.
16. SCHNATTER, R.A., T.W. ARMSTRONG, M.J. NICOLICH, *et al.* 1996. Lymphohaemotopoietic malignancies and quantitative estimates of exposure to benzene in Canadian petroleum distribution workers. Occup. Environ. Med. **53:** 773–781.
17. RUSHTON, L. & H. ROMANIUK. 1997. A case-control study to investigate the risk of leukaemia associated with exposure to benzene in petroleum marketing and distribution workers in the United Kingdom. Occup. Environ. Med. **54:** 152–166.
18. LAN, Q., L. ZHANG, G. LI, *et al.* 2004. Hematoxicity in workers exposed to low levels of benzene. Science **306:** 1774–1776.
19. HARRIS, N., E. JAFFE, H. STEIN, *et al.* 1994. A revised European-American classification of lymphoid neoplasms: a proposal from the International Lymphoma Study Group. Blood **84:** 1361–1392.
20. MILLER, B., W. FRANSMAN, D. HEEDERIK, *et al.* 2005. A review of the data quality and comparability of case-control studies of low-level exposure to benzene in the Petroleum Industry. Institute of Occupational Medicine. Edinburgh. Report No.: TM/05/04.
21. CHRISTIE, D., K. ROBINSON, I. GORDON, *et al.* 1990. Health Watch eighth annual report 1988-9. The University of Melbourne, Department of Community Medicine. Melbourne, Australia.
22. BISBY, J.A. & G.G. ADAMS. 1993. Health Watch ninth report 1992. The University of Melbourne, Department of Public Health and Community Medicine. Melbourne, Australia.

23. BISBY, J.A. & G.G. ADAMS. 1999. Health Watch tenth report 1998. The University of Melbourne, Department of Public Health and Community Medicine. Melbourne, Australia.
24. GUN, R.T., L. PILOTTO, P. RYAN, *et al.* 2000. Health Watch eleventh report 2000. Department of Public Health, Adelaide University. Adelaide, Australia.

Benzene Exposure and Multiple Myeloma

A Detailed Meta-analysis of Benzene Cohort Studies

PETER F. INFANTE

Environmental and Occupational Health, School of Public Health and Health Services, The George Washington University, Washington, DC 20037, USA

ABSTRACT: Case reports and epidemiological studies of workers exposed to benzene have demonstrated associations with a number of lymphohematopoietic diseases, but the association with multiple myeloma (MM) has been less apparent. Data from all of the "benzene cohort studies" conducted to date have been selected and evaluated for inclusion in a meta-analysis. The analysis demonstrates a significant excess in the relative risk (RR) of MM in relation to benzene exposure. Pooling the data from seven cohort studies, a meta-analysis yields a statistically significant weighted RR estimate of 2.13 (95% CI = 1.31–3.46). In the analysis of cohort data, an understanding of the cohort follow-up period in relation to benzene exposure and RR of MM is important. Exposure-related RRs of disease decline after the median latency periods are exceeded, particularly when exposure has terminated decades earlier. The positive epidemiological evidence for benzene as a cause of MM is supported by biological plausibility for such an effect from benzene exposure. Studies of refinery workers are difficult to interpret in relation to benzene exposure and risk of MM, but are limited in the study design and analysis. Nonetheless, they provide some support for an association between refinery work and MM.

KEYWORDS: benzene; myeloma; leukemia; cancer risk; occupational; follow-up period; petrochemicals

INTRODUCTION

Multiple myeloma (MM) is a neoplasm of plasma cells that are derived from B-lymphocytes located primarily in the bone marrow. For the past 40 years,

Address for correspondence: Peter F. Infante, D.D.S., Dr.P.H., 200 S. Oak Street, Falls Church, VA 22046. Voice: 571-641-3047; fax: 571-641-3407 (call in advance of faxing).
e-mail: pinfante@starpower.net

a secular increase in the incidence of MM is apparent in many industrialized countries. In the United States, the current death rate is about twice as high in men as compared to women and about twice as high in blacks as compared to whites. Case reports and epidemiological studies of workers exposed to benzene have demonstrated associations with a number of diseases of the lymphohematopoietic system, including aplastic anemia, various cytopenias, myelofibrosis, myelodysplastic syndrome, leukemia, and non-Hodgkin's lymphoma (NHL), but the association between benzene exposure and MM as demonstrated among cohorts of benzene-exposed workers has been less apparent. The suspicion of an association between benzene exposure and MM was first raised in case reports by Torres et al.[1] in 1970. Subsequently, Aksoy et al.[2] added four more cases of benzene-associated MM to the literature. An association was further supported in 1983 by DeCoufle et al.[3] who identified two cases of MM in a cohort of only 259 benzene-exposed chemical workers. In 1987, the update of mortality among Pliofilm workers by Rinsky et al.[4] provided much stronger evidence of a link between benzene exposure and MM. Since that time, elevated risks of MM have been identified in other benzene cohorts,[5-8] and investigators have reported a dose–response for MM in relation to cumulative dose, or in relation to the number of peak benzene exposures experienced by the cohort.[7] On the other hand, several studies of benzene cohorts have reported no elevated risk for MM.[9-11] The Sorahan et al. study,[11] however, has been criticized for significant under-ascertainment of cancer deaths.[12]

Epidemiological studies have demonstrated an elevated risk of MM associated with painting, or with paints containing organic solvents,[13-20] many of which have been contaminated with benzene, while other studies of painters have shown no association with an elevated risk of MM.[21,22] Chemical workers also have demonstrated an elevated risk of death from MM.[23]

Some studies of refinery workers have provided data indicating an elevated risk of MM,[24-29] while others have not.[30-32] Interpretation of the findings from the study of cohorts of petroleum refinery workers, however, is sometimes limited by the difficulty in determining benzene exposure to cohort members on a retrospective basis in the absence of exposure monitoring data, particularly when extrapolating exposures 30 or more years back in time. Mis-classification of benzene exposure among these workforces, however, will dilute the findings of cohort studies and usually bias results of dose–response analyses toward the null hypothesis of no association.

Another difficulty in identifying an association epidemiologically, particularly within the benzene cohort studies, per se, arises from the low mortality rate of MM in the general population, e.g., 3×10^{-5} and the fact that most benzene-exposed cohorts have been small. In addition, investigators have sometimes combined MM with other forms of lymphoma in their analyses. As a result, information from these studies related specifically to benzene and MM is sparse.

THE COHORT FOLLOW-UP PERIOD USED TO ESTIMATE THE RELATIVE RISK OF DISEASE SHOULD CONSIDER THE MEDIAN LATENCY PERIOD FOR THE DISEASE OF ETIOLOGICAL INTEREST

The follow-up periods of several of the benzene cohorts have been extended over time. Several reports reflecting these different periods of follow-up for the benzene cohorts have been published. As a result, the same cohort indicates different estimates in the relative risk (RR) of the diseases of interest depending upon the follow-up period. Silver *et al.*[33] state that a longer follow-up period is generally thought to result in a more precise estimate of RR for the exposed population, but that the effect of follow-up time, *per se*, on risk has not been given much attention. They expressed concern that summary estimates of risk may not be generalized to the same cohort, or to other cohorts followed for different lengths of time.

As an occupational cohort is followed over time, one should expect a rise and then a fall in the RR of disease related to the exposure associated with the disease. Silver *et al.*[33] cite several examples of exposure and disease where upon cessation of exposure, the RR of disease rises and then declines, e.g., atomic bomb survivors and RR of leukemia, uranium miners and RR of lung cancer, and asbestos exposure and RR of lung cancer. The discontinuation of cigarette smoking is also known to result in a decline over time in the RR of lung cancer. It is unlikely that one would determine the RR of lung cancer in relation to cigarette smoking by following a cohort of ex-smokers until their death and then base the estimate of RR on the final follow-up period. Likewise, it may be inappropriate to determine the risk of disease related to occupational exposures based on the final follow-up of an occupational cohort that may have received its last exposure decades earlier.

This raises the question, "which follow-up period should be selected for a qualitative estimate of the RR of disease?" In general, the period of follow-up which includes the median latency period for the disease of interest should be included, but the estimate of RR often declines when follow-up extends much beyond this period. Subsequent to this period, one should expect to observe a decline in the RR of the disease. A few of the factors responsible for the observation of this decline in disease risk over time are mentioned. First, for a "stop exposure cohort," e.g., one in which the cohort members are no longer exposed, the mortality experience of the cohort being followed in successive years of follow-up will reflect health risks related to successively lower exposures. Second, for a cohort that has new members added as the follow-up period is extended, risk of disease related to the exposure of interest also should decline, because newer cohort members, in general, will have been employed in more recent time periods when occupational exposures were lower. Third, it is likely that those more susceptible to disease risk related to the toxic exposure may die at relatively shorter latency periods than those less susceptible. Thus, the surviving cohort members as a group, may be less susceptible than the overall cohort, and therefore demonstrate a lower RR of disease in relation to the exposure. Therefore, if an occupational exposure response relationship exists, one should expect a decline in exposure to be accompanied by a decline in the RR of disease.

In the study of refinery workers in the United States, a fourth phenomenon is at play in relation to a reduction of RR of disease as new follow-up periods are added. Many of the jobs that entail high-level exposure to hydrocarbons have been performed over the past 20–30 years by contract employees. These employees are not included in the cohort study analyses of plant employees, because they do not work directly for the refinery. As a result, more recently hired refinery cohorts are comprised of workers, who in general have experienced lower exposures as compared to those hired in earlier times, even in situations where atmospheric exposure levels may have remained the same. For this reason alone, one should expect to see a decline in the risk of diseases related to refinery employment as cohorts are followed over time.

As a result of the interplay of these factors, RR estimates based on successive follow-up periods should be expected to result in successively lower estimates of RR of disease, and particularly for follow-up periods that extend beyond the median latency period for the disease being evaluated.

METHODOLOGY FOR DETERMINING SELECTION OF BENZENE COHORTS FOR ANALYSIS

In order to determine the extent to which benzene may be a cause of MM, the data from all of the "benzene cohort studies" conducted to date have been evaluated for inclusion in the analysis. These studies were chosen for evaluation with the assumption that the cohort members were selected on the basis of a determination that they were actually exposed to benzene. This is in contrast to the information available from the study of refinery workers. With the latter group of cohorts, it is difficult to determine which cohort members may have been exposed to measurable levels of benzene, when these exposures occurred, and their degree of benzene exposure. Thus, eight cohort studies of workers exposed to benzene were identified from the literature. These studies are listed in TABLE 1. In some cases, the cohorts have been updated several times and additional cohort members have been added over the years of follow-up. The strengths and weaknesses of individual studies have been evaluated and the basis for inclusion or exclusion of data from the studies is explained. The overall estimate for the RR of MM is then calculated.

TABLE 1. Options for selecting observed and expected deaths from multiple myeloma identified in the benzene cohort studies as published to date by author and year of publication

Cohort	Authors	Multiple myeloma			
		Year	Obs	Exp	SMR
1	DeCoufle et al.	1983	1	0.23	4.35
2	Rinsky et al.	1987	4	0.98	4.09
	Rinsky et al.	2002	5*	2.46*	2.04*
3	Wong	1987	2	0.56	3.57
			3*	0.10*	30.00*
4	Fu et al. (Florence)	1996	3	1.04	2.88
5	Yin et al.†	1996	1	2.50	0.40
6	Ireland et al.	1997	3*	0.93*	3.23*
	Collins et al.	2003	8	4.20	1.90
7	Bloemen et al.	2004	3	4.16	0.72
8	Sorahan et al.‡	2005	6	9.50	0.63

Obs=observed; Exp = expected.
*Alternate data points considered, but rejected for use in analysis.
†Calculated as RR using disease rates in non-benzene-exposed industrial workers.
‡Not selected because of evidence of significant under-ascertainment of cancer deaths.[12]

REVIEW OF STUDIES AND ESTIMATE OF RELATIVE RISK OF MYELOMA AMONG BENZENE COHORT STUDIES

In a small cohort study of white male chemical workers exposed to benzene, DeCoufle et al.[3] observed two "cases" of MM and one "death" from MM, e.g., one of the MM cases died from acute nonlymphocytic leukemia (ANLL) following treatment with radiotherapy and chemotherapy for MM. The authors did not calculate an expected number of deaths for MM alone, but rather estimated an expected number of 0.23 deaths for causes MM, polycythemia vera, and other neoplasms of the lymphoid tissue combined (ICD 8th Revision codes 202, 203, and 208). Thus, one can estimate that DeCoufle observed one "death" from MM versus 0.23 expected, standardized mortality ratio (SMR) = 4.35 [95% confidence interval (CI) = 0.1–24.2].

The update of the NIOSH Pliofilm study of benzene-exposed workers by Rinsky et al.[4] demonstrates a statistically significant excess of death from MM. Four deaths from MM were observed versus 0.98 expected, SMR = 4.09 (95% CI 1.10–10.47). The workers who died from MM all had a latency period of more than 20 years (ranging from 22 to 27 years). The median latency period was 25 years since initial exposure to benzene.

Rinsky et al.,[34] however, have provided more recent analyses for MM and leukemia among the Pliofilm cohort members. In the follow-up through December 31, 1996, there is now a total of five MM deaths and 2.35 expected among white males for an RR of 2.12. When the authors included white women,

TABLE 2. Effect of follow-up time on risk estimates for leukemia and myeloma in the NIOSH Pliofilm study of white males

Causes of death						
	All causes		Leukemia		Myeloma	
Authors	Obs	SMR	Obs	SMR	Obs	SMR
Infante et al.[35] 1950–1975 follow-up N = 748	140	0.75*†	7	5.60*	2	NE
Rinsky et al.[‡56] 1950–1975 follow-up N = 1006 (+258)	229	1.06	8	4.68*	2	NE
Rinsky et al.[4] 1950–1981 follow-up N = 1165 (+159)	330	0.99	9	3.37*	4	4.09*
Rinsky et al.[34] 1950–1996 follow-up N = 1165	656	0.99	15	2.56*	5	2.12

Cohort first exposed 1936, last exposed 1976. NE = expected not estimated; N = number in cohort; Obs = observed.
*$P < 0.05$.
†Follow-up only 75% completed.
‡Includes addition of workers first exposed >1950, "group 2."
†Additions to cohort during updated follow-up.

who were newly added to the cohort, the observed remained at five MM deaths, but the expected increased to 2.45, SMR = 2.04 (95% CI = 0.66–4.76).

The Rinsky et al.[34] update of the NIOSH Pliofilm benzene cohort is accompanied by another analysis by Silver et al.[33] This paper states that the RR for MM (and leukemia) in relation to benzene exposure rises and then declines with successive years of follow-up of the benzene cohort, whose last exposure occurred in 1976. The RRs for leukemia and MM by follow-up period are shown in TABLE 2. The data demonstrate a decline in the SMR for leukemia from 5.60 in the original report[35] to 2.56 in the last published follow-up, and a decline in the SMR for MM from 4.09 in the Rinsky 1987 publication to 2.12 in the most recent follow-up. As a possible interpretation for the fall in the RR of these diseases among the Pliofilm benzene cohort members over time, Silver et al.[33] offer two explanations. First, the 1996 cohort follow-up period resulted in 20 years passing between the last benzene exposure to the cohort and the end of the follow-up. Since the follow-up period in the most recent publication is likely to be well beyond the median latency period for benzene to induce leukemia in general (they acknowledge that different types of leukemia may have different median latencies, but the majority in this cohort were ANLL), following up the cohort over such a long period of time results in less expression of the disease and hence a lower RR. Second, they argue that workers, who were the most susceptible to leukemia from benzene, succumb relatively earlier to the disease, leaving a less susceptible population at risk for the majority of the follow-up period. Silver *et al.*[33] state that summary estimates are misleading and that time-specific estimates must be provided, or one may underestimate the risk to the exposed workers. Although the authors did not address MM in as much detail as they did for leukemia, the same phenomena is apparent.

As a result of reviewing the two above-mentioned papers by the NIOSH investigators, preference was given to the data on benzene exposure and RR of MM from the follow-up that was completed in 1981 by Rinsky *et al.*[4] This follow-up period was selected for the following reasons: (*a*) from the NIOSH Pliofilm cohort study, 25 years appears to be the median latency period for benzene exposure and death from MM as shown in the Rinsky *et al.* 1987 paper; therefore, one might expect the RR to begin to decline shortly after this period of time; and (*b*) benzene exposure to the cohort began in 1936 and ended in 1976—a 40-year maximum exposure period; therefore, 1956 is the most likely mid-point year of the exposure period. The addition of 25 years (estimated median latency for benzene exposure and MM in the study) to the 1956 mid-point cohort exposure year leads to 1981 as the follow-up year that might provide the most sensitive period for evaluating the RR of MM among cohort members. Based on the findings and discussion by Silver *et al.*,[33] one might also justify using data from the 1981 follow-up to estimate risk from MM and leukemia, because risk for both of these diseases declines after this period in the published studies.

In the benzene cohort study by Wong,[8] using the U.S. general population for comparison of mortality, the author only presented results for observed and expected deaths from MM for the group intermittently exposed to benzene. He observed two deaths from MM versus 0.56 expected (SMR = 3.57; 95% CI = 0.43–12.90). For the overall cohort continuously exposed to benzene, he observed three deaths from MM versus none in a comparison group comprised of workers at the same facilities not exposed to benzene. Wong[36] considered the internal comparison group the most appropriate in his study. Therefore, one could select the RR of 3.57 or a risk of 30 (3 versus 0.1 by arbitrarily assigning 0.1 to the expected for the internal comparison group). An SMR of 3.57 for MM was selected as the most conservative estimate of MM risk for this cohort.

Fu *et al.*[5] studied mortality among two cohorts of shoe manufacturers; one from three English towns and the other from Florence, Italy. The authors stated "the exposure information necessary to calculate time-related variables was not available for the English shoe workers" and "...only job title as indicated in the 1939 census was collected, and dates of employment were not available." They further indicated that "it is unclear how much exposure to benzene occurred among workers in the British shoe and boot industry...documentation on the specific solvents to which the workers in this study were exposed are not available" (p. 395). They considered 2.5% of the cohort having been highly exposed to "solvents," though they lacked information on the content of the solvents.

The authors stated that the Florence shoe workers began to use benzene in the early 1950s. They estimated that 70% of the glue consisted of benzene before 1960. By the end of 1963, however, a national law required that the benzene content of glues be limited to 2%. Exposure concentrations were not available, but the authors stated that atmospheric benzene exposures in the Pavia shoe industry ranged from 25 to 600 ppm. They further estimated that 26% of the cohort was highly exposed to "solvents." Since it is not known whether the English cohort was exposed to solvents containing benzene, data from the Florence cohort only were included in the analyses. The observed risks for leukemia and MM among these two cohorts are consistent with the view that there was little, or no use of benzene among the English shoeworkers, but that benzene was used by the Florence cohort members. For the Italian cohort, the SMRs for leukemia and MM were 214 (eight deaths observed) and 288 (three deaths observed), respectively. For the English cohort, the SMRs for leukemia and MM were 89 (16 deaths observed) and 104 (eight deaths observed), respectively.

The most recent update of the study of Chinese workers exposed to benzene by Yin et al.[9] does not indicate any excess of MM. There was one observed case of MM versus 2.50 expected based on comparison to rates among non-benzene-exposed workers, (RR = 0.40, 95% CI = 0.1–10.7). These data were used in the analyses. MM has a much less frequent occurrence in the Chinese population as compared to Western populations and the authors noted that this factor resulted in little statistical power to assess its risk as indicated by the wide confidence interval surrounding the estimate of risk.

Ireland et al.[6] evaluated mortality in a small cohort of benzene-exposed workers at the Monsanto Company in Sauget, Illinois. For MM among benzene-exposed production workers, three deaths were observed versus 0.93 expected (SMR = 3.23, 95% CI 0.7–9.4) For workers who had achieved 20+ years since first benzene exposure (latency), there were three deaths observed from MM versus 0.82 expected deaths (SMR = 3.66, 95% CI = 0.75–10.7). For workers who had estimated cumulative benzene exposure of between 1 ppm-year and >6 ppm-years, there were three observed deaths versus 0.56 expected (SMR = 5.36, 95% CI = 1.1–15.7).

The Ireland et al.[6] benzene cohort was followed up for 7 more years with the addition of 245 workers by Collins et al.[7] Overall, they observed eight MM deaths versus 4.2 expected (SMR = 1.9, 95% CI = 0.8–3.8). For those cohort members with estimated benzene exposures between 1 ppm-year and >6 ppm-years, there were six MM deaths versus 2.8 expected (SMR = 2.14, 95% CI = 0.8–4.67). The authors noted that 10 of 13 deaths from MM occurred among workers exposed to benzene (eight with known cumulative

TABLE 3. Cohort mortality study of hourly workers, who began employment at Monsanto Company in Sauget, Illinois between 1940–1977

	Myeloma		Leukemia (ANLL)	
	Obs	SMR	Obs	SMR
Ireland et al.[6] 1940–1991 follow-up $N = 4172$	3	3.23	2	2.78
Collins et al.[7] 1940–1997 follow-up $N = 4417$ (+245)	8	1.90	5	2.17

Myeloma by cumulative benzene dose (ppm-years)							
Nonexposed		< 1ppm-year		1–6 ppm-years		>6 ppm-years	
Obs	SMR	Obs	SMR	Obs	SMR	Obs	SMR
5	1.1	2	1.4	2	1.5	4	2.6

Myeloma by no. of days with peak exposures >100*							
None		<7		7–40		>40	
Obs	SMR	Obs	SMR	Obs	SMR	Obs	SMR
9	1.2	0	0.0	1	1.7	3	4.0

Collection of benzene exposure data began at 1980, cohort median cumulative benzene exposure = 3 ppm-year. Obs = observed.
*15-min exposure period.
+Additions to cohort during updated follow-up.

exposure) and they all achieved 20 or more years of latency, SMR = 1.80 (95% CI = 0.9–3.3).

The effect of follow-up time on the RR of MM and leukemia in the Monsanto cohort can be seen from the data in TABLE 3. Results from the two follow-up periods indicate that the SMR for ANLL was lowered from 2.78 (two deaths) to 2.17 (five deaths). The SMR for MM declined from 3.23 (three deaths) to 1.90 (eight deaths) between the follow-up periods. Therefore, the cohort shows a reduction in risk of MM and leukemia as the follow-up period is extended. The most recent overall results by Collins et al.[7] were included in the estimate of MM risk in the meta-analysis. This was done based on the following information. The cohort of Monsanto employees was followed up from 1940 to 1997, a 57-year period. The authors stated that all of the benzene-exposed MM deaths occurred more than 20 years after the initial exposure and one death is known to have occurred more than 30 years after the initial exposure, but they did not provide information on latency for specific cases. Therefore, a 25-year latency period was assumed based on this information along with the information provided in the NIOSH study. Since the mid-point exposure year for the cohort would be 1969 and the median latency for MM was assumed to be 25 years, a follow-up around the time of 1994, or a short time later may provide the most full expression of RR for MM among the cohort members. The earlier follow-up of the cohort was to 1991 and the latter was to 1997. Since 1994 is midway between these, and the latency might actually be longer than the assumed 25 years, the results based on the 1997 follow-up period were selected.

Though not a part of the overall benzene cohort analysis, Collins et al.[7] also evaluated lymphohematopoietic cancer by cumulative and peak benzene exposure. The authors concluded that they found increasing risk with cumulative benzene exposure and MM. They stated, "For MM, the SMRs were 1.1 (95% CI 0.3–2.5) in the nonexposed group, 1.4 (95% CI 0.2–5.1) in the <1 ppm-years, 1.5 (95% CI 0.2–5.4) in the 1–6 ppm-years, and 2.6 (95% CI 0.7–6.7) in the >6 ppm-year group" (p. 676). However, when peak exposures over 100 ppm for 40 or more days were considered, the SMR for MM was 4.0 (95% CI 0.8–11.7). The authors stated that a high number of peak exposures to benzene is a better predictor of risk than cumulative exposure (for MM and leukemia) and concluded that dose–rate of benzene exposure may be an important factor for evaluating benzene exposure and lymphohematopoietic cancers.

Ott et al.[37] followed up 594 workers employed at the Dow Chemical facility in Midland, Michigan from 1940 to 1973. These workers were exposed to benzene prior to 1940. Bond et al.[38] included an additional 362 cohort members for a total of 956 and extended the follow-up of the cohort till 1982. Bloemen et al.[10] combined the Midland, Michigan cohort previously studied by Ott et al.[37] and by Bond et al.[38] with a new group of 1310 workers characterized as having been employed in chlorobenzol area jobs with the lowest benzene exposures in unidentified Dow facilities in the United States. Cohort follow-up was extended till 1996. In the latter study, analyses were not presented separately for the previously studied Dow Midland facility cohort members and for the cohort additions from other facilities. As a result, it is not possible to determine the effect of follow-up time on risk estimates for MM and leukemia for the original Dow Midland cohort. The benzene exposures for the entire cohort[10] were terminated by 1978 as ethylcellulose production was discontinued in 1977, while the chlorobenzol and alkyl benzene operations ceased in 1978.

The results from the follow-up of the Dow chemical workers are shown in TABLE 4. In the most recent publication, Bloemen et al.[10] observed three deaths from MM versus 4.17 expected, the SMR = 0.72 (95% CI = 0.15–2.10). In the previous publication by Bond et al.,[38] one MM death had been observed, but an expected number of deaths was not provided. For death from ANLL among cohort members, there were four observed versus 3.6 expected, SMR = 1.11. Yet, in the previous two follow-up analyses of the Midland portion of the

TABLE 4. Results of Dow Midland benzene cohort mortality study by authors and follow-up period

	Causes of death							
	All causes		ANLL			Myeloma		
Authors	Obs	SMR	Obs	Exp	SMR	Obs	Exp	SMR
Ott et al.[37] 1940–1973 follow-up N = 594	102	0.80*	3	0.80	3.75[†]	0	NE	—
Bond et al.[38] 1940–1982 follow-up N = 956 (+362)	225	0.84*	4	0.90	4.44*	1	NE	—
Bloemen et al.[10] 1940–1996 follow-up N = 2266 (+1310)[‡]	972	0.90*	4	3.60	1.11	3	4.17	0.72
Within last follow-up period (1983–1996)			0	2.70	—	2	???	???

Benzene exposures to the cohort ceased in 1977 and 1978. NE = no expected provided; Obs = observed; Exp = expected.
*$P < 0.05$.
+Additions to cohort during updated follow-up.
[†]SIR was 3 "cases" versus 0.8 expected based on Third National Cancer Survey incidence data.
[‡]Number of very low benzene-exposed cohort members added from facilities other than Midland.
??? = not calculated.

cohort, the results for ANLL demonstrated a statistically significant excess. Ott et al.[37] reported a standardized incidence ratio (SIR) of 3.75 and Bond et al.[38] reported an SMR of 4.44. As shown in TABLE 4, within the new follow-up period of 1983–1996, no additional deaths from ANLL were observed, while 2.7 deaths from this cause were expected. Exposures to the cohort began prior to 1940 and by the end of the most recent follow-up period, 19 years have elapsed since the last exposure for those who remained on the job until the operations ceased in 1977–1978. Since the cohort selection began with those employed in 1938 and all benzene exposures were terminated by 1978, and assuming a 25-year median latency period for MM, the most sensitive follow-up period may have been around 1983. Therefore, the Bond et al.[38] study which followed the cohort through 1982 may have provided the best estimate of MM risk among the cohort members. For this follow-up, one death from MM was observed and an expected was not provided. Therefore, the study results based upon the Bloemen et al.[10] follow-up to 1996 were included in the meta-analysis as shown in TABLE 5.

Sorahan et al.[11] recently published the results of a cohort mortality and morbidity study of workers exposed to benzene in the United Kingdom. The authors observed a slight deficit for MM (SMR = 0.63, 95% CI = 0.2–1.4). The risk for total leukemia was slightly elevated (SMR = 1.37, 95% CI = 0.86–2.07) as was the risk for AML (SMR = 1.82, 95% CI = 0.94–3.18). In the study, the highest risk of cancer achieving statistical significance was "cancers of uncertain origin," SMR = 140, based on 68 cancer deaths, $P < 0.001$. Because of significant under-ascertainment of cancer deaths that occurred among the cohort members,[12] it is difficult to include this study in the analysis for MM. Furthermore, among cohort members who died in the earliest period of follow-up, e.g., 1968–1974, cancer registrations were missing for an estimated 50% (46/91) of the cancer deaths. As such, they likely represent deaths that occurred among cohort members who were exposed during earlier periods of employment, when benzene exposures were relatively higher. Since MM and other lymphohematopoietic cancer deaths are so rare, the identification of only a few additional deaths from these causes can make a significant difference in the estimate of RR. Other limitations of the study also have been noted.[12] Because of these limitations, the study results were not included in the meta-analyses.

SUMMARY OF DATA FROM THE BENZENE COHORT STUDIES IN RELATION TO RISK OF MYELOMA

The evaluation of the study design, methodology and data from the eight

TABLE 5. Summary of estimates of relative risk of myeloma identified in benzene cohort studies selected for inclusion in analysis by author and year of publication

	Estimates of risk of multiple myeloma				
Authors	Year	Obs	Exp	SMR	95% CI
DeCoufle et al.	1983	1	0.23	4.35	(0.1–24.2)
Rinsky et al.	1987	4	0.98	4.09	(1.1–10.5)
Wong	1987	2	0.56	3.57	(0.4–12.9)
Fu et al. (Florence)	1996	3	1.04	2.88	(0.6–8.4)
Yin et al.[†]	1996	1	2.50	0.40	(0.1–10.7)
Collins et al.	2003	8	4.20	1.90	(0.8–3.8)
Bloemen et al.	2004	3	4.16	0.72	(0.2–2.1)
All studies combined					
Overall pooled added numbers		22	13.67	1.61	(1.01–2.44)
Weighted (Poisson) RR*				2.13	(1.31–3.46)

Obs = observed; Exp = expected.

[†]Yin et al. calculated relative risks based on incidence rates (the others used SMRs) increasing the estimated standard errors.

*For the weighted estimate, all individual study weights were derived from standard errors calculated assuming a Poisson distribution (http:home.clara.net/sisa/smr.htm). Pooled weighted ln RR and RR estimates and confidence limits were derived using standard methods for meta-analysis, fixed effects model (X^2 test of homogeneity, $P > 0.05$).[39,40]

benzene cohort studies published to date resulted in the use of data from seven of the studies upon which to base an estimate of the RR of MM, as shown in TABLE 5. Using the overall pooled numbers, there are 22 observed MM deaths and 13.67 expected, SMR = 1.61 (95% CI = 1.01–2.44). For the weighted estimate, all individual study weights were derived from standard errors calculated assuming a Poisson distribution (http:home.clara.net/sisa/smr.htm.). Pooled weighted ln RR and RR estimates and confidence limits were derived using standard methods for meta-analysis, fixed effects model (X^2 test of homogeneity, $P > 0.05$).[39,40] The pooled weighted estimate of MM risks is shown in TABLE 5, RR = 2.13 (95% CI = 1.31–3.46). Of the studies included in the estimate, the RR of MM for five studies ranged from 1.90 to 4.35. In two of the studies, the RR for MM was 0.40 and 0.72. In one of these studies,[10] benzene exposure had terminated 20 years prior to the last year of follow-up, and was based on a follow-up period that was 13 years beyond the estimated most sensitive follow-up year to observe an elevation in the risk of MM in the study. This factor may have had some influence on the estimate of MM risk observed in the study.

BIOLOGICAL PLAUSIBILITY OF BENZENE AS A CAUSE OF MULTIPLE MYELOMA

In addition to the evaluation provided above, there is a biologically plausible basis for establishing benzene as a cause of myeloma[3,4,41]: (*a*) MM is a tumor of plasma cells within the bone marrow, which are derived from B-lymphocytes; (*b*) the bone marrow is a target organ for benzene toxicity causing aplastic anemia, various cytopenias (including depression of B-lymphocytes,[42] myelofibrosis, myelodysplastic syndrome, and leukemia; (*c*) benzene is associated with an increased risk of chromosomal damage to circulating lymphocytes[43–45]; (*d*) and, more recently to DNA damage to B-lymphocytes specifically[46]; (*e*) workers exposed to benzene also have demonstrated an elevated risk of chronic lymphocytic leukemia,[27,30,47] which is also a cancer of B cell lineage. Thus, benzene has shown very specific toxicity and genetic alteration not only to the target organ, the bone marrow, but also to the specific cells within the bone marrow from which plasma cells are derived, e.g., the B-lymphocytes.

The findings in humans are consistent with the results of the National Toxicology Program study of benzene in experimental animals.[48] This study demonstrated that benzene induced solid tumors at multiple sites in both rats and mice. The study also demonstrated a highly significant dose–response for lymphoma in male and female mice exposed to benzene.[48] Cronkite *et al.*[49] also have demonstrated the induction of lymphomas in mice exposed to benzene by inhalation. Therefore, at least six lines of evidence support the epidemiological finding that benzene exposure is associated with a significantly elevated risk of myeloma.

DOES EVIDENCE FROM TWO LARGE COHORT STUDIES OF REFINERY WORKERS SUPPORT OR DETRACT FROM THE EVIDENCE PROVIDED BY BENZENE COHORT STUDIES?

Although the purpose of the evaluation was to determine whether or not the benzene cohort studies provided evidence of an association between benzene exposure and myeloma, results from two of the largest cohort studies of petroleum refinery workers were also reviewed. This was done in order to put the findings of these studies into perspective in relation to those of the benzene cohort studies.

Australian Health Watch Study

The Australian Institute of Petroleum periodically publishes the results of its medical surveillance program referred to as Health Watch. Employees must have 5 years of employment to enter the program which was initiated in 1980. After termination of employment, the members of the cohort can choose to stay in the program or opt out. Between 1992 and 2004, six reports that represent the findings of morbidity and mortality among these workers have been reviewed.[24–27,30,31] The findings for the incidence and mortality of myeloma are presented in TABLE 6 and the results for leukemia incidence are shown in TABLE 7. In the overall analysis for myeloma, the SIR ranges from a high of 1.9 in the 10th Report[25] to 1.7 in the 11th Report.[26] Myeloma mortality ranges from an SMR of 2.6 in the 9th Report[24] to 1.7 in the 11th Report. As shown in TABLE 6, for the entire study population in the 11th Report, 15 cases of MM were observed, SIR = 1.7 (95% CI 0.9–2.8). Among workers employed in terminal work, a significant excess of myeloma was demonstrated, 10 cases of MM were observed, SIR = 2.50 (95% CI = 1.2–4.6). Among refinery workers, the risk of MM was essentially identical to the expected, SIR = 1.08 (95% CI = 0.4–2.8). The significant elevation in MM risk among terminal workers is noteworthy because as shown in TABLE 7, the highest RR of leukemia was also observed among terminal workers (16 observed versus 8.8 expected, SIR = 1.82, 95% CI = 1.04–2.95).[26] Gray et al.[27] concluded that the significantly elevated leukemia risk identified among terminal workers in the 11th Report "is probably explained by the historically higher exposures in terminals rather than any other site characteristic" (p. xiv). It is not clear from the study why this same comment could not be made in relation to the significant excess of MM that was identified among terminal workers. In the journal publication of this study,[31] the authors do not include the results for the analysis of MM among terminal workers. The reason for the omission of the results of this analysis in the journal publication is not provided.

In additional analyses of the Australian petroleum refinery workers, Gray et al.[27] were able to demonstrate a dose–response for benzene exposure and

TABLE 6. Health Watch: The Australian Institute of Petroleum Health Surveillance Program (began in 1980)

| | | Myeloma ||||||
| | | Incidence ||| Mortality |||
		Obs	SIR	(95% CI)	Obs	SMR	(95% CI)
Ref. 24	9th Report (1992)	6	1.8	(0.7–3.9)	6	2.6	(1.0–5.7)
Ref. 25	10th Report (1998)	13	1.9	(1.0–3.3)	8	1.6	(0.7–3.2)
Ref. 26	11th Report (2000)	15	1.7	(0.9–2.8)	11*	1.7	(0.9–3.1)
	Terminals	10	2.5	(1.2–4.6)			
	Refinery	4	1.08	(0.4–2.8)			
Ref. 31	Gun et al.	15	1.7	(0.96–2.84)	8*	1.6	(0.7–3.2)

Obs = observed.

*Difference in number of deaths is likely explained by Gun et al. following the cohort to the end of 1996, while the 11th Report followed the cohort to the end of 1998; no analysis was presented for terminal workers in Gun et al.

leukemia, but not for benzene exposure and MM. However, the dose–response analysis for benzene exposure and MM contained only six cases of MM spread over three dose groupings—far too small a number upon which to draw any meaningful conclusions about the lack of a dose–response. (Nine of the 15 MM cases in the study were used to set the RR to 1.0 for comparison.)

Wong and Raabe Meta-analysis of Myeloma Risk among Refinery Workers

Wong and Raabe[32] have evaluated mortality from MM through meta-analyses of petroleum-exposed cohorts. This study did not demonstrate an

TABLE 7. Health Watch: The Australian Institute of Petroleum Health Surveillance Program (began in 1980)

| | | Leukemia incidence |||
		Obs	SIR	(95% CI)
11th Report[26]	Terminals	16	1.8	(1.04–2.95)
	Refinery	11	1.3	(0.67–2.39)
	All workplaces	30*	1.5	(1.02–2.15)
Gun et al.[31]	All workplaces	27*	1.4	(0.91–2.02)

Obs = observed.

*Difference in number of cases is likely explained by Gun et al. using a shorter follow-up period. See footnote to Table 6.

elevation in risk for MM. For workers employed at U.S. refineries, the myeloma SMR = 0.97 (95% CI = 0.81–1.17). With data combined for workers employed in the United States, the United Kingdom, Canada, and Australia, the myeloma SMR = 0.95 (95% CI = 0.83–1.07). The authors concluded that petroleum workers are not at an increased risk of myeloma as a result of their exposure to benzene, benzene-containing liquids, or other petroleum products in their work environment. These conclusions, however, must be placed in perspective related to other causes of death among those included in the meta-analysis. Somewhat surprisingly, the authors do not provide information on the SMRs for the category "all causes of death," nor for the category "all cancer deaths." Therefore, it is not possible to evaluate the impact of the "healthy worker effect" (HWE) on the findings related to MM. Evaluation of the HWE in the study is particularly relevant because of the previously published findings regarding the U.S. gasoline distribution workers, who were also included in the Wong and Raabe meta-analysis for MM.[32]

For example, the Wong et al.[50] study of U.S. petroleum distribution workers contributed over 18,000 cohort members to the Wong and Raabe[32] meta-analyses for MM. For those who were considered land-based terminal workers, their overall mortality rate was only 50% of the expected based on death rates for the U.S. standard population. The overall mortality rate for marine workers was 77% of the standard population rates. Enterline[51] suggested that these findings may be due to selection bias in the Wong et al. study of gasoline distribution workers.[50] Concern about the methodology used in the study was also expressed by Infante[52] during the evaluation of the Wong et al. study.[50]

Further suspicion about selection bias in the Wong and Raabe studies of refinery workers[32,53,54] is provided by differences in the numbers of cohort members being followed up during identical time periods of follow-up, and also by different numbers of cohort members being followed up when the authors conducted meta-analyses for MM,[32] leukemia,[53] and NHL,[54] as presented in separate publications.

In terms of exposure to benzene among the cohorts evaluated by Wong and Raabe,[32,53,54] the authors provide little credible information. In the myeloma study, Wong and Raabe[32] state that the benzene component in the gasoline is highly correlated with total hydrocarbon (THC) exposure. Their source for this opinion is Smith et al.[55] Yet, Smith et al.[55] stated that THC was a reasonable surrogate for one or more of the major hydrocarbon components in the vapor mixture of gasoline based on sampling data after 1969, but that "benzene was an exception to this because the benzene content of gasoline has varied over time as a result of changes in gasoline blending practices by the refineries" (p. 14). Since Wong was a co-author of the THC exposure assessment study,[55] one would surmise that he might have known about the limitations of using gasoline as a surrogate for benzene exposure in MM study.[32] The use of a constant proportion of benzene in the THC vapor, as done by Wong and Raabe,[32] will result in mis-classification of benzene exposure.

In summary, the significant increase in MM incidence among terminal workers in the Health Watch cohort study[26] provides some evidence of an association between benzene exposure and MM. The Wong and Raabe meta-analysis of refinery workers[32] cohorts shows no association between refinery work exposures and RR of MM, but the analyses suffer from limitations in data presentation, benzene exposure mis-classification, and potential selection bias. As a result, they do not allow for any meaningful interpretation regarding benzene exposure and RR of MM. Nor do the findings detract from the evidence that benzene exposure is a cause of MM, as supported by the current meta-analysis of data from the benzene cohort studies in conjunction with other findings related to the toxicity of benzene on B-lymphocytes and lymphoma in experimental animals.

CONCLUSIONS

A meta-analysis of data from all well-conducted benzene cohort studies demonstrates a statistically significant elevation in the risk of death from MM.

Consideration of cohort follow-up period in relation to median latency period for the disease being evaluated is important in determining the RR of disease.

The positive epidemiological evidence for benzene and myeloma is supported by other study results related to the biological plausibility for such an effect from benzene exposure.

Cohort studies of refinery workers are difficult to interpret in relation to benzene exposure and risk of MM, because of limitations in exposure assessment, study design, and analysis. Yet, one large study of petroleum refinery workers provides suggestive additional evidence of an association between benzene exposure and myeloma.

ACKNOWLEDGMENT

The author thanks Dr. Steven Bayard, Occupational Safety and Health Administration, for providing statistical support for the analyses presented in this article.

REFERENCES

1. TORRES, A., M. GIRALT & A. RAICHS. 1970. Coexistancia de entecedentes benzolicos cronicos y plasmocitama mutiple. Presentacion de das Casos. Sangre **15:** 275–279.
2. AKSOY, M. *et al*. 1984. Clinical observations showing the role of some factors in the etiology of multiple myeloma. Acta Haematol. **71:** 116–120.
3. DECOUFLE, P., W.A. BLATTNER & A. BLAIR. 1983. Mortality among chemical workers exposed to benzene and other agents. Environ. Res. **30:** 16–25.

4. RINSKY, R.A. et al. 1987. Benzene and leukemia: an epidemiologic risk assessment. N. Eng. J. Med. **316:** 1044–1050.
5. FU, H. et al. 1996. Cancer mortality among shoe manufacturing workers: an analysis of two cohorts. Occup. Environ. Med. **53:** 394–398.
6. IRELAND, B. et al. 1997. Cancer mortality among workers with benzene exposure. Epidemiology **8:** 318–320.
7. COLLINS, J.J. et al. 2003. Lymphohematopoietic cancer mortality among workers with benzene exposure. Occup. Environ. Med. **60:** 676–679.
8. WONG, O. 1987. An industry wide study of chemical workers occupationally exposed to benzene. II. Dose response analyses. Br. J. Ind. Med. **44:** 382–395.
9. YIN, S-N. et al. 1996. A cohort study of cancer among benzene-exposed workers in China: overall results. Am. J. Ind. Med. **29:** 227–235.
10. BLOEMEN, L.J. et al. 2004. Lymphohematopoietic cancer risk among chemical workers exposed to benzene. Occup. Environ. Med. **61:** 270–274.
11. SORAHAN, T., L.J. KINLEN & R. DOLL. 2005. Cancer risks in a historical UK cohort of benzene exposed workers. Occup. Environ. Med. **62:** 231–236.
12. INFANTE, P.F. 2005. Cancer risks in a UK benzene exposed cohort. Occup. Environ. Med. **62:** 905.
13. BETHWAITE, P.B., N. PEARCE & J. FRASER. 1990. Cancer risks in painters: study based on the New Zealand Cancer Registry. Br. J. Ind. Med. **47:** 742–746.
14. DEMERS, P.A. et al. 1993. A case-control study of multiple myeloma and occupation. Am. J. Ind. Med. **23:** 629–639.
15. FIRTH, H.M. et al. 1993. Male cancer mortality by occupation: 1973–1986. N. Z. Med. J. **106:** 328–330.
16. FRIEDMAN, G.D. 1986. Multiple myeloma: relation to propoxyphene and other drugs, radiation and occupation. Int. J. Epidemiol. **15:** 424–426.
17. HEINEMAN, E.F. et al. 1992. Occupational risk factors for multiple myeloma among Danish men. Cancer Causes Control **3:** 555–568.
18. LA VECCHIA, C. et al. Occupation and lymphoid neoplasms. Br. J. Cancer **60:** 385–388.
19. LUNDBERG, I. & R. MILATOU-SMITH. 1998. Mortality and cancer incidence among Swedish paint industry workers with long-term exposure to organic solvents. Scand. J. Work Environ. Health **24:** 270–275.
20. MORRIS, P.D. et al. 1986. Toxic substance exposure and multiple myeloma: a case-control study. J. Natl. Cancer Inst. **76:** 987–993.
21. STEENLAND, K. & S. PALU. 1999. Cohort mortality study of 57,000 painters and other union members: a 15 year update. Occup. Environ. Med. **56:** 315–321.
22. TERSTREGGE, C.W. et al. 1995. Mortality patterns among commercial painters in The Netherlands. Int. J. Occup. Environ. Health **1:** 303–310.
23. MASSOUDI, B.L. et al. 1997. A case-control study of hematopoietic and lymphoid neoplasms: the role of work in the chemical industry. Am. J. Ind. Med. **31:** 21–27.
24. BISBY, J. 1992. Health watch: the Australian Institute of Petroleum Surveillance Program, Ninth Report, Univ Melbourne, Dept Pub Health Comm Med, Carlton, Victoria, 97 p.
25. BISBY, J.A. et al. 1998. Health Watch, The Australian Institute of Petroleum Health Surveillance Program, Tenth Report, Univ Melbourne, Dept General Practice and Public Health, 83 p.
26. GUN, R. et al. 2000. Health Watch. The Australian Institute of Petroleum Health Surveillance Program. Eleventh Report, University of Adelaide, Department of Public Health, 75 p.

27. GRAY, C. *et al*. 2001. Lympho-haematopoietic Cancer and Exposure to Benzene in the Australian Petroleum Industry. Monash University and Deakin University. 221 p.
28. SCHOTTENFELD, D. *et al*. 1981. A prospective study of morbidity and mortality in petroleum industry employees in the United States–a preliminary report. *In* Quantification of Occupational Cancer. R. PETO & M. SCHNEIDERMAN, Eds., 247–260. Banbury Report 9. Cold Spring Harbor Laboratory. New York.
29. THOMAS, T.L. *et al*. 1982. Mortality patterns among workers in three Texas oil refineries. J. Occup. Med. **24:** 135–141.
30. GLASS, D.C. *et al*. 2003. Leukemia risk associated with low-level benzene exposure. Epidemiology **14:** 569–577.
31. GUN, R.T. *et al*. 2004. Update of a prospective study of mortality and cancer incidence in the Australian petroleum industry. Occup. Environ. Med. **61:** 150–156.
32. WONG, O. & G.K. RAABE. 1997. Multiple myeloma and benzene exposure in a multinational cohort of more than 250,000 petroleum workers. Regul. Toxicol. Pharmacol. **26:** 188–199.
33. SILVER, S.R. *et al*. 2002. Effect of follow-up time on risk estimates: a longitudinal examination of the relative risks of leukemia and multiple myeloma in a rubber hydrochloride cohort. Am. J. Ind. Med. **42:** 481–489.
34. RINSKY, R.A. *et al*. 2002. Benzene exposure and hematopoietic mortality: a long-term epidemiologic risk assessment. Am. J. Ind. Med. **42:** 474–480.
35. INFANTE, P.F. *et al*. 1977. Leukaemia in benzene workers. Lancet **ii:** 76–78.
36. WONG, O. 1987. An industry wide mortality study of chemical workers occupationally exposed to benzene. I. General results. Br. J. Ind. Med. 44: 382–395.
37. OTT, M.G. *et al*. 1978. Mortality among individuals occupationally exposed to benzene. Arch. Environ. Health **33:** 3–10.
38. BOND, G.G. *et al*. 1986. An update of mortality among chemical workers exposed to benzene. Br. J. Ind. Med. **43:** 685–691.
39. ROTHMAN, K.J. & S. GREENLAND. 1998. Modern Epidemiology, 2nd ed., Lippicott, Williams, Wilkins. Philadelphia.
40. National Cancer Institute. 1993. Respiratory health effects of passive smoking: lung cancer and other disorders. The Report of the U.S. Environmental Protection Agency. Monograph 4. NIH Pub No. 93-3605.
41. GOLDSTEIN, B. 1990. Is exposure to benzene a cause of human multiple myeloma? Ann. N. Y. Acad. Sci. **609:** 225–230.
42. LAN, Q., L. ZHANG, G. LI, *et al*. 2004. Hematotoxicity in workers exposed to low levels of benzene. Science **306:** 1774–1776.
43. FORNI, A.M. *et al*. 1971. Chromosome changes and their evolution in subjects with past exposure to benzene. Arch. Environ. Health **23:** 385–391.
44. PICCIANO, D. 1979. Cytogenetic study of workers exposed to benzene. Environ. Res. **19:** 33–38.
45. SARTO, F. *et al*. 1984. A cytogenetic study on workers exposed to low concentrations of benzene. Carcinogenesis **5:** 827–832.
46. SUL, D. *et al*. 2002. Single strand DNA breaks in T- and B-lymphocytes and granulocytes in workers exposed to benzene. Toxicol. Lett. **134:** 87–95.
47. MCMICHAEL, A.J. *et al*. 1975. Solvent exposure and leukemia among rubber workers: an epidemiologic study. J. Occup. Med. **17:** 234–239.
48. HUFF, J.E. *et al*. 1989. Multiple-site carcinogenicity of benzene in Fisher rats and B6C3F1 mice. Environ. Health Perspect. **82:** 125–163.

49. CRONKITE, E.P. *et al*. 1984. Benzene inhalation produces leukemia in mice. Toxicol. Appl. Pharmacol. **75:** 358–361.
50. WONG, O., F. HARRIS & T.J. SMITH. 1993. Health effects of gasoline exposure. II. Mortality patterns of distribution workers in the United States. Environ. Health Perspect. **101**(Suppl 6): 63–76.
51. ENTERLINE, P.E. 1993. Review of new evidence regarding the relationship of gasoline exposure to kidney cancer and leukemia. Environ. Health Perspect. **101**(Suppl 6): 101–103.
52. INFANTE, P.F. 1993. State of the science on the carcinogenicity of gasoline with particular reference to recent cohort mortality study results. Environ. Health Perspect. **101**(Suppl 6): 105–109.
53. WONG, O. & G.K. RAABE. 1995. Cell-type-specific leukemia analyses in a combined cohort of more than 208,000 petroleum workers in the United States and the United Kingdom, 1937–1989. Regul. Toxicol. Pharmacol. **21:** 307–321.
54. WONG, O. & G.K. RAABE. 2000. Non-Hodgkin's lymphoma and exposure to benzene in a multinational cohort of more than 308,000 petroleum workeres, 1937 to 1996. J. Occup. Environ. Med. **42:** 554–568.
55. SMITH, T.J., S.K. HAMMOND & O. WONG. 1993. Health effects of gasoline exposure. I. Exposure assessment for U.S. distribution workers. Environ. Health Perspect. **101**(Suppl 6): 13–21.
56. RINSKY, R.A. *et al*. 1981. Leukemia in benzene workers. Am. J. Ind. Med. **2:** 217–245.

Dangerous and Cancer-Causing Properties of Products and Chemicals in the Oil Refining and Petrochemical Industries. Part XXX

Causal Relationship between Chronic Myelogenous Leukemia and Benzene-Containing Solvents

MYRON A. MEHLMAN

Department of Medicine, The Mount Sinai Medical Center, New York, New York and Preventive Medicine and Community Health, University of Texas Medical Branch at Galveston, Galveston, Texas, USA

ABSTRACT: Benzene and benzene-containing products and solvents have long been associated with bone marrow toxicity. Both animal studies and human epidemiological studies have shown statistically significant increases of leukemia and other lymphohematopoietic cancers in workers exposed to benzene. The most common leukemia that has been associated with benzene exposure, also called benzene poisoning, is acute myelocytic leukemia (AML). A review of the epidemiological literature on workers exposed to benzene or benzene-containing solvents and products shows, without question, that this exposure is significantly related to other types of leukemia and lymphoma. In this article, we review the literature on the relationship between benzene exposure and chronic myelogenous leukemia (CML) and find that benzene and benzene-containing products are significantly related to morbidity and mortality from CML.

KEYWORDS: benzene; leukemia; CML; chronic myelogenous leukemia; solvents; cancer

INTRODUCTION

Chronic myelogenous leukemia (CML) (ICD-9 code 204.1) can result from exposure to toxic substances known to cause damage to DNA in the bone marrow. As a result of this injury, uncontrolled growth of these altered white blood cells occurs. The incidence of CML in individuals up to 50 years of age is approximately 2 per 100,000. Between the ages of 50 and 70 years, the incidence is slightly over 2/100,000, and at 80 years of age and older, incidence increases to 8–13/100,000.

In over 90% of cases of CML, a cytogenetic abnormality known as the Philadelphia chromosome (Ph^1) is present. Nowell and Hungerford[1] identified the abnormality in chromosome 22q- in patients with CML. In 1973 Rowley[2] found that Ph^1 results from a reciprocal translocation between chromosomes 9 and 22. Ph^1 was later shown to carry a unique fusion gene, reflective of the translocation and termed BCR-ABL.[3] Ph^1 appears only in CML. It can be produced in tissue culture and in humans by ionizing radiation.

CML is thought to occur as a result of a genetic mutation in a single pluripotential hematopoietic stem cell that then proliferates in the bone marrow causing suppression of normal red blood cells, platelets, and white blood cells. Among the proteins encoded by this gene is a tyrosine kinase,[4,5] which "turns on" proliferation of the stem cell line and prevents its normal "programmed" cell death, leading to the marked increase in the numbers of myeloid cells and their precursors in the bone marrow and the circulation. Active research is now under way to block the activation of this protein, and some specific drugs have produced high rates of remission of CML. The cytogenetics of this process[6] and responses to the therapeutic agent that blocks production of the protein have been described in a recent edition of the *New England Journal of Medicine*.[7] Other cells in the bone marrow are either suppressed or are crowded out leading to a deficit in red blood cells, normal white blood cell lines, platelets with resultant anemia, infection and bleeding due to decreased clotting ability. Death is usually caused by a terminal infection, such as pneumonia.

REVIEW OF THE LITERATURE

Benzene—A Human Carcinogen

The hazards of benzene in gasoline have been recognized since at least 1928.[8] Early accounts describe aplastic anemia, a known precursor of leukemia; similar hazards of benzene in gasoline have documented.[9,10] Benzene has been shown to cause cancers in both animals and humans.[11–19] Benzene is currently classified by the Environmental Protection Agency (EPA), the American Conference of Governmental Industrial Hygienists (ACGIH), and the International Agency for Research against Cancer (IARC) as a human carcinogen. Benzene is a known carcinogen that causes nearly all varieties of lymphohematopoietic cancers in animals and humans.

CML was first described in 1845 as "splenomegaly with leukocytosis" or "splenic leukemia" because of the associated enlargement of the spleen.[20]

TABLE 1. Types of leukemias and lymphomas from benzene exposure in humans[a]

Acute myelogenous leukemia	Chronic lymphocytic leukemia
Acute lymphocytic leukemia	Hairy cell leukemia
Acute erythroleukemic leukemia	Hodgkin's lymphoma
Acute myelomonocytic leukemia	Non-Hodgkin's lymphoma
Acute promyelocytic leukemia	Lymphosarcoma
Acute undifferentiated leukemia	Multiple myeloma
Chronic myelogenous leukemia	Reticulum cell sarcoma

[a]Sources: Aksoy et al., 1989;[21] Bond et al., 1986;[22] Decouflé et al., 1983;[23] Delore et al., 1928;[24] Goguel et al., 1967;[25] Goldstein 1977;[26] Infante et al., 1985,[27] 1995;[28] McMichael et al., 1974, 1975, 1976;[29–31] Rinsky, 1981, 1987;[32,33] Savitz et al., 1997;[34] Schwartz, 1987;[35] Travis et al., 1994;[36] Vianna and Polan, 1979;[37] Vigliani and Forni, 1976;[38] Vigliani et al., 1976;[39] Wong, 1987;[40] Yin et al., 1987, 1989, 1994, 1996.[41–44]

Although acute myelocytic leukemia (AML) is the leukemia type most commonly associated with benzene exposure, many studies have shown that other types of leukemia, including CML, are caused by benzene (TABLE 1).[21–44] In 1967 Goguel et al.[25] reported 43 cases of leukemia associated with benzene exposure. In this study, there were more cases of CML (13 cases, 29%) than of AML (11 cases, 25%) (TABLE 2).[25]

A comparison of 74,838 benzene-exposed workers and 35,805 nonexposed workers[36] showed 82 cases of leukemia (9 cases of CML) in exposed workers and only 13 cases (2 cases of CML) in nonexposed workers (TABLE 3).[36] Smith et al.[45] reported 6 cases of leukemia in rubber workers exposed to benzene and benzene-containing products; 66% of leukemias were CML. Infante[28] looked at data from benzene-exposed workers in China between the years of 1972–1981 previously published by Yin et al.[42] and found a significant ($P < 0.001$) excess of CML in benzene-exposed workers. Rinsky[33] concluded "The epidemiological evidence linked benzene to leukemia, both acute and chronic lymphocytic and chronic myeloid leukemia [CML]." The link between benzene exposure and various leukemias, including CML, has been reported by many other investigators.[30,34,40,46,47]

TABLE 2. Benzene-induced leukemias in Paris between 1950 and 1967

Type of leukemia	Cases	%
Acute myelocytic leukemia (AML)	11	25
Acute lymphocytic leukemia (ALL)	2	4.5
Acute erythroleukemia	5	11.4
Acute undifferentiated leukemia	2	4.5
Acute promyelocytic leukemia	2	2.3
Chronic myelogenous leukemia	13	29
Chronic lymphocytic leukemia	8	18.2

Source: Goguel et al., 1967.[25]

TABLE 3. Leukemia in 74,828 benzene-exposed and 35,805 nonexposed workers in China

Type of leukemia	Exposed workers number (%)	Nonexposed workers number (%)
Aplastic leukemia	9 (11%)	0
Acute leukemia	32 (39%)	6 (46%)
Myelodysplastic syndrome	7 (9%)	0
Chronic granulocytic leukemia	9 (11%)	2 (15%)
Malignant lymphoma	20 (24%)	3 (23%)
Other	5 (6%)	2 (15%)

Source: Travis et al., 1994.[36]

In 1986 Wong[48] submitted a statement to OSHA during Benzene Hearings (page 2702). He reported 2 cases of CML, 1 case of unspecified leukemia, 2 cases of lymphocytic leukemia and 1 case of acute lymphocytic leukemia (ALL) and no cases of AML.

In 2003 Adegoke et al.[49] reported results of a case control study of 486 leukemia subjects and 502 controls in Shanghai, China between 1987 and 1989. They found adjusted odds ratios (OR) showed a statistically significant increased risk for all leukemias ($P < 0.01$), for AML ($P < 0.01$), and for CML ($P < 0.01$) in persons exposed to benzene compared to controls (TABLE 4).[49]

There is substantial peer-reviewed scientific epidemiological literature demonstrating that exposure to solvents or products containing benzene, is associated with increased risk of CML in workers exposed to these products. In 1963, Tareeff et al.[50] reported 10 cases of chronic leukemia and 6 cases of

TABLE 4. Leukemia and benzene exposure, 1987–1989, in Shanghai, China

Chemical/product	Disease	Exposure duration	Odds Ratio	95% CI	P for trend
Benzene	All leukemias	≥15 years	3.3*	1.9–6.91	0.01
	ALL	≥15 years	3.9*	1.3–11.8	0.12
	AML	≥15 years	2.9*	1.2–7.0	0.01
	CML	≥15 years	5.0*	1.8–13.9	0.01
Gasoline	CML	≤15 years	1.5	0.8–3.0	0.25
Paints	All leukemias	≤15 years	2.3*	1.2–4.7	0.09
	ALL	≤15 years	3.4*	1.2–9.5	0.15
	AML	≤15 years	1.8	0.7–4.2	0.92
	CML	≤15 years	3.7*	1.4–9.9	0.02

ALL = Acute lymphocytic leukemia; AML = acute myelogenous leukemia; CML = chronic myelogenous leukemia. CI = confidence interval.
*Statistically significant.
Source: Adegoke et al., 2003.[49]

acute leukemia associated with exposure to benzene in industrial workers in the former Soviet Union; 5 of these cases were CML. Browning[51] reported 65 cases of benzene-induced leukemia from the existing literature; 21 cases were CML. She concluded that leukemia in relation to benzene exposure was a fact and that all cell types of leukemia were related to benzene exposure. In 1967 Goguel et al.[25] reported 44 cases of benzene-induced leukemia between 1950 and 1965; 13 of these cases were CML. Gerard and Revol[52] reported 17 cases of AML, 4 cases of CML, and 9 cases of chronic lymphocytic leukemia (CLL) resulting from benzene exposure. Other investigators have reported increased relative risk of leukemia, including CML, in workers exposed to benzene.[42,44,53,54]

A number of other studies have shown increased risk of CML from exposure to petroleum products containing benzene. These include the recent studies by the Australian Institute of Petroleum Health Surveillance Program, Health Watch, Tenth Report;[5] Infante[56] on the carcinogenicity of gasoline; Jakobsson et al.;[57] Lindquist et al.;[58] Lumley et al.;[59] Naizi and Fleming, 1989;[60] Rushton;[61] and Schwartz.[36]

A population-based case-control study of 486 men and women was conducted in Shanghai, China from 1987–1989 to evaluate the association of selected occupational exposures with the risk of leukemia. The authors report that exposure to benzene was associated with elevated risk of CML. The odds ratio (OR) was statistically significant at 2.5 (95% confidence interval (CI) = 1.3–10.2). In essence, those individuals with occupational exposure to benzene demonstrated a 2.5-fold increase in the development of CML. In addition, the authors reported that those individuals occupationally exposed to organic solvents had an OR of 1.7, which was borderline statistically significant (95% CI=1.0–3.0).[49]

Savitz and Andrews[33] reported a statistically significant relative risk for non-AML leukemia, which would include CML, of 3.0 (95% CI = 1.3–5.8) based on 8 cases. The Chinese cohort study of Yin et al.[44] reported an imprecise 2.5 relative risk for CML and exposure to benzene (95% CI = 0.7–16.9), which is not statistically significant, but these findings are in agreement with other studies that showed statistically significant in risk.

In 1988 Linet et al.,[62] reviewing data from 1961 to 1979, found that 19% of 5,351 leukemia cases were CML. When the CML cases were cross-referenced with occupational information, a 1.5-fold excess of CML was revealed for motor mechanics, who were exposed to gasoline and its additives (standardized incidence ratio (SIR) = 1.5; $P < 0.05$).

In 1979 the Australian petroleum industry set up a health surveillance program (health watch) to assess the long-term health of its employees. A separate case-control study that focuses on the relationship of certain blood and bone marrow cancers and benzene exposure has also been established. From time to time, reports are issued by the Monash and Deakin Universities and are published in the medical and scientific literature. In 1991 this group published the incidence of cancer in the employees of the Australian petroleum industry from 1981 to 1989. The authors reported a fourfold excess of myeloid leukemia, which includes CML, over the expected number of cases.[55] This 4.0 SIR was statistically significant (95% CI = 1.6–8.2) for exposed persons.[63]

Presence of Benzene in Petroleum Solvents

Solvents that contain benzene and to which individuals are frequently exposed include Stoddard Solvent (Varsol), white spirits, Ligroin, mineral spirits, and more. The sources of exposure to Stoddard Solvent are shown in TABLE 5. Ligroin is also known as VP&M naphtha, refined solvent naphtha, solvent naphtha, varnish makers' and painters' naphtha; benzoline, canadol mineral spirits, petroleum, and benzene.

The Exxon Company, USA, MSDS data sheet dated 11/07/1988 shows the presence of benzene in SNG Feedstock Naphtha. This naphtha product contains 2% benzene or 20,000 ppm. The MSDS dated 11/07/1988 for C F 181 Naphtha, also manufactured by Exxon Company, USA, shows a concentration of 0.25% of benzene or 2,500 ppm. TABLE 6 shows the measured concentrations of benzene in solvents, such as toluene, xylene, and ethyl benzene. Other solvents that have been shown to contain substantial levels of benzene are shown in TABLE 7.

TABLE 5. Sources of exposure to stoddard solvent

Paint	Coatings
Paint thinner	Waxes
Dry cleaning fluid	Equipment cleaning fluid

Source: Toxicological profile, 1998.[65]

TABLE 6. Benzene levels in toluene, xylene, & ethyl benzene

Product name	Parts per million (ppm)	Information source
Toluene	2,000–10,000	Shell Oil Corp., 1977
Toluene	1,000	Shell Oil Corp., 1977
Toluene	1,000	General Electric Corp., 1977
Toluene	1,000	Dupont Corp., 1978
Toluene	1,000	Dupont Corp., 1978
Toluene	5,000	Dupont Corp., 1978
Toluene	5,000–50,000	Shell Oil Corp., 1977
Toluene	25,000	MSDS, CP Chem, 2003
Toluene	1,000	General Electric Corp., 1977
Toluene	25,000	Chevron MSDS, 2003
Toluene + Xylene	5,000–50,000	Boenhein and Person, 1973[65]
Xylene	1,000	Dupont Memo
Ethylbenzene	300	Shell Oil Corp., 1977
Ethylbenzene	887	General Electric Corp., 1977

TABLE 7. Solvents containing substantial levels of benzene

140° Flash aliphatic solvent	Hydraulic fluids	Shell Sol BJ-19EG
Asphalts	Kerosene	Shell Sol BJ-77EG
Butadiene	Lacquer thinner	Shell DAN
Butene	Liquid wrench	Shell rubber solvent
Calibration fluid	Mineral spirits	Slop oil
	Monochlorobenzene	Solvasol-1
Cyclodexanol	Naphthas	Solvasol-2
Dicyclopentadiene	Piperylene	Trimethyl benzene
Dichloropentadiene	Rubber cement	Vinyl thinner
Hexane	Rubber solvent	VM&P Naphtha

DISCUSSION/CONCLUSION

Numerous studies have shown that individuals exposed to solvents containing benzene have an increased risk of two or more of developing CML with statistical significance of 0.05 or lower. According to Bradford-Hill[66]:

"Before deducing 'causation' and taking action, we shall not invariably have to sit around awaiting the results of the research."

"The whole chain may have to be unraveled or a few links may suffice. It will depend on circumstances... All scientific work is incomplete, whether it be observation or experimental, and scientific work is liable to be repeated or modified by advancing knowledge. That does not confer upon us a freedom to ignore the knowledge we already have or to postpone the action that it appears to demand in a given time."

The science of information gathering and certainty determinations includes epidemiological and animal studies and takes into account target tissues or organs, metabolism, mechanisms, pathophysiology of disease as it relates to chemical exposure, biological plausibility, all used by credible scientists in these fields. The generally accepted standard of scientific probability showing that a probability value of less than 5% ($P \leq 0.05$) demonstrates 95% likelihood that the association shown is not due to chance. Based on the scientific evidence available, it is reasonable and prudent to conclude that benzene is causally related to CML.

REFERENCES

1. NOWELL, P.C. & D.A. HUNGERFORD. 1960. A minute chromosome in human chronic granulocytic leukemia. Science **132:** 1497.
2. ROWLEY, J.D. 1973. A new consistent chromosomal abnormality in chronic myelogenous leukaemia identified by quinacrine fluorescence and Giemsa staining. Nature **243:** 290–293.
3. SHTIVELMAN, E. *et al.* 1985. Fused transcript of abl and bcr genes in chronic myelogenous leukemia. Nature **315:** 550–554.

4. DALEY, G.Q. et al. 1990. Induction of chronic myelogenous leukemia in mice by the P210ber/abl gene of the Philadelphia chromosome. Science **247:** 824–830.
5. LUGO, T.G. et al. 1990. Tyrosine kinase activity and transformation potency of ber-abl oncogene products. Science **247:** 1079–1082.
6. GOLDMAN, J.M. & J.V. MELO. 2003. Chronic myeloid leukemia—Advances in biology and new approaches to treatment. N. Engl. J. Med. **349:** 1451–1464.
7. HUGHES, T.P. 2003. Frequency of major molecular responses to imatinib or interferon alfa plus cytarabine in newly diagnosed chronic myeloid leukemia. N. Engl. J. Med. **349:** 1423–1430.
8. ASKEY, J.M. 1928. Aplastic anemia due to benzol poisoning. Calif. West. Med. **29:** 262–263.
9. TONDEL, M. et al. 1995. Myelofibrosis and benzene exposure. Occup. Med. **45:** 51–52.
10. INFANTE, P.F. et al. 1990. Benzene in petrol: a continuing hazard. Lancet **335:** 814–815.
11. MEHLMAN, M.A. 1983. Advances in modern environmental toxicology. Vol. IV: Carcinogenicity and Toxicity of Benzene. Princeton Scientific Publishing. Princeton, NJ.
12. MEHLMAN, M.A. 1985a. Benzene: scientific update. Am. J. Indust. Med. **7:** 361–365.
13. MEHLMAN, M.A. 1985b. Advances in modern environmental toxicology. Vol. 14: Carcinogenicity and Toxicity of Benzene. Princeton Scientific Publishing. Princeton, NJ.
14. MEHLMAN, M.A. 1989. Advances in modern environmental toxicology. Vol. XVI: Benzene: occupational and environmental hazards—Scientific update. Princeton Scientific Publishing. Princeton, NJ.
15. MEHLMAN, M.A. 1990. Dangerous properties of petroleum refining products: carcinogenicity of motor fuels gasoline. Teratog. Carcinog. Mutag. **10:** 399–408.
16. MEHLMAN, M.A. 1991. Benzene health effects: unanswered questions still not addressed. Am. J. Ind. Med. **20:** 707–711.
17. U.S. ENVIRONMENTAL PROTECTION AGENCY. 1984. Regulatory strategies for the gasoline marketing industry. 49FR31706. Federal Register August 8.
18. U.S. ENVIRONMENTAL PROTECTION AGENCY. 1986. Evaluation of the potential carcinogenicity of benzene. Review draft. Carcinogen Assessment Group, Office of Health And Environmental Assessment. OHEA-073–29.
19. POKLIS, A. & C. BURKETT. 1977. Gasoline sniffing a review. Clin. Toxicol. **11:** 35–41.
20. GEARY, G.G. 2000. The story of chronic myeloid leukaemia. Br. J. Haematol. **110:** 2–11.
21. AKSOY, M. 1989. Hematotoxicity and carcinogenicity of benzene. Environ. Health Persp. **82:** 193–197.
22. BOND, G.G. et al. 1986. An update of mortality among workers exposed to benzene. Br. J. Ind. Med. **43:** 685–691.
23. DECOUFLÉ, P. et al. 1983. Mortality among chemical workers exposed to benzene and other agents. Environ. Res. **30:** 16–25.
24. DELORE, P. & C. BORGOMANO. 1928. Acute leukemia in the course of benzene poisoning: the toxic origin of certain acute leukemias and their relationship to severe anemia. J. Med. Lyon **9:** 227–233.
25. GOGUEL, A. et al. 1967. Les leucemies benzeniques de la region Parisienne entre, 1950 et 1965. Etude de 50 observations. Nouv. Rev. Fr. Hematol. **7:** 465–480.

26. GOLDSTEIN, B. 1977. Hematotoxicity in humans. J. Toxicol. Environ. Health (Suppl. 2): 69–105.
27. INFANTE, P.F. & M.C. WHITE. 1985. Projections of leukemia risk associated with occupational exposure to benzene. Am. J. Med. **7:** 403–413.
28. INFANTE, P.F. 1995. Benzene and leukemia: cell types, latency and amount of exposure associated with leukemia. *In* Update on benzene, advances in occupational medicine and rehabilitation. M. Imbriani, *et al.*, Eds.: 107–120. Fondazione Salvatore Maugeri Edizioni. Pavia, Italy.
29. MCMICHAEL, A.J. *et al.* 1974. An epidemiological study of mortality within a cohort of rubber workers 1964–1972. J. Occup. Med. **16:** 458–464.
30. MCMICHAEL, A.J. *et al.* 1975. Solvent exposure and leukemia among rubber workers: an epidemiologic study. J. Occup. Med. **17:** 234–239.
31. MCMICHAEL, A.J. *et al.* 1976. Cancer mortality among rubber workers: an epidemiologic study. Ann. N. Y. Acad. Sci. **271:** 125–137.
32. RINSKY, R.A. *et al.* 1981. Leukemia in benzene workers. Am. J. Ind. Med. **2:** 217–245.
33. RINSKY, R. 1987. Benzene and leukemia: an epidemiological risk assessment. N. Engl. J. Med. **316:** 1044–1050.
34. SAVITZ, D.A. & K.W. ANDREWS. 1997. Review of epidemiologic evidence on benzene and lymphatic and hematopoietic cancers. Am. J. Ind. Med. **31:** 287–295.
35. SCHWARTZ, E. 1987. Proportionate mortality ratio analysis of automobile mechanics and gasoline service station workers in New Hampshire. Am. J. Ind. Med. **12:** 91–99.
36. TRAVIS, L.B. *et al.* 1994. Hematopoietic malignancies and related disorders among benzene-exposed workers in China. Leuk. Lymphoma **14:** 91–102.
37. VIANNA, N.J. & A. POLAN. 1979. Lymphomas and occupational benzene exposure. Lancet **1:** 1394–1395.
38. VIGLIANI, E.C. 1976. Leukemia associated with benzene exposure. In: Occupational Carcinogenesis. Ann. N.Y. Acad. Sci. **271:** 143–151.
39. VIGLIANI, E.C. & A. FORNI. 1976. Benzene and leukemia. Environ. Res. **11:** 122–127.
40. WONG, O. 1987. An industry wide mortality study of chemical workers occupationally exposed to benzene. II. Dose response analysis. Br. J. Ind. Med. **44:** 365–395.
41. YIN, S.N. 1987. Leukemia in benzene workers: a retrospective cohort study. Br. J. Indust. Med. **44:** 124–128.
42. YIN, S.N. *et al.* 1989. A retrospective study of leukemia and other cancers in benzene workers. Environ. Health Persp. **82:** 207–214.
43. YIN, S.N. *et al.* 1994. Cohort study among workers exposed to benzene in China: general methods and resources. Am. J. Ind. Med. **26:** 383–400.
44. YIN, S-N. 1996. A cohort study of cancer among benzene-exposed workers in China: overall results. Am. J. Ind. Med. **29:** 227–235.
45. SMITH, A. *et al.* 1994. Assessment of cancer clusters using limited cohort data with spreadsheets: application to a leukemia cluster among rubber workers. Am. J. Indust. Med. **25:** 813–823.
46. BOUSSER, J. & S. TARA. 1951. Apropos de trois cas de leucemic myeloide chronique provoques par le benzol. Arch. Mal. Prof. **12:** 399–404.
47. LIAUDET, J. & M. COMBAZ. 1973. Meucemic myeloide chronique chez un chimiste petrolier de 35 ansn manipulant du benzene depuis l'age de 18 ams. J. Eur. Toxicol. **6:** 309–313.

48. WONG, O. 1986. Statement submitted to the OSHA benzene hearing. Environmental health associates I March 4, 1986.
49. ADEGOKE, O.J. *et al.* 2003. Occupational history and exposure and the risk of adult leukemia in Shanghai. Ann. Epidemiol. **13:** 485–494.
50. TAREEFF, E.M. *et al.* 1963. Benzene leukemias. Acta. Unio. Internat. Contra. Cancrum. **19:** 751–755.
51. BROWNING, E. 1965. Toxicity and metabolism of industrial solvents. Elsevier. New York. 3–65.
52. GIRARD, R. & L. REVOL. 1970. La fréquence d'une exposition benzénique au cours des hémopathies graves. Mouvelle Review Francaise d'Hematologie **10:** 477–484.
53. HAYES, R.B. *et al.* 2001. Benzene and lymphohematopoietic malignancies in humans. Amer. J. Indust. Med. **40:** 117–126.
54. OLUFEMI, J. 2002. Occupational history and exposure and the risk of adult leukemia in Shanghi. Abstr # 3809.
55. HEALTH WATCH, TENTH REPORT. 1998. The Australian Institute of Petroleum Health Surveillance Program. University of Melbourne. 83.
56. INFANTE, P.F. 1993. State of the science on the carcinogenicity of gasoline with particular reference to cohort mortality study results. Env. Health Perspect. **101**(Suppl 6): 105–109.
57. JAKOBSSON, R. *et al.* 1993. Acute myeloid leukemia among petrol station attendants. Arch. Environ. Health **48:** 255–259.
58. LINDQUIST, E. *et al.* 1991. Acute leukemia in professional drivers exposed to gasoline and diesel. Eur. J. Haematol. **47:** 98–103.
59. LUMLEY, M. *et al.* 1990. Benzene in petrol. Lancet **336:** 1318–1319.
60. NAIZI, G.A. & A.E. FLEMING. 1989. Blood dyscrasia in unofficial vendors of petrol and heavy oil and motor mechanics in Nigeria. Tropical Doctor **19:** 55–58.
61. RUSHTON, L.A. 1993. A 39-year follow-up of the UK refinery and distribution center studies: results for kidney cancer and leukemia. Environ. Health Perspect. **101**(Suppl 6): 77–84.
62. LINET, M.S. *et al.* 1988. Leukemias and occupation in Sweden: a registry-based analysis. Am. J. Indust. Med. **14:** 319–330.
63. CHRISTIE, D. *et al.* 1991. A prospective study in the Australian petroleum industry. II Incidence of cancer. Br. J. Ind. Med. **48:** 511–514.
64. BOENHEIM, A.F. & A.J. PEARSON. 1973. Petroleum hydrocarbon solvents. Modern Petroleum Technology. Fourth edition. Applied Science Publishers. Barking.
65. ATSDR. 1998. Toxicology Profiles – Benzene.
66. BRADFORD-HILL, A. 1965. The environment and disease: association or causation? Proc. Royal Soc. Med. **48:** 295–300.

Causal Relationship between Non-Hodgkin's Lymphoma and Exposure to Benzene and Benzene-Containing Solvents

MYRON A. MEHLMAN

Department of Community Medicine, The Mount Sinai Medical Center, New York, New York 10029, USA

Department of Preventive Medicine and Community Health, University of Texas Medical Branch, Galveston, Texas 77555, USA

ABSTRACT: Non-Hodgkin's lymphoma (NHL) is a malignant neoplasm of the lymphatic system made up of mainly B cell lymphocytes. A large number of studies have shown significant associations between NHL and benzene or benzene-containing solvents and products. This article summarizes studies detailing these associations and indicates those that are significant. Based on an analysis of the literature and the weight of evidence from numerous studies, it is reasonable to conclude that exposure to benzene or to solvents or products containing benzene is causally related to NHL.

KEYWORDS: benzene; lymphoma; non-Hodgkin's lymphoma; carcinogen; hematoreticular

INTRODUCTION

Non-Hodgkin's lymphoma (NHL) is a malignant neoplasm of cells of the lymphatic system. These cancers have been identified as B cell and T cell lymphomas and are distinguished from Hodgkin's lymphoma, which differs from other lymphomas in terms of unique cell type, sites involved, treatment, and prognosis. B cell lymphomas make up approximately 85–90% of NHL.

In addition to benzene and solvents containing benzene, NHL may be associated with chronic inflammatory diseases, such as celiac disease and rheumatic arthritis. Immune suppression has been reported in association with NHL. Ansell and Armitage[1] described the classification and chromosomal abnormalities in NHL. A large number of studies have shown statistically significant increased risk for NHL in persons who have been exposed to benzene.

Address for correspondence: Myron A. Mehlman, Ph.D., 7 Bouvant Drive, Princeton, NJ 08540. Voice: 609-683-1493; fax: 609-683-0838.
e-mail: mehlman@patmedia.net

RESULTS

Significantly increased risk for NHL has been associated with exposure to mineral spirits, aromatic hydrocarbons, and solvents containing benzene. Exposure to mineral spirits was shown to have statistically significant increases in risk for NHL by Hours et al.[2] with an odds ratio (OR) of 14.86 and a confidence interval (CI) of 2.76–80. Similar increases were shown by Hardell et al.,[3] Nordstrom et al.,[4] Pasqualetti et al.[5] (OR = 3.4, 95% CI = 1.65–7.4), Persson,[6] Persson et al.,[7] and Persson and Fredriksson[8] (TABLE 1).

The association between NHL and aromatic solvents is shown in TABLE 2.[2,5,9,10]

Statistically significant increases in NHL were found in workers exposed to solvents,[2,8,11,12] and white spirits.[3,13–16]

Adegoke et al.[17] showed statistically significant increases in acute lymphocytic leukemia (ALL), acute myelocytic leukemia (AML), and chronic myelogenous leukemia (CML) from exposure to benzene (TABLE 3).

Studies[18–20] show increased risk for NHL (TABLE 4) in employees exposed to solvents. Similar increased risk for NHL in employees exposed to benzene or benzene-containing solvents has been shown by Dryver et al.[9] (TABLE 5); Hardell et al.[21] (TABLE 6); Hardell et al.[3] (TABLE 7); Hayes et al.[14] (TABLE 8); Li et al.[22] (TABLE 9); Mao et al.[23] (TABLE 10); Nordstrom et al.[4] (TABLE 11); Olsson & Brandt[24] (TABLE 12); Persson et al.[7] (TABLE 13); Persson[25] (TABLE 14); Rego[26] (TABLE 15); Vianna & Polan[27] (TABLE 16); Wong[28] (TABLE 17); and Yin et al.[29] (TABLE 18). TABLE 19 shows some occupations in which workers are exposed to benzene.

DISCUSSION

The weight of evidence from numerous studies examined and the source of information for 95% certainty determinations include epidemiological and

TABLE 1. NHL and exposure to mineral spirits[a]

Author (Substance studied)	ICD-9 Code	OR[b]	95% CI[b]
Hardell et al.[3] (White spirits)	200, 202	3.2*	1.3–8.3
Hours et al.[2] (Mineral spirits)	200, 202	14.86*	
Nordstrom et al.[4] (White spirits)	200, 202	2.0*	
Pasqualetti et al.[5] (Mineral spirits, mineral ointment)	200, 202	3.4*	
Persson[6] (White spirits, >1 yr)	200	3.1	
Persson et al.[7] (White spirits)	200	1.6	
Persson & Fredriksson[8] (White spirits, >1 yr)	200	2.6*	

[a]Includes petroleum-derived solvent mixtures known as white spirits, Stoddard Solvent, rubber solvent, VM&P naphtha, benzene, and ligroin composed of 2–25% aromatics.
[b]OR = Odds Ratio. CI = Confidence Interval.
* Statistically significant at $P \leq 0.05$.

TABLE 2. NHL and aromatic solvents

Author (Substance studied)	ICD-9 Codes	Findings	Percent Increase
Dryver et al.[9]	200, 202	OR = 1.72*	72%
Hours et al.[2]	200, 202	OR = 2.1	110%
Pasqualetti et al.[5] (Aromatics)	200, 202	OR = 2.15**	115%
Wilcosky et al.[10] (Benzene)	200	RR = 3.0	200%
Wilcosky et al.[10] (Xylene)	200	RR = 3.7*	271%

*$P < 0.05$; OR = Odds Ratio. **$P < 0.001$. RR = relative risk.

TABLE 3. Hematolymphoreticular neoplasms and exposure to benzene

Chemical/Product	Disease	Exposure Duration	OR[a]	95% CI[a]	p for Trend
Benzene	All leukemias	≥ 15 years	3.3*	1.6–6.91	0.01
	ALL	> 15 years	3.9*	1.3–11.8	0.12
	AML	> 15 years	2.9*	1.2–7.0	0.01
	CML	> 15 years	5.0*	1.8–13.9	0.01
Gasoline	CML	> 15 years	1.5	0.8–3.0	0.25
Paints	All leukemias	< 15 years	2.3*	1.2–4.7	0.09
	ALL	< 15 years	3.4*	1.2–9.5	0.15
	AML	< 15 years	1.8	0.7–4.2	0.92
	CML	< 15 years	3.7*	1.4–9.9	0.02

[a]OR = Odds Ratio. CI = Confidence Interval.
* < 0.05.
FROM: Adegoke et al.[17]

TABLE 4. Increases in NHL and solvent exposure

	OR[a]	95% CI[a]
Berlin et al.[18]	1.9	0.8–4.0
Blair et al.[19]	1.9	0.9–3.8
Brandt et al.[20]	3.3	1.9–5.8

[a]OR = Odds Ratio. CI = Confidence Interval.

TABLE 5. NHL

Compound	OR[a]	95% CI[a]
Aromatic hydrocarbons	1.72	1.1–2.7
Gasoline	1.59	1.04–2.05
Solvents > 5 years exposure	1.59	1.11–2.28
Painters	1.77	1.13–2.76

[a]OR = Odds Ratio. CI = Confidence Interval.
FROM: Dryver et al.[9]

TABLE 6. Exposure to organic solvents and malignant lymphoma

Compound	OR[a]	95% CI[a]
Styrene, trichloroethylene, perchloroethylene, benzene	4.6	1.9–11.4
Other Solvents	2.8	1.6–4.8
Solvents, Phenoxy acids and chlorophenols	8.5	4.3–17.2
Phenoxy acids and organic solvents	11.2	3.2–39.7
Chlorophenols	5.7	2.7–12.2

[a]OR = Odds Ratio. CI = Confidence Interval.
See Tables V, VI, and VII, p. 173, Hardell et al.[21]

TABLE 7. Exposure to organic solvents and NHL

Organic Solvents	OR[a]	95% CI[a]
All	2.4	1.4–3.9
High grade	2.9	1.6–5.6
Low grade	1.8	0.8–3.8
Benzene	28.0	18–73.0
Thinner	3.4	1.4–10.0
Turpentine	3.3	0.9–17
White spirits	3.2	1.3–8.3

[a]OR = Odds Ratio. CI = Confidence Interval.
See Table 2, p. 2387, Hardell et al.[3]

TABLE 8. Relative risk and extent of exposure to benzene

	Cumulative ppm-years			
Neoplasm	<40	40–99	>100	P Value
All hematological neoplasms				
RR	2.2	2.9	2.9	0.04*
95% CI	1.1–4.5	2.9	1.4–5.2	
NHL				
RR	3.3	1.1	3.5	0.02*
95% CI	0.8–13.9	1.1	0.1–11.1	

*Statistically significant. RR = relative risk
See Table 2, p. 1068, Hayes et al.[14]

TABLE 9. Incidence and deaths in male painters and paint manufacturers exposed to benzene

Type of Risk	Exposed/Unexposed	Relative Risk	95% CI[a]
All causes, mortality risk	622/485	1.22	1.08–1.37
All neoplasms, mortality risk	250/176	1.39	1.14–1.69
All hematopoietic and lymphoproliferative disorders, incidence risk	24/10	2.71	1.22–5.96
Leukemia, incidence risk	15/7	2.43	1.02–6.0

[a]CI = Confidence Interval.
FROM: Li et al.[22]

TABLE 10. NHL and occupational exposure to chemicals in Canada

Agent	Number	OR[a]	95% CI[a]
Benzidine	21	1.9	1.1–1.5
Benzidine, >4 years exposure	15	2.2	1.2–4.1*
Benzene	36	1.2	0.8–1.9
Mineral cutting or lubricating oil	177	1.3	1.0–1.5

[a]OR = Odds Ratio. CI = Confidence Interval.
*$P = 0.02$ for trend.
FROM: Mao et al.[23]

TABLE 11. NHL

Solvent	Diseased/Total	OR[a]	95% CI[a]
All solvents	51/143	1.5	0.99–2.3
White spirits	33/69	2.0	1.2–3.4
Paints	1.1/11	4.3	1.8–10.3
Turpentine	5/11	2.0	0.7–5.9
Other solvents	5/9	2.4	0.8–7.4

[a]OR = Odds Ratio. CI = Confidence Interval.
FROM: Nordstrom et al.[4]

TABLE 12. NHL

NHL Location at Presentation[a]	OR[a]	95% CI[a]
NHL, supradiaphragmatic location	6.5	3.2–13.3
NHL, below the diaphragm or generalized	2.3	1.3–4.3

Dose Response[b]		
Solvent exposure ≤ 10 years	1.8	1.2–2.7
Solvent exposure ≤ 20 years	3.3	1.5–7.1
Solvent exposure ≤ 30 years	6.0	1.9–19.0*

[a]OR = Odds Ratio. CI = Confidence Interval.
[b]In workers exposed to organic solvents.
*$\chi^2 = 17.7, P = 0.0001$.
FROM: Olson & Brandt.[24]

TABLE 13. Occupational risk factors for development of malignant lymphoma

Agent	Patients/Controls	OR[a]	95% CI[a]
Solvent exposure for 2–5 years vs. 0–1 year	24/34	1.7	0.6–2.2
Gasoline	10/11	2.1	
Aviation gasoline	4/3	3.0	

[a]OR = Odds Ratio. CI = Confidence Interval.
FROM: Persson et al.[7]

TABLE 14. Malignant lymphoma

Compound	OR[a]	95% CI[a]
Solvents (occupational)	1.9*[b]	0.1.1–3.2
Thinner	1.9*	1.0–3.3
White spirits	2.6*	1.3–4.7
Painters	2.5[c]	0.5–9.6

*Statistically significant.
[a] OR = Odds Ratio. CI = Confidence Interval.
[b] Logistics odds ratio (LOR).
[c] 150% increase n.s.
FROM: Persson.[25]

TABLE 15. NHL and organic solvents

Condition	OR[a]	95% CI[a]
All NHL & solvents	1.67	0.97–2.87
All NHL & solvents, > 5 years	1.87	1.05–3.34
Painting/paper hanging	1.9	0.9–3.8
Petroleum refining	1.6	0.5–5.8

[a] OR = Odds Ratio. CI = Confidence Interval.
FROM: Rego.[26]

TABLE 16. NHL

	Odds Ratio
Occupational benzene exposure	1.6–2.0
Dose response	
For workers < 45 years old	0.9–1.3
For workers > 45 years old	1.7–2.1*

*$P < 0.01$.
FROM: Vianna and Polan.[27]

TABLE 17. NHL and leukemia in benzene-exposed workers

A benzene-exposed group of 3074 chemical workers was compared with a group of 3074 nonexposed controls

The exposed group experienced 7 leukemias (RR = 2.9); the control group had no leukemias

The RR for non-Hodgkin's lymphopoietic cancer for the continuously exposed group was 3.77 ($P < 0.05$)

Wong's data demonstrate statistically significant dose–response relationships between exposure to benzene and lymphopoietic cancers.

FROM: Wong.[28] RR = relative risk.

TABLE 18. Malignant lymphoma in benzene-exposed workers in China

Condition	RR[a]	95% CI[a]
Malignant lymphoma and related disorders (codes 200-202)	4.5	1.3–28.4
NHL	3.0	
Nodal and extranodal NHL	4.0	1.1–25.7

[a]RR = relative risk. CI = Confidence Interval.
FROM: Yin et al.[29]

TABLE 19. Occupations with exposures to benzene

Adhesive makers	Dry cleaning workers	Perfume makers
Airplane mechanics	Dye makers	Petrochemical workers
Alcohol production workers	Electroplaters	Petroleum refinery workers
Art glass workers	Enamel workers	Pharmaceutical workers
Artificial leather makers	Engravers	Phenol makers
Asbestos product impregnators	Ethylbenzene workers	Photographic chemical makers
Asphalt workers	Explosive makers	Picric acid makers
Automotive workers	Fuel oil handlers	Polish makers
Battery makers, dry	Fumigant makers	Pottery decorators
Belt scourers	Fungicide makers	Printers
Benzene hexachloride makers	Furniture finishers	Putty makers
Benzene workers	Gas workers	Resin makers
Biphenyl workers	Glue makers	Rotogravure printers
Brake lining makers	Herbicide makers	Rubber cementers
Bronze workers	Hydrochloric acid workers	Rubber gasket makers
Burnishers	Ink makers	Rubber makers
Can makers	Insecticide makers	Rubber reclaimers
Carbolic acid workers	Lacquer makers	Shellac makers
Cast scrubbers, electroplating	Leather makers	Shoe factory workers
Chemical synthesis	Linoleum makers	Shoe finishers
Chlorobenzene workers	Lithographers	Solvent makers
Chlorodiphenyl workers	Maleic acid makers	Stain makers
Clutch disc impregnators	Mirror silverers	Stainers
Coal tar refiners	Nitrobenzene makers	Styrene makers
Coal tar workers	Nitrocellulose workers	Synthetic fiber makers
Coatings workers	Oil processors	Trinitrotoluol makers
Cobblers	Oilcloth makers	Type cleaners
Coke oven workers	Organic chemical synthesizers	Varnish makers
DDT makers	Paint makers	Vulcanizers
Degreasers	Painters	Wax makers
Detergent makers	Paraffin processors	Welders
Dichlorobenzene workers	Pencil makers	Window shade makers
Disinfectant workers		Wire insulators

animal studies and take into consideration target tissues, metabolism, mechanism, and pathophysiology, which are standard methodology used by scientists in the field. The generally accepted standard of scientific probability of data showing a P value of 0.05 or less is generally accepted as demonstrating a likelihood of 95% that the association is not due to chance.

Thus, it is reasonable and prudent to conclude that exposure to benzene or to solvents or products containing benzene is causally related to NHL.

REFERENCES

1. Ansell, S.M. & J. Armitage. 2005. Non-Hodgkin lymphoma: diagnosis and treatment. Mayo Clin. Proc. **80:** 1087–1097.
2. Hours, M. *et al.* 1995. Occupational exposure and malignant hemopathies: a case-control study in Lyon (France). Rev. Epidemiol. Sante Publique **43:** 231–241.
3. Hardell, L. *et al.* 1994. Exposure to phenoxyacetic acids, chlorophenols, or organic solvents in relation to histopathology, stage, and anatomical location of non-Hodgkin's lymphoma. Cancer Res. **54:** 2386–2389.
4. Nordstrom, M. *et al.* 1998. Occupational exposures, animal exposure and smoking as risk factors for hairy cell leukaemia evaluated in a case-control study. Br. J. Cancer **77:** 2048–2052.
5. Pasqualetti, P. *et al.* 1991. Occupational risk for hematological malignancies. Am. J. Hematol. **38:** 147–149.
6. Persson, B. 1989. Malignant lymphomas and occupational exposures. Br. J. Indust. Med. **46:** 516–520.
7. Persson, B. *et al.* 1993. Some occupational exposures as risk factors for malignant lymphomas. Cancer **72:** 1773–1778.
8. Persson, B. & M. Fredriksson. 1999. Some risk factors for non-Hodgkin's lymphoma. Int. J. Occup. Med. Environ. Health. **12:** 135–142.
9. Dryver, E. *et al.* 2002. Exposures, hormones, heredity and the risk of non-Hodgkin's lymphoma: results from a case control study in Southern Sweden. Blood **100:** 770a.
10. Wilcosky, T.C. *et al.* 1984. Cancer mortality and solvent exposure in the rubber industry. Am. Ind. Hyg. Assoc. J. **45:** 809–811.
11. Rego, M.A. *et al.* 2002. Non-Hodgkin's lymphoma and organic solvents. J. Occup. Environ. Med. **44:** 874–881.
12. Fabbro-Peray, P. *et al.* 2001. Environmental risk factors for non-Hodgkin's lymphoma: a population based case-control study in Languedoc-Roussillon, France. Cancer Causes Control **12:** 201–212.
13. Nilsson, R. *et al.* 1998. Leukemia, lymphoma and multiple myeloma in seamen on tankers. Occup. Med. **55:** 517–522.
14. Hayes, R.B. *et al.* 1997. Benzene and the dose-related incidence of hematologic neoplasms in China. For the Chinese Academy of Preventive Medicine–National Cancer Institute Benzene Study Group. J. Natl. Cancer Inst. **89:** 1065–1071.
15. Tatham, L. *et al.* 1997. Occupational risk factors for subgroups of non-Hodgkin's lymphoma. Epidemiology **8:** 551–558.
16. Weisenberger, D.D. 1994. Epidemiology of non-Hodgkin's lymphoma: recent findings regarding an emerging epidemic. Ann. Oncol. **5** (Suppl 1): SI9–S24.

17. ADEGOKE, O.J. et al. 2003. Occupational history and exposure and the risk of adult leukemia in Shanghai. Ann. Epidemiol. **13:** 485–494.
18. BERLIN, K. et al. 1995. Cancer incidence and mortality of patients with suspected solvent-related disorders. Scand. J. Work Environ. Health **21:** 362–367.
19. BLAIR, A. et al. 1993. Evaluation of risks for non-Hodgkin's lymphoma by occupation and industry exposures from a case-control study. Am. J. Indust. Med. **23:** 301–312.
20. BRANDT, L. 1992. Exposure to organic solvents and risk of haematological malignancies. Leukemia Res. **16:** 67–70.
21. HARDELL, L. et al. 1981. Malignant lymphoma and exposure to chemicals, especially organic solvents, chlorophenols and phenoxy acids: a case control study. Br. J. Cancer **43:** 169–176.
22. LI, G.-L. et al. 1994. Gender differences in hematopoietic and lymphoproliferative disorders and other cancer risks by major occupational group among workers exposed to benzene in China. J. Occup. Med. **36:** 875–881.
23. MAO, Y. et al. 2000. Non-Hodgkin's lymphoma and occupational exposure to chemicals in Canada. Ann. Oncol. **11** (Suppl 1)**:** 69–73.
24. OLSSON, H. & L. BRANDT. 1988. Risk of non-Hodgkin's lymphoma among men occupationally exposed to organic solvents. Scand. J. Work Environ. Health **14:** 246–251.
25. PERSSON, B. 1995. Occupational exposures and malignant lymphoma. Linköping University Medical Dissertation No. 475. Dept. of Occupational and Environmental Medicine, Linköping, Sweden.
26. REGO, M.A. 1998. Non-Hodgkin's lymphoma risk derived from exposure to organic solvents: a review of epidemiologic studies. Cad Saúde Publica, Rio de Janeiro **14** (Sup 3)**:** 41–66.
27. VIANNA, N. J. & A. POLAN. 1979. Lymphomas and occupational benzene exposure. Lancet **1:** 1394–1395.
28. WONG, O. 1986. Testimony submitted to OSHA Benzene Hearing, March 4, 1986.
29. YIN, S-N. et al. 1996. A cohort study of cancer among benzene-exposed workers in China: Overall results. Am. J. Ind. Med. **29:** 227–235.

Genetic Polymorphism of Toxicant-Metabolizing Enzymes and Prognosis of Chinese Workers with Chronic Benzene Poisoning

J-X. WAN,[a] Z-B. ZHANG,[a] J-R. GUAN,[b] D-Z. CAO,[c] R. YE,[b] X-P. JIN,[a] AND Z-L. XIA,[a]

[a]*Department of Occupational Health and Toxicology, School of Public Health, Fudan University, Shanghai, 200032, China*

[b]*Hangzhou Centers for Disease Control and Prevention*

[c]*Institute of Occupational Health attached to Maanshan Steel & Iron Group*

ABSTRACT: Workers with chronic benzene poisoning (CBP) sometimes have a white blood cell count (WBC) below 4×10^9/L even after cessation of workplace exposure to benzene for years. In order to explore this phenomenon, 120 workers with CBP were divided into two groups depending on the WBC, the mean diagnostic age of CBP, benzene exposure duration, and body mass index (BMI). The proportion of genotypes of cytochrome P450 2E1 (CYP2E1), glutathione-S-transferase mu-1 (GSTM1), glutathione-S-transferase theta-1 (GSTT1), myeloperoxidase (MPO), and NAD(P)H, quinone oxidoreductase 1 (NQO1) were compared between workers with WBC $<4 \times 10^9$/L and those with WBC $\geq 4 \times 10^9$/L. With methods of logistic regression, a risk model was set up to predict the prognosis of CBP workers. The results indicated that the BMI of workers with WBC $<4 \times 10^9$/L was lower than that of workers with WBC of $\geq 4 \times 10^9$/L (21.40 ± 2.76 versus 23.09 ± 3.36, $P = 0.01$), and the logistic regression model suggested there was a 4.5-fold increased risk among workers carrying GSTT1 null genotype (95% CI = 1.13–17.54) compared with workers with GSTT1 non-null genotype. Our findings suggest that benzene exposure duration, BMI, and GSTT1 genotype may impact prognosis of the CBP workers.

KEYWORDS: chronic benzene poisoning; prognosis; genetic polymorphism; lifestyle; late effects

INTRODUCTION

Exposure to benzene produces hematotoxicity, including pancytopenia, aplastic anemia, myelodysplastic, and acute myeloid leukemia.[1–4] Humans are

Address for correspondence: Prof. Z-L. Xia, Department of Occupational Health and Toxicology, School of Public Health, Fudan University, Shanghai, 200032. Voice: +86-21-64041900; fax: +86-21-64178160.
 e-mail: xzl580428@yahoo.com

Ann. N.Y. Acad. Sci. 1076: 129–136 (2006). © 2006 New York Academy of Sciences.
doi: 10.1196/annals.1371.041

exposed to benzene via occupational exposure or environmental via contact with cigarette smoke, gasoline emissions, or products of incomplete combustion. Substantive research has been performed to elucidate the mechanism of benzene poisoning (BP).[5–7] To date, there have been few studies of the recovery status of the chronic benzene poison (CBP) workers after cessation of occupational benzene exposure.[8,9] We observed that the white blood cell (WBC) count of some workers remained below 4×10^9/L even after they have been away from workplace benzene exposure for years. With methods of logistic regression, we analyzed the relationship between the prognosis and the diagnostic age of CBP, benzene exposure duration, body mass index (BMI) and the proportion of genotypes of cytochrome P450 2E1 (CYP2E1), glutathione-S-transferase mu-1 (GSTM1), glutathione-S-transferase theta-1 (GSTT1), myeloperoxidase (MPO), and NAD(P)H, quinone oxidoreductase 1 (NQO1) were compared between WBC $<4 \times 10^9$/L group and WBC $\geq 4 \times 10^9$/L group.

MATERIALS AND METHODS

Subjects

One hundred and twenty BP workers were recruited from Maanshan and Hangzhou, China, and included 46 males and 74 females. BP was diagnosed from 1980 to 1998 by the locally authorized Occupational Disease Diagnostic Team, and patients were registered in the hospitals of prevention and treatment for occupational diseases, which cooperated with us. The diagnostic criteria for occupational BP, according to the Ministry of Health, China, include: (*a*) total WBC $< 4,000/\mu$L or WBC between 4,000 and 4,500/μL and platelet count $< 80,000/\mu$L, with repeated confirmation of this count in a few months by a peripheral blood examination; (*b*) individuals with documented benzene exposure who had been employed for at least 6 months in the factory; and (*c*) exclusion of other causes of abnormal blood counts such as chloromycetin use and ionizing radiation. The medical records of patients were reviewed independently. Those with WBC $>3500/\mu$L were evaluated to confirm the BP diagnosis. There were 44 workers whose WBC were below 4×10^9/L.

WBC Count

Routine blood examination and assay of alanine transaminase (ALT) were completed in the Institute of Occupational Health attached to Maanshan Steel & Iron Group and Hangzhou Hospital for Occupational Diseases. The method used in ALT determination was Elisa-linked continuously ultraviolet monitoring assay.

Benzene Exposure Duration and Diagnostic age of BP

Benzene exposure duration means the actual time that workers were engaged in benzene exposure work, and the diagnostic age of BP denotes that the age at which the workers started to get benzene poisoning (the date of diagnosis minus that of their birth).

Calculation of BMI

BMI equals the body weight (kg) divided by square of the body height (m^2).

Determination of Genetic Polymorphisms of CYP2E1, NQO1, MPO, GSTM1 and GSTT1

The single nucleotide polymorphisms (SNPs) in the promotors and the complete coding regions of CYP2E1, NQO1, and MPO were determined with methods of polymerase chain reaction (PCR), sequencing and DHPLC (denaturing high-performance liquid chromatography).[10,11] The genotypes of GSTT1 and GSTM1 were determined with method of PCR.[12]

Statistical Analysis

A database was established and the data were analyzed with SPSS 10.0. Diagnostic age, BP duration, and BMI in WBC $<4 \times 10^9$/L group and WBC $\geq 4 \times 10^9$/L group were analyzed (*t-test*). The χ^2 *test* was adopted to compare the difference of the lifestyle (smoking and alcohol consumption) and the proportion of genotypes of CYP2E1, GSTM1, GSTT1, MPO, and NQO1 in the two groups. Multiple factors analysis was implemented with method of logistic regression.

RESULTS

Distribution of Gender, Age of Diagnosis and Years of Benzene Exposure

The distributions of gender, age of diagnosis, years of BP, and ALT in subjects are shown in TABLE 1.

Comparison of Age of Diagnosis, Benzene Exposure Duration, and BMI

There was no significant difference in age of diagnosis and BP duration between the two groups, but the BMI of the WBC $<4 \times 10^9$/L group was lower than that of WBC $\geq 4 \times 10^9$/L group ($P = 0.01$) (TABLE 2).

TABLE 1. Demographics of subjects with WBC $<4 \times 10^9$/L and WBC $\geq 4 \times 10^9$/L

Items	WBC $<4 \times 10^9$/L		WBC $\geq 4 \times 10^9$/L	
	n	%	n	%
Gender				
Male	14	1.82	32	42.11
Female	30	68.18	44	57.89
Age of diagnosis				
<30	7	15.91	18	23.68
~35	14	31.82	20	26.32
~40	13	2954	17	22.37
~45	7	15.91	11	14.47
>45	3	6.82	10	13.16
Years of benzene exposure				
<5	7	15.91	13	17.11
~10	12	27.27	23	30.26
~15	11	25.00	14	18.42
~20	5	11.36	15	19.74
>20	9	20.46	11	14.47
ALT				
≤40	44	100.00	76	100.00
>40	0	0.00	0	0.00

Relationship Between Lifestyle, Polymorphisms in Toxicant Metabolizing Genes and WBC Recovery Level of BP Workers

The proportion of null genotype of GSTT1 in the WBC $<4 \times 10^9$/L group was higher than that of WBC $\geq 4 \times 10^9$/L group (64.29% versus 46.67%), but there was no significant difference ($P = 0.07$), so did other factors between the two groups (TABLE 3).

Logistic Analysis of WBC Recovery Level of BP Workers

Logistic analysis of WBC recovery level of BP workers was implemented with WBC level (WBC $\geq 4 \times 10^9$/L = 0; WBC $<4 \times 10^9$/L = 1) as the depen-

TABLE 2. Comparison of age of diagnosis, benzene exposure duration and BMI in WBC $<4 \times 10^9$/L group and WBC $\geq 4 \times 10^9$/L group

Items	n	WBC $<4 \times 10^9$/L	WBC $\geq 4 \times 10^9$/L	t	P
Diagnosis of age year	120	35.59 ± 6.98	36.39 ± 7.95	0.557	0.578
Years of benzene exposure	120	13.85 ± 8.63	12.64 ± 7.09	–0.830	0.408
BMI (kg/m^2)	99	21.40 ± 2.76	23.09 ± 3.36	2.636	0.010

TABLE 3. Effects of gender, lifestyle, and genetic polymorphisms of toxicant-metabolizing enzymes on WBC recovery level of BP workers

Items	WBC <4×10^9/L (%)	WBC $\geq 4 \times 10^9$/L (%)	OR (95% CI)
Gender			
Male	14 (31.82)	32 (42.11)	0.64 (0.27–1.50)
Female	30 (68.12)	44 (57.89)	1.00
Lifestyle			
Smoking			
Yes	5 (11.90)	9 (12.33)	0.96 (0.26–3.47)
No	37 (88.10)	64 (87.67)	1.00
Alcohol consumption			
Yes	4 (9.30)	10 (13.51)	0.66 (0.16–2.50)
No	39 (90.70)	64 (86.49)	1.00
Genetic factors			
NQO1			
c.559C > T			
T/T	10 (28.57)	19 (26.03)	1.14 (0.42–3.05)
C/T and C/C	25 (71.43)	54 (73.97)	1.00
CYP2E1			
96-bp insertion			
Ins$_{96}$–/–	26 (63.41)	46 (63.01)	1.02 (0.43–2.43)
Ins$_{96}$–/+ and +/+	15 (36.59)	27 (36.99)	1.00
c. –1293G > C			
G/G	24 (54.55)	46 (60.53)	0.78 (0.35–1.77)
G/C and C/C	20 (45.45)	30 (39.47)	1.00
c.1263C > T			
C/C	26 (74.29)	43 (74.14)	1.01 (0.35–2.93)
C/T	9 (25.71)	5 (25.86)	1.00
MPO			
c. –463G > A			
G/G	37 (86.05)	60 (81.08)	1.44 (0.46–4.64)
G/A	6 (13.95)	14 (18.92)	1.00
IVS8 +19G > A			
G/G	7 (19.44)	16 (21.92)	0.86 (0.28–2.55)
G/A	29 (80.56)	57 (78.08)	1.00
GSTM1			
Null	21 (51.22)	34 (50.70)	1.02 (0.44–2.39)
Non-null	20 (48.78)	33 (49.25)	1.00
GSTT1			
Null	27 (64.29)	35 (46.67)	2.06 (0.88–4.83)
Non-null	5 (35.71)	40 (53.33)	1.00

dent variable and gender, age of diagnosis, benzene exposure duration, BMI, and polymorphisms of benzene toxicant-metabolizing genes as covariables (logistic methods of Forward Wald). The results suggested that benzene exposure duration, BMI, and genetic polymorphisms of GSTT1 may affect the WBC recovery of BP workers (TABLE 4).

DISCUSSION

Previous studies have shown that the toxicity of benzene derives from its metabolites. Once absorbed, benzene is metabolized by CYP2E1 to yield phenol, hydroquinone (HQ), catechol (CAT), and 1,2,4-benzenetriol.[13] These metabolites accumulate in the bone marrow[14] where they undergo autoxidation or activation by peroxidases to yield the corresponding quinones,[15–17] which are believed to be among the ultimate toxic metabolites of benzene.[18] In theory, when BP workers cease occupational exposure to benzene, the metabolites of benzene in their bodies will decrease gradually, so that the effects of metabolites of benzene will diminish over time. If there is no significant damage in the function of DNA, the cellular functions should recovery gradually, and the WBC will recover to normal.

However, in the study of the association of genetic polymorphisms of toxicant-metabolizing enzymes with susceptibility to BP, we found that not all the BP workers' WBC quantities exceeded 4×10^9/L even after their cessation of occupational benzene exposure for years. We found that 36.7% of BP workers' peripheral WBC remained below 4×10^9/L. The study of susceptibility to BP indicates that there are effects between the lifestyle such as smoking, alcohol consumption, and polymorphisms in toxicant-metabolizing genes such as *NQO1 c.559C > T, CYP2E1 c.−1293G > C*, which lead to a supposition that lifestyle and polymorphisms in toxicant-metabolizing genes may affect the prognosis of BP workers. Our results suggested that there was a significant difference for BMI in the WBC $<4 \times 10^9$/L and the WBC $\geq 4 \times 10^9$/L groups (P = 0.01). Results of logistic regression also indicated that there would be a 0.2771-fold decrease in the risk of CBP for individuals exposed to benzene who had 0.01 kg/m^2 increase in BMI; this may be related to nutritional conditions of the BP workers. However, it has been reported that the mean blood pressure, blood sugar, and triglycerides in Chinese whose BMI exceeds 22.6 are higher than those in Chinese whose BMI is 22.6. Thus, the recovery of the BP workers needs appropriate instead of excessive nutrition. In the model of logistic regression, the risk of BP for individuals with null GSTT1 gene was 4.4553 times higher than for individuals with non-null GSTT1 gene (95% CI: 1.1313–17.5459). Moreover, individuals with null GSTT1 gene and NQO1 c.559C > T at the same time were susceptible to BP.[12] Our results suggest that individuals with null *GSTT1* gene should avoid being engaged in work exposure to benzene.

Benzene is a confirmed carcinogen, and previous studies have indicated that excess risk of leukemia is associated with cumulative benzene exposures.[19–22] In a retrospective cohort study conducted in China, some cases of leukemia also had a history of CBP before the leukemia developed.[21] The factors involved in the prognosis of CBP may also contribute to the development of leukemia. However, this study was not designed to explore the impact factors in the prognosis of BP workers, but it was merely an initial exploration of

TABLE 4. Logistic analysis of WBC recovery level of CBP workers

Variables	ß	Wald	P	OR (95% CI)
Years of benzene exposure[a]	0.0907	4.6626	0.0308	1.0949 (1.0084 –1.1888)
$GSTT1$[b]	1.4941	4.5640	0.0327	4.4553 (1.1313– 17.5459)
BMI[c]	–0.3245	5.4900	0.0191	0.7229 (0.5511 0.9483)
Constant	4.4734	2.6245	0.1052	

[a] Benzene exposure duration as a continuous variable 19,20,21......
[b] $GSTT1$ null genotype vs. non-null genotype.
[c] BMI as a continuous variable kg/m^2 16.81,16.82,16.83......

factors affecting the prognosis of BP workers. Other factors that may affect the prognosis of BP workers, such as therapy scheme and nutrition, are subjects of further research.

ACKNOWLEDGMENTS

This study was supported by National Natural Science Foundation of China (30271113) and 973 Program (2002CB512902).

REFERENCES

1. AKSOY, M., K. DINCOL, T. AKGUN, *et al.* 1972. Details of blood changes in 32 patients with pancytopenia associated with long-term exposure to benzene. Br. J. Ind. Med. **29:** 56–64.
2. YIN, S.N., G. LI, *et al.*1987. Leukemia in benzene workers, a retrospective cohort study. Br. J. Ind. Med. **44:** 124–128.
3. LINET, M.S., S.-N. YIN, L.B. TRAVIS, *et al.* 1996. Clinical features of hematopoietic malignancies and related disorders among benzene-exposed workers in China. Environ. Health Perspect. **104:** 1353–1364.
4. YARDLEY-JONES, A. D. ANDERSON & D.V. PARKE. 1991. The toxicity of benzene and its metabolism and molecular pathology in human risk assessment. Br. J. Ind. Med. **48:** 437–444.
5. YOON, B.I., Y. HIRABAYASHI, Y. KAWASAKI, *et al.* 2001. Mechanism of action of benzene toxicity: cell cycle suppression in hemopoietic progenitor cells (CFU-GM). Exp. Hematol. **29:** 278–85.
6. HIRAKU, Y. & S. KAWANISHI. 1996. Oxidative DNA damage and apoptosis induced by benzene metabolites. Cancer Res. **56:** 5172–5178.
7. TUO, J., S.P. WOLFF, S. LOFT, *et al.* 1998. Formation of nitrated and hydroxylated aromatic compounds from benzene and peroxynitrite, a possible mechanism of benzene genotoxicity. Free Radic. Res. **28:** 369–375.
8. GUBERAN, E. & P. KOCHER. 1971. Long-term prognosis of chronic benzene poisoning checking of a population ten years after exposure. Schweiz. Med. Wochenschr. **101:** 1789–1790.

9. DROGICHINA, E.A., L.A. ZORINA & I.A. GRIBOVA. 1971. Clinical picture and prognosis of nervous system changes in chronic benzene poisoning. Gig. Tr. Prof. Zabol. **15:** 18–21.
10. LIU, W., D.I. SMITH, K.J. RECHTZIGEL, et al. 1998. Denaturing high performance liquid chromatography (DHPLC) used in the detection of germline and somatic mutations. Nucleic Acids Res. **26:** 1396–1400.
11. KUKLIN, A., K. MUNSON, D. GJERDE, et al. 1997–98. Detection of single-nucleotide polymorphisms with the WAVE DNA fragment analysis system. Genet. Test **1:** 201–206.
12. WAN, J., J. SHI, L. HUI, et al. 2002. Association of genetic polymorphisms in CYP2E1, MPO, NQO1, GSTM1, and GSTT1 genes with benzene poisoning. Environ. Health Perspect. **110:** 1213–1218.
13. SEATON, M.J., P.M. SCHLOSSER, J.A. BOND, et al. 1994. Benzene metabolism by human liver microsomes in relation to cytochrome P450 2E1 activity. Carcinogenesis **15:** 1799–1806.
14. RICKERT, D.E., T.S. BAKE, J.S. BUS, et al. 1979. Benzene disposition in the rat after exposure by inhalation. Toxicol. Appl. Pharmacol. **49:** 417–423.
15. GREENLEE, W.F., J.D. SUN & J.S. BUS. 1981. A proposed mechanism of benzene toxicity: formation of reactive intermediates from polyphenol metabolites. Toxicol. Appl. Pharmacol. **59:** 187–195.
16. THOMAS, D.J., A. SADLER, V.V. SUBRAHMANYAM, et al. 1990. Bone marrow stromal cell bioactivation and detoxification of the benzene metabolite hydroquinone: comparison of macrophages and fibroblastoid cells. Mol. Pharmacol. **37:** 255–262.
17. LEVAY, G.D., W. ROSS & J. BODELL. 1993. Peroxidase activation of hydroquinone results in the formation of DNA adducts in HL-60 cells, mouse bone marrow macrophages and human bone marrow. Carcinogenesis **14:** 2329–2334.
18. IRONS, R.D. 1985. Quinones as toxic metabolites of benzene. J. Toxicol. Environ. Health **16:** 673–678.
19. GLASS, D.C., C.N. GRAY, D.J. JOLLEY, et al. 2003. Leukemia risk associated with low-level benzene exposure. Epidemiology **14:** 569–577.
20. RINSKY, R.A., A.B. SMITH, R. HORNUNG, et al. 1987. Benzene and leukemia. An epidemiologic risk assessment. N. Engl. J. Med. **316:** 1044–1050.
21. YIN, S.-N., G.L. LI, F.D. TAIN, et al. 1987. Leukaemia in benzene workers: a retrospective cohort study. Br. J. Ind. Med. **44:** 124–128.
22. HAYES, R.B., S.-N. YIN, M. DOSEMECI, et al. 2001. Benzene and lymphohematopoietic malignancies in humans. Am. J. Ind. Med. **40:** 117–126.

Gene–Environment Interactions in Cancer

Do They Exist?

KARI HEMMINKI,[a,b] ASTA FÖRSTI,[a,b] AND JUSTO LORENZO BERMEJO[a]

[a]*Division of Molecular Genetic Epidemiology, German Cancer Research Center (DKFZ), Im Neuenheimer Feld 580, D-69120 Heidelberg, Germany*

[b]*Center for Family Medicine, Karolinska Institute, 141 83 Huddinge, Sweden*

> ABSTRACT: Single nucleotide polymorphisms (SNPs) are extensively used in case–control studies of practically all cancer types. In addition to the pure genetic studies, gene–environment studies, which simultaneously consider environmental factors, have been increasingly conducted. All SNP studies aim at the identification of the role of inherited cancer susceptibility genes. However, being genetic markers, they are applicable only on heritable conditions, which is often a neglected fact. Based on the data on the heritability of cancer and the importance of environmental factors in cancer etiology, we discuss the likelihood of successful gene–environment studies. The available evidence is not conclusive, but it consistently points to a minor heritable etiology in cancer, which will hamper the success of SNP-based association studies. We use simulation techniques to examine which situations would favor the application of a gene–environment approach instead of the traditional environmental approach in case–control studies. The results show that well-chosen candidate gene with a relatively low allele frequency may improve the power to detect environmental determinants of a disease. However, this advantage is lost when the number of underlying genes increases. We are concerned about an indiscriminate use of genetic tools for cancers, which are mainly environmental in origin. The likelihood of success for SNP-based gene–environment studies increases if established environmental risk factors are tested on proven candidate genes. Enhancing the likelihood that the disease causation is genetic, for example, by selecting familial cases, may increase the power of the studies, and the rareness of those cases calls for collaborative networks.
>
> KEYWORDS: SNP; case–control study; statistical power; heritability; twin

Address for correspondence: Kari Hemminki, Division of Molecular Genetic Epidemiology, German Cancer Research Center (DKFZ), Im Neuenheimer Feld 580, D-69120 Heidelberg, Germany. Voice: 49-6221-421800; fax: 49-6221-421810.
e-mail: k.hemminki@dkfz.de

INTRODUCTION

Genetic association studies on cancer apply usually a case–control design and use single nucleotide polymorphisms (SNPs) in candidate genes or adjacent loci as markers. In the simplest form, a SNP with a known or assumed function is selected, and the genotypes are determined in cases and controls to test for association. In the more complex variant, the "gene–environment study," genotypes are tested with various environmental factors for possible interactions.[1] Gene–environment studies may search for new etiologies (when the effects of the tested genes or environmental exposures, or both, are unknown), or they may explore the mechanisms of cellular action of known environmental factors, such as smoking. Any application of SNPs assumes that heritable variations in gene functions convey cancer risk, that is, they assay for heritable effects.[2,3] Heritable etiology of cancer may not be overwhelming, and the use of genetic tools to dissect causation of a primarily environmental disease may not be successful. This may be one of the fundamental reasons for the lack of proof-of-principle results on gene–environment interactions in cancer; however, large case–control studies may help to offer some evidence.[4] To define the terms, we use "genetic" synonymously with "heritable," for genetic traits transmitted in the germ line. In the literature, "genetic" often refers to any changes in DNA, including somatic events, which are important in cancer. Whether heritable events are also important in cancer will be discussed in the next section.

The fact that "only 10% of smokers are diagnosed with lung cancer" is often cited as evidence for inherited differences in susceptibility to tobacco carcinogenesis.[5] This figure is lower than the 24% estimated for male smokers by an IARC working group and it fails to consider the stochastic nature of the carcinogenic process.[5] Time-dependent stochastic effects and epigenetic changes probably play an important role in long-term animal bioassays, in which only a fraction of inbred animals contracts cancer when exposed to a constant level of carcinogen. Recent data show interesting epigenetic differences in monozygotic twins.[6] Studying methylation patterns and histone acetylation, the authors found an age-dependent divergence in these indices. The patterns were identical at birth, but they diverged at advanced ages to levels that distinguished gene expression levels in twin pairs.

In the following, we discuss first, heritability as an underpinning for gene–environment studies. We use then simulation techniques to investigate under which situations gene–environment studies may exceed the power of mere environmental studies. Feasibility of gene–environment studies should not be discussed without considering the unsolved, two-dimensional (selection of genes and selection of environmental factors) multiple comparisons problems. However, we refer to a recent paper with a short introduction to the problem.[7]

HERITABLE VS. ENVIRONMENTAL CAUSES OF CANCER

The heritability data of human diseases is often based on twin studies, which aim at estimating the proportion of the total phenotypic variance, which is attributable to genetic effects.[8] If the disease is analyzed as a binary trait (affected with cancer/unaffected), the individual probability to develop cancer (liability) is usually invoked to calculate heritability. Genetic, shared environmental and nonshared environmental effects are then estimated from concordance of cancer in monozygotic and dizygotic twins.[3,9] There are two important qualifications to the obtained results. First, the models applied do not allow interactions; this condition is probably violated to a variable degree in all twin studies. Second, as specified in the definition of heritability, the method only assesses the proportion of total phenotypic variation due to genetic effects; that is, the heritability relies on the phenotypic variation in the population of twins and may be different for the general population.

Twin studies, albeit their limitations, are nevertheless the most direct way to examine disease etiology. According to the Nordic twin study, the estimated heritabilities for site-specific cancers were only significant for the colorectum (35%), breast (27%), and prostate (42%).[10] Nonshared, random environment was the major contributor to all cancers, in line with the other evidence, discussed below, on the importance of environmental effects in cancer. The estimates were derived from concordance rates for cancer in twins according to zygosity. Twin concordance by the age of 75 years was only 11% for colorectal, 13% for breast, and 18% for prostate cancer in monozygotic twins, but these were higher than the corresponding percentages in dizygotic twins (5%, 9%, and 3%, respectively).[10] For other cancer sites concordance rates were lower, but the number of affected twin pairs was low. Monozygotic twins share 100% of their genes and they are exposed to common environmental factors, particularly early in life, more than any other pair of human beings. Even if the heritability estimates are imprecise, these data provide little support to strong heritable effects or gene–environment interactions in cancer. Similar modeling has been applied to other types of familial relationships, for example, to sibling pairs affected by cancer; the analyses resulted in even lower heritability estimates.[11]

Inherited cancer syndromes of high penetrance and probably no appreciable environmental influence are thought to account for 1–2% of cancers, at most.[12] Low penetrant familial cancers may amount to 10% of cancers, but the proportion depends on the definition, for example, whether only first-degree or more distant relatives are considered. Familial risks for most cancers are around two and familial attributable fractions range from 9.1% for prostate cancer to 0.2% for connective tissue tumors.[13] Thus, only a small fraction of cancer would be prevented if all familial cancers were avoided. If environmental effects were excluded or quantified, the attributable fractions from family studies would reflect the genetic determination of cancer. However, because of low penetrance, the results underestimate true heritable effects. The available data suggest that apart from prostate, breast, and colorectal cancer, the contribution of the familial risk observed in nuclear families to the etiology of cancer is limited. Familial aggregation of cancer has been observed over many generations in Icelandic and Utah pedigrees.[14,15] However, in Iceland the risk for spouses exceeded those for second-degree relatives in lung, stomach, and colon cancer, implying contribution of environmental effects to the familial clustering.

The familial aggregation of cancer is a direct measure of a possible heritable cause and, if it is lacking or small, the likelihood of a heritable influence is also small.[3] There are direct indicators of the importance of environmental factors, which, however, may not rule out the operation of gene–environment interactions, but they should be at least considered before studies are launched. Incidence changes upon time in migration studies point to a predominant environmental contribution to cancer causation.[16–19] These data are in line with large international differences in the incidence of cancer, which for almost any site are at least 10-fold.[20] Moreover, there have been major incidence changes in single regions. For example, during the operation of Swedish Cancer Registry, from 1958 to 2003, the incidence of male melanoma has increased 7.7-fold, squamous cell skin cancer 4.1-fold, prostate cancer and non-Hodgkin's lymphomas both 3.2-fold, and breast cancer 2.2-fold. In the same period, the incidence of male gastric cancer has decreased 3.4-fold. The incidence changes in Sweden are not unique and they are found in other registration systems of long follow-up.[21]

WHEN DO POWER CONSIDERATIONS FAVOR A GENE–ENVIRONMENT APPROACH

Being popular, gene–environment studies are likely to offer some advantages over environmental studies, more than just the hype with genetic markers. Here we address this question by comparing the statistical power of a study assessing the effect of an environmental risk factor (environmental study) with the power of a study that simultaneously considers a susceptibility gene (gene–environment study). FIGURE 1 shows the population stratification by exposure status and genotype, assuming exposure and allele frequencies of 0.1. For simplicity, we assume that the exposure only increases the risk of cancer among carriers of variant forms of the gene (A). In addition, we assume a dominant genetic effect, that is, exposed genotypes Aa and AA are at increased risk of cancer. In the environmental study, individuals are classified as exposed (10%) or unexposed (90%). In the gene–environment study, the proportion of individuals at high risk is $0.1 \cdot (1-(0.9 \cdot 0.9)) = 1.9\%$. FIGURE 1 shows that rare risk genotypes limit the risk population, which may be a statistical disadvantage.

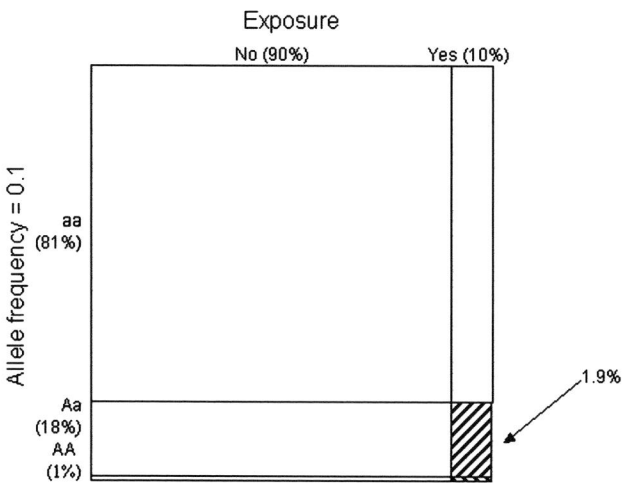

FIGURE 1. A scheme showing how a population is divided when the exposure frequency is 10% and allele frequency is also 10%. These frequencies are used in the later examples.

On the other hand, cases preferentially enrich in these genotypes, which is a statistical advantage.

We consider first the likelihood of identifying the effect of the environmental exposure E on cancer risk. Let p_E represent the exposure frequency, OR the odds ratio for exposed versus unexposed individuals, and f_0 the disease penetrance among unexposed individuals. In order to estimate the power to detect the exposure effect, exposure statuses were simulated for 500 cases and 500 controls under a range of parameter settings (any p_E; OR = 1.2, 1.5, or 2.0; $f_0 = 5\%$). For each parameter combination, 100,000 data sets were generated using the SAS function RANUNI. The simulated data were analyzed using the case–control study design. The power to detect an association between environmental exposure and cancer risk was estimated by the proportion of studies, which resulted in a significant exposure effect at the 5% confidence level. The results are presented in FIGURE 2. When 10% of the individuals were exposed, the power to detect a significant effect was 95% for an OR of 2, 50% for an OR of 1.5, and 12% for an OR of 1.2. At an exposure frequency of 0.5, an OR of 1.2 attained a power of 30%.

Similar calculations were carried out for a hypothetical susceptibility allele A with frequency p_A and genotype relative risk GRR. We considered two penetrance models: all carriers of the allele A were at increased risk (dominant model), or only AA homozygotes showed increased susceptibility (recessive model). Genotypes (carrier/noncarrier) were simulated for 500 cases and 500 controls under several scenarios (any p_A; GRR = 1.2, 1.5, or 2.0; $f_0 = 5\%$; dominance or recessiveness). Hardy–Weinberg equilibrium was assumed. The

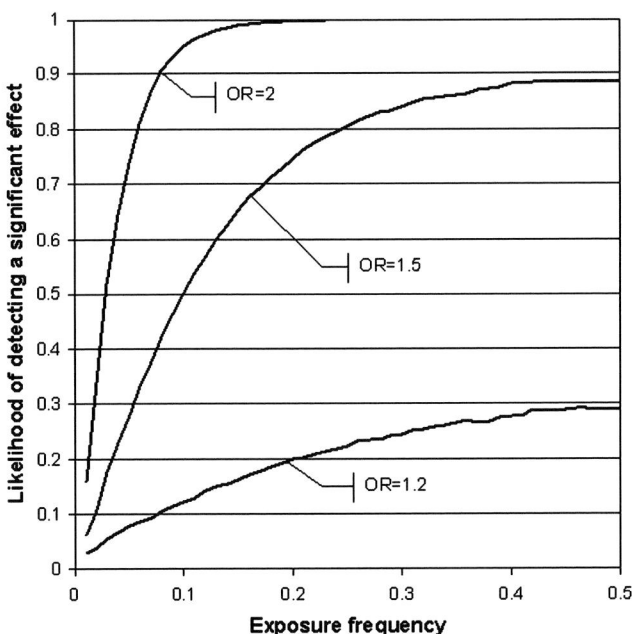

FIGURE 2. Power of a case–control study (probability of detecting the effect of an environmental risk factor) according to exposure frequency and odds ratio for environmental exposure (OR). The assumed sample size was 500 cases plus 500 controls. The type I error probability was fixed at 0.05.

power to identify an association between the allele A and cancer risk was calculated as the proportion of the 100,000 simulated data sets that resulted in a significant gene effect at the 5% confidence level. Results are shown in FIGURE 3. The highest power was reached for intermediary allele frequencies, 100% power for a wide range of allele frequencies when the GRR equaled 2. For a GRR of 1.2, an allele frequency of 0.3 attained a 32% power. The dependence of power on the penetrance model of the allele is also shown in FIGURE 4. When the GRR was 1.5, a power higher than 90% was reached for allele frequencies between 0.2 and 0.34 (dominant allele), and for allele frequencies between 0.59 and 0.75 (recessive allele) (FIG. 4).

We also explored the power of case–control studies when the environmental exposure only increases the risk of cancer among certain genotypes. A dominant penetrance model was assumed, that is, exposed carriers of the allele A were at increased risk; exposed noncarriers and unexposed individuals developed cancer with probability $f_0 = 5\%$. Realistic data on the odds ratio for high-risk versus low-risk individuals (ORr) are sparse in the literature. In contrast, there is increasing information about the magnitude of environmental exposures (OR). To make use of available information, we fixed the OR and

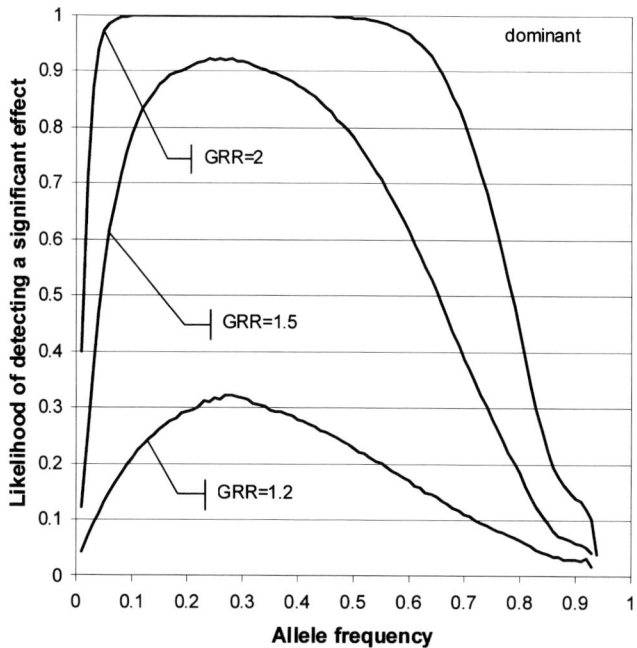

FIGURE 3. Power of a case–control study (probability of detecting the effect of a susceptibility allele) according to allele frequency and genotype relative risk (GRR). The assumed sample size was 500 cases plus 500 controls. The type I error probability was fixed at 0.05.

examined the relationship between allele frequency and ORr. The environmental effect was assumed to be relatively weak (OR = 2, 1.5, or 1.2), since environmental exposures with strong effects would result in a high power and less interesting scenarios. FIGURE 5 shows the results. If only exposed carriers of the allele A with frequency $p_A = 0.1$ were at increased risk of cancer, the ORr = 7.7 resulted in a marginal exposure effect OR = 2; ORr = 3.9 resulted in OR=1.5 and ORr = 2.1 resulted in OR = 1.2.

In order to explore the relevance of consideration of gene–environment interactions, we simulated the exposure statuses and genotypes of 500 cases and 500 controls according to reasonable parameter settings (any p_A; $p_E = 10\%$; OR = 1.2, 1.5, or 2.0; $f_0 = 5\%$). Dominance and Hardy–Weinberg equilibrium were assumed. For each of the 100,000 simulated data sets, the exposure odds ratio among susceptible genotypes was compared with the exposure odds ratio among noncarriers using the Breslow–Day test of homogeneity.[22] The power to detect a modulation of the environmental effect by the individual genotype was estimated as the percentage of tests, which resulted in rejection of the null hypothesis "the two odds ratios are homogeneous" at the 5% confidence level. The results are presented in FIGURE 6. When the OR was 1.5 and the allele

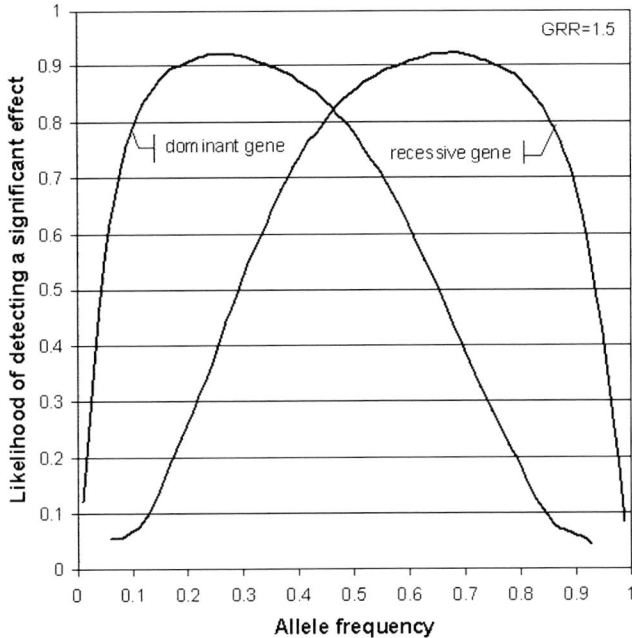

FIGURE 4. Power of a case–control study (probability of detecting the effect of a susceptibility allele) according to allele frequency, genotype relative risk (GRR), and inheritance mode, according to a dominant and recessive mode of inheritance. The assumed sample size was 500 cases plus 500 controls. The type I error probability was fixed at 0.05.

frequency was 0.2, the power to detect the modulation of the exposure effect by genotype was about 96%. Under similar conditions, the power to detect a significant exposure effect was 50% (see arrow, which is obtained from FIG. 2). This analysis shows that gene–environment study is more powerful than the purely environmental study when allele frequency of the tested candidate gene is low.

The present exercise shows that the consideration of candidate genes may, under optimal conditions, boost the power to detect environmental determinants of a disease. However, our example is based on the unlikely assumption that the environmental effect is modified by a single gene that the scientist luckily was able to select. Under more likely situations, tens or hundreds of genes may interact and modify the carcinogenic effect of the environmental exposure. That may be the reason why tobacco smoking still has not been linked with interacting genes. Logistics of sample collection and SNP analysis are also a budgetary issue of major concern. We used a simple model, but variations in our model had relatively little impact compared to the effects of multiple genes.

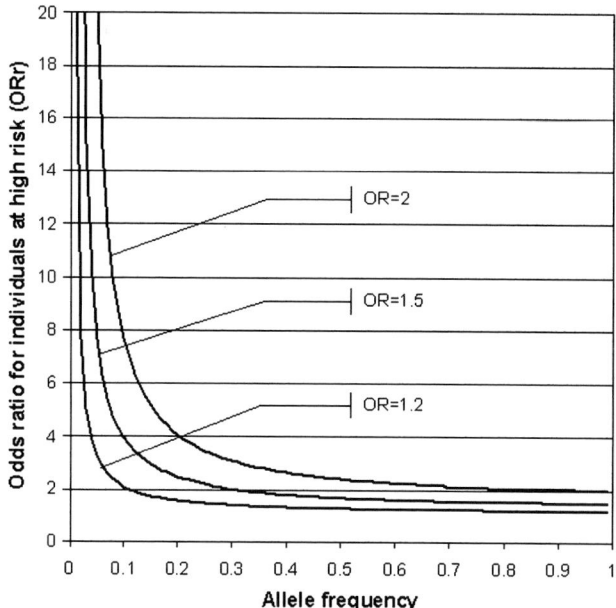

FIGURE 5. Odds ratio (ORr) for individuals at high risk (genotypes Aa and AA) that results in an odds ratio for environmental exposure (OR) equal to 2, 1.5, or 1.2, according to the allele frequency. The disease prevalence among individuals at low risk was fixed at 5%.

CONCLUSIONS

Data based on regional differences in cancer incidence, changes in incidence upon migration, and correlations of risk between twins and other family members consistently point to a major environmental etiology in all common cancers.[10,11,23] We find little evidence that "gene–environmental interactions underlie almost all cancers" and even in some assumed gene–environmental interactions, there is a tendency to ascribe them to genetic rather than to environmental causes.[24,25] The large shifts in cancer incidence are clearly driven by mostly undefined and probably numerous environmental factors. If genes interact with the environment, it is likely that a plethora of genes is involved, posing a research task of enormous complexity, probably remaining largely unresolved.[7] The practical implication is that many gene–environment interactions will not be detected, even if they existed. This may not be a great surprise to people working in the field and trying to encourage students who are disappointed because of negative results. Gene–environment interactions should be primarily attempted with known environmental causes or known genes, to establish credibility in the field. Gene–environment studies should not be used as a temporary escape because of the problems faced by etiological

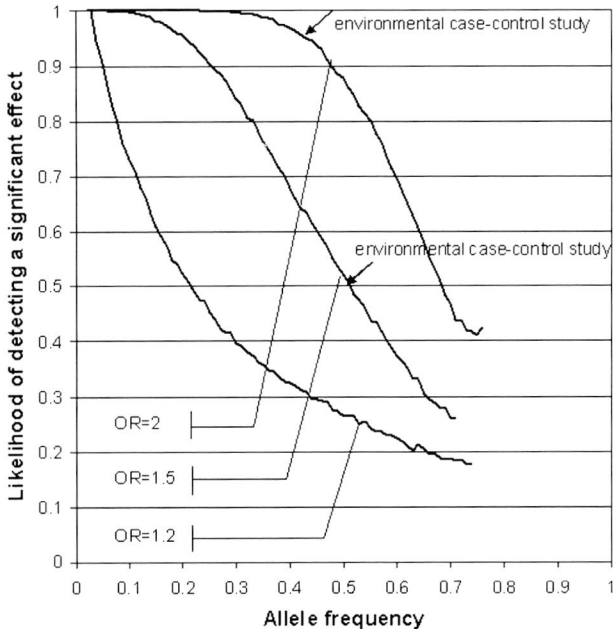

FIGURE 6. Power of a case–control study to detect a gene–environment interaction (heterogeneity of environmental effects by genotype), according to allele frequency and odds ratio for environmental exposure (OR). The assumed parameters were: sample size 500 cases plus 500 controls, dominant gene, exposure frequency 10%, and disease prevalence among individuals at low risk 5%. The type I error probability was fixed at 0.05. The arrows show the power of a case–control study to identify the environmental effect (cf. FIG. 2). The arrow to the bottom curve would be close to the lower right corner.

cancer epidemiology and classical gene identification approaches. Enhancing the likelihood that the disease is genetic, for example, by analyzing familial, or even more preferably early onset familial cases, may increase the power of the studies.[26,27] But those cases are rare and collaborative efforts should be encouraged. We believe that a balanced assessment of the causes of cancer will permit to advance in etiological understanding, to target optimally research, and to help choosing correct research tools. In spite of the employed technological power, the success of using a genetic tool for a disease, which is mainly environmental may not be optimal and it needs to prove itself while funding bodies are still favorable.

ACKNOWLEDGMENTS

This work was supported by Deutsche Krebshilfe the Swedish Cancer Society and EU LSHC-CT-2004-503465.

REFERENCES

1. BRENNAN, P. 2002. Gene-environment interactions and aetiology of cancer: what does it mean and how can we measure it. Carcinogenesis **23:** 381–387.
2. HEMMINKI, K., A. FÖRSTI & J.L. BERMEJO. 2005. Single nucleotide polymorphisms (SNPs) are inherited from parents and they measure heritable events. J. Carcinog. **4:** 2.
3. BURTON, P., M. TOBIN & J. HOPPER. 2005. Key concepts in genetic epidemiology. Lancet **366:** 941–951.
4. Garcia-Closas, M. et al. 2005. NAT2 slow acetylation, GSTM1 null genotype, and risk of bladder cancer: results from the Spanish Bladder Cancer Study and meta-analyses. Lancet **366:** 649–659.
5. IARC. 2004. IARC Monographs on the Evaluation of carcinogenic Risk to Numans, Vol. 83 Tobacco Smoke and Involuntary Smoking. IARC. Lyon.
6. FRAGA, M.F. et al. 2005. Epigenetic differences arise during the lifetime of monozygotic twins. Proc. Natl. Acad. Sci. USA **102:** 10604–10609.
7. HEMMINKI, K. & J.L. BERMEJO. 2005. Relationships between familial risks of cancer and the effects of heritable genes and their SNP variants. Mut. Res. **592:** 6–17.
8. SHAM, P. 1998. Statistics in Human Genetics. John Wiley & Sons. New York.
9. RISCH, N. 2001. The genetic epidemiology of cancer: interpreting family and twin studies and their implications for molecular genetic approaches. Cancer Epidemiol. Biomarkers Prev. **10:** 733–741.
10. LICHTENSTEIN, P. et al. 2000. Environmental and heritable factors in the causation of cancer. N. Engl. J. Med. **343:** 78–85.
11. CZENE, K., P. LICHTENSTEIN & K. HEMMINKI. 2002. Environmental and heritable causes of cancer among 9.6 million individuals in the Swedish Family-Cancer Database. Int. J. Cancer **99:** 260–266.
12. PONDER, B. 2001. Cancer genetics. Nature **411:** 336–341.
13. HEMMINKI, K. & K. CZENE. 2002. Attributable risks of familial cancer from the family-cancer database. Cancer Epidemiol. Biomarkers Prev. **11:** 1638–1644.
14. AMUNDADOTTIR, L.T. et al. 2004. Cancer as a complex phenotype: pattern of cancer distribution within and beyond the nuclear family. PLoS Med. **1:** e65.
15. KERBER, R.A. & E. O'BRIEN. 2005. A cohort study of cancer risk in relation to family histories of cancer in the Utah population database. Cancer **103:** 1906–1915.
16. DOLL, R. & R. PETO. 1981. The causes of cancer. J. Natl. Cancer Inst. **66:** 1191–1309.
17. PARKIN, D.M. & M. KHLAT. 1996. Studies of cancer in migrants: rationale and methodology. Eur. J. Cancer **32A:** 761–771.
18. HEMMINKI, K., X. LI & K. CZENE. 2002. Cancer risks in first generation immigrants to Sweden. Int. J. Cancer **99:** 218–228.
19. HEMMINKI, K. & X. LI. 2002. Cancer risks in second-generation immigrants to Sweden. Int. J. Cancer **99:** 229–237.
20. IARC. 2002. Cancer Incidence in Five Continents. D.M. Parkin, et al. Eds.: 781 pages. IARC Scientific Publication No. 155. LYON, France.
21. POLEDNAK, A. 1994. Trends in cancer incidence in Connecticut, 1935–1991. Cancer **74:** 2863–2872.
22. DOS SANTOS SILVA, I. 1999. Cancer Epidemiology: Principles and Methods. IARC. Lyon.

23. PETO, J. 2001. Cancer epidemiology in the last century and the next decade. Nature **411:** 390–395.
24. WILLETT, W. 2002. Balancing life-style and genomics research for disease prevention. Science **296:** 695–698.
25. VINEIS, P. 2004. A self-fulfilling prophecy: are we underestimating the role of environment in gene-environment interaction research? Int. J. Epidemiol. **33:** 945–946.
26. HOULSTON, R.S. & J. PETO. 2003. The future of association studies of common cancers. Hum. Genet. **112:** 434–435.
27. ANTONIOU, A.C. & D.F. EASTON. 2003. Polygenic inheritance of breast cancer: implications for design of association studies. Genet. Epidemiol. **25:** 190–202.

Manufactured Uncertainty

Protecting Public Health in the Age of Contested Science and Product Defense

DAVID MICHAELS

Department of Environmental and Occupational Health, The George Washington University School of Public Health and Health Services, Washington DC 20037, USA

ABSTRACT: The strategy of "manufacturing uncertainty" has been used with great success by polluters and manufacturers of dangerous products to oppose public health and environmental regulation. This strategy entails questioning the validity of scientific evidence on which the regulation is based. While this approach is most identified with the tobacco industry, it has been used by producers of asbestos, benzene, beryllium, chromium, diesel exhaust, lead, plastics, and other hazardous products to avoid environmental and occupational health regulation. It is also central to the debate on global warming. The approach is now so common that it is unusual for the science *not* to be challenged by an industry facing regulation. Manufacturing uncertainty has become a business in itself; numerous technical consulting firms provide a service often called "product defense" or "litigation support." As these names imply, the usual objective of these activities is not to generate knowledge to protect public health but to protect a corporation whose products are alleged to have toxic properties. Evidence in the scientific literature of the funding effect—the close correlation between the results of a study desired by a study's funder and the reported results of that study—suggests that the financial interest of a study's sponsors should be taken into account when considering the study's findings. Similarly, the interpretation of data by scientists with financial conflicts should be seen in this light. Manufacturing uncertainty is antithetical to the public health principle that decisions be made using the best evidence currently available.

KEYWORDS: uncertainty; certainty; doubt; regulation; product defense; litigation support; funding effect; public health

Address for correspondence: David Michaels, Ph.D., M.P.H., Department of Environmental and Occupational Health, The George Washington University School of Public Health and Health Services, 2100 M St. NW, Suite 203, Washington DC, 20037. Voice: 202-994-2461; fax: 202-994-0011.
e-mail: eohdmm@gwumc.edu

INTRODUCTION

In March 2002, a nuclear reactor near Toledo, OH, came within a quarter inch of a major radiation release, possibly the worst accident of this type in U.S. history. Water mixed with boric acid had eaten through six inches of carbon steel, leaving only a thin layer of stainless steel to contain the water in the Davis–Besse nuclear reactor's vessel head. When finally seen by safety inspectors, that last steel layer was bulging, barely able to contain the highly pressurized coolant.

Three months earlier, two other reactors had developed similar cracks. After studying the situation, experts at the U.S. Nuclear Regulatory Commission (NRC) predicted a high probability of finding cooling system breaches at the Ohio plant and asked operators of all similar reactors to shut down voluntarily and inspect for damage. The operator of Davis–Besse refused and NRC staff prepared an order demanding that the reactor be shut down and inspected. But that order was never issued. Desiring to protect the financial health of the operator, the NRC manager demanded "absolute proof" that the vessel head was damaged before he would order a shut down and inspection, proof that could only be obtained with the shut down and inspection.[1]

Absolute certainty in the realm of medicine and public health is rare. Our public health programs will not be effective if absolute proof is required before we act; the best available evidence must be sufficient. Yet we see a growing trend that demands *proof* over precaution in the realm of public health.[2]

Few scientific challenges are more complex than understanding the cause of disease in humans. Scientists cannot feed toxic chemicals to people to see what dose causes cancer. Instead, we must harness the "natural experiments" where exposures have already happened in the field. In the laboratory, we can use only animals. Both epidemiologic and laboratory studies have many uncertainties, and scientists must extrapolate from study-specific evidence to make causal inferences and recommend protective measures. Absolute certainty is rarely an option. Our regulatory programs will not be effective if such proof is required before we act; the best available evidence must be sufficient.

THE TOBACCO ROAD

Years ago, a tobacco executive unwisely committed to paper the perfect slogan for his industry's disinformation campaign: *"Doubt is our product."*[3] With tobacco, doubt turned out to be less addictive for the public than the leaf itself, and the industry finally abandoned its strategy.

I call this strategy "manufacturing uncertainty,"[4] and no industry manufactured more uncertainty over a longer period than the tobacco companies. Following a strategic plan developed in the mid 1950s by the public relations firm Hill and Knowlton, a firm that manufactured uncertainty on behalf of various industries over the course of decades, Big Tobacco hired scientists to challenge the growing consensus linking cigarette smoking with lung cancer and other adverse health effects. This industry campaign had three basic messages: Cause and effect relationships have not been established in any way; statistical data do not provide the answers; more research is needed. As recently as 1989, a spokeswoman appearing on national television dismissed claims that tobacco caused lung cancer as "...just statistics. The causal relationship between smoking and cancer has not yet been established."[5]

The industry even started its own "scientific" publication, *Tobacco and Health Research*, for which the main criterion for articles was straightforward: "...the most important type of story is that which casts doubt on the cause and effect theory of disease and smoking." Editorial guidelines stated that headlines "should strongly call out the point—Controversy! Contradiction! Other Factors! Unknowns!"[6]

Doubt turned out to be less addictive for the public than tobacco, and the industry finally abandoned its strategy. Thanks to its efforts, however, public health protections and compensation for tobacco's victims were delayed for decades. The practices perfected by tobacco executives and public relations are alive and well.

Learning from tobacco, other industries have discovered that debating the science is much easier and more effective than debating the policy. Witness the debate over global warming. Many studies link human activity, especially burning of carbon fuels, with global warming.[7] Waiting for absolute certainty that the accumulation of greenhouse gases will result in dramatic changes in the climate seems far riskier and potentially far more expensive to address than acting now to control the causes of global warming. Opponents of preventive action, led by the fossil fuels industry, avoid this policy debate by challenging the science instead with a classic uncertainty campaign. I need only cite a memo from the political consultant Frank Luntz, delivered to his clients in early 2003. In "Winning the Global Warming Debate," Luntz wrote:

> Voters believe that there is *no consensus* about global warming within the scientific community. Should the public come to believe that the scientific issues are settled, their views about global warming will change accordingly. Therefore, *you need to continue to make the lack of scientific certainty a primary issue in the debate*... The scientific debate is closing [against us] but not yet closed. There is still a window of opportunity to challenge the science. (emphasis in original)[8]

There has been substantial media coverage of the political machinations behind the global warming debate, and we all know about the behavior of the tobacco industry. Less well known are the campaigns mounted to question studies documenting the adverse health effects of exposure to beryllium, lead, mercury, vinyl chloride, chromium, benzene, benzidine, nickel, and a long list of other toxic chemicals and pharmaceuticals. In fact, it is unusual for the science behind any proposed public health or environmental regulation *not* to be challenged, no matter how powerful the evidence.

How ridiculous can it get? There is widespread agreement in the scientific community that broad-spectrum ultraviolet (UV) radiation, whether from sunlight or from tanning lamps, causes skin cancer. Yet trade associations representing the indoor tanning industry have attempted to derail the "cancer-causing" designation by questioning the scientific evidence.[9]

Manufacturing uncertainty on behalf of big business has become a big business in itself. The "product defense" firms have become experienced, adept, and successful consultants in epidemiology, biostatistics, and toxicology. The work of these product defense firms bears the same relationship to science as the Arthur Andersen Company does to accounting—or did before it went bankrupt following the Enron debacle.

BERYLLIUM: NATIONAL DEFENSE OR "PRODUCT DEFENSE"?

The metal beryllium is extremely useful and almost unimaginably toxic. Breathing the tiniest amount of this lightweight metal can cause disease and death. As a neutron moderator that increases the yield of nuclear explosions, beryllium is vital to the production of weapon systems. Throughout the cold war, the U.S. nuclear weapons complex was the nation's largest consumer of the substance. As a result, however, hundreds of weapons workers have developed chronic beryllium disease (CBD). It is not just machinists who work directly with the metal who develop CBD, but also others simply in the vicinity of the milling and grinding processes, often for very short periods of time, and even people living near beryllium factories.

As Assistant Secretary of Energy for Environment, Safety, and Health from 1998 to 2001, I was the chief safety officer for the nuclear weapons complex, responsible for protecting the health of workers, the communities, and the environment around the production and research facilities. In 1998 the Department of Energy's (DOE) exposure standard had been unchanged for almost 50 years, and there were hundreds of cases of beryllium disease in the nuclear weapons complex and in factories that supplied beryllium products.

The history of this original DOE beryllium standard is legendary. It was developed in a 1948 discussion held in the back seat of a taxi by Merril Eisenbud, an Atomic Energy Commission (AEC) industrial hygienist, and Willard Machle, a physician who was a consultant to the firm building the Brookhaven Laboratory in Long Island, New York. Eisenbud discusses this history in his autobiography, noting that they selected the exposure limit "in the absence of an epidemiological basis for establishing a standard"[10] (p. 55). The AEC "tentatively" adopted a standard of 2 ug/m^3 in 1949, and then reviewed it annually for 7 years before permanently accepting it.

When first implemented, the 2 ug/m^3 standard resulted in a dramatic decrease in new beryllium disease cases. But by 1951, Eisenbud recognized that the the distribution of the chronic form of beryllium disease did not follow the usual exposure-response model seen for most toxic substances and hypothesized an immunological susceptibility.[11] It was not long before CBD was seen among workers hired after the 1949 standard went into effect, and whose exposure appeared to be below the 2 ug/m^3 standard.[12] Moreover, CBD had been diagnosed in persons with no workplace exposure to the metal, including individuals who simply laundered the clothes of workers, drove a milk delivery truck with a route near a beryllium plant, or tended cemetery graves near a beryllium factory.[12]

When the Occupational Safety and Health Administration (OSHA) was established in 1971 to protect the health of workers in the private sector, it simply adopted the taxicab standard. By the 1980s, however, it was clear that workers exposed to beryllium levels well below the standard were developing disease. As both the DOE and OSHA began the time-consuming legal process of changing their standards, the beryllium industry objected. At one public meeting, the Director of Environmental Health and Safety of Brush Wellman, the leading U.S. producer of beryllium products asserted (according to DOE's minutes of the meeting): "Brush Wellman is unaware of any scientific evidence that the standard is not protective. However, we do recognize that there have been sporadic reports of disease at less than 2 ug/m^3. Brush Wellman has studied each of these reports and found them to be scientifically unsound."[13]

In 1991, Brush managers were told that if they were "asked in some fashion whether or not the 2 ug/m^3 standard is still considered by the company to be reliable," they should answer "In most cases involving our employees, we can point to circumstances of exposure (usually accidental), higher than the standard allows. In some cases, we have been unable (for lack of clear history) to identify such circumstances. However, in these cases we also cannot say that there was not excessive exposure."[14]

This was the industry's primary argument, and it was based on a flawed logic. It was not difficult to go back into the work history of anyone with CBD and estimate that at some point in time, the airborne beryllium level must have exceeded the standard. Brush did this and then reasoned that the 2 ug/m^3 must be fully protective since most people who had CBD had, at some point, been exposed to levels above the standard.

The ever-increasing number of CBD cases identified at facilities across the nuclear weapons complex as well as in the beryllium industry's own factories rendered less plausible the claim that the old standard was safe. In September 1999 Brush Wellman sponsored a conference, in collaboration with the American Conference of Governmental Industrial Hygienists, to bring "leading scientists together to present and discuss the current information and new research on the hazards posed by beryllium"[15] (p. 527). The papers were subsequently published together in an industrial hygiene journal. Clearly, one purpose of the conference was to influence the government standard setting on beryllium: at the time of the conference, DOE was a few months away from issuing its final rule and OSHA had signaled its intention to revise its outdated standard.

Several papers were presented by scientists employed by Exponent, Inc., the beryllium industry's product defense consultant. These included a paper entitled "Identifying an Appropriate Occupational Exposure Limit (OEL) for Beryllium: Data Gaps and Current Research Initiatives" that promoted the industry's new rationale for opposing a new, stronger beryllium standard: that more research is needed on the effects of particle size, of exposure to beryllium compounds and of skin exposure to CBD risk. The paper concluded: "At this time, it is difficult to identify a single new TLV [threshold limit value] for all forms of beryllium that will protect nearly all workers. It is likely that within three or four years, a series of TLVs might need to be considered. ... In short, the beryllium OEL could easily be among the most complex yet established"[15] (p. 536).

After reviewing the public comments and the literature on beryllium's health effects, the DOE health and safety office concluded that, while more research is always desirable, we had more than enough information to warrant immediate implementation of a stronger beryllium disease prevention standard. Over the industry's objections, we issued a new rule, reducing the acceptable workplace exposure level by a factor of 10.

Simultaneously, OSHA also recognized the inadequacy of its own standard[16] and announced its commitment to issuing a stronger one.[17] However, when the George W. Bush Administration took office in 2001, the commitment to strengthening its beryllium rule was dropped from the agency's formal regulatory agenda.

In November 2002 OSHA implicitly accepted the industry's approach by issuing a call for additional data on the relationship of beryllium disease to, among other things, particle size, particle surface area, particle number, and skin contact.[18] In the few years since DOE issued its standard, however, researchers have published several epidemiologic studies that demonstrate that the 2.0 $\mu g/m^3$ standard does not prevent the occurrence of CBD.[19–22]

In addition to CBD, the scientific community widely recognizes that beryllium also increases the risk of lung cancer[23,24]; several studies conducted by epidemiologists at the Center for Disease Control (CDC) support this conclusion.[25–27] In 2002, however, scientists at a product defense firm published a 10-year-old reanalysis of one of the CDC studies.[28] By changing some parameters, the statistically significant elevation of lung cancer rates was no longer statistically significant. (Such alchemy is rather easily accomplished, of course, while the opposite—turning insignificance into significance, is extremely difficult.) Not coincidentally, this particular firm had done extensive work for the tobacco industry.[2] The new analysis was published in a peer-reviewed journal—not one with much experience in epidemiology, but peer-reviewed nevertheless, and the industry now touts its study as evidence that everyone else is wrong.

And so it goes today, in industry after industry, with study after study, year after year. Data are disputed, data have to be reanalyzed. Animal data are deemed not relevant, human data not representative, exposure data not reliable. More research is always needed. Uncertainty is manufactured. Its purpose is always the same: shielding corporate interests from the inconvenience and economic consequences of public health protections.

PPA: THE TRICKS OF THE TRADE

In order to attract new clients, some of these firms even brag about their successes. Until I wrote about it in *Scientific American*,[2] the Weinberg Group (another firm that had worked extensively for the tobacco industry) advertised on its web site its contribution to the effort to oppose the Food and Drug Administration's (FDA) belated clampdown on phenylpropanolamine (PPA), the over-the-counter drug that was widely used for decades as a decongestant and appetite suppressant.

Here is a short version of the history of PPA. Reports of hemorrhagic strokes in young women who had taken a PPA-containing drug began circulating in the early 1970s. Twenty years later, when the FDA finally raised official questions about the safety of PPA, the manufacturers rejected them. Eventually, a compromise was reached. The drug manufacturers would select an investigator and fund an epidemiologic study whose design would be jointly approved by the FDA. They chose the Yale University School of Medicine. In October 1999 the manufacturers and the FDA learned that the study confirmed the causal relationship between PPA and hemorrhagic stroke.[29] The study was published the following year in the *New England Journal of Medicine*.[30]

When they initially learned of the study's findings, did the manufacturers immediately withdraw this drug, which by then had annual sales of more than $500 million, but was responsible, according to an FDA analysis, of between 200 and 500 strokes per year among 18-to-49-year-olds?[31] No. Instead, they turned to the Weinberg Group to attack the study itself, focusing on "bias and areas of concern."[32] The manufacturers recognized that the FDA would eventually force the drug off the market, but they stalled for almost a year, enough time to reformulate their products. And when the FDA finally requested manufacturers to stop marketing PPA in November 2000, the industry was prepared to ship reformulated products immediately.[29]

Here is the full text of the web page on the Weinberg Group's work on PPA:

ADVERSE EVENT LINKED TO OTC PRODUCT

A pharmaceutical company retained THE WEINBERG GROUP to audit the results of a FDA-requested, industry-sponsored case-control study that linked their over-the-counter (OTC) product and several others with a serious, life-threatening adverse event. There was a substantial concern from the FDA based on reports of adverse events that use of these OTC products would present a public health problem. The study was commissioned to answer the question of risk with a controlled investigation. According to the study investigators, the results of the study showed a strong association between these products and a severe, life-threatening adverse event. Epidemiologists at THE WEINBERG GROUP led experts and consultants to some of the other affected OTC companies, in an effort that included a reanalysis of the raw data from the case-control study, and an assessment of the study's methodological flaws. The unique ability of the experts at THE WEINBERG GROUP to combine their expertise in epidemiology and biostatistics with strategic thinking enabled them to lead the pharmaceutical company's effort in their dispute with the FDA.[33]

THE FUNDING EFFECT

The biomedical literature extensively discusses the "funding effect," a term used to describe the close correlation between the results of a study desired by a study's funder and the reported results of that study.[34–36] Recent reviews in leading biomedical journals found that industry sponsorship was strongly associated with proindustry conclusions.[37,38]

The funding effect has also been seen in studies that look at the toxic effects of chemical exposures. The disparity between the results of studies examining the risk of lung cancer among beryllium-exposed workers discussed above is an example of the funding effect: Three government-funded analyses find an elevated risk while the one industry-funded analysis (actually reanalysis) does not.

An even more striking example in the toxicology literature is the debate over the effects of low-dose exposure to bisphenol A (BPA), an environmental estrogen used in the manufacture of polycarbonate plastic, a resin widely used in food cans and dental sealants. Exposure to BPA had been found in some studies to alter endocrine function at very low doses. In response, the American Plastics Council hired the Harvard Center for Risk Analysis (HCRA) to conduct a weight-of-the-evidence review of the toxicology. The HCRA panel reviewed 19 animal studies and reported that it found no consistent affirmative evidence of low-dose BPA effects.[39]

This conclusion was challenged by scientists who felt that the HCRA had chosen to examine only a minority of the 47 studies available at the time. These scientists reviewed the 115 published that had been published through December 2004 and found results that differed markedly with the HCRA analysis.[40]

As can be seen in TABLE 1, 90% (94 of 104) of the studies paid for with government funds reported an effect associated with BPA exposure; not a

TABLE 1. Biased outcome due to source of funding in low-dose *in vivo* BPA research as of December 2004

Source of funding	Number of studies & effect reported	
	Harm	No harm
Government	94	10
Chemical corporations	0	11
Total	94	22

Adapted from: Vom Saal, & Hughes.[40]

single one of the 11 corporate funded studies found an effect. The correlation between sponsor and result requires no test of statistical significance beyond Joseph Berkson's test of "interocular traumatic impact"—the results hit you right between the eyes.

VIOXX: CONFLICTED SCIENCE AND ITS CONSEQUENCES

I am not presuming here that the scientists involved in "manufacturing uncertainty" knowingly promote deadly products. More likely, scientists, along with the corporate executives and attorneys who hire them, convince themselves that the products they are defending are safe and that the evidence of harm is inaccurate, misleading, or trivial.

This can be seen in the recent evidence on the cardiac effects of Vioxx (rofecoxib), Merck & Co., Inc.'s blockbuster pain reliever that was taken off the market in November 2004, accompanied by headlines around the world. Even before the FDA approved Vioxx in May 1999, the agency reviewed data that suggested Vioxx could increase heart disease risk. Several independent scientists (i.e., not on Merck's payroll) also raised red flags, but for the most part, they were ignored by the FDA. Then the results of a clinical trial appeared in early 2000, just a few months after the drug was put on the market, linking Vioxx with an increased risk of heart attack. Merck had chosen naproxen (sold under the brand name Aleve) as the comparison treatment in the trial because aspirin, perhaps a more obvious choice, was known to lower cardiovascular disease risk, and the company did not want its trial to show more heart attacks among the study participants who took Vioxx. But the results showed that participants who took Vioxx for more than 18 months had five times the risk of heart attack as those taking naproxen.[41]

Merck's scientists faced a dilemma. They could interpret this finding to mean either that Vioxx increased heart attack risk by 400% or that naproxen was, like aspirin, beneficial in reducing the risk of heart attack by an astounding 80%. When a double-blind trial using a placebo control found seven excess heart attacks per every 1000 users per year, the correct interpretation was clear: Vioxx causes heart attacks. One FDA analysis estimates that Vioxx caused between 88,000 and 139,000 heart attacks, 30–40% of which were fatal, in the 5 years the drug was on the market.[42]

Subsequent litigation has uncovered memos documenting that Merck executives were concerned about the increased risk of heart attacks associated with Vioxx, but that they downplayed these concerns in their communications with physicians and resisted the FDA's efforts to add warnings to Vioxx's label.[43] It is hard to imagine that the drug maker's scientists were consciously promoting a product they knew would result in disease and death. At the same time, it is hard to imagine they honestly thought naproxen reduced the risk of heart attack by 80%. It seems more likely that their allegiances were so tightly linked with the products they have worked on, as well as the financial health of their employers, that their judgment became fatally impaired.

A NEW REGULATORY PARADIGM

The lessons of the past 40 or 50 years and the import of the government's actions over the past 4 years are clear. A new regulatory paradigm is needed. Federal agencies must ensure that data and scientific analyses provided by manufacturers are independently verified. Opinions submitted to regulatory agencies by corporate scientists and, especially, the product defense industry must be taken as advocacy primarily, not as science. Below are a few steps that begin to approach this new paradigm.

It has become apparent that some industry-supported research is never published because the sponsor did not like the results. Following a series of alarming instances in which the sponsors of research used their financial control to the detriment of the public's health, a group of leading biomedical journals established policies that make their published articles transparent to commercial bias and that require authors to accept full control and responsibility for their work.

These journals will now only publish studies done under contracts in which the investigators had the right to publish the findings without the consent or control of the sponsor. In a joint statement, the editors of the journals asserted that contractual arrangements allowing sponsor control of publication "erode the fabric of intellectual inquiry that has fostered so much high-quality clinical research"[44] (p. 1233).

But the federal regulatory agencies charged with protecting our health and environment have no similar requirements. When studies are submitted to the EPA or OSHA, for example, the agencies do not have the authority to inquire who paid for the studies or whether these studies would have seen the light of day if the sponsor did not approve the results.

Federal agencies should adopt, at a minimum, requirements for "research integrity" comparable to those used by biomedical journals: Parties submitting data from research they have sponsored must disclose if the investigators had the right to publish their findings without the consent or influence of the sponsor.[45]

It is also important to recognize that the opinions of virtually any scientist can be clouded by conflict of interest, even if it is not apparent to the scientist. Conflict of interest inevitably shapes judgment, and this must be factored into the consideration of the analyses and opinions of scientists employed by industry.

Public health is not well served by the unequal treatment of public and private science. While raw data from government-funded studies are generally available to private parties for inspection and reanalysis, enabling product defense experts to conduct *post hoc* analyses that challenge troubling findings, industry is under no obligation to release comparable raw data from their own studies. When private sponsors conduct research to influence public regulatory proceedings, these studies should be subject to the same access and reporting provisions as those applied to publicly funded science.[46]

Apologists for polluters and manufacturers of dangerous products commonly complain about government regulation, asserting that the agencies are not using "sound science." In fact, many of these manufacturers of uncertainty do not want "sound science"; they want something that sounds like science but lets them do exactly what they want to do.

We all recognize that the science is just one part of policy making. In shaping rules and programs to protect the public health and environment, decision makers also have to consider economic issues, values, and a host of other factors. In our current regulatory system, debate over science has become a substitute for debate over policy and the values upon which policy should be based.

Opponents of regulation use the existence of uncertainty, no matter its magnitude or importance, as a tool to counter imposition of public health protections that may cause them financial difficulty. It is important that those charged with protecting the public's health recognize that the desire for absolute scientific certainty is both counterproductive and futile. This recognition underlies the wise words of Sir Austin Bradford Hill delivered in an address to the Royal Society of Medicine in 1965:

> All scientific work is incomplete—whether it be observational or experimental. All scientific work is liable to be upset or modified by advancing knowledge. That does not confer upon us a freedom to ignore the knowledge we already have, or to postpone action that it appears to demand at a given time.
>
> Who knows, asked Robert Browning, but the world may end tonight? True, but on available evidence most of us make ready to commute on the 8:30 next day[47] (p. 300).

REFERENCES

1. OFFICE OF THE INSPECTOR GENERAL US NUCLEAR REGULATORY COMMISSION. NRC's Regulation of Davis-Besse Regarding Damage to the Reactor Vessel Head. Case No. 02-03S, Dec 30, 2002.
2. MICHAELS, D. 2005. Doubt is their product. Sci. Am. **292:** 96–101.
3. Smoking and health proposal. Brown & Williamson Document No. 332506. Available at http://legacy.library.ucsf.edu/tid/rgy93foo, accessed July 5, 2006.
4. MICHAELS, D. & C. MONFORTON. 2005. Manufacturing uncertainty: contested science and the protection of the public's health and environment. Am. J. Publ. Health **95:** S39–S48.
5. BRENNAN, D. Tobacco Institute, in a 1989 interview on ABC television's Good Morning America.
6. THOMPSON, C. Memorandum to William Kloepfer, Jr. and the Tobacco Institute, Inc. from Hill and Knowlton, Inc.; October 18, 1968. Tobacco Institute Document TIMN0071488-1491.
7. NATIONAL ACADEMY OF SCIENCES. 2003. Planning climate and global change research: a review of the draft U.S. climate change science program strategic plan. National Academies Press. Washington DC.
8. LUNTZ, F. Memo on the environment. Available at Environmental Working Group: http://www.ewg.org:16080/briefings/luntzmemo/pdf/LuntzResearch_environment.pdf, accessed July 5, 2006.
9. NATIONAL TOXICOLOGY PROGRAM, BOARD OF SCIENTIFIC COUNSELORS. 2000. Summary of minutes from the report on Carcinogens Subcommittee Meeting, December 13-15. Available at: http://ntp-server.niehs.nih.gov/ntp/htdocs/Liason/121300.pdf, accessed July 5, 2006.
10. EISENBUD, M. 1990. An Environmental Odyssey: people, Pollution, and Politics in the Life of a Practical Scientist. University of Washington Press. Seattle, WA.
11. STERNER, J.H. & M. EISENBUD. 1951. Epidemiology of beryllium intoxication. Arch. Industr. Hyg. Occup. Med. **4:** 123–151.
12. NATIONAL INSTITUTE FOR OCCUPATIONAL SAFETY AND HEALTH. Criteria for a recommended standard to beryllium exposure. DHEW HSM 72-10268 **1972:** IV–21.
13. January 1997 Beryllium Public Forum Albuquerque, NM. See www.eh.doe.gov/be/forumal.htm
14. Efficacy of the 2 ug/m^3 Standard. Exhibit B, Document ID CB053353. December 1991.
15. PAUSTENBACH, D.J., A.K. MADL & J.F. GREENE. 2001. Identifying an appropriate occupational exposure limit (OEL) for beryllium: data gaps and current research initiatives. Appl. Occup. Environ. Hyg. **16:** 527–538.
16. JEFFRESS, C.N. Letter to Peter Brush, Acting Assistant Secretary, DOE. August 27, 1998.
17. U. S. DEPARTMENT OF LABOR. 1998. Unified Regulatory Agenda. 63 Federal Register 22218, April 27.
18. OCCUPATIONAL SAFETY AND HEALTH ADMINISTRATION. 2002. Occupational exposure to beryllium: request for information. 67 Federal Register 228: 70707, November 26.
19. STANGE, A.W., D.E. HILMAS, F.J. FURMAN, *et al.* 2001. Beryllium sensitization and chronic beryllium disease at a former nuclear weapons facility. Appl. Occup. Environ. Hyg. **16:** 405–417.

20. HENNEBERGER, P.K., D. CUMRO, D.D. DEUBNER, *et al.* 2001. Beryllium sensitization and disease among long-term and short-term workers in a beryllium ceramics plant. Int. Arch. Occup. Environ. Health **74:** 167–176.
21. KELLEHER, P.C., J.W. MARTYNY, M.M. MROZ, *et al.* 2001. Beryllium particulate exposure and disease relations in a beryllium machining plant. J. Occ. Environ. Med. **43:** 238–249.
22. ROSENMAN, K., V. HERTZBERG, C. RICE, *et al.* 2005. Chronic beryllium disease and sensitization at a beryllium processing facility. Environ. Health Perspect. **113:** 1366–1372.
23. INTERNATIONAL AGENCY FOR RESEARCH ON CANCER. 1993. Beryllium, cadmium, mercury, and exposures in the glass manufacturing industry. Beryll. Beryll. Comp. **58:** 41–117.
24. NATIONAL TOXICOLOGY PROGRAM. 2002. 10th Report on Carcinogens. Available at: http://ntp.niehs.nih.gov/ntp/roc/eleventh/profiles/s022bery.pdf, accessed July 5, 2006.
25. STEENLAND, K. & E. WARD. 1991. Lung cancer incidence among patients with beryllium disease: a cohort mortality study. J. Natl. Cancer Inst. **83:** 1380–1385.
26. WARD, E., A. OKUN, A. RUDER, *et al.* 1992. A mortality study of workers at seven beryllium processing plants. Am. J. Ind. Med. **22:** 885–904.
27. SANDERSON, W.T., E.M. WARD, K. STEENLAND, *et al.* 2001. Lung cancer case-control study of beryllium workers. Am. J. Ind. Med. **39:** 133–144.
28. LEVY, P.S., H.D. ROTH, *et al.* 2002. Beryllium and lung cancer: a reanalysis of a NIOSH cohort mortality study. Inhal. Toxicol. **14:** 1003–1015.
29. SACK, K. & A. MUNDY. A dose of denial. *Los Angeles Times*. March 28, 2004. Available at: http://www.latimes.com/news/nationworld/nation/la-na-ppa28mar28-1,1,4339482,print.htmlstory?coll=la-home-headlines&ctrack=1&cset=true. Accessed October 10, 2005.
30. KERNAN, W.N., C.M. VISCOLI, *et al.* 2000. Phenylpropanolamine and the risk of hemorrhagic stroke. N. Engl. J. Med. **343:** 1826–1832.
31. LAGRENADE, L., P. NOURJAH, *et al.* 2001. Estimating public health impact of adverse drug events in pharmacoepidemiology: phenylpropanolamine and hemorrhagic stroke. Poster presentation at the 2001 FDA Science Forum: Science across the boundaries. Washington DC, February 15–16.
32. KIRTON, W. Email to Bayer representatives, SUBJECT: CHPA Yale Study Meeting, 1/21/99. Available through the *Los Angeles Times*, "A Dose of Denial" at http://www.latimes.com/news/nationworld/nation/la-na-ppa28mar28-1,1,2552623.htmlstory?coll=la-home-headlines. Accessibility verified October 10, 2005.
33. WEINBERG GROUP. Adverse event linked to OTC product. was available at: http://www.weinberggroup.com. Accessibility verified: July 30, 2004.
34. KRIMSKY, S. 2003. Science in the private interest: has the lure of profits corrupted the virtue of biomedical research? Rowman-Littlefield Publishing. Lanham, MD.
35. KRIMSKY, S. 2005. The funding effect in science and its implications for the judiciary. J. Law Pol. **8:** 43–68.
36. SMITH, R. 2005. Medical journals are an extension of the marketing arm of pharmaceutical companies. PLoS Med. **2:** e138.
37. BEKELMAN, J.E., Y. LI & C.P. GROSS. 2003. Scope and impact of financial conflicts of interest in biomedical research: asystematic review. *JAMA* **289:** 454–465.

38. LEXCHIN, J., L.A. BERO, *et al.* 2003. Pharmaceutical industry sponsorship and research outcome and quality. BMJ **326:** 1167–1170.
39. GRAY, G.M., J.T. COHEN, *et al.* 2004. Weight of the evidence evaluation of low-dose reproductive and developmental effects of bisphenol A. Hum. Ecol. Risk Assess. **10:** 875–921.
40. VOM SAAL, F.S. & C. HUGHES. 2005. An extensive new literature concerning low-dose effects of bisphenol A shows the need for a new risk assessment. Environ. Health Perspect. **113:** 926–933.
41. BOMBARDIER, C., L. LAINE, *et al.* 2000. Comparison of upper gastrointestinal toxicity and rofecoxib and naproxen in patients with rheumatoid arthritis. N. Engl. J. Med. **343:** 1520–1528.
42. GRAHAM, D. Testimony before the U.S. Senate Finance Committee, November 18, 2004. Available at: http://finance.senate.gov/hearings/testimony/2004test/111804dgtest.pdf. Accessibility verified: October 10, 2005.
43. BERENSON, A. 2005. For Merck, Vioxx paper trail won't go away. New York Times. August 21.
44. DAVIDOFF, F., C.D. DEANGELIS, J.F. DRAZEN, *et al.* 2001. Sponsorship, authorship, and accountability. JAMA **286:** 1232–1234.
45. MICHAELS, D. & W. WAGNER. 2003. Disclosures in regulatory science. Science **302:** 2073.
46. WAGNER, W. & D.M. MICHAELS. 2004. Equal treatment for regulatory science: extending the controls governing public research to private research. J. Law Med. **30:** 119–154.
47. HILL, A.B. 1965. The environment and disease: association or causation? Proc. Royal Soc. Med. **58:** 295–300.

Misuse of Genetic Data in Environmental Epidemiology

PAOLO VINEIS*

Department of Epidemiology and Public Health, Imperial College London, W2 1PG London, United Kingdom

Department of Epidemiology, University of Torino, 10123 Torino, Italy

> ABSTRACT: The implications of attributing health and diseases more to "nature" (genes) or "nurture" (the environment) have been debated for a long time. Although considerable advancements have been made both in theoretical clarification of concepts, and in the study of the origins of disease, there is still much confusion, for example, in the press and in the beliefs of the population. There is a large consensus, among scientists, that only a small fraction of diseases is due to genes in the usual meaning, that is, according to mendelian inheritance (say around 5% of all diseases), whereas the vast majority of cases are due to environmental exposures or to "gene–environment interactions." In this article I will briefly discuss a model for gene–environment interactions, and I will recall an important discussion that took place decades ago around the mistakes related to attributing diseases or other traits to inheritance. Finally, I will describe a specific example of potential misuse of genetic information.
>
> KEYWORDS: gene–environment interactions; epidemiology; genetic screening

A MODEL OF GENE–ENVIRONMENT INTERACTIONS

Gene–environment interactions are probably the most important explanation for disease onset, meaning that most diseases do not develop without an environmental component, but the onset in a specific individual depends on genetic background. An effective model to describe gene–environment interactions has been provided by Ottman (TABLE 1).[1] This is quite useful because it is based on concrete biological knowledge.

In Model A, the effect of G (genotype) is to produce or increase expression of a risk factor (E) that can also be produced (or perhaps eliminated) nongenetically. For example, homozygotes for the phenylketonuria (PKU) mutation have

---

Address for correspondence: Paolo Vineis, Department of Epidemiology and Public Health, Imperial College London, Norfolk Place W2 1PG, London, United Kingdom. Voice: 0044 20 75943372; fax: 0044 20 75943196.

e-mail: p.vineis@imperial.ac.uk

TABLE 1. Ottman's models of gene–environment interactions

Model A: the effect of G is to produce or increase expression of a risk factor (E) that can also be produced nongenetically (e.g., PKU)
Model B: G exacerbates the effect of E (e.g., UV and skin cancer)
Model C: E exacerbates the effect of G but there is no effect in persons with the low-risk genotype (e.g., porphyria variegata)
Model D: both G and E are required to obtain the effect (e.g., G6PD deficiency)
Model E: G and E both have a separate effect, but when they occur together the effect is much higher (e.g., α-1-antitrypsin and COPD)

NOTE: G = genotype, E = environment.

a deficiency in the enzyme required to convert phenylalanine to tyrosine. If untreated, they will cumulate phenylalanine in the blood and develop mental retardation, although dietary restriction keeps such levels low and prevents retardation. In Model B, G exacerbates the effect of E. For example, Xeroderma Pigmentosum (XP) is an autosomal recessive disorder in which exposure to UV light causes a large number of skin cancers because of a defect of DNA repair enzymes. However, skin cancer is associated with UV exposure also in people without XP. In Model C, E exacerbates the effect of G but there is no effect in persons with the low-risk genotype. For example, an autosomal dominant disorder, porphyria variegata, is characterized by severe skin problems. Exposure to barbiturates strongly exacerbates the symptoms and can lead to death. In Model D, both G and E are required to obtain the effect. G6PD is an X-linked recessive disorder: individuals are asymptomatic unless they eat fava beans; in which case they develop severe hemolytic anemia. Fava beans do not produce any symptoms in normal individuals. Finally, in Model E, G and E both have a separate effect, but when they occur together the effect is much higher. For example, chronic obstructive pulmonary disease (COPD) risk is increased in smokers without α-1-antitrypsin deficiency and in nonsmokers with the deficiency, but risk is increased greatly in smokers with the deficiency. I suspect that most cases of gene–environment interactions relevant to common diseases belong to category E or B.

A CRITIQUE OF THE USUAL APPROACH

It is worth mentioning an important misunderstanding about the role of genes in causing disease. This issue was clarified in an important paper by Richard Lewontin[2] many years ago, but it is still a matter of confusion. The main idea of Lewontin's paper is that when we evaluate gene–environment interactions we use the "analysis of variance" paradigm, that is, we try to combine the two main effects (genes versus environment), plus their interactive term, in a linear model. Causal models presuppose a linear combination of factors as the base line, variances are then computed and the role of the two main effects (or their interaction) is apportioned accordingly. But, Lewontin argues, the analysis-of-variance approach is misleading. There is no theoretical justification for the presumption of a linear explanation (this is done for the sake of simplicity but does not correspond to any reasonable biological reason). By contrast, all the experiments done with, for example, *Arabidopsis* (a plant) or *Drosophila* (based for example on radiation-induced mutations) show that mutations cause a change in what is called the "norm of reaction," that is, the ability of the organism to react to different environmental conditions. The way in which the mutant strain will react, say, to different temperatures, is not predictable if the environmental conditions are not specified. Usually what happens is "canalization," that is, under "normal" conditions there is a certain norm of reaction that is the same for the wild type and the mutants, whereas in changing environments the wild type and the mutant differ in the norm of reaction. What this suggests is that in at least some cases a nonlinear explanation is going to be required. In practical terms, it means that all attempts to explain disease on the basis of either the environment or genes (or their interaction) are in fact doomed to fail, because two organisms with different gene variants will have exactly the same response in a normal environment, and a totally different response in an abnormal environment.

A second confusion, still largely present in the press, concerning heredity versus genetic causation, is related to the debate that took place around IQ a few years ago, after the publication of "The Bell Curve" by Herrnstein and Murray (1994). As it was pointed out,[3] the basic confusion in this book and in many other similar papers was between heritability and genetic determination. Heritability has to do with similar patterns of observable traits between parents and the offspring, whereas a characteristic is "genetically determined" if it is coded in and caused by the genes in a normal environment. Two extreme examples are the following: (*a*) the number of fingers in humans is totally genetically determined and the rare deviations from five fingers are caused by defects of development, for example, from thalidomide, and therefore are not heritable; and (*b*) wearing skirts among European populations has a very strong heritability (it occurs only in women, with the exception of the odd Scotsmen): it is related to having XX versus XY, however, it is not genetically determined.[3] Such misconceptions are clearly relevant to the discussion about the heritability versus genetic determination of cancer. For example, the study of twins, which often does not rule out similar environmental exposures for the pair, cannot be used to infer that cancer or schizophrenia are due to inherited changes in DNA. Another way of making the same point would be to point out that the environment itself is inherited (for example in the case of the propensity to wear skirts). The same applies to claims that IQ has 60% heritability, academic performance 50%, and occupational status 40%: these figures do not mean that such characteristics are inherited through genes (DNA), that is, that there is genetic determination, but only that there is a strong association between the characteristic in the index subject and the same characteristic in the parents (e.g., voting behavior).

THE RISKS OF GENETIC SCREENING: THE EXAMPLE OF NICO TEST

Let us examine a recent example of how predictive medicine could look in future years. The English newspaper "The Telegraph" on December 2, 2004 has given the following news, under the headline "DNA test can identify the smoker's gene:" a private firm from Oxford, UK (G-Nostics), which is one of the "spin-off" companies of the Oxford University, has put on the market a kit for the identification of the carriers of a genetic variant of the gene *DRD2*, involved in the syndrome of nicotine addiction. The carriers of the variant would be more susceptible to developing addiction, but also to responding to a treatment with nicotine patches, and then should be treated as really "sick" people. The offer of the test is based on the results of a series of genetic analyses and of a randomized experiment (the most persuasive type of clinical investigation). This consisted in enrolling 1532 heavy smokers, who were randomly allocated to two arms, the first receiving the nicotine patch, and the second other types of antismoking treatments. Subsequently (after 8–10 years) 755 of these subjects accepted to donate a blood sample to determine their genetic characteristics. At this point the researchers observed that those who quit smoking with the use of a patch were predominantly subjects with the genetic variant of *DRD2*, who therefore turned out to be the best target for the this kind of dissuasion intervention. The difference between carriers of the variant and "normal" subjects was strong (a likelihood of quitting 2–3 times higher) and statistically significant. However, the effectiveness of the dissuasion intervention was at short time (between 1 and 12 weeks). The mentioned article is substantially correct and shows a sound scientific background.

The same authors from the Oxford University have then published a systematic review on the association between genes and nicotine addiction, claiming that the scientific evidence was at the moment rather weak and that studies show great heterogeneity of results. These same researchers, however, seem to be responsible for the spin-off mentioned above, that is, marketing of a kit for general practitioners (G-Nostics). Leaving the comments on the conflict of interest apart, it is worth noting the slippery slope that is thus created on the basis of premature introduction into practice of a test like this. It is not unrealistic to imagine the following potential scenarios: (*a*) there will be probably a category of people who, having a greater genetic resistance to the effectiveness of the nicotine patch and other antismoking devices, will end up with thinking they have no hope of quitting (or, even worse, could think they are protected from cancer thanks to some other gene variant like *CYP2D6*, also analyzed by G-Nostics); and (*b*) what will happen when similar genes will be tested to prevent obesity and its consequences? And what about some forms of social deviance (antisocial behaviors)?

The slippery slope does not consist only in the fact that an approach that is typical of clinical medicine (pharmacogenetics) is extended to a behavior (smoking), thus suggesting that complex behaviors not only belong to the category of disease, but also in the fact that for these behaviors it is literally impossible to disentangle what is due to social relationships and interactions and what to the individual history, including genetic characteristics. On scientific grounds it is quite likely that addictions and behaviors that are hazardous to health, and even antisocial, derive from an interaction between the environment and individual susceptibility, including the genetic form. But the two aspects cannot be separated, and to try to separate them may lead to a very dangerous slope: a stigma toward minorities (for carriers of the susceptibility genes); increasing conflict between carriers of such genes and insurances, in the countries where the latter are powerful; a widespread climate of weakening of responsibility, since the culprit of diseases and behaviors is attributed to genes; and the diffusion of a model of causality more and more influenced by hard natural sciences rather than social and political determinants of the most significant events in people's lives.

REFERENCES

1. OTTMAN, R. 1996. Gene-environment interaction: definitions and study designs. Prev. Med. **25:** 764–770.
2. LEWONTIN, R. 1972. The analysis of variance and the analysis of causes. Am. J. Hum. Genet. **26:** 400–411.
3. BLOCK, N. 1995. How heritability misleads about race. Cognition **56:** 99–128.

Collision of Basic and Applied Approaches to Risk Assessment of Thyroid Toxicants

R. THOMAS ZOELLER

Biology Department and Molecular and Cellular Biology Program, Morrill Science Center, University of Massachusetts, Amherst, Massachusetts 01003, USA

ABSTRACT: Thyroid hormone (TH) is essential for normal brain development; therefore, any environmental chemical that interferes sufficiently with thyroid function, TH metabolism, or TH action may exert adverse effects on brain development. Important known differences in aspects of thyroid endocrinology between the fetus, infant, and adult allow us to identify age-dependent vulnerabilities to thyroid toxicants with some confidence. These differences include the size of the hormone pool stored in the thyroid gland at different ages as well as the age-dependent sensitivity to mild TH insufficiency. Several recent studies that describe risk assessments of the environmental contaminant, ammonium perchlorate, provide good examples of conclusions based on the selective consideration of these known aspects of the thyroid system. Specifically, authors who consider age-dependent differences in thyroid endocrinology suggest that safe levels of perchlorate should be set at relatively low levels (low parts per billion). In contrast, authors who do not consider these known age-dependent differences in thyroid endocrinology recommend safe levels of perchlorate at high (hundreds) parts per billion to parts per million. Emerging evidence indicates that a variety of high production volume chemicals can directly interact with the TH receptor. As testing paradigms are designed by regulatory agencies, these age-dependent differences in thyroid endocrinology must be considered.

KEYWORDS: thyroid hormone; brain development; perchlorate; PCB; thyroid receptors

INTRODUCTION

Thyroid hormone (TH) is essential for brain development, but the symptoms of hypothyroidism at birth are not readily or uniformly apparent. Before universal neonatal screening for congenital hypothyroidism (CH), these children

Address for correspondence: Professor R. Thomas Zoeller, Biology Department, Morrill Science Center, 611 North Pleasant Street, University of Massachusetts, Amherst, MA 01003.
e-mail:tzoeller@bio.umass.edu

often were not identified for some months after birth.[1,2] The consequences of this delay in diagnosis and treatment were disastrous. The mean full-scale intelligence quotient (IQ) of CH infants was found to be 76 before universal screening,[3] but when categorized according to the timing of diagnosis, it became evident that the older the infants were before they were diagnosed with CH (and treated), the more severe were the intellectual and developmental deficits.[3,4] Thus, from a public health perspective, TH dysfunction is important to discover without reliance solely on clinical presentation.

Taken together, these studies demonstrate that TH is essential for brain development from early fetal development (first trimester) to at least 2 years after birth. Studies of the human fetal and neonatal brain demonstrate that thyroid hormone receptors (TRs) are expressed early in development,[5–8] and that different TRs are selectively expressed in very specific temporal and spatial pattern, much like that described for the developing rat brain.[9] Thus, there are two critical issues to address when evaluating the mechanisms by which environmental chemicals might interfere with TH signaling in the developing brain. The first is that if a chemical can cause a reduction in circulating levels of TH, it is important to determine empirically whether the reduction in TH is sufficient to cause TH insufficiency. Second, if a chemical can directly interfere with TH signaling at the receptor, it is important to determine empirically whether the effect is isoform specific. These two topics will be addressed in sequence.

THYROID HORMONE INSUFFICIENCY—TO WHAT EXTENT CAN COMPENSATORY MECHANISMS AMELIORATE THE CONSEQUENCES OF LOW THYROID HORMONE?

The thyroid endocrine system is governed by mechanisms that are in many ways different from those governing the sex hormones, despite the fact that both are controlled by classic neuroendocrine systems. These mechanisms appear to allow a constant supply of TH to cells despite fluctuations in TH synthesis or in circulating levels of TH.

The thyroid system is a classic neuroendocrine axis (FIG. 1); the hypothalamus controls the pituitary gland, which in turn controls the thyroid, and feedback mechanisms between thyroid secretions and the hypothalamus and pituitary maintain the activity of this axis within relatively narrow limits.[10] The active THs, thyroxine (T_4), and triiodothyronine (T_3) are formed in the thyroid gland. These hormones are synthesized in an unusual way in that they are derived from coupling two iodinated tyrosyl residues that make up the larger hormone precursor, thyroglobulin (Tg). Tg is a large glycoprotein containing two identical subunits each of nearly 3000 amino acids, creating a 660 kDa mature protein.[11] Following iodination, the protein is stored in the

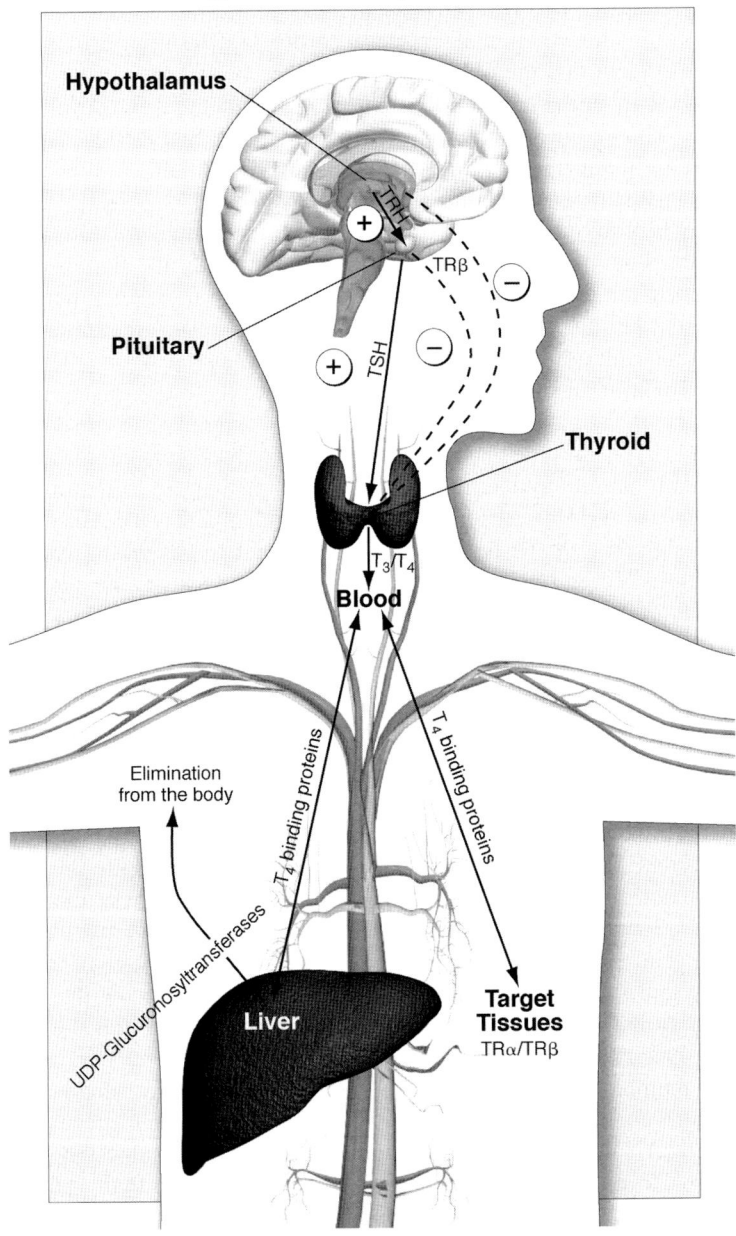

FIGURE 1. The hypothalamic-pituitary-thyroid (HPT) Axis. Neurons whose cell bodies reside in the hypothalamic paraventricular nucleus (PVN) synthesize the tripeptide TRH.[123,124] Although TRH-containing neurons are widely distributed throughout the brain,[125,126] TRH neurons in the PVN project uniformly to the median eminence,[127,128] a neurohemal organ connected to the anterior pituitary gland by the hypothalamic-pituitary-portal vessels,[129] and are the only TRH neurons to regulate the pituitary-thyroid axis.[21,130] TRH is delivered by the pituitary-portal vasculature to the anterior pituitary gland to stimulate the synthesis and release of TSH.[131] TRH stimulates the synthesis of the TSH beta subunit.[131] However, TRH also affects the post-translational glycosylation of TSH which affects its biological activity.[132–137] Interestingly, a recent report by Nikrodhanond et al.[138] demonstrates that the role of TRH in regulating the pituitary-thyroid axis is stronger than the role of TH negative feedback.

Pituitary TSH is one of three glycoprotein hormones of the pituitary gland and is composed of an alpha and a beta subunit.[139] All three pituitary glycoproteins (luteinizing hormone, LH; follicle stimulating hormone, FSH; and TSH) share the same alpha subunit.[140] Pituitary TSH binds to receptors on the surface of thyroid follicle cells stimulating adenylate cyclise.[141,142] The effect of increased cAMP is to increase the uptake of iodide into thyroid cells, increase iodination of tyrosyl residues on TG by thyroperoxidase, increase the synthesis and oxidation of TG, TG uptake from thyroid colloid, and production of the iodothyronines T_4 and T_3. T_4 is by far the major product released from the thyroid gland.[142] Recent anatomical studies have shown that human pituitary thyrotropes express the mRNA encoding the TSH receptor,[143,144] which may represent a negative feedback loop accounting for the fact that serum TSH is reduced in some Gravesatients with normal levels of TH.[145]

THs are carried in the blood by specific proteins. In humans, about 75% of T_4 is bound to thyroxine-binding globulin (TBG), 15% is bound to transthyretin (TTR) and the remainder is bound to albumin.[146] TBG, the least abundant but most avid T_4 binder, is a member of a class of proteins that includes cortisol binding protein (CBP) and is cleaved by serine proteases in serum.[147] These enzymes are secreted into blood during inflammatory responses and, in the case of CBP, can induce the release of cortisol at the site of inflammation. The physiological significance of this observation is presently unclear for TBG.[146] However, it is clear that T_4 is release from TBG upon cleavage.[148,149] The presence and abundance of the different binding proteins varies among the vertebrates and may be developmentally regulated in a generalized manner. In the rat, high serum levels of TBG are found in the fetus and the early postnatal pup,[150,151] adult levels of TBG are low, but low serum T_4 appears to increase both serum TBG and liver biosynthesis in the rodent.[151] Interestingly, organisms across taxa appear to have the greatest carrying capacity for T_4 in serum during development compared to their respective adult forms.[152] This may be a mechanism by which T_4 can be adequately maintained at a developmental time when it is uniformly important.

THs (T_4 and T_3) exert a negative feedback effect on the release of pituitary TSH and on the activity of hypothalamic TRH neurons.[18,19,124] Although it is clear that TH regulates the expression of TSH[153–155] and TRH[18,123,124,156] in a negative feedback manner, it is also clear that the functional characteristics of negative feedback must include more than simply the regulation of the gene encoding the secreted protein/peptide. In addition, fasting suppresses the activity of TRH neurons by a neural mechanism that may involve leptin.[157,158] This fasting-induced suppression of TRH neurons results in the reduction of circulating levels of thyroid hormone. In humans and perhaps in rodents, circulating levels of T_4 and of T_3 fluctuate considerably within an individual; therefore, TSH measurements are considered to be diagnostic of thyroid dysfunction.[15,159,160] However, individual T_4 levels in humans vary within far narrower limits than the population limits (i.e., the population reference range).[10,161] In addition, variance in serum T_4 in pairs of monozygotic twins is far more correlated than that in pairs of dizygotic twins or the general population.[162] Thus, the set-point around which negative feedback appears to function has a very strong genetic component in humans and perhaps in other animals.

T_4 and T_3 are actively transported into target tissues.[163–170] T_4 can be converted to T_3 by the action of outer-ring deiodinases (ORD, type I and type II).[22] Peripheral conversion of T_4 to T_3 by these ORDs accounts for nearly 80% of the T_3 found in the circulation.[160] Thyroid hormones are cleared from the blood in the liver following sulfation or sulfonation by sulfotransferases, or following glucuronidation by UDP-glucuronosyl transferase.[171,172] These modified THs are then eliminated through the bile.

T_4 and/or T_3 are actively concentrated in target cells about 10-fold over that of the circulation, although this is tissue-dependent. The receptors for T_4 and T_3 (TRs) are nuclear proteins that bind to DNA and regulate transcription.[173–177] There are two genes that encode the TRs, c-erbA-alpha (TRa) and c-erbA-beta (TRb). Each of these genes is differentially spliced, forming 3 separate TRs, TRa1, TRb1, and TRb2. The effects of TH are quite tissue-, cell-, and developmental stage-specific and it is believed that the relative abundance of the different TRs in a specific cell may contribute to this selective action.

colloid, the fluid filling the central core of the thyroid follicle. At the time of hormone release, iodinated Tg is taken up into the cell from the colloid, digested by lysosomal enzymes, liberating T_3 and T_4 into the blood.[12] T_4 is the predominant iodothyronine released by the thyroid gland; circulating T_3 is formed largely from peripheral deiodination of T_4.[13] The pituitary glycoprotein hormone, thyrotropin (TSH),[14] regulates the synthesis and secretion of THs by activating adeylate cyclase in thyroid follicular cells. However, there are a number of important extrathyroidal processes that combine to maintain circulating THs within a relatively narrow concentration range. Although T_4 is the predominant form of TH in the serum, T_3 is the active hormone at the receptor. The term *thyroid hormone* will be abbreviated in the remainder of this paper to "TH" to include both T_4 and T_3, recognizing the differences between the two.

Normal variation in circulating concentrations of T_4 reflects short-term pulsatile and diurnal variation.[15] THs exert a negative feedback effect on pituitary secretion of TSH,[16,17] and on the hypothalamic secretion of the releasing factor, thyrotropin-releasing hormone (TRH).[18,19] Although it is clear that TRH is a major factor regulating TSH secretion, several hypothalamic factors contribute to TSH regulation, including somatostatin, dopamine, and norepinephrine.[16] Moreover, some investigators suggest that the primary role of TRH in the regulation of TSH secretion is to modulate the set-point around which TH act on the pituitary.[20,21] Thus, circulating levels of TH, and the balance between different forms of these hormones, are controlled by a number of processes.

Once inside the cell, T_4 must be converted to T_3 before it can exert an action on the receptor. This conversion from T_4 to T_3 is mediated by a deiodinase enzyme, the type 1 or type 2 deiodinase (D1 or D2).[22] Moreover, a type 3 deiodinase (D3) can inactivate T_3. Finally, there are cellular binding proteins that may play a role in maintaining intracellular TH levels, or delivering the hormone to the nucleus.[23,24] However, in the brain, this may be cell-type specific (FIG. 2).[25,26] That is, T_4 may be converted to T_3 in glial cells before T_3 can act in neurons. This would require a T_4 transporter on glial cells and a T_3 transporter on neurons. This appears to be the case inasmuch as the organic anion transporter 14 (OATP-14, a T_4-selective transporter) is expressed in areas of glia and the monocarboxylate transporter 8 (MCT8, a T_3 selective transporter) is expressed in areas of neurons.[27]

Thus, the thyroid signaling system appears on the surface to be optimized for maintaining a continuous supply of hormone to cells. Specifically, one can imagine that as T_4 levels begin to decline, a number of compensatory mechanisms respond to maintain tissue levels of T_3. Specifically, TSH levels would increase as a consequence of the negative feedback system. In addition, tissue levels of D2 might increase to more effectively convert T_4 to T_3. Finally, transporters may respond in a manner that would cause a compensatory increase in the ability of cells to sequester T_4 and/or T_3. In addition, the thyroid glanditself appears to be very capable of responding to low iodide with mechanisms

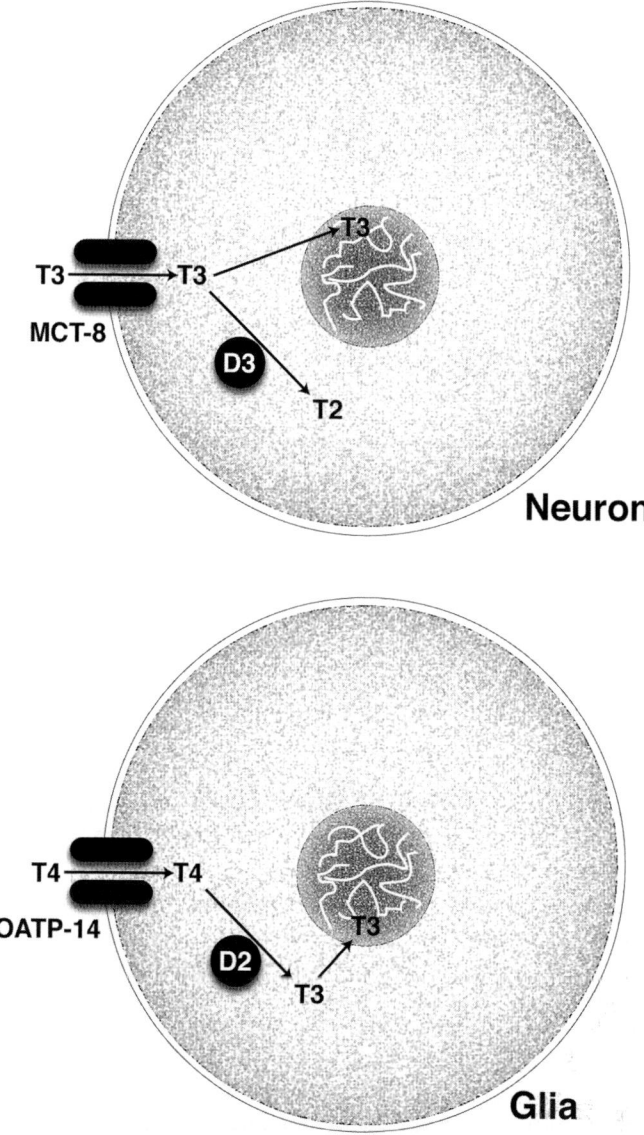

FIGURE 2. Transport and cellular uptake of THs. The current model is that glial cells (astrocytes) take up T_4 through a specific transporterhe organic anion transport protein-14 (in the rat),178 after which T_4 is converted to T_3 by the outer ring deiodinase, D2. T_3 then leaves the glial cell and is transported into neurons through the action of the monocarboxylate transporter 8 (MCT8).[27] T_3 can act in the neuron to regulate gene expression, and it is degraded by the action of the inner ring deiodinase, D3. This two-step process may also occur in the pituitary gland.[179]

that increase iodide uptake. For example, the normal percentage uptake of radioactive iodide in the euthyroid population is about 15–20% (e.g., see Ref. 28). In contrast, patients with Graves' disease exhibit a percentage uptake of radioactive iodide of over 40%.[29] Thus, even the thyroid gland appears to be capable of mounting compensatory responses to situations that would tend to cause a reduction in circulating levels of TH.

Although compensation doubtlessly occurs to some extent, there is little or no empirical experimental evidence for compensation. For example, if circulating levels of TSH are elevated but TH levels are not different from controls (something not often seen in rodent studies), one must capture events downstream of TH action in tissues before concluding that compensation has occurred. Moreover, additional measures of compensatory mechanisms cannot make up for the lack of information about TH action. For example, T_4 suppresses D2 expression in the brain[30] and while this observation is certainly important and is consistent with the hypothesis that compensation has occurred, it does not directly test the hypothesis that these apparently compensatory changes protect the tissue from damage due to low TH. Likewise, measurements of tissue levels of T_4 or T_3 do not directly test the hypothesis that putative compensatory mechanisms have ameliorated effects of low T_4 on TH action for the same reasons. Therefore, conclusions about whether compensation has occurred must be based on direct measurement of TH actions rather than on indirect measures (e.g., TSH, sodium/iodide symporter [NIS], etc.). This has rarely if ever been performed.

CONSEQUENCES FOR RISK ASSESSMENT—THE CASE OF PERCHLORATE

Ammonium perchlorate is the principal oxidant for solid propellants in the defense industry.[31,32] Perchlorate contamination of ground water across the United States has recently become apparent[33] and, therefore, it is important to determine a "safe" level of perchlorate to which humans can be exposed. Perchlorate inhibits iodide uptake into the thyroid gland by the NIS.[34] This action of perchlorate can lead to an inhibition of TH synthesis, and thus a reduction in circulating levels of TH, and was the basis of both its prior clinical use in the treatment of hyperthyroidism, and its potential toxicity as an environmental contaminant. To establish the dose–response in humans for perchlorate inhibition of thyroidal iodide uptake, and short-term effects on circulating TH, Greer et al.[35] gave perchlorate in drinking water at 0.007, 0.02, 0.1, or 0.5 mg/kg per day to 37 male and female volunteers for 14 days. In 24 subjects, 8- and 24-hour measurements of thyroidal ^{123}I uptake (RAIU) were performed before exposure, on exposure days 2 and 14 (E2 and E14), and 15 days postexposure (P15).

In general, this study allowed the estimation of a "no-effect level" of perchlorate of 5.2 or 6.4 μg/kg per day.[35] This value was calculated based on the dose–response of perchlorate consumption on the ability of the thyroid gland to take up iodide. Thus, some have considered this value (5.2 μg/kg per day) to be a threshold below which perchlorate would have no effect. The concentration of perchlorate in drinking water required to deliver this dose to a healthy adult is approximately 200 ppb.

Based on this information, Greer *et al.* concluded that perchlorate concentrations of about 200 ppb (and possibly higher) should be of no health concern in an iodine-sufficient population.[35] Although this conclusion is marginally defensible for normal, euthyroid adults, several key aspects of the normal adult thyroid system are significantly different from that of infants, which makes this conclusion inappropriate for a significant proportion of the normal human population—fetuses, neonates, and infants.

Specifically, Greer *et al.*[35] postulated that 0.5 mg/kg per day of perchlorate failed to influence circulating levels of TH in healthy adults in 14 days of exposure because the normal adult thyroid gland contains a very large storage capacity of unreleased TH. In fact, these authors estimate that there should be sufficient hormone stored in the adult thyroid gland to last for several months. Thus, no concentration of perchlorate should cause a reduction in circulating levels of TH of a normal euthyroid adult within 14 days. The case is quite different for a late gestation fetus or neonate. Specifically, Vulsma *et al.*[36] (cited in Ref. 37) estimated that the neonatal thyroid gland contains TH equivalent to only a single day's secretion. This estimate was revised by van den Hove *et al.*[37] and Savin *et al.*[38] who empirically measured intrathyroidal stores of TH in human fetuses and neonates and found that the amount of hormone stored in the colloid is less than that required for a single day. Thus, inhibiting TH synthesis in a fetus/neonate would immediately be manifested as a decrease in circulating levels of TH.

Three additional characteristics of a child's thyroid system should be considered performing risk analysis of a thyroid toxicant such as perchlorate. First, neonates have sole source of nutrition—milk—and perchlorate may be concentrated in milk. Perchlorate acts on the NIS,[34] a protein that is induced in lactating breast tissue by prolactin.[39–42] Thus, it is possible that perchlorate is concentrated in milk.[43,44] In support of this, Kirk *et al.* have found some relatively high levels of perchlorate in human milk.[45] These authors report that perchlorate is found in a number of human milk samples and that the level of perchlorate in milk was not predicted by the level of perchlorate in their drinking water. Thus, it is not possible to conclude from this study that milk concentrates perchlorate, but it does indicate that perchlorate exposure is more widespread than is predicted by contaminated drinking water.

Second, a short period of TH insufficiency may produce permanent neurological deficits in children. van Vliet reviewed the evidence that a period of TH insufficiency of as little as 14 days may be long enough to produce permanent neurological deficits in neonates.[46] Finally, infants drink six times as much fluid per unit body weight as adults. Therefore, in the example of perchlorate, a drinking water concentration of 200 ppb would deliver about 129 μg perchlorate to a 4 kg infant being fed a formula prepared with tap water. This would be a dose of about 32.2 μg/kg, which would undoubtedly inhibit iodide uptake in this infant.

Considering these differences between the thyroid system of adults and infants, it is predictable that any chemical, such as perchlorate, that reduces TH synthesis will reduce circulating levels of TH. However, it is not known to what extent iodide uptake must be inhibited by perchlorate to cause a reduction in TH synthesis. Some authors have suggested that as much as 75% iodide uptake inhibition would be required for deleterious effects to be produced by perchlorate.[47] This estimate, as articulated in the report of the National Academy Report is without citation. To our knowledge, there are no quantitative studies defining the degree to which iodide uptake inhibition must occur to inhibit TH synthesis. Thus, for human infants, it remains to be determined the degree to which they can tolerate iodide uptake inhibition.

A significant issue is also that infants cannot tolerate TH insufficiency for many days. For example, long-term studies of children with CH that have been treated with T_4 replacement indicate that very subtle differences in circulating levels of T_4 are associated with significant differences in intellectual performance later in life.[48–51] Thus, the consequences of perchlorate exposure to infants could be the production of a lifetime of cognitive deficit, the severity of which would be proportional to the severity of TH insufficiency.

Considering the difference between adults and infants in their ability to compensate for perchlorate exposure and tolerate TH insufficiency, it is remarkable that recent reviews of perchlorate risk assessment fail to mention the infant. For example, Fields et al.[52] conclude that, "Of more general concern is whether the 10-fold default uncertainty factor is needed for intraspecies (i.e., within human) variability to protect such hypothetical susceptible sub-populations" (p. 52). Although their focus was on vegetarians who may (or may not) be iodide insufficient, their presumptive concern about the 10-fold uncertainty factor to protect vegetarians failed to mention infants as a subpopulation of humans with known vulnerabilities to TH insufficiency. Perhaps even more remarkably, Crump and Gibbs[53] performed benchmark dose analysis using changes in serum free T_4 and TSH (which are not changed in the three studies used in their analysis). Thus for the Greer study, which could not have identified reduced T_4 or TSH in a 14-day study for reasons described above, the benchmark dose levels ranged from about 50 to 60 mg per day (around 850 μg/kg per day). Again for this analysis, there is no mention of the differences in the thyroid system between adults and infants or the differences in consequences of TH insufficiency between adults and infants. Moreover, 850 μg/kg per day is more than 1200 times the reference dose (RfD) proposed by the NAS committee.

In contrast, Ginsberg and Rice[54] come to a conclusion that is the opposite of that proposed by Fields et al.[52] Specifically, they argue that additional uncertainty factors may be considered by risk assessors precisely to protect the fetus and nursing infants.[54] Thus, the estimated "safe" level of perchlorate consumption proposed by different authors appears to be dependent upon whether infants are considered in the analysis. However, because most infants are normal and all normal infants have very well-characterized differences in their thyroid systems compared to normal adults, it is unclear how any analysis could fail to include infants as a vulnerable subpopulation worthy of protection.

ENVIRONMENTAL CHEMICALS AS THYROID HORMONE MIMICS: A CHALLENGE FOR RISK ASSESSMENT

Despite early speculations that environmental chemicals may act as imperfect TH analogs,[55,56] few studies had tested this hypothesis until recently. Now, several recent reports show that a broad range of chemicals to which humans are routinely, and inadvertently, exposed can bind to TRs and may produce complex effects on TH signaling. Perhaps the best example is that of polychlorinated biphenyls (PCBs)—industrial chemicals consisting of paired phenyl rings with various degrees of chlorination.[57] Although the production of PCBs was banned in the mid 1970s, these contaminants are routinely detected in the environment[58] and in human tissues[59] at high concentrations. PCB body burden is associated with lower full-scale IQ, reduced visual recognition memory, attention, and motor deficits,[60–65] and structural differences in areas of white matter.[66,67]

PCBs reduce circulating levels of T_4 in animals,[68–72] and some authors propose that PCBs exert neurotoxic effects on the developing brain by causing a state of relative hypothyroidism.[73–75] This concept is supported by the observations that the ototoxic effect of PCB exposure can be partially ameliorated by T_4 replacement,[76] and that the cerebellum, a tissue highly sensitive to TH insufficiency,[77–79] is targeted by PCB exposure. PCB exposure alters motor behavior associated with cerebellar function,[80,81] as well as cerebellar anatomy.[81] Interestingly, PCB exposure is associated with an increase in expression of glial fibrillary acidic protein,[81] which is also increased by TH insufficiency.[82] Finally, in young children, the association between PCB body burden and behavioral measures of response-inhibition is stronger in those children that have a smaller corpus callosum,[66] an area of the brain affected by TH.[83–85] Thus, it is possible that PCBs exert at least some neurotoxic effects on the developing cerebellum by causing a state of relative hypothyroidism.

However, PCB exposure does not produce effects on animals that are fully consistent with effects of TH insufficiency, such as body weight gain during development[68,86,87] or the timing of eye opening.[69] In addition, despite the reduction in serum T_4, PCB exposure increases the expression of several TH-responsive genes in the fetal[86,87] and neonatal[68] brain, indicating that at least some individual PCB congeners, or their metabolites, can act as TR agonists *in vivo*. Recently, Kitamura *et al.*[88] reported that nine separate hydroxylated PCB congeners can bind to the rat TR with an IC_{50} as low as 5 μM. In addition, using a human neuroprogenitor cell line, Fritsche *et al.*[89] found that a specific PCB congener could mimic the ability of T_3 in increasing oligodendrocyte differentiation, and that this effect was blocked by the selective TR antagonist NH3. Finally, Arulmozhiraja and Morita[90] have identified several PCB congeners that exhibit weak TH activity in a yeast two-hybrid assay optimized to identify such activity.

However, not all recent reports indicate that PCBs act as agonists on the TR. Kimura-Kuroda *et al.*[91] found that two separate hydroxylated PCBs interfere with T_3-dependent neurite outgrowth in mouse cerebellar granule cell primary cultures. In addition, Bogazzi *et al.*[92] found that a commercial mixture of PCBs (Aroclor 1254) exhibited specific binding to the rat TRβ at approximately 10 μM. This concentration inhibited TR action on the malic enzyme (ME) promotor in a CAT assay and this effect required an intact TR element (TRE). However, the PCB mixture did not alter the ability of TR to bind to the ME TRE in a gel shift assay. In contrast, Iwasaki *et al.*[93] found that a specific hydroxylated PCB congener inhibits TR-mediated transcriptional activation in a luciferase assay at concentrations as low as 10^{-10} M. This effect was observed in several cell lines, but was not observed using a glucocorticoid response element. Miyazaki *et al.*[94] followed this report by showing that PCBs can dissociate TR:RXR heterodimers from a TRE.

It is clear that PCBs are neurotoxic in humans and animals, and that they can interact directly with the TR. However, the consequences of PCB exposure on TR action appear to be quite complex. This complexity includes acting as an agonist or antagonist and may include TR isoform selectivity inasmuch as most studies have been performed using the TRβ, leaving the TRα relatively unstudied in this context. In addition, considering that there are 209 different chlorine substitution patterns on the biphenyl backbone and that these can be metabolized (hydroxyl- and methylsulfonyl-metabolites[95]), it is possible that different chemical species exert different effects. Finally, PCBs may exert different actions on TRs depending on associated heterodimer partners, promotor structure, or different co-factors. This complexity will be important to pursue because the effects of PCB exposure in humans is far better studied than for structurally related compounds such as polybrominated biphenyls (PBBs) and polybrominated diphenyl ethers (PBDEs). Thus, mechanistic studies on PCBs can be more easily and effectively coupled to specific human health outcomes.

Bisphenol-A (BPA, 4,4′-isopropylidenediphenol) also represents another example of an environmental chemical to which humans are exposed that may produce complex effects on TH action. BPA is produced at a rate of over

800 million kg annually in the United States alone,[96] and is used primarily in the manufacture of plastics including polycarbonate plastics, epoxy resins that coat food cans, and in dental sealants.[97,98] Howe et al.[97] estimated human consumption of BPA from expoxy-lined food cans alone to be about 6.6 µg/person per day. BPA has been reported in concentrations of 1–10 ng/mL in serum of pregnant women, in the amniotic fluid of their fetus, and in cord serum taken at birth.[99,100] Moreover, BPA concentrations of up to 100 ng/g were reported in placenta.[99] BPA is also halogenated (brominated or chlorinated) to produce flame-retardants. Tetrabromobisphenol-A (TBBPA) is the most commonly used with over 60,000 tons produced annually.[101,102] Thomsen et al.[103] recently reported that brominated flame retardants, including TBBPA, have increased in human serum from 1977 to 1999 with concentrations in adults ranging from 0.4 to 3.3 ng/g serum lipids. However, infants (0–4 years) exhibited serum concentrations that ranged from 1.6 to 3.5 times higher.[103]

Considering this pattern of human exposure, it is potentially important that BPA has been shown to bind to the TR.[104] Best characterized as a weak estrogen,[105] binding to the estrogen receptor (ER) with a K_i of approximately 10^{-5} M,[106,107] BPA binds to and antagonizes T_3 activation of the TR[108,109] with a K_i of approximately 10^{-4} M, but as little as 10^{-6} M BPA significantly inhibits TR-mediated gene activation.[109] Moreover, Moriyama et al. found that BPA reduced T_3-mediated gene expression in culture by enhancing the interaction with the co-repressor N-CoR.[104] Interestingly, we have found that developmental exposure to BPA in rats produces an endocrine profile similar to that observed in thyroid resistance syndrome.[110] Specifically, T_4 levels were elevated during development in the pups of BPA-treated animals, but TSH levels were not different from controls.[111] This profile is consistent with BPA inhibition of TRβ-mediated negative feedback. However, the TH-response gene, RC3, was elevated in the dentate gyrus of these BPA-treated animals.[111] Because the TRα isoform is expressed in the dentate gyrus, we concluded that BPA may be a selective TRβ antagonist *in vivo*.

If BPA acts as a TR antagonist *in vivo*, it is predictable that specific developmental events and behaviors would be affected by developmental exposure to BPA. In this regard, Seiwa et al.[112] have shown that BPA blocks T_3-induced oligodendrocyte development from precursor cells (OPCs). In addition, it is potentially important that BPA-exposed rats exhibit ADHD-like symptoms.[113] This is important because the apparent endocrine profile induced by BPA, in which serum T_4 is elevated despite normal TSH levels, and in which the TRα receptor appears to be overstimulated, is similar to the endocrine profile in humans caused by a mutation in the TRβ receptor. The subsequent syndrome, thyroid resistance syndrome,[114] is associated with a number of symptoms including attention deficit disorders.[110,115] Thus, it is possible that the endocrine profile produced by BPA could be associated with adverse neurodevelopmental effects.

Despite the antagonistic effects of BPA on the TRβ, halogenated BPAs appear to act as TR agonists.[108] Both TBBPA and tetrachlorobisphenol A (TCBPA) can bind to the TR and induce GH3 cell proliferation and growth hormone production.[108] Thus, these compounds may exert agonistic effects on the TR and this could be important during early brain development. For example, TH of maternal origin can regulate gene expression in the fetal brain,[116–118] one of these genes codes for Hes1.[87] Considering the role of HES proteins in fate specification in the early cortex,[119–121] the observation that industrial chemicals can activate the TR and increase HES expression[87] may indicate that these chemicals can exert subtle effects on early differentiative events. In addition, TH exerts equal and opposite effects on the numbers of oligodendrocytes and astrocytes found in the corpus callosum and anterior commissure,[122] indicating that TH controls the balance of production of these two cell types in the postnatal brain. Similarly to events in the early fetal brain controlled by Hes1, inappropriate TH stimulation may have long-term consequences on the balance of glial cell types that make up major bridging structures in the brain.

CONCLUSIONS

The human population is exposed to a large number of chemicals that can influence thyroid function and, perhaps, TH action. Current screens and tests for thyroid toxicants focus primarily—if not exclusively—on evaluating effects on thyroid function as indicated by changes in hormone levels. Mechanistic studies will be required to fully characterize thyroid toxicants because it is becoming clear that some compounds can exert direct effects on TH signaling and may be having important public health consequences that are currently not appreciated. Moreover, it is important to focus on the potential effects of thyroid toxicants at different life stages because both the sensitive to various toxicants and the ultimate consequences of exposures will differ among individuals of different ages.

REFERENCES

1. JACOBSEN, B.B. & N.J. BRANDT. 1981. Congenital hypothyroidism in Denmark: incidence, types of thyroid disorders and age at onset of therapy in children: 1970-1975. Arch Dis Child. **56:** 134–136.
2. ALM, J. *et al.* 1984. Incidence of congenital hypothyroidism: retrospective study of neonatal laboratory screening versus clinical symptoms as indicators leading to diagnosis. Br. Med. J. **289:** 1171–1175.
3. KLEIN, R. 1980. History of congenital hypothyroidism. *In* Neonatal Thyroid Screening. G.N. Burrow & J.H. Dussault, Eds.: 51–59. Raven Press. New York.
4. KLEIN, R.Z. & M.L. MITCHELL. 1996. Neonatal Screening for Hypothyroidism. *In* The Thyroid: a Fundamental and Clinical Text. L.E. Braverman & R.D. Utiger, Eds.: 984–988. Lipponcott-Raven. Philadelphia.

5. CHAN, S. & M.D. KILBY. 2000. Thyroid hormone and central nervous system development. J. Endocrinol. **165:** 1–8.
6. KILBY, M.D. *et al.* 2000. Expression of thyroid receptor isoforms in the human fetal central nervous system and the effects of intrauterine growth restriction. Clin Endocrinol. (Oxf.) **53:** 469–477.
7. CHAN, S. *et al.* 2002. Early expression of thyroid hormone deiodinases and receptors in human fetal cerebral cortex. Brain Res. Dev. Brain Res. **138:** 109–116.
8. KILBY, M.D. 2003. Thyroid hormones and fetal brain development. Clin. Endocrinol (Oxf). **59:** 280–281.
9. BRADLEY, D.J., H.C. TOWLE & W.S. YOUNG. 1992. Spatial and temporal expression of alpha- and beta-thyroid hormone receptor mRNAs, including the beta-2 subtype, in the developing mammalian nervous system. J. Neurosci. **12:** 2288–2302.
10. ANDERSEN, S. *et al.* 2002. Narrow individual variations in serum T(4) and T(3) in normal subjects: a clue to the understanding of subclinical thyroid disease. J. Clin. Endocrinol. Metab. **87:** 1068–1072.
11. SPENCER, C.A. 2000. Thyroglobulin. *In* Werner and Ingbar's The Thyroid: a Fundamental and Clinical Text. L.E. Braverman & R.D. Utiger, Eds.: 402–413. Lippincott Williams and Wilkins. Philadelphia, PA.
12. TAUROG, A. 2004. Hormone synthesis: thyroid iodine metabolism. *In* The Thyroid: a Fundamental and Clinical Text. L.E. Braverman & R.D. Utiger, Eds.: 61–85. Lippincott-Raven. Philadelphia.
13. LEONARD, J.L. & J. KOEHRLE. 1996. Intracellular pathways of iodothyronine metabolism. *In* The Thyroid: a Fundamental and Clinical Text. L.E. Braverman & R.D. Utiger, Eds.: 125–161. Lippincott-Raven. Philadelphia.
14. SPAULDING, S.W. 2000. Biological actions of thyrotropin. *In* The Thyroid: a Fundamental and Clinical Text. L.E. Braverman & R.D. Utiger, Eds.: 227–233. Lippincott Williams and Wilkins. Philadelphia.
15. STOCKIGT, J.R. 2000. Serum thyrotropin and thyroid hormone measurements and assessment of thyroid hormone transport. *In* The Thyroid: a Fundamental and Clinical Text. L.E. Braverman & R.D. Utiger, Eds.: 376–392. Lippincott-Raven. Philadelphia.
16. MORLEY, J.E. 1981. Neuroendocrine control of thyrotropin secretion. Endocrine Rev. **2:** 396–436.
17. SCANLON, M.F. & A.D. TOFT. 2000. Regulation of Thyrotropin Secretion. *In* The Thyroid: a Fundamental and Clinical Text, 8th Edition. L.E. Braverman & R.D. Utiger, Eds.: 234–253. Lippincott William & Wilkins. Philadelphia.
18. KOLLER, K.J. *et al.* 1987. Thyroid hormones regulate levels of thyrotropin-releasing hormone mRNA in the paraventricular nucleus. Proc. Natl. Acad. Sci. USA. **84:** 7329–7333.
19. RONDEEL, J.M.M. *et al.* 1989. In vivo hypothalamic release of thyrotropin-releasing hormone after electrical stimulation of the paraventricular area: comparison between push-pull perfusion technique and collection of hypophysial portal blood. Endocrinology **125:** 971–975.
20. GREER, M.A. *et al.* 1993. Evidence that the major physiological role of TRH in the hypothalamic paraventricular nuclei may be to regulate the set-point for thyroid hormone negative feedback on the pituitary thyrotroph. Neuroendocrinology. **57:** 569–575.

21. TAYLOR, T. et al. 1990. The paraventricular nucleus of the hypothalamus has a major role in thyroid hormone feedback regulation of thyrotropin synthesis and secretion. Endocrinology **126:** 317–324.
22. ST. GERMAIN, D.L. & V.A. GALTON. 1997. The deiodinase family of selenoproteins. Thyroid **7:** 655–668.
23. NISHII, Y. et al. 1993. Induction of cytosolic triiodo-L-thyronine (T3) binding protein (CTBP) by T3 in primary cultured rat hepatocytes. Endocr. J. **40:** 399–404.
24. MORI, J. et al. 2002. Nicotinamide adenine dinucleotide phosphate-dependent cytosolic T(3) binding protein as a regulator for T(3)-mediated transactivation. Endocrinology **143:** 1538–1544.
25. BERNAL, J., A. GUADANO-FERRAZ & B. MORTE. 2003. Perspectives in the study of thyroid hormone action on brain development and function. Thyroid. **13:** 1005–1012.
26. BERNAL, J. 2005. The significance of thyroid hormone transporters in the brain. Endocrinology. **146:** 1698–1700.
27. HEUER, H. et al. 2005. The monocarboxylate transporter 8 linked to human psychomotor retardation is highly expressed in thyroid hormone-sensitive neuron populations. Endocrinology **146:** 1701–1706.
28. BRAVERMAN, L.E. et al. 2005. The effect of perchlorate, thiocyanate, and nitrate on thyroid function in workers exposed to perchlorate long-term. J. Clin. Endocrinol. Metab. **90:** 700–706.
29. MESTMAN, J.H. 1999. Diagnosis and management of maternal and fetal thyroid disorders. Curr. Opin. Obstet. Gynecol. **11:** 167–175.
30. BURMEISTER, L.A., J. PACHUCKI & D.L. ST. GERMAIN. 1997. Thyroid hormones inhibit type 2 iodothyronine deiodinase in the rat cerebral cortex by both pre- and posttranslational mechanisms. Endocrinology **138:** 5231–5237.
31. EPA, U. S. 1998. Perchlorate environmental contamination: toxicological review and risk characterization based on emerging information. http://www.epa.gov/ncea/perch.htm.
32. EPA, U.S. 1999. Perchlorate. http://www.epa.gov/OGWDW/ccl/perchlor/perchlo.html.
33. URBANSKY, E.T. 1998. Perchlorate chemistry: implications for analysis and remediation. Bioremed. J. **2:** 81–95.
34. WOLFF, J. 1998. Perchlorate and the thyroid gland. Pharmacol. Rev. **50:** 89–105.
35. GREER, M.A. et al. 2002. Health effects assessment for environmental perchlorate contamination: the dose response for inhibition of thyroidal radioiodine uptake in humans. Environ. Health Perspect. **110:** 927–937.
36. VULSMA, T. 1991. Etiology and Pathogenesis of Congenital Hypothyroidism. Evaluation and Examination of Patients Detected by Neonatal Screening in the Netherlands. Academisch Proefschrift, Amsterdam.
37. VAN DEN HOVE, M.F. et al. 1999. Hormone synthesis and storage in the thyroid of human preterm and term newborns: effect of thyroxine treatment. Biochimie. **81:** 563–570.
38. SAVIN, S. et al. 2003. Thyroid hormone synthesis and storage in the thyroid gland of human neonates. J. Pediatr. Endocrinol. Metab. **16:** 521–528.
39. SPITZWEG, C. et al. 1998. Analysis of human sodium iodide symporter gene expression in extrathyroidal tissues and cloning of its complementary deoxyri-

bonucleic acids from salivary gland, mammary gland, and gastric mucosa. J. Clin. Endocrinol. Metab. **83:** 1746–1751.
40. RILLEMA, J.A. & D.L. ROWADY. 1997. Characteristics of the prolactin stimulation of iodide uptake into mouse mammary gland explants. Proc. Soc. Exp. Biol. Med. **215:** 366–369.
41. RILLEMA, J.A., T.X. YU & S.M. JHIANG. 2000. Effect of prolactin on sodium iodide symporter expression in mouse mammary gland explants. Am. J. Physiol. Endocrinol. Metab. **279:** E769–772.
42. PERRON, B. *et al.* 2001. Cloning of the mouse sodium iodide symporter and its expression in the mammary gland and other tissues. J. Endocrinol. **170:** 185–196.
43. HOWARD, B.J. *et al.* 1996. A review of countermeasures to reduce radioiodine in milk of dairy animals. Health Physics. **71:** 661–673.
44. MOUNTFORD, P.J. *et al.* 1986. Transfer of radioiodide to milk and its inhibition. Nature **322:** 600.
45. KIRK, A.B. *et al.* 2005. Perchlorate and Iodide in dairy and breast milk. Environ. Sci. Technol. **39:** 2011–2017
46. VAN VLIET, G. 1999. Neonatal hypothyroidism: Treatment Outcome. Thyroid **9:** 79–84.
47. NAS. 2005. Health Implications of Perchlorate Ingestion.
48. HEYERDAHL, S., F.B. KASE & S.O. LIE. 1991. Intellectual development in children with congenital hypothyroidism is relation to recommended thyroxine treatment. J. Pediatr. **118:** 850–857.
49. HEYERDAHL, S. 2001. Longterm outcome in children with congenital hypothyroidism. Acta. Paediatr. **90:** 1220–1222.
50. HEYERDAHL, S. & B. OERBECK. 2003. Congenital hypothyroidism: developmental outcome in relation to levothyroxine treatment variables. Thyroid **13:** 1029–1038.
51. OERBECK, B. *et al.* 2003. Congenital hypothyroidism: influence of disease severity and L-thyroxine treatment on intellectual, motor, and school-associated outcomes in young adults. Pediatrics. **112:** 923–930.
52. FIELDS, C., M. DOURSON & J. BORAK. 2005. Iodine-deficient vegetarians: a hypothetical perchlorate-susceptible population? Regul. Toxicol. Pharmacol. **42:** 37–46.
53. CRUMP, K.S. & J.P. GIBBS. 2005. Benchmark calculations for perchlorate from three human cohorts. Environ. Health Perspect. **113:** 1001–1008.
54. GINSBERG, G. & D. RICE. 2005. The NAS Perchlorate Review: questions remain about the perchlorate RfD. Environ. Health Perspect. **113:** 1117–1119.
55. MCKINNEY, J.D. & C.L. WALLER. 1994. Polychlorinated biphenyls as hormonally active structural analogues. Environ. Health Perspect. **102:** 290–297.
56. MCKINNEY, J.D. & C.L. WALLER. 1998. Molecular determinants of hormone mimicry: halogenated aromatic hydrocarbon environmental agents. J. Toxicol. Environ. Health B. Crit. Rev. **1:** 27–58.
57. CHANA, A. *et al.* 2002. Computational studies on biphenyl derivatives. Analysis of the conformational mobility, molecular electrostatic potential, and dipole moment of chlorinated biphenyl: searching for the rationalization of the selective toxicity of polychlorinated biphenyls (PCBs). Chem. Res. Toxicol. **15:** 1514–1526.

58. BREIVIK, K. et al. 2002. Towards a global historical emission inventory for selected PCB congeners—a mass balance approach. 1. Global production and consumption. Sci. Total. Environ. **290:** 181–198.
59. FISHER, B.E. 1999. Most unwanted. Environ. Health Perspect. **107:** A18–23.
60. HUISMAN, M. et al. 1995. Neurological condition in 18-month-old children perinatally exposed to polychlorinated biphenyls and dioxins. Early Hum.f Dev. **43:** 165–176.
61. JACKSON, T.A. et al. 1997. The partial agonist activity of antagonist-occupied steroid receptors is controlled by a novel hinge domain-binding coactivator L7/SPA and the corepressors N-CoR or SMRT. Mol. Endocrinol. **11:** 693–705.
62. OSIUS, N. et al. 1999. Exposure to polychlorinated biphenyls and levels of thyroid hormones in children. Environ. Health Perspect. **107:** 843–849.
63. KORRICK, S.A. & L. ALTSHUL. 1998. High breast milk levels of polychlorinated biphenyls (PCBs) among four women living adjacent to a PCB-contaminated waste site. Environ. Health Perspect. **106:** 513–518.
64. AYOTTE, P. et al. 2003. Assessment of pre- and postnatal exposure to polychlorinated biphenyls: lessons from the inuit cohort study. Environ. Health Perspect. **111:** 1253–1258.
65. WALKOWIAK, J. et al. 2001. Environmental exposure to polychlorinated biphenyls and quality of the home environment: effects on psychodevelopment in early childhood. Lancet **358:** 1602–1607.
66. STEWART, P. et al. 2003. Prenatal PCB exposure, the corpus callosum, and response inhibition. Environ. Health Perspect. **111:** 1670–1677.
67. STEWART, P. et al. 2005. Response inhibition at 8 and 9 1/2 years of age in children prenatally exposed to PCBs. Neurotoxicol. Teratol. **27:** 771–780.
68. ZOELLER, R.T. A.L. DOWLING & A.A. VAS. 2000. Developmental exposure to polychlorinated biphenyls exerts thyroid hormone-like effects on the expression of RC3/neurogranin and myelin basic protein messenger ribonucleic acids in the developing rat brain. Endocrinol. **141:** 181–189.
69. GOLDEY, E.S. et al. 1995. Developmental exposure to polychlorinated biphenyls (Aroclor 1254) reduces circulating thyroid hormone concentrations and causes hearing deficits in rats. Toxicol. Appl. Pharmacol. **135:** 77–88.
70. BASTOMSKY, C.H. 1977. Goitres in rats fed polychlorinated biphenyls. Can. J. Physiol. Pharmacol. **55:** 288–292.
71. BASTOMSKY, C.H. 1977. Enhanced thyroxine metabolism and high uptake goiters in rats after a single dose of 2,3,7,8-tetrachlorodibenzo-p-dioxin. Endocrinology. **101:** 292–296.
72. BASTOMSKY, C.H., P.V.N. MURTHY & K. BANOVAC. 1976. Alterations in thyroxine metabolism produced by cutaneous application of microscope immersion oil: effects due to polychlorinated biphenyls. Endocrinology. **98:** 1309–1314.
73. CROFTON, K.M. 2004. Developmental disruption of thyroid hormone: correlations with hearing dysfunction in rats. Risk Anal. **24:** 1665–1671.
74. CROFTON, K.M. et al. 2000. PCBs, thyroid hormones, and ototoxicity in rats: cross-fostering experiments demonstrate the impact of postnatal lactation exposure. Toxicol. Sci. **57:** 131–140.
75. BROUWER, A. et al. 1999. Characterization of potential endocrine-related health effects at low- dose levels of exposure to PCBs. Environ. Health Perspect. **107** (Suppl 4): 639–649.

76. GOLDEY, E.S. & K.M. CROFTON. 1998. Thyroxine replacement attenuates hypothyroxinemia, hearing loss, and motor deficits following developmental exposure to Aroclor 1254 in rats. Toxicol. Sci. **45:** 94–105.
77. KOIBUCHI, N. & W.W. CHIN. 2000. Thyroid hormone action and brain development. Trend. Endocrinol. Metab. **11:** 123–128.
78. LI, G.H. et al. 2004. Impact of thyroid hormone deficiency on the developing CNS: cerebellar glial and neuronal protein expression in rat neonates exposed to antithyroid drug propylthiouracil. Cerebellum **3:** 100–106.
79. YOUSEFI, B. et al. 2005. Postnatal changes of steroid receptor coactivator-1 immunoreactivity in rat cerebellar cortex. Thyroid **15:** 314–319.
80. ROEGGE, C.S. et al. 2004. Motor impairment in rats exposed to PCBs and methylmercury during early development. Toxicol. Sci. **77:** 315–324.
81. NGUON, K., M.G. BAXTER & E.M. SAJDEL-SULKOWSKA. 2005. Perinatal exposure to polychlorinated biphenyls differentially affects cerebellar development and motor functions in male and female rat neonates. Cerebellum **4:** 112–122.
82. GRANHOLM, A.C. 1985. Effects of thyroid hormone deficiency on glial constituents in developing cerebellum of the rat. Exp. Brain Res. **59:** 451–456.
83. TRAGGIAI, C. & R. STANHOPE. 2004. Body mass index and hypothalamic morphology on MRI in children with congenital midline cerebral abnormalities. J. Pediatr. Endocrinol. Metab. **17:** 219–221.
84. SCHOONOVER, C.M. et al. 2004. Thyroid hormone regulates oligodendrocyte accumulation in developing rat brain white matter tracts. Endocrinology **145:** 5013–5020.
85. BERBEL, P. et al. 1994. Role of thyroid hormones in the maturation of interhemispheric connections in rats. Behav. Brain. Res. **64:** 9–14.
86. GAUGER, K.J. et al. 2004. Polychlorinated biphenyls (PCBs) exert thyroid hormone-like effects in the fetal rat brain but do not bind to thyroid hormone receptors. Environ. Health Perspect. **112:** 516–523.
87. BANSAL, R. et al. 2005. Maternal thyroid hormone increases HES expression in the fetal rat brain: An effect mimicked by exposure to a mixture of polychlorinated biphenyls (PCBs). Brain Res. Dev. Brain Res. **156:** 13–22.
88. KITAMURA, S. et al. 2005. Thyroid hormone-like and estrogenic activity of hydroxylated PCBs in cell culture. Toxicology **208:** 377–387.
89. FRITSCHE, E. et al. 2005. Polychlorinated biphenyls disturb differentiation of normal human neural progenitor cells: clue for involvement of thyroid hormone receptors. Environ. Health Perspect. **113:** 871–876.
90. ARULMOZHIRAJA, S. & M. MORITA. 2004. Structure-activity relationships for the toxicity of polychlorinated dibenzofurans: approach through density functional theory-based descriptors. Chem. Res. Toxicol. **17:** 348–356.
91. KIMURA-KURODA, J., I. NAGATA & Y. KURODA. 2005. Hydroxylated metabolites of polychlorinated biphenyls inhibit thyroid-hormone-dependent extension of cerebellar Purkinje cell dendrites. Brain Res. Dev. Brain Res. **154:** 259–263.
92. BOGAZZI, F. et al. 2003. Effects of a mixture of polychlorinated biphenyls (Aroclor 1254) on the transcriptional activity of thyroid hormone receptor. J. Endocrinol. Invest. **26:** 972–978.
93. IWASAKI, T. et al. 2002. Polychlorinated biphenyls suppress thyroid hormone-induced transactivation. Biochem. Biophys. Res. Commun. **299:** 384–388.
94. MIYAZAKI, W. et al. 2004. Polychlorinated biphenyls suppress thyroid hormone receptor-mediated transcription through a novel mechanism. J. Biol. Chem. **279:** 18195–18202.

95. KATO, Y. *et al.* 1998. Reduction of thyroid hormone levels by methylsulfonyl metabolites of polychlorinated biphenyl congeners in rats. Arch. Toxicol. **72:** 541–554.
96. REPORTER, C.M. 1999. ChemExpo Chemical Profile: Bisphenol-A. Schnell Publishing Company. Sutton, England.
97. HOWE, S.R., L. BORODINSKY & R.S. LYON. 1998. Potential exposure to bisphenol A from food-contact use of epoxy coated cans. J. Coatings Technology **70:** 69–74.
98. LEWIS, J.B. *et al.* 1999. Identification and characterization of estrogen-like components in commercial resin-based dental restorative materials. Clin. Oral Investig. **3:** 107–113.
99. SCHONFELDER, G. *et al.* 2002. Parent bisphenol A accumulation in the human maternal-fetal-placental unit. Environ. Health Perspect. **110:** A703–A707.
100. IKEZUKI, Y. *et al.* 2002. Determination of bisphenol A concentrations in human biological fluids reveals significant early prenatal exposure. Hum. Reprod. **17:** 2839–2841.
101. WHO, E. H. C. 1995. Tetrabromobisphenol A and derivatives. World Health Organization. Brussels, Belgium.
102. WHO, E. H. C. 1997. Flame-retardants: a general introduction. World Health Organization. Brussels, Belgium.
103. THOMSEN, C., E. LUNDANES & G. BECHER. 2002. Brominated flame retardants in archived serum samples from Norway: a study on temporal trends and the role of age. Environ. Sci. Technol. **36:** 1414–1418.
104. MORIYAMA, K. *et al.* 2002. Thyroid hormone action is disrupted by bisphenol A as an antagonist. J. Clin. Endocrinol. Metab. **87:** 5185–5190.
105. STAPLES, C.A. *et al.* 1998. A review of the environmental fate, effects, and exposures of bisphenol A. Chemosphere **36:** 2149–2173.
106. KRISHNAN, A.V. *et al.* 1993. Bisphenol A: an estrogenic substance is released from polycarbonate flasks during autoclaving. Endocrinology **132:** 2279–2286.
107. GAIDO, K.W. *et al.* 1997. Evaluation of chemicals with endocrine modulating activity in yeast-based steroid hormone receptor gene transcription assay. Toxicol. Appl. Pharmacol. **143:** 205–212.
108. KITAMURA, S. *et al.* 2002. Thyroid hormonal activity of the flame retardants tetrabromobisphenol A and tetrachlorobisphenol A. Biochem. Biophys. Res. Commun. **293:** 554–559.
109. MORIYAMA, K. *et al.* 2002. Thyroid hormone action is dusrupted by bisphenol A as an antagonist. J. Clin. Endocrinol. Metab. **87:** 5185–5190.
110. CHENG, S.Y. 2005. Thyroid hormone receptor mutations and disease: beyond thyroid hormone resistance. Trends Endocrinol. Metab. **16:** 176–182.
111. ZOELLER, R.T., R. BANSAL & C. PARRIS. 2005. Bisphenol-A, an environmental contaminant that acts as a thyroid hormone receptor antagonist *in vitro*, increases serum thyroxine, and alters RC3/neurogranin expression in the developing rat brain. Endocrinology **146:** 607–612.
112. SEIWA, C. *et al.* 2004. Bisphenol A exerts thyroid-hormone-like effects on mouse oligodendrocyte precursor cells. Neuroendocrinology **80:** 21–30.
113. ISHIDO, M. *et al.* 2004. Bisphenol A causes hyperactivity in the rat concomitantly with impairment of tyrosine hydroxylase immunoreactivity. J. Neurosci. Res. **76:** 423–433.
114. REFETOFF, S. *et al.* 1994. The syndromes of resistance to thyroid hormone: update 1994. *In* Clinical and Molecular Aspects of Diseases of the Thyroid, Vol. 3.

L.E. Braverman & S. Refetoff, Eds.: 336–343. Endocrine Society. Bethesda, MD.
115. TOGASHI, M. et al. 2005. Rearrangements in thyroid hormone receptor charge clusters that stabilize bound 3,5',5-triiodo-L-thyronine and inhibit homodimer formation. J. Biol. Chem. **280:** 25665–25673.
116. DOWLING, A.L.S. et al. 2000. Acute changes in maternal thyroid hormone induce rapid and transient changes in specific gene expression in fetal rat brain. J. Neurosci. **20:** 2255–2265.
117. DOWLING, A.L.S., E.A. IANNACONE & R.T. ZOELLER. 2001. Maternal hypothyroidism selectively affects the expression of neuroendocrine-specific protein-A messenger ribonucleic acid in the proliferative zone of the fetal rat brain cortex. Endocrinology **142:** 390–399.
118. DOWLING, A.L.S. & R.T. ZOELLER. 2000. Thyroid hormone of maternal origin regulates the expression of RC3/Neurogranin mRNA in the fetal rat brain. Brain Res. **82:** 126–132.
119. WU, Y. et al. 2003. Hes1 but not Hes5 regulates an astrocyte versus oligodendrocyte fate choice in glial restricted precursors. Dev. Dyn. **226:** 675–689.
120. SCHUURMANS, C. & F. GUILLEMOT. 2002. Molecular mechanisms underlying cell fate specification in the developing telencephalon. Curr. Opin. Neurobiol. **12:** 26–34.
121. GAIANO, N. & G. FISHELL. 2002. The role of notch in promoting glial and neural stem cell fates. Annu. Rev. Neurosci. **25:** 471–490.
122. SHARLIN, D.S., R. BANSAL & R.T. ZOELLER. 2006. Polychlorinated biphenyls exert selective effects on cellular composition of white matter in a manner inconsistent with thyroid hormone insufficiency. Endocrinology **147:** 846–858.
123. SEGERSEN, T.P. et al. 1987. Localization of thyrotropin-releasing hormone prohormone messenger ribonucleic acid in rat brain by in situ hybridization. Endocrinology **121:** 98–107.
124. SEGERSEN, T.P. et al. 1987. Thyroid hormone regulates TRH biosynthesis in the paraventricular nucleus of the rat hypothalamus. Science **238:** 78–80.
125. LECHAN, R.M., P. WU & I.M.D. JACKSON. 1986. Immunolocalization of the thyrotropin-releasing hormone prohormone in the rat central nervous system. Endocrinology **119:** 1210–1216.
126. JACKSON, I.M.D., P. WU & R.M. LECHAN. 1985. Immunohistochemical localization in the rat brain of the precursor for thyrotropin releasing hormone. Science **229:** 1097–1099.
127. ISHIKAWA, K. et al. 1988. Immunocytochemical delineation of the thyrotrophic area: origin of thyrotropin-releasing hormone in the median eminence. Neuroendocrinology **47:** 384–388.
128. MERCHENTHALER, I. & Z. LIPOSITS. 1994. Mapping of thyrotropin-releasing hormone (TRH) neuronal systems of rat forebrain projecting to the median eminence and the OVLT. Immunocytochemistry combined with retrograde labeling at the light and electron microscopic levels. Acta. Biol. Hung. **45:** 361–374.
129. MARTIN, J.B. & S. REICHLIN. 1987. Clinical Neuroendocrinology. F.A. Davis Company. Philadelphia, PA.
130. AIZAWA, T. & M.A. GREER. 1981. Delineation of the hypothalamic area controlling thyrotropin secretion in the rat. Endocrinology **109:** 1731–1738.
131. HAISENLEDER, D.J. et al. 1992. Differential actions of thyrotropin (TSH)-releasing hormone pulses in the expression of prolactin and TSH subunit messenger

ribonucleic acid in rat pituitary cells *in vitro*. Endocrinology **130:** 2917–2923.
132. HAREL, G. *et al.* 1993. Effect of thyroid hormone deficiency on glycosylation of rat TSH secreted *in vitro*. Horm. Metab. Res. **25:** 278–280.
133. LIPPMAN, S.S., S. AMR & B.D. WEINTRAUB. 1986. Discordant effects of thyrotropin (TSH)-releasing hormone on pre- and posttranslational regulation of TSH biosynthesis in rat pituitary. Endocrinology **119:** 343–348.
134. MAGNER, J.A., J. KANE & E.T. CHOU. 1992. Intravenous thyrotropin (TSH)-releasing hormone releases human TSH that is structurally different from basal TSH. J. Clin. Endocrinol. Metab. **74:** 1306–1311.
135. TAYLOR, T. & B.D. WEINTRAUB. 1985. Thyrotropin (TSH)-releasing hormone regulation of TSH subunit biosynthesis and glycosylation in normal and hypothyroid rat pituitaries. Endocrinology **116:** 1968–1976.
136. TAYLOR, T., N. GESUNDHEIT & B.D. WEINTRAUB. 1986. Effects of *in vivo* bolus versus continuous TRH administration on TSH secretion, biosynthesis, and glycosylation in normal and hypothyroid rats. Mol. Cell. Endocrinol. **46:** 253–261.
137. WEINTRAUB, B.D. *et al.* 1989. Effect of TRH on TSH glycosylation and biological action. Ann N. Y. Acad. Sci. **553:** 205–213.
138. NIKRODHANOND, A.A. *et al.* 2006. Dominant role of thyrotropin-releasing hormone in the hypothalamic-pituitary-thyroid axis. J. Biol. Chem. **281:** 5000–5007.
139. WONDISFORD, F.E., J.A. MAGNER & B.D. WEINTRAUB. 1996. Chemistry and biosynthesis of thyrotropin. *In* The Thyroid: a Fundamental and Clinical Text. L.E. Braverman & R.D. Utiger, Eds.: 190–206. Lippincott and Raven. Philadelphia, PA.
140. HADLEY, M.E. 2000. Endocrinology. Prentice Hall. Upper Saddle River, NJ.
141. WONDISFORD, F.E., J.A. MAGNER & B.D. WEINTRAUB. 1996. Thyrotropin. *In* The Thyroid: a Fundamental and Clinical Text. L.E. Braverman & R.D. Utiger, Eds.: 190–206. Lippincott-Raven. Philadelphia, PA.
142. TAUROG, A., M.L. DORRIS & D.R. DOERGE. 1996. Minocycline and the thyroid: antithyroid effects of the drug, and the role of thyroid peroxidase in minocycline-induced black pigmentation of the gland. Thyroid. **6:** 211–219.
143. PRUMMEL, M.F. *et al.* 2000. Expression of the thyroid-stimulating hormone receptor in the folliculo-stellate cells of the human anterior pituitary. J. Clin. Endocrinol. Metab. **85:** 4347–4353.
144. THEODOROPOULOU, M. *et al.* 2000. Thyrotrophin receptor protein expression in normal and adenomatous human pituitary. J. Endocrinol. **167:** 7–13.
145. PRUMMEL, M.F., L.J. BROKKEN & W.M. WIERSINGA. 2004. Ultra short-loop feedback control of thyrotropin secretion. Thyroid **14:** 825–829.
146. SCHUSSLER, G.C. 2000. The thyroxine-binding proteins. Thyroid **10:** 141–149.
147. FINK, I.L. *et al.* 1986. Complete amino acid sequence of human thyroxine-binding globulin deduced from cloned DNA: close homology to the serine antiproteases. Proc. Natl. Acad. Sci. USA. **83:** 7708–7712.
148. JANSSEN, O.E. *et al.* 2002. Characterization of T(4)-binding globulin cleaved by human leukocyte elastase. J. Clin. Endocrinol. Metab. **87:** 1217–1222.
149. GRASBERGER, H. *et al.* 2002. Loop variants of the serpin thyroxine-binding globulin: implications for hormone release upon limited proteolysis. Biochem. J. **365:** 311–316.

150. VRANCKX, R. et al. 1990. The hepatic biosynthesis of rat thyroxine binding globulin (TBG): demonstration, ontogenesis, and up-regulation in experimental hypothyroidism. Biochem. Biophys. Res. Commun. **167:** 317–322.
151. VRANCKX, R. et al. 1994. Regulation of rat thyroxine-binding globulin and transthyretin: studies in thyroidectomized and hypophysectomized rats given tri-iodothyronine or/and growth hormone. J. Endocrinol. **142:** 77–84.
152. RICHARDSON, S.J. et al. 2005. Developmentally regulated thyroid hormone distributor proteins in marsupials, a reptile, and fish. Am. J. Physiol. Regul. Integr. Comp. Physiol. **288:** R1264–R1272.
153. FRANKLYN, J.A. et al. 1987. Effect of hypothyroidism and thyroid hormone replacement *in vivo* on pituitary cytoplasmic concentrations of thyrotropin-ß and alpha-subunit messenger ribonucleic acids. Endocrinology **120:** 2279–2288.
154. MIRELL, C.J. et al. 1987. Influence of thyroidal status on pituitary content of thyrotropin ß- and alpha-subunit, growth hormone, and prolactin messenger ribonucleic acids. Mol. Endocrinol. **1:** 408–412.
155. SHUPNIK, M.A. & E.C. RIDGWAY. 1987. Thyroid hormone control of thyrotropin gene expression in rat anterior pituitary cells. Endocrinology **121:** 619–624.
156. ZOELLER, R.T., R.S. WOLFF & K.J. KOLLER. 1988. Thyroid hormone regulation of messenger ribonucleic acid encoding thyrotropin (TSH)-releasing hormone is independent of the pituitary gland and TSH. Mol. Endocrinol. **2:** 248–252.
157. LAGRADI, G. et al. 1997. Leptin prevents fasting-induced suppression of prothyrotropin-releasing hormone messenger ribonucleic acid in neurons of the hypothalamic paraventricular nucleus. Endocrinology **138:** 2569–2576.
158. FEKETE, C. et al. 2000. Association of cocaine- and amphetamine-regulated transcript-immunoreactive elements with thyrotropin-releasing hormone-synthesizing neurons in the hypothalamic paraventricular nucleus and its role in the regulation of the hypothalamic-pituitary-thyroid axis during fasting. J. Neurosci. **20:** 9224–9234.
159. ROTI, E. et al. 1993. The use and misuse of thyroid hormone. Endocr. Rev. **14:** 401–423.
160. CHOPRA, I.J. 1996. Nature, source, and relative significance of circulating thyroid hormones. *In* The Thyroid: a Fundamental and Clinical Text. L.E. Braverman & R.D. Utiger, Eds.. 111–124. Lippincott-Raven. Philadelphia, PA.
161. ANDERSEN, S. et al. 2003. Biologic variation is important for interpretation of thyroid function tests. Thyroid **13:** 1069–1078.
162. HANSEN, P.S. et al. 2004. Major genetic influence on the regulation of the pituitary-thyroid axis: a study of healthy Danish twins. J. Clin. Endocrinol. Metab. **89:** 1181–1187.
163. OPPENHEIMER, J.H. 1983. The nuclear receptor-triiodothyronine complex: relationship to thyroid hormone distribution, metabolism, and biological action. *In* Molecular Basis of Thyroid Hormone Action. J.H. Oppenheimer & H.H. Samuels, Eds.: 1–35. Academic Press. New York, NY.
164. EVERTS M.E. et al. 1994. Uptake of triiodothyroacetic acid and its effect on thyrotropin secretion in cultured anterior pituitary cells. Endocrinology **135:** 2700–2707.
165. EVERTS M.E. et al. 1994. Uptake of thyroxine in cultured anterior pituitary cells of euthyroid rats. Endocrinology **134:** 2490–2497.

166. EVERTS M.E. *et al.* 1995. Uptake of 3,5',5,5'-tetraiodothyroacetic acid and 3,3',5'-triiodothyronine in cultured rat anterior pituitary cells and their effects on thyrotropin secretion. Endocrinology **136:** 4454–4461.
167. KRAGIE, L. 1996. Membrane Iodothyronine transporters, Part II: review of protein biochemistry. Endocr. Res. **22:** 95–119.
168. DOCTER, R. *et al.* 1997. Expression of rat liver cell membrane transporters for thyroid hormone in Xenopus laevis oocytes. Endocrinology **138:** 1841–1846.
169. MOREAU, X., P.J. LEJEUNE & R. JEANNINGROS. 1999. Kinetics of red blood cell T3 uptake in hypothyroidism with or without hormonal replacement, in the rat. J. Endocrinol. Invest. **22:** 257–261.
170. FRIESEMA, E.C.H. *et al.* 1999. Identification of thyroid hormone transporters. Biochem. Biophys. Res. Commun. **254:** 497–501.
171. HOOD, A. & C.D. KLAASSEN. 2000. Differential effects of microsomal enzyme inducers on in vitro thyroxine (T(4)) and triiodothyronine (T(3)) glucuronidation. Toxicol. Sci. **55:** 78–84.
172. HOOD, A. & C.D. KLAASSEN. 2000. Effects of microsomal enzyme inducers on outer-ring deiodinase activity toward thyroid hormones in various rat tissues. Toxicol. Appl. Pharmacol. **163:** 240–248.
173. LAZAR, M.A. 1993. Thyroid hormone receptors: multiple forms, multiple possibilities. Endocr. Rev. **14:** 184–193.
174. LAZAR, M.A. 1994. Thyroid hormone receptors: update 1994. Endocr. Rev. Monogr. **3:** 280–283.
175. OPPENHEIMER, J.H., H.L. SCHWARTZ & K.A. STRAIT. 1994. Thyroid hormone action 1994: the plot thickens. Eur. J. Endocrinol. **130:** 15–24.
176. MANGELSDORF, D.J. & R.M. EVANS. 1995. The RXR heterodimers and orphan receptors. Cell **83:** 841–850.
177. OPPENHEIMER, J.H. & H.L. SCHWARTZ. 1997. Molecular basis of thyroid hormone-dependent brain development. Endocr. Rev. **18:** 462–475.
178. FRIESEMA, E.C. *et al.* 2005. Thyroid hormone transporters. Vitam. Horm. **70:** 137–167.
179. ALKEMADE, A. *et al.* 2006. Novel neuroanatomical pathways for thyroid hormone action in the human anterior pituitary. Eur. J. Endocrinol. **154:** 491–500.

Science and Policy in Risk Assessments of Chlorinated Ethenes

CHRISTINA RUDÉN

Department of Philosophy and the History of Technology Royal Institute of Technology, SE-100 44 Stockholm, Sweden

ABSTRACT: In this article the use of data obtained from standardized experimental methods, for example, as specified in OECD guidelines for the testing of chemicals, epidemiology data, and mechanism data obtained from nonstandardized experimental methods in carcinogen risk assessment is scrutinized using the most recent risk assessments made by International Agency for Research on Cancer (IARC), the MAK(MAK)-Kommission, World Health Organization (WHO), European Centre for Ecotoxicology and Toxicology of Chemicals (ECETOC), and American Conference of Governmental Industrial Hygienists (ACGIH) for the four chlorinated ethenes as examples. The analysis shows that there was little controversy among these risk assessors about the interpretation of standardized animal data. On the other hand, they differ in their interpretation of epidemiology data, in particular in their assessment of statistical significance including the use of meta-analyses, and in quality evaluation of studies initiated on the basis of *a priori* concerns for carcinogenicity. The selection of mechanism data for species extrapolation is diverse among these risk assessors. Furthermore, in some cases they refrain from transparently motivating significant claims about mechanisms of toxicity by avoiding to give (explicit) references to the sources of information forming the basis of these claims or conclusions. This practice is not according to the scientific standards that should be required of a risk assessment document, and it makes it difficult to follow the argumentation and consequently to scrutinize the scientific accuracy of the conclusions drawn. In this article it is concluded that in some of these risk assessment documents, the use of mechanism data is not according to the scientific standards that should be required. It is furthermore concluded that if the use of mechanism data in these documents are representative of risk assessments in general, then there is an urgent need for further development and implementation of quality criteria for the use of mechanism data in species extrapolation.

KEYWORDS: regulatory toxicology; health risk assessment; epidemiology; mechanism of toxicity; carcinogenicity; chlorinated ethenes

Address for correspondence: Christina Rudén, Department of Philosophy and the History of Technology, Royal Institute of Technology, Teknikringen 78B, SE-100 44 Stockholm, Sweden. Voice: +46-8-790-95-87; fax: +46-8-790-95-17.
 e-mail: cr@infra.kth.se

INTRODUCTION AND BACKGROUND

Several ambitious measures have been taken to make a systematic account of the regulatory process,[1,2] and attempts have even been made to harmonize risk assessment procedures as in the IPCS harmonization project (www.who.int/ipcs).[3] However, the combined effect of scientific uncertainties and a significant reliance on case-by-case assumptions made by individual experts has made it difficult to achieve a risk assessment process that is fully predictable and systematic in all aspects. It is therefore essential to continuously scrutinize and evaluate this process, and to provide systematic feedback from its practical performance. This requires careful systematic studies of the actual workings of the process.

A generally accepted principle in toxicological risk assessment is that adverse effects seen in animal studies indicate that the chemical under study will cause a similar effect in humans, and standardized animal experiments are typically used for the purpose of effect identification. The design and procedures of standardized experiments are laid down in detailed guidelines, such as the OECD Testing Guidelines and the principles of Good Laboratory Practices (GLP). Guideline studies are readily accepted in the risk assessment process,[4] and the interpretation of standardized data is furthermore aided by the classification and labeling criteria (European Directive 67/548). In combination, the test guidelines and the classification and labeling criteria provide a rather detailed and generally accepted set of evaluation and interpretation criteria for standardized animal experiments.

If data from exposed humans are available, no species extrapolation is necessary and therefore, high quality epidemiological data are usually assigned significant weight in the risk assessment process. The quality and strength of the epidemiological evidence are determined by the overall design of the study (accuracy of exposure assessment and disease characterization, control for confounding, and lack of selection biases, etc.), and on the resulting power of the study to detect an effect. The performance of an epidemiological study is dependent on the identification and enrolment of individual human beings, and thus dependent on the prerequisites available. Therefore, guidelines for evaluation of epidemiological study design cannot be as detailed as the evaluation guidelines for standardized animal experiments.[4]

Standardized animal experiments produce little information on the exact mechanism of toxicity. Instead, the use of nonstandardized experimental methods is often needed to obtain this kind of data. These methods include data on metabolism, absorption, and various aspects of toxicity, for example, cytotoxicity, macromolecule binding, etc.[4] (p. 89). The use of mechanism data in risk assessment has been established in, for example, the criteria for classification of carcinogenicity, that state that a substance should not be classified as a carcinogen if "the mechanism of experimental tumor formation is clearly identified, with good evidence that this process cannot be extrapolated to man"

(Directive 67/548/EEC, par. 4.2.1.2). However, how much and what kind of data are needed to consider a mechanism "clearly identified," and what is meant by "good evidence" is not further specified in this directive. According to the EC TGD, data obtained from nonstandardized methods should be evaluated on a case-by-case basis using expert judgment[4] (p. 87). The availability of detailed evaluation and interpretation criteria are thus limited also for nonstandardized (mechanism) data.

In this article the use of animal data, epidemiology, and mechanism data in risk assessment is scrutinized using examples from carcinogen risk assessments of chlorinated ethenes. It is hypothesized that the use of animal data in risk assessment is more stringent and predictable compared to the use of epidemiology and especially mechanism data.

MATERIALS AND METHODS

The Riskline database (available at www.kemi.se) was searched for carcinogen risk assessments made of the chlorinated ethenes: vinylchloride (Cas 75-01-4), dichloroethene (Cas 75-35-4), trichloroethene (Cas 79-01-6), and tetrachloroethene (Cas 127-18-4). Risk assessments documents were selected with the objective to obtain a similar set of highly influential risk assessment documents, published as recently as possible for each chemical. On the basis of these criteria the risk assessment documents made by the International Agency for Research on Cancer (IARC), The German MAK(MAK)-Kommission, the American Conference of Governmental Industrial Hygienists (ACGIH), the World Health Organization (WHO), and the European Centre for Ecotoxicology and Toxicology of Chemicals (ECETOC) were collected.

In each risk assessment document, the conclusions drawn were identified and described. A standardized method for this purpose was used, namely a cancer risk assessment index (CRAI) that has three parameters. The first parameter describes the conclusion on animal carcinogenicity based on animal data. The second parameter refers to the conclusions drawn from epidemiology, and the third parameter describes the overall conclusion regarding cancer risk to humans. The parameters are assigned a "+" if the conclusion drawn by the risk assessors indicates a carcinogenic potential, and if carcinogenicity is described as implausible, then a "−" is assigned.[5]

The CRAI index has a theoretical maximum of eight groups, but all risk assessments that I have studied to date belong to one of the following four groups:

(1) − − − not carcinogenic in animals, negative epidemiology, no/implausible human cancer risk.
(2) + − − carcinogenic in animals, negative epidemiology, no/implausible human cancer risk.

(3) + − + carcinogenic in animals, negative epidemiology, a plausible human cancer risk.
(4) + + + carcinogenic in animals, positive epidemiology, a plausible human cancer risk.

Result of the CRAI Categorization

The results of the CRAI categorization of these 20 risk assessment documents are shown in TABLE 1. The CRAI categorization was used to identify differences in how these expert groups have assessed the carcinogen risks with these substances. In particular, the CRAI categorization gives some indications on whether epidemiological data were important for the overall conclusion (as in the + + + group), or if animal data were extrapolated to assess the human risk (as in the + − + group), or if positive animal data were considered irrelevant to human risk (the + − − group). Thereafter the key data and assumptions used to arrive at different conclusions were identified and compared to enable a discussion of how these differences were scientifically motivated.

It should be noted that this is a study of risk assessments made in the past. The conclusions drawn in the respective risk assessment documents do not necessarily reflect the current views of the expert groups.

Analysis of the Reasons for Diverging Conclusions

Vinylchloride

There is consensus among the vinylchloride risk assessors that the epidemiological data and the animal data indicate an increased risk of liver tumors after exposure to this substance. Regarding the mechanism behind the carcinogenicity the conclusions differ, however. In four of the five risk assessments it is concluded that vinylchloride has a mutagenic/genotoxic potential

TABLE 1. CRAI categorization and year of publication for the 20 selected carcinogen risk assessments

	IARC	MAK	WHO	ACGIH	ECETOC
Vinylchloride	+ + +	+ + +	+ + +	+ + +	+ + +
	(1987)	(1993)	(1999)	(1999)	(1988)
Dichloroethene	+ − −	+ − +	+ − +	+ − −	+ − −
	(1999)	(1997)	(1990)	(1999)	(1985)
Trichloroethene	+ + +	+ + +	+ − −	− − −	+ − −
	(1995)	(1996)	(1985)	(1993)	(1994)
Tetrachloroethene	+ + +	+ − +	+ − −	+ − −	+ − −
	(1995)	(1992)	(1984)	(1993)	(1999)

(WHO,[6] MAK,[7] IARC,[8] ECETOC,[9]). In the ACGIH risk assessment, on the other hand, it is concluded that "cell proliferation" is the main mechanism behind vinylchloride-induced hepatic angiosarcoma, and the ACGIH furthermore claim that to the extent that vinylchloride-DNA adduct formation occurs, these will be "routinely repaired prior to cell replication, and only when DNA repair mechanisms are overwhelmed does genetic damage lead to the development of cancer"[10] (p. 11).

The conclusion about carcinogenic mechanism has direct implications for risk management of mutagenic or genotoxic chemicals. Since for this mechanism the generally accepted default assumption is that the cancer risk will increase with any exposure above zero, that is, there is no threshold dose under which exposure is considered safe. Along this line, the MAK-Kommission[7] and WHO[6] conclude that a safe limit cannot be determined for this chemical. But, not all risk assessors adhere to this assumption. The ECETOC explicitly concludes that vinylchloride is "mutagenic *in vitro* and *in vivo*"[9] (p. 95) but that this effect is seen only at high doses, and the ECETOC concludes that: "although it is not possible to set definitely safe levels of exposure for genotoxic carcinogens, the evidence presented in this report does not suggest that occupational exposure at current levels [≤ 3 ppm] presents any significant risk"[9] (p. 96).

Similarly, the ACGIH conclude that a safe level of exposure can be determined, and proposes that this level be set at 1 ppm.[10] This conclusion is based on the identification of: "[A]n approximate 6.5 ppm lowest-observed-effect-level (LOEL) for vinylchloride in a strong epidemiological study (Simonato *et al.*, 1991) and the demonstrated effectiveness of a 1 ppm occupational exposure level in preventing the occurrence of angiosarcoma in exposed workers"[10] (p. 12). IARC draws no explicit conclusion about the dose–response relationship.[8] For a summary of conclusions and the references used to substantiate them, see TABLE 2.

Dichloroethene (Vinylidenechloride)

The dichloroethene database includes little evidence from epidemiological studies, and the available human data has thus not been considered conclusive or positive in any of these risk assessments.

There are 18 long-term animal studies of dichloroethene covering rat, mouse, and hamster, as well as inhalation, oral, dermal, and subcutaneous routes of exposure. Seventeen of these animal studies are negative, while one study shows evidence of increased incidences of kidney adenocarcinoma in the mouse.

The 18 animal studies are of variable quality and many of them are not performed according to OECD guidelines, either due to an insufficient number of animals or because of the exposure duration being too short.[11] (pp. 123–127)

TABLE 2. A summary of the conclusions that these risk assessors draw about the vinylchloride mutagenic or genotoxic potential, their conclusions on dose–response, and the references to substantiate these conclusions

Risk assessment document	Conclusions on mutagenicity	Conclusion dose–response	Risk management implication	References used explicitly to depart from default assumptions
IARC 1987	Genotoxic and mutagenic; refers to positive data for: chromosomal aberrations and SCE in humans and rodents *in vivo*. Alkylation of DNA in rats and mice *in vivo*. SCE, mutations, USD, cell transformation *in vitro*. Sex linked recessive lethal mutations in Drosophila. Mutagenicity in plants and fungi. Gene conversion in yeast. DNA damage and mutations in bacteria binding to isolated DNA in the presence of a metabolizing system.	No explicit dose–response assessment.	No safe exposure level can be determined.	N.a.
MAK 1993	Genotoxic and mutagenic; VC metabolites chloroethylene oxide and chloroacetaldehyde, are reactive mutagenic compounds, which can cause chemical modification of cellular macromolecules, such as proteins and nucleic acids.	Linear and increasing from zero exposure.	No safe exposure level can be determined.	N.a.
WHO 1999	Genotoxic and mutagenic; VC is mutagenic and clastogenic in humans.	No explicit conclusion on dose–response.	? The exposures should be kept "as low as possible."	?
ACGIH 1999	(Mainly) nongenotoxic; cell proliferation...plays a significant role in the production of vinylchloride-induced hepatic angiosarcoma. Since DNA adducts are routinely repaired they will not lead to cancer at low doses.	Threshold effect	Exposure ≤ 1 ppm is safe	Maltoni *et al.*, 1981 (animal study) Simonato *et al.*, 1991 (epidemiology)
ECETOC 1988	Genotoxic and mutagenic; "vinylchloride causes chromosomal aberrations in human beings" and it is "mutagenic *in vitro* and *in vivo*."	Threshold effect	Exposure ≤ 3 ppm is safe	None

N.a. = not applicable; ? = no explicit comment; SCE = sister chromatid exchange; USD = unscheduled DNA synthesis; VC = vinylchloride. This is since the default assumptions (*a*) that animal data are relevant to assess human risk, and (*b*) that there is no dose–response threshold for genotoxic carcicogens, are adhered to.

Among the negative studies there are two studies from the National Toxicology Programme (gavage studies in the rat and the mouse) using 50 animals per group and sex, and exposing the animals for 24 months.

The positive study was performed by Maltoni and co-workers. In this study 30–60 animals per group were used and exposure duration was 12 months (4 h per day inhalation) (Maltoni et al.,1977; 1984, as cited in MAK[11]). The positive mouse experiment is acknowledged in all risk assessments under study.[11,15] Regarding the mechanism behind the mouse kidney tumors, the MAK-Kommission noted that the carcinogenic effect in the mouse kidney was seen in combination with marked nephrotoxicity, but still this effect is considered potentially relevant to humans and dichloroethene is classified by MAK-Kommission as a suspected human carcinogen (MAK category IIIB).[11]

The WHO,[12] ECETOC,[13] and ACGIH[14] also conclude that toxicity and carcinogenicity were associated in this case. The WHO concludes that this is not sufficient to dismiss these data in the human risk assessment, in particular in the light of the (nonsignificant) increase in cancer risks seen in epidemiological studies[12] (p. 129f.). The ECETOC, on the other hand, conclude that the effect is considered species specific and probably not relevant to humans, and the ACGIH conclude that if exposures are kept low the "theoretical cancer risk" will be "minimize[d] if not eliminate[d] for all practical purposes"[14] (p. 4). Dichloroethene is considered by the ACGIH to be not classifiable as a human carcinogen (ACGIH group A4).

The IARC does not comment on the possible mechanism behind this effect and dichloroethene is considered not classifiable as a human carcinogen (IARC group 3).[15]

The overall conclusions on the carcinogenic potential of dichloroethene to humans among these risk assessors thus range from "not carcinogenic" or "not classifiable" (ECETOC,[13] ACGIH,[14] IARC[15]) to "suspected human carcinogen" (MAK,[11] WHO[12]). For a summary of conclusions and the references used to substantiate them, see TABLE 3.

Trichloroethene

In the trichloroethene case, two of the risk assessments under study have interpreted epidemiology to be positive, namely IARC[16] and MAK,[17] and three risk assessments concluded that epidemiology is negative (ACGIH,[18] ECETOC,[19] WHO[20]).

The IARC classification of trichloroethene as a "probable human carcinogen" is mainly based on a meta-analysis of the results reported in three epidemiological studies: Spirtas et al.,[21] Axelson et al.,[22] and Anttila et al.[23] (together with the data on animals, which were considered to be "sufficient"). The results in these epidemiological studies taken individually indicate no significantly increased incidences. A meta-analysis of the data, however, shows

TABLE 3. A summary of the conclusions that these risk assessors draw about the dichloroethene mechanism of action, human relevance of animal data, and the references explicitly used to substantiate these conclusions

Risk assessment document	Conclusion mechanism	Human relevance of animal data	References used explicitly to depart from default assumptions
IARC 1987	None explicit	No comment; not classifiable as a human carcinogen	N.a.
MAK 1993	Nongenotoxic; kidney toxicity and associated cell replication	Potentially relevant; suspected human carcinogen	Maltoni et al., 1977, 1984
WHO 1999	Nongenotoxic; or through "repeated kidney damage" that "facilitates the expression of the genotoxic potential of metabolites."	Potentially relevant; the consistency of excess cancer deaths in epidemiology (even though not statistically significant) "is worth mentioning."	None explicit.
ACGIH 1999	Nongenotoxic	Animal tumors not considered relevant to humans. Not classifiable as a human carcinogen	Reitz et al., 1980; Andersen et al., 1980; D'Souza et al., 1988
ECETOC 1988	Nongenotoxic; or through "repeated kidney damage" that "facilitates the expression of the genotoxic potential of metabolites."	Animal tumors probably not relevant to humans	None

N.a. = not applicable.

TABLE 4. The risk assessors' conclusions regarding epidemiology data for trichloroethene, and the references used to substantiate these conclusions

Risk assessment document	Conclusion on epidemiology	Key epidemiology studies
MAK 1996	Positive (renal cell tumors)	Henschler et al., 1995
IARC 1995	Positive (liver and biliary tract, and NHL)	Spirtas et al., 1991; Axelson et al., 1994; Anttila et al., 1995
ECETOC 1994	Negative	Spirtas et al., 1991; Axelson et al., 1994
ACGIH 1993	Negative	Spirtas et al., 1991
WHO 1985	Negative	N.a.

N.a. = not applicable.

an increased relative risk of cancer of the liver and biliary tract, and for non-Hodgkin's lymphoma (NHL)[16] (p. 135).

The MAK-Kommission[17] motivates their classification of trichloroethene as a human carcinogen by the German epidemiology data reported by Henschler et al.,[24] in combination with "the known mechanism of action of TCE"[17] (p. 202).

The epidemiological studies performed by Spirtas et al.,[21] Axelson et al.,[22] Anttila et al.,[23] and Henschler et al.[24] were thus considered key studies by the IARC and the MAK- Kommission. In TABLE 4, the use of these particular studies in the selected risk assessments are summarized. It could be noted that the two risk assessors concluding that trichloroethene epidemiology is positive, base their conclusion on different studies, reporting carcinogenic effects in different organs.[5]

Trichloeroethene bioassays have, for example, provided evidence of liver tumors in the mouse, and kidney tumors in the rat after trichloroethene exposure. IARC,[16] MAK,[17] ECETOC,[19] and WHO[20] have all considered the trichloroethene animal data to be positive. In the ACGIH risk assessment, on the other hand, it is concluded that the mouse liver tumors were a result of using very high doses resulting in liver toxicity and that "tumours are not expected if liver injury does not occur"[18] (p. 2). (Data reporting kidney tumors in the rat are not referred to by the ACGIH).

The mouse liver tumors were by MAK and ECETOC considered to be causally connected to peroxisome proliferation and therefore not relevant to humans.[17,19] The IARC, on the other hand, concludes that the hypothesis that peroxisome proliferation caused the liver tumors seen in mice is plausible, however, they note that the epidemiological data corroborate an increased risk of cancer in the liver and biliary tract.[16] WHO has not included a discussion on carcinogenic mechanisms of the liver tumors in their assessment.[20]

Regarding the rat kidney tumors, the MAK-Kommission states that they are caused by genotoxic metabolites formed locally in the kidney and that this mechanism is corroborated by both animal and epidemiology data.[17] IARC[16] and WHO[20] make no explicit comment on the potential carcinogenic mechanism. (IARC classified trichloroethene as a probable human carcinogen, while WHO only gives very vague statements on the potential carcinogenicity of trichloroethene). The ECETOC,[19] on the other hand, concludes that the mechanism is unspecific toxicity due to high experimental doses, and finally, as already mentioned, the ACGIH does not refer to any of the data indicating an increase risk of kidney tumors even though these data were published well before the ACGIH risk assessment.[18,25,26] ACGIH considered trichloroethene "Not suspected as a human carcinogen." The motivation given for this is that "the substance has been demonstrated by well-controlled epidemiological studies not to be associated with any increased risk of cancer in humans"[18] (p. 4). For a summary of conclusions and the references used to substantiate them, see TABLE 5.

Tetrachloroethene (tetra)

Epidemiology has been considered positive in one risk assessment of tetra (IARC[27]), and negative in the others (ECETOC,[28] MAK,[29] ACGIH,[30] WHO[31]). The IARC refers to five cohort studies, namely Olsen et al.,[32] Blair et al.,[33] Spirtas et al.,[21] Anttila et al.,[23] and Ruder et al.,[34] and the IARC conclusion on epidemiology is based on findings of elevated relative risk for NHL in all three cohort studies in which such results were reported, that is, Blair et al.,[33] Spirtas et al.,[21] and Anttila et al.[23]

The ECETOC[28] refers to all these five cohort studies. ECETOC points at the studies published by Blair, Ruder, and Anttila as being the key studies since these three studies contain specific exposure information on tetra as well as subcohorts with main exposure to tetra. The ECETOC acknowledges the increased incidence of NHL in Spirtas et al.,[21] but this study is given little weight since the study was initiated because of concern for an excess risk of lymphatic and hematopoietic cancers[28] (p.139). The Anttila study is described by the ECETOC as positive on p. 143 ("Anttila et al. reported an excess of NHL in the subcohort of workers monitored for PER in blood"), and as negative on p.136 ("increased risks were seen for cancers at some sites, but none was significant").[28] Regarding the Blair study, ECETOC states that it reported an increased risk of "lymphosarcoma and reticulosarcoma," but that it did "not provide results for NHL as a cause of death." Furthermore, the ECETOC concludes that there was "no excess of deaths due to lymphosarcoma or reticulosarcoma in the Ruder study[34] and no excess of deaths due to other lymphatic or hematopoietic cancers." The ECETOC concludes that "available epidemiological studies were either negative or were not sufficient to provide evidence of a relationship between exposure to [tetra] and cancer in humans"[28] (p. 166). The use of these data in the risk assessment documents under study are summarized in TABLE 6.

TABLE 5. A summary of the trichloroethene risk assessments' conclusions about the mechanism of potential carcinogenicity including, the relevance of animal data and the references explicitly used to substantiate these claims

Risk assessment document	Conclusion mechanism	Human relevance of animal data	References used explicitly to depart from default assumptions
MAK 1996	Liver: peroxisome proliferation. Kidney: genotoxicity	Liver: not relevant. Kidney: relevant	Liver: none explicit. Kidney: N.a. (Pos epidemiology)
IARC 1995	Liver: peroxisome proliferation? Kidney: No comment	Liver: unknown. Kidney: relevant.	Liver: N.a. (Pos epidemiology). Kidney: none explicit
ECETOC 1994	Liver: peroxisome proliferation. Kidney: cytotoxicity	Liver: not relevant. Kidney: not relevant at low exposures	Liver: Bull et al., 1990, 1993; DeAngelo et al., 1992; Dees and Travis, 1993; Dekant et al., 1984, 1986; Elcombe et al., 1985; Green and Prout, 1985; Herren-Freund et al., 1987; Klaunig et al., 1989; Knadle et al., 1990; Larson and Bull, 1992a, b; Mirsalis et al., 1985; Randall and Sipes, 1984; Reddy et al., 1980; Stott et al., 1982; Watson et al., 1993. Kidney: Birner et al., 1993; Dekant et al., 1986b; Goldsworthy et al., 1988; Green et al., 1988, 1990; Seldén et al., 1993; Terracini and Parker, 1965
ACGIH 1993	Liver: cytotoxicity. Kidney: data not included	Liver: not relevant at low exposures. Kidney: none	Liver: Stott et al., 1982. Kidney: None
WHO 1985	Liver: no comment. Kidney: no comment	Liver: unknown. Kidney: unknown	Liver: none. Kidney: none

N.a. = not applicable.

TABLE 6. Interpretation of tetrachloroethene epidemiological data on NHL

	Anttila et al., 1995	Ruder et al., 1994	Spirtas et al., 1991	Blair et al., 1990
ECETOC 1999	?*	Neg.	(Pos.).**	Neg.
IARC 1995	Pos.	Not reported	Pos.	Pos.
ACGIH 1993	Not available	Not available	Not referred to	(Pos.)***
MAK 1992	Not available	Not available	Not referred to	(Pos.)***
WHO 1984	Not available	Not available	Not available	Not available

* This study is described by the ECETOC as positive on p. 143 ("Anttila et al. reported an excess of NHL..."), and as negative on p.136 ("Increased risks...but none was significant").
** Not considered a key study by the ECETOC since it was initiated due to *a priori* concerns about cancers in the lymphatic and hematopoietic tissues.
*** Not considered a key study by these risk assessors, due to mixed exposures.
? = no explicit comment.

All these risk assessors have concluded that tetrachloroethene has a carcinogenic potential in animals. The positive finding that has been most extensively discussed is hepatocellular carcinomas in mice. The MAK-Kommission[29] and IARC[27] considered tetrachloroethene as a "suspected" and "probable" human carcinogen respectively, on the basis of animal data. The ECETOC,[28] ACGIH,[30] and WHO,[31] on the other hand, have concluded that the effects seen in animal experiments are probably not relevant to humans.

The ECETOC[28] concludes that the occurrence of hepatocellular carcinomas in mice is causally connected to peroxisome proliferation, and since humans are not susceptible to peroxisome proliferation, this effect is not relevant to humans. This line of arguments is substantiated by four references namely Birner et al., 1998; Herren-Freund et al., 1987; Ikeda et al., 1972 (abstract); Ohtsuki et al., 1983 (as cited in ECETOC[28]). The MAK-Kommission argues a number of reasons why the animal data are not directly applicable to humans, for example, no clear dose–response, extremely high exposures (one reference is given to a U.S. EPA draft report on PBPK modeling), no direct pathological–anatomical equivalent in humans for the tumors seen in animals, and lack of genotoxicity. Besides the U.S. EPA report no explicit references are given to substantiate these arguments. The MAK-Kommission concludes that "the currently available data do not permit an unambiguous assessment of the carcinogenic potential of tetrachloroethene in man," but despite this, they acknowledge a "theoretical possibility that under certain conditions in certain organs genotoxic effects could also become manifest in man" (p. 4), and it is concluded that tetra is a suspected human carcinogen.[29]

The ACGIH concludes that "little or no genotoxicity has been reported from studies with Tetra," and the mouse liver tumors are considered irrelevant to humans since they are the result of peroxisome proliferation. No particular references are given that can be directly connected to this conclusion. ACGIH concludes that "Because Tetra was carcinogenic in mice and rats at relatively

TABLE 7. A summary of the tetra risk assessments' conclusions about the mechanism of potential carcinogenicity including the relevance of animal data and the references explicitly used to substantiate these claims

Risk assessment document	Conclusion mechanism	Human relevance of animal data	References used explicitly to depart from default assumptions
MAK 1992	Non-genotoxic?	Relevant "...theoretical possibility that under certain conditions in certain organs genotoxic effects could also become manifest in man."	Reitz and Nolen, 1986
IARC 1995	Not PP	Relevant poor correlation between PP and tumor formation in the mouse liver. Different mutation spectrums in tetrachloroethene-induced mouse liver tumors compared to those induced by trichloroethene.	Odum et al., 1988, Goldsworthy and Popp, 1987; Anna et al., 1994
ECETOC 1999	PP	Not relevant	Birner et al., 1998; Herren-Freund et al., 1987; Ohtsuki et al., 1983; Ikeda et al., 1972
ACGIH 1993	PP	Not relevant	No (explicit) references
WHO 1984	No comment	No comment	N.a.

N.a. = not applicable; PP = peroxisome proliferation; ? = no explicit comment.

high doses, an A3, Confirmed Animal Carcinogen with Unknown Relevance to Humans, notation is assigned to the substance"[30] (p. 4).

WHO concludes that "Tetra was found to be carcinogenic in mice but not in rats. Evidence from epidemiological studies...is insufficient to conclude that exposure to Tetra causes cancer in human beings" (p. 10). No specific comments (or references) regarding species specificity or carcinogenic mode of action are given.[31]

IARC concludes that the animal data cannot be dismissed by peroxisome proliferation since there is a poor correlation between peroxisome proliferation and tumor formation in the mouse liver.[27] For a summary of conclusions and references, see TABLE 7.

DISCUSSION AND CONCLUSIONS

The CRAI categorization shows that the interpretation of standardized animal data is similar for all these chemicals and risk assessment documents. One single exemption is the ACGIH interpretation of trichloroethene-induced mouse liver tumors that are considered not to be a direct effect of trichloroethene. There was thus little controversy among these risk assessors about the interpretation of standardized animal data.

These risk assessors differ in their interpretation of epidemiology data. In particular, in the assessment of statistical significance and the use of meta-analyses, and in data quality evaluation of studies initiated on the basis of *a priori* concerns for carcinogenicity. The results of this investigation furthermore suggest that the relative importance of epidemiology data, to animal data differ among these risk assessors. The ECETOC and the ACGIH seem less prone to consider a chemical carcinogenic to humans on the basis of animal data only.

The relevance of animal data to human risk assessment is often uncertain, since detailed knowledge about species differences is usually not available for individual chemicals. There may be either quantitative or qualitative differences in sensitivity or metabolism, and this may limit the relevance of interspecies comparisons. However, basing human health risk assessment on animal data is a standard procedure (since we do not perform toxicological experiments on human beings), based on the assumption that toxic and carcinogenic effects seen in mammalian experimental species are relevant to humans. This a generally accepted and scientifically motivated assumption since the physiology of the common test species is similar to that of humans.

Obviously, both under- and overestimation of human risk can result from interspecies differences, and in the absence of detailed knowledge about potential species differences it runs counter to generally accepted concepts of health protection to assume, without any supporting evidence, that humans are much less sensitive than the experimental species.

High quality human data therefore provide important information to a health risk assessment. A major difference between epidemiology and animal models is that epidemiology is predominantly an observational and not an experimental science. Epidemiologists study exposures and disease occurrence in a real-life setting, and are thus depending on a multitude of influences (a myriad of exposures, genetic aspects, human behavior, and lifestyle factors, etc.), many of which are interrelated and have strong confounding potential.

As a rough rule of thumb, epidemiological studies cannot reliably detect excess relative risks about 10% or smaller, and in many cases excess risk much higher than 10% may go undetected. For the more common types of cancer in industrial countries, lifetime risks are between 1% (leukemia) and 10% (breast cancer in Swedish women). Therefore, even in the more sensitive studies, the limits of an observable excess lifetime risk are in the order of 1/100 or 1/1000. These are risks that in many cases are considered unacceptable.[35]

Due to the obstacles in designing epidemiology studies, we can expect conclusive epidemiological data to become available only for a limited number of substances. It is furthermore clear that it should not be expected that all human carcinogens have sufficiently large effects to be detectable in epidemiological studies. Therefore, health risk assessment (and risk management) of chemicals cannot depend on the availability of epidemiological data.

Mechanism data used for species extrapolation are obtained from a wide range of methods and models. Furthermore, the selection of mechanism data by these risk assessors is also diverse and different risk assessors use different sets of data to substantiate their mechanistic arguments in species extrapolation. Second, there is diversity in how these references are incorporated and presented in the risk assessment document. Sometimes the mechanistic arguments and the corresponding scientific references are given in the same section, and in these cases it is relatively easy to follow the argumentation and to identify the scientific basis of it. In other cases, the mechanistic discussions and species extrapolations are presented in separate sections. This makes it sometimes difficult to follow the argumentation and to scrutinize the scientific accuracy of it. Avoiding to give (transparent) references to significant claims in a risk assessment document is not according to the scientific standards that should be required. This requirement is particularly important if the use of references aim at departuring from generally accepted default assumptions. If the use of mechanism data in these risk assessment documents is representative for risk assessments in general, then there is need for further development and implementation of quality criteria for the use of mechanism data in species extrapolation.

REFERENCES

1. EUROPEAN COMMISSION. 2000. First report on the harmonisation of risk assessment procedures, part 1, 26–27.
2. NRC, NATIONAL RESEARCH COUNCIL. 1994. Science and judgement in risk assessment. Committee on Risk Assessment of Hazardous Air Pollutants, Commission on Life Sciences. National Academy Press.Washington, DC
3. SONICH-MULLIN, C. R. et al. 2001. IPCS conceptual framework for evaluating a mode of action for chemical carcinogenesis. Reg. Tox. Pharmacol. **34:** 146–152.
4. EUROPEAN COMMISSION. 2003. Technical guidance document in support of the commission directive 93/67/EEC in risk assessment for new notified substances and the Commission regulation (EC) 1488/94 on risk assessment for existing substances. Available online at: www.ecb.it.
5. RUDÉN, C. 2001. Interpretations of primary carcinogenicity data in 29 trichloroethylene risk assessments. Toxicology **169:** 209–225.
6. WHO. 1999. Vinyl chloride. Environmental Health Criteria **215**.
7. MAK-KOMMISSION. 1993. Vinyl chloride. Toxikologisch-arbeitsmedizinische Begrundung von MAK-Werten.
8. IARC. 1987. Vinyl chloride. Suppl. 7 pp 373–374.
9. ECETOC. 1988. Vinyl chloride. Technical Report Vol 31.
10. ACGIH. 1999. Vinyl chloride. Documentation of the TLVs. Seventh edition.
11. MAK. 1997. Vinylidene chloride. Toxikologisch-arbeitsmedizinische Begrundung von MAK-Werten.
12. WHO. 1990. Vinylidene chloride. Environmental Health Criteria **100**.
13. ECETOC. 1985. Vinylindene chloride [sic]. Joint Assessment of Commodity Chemicals Vol 5.
14. ACGIH. 1999. Vinylidene chloride. Documentation of the TLVs. Seventh Edition.
15. IARC. 1999. Vinylidene chloride. Monographs Vol. **71**.
16. IARC. 1995. Trichloroethylene. Monographs Vol. **63**.

17. MAK. 1996. Trichloroethene. Toxikologisch-arbeitsmedizinische Begrundung von MAK-Werten.
18. ACGIH. 1993. Trichloroethylene. Documentation of the TLVs. Seventh edition.
19. ECETOC. 1994. Trichloroethylene. Technical report No 60.
20. WHO. 1985. Trichloroethene. Environmental Health Criteria **50**.
21. SPIRTAS, R. *et al.* 1991. Retrospective cohort mortality study of workers at an aircraft maintenance facility I. Epidemiological results. Br. J. Ind. Med. **48:** 515–530.
22. AXELSON, O. *et al.* 1994. Updated and expanded Swedish cohort study on trichloroethylene and cancer risk. J. Occup. Med. **36:** 556–562.
23. ANTTILA, A. *et al.* 1995. Cancer incidence among Finnish workers exposed to halogenated hydrocarbons. J. Occup. Environ. Med. **37:** 797–806.
24. HENSCHLER, D. *et al.* 1995. Increased incidence of renal cell tumors in a cohort of cardboard workers exposed to trichloroethylene. Arch. Toxicol. **69:** 291–299.
25. RUDÉN, C. 2003. Scrutinizing ACGIH risk assessments—the trichloroethylene case. Am. J. Ind. Med. **44:** 207–213.
26. Rudén, C. 2005. Re: Am J Ind Med **44:** 204–213, 2003. Response to the Letter to the Editor from Vickie L. Wells (ACGIH). Am. J. Ind. Med. 44: 464–466.
27. IARC. 1995. Tetrachloroethene. Monographs Vol. **63**.
28. ECETOC. 1999. Tetrachloroethylene. Joint Assessment of Commodity Chemicals **39**.
29. MAK. 1992. Tetrachloroethylene. Toxikologisch-arbeitsmedizinische Begrundung von MAK-Werten.
30. ACGIH. 1993. Tetrachloroethylene. Documentation of the TLVs. seventh edition.
31. WHO. 1984. Tetrachloroethylene. Environmental Health Criteria **31.**
32. OLSEN, G.W. *et al.* 1989. Mortality experience of a cohort of Louisiana chemical workers. J. Occup. Med. **31:** 32–34.
33. BLAIR, A. *et al.* 1990. Cancer and other causes of death among a cohort of dry cleaners. Br. J. Ind. Med. **47:** 162–168.
34. RUDER, A.M. *et al.* 1994. Cancer mortality in female and male dry-cleaning workers. J. Occup. Med. **36:** 867–874.
35. VAINIO, H. & L. TOMATIS. 1985. Exposure to carcinogens: scientific and regulatory aspects. Ann. Am. Con. Govt. Indus. Hygien. **2:** 135–143.

Potential Health Effects of Occupational Chlorinated Solvent Exposure

AVIMA M. RUDER

National Institute for Occupational Safety and Health, Centers for Disease Control and Prevention, Cincinnati, Ohio 45226, USA

ABSTRACT: Based on toxicology, metabolism, animal studies, and human studies, occupational exposure to chlorinated aliphatic solvents (methanes, ethanes, and ethenes) has been associated with numerous adverse health effects, including central nervous system, reproductive, liver, and kidney toxicity, and carcinogenicity. However, many of these solvents remain in active, large-volume use. This article reviews the recent occupational epidemiology literature on the most widely used solvents, methylene chloride, chloroform, trichloroethylene, and tetrachloroethylene, and discusses other chlorinated aliphatics. The impact of studies to date has been lessened because of small study size, inability to control for confounding factors, particularly smoking and mixed occupational exposures, and the lack of evidence for a solid pathway from occupational exposure to biological evidence of exposure, to precursors of health effects, and to health effects. International differences in exposure limits may provide a "natural experiment" in the coming years if countries that have lowered exposure limits subsequently experience decreased adverse health effects among exposed workers. Such decreases could provide some evidence that higher levels of adverse health effects were associated with higher levels of solvent exposure. The definitive studies, which should be prospective biomarker studies incorporating body burden of solvents as well as markers of effect, remain to be done.

KEYWORDS: chlorinated solvents; occupational exposure; trichloroethylene; tetrachloroethylene; health effects; cancer

BACKGROUND

Chlorinated methanes, ethanes, and ethenes are among the most widely used and useful chemical compounds. Chlorinated methanes are methyl chloride (CAS 74-87-3), methylene chloride (dichloromethane, CAS 75-09-2), chloroform (trichloromethane, CAS 67-66-3), and carbon tetrachloride (CAS 56-23-5), with 1–4 chlorines, respectively, substituted for hydrogens. The main use of

Address for correspondence: Avima Ruder, Ph.D., National Institute for Occupational Safety and Health, Mailstop R–16, 4676 Columbia Parkway, Cincinnati, OH 45226. Voice: 513-841-4440; fax: 513-458-7102.
  e-mail: amr2@cdc.gov

Ann. N.Y. Acad. Sci. 1076: 207–227 (2006). © 2006 New York Academy of Sciences.
doi: 10.1196/annals.1371.050

methyl chloride is in the manufacture of silicone; the other three compounds have been used as solvents, paint removers, degreasers, cleaning compounds, and chemical intermediates.

There are nine chlorinated ethanes. Most are used primarily to produce other chemicals rather than as solvents. The chief use of ethyl chloride (75-00-3) has been in the production of the gasoline fuel additive tetraethyl lead. As leaded gasoline use has declined, so has the use of ethyl chloride.[1] Ethylidene dichloride (1,1-dichloroethane, CAS 75-34-3) and ethylene dichloride (1,2-dichloroethane, CAS 107-06-2) are chemical intermediates in the production of chlorinated ethenes. Methyl chloroform (1,1,1-trichloroethane, CAS 71-55-6) was widely used as a solvent but is being phased out, except for use in closed systems.[2] The other trichloroethane, vinyl trichloride (1,1,2-trichloroethane, CAS 79-00-5), and the four higher chlorinated compounds (1,1,1,2-tetrachloroethane, CAS 630-20-6; 1,1,2,2-tetrachloroethane, CAS 79-34-5; pentachloroethane, CAS 76-01-7; and hexachloroethane, CAS 67-72-1) are all used mainly as chemical intermediates.

The five chlorinated ethenes include vinyl chloride (CAS 75-01-4) and vinylidene chloride (1,1-dichloroethylene, CAS 75-35-4), used primarily as precursors for polyvinyl chloride and polyvinylidene chloride (plastic wrap) respectively, and three solvents/degreasers: 1,2-dichloroethylene (CAS 540-59-0), trichloroethylene (TCE, CAS 79-01-6), and tetrachloroethylene (perchloroethylene, PCE, CAS 127-18-4). The latter two are widely used, especially for metal degreasing and clothes cleaning. Vinyl chloride has been classified as a human liver carcinogen, and occupational exposure has been limited by regulatory bodies in many countries. Exposure regulations for the other chlorinated ethenes are less restrictive.

Based on toxicology, metabolism, animal studies, and human studies, occupational exposure to chlorinated aliphatic solvents (methanes, ethanes, and ethenes) has been associated with a number of adverse health effects including central nervous system, reproductive, liver, and kidney toxicity, and carcinogenicity. However, many of these solvents remain in active and even in large-volume use. There has been much discussion in the literature as to whether toxicity and carcinogenicity of these solvents has been established, disproved, or is still in doubt.[3–10]

The intent of this article is to present the most recent field and clinical studies on solvent-exposed workers and to suggest what studies are needed to resolve ambiguities regarding the toxicity and carcinogenicity of these solvents.

Additional information on these compounds can be found in a supplementary table available from the author or at http://www.cdc.gov/niosh/ext-suppmat/Chlorinated-Solvents/. TABLE 1 lists all the compounds and their toxicity ratings by the International Agency for Research on Cancer (IARC) and other organizations; TABLE 2 gives volumes of use and numbers of workers potentially exposed for the same compounds.

TABLE 1. Chlorinated solvents and related compounds: toxicity assessments

Common/generic names	Composition	CAS number	IARC group[a]	IARC– human	IARC– animal	NTP·11[b]	WHO[c]	EPA IRIS[d]	U.S. NIOSH REL[e]
Chlorinated methanes									
Methyl chloride, chloromethane	CH_3Cl	74-87-3	3	I	I	N/A	Possible central nervous system (CNS) and reproductive toxin (ICSC 419)	D	Lowest feasible[f]
Methylene chloride, dichloromethane	CH_2Cl_2	75-09-2	2B	I	S	R	Possible CNS and liver toxin, carcinogen (ICSC 58)	B2	Lowest feasible[f]
Chloroform, trichloromethane	$CHCl_3$	67-66-3	2B	I	S	R	Possible CNS, liver, kidney toxin, carcinogen (ICSC 27, CICAD 58)	B2	2 ppm; lowest feasible[f]
Carbon tetrachloride	CCl_4	56-23-5	2B	I	S	R	Possible CNS, liver, kidney toxin, carcinogen (ICSC 24)	B2	2 ppm; lowest feasible[f]
Chlorinated ethanes									
Ethyl chloride, 1-chloroethane	C_2H_5Cl	75-00-3	3	NC	L	Nominated	Possible CNS toxin (ICSC 132)	N/A	Caution[g]
Ethylidene dichloride, 1,1-dichloroethane	$C_2H_4Cl_2$	75-34-3	N/A	—	—	N/A	Possible CNS, liver, kidney toxin (ICSC 249)	C	100 ppm; caution[g]
Ethylene dichloride, 1,2-dichloroethane	$C_2H_4Cl_2$	107-06-2	2B	I	S	R	Possible CNS, liver, kidney toxin, probable carcinogen (ICSC 250, CICAD 1)	B2	1 ppm; lowest feasible[f]; caution[g]

Continued.

TABLE 1. Continued.

Common/generic names	Composition	CAS number	IARC group[a]	IARC–human	IARC–animal	NTP_11[b]	WHO[c]	EPA IRIS[d]	U.S. NIOSH REL[e]
Methyl chloroform, 1,1,1-trichloroethane	$C_2H_3Cl_3$	71-55-6	3	I	I	Nominated	Possible cardiac, CNS, liver, kidney toxin (ICSC 79)	D	350 ppm; caution[g]
Vinyl trichloride, 1,1,2-trichloroethane	$C_2H_3Cl_3$	79-00-5	3	ND	L	N/A	Possible CNS, liver, kidney toxin (ICSC 80)	C	10 ppm; lowest feasible[f]; caution[g]
1,1,1,2-tetrachloroethane	$C_2H_2Cl_4$	630-20-6	3	ND	L	Nominated	Possible CNS toxin (ICSD 1486)	C	Caution[g]
1,1,2,2-tetrachloroethane	$C_2H_2Cl_4$	79-34-5	3	I	L	Nominated	Possible CNS, liver, kidney toxin (ISCS 332, CICAD 3)	C	1 ppm; lowest feasible[f]; caution[g]
Pentachloroethane	C_2HCl_5	76-01-7	3	ND	L	Nominated	Possible CNS toxin (ICSD 1394)	N/A	Caution[g]
Hexachloroethane	C_2Cl_6	67-72-1	2B	I	S	Nominated	Possible CNS, liver, kidney toxin (ICSC 51)	C	1 ppm; lowest feasible[f]; caution[g]
Chlorinated ethenes Vinyl chloride, chloroethylene	C_2H_3Cl	75-01-4	1	S	S	K	Possible circulatory system, liver, spleen toxin, known carcinogen (ICSC 82)	A	Lowest feasible[f]
1,1-dichloroethylene, vinylidene chloride	$C_2H_2Cl_2$	75-35-4	3	I	L	Nominated	Possible liver, kidney toxin (ICSC 83, CICAD 51)	C	Lowest feasible[f]

Continued.

TABLE 1. *Continued.*

Common/generic names	Composition	CAS number	IARC group[a]	IARC– human	IARC– animal	NTP_11[b]	WHO[c]	EPA IRIS[d]	U.S. NIOSH REL[e]
1,2-dichloroethylene, dichloroethene	$C_2H_2Cl_2$	(m,x) 540-59-0 (cis) 156-59-2 (trans) 156-60-5	N/A	—	—	N/A	Possible liver toxin (ICSC 436)	N/A	200 ppm
TCE, 1,1,2-trichloroethylene	C_2HCl_3	79-01-6	2A	L	S	R	Possible CNS, liver, kidney toxin, probable carcinogen (ICSC 81)	N/A	Lowest feasible[f]
Perchloroethylene, tetrachloroethylene	C_2Cl_4	127-18-4	2A	L	S	R	Possible CNS, liver, kidney toxin, probable carcinogen (ICSC 76)	N/A	Lowest feasible[f]

NOTE: N/A (not assessed).
[a]IARC = International Agency for Research on Cancer; Groups: 1 (human carcinogen), 2A (probable human carcinogen), 2B (possible human carcinogen), 3 (not classifiable as to human carcinogenicity). Evidence for carcinogenicity: S (sufficient), L (limited), I (inadequate), ND (no data as to carcinogenicity).
[b]NTP'11 = National Toxicology Program Report on Carcinogens, 11th edition; K = known to be a human carcinogen; R = reasonably anticipated to be a human carcinogen.
[c]WHO = World Health Organization; CICAD = Concise International Chemical Assessment Document; ICSC = International Chemical Safety Card.
[d]EPA IRIS = Integrated Risk Information System; weight of evidence characterization: A = human carcinogen; B2 = probable human carcinogen based on sufficient evidence of carcinogenicity in animals; C = possible human carcinogen; D = not classifiable as to human carcinogenicity.
[e]NIOSH Recommended Exposure Limits.[65]
[f]Potential occupational carcinogen, NIOSH Pocket Guide Appendix A.[66]
[g]All chloroethanes are given a "caution" rating because of their structural similarity to the four chloroethanes shown to be carcinogenic in animals. NIOSH Pocket Guide Appendix C.[67]

TABLE 2. Chlorinated solvents and related compounds: volume of use and regulations

Common/generic names	CAS number	Uses	Volume of use in the United States[a] (year)	Estimated no. of U.S. workers potentially exposed[b]	Volume of use in Europe[a] (year)	Estimated no. of European workers potentially exposed[c]	U.S. OSHA PELs[d]	Other PELs[e]
Chlorinated methanes								
Methyl chloride, chloromethane	74-87-3	Manufacture of silicone	347 (1998)[C]	10,000	417 (1998)[C]	—	100 ppm	50 ppm (Germany, Spain, United Kingdom)
Methylene chloride, dichloromethane	75-09-2	Solvent, paint remover, degreaser	231 (1998)[C]	1,400,000	346 (1998)[C]	285,000	25 ppm	50 ppm (Spain); 3A (Germany)
Chloroform, trichloromethane	67-66-3	Solvent, chemical intermediate	204 (1998)[C]	96,000	273 (1998)[C]	—	50 ppm	0.5 ppm (Germany); 2 ppm (Spain, United Kingdom)
Carbon tetrachloride	56-23-5	Degreaser, chemical intermediate	50 (1998)[C]	104,000	26 (1998)[C]	77,000	10 ppm	0.5 ppm (Germany); 2 ppm (United Kingdom); 5 ppm (Spain)
Chlorinated ethanes								
Ethyl chloride, 1-chloroethane	75-00-3	Largest use production of tetraethyl lead, refrigerant, solvent	49 (1986)[B]	50,000	—	—	1000 ppm	3B (Germany); 100 ppm (Spain); 1000 ppm (United Kingdom)
Ethylidene dichloride, 1,1-dichloroethane	75-34-3	Solvent, chemical intermediate	11,922 (1998)[G]	2,000	9,495 (1998)[G]	—	100 ppm	100 ppm (Germany, Spain, United Kingdom)
Ethylene dichloride, 1,2-dichloroethane	107-06-2	Primarily used to produce vinyl chloride	8,468 (1992)[A]	83,000	—	—	50 ppm	C2 Germany; C2, 5 ppm (Spain)
Methyl chloroform, 1,1,1-trichloroethane	71-55-6	Industrial cleaner/degreaser	91 (1997)[D]	2,500,000	30 (1997)[D]	—	350 ppm	100 ppm (Spain); 200 ppm (Germany)
Vinyl trichloride, 1,1,2-trichloroethane	79-00-5	Chemical intermediate in production of vinylidene chloride	186 (1980s)[E]	1,000	—	—	10 ppm	10 ppm (Germany, Spain)
1,1,1,2-tetrachloroethane	630-20-6	Solvent, chemical intermediate	—	—	—	—	—	—

Continued.

TABLE 2. *Continued.*

Common/generic names	CAS number	Uses	Volume of use in the United States[a] (year)	Estimated no. of U.S. workers potentially exposed[b]	Volume of use in Europe[a] (year)	Estimated no. of European workers potentially exposed[c]	U.S. OSHA[d]	Other PELs[e]
1,1,2,2-tetrachloroethane	79-34-5	Solvent, chemical intermediate	—	4,000	—	—	5 ppm	1 ppm (Germany)
Pentachloroethane	76-01-7	Solvent, chemical intermediate	—	200	—	—	—	5 ppm (Germany)
Hexachloroethane	67-72-1	Moth repellent, lubricating oil ingredient	—	8,500	—	—	1 ppm	1 ppm (Germany, Spain); 5 ppm (United Kingdom)
Chlorinated ethenes								
Vinyl chloride, chloroethylene	75-01-4	Primarily used to produce polyvinyl chloride	6741 (1997)[F]	81,000	8751 (1997)[F]	41,300	1 ppm	C1 (Germany); 3 ppm (Spain)
1,1-dichloroethylene, vinylidene chloride	75-35-4	Primarily used to produce poly (vinylidene chloride) copolymers	79 (1992)[A]	3,000	—	—	—	3B, 2 ppm (Germany); 5 ppm (Spain)
1,2-dichloroethylene, dichloroethene	(mix) 540-59-0 (cis) 156-59-2 (trans) 156-60-5	Degreaser	—	200	—	—	200 ppm	200 ppm (Germany, Spain, United Kingdom)
TCE, 1,1,2-trichloroethylene	79-01-6	Degreaser	89 (1997)[F]	401,000	108 (1997)[F]	284,000	100 ppm	C1 (Germany); 80 mg/L TCA in urine end of work week (Spain)
Perchloroethylene, tetrachloroethylene	127-18-4	Dry-cleaning solvent, degreaser	136 (1997)[F]	688,000	127 (1997)[F]	870,000	100 ppm	3B (Germany) 25 ppm (Spain); 50 ppm (United Kingdom)

NOTE: [110]Use data from A—Reed 1993[106], B—Miller 1993[1], C—Chemical Economics Handbook 1999[107], D—Leder *et al*, 1999[108], E—IARC 1999[109], F—Cowfer and Gorensek 1997[110], G—Jebens 1999.[111]

[a] in thousands of metric tons.
[b] from NIOSH 1990.[112]
[c] From CAREX statistics for 19 European Union countries.[113]
[d] U.S. Occupational Safety and Health Administration: permissible Exposure Limit (time-weighted average) for an 8-h day. From U.S. ICSC.[114]
[e] Permissible Exposure Limit (time-weighted average) for an 8-h day. From EC ICSC and country-specific regulations[114–118] German categories: C1 (human carcinogen), C2 (animal carcinogen), 3A (probably category 4 or 5 but insufficient data for occupational exposure limit [Maximale Arbeitsplatz Konzentrationen, MAK] value), 3B (insufficient evidence for classification, MAK possible if not genotoxic), MAKs are not assigned to C1 and C2 chemicals.

METHODS

We focus on the most widely used of these solvents (methylene chloride, chloroform, TCE, PCE) and discuss other chlorinated aliphatic solvents and related compounds, such as vinyl chloride. A PubMed search in July 2005 on solvents, health effects, and occupational exposure found 3193 citations. Within this group, the epidemiology literature published since 1990 on health effects of occupational exposure was reviewed and summarized (TABLE 3; available from author or at http://www.cdc.gov/niosh/ext-supp-mat/Chlorinated-Solvents/). When two or more updates were published on the same study population, the most recent is cited. Case reports and the literature on environmental exposure were excluded. Meta-analyses and reviews are not included in the tables of studies. Because abstracts do not provide enough information to assess a study, only papers that could be obtained were included.

In cohort mortality studies, the common metrics are the standardized mortality ratio (SMR), comparing mortality in a cohort with mortality in the general population, adjusted for age, sex, gender, and calendar period, and the rate ratio (RR, also known as relative risk), comparing mortality in one subcohort with that in another. Similarly, cancer incidence studies use the standardized incidence ratio (SIR) and or the RRs. Because cancer is the only chronic disease group for which mandatory reporting and state registries are widespread, SIR studies for cardiovascular, respiratory, and other chronic diseases are rarely feasible.

Case–control and exposed-referent cross-sectional studies compare exposure among individuals with and without disease or physiological changes among exposed and unexposed individuals, using unadjusted and adjusted odds ratios (OR).

RESULTS

Reviews of the levels of occupational exposure to chlorinated solvents over time have documented generally decreasing exposures from the 1940s for TCE, PCE, and other solvents.[11–14] Epidemiological studies of solvent-exposed cohorts generally have found higher risks of adverse health effects among those exposed before 1970. However, studies of recently exposed workers have continued to document adverse health effects. These include neurological effects, kidney and liver damage, reproductive effects, cardiovascular effects, and cancer. These studies are cited below in system/disease groupings and then by solvent.

The short-term narcotic effects of chlorinated solvents have been known since the mid-19th century. Several chlorinated solvents, especially chloroform and TCE, were used for anesthesia until fairly recently.[15,16] The literature on long-term neurological effects of solvent exposure has focused on tests of sensory abilities, such as failure in blue-yellow color discrimination or pattern recognition.[17–21]

Exposure to chlorinated solvents has been associated with kidney and liver damage in case reports for 80 years and has been studied more systematically for about 20 years.[22,23] Case–control and cross-sectional exposed-referent studies have examined somatic cell mutations and kidney proteins in the urine of individuals exposed and unexposed to chlorinated solvents; however, differences between solvent-exposed and -unexposed individuals in levels of various proteins in the urine have not been related to clinical signs of kidney damage.[24–40]

Although inhalation is the main route of occupational exposure to chlorinated solvents, there is very little literature on short- or long-term effects on the respiratory system. Elevated SMRs for respiratory disease mortality not associated with duration of employment have been seen in cohort studies,[41–43] but lung cancer has not been associated with solvent exposure.

Cardiac arrhythmias often occur when TCE was used as an anesthetic.[44] Arrhythmia and other transient adverse cardiac effects have been associated with occupational exposure to methylene chloride, chloroform, and TCE.[43,45–49] The "healthy worker" effect, which typically leads to a reduction in observed numbers of cardiovascular deaths in occupational cohorts, could mask a low-to-moderate increase in cardiovascular mortality.[50]

Reproductive effects, including both gamete damage and developmental damage to the embryo and fetus, have been reported in a number of studies of solvent-exposed workers.[51–59] Spontaneous abortions or delays in becoming pregnant among solvent-exposed women show consistency for PCE but not TCE.[60–63]

There have been a few studies of immunological effects, including the occurrence of autoimmune diseases, in solvent-exposed workers, some reporting increased risk, others reporting lack of risk.[64–69]

Based on animal studies and human case reports of damage to the liver, biliary tract, and urinary organs from chlorinated solvent exposure, these organs have become a focus of epidemiology investigations. Many case–control cancer studies have targeted the liver, biliary tract, and urinary organs[70–73]; others have investigated breast and nervous system cancer.[74–76] In the Scandinavian countries, groups of workers monitored for acute or long-term solvent exposure or known from census records to have worked in certain industries have been linked to cancer registries.[48,77–80]

Occupational cohort studies have been concentrated in a few industries: cellulose triacetate or cellulose fiber production[81–86]; aircraft manufacturing or maintenance[43,87,88]; dry cleaning[41,42]; and various industries (mainly metalworking, electronics, painting, printing, and dry cleaning) using TCE.[89–91] In general, findings in the cohort studies have not been consistent, even within the same industries. For example, in the aerospace industry, Blair et al.[43] found increasing risk for colon and liver cancer and for multiple myeloma with increasing TCE exposure, whereas Boice et al.[87] and Morgan et al.[88] saw no increases for those sites. However, both Blair et al.[41] and Ruder et al.[42] found elevated SMRs for bladder, cervical, and esophageal cancer, and both found increased respiratory disease mortality among PCE-exposed dry-cleaning workers. There are several possible explanations for inconsistent results among cohorts in the same industry. The cohorts may actually have had different combinations of exposures. A carcinogen might be associated with cancer at more than one site, and by chance one site might have been in excess in one cohort, another site in a second cohort. Individuals exposed to several carcinogens might develop a cancer associated with any one of them, and the distribution of cancers could vary by chance from cohort to cohort.

Methylene Chloride

Six case–control studies in populations with multiple chemical exposures have evaluated associations with methylene chloride exposure for biliary and liver cancer or renal cell carcinoma (RCC) and no increase was found.[70,71] For breast, brain, and rectal cancers, associations were found between level of exposure and cancer OR.[74,75,92,93] A case–control study of spontaneous abortion in exposed women found no association.[62] Mortality studies conducted in cellulose triacetate manufacturing cohorts exposed to acetone and methanol as well as methylene chloride show divergent findings of excess prostate and cervical cancer,[81] brain cancer and Hodgkin lymphoma,[82] biliary and liver cancer,[83] colon cancer,[85,86] and ischemic heart disease.[84] Within an aircraft manufacturing cohort exposed to many chemicals, any methylene chloride exposure was associated with increased risk of non-Hodgkin's lymphoma (NHL), multiple myeloma, and breast cancer.[43] One exposure assessment, also in triacetate production, found a correlation between the level of methylene chloride and level of carboxyhemoglobin.[49]

Chloroform

Three case–control studies evaluated risks of biliary and liver cancer, RCC, and astrocytomas and chloroform exposure and found some evidence of an effect on RCC risk.[70,71,75] Two cross-sectional exposed-referent studies looked at pregnancy outcomes among women who worked during their pregnancies. Dentists who used chloroform-based root canal sealers, among other chemicals, had longer time to pregnancy leading to live birth than did nonexposed women.[52] Chloroform-exposed lab workers were more likely than the unexposed to experience spontaneous abortions.[94] A greater percentage of exposed than unexposed workers had low serum prealbumin and transferrin, lower scores for attention, memory, and perception, and higher scores for fatigue, depression, and anger.[95]

Carbon Tetrachloride

Five case–control studies found no association between carbon tetrachloride exposure and biliary and liver cancer or RCC[70,71]; some evidence for breast cancer, but not increasing risk with increasing exposure[74]; some effect for colon and rectal cancer[96]; and a dose–response relationship for brain cancer.[75] Exposure was associated with increased OR of scleroderma in men but not women.[97] Exposures to women working in their second or third trimester appeared to increase the risk of a small-for-gestational-age liveborn baby.[51] A study of hepatic proteins in the urine found significant differences between carbon tetrachloride-exposed men and the unexposed and also between the unexposed and "normal ranges."[32] Within an aircraft manufacturing cohort exposed to many chemicals, any carbon tetrachloride exposure was associated with increased risk of NHL and myeloma (especially in women), and breast cancer.[43]

Tetrachloroethylene

Since 1990, there have been seven case–control, two cross-sectional, twelve cross-sectional exposed-referent, three linkage, and four cohort studies of PCE published; all but five of these focused on workers in the dry-cleaning industry with few or no other chemical exposures. Within the cohorts or study populations with multiple exposures, three studies found some increased risk of scleroderma and of biliary and liver or brain cancer[64,70,75]; others found no increased risk for RCC[71] or for any cancer.[87]

Neurological studies found exposure-correlated differences between groups (any versus none or high versus low) for pattern memory and recognition and blue-yellow color confusion.[17,19,20] Higher rates of reproductive failure among PCE-exposed women and of eccentric sperm morphology and motility among PCE-exposed men; suggestions of longer times to pregnancy among PCE-exposed women and the partners of PCE-exposed men have been reported.[54,55,58,63] Three studies of spontaneous abortion in exposed women have shown increased risks, especially for dry-cleaning machine operators.[60–62]

Four studies of kidney proteins in urine have been done in dry-cleaner populations; there is no consistency between studies, not even for the referent ranges.[36–39] Studies of hepatic or immunological changes found differences between dry cleaners and referents but could not correlate those changes with clinical effects.[33,35,68,69]

Three studies linked census records for those working as "dry cleaner or laundry worker" or a database of those being monitored for PCE exposure to cancer registries, finding statistically significant increased risks for liver and pancreatic cancer, Hodgkin's disease, and leukemia.[78,79,98] As noted above, updates of two cohorts of dry-cleaning workers found elevated SMRs for bladder, cervical, and esophageal cancer, and both found increased respiratory disease mortality.[41,42]

Trichloroethylene

Since 1990, there have been sixteen case–case or case–control, three cross-sectional, five cross-sectional exposed-referent, two linkage, and seven cohort studies published. Many of these studies did not report contemporaneous exposure to other chemicals, but it is likely that such exposures occurred.

A total of 40% of 263 TCE-exposed workers developed severe generalized dermatitis within 3 months of their first exposure; unaffected co-workers with longer durations of exposure differed significantly only in their tumor necrosis factor-α (TNF-α) genotypes which could not independently differentiate cases from controls.[65] TCE-exposed and -unexposed co-workers in a printing factory differed in serum levels of cytokines but not in clinical signs or symptoms[67]; another study of printers found no differences in nerve function between TCE exposed and unexposed[21]; and a study of semen parameters and endocrine profiles in a TCE-exposed factory (no referents) saw deviations from the World Health Organization normal parameters for semen and a correlation between dehydroepiandrosterone sulphate and years of TCE exposure.[56,99] Time to pregnancy was delayed for women with high exposures.[63] Two studies of spontaneous abortion in exposed women had conflicting results of no increased risk[61] and a threefold increased risk.[62] Case–control studies of scleroderma (systemic sclerosis) and oral cleft defects in babies of women who worked in their first trimester found increased risk for TCE exposure.[57,64,66,100]

Studies of liver and kidney proteins in urine or somatic mutations have been done in RCC case–case, RCC case–control, and cross-sectional exposed-referent populations. Statistically significant findings of "urinary protein patterns indicative of kidney damage" or of somatic mutations could not be related causally to TCE exposure in cases or to clinical effects among TCE-exposed workers.[24,26–29,31,40] A weak relationship was seen between levels of urinary N-acetyl-β-gluoosaminidase and concurrent hours per week exposed to TCE.[101] Bolt et al.[24] recently studied excretion of α_1-microglobulin among RCC cases and controls, with some cases and some controls having been exposed to TCE. Among the RCC cases but not among the controls there was a significant difference in α_1-microglobulin levels in the urine.[24] Excess RCC among those exposed to TCE has been reported in a number of studies.[30,71–73,91]

Other studies of cancer among those exposed to TCE have reported increased risks for cancer overall, for liver and brain cancer in several studies, and for esophageal, stomach, biliary, pancreatic, colon, cervical, and prostate cancer, NHL, and multiple myeloma.[43,75,78,80,89,90] Other studies have reported no increased cancer risk among the TCE exposed.[48,87,88]

Other Chlorinated Solvents

Dosemeci et al.[71] examined RCC risk among those exposed to methyl chloride, 1,2-dichloroethane, 1,1,1-trichloroethane, or 1,1,2-trichloroethane and found an increase risk for dichloroethane-exposed women. A follow-up years later on seamen previously poisoned in a methyl chloride accident found a greater risk of cardiovascular death than among referent seamen not exposed to methyl chloride.[77]

Two 1,1,1-trichlorethane case–control studies looked at scleroderma risk, one finding an elevation associated with self-reported exposure, but no elevation associated with exposure reevaluated by an expert[64]; the other study showed elevated risk for men but not women.[97] An innovative study linked a register of workers monitored for serum 1,1,1-trichloroethane levels with subsequent cancer registry records. Several SIRs were elevated, especially those for cancer of the central nervous system, NHL, and myeloma.[78] A retrospective cohort study of aircraft maintenance workers with numerous chemical exposures found elevated RRs for NHL in men and myeloma and breast cancer in women.[43] A case–control study found nonlinear increases in astrocytoma risk with increasing cumulative exposure.[75] Among women reporting 1,1,1-trichloroethane exposure before and during pregnancy, two studies found increased odds of spontaneous abortion[61,62]; while in another study time to pregnancy was not delayed.[63]

DISCUSSION

The impact of studies to date has been limited by small study size and/or inability to control for confounding factors, particularly smoking and mixed occupational exposures, the lack of evidence linking occupational exposure to biological evidence of exposure, to precursors of health effects, to health effects, and varying, inconsistent outcomes. However, a number of studies consistently have reported increases in bladder and cervical cancer among those exposed to PCE[41,42,78] and of kidney, liver, and brain cancer among those exposed to TCE.[30,43,71–73,75,80,91,98] Among PCE-exposed men and women, there have also been consistent reports of spontaneous abortion or other fertility problems[54,55,60–63] and neurological problems.[17,19,20]

Carcinogenic metabolites may be formed via pathways that exhibit a high degree of genetic polymorphisms. For example, trichloroacetic acid (TCA) forms from PCE and TCE via CYP2E1-mediated metabolism, whereas mutagenic metabolites S-(l,2-dichlorovinyl)-L-cysteine and S-(l,2,2-trichlorovinyl)-L-cysteine form via the glutathione pathway and are deactivated via the *N*-acetyltransferase pathway.[102,103] All three pathways are highly polymorphic, and risk does depend on genotypes as well as on exposure.[104] Most of the biomarker studies have not evaluated variation in susceptibility by genotype.

For some of these solvents (methyl chloride, ethyl chloride, 1,1-dichloroethane, tetra-, penta-, and hexachloroethane, and dichloroethylene) there is little or no literature on human health effects. Large cohorts of workers exposed to solvents mainly used as chemical intermediates in closed systems might be difficult to assemble, but the number of study participants needed for cross-sectional biomarker studies is much smaller.

The biomarker studies to date, although showing some significant differences between exposed and referent groups, or between an exposed group and laboratory reference values, have for the most part not demonstrated any link between these changes in immunological parameters or levels of kidney and liver proteins in the urine and health effects. This is at least partially due to the retrospective nature of many studies. High or low levels of urinary proteins in tests conducted after an exposure has occurred may have been out of range before the exposure began; high or low levels of urinary proteins in tests conducted after a diagnosis has been made may not have been out of range before the diagnosis. In addition, the connections between, for example, unusual levels of urinary proteins and kidney damage have not been explored in prospective studies.

The cohort studies lack information on nonoccupational exposures or exposures in other jobs. In most industries, solvent-exposed workers are exposed to a number of solvents as well as to other substances (e.g., heavy metals) that have been associated with the development of cancer and other chronic diseases.

For some solvents, TCE in particular, the literature reexamining and reappraising field studies and clinical studies is as extensive or more extensive than the primary literature. Although these studies can be quite useful, particularly in calling attention to overlooked research or in meta-analysis of studies that can be combined, they cannot replace the primary studies. Some of the effort going into these reviews and meta-analyses could be redirected to filling in the gaps in the investigative studies.

In the future, international differences in exposure limits may provide a "natural experiment" if countries such as Germany and Spain, which have limited exposure, subsequently experience decreased adverse health effects among exposed workers. Such decreases could provide some evidence that higher levels of health effects (previously in those countries or concurrently in countries such as the United States with higher exposure limits) were associated with higher levels of solvent exposure.

CONCLUSION

The definitive studies will be those whose results and conclusions are accepted universally, whether or not those results show an association of solvent exposure with a health effect. To achieve consensus, those studies should incorporate body burden, not merely ambient levels, of solvents as well as biological markers of effect. Ideally the studies would be prospective, to establish preexposure base lines for physiological outcomes and background levels for solvent body burden, would genotype participants for variations in genes coding for enzymes in relevant metabolic pathways, and would find methods to clarify the association between subclinical physiological changes and health effects. If possible, exposed individuals studied would be exposed predominantly to a single solvent. This last condition could be met for PCE with the participation of dry-cleaning workers (NIOSH [National Institute for Occupational Safety and Health] has conducted a pilot study fulfilling most of the conditions listed above).[105] There may not be a working population exposed only to TCE or only to one of the other chlorinated solvents. For all of the chlorinated aliphatic solvents, the definitive studies remain to be done.

ACKNOWLEDGMENTS

The assistance of Lucy Schoolfield, late NIOSH Hamilton Librarian, in assembling relevant literature, and the helpful comments of Lynne Pinkerton, Elizabeth Whelan, and anonymous reviewers are gratefully acknowledged.

The findings and conclusions in this report are those of the author and do not necessarily represent the views of the National Institute for Occupational Safety and Health.

SUPPLEMENTAL TABLES

Supplemental tables are available from author or at http://www.edc.gov/niosh/ext-supp-mat/Chlorinated-Solvents/.

REFERENCES

1. MILLER, M.C. 1993. Ethyl chloride. *In* Kirk-Othmer Encyclopedia of Chemical Technology, Vol. 6. J.I. Kroschwitz & M. Howe-Grant, Eds.: 1–10. John Wiley & Sons. New York, NY.
2. SNEDECOR, G. 1993. Other chloroethanes. *In* Kirk-Othmer Encyclopedia of Chemical Technology, Vol. 6. J.I. Kroschwitz & M. Howe-Grant, Eds.: 11–36. John Wiley & Sons. New York, NY.

3. RAMLOW, J.M. 1995. Critique of review of chlorinated solvents epidemiology. Am. J. Ind. Med. **27:** 313–316.
4. WARTENBERG, D., D. REYNER & C.S. SCOTT. 2000. Trichloroethylene and cancer: epidemiologic evidence. Environ. Health Perspect. **108:** 161–176.
5. WARTENBERG, D., D. REYNER & C.S. SCOTT. 2001. Errors in TCE analysis: response. Environ. Health Perspect. **109:** A108–A109.
6. WARTENBERG, D. & C.S. SCOTT. 2002. Carcinogenicity of trichloroethylene. Environ. Health Perspect. **110:** A13–A14.
7. NORMAN, W.C. 2005. Trichloroethylene as human carcinogen. Toxicology **208:** 171–172.
8. NORMAN, W.C. 3rd & P. BOGGS. 1996. Flawed estimates of methylene chloride exposures. Am. J. Ind. Med. **30:** 504–505.
9. SASS, J.B., B. CASTLEMAN & D. WALLINGA. 2005. Vinyl chloride: a case study of data suppression and misrepresentation. Environ. Health Perspect. **113:** 809–812.
10. BOFFETTA, P., K.A. MUNDT & L.D. DELL. 2005. Epidemiologic evidence for the carcinogenicity of vinyl chloride monomer. Scand. J. Work Environ. Health **31:** 236.
11. RAASCHOU-NIELSEN, O. *et al.* 2002. Exposure of Danish workers to trichloroethylene, 1947–1989. Appl. Occup. Environ. Hyg. **17:** 693–703.
12. EARNEST, G.S. 2002. A control technology evaluation of state-of-the-art, perchloroethylene dry-cleaning machines. Appl. Occup. Environ. Hyg. **17:** 352–359.
13. VON GROTE, J. *et al.* 2003. Reduction of occupational exposure to perchloroethylene and trichloroethylene in metal degreasing over the last 30 years: influences of technology innovation and legislation. J. Expo. Anal. Environ. Epidemiol. **13:** 325–340.
14. LUDWIG, H.R. *et al.* 1983. Worker exposure to perchloroethylene in the commercial dry cleaning industry. Am. Ind. Hyg. Assoc. J. **44:** 600–605.
15. OSTLERE, G. 1953. The history of trichlorethylene anaesthesia. Anaesthesia **8:** 21–25.
16. WEDGWOOD, J.J. & A.A. SPENCE. 1997. The rise and fall of chloroform anaesthesia. Proc. R. Coll. Physicians Edinb. **27:** 575–591.
17. CAVALLERI, A. *et al.* 1994. Perchloroethylene exposure can induce colour vision loss. Neurosci. Lett. **179:** 162–166.
18. BRUNETTI, D. *et al.* 1997. Cancer risk in first-degree relatives of children with malignant tumours (province of Trieste, Italy). Int. J. Cancer **73:** 822–827.
19. ECHEVERRIA, D., R.F. WHITE & C. SAMPAIO. 1995. A behavioral evaluation of PCE exposure in patients and dry cleaners: a possible relationship between clinical and preclinical effects. J. Occup. Environ. Med. **37:** 667–680.
20. SHARANJEET-KAUR *et al.* 2004. Effect of petroleum derivatives and solvents on colour perception. Clin. Exp. Optom. **87:** 339–343.
21. RUIJTEN, M.W., M.M. VERBERK & H.J. SALLE. 1991. Nerve function in workers with long-term exposure to trichloroethene. Br. J. Ind. Med. **48:** 87–92.
22. PHILLIPS, S.D. & J.C. WAKSMAN. 2004. Hepatorenal solvent toxicology. Clin. Occup. Environ. Med. **4:** 731–740.
23. LASH, L.H. & J.C. PARKER. 2001. Hepatic and renal toxicities associated with perchloroethylene. Pharmacol. Rev. **53:** 177–208.

24. BOLT, H.M. et al. 2004. Urinary α_1-microglobulin excretion as biomarker of renal toxicity in trichloroethylene-exposed persons. Int. Arch. Occup. Environ. Health **77:** 186–190.
25. BRAUCH, H. et al. 1999. Trichloroethylene exposure and specific somatic mutations in patients with renal cell carcinoma. J. Natl. Cancer Inst. **91:** 854–861.
26. BRAUCH, H. et al. 2004. VHL mutations in renal cell cancer: does occupational exposure to trichloroethylene make a difference? Toxicol. Lett. **151:** 301–310.
27. BRÜNING, T. et al. 1996. Preexistence of chronic tubular damage in cases of renal cell cancer after long and high exposure to trichloroethylene. Arch. Toxicol. **70:** 259–260.
28. BRÜNING, T. et al. 1999. Pathological excretion patterns of urinary proteins in renal cell cancer patients exposed to trichloroethylene. Occup. Med. (Lond.) **49:** 299–305.
29. BRÜNING, T. et al. 1999. Glutathione transferase alpha as a marker for tubular damage after trichloroethylene exposure. Arch. Toxicol. **73:** 246–254.
30. BRÜNING, T. et al. 2003. Renal cell cancer risk and occupational exposure to trichloroethylene: results of a consecutive case-control study in Arnsberg, Germany. Am. J. Ind. Med. **43:** 274–285.
31. GREEN, T. et al. 2004. Biological monitoring of kidney function among workers occupationally exposed to trichloroethylene. Occup. Environ. Med. **61:** 312–317.
32. TOMENSON, J.A. et al. 1995. Hepatic function in workers occupationally exposed to carbon tetrachloride. Occup. Environ. Med. **52:** 508–514.
33. BRODKIN, C.A. et al. 1995. Hepatic ultrasonic changes in workers exposed to perchloroethylene. Occup. Environ. Med. **52:** 679–685.
34. FRANCHINI, I. et al. 1983. Early indicators of renal damage in workers exposed to organic solvents. Int. Arch. Occup. Environ. Health **52:** 1–9.
35. GENNARI, P. et al. 1992. gamma-Glutamyltransferase isoenzyme pattern in workers exposed to tetrachloroethylene. Am. J. Ind. Med. **21:** 661–671.
36. MUTTI, A. et al. 1992. Nephropathies and exposure to perchloroethylene in drycleaners. Lancet **340:** 189–193.
37. SOLET, D. & T.G. ROBINS. 1991. Renal function in dry-cleaning workers exposed to perchloroethylene. Am. J. Ind. Med. **20:** 601–614.
38. VERPLANKE, A.J., M.H. LEUMMENS & R.F. HERBER. 1999. Occupational exposure to tetrachloroethene and its effects on the kidneys. J. Occup. Environ. Med. **41:** 11–16.
39. VYSKOCIL, A. et al. 1990. Study on kidney function in female workers exposed to perchlorethylene. Hum. Exp. Toxicol. **9:** 377–380.
40. SELDEN, A. et al. 1993. Trichloroethylene exposure in vapour degreasing and the urinary excretion of N-acetyl-beta-D-glucosaminidase. Arch. Toxicol. **67:** 224–226.
41. BLAIR, A., S.A. PETRALIA & P.A. STEWART. 2003. Extended mortality follow-up of a cohort of dry cleaners. Ann. Epidemiol. **13:** 50–56.
42. RUDER, A.M., E.M. WARD & D.P. BROWN. 2001. Mortality in dry-cleaning workers: an update. Am. J. Ind. Med. **39:** 121–132.
43. BLAIR, A. et al. 1998. Mortality and cancer incidence of aircraft maintenance workers exposed to trichloroethylene and other organic solvents and chemicals: extended follow up. Occup. Environ. Med. **55:** 161–171.

44. DHUNÉR, K.G., P. NORDQVIST & B. RENSTROM. 1957. Cardiac irregularities in trichloroethylene poisoning. Acta. Anesth. Scand. **2:** 121–135.
45. LOWRY, L.K., R. VANDERVORT & P.L. POLAKOFF. 1974. Biological indicators of occupational exposure to trichloroethylene. J. Occup. Med. **16:** 98–101.
46. LLOYD, J.W., R.M. MOORE JR. & P. BRESLIN. 1975. Background information on trichloroethylene. J. Occup. Med. **17:** 603–605.
47. SODEN, K.J. 1993. An evaluation of chronic methylene chloride exposure. J. Occup. Med. **35:** 282–286.
48. AXELSON, O. et al. 1994. Updated and expanded Swedish cohort study on trichloroethylene and cancer risk. J. Occup. Med. **36:** 556–562.
49. SODEN, K.J., G. MARRAS & J. AMSEL. 1996. Carboxyhemoglobin levels in methylene chloride-exposed employees. J. Occup. Environ. Med. **38:** 367–371.
50. CHOI, B.C. 2001. Mathematical procedure to adjust for the healthy worker effect: the case of firefighting, diabetes, and heart disease. J. Occup. Environ. Med. **43:** 1057–1063.
51. SEIDLER, A. et al. 1999. Maternal occupational exposure to chemical substances and the risk of infants small-for-gestational-age. Am. J. Ind. Med. **36:** 213–222.
52. DAHL, J.E. et al. 1999. Dental workplace exposure and effect on fertility. Scand. J. Work. Environ. Health **25:** 285–290.
53. WENNBORG, H. et al. 2005. Congenital malformations related to maternal exposure to specific agents in biomedical research laboratories. J. Occup. Environ. Med. **47:** 11–19.
54. ESKENAZI, B. et al. 1991. A study of the effect of perchloroethylene exposure on the reproductive outcomes of wives of dry-cleaning workers. Am. J. Ind. Med. **20:** 593–600.
55. ESKENAZI, B. et al. 1991. A study of the effect of perchloroethylene exposure on semen quality in dry cleaning workers. Am. J. Ind. Med. **20:** 575–591.
56. CHIA, S.E. et al. 1996. Semen parameters in workers exposed to trichloroethylene. Reprod. Toxicol. **10:** 295–299.
57. LORENTE, C. et al. 2000. Maternal occupational risk factors for oral clefts. Occupational Exposure and Congenital Malformation Working Group. Scand. J. Work. Environ. Health **26:** 137–145.
58. OLSEN, J. et al. 1990. Low birthweight, congenital malformations, and spontaneous abortions among dry-cleaning workers in Scandinavia. Scand. J. Work. Environ. Health **16:** 163–168.
59. AHLBORG, G. JR. 1990. Pregnancy outcome among women working in laundries and dry-cleaning shops using tetrachloroethylene. Am. J. Ind. Med. **17:** 567–575.
60. DOYLE, P. et al. 1997. Spontaneous abortion in dry cleaning workers potentially exposed to perchloroethylene. Occup. Environ. Med. **54:** 848–853.
61. LINDBOHM, M.L. et al. 1990. Spontaneous abortions among women exposed to organic solvents. Am. J. Ind. Med. **17:** 449–463.
62. WINDHAM, G.C. et al. 1991. Exposure to organic solvents and adverse pregnancy outcome. Am. J. Ind. Med. **20:** 241–259.
63. SALLMEN, M. et al. 1995. Reduced fertility among women exposed to organic solvents. Am. J. Ind. Med. **27:** 699–713.
64. GARABRANT, D.H. et al. 2003. Scleroderma and solvent exposure among women. Am. J. Epidemiol. **157:** 493–500.

65. DAI, Y. *et al.* 2004. Genetic polymorphisms of cytokine genes and risk for trichloroethylene-induced severe generalized dermatitis: a case-control study. Biomarkers **9:** 470–478.
66. DIOT, E. *et al.* 2002. Systemic sclerosis and occupational risk factors: a case-control study. Occup. Environ. Med. **59:** 545–549.
67. IAVICOLI, I., A. MARINACCIO & G. CARELLI. 2005. Effects of occupational trichloroethylene exposure on cytokine levels in workers. J. Occup. Environ. Med. **47:** 453–457.
68. TORAASON, M. *et al.* 2003. Effect of perchloroethylene, smoking, and race on oxidative DNA damage in female dry cleaners. Mutat. Res. **539:** 9–18.
69. ANDRYS, C. *et al.* 1997. Immunological monitoring of dry-cleaning shop workers—exposure to tetrachloroethylene. Cent. Eur. J. Public Health **5:** 136–142.
70. BOND, G.G. *et al.* 1990. Liver and biliary tract cancer among chemical workers. Am. J. Ind. Med. **18:** 19–24.
71. DOSEMECI, M., P. COCCO & W.H. CHOW. 1999. Gender differences in risk of renal cell carcinoma and occupational exposures to chlorinated aliphatic hydrocarbons. Am. J. Ind. Med. **36:** 54–59.
72. HENSCHLER, D. *et al.* 1995. Increased incidence of renal cell tumors in a cohort of cardboard workers exposed to trichloroethene. Arch. Toxicol. **69:** 291–299.
73. VAMVAKAS, S. *et al.* 1998. Renal cell cancer correlated with occupational exposure to trichloroethene. J. Cancer Res. Clin. Oncol. **124:** 374–382.
74. CANTOR, K.P. *et al.* 1995. Occupational exposures and female breast cancer mortality in the United States. J. Occup. Environ. Med. **37:** 336–348.
75. HEINEMAN, E.F. *et al.* 1994. Occupational exposure to chlorinated aliphatic hydrocarbons and risk of astrocytic brain cancer. Am. J. Ind. Med. **26:** 155–169.
76. COCCO, P., E.F. HEINEMAN & M. DOSEMECI. 1999. Occupational risk factors for cancer of the central nervous system (CNS) among U.S. women. Am. J. Ind. Med. **36:** 70–74.
77. RAFNSSON, V. & G. GUDMUNDSSON. 1997. Long-term follow-up after methyl chloride intoxication. Arch. Environ. Health **52:** 355–359.
78. ANTTILA, A. *et al.* 1995. Cancer incidence among Finnish workers exposed to halogenated hydrocarbons. J. Occup. Environ. Med. **37:** 797–806.
79. TRAVIER, N. *et al.* 2002. Cancer incidence of dry cleaning, laundry and ironing workers in Sweden. Scand. J. Work. Environ. Health **28:** 341–348.
80. HANSEN, J. *et al.* 2001. Cancer incidence among Danish workers exposed to trichloroethylene. J. Occup. Environ. Med. **43:** 133–139.
81. GIBBS, G.W., J. AMSEL & K. SODEN. 1996. A cohort mortality study of cellulose triacetate-fiber workers exposed to methylene chloride. J. Occup. Environ. Med. **38:** 693–697.
82. HEARNE, F.T. & J.W. PIFER. 1999. Mortality study of two overlapping cohorts of photographic film base manufacturing employees exposed to methylene chloride. J. Occup. Environ. Med. **41:** 1154–1169.
83. LANES, S.F. *et al.* 1993. Mortality update of cellulose fiber production workers. Scand. J. Work. Environ. Health **19:** 426–428.
84. TOMENSON, J.A. *et al.* 1997. Mortality of workers exposed to methylene chloride employed at a plant producing cellulose triacetate film base. Occup. Environ. Med. **54:** 470–476.

85. GOLDBERG, M.S. & G. THÉRIAULT. 1994. Retrospective cohort study of workers of a synthetic textiles plant in Quebec: II. Colorectal cancer mortality and incidence. Am. J. Ind. Med. **25:** 909–922.
86. GOLDBERG, M.S. & G. THÉRIAULT. 1994. Retrospective cohort study of workers of a synthetic textiles plant in Quebec: I. General mortality. Am. J. Ind. Med. **25:** 889–907.
87. BOICE, J.D. JR. et al. 1999. Mortality among aircraft manufacturing workers. Occup. Environ. Med. **56:** 581–597.
88. MORGAN, R.W. et al. 1998. Mortality of aerospace workers exposed to trichloroethylene. Epidemiology **9:** 424–431.
89. GREENLAND, S. et al. 1994. A case-control study of cancer mortality at a transformer-assembly facility. Int. Arch. Occup. Environ. Health **66:** 49–54.
90. RITZ, B. 1999. Cancer mortality among workers exposed to chemicals during uranium processing. J. Occup. Environ. Med. **41:** 556–566.
91. RAASCHOU-NIELSEN, O. et al. 2003. Cancer risk among workers at Danish companies using trichloroethylene: a cohort study. Am. J. Epidemiol. **158:** 1182–1192.
92. COCCO, P., M. DOSEMECI & E.F. HEINEMAN. 1998. Occupational risk factors for cancer of the central nervous system: a case-control study on death certificates from 24 U.S. states. Am. J. Ind. Med. **33:** 247–255.
93. DUMAS, S. et al. 2000. Rectal cancer and occupational risk factors: a hypothesis-generating, exposure-based case-control study. Int. J. Cancer **87:** 874–879.
94. WENNBORG, H. et al. 2000. Pregnancy outcome of personnel in Swedish biomedical research laboratories. J. Occup. Environ. Med. **42:** 438–446.
95. LI, L.H. et al. 1993. Studies on the toxicity and maximum allowable concentration of chloroform. Biomed. Environ. Sci. **6:** 179–186.
96. SIEMIATYCKI, J. et al. 1991. Associations between occupational circumstances and cancer. In Risk Factors for Cancer in the Workplace. J. Siemiatycki, Ed.: 141–295. CRC Press. Boca Raton, FL.
97. NIETERT, P.J. et al. 1998. Is occupational organic solvent exposure a risk factor for scleroderma? Arthritis Rheum. **41:** 1111–1118.
98. LYNGE, E. & L. THYGESEN. 1990. Primary liver cancer among women in laundry and dry-cleaning work in Denmark. Scand. J. Work. Environ. Health **16:** 108–112.
99. CHIA, S.E., V.H. GOH & C.N. ONG. 1997. Endocrine profiles of male workers with exposure to trichloroethylene. Am. J. Ind. Med. **32:** 217–222.
100. NIETERT, P.J. & R.M. SILVER. 2000. Systemic sclerosis: environmental and occupational risk factors. Curr. Opin. Rheumatol. **12:** 520–526.
101. RASMUSSEN, K., C.H. BROGREN & S. SABROE. 1993. Subclinical affection of liver and kidney function and solvent exposure. Int. Arch. Occup. Environ. Health **64:** 445–448.
102. BERNAUER, U. et al. 1996. Biotransformation of trichloroethene: dose-dependent excretion of 2,2,2-trichloro-metabolites and mercapturic acids in rats and humans after inhalation. Arch. Toxicol. **70:** 338–346.
104. THIER, R. et al. 2003. Markers of genetic susceptibility in human environmental hygiene and toxicology: the role of selected CYP, NAT, and GST genes. Int. J. Hyg. Environ. Health **206:** 149–171.
105. RUDER, A.M. et al. 1998. A pilot study: biological markers for perchloroethylene exposures and possible health effects related to that exposure; a project to test feasibility and refine methods. Protocol HSRB 98-DSHEFS-03. 1–151. National Institute for Occupational Safety and Health. Cincinnati, OH.

106. REED, D.J. 1993. Chlorocarbons and chlorohydrocarbons: survey. *In* Kirk-Othmer Encyclopedia of Chemical Technology, Vol. 5. Kroschwitz, J.I. & M. Howe-Grant, Eds.: 1017–1028. John Wiley & Sons. New York, NY.
107. CHEMICAL ECONOMICS HANDBOOK PROGRAM. 1999. Chlorinated methanes. *In* Chemical Economics Handbook. Chemical Economics Handbook Program, Ed.: 635.2000D-635.2002R. SRI International. Menlo Park, CA.
108. LEDER, A.E., W. BLYTH & M. ISIKAWA-YAMAKI. 1999. C_2 chlorinated solvents. *In* Chemical Economics Handbook. Chemical Economics Handbook Program, Ed.: 632.3000A-632.3002L. SRI International. Menlo Park, CA.
109. INTERNATIONAL AGENCY FOR RESEARCH ON CANCER. 1999. 1,1,2-Trichloroethane IARC Monogr Eval Carcinog Risks Hum **71(Pt 3):** 1153–1161.
110. COWFER, J.A. & M.B. GORENSEK. 1997. Vinyl chloride. *In* Kirk-Othmer Encyclopedia of Chemical Technology, Vol. 24. J.I. Kroschwitz & M. Howe-Grant, Eds.: 851–882. John Wiley & Sons. New York, NY.
111. JEBENS, A.M. 1999. Ethylene dichloride. *In* Chemical Economics Handbook. Chemical Economics Handbook Program, Ed.: 651.5000A-651.5001T. SRI International. Menlo Park, CA.
112. NATIONAL INSTITUTE FOR OCCUPATIONAL SAFETY AND HEALTH. 1990. National Occupational Exposure Survey.
113. EUROPEAN UNION. 1998. Carcinogen Exposure: Occupational Exposure to Carcinogens in the European Union in 1990-93. 2006. http://www.ttl.fi/Internet//English/Organization/Collaboration/Carex/carex_desc.htm
114. UNITED NATIONS ENVIRONMENTS PROGRAMME, INTERNATIONAL LABOUR OFFICE & WORLD HEALTH ORGANIZATION. 2005. International Chemical Safety Cards (ICSCs): International Programme on Chemical Safety (U.S. OSHA limits).
115. DEUTSCHE FORSCHUNGSGEMEINSCHAFT. 2003. List of MAK and BAT Values 2003: Maximum Concentrations and Biological Tolerance Values at the Workplace. Wiley. New York.
116. INTERNATIONAL OCCUPATIONAL SAFETY AND HEALTH INFORMATION CENTRE. 2005. Internationl Chemical Safety Cards (ICSCs): International Programme on Chemical Safety (EC limits).
117. INSTITUTO NACIONAL DE SEGURIDAD E HIGIENE EN EL TRABAJO, ESPANA. 2005. Occupational Exposure Limits (Spain).
118. HEALTH AND SAFETY EXECUTIVE. 2005. Occupational Exposure Standards (United Kingdom).

Endocrine Disruptors

Challenges for Environmental Research in the 21st Century

MARY S. WOLFF

Department of Community and Preventive Medicine, Mount Sinai School of Medicine, New York, New York 10029, USA

ABSTRACT: During the past 10 years, there has been a worldwide decline in the use and human exposure to many chemicals, including pesticides and persistent organic pollutants (POPs). However, a new generation of chemicals that have endocrine disrupting (ED) potential have emerged. Their presence in the environment and concomitant levels in humans are prevalent, although the sources of these contemporary-use industrial chemicals are not entirely identified. They include the phthalates, alkylphenols, brominated diphenyl ethers, and perfluorinated organics (PFOCs). The alkylphenols, especially bisphenol A, are potent EDs. Levels vary by geography, race/ethnicity, age and gender, and human health effects are just beginning to be assessed. This article discusses the toxicology, human exposure, and potential health effects of EDs that are likely to be important in the 21st century.

KEYWORDS: endocrine disruptors; phthalates; organochlorines; phenols; toxicology

INTRODUCTION

The concept of environmental endocrine disruptors (EDs), defined as hormonally active xenobiotics, is barely two decades old: the first citation in the *PubMed* was listed in 1987 followed by the seminal publications by Colborn and colleagues that appeared in the early 1990s.[1,2] Early on, the list of chemical suspects was long,[2,3] and much toxicologic research then ensued regarding well-known and newer chemicals. Research after this time became less centered on noncancer end points, shifting to receptor-based exper-

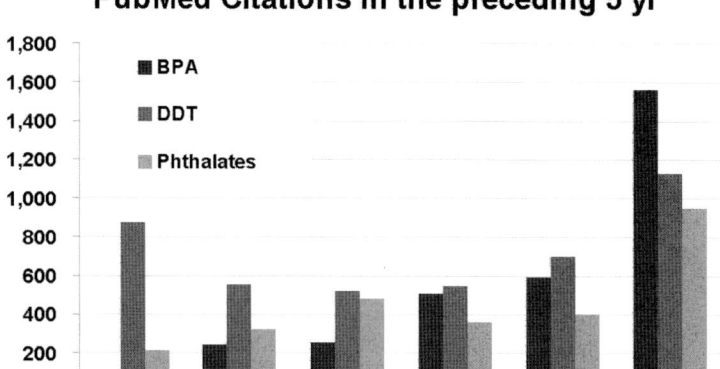

FIGURE 1. Number of the *PubMed* citations for three EDs from 1975–2004 by 5-year intervals: BPA, DDT, and phthalates.

iments, and was aptly termed *functional toxicology*.[4] The National Academy of Sciences completed a health evaluation of EDs in 1999, and the European Union established a special Directorate to monitor EDs around the same time.[5]

For many years before the ED phenomenon, epidemiologic research had been undertaken regarding more well-known exposures such as dioxins, bis (4-chlorophenyl)-1,1,1-trichloroethane (DDT), polychlorinated biphenyls (PCBs) (so-called persistent organic pollutants, or POPs), and lead (Pb). These exposures were better understood than some of the newly identified industrial chemicals, and methods were available to quantify them in the ambient environment and in human tissues. Today, research continues on these exposures, which can still be found at significant levels in developing countries, especially among children, who are a vulnerable population.[6] At the same time, biologic evidence has accumulated about a new generation of EDs, illustrated by a dramatic increase in citations about less-studied chemicals, including phthalates and alkylphenols (FIG. 1). *bis*PhenolA (BPA), now a topic of vigorous research, was virtually unknown before 1980, while health effects of DDT have been numerous and continue to be done. Toxicologic but little epidemiologic research exists for phthalates. Recently, a number of well-designed reproductive studies about phthalates in humans have been published in the United States and Europe. Determinations of biomarkers in humans have just begun for some of the newer ED chemicals, but it has become apparent that the 21st century brings a different range of chemical exposures to the forefront.

TOXICOLOGY

Earliest research focused on estrogenic effects of EDs, but these assessments were quickly extended in the 1990s to other end points, including androgen, thyroid, and orphan receptors.[7–9] Furthermore, whereas EDs are almost always weaker hormone agonists than steroids, their responses in nonhormonal assays may be more potent than steroids. For example, nonylphenol is a much weaker estrogen receptor (ER) agonist than diethystilbestrol (DES), but its effect on telomeres is more potent.[10] Evidence about multiple chemicals has shown that EDs may have additive or synergistic effects in chemical combinations[11–14] or together with endogenous hormones.[15] A further consideration is that dose–response curves are not always monotonic, and so low-dose effects do not necessarily extrapolate directly from high-dose experiments.[16,17] Indeed, as discussed in this symposium dose response at low levels can be supralinear (e.g., benzene and leukemia, as reported by Hayes at this symposium).

For some time, risk assessments have recognized the need to incorporate the relative potency of different environmental agents. The best example is the application of toxic equivalent factors (TEF) that were developed for dioxin and related compounds. TEFs allow a summation of individual compound concentrations based on their biologic activity.[18] In a discussion of estrogenic effects of environmental agents, Safe in 1995 applied the principle of estrogenic equivalents for organochlorine POPs in humans (concentration × relative activity, based on ER-binding or cell proliferation bioassays), concluding that biomarker levels were so low as to be swamped by natural "estrogens" such as phytoestrogens and hormone replacement (TABLE 1).[19,20] This particular approach neglected several factors, for example, the persistence of organochlorines compared to estrogen and the effect of EDs on nonhormonal end points.[21] But the principle of effective biologic ED levels remains important[4] and supports the need to examine closely the new-age EDs that are found at high levels in humans with respect to their relative biologic potency and the context of other exposures. The examples cited by Safe, updated with additional information in TABLE 1, suggest that BPA but not phthalates may be able to exert an estrogenic effect *in vivo* at levels currently observed in people.

Research on human health effects shifted over the past two decades from investigation mainly of cancer to other hormonally related outcomes. Studies have addressed fetal and child development, male and female reproductive function, respiratory function, and cognitive function. Such effects may differ markedly depending on the timing of ED exposure during an organism's lifetime.[22] Therefore, susceptible windows of effect have also received attention, as pointed out in this symposium,[23] for a variety of reasons. "Critical windows" may have very different endogenous hormone levels that enhance or reduce toxicity of EDs. Intraindividual susceptibility factors may differ throughout the life span, including variability in absorption, elimination, and biologic response to EDs. Genetic variation in metabolism and disposition of EDs has been characterized for some enzyme systems related to polycyclic aromatic hydrocarbons (PAH) and organochlorines (e.g., the cytochrome P450 and glutathione transferase coding genes) and to organophosphate (OP) pesticides (oxonases, esterases).[24] Recent studies have found individual variability in glucuronidation capacity for phthalates and phenols.[25–27] Less than 50% of people are equol producers, an important hormonal metabolism pathway for phytoestrogens.[28] At the end of the exposure–response continuum, target organ sensitivity to EDs may also vary among individuals, for example, with different variants in hormone receptors.[29]

MAGNITUDE OF EXPOSURE TO EDS

The extent of ED exposures can be appreciated in the levels of biomarkers in populations with and without known exposures. In North America and Europe, persistent organics and Pb have declined dramatically over the past few decades, as shown by the levels of organochlorine pesticides in breast milk. DDT declined more than 20-fold from 1965 to 1995 in both Canada and Sweden (FIG. 2 A). Other pesticides (dieldrin, heptachlor) showed identical reductions.[30] The drop in PCB levels was less dramatic over this period, perhaps because of some continuing low-level exposures (FIG. 2 B). However, recent data suggest a decline for PCBs by 2001 in the United States,[31,32] with the same half-life as the organochlorine pesticides (FIG. 2). A rule of thumb is that three half-lives are needed to eliminate a substance in the absence of additional exposure ($0.5^{3.3}$ or 3.3 half-lives = 10% left, or 90% clearance). For the data in FIGURE 2, the clearance time is thus about 19 years. As a result, levels of persistent organochlorines today are largely undetected in children and are increasingly so in adults (TABLE 2). Similarly, elemental Pb has declined more than 10-fold since the 1960s, and it was lower in the 2000–2001 than in the 1999 NHANES data.[31] Other pesticides, such as the OPs, have also become

TABLE 1. Net effect of environmental estrogens (dose × relative potency)

	"Estrogen Equivalents"
Morning-after pill	333×10^3
Birth control	17×10^3
Hormone replacement	3×10^3
Dietary flavonoids	100
*BPA	~1
+Organochlorines	2×10^{-6}
*Phthalates	$<10^{-6}$

NOTE: Adapted from Safe, 1995 (Refs. 19–20).
*Not assessed by Safe, 1995.
+Persistent.

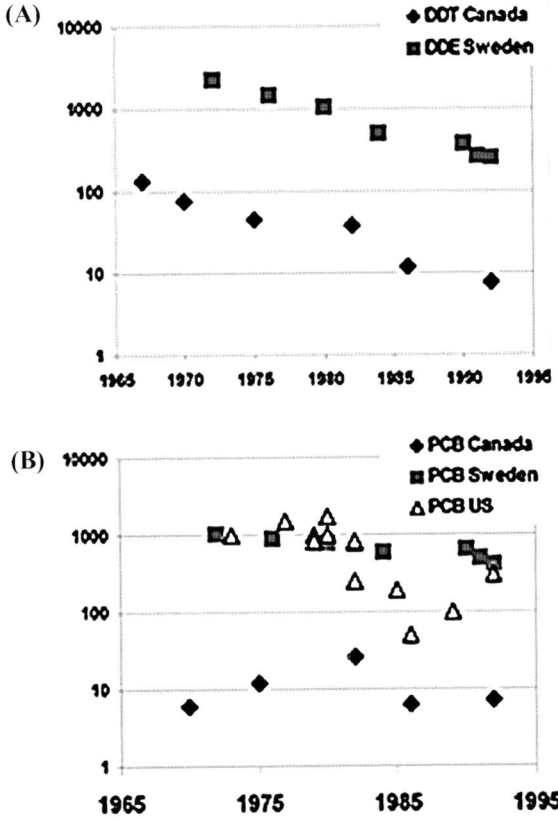

FIGURE 2. Decline in persistent organochlorines during the late 20th century. Slopes of the DDE regression models were -0.11 ln-ng/g.year for both curves (ln-DDE versus year; $t_{\frac{1}{2}} = 6.3$ years). For the PCB data shown in panel **B** from the U.S. the beta was -0.16 ($t_{\frac{1}{2}} = 4.3$ years); with added data from 1973 and 2003 the slope was -0.11 ($t_{\frac{1}{2}} = 6.3$ years). The PCB data from Sweden had a slope of -0.04; the Canadian PCB slope was not computed. (Data from Refs. 54–56.)

less commonly used, but their levels have not yet disappeared. Measurements in NHANES showed a decline in recent years of the alkylphosphates which are generic metabolites of OPs.[31] Specific metabolites for OPs (e.g., malathion dicarboxylic acid) were also measured by the Centers for Disease Control and Prevention (CDC); medians for five of the six were below the detection limits. Another specific OP biomarker, 3,5,6-trichloro-2-pyridinol, was detected (median 2.2 μg/L in the 2000–2001 samples), and these levels were similar to values reported in Europe and elsewhere.

In contrast, other, newer ED exposures have risen or remained stable. Typical 95th percentiles of the NHANES results are shown in TABLE 2, with two

TABLE 2. Levels of selected environmental biomarkers, 95th percentiles, NHANES 2001–2002

	Descending order of median biomarkers		
	6–11 years	≥20 years	unit+
Phytoestrogens (sum of 6 metabolites)	4,260	5,260	μg/L U
Phthalates (sum of 12 metabolites)	1,630	3,280	μg/L U
Alkylphosphates (sum of 6 metabolites)	96	62	μg/L U
Pb	37	46	μg/L B
PAH-OH (sum of 22 metabolites)	24	46	μg/L U
3 chlorophenols (Cl_5,Cl_3)	23	18	μg /L U
DDE + DDT**	2.3	17	μg /L B
Alkylphenols (BPA)*	2	1.3	μg/L U
Hg (mercury)	1.9	4.6	μg/L B
34 PCB, 17 PCDD/Fs**	0.5	3.5	μg/L B
Cotinine	3.2	1.4	μg/L B

NOTE: +U = urine, B = blood.
**12–19 years; not done in 6–11 years.
*Ref. 34 and unpublished data. Other data from cdc.gov/exposurereport/
Medians were approximately half the 95th percentiles for phytoestrogens, phthalates, DDE/T, and less than one-tenth for alkylphosphates.
DDE = dichlorodiphenyldichloroethylene; DDT = dichlorodiphenyltrichloroethane.

families of chemicals having >1000 μg/L, four groups being above 10 μg/L, while others were below 5 μg/L. The highest levels are the phthalate metabolites that on average are >200 μg/L in children and adults, with the 95th percentile above 1 mg/L (TABLE 2). Children typically had higher levels of many biomarkers than adults, attributable to exposure levels and to smaller body size. Findings are similar in other countries.[33] Among Blacks, patterns of exposure in the U.S. NHANES data find higher phthalates and certain phenols and lower phytoestrogens than in Hispanics and Whites. BPA levels have generally been reported to be below 10 μg/L, and nonylphenol and related alkylphenols are lower.[34] Even occupationally exposed populations do not have high alkylphenol biomarkers.[35] While blood biomarker levels are not directly comparable to urine concentrations for all chemicals, the pattern of exposures in TABLE 2 underscores the dominance of new-age chemicals including the phytoestrogens and phthalates, whereas Pb remains intermediate in level. Still emerging as significant contaminants are the polybrominated biphenylethers (PBDEs) and perfluorinated organics (PFOCs). Population-based research or large-scale studies are not yet available, but preliminary data have been reported in small samples that are not necessarily generalizable (e.g., age, sex, race). These findings show PBDE levels of 15–31 ng/g lipid in blood and milk,[36–38] which were less than one-fifth the level of PCBs in New York women in 2002[39] and California women in 1999.[37] However, there may be wide geographic variability: serum pools created in Dallas in 2003 had PBDE levels (61 ng/g lipid) that were more than twice that of PCBs.[32] These reported median PFOC concentrations in blood range from 8 μg/L in Japan[40] to 28–73 μg/L in Canada[41] and the United States.[42]

Environmental sources of ED exposures are not entirely known, although a number of chemicals have been traced to dietary and air contamination. ED exposures are likely to be significant in the indoor environment, where they are pervasive at levels (i.e., micrograms) sufficient to exert an effect on hormone balance.[14,43] Contact with certain products, such as plastics, detergents, personal care products (lotion, perfume, shampoo, spermicides, makeup), dental sealants, and epoxy resins, contribute to phthalate and BPA exposure.[44–47] PAH are hydrocarbon combustion products that arise from air pollution, food, and tobacco smoke. The highest PAH phenols found in NHANES samples were 1- and 2-naphthol (TABLE 2), which may come from mothballs. Some studies suggest that intake of PAH, pesticides, and phthalates is greater from diet than other sources.[48,49] Limited data also suggest that the tetrabrominated diphenyl ethers may come from diet while higher-brominated exposures arise from dust.[50,51] Occupational exposures to EDs may provide further clues to sources, and such populations also afford the opportunity to examine potential health effects.[52]

FUTURE DIRECTIONS AND RESEARCH NEEDS

As research continues on health effects of EDs, significant gaps in research remain to be filled. We need population data on the newer EDs, such as PBDEs and PFOCs. At the same time, basic knowledge about fate and transport in the body is very sketchy for many chemicals: where do they come from and how long do they last in the body? Population levels of PBDEs are increasing, but we still know little about the origin and persistence in humans of this complex family of chemicals although they have been used as fire retardants for at least three decades.[53] Individual susceptibility factors have not been well characterized, and they span the entire range of exposure and disease, from chemical absorption and metabolism to target organ sensitivity and repair of tissue damage. The breadth and combination of these variations pose a singular challenge to scientific investigation. Finally, as we learn more about single chemicals and distinct families of environmental agents, we are faced with the need to address the complexity of ordinary life: we are exposed not to one but to many EDs, and it will require new insights and clever methodologic approaches to unravel their combined effects.

REFERENCES

1. COLBORN, T. & C. CLEMENT. 1992. Chemically-Induced Alterations in Sexual and Functional Development: The Wildlife Human Connection. Princeton Scientific Publishing. Princeton, NJ.

2. COLBORN, T., F.S. VOM SAAL & A.M. SOTO. 1993. Developmental effects of endocrine-disrupting chemicals in wildlife and humans. Environ. Health Perspect. **101:** 378–384.
3. KAVLOCK, R.J., G.P. DASTON, C. DE ROSA, *et al.* 1996. Research needs for the risk assessment of health and environmental effects of endocrine disruptors: a report of the U.S. EPA-sponsored workshop. Environ. Health Perspect. **104**(suppl. 4): 715–740.
4. MCLACHLAN, J.A. 1993. Functional toxicology: a new approach to detect biologically active xenobiotics. Environ. Health Perspect. **101:** 386–387.
5. ENVIRONMENT, D.G. 2005. Information Brochure. European Commission. available at: http://europa.eu.int/comm/dgs/environment/index-en.htm
6. ROGAN, W.J. & A. CHEN. 2005. Health risks and benefits of bis(4-chlorophenyl)-1,1,1-trichloroethane (DDT). Lancet **366:** 763–773.
7. ZACHAREWSKI, T. 1998. Identification and assessment of endocrine disruptors: limitations of *in vivo* and *in vitro* assays. Environ. Health Perspect. **106**(Suppl. 2): 577–582.
8. MCLACHLAN, J.A. 2001. Environmental signaling: what embryos and evolution teach us about endocrine disrupting chemicals. Endocr. Rev. **22:** 319–341.
9. SOHONI, P. & J.P. SUMPTER. 1998. Several environmental oestrogens are also anti-androgens. J. Endocrinol. **158:** 327–339.
10. ROY, D., J.B. COLERANGLE & K.P. SINGH. 1998. Is exposure to environmental or industrial endocrine disrupting estrogen-like chemicals able to cause genomic instability? Front. Biosci. **3:** d913–d921.
11. BERGERON, J.M., D. CREWS & J.A. MCLACHLAN. 1994. PCBs as environmental estrogens: turtle sex determination as a biomarker of environmental contamination. Environ. Health Perspect. **102:** 780–781.
12. RAJAPAKSE, N., E. SILVA & A. KORTENKAMP. 2002. Combining xenoestrogens at levels below individual no-observed-effect concentrations dramatically enhances steroid hormone action. Environ. Health Perspect. **110:** 917–921.
13. SOTO, A.M., K.L. CHUNG & C. SONNENSCHEIN. 1994. The pesticides endosulfan, toxaphene, and dieldrin have estrogenic effects on human estrogen-sensitive cells. Environ. Health Perspect. **102:** 380–383.
14. SILVA, E., N. RAJAPAKSE & A. KORTENKAMP. 2002. Something from "nothing"—eight weak estrogenic chemicals combined at concentrations below NOECs produce significant mixture effects. Environ. Sci. Technol. **36:** 1751–1756.
15. WELSHONS, W. 2006. Evaluation and prediction of risks of environmental estrogens. Ann. N. Y. Acad. Sci.
16. ZOELLER, T. 2006. Collision of basic and applied approaches to risk assessment of thyroid toxicants. Ann. N. Y. Acad. Sci.
17. VOM SAAL, F.S. & C. HUGHES. 2005. An extensive new literature concerning low-dose effects of bisphenol A shows the need for a new risk assessment. Environ. Health Perspect. **113:** 926–933.
18. WALKER, N.J., P.W. CROCKETT, A. NYSKA, *et al.* 2005. Dose-additive carcinogenicity of a defined mixture of "dioxin-like compounds". Environ. Health Perspect. **113:** 43–48.
19. SAFE, S. & V. KRISHNAN. 1995. Chlorinated hydrocarbons: estrogens and antiestrogens. Toxicol. Lett. **82–83:** 731–736.
20. SAFE, S.H. 1995. Environmental and dietary estrogens and human health: is there a problem? Environ. Health Perspect. **103:** 346–351.

21. WOLFF, M.S. & P.J. LANDRIGAN. 1994. Environmental estrogens. Science **266:** 526–527.
22. LAMARTINIERE, C.A. 2002. Timing of exposure and mammary cancer risk. J. Mammary. Gland. Biol. Neoplasia. **7:** 67–76.
23. MANTOVANI, A. 2006. Risk assessment of endocrine disruptors: the role of toxicological studies. Ann. N. Y. Acad. Sci.
24. BRYK, B., L. MOYAL-SEGAL, E. PODOLY, et al. 2005. Inherited and acquired interactions between ACHE and PON1 polymorphisms modulate plasma acetylcholinesterase and paraoxonase activities. J. Neurochem. **92:** 1216–1227.
25. SILVA, M.J., D.B. BARR, J.A. REIDY, et al. 2003. Glucuronidation patterns of common urinary and serum monoester phthalate metabolites. Arch. Toxicol. **77:** 561–567.
26. YE, X., Z. KUKLENYIK, L.L. NEEDHAM & A.M. CALAFAT. 2005. Quantification of urinary conjugates of bisphenol A, 2,5-dichlorophenol, and 2-hydroxy-4-methoxybenzophenone in humans by online solid phase extraction-high performance liquid chromatography-tandem mass spectrometry. Anal. Bioanal. Chem. **4:** 638–644.
27. VOLKEL, W., T. COLNOT, G.A. CSANADY, et al. 2002. Metabolism and kinetics of bisphenol a in humans at low doses following oral administration. Chem. Res. Toxicol. **15:** 1281–1287.
28. SETCHELL, K.D., N.M. BROWN & E. LYDEKING-OLSEN. 2002. The clinical importance of the metabolite equol-a clue to the effectiveness of soy and its isoflavones. J. Nutr. **132:** 3577–3584.
29. KUIPER, G.G., J.G. LEMMEN, B. CARLSSON, et al. 1998. Interaction of estrogenic chemicals and phytoestrogens with estrogen receptor beta. Endocrinology **139:** 4252–4263.
30. WOLFF, M.S. 1999. Half-lives of organochlorines (OCs) in humans. Arch. Environ. Contam. Toxicol. **36:** 504.
31. Available at www.cdc.gov/exposurereport/. 2005.
32. SCHECTER, A., O. PAPKE, K.C. TUNG, et al. 2005. Polybrominated diphenyl ether flame retardants in the U.S. population: current levels, temporal trends, and comparison with dioxins, dibenzofurans, and polychlorinated biphenyls. J. Occup. Environ. Med. **47:** 199–211.
33. KOCH, H.M., R. PREUSS, H. DREXLER & J. ANGERER. 2005. Exposure of nursery school children and their parents and teachers to di-n-butylphthalate and butylbenzylphthalate. Int. Arch. Occup. Environ. Health **78:** 223–229.
34. CALAFAT, A.M., Z. KUKLENYIK, J.A. REIDY, et al. 2005. Urinary concentrations of bisphenol A and 4-nonylphenol in a human reference population. Environ. Health Perspect. **113:** 391–395.
35. YE, X. & N.L.C.A KUKLENYIK Z. 2005. Automated on-line column-switching HPLC-MS/MS method with peak focusing for the determination of nine environmental phenols in urine. Anal Chem. **77:** 5407–5413.
36. FOCANT, J.F., A. SJODIN, W.E. TURNER & D.G. PATTERSON JR. 2004. Measurement of selected polybrominated diphenyl ethers, polybrominated and polychlorinated biphenyls, and organochlorine pesticides in human serum and milk using comprehensive two-dimensional gas chromatography isotope dilution time-of-flight mass spectrometry. Anal. Chem. **76:** 6313–6320.
37. PETREAS, M., J. SHE, F.R. BROWN, et al. 2003. High body burdens of $2,2',4,4'$-tetrabromodiphenyl ether (BDE-47) in California women. Environ. Health Perspect. **111:** 1175–1179.

38. MORLAND, K.B., P.J. LANDRIGAN, A. SJODIN, et al. 2005. Body burdens of polybrominated diphenyl ethers among urban anglers. Environ. Health Perspect. **113:** 1689–1692.
39. WOLFF, M.S., S.L. TEITELBAUM, P.J. LIOY, et al. 2005. Exposures among pregnant women near the World Trade Center site on 11 September 2001. Environ. Health Perspect. **113:** 739–748.
40. INOUE, K., F. OKADA, R. ITO, et al. 2004. Perfluorooctane sulfonate (PFOS) and related perfluorinated compounds in human maternal and cord blood samples: assessment of PFOS exposure in a susceptible population during pregnancy. Environ. Health Perspect. **112:** 1204–1207.
41. KUBWABO, C., N. VAIS & F.M. BENOIT. 2004. A pilot study on the determination of perfluorooctanesulfonate and other perfluorinated compounds in blood of Canadians. J. Environ. Monit. **6:** 540–545.
42. KUKLENYIK, Z., J.A. REICH, J.S. TULLY, et al. 2004. Automated solid-phase extraction and measurement of perfluorinated organic acids and amides in human serum and milk. Environ. Sci. Technol. **38:** 3698–3704.
43. RUDEL, R.A., J.G. BRODY, J.D. SPENGLER, et al. 2001. Identification of selected hormonally active agents and animal mammary carcinogens in commercial and residential air and dust samples. J. Air Waste Manag. Assoc. **51:** 499–513.
44. KAVLOCK, R., K. BOEKELHEIDE, R. CHAPIN, et al. 2002. NTP Center for the Evaluation of Risks to Human Reproduction: phthalates expert panel report on the reproductive and developmental toxicity of di-n-octyl phthalate. Reprod. Toxicol. **16:** 721–734.
45. MATSUMOTO, A., N. KUNUGITA, K. KITAGAWA, et al. 2003. Bisphenol A levels in human urine. Environ. Health Perspect. **111:** 101–104.
46. TEITELBAUM, S.L., A.M. CALAFAT, J.A. BRITTON, et al. 2005. Temporal variability in urinary phthalate metabolites, phenols and phytoestrogens among children. Epidemiology **16:** S41.
47. DUTY, S.M., R.M. ACKERMAN, A.M. CALAFAT, et al. 2005. Personal care product use predicts urinary concentrations of some phthalate monoesters. Environ. Health Perspect. **113:** 1530–1535.
48. FROMME, H., T. LAHRZ, M. PILOTY, et al. 2004. Occurrence of phthalates and musk fragrances in indoor air and dust from apartments and kindergartens in Berlin (Germany). Indoor Air **14:** 188–195.
49. WILSON, N.K., J.C. CHUANG & C. LYU. 2001. Levels of persistent organic pollutants in several child day care centers. J. Expo. Anal. Environ. Epidemiol. **11:** 449–458.
50. SCHECTER, A., O. PAPKE, J.E. JOSEPH & K.C. TUNG. 2005. Polybrominated diphenyl ethers (PBDEs) in U.S. computers and domestic carpet vacuuming: possible sources of human exposure. J. Toxicol. Environ. Health A **68:** 501–513.
51. JONES-OTAZO, H.A., J.P. CLARKE, M.L. DIAMOND, et al. 2005. Is house dust the missing exposure pathway for PBDEs? An analysis of the urban fate and human exposure to PBDEs Environ. Sci. Technol. **39:** 5121–5130.
52. KUKLENYIK, Z., J. EKONG, C.D. CUTCHINS, et al. 2003. Simultaneous measurement of urinary bisphenol A and alkylphenols by automated solid-phase extractive derivatization gas chromatography/mass spectrometry. Anal. Chem. **75:** 6820–6825.
53. NORRIS, J.M., R.J. KOCIBA, B.A. SCHWETZ, et al. 1975. Toxicology of octabromobiphenyl and decabromodiphenyl oxide. Environ. Health Perspect. **11:** 153–161.
54. LAKIND, J.S., C.M. BERLIN & D.Q. NAIMAN. 2001. Infant exposure to chemicals in breast milk in the United States: what we need to learn from a breast milk monitoring program. Environ. Health Perspect. **109:** 75–88.

55. CRAAN, A.G. & D.A. HAINES. 1998. Twenty-five years of surveillance for contaminants in human breast milk. Arch. Environ. Contam. Toxicol. **35:** 702–710.
56. NOREN, K., A. LUNDEN, E. PETTERSSON & A. BERGMAN. 1996. Methylsulfonyl metabolites of PCBs and DDE in human milk in Sweden, 1972–1992. Environ. Health Perspect. **104:** 766–772.

Risk Assessment of Endocrine Disrupters

The Role of Toxicological Studies

ALBERTO MANTOVANI

Department of Food Safety and Veterinary Public Health, Istituto Superiore di Sanità, 00161 Rome, Italy

> ABSTRACT: Endocrine disrupters (ED) represent a good challenge for experimental toxicology. In order to deal with several critical points relevant to risk assessment: (*a*) ED may induce long-term effects upon exposure in susceptible developmental phases, including postnatal life up to puberty; thus, efforts are required to refine testing strategies, for example, by supporting the two-generation rodent study with a comprehensive *in vitro/in vivo* screening battery; (*b*) due to the regulatory role of endocrine homeostasis, mechanisms of endocrine disruption may impact on immune, neurobehavioral, and reproductive development, as well as on susceptibility to cancer; (*c*) the potential multiple exposure to ED with common targets through diet and/or living environment calls for the development of models to understand mechanisms of interactions and effects of mixtures; and (*d*) last but not least, ED may interact with a number of factors related to differential vulnerability of individuals or population subgroups, including the intake of nutrients or bioactive food components. Besides reducing the chance for noxious chemicals to enter our life, toxicological research on mechanisms may also lead to the definition of possible biomarkers of exposure, effect, and susceptibility that may be further exploited in human health surveillance.
>
> KEYWORDS: development; prenatal development; prepubertal stage; fertility; biomarkers; two-generation study; mechanisms of action

INTRODUCTION

In the last decade toxicological research and regulatory agencies are paying attention toward the risk assessment of xenobiotics with potential endocrine activities, the so-called endocrine disrupters (ED).[1] ED are a heterogeneous ensemble of substances that can interfere with the function of the endocrine system through diverse mechanisms, such as agonism/antagonism

---

Address for correspondence: Alberto Mantovani, Department of Food Safety and Veterinary Public Health, Istituto Superiore di Sanità, viale Regina Elena, 299, 00161 Rome, Italy. Voice: ++39-06-4990-2815; fax: ++39-06-4990-2658.
  e-mail: alberto@iss.it

with nuclear receptors, inhibition of hormone biosynthesis, and interference with the hypothalamus–hypophysis axis. On the basis of available knowledge, the homeostasis of sex steroids and thyroid are the main, but not exclusive, targets of ED effects.[1,2] Thus, the topic of reproductive health, including the continuum from gamete production and fertilization through to intrauterine and postnatal development of progeny, is recognized as especially vulnerable to endocrine disruption. In particular, development and differentiation of target tissues may be critically sensitive phases because hormones can play crucial regulatory roles as shown, for example, for gonadal steroids in sexual differentiation.[3,4]

ED include such diverse compounds as persistent organic pollutants, for example, dioxins and polychlorinated biphenyls (PCB),[5] several groups of chemicals used in plant and/or farm animal production, such as chlorinated insecticides and imidazole fungicides,[6,7] a number of widespread industrial chemicals, such as some phthalates and brominated flame retardants,[8,9] and heavy metals, such arsenic and cadmium.[10,11] A peculiar group of endocrine-active compounds are the so-called "phytoestrogens"; these include such compounds as isoflavones, lignans, coumestans, etc., that are naturally present in vegetable food commodities, such as soy, and act as modulators of estrogen receptors.[12] A good dietary intake of phytoestrogens might be a protective factor toward several cancers, for example, breast, prostate, and postmenopausal diseases, such as osteoporosis; however, possible concerns exist regarding the exposure to high doses during pregnancy or early infancy, for example, through the use of dietary integrators or soy-based milk.[13,14]

Due to their diversity, ED have different targets. Examples of enzyme inhibitors include ethylenethiourea, a byproduct of thiocarbammates that targets thyroid peroxidase,[15] or azole fungicides that impair the steroid biosynthesis.[16] Among compounds interfering with nuclear receptors, vinclozolin and other dicarboximide pesticides are antagonists of androgen receptor,[17] whereas "xenoestrogens," such as *o,p'* DDT and the industrial chemicals bisphenol A and nonylphenol act as agonists of estrogen receptor alpha.[18] Other mechanisms might deserve more attention, such as the interactions with the regulatory hypothalamus–pituitary network as by the triazine herbicides in rats,[19] or with hormone transport as shown by the hydroxylated PCB metabolites inhibiting human estrogen sulfotransferase.[20]

Such target specificity should not be taken in a too rigid way: bisphenol A and nonylphenol may act also as antiandrogens.[21] Endocrine homeostasis is a complex system with a high degree of interaction and interdependency. Moreover, normal endocrine function is not steady state, but depends on cyclical patterns, age, and developmental stage. A major example is represented by dioxins and dioxin-like compounds; these contaminants activate a family of transcription factors, the arylhydrocarbon receptor and the arylhydrocarbon receptor nuclear translocator, resulting in multifaceted effects through a complex cross-talk with the estrogen receptors and other nuclear receptors as well.[22] A major issue is whether exposure to ED actually leads to adverse effects on human health. Whereas high exposures to specific ED may occur in some occupational settings, especially in developing countries,[23] the general population likely experiences a low-level intake of multiple ED that may contaminate the food chain and environment through many different usage patterns and exposure pathways.[24] A number of studies link increased exposure to persistent organic pollutants and other ED to male infertility, malformations of the male reproductive system, for example, hypospadia and cryptorchidism, gynecological problems (recurrent pregnancy loss, endometriosis), increased susceptibility to cancer of the testis and other target tissues, as well as developmental delays in children.[25,26] Nevertheless, the actual role of ED as risk factors for human health is still a matter of debate.[27]

In such a controversial field, toxicology plays a major role in order both to set policies for regulation and risk management of chemicals and to support the assessment of issues potentially relevant to public health.

TOXICOLOGICAL TESTING AND THE RISK ASSESSMENT OF ED

Whereas the available knowledge is steadily increasing, ED still are a challenge for toxicological research, particularly for hazard identification and characterization components of the risk assessment process.[4]

Hazard Identification

A critical issue for toxicological studies on ED is the identification of long-term effects upon exposure in susceptible developmental phases, including postnatal life, at dose levels well below those inducing evident birth defects and/or fetal growth retardation. Examples of subtle effects elicited by exposure during organogenesis in rodents include: impaired spermatogenesis induced by the synthetic estrogen diethylstilboestrol or the chlorinated insecticide lindane,[28] altered estrous cyclicity by the major PCB metabolite 4-OH-CB-107,[29] and impaired histogenesis of thyroid and adrenals by the fungicide methyl thiophanate.[30]

Current regulatory tests are relevant to the hazard identification of ED. An effort toward refinement of the testing battery is required in order to improve the ability to catch the actually relevant end points. The prenatal developmental toxicity studies may identify effects upon exposure during pregnancy, including morphological abnormalities of target tissues and/or organs, for example, hypospadias,[31] which are relevant to ED toxicity. However, to achieve a proper hazard identification of ED, histopathological examinations and/or prolonged postnatal observation could be included, based on the available information on targets most relevant for a given chemical or group of chemicals.[32]

Hazard Characterization

Besides identifying effects, understanding mechanisms may be important for the adequate understanding of hazards. Although most attention has focused initially on the reproductive impact, the effects of ED may be quite far reaching. The later outcomes of altered programming *in utero* and during early childhood are particularly important for body systems with prolonged maturation, such as the immune and nervous systems. They may be considered as communicating networks of signaling pathways that are critical for interaction with the environment and are regulated by the endocrine system. For instance, the putative xenoestrogen bisphenol A induces neurobehavioral and immune alterations[33,34]; persistent organic pollutants may also affect the functional development of immune and nervous systems, possibly acting through agonism of the arylhydrocarbon receptor,[35] and thyroid interference,[36] respectively. Subtle effects may have important long-term consequences. Impaired differentiation of target tissues might be linked to increased susceptibility to cancer later in life; examples are alterations of mammary tissue morphogenesis induced by bisphenol A and tetrachloro-p-dibenzo-dioxin[37,38] and testicular morphogenesis by di-n-butyl phthalate.[39]

Two-Generation Study

This protocol (OECD testing guideline 416) is a key component of testing chemicals to which humans may be exposed through food and living environment, such as many ED.[40] The test substance is administered continuously through diet or, sometimes drinking water, to parental (P) and subsequent offspring generations (F1, F2) with one or two litters per generation; the study is normally conducted with the rat. The two-generation test provides information about effects on male and female reproductive performances, pregnancy outcomes, maternal lactation and offspring care, prenatal and postnatal survival, growth and development of the offspring as well as on their reproductive capacity; in particular, it also provides a reliable assessment of exposure of newborn and peripubertal animals.[41] Most important, the two-generation study is the only one regulatory test where a group of animals of the F1 generation is exposed to a substance throughout the entire development, that is, from gamete stage until sexual maturity and production of offspring; accordingly, the two-generation test is recognized as the key study for hazard identification of ED.[4] Moreover, it is a relatively flexible protocol, where several additional parameters and/or satellite groups may be fed-in to investigate long-term effects on the prenatal and postnatal development of the immune, nervous, and reproductive systems.[42–44] A potential drawback of multigeneration tests is their complexity. The investigation of all parameters relevant to reproductive function as well as development will result in a huge quantity of data, since the combined evaluation of a number of parameters is required for a sensitive identification of certain critical effects. For example, indicators of impaired male reproductive development include: anogenital distance, hypospadias, age and weight of offspring on the day of balanopreputial separation, weight of testis and accessory glands, and sperm parameters (count, motility, morphology, genetic integrity). In rodents, fertility is quite an insensitive parameter, since severe damage to reproductive functional development is required to detect an appreciable effect; on the contrary, the qualitative and quantitative histological examination of reproductive tissues is likely the most sensitive marker.[44] Thus, the OECD recommends a stepwise approach to the toxicological testing of ED, where screening *in vitro/in vivo* assays should identify the most relevant mechanisms/end points to be investigated; as a result, targeted two-generation assays would allow to follow up late outcomes and identify dose–response curves for critical effects.[45]

DEVELOPMENT OF A SCREENING BATTERY FOR ED

In vitro or short-term *in vivo* assays tend to focus on a narrow set of effects/mechanisms, for example, receptor binding and/or transactivation, whereas interaction with endocrine homeostasis is a complex event targeting a network of signals among different tissues: thus, no such test can yet be regarded as a comprehensive screen for ED.[46] Nevertheless, several screening assays for the most relevant mechanisms have been developed and are in course of validation. *In vivo* short-term assays have received much interest since they appear to mimic the complexity of the endocrine system in a more reliable way. The uterotrophic assay and Hershberger assay are developed to identify interferences with estrogen- or androgen-related pathways, respectively; both are based on the weight changes of reproductive tissues, as well as having a time-honored history in endocrinology.[45,46] An enhanced version of the subacute 28-day rat toxicity test, the most common test applied to comprehensively test chemical toxicity worldwide, is also internationally recommended as a first-tier assay for ED. Additional parameters include serum levels of LH, thyroid, and steroid hormones, assessment of the estrous cycle, and detailed examination of reproductive tissues, thyroid, and pituitary, as well as of spermatogenesis.[47,48]

A potential shortcoming of the 28-day test concerns the use of young adult rodents that might be not as sensitive as animals at earlier life stages. The U.S. Environmental Protection Agency endorses the uses of the 20-day female pubertal assay as an integration of a first-tier *in vivo* testing battery including the uterotrophic and Hershberger assays: the pubertal assay could identify effects on steroidogenesis, antithyroid, and antiestrogenic activities as well as perturbances of hypothalamus-pituitary-gonadal axis.[49] *In vitro* tests include the activation of estrogen and androgen receptors for binding/transcriptional activity and assays for steroidogenesis; however, few or no reliable assays are yet available for such important actions as interference with thyroid function or with hypothalamic–pituitary axis.[50] Combinations of assays to compose testing strategies is a promising way to implement an intelligent, efficient, mechanism-driven approach to ED hazard identification.[50,51] Therefore, the validation of integrated testing batteries emerges as a critical issue, together and beyond validation of individual assays. Implementation of batteries entails specific challenges, such as including more mechanisms besides estrogen/androgen agonism/antagonism and strengthening the sensitivity of the assays by means of metabolic activation systems, molecular biology tools, etc. Such problems are outlined, for example, in the objectives of ReProTect, an Integrated Project funded within the European 6th Framework Programme.[51]

TOXICOLOGICAL RESEARCH AND NEW CHALLENGES IN THE RISK ASSESSMENT OF ED

Besides reducing the chance for noxious chemicals to enter our life through the identification of ED and associated hazards, toxicological assays may also provide important information for other components of risk assessment. Examples shall be given concerning the elucidation of the full spectrum of effects.

Elucidation of the Spectrum of ED-Induced Effects

ED toxicology cannot be simplistic. A different susceptibility of tissues to certain mechanisms or chemicals may be observed not only according to the pharmacokinetics but also to the tissue-specific patterns of cross-talk among nuclear receptors.[52] An example is the different susceptibility of reproductive tissues, hypothalamus, and pituitary as regards estrogen receptor alpha expression following neonatal exposure of rodents to three compounds that are all labeled as "estrogen agonists," that is, diethylstilboetrol, bisphenol A, and octylphenol.[53] Basic research on mechanisms is also expanding the range of cellular targets of ED, in particular with regard to interference with nuclear receptors. The progesterone receptor may be a target equally or even more sensitive than estrogen receptor alpha[54] for several putative "estrogenic" chlorinated compounds. Arsenic interacts with the glucocorticoid receptor[55]; arsenic intake is associated with an increased risk of diabetes in both rodents and humans.[10] Phthalates may activate pregnane-X-receptor that mediate the induction of enzymes involved in steroid metabolism and xenobiotic detoxification.[56] Parallel to the new data on mechanisms, experimental studies increasingly point to effects other than on reproductive function, such as immune modulation[34,35] and neurobehavioral development.[26,32,36] Under this respect, pubertal assays are a resource that could be increasingly exploited to investigate effects in juvenile stages[57] where major events involving the whole endocrine network as well as growth, metabolism, and immunity do occur.[58] For instance, delayed preputial separation, elevated TSH, and reduced T3 and T4 were the most evident effects in a rat pubertal assay on a commercial mixture of polybrominated flame retardants. Reduced serum T3 in males appeared as the critical effect.[59] Experimental studies also point to further targets of xenobiotics that might deserve greater attention, for example, pancreatic islets,[60] which may point to a role of chemicals in a major public health issue as diabetes, or adrenals,[30] that may be relevant to such multiple end points as immunity, electrolyte and glucose metabolism, and steroid biosynthesis.

"Real-Life" Toxicology

Experimental models cannot mimic the complexity of real-life situations; nevertheless, experimental toxicology may both provide possible tools to be used in public health research and surveillance (e.g., biomarkers) and support the clarification of what may occur in "real life," for example, where organisms of different ages are exposed to a mixture of low-level ED residues in diet. Models that may lead to a better understanding (hence, to a better management) of what may occur in actual exposure situations do need facing several major issues, still under debate. Effects of *mixtures* of chemicals sharing the same cellular targets are under investigation. Whereas it is generally assumed that effects of combined exposures to xenobiotics, including ED, unlikely depart from dose additivity,[61] it is worth mentioning that some studies indicate possible exceptions. For instance, PCB and dioxins may show a greater-than-additive effects on thyroid function markers in rats, albeit only at high dose levels[62] whereas mixtures of estrogen agonists, such as alkylphenols and genistein showed less-than-additive effects on estrogen-dependent MCF-7 human breast cancer cells, possibly indicating weak antagonism.[63] Therefore, ED toxicology needs to take into account the cross-talk among different signaling pathways. Another relevant issue in risk assessment are potential adverse health effects of combined exposures to chemicals and nonchemical stressors.[64] In the field of food safety it is especially interesting to obtain more data on the *interactions between ED and bioactive food components* that may modulate endocrine homeostasis, either directly or indirectly. Indeed, a number of data already point out possible protective effects toward PCB of, for example, antioxidants, such as vitamins E and C.[65,66] The case of endocrine-active substances in vegetables, so-called, "phytoestrogens," is less clear-cut. Some experimental studies suggest a protective effect toward oxidative stress[67] and chemically induced mammary tumors.[68] However, it cannot be ruled out altogether that phytoestrogens may add to ED effects: for instance, complex interactions, including the potentiation of some effects, have been elicited by the combined exposure to soy-phytoestrogen genistein and the estrogen-alpha agonist methoxychlor on the pre–postnatal sexual development in rats.[69]

Experimental studies may identify *possible biomarkers* to be exploited both as early signals of effect in experimental studies and, even more important, as biomarkers of effective dose in health surveillance. The standardization of *in vitro* receptor binding and transactivation assays may lead to the development of total exposure biomarkers, such as the "total estrogen effective burden" that may be implemented to screen estrogenicity in a series of biological samples, thus strengthening exposure assessment in epidemiological studies.[70] The development of possible biomarkers is supported by the rapid spread among toxicologists of new tools provided by molecular biology (e.g., transcriptomics, proteomics) that allow to investigate the patterns of modulation of gene transcripts and protein synthesis.[71] In particular, DNA microarrays in ED-exposed *in vitro* systems have been viewed as a rapid and straightforward approach to provide a "signature" of transcriptional activity in tissues at one point in time.[72] Moreover, the development of study protocols to investigate gene expression in relation to dose and timing of exposure has to be envisaged.[73] Novel targets may be revealed from an increased output of toxicogenomic data on ED; for instance, the xenoestrogen bisphenol regulated a number of growth- and development-related genes in endometrium- and breast-derived cells lines (Ishikawa and MCF-7, respectively), suggesting a possible action on early developmental stages of such tissues.[74] Array technologies might also support comparative investigation on strain- and/or species-related susceptibility to a given mechanism, as in a recent paper on AhR polymorphism in rat.[75] Thus, integrated transcript and protein profiling looks as a promising, indeed fascinating strategy toward biomarker discovery, even though only some initial steps have been undertaken till now.[76] Nevertheless, however fascinating and innovative, the "omics" have to be handled with care. Important issues include: large amounts of bioinformatic data that should be interpretable, or the lack of harmonized quality standards established through interlaboratory validation studies.[77] Such issues should be carefully assessed in order to introduce the new biomolecular tools in toxicological testing.

CONCLUSIONS

Hazard identification of ED can be reliably based on a multitiered strategy with two-generation studies, supported and targeted by a battery of *in vitro/in vivo* screening assays, as the tests of choice to assess dose-response and comprehensive effect patterns.[4,45] Assays targeting specific susceptible phases of development may usefully integrate the hazard identification process; for instance, peripubertal assays[57] may support risk assessment of exposures occurring during infancy/childhood through food commodities, living environment, etc.[78] Nevertheless, some major issues for research in the hazard identification of ED are also apparent, such as:

- more comprehensive *in vitro* batteries,[50,51] making full avail of knowledge of ED mechanisms of action, thus including more receptor-mediated as well as nonreceptor-mediated mechanisms[16,20,53–56];
- experimental models and end points, including new tools provided by molecular biology, to evaluate: (*a*) effects on the functional development of complex systems other than the reproductive one, such as the nervous and immune systems[26,32,34–36]; and (*b*) long-term effects of exposures during early life, such as reproductive senescence,[29] gender-related autoimmunity,[79] or enhanced predisposition to cancer.[37–39]

Experimental models of toxicological research may also support the real-life evaluation of the ED intake, such as exposures to mixtures,[63,64] combined exposures to ED, and other factors (nonchemical stressors, nutrients, etc.[64–69]), and identification of possible biomarkers for health surveillance.[70,71,73] Up-to-date approaches, such as toxicogenomics may, indeed, be useful in dissecting complex mechanisms.[72–76] More data on dose- and time-related patterns of gene expression changes are needed,[73] and issues concerning validation have to be taken into account carefully.[77] The development of real-life approaches are among the most enticing issues in toxicological research; the predictive power of approaches developed by ED toxicology may substantially benefit from cross-talk with other research fields, such as biology of nuclear receptors and transcription factors,[52,80] nutrigenomics,[81] etc. The aim is to underpin critical mechanisms within multifactorial events (such as, endocrine-related health disorders) in order to eventually strengthen the scientific bases of strategies for risk management and prevention.

ACKNOWLEDGMENTS

The present article has been elaborated within the frame of the EU FP6 Project REPROTECT (Development of a novel approach in hazard and risk assessment or reproductive toxicity by a combination and application of *in vitro*, tissue and sensor technologies, LSHB-CT-2004-503257Q) as well as of the following research grants supported by the Italian Health Ministry: *"Risk analysis of the presence of residues in foods of animal origin"* (grant 4AF`F3); *"Multi-center study on the relationship between endocrine diseases and environmental and occupational exposures"* (grant M6B).

The contribution in the elaboration of the manuscript by Mrs. Francesca Baldi is gratefully acknowledged.

REFERENCES

1. NEUBERT, D. 1997. Vulnerability of the endocrine system to xenobiotic influence. Regul. Toxicol. Pharmacol. **26:** 9–29.
2. MANTOVANI, A. *et al.* 1999. Problems in testing and risk assessment of endocrine disrupting chemicals with regard to developmental toxicology. Chemosphere **39:** 1293–1300.
3. PRYOR, J.L. *et al.* 2000. Critical windows of exposure for children's health: the reproductive system in animals and humans. Environ. Health Perspect. **108:** 491–503.
4. MANTOVANI, A. 2002. Hazard identification and risk assessment of endocrine disrupting chemicals with regard to developmental effects. Toxicology **181/182:** 367–370.
5. ROSS, G. 2004. The public health implications of polychlorinated biphenyls (PCBs) in the environment. Ecotoxicol. Environ. Saf. **59:** 275–291.
6. BURANATREVEDH, S. & D. ROY. 2001. Occupational exposure to endocrine-disrupting pesticides and the potential for developing hormonal cancers. J. Environ. Health **64:** 17–29.
7. MANTOVANI, A. & A. MACRÌ. 2002. Endocrine effects in the hazard assessment of drugs used in animal production. J. Exp. Clin. Cancer Res. **21:** 445–456.
8. LATINI, G. *et al.* 2004. Plasticizers, infant nutrition and reproductive health. Reprod.Toxicol. **19:** 27–33.
9. GILL, U. *et al.* 2004. Polybrominated diphenyl ethers: human tissue levels and toxicology. Rev. Environ. Contam. Toxicol. **183:** 55–97.
10. TSENG, C.H. 2004. The potential biological mechanisms of arsenic-induced diabetes mellitus. Toxicol. Appl. Pharmacol. **197:** 67–83.
11. HENSON, M.C. & P.J. CHEDRESE. 2004. Endocrine disruption by cadmium, a common environmental toxicant with paradoxical effects on reproduction. Exp. Biol. Med. (Maywood) **229:** 383–392.
12. STARK, A. & Z. MADAR. 2002. Phytoestrogens: a review of recent findings. J. Pediatr. Endocrinol. Metab. **15:** 561–572.
13. BRANCA, F. & S. LORENZETTI. 2005. Health effects of phytoestrogens. Forum Nutr. **57:** 100–111.
14. TUOHY, P.G. Soy infant formula and phytoestrogens. J. Paediatr. Child Health **39:** 401–405.
15. BRUCKER-DAVIS, F. 1998. Effects of environmental synthetic chemicals on thyroid function. Thyroid **8:** 827–856.
16. ZARN, J.A. *et al.* 2003. Azole fungicides affect mammalian steroidogenesis by inhibiting sterol 14 alpha-demethylase and aromatase. Environ. Health Perspect. **111:** 255–261.
17. GRAY, L.E. *et al.* 2001. Effects of environmental antiandrogens on reproductive development in experimental animals. Hum. Reprod. Update **7:** 248–264.
18. WOZNIAK, A.L. *et al.* 2005. Xenoestrogens at picomolar to nanomolar concentrations trigger membrane estrogen receptor-alpha-mediated Ca2+ fluxes and prolactin release in GH3/B6 pituitary tumor cells. Environ. Health Perspect. **113:** 431–439.
19. MCMULLIN, T.S. *et al.* 2004. Evidence that atrazine and diaminochlorotriazine inhibit the estrogen/progesterone induced surge of luteinizing hormone in female

Sprague-Dawley rats without changing estrogen receptor action. Toxicol. Sci. **79:** 278–286.
20. KESTER, M.H. *et al.* 2000. Potent inhibition of estrogen sulfotransferase by hydroxylated PCB metabolites: a novel pathway explaining the estrogenic activity of PCBs. Endocrinology **141:** 1897–1900.
21. LEE, H.J. *et al.* 2003. Antiandrogenic effects of bisphenol A and nonylphenol on the function of androgen receptor. Toxicol. Sci. **75:** 40–46.
22. HOMBACH-KLONISCH, S. *et al.* 2005. Molecular actions of polyhalogenated arylhydrocarbons (PAHs) in female reproduction. Curr. Med. Chem. **12:** 599–616.
23. EJAZ, S. *et al.* 2004. Endocrine disrupting pesticides: a leading cause of cancer among rural people in Pakistan. Exp. Oncol. **26:** 98–105.
24. RUDEL, R.A. *et al.* 2003. Phthalates, alkylphenols, pesticides, polybrominated diphenyl ethers, and other endocrine-disrupting compounds in indoor air and dust. Environ. Sci. Technol. **37:** 4543–4553.
25. AMARAL MENDES, J.J. 2002. The endocrine disrupters: a major medical challenge. Food Chem. Toxicol. **40:** 781–788.
26. COLBORN, T. 2004. Neurodevelopment and endocrine disruption. Environ. Health Perspect. **112:** 944–949.
27. SAFE, S. 2005. Clinical correlates of environmental endocrine disruptors. Trends Endocrinol. Metab. **16:** 139–144.
28. TRAINA, M.E. *et al.* 2003. Long-lasting effects of lindane on mouse spermatogenesis induced by in utero exposure. Reprod. Toxicol. **17:** 25–35.
29. MEERTS, I.A. *et al.* 2004. Effects of in utero exposure to 4-hydroxy-2,3,3',4',5-pentachlorobiphenyl (4-OH-CB107) on developmental landmarks, steroid hormone levels, and female estrous cyclicity in rats. Toxicol. Sci. **82:** 259–267.
30. MARANGHI, F. *et al.* 2003. Histological and histomorphometric alterations in thyroid and adrenals of cd rat pups exposed in utero to methyl thiofanate. Reprod. Toxicol. **17:** 617–623.
31. GRAY, L.E. JR. *et al.* 2004. Toxicant-induced hypospadias in the male rat. Adv. Exp. Med. Biol. **545:** 217–241.
32. MANTOVANI, A. & G. CALAMANDREI. 2001. Delayed developmental effects following prenatal exposure to drugs. Curr. Pharm. Des. **7:** 859–880.
33. ADRIANI, W. *et al.* 2003. Altered profiles of spontaneous novelty seeking, impulsive behavior, and response to D-amphetamine in rats perinatally exposed to bisphenol A. Environ. Health Perspect. **111:** 395–401.
34. YOSHINO, S. *et al.* 2004. Prenatal exposure to bisphenol A up-regulates immune responses, including T helper 1 and T helper 2 responses, in mice. Immunology **112:** 489–495.
35. NOHARA, K. *et al.* 2000. The effects of perinatal exposure to low doses of 2,3,7,8-tetrachlorodibenzo-p-dioxin on immune organs in rats. Toxicology **154:** 123–133.
36. BOWERS, W.J. *et al.* 2004. Early developmental neurotoxicity of a PCB/organochlorine mixture in rodents after gestational and lactational exposure. Toxicol. Sci. **77:** 51–62.
37. MARKEY, C.M. *et al.* 2001. In utero exposure to bisphenol A alters the development and tissue organization of the mouse mammary gland. Biol. Reprod. **65:** 1215–1223.

38. FENTON, S.E. *et al*. 2002. Persistent abnormalities in the rat mammary gland following gestational and lactational exposure to 2,3,7,8-tetrachlorodibenzo-p-dioxin (TCDD). Toxicol. Sci. **67:** 63–74.
39. MAHOOD, I.K. *et al*. 2005. Abnormal Leydig Cell aggregation in the fetal testis of rats exposed to di (n-butyl) phthalate and its possible role in testicular dysgenesis. Endocrinology **146:** 613–623.
40. BARLOW, S.M. *et al*. 2002. Hazard identification by methods of animal-based toxicology. Toxicol. Food Chem. **40:** 145–149.
41. LAMB, J.C. & S.M. BROWN. 2000. Chemical testing strategies for predicting health hazards to children. Reprod. Toxicol. **14:** 83–94.
42. OGATA, R. *et al*. 2001. Two-generation reproductive toxicity study of tributyltin chloride in female rats. J. Toxicol. Environ. Health A. **63:** 127–144.
43. OMURA, M. *et al*. 2001. Two-generation reproductive toxicity study of tributyltin chloride in male rats. Toxicol. Sci. **64:** 224–132.
44. MANTOVANI, A. & F. MARANGHI. 2005. Risk assessment of chemicals potentially affecting male fertility. Contraception **72:** 308–313.
45. GELBKE, H.P. *et al*. 2004. OECD test strategies and methods for endocrine disruptors. Toxicology **205:** 17–25.
46. CLODE, S.A. 2006. Assessment of in vivo assays for endocrine disruption. Best Pract. Res. Clin. Endocrinol. Metab. **20:** 35–43.
47. MELLERT, W. *et al*. 2003. Detection of endocrine-modulating effects of the antithyroid acting drug 6-propyl-2-thiouracil in rats, based on the "Enhanced OECD Test Guideline 407". Regul. Toxicol. Pharmacol. **38:** 368–377.
48. WASON, S. *et al*. 2003. 17alpha-methyltestosterone: 28-day oral toxicity study in the rat based on the "Enhanced OECD Test Guideline 407" to detect endocrine effects. Toxicology **192:** 119–137.
49. GRAY, L.E. JR. *et al*. 2004. Use of the laboratory rat as a model in endocrine disruptor screening and testing. ILAR J. **45:** 425–437.
50. BREMER, S. *et al*. 2005. Reproductive and developmental toxicity. Altern. Lab Anim. **33:** 183–209.
51. HARENG, L. *et al*. 2005. The integrated project ReProTect: a novel approach in reproductive toxicity hazard assessment. Reprod. Toxicol. **20:** 441–452.
52. GRONEMEYER, H. *et al*. 2004. Principles for modulation of the nuclear receptor superfamily. Nat. Rev. Drug Discov. **3:** 950–964.
53. KHURANA, S. *et al*. 2000. Exposure of newborn male and female rats to environmental estrogens: delayed and sustained hyperprolactinemia and alterations in estrogen receptor expression. Endocrinology **141:** 4512–4517.
54. VILLA, R. *et al*. 2004. Target-specific action of organochlorine compounds in reproductive and nonreproductive tissues of estrogen-reporter male mice. Toxicol. Appl. Pharmacol. **201:** 137–148.
55. BODWEL, l.J.E. *et al*. 2004. Arsenic at very low concentrations alters glucocorticoid receptor (GR)-mediated gene activation but not GR-mediated gene repression: complex dose-response effects are closely correlated with levels of activated GR and require a functional GR DNA binding domain. Chem. Res. Toxicol. **17:** 1064–1076.
56. HURST, C.H. & D.J. WAXMAN. 2004. Environmental phthalate monoesters activate pregnane X receptor-mediated transcription. Toxicol. Appl. Pharmacol. **199:** 266–274.

57. KIM, H.S. *et al.* 2002. Evaluation of the 20-day pubertal female assay in Sprague-Dawley rats treated with DES, tamoxifen, testosterone, and flutamide. Toxicol. Sci. **67:** 52–62.
58. GOLUB, M.S. *et al.* 2004. Endocrine disruption in adolescence: immunologic, hematologic, and bone effects in monkeys. Toxicol. Sci. **82:** 598–607.
59. STOKER, T.E. *et al.* 2004. Assessment of DE-71, a commercial polybrominated diphenyl ether (PBDE) mixture, in the EDSP male and female pubertal protocols. Toxicol. Sci. **78:** 144–155.
60. NOVELLI, M. *et al.* 2005. 2,3,7,8-Tetrachlorodibenzo-p-dioxin-induced impairment of glucose-stimulated insulin secretion in isolated rat pancreatic islets. Toxicol. Lett. **156:** 307–314.
61. GROTEN, J.P. 2000. Mixtures and interactions. Food Chem. Toxicol. **38:** S65–S71.
62. CROFTON, K.M. *et al.* 2005. Thyroid-hormone-disrupting chemicals: evidence for dose-dependent additivity or synergism. Environ. Health Perspect. **113:** 1549–1554.
63. RAJAPAKSE, N. *et al.* 2004. Deviation from additivity with estrogenic mixtures containing 4-nonylphenol and 4-tert-octylphenol detected in the E-SCREEN assay. Deviation from additivity with estrogenic mixtures containing 4-nonylphenol and 4-tert-octylphenol detected in the E-SCREEN assay. Environ. Sci. Technol. **38:** 6343–6352.
64. JONKER, D. *et al.* 2004. Safety evaluation of chemical mixtures and combinations of chemical and non-chemical stressors. Rev. Environ. Health. **19:** 83–139.
65. SENTHIL KUMAR, J. *et al.* 2004. Effects of Vitamin C and E on PCB (Aroclor 1254) induced oxidative stress, androgen binding protein and lactate in rat Sertoli cells. Reprod. Toxicol. **19:** 201–208.
66. HENNIG, B. *et al.* 2005. Modification of environmental toxicity by nutrients: implications in atherosclerosis. Cardiovasc. Toxicol. **5:** 153–160.
67. FAROMBI, E.O. *et al.* 2004. Commonly consumed and naturally occurring dietary substances affect biomarkers of oxidative stress and DNA damage in healthy rats. Food Chem. Toxicol. **42:** 1315–1322.
68. SAARINEN, N.M. *et al.* 2002. Enterolactone inhibits the growth of 7,12-dimethylbenz(a)anthracene-induced mammary carcinomas in the rat. Mol. Cancer Ther. **1:** 869–876.
69. YOU, L. 2004. Phytoestrogen genistein and its pharmacological interactions with synthetic endocrine-active compounds. Curr. Pharm. Des. **10:** 2749–2757.
70. FERNANDEZ, M.F. *et al.* 2004. Assessment of total effective xenoestrogen burden in adipose tissue and identification of chemicals responsible for the combined estrogenic effect. Anal. Bioanal. Chem. **379:** 163–170.
71. KUSUMEGI, T. *et al.* 2004. BMP7/ActRIIB regulates estrogen-dependent apoptosis: new biomarkers for environmental estrogens. J. Biochem. Mol. Toxicol. **18:** 1–11.
72. NACIFF, J.M. & G.P. DASTON. 2004. Toxicogenomic approach to endocrine disruptors: identification of a transcript profile characteristic of chemicals with estrogenic activity. Toxicol. Pathol. **32:** 59–70.
73. KATO, N. *et al.* 2004. Gene expression profile in the livers of rats orally administered ethinylestradiol for 28 days using a microarray technique. Toxicology **200:** 179–192.
74. SINGLETON, D.W. *et al.* 2006. Gene expression profiling reveals novel regulation by bisphenol-A in estrogen receptor-alpha-positive human cells. Environ. Res. **100:** 86–92.

75. OKEY, A.B. *et al.* 2005. Toxicological implications of polymorphisms in receptors for xenobiotic chemicals: the case of the aryl hydrocarbon receptor. Toxicol. Appl. Pharmacol. **207:** 43–51.
76. MERRICK, B.A. & J.H. MADENSPACHER. 2005. Complementary gene and protein expression studies and integrative approaches in toxicogenomics. Toxicol. Appl. Pharmacol. **207:** 189–194.
77. REYNOLDS, V.L. 2005. Applications of emerging technologies in toxicology and safety assessment. Int. J. Toxicol. **24:** 135–137.
78. GARRY, V.F. 2004. Pesticides and children. Toxicol. Appl. Pharmacol. **198:** 152–163.
79. PEEVA, E. & M. ZOUALI. 2005. Spotlight on the role of hormonal factors in the emergence of autoreactive B-lymphocytes. Immunol. Lett. **101:** 123–143.
80. KALAITZIDIS, D. & T.D. GILMORE. 2005. Transcription factor cross-talk: the estrogen receptor and NF-kappaB. Trends Endocrinol. Metab. **16:** 46–52.
81. BAUER, M. *et al.* 2004. Linking nutrition to genomics. Biol. Chem. **385:** 593–596.

Diesel Exhaust and Coal Mine Dust

Lung Cancer Risk in Occupational Settings

BARBARA HOFFMANN AND KARL-HEINZ JÖCKEL

Institute for Medical Informatics, Biometry and Epidemiology, University Clinics of Essen, 45122 Essen, Germany

ABSTRACT: Conflicting evidence on the carcinogenicity of diesel exhaust (DE) and coal mine dust in occupational settings exist. Exposure measurement in most studies is inferred on the basis of job classifications and may lead to misclassification. Confounding behavioral factors (i.e., smoking) and occupational risk factors (exposure to asbestos, arsenic, radon) need to be considered. We evaluated the epidemiological evidence and current findings of the carcinogenicity of DE and coal mine dust in occupational settings. Pertaining literature was identified through Medline search and recent review articles. Strengths and limitations of recent approaches are discussed. Many epidemiological studies have addressed the question of carcinogenicity in workers exposed to DE, and most showed a low-to-medium increase in the risk of bronchial carcinoma. The pooled relative risk (RR) estimates lie between 1.33 and 1.47, and a consistent rise in risk across various job categories and study designs point to a causal relationship. Data on the carcinogenicity of coal mine dust are less consistent and the potential for confounding by unmeasured risk factors (arsenic, radon, DE) are higher. While silica as one of its components has been evaluated as carcinogenic, there is inadequate evidence for the carcinogenicity of pure coal dust according to the International Agency for Research on Cancer (IARC). There is sufficient evidence for a causal relationship between DE and lung cancer in occupational settings. The evidence for coal mine dust is less convincing, but individual studies show an increase in risk of lung cancer in exposed workers.

KEYWORDS: diesel exhaust; coal mine dust; lung cancer; occupational epidemiology

HEALTH EFFECTS OF DIESEL EXHAUST

As early as 1955, diesel exhaust (DE) extracts had been shown to cause skin papillomas in mice. It took another 22 years to recognize the mutagenicity of

Address for correspondence: Barbara Hoffman, Institute for Medical Informatics, Biometry and Epidemiology, University Clinics of Essen Hufelandstr. 55, 45122 Essen, Germany.
Voice: 0049-201-7234514; fax: 0049-201-7235933.
e-mail: Barbara.Hoffmann@medizin.uni-essen.de

DE extracts in animal bioassays. Evidence was found that components of diesel extract act as weak tumor promoters.[1] The increasing replacement of gasoline engines with the more efficient diesel engines in industry and traffic led to increased exposure of the general population and to high-exposure situations of workers in occupational settings, such as railroad workers and truck drivers. The early findings on carcinogenicity and mutagenicity and the increasing exposure of the general population and special occupational groups entailed increasing public health alertness.

The rising concern about adverse health effects of DE initiated the planning and conduction of epidemiological studies and long-term animal exposure experiments during the 1980s. Long-term inhalation bioassays with DE were carried out in the United States, Germany, Switzerland, and Japan, and tested responses of rats, mice, hamsters, cats, and monkeys.[2–4] Adequate and consistent evidence for the induction of lung cancer and hyperplasia and metaplasia of lung epithelium was found in rats. There was only equivocal evidence for carcinogenicity in mice and no evidence in hamsters, cats, and monkeys, but these studies lack power to rule out such an effect because of study design and small numbers of animals.[5]

According to our current understanding of cancer mechanisms, carcinogenesis in animals and humans is a multistage, multifactorial process that involves alterations in genes controlling cell proliferation, cell growth, and programmed cell death. Genotoxic and nongenotoxic mechanisms are both biologically plausible alternatives to explain the development of lung cancer in animals exposed to DE.[5]

Epidemiology

Much epidemiological evidence exists on the association between DE and lung cancer risk. A detailed discussion of individual studies has been performed by several authoritative agencies,[1,5–7] and most of the epidemiological studies were conducted on occupational cohorts.[1] Additional evidence is available from population-based studies examining the effect of ambient particulate matter.[8] The preferred use of occupational cohorts stems from the fact that workers under specific working conditions are highly exposed to DE, and job stability allows a more accurate description of lifetime DE exposure. Several methodological issues need to be discussed when evaluating the existing evidence.

One major challenge in epidemiological studies is exposure assessment. Due to their retrospective design, almost all epidemiological studies lack concurrent personal exposure measurements. Instead, exposure assessment is usually based on the affiliation to a special industry or occupational group (railroad workers, truck and bus drivers, garage workers, heavy equipment operators, etc.), reconstructed from self-reported work histories or company records.

Probability and intensity of exposure to DE can be determined by expert ratings or with the help of a job-exposure matrix. Even though these indirect approaches are able to identify individuals with high exposure, misclassification of exposure cannot be ruled out as a cause of bias. In most cases, misclassification will occur in both directions (nondifferential misclassification) and usually leads to an underestimation of the true effect.[9]

One study that incorporates personal exposure measurements and area measurements of total carbon as a surrogate for DE exposure is a cohort mortality study conducted in 5536 male potash miners in Germany.[10] Exposure was assessed from concentration measurements of total carbon in personal dust samples and multiplied by years of exposure to give a quantitative exposure measure. After a cumulative exposure of 4.9 years $\times$ mg/m^3, corresponding to 20 years of exposure in the highest exposure category, a relative risk (RR) for lung cancer of 1.7 (0.5–5.8) was found. While the main strength of this study is the availability of personal exposure measurements, the relatively small cohort, the lack of a latency analysis, and the possibility of confounding by smoking potentially limit the explanatory power of the study. Nevertheless, this is the first study with direct exposure measurements, and its results are in agreement with prior studies that use only an indirect exposure assessment.

The choice of an adequate reference group also poses some degree of difficulty in epidemiological studies. The use of an external reference group (i.e., general population and national mortality rates) in the calculation of the age-standardized mortality rate (SMR) of an employed population leads to an underestimation of the true effect, because the economically active group has, on average, a better health status than the general population (healthy worker effect). Cohort studies using an internal reference group and case–control studies might result in a less biased estimate. Not surprisingly, the early cohort studies that use an external reference group show only equivocal evidence,[11–17] while the majority of the cohort studies using an internal comparison group[10,18–24] and almost all case–control studies[20,25–41] conducted within the last 25 years show an increased lung cancer risk in DE-exposed workers with an elevation of risk between 20% and 50% (FIGS. 1 and 2).

A third relevant issue in epidemiological studies consists of an adjustment for confounding exposures, such as smoking, exposure to asbestos, and other particles or carcinogens, which is often lacking in cohort studies. Many recent case–control studies concerned with detailed assessment of potential confounding risk factors with the help of personal interviews, found no qualitative difference between the adjusted and the unadjusted results.[42] This is exemplified by two recent, large, case–control studies on occupational exposures in Germany.[43,44] The two case–control studies conducted between 1988 and 1996 were pooled for a joint analysis.[38,39,45] A total of 3498 incident male lung cancer cases and 3541 male population controls were included in the pooled analysis. Face-to-face interviews allowed the construction of detailed life-long work histories and the assessment of possible confounding factors. Occupational

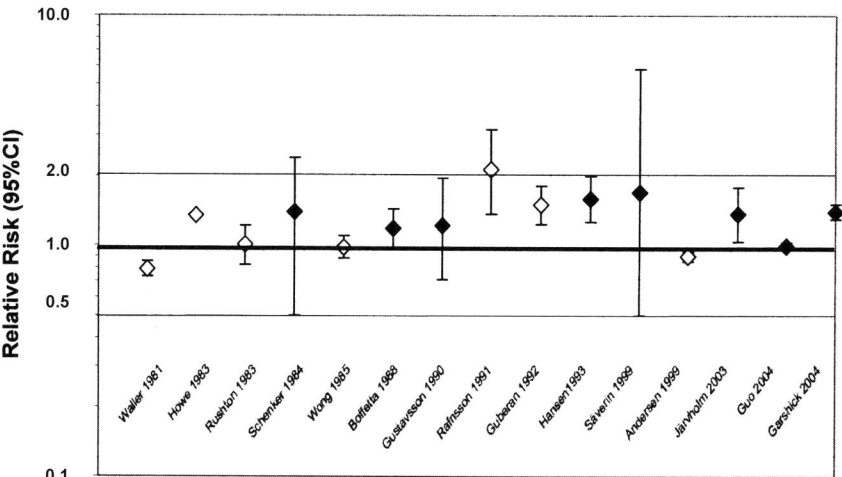

FIGURE 1. Cohort studies on occupational exposure to DE and lung cancer risk since 1981. Open marks indicate cohort studies with external reference group, filled marks represent cohort studies with internal reference group.

exposures were evaluated by expert rating. Cumulative personal exposure was calculated as duration of exposure to DE. The evaluation of lung cancer risk for all jobs with DE exposure combined showed a crude odds ratio (OR) of 1.91. Adjustment for smoking and asbestos exposure led to a decrease in the effect to 1.43 (95% CI: 1.23–1.67). Lung cancer risk increased up to the category of 10–20 years of exposure and decreased thereafter. Heavy equipment

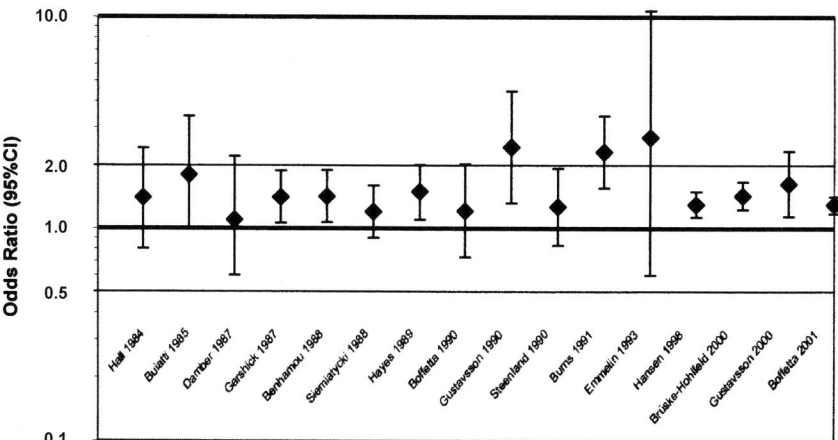

FIGURE 2. Case–control studies in occupational exposure to DE and lung cancer since 1984. ORs are adjusted for smoking.

operators showed the highest increase in risk (OR 2.31, 95% confidence interval [CI]: 1.44–3.70), followed by workers in traffic-related jobs (OR 1.53), drivers of farming tractors (OR 1.29), and professional drivers (OR 1.25). A recent study in Sweden, conducted with a similar design including detailed confounder assessment, is in close agreement with the above mentioned results.[40] For workers within the highest quartile of cumulative exposure versus no exposure, the risk of lung cancer was elevated to 1.63 (95% CI: 1.14–2.33), and a positive exposure–response curve was noted.

Two occupational groups that have received special attention are railroad workers and truck drivers. The results of occupational cohort and case–control studies of railroad workers and truck drivers are generally consistent in showing a weak association between exposure to DE and lung cancer. They suggest that prolonged exposure to DE over many years is associated with a 1.2–1.5 times increase in the RR of lung cancer incidence or mortality in male workers. The most recent extended analysis of the U.S. American railroad workers,[24] including 38 years of follow-up and 4351 lung cancer cases, obtained a RR of 1.40 (95% CI: 1.30–1.51) for workers in jobs associated with operating trains powered by diesel engines.

Extensive criticism has been raised regarding several aspects of study design in the two major studies on railroad workers[28,29] and truck drivers [34,46,47]; these criticisms include validity of exposure measurements, lack of consideration of a latency period, and adequate adjustment for age.[48,49] Several other independent studies however have found similar results,[16,22,31,33,35,37,38,50,51] strengthening the findings of Garshick et al.[24] and Steenland et al.[47]

Reviews and Meta-Analyses

Several reviews and two formal meta-analyses have been carried out within the last 10 years.[1,6,42,52–56] The two formal meta-analyses find that the majority of the included studies show risk estimates greater than 1. The pooled RR in Bhatia et al. is 1.33 (95% CI: 1.24–1.44)[42] and in Lipsett and Campleman 1 47 (95% CI: 1.29–1.67).[53] Increased risk estimates are also seen when stratifying by study design (cohort versus case–control, cohort with external comparison group versus cohort with internal comparison group), by adjustment for smoking, and by occupational group. The authors of the meta-analyses and most of the reviewers come to the conclusion that the evidence supports a causal association between diesel exposure and lung cancer and that the observed effect is unlikely to be due to residual confounding by smoking.

Other Important Evidence

In addition to the data from animal experiments and the epidemiological evidence, other key data are judged to be supportive of carcinogenicity of

DE. DE is a complex mixture of hundreds of constituents in the gaseous and particulate phases. Several organic compounds are known to be mutagenic or carcinogenic. Moreover DE, diesel particulate matter, and diesel particulate matter extracts have been found to cause chromosomal aberrations. Elevated levels of DNA adducts in lymphocytes have been shown in exposed workers.

Summary of Causal Evidence on Carcinogenicity of DE

There is biologic plausibility for the carcinogenicity of DE, supported by the induction of cancer in laboratory animals, the existence of mutagenic substances, and human carcinogens in the complex mixture of DE, as well as the formation of increased levels of DNA adducts in exposed workers. Findings are consistent in so far as the majority of epidemiological studies with different designs and conducted in various occupational groups show an increased risk in lung cancer. Even though almost all studies lack personal exposure measurements, surrogate measures for cumulative exposure, such as duration and intensity of exposure, reveal positive exposure–response relationships in several studies.

The smoking-adjusted pooled overall risk in meta-analysis lies between 1.3 and 1.5, conveying a small-to-moderate increase in risk. Several studies though show higher risk estimates for subgroups of highly exposed individuals. The possibility of residual confounding is low, considering that risk increases in studies with a more carefully conducted smoking adjustment are not substantially lower than in studies without smoking adjustment.

In 1989 the body of evidence led the International Agency for Research on Cancer (IARC) to classify exposure to DE as "probably carcinogenic to humans" (Group 2a), on the basis of sufficient evidence of carcinogenicity in experimental animals and limited evidence of carcinogenicity in humans.[6] Since then, many additional epidemiological studies and two meta-analyses have been conducted. In 2002, the U.S. Environmental Protection Agency concluded that DE is a probable human carcinogen.[1] Further epidemiological evidence that has accumulated since then strengthens the above mentioned conclusions. The underlying evidence is consistent with a causal association.

Open Questions on DE

Even though the evidence for a causal relationship between DE exposure and lung cancer is strong, several questions remain. The role of the various DE constituents in mediating the carcinogenic effect is still unclear. There are only experimental data from rat inhalation studies on the mode of tumor induction and the possible existence of a threshold. Only limited information exists on the size and relevance of carcinogenic effects of DE at environmental levels. Quantitative risk assessment has been hampered by a lack of actual exposure data in the retrospective studies. The Health Effects Institute has recently published recommendations for further research strategies addressing this problem.[57] Finally, the relevance of historical data for the present exposure situations needs to be studied thoroughly.

Coal Mine Dust

Coal dust is a heterogeneous byproduct of coal mining and the subsequent use of coal. The composition of coal mine dust varies according to coal type as well as extent and nature of the embedding rock, including carbonaceous material from the coal itself, minerals (e.g., clays, sulphide ores, phosphates, and quartz), metals, and organic compounds. The airborne respirable dust contains a variable amount of coal dust (40–95%) and quartz (1–10%). Due to the widespread use of diesel-powered equipment in underground mines, an additional exposure to DE is probable. Because of the different extent and pace of dieselization in different countries and mines, a quantitative comparison of DE in coal mine dust is difficult.

IARC evaluated inhaled crystalline silica in the form of quartz or cristobalite from occupational sources as carcinogenic to humans (Group 1) in 1997. The evidence on the carcinogenicity of coal mine dust and of coal dust itself is still under debate.[58]

Epidemiology

Epidemiological studies, evaluating health effects where study subjects are typically exposed to a complex mixture of several possible and/or ascertained carcinogens, must deal with a combination of effects. A coal miner is exposed to coal mine dust that consists of various components with or without known carcinogenic effects, that is, quartz, DE, and pure coal dust. Other work-related or behaviorally determined exposures exert potential health effects on the miner and include smoking, asbestos, and radon exposures. To further complicate the situation, several selection machanisms take place on the group of miners, who carry out highly exposed under-around work, leading to changes in individual exposure status (healthy worker and healthy worker survivor effect). Therefore it is often not possible to deduce the specific action of one single pollutant from an epidemiological study, but rather to investigate the adverse health effects of a complex occupational exposure situations, characterized by a mixture of pollutants and effects.

Cohort Mortality Studies

A large body of literature exists concerning cohort mortality studies in coal miners. In most of them, all-cause, all-cancer, and lung-cancer specific mortality rates were lower than national rates, exhibiting the well-known healthy worker effect. Further bias results from the lack of individual information on other risk factors, possibly confounding the association under question.

In addition, the aforementioned studies provide conflicting results on the relationship between the degree of coal workers' pneumoconiosis and lung cancer-specific SMR. While two studies found a negative association,[59,60] others found no certain exposure–response relationship[61,62] or higher SMR in silicotics than in nonsilicotics.[63] Selection mechanisms, like the healthy worker survivor effect, resulting in the exclusion of workers with early signs of exposure-related disease from the highly exposed underground work, may be responsible for these differences. To our knowledge, no cohort studies with a well-conducted adjustment for confounding factors and intense scrutiny of selection mechanisms acting on the group of underground workers exist.

Case–Control Studies

Six case–control studies investigated the association between working as a coal miner or other coal dust exposure and lung cancer (FIG. 3).[43,51,64–67] All show an elevated risk; three indicate a positive exposure–response relationship.[64,66,68] ORs are adjusted for smoking and partly for other possible confounders such as asbestos exposure.

The intense selection mechanisms among coal miners at all stages before and during employment are possibly responsible for the differences between cohort mortality and case–control studies.[63] Case–control studies, incorporating workers who were excluded from underground work from the start or early

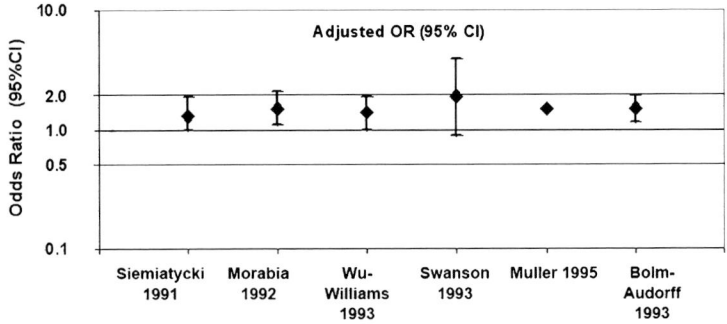

FIGURE 3. Case–control studies on the association between work as a coal miner and lung cancer. Results are adjusted for smoking.

after starting underground work, seems to be a better design to detect adverse health effects.[43]

Summary of Evidence on Carcinogenicity of Coal Mine Dust

Coal mine dust is a complex mixture of constituents, some of which exert known, probable, or suspected carcinogenicity. Inhalation of quartz has been recognized as carcinogenic by several authoritative agencies. While almost all cohort mortality studies show no definite increase in lung cancer risk, several aspects of study design pose great problems in their evaluation. Selection mechanisms lead to a pronounced healthy worker and healthy worker survivor effect, which are visible in the low SMR estimates. The case–control study design is not as strongly affected by the healthy worker effect. All case–control studies examining the effect of exposure to coal mine dust on lung cancer risk have observed a consistent small-to-moderate increase in risk, which remains elevated after adjustment for other exposures and lifestyle factors. Due to missing personal exposure measurements, it is not possible to give a definite statement on the carcinogenicity of single components.

CONCLUSIONS

The causal evidence on occupational exposure to DE and lung cancer risk has been strengthened by recent studies. Regarding occupational exposure to coal mine dust, an increase in lung cancer risk can be consistently observed in case–control studies, which probably show less biased estimates than retrospectively performed cohort studies. However, it is not possible to determine the responsible constituent(s) of the complex mixture encountered in underground coal mines.

REFERENCES

1. U.S. ENVIRONMENTAL PROTECTION AGENCY. 2002. Health Assessment Document for Diesel Engine Exhaust. Office of Research and Development. Washington, DC.
2. MAUDERLY, J.L., R.K. JONES, W.C. GRIFFITH, *et al.* 1987. Diesel exhaust is a pulmonary carcinogen in rats exposed chronically by inhalation. Fundam. Appl. Toxicol. **9:** 208–221.
3. HEINRICH, U., H. MUHLE, S. TAKENAKA, *et al.* 1986. Chronic effects on the respiratory tract of hamsters, mice and rats after long-term inhalation of high concentrations of filtered and unfiltered diesel engine emissions. J. Appl. Toxicol. **6:** 383–395.

4. IWAI, K., T. UDAGAWA, M. YAMAGISHI, *et al.* 1986. Long-term inhalation studies of diesel exhaust on F344 SPF rats. Incidence of lung cancer and lymphoma. Dev. Toxicol. Environ. Sci. **13:** 349–360.
5. HEI. 1995. Diesel Exhaust: A Critical Analysis of Emissions, Exposure, and Health Effects. Health Effects Institute. Cambridge, MA.
6. INTERNATIONAL AGENCY FOR RESEARCH ON CANCER. 1989. Diesel and gasoline engine exhausts and some nitroarenes. [46]. IARC Monographs On The Evaluation Of Carcinogenic Risks To Humans. Lyon, France.
7. HEI. 1999. Diesel emissions and lung cancer: epidemiology and quantitative risk assessment. A Special Report of the Institute's Diesel Epidemiology Expert Panel. Health Effects Institute. Cambridge, MA.
8. POPE, C.A. III, R.T. BURNETT, M.J. THUN, *et al.* 2002. Lung cancer, cardiopulmonary mortality, and long-term exposure to fine particulate air pollution. JAMA **287:** 1132–1141.
9. SILVERMAN, D.T. 1998. Is diesel exhaust a human lung carcinogen? Epidemiology **9:** 4–6.
10. SÄVERIN, R., A. BRÄNLICH, D. DAHMANN, *et al.* 1999. Diesel exhaust and lung cancer mortality in potash mining. Am. J. Ind. Med. **36:** 415–422.
11. WALLER, R.E. 1981. Trends in lung cancer in London in relation to exposure to diesel fumes. Environ. Int. **5:** 479–483.
12. HOWE, G.R., D. FRASER, J. LINDSAY, *et al.* 1983. Cancer mortality (1965–1977) in relation to diesel fume and coal exposure in a cohort of retired railway workers. J. Natl. Cancer Inst. **70:** 1015–1019.
13. RUSHTON, L., M.R. ALDERSON & C.R. NAGARAJAH. 1983. Epidemiological survey of maintenance workers in london-transport executive bus garages and chiswick-works. Br. J. Ind. Med. **40:** 340–345.
14. WONG, O., R.W. MORGAN, L. KHEIFETS, *et al.* 1985. Mortality among members of a heavy construction equipment operators union with potential exposure to diesel exhaust emissions. Br. J. Ind. Med. **42:** 435–448.
15. RAFNSSON, V. & H. GUNNARSDOTTIR. 1991. Mortality among professional drivers. Scand. J. Work Environ. Health **17:** 312–317.
16. GUBERAN, E., M. USEL, L. RAYMOND, *et al.* 1992. Increased risk for lung cancer and for cancer of the gastrointestinal tract among Geneva professional drivers. Br. J. Ind. Med. **49:** 337–344.
17. ANDERSEN, A., L. BARLOW, A. ENGELAND, *et al.* 1999. Work-related cancer in the Nordic countries. Scand. J. Work Environ. Health **25:** 1–114.
18. SCHENKER, M.B., T. SMITH, A. MUNOZ, *et al.* 1984. Diesel exposure and mortality among railway workers: results of a pilot study. Br. J. Ind. Med. **41:** 320–327.
19. BOFFETTA, P., S.D. STELLMAN & L. GARFINKEL. 1988. Diesel exhaust exposure and mortality among males in the american-cancer-society prospective-study. Am. J. Ind. Med. **14:** 403–415.
20. GUSTAVSSON, P., N. PLATO, E.B. LIDSTROM, *et al.* 1990. Lung-cancer and exposure to diesel exhaust among bus garage workers. Scand. J. Work Environ. Health **16:** 348–354.
21. HANSEN, E.S. 1993. A follow-up study on the mortality of truck drivers. Am. J. Ind. Med. **23:** 811–821.
22. JÄRVHOLM, B. & D. SILVERMAN. 2003. Lung cancer in heavy equipment operators and truck drivers with diesel exhaust exposure in the construction industry. Occup. Environ. Med. **60:** 516–520.

23. GUO, J., T. KAUPPINEN, P. KYYRONEN, *et al*. 2004. Occupational exposure to diesel and gasoline engine exhausts and risk of lung cancer among Finnish workers. Am. J. Ind. Med. **45:** 483–490.
24. GARSHICK, E., F. LADEN, J.E. HART, *et al*. 2004. Lung cancer in railroad workers exposed to diesel exhaust. Environ. Health Perspect. **112:** 1539–1543.
25. HALL, N.E.L. & E.L. WYNDER. 1984. Diesel exhaust exposure and lung cancer—a case–control study. Environ. Res. **34:** 77–86.
26. BUIATTI, E., D. KRIEBEL, M. GEDDES, *et al*. 1985. A case control study of lung cancer in Florence, Italy. I. Occupational risk factors. J. Epidemiol. Comm. Health **39:** 244–250.
27. DAMBER, L.A. & L.G. LARSSON. 1987. Occupation and male lung cancer—a case–control study in Northern Sweden. Br. J. Ind. Med. **44:** 446–453.
28. GARSHICK, E., M.B. SCHENKER, A. MUNOZ, *et al*. 1987. A case–control study of lung cancer and diesel exhaust exposure in railroad workers. Am. Rev. Respir. Dis. **135:** 1242–1248.
29. GARSHICK, E., M.B. SCHENKER, A. MUNOZ, *et al*. 1988. A retrospective cohort study of lung-cancer and diesel exhaust exposure in railroad workers. Am. Rev. Respir. Dis. **137:** 820–825.
30. BENHAMOU, S., E. BENHAMOU & R. FLAMANT. 1988. Occupational risk-factors of lung-cancer in a French case–control study. Br. J. Ind. Med. **45:** 231–233.
31. SIEMIATYCKI, J., M. GERIN, P. STEWART, *et al*. 1988. Associations between several sites of cancer and 10 types of exhaust and combustion products—results from a case-referent study in Montreal. Scand. J. Work Environ. Health **14:** 79–90.
32. HAYES, R.B., T. THOMAS, D.T. SILVERMAN, *et al*. 1989. Lung-cancer in motor exhaust-related occupations. Am. J. Ind. Med. **16:** 685–695.
33. BOFFETTA, P., R.E. HARRIS & E.L. WYNDER. 1990. Case–control study on occupational exposure to diesel exhaust and lung-cancer risk. Am. J. Ind. Med. **17:** 577–591.
34. STEENLAND, N.K., D.T. SILVERMAN & R.W. HORNUNG. 1990. Case–control study of lung cancer and truck driving in the teamsters-union. Am. J. Public Health **80:** 670–674.
35. BURNS, P.B. & G.M. SWANSON. 1991. The Occupational Cancer Incidence Surveillance Study (OCISS): risk of lung cancer by usual occupation and industry in the Detroit metropolitan area. Am. J. Ind. Med. **19:** 655–671.
36. EMMELIN, A., L. NYSTROM & S. WALL. 1993. Diesel exhaust exposure and smoking—a case-referent study of lung cancer among Swedish dock workers. Epidemiology **4:** 237–244.
37. HANSEN, J., O. RAASCHOU-NIELSEN & J.H. OLSEN. 1998. Increased risk of lung cancer among different types of professional drivers in Denmark. Occup. Environ. Med. **55:** 115–118.
38. BRUSKE-HOHLFELD, I., M. MOHNER, W. AHRENS, *et al*. 1999. Lung cancer risk in male workers occupationally exposed to diesel motor emissions in Germany. Am. J. Ind. Med. **36:** 405–414.
39. BRUSKE-HOHLFELD, I., M. MOHNER, H. POHLABELN, *et al*. 2000. Occupational lung cancer risk for men in Germany: results from a pooled case–control study. Am. J. Epidemiol. **151:** 384–395.
40. GUSTAVSSON, P., R. JAKOBSSON, F. NYBERG, *et al*. 2000. Occupational exposure and lung cancer risk: a population-based case-referent study in Sweden. Am. J. Epidemiol. **152:** 32–40.

41. BOFFETTA, P., M. DOSEMECI, G. GRIDLEY, *et al.* 2001. Occupational exposure to diesel engine emissions and risk of cancer in Swedish men and women. Cancer Causes Control **12:** 365–374.
42. BHATIA, R., P. LOPIPERO & A.H. SMITH. 1998. Diesel exhaust exposure and lung cancer. Epidemiology **9:** 84–91.
43. JÖCKEL, K.H., W. AHRENS, I. JAHN, *et al.* 1998. Occupational risk factors for lung cancer: a case–control study in West Germany. Int. J. Epidemiol. **27:** 549–560.
44. KREIENBROCK, L., M. KREUZER, M. GERKEN, *et al.* 2001. Case–control study on lung cancer and residential radon in western Germany. Am. J. Epidemiol. **153:** 42–52.
45. JÖCKEL, K.-H., I. BRÜSKE-HOHLFELD & E. WICHMANN. 1998. Lungenkrebsrisiko Durch Berufliche Exposition. Ecomed. Landsberg. Germany.
46. STEENLAND, K., D. SILVERMAN & D. ZAEBST. 1992. Exposure to diesel exhaust in the trucking industry and possible relationships with lung cancer. Am. J. Ind. Med. **21:** 887–890.
47. STEENLAND, K., J. DEDDENS & L. STAYNER. 1998. Diesel exhaust and lung cancer in the trucking industry: exposure-response analyses and risk assessment. Am. J. Ind. Med. **34:** 220–228.
48. BUNN, W.B. III, P.A. VALBERG, T.J. SLAVIN, *et al.* 2002. What is new in diesel. Int. Arch. Occup. Environ. Health **75**(Suppl.): S122–S132.
49. BUNN, W.B., T.W. HESTERBERG, P.A. VALBERG, *et al.* 2004. A reevaluation of the literature regarding the health assessment of diesel engine exhaust. Inhal. Toxicol. **16:** 889–900.
50. JAKOBSSON, R., P. GUSTAVSSON & I. LUNDBERG. 1997. Increased risk of lung cancer among male professional drivers in urban but not rural areas of Sweden. Occup. Environ. Med. **54:** 189–193.
51. SWANSON, G.M., C.S. LIN & P.B. BURNS. 1993. Diversity in the association between occupation and lung cancer among black-and-white men. Cancer Epidemiol. Biomarkers Prev. **2:** 313–320.
52. CALIFORNIA ENVIRONMENTAL PROTECTION AGENCY. 1998. Part B: Health Risk Assessment for Diesel Exhaust. Proposed Identification of Diesel Exhaust as a Toxic Air Contaminant. CalEpa, Office of Environmental Health Hazard Assessment, Sacramento, CA.
53. LIPSETT, M. & S. CAMPLEMAN. 1999. Occupational exposure to diesel exhaust and lung cancer: a meta-analysis. Am. J. Public Health **89:** 1009–1017.
54. COHEN, A.J. & M.W.P. HIGGINS. 1995. Health Effects of Diesel Exhaust: Epidemiology. Health Effects Institute. Cambridge, MA.
55. STÖBER, W. & U.R. ABEL. 1996. Lung cancer due to diesel soot particles in ambient air? A critical appraisal of epidemiological studies addressing this question. Intern. Arch. Occup. Environ. Health **68:** S3–S61.
56. MUSCAT, J.E. 1996. Carcinogenic effects of diesel emissions and lung cancer: the epidemiologic evidence is not causal. J. Clin. Epidemiol. **49:** 891–892.
57. THE HEALTH EFFECTS INSTITUTE. 2002. Executive summary: research directions to improve estimates of human exposure and risk from diesel exhaust. A Special Report of the Diesel Epidemiology Working Group. Cambridge, MA.
58. INTERNATIONAL AGENCY FOR RESEARCH ON CANCER. 1997. Silica, some silicates, coal dust and para-aramid fibrils. World Health Organization. [68]. World Health Organisation. IARC Monographs on the Evaluation of Carcinogenic Risks to Humans. Lyon, France.

59. ROOKE, G.B., F.G. WARD, A.N. DEMPSEY, et al. 1979. Carcinoma of the lung in Lancashire coalminers. Thorax **34:** 229–233.
60. MILLER, B.G. & M. JACOBSEN. 1985. Dust exposure, pneumoconiosis, and mortality of coalminers. Br. J. Ind. Med. **42:** 723–733.
61. GOLDMAN, K.P. 1965. Mortality of coal-miners from carcinoma of the lung. Br. J. Ind. Med. **22:** 72–77.
62. COCHRANE, A.L., T.J. HALEY, F. MOORE, et al. 1979. The mortality of men in the Rhondda Fach, 1950–1970. Br. J. Ind. Med. **36:** 15–22.
63. MORFELD, P., K. LAMPERT, H. ZIEGLER, et al. 1997. Overall mortality and cancer mortality of coalminers: attempts to adjust for healthy worker selection effects. Ann. Occup. Hyg. **1**(Suppl. 41): 346–351.
64. MORABIA, A., S. MARKOWITZ, K. GARIBALDI, et al. 1992. Lung cancer and occupation: results of a multicentre case–control study. Br. J. Ind. Med. **49:** 721–727.
65. SIEMIATYCKI, J. 1991. Risk factors for cancer in the workplace. CRC Press. Boca Raton, FL.
66. WU-WILLIAMS, A.H., Z.Y. XU, W.J. BLOT, et al. 1993. Occupation and lung cancer risk among women in northern China. Am. J. Ind. Med. **24:** 67–79.
67. MULLER, M., P. BARTSCH, A. ALBERT, et al. 1995. Epidemiological study about occupational risk factors of lung cancer in the province of Liege. 15th International Conference of environmental and occupational lung diseases. Orlando, FL.
68. BOLM-AUDORFF, U., K.-H. JÖCKEL, B. KILGUSS, et al. 1993. Bösartige Tumoren der ableitenden Harnwege und Risiken am Arbeitsplatz. [Fb 697]. Bremerhaven, Wirtschaftsverlag NW, Verlag für neue Wissenschaft. Schriftenreihe der Bundesanstalt für Arbeitsschutz–Forschung. Bundesanstalt für Arbeitsschutz.

Lung Tumor Risk Estimates from Rat Studies with Not Specifically Toxic Granular Dusts

MARKUS ROLLER[a] AND FRIEDRICH POTT[b]

[a]*Advisory Office for Risk Assessment, 44229 Dortmund, Germany*

[b]*Heinrich-Heine-University Düsseldorf, Düsseldorf, Germany*

ABSTRACT: Since 1985 several carcinogenicity studies have been published about lung tumors in rats after exposure to *respirable granular biodurable particles without known significant specific toxicity* (abbreviation of this complex definition by the three letters GBP to substitute the former term *inert dusts*). During this time, the relevance of the carcinogenicity of GBP in rats was questioned, for example, because no lung tumors from GBP were found in hamsters and carcinogenicity in mice was questionable. However, the carcinogenesis and the tumor risk from quartz appear similar in men and rats, and the effects of GBP in rats appear not to differ, on principle, from that of quartz, but at a much higher dose level. We calculated the excess risk (ER) of GBP in rats from the final results of an instillation study with 16 GBP types in connection with results of inhalation experiments with carbon black, titanium dioxide, and diesel particles. Retained particle volume together with some indicator of particle size was identified as the best suitable dose metric and the dose-response relationships were analyzed on the basis of the multistage model. By relating the results to the available dose–response slopes after inhalation, ER for workplace-like exposure were calculated for three particle size classes and an exposure to 0.3 mg/m^3 (density 2–2.5 g/mL); mean diameter 1.8–4 μm (*GBP-fine-large*): ER 0.1%; 0.09–0.2 μm (*GBP-fine-small*): ER 0.2%; 0.01–0.03 μm (*GBP-ultra-fine*): ER 0.5%.

KEYWORDS: lung cancer; particles; risk assessment

INTRODUCTION

Since 1985, several carcinogenicity studies with chronic inhalation exposure to diesel engine exhaust, carbon black, and titanium dioxide have resulted in benign and malignant lung tumors in rats but not in hamsters[1–3]; in mice,

Address for correspondence: Dr. Markus Roller, Advisory Office for Risk Assessment, Doldenweg 14, 44229 Dortmund, Germany. Voice and fax: +49-231-79-79-489.
e-mail: Markus.Roller@t-online.de

there was no or just a questionable tumor response. Repeated intratracheal instillation of diesel soot and other granular (nonfibrous) biodurable dusts of different chemical compositions without known significant specific toxicity also induced lung tumors in rats.[4,5]

Regarding the analogy between diesel exhaust and other particle types which are not specifically toxic, for example, carbon black and titanium dioxide, the interpretation of the data is not uniform. For risk assessment of lung cancer from biodurable particles it is important to answer the question, whether the effect from diesel particles is largely caused by a pure particle effect or not. The opinion maintains that there is a significant difference between the carcinogenesis of diesel and other particles because the organic substances with numerous well-known carcinogenic polycyclic aromatic hydrocarbons (PAH) are adsorbed at the elemental carbon core of diesel particles.[6] The results of Dasenbrock et al.[5] seem to support this conclusion at the first look, because the group of rats which received 15 mg extracted diesel particles per instillations produced lung tumors in only 4% compared to 17% by 15 mg original diesel soot, which contained 43% organic material. However, the results of other groups, especially extracted diesel particles with adsorbed benzo[a]pyrene (BaP) show inconsistencies which cannot be explained reasonably. Furthermore, two inhalation experiments with rats can be well compared: pyrolyzed tar pitch aerosol which contained about 90 μg BaP per m^3 as a reference substance for PAH and a minimal concentration of elemental carbon resulted in a tumor response of 18%[7]; inhalation exposure to 4.2 mg diesel exhaust per m^3 with 12 ng BaP induced lung tumors in 16%.[8] The relationship between the BaP concentrations per m^3 in these two experiments amounts to about 7500 to 1. Also another inhalation experiment with pyrolyzed tar pitch and carbon black showed that a very high PAH concentration (20 μg BaP/m^3, total aerosol 1 mg/m^3) resulted in 4% lung tumors, 6 mg carbon black/m^3 (BaP < 0.1 ng/m^3) in 18%; the simultaneous exposure to carbon black and tar pitch can be interpreted as a more than additive effect.[9] Assuming that the PAH-composition with BaP as a reference substance was at a factor of 10 more potent in diesel exhaust than in tar pitch aerosol and the adsorption of PAH at the surface of elemental carbon core enhances the carcinogenic potency of PAH again by a factor of 10, then little more than 1% of the carcinogenicity of diesel exhaust in rat inhalation studies can be explained by PAH. The most reasonable explanation for the carcinogenicity of diesel soot in rats and humans is the activity of its content of respirable granular biodurable particles (GDP) without specific toxicity.[3,10] Remarkably, no lung tumors from diesel particles and carbon black were observed in hamsters. However, some other substances that are human carcinogens did not show a lung tumor response in hamsters either. These are asbestos fibres,[11] PAH-rich pyrolysis exhaust,[7] cadmium compounds,[12] quartz,[13] and nickel compounds.[14]

In 1995, the state of knowledge of dust carcinogenicity had been described and interpreted by several authors in the proceedings of the conference *Particle*

Overload in the Rat Lung and Lung Cancer.[15] In general, it was concluded that the rat lung tumors are rat-specific and occur only under so-called overload conditions of the lung. However, epidemiologic studies of the last 10 years confirmed the carcinogenicity of quartz with a similar risk as in rats.[16] Moreover, the mode of action does not differ essentially between both species and also the effects of not specifically toxic dusts in rats appear not to differ, on principle, from that of quartz, but at a much higher dose level. Furthermore, epidemiologic studies with diesel exhaust led to higher lung cancer risk estimates, related to the long-term exposure concentration, than the rat inhalation studies.[3,17,18] However, simultaneous exposure to not specifically toxic dusts has not been taken into consideration in the epidemiologic studies up to now. Also coal dust studies are not consistently "negative" as in older studies referred by Mauderly.[19] A healthy worker selection effect was substantiated.[20–22] Therefore, it is justified to assume that the analogy between the lung of rats and humans exists to a certain extent not only qualitatively but also quantitatively. Therefore, an estimation of the lung tumor risk in rats from chronic inhalation exposure to diesel and other not significantly specifically toxic dusts can be helpful for risk assessment of workers. Finally, the concentrations and the currently used unit risks for arsenic, cadmium, chromium, nickel, and PAH are too low to explain the increased lung cancer risks attributed to relatively low environmental concentrations of particulate matter ($PM_{2.5}$) in more recent epidemiological studies, for example, by Pope *et al*.[23] It has been calculated, that the increased lung cancer risks in these studies can only be plausibly explained in the sense of a cause-and-effect relationship if substantial carcinogenic potency is also attributed to the GBP within the environmental $PM_{2.5}$ fraction.[24]

In the past, the simple name *inert dust* was used as a general term for dusts without specific toxicity. In the last years, the abbreviation *PSP* was also used for such particles. It was defined as *nonfibrous poorly soluble particles of low acute toxicity, chemically distinct*.[25] However, the restriction to *low acute toxicity* is not sufficient, because chronically toxic dusts should be excluded too. Therefore, we developed a more precise definition after the end of an instillation study with 19 dusts.[10,26] We classified 16 of the 19 tested dusts as *respirable granular biodurable particles without known significant specific toxicity*. The chosen abbreviation *GBP* means only a *special selection* of the uncounted number of *granular biodurable dust types* which includes a lot of specifically toxic dusts like quartz or genotoxic particles like nickel oxide. So, the abbreviation *GBP* should be reserved for the nine-word definition.

DATABASE FOR THE CANCER RISK ESTIMATE OF GBP

There are several experimental data that can be included into a quantitative risk assessment for GBP. More dusts have been tested by instillation than by inhalation. To make use of the comprehensive database of the 19-dust instillation study we compare the carcinogenic potencies related to long-term retention of dust volume with dose–response relationships after inhalation as far as comparable data are available.

(1) TABLE 1 contains a summary of inhalation carcinogenicity studies with GBP in which the dust mass retained in the lung was measured at some time points after start of exposure. Inhalation studies without such lung burden data are not listed in TABLE 1.
(2) TABLE 2 summarizes the 19-dust study with instillation of 16 GBP types, which has just been finished by histological tumor diagnostics.[10] These data are used for calculation of dose–response relationships and for comparison with the inhalation experiments. Further instillation studies with different experimental design and less experimental groups were not included in the risk estimates.

One of the main questions of the 19-dust study was "which physicochemical dust characteristics determine carcinogenic potency?" or in other words "which is the appropriate dose metric?" The primary dose measure, of course, is the instilled dust mass. The long-time lung dust burden after the last instillation is estimated to amount to two-thirds of the dose instilled. This value is in the upper range of many reported retention data.[10,42] The retained dust mass and the density of the material determine the retained volume, the retained dust mass and the specific surface area of the dust determine the retained surface area.

The nonlinear regression analysis of the dose–response relationships of our 16 GBP relatively clearly led to the conclusion that the retained dust volume in combination with some information about (mean) particle size is the best suitable dose metric to date. Goodness-of-fit was about the same when the information about particle size was expressed in terms of three size classes and when a continuous function was used. The continuous function describes an increase of carcinogenic potency with decreasing (mean) particle size— and a decrease of carcinogenic potency with increasing (mean) particle size, respectively. This seems biologically more plausible than a sharp borderline between different particle sizes; for practical purposes, however, use of three size classes seems to be appropriate. FIGURE 1 shows the dose–response relationships for the three size classes *GBP-ultra-fine* (mean diameter 0.01– 0.03 μm), *GBP-fine-small* (0.09–0.2 μm), and *GBP-fine-large* (1.8–4 μm). Using surface area as a dose metric only makes sense when the different carcinogenic potencies of various dusts can be explained by their surface areas and particle sizes have not to be considered separately—this is the rationale of the *surface area concept*: "smaller particles are more effective *because* of their larger surface area." However, no plausible dose–response relationship was obtained in our regression analysis when just retained surface area was used as the dose metric—as can easily be seen on FIGURE 2. This contrasts with the reasonable volume-size-response curves on FIGURE 1.

TABLE 1. Inhalation carcinogenicity studies with GBP-F and GBP-UF in different rat strains

Dust	Diameter [μm][a]	Specif. surface area (BET) [m²/g]	Density [g/mL]	Exposure [mg/m³]	Exposure [h/wk, 24 mth]	GBP dose per lung after ca 1 yr exposure[b] [mg]	[μL][c]	[μL]/g[d]	after ~2 yr [mg][e]	Rats with lung tumor(s) [abs../at risk]	%	Tum./GBP in lung [%/μL][f]	Rat strain	References
Coal	—	—	1.4[g]	200	5 × 5	♀43	31	18	96	4/36	11	0.35	Sprague-Dawley, female	27
control	—	—	—	0	—	♀—	—	—	—	0/6	0[h]	—		
TiO₂ rutile	MMAD 1.5-1.7 (equival. to ~0.8 geom. diam.)	—	4.26[i]	10 ~84% resp.	5 × 6	♀8.7 ♂10.1	2.0 2.4	0.87 0.73	32 21	1/75 2/71	1.3 2.8	—	Sprague-Dawley, female and male	24
				50	5 × 6	♀60 ♂76	14 18	6.0 5.5	130 118	0/74 1/75	0 1.3	0		
				250	5 × 6	♀382 ♂362	90 85	38 26	546 785	26/74 13/77	35 17	0.39 0.17		
control	—	—	—	0	—	♀— ♂—	—	—	—	—	—	—		
Diesel engine exhaust, ~12% organic substances	MMAD ~0.25, 0.01-10	—	1.85[k]	0.35	5 × 7	♀♂0.24	0.13	0.09	0.6	3/223	1.3	—	F344/Crl, female and male	29
				3.5	5 × 7	♀♂2.18	1.18	0.79	11.5	8/221	3.6[m]	2.3		
				7.1	5 × 7	♀♂7.29	3.94	2.63	20.5	29/227	12.8	3.0		
control	—	—	—	0	—	♀♂—	—	—	—	0/77 2/79	0 2.5	—		
Toner	MMAD ~4, geom. SD 1.5; mean 3.5	3.6	1.2	0.35 respirable	5 × 6	♀0.12[n] ♂0.16[n]	0.1 0.13	0.1 0.09	0.19 0.24	2/230 1/112	0.9 0.9	—	F344 female and male	30,31
				1.5	5 × 6	♀0.69[n] ♂1.03[n]	0.57 0.85	0.55 0.60	1.41 2.05	0/114	0	0 0		
				5.4	5 × 6	♀5.32[n] ♂8.7[n]	4.4 7.3	4.3 5.1	13.0 18.1	5/114	4.4[p]	p		
TiO₂ rutile	MMAD ~1.1	—	4.3	3.9 respirable	5 × 6	♀1.4[n] ♂1.8[n]	0.33 0.42	0.32 0.30	2.24 3.20	2/113	1.8	—		
Quartz DQ12 (pos. control)	MMAD ~1.4	—	2.6	0.74 respirable	5 × 6	♀0.57[n] ♂0.75[n]	0.22 0.29	0.15 0.19	0.79 1.03	20/113	17.7	—		
control	—	—	—	0	—	♀♂—	—	—	—	3/111	2.7	—		
Talc	MMAD 2.7-3.2	—	2.8[q]	6	5 × 6	♀4.7[r] ♂4.4[r]	1.7 1.6	1.7 1.6	9.1[r] 10.5[r]	0/48 1/50	0 2	0	F344/N, female and male	32
				18	5 × 6	♀14[r] ♂21[r]	5.1 7.5	5.1 7.5	29.4[r] 24.2[r]	13/50 1/50	26 2	4.7[r]		
control	—	—	—	0	—	♀— ♂—	—	—	—	1/50 0/49	2 0	—		

continued.

TABLE 1. Continued.

Dust	Diameter [μm][a]	Specif. surface area (BET) [m²/g]	Density [g/mL]	Exposure [mg/m³]	Exposure [h/wk, 24 mth]	GBP dose per lung after ca 1 yr exposure[b] [mg]	GBP dose per lung after ca 1 yr exposure[b] [μL][c]	GBP dose per lung after ca 1 yr exposure[b] [μL/g][d]	GBP dose per lung after ~2 yr [mg][e]	Rats with lung tumor(s) [abs./at risk,%]	Tum./GBP in lung [%/μL][f]	Rat strain	References
Carbon black Printex 90	0.014 MMAD 1.1	230	1.85	6	10 mth[v] 20 mth[v]	15.4 15.4	8.3 8.3	5.5 5.5		12/72 17 7/72 9.7	2.2 1.0	Wistar, female and male	9
$NO_2 + SO_2$ + formaldehyde	—	—	—	5 + 5 + 3 ppm	10 mth[v] 20 mth[v]					1/72 1.4 0/72 0	— —		
control		—	—	0						0/72 0	—		
Carb. bl. Pr. 90[f]	0.014[u]	227	1.85	7-12[v]	5 × 18	38	20.5	14	44	39/100 39	1.9	Wistar, female and male	33
TiO_2 P25	0.021[u]	48	3.8	7-15[w]	5 × 18	35	9.2	6.1	39	32/100 32	3.4		
Diesel engine emissions, ~40 % org. substances	MMAD 0.25, 0.015-16	Native 8, extract 96-130	1.85[k]	0.8 2.5 7.0	5 × 18 5 × 18 5 × 18	2.8 11 36	1.5 5.9 19	1.0 4.0 13	6.3 24 64	0/198 0 11/200 5.5 22/100 22	0 0.8 1.1		
control		—	—	0		—	—	—	—	1/217 0.5	—		
Carbon black, furnace black Elftex-12	< 0.05[x]		1.85[y]	2.5 2.5 6.6 6.6	5 × 16 5 × 16 5 × 16 5 × 16	♀6.2 ♂7.9 ♀12 ♂15	3.4 4.3 6.5 8.1	2.2 2.8 4.3 5.4	17 25 37 40	8/107 7.5 2/106 1.9 28/105 27 4/106 3.8	2.2 — — 4.1	F344/N, female and male	34
Diesel engine emissions			1.85[k]	2.4 2.4 6.3 6.3	5 × 16 5 × 16 5 × 16 5 × 16	♀9.8 ♂12 ♀21 ♂28	5.3 6.5 11 15	3.5 4.3 7.6 10	36 45 81 90	8/105 7.6 5/105 4.7 29/106 27 9/106 8.5	1.4 — 2.5 —		
control	—		—	0 0	— —	♀— ♂—	— —	— —	— —	0/105 0 3/109 2.8	— —		

continued.

TABLE 1. Continued.

Dust	Diameter [μm][a]	Specif. surface area (BET) [m²/g]	Density [g/mL]	Exposure [mg/m³]	Exposure [h/wk, 24 mth]	GBP dose per lung after ca 1 yr exposure[b] [mg]	[μL][c]	[μL]/g[d]	after ~2 yr [mg][e]	Rats with lung tumor(s) [abs./at risk,%]	Tum./GBP in lung [%/μL][f]	Rat strain	References
Diesel engine emissions		1.85[y]		3.5 ± 1.4	3 × 17							F344, female and male	35
					3 mth	1.35	0.7	0.49		0/48[c]	0		
					6 mth	2.3	1.2	0.83		6/43[c]	14.0		
					9 mth					19/47[c]	40.4		
					12 mth	4.5	2.4	1.6		10/44[c]	22.7		
control	—	—	—	0		—	—	—	—	1/48	2.1	—	

[a] Either mass median aerodynamic diameter (MMAD, partly with range) or mean diameter of primary particles.
[b] Probably, this dose represents the best effective dose of the existing data. Therefore, only those experiments could be included into the study which contained such a value or it could be interpolated.
[c] Calculated from the measured retained particle mass and density.
[d] The dust volume retained in the lung is important in connection with the limit value of 1 μL dust per g control lung of the German MAK-Kommission.[36] In most experiments, the lung wet weight of the control group is not given but the assumption of 1.5 g was used according to Greim,[36] with the following exceptions: coal 1.68 g,[27] TiO₂ Sprague-Dawley rats 2.35 g, male 3.25,[28] talc experiment (see footnote f); for the study with toner and TiO₂, the arithmetic means of the given lung wet weights after 9 and 15 months were calculated.
[e] Additional information, but inhalation studies with asbestos dusts with one and two years of exposure did not show participation of the second year of exposure for the tumour response.
[f] The tumour response in % per μL GBP retained in the lung one year after start can be used as a measure for the carcinogenic potency of a dust in the exposed group. The quotient may help to compare the carcinogenic potencies of different groups if the background tumour incidence is far below 1% like in female Wistar rats. For its calculation, animals with tumours of the respective control group were subtracted if the percentage was small in relation to the tumour rate of the exposed group.
[g] Concluded by analogy with coal dust.
[h] Only 6 lungs were examined histologically, therefore no statistically significantly higher tumour response in the exposed group. Among 485 control rats, no lung tumour observed macroscopically.
[i] According to Weast et al.[37]
[k] Assumption of the value for the calculation in analogy to carbon black; the influence of the organic part of diesel particles on the density is not known.
[m] Significant higher than control (p <0.05); approximately equal numbers of males and females in each group.
[n] Mean value of the measurements after 9 and 15 months.
[p] p ~ 0.05 versus 6 rats with tumour of 450 (1.33%) not or lowly exposed.
[q] According to Ramdohr & Strunz.[38]
[s] Exposure time 5 days per week, 18 h per day.
[t] 0.04% of the particle mass could be extracted as organic substances (diesel 40% with the same method). Mean values per mg Printex 90: 0.6 pg benzo[a]pyrene (diesel soot 3.9 ng), 1-nitropyrene <0.5 ng (diesel soot 19.1 ng).
[u] MMAD 0.64 μm for carbon black and 0.8 μm for TiO₂ after removing the large agglomerates of primary particles using a cyclone.
[v] Mean concentrations: 7.4 mg/m³ 4 months, 12.2 mg/m³ 20 months.
[w] Mean concentrations: 7.2 mg/m³ 4 months, 14.8 mg/m³ 4 months, 7.2 mg/m³ 16 months.
[x] Particle size not given; value concluded by analogy with furnace blacks produced by Degussa between 0.014 and 0.56 μm.
[y] Density not given; densities of carbon blacks produced by Degussa 1.8 – 1.9 g/mL.
[z] 182 of 192 rats exposed in four groups survived longer than 18 months, 35 of them (19%) developed at least 1 lung tumour. The given numbers of examined animals per group refer to rats which survived at least 18 months.

QUANTITATIVE RISK ASSESSMENT FOR GBP BASED ON RAT DATA

FIGURE 3 shows dose–response relationships of inhalation studies with diesel, ultrafine carbon black, and TiO_2, for which lung burden had been measured (TABLE 1). Compared to the instillation data, the slopes of the dose–response relationships differ by a factor between about 2 and 9; on average, some factor between 5 and 6 seems to be suitable. The slopes for the excess risks (ER) after instillation were calculated as 25.9%, 9.6%, and 4.6% ER per μL/g control lung for *GBP-ultra-fine*, *GBP-fine-small*, and *GBP-fine-large*, respectively.[10]

Greim[36] assumed that an exposure concentration of 1.2 μL/m³ (corresponding to a mass concentration of 3 mg/m³ for particles of the density 2.5 g/mL) leads to a lung burden of 1 μL/g lung after long-term (occupational) exposure. If this conversion procedure and additionally a factor between 5 and 6 to convert from instillation to inhalation are applied, the following ER for dusts with density of 2–2.5 g/mL are obtained: ER of about 1% per 3 mg/m³ for large-fine GBP, 0.2% and 0.5% per 0.3 mg/m³ for small-fine and ultra-fine GBP, respectively.

These exposure-specific risks can be compared with further data. For example, a unit risk for environmental exposure to diesel engine emissions was calculated by Umweltbundesamt (UBA).[3] The unit risk value of 1×10^{-4} per μg/m³ (related to the elemental carbon core of diesel particles) may be converted to the exposure time pattern of the workplace situation by using

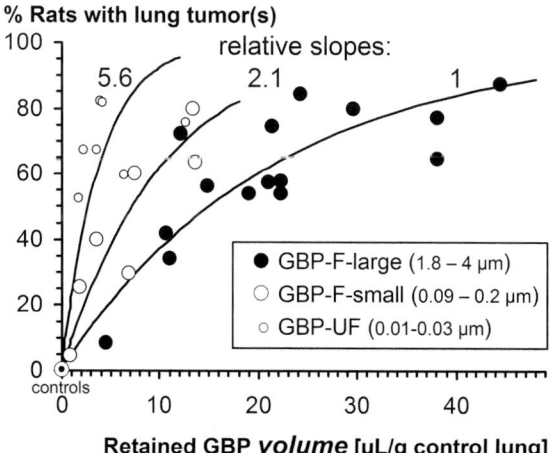

FIGURE 1. Dose–response relationships after instillation of GBP in rats. Dose metric: retained *dust volume* (two-thirds of the dose instilled are estimated to be retained in the lung for a long time); nonlinear regression for three particle size categories.

TABLE 2. Instillation carcinogenicity study with 19 respirable granular dusts in female Wistar rats (HsdCpb:WU)[10]

Dust	Diam.[a] [μm] GBP[b] size class	Specific surface area[c] (BET) [m²/g]	Density [g/mL]	Dose i.tr. weekly instill. × mg	Volume[d] [μL]	Lung burden of GBP volume total[e] [μL]		Rats with ≥ 1 primary lung tumour (no. with ben.tum. only + no. with mal. tum./at risk = %)	Survival 50 % [weeks]	Tum./total lung burden of dust[g] [%/μL]	Lungs with metastases of tumours from other sides
							[μL/g lung][f]				
Part 1: mining dusts (coal dust study)											
Carrier fluid[h]	—	—	—	20 × 0.4 mL	—	—	—	0 + 0/47 = 0.0 %	110	—	17.0 %
Lean coal < 0.1 % SiO₂	4.0 GBP-F-la	4.1	1.4	11 × 6	47	31	21	4 + 23/47 = 57.4 %	109	1.8	17.0 %
				20 × 6	86	57	38	1 + 30/48 = 64.6 %	101	1.1	20.8 %
Lower rich coal < 0.1 % SiO₂	1.8 GBP-F-la	9.9	1.4	10 × 6	43	29	19	10 + 16/48 = 54.2 %	108	1.9	10.4 %
				20 × 6	86	57	38	2 + 32/44 = 77.3 %	106	1.4	13.6 %
Rich coal, mine 1.3 % SiO₂	3.4 GBP-F-la	6.4	1.8	10 × 6	33	22	15	5 + 22/48 = 56.3 %	106	2.5	14.6 %
				20 × 6	67	44	30	10 + 26/45 = 80.0 %	99	1.8	24.4 %
Steam coal, mine 9.0 % SiO₂	2.4 GBP-F-la	10.9	2.2	10 × 6	27	18	12	5 + 26/43 = 72.1 %	108	4.0	14.0 %
				20 × 6	55	36	24	8 + 30/45 = 84.4 %	95	2.3	8.9 %
Rock, coal mine 16.7 % SiO₂	2.3 GBP-F-la	17.6	2.4	10 × 6	25	17	11	3 + 13/47 = 34.0 %	102	2.0	17.0 %
				20 × 6	50	33	22	5 + 21/45 = 57.8 %	105	1.7	15.6 %
Quartz DQ12 99.1 % SiO₂	1.1 (no GBP)	8.8	2.6	5 × 1[i]	1.9	0.85	0.57	6 + 17/35 = 65.7 %	103	> 5[j]	22.9 %
				10 × 1[i]	3.8	1.7	1.1	5 + 20/35 = 71.4 %	106	> 28[j]	11.4 %
				10 × 2[i]	7.7	3.4	2.3	8 + 20/36 = 77.8 %	100	> 15[j]	5.6 %
Part 2: non-mining dusts – series (a)											
Carbon black, lamp black 101	0.095 GBP-F-sm	18.4	1.85	5 × 6[k]	18	11	7.5	15 + 12/45 = 60.0 %	106	5.5	15.6 %
				10 × 6[m]	34	22	15	12 + 17/46 = 63.0 %	104	2.9	10.9 %
				20 × 6[n]	68	44	30	histology not performed	108	—	—
Carbon black, furnace black Printex 90	0.014 GBP-UF	337	1.85	5 × 1.5[p]	5	3	2.2	14 + 17/46 = 67.4 %	110	22.5	13.0 %
				5 × 3[q]	10	6	4.0	4 + 12/18 = 88.9 %	112	14.8	11.1 %
				5 × 3[r]	8	5	3.3	6 + 15/27 = 77.8 %	107	15.6	22.2 %
				5 × 3[s]	9	5.8	3.9	10 + 27/45 = 82.2 %		14.9	17.8 %
				5 × 6	16	11	7.2	7 + 33/48 = 83.3 %	108	7.6	10.4 %
				10 × 6	32	22	14	histology not performed	100	—	—
Aluminum oxide C [8]	0.013 GBP-UF	124	3.2	5 × 6	9	6	4	7 + 29/44 = 81.8 %	111	13.6	15.9 %
				10 × 6	19	12	8	12 + 22/47 = 72.3 %	97	6.0	10.6 %
Aluminum silicate P 820	0.015 GBP-UF	62.9	2.1	5 × 6	14	10	6	10 + 18/47 = 59.6 %	107	6.0	23.4 %
				10 × 6	29	19	13	15 + 19/45 = 75.6 %	108	4.0	22.2 %
Kaolin ~Al₂Si₂O₅(OH)₄	2.0 GBP-F-la	19	2.5	10 × 6	24	16	11	8 + 12/48 = 41.7 %	115	2.6	8.3 %
				20 × 6	48	32	21	7 + 28/47 = 74.5 %	121	2.3	4.3 %
No treatment (1)	—	—	—	—	—	—	—	1 + 0/46 = 2.2 %	124	—	4.3 %
Part 2: non-mining dusts – series (b)											
Diesel soot, lorry	0.2 GBP-F-sm	12.9 native, 34.5[t] extracted	1.85[u]	3 × 2.5	> 4.1	1.4[v]	0.9	1 + 1/45 = 4.4 %	117	3.1	20.0 %
				5 × 3	> 8.1	2.7	1.8	7 + 5/47 = 25.5 %	115	9.4	21.3 %
				5 × 6	> 16.2	5.4	3.6	12 + 6/45 = 40.0 %	108	7.4	22.2 %
TiO₂, P 25, hydrophilic	0.025 GBP-UF	52	3.8	5 × 3	3.9	2.6	1.7	9 + 13/42 = 52.4 %	114	20.2	14.3 %
				5 × 6	7.9	5.3	3.5	8 + 23/46 = 67.4 %	114	12.7	15.2 %
				10 × 6	16	11	7.0	11 + 21/46 = 69.6 %	104	6.3	15.2 %

continued.

TABLE 2. Continued.

Dust	Diam.[a] [μm] GBP[b] size class	Specific surface area[c] (BET) [m²/g]	Density [g/mL]	Dose i.tr. weekly instill. × mg	Volume[d] [μL]	Lung burden of GBP volume total[e] [μL]	[μL·g lung][f]	Rats with ≥ 1 primary lung tumour (no. with ben.tum. only + no. with mal. tum./at risk = %)	Survival 50 % [weeks]	Tum./total lung burden of dust[g] [%/μL]	Lungs with metastases of tumours from other sides
TiO₂, P 805, hydrophobic	0.021 (no GBP)	32.5	3.8	15 × 0.5 30 × 0.5	2.0 3.9	"low" (lipophilic) "low" (lipophilic)		0 + 0/11 = 0.0 % 1 + 0/15 = 6.7 %	86 114		9.1 % 6.7 %
Test Toner	3.5 GBP-F-la	3.6	1.2	10 × 6 20 × 6	50 100	33 67	22 44	6 + 7/24 = 54.2 % 7 + 14/24 = 87.5 %	111 101	1.6 1.3	0.0 % 16.7 %
TiO₂, anatase[x]	0.2 GBP-F-sm	9.9	3.9	10 × 6 20 × 6	15 31	10 21	6.8 14	7 + 6/44 = 29.5 % 17 + 11/44 = 63.6 %	108 113	3.0 3.0	11.4 % 2.3 %
ZrO₂	< 5; ~2[w] GBP-F-la	4.4	5.85	10 × 6	10	6.8	4.6	4 + 0/47 = 8.5 %	115	1.3	10.6 %
Lung dust, coal miner, silicosis III	0.2 GBP-F-sm	12.2	~2	10 × 6 20 × 6	30 60	20 40	13 27	5 + 27/40 = 80.0 % histology not performed	117 107	4.0 –	20.0 % –
SiO₂ amorphous (silica fumed)	0.014 (no GBP)	210	2.2	5 × 3 10 × 3				2 + 0/35 = 5.7 % histology not performed	113 112	–	14.3 %
No treatment (2)	–	–	–	–	–	–	–	0 + 0/46 = 0.0 %	113	–	13.0 %

[a] Mostly average or mean diameter of primary particle size. In the case of diesel soot, the results of the particle size measurements 39,40 led to the conclusion that the average size of the mostly aggregated ultrafine primary particles is in the size class of 0.9-0.2 μm. This is supported by other evaluations.³ The bio-durability of diesel soot agglomerates is not known.
[b] GBP = respirable granular bio-durable particles without known significant toxicity. The sequence of the experimental groups is analogous as in the experiment. For evaluation of the carcinogenic potency, the tested GBP were divided up into three size classes: ultrafine GBP = GBP–UF (mean diameter 0.01-0.03 μm); fine str all GBP = GBP-F-sm (mean diameter 0.09-0.2 μm; fine large GBP = GBP-F-la (mean diameter 1.8-4 μm).
[c] According to Eickhoff.⁴¹ These values were used for further calculations.
[d] Calculated from particle mass and density.
[e] With the exception of quartz (see footnote ʲ), diesel soot (see footnote ᵛ), hydrophobic TiO₂, and amorphous silica, the dust part retained in the lung at long term is consistently assumed with a mean of two-thirds of the dose instilled; this is rather in the upper range of the published data. 10,42
[f] Calculated under the precondition that the wet weight of the control rats amounts to 1.5 g. The German MAK standard for dust volume burden is 1 μL per g lung.³⁶
[g] Measure of the carcinogenic potency of a dust in the respective group for comparison with other groups of the experiment.
[h] 0.9 % NaCl solution, phosphate buffered, by adding Tween 80: 0.5 % for control, coal dusts, diesel soot and carbon black, 1% for hydrophobic TiO₂ and toner.
[i] Due to the known tendency of quartz to migrate from the lung to the lymph nodes, the percentage of quartz retained in the lungs is expected to decrease more than the lung burden of GBP. In earlier instillation experiments with quartz DQ12, bronchial clearance and lymphotropy reduced the SiO₂ content in the lung to about one-third of the instilled mass 10 months after instillation (Brockhaus A., Pott, F., 1968/69, unpublished). This corresponds roughly with retention data of the inhalation experiment of Bellmann et al.³¹ Hence, an additional reduction factor of two-thirds was regarded for DQ12 (estimated retention = DQ12 instilled × 2/3 × 2/3). As a general example: Instillation of 10 mg quartz with a density of 2.6 g/mL result in 1.7 μL retained dust. However, the values in the table given for % lung tumours per μL quartz per lung are calculated as if 2/3 of the instilled quartz dose persisted in the lung like GBP. The symbol > which precedes the figures written in italics indicates the existing uncertainties.
[k-q] One additional instillation by error. The dust volume of this instillation is included in the calculation of the totally instilled volume:
[k] 1 × 2.5 mg diesel soot.
[m] 1 × 3 mg diesel soot.
[n] 1 × 6 mg diesel soot.
[p] 1 × 3 mg TiO₂-UF hydrophilic.
[q] First subgroup of the 15 mg group: 1 × 6 mg TiO₂-UF hydrophilic.
[r] Second subgroup of the 15 mg group: no additional instillation.
[s] Results of the two subgroups are combined in this line; they are the basis for further calculations.
[t] Specific surface area of extracted soot was used for statistical calculations
[u] Value used for the carbonaceous particle core of diesel soot in analogy with carbon black; because the lower density of the organic substances, the volume of the instilled dose is higher than the volume calculated from mass and density of elemental carbon.
[v] According to UBA³ it is assumed that 50 % of the native diesel soot of lorries are organic substances and that they will be dissolved in the lung. Hence, 7.5 mg instilled total particle mass of diesel soot is reduced in the lung by two mechanisms: 2/3 of this mass (5 mg or 2.7 μL) is estimated to be retained in the lung for a longer period according to footnote ᵉ, but it is considered to be reduced by dissolution of the organic part at 50 % to 2.5 mg or 1.35 μL elemental carbon.
[w] Particle size < 5 μm described by the supplier. Rödelsperger (personal communication, 2005) found a heterogeneous material with 2 components containing ultrafine particles and much larger particles up to 10 μm by scanning electron microscopy. However, it was concluded from the relatively small specific surface area (4.4 m²/g) that the dust is adequately sorted into the size class GBP-F-large.
[x]

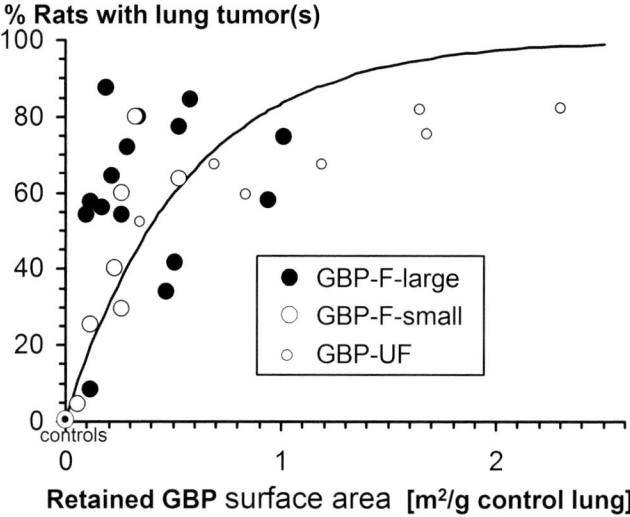

FIGURE 2. Dose–response relationships after instillation of GBP in rats. Dose metric: retained *dust surface area*; nonlinear regression for all dusts together (different dot symbols for illustrative purposes only).

the factor of 6. Related to a long-term exposure concentration of 0.3 µg/m^3, this corresponds to an ER of 0.5%. Today, we consider it more appropriate to classify diesel particles as *small-fine* (mean diameter around 0.2 µm). But, the unit risk of UBA was derived under the assumption that diesel particles are ultrafine, and data of ultrafine carbon black and TiO$_2$ dusts were included in the calculation. If, therefore, the UBA unit risk is used as a measure for the carcinogenic potency of ultrafine dusts after inhalation, then the estimate agrees with the value derived above very well.

The inhalation study with toner dust was evaluated by the authors as *negative*.[30,31] But it has been described that an exposure-related increase in the highest exposure group has to be taken into consideration due to a marginally positive trend test, due to a dose–response relationship for inflammatory effects, and due to the general finding of GBP carcinogenicity.[10] The exposure concentration in the "highest" exposure group of this large-fine dust type was not actually "high," the time-weighted average (TWA) concentration was only 4.4 mg/m^3, which is not much higher than the general dust limit value in Germany. The exposure concentrations of the two lower dose groups were so low that it is reasonable to take the tumor frequencies of these groups as information about the background risk of the rat strain. The lung tumor frequency of these two groups together with the clean air control was 4/337 = 1.2%. The lung tumor frequency in the "highest" dose group was 5/114 = 4.4%, the ER is then 4.4% − 1.2% = 3.2%. If this is related to the TWA concentration of 4.4 mg/m^3 an ER of 2% per 3 mg/m^3 for *GBP-fine-large* is obtained. If it is

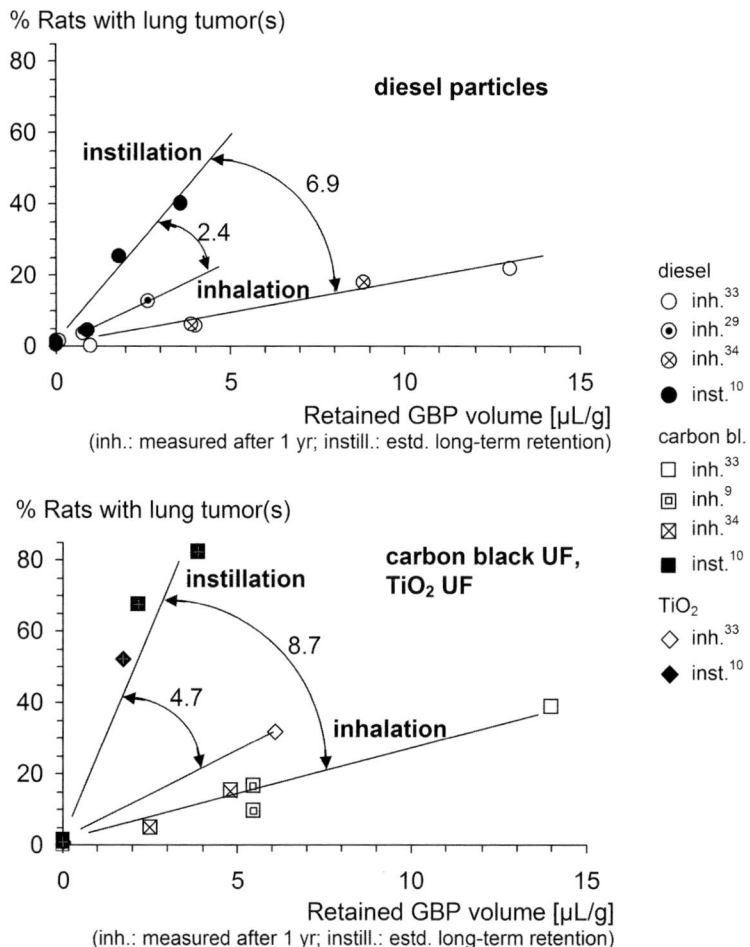

FIGURE 3. Comparison of dose–response relationships of carcinogenicity studies with inhalation [9,29,33,34] and with instillation.[10]

taken into consideration that the density of the toner dust is only 1.2 g/mL, then this estimate is also in good agreement with the one derived above.

A quantitative risk assessment for quartz dust has led to an estimate of an ER of 1% after long-term occupational exposure to 0.05 mg SiO_2/m^3.[16] The estimate was particularly based on an inhalation study with quartz DQ12. The logistic regression analysis of the 19-dust study led to an odds ratio of 30 for pure quartz DQ12 in comparison with *GBP-fine-large*.[10] Application of the factor 30 to the specific risk for DQ12 gives an estimate of the specific risk for large-fine GBP of 2% per 3 mg/m^3 (3 mg/m^3/0.05 mg/m^3 = 60, reduced by a factor of 30). This ER value for *GBP-fine-large* is also relatively close to the estimates derived above.

REFERENCES

1. IARC MONOGRAPHS ON THE EVALUATION OF CARCINOGENIC RISKS TO HUMANS. 1989. Vol. 46. Diesel and gasoline engine exhaust and some nitroarenes. World Health Organization, International Agency for Research on Cancer. Lyon, France.
2. IARC MONOGRAPHS ON THE EVALUATION OF CARCINOGENIC RISKS TO HUMANS. 1996. Vol. 65. Printing processes and printing inks, carbon black and some nitro compounds. Carbon black. pp. 149–262. World Health Organization, International Agency for Research on Cancer. Lyon, France.
3. UBA. UMWELTBUNDESAMT. 1999. Ed. Durchführung eines Risikovergleichs zwischen Dieselmotoremissionen und Ottomotoremissionen hinsichtlich ihrer kanzerogenen und nicht-kanzerogenen Wirkungen. Berichte 2/99—Forschungsbericht 297 61 001/01. Report Nr. UBA FB 99–033. Erich Schmidt Verlag. Berlin.
4. POTT, F., D.L. DUNGWORTH, U. HEINRICH, *et al.* 1994. Lung tumours in rats after intratracheal instillation of dusts. (Inhaled Particles VII). Ann. Occup. Hyg. **38** (Suppl 1): 357–363.
5. DASENBROCK, C., L. PETERS, O. CREUTZENBERG, *et al.* 1996. The carcinogenic potency of carbon particles with and without PAH after repeated intratracheal administration in the rat. Toxicol. Lett. **110:** 1–7.
6. Report on Carcinogens, Tenth Edition; U.S. Department of Health and Human Services, Public Health Service, National Toxicology Program, December 2002.
7. HEINRICH, U., F. POTT, U. MOHR, *et al.* 1986. Lung tumours in rats and mice after inhalation of PAH-rich emissions. Exp. Pathol. **29:** 29–34.
8. HEINRICH, U., H. MUHLE, S. TAKENAKA, *et al.* 1986. Chronic effects on the respiratory tract of hamsters, mice and rats after long-term inhalation of high concentrations of filtered and unfiltered diesel engine emissions. J. Appl. Toxicol. **6:** 383–395.
9. HEINRICH, U., D.L. DUNGWORTH, F. POTT, *et al.* 1994. The carcinogenic effects of carbon black particles and tar-pitch condensation aerosol after inhalation exposure of rats. Inhaled particles VII. Ann. Occup. Hyg. **38**(Suppl 1): 351–356.
10. POTT, F. & M. ROLLER. 2005. Carcinogenicity study with nineteen granular dusts in rats. Eur. J. Oncol. **10:** 249–281.
11. WHO (WORLD HEALTH ORGANIZATION). 1986. Asbestos and other natural mineral fibres, IPCS International Programme on Chemical Safety. Environmental Health Criteria 53, World Health Organization. Geneva.
12. IARC MONOGRAPHS ON THE EVALUATION OF CARCINOGENIC RISKS TO HUMANS. 1993. Vol.58: Beryllium, Cadmium, Mercury, and Exposures in the Glass Manufacturing Industry. Cadmium and Cadmium Compounds. pp. 119–237. World Health Organization, International Agency for Research on Cancer. Lyon, France.
13. IARC Monographs on the Evaluation of Carcinogenic Risks to Humans. 1997. Vol. 68: Silica, Some Silicates, Coal Dust and para-Aramid Fibrils. Silica. pp. 41–242. World Health Organization, International Agency for Research on Cancer. Lyon, France.
14. IARC Monographs on the Evaluation of Carcinogenic Risks to Humans. 1990. Vol. 49: Chromium, Nickel and Welding. Nickel and Nickel Compounds. pp. 257–445. World Health Organization, International Agency for Research on Cancer. Lyon, France.
15. MAUDERLY, J.L. & R.J. MCCUNNEY, Eds. 1996. Particle overload in the rat lung and lung cancer: implications for human risk assessment. Proceedings of a conference

held in the Massachusetts Institute of Technology on March 29–30, 1995. Taylor and Francis. Bristol, PA.
16. SCHLÜTER, G. 2003. Stellungnahme des Beraterkreises Toxikologie des AGS zur Frage der Einstufung von kristallinem Siliciumdioxid (Quarz [14808-60-7] und Cristobalit [14464-46-1]) in Form von alveolengängigen Stäuben. *In* Bundesanstalt für Arbeitsschutz und Arbeitsmedizin BAuA. Quarz - Einstufung & Dosis-Wirkungs-Beziehungen, Eds.: 109–133. Workshop vom 07./08. März 2002 in Berlin. Schriftenreihe der Bundesanstalt für Arbeitsschutz und Arbeitsmedizin. Tb131. Bremerhaven, Wirtschaftsverlag NW.
17. DAWSON, S.V. & G.V. ALEXEEFF. 2001. Multi-stage model estimates of lung cancer risk from exposure to diesel exhaust, based on a U.S. railroad worker cohort. Risk Anal. **21:** 1–18.
18. STAYNER, L., D. DANKOVIC, R. SMITH, *et al.* 1998. Predicted lung cancer risk among miners exposed to diesel exhaust particles. Am. J. Ind. Med. **34:** 207–219.
19. MAUDERLY, J.L. 1994. Contribution of inhalation bioassays to the assessment of human health risks from solid airborne particles. *In* Toxic and carcinogenic effects of solid particles in the respiratory tract. ILSI-Monographs.Mohr, U. *et al.* Eds.: 355–365. ILSI Press.Washington, DC.
20. BECKLAKE, M.R. 1998. Workplace pollution and airway disease: evidence from community and workforce based studies. *In* Relationships Between Respiratory Disease and Exposure to Air Pollution. ILSI monographs. Mohr, U. *et al.* Eds.: 133–145. ILSI Press.Washington, DC.
21. BOLM-AUDORFF, U., M. MÖHNER, P. MORFELD, *et al.* 1998. Lungenkrebsrisiko durch berufliche Exposition—Quarzstäube. *In* Lungenkrebsrisiko durch berufliche Exposition. Fortschritte in der Epidemiologie K.H. Jöckel, *et al.* Eds.: 186–209. Ecomed Verl. Ges., Landsberg/Lech.
22. MORFELD, P., K. LAMPERT, H. ZIEGLER, *et al.* 1997. Längsschnittstudie zum Einfluß der Bergarbeiterpneumokoniose auf die Lungenkrebsmortalität von unter Tage tätigen Bergleuten im Deutschen Steinkohlenbergbau. *In*: Psychomentale Belastungen und Beanspruchungen im Wandel von Arbeitswelt und Umwelt. Kanzerogenese und Synkanzerogenese. E. Borsch-Galetke & F. Struwe, Eds.: 133–137. Rindt-Druck, Fulda.
23. POPE, C.A., R.T. BURNETT, M.J. THUN, *et al.* 2002. Lung cancer, cardiopulmonary mortality, and long-term exposure to fine particulate air pollution. JAMA **287:** 1132–1141.
24. ROLLER, M. 2005. Die Risikoabschätzungen des LAI vor dem Hintergrund der Feinstaubdiskussion im Jahre 2005. Gefahrstoffe – Reinhalt. Luft. **65:** 425–434.
25. ILSI RISK SCIENCES INSTITUTE WORKSHOP PARTICIPANTS. 2000. The relevance of the rat lung response to particle overload for human risk assessment: a workshop consensus report. Inhal. Toxicol.**12:** 1–17.
26. POTT, F. & M. ROLLER. 2003. Untersuchungen zur Kanzerogenität granulärer Stäube an Ratten — Ergebnisse und Interpretationen. Kurzbericht über das Projekt F1843 der Bundesanstalt für Arbeitsschutz und Arbeitsmedizin, veröffentlicht am 28.08.2003 im Internet unter. Available at http://www.baua.de/fors/f1843.htm
27. MARTIN, J.C., H. DANIEL, & L. LE BOUFFANT. 1977. Short- and long-term experimental study of the toxicity of coal-mine dust and of some of its constituents. *In*: Inhaled Particles IV, Vol. 1. Walton, W., Ed.: 361–371. Pergamon Press. Oxford.

28. LEE, K.P., H.J. TROCHIMOWICZ & C.F. REINHARDT. 1985. Pulmonary response of rats exposed to titanium dioxide (TiO_2) by inhalation for two years. Toxicol. Appl. Pharmacol. **79:** 179–192.
29. MAUDERLY, J.L., R.K. JONES, W.C. GRIFFITH, *et al*. 1987. Diesel exhaust is a pulmonary carcinogen in rats exposed chronically by inhalation. Fundam. Appl. Toxicol. **9:** 208–221.
30. MUHLE, H., B. BELLMANN, O. CREUTZENBERG, *et al*. 1991. Pulmonary response to toner upon chronic inhalation exposure in rats. Fundam. Appl. Toxicol. **17:** 280–299.
31. BELLMANN, B., H. MUHLE, O. CREUTZENBERG, *et al*. 1991. Lung clearance and retention of toner, utilizing a tracer technique, during chronic inhalation exposure in rats. Fundam. Appl. Toxicol. **17:** 300–313.
32. NATIONAL TOXICOLOGY PROGRAM (NTP). 1993. Toxicology and carcinogenesis studies of talc in F344/N Rats and $B6C3F_1$ Mice. Technical Report Series No. 421, NIH Publ. No. 93-3152.
33. HEINRICH, U., R. FUHST, S. RITTINGHAUSEN, *et al*. 1995. Chronic inhalation exposure of Wistar rats and two different strains of mice to diesel engine exhaust, carbon black and titantium dioxide. Inhal. Toxicol. **7:** 533–556.
34. NIKULA, K.J., M.B. SNIPES, E.B. BARR, *et al*. 1995. Comparative pulmonary toxicities and carcinogenicities of chronically inhaled diesel exhaust and carbon black in F344 rats. Fundam. Appl. Toxicol. **25:** 80–94.
35. IWAI, K., T. UDAGAWA, M. YAMAGISHI, *et al*. 1986. Long-term inhalation studies of diesel exhaust on F344 SPF rats. Incidence of lung cancer and lymphoma. *In*: Carcinogenic and Mutagenic Effects of Diesel Engine Exhaust. N. Ishinishi, *et al*. Eds.:349–360. Oxford: Elsevier Sci. Publ. (Biomedical Div.). Amsterdam, New York. (Developments in Toxicology and Environmental Science. Vol. 13).
36. GREIM, H. 1997. Ed. Toxikologisch-arbeitsmedizinische Begründungen von MAK-Werten (Maximale Arbeitsplatzkonzentrationen) der Senatskommisssion zur Prüfung gesundheitsschädlicher Arbeitsstoffe der Deutschen Forschungsgemeinschaft. Allgemeiner Staubgrenzwert. 25. Lieferung 1997. Wiley-VCH, Weinheim.
37. WEAST, R.C. *et al*. Ed. 1989. CRC Handbook of Chemistry and Physics. CRC Press. Boca Raton, FL.
38. RAMDOHR, P. & H. STRUNZ. 1967. Klockmann's Lehrbuch der Mineralogie. Ferd. Enke Verlag. Stuttgart.
39. KLINGENBERG, H., D. SCHÜRMANN & K.-H. LIES. 1991. Dieselmotorabgas—Entstehung und Messung. *In* Krebserzeugende Stoffe in der Umwelt—Herkunft, Messung, Risiko, Minimierung. VDI-Berichte 888. VDI-Verlag. Düsseldorf.
40. RÖDELSPERGER, K., B. BRÜCKEL, S. PODHORSKY, *et al*. 2002. Characterisation of ultrafine particles by electron microscopy. *In* Crucial Issues in Inhalation Research—Mechanistic, Clinical and Epidemiologic. Heinrich, U. & U. Mohr, Eds.: 221–232. Fraunhofer IRB Verlag. Stuttgart.
41. EICKHOFF, H.-P. 2001. Bestimmung von Dichte, spezifischer Partikeloberfläche und Partikelgröße Bericht-Nummer B0104014 – Teil 1.1. BET-Messwerte. Gesellschaft für Oberflächen- und Festkörperuntersuchung mbH Hamburg. Auftrags-Nr.Z2.2-14120 F1843. Supported by the Federal Institute for Occupational Safety and Health.
42. DRISCOLL, K.E., D.L. COSTA, G. HATCH, *et al*. 2000. Intratracheal instillation as an exposure technique for the evaluation of respiratory tract toxicity: uses and limitations. Toxicol. Sci. **55:** 24–35.

Measurements of Asbestos Burden in Tissues

RONALD F. DODSON[a] AND MARK A.L. ATKINSON[b]

[a] *ERI Consulting, Inc., Tyler, Texas 75701, USA*

[b] *The University of Texas Health Center at Tyler, Tyler, Texas 75708-3154, USA*

ABSTRACT: Asbestos inhaled into the lung is recognized as a potential causal agent for the development of diseases in man. The diseases induced by asbestos include lung cancer, fibrosis of the lung (asbestosis), and extrapulmonary tumors including mesothelioma (a tumor of the serosal membrane), as well as fibrosis and other changes in the pleura linings. The cause of these diseases can often be more specifically linked to asbestos exposure once tissue burden of asbestos is established. The asbestos burden in tissue can be defined as the number of asbestos bodies and/or the numbers and types of asbestos fibers found in the tissue. In either of these cases the quality of information is directly dependent on the preparative techniques and instrumentation used in the analysis. The present article will discuss the significance of findings of tissue burden based on both these variables.

KEYWORDS: asbestos; electron microscopy; ferruginous bodies; light microscopy

INTRODUCTION

Asbestos minerals have been reported as having been used in over 3,000 commercial applications in the United States alone.[1] There are potentially an equal number of products through which asbestos exposure may occur by virtue of asbestos being a component of the primary mineral mined for a given commercial application. Often these products have not been recognized by either consumers or workers as containing asbestos. An example of such products where asbestos exposures might occur include exposures to products containing vermiculite[2–4] and talc.[5–7]

It is well established that asbestos exposure carries the risk of inducing malignant and nonmalignant diseases of the respiratory system as well as pathological responses in extrapulmonary sites in the body.[8] Historical data concerning occupational exposure to asbestos provide important information

Address for correspondence: Ronald F. Dodson, ERI Consulting, Inc., 2026 Republic Drive, SteA, Tyler, TX 75701. Voice: 903-534-5001; fax: 903-534-8701.
 e-mail: Ron@ericonsulting.com

relating the potential for past exposure to asbestos leading to these diseases. However, possible sources of secondary or environmental exposures are often overlooked because the exposed individual either has forgotten such exposure or had no knowledge at the time that asbestos was a component of the dust to which they were exposed in the past.[9]

A careful and appropriately conducted analysis of tissue samples for both the presence and amount of asbestos, irrespective of fiber size, offers an insight into prior exposure. However, there are significant variables that dictate the quality of information that is to be gained from tissue analysis. These include the technique used for preparation of the tissue, the amount of tissue analyzed, and the instrument used for the analysis. Further, when results are presented the method of analysis in a given count scheme does not disclose what is included and, perhaps more important, what is excluded in the analysis. The following article describes what can reasonably be expected from the various analytical procedures used to characterize tissue burden of asbestos and the limits associated with the various analytical applications now in use.

Data Obtained from Analysis of Ferruginous Bodies by Light Microscopy

A ferruginous body is simply an iron-coated structure that forms when pulmonary macrophages interact over time with an inhaled particle. If these structures are formed on an asbestos fiber the entity can appropriately be termed an *asbestos body*.[10,11] There are other inhaled structures, both fibrous and nonfibrous, that may also stimulate the deposition of a ferruginous coating.[12–17] It is therefore important that a ferruginous body seen in a tissue section conforms to the definition of "an elongated structure formed on a clear, elongated core" if the term *asbestos body* is to be applied. Meeting these morphological criteria is important because the presence of these structures in combination with areas of fibrosis in tissue sections suffice for the pathological diagnosis of asbestosis.[8,18] It should be noted that there are significant limitations to recognizing asbestos bodies in tissue sections. These limitations include the fact that the structures have to be appropriately orientated in the plane of section to permit identification based on the previously mentioned morphological definition and that tissue sections are very insensitive sources for determining the presence of ferruginous bodies due to the limited amount of tissue sampled, the orientation of the bodies in relation to the plane of the section, and the associated random sampling error due to the amount of tissue evaluated.[18,19]

These limitations can be circumvented by sampling schemes that involve the use of techniques by which larger pieces of tissue can be analyzed for the presence of ferruginous bodies. These techniques involve dissolving the tissue and collecting the particulates, including ferruginous bodies, on filters. It is critical that such procedures cause the least disruption of particulates collected from the tissue and collect them in a state as similar as possible

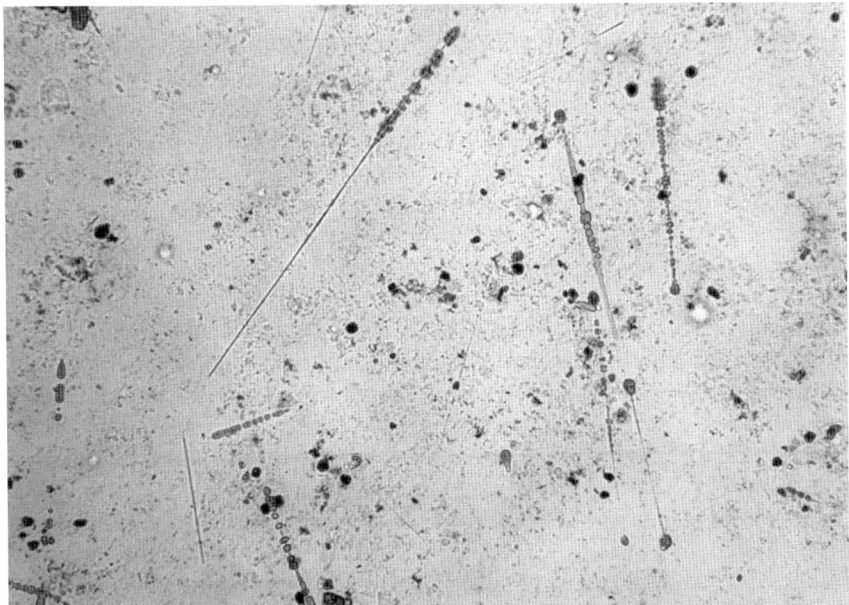

FIGURE 1. The ferruginous bodies in this field have been isolated from human lung tissue via a digestion technique. The bodies are in a flat plain of orientation on the cleared filter that permits the morphological identification of their features thus permitting their designation as "asbestos bodies."

to that in which they exist in the tissue. This can be best achieved by using the most direct method of isolation. Tissue preparations involving appreciable manipulations of the tissue can potentially disrupt ferruginous coatings as well as fragment uncoated asbestos components within the tissue. The procedure used in our laboratory is a digestion technique that involves dissolution of the tissue components by a modified bleach digestion technique[19] followed by filtration resulting in a "clean" preparation. Subsequently, the filter matrix can be made transparent permitting an unobstructed view by the light microscope of the undissolved material and consequently optimum resolution while screening for ferruginous bodies.[20] The ferruginous coating formed along the central fiber makes the asbestos body recognizable in the light microscope even if the asbestos fiber in the central core is too thin to be easily resolved (FIG. 1). The ferruginous bodies lie flat on the surface of the filter thus eliminating the limitations imposed by orientation that are encountered in tissue sections. The trained analyst can then easily identify whether or not a ferruginous body morphologically conforms to the definition of an asbestos body or if it is formed on a nonasbestos entity.

Another limitation that should be recognized is that the presence of asbestos bodies indicates that an exposure to longer asbestos fibers has occurred in the past. Asbestos bodies are formed primarily on asbestos fibers longer than 8–10 μm and often on fibers with thicker diameters than many equivalent length uncoated fibers. Thus asbestos bodies represent a selected population of the longer fibers in the lung at the time of sampling. There are further compounding issues to be considered in weighing the assessment of the presence or absence of ferruginous bodies in digested materials, namely that some animal species and some humans are "poor producers" of ferruginous coatings even when long fibers are present in the lung tissue.[10,21–23] There is also a wide individual variation of the ratio of uncoated asbestos fibers in tissue to the number of ferruginous bodies.[21] Thus it is not possible to extrapolate the number of uncoated asbestos fibers based on the number of asbestos bodies found. The number of asbestos bodies per gram of tissue (determined from a digested sample) can, however, be used to correlate with types of past exposure. Several laboratories have concluded that the burden of asbestos bodies in lung tissue of the general population is represented as 0–20 asbestos bodies per gram of wet tissue.[24–28]

The definitive identification of the type of asbestos in the core of a ferruginous body, as with uncoated asbestos fibers, can only be achieved by analytical transmission electron microscopy ideally in conjunction with X ray energy dispersive analysis and selected area diffraction techniques. Finally, it is highly unusual for chrysotile fibers to be found as the core of asbestos bodies. This is explained by the fact that this form of asbestos is usually inhaled as a fiber with lengths shorter than 8 μm and, even when longer, they are usually thin single fibrils.[28] Thus it is an exception when the cores of asbestos bodies are not formed on amphibole fibers and a ferruginous body count provides little or no information about chrysotile exposure.

Analysis of Uncoated Asbestos Fibers

The use of light microscopy to detect uncoated asbestos fibers in tissue sections is an exercise in futility. The background presence of tissue components compromise resolution to such an extent that it is impossible, with only rare exceptions, to detect uncoated fibers. Fibers that are seen are thicker and usually nonasbestos fibers. It is highly unlikely that uncoated asbestos structures could be detected. A further limitation in the use of the light microscope to detect even isolated uncoated fibers is its inherent resolution. Langer *et al.*[29] correctly pointed out that "the optical microscope delivers a select, biased population. First we can only study what the microscope sees—and it only sees large fibers, those thicker than 0.5 μm in diameter." (p. 360). Rood and Streeter[30] "assumed a limit of 0.3 μm" (p. 333). as the limit of resolution. Uncoated fibers in tissue are usually present as individual "fibrils" with diameters as small as 50 nm for chrysotile and 500 nm for amphiboles[31] and would thus be "invisible" to the light microscope. This makes the phase contrast light microscope of no

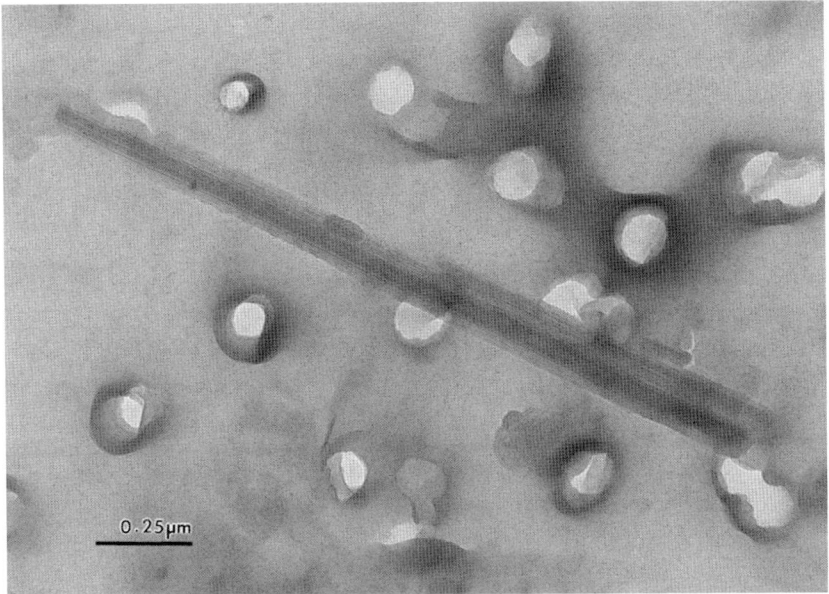

FIGURE 2. This chrysotile "bundle" was isolated via a digestion technique from omental tissue of an individual with mesothelioma. The replica permits comparison of the 0.2 micron pore sizes in the filter with the diameter of the bundle. This size fiber represents the more commonly found sizes in extrapulmonary sites and, due to either parameters of diameter and/or length, would not be included in a light microscopy based count of an environmental sample.

value in fiber counting for determination of tissue burden even if the sample is prepared via a digestion procedure that permits the maximum resolution attainable by the light microscope on the cleared filter.

Further confusion in assessing the data of fiber burden in tissue is contained in the guidelines for determination of fiber burden in the work place (OSHA–NIOSH counting scheme) that uses a fiber definition of an elongated structure that has greater than a 3:1 aspect ratio and equal to or longer than 5 μm length. This definition is based on a scheme using the phase contrast light microscopy. This counting scheme for fiber burden in air samples has been described by Langer et al.[32] as based on an emphasis on "practicality and theoretical considerations" (p. 254). not in an effort to necessarily target a population of fibers relating to specific health risks nor to include all of the fibers present. The purpose is simply to standardize the procedure to a particular subset of fibers. In addition, the counting scheme counts any "fiber" and does not distinguish the type of fiber. In essence, the majority of uncoated asbestos fibers in tissue (FIGS. 2 and 3) and in many environmental samples (FIG. 4) are either not resolvable (due to their thin diameter) by the light microscope and/or would be excluded from a count that only includes 5 μm and longer fibers.[33]

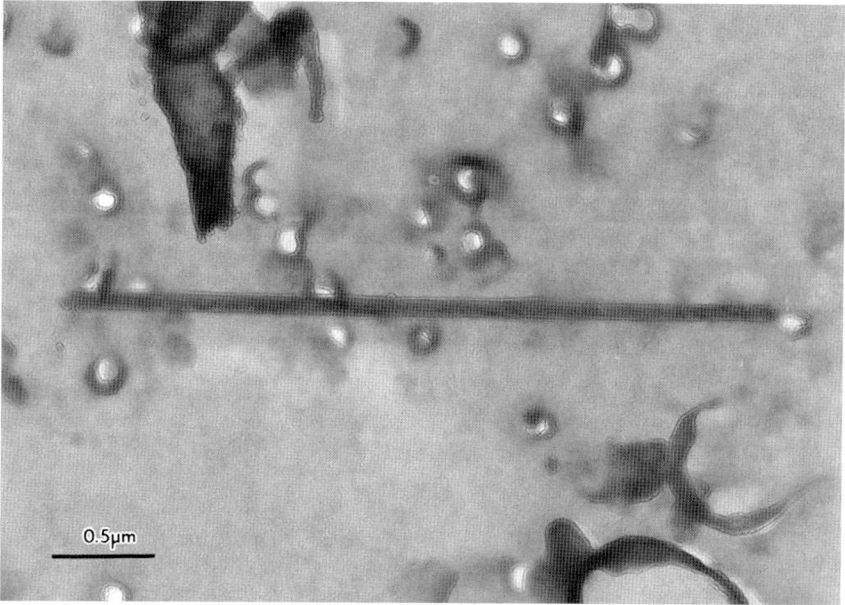

FIGURE 3. The amosite fiber in this micrograph was isolated from mesentary tissue in an individual with mesothelioma. The size of this fiber indicates that short amphibole fibers as well as short chrysotile fibers do reach extrapulmonary sites. A comparison of the diameter of the fiber with the adjacent 0.2 micron pores emphasize the necessity of using the resolution afforded in an analytical transmission electron microscope if they are to be detected in a sample.

Instruments with the greatest ability to detect uncoated fibers from tissue and to determine their composition are the electron microscopes. The scanning electron microscope permits better quantitative analysis than the light microscope due to its better resolution. However, the scanning electron microscope is less desirable than the analytical transmission electron microscope (TEM)[34] for the analysis of uncoated fibers from tissue samples because the TEM cannot only detect the smallest fiber but also identify the mineral type based on a combination of morphological appearance, elemental composition determined by X ray energy dispersive analysis and crystalline orientation by selected area diffraction. As with use of the light microscope, the use of either instrument is only as accurate as the tissue preparation and the magnification permit. It is imperative that the data provided from a tissue analysis define what is included in a count so that the reader understands what is excluded. For example, a count including only 5 μm and longer fibers may well exclude the vast majority of the fibers present in a tissue sample even if the analytical transmission electron microscope is used. Furthermore, using an analytical transmission electron in a count scheme at very low magnification will exclude appreciable populations

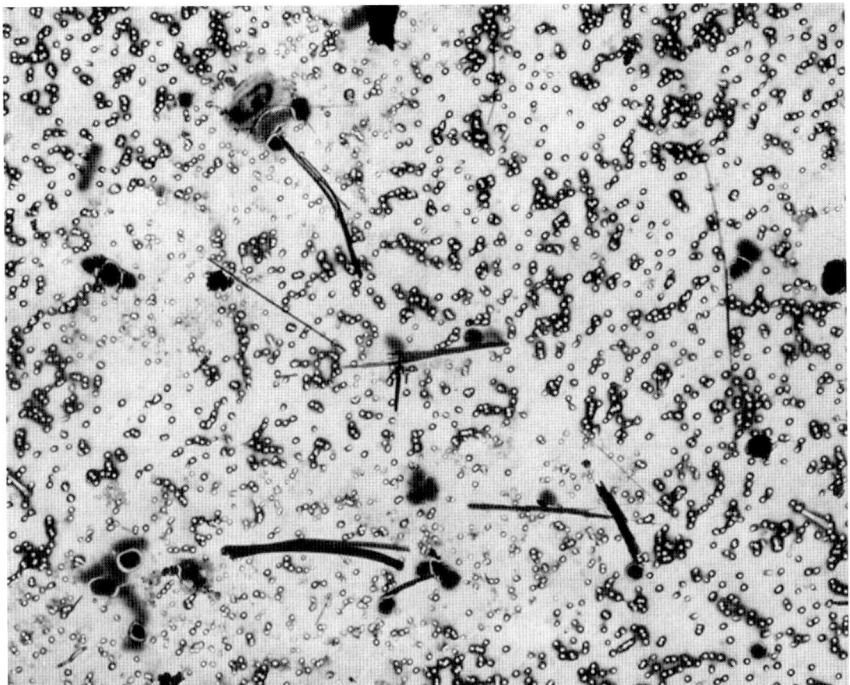

FIGURE 4. The chrysotile fibers and bundles shown in this micrograph are from materials collected by rinsing dust from the surface of unused brake linings. The asbestos structures seen in the field would not be counted in the phase contrast light microscopy count scheme that includes "fibers" longer than 5 μm, because the structures are shorter than 5 μm, or of insufficient diameter, to be detected by light microscopy.

of fibers because these are no longer "visible." Therefore, to interpret analytical data, it is critical to know the physical limits of fibers, both diameter and length, that are theoretically resolvable as well as those arbitrarily included in the count scheme. The preparation procedure of a sample is also of critical importance to ensure that the collected particulates have minimal alteration from their morphological status in the tissue. Safeguards should include using a "more direct method of preparation" (p. 226), with avoidance of more traumatic procedures for isolation of particulates, such as drying the tissue during the isolation process,[35] which may crack ferruginous bodies or even cause separation of fiber bundles into small units.

In reviewing asbestos fibers detected by tissue analysis that included both short (<5 μm) and thin fibers, Pooley and Ranson[36] have indicated that the light microscope would only "visualize 5% of the crocidolite, 26.5% of the amosite, and 0.14% of the chrysotile present in the lung tissue" (p. 1). Looking at this in a different perspective, Churg and Warnock[37] have indicated that most asbestos fibers in lung tissue are too short to form ferruginous bodies or to be visible by the light microscope. Data from our own laboratory have also found that the vast majority of uncoated asbestos fibers found in tissue are less than 5 μm with even most longer fibers being below the limit of delectability (due to their thin diameter) using the light microscope.[38–41]

Several additional issues concerning uncoated fibers are of appreciable importance when reviewing data from tissue analysis. The shorter uncoated fibers are the population of fibers that are more readily cleared from the lung during the time from last exposure. However, these are the same fibers that are more easily relocated to the extrapulmonary sites where asbestos-induced diseases occur (FIGS. 3 and 4).[40–44] In this context Sebastien et al.[42] actually indicated that "lung parenchymal retention is not a good indicator of pleural retention: indeed, there was no relationship between parenchymal and pleural concentrations" (p. 242). They concluded that "asbestos dusts in the parietal pleura were related to type and size: most of the fibres were short chrysotile fibres" (p. 242).

DISCUSSION

Data obtained from tissue regarding asbestos burden are dependent on both the preparation and the analytical procedures used to obtain the data. Asbestos bodies found in tissue sections are a useful indicator of past exposure to longer fibers of asbestos. However, tissue sections are a very insensitive media for identification of asbestos bodies. Asbestos bodies are, with rare exception, formed on amphibole asbestos fibers, and their presence provides no information regarding the overall tissue burden of uncoated asbestos fibers or chrysotile exposure. The assessment of larger pieces of tissue, such as is possible with digestion procedures, in lieu of screening tissue sections; permits a better perspective of actual tissue burden of ferruginous bodies.

The light microscope offers very limited information regarding uncoated fiber burden because the majority of uncoated fibers in tissue are below the limit of resolution of the light microscope and does not provide a means of identification of the type of fibers seen. The scanning electron microscope offers somewhat more information regarding fiber content in tissue but is limited in resolution and the ability to specifically identify the composition of uncoated fibers compared with the capabilities of the analytical transmission electron microscope.

The value of information of fiber burden in tissue is directly related to the quality of the procedures used in fiber isolation and the necessity for the inclusion of short fibers (<5 μm) in the data presentation if overall burden is to be accurately represented. It is evident from the data presented that the short fibers of asbestos make up the majority of fibers in tissue and are the fiber types more likely to be relocated to the extrapulmonary sites where asbestos-induced diseases occur. Their exclusion from any count scheme creates a biased base of information concerning tissue burden because the results preferentially ignore most chrysotile fibrils and all but the thicker and longer amphibole asbestos fibers.

REFERENCES

1. CRAIGHEAD, J.E. & B.T. MOSSMAN. 1982. The pathogenesis of asbestos-associated diseases. N. Engl. J. Med. **306:** 1446–1455.
2. MURAVOV, O.I., W.E. KAYE, M. LEWIN, *et al*. 2005. The usefulness of computed tomography in detecting asbestos-related pleural abnormalities in people who had indeterminate chest radiographs: the Libby, MT, experience. Int. J. Hyg. Environ. Health **208:** 87–99.
3. MCDONALD, J.C., A.D. MCDONALD, P. SEBASTIEN & K. MOY. 1988. Health of vermiculite miners exposed to trace amounts of fibrous tremolite. Brt. J. Ind. Med. **45:** 630–634.
4. ANDERSON, B.A., S.M. DEARWENT, J.T. DURANT, *et al*. 2005. Exposure pathway evaluation for sites that processed asbestos-contaminated vermiculite. Int. J. Hyg. Environ. Health. **208:** 55–65.
5. GAMBLE, J.F., W. FELLNER & M.J. DIMEO. 1979. An epidemiologic study of a group of talc workers. Am. Rev. Respir. Dis. **119:** 741–753.
6. DE VUYST, P., P. DUMORTIER, P. LEOPHONTE, *et al*. 1987. Mineralogical analysis of bronchoalveolar lavage in talc pneumoconiosis. Eur. J. Respir. Dis. **70:** 150–156.
7. ROGGLI, V.L., R.T. VOLLMER, K.J. BUTNOR & T.A. SPORN. 2002. Tremolite and mesothelioma. Ann. Occup. Hyg. **46:** 447–453.
8. CRAIGHEAD, J.E., J.L. ABRAHAM, A. CHURG, *et al*. 1982. The pathology of asbestos-associated disease of the lungs and pleural cavities: diagnostic criteria and proposed grading schema. Arch. Pathol. Lab. Med. **106:** 544–596.
9. TOSSAVAINEN, A., D.W. HENDERSON, J. RANATANEN, *et al*. 1997. Asbestos, asbestosis, and cancer. Scand. J. Work. Environ. Health. **23:** 311–316.
10. DAVIS, J.M.G. 1972. Further observations on the ultrastructure and chemistry of the formation of asbestos bodies. Exp. Mol. Pathol. **13:** 346–358.
11. GOVERNA, M. & C. ROSANDA. 1972. A histochemical study of the asbestos body coating. Br. J. Ind. Med. **29:** 154–159.
12. GROSS, P., J. TUMA & R.T.P. DETREVILLE. 1971. Unusual ferruginous bodies. Arch. Environ. Health **22:** 534–537.
13. GROSS, P., R.T.P. DETREVILLE, L.J. CRALLEY & J.M.G. DAVIS. 1968. Pulmonary ferruginous bodies. Arch. Pathol. **85:** 539–546.
14. HOLMES, A., A. MORGAN & W. DAVISON. 1968. Formation of pseudo-asbestos bodies on sized glass fibres in the hamster lung. Ann. Occup. Hyg. **27:** 301–313.
15. DODSON, R.F., M.F. O'SULLIVAN, C.J. CORN, *et al*. 1985. Ferruginous body formation on a nonasbestos mineral. Arch. Pathol. Lab. Med. **109:** 849–852.
16. DODSON, R.F., M.F. O'SULLIVAN, C.J. CORN, *et al*. 1993. Analysis of ferruginous bodies in bronchoalveolar lavage from foundry workers. Br. J. Ind. Med. **50:** 1032–1038.
17. DODSON, R.F., M.F. O'SULLIVAN, C.J. CORN & S.P. HAMMAR. 1995. Quantitative comparison of asbestos and talc bodies in an individual with mixed exposure. Am. J. Ind. Med. **27:** 207–215.

18. CROUCH, E. & A. CHURG. 1984. Ferruginous bodies and the histological evaluation of dust exposure. Am. J. Surg. Pathol. **8:** 109–116.
19. WILLIAMS, M.G., R.F. DODSON, C.J. CORN & G.A. HURST. 1982. A procedure for isolation of amosite asbestos and ferruginous bodies from lung tissue and sputum. J. Toxicol. Environ. Health **10:** 627–638.
20. DODSON, R.F., M.F. O'SULLIVAN, M.G. WILLIAMS & G.A. HURST. 1982. Analysis of cores of ferruginous bodies from former asbestos workers. Environ. Res. **28:** 171–178.
21. DODSON, R.F., M.F. O'SULLIVAN & C.J. CORN. 1996. Relationships between ferruginous bodies and uncoated asbestos fibers in lung tissue. Arch. Environ. Health **16:** 637–647.
22. CHURG, A. 1989. The diagnosis of asbestosis. Hum. Pathol. **20:** 97–99.
23. DODSON, R.F., M.G. WILLIAMS, J. HUANG & J.R. BRUCE. 1999. Tissue burden of asbestos in nonoccupationally exposed individuals from East Texas. Am. J. Ind. Med. **35:** 281–286.
24. DODSON, R.F., M.F. O'SULLIVAN, D.R. BROOKS & J.R. BRUCE. 2001. Asbestos content of omentum and mesentery in nonoccupationally exposed individuals. Toxicol. Ind. Health **17:** 138–143.
25. DODSON, R.F., S.D. GREENBERG, M.G. WILLIAMS, et al. 1984. Asbestos content in lungs of occupationally and nonoccupationally exposed individuals. J. Am. Med. Assoc. **252:** 68–71.
26. BREEDIN, P.H. & D.H. BUSS. 1976. Ferruginous (asbestos) bodies in the lungs of rural dwellers, urban dwellers, and patients with pulmonary neoplasms. South Med. J. **69:** 401–404.
27. ROGGLI, B.L., P.C. PRATT & A.R. BRODY. 1986. Asbestos content of lung tissue in asbestos associated diseases: a study of 110 cases. Br. J. Ind. Med. **43:** 18–28.
28. LEVIN, J.L., M.F. O'SULLIVAN, C.J. CORN & R.F. DODSON. 1995. An individual with a majority of ferruginous bodies formed on chrysotile cores. Arch. Environ. Health **59:** 462–465.
29. LANGER, A.M., I.J. SELIKOFF & A. SASTRE. 1971. Chrysotile asbestos in the lungs of persons in New York City. Arch. Environ. Health **22:** 348–361.
30. ROOD, A.P. & R.R. STREETER. 1984. Size distributions of occupational airborne asbestos textile fibres as determined by transmission electron microscopy. Ann. Occup. Hyg. **28:** 333–339.
31. MUELLER, P.K. & R.L. STANLEY. 1975. Asbestos Fiber Atlas 9–19. (Anonymous) Washington DC, US, Environmental Protection Agency.
32. LANGER, A.M., R.P. NOLAN & J. ADDISON. 1991. Distinguishing between amphibole asbestos fibers and elongate cleavage fragments of their non-asbestos analogues. *In* Mechanisms of Fibre Carcinogenesis. R.C. Brown, J.A. Hoskins & N.F. Johnson, Eds.: 253–267. Plenum. New York.
33. DODSON, R.F., M.A.L. ATKINSON & J.L. LEVIN. 2003. Asbestos fiber length as related to potential pathogenicity: a critical review. Am. J. Ind. Med. **44:** 291–297.
34. UPTON, A.C., J.C. BARRETT, M.R. BECKLAKE, et al. 1991. Health Effects Institute-Asbestos Research: Asbestos in Public and Commercial Buildings: A Literature Review and Synthesis of Current Knowledge. 4–20. (Anonymous) Health Effect Institute. Cambridge Ma. section 4.4.2.4
35. ASHCROFT, T. & A.G. HEPPLESTON. 1973. The optical and electron microscopic determination of pulmonary asbestos fiber concentrations and its relationship to human pathological reaction. J. Clin. Pathol. **26:** 224–234.

36. POOLEY, F.D. & D.L. RANSON. 1986. Comparison of the results of asbestos fibre dust counts in the lung tissue obtained by analytical electron microscopy and light microscopy. J. Clin. Pathol. **39:** 313–317.
37. CHURG, A. & M.L. WARNOCK. 1980. Asbestos fibers in the general population. Am. Rev. Respir. Dis. **122:** 669–677.
38. DODSON, R.F., M.F. O'SULLIVAN & C.J. CORN. 1993. Technique dependent variations in asbestos burden as illustrated in a case of nonoccupational exposed mesothelioma. Am. J. Ind. Med. **24:** 235–240.
39. DODSON, R.F., M.F. O'SULLIVAN, C.J. CORN, et al. 1997. Analysis of asbestos fiber burden in lung tissue from mesothelioma patients. Ultrastruct. Pathol. **212:** 321–336.
40. DODSON, R.F., M.G. WILLIAMS, C.J. CORN, et al. 1990. Asbestos content of lung tissue, lymph nodes, and pleural plaques from former shipyard workers. Am. Rev. Respir. Dis. **142:** 843–847.
41. DODSON, R.F., M.F. O'SULLIVAN, J. HUANG, et al. 2000. Asbestos in extrapulmonary sites, omentum and mesentery. Chest **117:** 486–493.
42. SEBASTIEN, P., X. JANSON, A. GAUDICHET, et al. 1980. Asbestos retention in human respiratory tissues: comparative measurements in lung parenchyma and in parietal pleura. In Biological Effects of Mineral Fibers. J.C. Wagner, Ed.: 237–246. IARC. Lyon, France.
43. SUZUKI, Y. & S.R. YUEN. 2001. Asbestos tissue burden study on human malignant mesothelioma. Ind. Health **39:** 150–160.
44. SUZUKI, Y. & S.R. YUEN. 2002. Asbestos fibers contributing to the induction of human malignant mesothelioma. Ann. N. Y. Acad. Sci. **982:** 160–176.

Asbestos Ban in India

Challenges Ahead

TUSHAR KANT JOSHI, UTTPAL B. BHUVA, AND PRIYANKA KATOCH

Centre for Occupational and Environmental Health, Lok Nayak Hospital, New Delhi 110002, India

ABSTRACT: **Rapidly industrializing India is described by the International Monetary Fund as a young, disciplined, and vibrant economy with a projected growth of 6.7% for 2005. The total workforce of 397 million has only 7% of workers employed in the organized sector with construction, where asbestos exposure is prevalent, employing 4.4%. The domestic production of asbestos declined from 20,111 tons in 1998–1999 to 14,340 tons in 2002–2003. The imports from Russia and Canada increased from 61,474 tons in 1997–1998 to 97,884 tons in 2001–2002. The production of asbestos cement products went up from 0.68 million tons in 1993–1994 to 1.38 million tons in 2002–2003. The asbestos industry has been delicensed since March 2003. The number of asbestos-based units stood at 32, with the western state of Maharashtra having the largest number. According to official figures, the industry employs 8000 workers. The occupational exposure standard is still 2 fibers/mL, worse still, mesothelioma is not recognized as an occupational disease. The latest cancer registry data have no information on mesothelioma. The health and safety legislation does not cover 93% of workers in the unorganized sector where asbestos exposures are extremely high. Workers remain uninformed and untrained in dealing with asbestos exposure. Enforcement agencies are not fully conscious of the risks of asbestos exposure. Industrial hygiene assessment is seldom carried out and pathologists do not receive training in identifying mesothelioma histopathologically. The lack of political will and powerful influence of the asbestos industry are pushing India toward a disaster of unimaginable proportion.**

KEYWORDS: **asbestos; mesothelioma; occupational safety**

INTRODUCTION

Throughout the world, many adults, and some children, spend most waking hours at work facing a variety of hazards almost as numerous as the different

---

Address for correspondence: Tushar Kant Joshi, Head of Occupational and Environmental Medicine Programme, Centre for Occupational and Environmental Health, Ground Floor B.L. Taneja Block, Lok Nayak Hospital, MAMC New Delhi 110002, India. Voice: 0091/011/23214731; fax: 0091/011/23235574.

e-mail: joshi˙tk@rediffmail.com

types of work. The disease burden from these selected occupational risks amounts to 1.5% of the global burden in terms of disability adjusted life years. The workplace risks are almost entirely preventable. Occupational fatality rates reported in industrializing countries are at least two to five times higher than the rates reported in industrialized countries.[1]

Work-related morbidity and mortality not only result in suffering and hardship for the worker and the family but also add to the overall cost to society through lost productivity and increased use of medical and welfare services. The cost to the society has been estimated at 2–14% of the gross national product in different studies in different countries.[2] India is poised for a rapid growth and the state of occupational health raises some questions about the country's willingness to address issues of serious concern, including those created by the continued growth of asbestos industry.

New Economy and Challenges to Occupational Safety and Health

The New Economy is global, fast, cyber driven, dynamic, networked, and efficient. What makes the New Economy stand out from the past is that it is English dominated, U.S. controlled, sophisticated, and strong. In the middle of all that, India stands out with a very special place in the world! India is young, vibrant, disciplined, and ready to perform in the New Economy. And that is why outsourcing will drive the Indian economy upward according to International Monetary Fund. The International Monetary Fund has projected a 6.7% growth for India in 2005, while slightly trimming its forecast for the current year to 6.4%.[3]

To combat the occupational safety and health (OSH) challenges created by the economic liberalization that started in the early 1990s, the Planning Commission of India set up a working group in the Ministry of Labour for the current 10th Five-Year Plan (2002–2007). This was to demonstrate the willingness of the Indian government to play a major role in improving OSH of the workers in the industries, ports, mines, and in the unorganized sectors. The experts admitted that there was a certain degree of compromise and laxity in overseeing the rules and regulations already framed under various Acts to take care of OSH at the national level, the state level, and at the organizational level. The need for better training of the workers and management was highlighted to improve the OSH management system.[4]

The Union Ministry of Labour, Government of India, in its meeting on June 4, 2001, emphasized the need to integrate existing resources in the area of OSH by focusing on the protection of the workers from the hazards as well as for the elimination of work-related injuries, ill health, diseases, accidents, and deaths. Further, in view of the present scenario of OSH, there was need for formulating, implementing, and for periodical review of the coherent occupational health and safety management policy at the national level for the establishment and promotion of OSH management system in the organization.

Three working subgroups were constituted to prepare a paper for the 10th Five-Year Plan on OSH. The working subgroups were as follows.[5]

Mines

The subgroup headed by Director General of Mines Safety included members from Coal India Ltd., Ministry of Mines, Indian Institute of Miners Health, National Mining Development Corporation, and other organizations.

Industry and Port

The subgroup headed by Director General of Factory Advice Service and Labour Institutes included Director General, National Safety Council, Ministry of Surface Transport, Ministry of Environment and Forests, National Institute of Occupational Health, and State Labour departments.

Unorganized Sector

This sector included occupations, such as agriculture, construction, beedi (rolled tobacco leaf) workers, fish processing, and cashew cracking. This subgroup was headed by Joint Secretary (Industrial Safety and Health) and included members from safety and health institute and state labour departments.

STATE OF OCCUPATIONAL SAFETY AND HEALTH IN INDIA

There is reason to believe that the lack of adequate concern among Indian authorities is responsible for gross underestimation of occupational illness including asbestos-related health problem in India. This is evidenced by the few diseases reported against a much large number expected, as estimated by WHO and independent researchers. The International Labour Office, which collects and publishes global accident rates and supports member states to enhance their recording and notification systems for occupational accidents and diseases, finds the figures on both China and India less reliable, and only limited information is available for the two most important sectors in developing countries: the agriculture and informal sectors.[6] This is despite the fact that India has one of the oldest systems of factory and mine inspectorates dating back to the nineteenth century. The authorities are quick to cite the following laws in defense of Government of India's commitment to protect worker health and safety.

Legislation Relating to OSH in India[7]

1. The Factories Act, 1948, and the State Factories Rules framed thereunder.
2. The Dock Workers (Safety, Health & Welfare) Act, 1986, and the Regulations framed thereunder.

3. The Mines Act, 1952, and the Rules framed thereunder.
4. The Plantation Labour Act, 1951.
5. The Shop & Establishments Act.
6. The Explosives Act, 1884, and the Rules framed thereunder.
7. The Petroleum Act, 1934, and the Rules framed thereunder.
8. The Insecticides Act, 1968, and the Rules framed thereunder.
9. The Indian Electricity Act, 1910, and the Indian Electricity Rules, 1956.
10. The Indian Boilers Act, 1923, and the Indian Boilers Regulations.
11. The Dangerous Machines (Regulation) Act, 1983.
12. The Environment (Protection) Act, 1986, and the Manufacture, Storage & Import of Hazardous Chemicals Rules, 1989 and other Rules framed thereunder.
13. The Indian Atomic Energy Act, 1962.

The plethora of rules have only created confusion, since many do not incorporate elements of OSH, and safety such as exposure standards, and the ones prescribed are either outdated or not relevant to the conditions of work. The nonimplementation of preventive and protective measures has resulted in the perpetuation of hazardous conditions of work and unsafe work practices. An example is that of the construction sector, which is experiencing an unprecedented boom and employs millions of workers. The standards prescribed under the construction workers' act are inadequate to create safe and healthy conditions of work. TABLE 1 refers to Schedule IX of the Indian Building and Construction Workers Act, which includes hazardous processes.[8]

The list does not include processes involving handling, sawing, or repairing of asbestos cement products such as pipes or sheets, etc., which are known to create dangerous conditions of exposures. Damage to asbestos-containing material can result in the release of asbestos fibers that become airborne and are readily inhaled. These fibers can remain in the lungs for long periods and can cause serious lung diseases.[9]

The limit of exposure set in Schedule XII is 2 fibers/mL. For fibers greater than 5 μm in length and less than 3 μm in breadth, with length-to-breadth ratio equal to or greater than 3:1. The Schedule II of the Act lists asbestosis alone, as the notifiable occupational disease and does not include lung cancer and

TABLE 1. Schedule IX hazardous process (Indian Building and Construction Worker Act, 1996)

Works
Roof work
Steel erection
Work under and over water
Demolition
Work in confined space

mesothelioma despite the increasing use of asbestos in construction activities. This works to the advantage of employers who would not accept liability when a construction worker seeks compensation for asbestos-related disability.

The Factories Act, 1948, applicable to the manufacturing units, does slightly better and the Schedule II includes the following exposure limits: Amosite: 0.5 fibers/mL; Chrysotile: 1.0 fibers/mL; Crocidolite: 0.2 fibers/mL.[10]

Why exposure limits for crocidolite are given when the material was banned in 1993 is anybody's guess. The Schedule III on notifiable diseases (occupational diseases) includes asbestosis but not mesothelioma. The Mines Act, 1952, alone recognizes asbestosis, and cancer of the lung or stomach or pleura and peritoneum, i.e., mesothelioma, as a notifiable and occupational disease.

Unless Government of India brings down the exposure limits to 0.1 fibers/mL and evolves a strategy to strictly monitor the workplace and the health of the exposed workers, and takes action against the violators till a ban on asbestos comes into effect, serious damage to health of the workers and the environment would continue to occur.

ASBESTOS PRODUCTION AND USE IN INDIA

The information provided by the Indian parliament about asbestos import, export, and the production of asbestos-based products in India is given in TABLES 2–5.[11]

From the above tables, it is clear that domestic asbestos production is declining but the imports are increasing. It is likely that the tariff reduction on imports has rendered indigenous production uneconomical. It is also obvious that the export of asbestos products amounts to a fraction of imports, suggesting that the bulk of imports are consumed domestically.

As shown in TABLE 4, the production of asbestos cement products is increasing every year in India. The asbestos cement manufacturers are unable to provide any statistics on the morbidity and compensation. This gives the industry an unfair advantage, as the huge profits are diverted to influence decision makers and undertake misinformation campaign describing chrysotile to be safe under conditions of "controlled use."

TABLE 5 includes the information which the minister of Commerce and Industry, Government of India, provided regarding the number of asbestos cement plants in the country in 2003.[12]

TABLE 2. Asbestos production, export, and import in India

Year	1998–1999	1999–2000	2000–2001	2001–2002	2002–2003
Production	20,111	18,550	15,397	11,148	14,340
Export	264	166	391	809	—
Import	76,094	60,583	60,625	97,884	—

NOTE: Figures given are in metric tons.

TABLE 3. Asbestos import in 2001–2002 from various countries

Countries	Import in tons
Russia	36,994
Canada	32,180
Zimbabwe	11,933
Brazil	7,126
Kazakhstan	5,842
Others	3,809
Total	97,884

[Reply to Lok Sabha (lower house of Indian Parliament) starred question No. 685 for answer on May 9, 2003]

(a) The minister admitted no separate information is collected and maintained on the incidence of lung cancer in asbestos industries centrally.
(b) Asbestos and asbestos products manufacturing industries have been delicensed with effect from 11/3/1998.

This has implications for worker safety and health as delicensing results in bypassing rules and regulations. However, asbestos-based industries require prior environmental clearance, irrespective of the size of the plan and the level of investment. The provision of Environmental Impact Assessment Notification, 1994, relating to medical health care of the workers, occupational health, and hygiene have to be adhered to by all asbestos-based units. Further, these units have to comply with various Bureau of Indian Standards safety and health standards for use of asbestos and asbestos products.[13]

The reply of the Government was not convincing when information was sought on its preparedness to handle asbestos-related issues. The concerned minister provided information as given in Box 1.[13]

Box 1: Information provided in the Indian House of Parliament on asbestos by the Minister of State in the Ministry of Labour

TABLE 4. Production of asbestos cement products in India from 1993 to 2003

Year	Production in metric tons
1993–1994	681
1994–1995	758
1995–1996	980
1996–1997	855
1997–1998	924
1998–1999	976
1999–2000	1,302
2000–2001	1,211
2001–2002	1,333
2002–2003	1,387

TABLE 5. State-wise distribution of asbestos cement plant in India

Name of States	No. of chrysotile plants
Assam	1
Andhra Pradesh	3
Gujarat	1
Jharkhand	1
Haryana	1
Karnataka	1
Kerala	1
Madhya Pradesh	2
Maharasthra	9
Orissa	1
Tamil Nadu	6
Uttar Pradesh	1
West Bengal	2
Rajasthan	1
Union Territory of Dadra and Nagar Haveli	1
Total	32

(a) The Government has not received any report from the International Labour Organization (ILO) relating to a recent survey conducted by them, wherein asbestos is believed to be responsible for death of a hundred thousand workers annually across the world.

(b) As per the details collected by Labour Bureau, Shimla, about 8321 workers are engaged in the manufacture of asbestos cement and other cement products.

(c) The manufacture, handling, and processing of asbestos and its products have been one of the industries involving hazardous processes as given in the First Schedule under the Factories Act, 1948. The provision under Chapter IV-A of the Factories Act relating to hazardous processes therefore becomes applicable to this class of industry. The permissible threshold limit of asbestos has also been indicated in the Second Schedule of the Factories Act.

(d) Under Section 87, concerning dangerous operation, the State Governments are empowered to make rules applicable to any factory or call or description of factories on dangerous operation where the State Government is of the opinion that any manufacturing process or operation carried on in a factory exposes any person employed in it to a serious risk of bodily injury, poisoning, or disease.

The supervision of OSH is hampered due to shortage of trained inspectors and lack of industrial hygiene surveillance. Health and safety is a state subject, but as the states are evading tough action against the asbestos-based establishments, the central government has to intervene to protect a large number of workers employed in construction and asbestos units. These workers are at risk of developing asbestos-related disease including mesothelioma and lung cancer leading to an epidemic as is predicted by investigators for the countries that used asbestos in the past. When the various estimates from the studies are extrapolated to include the world population, they project that the asbestos cancer epidemic will cause 5–10 million deaths, past and present.[14] In this conservative estimate, it is assumed that asbestos exposures are going to cease and that the epidemic will run itself out, but the world's production of asbestos, which went down by half in the 1990s, seems to have stabilized at around 2 million tons/year in 2001–2002, and further progress is far from assured.

Employment and Risk from Asbestos Establishments in India

A serious concern in India is the lack of OSH arrangements available to the majority of workers employed in the unorganized sector. The information available through national sample survey for the year 1999–2000 revealed that total employment in the both organized and unorganized sectors in the country was 397 million. Of this, about 28 million were in the organized sector and the balance 369 million in the unorganized sector. According to the information provided by the Registrar General of India, the work participation rate for women was 25.68% in 2001.[15]

According to the report of the Ministry of Labour, Government of India, the employment of workers in the asbestos industries for the years 2001–2003 is shown in TABLE 6.

Asbestos, in India, occurs in the states of Andhra Pradesh, Rajasthan, Bihar, Karnataka, Tamil Nadu, and Manipur. In Bihar, chrysotile asbestos occurs in Singhbhum districts associated with serpentinized dunites and peridodites and is usually between 3.1 and 6.2 mm long. Chrysotile asbestos fibers are short and are between 9.4 and 15.75 mm in length. In Lakshmana mines in Cuddapah district in Andhra Pradesh, chrysotile fibers are between 0.076 and 0.152 mm in length. In India, asbestos as a raw material is received from Canada without any warning and India sends back the finished product to them along with the warning "hazardous product."[16]

At present in India, more than 30 mines are in operation, producing mainly chrysotile and tremolite. The mining and milling and related processes expose the people to the risk of cancer. Women are more affected by the exposure in processing units as compared to men who generally work in mines. Direct and indirect employment in asbestos-related industry and mines constitute a work-

TABLE 6. Employment in asbestos industries for the year 2001–2003

Employment	2001–2002	2002–2003
Per day	215	196
Percentage share of total mines	0.16	0.16

force of around 100,000. Latency period (length of the time between exposure and the onset of diseases) in India is estimated to be 20–37 years. The causes for lung and breathing problems are mainly due to obsolete technology and direct contact with the asbestos products without proper precaution, because in India asbestos are sold without statutory warning.[17]

In another report, the State Pollution Board of Rajasthan admitted that there were 33 asbestos mines in the state and the Board had issued directions under Section 31A of Air Act, 1981 to 29 units. Presently, 25 units were closed or were nonoperational and only 8 units were in operation, out of which 5 were operating after obtaining stay orders from the High Court.[18]

ASBESTOS EXPOSURE IN INDIA

It has been known that whereas asbestosis is typically confined to the workplace, mesothelioma could be contracted from nonoccupational and environmental exposures, making exposure to the asbestos-based products potentially hazardous.[19] The widespread use of asbestos products in India without any awareness and warning is a cause of serious concern.

Another contentious issue has been the inclusion of chrysotile in the list of chemicals requiring Prior Informed Consent (PIC). The 11th session of the Intergovernmental Negotiating Committee (INC-11) for an International Legally Binding Instrument for the Application of the PIC Procedure for Certain Hazardous Chemicals and Pesticides in International Trade met on Saturday, September 18, 2004 in Geneva. The delegates discussed the inclusion of additional chemicals in the interim PIC Procedure. The INC agreed to make parathion and tetraethyl and tetramethyl lead subject to the interim PIC Procedure, but failed to reach consensus on the addition of chrysotile asbestos. India and China indicated their opposition to including chrysotile.[20]

A few studies on asbestos-exposed workers have been undertaken in India. The exposure was studied in mining, milling, and asbestos cement factories. It was found that the exposure was below the limit of 2 fibers/mL in asbestos cement factories and in underground mines where wet drilling was carried out. In open-cast mining also, the fiber lends were found to be lower than 2 fibers/mL. But exposure was very high in milling of asbestos fibers (20–222 fibers/mL).[21]

In another study, airborne asbestos dust concentration and occupational health environment of workers in asbestos products manufacturing unit were monitored, and compared with the standards. Study reveals that overall airborne asbestos concentration in the unit is well within the limit, but the workers, who were exposed to airborne asbestos dust, showed a marked increase in deterioration of lung function as compared to the control population, which was not exposed to this dust. Furthermore, the population, which was exposed to airborne asbestos dust along with other predisposing factors like cigarette smoking, showed a marked deterioration of lung function as compared to the population exposed only to airborne asbestos dust.[22]

The national program for control and treatment of occupational diseases in India found that the prevalence of silicosis was 6.2–34% in mica miners, 4.1% in manganese miners, 30.4% in lead and zinc miners, 9.3% in deep and surface coal miners, 27.2% in iron foundry workers, and 54.6% in slate-pencil workers. Prevalence of asbestosis was extended from 3% in asbestos miners to 21% in mill workers. In textile workers, Byssinosis was as common as 28–47%.[23]

Asbestos-Related Illness: Global and Indian Profile

Dust exposure at global scale still constitutes the most serious occupational exposure. Millions of workers in a variety of occupations are exposed to microscopic airborne particles of silica, asbestos, and coal dust. These particles may not only cause cancer of the lung, trachea, and bronchus but also nonmalignant respiratory diseases, such as silicosis, asbestosis, and pneumoconiosis ("dusty lung").[24] Rate trends in developing countries are mostly unknown, but the magnitude of the problem is substantial (FIG. 1).[25,26]

Asbestos-related diseases, especially mesothelioma, are on the rise in industrialized countries. According to a recent report, deaths from exposure to asbestos are increasing and will peak in the next decade. Annual deaths from mesothelioma among men in Britain will rise to between 1950 and 2450 a year, between 2011 and 2015, compared with 153 deaths in 1968, say researchers writing in the *British Journal of Cancer* (http://www.nature.com/bjc doi:10.1038/sj.bjc.6602307).

In the United States, a peak of about 2000 cases has been estimated. In Australia, the incidence of mesothelioma is expected to peak at about 700 cases a year in 2010, and in the Netherlands, it has been predicted that pleural mesothelioma will peak at about 2028, with up to 900 cases a year. In France, the number of deaths is predicted to reach a peak at 2200 cases a year, sometime after 2020.[77]

There has been anecdotal reporting of asbestos-related malignancies in India. A study reported that in the 25-year period from April 1, 1966 to March 31, 1991, a total of 76,239 surgical biopsies were conduced, 234 of which were from the pleura. Fifteen pleural biopsies of the 234 pleural specimens received were diagnosed as primary pleural mesothelioma, giving an incidence of 0.02% of all biopsy materials and 6.4% of pleural biopsies. The patients were in the 29–65 years age group (mean: 46.5 years); all were males except three, the male-to-female ratio being 4:1. All patients had lived in urban environment, with only two coming from rural areas. The author reported that there was no contact with asbestos.[28] However, this assumption is questionable, since there is nothing to suggest that detailed occupational histories were recorded.

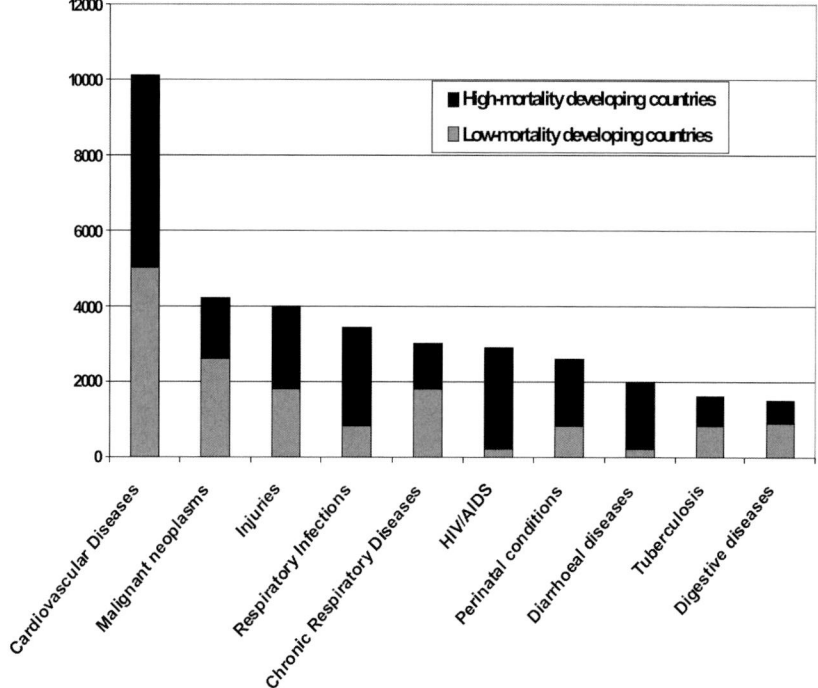

FIGURE 1. Deaths attributable to 10 leading causes in developing countries, 2001.[26]

The latest Indian cancer registry by the Indian Council of Medical Research states that among males, cancers of sites associated with the use of tobacco are the most frequent. Cancer of the lung is numerically the number one cancer. It is the leading site in the metropolitan cities of Delhi and Mumbai, and Bhopal, wherein it constituted about 10% of cancers of all sites. It is the second and third leading site among males in Bangalore and Chennai, respectively. In females also, cancer of the lung is 1 of 10 leading sites in 4 of the 6 registries at Bhopal, Chennai, Delhi, and Mumbai. However, the registry does not provide estimates for pleural mesothelioma.[29] In the absence of a surveillance system for asbestos-related cancers, it is difficult to construct a cancer burden of asbestos exposure.

Like many other countries, India is also plagued by inadequacies in primary reporting, collating, classifying, and publishing of data relating to occupationally caused diseases and injury. From the beginning, there have been opposing forces at work when employer responsibility for injury or disease is an issue. The history of occupational health has been a long struggle by workers to establish accountability against employers who have sought to deny or reduce their liability. An additional problem in India is that of identification of occupational illness due to paucity of trained occupational health physicians. Against

TABLE 7. Notifiable diseases in India as reported by States

States/Uts	Occupational diseases
Assam	–
Gujarat	52
Haryana	–
Himachal Pradesh	–
Madhya Pradesh	–
Orissa	1,888
Punjab	–
Rajasthan	–
West Bengal	23
Chandigarh	–
Delhi	–
Total	1,963

NOTE: Information in respect of other States/Union territories (Uts) is not available.

an estimated requirement of 8000 occupational health physicians, only a few hundred are available in the country at present, raising serious question about the intentions of the government to take on the challenge of preventing occupational diseases seriously. According to WHO and ILO, the problem is that many countries lack the technical and social infrastructure to provide protection for their working and nonworking populations against the hazards of physical, chemical, biological, psychosocial, or ergonomic character. As a result, what is an economic blessing today, may lead to considerable deterioration in the health status of working populations of the developing world tomorrow.[30] The available statistics on occupational diseases (notifiable diseases), compiled by the office of the Director General of Factory Advice Service and Labour Institutes, Mumbai, are presented in TABLE 7. Only three states have reported such incidents.[4]

It is apparent that no serious efforts have been made in India to identify and report occupational diseases. Not surprisingly, the principal argument put forward by the asbestos industries is that asbestos has not caused any disease in India.

According to researchers, the reporting of occupational diseases varies widely and not all reported diseases are compensated. The estimated number of work-related diseases (incidence and mortality) for different regions is given in TABLE 8. These countries are predominantly from the Established Market Economies region. These estimates are based on direct method of calculation. However, indirect methods indicate that the number of new cases of occupational diseases could be approximately 10.7 million per year.[31]

The position statement by the Collegium Ramazzini calling for an international ban on asbestos highlights the risk prevailing in a newly industrializing country like India. According to the statement, about 90% of global asbestos' use today is in asbestos cement construction materials; mainly flat sheet corrugated roofing panels and pipes. Installation, renovation, maintenance,

TABLE 8. Occupational diseases (direct method)

Region	Incidence/year	Mortality/year
EME	837,400–895,500	109,800–113,400
FSE	201,000–206,700	46,800
IND	924,700–1,902,300	121,000
CHN	88,300–2,537,900	8,600–161,500
OAI	711,500–1,463,700	93,100
SSA	537,400–1,105,600	70,400
LAC	407,000–803,000	64,200
MEC	533,000–1,906,600	69,800
WORLD	4,240,700–10,010,800	583,700–704,200

EME = Established Market Economies; FSE = Former Socialist Economies of Europe; IND = India; CHN =China; OAI = other Asia and islands; SSA = sub-Saharan Africa; LAC = Latin America and Caribbean; MEC = Middle Eastern Crescent.

and demolition of these materials give rise to very high exposures for millions of workers and members of the general public every day, all over the world (TABLE 9).[32,33]

ASBESTOS AND SHIPBREAKING INDUSTRY

Of the approximate 45,000 ocean-going ships in the world, about 700 are taken out of service every year. In the early 1970s, shipbreaking was a highly mechanized industrial operation carried out in the shipyards of Great Britain, Taiwan, Mexico, Spain, and Brazil. The rigorous environmental and health and safety standards have shifted it to poorer Asian states. It is estimated that over 100,000 workers are employed at shipbreaking yards worldwide. By the end of the 1990s, 70% were being scrapped in India (FIG. 2).

Most ships being dismantled today were built in the 1970s, prior to the banning of many hazardous substances including asbestos. In Asia, old ships

TABLE 9. Estimated annual incidence of occupational injury and disease in the world (indirect method)[32]

	Number of new cases per year
Injuries	100,688,000
Diseases	–
Pesticide poisoning	109,000
Other poisoning	122,000
Cancer	191,000
Mental disorders	318,000
Pneumoconiosis	453,000
Noise-induced hearing loss	1,628,000
Skin disorders	1,895,000
Chronic respiratory diseases	2,631,000
Musculoskeletal disorders	3,337,000

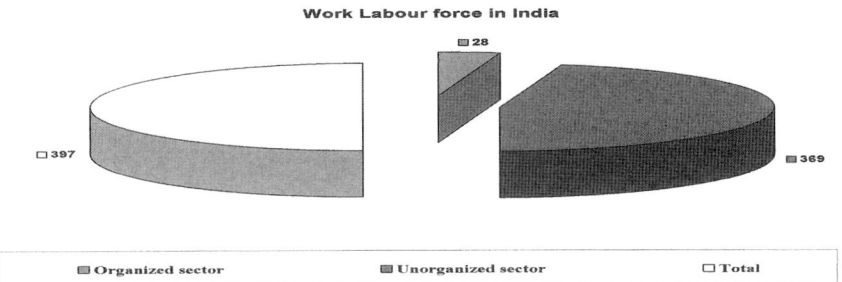

FIGURE 2. Work Labour Force in India.

containing these materials are being cut up by hand, on open beaches under inhumane working conditions. In 1998, Greenpeace sent investigators to the world's largest scrapping site for ocean-going ships in Alang, India. They witnessed appalling worker conditions and mass environmental pollution: workers routinely had to remove carcinogenic asbestos with their bare hands; materials and objects containing asbestos are widely distributed around the country, both as waste and for reuse.[34]

CONCLUSION

OSH in India has come a long way. With the amendment of The Factories Act in 1987, and the addition of a schedule on hazardous processes and hazardous substances, a major step forward was taken. The Chapter IV-A of The Factories Act, 1948 lays down comprehensive provisions for workers' health protection. It was expected to promote identification of occupational illness leading to the creation of a broad canvas of occupational diseases.[35] However, the Indian Ministry of Labour is still not able to provide reliable data on occupational diseases including those caused by asbestos exposure.

The Ministry of Health and Family Welfare, Government of India, has promulgated the National Health Policy (NHP) after a gap of 18 years, ever since the first attempt was made in 1983.[36] Occupational health was one of the components of the NHP 1983 and now also included in the NHP 2002, but very little attention has been paid to mitigate the effect of occupational disease through proper program.[23]

There is no indication at this time that a global ban on asbestos is likely to be accepted by all countries, and international enforcement of a ban on asbestos is unlikely to occur. In developing countries, where little or no protection of workers and communities is taking place, the asbestos cancer epidemic may be even more devastating and may continue indefinitely, which Europe and the other industrialized countries are experiencing now.[37]

The challenges in realizing a ban on asbestos are profound in India. A well-entrenched asbestos lobby with strong links with powers that be, is working hard to discredit the scientists, activists, and nongovernmental organizations (NGOs). Asbestos industry makes huge profits and spends a pittance on worker health protection. The lack of support from trade unions, weak enforcement of existing regulations, and shortage of trained hygienists, and occupational health physicians hamper in determining the asbestos exposure, and identifying morbidity and mortality attributable to such exposure. Even if a ban is effected in the coming few years, the exposure of the past would still result in an epidemic in the future.

Globalization is generating a new hope in India. Many Indians understand that retreating from their globalizing strategy now would be a disaster and result in India's neighborhood rival, China, leaving India in the dust. With prosperity coming to a few, the great majority are simply spectators to this drama. India's problem is not too much globalization, but too little good governance. India needs a political reform revolution to go with its economic one.[38]

REFERENCES

1. World Health Report. 2002. Reducing Risks, Promoting Healthy life. World Health Organization. Geneva.
2. MIKHEEV, M. 1994. New epidemics: the challenge for international health work. *In* New Epidemics in Occupational Health. 27–33. Finnish Institute of Occupational Health. Helsinki.
3. REDDY B. 2004. India's growth projection from IMF is hinged on "New Economy" revolution. India Daily, September 30.
4. Planning Commission, Government of India, Parliament Street New Delhi. Order no. M-13015/9/2000-LEM/LP, April 27, 2001.
5. Minutes of the meeting of the working group on occupational safety and health. Union Ministry of Labour, Government of India, Shram-Shakti Bhawan, New Delhi, June 4, 2001.
6. TAKALA, J. Global estimate of fatal occupational accidents. Sixteenth International Conference of Labour Statisticians, International Labour Office, Geneva, October 1998.
7. The Directorate General of Factory Advice Service and Labour Institutes. 2003. Central Labour Institute, Mankikar Marg, Sion, Mumbai. www.dgfasli.nic.in/working group/intro.htm
8. The Building and Other Construction Workers (Regulation of Employment and Working Conditions) Act 1996.
9. DAVE, S.K. 1993. Asbestosis: epidemiology, clinical manifestations, diagnosis and treatment. Indian J. Clin. Pract. **3:** 40–49.
10. The Factories Act. 1948. Union Ministry of Labour, Government of India, Shram-Shakti Bhawan, New Delhi – 110 001.
11. Members Reference service. Lok Sabha Secretariat, Parliament library and reference and research, documentation and information service, Government of India. Ref. No. 10/Ref./2005; subject-asbestos industry, January 4, 2005.
12. Starred Question No. 685. Asbestos cement plants in the country. Government of India, Ministry of Commerce and Industry, Lok Sabha, May 9, 2003.

13. Starred Question No. 3634. Workers in asbestos cement products. Government of India, Ministry of Labour, Lok Sabha, April 4, 2003.
14. LEIGH, J. 2001. Asbestos-related diseases: international estimates of future liability. In Working Safely in a Changing World. Proceedings of the 5th International Congress on Work Injuries Prevention, Rehabilitation, and Compensation and 2nd Australian National Workers Compensation Symposium (Workcongress 5), March 18–21, 2001, Adelaide, Australia. p. 73. Workcover Corporation South Australia. Adelaide.
15. Ministry of Labour and Employment. Government of India, Annual Report 2004–2005. www.labour.nic.in.
16. SUBRAMANIAN, V. & N. MADHVAN. 2005. Asbestos problem in India. Lung Cancer **49** (Suppl. 1): S9–S12.
17. RAMANATHAN, A.L., V. SUBRAMANIAN. 2001. Present status of asbestos mining and related health problems in India—a survey. Ind. Health **39**: 309–315.
18. Sixth meeting of the Task Force for Cement Industries and Third meeting of the Task Force for Asbestos Based Industries. Central Pollution Control Board, Ministry of Environment and Forests, Government of India, July 12, 2005.
19. TWEEDALE, G. & J. MCCULLOCH. 2004. Chysophiles versus chrysophobes: the white asbestos controversy, 1950s–2004. The history of Science Society. Isis **95**: 239–259.
20. First meeting of the Conference of the Parties to the Rotterdam Convention on the Prior Informed Consent Procedure for Certain Hazardous Chemicals and Pesticides in International Trade, Geneva, Switzerland, September 20–24, 2004.
21. National Institute of Occupational Health Department of Occupational Hygiene. http://www.nioh.org/niohdeptocchygin.htm.
22. GAUTAM, A.K. et al. 2003. Environmental monitoring of asbestos products manufacturing units—a case study. Indian J. Environ. Health **45**: 289–292.
23. National Programme for Control and Treatment of Occupational Diseases. http://www.ndc-nihfw.org/html/Programmes/NationalProgrammeForControl Treatment.htm.
24. National Institute for Occupational Safety and Health. Work-Related Lung Disease Surveillance Report 1999. Division of Respiratory Disease Studies, National Institute for Occupational Safety and Health (NIOSH), Cincinnati, OH, 1999.
25. CHEN, W. et al. 2001. Exposure to silica and silicosis among tin miners in China: exposure-response analyses and risk assessment. Occup. Environ. Med. **58**: 31–37.
26. World Health Report. 2003. Shaping the Future. World Health Organization. Geneva.
27. ABERGAVENNY, R. & B.M.J. DOBSON. 2005. Asbestos related cancer deaths set to rise. Br. Med. J. **330**: 327.
28. KINI, U., S. SHARRIF & J.A. THOMAS. 1992. Primary pleural mesothelioma in south India: a 25-year study. J. Surg. Oncol. **49**: 196–201.
29. Population based Cancer registries. Indian Council of Medical Research, New Delhi, Consolidated Report, 1990–1996. http://icmr.nic.in/ncrp/cancer_regoverview.htm.
30. WHO Press Release. WHO/31, June 8, 1999. WHO. Geneva.
31. JAMES, Leigh, P. MACASKILL, E. KUOSMA & J. MANDRYK. Global burden of disease and injury due to occupational factors. From the National Institute of Occupational Health and Safety, Sydney, Australia, and Finnish Institute of Occupational Health, Helsinki, Finland.

32. CASTLEMAN, B. 2003. "Controlled use" of asbestos. Int. J. Occup. Environ. Health **9:** 294–298.
33. 2005. Eleventh Collegium Ramazzini statement; call for an international ban on asbestos; statement update http://www.collegiumramazzini.org/download/11_EleventhCRStatement(2004).pdf.
34. Shipbreaking: a global environmental, health and labour challenge. A Greenpeace Report for IMO MEPC 44th session, March 2000.
35. JOSHI, T.K. 2005. Occupational health in India—is anyone watching? Bull. Occup. Environ. Health: 6–7.
36. National Health Policy. unpan1.un.org/intradoc/groups/public/documents/APCITY/UNPAN 009630.pdf.
37. LADOU, J. 2004. The asbestos cancer epidemic—commentary. Environ. Health Perspect. **112**(3): 285–290.
38. GEORGE, A.M. 2004. India untouched: the forgotten face of rural poverty. New York Times, Library of Congress Control Number: 2004095498, printed in the United States of America 10 9 8 7 6 5 4 3 2 1.

Ionizing Radiation and Cardiovascular Disease

DAVID G. HOEL

Medical University of South Carolina, Charleston, South Carolina 29401, USA

ABSTRACT: For more than 15 years the A-bomb survivor studies have shown increased noncancer mortality due to radiation exposures. The most prominent cause of this increase is circulatory disease mortality. Although the estimated relative risk is less than for solid cancers (1.2 versus 1.6 per Sv), there are measurable increases in cardiovascular disease mortality at doses greater than 0.5 Sv. The evidence for circulatory diseases in mortality studies of occupational cohorts exposed to external radiation is less compelling. It is generally accepted that atherosclerosis is an inflammatory disease of the arteries and a risk factor for myocardial infarction. Immunological markers for inflammatory disease have been shown to be dose related in A-bomb survivors. Evidence from animal studies reveals increased cardiovascular mortality and arterial endothelial damage from both neutron and, to a lesser extent, gamma exposures.

KEYWORDS: ionizing radiation; neutron; gamma; A-bomb; cardiovascular disease; atherosclerosis; immunological markers; inflammatory disease

INTRODUCTION

Occupational studies of workers exposed to external radiation have focused primarily on cancer effects. Recently, studies of the A-bomb survivors have detected increased mortality from noncancer diseases.[1] This is not surprising since it has been known for some time that high radiation doses used in medical therapy have caused increased risk for myocardial infarction. The earliest occupational studies that included information on circulatory disease mortality were those focusing on medical radiologists. There were increases in cardiovascular disease (CVD) mortality among U.S. radiologists,[2,3] although no effects were seen in studies in the United Kingdom.[4] There were no time trends in cardiovascular mortality among the U.S. radiologists, which would be expected because of the decreasing exposures over time. Recently, U.S. radiology technicians showed a time trend in both stroke and coronary heart disease mortality.[5]

Address for correspondence: David G. Hoel, Medical University of South Carolina, Charleston, SC 29401. Voice: 843-876-1109; fax: 843-876-1126.
e-mail: hoel@musc.edu

TABLE 1. Occupational studies and circulatory disease mortality

Study	Workers (circulatory deaths)	ERR* per Sv	Comments
Medical workers			
U.K. radiologists (Berrington 2001)[4]	2,698 (514)	<0	Time trend in cancer but not in CVD
U.S. radiologists (Matanoski 1975)[2]	30,084	0.2	Time trend in cancer but not in CVD
U.S. radiology techs (Hauptmann 2003)[5]	90,284 (1,070)	0.01–0.42	Time trend in both stroke and CHD
Nuclear workers study			
IARC 3 country study (Cardis 1995)[6]	95,673 (7,885)	0.26	5% workers > 0.2 Sv; 2% workers > 0.4 Sv
U.S. power reactors (Howe 2004)[7]	53,698 (350)	8.3	95% CI: (2.3, 18.2)
Mayak workers (Bolotnikova 1994)[9]	9,373 (749)	0.01	
Chernobyl emergency (Ivanov 2001)[8]	65,905 (1,728)	0.79	Exposures 0 to 0.35 Sv

*Excess relative risk.

The International Agency for Research on Cancer (IARC) study of nuclear workers is the most widely known survey.[6] This mortality study of nuclear workers included almost 8000 deaths due to circulatory disease. The deaths showed a significant association with radiation exposure. Recently, a mortality study of workers at 15 nuclear power reactors in the United States revealed a large relative risk from circulatory disease mortality.[7] The Chernobyl emergency workers[8] were also found to have increased CVD mortality. Mayak workers did not exhibit such an increase.[9] The results from these studies are mixed (see TABLE 1) and have, along with many other studies, been carefully reviewed by McGale and Darby.[10] In their analysis they concluded that there was not yet sufficient evidence to conclude that occupational levels of radiation are causally associated with increased cardiovascular mortality.

CIRCULATORY DISEASE IN A-BOMB STUDIES

Cancer has been the primary focus of the A-bomb studies. This is natural since radiation-induced cancer mortality is the most common late effect of radiation in a survivor population. More recently increased noncancer mortality, attributed to radiation, has been observed (see TABLE 2). Of particular interest are circulatory effects that account for the majority of radiation-induced noncancer mortality.

The adult health study of the A-bomb survivor cohort involves biennial medical evaluations that are conducted on about 20,000 individuals. They

TABLE 2. A-bomb life span studies mortality 1968–1997

Cause of death	ERR per Sv	Total	Excess
Solid cancers*	0.62 (0.49, 0.71)**	6,778	406***
All noncancer diseases	0.14 (0.08, 0.20)	14,459	273
Heart disease	0.17 (0.08, 0.26)	4,477	101
Stroke	0.12 (0.02, 0.22)	3,954	64
Respiratory disease	0.18, 0.06, 0.32)	2,266	57
Digestive disease	0.15 (0.00, 0.32)	1,292	27
Infectious disease	−0.02 (<−0.2, 0.25)	397	−1
All other diseases	0.08 (−0.04, 0.23)	2,073	24

* Recalculated from Life Span Study (LSS) data.
** Excess relative risk (90% confidence interval).
*** Estimated radiation associated deaths.
Source: Preston et al.[1]

were evaluated[11] for disease incidence during the period 1968–1998. Among those individuals exposed at less than 40 years of age, the excess relative risk of a myocardial infarction was estimated to be 0.25 ($P < 0.05$). For subjects examined at younger ages the estimated excess relative risks were greater (1.56 at 30 years, 0.83 at 40 years, and 0.44 at 50 years). The authors presented a quadratic fit to the myocardial infarction data ($P = 0.05$) but indicated that a simple linear response was not significant ($P = 0.10$). They also presented data showing a significant quadratic relationship with hypertension and radiation exposure ($P = 0.03$) and discussed a previous study showing a significantly higher level of total cholesterol among the exposed subjects.

Heart disease mortality for essentially the same period, 1968–1997, also was significantly increased in dose, with TABLE 3 showing that the estimated relative risk for individuals exposed at ages less than 40 years to be 0.19 (90% CI 0.06, 0.34). The risk estimates given in the table indicate that women appear to be more susceptible than men, and those exposed at ages less than 40 years are also more susceptible. Mortality due to stroke yielded similar results.

Heart disease mortality in the A-bomb survivors was further analyzed for a threshold-like response. Although the data was not sufficient for firm conclusions, it appeared that the inclusion of a threshold term in the linear dose response did not at all improve the statistical fit to the data. Also, neither did the inclusion of a quadratic term improve the fit of the dose–response function. In contrast to the incidence data, a simple linear response was better than a purely quadratic response, although not significantly so.

Several known risk factors for circulatory disease have been measured in the A-bomb survivors. Logistic regression analysis of the prevalence of pulse wave velocity (PWV) abnormalities (i.e., PWV > 9 m/s) estimated the odds ratio to be 1.28 ($P = 0.06$) for an exposure of 1 Sv.[12] This noninvasive measure of large artery elasticity and stiffness is a

TABLE 3. A-bomb life span studies: mortality 1968–1997

Group	HEART	
	All Doses	Doses less than 1 Sv
Total	0.17 (0.08, 0.26)*	0.21 (0.06, 0.38)
Total exposed < age 40 years	0.19 (0.06, 0.34)	0.24 (0.01, 0.51)
Female	0.20 (0.07, 0.33)	0.23 (0.04, 0.44)
Female exposed < age 40 years	0.29 (0.09, 0.51)	0.36 (0.04, 0.72)
Male	0.14 (0.01, 0.28)	0.20 (−0.04, 0.46)
Male exposed < age 40 years	0.10 (−0.07, 0.30)	0.09 (−0.24, 0.49)
	STROKE	
	All Doses	Doses less than 1 Sv
Total	0.12 (0.02, 0.22)*	0.10 (−0.05, 0.26)
Total exposed < age 40 years	0.18 (0.04, 0.34)	013 (−0.11, 0.40)
Female	0.13 (0.00, 0.27)	0.17 (−0.03, 0.39)
Female exposed < age 40 years	0.16 (−0.02, 0.37)	0.10 (−0.20, 0.45)
Male	0.11 (−0.03, 0.26)	0.00 (−0.22, 0.25)
Male exposed < age 40 years	0.22 (0.01, 0.49)	0.17 (−0.21, 0.62)

*Excess relative risk (90% likelihood bounds).

predictor of CVD risk whether it is causative or a marker of disease already present.[13] Another CVD risk measure is aortic arch calcification with an estimated relative risk of 1.17 ($P < 0.01$) at 1 Sv with the relative risk being greater in women than in men.[14]

Atherosclerosis is an inflammatory disease of the arteries that can lead to ischemia of the heart and brain resulting in infarction.[15] The chronic inflammation of the artery is initiated by endothelial dysfunction that may be caused, for example, by elevated low-density lipoproteins, free radicals from cigarette smoking, hypertension, diabetes, and infection (e.g., herpes viruses). The chronic inflammation results in increased levels of macrophages and lymphocytes, whose activation causes the release of cytokines and growth factors leading to increased damage, local necrosis, and altered blood flow.

The macrophages and lymphocytes involved in chronic inflammation in atherosclerosis have been shown to be regulated by T cell cytokines (interferon-γ [IFN-γ], tumor necrosis factor-α [TNF-α], interleukin-2 [IL-2], -4, and -10). Also, the inflammatory cytokine IL-6 induces the liver's synthesis of acute-phase plasma, proteins such as C-reactive protein (CRP), which is thus a surrogate for IL-6. Increases in CRP are a risk factor for inflammation and useful in predicting susceptibility to myocardial infarctions, stroke, and peripheral arterial disease. In a prospective study of baseline values of IL-6, men in the highest quartile had a relative risk of 2.3 (95% CI 1.3, 4.3) for a subsequent myocardial infarction compared with those in the lowest quartile.[16]

TABLE 4. Radiation associated immunological markers of inflammation: A-bomb survivor clinical studies

- Leukocyte counts
- T cell cytokines
   IL-6, IL-10, IFN-γ,
   TNF-α
- Acute phase proteins
   α1-globulin, α2-globulin, CRP
- B cell antigen receptor
   IgA, IgM
- Other markers
   Erythrocyte sedimentation rate, sialic acid

Several studies of the late effects of radiation and levels of inflammatory markers have been reported for the A-bomb survivors. It has been shown that IL-6 levels were increased by about 10% per Sv and CRP levels by 28% per Sv after adjusting for known risk factors (age, gender, smoking, body mass, cholesterol, high-density lipoprotein cholesterol, triglycerides, systolic blood pressure, etc.)[17]

In a study of persistent subclinical inflammation, leukocyte counts, erythrocyte sedimentation rates, α1-globulin, α2-globulin, and sialic acid levels were measured.[18] All showed a positive association with radiation exposure. The only other marker measured in the study was the neutrophil count that was increased with radiation dose, but not significantly so. Recently, inflammatory biomarker levels (TNF-α, IL-10, IgG, IgM, IgA, IgE, IFN-γ) as well as erythrocyte sedimentation rates were measured among the A-bomb survivors.[19] All of these markers were positively associated with radiation dose except for IgE (see TABLE 4) and IgG. Both Neriishi and Hayashi, who carried out these studies, estimate that the effect of 1 Sv radiation exposure is equivalent to an increase in immunological age of about 9–10 years.

Neriishi[18] further hypothesizes that the presence of auto-antibodies and T cell immune impairment[20] may have led to persistent inflammation. Thus immunological damage by radiation may be the cause of late CVD effects. Both bone marrow and the thymus are quite radiogenic and the observed immunological effects were being measured 40–50 years after the radiation exposure. This suggests that permanent genetic damage to lymphocyte progenitor cells may have occurred.

ANIMAL STUDIES

There have been relatively few animal studies on the effects on radiation-induced circulatory disease (e.g., see O'Connor[21] for a review of cerebral vasculature). Some of these studies involve very high doses, such as 10 Gy and greater, which are relevant to radiation exposures used in medical therapy, but not for occupational exposures.[22] There was a series of studies that evaluated the pathology of the coronary arteries and heart in female B6CF$_1$ mice after receiving whole body exposures to gamma rays or fission neutrons.[23–26] In one of these studies[25] degenerative changes in the coronary arteries or aorta began appearing 3–6 months after irradiation and became more severe at 2 years after exposure. The effects were basically smooth muscle degeneration in the medial layer. Some arterial plaques were observed that appeared to be similar to lesions in humans with CVD. However, these lesions in the mice contained no lipids. For mice that received a total of 2.4 Gy of whole body neutron exposure in 24 equal weekly fractions, there was extensive degeneration of the coronary arteries. What was especially interesting was that a second group of mice receiving the same total exposure, but as a single dose, had less tissue damage. The acute 2.4 Gy-exposed mice were not as severely affected as another experimental group that received 0.8 Gy of fractionated neutron exposures. There was relatively little smooth muscle degeneration in mice in the 0.8 Gy acute neutron exposure group. The lowest exposure group of 0.2 Gy of fractionated neutron exposure showed effects that were also greater than those of the 0.8 Gy acutely exposed group. For gamma exposures at 7.88 Gy, there was evidence of effects of acute exposures, but in contrast to the neutron exposures, there were only minimal effects with fractionation of these doses. This evidence results in a very high estimated relative biological effectiveness (RBE) for fractionated exposures. The authors estimated that for a fractionated neutron dose of 0.2 Gy, the RBE would range between 40 and 130. For acute doses, they estimated the RBE to be less than 10 at the 0.8 Gy level of neutron exposure.

A second experiment[24] measured the volume of degenerated smooth muscle cells of coronary arteries of mice 15 months after exposure to gamma rays and several types of high-energy particles. The average linear energy transfer (LET) estimates were 0.8, 80, 150, and 600 keV/μM for ^{60}Co, ^{12}C, ^{20}Ne, and ^{40}Ar, respectively. The data are limited for dose–response analysis but a simple additive to background linear fit with an experimental cell-killing term (see FIG. 1) indicated that the high-energy particles have a similar estimated linear effect. The ratio of the linear terms, which corresponds to a low-dose RBE estimates, is about 4 for this example (see FIG. 1 for the parameter values). Using a simpler and more realistic dose–response function, such as the square root of dose, the fit to the data is poorer and the estimated RBE values range from about 2.5 to 6.5 with the larger data set (^{20}Ne) giving an estimate of 4. Now, from the neutron study using acute and fractionated exposures one must assume that the fractionated exposures of high-energy particles may cause considerably more damage than the acute exposures in this study.

The lifetime studies of the same B6CF$_1$ mouse showed CVD as a contributing factor to mortality.[27] These studies show increased effects with both neutron and gamma exposures with a greater relative risk in female mice and for neutron exposures. The estimated RBE for neutrons in these studies have yet to be estimated but are likely to be similar to the estimated RBE of 4

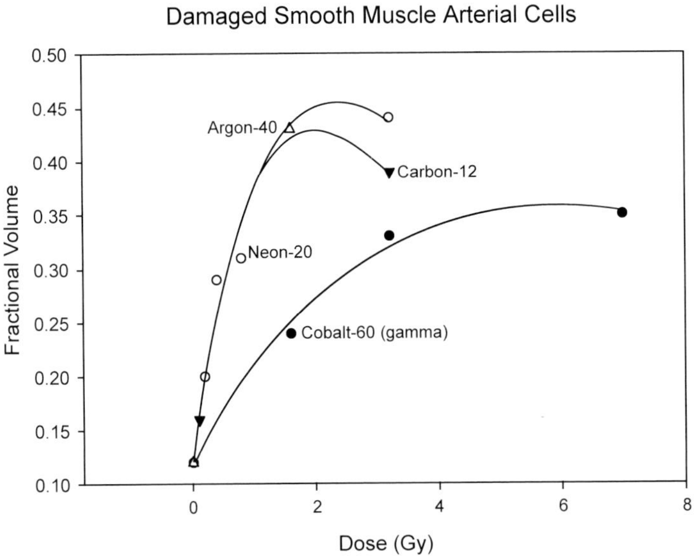

Damaged Smooth Muscle Arterial Cells

FIGURE 1. Plot of the volume of coronary artery degeneration in B6CF$_1$ mice 15 months after irradiation by charged particles or ^{60}Co gamma photons. The fitted curves are $0.12 + \alpha \cdot d \cdot \exp(-\beta \cdot d)$ where d is the exposure in Gy. The estimated parameters α and β are (Cobalt-60: 0.11, 0.17; Neon-20: 0.39, 0.43; Carbon-12: 0.42, 0.50). The data are from the paper by VV Yang and EJ Ainsworth.[24]

observed in the previous study of smooth muscle damage using other high-energy particles.

CONCLUSIONS

There is reasonably good evidence that at doses greater than 0.5 Sv, ionizing radiation is a risk factor for circulatory disease in humans. For animals, the experimental doses at which CVD effects have been observed are generally greater. The mechanism of this effect is not clear. It may simply involve direct damage to the arterial endothelial cells. If this is the case then it could be that the effect does not occur at low doses. Also, in using animal models for humans it must be noted that the pathophysiology of artherosclerosis may differ particularly due to the effects of lipids in the diet.[27]

The more interesting possibility is that damage to the immune system results in late inflammatory reactions leading eventually to atherosclerosis. If this is the case, then the effects of radiation may occur at lower doses, and there may be increased risks for other inflammatory conditions besides circulatory disease.

The difficulty in determining the likelihood of low dose cardiovascular effects in contrast to cancer is that the relative effect is less than that of radiation-induced cancer and the latency period in both animals and humans appears to exceed that of cancer. What is clear is that high LET radiation is likely a greater risk for cardiovascular mortality than either gamma or X ray. In the case of neutrons, the effects of fractionated exposures are greater than those of acute exposures. This leads us to conclude that prolonged exposure to high-energy particles will likely well exceed the risk of acute gamma exposures observed in the A-bomb survivors for CVDs.

ACKNOWLEDGMENTS

This work was supported by NASA grant NAG9-1518.

This report makes use of data obtained from the Radiation Effects Research Foundation (RERF) in Hiroshima, Japan. RERF is a private foundation funded equally by the Japanese Ministry of Health and Welfare and the U.S. Department of Energy through the U.S. National Academy of Sciences. The conclusions in this report are those of the author and do not necessarily reflect the scientific judgment of RERF or its funding agencies.

REFERENCES

1. PRESTON, D.L., Y. SHIMIZU & D.A. PIERCE. 2003. Studies of mortality of atomic bomb survivors. Report 13: solid cancer and noncancer disease mortality: 1950–1997. Rad. Res. **160:** 381–407.
2. MATANOSKI, G.M., R. SELTSER, P.E. SARTWELL, et al. 1975. The current mortality rates of radiologists and other physician specialists: specific causes of death. Am. J. Epidemiol. **101:** 199–210.
3. MATANOSKI, G.M. 1981. Risk of cancer associated with occupational exposure in radiologists and other radiation workers. *In* Cancer Achievements, Challenges, and Prospectives for the 1980's. J.H. Burchenal & H.F. Oettgen, Eds.: 241–254. Grune and Stratton. New York, NY.
4. BERRINGTON, A., S.C. DARBY, H.A. WEISS, et al. 2001. 100 years of observation on British radiologists: mortality from cancer and other causes 1897–1997. Br. J. Rad. **74:** 507–519.
5. HAUPTMANN, M., A.K. MOHAN, M.M. DOODY, et al. 2003. Mortality from diseases of the circulatory system in radiologic technologists in the United States. Am. J. Epidemiol. **157:** 239–248.
6. CARDIS, E., E.S. GILBERT, L. CARPENTER, et al. 1995. Effects of low doses and low dose rates of external ionizing radiation: cancer mortality among nuclear industry workers in three countries. Rad. Res. **142:** 117–132.
7. HOWE, G.R., L.B. ZABLOTSKA, J.J. FIX, et al. 2004. Analysis of the mortality experience amongst U.S. nuclear power industry workers after chronic low-dose exposure to ionizing radiation. Rad. Res. **162:** 517–526.
8. IVANOV, V.K., A.I. GORSKI, M.A. MAKSIOUTOV, et al. 2001. Mortality among the Chernobyl emergency workers: estimation of radiation risks (preliminary analysis). Health Phy. **81:** 514–521.

9. BOLOTNIKOVA, M.G., N.A. KOSHURNIKOVA, N.S. KOMLEVA, *et al.* 1994. Mortality from cardiovascular diseases among male workers at the radiochemical plant of the 'Mayak' complex. Sci. Total Environ. **142:** 29–31.
10. MCGALE, P. & S.C. DARBY 2005. Low doses of ionizing radiation and circulatory diseases: a systematic review of the published epidemiological evidence. Radiat. Res. **163:** 247–257.
11. YAMADA, M., F.L. WONG, S. FUJIWARA, *et al.* 2004. Noncancer disease incidence in atomic bomb survivors, 1958–1998. Radiat. Res. **161:** 622–632.
12. UEDA, H. 1995. Arteriosclerosis in the atomic-bomb survivors. RERF Update **7:** 6–7.
13. SAFAR, H., J.J. MOURAD, M. SAFAR, *et al.* 2002. Aortic pulse wave velocity, an independent marker of cardiovascular risk. Arch Mal Cover Vaiss **95:** 1215–1218.
14. KAWAMURA, S., K. KODAMA, Y. SHIMIZU, *et al.* 1992. Prevalence of aortic arch calcification in the AHS population. Nagasaki Med. J. **67:** 474–478.
15. ROSS, R. 1999. Atherosclerosis—an inflammatory disease. N. Engl. J. Med. **340:** 115–126.
16. RIDKER, P.M., N. RIFAI, M.J. STAMPFER, *et al.* 2000. Plasma concentration of interleukin-6 and the risk of future myocardial infarction among apparently healthy men. Circulation **101:** 1767–1772.
17. HAYASHI, T., Y. KUSUNOKI, M. HAKODA, *et al.* 2003. Radiation dose-dependent increases in inflammatory response markers in A-bomb survivors. Int. J. Radiat. Biol. **79:** 129–136.
18. NERIISHI, K., E. NAKASHIMA & R.R. DELONGCHAMP. 2001. Persistent subclinical inflammation among A-bomb survivors. Int. J. Radiat. Biol **77:** 475–482.
19. HAYASHI, T., Y. MORISHITA, Y. KUBO, *et al.* 2005. Long-term effects of radiation dose on inflammatory markers in atomic bomb survivors. Am. J. Med. **118:** 83–86.
20. AKIYAMA, M. 2000. Late effects of radiation on the human immune system: an overview of immune response among the atomic-bomb survivors. Int. J. Radiat. Biol. **68:** 497–508.
21. O'CONNOR, M.M. & M.R. MAYBERG. 2000. Effects of radiation on cerebral vasculature: a review. Neurosurgey **46:** 138–151.
22. LAUK, S., Z. KISZEL, J. BUSCHMANN, *et al.* 1985. Radiation-induced heart disease in rats. Int. J. Radiat. Oncol. Biol. Phys. **11:** 801–808.
23. STEARNER, S.P., V.V. YANG & R.L. DEVINE. 1979. Cardiac injury in the aged mouse: comparative ultrastructural effects of fission spectrum neutrons and gamma rays. Radiat. Res. **78:** 429–447.
24. YANG, V.V. & E.J. AINSWORTH. 1982. Late effects of heavy charged particles on the fine structure of the mouse coronary artery. Radiat. Res. **91:** 135–144.
25. YANG, V.V., S.P. STEARNER & E.J. AINSWORTH. 1978. Late ultrastructural changes in the mouse coronary arteries and aorta after fission neutron or 60Co gamma irradiation. Radiat. Res. **74:** 436–456.
26. YANG, V.V., S.P. STEARNER & S.A. TYLER. 1976. Radiation-induced changes in the fine structure of the heart: comparison of fission neutrons and 60Co gamma rays in the mouse. Radiat. Res. **67:** 344–360.
27. MURROS, K.E. & J.F. TOOLE. 1989. The effect of radiation on carotid arteries. A review article. Arch. Neurol. **46:** 449–455.

Power-Frequency Electric and Magnetic Fields in the Light of Draper *et al.* 2005

JOHN SWANSON,[a] TIM VINCENT,[b] MARY KROLL,[b] AND GERALD DRAPER[b]

[a]*National Grid, London, WC2N 5EH, United Kingdom*

[b]*Childhood Cancer Research Group, Oxford, OX2 6HJ, United Kingdom*

ABSTRACT: Power-frequency electric and magnetic fields are produced wherever electricity is used; exposure is ubiquitous. Epidemiologic studies find an association between children living in homes with the highest magnetic fields and childhood leukemia, but bias is a possible alternative to a causal explanation. A new study, Draper *et al.*, looks at residence close to high-voltage power lines, one source of exposure to such fields, and its design avoids any obvious bias. It finds elevated childhood leukemia rates, but extending too far from the power lines to be straightforwardly compatible with the existing literature. This leads to an examination of alternative explanations: magnetic fields, other physical factors, such as corona ions, the characteristics of the areas power lines pass through, bias, and chance. The conclusion is that there is currently no single preferred explanation, but that this is a serious body of science that needs further work until an explanation is found.

KEYWORDS: childhood leukemia; magnetic fields; power lines; epidemiology

INTRODUCTION

There are serious health concerns about the electric and magnetic fields produced by the distribution and use of electricity. These fields are usually categorized as a form of radiation, nonionizing radiation. Most of the radiological community is concerned with ionizing radiation and regards nonionizing radiation as something of a fringe subject. This article surveys the existing knowledge on power-frequency electric and magnetic fields (EMFs) and in particular updates that evidence in the light of a new study by Draper *et al.*[1] published in the *British Medical Journal* in June 2005. A focus of the present article is to address whether the subject of power-frequency fields deserves serious scientific attention or not.

Address for correspondence: John Swanson, National Grid 1-3 Strand, London, WC2N 5EH, UK. Voice: +44 20 7004 3134; fax: +44 20 70043131.
 e-mail: john.swanson@physics.org

The evidence that there may be health effects is driven by epidemiology and this article therefore concentrates on epidemiology. The agent under consideration is power-frequency electric and magnetic fields, principally the magnetic fields. These fields are often collectively referred to as EMFs, but are in fact separate entities that, while both originating from the generation, distribution, and use of electricity, are not particularly well correlated and need treating separately. The common categorization as nonionizing radiation reflects the fact that these fields are part of the electromagnetic spectrum. However, at these extremely low frequencies, radiation in the scientific sense, where the electric and magnetic fields are coupled together in a way that propagates through space, is in fact negligible.

Exposure comes from many sources, but most epidemiologic studies of childhood leukemia have concentrated on assessing the fields in the home because this is the major source of exposure for young children. In the general volume of the home, fields come mainly from low-voltage electricity supply wiring (sometimes including house wiring but usually the distribution circuits supplying the home) and, in a minority of homes, higher-voltage power lines outside the home. Magnetic fields are also produced by domestic electrical appliances, but these fields are generally localized to less than a meter from the appliance.

THE PRESENT POSITION

Electric and magnetic fields, at values above a few tens of kilovolts per meter (kV/m) and above a few millitesla (mT), have recognized acute effects; they induce fields in the body, which can interfere with nerves.[2] Exposure limits such as International Commission on Non-Ionizing Radiation Protection (ICNIRP)[3] are designed to protect the population from these effects. This article concerns the possibility that much lower levels of magnetic field, of order 1 μT, might cause disease. The evidence is strongest for magnetic fields causing childhood leukemia; if they cause that they may cause other health effects as well, but if the evidence for causation of childhood leukemia is judged too weak, then the evidence currently available on other health effects is likely also to be judged too weak.

The first epidemiologic study of childhood cancer and EMFs was published in 1979.[4] Since then there have been over 20 studies, of variable quality. Early studies are now considered to be weak in terms of either case and control selection, exposure assessment, size, or all three. Later studies, however, while not without problems, are considerably better.

The present position on magnetic fields and the epidemiology of childhood leukemia is best summarized by the pooled analysis of Ahlbom et al.[5] (TABLE 1). This included the data from only those nine studies that met specified quality criteria, principally that they had exposure assessments that involved a 24 h or longer assessment of the field in the home. A different pooled analysis

TABLE 1. Main results from Ahlbom et al.[5] (adapted from table 3 in that paper). Relative risks and numbers of subjects by exposure category, leukemia. Reference level <0.1 μT

Study	Relative risks (and 95% confidence intervals)			Numbers of subjects ≥0.4 μT	
	0.1–<0.2 μT	0.2–<0.4 μT	≥0.4 μT	Observed	Expected
Measurement studies					
Canada	1.29	1.39	1.55	13	10.3
Germany	1.24	1.67	2.00	2	0.9
New Zealand	0.67	4 cs/0 ct	0 cs/0 ct	0	0
UK	0.84	0.98	1.00	4	4.4
USA	1.11	1.01	3.44	17	4.7
Calculated field studies					
Denmark	2.68	0 cs/8 ct	2 cs/0 ct	2	0
Finland	0 cs/19 ct	4.11	6.21	1	0.2
Norway	1.75	1.06	0 cs/10 ct	0	2.7
Sweden	1.75	0.57	3.74	5	1.5
Summary					
Measurements studies	1.05 (0.9–1.3)	1.15 (0.9–1.5)	1.87 (1.1–3.2)	36	20.1
Calculated field studies	1.58 (0.8–3.3)	0.79 (0.3–2.3)	2.13 (0.9–4.9)	8	4.4
All studies	1.08 (0.9–1.3)	1.11 (0.8–1.5)	2.00 (1.3–3.1)	44	24.2

cs = cases; ct = controls.

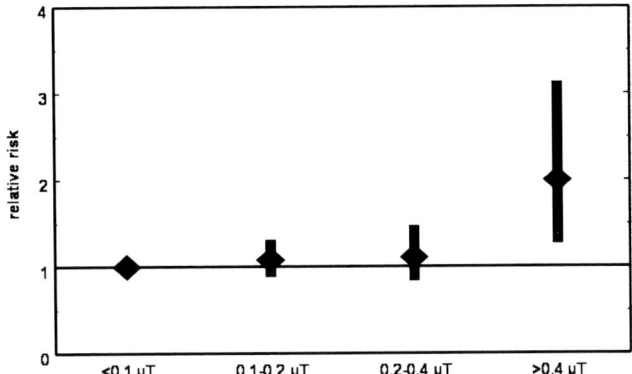

FIGURE 1. Relative risks for leukemia from Ahlbom et al.[5] with 95% confidence intervals.

conducted at a similar time,[6] including all studies for which data were available regardless of quality, found similar results.

The reference category in Ahlbom et al. is a field in the home of less than 0.1 µT, and for the top exposure category, greater than 0.4 µT, there is a relative risk of 2.00 (FIG. 1). This is statistically significant ($P = 0.002$) and based on 44 exposed cases (the study had 3247 cases in total), so criticisms of being based on small numbers would seem unfounded. The intermediate exposure categories have relative risks around 1.1, nonsignificant, and are usually described as being near the no-effect level, implying that the risk appears only in the highest exposure category. In fact, the data also appear to be consistent with a progressive increase in risk, which is probably more plausible.

Since that pooled analysis, one major further study[7] and some smaller studies have appeared with similar results, strengthening the conclusion that there is an association in the study populations between childhood leukemia and EMF exposure. However, this does not establish causation; alternative explanations need to be considered.

Chance, as already discussed, seems highly unlikely, but logically is always possible, and the less credible any alternative explanations are, the more credible chance becomes.

Confounding is a possibility. However, no meaningful confounding has yet been identified in the studies. Most have adjusted for factors such as socioeconomic status (SES), but usually without greatly changing the result. In the mid-1990s, traffic density (which is associated with exposure: magnetic fields are higher on busier roads) was investigated, but without establishing that it explains the results. A recent research area is the small contact voltages in homes, which are produced by similar sources to magnetic fields and are therefore associated with them.[8]

The most likely noncausal explanation would be bias. The individual studies in the Ahlbom et al. pooled analysis fall into two categories: four studies in the Nordic countries, which calculated exposure from external power lines, and five studies in the rest of the world, which required entry into the home to measure the field. The Nordic studies are based on national population registries so were able to select seemingly unbiased control groups, and did not require any subject participation; they are therefore probably free from control bias. The other studies, however, which do require subject participation, suffer from incomplete and differential case and control participation, generally biasing the control group to higher SES. In addition, while some of these studies used unbiased sources to select the control group, others used random-digit telephone dialing or insurance records, potentially creating a socio-economic bias even before participation is considered.

It is undisputed that the control groups in this category of study are biased. It is also undisputed that exposure can vary with SES (usually being lower in more affluent households). Opinions differ as to the extent to which adjusting for some measure of SES (which most studies have done) adjusts adequately for any bias in exposure. The limited evidence available[9] suggests that the direction of the bias is such as to produce an exaggerated estimate of the relative risk; when some of the subjects who had been excluded from the original study were included in a subsequent partial reanalysis, the relative risk reduced. However, there is no proof that this bias actually accounts for the finding of the pooled analysis (or accounts for enough of it to suggest that the remainder could more plausibly be due to chance). Further, this bias does not affect the Nordic studies. When pooled, these show similar results to the other studies (see TABLE 1), an argument against bias as the explanation. However, these Nordic studies as a group on their own are less statistically significant (in fact $P = 0.04$). So if the unbiased Nordic studies are attributed to chance, and bias is accepted as the explanation of the other studies, a noncausal explanation is possible.

At first sight, this is not as natural an explanation of the epidemiology as a causal explanation. However, the causal explanation is not without problems as well. One is the lack of support from laboratory experiments.[10] The other is the absence of a plausible mechanism. These factors make a causal explanation less likely, and it is largely this that motivates the search for noncausal explanations, such as bias. An extreme view is that mechanistic arguments show that any effects of fields at these relatively low levels are impossible; the experimental evidence confirms the absence of effects; and therefore the epidemiologic evidence must be flawed, it is just a case of identifying the flaw.

A more sensible view is that the epidemiology does not establish a risk, but it certainly poses a serious question. It was the epidemiologic evidence, specifically the Ahlbom *et al.* pooled analysis, that led the International Agency for Research on Cancer (IARC) to classify magnetic fields as "possibly carcinogenic" in 2001.[10] Specifically, IARC classified the epidemiologic evidence on childhood leukemia as "limited" but the epidemiologic evidence on all other cancers and the animal evidence as "inadequate." If there were robust supporting evidence from the laboratory or if there were an accepted mechanism, it is arguable that magnetic fields could be regarded as an established carcinogen.

If there were a causal effect for childhood leukemia in homes with magnetic-field exposures above 0.4 µT, the number of attributable cases would be small, because the prevalence of such homes is low. In the UK, where only 0.4% of homes have that field, the attributable cases from fields above 0.4 µT are estimated to be two per year or around half a percent of all cases. If the much smaller relative risks in the intermediate exposure categories in the Ahlbom *et al.* pooled analysis are also included, the number of attributable cases would roughly double; and if exposure misclassification means any true relative risk is diluted in the studies available, it could be larger again but would still account for only a small number of the over 400 cases of childhood leukemia that occur per year in the UK.

However, for those children who are exposed, a relative risk of 2 (from any source) for childhood leukemia would be an absolute attributable risk of five per hundred thousand per year. In the UK, a general framework for society's toleration of risk is set by the Health and Safety Executive.[11] In this framework, a risk of death of 5/100,000 per year would be firmly in the "reduce if possible" category and close to the upper limit of tolerability for the general public. Not all cases of childhood leukemia are fatal (the current 5-year survival rate is about 80%) so the figures are not directly comparable. Nonetheless, for a disease as serious as childhood leukemia, the Health and Safety Executive framework may give a guide as to society's likely toleration of any risk.

DRAPER *et al.* 2005

Into this existing background of epidemiologic evidence relating to magnetic fields comes a new study by Draper *et al.*[1] published in the *British Medical Journal* in June 2005 to the accompaniment of intensive though short-lived media attention. This study was led by the Childhood Cancer Research Group (CCRG) at the University of Oxford with collaboration from National Grid, the company that owns and operates the electricity transmission system. Because of the public scepticism of research involving industry, elaborate lengths were taken to protect the independence of the study. In the event, considering that the study was a positive one, "industry fix" was one accusation not levelled.

The study concerned domicile at birth in relation to power lines. It included cases of childhood cancer diagnosed in England and Wales from 1962 to 1995 from the National Registry of Childhood Tumors held by the CCRG. The National Registry contains 33,000 cases for the relevant period, 29,081 of which were available for these analyses, including 9700 cases of leukemia (some cases could not be included for various reasons such as adoption or insufficient information on place of birth). One control per case was selected from birth registers, matched on age, sex, and birth registration district. Registration for childhood cancer, and leukemia in particular, is virtually complete, certainly for the later part of the period, as is birth registration. Cases and controls were then compared for the proximity of the address at birth to overhead transmission lines at 275 kV and 400 kV using grid references.

The study was originally conceived as a test of whether magnetic fields cause childhood cancer. The principle in this study, just as in the Nordic studies, is that in a home close enough to a power line, that power line is the dominant source of magnetic field. Therefore, exposure can be estimated from the power line alone, and can be calculated from appropriate records. Measurements, with the need for subject participation and hence the scope for bias, are unnecessary. The average distance for the magnetic field from a power line to fall to 0.1 μT (the reference category in the Ahlbom *et al.* pooled analysis, which was also chosen as the reference category in Draper *et al.*) is 100–200 m. But it is theoretically possible to get 0.1 μT as far as 400 m under unusual circumstances. Draper *et al.* allowed for a further margin of uncertainty and examined all subjects within 600 m. In practice though, in homes at these greater distances the field from the power line is much less than the field from other sources; hence the assumption that the power line is the dominant source of exposure breaks down. Thus, a study conceived as a test of magnetic fields actually included homes so far from the power lines that it is hard to see how the fields from the line could be significant.

In Draper *et al.*, for the analysis of proximity to power lines, the reference category became greater than 600 m, and this analysis of proximity is presented as an investigation in its own right, not just as a surrogate for exposure to fields. Only this proximity analysis has been presented so far. Publication of the calculated fields is delayed while awaiting cross-checking with the only directly comparable study.[12]

The results of the study are summarized in Table 2 and FIGURE 2. For central nervous system tumors and "other" tumors the Draper *et al.* results did not suggest any effect. But for childhood leukemia, the relative risk in every 100-m distance band within 600 m was elevated. The *P* value of a trend for risk with the reciprocal of distance was statistically significant at 0.01. This *P*-for-trend is the preferred method of judging the statistical significance of these results, rather than confidence intervals for the risk estimate in each separate band, which could be made arbitrarily larger or smaller by choosing more or fewer bands.

POSSIBLE EXPLANATIONS

Just as with the previous epidemiology, the results given by Draper *et al.* could be due to chance. The *P* value of 0.01 makes this superficially unlikely. However, Draper *et al.* point out (and other commentators have subsequently reinforced) that the leukemia controls differ in their distribution of distance to power lines from the remaining controls. If the leukemia cases are compared

TABLE 2. Main results from Draper et al.[1] (adapted from table 2 in that paper). Distance of birth address from nearest National Grid line for cases and controls in each diagnostic group, and estimated relative risk (RR)

Distance (m)	Leukemia			CNS/brain tumors			Other diagnoses		
	Cases	Controls	RR	Cases	Controls	RR	Cases	Controls	RR
0–49	5	3	1.67	3	7	0.44	7	6	1.17
50–99	19	11	1.79	4	6	0.69	15	16	0.91
100–199	40	25	1.64	26	32	0.82	37	45	0.81
200–299	44	39	1.16	38	28	1.35	66	76	0.87
300–399	61	54	1.15	35	30	1.19	79	65	1.21
400–499	78	65	1.23	40	42	0.96	80	97	0.82
500–599	75	56	1.36	54	41	1.33	86	85	1.01
600+(reference)	9,378	9,447	1.00	6,405	6,419	1.00	12,406	12,386	1.00
Total	9,700	9,700		6,605	6,605		12,776	12,776	

with their matched controls, the results described above are obtained; if they are instead compared to all the controls, the positive results largely disappear. This difference between control sets could be a chance event, in which case the whole result is more plausibly attributable to chance; or it could be a consequence of the matching factors used to choose the controls, and represent a genuine difference between children who get leukemia and other childhood cancers, in which case the result stands. The original authors recognize both possibilities, but feel that because controls were selected using a clearly specified procedure that has not been questioned, it is deeply unsatisfactory to abandon those and use alternative controls just because the results obtained using the original controls are deemed surprising.

If the result stands, an explanation must be sought. Having set up the study motivated by the evidence on magnetic fields, Draper et al. found an effect

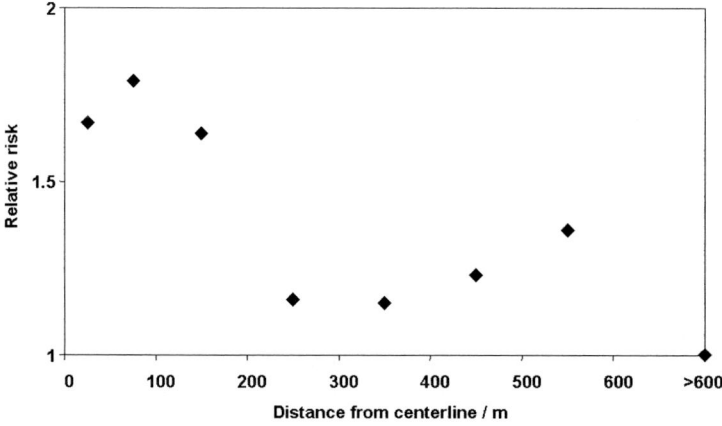

FIGURE 2. Relative risks for leukemia from Draper et al.[1]

that extends far too far from the power lines to be compatible with the existing evidence on magnetic fields. The elevated risks within, say, 100 m could be a consequence of the same magnetic-field effect suggested in other studies, with a different explanation of the results from 100 m to 600 m. Alternatively, the whole of the results could have a different explanation. Most commentators agree it is not a straightforward confirmation of the results. Paradigm shifts take time, however, and understandably not everyone wishes to abandon previous convictions on the basis of a single study. Some commentary on Draper *et al.* has struggled to explain it in terms of magnetic fields, even to the point of recognizing the incompatibility beyond 100 m but using this to suggest that the study is flawed rather than look for an alternative causal explanation.

Of course, any study might be flawed. But, for a research scientist, it is surely better to look at new data for any pointers to new understanding they might give you, rather than for how to force them into the existing paradigm. What explanations could there be for an effect at distances up to 600 m from power lines? Indeed, the effect may extend even further; Draper *et al.* did not investigate beyond 600 m.

One hypothesis already exists: "corona ions," advanced by Professor Henshaw and co-workers at Bristol University in 1999 and since.[13] The very high electric fields on the surface of power line conductors ionize the surrounding air and produce corona ions. These are blown away from the line by the wind. The suggestion is that they attach themselves to existing airborne carcinogenic pollutants, increasing the charge on them; and that when these pollutants are inhaled they have a higher probability of retention within the body because of the increased charge. There has been a dispute as to whether the magnitude of these effects is large enough to be significant. The then National Radiological Protection Board (NRPB) set up an *ad hoc* expert group to examine the question. They concluded,[14] "it seems unlikely that corona ions would have more than a small effect on the long-term health risks associated with particulate air pollutants, even in the individuals who are most affected," but this was without performing the necessary detailed modeling, which they decided was too difficult. There is also no agreement on whether, or to what extent, airborne pollutants cause childhood leukemia. But whatever theoretical concerns there may be, there is no substitute for an experimental test. So Draper *et al.* tested this idea on their data, specifically testing whether the excess of cases over controls was larger downwind than upwind of power lines.

The model used to test this was deliberately the same one as used by another group at Bristol University which had previously reported (in conference presentations and in the media) that there was a highly significant excess of adult cancers downwind of power lines. This was a highly simplified model where all winds are deemed to come from the prevailing south west and only the point of the line nearest to the home in question is considered as a source of corona ions. Using this model, there was actually a slightly greater risk of childhood leukemia upwind than downwind of lines, the reverse of what would be predicted, and therefore there is no evidence that corona ions are the explanation of the results. But the model was simplistic. Moreover, "no evidence of risk" is not the same as "evidence of no risk." Final resolution of this awaits a more sophisticated test including measurements of actual wind direction from meteorological stations round the country.

Other possible physical effects of power lines have been suggested: chemicals washed off either the pylons or conductors; nitrous oxides, ozone, or other chemicals produced by corona on the conductors; herbicides used to control vegetation near them; or the idea that residence at some distance from a power line is a marker for increased time spent close to the power line where the magnetic field is greater. None seems particularly likely.

The other main candidate causal explanation is that the higher leukemia rates stem from the characteristics of the people who live near to power lines or the areas through which power lines pass. There is evidence[15] that leukemia rates vary with SES; in contrast to most childhood diseases, they increase with increasing SES. There is also growing evidence that leukemia rates are affected, among other things, by various factors linked in different ways to exposure to infections. For example, childhood leukemia appears to be caused by exposure to previously unencountered infectious agents through influxes of urban populations into rural areas,[16] and appears to be prevented by early exposure to infection generally, for instance through attendance at childcare.[17] There is also evidence that many socio-economic factors vary around power lines in the UK. This is not as straightforward as might be expected. In the urban areas of highest population density, deprivation increases close to power lines. In areas of lower population density, however, the reverse effect occurs, and deprivation decreases. Both of these probably stem from how power lines are routed. In England and Wales it is the latter effect that is greater for 400 and 275 kV lines, and overall, deprivation decreases closer to these power lines.

It would not be at all surprising if other social or demographic factors varied round power lines as well. SES as such does not appear to explain the findings of Draper et al.; adjusting for it using Carstairs' Index, a measure of the SES of a small area containing the address, affected only the second decimal place of the risk ratios, despite the study exhibiting a clear association between SES and leukemia rates. But the evidence on which specific social factors are linked to leukemia, and exactly which vary round power lines, is currently inadequate. For the moment, it remains an intriguing but unproven possibility that Draper et al. could ultimately come to be seen as another in the sequence of studies which show social risk factors for childhood leukemia.

There is an interesting if arcane debate as to whether such an effect would be "causal." In a scientific sense, if power lines are routed through areas that already have the pertinent population characteristics, this would not be a casual effect of the power line; but if it is the presence of the power line which affects the area and influences the population characteristics, this could be regarded as a causal effect albeit indirect.

If the results of Draper *et al.* are hypothesized as suggesting the raised leukemia rates round power lines are attributable to characteristics of the population who live there, there is a further speculation: that this could also explain the raised leukemia rates round power lines in the methodologically similar Nordic studies. This would then leave the evidence for a magnetic-field effect coming only from those other studies, the measured-field studies in Ahlbom *et al.*, which are known to have bias. In this hypothesis, Draper *et al.* is seen as weakening rather than strengthening the epidemiologic evidence for magnetic fields as a cause of childhood leukemia.

There is one other difference between Draper *et al.* and the preceding literature: Draper *et al.* looked at address at birth (because it was only for birth that a suitable control set was available), whereas most previous studies, particularly those with measurements, have looked at address immediately prior to diagnosis (because that is the easiest address to assess retrospectively). This could be invoked as a reason for different findings; perhaps perinatal exposure to magnetic fields has effects at lower levels (and hence at greater distances from the power line) than exposure later in life. But given that half of leukemia cases have the same address at birth and diagnosis, it seems unlikely that looking at birth rather than diagnosis exposure is likely to make much difference to the findings.

OVERALL COMMENTS

It is perfectly possible that either, or both, the existing epidemiology concerning childhood cancer and magnetic fields, as summarized in Ahlbom *et al.*, and the new results from Draper *et al.* are merely chance findings; but this does not seem very likely. It is more natural to regard these studies as fairly strong evidence that something is happening, and to ask what that "something" might be. The main candidate explanations for the epidemiology of magnetic fields generally are bias, or a direct causal effect of magnetic fields. Neither seems to explain Draper *et al.* It seems we must either have a magnetic-field effect plus something extra (social factors or chance) in Draper *et al.* or a single effect (presumably social factors) in Draper *et al.* and the Nordic studies plus bias in the other studies; or two different causal factors (magnetic fields and corona ions).

What is clear is that the evidence associating childhood leukemia with magnetic fields and power lines is substantial and deserves further research until an explanation(s) is found. The number of attributable cases, if there is a causal effect, may not be large. But aside from the worthy goal of removing, if possible, the cause of even just a few cases of leukemia, research into EMFs offers an extra handle on the etiology of childhood leukemia generally, whether that be insights into the physical mechanism if there is a magnetic-field explanation, the effect of pollution in a corona-ion explanation, or exactly which population characteristics are relevant if that is the explanation. It is not as if the causes of childhood leukemia are completely understood; in fact, it could be argued that, set alongside the dramatic improvement in treatment over the last few decades, the slower progress in identifying causes is disappointing. We cannot afford to ignore any avenue for further investigation.

ACKNOWLEDGMENTS

CCRG receives funding from the Department of Health, and the Scottish Ministers. The views expressed are those of the authors and not necessarily those of National Grid, the Department of Health, and the Scottish Ministers.

COMPETING INTERESTS

The authors declare competing financial interest. JS is employed by National Grid and worked on this article with their permission.

REFERENCES

1. DRAPER, G. et al. 2005. Childhood cancer in relation to distance from high voltage power lines in England and Wales: a case-control study. Br. Med. J. **330:** 1290–1293.
2. REILLY, J.P. 2002. Neuroelectric mechanisms applied to low frequency electric and magnetic field exposure guidelines—Part 1: sinusoidal waveforms. Health Phys. **83:** 341–355.
3. INTERNATIONAL COMMISSION ON NON-IONIZING RADIATION PROTECTION. 1998. Guidelines for limiting exposure to time-varying electric, magnetic, and electromagnetic fields (up to 300 GHz). Health Phys. **74:** 494–522.
4. WERTHEIMER, N. & E. LEEPER. 1979. Electrical wiring configurations and childhood cancer. Am. J. Epidemiol. **109:** 273–284.
5. AHLBOM, A. et al. 2000. A pooled analysis of magnetic fields and childhood leukaemia. Br. J. Cancer **83:** 692–698.
6. GREENLAND, S. et al. 2000. A pooled analysis of magnetic fields, wire codes, and childhood leukemia. Epidemiology **11:** 624–634.
7. SCHUZ, J. et al. 2001. Residential magnetic fields as a risk factor for childhood acute leukaemia: results from a German population-based case-control study. Int. J. Cancer **91:** 728–735.
8. KAVET, R. 2005. Contact current hypothesis: summary of results to date. Bioelectromagnetics **26:** S75–S85.
9. HATCH, E.E. et al. 2000. Do confounding or selection factors of residential wiring codes and magnetic fields distort findings of electromagnetic fields studies? Epidemiology **11:** 189–198.
10. IARC. 2002. Monographs on the evaluation of carcinogenic risks to humans. Volume 80 Non-ionizing radiation, part 1: static and extremely low-frequency (ELF) electric and magnetic fields. IARC Press. Lyon.

11. HSE. 2001. Reducing risks, protecting people, HSE's decision making process. HMSO. Norwich, UK.
12. UK CHILDHOOD CANCER STUDY INVESTIGATORS. 2002. Childhood cancer and residential proximity to power lines. Br. J. Cancer **83:** 1573–1580.
13. FEWS, A.P. *et al.* 1999. Corona ions from power lines and increased exposure to pollutant aerosols. Int. J. Radiat. Biol **75:** 1523–1531.
14. NRPB. 2004. Particle deposition in the vicinity of power lines and possible effects on health. Documents of the NRPB **15:** 3–62.
15. DRAPER, G.J. *et al.* 1991. The geographical epidemiology of childhood leukaemia and non-Hodgkin lymphomas in Great Britain, 1966–1983, OPCS studies on medical and population subjects no. 53. OPCS. London.
16. KINLEN, L. & R. DOLL. 2004. Population mixing and childhood leukaemia: fallon and other US clusters. Br. J. Cancer **91:** 1–3.
17. GILHAM, C. *et al.* 2005. Day care in infancy and risk of childhood acute lymphoblastic leukaemia: findings from UK case-control study. Br. Med. J. **330:** 1294–1297.

Getting It Right the First Time

Developing Nanotechnology while Protecting Workers, Public Health, and the Environment

JOHN M. BALBUS, KAREN FLORINI, RICHARD A. DENISON, AND SCOTT A. WALSH

Environmental Defense, Washington, DC 20009

ABSTRACT: Nanotechnology, the design and manipulation of materials at the atomic scale, may well revolutionize many of the ways our society manufactures products, produces energy, and treats diseases. Innovative nanotechnology products are already reaching the market in a wide variety of consumer products. Some of the observed properties of nanomaterials call into question the adequacy of current methods for determining hazard and exposure, and for controlling resulting risks. Given the limitations of existing regulatory tools and policies, two distinct kinds of initiatives are urgently needed: first, a major increase in the federal investment nanomaterial risk research, and second, rapid development and implementation of voluntary standards of care pending development of adequate regulatory safeguards. The U.S. government should increase federal funding for nanomaterial risk research under the National Nanotechnology Initiative to at least $100 million annually for the next several years. Several voluntary programs are currently at various stages of evolution, though the eventual outputs of each of these are still far from clear. Ultimately, effective regulatory safeguards, harmonized globally, are necessary to provide a level playing field for industry while adequately protecting human health and the environment.

KEYWORDS: nanotechnology; nanoparticles; nanotoxicology; risk assessment; risk management; occupational health; environmental health

INTRODUCTION

Nanotechnology, the design and manipulation of materials at the atomic scale, may well revolutionize many of the ways our society manufactures products, produces energy, and treats diseases. Hundreds of large and small nanotechnology companies are developing a wide variety of materials for use

Address for correspondence: John M. Balbus, M.D., M.P.H., 1875 Connecticut Avenue, N.W. Suite 600, Washington, DC 20009. Voice: 202-387-3500; fax: 202-234-6049.
e-mail: jbalbus@environmentaldefense.org

in electronics, medical diagnostic tools and therapies, construction materials, personal care products, paints and coatings, environmental cleanup, energy production and conservation, environmental sensors, and many other important applications.

Deliberate exploitation of properties that only become evident at the nanoscale is central to these applications. Such properties include highly specific binding over a huge surface that arises from tiny particle size, absorption, and radiation of specific wavelengths of light, penetration of cellular barriers, and high tensile strength and durability. Carefully controlled, these properties may provide highly beneficial products. But these new and enhanced properties also raise the possibility of unintended adverse consequences for human health and the environment. The same binding properties that deliver therapeutics to cancer cells might also, for example, deliver toxic substances to aquatic organisms if similar materials are released or used in the ambient environment. The electrical properties that drive applications in computers may lead to oxidative damage in living tissues. It is essential that these potential harms are identified prospectively and mitigated—ideally through material design, or where that is not feasible through use restrictions.

NANOTECHNOLOGY EXPOSURES AND RISKS–A LIFE CYCLE VIEW

Innovative nanotechnology products are already reaching the market in a wide variety of consumer products, and the National Science Foundation predicts that the global market for nanomaterials could reach $1 trillion within a decade.[1] Some of these products, and others now in the pipeline, will result in human and environmental exposures to nanoparticles. Examples include drugs and cosmetics, and uses for remediation of groundwater contamination. Other products may also entail substantial exposures, though not necessarily during a product's useful life. For example, tennis rackets, automobile running boards, and other products contain carbon nanotubes embedded within resins or other matrices. While exposure to individual nanoparticles during the product's intended use seems unlikely, a product's life cycle includes not just the product's useful life, but also its manufacture (and manufacture of its components) and its disposal or recycling/reclamation. Human or environmental exposures during these other stages may be substantial. For instance, although computer users are highly unlikely to inhale carbon nanotubes bound in their computer screen, exposure potential may dramatically increase when recyclers ultimately grind up those screens for use as road aggregate. Human exposures are most obvious for the workers doing the grinding, but may also occur in road construction workers, and perhaps to travelers and neighbors as the road surface weathers with time and traffic. One study has shown that finely ground carbon nanotubes can damage lung tissues,[2] illustrating the importance of considering a product's complete life cycle to understand exposures, tailor toxicity testing, and thus address risks effectively.

With commercialization of more products containing or comprising nanomaterials comes growing opportunities for human and environmental exposure, lending urgency to the need to understand the potential hazards of nanomaterials. It also raises the question of whether and how carefully regulators are reviewing these new materials before they reach the market.

Numerous examples demonstrate that failure to sufficiently consider the potential adverse effects of technological advances can lead to immense costs, from human and environmental as well as financial perspectives. The widespread use of tetraethyl lead in motor fuels has left a legacy of impaired cognitive function and shortened life span,[3] as well as persistent environmental contamination. Similarly, widespread use of asbestos has created a tremendous human burden of lung disease and mesothelioma—and has also resulted in massive litigation and cleanup costs for many companies that mined asbestos and manufactured asbestos products, as well as for building owners that installed such products. The total cost of liability for asbestos-related losses could reach $200 billion.[4] Finally, failure to address potential harms proactively could lead to a repeat of what is occurring in the biotechnology sector, where European consumers' resistance to genetically modified foodstuffs is said to cost the U.S. agricultural sector $300 million annually in lost crop export revenues.[5]

HOW DO NANOMATERIALS DIFFER FROM CONVENTIONAL SUBSTANCES?

In some cases, the very properties that make nanomaterials uniquely useful in biomedical or other commercial applications also raise the potential for novel mechanisms and targets of toxicity. For example, the ability of certain nanoparticles to penetrate cell membranes, which new applications to deliver targeted therapies exploit, suggests that nanoparticles will also be able to cross physiologic barriers and enter body compartments that larger particles and smaller molecules do not readily access. Particles of different sizes gain entry into cells via very different mechanisms. Those larger than 500 nm primarily gain entry through active endocytosis; those smaller than 200 nm gain entry through a variety of active and nonactive mechanisms.[6] One study of 20 nm polystyrene beads suggests that they enter cells by passing directly through membranes without requiring specific transport mechanisms. Once inside the cells, the nanoparticles distributed throughout the cytoplasm and appeared to bind to a variety of structures, including endosomes and cytoskeletal elements. Aggregation after entry occurred and was inhibited by blockers of microtubules, suggesting a role for active transport processes intracellularly.[7]

The manner in which different individual and aggregated nanoparticles may interact with critical cellular substructures, such as cytoskeletal and motile elements is poorly understood, and cannot be inferred from studies of chemical agents or randomly generated nanoparticles. Surface modifications may allow nanoparticles to bind to cell-surface receptors and avoid internalization[8] or be taken up by specific transport mechanisms, allowing cell targeting for therapeutic agents. It is clear that subtle variations in nanoparticle surfaces, whether due to intentional coating prior to entry into the body or unintentional surface binding or coating degradation once inside the body, can have dramatic impacts on where and how nanoparticles gain entry into cells, as well as where and how they are transported within cells after entry. Understanding the implications of such transport as well as assuring the stability of surface properties throughout the life span of manufactured nanoparticles will be critical to assuring safety.

Preliminary efforts to use nanoparticles for therapeutic interventions indicate that at least some nanomaterials have unanticipated toxic effects—effects that have been detected only because of the testing that routinely occurs in the course of drug development. In one example, researchers developing nanoparticles designed to target gliosarcoma tumor cells noted that, of 20 such materials, all caused adverse effects on the reticular endothelial system and the kidneys.[9] No papers documenting this example of unintended consequences have been published as yet; study of these events is crucial to understanding potential structure–activity relationships for nanomaterials.

Understanding the behavior of nanoparticles requires careful characterization of their surface properties. For a given mass of particles, surface area increases exponentially with decreasing diameter (and increasing number). This increased surface-area-to-mass ratio may be a critical feature in understanding the toxicity of nanomaterials. For example, in a study comparing the toxicity of conventional versus nano-sized particles of titanium dioxide, the nanoparticles appeared significantly more toxic when the dose was reported on a mass basis, but the distinction essentially disappeared when the dose was reported on a surface area basis.[10] The higher surface area to volume ratio also leads to higher particle surface energy, which may translate into higher reactivity.[11] Last, the combination of high surface area and small size may give nanoparticles unusual catalytic reactivity due to quantum effects, such as those seen with gold nanoparticles.[12] This combination of enhanced surface area and enhanced surface activity lends far greater complexity to the characterization of nanoparticles, and also precludes easy extrapolation about potential toxicity.

No studies on reproductive toxicity, immunotoxicity, or chronic health effects, such as cancer or developmental toxicity, have yet been published.[13] Of the limited number of short-term studies completed to date, several have found a variety of adverse effects. Studies in which single-walled carbon nanotubes (SWCNTs) were instilled into the lungs of rodents have consistently demonstrated that SWCNTs cause unusual lung granulomas and other signs of lung inflammation,[14–16] and one[16] found that SWCNTs also cause dose-dependent, diffuse interstitial fibrosis. One study of multiwalled carbon nanotubes (MWCNTs) showed similar lung toxicity, especially after the MWCNTs were finely ground.[2] SWCNT and MWCNT have also been shown to induce oxidative stress in skin cells.[17–19] These studies raise questions of potential toxicity at the beginning or end of the carbon nanotube (CNT) life cycle, through workplace exposures or if CNT-containing products undergo weathering, erosion, or grinding during recycling or disposal.

C_{60} fullerenes (commonly known as buckyballs) have been less well studied in mammalian models. They have been shown to be potent bactericides in water[20] as well as capable of being transported via the gills from water to the brains of fish, where they can cause oxidative damage to brain cell membranes.[21] Buckyballs also have caused oxidative stress in *in vitro* testing systems.[22]

Quantum dots can be made of a variety of inherently toxic materials, including cadmium and lead. As some of the key applications of quantum dots include diagnostic imaging and medical therapeutics, quantum dots have been studied relatively extensively in biological systems, although only a small portion of this research has focused on potential toxicity. Studies performed to date have mainly been *in vitro* cytotoxicity assays. While results have been somewhat inconsistent, studies that used longer exposure times were more likely to demonstrate significant toxicity.[23] Quantum dots typically have a core made of inorganic elements, but they are generally coated with organic materials, such as polyethylene glycol to enhance their biocompatibility or target them to specific organs or cells. Some coatings initially decrease toxicity by one or more orders of magnitude, but the coatings are known to degrade when exposed to air or ultraviolet light, after which toxicity increases. While the presumption has been that this cytotoxicity was caused by leakage of cadmium or selenium from the core, there is evidence that some of the molecules used as coatings may have independent toxicity.[23] Significant questions remain about the safety of quantum dots based on the available *in vitro* studies.

HOW WELL WILL CURRENT REGULATORY FRAMEWORKS PROTECT WORKERS, THE PUBLIC, AND THE ENVIRONMENT FROM NANOMATERIAL RISKS?

Effectively managing nanomaterials' potential risks will prove a challenge for existing occupational and environmental regulatory frameworks for at least four reasons. First, in most current regulatory programs, standards (and exemptions from them) are based on mass and mass concentration. Because of their high surface-area-to-mass ratios and enhanced surface activity, nanomaterials are likely to prove potent at far lower concentration levels than those envisioned when threshold standards were initially set. Second, although regulators can often reasonably predict at least some types of toxicity for new conventional materials based on extrapolation from related materials, too little is currently known about nanomaterials to enable such extrapolation.

Third, it appears many nanomaterials are being developed in a decentralized fashion, with a significant percentage of production coming from small, dispersed facilities. As a result, obtaining information on which materials are being produced and used, by what processes and for what applications—and directing any compliance and enforcement efforts to where they are needed—will be hampered by the sheer number of facilities involved. By the same token, a great deal of production, processing, and use will take place in facilities that may lack expertise and resources to understand and comply with environmental and occupational safeguards.

Last, the pace of the regulatory process lags far behind the speed with which nanomaterials are being brought to market. While substances marketed as drugs, food additives, and pesticides regularly receive significant scrutiny when first brought to market, most others do not. As a result, occupational and environmental protections generally must be developed after problems are identified or suspected, and then in regulatory proceedings that typically take several years to complete. The opportunity exists to recognize and control problems more proactively with nanotechnology. A more detailed discussion of specific regulatory issues under two key U.S. laws follows.

OCCUPATIONAL SAFETY AND HEALTH ACT

Under the Occupational Safety and Health Act (OSHAct), four types of regulatory mechanisms are most likely relevant for protecting workers from overexposure to nanomaterials: substance-specific standards, general respiratory protection standards, the hazard communication standard, and the "general duty clause." Each is examined below.

As a practical matter, substance-specific occupational standards are unlikely to be set in the absence of extensive toxicology data. The majority of standards have been based on findings of human epidemiology studies, which by definition follow widespread exposure and take even more time to conduct. Given the relative paucity of health data on nanoparticles, it is unlikely that any nanoparticle-specific standards will be put in place in the reasonable future. In their absence, inhalable nanoparticles will automatically be covered by the 5 mg/m^3 standard that applies to "particulates not otherwise regulated," sometimes called "nuisance dust" (29 CFR 1910.1000 Table Z-1). These mass-based standards, developed for conventional particles, are unlikely to protect workers from adverse effects of nanoparticle exposures: One author [14] has suggested that exposures to carbon nanotubes at 5 mg/m^3 for several weeks would be analogous to exposure levels he found to cause lung granulomas and inflammation in rats.

Second, the respiratory protection standard (CFR 1910.134) requires employers to provide workers with respirators or other protective devices when engineering controls are not adequate to protect health. The standard provides guidance in selecting specific personal protective equipment and in implementing workplace respiratory protection programs. Only respirators certified by the National Institute of Occupational Safety and Health (NIOSH) may be used, and employers must assess the effectiveness of the respirators they supply. The current lack of validated means to measure and characterize the form and size of nanoparticles in the air, as well as uncertainties regarding respirator performance with particles less than 30 nm and potential agglomerates around 300 nm,[24] will complicate implementation of this standard.

Third, OSHA's hazard communication standard (CFR 1910.1200) stipulates that all producers or importers of chemicals are obligated to develop material safety data sheets ("MSDSs"), which are intended to provide workers with available information on hazardous ingredients in products they handle and educate them on safe handling practices. However, even when accurate and up-to-date, MSDSs have significant limitations—most notably, there is no requirement to either generate data on potential hazards, or to disclose the absence of data. Moreover, in some instances a nanomaterial's MSDS has simply adopted the hazard profile for a supposedly related bulk material. For example, an MSDS for carbon nanotubes identifies the primary component as graphite, and goes on to cite information on the hazards of graphite without acknowledging any dissimilarity between the two substances.[25]

Finally, OSHAct's general duty clause [Section 5(a)(1)] is intended as a backstop to protect workers from exposures that are widely known to result in toxic effects but are not addressed specifically by an OSHA standard. The general duty clause, however, applies only to "recognized" hazards, a difficult criterion to meet in light of the current paucity of toxicity data on specific nanomaterials.

TOXIC SUBSTANCES CONTROL ACT

Beyond the occupational realm, the array of potential environmental regulatory authorities initially appears impressive, including the Clean Air Act, Clean Water Act, Resources Conservation and Recovery Act (RCRA, which addresses management of hazardous and other solid wastes), and the Toxic Substances Control Act (TSCA, which covers chemicals other than drugs, food additives, cosmetics, and pesticides). However, with the exception of some provisions of TSCA, existing regulations under these statutes are not directly relevant to nanomaterials, and adopting such standards would require the Environmental Protection Agency (EPA) to launch a lengthy, data-intensive rulemaking process that would take years to complete.[26]

Certain provisions of TSCA, however, do currently apply. Specifically, section 5 of TSCA requires the producer of a "new" chemical substance to send EPA a "Pre-Manufacture Notification" (PMN) before beginning to produce a substance. Unfortunately, there are no baseline data requirements for PMNs, and 85% of PMNs are submitted without any health data.[27] EPA can request additional data, but rarely does so; it typically conducts its review based on use of structure–activity relationship models, through which toxicological properties of an unstudied substance are estimated based on the extent of molecular structural similarity to substances with known toxicological properties. Because the models are based on the properties of bulk forms of conventional chemical substances, and because nanomaterials' novel and enhanced properties result from characteristics (e.g., size, shape) in addition to their molecular structure, existing models have little applicability to nanomaterials. It remains to be seen whether EPA will make more vigorous use of its authority to require actual toxicity data on nanomaterials to be generated and included in PMNs.

Other key questions also remain unresolved, including the extent to which nanomaterials qualify as "new" chemicals (thereby triggering PMN requirements). Under TSCA, a "new" chemical is one that is not already listed on the TSCA Inventory of chemicals in commerce, and a chemical is defined as a substance with "a particular molecular identity" [TSCA section 3, 15 USC section 2602(2)]. While nanomaterials whose molecular formula is not already included on the TSCA Inventory obviously constitute "new" materials, some parties appear to be assuming that other nanomaterials—those with a molecular structure identical to a substance already on the Inventory—do not qualify as new.

In October 2005, EPA announced plans to issue guidance on distinguishing "new" from "existing" nanomaterials. Environmental Defense has urged EPA to clarify that nanomaterials constitute "new" substances unless their chemical and physical properties are demonstrably identical to those of the conventional substance, on the grounds that only substances with the same properties *as well as* the same molecular structure share "a particular molecular identity." Environmental Defense also urged EPA not to apply mass based and other exemptions in the PMN program unless the underlying scientific rationale is appropriate when applied to nanomaterials.

TSCA also provides certain information gathering authorities. Under Section 8(a), EPA can require manufacturers to provide certain use and exposure information. Section 8(e) requires manufacturers to submit any information indicating that a substance poses a "significant risk" to health or the environment, while Section 8(d) allows EPA to require manufacturers to submit all toxicity-related data already in their possession. As further discussed below, EPA is currently conducting a multistakeholder process that is both designing a voluntary initiative to address nanomaterial risks and considering possible use of TSCA authorities.

ADDRESSING NANOMATERIAL RISKS: NEXT STEPS

Given the limitations of existing regulatory tools and policies, two distinct kinds of initiatives are urgently needed: first, a major increase in the federal investment nanomaterial risk research, and second, rapid development and implementation of voluntary standards of care pending development of adequate regulatory safeguards. A wide array of stakeholders must be involved in all components of the latter process, not only large and small businesses and the academic community, but also labor groups, health organizations, consumer advocates, community groups, and environmental organizations.

INCREASE GOVERNMENTAL INVESTMENT IN RISK RESEARCH

The U.S. government, as the largest single investor in nanotechnology research and development, needs to spend more to assess the health and environmental implications of nanotechnology and ensure that the critical research needed to identify potential risks is done expeditiously. Through the National Nanotechnology Initiative, the federal government spends roughly $1 billion annually on nanotechnology research and development. Of this, environmental and health implications research accounted for only $8.5 million (<1%) in FY 2004, and is expected to increase to only $38.5 million (<4%) in FY 2006.

The U.S. government should spend at least $100 million annually on hazard and exposure research for the next several years. While an annual expenditure of $100 million represents a significant increase over current levels, it is still less than 10% of the overall federal budget for nanotechnology development. Moreover, it is a modest investment compared to the potential benefits of risk avoidance and to the $1 trillion role that nanotechnology is projected to play in the world economy by 2015. The call for greatly expanded health and environmental research spending is buttressed by experts' assessments, as well as by testing costs associated with hazard characterization programs for conventional chemicals, and the research budgets for a roughly analogous risk characterization effort on risks of airborne particulate matter.[28]

But the U.S. government should not be the sole, or even the principal, funder of nanomaterial risk research. Other governments are also spending heavily to promote nanotechnology research and development, and they too should allocate some portion of their spending to address nanotechnology risks. And although government risk research has a critical role to play in developing the basic knowledge and methods to characterize and assess the risks of nanomaterials, private industry should fund the majority of the research and testing on the products they are planning to bring to market. Clearly, all parties will benefit if governments and industry coordinate their research to avoid redundancy and optimize efficiency.

DEVELOP VOLUNTARY STANDARDS OF CARE

Because federal agencies are unlikely to be able to put into place adequate provisions for nanomaterials quickly enough to address the products now entering or poised to enter the market, voluntary "standards of care" for nanomaterials must play a role in guiding the safe use of nanomaterials in the meantime. These standards should include a framework and a process by which to identify and manage nanomaterials' risks across a product's full life cycle, taking into account worker safety, manufacturing releases and wastes, product use, and product disposal. Such standards should be developed and implemented in a transparent and accountable manner, including public disclosure of the assumptions, processes, and results of the risk identification and risk management systems.

Several voluntary programs are currently at various stages of evolution, though the eventual outputs of each of these are still far from clear. In October 2005, a workgroup of an EPA advisory committee proposed a framework for a voluntary program aimed at producers, processors, and users of nanomaterials. The group also recommended using certain TSCA regulatory authorities to address nanomaterial risks.[29] In addition, Environmental Defense is working directly with industry to develop a framework for the responsible development, production, use, and disposal of nanoscale materials. Once developed, the framework will be pilot tested on specific nanoscale materials or applications of commercial interest to the company. Other U.S. multistakeholder efforts to develop voluntary standards are also under way through ASTM International[30] and the American National Standards Institute.[31] In addition, the International Standards Organization is convening a new Technical Committee on Nanotechnologies.[32]

Regulatory programs are essential to securing long-term public confidence in and support for nanotechnology.[33] In an ideal world, ample data on nanomaterials toxicity and exposures would already exist, allowing governments to establish appropriate safeguards. In reality such data are extremely limited and regulatory programs almost nonexistent. Significantly more federal support for research into the health and environmental effects of nanomaterials is urgently needed, along with rapid development of voluntary standards of care that can help provide interim protection for workers, the general public, and the environment until meaningful regulations can be put into place.

REFERENCES

1. Roco, M. The future of the National Nanotechnology Initiative. Available at http://www.nano.gov/html/res/slides.pdf. (Accessed on November 4, 2005).
2. MULLER, J. *et al*. 2005. Respiratory toxicity of multi-wall carbon nanotubes. Toxicol. Appl. Pharmacol. **207:** 221–231.

3. LUSTBERG, M. et al. 2002. Blood lead levels and mortality. Arch. Intern. Med. **162:** 2443–2449.
4. SEIFERT, C. 2004. July 15. Industry surveys. Insurance: Property-Casualty. Standard & Poor: NY.
5. HASTERT, D. 2003. Testimony of J. Dennis Hastert before the U.S. House of Representatives, Committee on Science. June 12, 2003. Available at: http://www.house.gov/science/hearings/research03/jun12/hastert.htm. (Accessed on November 4,2005).
6. REJMAN, J. et al. 2004. Size-dependent internalization of particles via the pathways of clathrin- and caveolae-mediated endocytosis. Biochem. J. **377:** 159–169.
7. EDETSBERGER, M. et al. 2005. Detection of nanometer-sized particles in living cells using modern fluorescence fluctuation methods. Biochem. Biophys. Res. Commun. **332:** 109–116.
8. GUPTA, A. et al. 2004. Lactoferrin and ceruloplasmin derivatized superparamagnetic iron oxide nanoparticles for targeting cell surface receptors. Biomaterials **25:** 3029–3040.
9. INSTITUTE OF MEDICINE OF THE NATIONAL ACADEMIES. 2005. Implications of nanotechnology for environmental health research. National Academic Press: Washington, DC.
10. OBERDORSTER, G. et al. 2005. Nanotoxicology: an emerging discipline evolving from studies of ultrafine particles. Environ. Health Perspect. **113:** 823–839.
11. OBERDORSTER, G. et al. 2005. Principles for characterizing the potential human health effects from exposure to nanomaterials: elements of a screening strategy. Part. Fibre Toxicol. **2:** 8.
12. DANIEL, M. et al. 2004. Gold nanoparticles: assembly, supramolecular chemistry, quantum-size-related properties, and applications toward biology, catalysis, and nanotechnology. Chem. Rev. **104:** 293–346.
13. WOODROW WILSON CENTER PROJECT ON EMERGING NANOTECHNOLOGIES 2005. "Nanotechnology. Environmental and Health Implications. A database of current research." Available at www.nanotechproject.net.
14. LAM, C. et al. 2003. Pulmonary toxicity of single-wall carbon nanotubes in mice 7 and 90 days after intratracheal instillation. Toxicol. Sci. **77:** 126–134.
15. WARHEIT, D. et al. 2004. Comparative pulmonary toxicity assessment of single-wall carbon nanotubes in rats. Toxicol. Sci. **77:** 117–125.
16. SHVEDOVA, A. et al. 2005. Unusual inflammatory and fibrogenic pulmonary responses to single-walled carbon nanotubes in mice. Am. J. Physiol. Lung Cell. Mol. Physiol. **289:** L698–L708.
17. MONTEIRO-RIVIERE, N. et al. 2005. Multi-walled carbon nanotube interactions with human epidermal keratinocytes. Toxicol. Lett. **155:** 377–384.
18. MANNA, S. et al. 2005. Single-walled carbon nanotube induces oxidative stress and activates nuclear transcription factor- kb in human keratinocytes. Nano Lett. **5(a)** : 1676–1684.
19. SHVEDOVA, A. et al. 2003. Exposure to carbon nanotube material: assessment of nanotube cytotoxicity using human keratinocyte cells. J. Toxicol. Environ. Health A. **66:** 1909–1926.
20. FORTNER, J. et al. 2005. C60 in water: nanocrystal formation and microbial response. Environ. Sci. Technol. **39:** 4307–4316.
21. OBERDORSTER, E. 2004. Manufactured nanomaterials (fullerenes, C60) induce oxidative stress in the brain of juvenile largemouth bass. Environ. Health Perspect. **112:** 1058–1062.

22. SAYES, C. et al. 2004. The differential cytotoxicity of water-soluble fullerenes. Nano Lett. **4:** 1881–1887.
23. HARDMAN, R. 2005. A toxicological review of quantum dots: toxicity depends on physico-chemical and environmental factors. Environ. Health Persp. Nat. Inst. Environ. Health Sci. doi: 10.1289/ehp.8284. Available at: http://dx.doi.org. (Accessed on November 4, 2005).
24. TSI. 2005. Mechanisms of filtration for high efficiency fibrous filters. Application Note ITI – 041, TSI Incorporated. Available at www.tsi.com/AppNotes/appnotes.aspx?Cid=24&Cid2=195&Pid=33&lid=439&file=iti_041#mech.
25. CARBON NANOTUBES, INC. UNDATED. MATERIAL SAFETY DATA SHEET – CNI CARBON NANOTUBES. Available at http://www.cnanotech.com/download_files/MSDS%20CNI%20Nanotubes.pdf. (Accessed on November 4, 2005).
26. ENVIRONMENTAL LAW INSTITUTE. 2005. Securing the Promise of Nanotechnology: Is U.S. Environmental Law Up to the Job? Available at: http://www.elistore.org/reports_detail.asp?ID=11116. (Accessed on November 4, 2005).
27. GOVERNMENT ACCOUNTABILITY OFFICE. 2005.Options exist to improve EPA's ability to assess health risks and manage its chemical review program, GAO-05-458, June 12. 2005.
28. DENISON, R. 2005. A proposal to increase federal funding of nanotechnology risk research to at least $100 million annually. Available at http://www.environmentaldefense.org/documents/4442_100milquestionl.pdf. (Accessed on November 4, 2005).
29. U.S. ENVIRONMENTAL PROTECTION AGENCY. Interim Ad Hoc Work Group on Nanoscale Materials, National Pollution Prevention and Toxics Advisory Committee (NPPTAC). 2005. Overview of Issues for Consideration by NPPTAC. October 8. 2005. Available at www.epa.gov/oppt/npptac/nanowgoverviewdraft051011final.doc. (Accessed on November 4, 2005).
30. ASTM INTERNATIONAL. 2005. Committee E56 on nanotechnology. Available at: http://www.astm.org/COMMIT/COMMITTEE/E56.htm. (Accessed on November 4, 2005).
31. AMERICAN NATIONAL STANDARDS INSTITUTE. 2005. Nanotechnology Standards Panel. Available at: http://www.ansi.org/standards_activities/standards_boards_panels/nsp/overview.aspx?menuid=3. (Accessed on November 4, 2005).
32. INTERNATIONAL STANDARDS ORGANIZATION. 2005. Nanotechnologies Technical Committee. Available at: http://www.iso.org/iso/en/stdsdevelopment/tc/tclist/TechnicalCommitteeDetailPage.TechnicalCommitteeDetail?COM MID=5932. (Accessed on November 4, 2005).
33. MACOUBRIE, J. 2005. Informed public perceptions of nanotechnology and trust in government. Washington, DC: Woodrow Wilson International Center for Scholars. Available at http://www.wilsoncenter.org/news/docs/macoubriereport.pdf. (Accessed on November 4, 2005).

Pesticides and Adult Respiratory Outcomes in the Agricultural Health Study

JANE A. HOPPIN,[a] DAVID M. UMBACH,[b] STEPHANIE J. LONDON,[a] CHARLES F. LYNCH,[c] MICHAEL C.R. ALAVANJA,[d] AND DALE P. SANDLER[a]

[a]*Epidemiology Branch, National Institute of Environmental Health Sciences, NIH, DHHS, Research Triangle Park, North Carolina 27709-2233, USA*

[b]*Biostatistics Branch, National Institute of Environmental Health Sciences, NIH, DHHS, Research Triangle Park, North Carolina 27709-2233, USA*

[c]*Department of Epidemiology, University of Iowa, Iowa City, Iowa 52242-5000, USA*

[d]*Occupational Epidemiology Branch, National Cancer Institute, NIH, DHHS, Rockville, Maryland 20852-7240, USA*

ABSTRACT: In the 1700s, Bernardino Ramazzini was among the first to describe respiratory disease among agricultural workers. Since then, farmers continue to have higher rates of respiratory illnesses, even as changes occur in occupational and environmental exposures on farms. While grain and animal exposures have been well studied for their role in agricultural lung diseases, pesticide exposures have not. Using the Agricultural Health Study, a prospective cohort study of ~89,000 licensed pesticide applicators and their spouses in Iowa and North Carolina, we are currently assessing the association of pesticides with respiratory outcomes, including wheeze, adult asthma, farmer's lung, and chronic bronchitis. At enrollment (1993–1997), 19% of farmers and 22% of commercial pesticide applicators reported wheeze in the previous year. Using logistic regression models adjusted for age, state, smoking status, and body mass index, we evaluated the association of 40 individual pesticides with wheeze within these two groups separately. In both groups, we observed strong evidence of an association of organophosphates with wheeze. For farmers, the organophosphates chlorpyrifos, malathion, and parathion were positively associated with wheeze; for the commercial applicators, the organophosphates chlorpyrifos, dichlorvos, and phorate were positively associated with wheeze. Chlorpyrifos was strongly associated with wheeze in a dose-dependent manner in both groups; use of chlorpyrifos for at least 20 days per year had an odds ratio of 1.48 (95% confidence interval [CI] = 1.00–2.19) for farmers and 1.96 (95% CI =

Address for correspondence: Jane A. Hoppin, ScD., NIEHS, Epidemiology Branch, MD A3-05, PO Box 12233, Research Triangle Park, NC 27709-2233. Voice: 919-541-7622; fax: 919-541-2511.
e-mail: hoppin1@niehs.nih.gov

1.05–3.66) for commercial applicators. Our wheeze results are consistent with recent animal models that support a role for organophosphates and respiratory outcomes.

KEYWORDS: farmers; occupational exposure; organophosphates; pesticide applicators

INTRODUCTION

Bernardino Ramazzini was among the first to identify that farmers were at higher risk of respiratory disease compared to the general population; Ramazzini's syndrome, an acute pulmonary disease of farmers associated with threshing grain, is named in his honor. Many aspects of farming have changed since the 1700s, and yet farmers continue to have higher rates of respiratory disease.[1] Pesticides may contribute to respiratory symptoms and disease. Surveys of farmers and rural residents have linked pesticides in general and particular pesticide classes with respiratory symptoms, but few have identified specific pesticides.

Evaluating the respiratory health effects of pesticides among farmers presents a challenge owing to high exposures to many other respiratory irritants and toxicants, such as animals and grains. Large sample sizes and good data on exposures to other respiratory hazards are necessary. The Agricultural Health Study (AHS), a prospective cohort study of private pesticide applicators (primarily farmers), their spouses, and commercial pesticide applicators, provides a unique opportunity to assess the respiratory hazards of pesticides. By having both farmers and commercial applicators, we can evaluate different patterns of pesticide use. To explore potential associations of pesticides with respiratory outcomes, we evaluated the common respiratory symptom wheeze and pesticide use in the past year in these two groups of pesticide applicators.

MATERIALS AND METHODS

We analyzed data from two populations participating in the AHS: private pesticide applicators (farmers) and commercial pesticide applicators. These two groups provide different perspectives on the potential impact of pesticides on respiratory health. Farmers may have exposure to other respiratory toxicants as a result of farm work, and pesticide applicators may have more days of pesticide exposure because pesticide application is their occupation, not just one aspect of their livelihood. The data from farmers were analyzed first and published in 2002,[2] while the data from commercial applicators was only recently analyzed to follow up on the associations observed among farmers.[3] Farmers and commercial applicators completed identical questionnaires with regard to their pesticide use and medical history. Here, we will summarize and compare the results from both populations.

The AHS is a large prospective cohort study of licensed pesticide applicators and their families in Iowa and North Carolina enrolled between 1993 and 1997.[4] AHS participants are primarily farmers and their spouses, but almost 5000 commercial pesticide applicators in Iowa enrolled in the same manner as the farmers. Commercial applicators apply pesticides for hire to land, plants, seed, animals, waters, and structures. Details of enrollment are described elsewhere.[4] Briefly, pesticide applicators enrolled by completing the enrollment questionnaire at the time of pesticide licensing; more than 80% of private licensed applicators and 47% of commercial pesticide applicators enrolled. Approximately 40% returned a second, more detailed questionnaire, which included items regarding respiratory outcomes. The analysis is limited to individuals who returned both questionnaires. Farmers who did and did not return the take-home questionnaire are similar with regard to demographic characteristics, farming practices, and medical history.[5] Among the commercial applicators, however, individuals who returned the take-home questionnaire were more likely to apply and handle pesticides than those who did not; we saw no difference between the groups with regard to medical history, smoking, and demographic factors.

Exposure assessment and statistical methods were similar for both analyses, and are presented in detail elsewhere.[2,3] Briefly, we assessed exposure and outcome using two self-administered questionnaires. These questionnaires provided information regarding 40 specific chemicals used in the year before enrollment, pesticide application methods, current agricultural activities, smoking history, medical history, and demographics. Wheeze in the past year was based on the response to the question: "How many episodes of wheezing or whistling in your chest have you had in the past 12 months?" Any positive response (one or more episodes) was defined as a wheeze case.

We focused on pesticide use in the year before enrollment because this was temporally relevant for the wheeze outcome. The questionnaire obtained information on lifetime pesticide use as well as current use. We created three level variables to evaluate lifetime use of individual chemicals: never use, use but not in the past year (former use), and current use (used in the past year). In our farmer paper,[2] we modeled pesticide use in a slightly different fashion; however, to ease comparison of results, we will present all results modeled as described here. For each pesticide, we evaluated both ever use in the past year and frequency of application. We used the questionnaire information on frequency of application to assess dose response.

Pesticides were evaluated using a common base logistic regression model controlling for potential confounders. Odds ratios (ORs) were not estimated for pesticides with fewer than five exposed cases. The base models included 10-year age categories, smoking history (current, past, never), body mass index (BMI, <25, ≥ 25 kg/m^2), and asthma/atopy status in four levels (both, asthma

TABLE 1. Demographic, medical, and farming characteristics of farmers and commercial pesticide applicators in the Agricultural Health Study at enrollment (1993–1997)

	Farmers ($n = 17{,}920$) (%)	Commercial applicators ($n = 2{,}255$) (%)
Wheeze in past year	19	22
Age (years)		
<31	8	21
31–40	22	34
41–50	26	27
51–60	23	12
>60	21	6
Sex		
Female	3	5
Male	97	95
Race		
White	98	99
Other race	2	1
State		
Iowa	69	100
North Carolina	31	0
BMI (kg/m^2)		
<23	11	13
23–24.9	16	17
25–26.9	23	22
27–31	33	30
>31	18	18
Smoke status		
Never	55	48
Past	32	30
Current	13	23
Drink alcohol in past year	68	86
Asthma/atopy status		
Neither	86	86
Atopy only	9	9
Asthma only	3	2
Both	2	3
Current farm activities		
Own or work on farm	96	31
Raise animals	60	18
Raise crops	91	25
Pesticide use history		
Years applied pesticides		
0	1	5
<1	2	8
2–5	10	25
6–10	14	22
11–20	33	27
21–30	26	11
>30	15	4
Days applied per year		
None	1	5

Continued.

TABLE 1. Continued

	Farmers ($n = 17{,}920$) (%)	Commercial applicators ($n = 2{,}255$) (%)
<5	17	11
5–9	25	11
10–19	31	15
20–39	18	22
40–59	4	13
60–150	3	19
>150	0	5

alone, atopy alone, neither). State was included in models for farmers, but not for commercial applicators because all lived in Iowa. Previously published models for farmers did not control for BMI; models reported here include BMI. As a result of including BMI as a covariate, our sample size is reduced to 17,920 farmers compared to the 20,468 presented previously.[2]

Pesticides are commonly used together and can be applied to crops and animals that may independently trigger asthma symptoms. For chemicals with evidence of confounding by other pesticides, the models were run again containing the chemical of interest and the use status for the potential confounder (current, former, never). As a result of this analysis, all models for commercial applicators control for use of chlorimuron-ethyl.

The farmer analysis used the prerelease data set dated February 2000, and the commercial applicator analysis used P1REL0310 release of the AHS data set, and all statistical analysis was done using SAS (Cary, NC).

RESULTS

The association of pesticides and wheeze was evaluated among 17,920 farmers and 2,255 commercial applicators. Wheeze is a common respiratory symptom of both farmers and commercial pesticide applicators; 19% of farmers and 22% of commercial applicators reported wheezing at least once in the year before enrollment (TABLE 1). Participants ranged in age from 18 to 88, with farmers being older on average than commercial applicators. Commercial applicators were only enrolled in Iowa, while farmers were recruited from both North Carolina and Iowa. Commercial applicators were more likely to smoke than farmers, with 23% being current smokers compared to 13% of farmers. Farmers had more years of pesticide application, but fewer annual days of pesticide application, compared with the commercial applicators. Over 96% of farmers reported living or working on a farm at the time of enrollment, while only 31% of commercial applicators did.

Pesticide use by farmers and commercial applicators was similar with regard to the chemicals used (TABLE 2). However, some pesticides, particularly herbicides, were more likely to be used by commercial applicators (e.g., chlorimuron-ethyl 24% versus 12%) than farmers. Farmers were slightly more likely to report some insecticides (dichlorvos, terbufos), but not all; diazinon use was more common among commercial applicators. Even among chemicals used more by farmers, commercial applicators who used these chemicals applied them more often.

Both herbicides and insecticides were associated with wheeze among farmers and commercial applicators (TABLE 3). For farmers, 5 of 40 pesticides used in the past year were significantly associated with wheeze. This is fewer chemicals than previously reported owing to the reduction in sample size by including BMI in the models; however, the OR estimates were essentially unchanged, with all but trichlorfon changing less than 2% from those reported previously. For commercial applicators, 3 of 36 pesticides (chlorimuron-ethyl, dichlorvos, and phorate) were associated with wheeze. Among commercial applicators, seven other herbicides were significantly associated with wheeze prior to adjustment for chlorimuron-ethyl use; inclusion of chlorimuron-ethyl in these models attenuated these observed OR to the null value, 1.0.[3] We saw no evidence of confounding by chlorimuron-ethyl (or any other chemical) for farmers; thus, these models do not adjust for other chemical use. In general, the significant OR estimates for farmers were much lower (ORs range from 1.13 to 1.37) than commercial applicators (ORs range from 1.62 to 2.48). The pesticides with the highest ORs were organophosphates, parathion for farmers (OR = 1.37, 95% confidence interval [CI] = 0.93, 2.03) and dichlorvos (OR = 2.48, 95% CI = 1.08, 5.66), and phorate (OR = 2.35, 95% CI = 1.36, 4.06) for commercial applicators. Chlorimuron-ethyl was significantly associated with wheeze among commercial applicators but not farmers with the OR much greater among commercial applicators (OR = 1.62 versus 1.08).

We assessed dose response for chlorimuron-ethyl and two organophosphate insecticides (chlorpyrifos and phorate) with sufficient use by both farmers and commercial applicators to allow for dose-response modeling (TABLE 4). We observed significant dose-response trends for both chlorimuron-ethyl and chlorpyrifos among both groups, while phorate was only significant among commercial applicators. Use of chlorimuron-ethyl was much more frequent among commercial applicators, as a result, we were able to estimate five exposure levels (<5 to 40+ days) compared to three levels for farmers. Increased wheeze was seen at all levels of chlorimuron-ethyl use among commercial applicators, but limited to farmers using it five or more days a year. For chlorpyrifos, the increased odds of wheeze was associated with use of 20+ days per year; the commercial applicator data allow us to assess the monotonic increase at higher dose levels (40+ days/year). Commercial applicators using chlorpyrifos more than 40 days a year had an OR of 2.40 (95% CI = 1.24, 4.65). Other dose-response models can be found in our previous papers.[2,3]

TABLE 2. Pesticide use patterns among farmers and commercial applicators in the Agricultural Health Study at enrollment (1993–1997)

	Farmers			Commercial applicators*		
	Never %	Former %	Current use** %	Never %	Former %	Current use** %
Herbicides						
2,4-D	20	44	36	24	33	43
Alachlor	44	47	9	55	27	19
Atrazine	26	44	30	48	26	26
Butylate	73	26	1	74	22	4
Chlorimuron-ethyl	68	20	12	62	14	24
Cyanazine	56	34	10	56	22	22
Dicamba	46	32	22	40	26	34
EPTC	79	18	3	66	21	14
Glyphosate	22	37	42	20	30	50
Imazethapyr	54	15	30	61	12	27
Metolachlor	52	29	19	54	19	26
Metribuzin	62	34	4	61	23	16
Paraquat	84	12	3	77	16	7
Pendimethalin	63	24	13	53	21	26
Petroleum oil	78	14	7	73	14	13
Trifluralin	45	37	18	51	25	24
Insecticides						
Carbamates						
Aldicarb	92	5	3	97	2	1
Carbaryl	57	31	12	60	30	11
Carbofuran	71	27	2	83	11	5
Organophosphates						
Chlorpyrifos	56	27	16	66	20	15
Coumaphos	90	7	3	97	3	1
Diazinon	79	15	6	73	17	10
Dichlorvos	88	9	3	92	7	1
Fonofos	77	19	4	86	10	4
Malathion	35	48	17	44	40	16
Parathion	92	7	1	96	3	1
Phorate	69	28	3	84	13	3
Terbufos	60	28	12	82	12	6
Trichlorfon	99	1	0	94	3	3
Other insecticides						
Lindane	86	13	1	92	7	1
Permethrin (animals)	87	9	5	92	5	3
Permethrin (crops)	86	9	5	76	12	12
Fungicides						
Benomyl	92	6	2	89	8	3
Captan	88	5	7	94	4	3
Chlorothalonil	92	4	3	88	5	6
Maneb	92	6	2	94	4	2
Metalaxyl	81	12	7	90	6	4
Ziram	99.5	0.3	0.2	99.7	0.2	0.1
Fumigants						
Aluminum phosphide	97	3	1	86	8	6
Methyl Bromide	85	11	4	96	2	1

*Results from commercial applicators previously presented in Hoppin et al.[3]
**Current use defined as ever use in the year prior to enrollment.

TABLE 3. Chemical specific odds ratio for wheeze in the past year for farmers and commercial pesticide applicators in the Agricultural Health Study Cohort, 1993–7

Chemical	Farmers ($n = 17{,}920$)			Commercial applicators ($n = 2{,}255$)		
	Odds ratio*	95% Confidence interval		Odds ratio**	95% Confidence interval	
Herbicides						
2,4-D	0.97	0.86	1.10	0.99	0.73	1.34
Alachlor	1.23	1.06	1.41	0.81	0.55	1.18
Atrazine	1.18	1.05	1.32	0.91	0.63	1.31
Butylate	1.19	0.81	1.75	1.08	0.62	1.88
Chlorimuron-ethyl	1.08	0.96	1.23	1.62	1.25	2.10
Cyanazine	1.05	0.91	1.21	0.80	0.55	1.18
Dicamba	1.05	0.93	1.18	0.78	0.58	1.07
EPTC	1.37	1.08	1.73	0.90	0.61	1.32
Glyphosate	1.05	0.94	1.17	1.14	0.83	1.57
Imazethapyr	1.04	0.93	1.15	1.03	0.71	1.50
Metolachlor	1.09	0.97	1.21	1.01	0.71	1.43
Metribuzin	1.07	0.88	1.31	0.99	0.66	1.48
Paraquat	1.22	0.98	1.51	0.93	0.59	1.47
Pendimethalin	1.02	0.91	1.16	0.99	0.70	1.41
Petroleum oil	1.26	1.09	1.47	1.18	0.83	1.66
Trifluralin	1.15	1.02	1.30	0.85	0.60	1.20
Insecticides						
Carbamates						
Aldicarb	0.99	0.78	1.24	–		
Carbaryl	1.13	0.99	1.29	1.11	0.78	1.58
Carbofuran	1.14	0.88	1.48	0.77	0.46	1.30
Organophosphates						
Chlorpyrifos	1.09	0.97	1.23	1.27	0.92	1.74
Coumaphos	0.95	0.75	1.22	1.94	0.63	5.96
Diazinon	1.10	0.93	1.31	0.84	0.57	1.23
Dichlorvos	1.13	0.88	1.46	2.48	1.08	5.66
Fonofos	1.12	0.91	1.38	1.46	0.86	2.46
Malathion	1.13	1.00	1.27	0.95	0.69	1.31
Parathion	1.37	0.93	2.03	–		
Phorate	1.02	0.80	1.31	2.35	1.36	4.06
Terbufos	1.10	0.96	1.25	1.36	0.87	2.12
Trichlorfon	2.40	0.82	6.98	0.63	0.31	1.29
Other Insecticides						
Lindane	0.85	0.59	1.21	0.67	0.22	2.02
Permethrin (crops)	1.08	0.89	1.31	0.89	0.61	1.29
Permethrin (animals)	1.28	1.06	1.55	1.37	0.70	2.68
Fungicides						
Benomyl	1.09	0.83	1.44	1.06	0.57	1.97
Captan	1.03	0.88	1.22	0.83	0.38	1.79
Chlorothalonil	0.88	0.70	1.11	0.97	0.61	1.54
Maneb/Mancozeb	1.11	0.86	1.43	0.61	0.24	1.55
Metalaxyl	1.15	0.98	1.35	0.71	0.39	1.30
Ziram	1.06	0.48	2.36	–		
Fumigants						
Aluminum phosphide	0.71	0.40	1.28	1.32	0.84	2.07
Methyl bromide	1.14	0.92	1.41	–		

*Odds ratios adjusted for age, BMI, smoking, asthma/atopy and previous use of pesticide. Commercial applicators models include adjustment for chlorimuron-ethyl; farmer models include state.

Odds ratios were not estimated for pesticides with <5 exposed cases.

**Selected odds ratios for commercial applicators were presented previously.[3]

TABLE 4. Selected pesticide-specific dose-response models for wheeze among pesticide applicators from the Agricultural Health Study, 1993–1997.

	Farmers ($n = 17{,}920$)				Commercial applicators ($n = 2{,}255$)			
Pesticide	Odds Ratio	95% Confidence interval		P-trend**	Odds Ratio	95% Confidence interval		P-trend**
Herbicides								
Chlorimuron-ethyl				<0.01				0.01
None	1.00				1.00			
<5 days	0.95	0.80	1.12		1.87	0.96	3.63	
5–9 days	1.26	1.02	1.56		1.38	0.75	2.52	
10–19 days	1.39	1.03	1.88		1.97	1.34	2.90	
20–39 days	–				1.41	0.94	2.13	
40+ days	–				1.50	0.81	2.79	
Organophosphate insecticides								
Chlorpyrifos				0.01				<0.01
None	1.00				1.00			
<5 days	1.00	0.84	1.18		1.00	0.56	1.80	
5–9 days	1.27	1.06	1.51		1.10	0.58	2.08	
10–19 days	0.91	0.71	1.17		0.77	0.39	1.49	
20–39 days	1.48	1.00	2.19		1.96	1.05	3.66	
40+ days	–				2.40	1.24	4.65	
Phorate				0.72				0.01
None	1.00				1.00			
<5 days	1.31	0.86	2.00		2.01	0.64	6.32	
5–9 days	0.73	0.47	1.16		3.67	1.25	10.80	
10+ days	1.09	0.71	1.66		2.10	1.04	4.25	

All models adjusted for: age, BMI, smoking status, and asthma-atopy status.
Commercial applicator models include adjustment for chlorimuron-ethyl.
Farmer models are adjusted for state.
*P-trend based on χ^2 test of trend using the categories presented.

DISCUSSION

Epidemiologic and animal data suggest that organophosphate insecticides have a role in respiratory outcomes.[6–9] Self-reported asthma was associated with organophosphate and carbamate use among Canadian farmers, with the stronger association for carbamate insecticides.[7] Respiratory symptoms were four times more common among Kenyan farm workers with acetylcholinesterase inhibition, a marker of organophosphate and carbamate exposure.[8] The AHS provides the first opportunity to look at individual pesticides in detail. Among the ~20,000 AHS farmers, increased wheeze was associated with three organophosphate insecticides (parathion, chlorpyrifos, and malathion); parathion had the highest odds of wheeze among all pesticides studied.[2] Among the ~2000 AHS commercial applicators, increased wheeze was associated with three organophosphates (chlorpyrifos, dichlorvos, and phorate); phorate and chlorpyrifos demonstrated dose-response trends.

Pesticides have not been well characterized with regard to their respiratory toxicity. In guinea pigs, organophosphate insecticides cause airway hyperreactivity.[6,9,10] This reaction may be associated with acetylcholinesterase inhibition of the vagal nerve.[7,10] A recent animal model suggests, however, that airway hyperreactivity may occur at doses below those that inhibit acetylcholinesterase.[6,9,10] This new model suggests that organophosphate-induced airway hyperreactivity results from effects of autoinhibitory M2 muscarinic receptors on the parasympathetic nerves in the lung and not from acetylcholinesterase inhibition nor from dysfunction of M3 muscarinic receptors in airway smooth muscle.[6] If this model is appropriate for humans, it suggests that systemic exposure to organophosphate insecticides results in increased airway responsiveness at levels below those that cause acetylcholinesterase inhibition.

Chlorimuron-ethyl was strongly associated with wheeze among commercial applicators, and it confounded the associations of all other herbicides and attenuated the estimates for many of the organophosphate insecticides. There is little previous evidence of an association of chlorimuron-ethyl, or herbicides in general, with respiratory outcomes. Among farmers, we observed increased wheeze with six herbicides, including chlorimuron-ethyl; we also observed a dose-response trend for chlorimuron-ethyl. However, we saw no evidence for confounding by chlorimuron-ethyl, other pesticides, or other agricultural exposures among farmers. Chlorimuron-ethyl use was almost twice as common among commercial applicators as that among farmers and the frequency of use was greater as well. Chlorimuron-ethyl is a sulfonylurea postemergent herbicide and is only available as a dry formulation,[11] which may make it more likely to result in exposure via the respiratory route. Because pesticide products contain both the pesticide active ingredient as well as a large percentage of "other ingredients," we are unable to determine whether chlorimuron-ethyl itself or one of the other ingredients in the pesticide product was responsible for the association with wheeze.

We observed higher ORs among commercial applicators than farmers. This observation may be due to a number of factors including (*a*) more frequent pesticide use among commercial applicators, (*b*) lower background exposures among commercial applicators to respiratory toxicants from the farm, and (*c*) a healthy worker effect among the farmers. Commercial pesticide application is a transient occupation, while farming is more often a lifelong commitment; thus, individuals who farm are possibly self-selected with regard to exposures, with sensitive individuals leaving farming or altering their farming practices to minimize exposure to respiratory triggers. Our cross-sectional analysis cannot rule out this phenomenon, which has been observed in other respiratory studies of farmers.[12–14] While the commercial applicators smoked more, it is unlikely that smoking influenced the higher ORs observed because commercial applicators were compared to other commercial applicators, and we made extensive efforts to control for the potential respiratory consequences of smoking. Smoking was unrelated to specific pesticide use among both farmers and commercial applicators.

The AHS provides a unique opportunity to explore associations between pesticides and respiratory outcomes. By studying both farmers and commercial applicators, we are able to explore the role of pesticides and wheeze in populations with similar but different patterns of exposure. The associations were stronger in commercial applicators than farmers potentially because of the increased frequency of pesticide use. Our results add to the emerging literature linking organophosphate insecticides to adverse respiratory outcomes and suggest a role for chlorimuron-ethyl.

ACKNOWLEDGMENTS

This research was supported by the Intramural Research Program of the National Institute of Environmental Health Sciences and the National Cancer Institute, National Institutes of Health.

REFERENCES

1. SCHENKER, M.B. et al. 1998. Respiratory health hazards in agriculture. Am. J. Respir. Crit. Care Med. **158:** S1–S76.
2. HOPPIN, J.A. et al. 2002. Chemical predictors of wheeze among farmer pesticide applicators in the Agricultural Health Study. Am. J. Respir. Crit. Care Med. **165:** 683–689.
3. HOPPIN, J.A. et al. 2006. Pesticides associated with wheeze among commercial pesticide applicators in the Agricultural Health Study. Am. J. Epidemiol. **163:** 1129–1137.
4. ALAVANJA, M.C. et al. 1996. The Agricultural Health Study. Environ. Health Perspect. **104:** 362–369.
5. TARONE, R.E. et al. 1997. The Agricultural Health Study: factors affecting completion and return of self-administered questionnaires in a large prospective cohort study of pesticide applicators. Am. J. Ind. Med. **31:** 233–242.
6. FRYER, A.D. et al. 2004. Mechanisms of organophosphate insecticide-induced airway hyperreactivity. Am. J. Physiol.Lung Cell. Mol. Physiol. **286:** L963–L969.
7. SENTHILSELVAN, A., H. MCDUFFIE & J.A. DOSMAN. 1992. Association of asthma with use of pesticides—results of a cross-sectional survey of farmers. Am. Rev. Respir. Dis. **146:** 884–887.
8. OHAYO-MITOKO, G.J.A. et al. 2000. Self reported symptoms and inhibition of acetylcholinesterase activity among Kenyan agricultural workers. Occup. Environ. Med. **57:** 195–200.
9. LEIN, P.J. & A.D. FRYER. 2005. Organophosphorus insecticides induce airway hyperreactivity by decreasing neuronal M2 muscarinic receptor function independent of acetylcholinesterase inhibition. Toxicol. Sci. **83:** 166–176.
10. SEGURA, P. et al. 1999. Identification of mechanisms involved in the acute airway toxicity induced by parathion. Naunyn Schmiedebergs Arch. Pharmacol. **360:** 699–710.
11. MEISTER, R.T., Ed. 2005. Crop Protection Handbook. Meister Media Worldwide. Willoughby, Ohio.

12. TUPI, K. *et al*. 1987. Effects of respiratory morbidity on occupational activity among farmers. Eur. J. Respir. Dis. Suppl. **152:** 206–211.
13. VOGELZANG, P.F. *et al*. 1999. Health-based selection for asthma, but not for chronic bronchitis, in pig farmers: an evidence-based hypothesis. Eur. Respir. J. **13:** 187–189.
14. POST, W., D. HEEDERIK & R. HOUBA. 1998. Decline in lung function related to exposure and selection processes among workers in the grain processing and animal feed industry. Occup. Environ. Med. **55:** 349–355.

The Assessment of Occupational Exposure to Diazinon in Nicaraguan Plantation Workers Using Saliva Biomonitoring

CHENSHENG LU,[a] TERESA RODRÍGUEZ,[b] AURA FUNEZ,[b] RENE S. IRISH,[c] AND RICHARD A. FENSKE[c]

[a]*Department of Environmental and Occupational Health, Rollins School of Public Health, Emory University, Atlanta, Georgia, 30322, USA*

[b]*Programa de Salud Ocupacional y Ambiental, Universidad Nacional Autónoma de Nicaragua, León (UNAN-León), Nicaragua*

[c]*Department of Environmental and Occupational Health Sciences, University of Washington, Seattle, Washington, 98195, USA*

ABSTRACT: A cross-sectional study with repeated sample collection in multiple days was conducted to assess diazinon exposures. Saliva and limited blood samples were collected from 10 banana plantation workers involved with diazinon application and their children aged 2–12 years living in Chinandega, Nicaragua. Diazinon concentration-time profiles in saliva varied between two plantations, which reflects the differences of work practices in each plantation. Salivary concentrations of diazinon measured in Plantation 1 applicators continued to increase 2 days after self-reported diazinon application, suggesting an ongoing exposure among these workers. However, salivary diazinon concentrations measured in Plantation 2 applicators were peaked 12 h prior to the first application, and then decreased 36 h post the first application. Diazinon concentrations in saliva were significantly correlated with the time-matched plasma samples collected from the same workers, which is in agreement with the previous published data from animal models. Children's exposure to diazinon through take-home pathway does not exist, as evident by the majority of nondetected saliva samples, and this finding was confirmed by the results from the urine samples. Severe dehydration was observed in many plantation workers and their children, resulting in the loss of some saliva samples, which no doubt have impaired the overall quality of the study results. Regardless, this article has demonstrated that saliva can be used to assess exposures to diazinon in pesticide applicators and children.

Address for correspondence: Chensheng Lu, Department of Environmental and Occupational Health, Rollins School of Public Health, Emory University, Room 226, 1518 Clifton Road, NE, Atlanta, GA 30322. Voice: 404-727-2131; fax: 404-727-8744.
e-mail: clu2@sph.emory.edu

Ann. N.Y. Acad. Sci. 1076: 355–365 (2006). © 2006 New York Academy of Sciences.
doi: 10.1196/annals.1371.057

KEYWORDS: saliva; diazinon; organophosphorus; pesticide exposure; biological monitoring

INTRODUCTION

The use of saliva in the biological monitoring program has recently been suggested as a possible tool for measuring pesticide exposure.[1,2] Saliva has been used widely in monitoring therapeutic drugs[3–5] and hormones,[6,7] and has only recently been used to measure pesticides in animals[8–11] and the farm worker population.[12,13] Other biological specimen samples, such as umbilical cord blood, meconium, and amniotic fluid, are being explored for monitoring pesticide exposure[14–17] in various populations for different purposes. However, the procedures for collecting these specimen samples can be invasive or inconvenient.

Saliva sampling, in comparison to other specimen sample collection methods, is convenient and noninvasive. Unlike blood drawing, saliva collection does not require medical personnel with special training. Saliva can be collected from a wide range of age groups with minimum discomfort. Saliva samples, however, can be contaminated with chemicals in the oral cavity. The measurements of chemical concentrations in saliva have the potential of reliability to reflect levels in plasma and, therefore, can be used as the indicator of body burden. This advantage will allow saliva measurements not only to be used for the purpose of exposure assessment but also for dose estimation as well.

Diazinon in saliva in an animal study,[11] and its salivary concentrations were highly correlated with plasma concentration under a control dosing study design. The pharmacokinetics of diazinon in saliva are not significantly different from those in arterial plasma, and diazinon concentration in saliva is significantly correlated with that of plasma. The saliva/plasma (S/P) concentration ratio of diazinon would be close to unity if the protein-unbound fraction of diazinon in plasma were used to calculate the S/P ratio. These findings suggest that the salivary concentration of diazinon can accurately predict diazinon level in the plasma, and the amount of diazinon measured in saliva reflects the unbound fraction of diazinon in plasma. The objective of this article was to validate salivary excretion of diazinon in a cohort of plantation applicators and their children in Nicaragua.

METHOD

Altogether 10 pesticide applicators employed in two banana plantations in Chinandega, Nicaragua and their children (ages 2–12 years) participated in this cross-sectional study with repeated sample collection for 3 days between

April 2003 and June 2003. Initial contact with banana plantation administrators was made via a plantation clinic doctor who had previously collaborated with the Universidad Nacional Autónoma de Nicaragua, León, Nicaragua (UNAN-Leon). Pesticide applicators who participated in this study were selected by the plantation administrators. The Human Subjects Committee at the University of Washington (03-5936-E-01), Seattle, Washington and the Ethics Committee at UNAN-Leon in Nicaragua approved the study protocol. The consent form was read to illiterate subjects. Subjects were compensated monetarily for the inconvenience and their time.

Each of the two research teams consisted of two field staffs. One team traveled to the applicators' houses to sample their children, while the other team, staffed with at least one phlebotomist/nurse, collected samples from the applicators at the banana plantation. All samples were collected in the presence of the staff who also recorded observations and obtained information from the applicators regarding pesticide use.

A complete set of saliva samples for the applicators and their children included three voids; right before application (approximately 5:30 AM), lunch break (11 AM), and the end of shift (4:30 PM) on the day of the application and the following day. A saliva sample was also collected a day prior to the scheduled application. Saliva samples were collected using a commercially available device, Salivette (Sarstadt Inc., Newton, NC), consisting of a cotton roll with its own centrifuge tube and cap. Saliva is confined within the sampling device, which then can be used for transportation and storage. Applicators and their children were asked, at each sampling time, to rinse their mouths thoroughly with bottled water twice, wait for 5 min, and then put the cotton roll into their mouths. They were asked to avoid touching the lip of the plastic tube or the cotton roll with their fingers or lips in order to minimize exogenous contamination. Subjects left the cotton roll inside their mouths for 2 min and were told to gently hold the saturated cotton roll with their teeth, guide it back into the tube, and recap the tube tightly.

Venous blood, taken by venipuncture (Green-top Vacutainer; BD, Franklin Lakes, NJ), was collected from the applicator by a phlebotomist/nurse a day before pesticide application, at the end of shift on Day 1, and at the beginning of the shift and the end of shift on Day 2. Subject's arm was carefully cleaned with alcohol pads several times before each time blood was drawn. Immediately after the blood was drawn, the Vacutainer tube was mixed by inverting the tube 10 times in order to distribute the anticoagulant evenly.

Upon collection of saliva and venous blood samples, both Salivette and Vacutainer were labeled and placed in a separated cooler with blue ice packs during transportation. After returning to the lab, saliva samples were was centrifuged at 3000 rpm for 10 min. The cotton roll and inner plastic tube were then removed and discarded, leaving the ultrafiltrate in the centrifuge tube. Salivette collects approximately 2 mL of saliva in 2 min. Vacutainer tubes were centrifuged at 5000 rpm for 10 min before removing the rubber cap from the tube. Plasma was then transferred from the Vacutainer tube to a prelabeled cryogenic vial. Saliva and venous plasma samples were stored in the freezer at $-10°C$ to $-20°C$ until shipping with dry ice to the University of Washington in Seattle, Washington for analysis.

Blank human saliva and venous plasma samples collected from volunteers without known exposure to diazinon were fortified with diazinon dissolved in methanol for the purpose of quality assurance and quality control (QA/QC). Blank specimen samples were split into two groups, one for diazinon fortification and the other one for determining the background diazinon level in the samples. Blank saliva and plasma samples were fortified with diazinon at the level of 12.2 μg/L in the lab on UNAN campus daily, carried in a cooler with blue ice packs to the field, and brought back to the lab with specimen samples collected from study subjects. Fortified specimen samples were then labeled and stored in the freezer as other field specimen samples until analysis.

Saliva and plasma samples were analyzed for diazinon using an enzyme-linked immunosorbant assay (ELISA) (EnviroGard 72700; Strategic Diagnostics Inc., Newark, DE) as described previously.[11] This ELISA kit is specific to diazinon with the limit of detection of 0.022 μg/L and some cross-activity with its oxon-metabolite and methyl pirimiphos at much higher levels of 0.2–900 μg/L, respectively.

RESULTS

Both plantations used Diazinon 60% and diluted 2 L of formula in 480 L of water. Eight and 12 20-L backpacks were applied at Plantations 1 and 2, respectively. Diazinon was sprayed approximately once a month on the ornamental plants around the perimeter of the plantation to avoid the propagation of insects from one block to another. Plantation management reported that diazinon had not been sprayed for at least 15 days prior to sampling. Applications at Plantation 1 took place between 5:30–10 AM and 1–3 PM, and from 6–10 AM and 1–3 PM at Plantation 2. Applicators wore protective equipment: coveralls, shin guards, impermeable capes, gloves, boots, and hats, but not masks. Some protective equipment used in Plantation 2 was in questionable condition as observed by the researchers.

Recoveries of diazinon from the fortified samples were comparable between saliva and plasma samples; greater variation was found in the fortified plasma samples (TABLE 1). There is approximately a 20% degradation of diazinon in saliva and plasma samples during the course of sample transportation and freezer storage. No adjustment for the degradation was made for the final concentration data.

A total of 37 blood and 62 saliva samples were collected from 10 applicators, and 62 saliva samples were collected from their children (TABLE 2). Altogether 14 and 13 saliva samples collected from applicators and children, respectively,

TABLE 1. Recovery efficiencies of diazinon from field fortified human venous plasma and saliva samples using the ELISA

Sample type	n	Fortify level (μg/L)	Observed level (μg/L)			Average Recovery [a] (%)
			Mean	c.v.	Range	
Venous plasma	6	0 [b]	0.7	84.5	0–1.6	
	6	12.2	10.3	54.4	4.5–14.3	84.3
Saliva	6	0 [b]	0.1	142.9	0–0.7	
	6	12.2	9.8	17.3	7.5–12.6	80.6

[a] Average recovery (%) = (observed level/fortify level) × 100%.
[b] Blank specimen samples.
c.v. = Coefficient of variations.

did not contain saliva after centrifugation. Seven saliva samples collected from one applicator at Plantation 2 were excluded from data analysis due to foreign objects found in all the saliva samples. Two of the subjects at Plantation 1 did not apply diazinon; one mixed the pesticide, and the other was a supervisor. Their data were also excluded from the analysis.

All the plasma samples, including those collected 1 day before the application from applicators contained detectable diazinon concentrations. Diazinon was found in 22 of 41 (54%) saliva samples collected from applicators and in 20 of 49 (41%) saliva samples collected from children. Diazinon concentrations measured in children's saliva samples were mostly low, and were not included in the data analysis. This decision is supported by the measurement of diazinon metabolite concentrations in the urine samples in which only 5% of children's urine samples contained detectable diazinon metabolite.[18]

Data were analyzed separately for Plantations 1 and 2 due to different diazinon application rates. Diazinon was detectable in the plasma samples taken from applicators at Plantation 1 1 day before their scheduled spray (TABLE 3), and this level remained relatively constant for the following 2 days, including the spray day (FIG. 1). Diazinon concentrations in saliva collected from the same group of applicators were not detectable before the scheduled spray,

TABLE 2. Numbers of venous blood and saliva samples collected from applicators and their children, and the ranges of diazinon concentrations in each type of specimen samples

Sample type	Total number of sample collected	Number of sample contained no. volume	Number of sample below LOD[a]	Number of samples above LOD[a]	Range of diazinon concentration
Venous plasma	37	0	0	37	0.05–0.28
Saliva (applicator)	55 [b]	14	19	22	0–0.15
Saliva (children)	62	13	29	20	0–0.11

[a] LOD=limit of detection; 0.022 μg/L.
[b] Seven saliva samples from one applicator at Plantation 2 excluded.

TABLE 3. Average diazinon concentrations in saliva and venous plasma collected from applicators at Plantations 1 and 2 during the 3-day study period

Plantation	Hours after 1st diazinon application [a]	Average diazinon concentration in saliva (μg/L) [b]	Average diazinon concentration in plasma (μg/L) [b]	S/P diazinon concentration ratio [c]
1	-12 [d]	0	0.159 (12)	0
	0	0 (0)		
	6	0.044 (45)		
	12	0.047 (43)	0.175 (30)	0.27
	24	0.041 (n.a.) [e]	0.171 (42)	0.24
	30	0.061 (25)		
	36	0.045 (5)	0.178 (23)	0.25
2	-12 [d]	0.046 (23)	0.119 (25)	0.39
	0	0.039 (36)		
	6	0.028 (37)		
	12	0.020 (88)	0.094 (22)	0.22
	24	0.011 (173)	0.068 (19)	0.16
	30	0.010 (173)		
	36	0.004 (173)	0.070 (20)	0.06

[a] The first diazinon application began at 5:30 AM at Plantation 1 and 6:00 AM at Plantation 2 and ended at 10:30 AM for both plantations. The second diazinon application began at 1:00 PM and ended at 3:00 PM for both plantations.
[b] Coefficient of variations (c.v.) in parentheses.
[c] S/P concentration ratio = diazinon concentration in saliva per diazinon concentration in plasma.
[d] Samples collected approximately 12 h prior to diazinon application.
[e] c.v. was not available.

increased as diazinon application started, and then gradually decreased for the rest of Day 1. Diazinon concentrations in saliva in Day 2 followed the same pattern as in Day 1, except the average peak concentration was higher on Day 2 than on Day 1. It appeared that applicators continued to be exposed to diazinon after the application on Day 1. Concentration-time profiles for diazinon corresponded better between plasma and saliva samples collected from applicators at Plantation 2 (FIG. 2). Diazinon concentrations in plasma and saliva samples were already at the peak levels (0.046 and 0.019 μg/L for saliva and plasma samples, respectively) before the scheduled spray, and then subsequently decreased throughout the study period (TABLE 3).

Diazinon concentrations in plasma samples were significantly correlated with those in saliva for the 21 pairs of time-matched saliva and plasma samples (Pearson test, $r = 0.521$, $P = 0.015$) (FIG. 3). Data not included in this analysis were 12 missing saliva samples and 4 questionable saliva samples with foreign objects collected from applicator at Plantation 2. The individual S/P concentration ratio of diazinon, ranging from 0.06 to 0.39 (TABLE 3), is not significantly different from the average S/P ratio (one-sample t-test), which is 0.23, and between two plantations (one-way ANOVA).

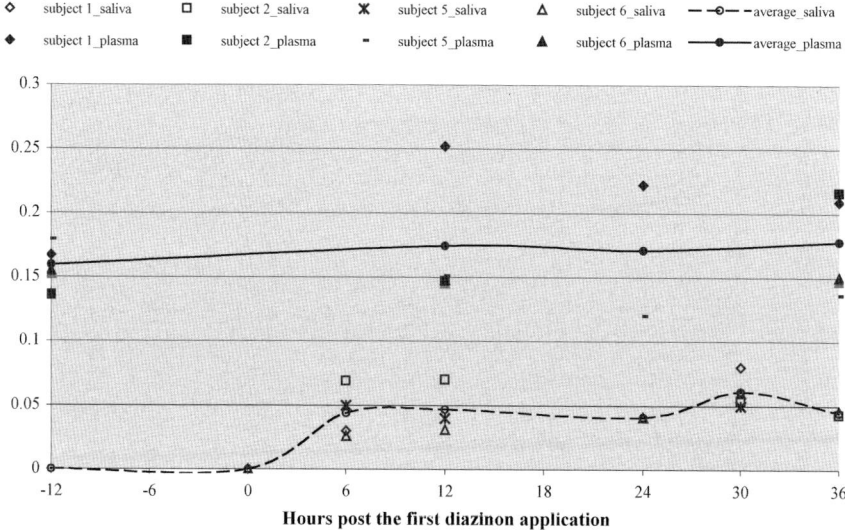

FIGURE 1. Individual and the average venous plasma and saliva concentration-time profiles of diazinon for applicators at Plantation 1.

DISCUSSION

This study presents the first human data to validate the feasibility of using saliva samples as tools to assess pesticide exposure in humans. A previous study demonstrated the feasibility of saliva biomonitoring for assessing commercial herbicide applicators' exposure to atrazine[13]; but no blood samples were taken in that study to confirm the presence of atrazine in saliva.

Applicators exhibited a very different exposure profile between the two plantations, and the most striking differences were the timing and the degree of exposure. It appears that the applicators at Plantation 1 exposed to diazinon as the result from the two applications, and the exposure continued in Day 2 after the termination of application in Day 1. Saliva samples collected prior to the scheduled application indicated that applicators at Plantation 1 were not exposed to diazinon but their plasma diazinon concentrations contradicted this finding. Unfortunately, there is no plausible explanation for this discrepancy. Data collected from applicators at Plantation 2 also suggested that diazinon has been used prior to the scheduled applicator day. Although plantation managers assured that the plantations had not been sprayed with diazinon for at least 15 days, it appears that applicators at both plantations were exposed to diazinon through different pathways or tasks that were not identified.

The findings from this study are supported by the urinary results, as published separately,[18] which showed the diazinon metabolite, 2-isopropxy-4-methyl-pyrimidinol (IMPY), was measured in urine samples collected from

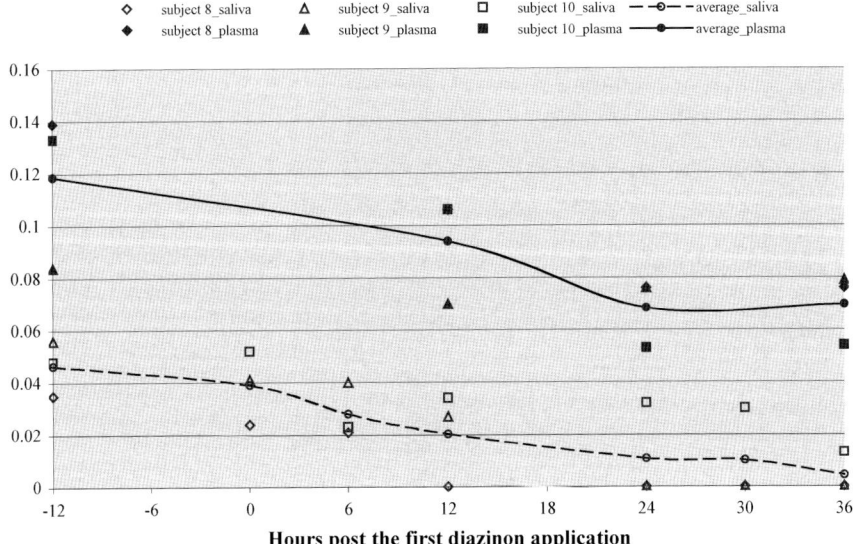

FIGURE 2. Individual and the average venous plasma and saliva concentration-time profiles of diazinon for applicators at Plantation 2.

applicators and their children. At both plantations, peak IMPY levels were reached after the second diazinon application, approximately 6 h after first diazinon application. However, it seems evident that the peak IMPY concentrations resulted from previous diazinon application that was not reported to us or other activities associated with contacting with diazinon before the scheduled spray date. The concentration-time profiles of IMPY excretion for applicators at both plantations approximately match the concentration-time profiles of plasma and saliva over the 3-day study period.

Diazinon concentrations measured in saliva collected from applicators were consistently lower than those in plasma, and on average the amount of diazinon in saliva only accounts for 23% of diazinon in plasma. This finding is consistent with the previous results published using animal models[11,19] except that the average S/P concentration ratio of diazinon was higher in humans than in rats. The probable explanation for this discrepancy is the different degree of protein binding of diazinon between human and rat. Although limited by a relatively small sample size, it is prudent to suggest that salivary excretion of diazinon in human, as well as in rat, is constant and mainly occurs by passive intracellular diffusion. Therefore, the fraction of protein-bound diazinon would not be able to cross the cell membrane of salivary glands because of the larger molecular size, which leads to a much lower concentration in saliva. In other words, by taking into account the protein binding of diazinon in plasma, the level of diazinon measured in saliva may represent the protein-unbound fraction of diazinon in plasma.

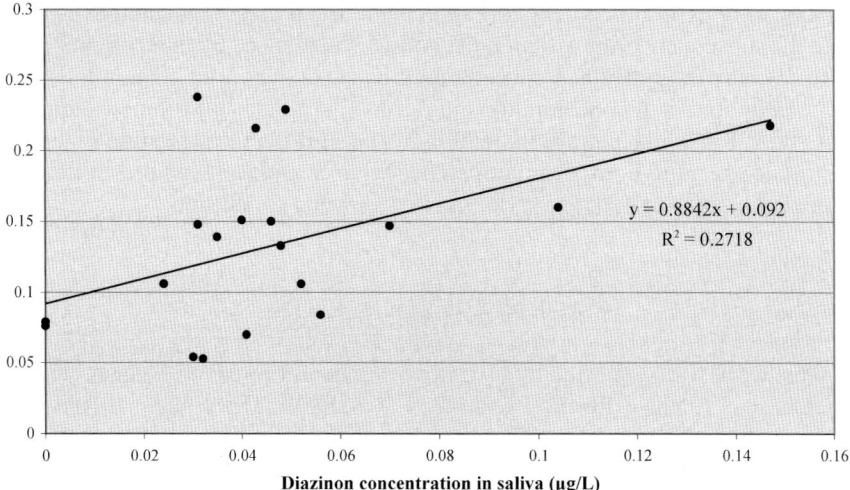

FIGURE 3. Regression model for venous plasma and saliva concentrations of diazinon measured in applicators at Plantations 1 and 2.

Several logistic reasons in the field encumbered the conduction of this study and limited the interpretation of the study results. First of all, although we were able to recruit study subjects from the plantations, the accessibility to the plantation applicators by the field staff during the working hours was somewhat restricted. The information regarding the tasks of those applicators before and after the diazinon application was not available, and therefore, limited the interpretation of several confounding results. Second, in the areas where the families lived, there was no electricity at night, which made sample collection during the evening hours impossible. One of the advantages of saliva biomonitoring is that sample can be collected from individuals repeatedly without ethical or medical consideration. Unfortunately, this limitation constrains the timing and the frequency of saliva sample collection. As shown in FIGURE 1, additional two saliva samples collected during lunch breaks in Days 1 and 2 provided more useful pharmacokinetic information than the three plasma samples collected over the 30-h study period. Last, approximately 20% of saliva samples from applicators contained no saliva. Dehydration seems to be the logical explanation because most of the missing saliva samples from applicators were collected at the end of work shift and to a less degree, during the early morning hours (6 AM) from both the applicators and their children. Information regarding the distribution and the subsequent excretion of diazinon as measured in saliva after exposure is therefore not sufficient for further pharmacokinetic analysis.

Regardless of those limitations, results from this study demonstrate that saliva can be used as an alternative specimen sample to blood for the purpose of assessing exposure to diazinon. In general, trend due to diazinon exposure and the subsequent elimination in the body are evidenced by the temporal pattern of decreasing diazinon concentration in saliva over time. Such trends are echoed by diazinon concentrations in plasma, particularly for blood samples collected from applicators at Plantation 2. The significant correlation of diazinon concentrations between time-matched saliva and plasma samples collected from applicators at both plantations confirmed this conclusion.

ACKNOWLEDGMENTS

This work was supported by U.S. Environmental Protection Agency STAR (Science to Achieve Results) Grant No. R-828606. The content of this article does not necessarily reflect the views of the Agency, and no official endorsement should be inferred. Additional support was provided through the National Institute of Health-Fogarty program for International Scholars in Occupational and Environmental Health (5 D43 TW00642), and the Research Department of the Swedish International Development Agency (SIDA-SAREC). We thank Sarah Weppner, Lisa Younglove, and Maria Tchong's assistance either in the field where this work was conducted in Nicaragua, or in the lab where the sample analysis was taken place in University of Washington, Seattle, WA. We also thank Nicaraguan families, who participated in this study and Dr. José René Gasteazoro H. in Nicaragua who assisted us in subject recruitment.

REFERENCES

1. BARR, D.B. *et al*. 2005. Biological monitoring of exposure to environmental chemicals throughout the life stages: requirements and issues for consideration for the Nacional Children's Study. Environ. Health Perspect. **113:** 1083–1091.
2. BRADMAN, A. & R.M. WHYATT. 2005. Characterizing exposure to nonpersistent pesticidas during pregnancy and early childhood in the Nacional Children's Study: a review of monitoring and measurement methodologies. Environ. Health Perspect. **113:** 1092–1099.
3. DORBITCH, R.K. & C.K. SVENSSON. 1992. Therapeutic drug monitoring in saliva—an update. Clin. Pharmacol. **23:** 365–379.
4. SCHRAMM, W. *et al*. 1992. Drugs of abuse in saliva: a review. J. Anal. Toxicol. **16:** 1–9.
5. NAVARRO, M. *et al*. 2001. Usefulness of saliva for measurement of 3,4-methylenedioxymethamphetamine and its metabolites: correlation with plasma drug concentrations and effect of salivary pH. Clin. Chem. **47:** 1788–1795.
6. VINING, R.F. *et al*. 1983. Saliva estriol measurements: an alternative to the assay of serum unconjugated estriol in assessing fetoplacental function. J. Clin. Endocrinol. Metab. **56:** 454–460.

7. MCGREGOR, J.A. *et al*. 1995. Salivary estriol as risk assessment for preterm labor: a prospective trial. Am. J. Obstet. Gynecol. **173:** 1337–1342.
8. LU, C. *et al*. 1997a. Determination of atrazine levels in whole saliva and plasma in rat: potential for salivary monitoring for occupational exposure. J. Toxicol. Environ. Health. **50:** 101–111.
9. LU, C. *et al*. 1997b. Correspondence of salivary and plasma concentrations of atrazine in rats under variable salivary flow rate and plasma concentration. J. Toxicol. Environ. Health **52:** 317–329.
10. LU, C. *et al*. 1998. Salivary concentrations of atrazine reflect free atrazine plasma levels in rats. J. Toxicol. Environ. Health **53:** 283–292.
11. LU, C. *et al*. 2004. Biological monitoring of diazinon exposure using saliva in an animal model. J. Toxicol. Environ. Health **66:** 2315–2325.
12. NIGG, H.N. *et al*. 1993. Quantification of human exposure to ethion using saliva. Chemosphere **26:** 897–906.
13. DENOVAN, L.A. *et al*. 2000. Saliva biomonitoring of atrazine exposure among commerical herbicide applicators. Intl. Arch. Occup. Environ. Health **73:** 457–462.
14. FOSTER, W. *et al*. 2000. Detection of endocrine disrupting chemicals in samples of second trimester human amniotic fluid. Clin. Endocrinol. Metab. **85:** 2954–2957.
15. MUCKLE, G. *et al*. 2001. Prenatal exposure of the northern Quebec Inuit infants to environmental contaminants. Environ. Health Perspect. **109:** 1291–1299.
16. WHYATT, R.M. & D.B. BARR. 2001. Measurement of organophosphate metabolites in postpartum meconium as a potential biomarker of prenatal exposure: a validation study. Environ. Health Perspect. **109:** 417–420.
17. BRADMAN, A. *et al*. 2003. Measurement of pesticidas and other toxicants in amniotic fluid as a potencial biomarker of prenatal exposure: a validation study. Environ. Health Perspect. **111:** 1779–1782.
18. RODRÍGUEZ, T. *et al*. 2006. Biological monitoring of pesticide exposure among applicators and their children in Nicaragua. Int. J Occup. Environ. Health (In press).
19. WU, H.X. *et al*. 1996. Diazinon toxicokinetics, tissue distribution and anticholinesterase activity in the rat. Biomed. Environ. Sci. **9:** 359–369.

Cancer and Pesticides

An Overview and Some Results of the Italian Multicenter Case–Control Study on Hematolymphopoietic Malignancies

LUCIA MILIGI,[a] ADELE SENIORI COSTANTINI,[a] ANGELA VERALDI,[a] ALESSANDRA BENVENUTI,[a] WILL,* AND PAOLO VINEIS[b]

[a]*Environmental and Occupational Epidemiology Unit, Centre for Study and Prevention of Cancer, Istituto Toscano Tumori, 50135 Florence, Italy*

[b]*University of Turin, 10126 Turin, Italy, and Imperial College London, London W2 1PG UK*

ABSTRACT: Exposure to pesticides is recognized as an important environmental risk factor associated with development of cancer. Epidemiological studies, although sometimes contradictory, have linked phenoxy acid herbicides with non-Hodgkin's lymphoma (NHL) and Soft Tissue Sarcoma (STS); organochlorine insecticides with STS, NHL, and leukemia; organophosphorous compounds with NHL and leukemia; and triazine herbicides with ovarian cancer. Exposure assessment is a crucial point in studying the association between cancer and pesticides. In order to investigate the association between hematolymphopoietic malignancies and occupational exposures, including pesticides, a population-based case–control study was carried out in Italy in 11 areas, 9 of which are agricultural or mixed areas. All newly diagnosed cases of hematolymphopoietic malignancies were collected in a 3-year period (1991–1993). The control group consisted of a random sample of the population residing in each area. The approach to infer exposures in agriculture was based on: the use of an agricultural questionnaire with 24 crop-specific questionnaires; expert agronomists who reviewed the collected information for each subject and translated it into pesticides histories. In total, 1925 cases and 1232 controls were interviewed in the nine agricultural areas. Increased risk was observed for some specific classes of pesticides. Furthermore, a nonstatistically significant increased risk of NHL was observed for subjects who were exposed to phenoxy herbicides not using protective equipment and a significant increased risk for exposure to 2, 4-dichlorophenoxy acetic acid (2,4-D).

*WILL: Italian Working Group Leukemia Lymphomas (Oriana Nanni, Forlì; Valerio Ramazzotti, Rome; Rosario Tumino, Ragusa; Emanuele Stagnaro, Genova; Stefania Rodella, Verona and Florence; Arabella Fontana, Novara; Carla Vindigni, Siena)

Address for correspondence: Lucia Miligi, DSc, Istituto Toscano Tumori, Environmental and Occupational Epidemiology Unit, Centre for Study and Prevention of Cancer; via di S.Salvi 12, 50135 Florence, Italy. Voice: 0039-055-6268348; fax: 0039-055-679954.
 e-mail: l.miligi@cspo.it

KEYWORDS: hematolymphopoietic malignancies; non-Hodgkin's lymphoma; leukemia; multiple myeloma; pesticides; exposure assessment; phenoxy herbicides

INTRODUCTION

The term *pesticides* encompass a heterogeneous group of chemicals developed to control a variety of pests. Pesticides are generally categorized according to the type of pest for which they have shown efficacious action. The primary categories are: insecticides, herbicides, and fungicides. All are broad categories with different chemical classes with a wide diversity of chemicals having different properties and different effects.

Pesticides have been employed since ancient time to control insects. Inorganic compounds, in particular arsenic compounds, were the first to be used. The discovery of insecticidal dichlorodiphenyl trichloroethane (DDT) during 1940s in Switzerland was the first step of the modern chemical age of pesticides. Organophosphates insecticides and phenoxy herbicides came approximately in use during this period. The production of carbamates and organochlorines started approximately during this time, but, during the 1960s and 1970s, many other chemicals were developed and introduced in the market. Finally in the last 20 years new generations of pesticides, such as pyrethroids, from naturally occurring pyrethrins, were developed.[1]

Meanwhile the use of pesticides has increased in both developed and developing countries during last decades, there has been a concern about their long-term effects on human health. Exposure to pesticides can lead to problems ranging from acute illness to chronic effects, such as reproductive effects, neurological effects, and cancers.

This article presents a brief review on the carcinogenicity of pesticides. Some results of the Italian multicenter case–control study on hematolymphopoietic malignancies are also presented.

Pesticides and Cancer

A number of pesticides or groups of pesticides have been identified as certain, probable, and possible carcinogens to humans by several national and international institutions.[2,3] Some pesticides have been banned from the market in the United States and in the European Community. Some pesticides are known to be genotoxic or tumor promotive, whereas others possess hormonal, immunotoxic, or hematotoxic properties.[1]

Farmers and industrial workers employed in chemical plants for manufacture of pesticides are considered at high-risk group to developing cancer following pesticide exposure. Considering farmers, several agricultural tasks may involve exposure to pesticides including mixing, loading, distribution, maintenance, repair of machinery and tools, and reentry into treated areas.

Several mortality studies were conducted in different countries with the aim to evaluate the risk of cancer among farmers. Farmers are at lower risk for most major causes of death than general population particularly for total mortality, total cancer, heath diseases and some specific cancers, such as lung cancer, but, in spite of the favorable overall cancer risk, this occupational group tends to experience increase risks for selected types of cancers especially Soft Tissue Sarcomas (STSs), non-Hodgkin lymphomas (NHLs), Hodgkin's disease (HD), leukemia, multiple myeloma (MM), and cancer of the skin and prostate.[1,4] Cancers of the breast, testis, endometrium, ovary, liver, and bladder have also been occasionally associated with pesticide exposure.[1] A number of studies have specifically considered pesticides applicators and manufacturers characterized by longer and more intense exposure.[5]

Most of the information on the carcinogenic risk of pesticides derives from case–control studies of selected cancer sites and, although they are limited by the difficulty in identifying the association with individual pesticides, these studies have suggested that some classes of pesticides are linked with cancer. In particular some studies, have linked phenoxy herbicides with STS, NHL, MM, and leukemia; organophosphorous compounds with NHL and leukemia; and triazine herbicides with ovarian cancer.[1]

Furthermore organochlorine compounds were linked with breast cancers and prostate cancer was associated mainly with organophosphates and organochlorine compounds.[6]

Exposure to different classes of pesticides has been linked with hematolymphopoietic malignancies and attention has been particularly focused on NHL. The incidence of this cancer has been increasing for the last 40 years in the United States and in other industrialized countries.[7] The possible role of pesticides in this increase has been suggested.

NHL and Pesticides

The first observations come from cohort studies on farmers in which increased risk of NHL was suggested,[4,8] but the role of specific pesticides was mainly investigated in the contest of case–controls studies. Particular attention has been given to the possible role of phenoxy herbicides. After the first study conducted in Sweden found increased risk of malignant lymphomas among agricultural workers exposed to phenoxy herbicides,[9] other case–control studies in New Zealand,[10] United States,[11–14] Sweden,[15,16] Canada,[17] and Australia[18] have focused on this topic. The observed results of these studies showed variations in relative risk estimates; the first Sweden study observed a significant increased RR of 4.7,[9] also confirmed in the following study by Persson,[15] but the magnitude of the risk was not replicated in the other following Swedish studies.[16] The U.S. study conducted in Kansas[11] showed a twofold increased risk and a study in Nebraska[12] observed a nonsignificant risk of 1.5, but others U.S. studies[13,14] showed only slight increased risks. Also a study in New Zealand found little evidence of increased risk for this exposure and NHL.[10] Finally, a Canadian study showed increased risk[17] for a broad range of chemical classes of pesticides including phenoxy herbicides.

Among cohort studies, the International Agency for Research on Cancer collaborative study among production workers and professional sprayers deserves attention. This study, that has brought together 21,863 workers in 36 cohorts in 12 countries,[19,20] showed a slight increased risk for exposure to phenoxy herbicides.

Differences in study results may depend on types of specific phenoxy herbicides used (e.g., 2,4-dichlorophenoxyacetic acid [2,4-D] and 2,4,5-trichlorophenoxyacetic acid [2,4,5-T]), different levels or absence of contaminations[1] of 2,3,7,8-tetrachlorodibenzo-p-dioxin (TCDD) a possible contaminant of 2,4,5-T[21]; differences in agricultural practices, determined also by climate, could also be considered.

Increased risk of NHL was also associated with different types of pesticides namely organophosphates and organochlorine insecticides, triazines herbicides.[9,12,13,22,23]

The Italian Multicenter Case–Control Study on Hematolymphopoietic Malignancies

We conducted a population-based case–control study to investigate, among others, the association between pesticide exposure and hematolymphopoietic malignancies. The study was conducted in 11 different areas in Italy, 5 of which were areas with a high prevalence of agricultural activities, and 4 were mixed areas. The predominant classes of pesticides used in the agricultural or mixed areas, due to the prevalent crops, were insecticides and fungicides. Results concerning occupations[24,25] and overall results on pesticide exposures by gender have been published elsewhere.[26]

Methods

We identified all incident cases of malignancies of the hematolymphopoietic system among residents of both sexes, aged 20–74 years. The control group was formed by a random sample of the general population resident in each of the areas under study, stratified by sex and 5-year age groups. We conducted person-to-person interviews to obtain information on a wide range of known or suspected risk factors. The association between pesticides, specifically phenoxy herbicides, and NHL was of major concern. For this reason a specific section of the questionnaire was devoted to agricultural work. A general agricultural questionnaire and 24 crop-specific questionnaires were explicitly designed for the crops commonly grown in the study areas (an example of crop-specific questionnaires is available on request). The subjects were also interviewed about specific crop diseases and whether pesticides were used. Detailed data were also collected on protective equipments used, period, and frequency of treatments, on means of application and reentry into the fields after treatment.

Exposure assessment was performed by experienced agronomists (one for each area). They reviewed blindly the information collected in the agricultural section of the questionnaire for each subject and translated it into pesticide exposure histories in terms of: type of treatment (e.g., herbicides), chemical families used (e.g., phenoxy acid), and active ingredients used (e.g., 2,4-D). A probability of usage in terms of chemical families and active ingredients was assigned according to an ordinal scale (low, medium, and high) considering: time period, crops and crop diseases or treatments applied, and area.

Data analyses were performed with the SAS package for the personal computer. Point estimates of odds ratios (OR) and the corresponding confidence intervals (95% CI) were calculated adjusting for age in five categories. We fitted logistic regression analyses for the variables of interest controlling for area and age. Cases of NHL and chronic lymphocytic leukemia (CLL) were considered together in the analyses, due to substantial biological similarities between the two diseases. For exposure in agriculture, the data for men and women, when it was possible in terms of number of exposed subjects, were analyzed separately for the nine prevalent and mixed agricultural areas. Analyses of data were performed at different levels considering crops, treatments, chemical classes, and individual pesticides. The analysis of the chemical classes of pesticides and specific pesticide was limited to the subjects to whom experts assigned medium or high probability of usage. Those who had never worked in agriculture formed the reference category.

Results

In total 1145 cases of NHL and CLL, 430 of leukemia, 258 of HD, 205 MM, and 1232 controls were interviewed in the nine agricultural or mixed areas. In a previous article we have presented results by types of crops grown and chemical classes for NHL and leukemia.[26] Concerning crops grown, we have found nonsignificant increased risk of leukemia for women in flower growing, for men in vegetables greenhouses, and for both genders engaged in orchards. A slight increasing risk of NHL for both genders that had worked in orchards was also observed. Data on MM showed increased risk for men and women working in orchards (men: OR 2.1; 95% CI 0.9–4.8, 12 exposed cases; women OR 2.2; 95% CI 0.7–6.6, 7 exposed cases; men and women OR 2.1; 95% CI 1.1–4.0, 19 exposed cases). Concerning HD, due to the small numbers of

TABLE 1. Exposed cases, OR, and 95%CI for NHL (ICD IX: 200,202) and CLL (ICD IX: 204.1) by type of treatment

	Men			Women			Men and women		
	Exp. cases	OR*	95%CI	Exp. cases	OR*	95%CI	Exp. cases	OR**	95%CI
Fungicides	134	0.8	0.6–1.1	53	0.8	0.5–1.2	187	0.8	0.6–1.0
Herbicides	49	0.8	0.5–1.3	24	1.3	0.7–2.5	73	1.0	0.7–1.4
Fumigants	7	1.0	0.2–3.7	0			7	0.6	0.2–1.9
Insecticides	104	0.8	0.6–1.1	43	0.9	0.5–1.4	147	0.8	0.6–1.1
Molluskicides	13	1.2	0.5–3.1	4	0.8	0.2–3.8	17	1.1	0.5–2.4
Rodenticides	20	0.7	0.4–1.3	6	1.2	0.4–4.1	26	0.8	0.5–1.5
For animal breeding	21	0.7	0.5–1.0	5	0.7	0.4–1.3	26	0.7	0.5–1.0
Seed treatments	50	0.8	0.5–1.3	11	0.9	0.4–2.2	61	0.9	0.6–1.3

*OR adjusted for age and area.
**OR adjusted for sex, age, and area.

exposed subjects, results are presented not separately by gender, but for this lymphoma we did not find increased risk by crop grown except for vegetables growing (OR 1.6; 95% CI 0.6–4.3, 6 exposed cases).

Results by types of treatments are presented for the pathologies under investigation in TABLES 1, 2, 3, and 4. No exposures to any type of treatment were found to be significantly associated with the pathologies under investigation. Increased risk of leukemia was observed for exposure to herbicides in both genders and for insecticides in women. Subjects exposed to herbicides were also at risk for MM.

Considering chemical classes of pesticides or individual pesticides, some increased risks of NHL and leukemia were found.[26] Among fungicides we

TABLE 2. Exposed cases, OR, and 95%CI for leukemias (ICD IX: 204–208) by type of treatment

	Men			Women			Men and women		
	Exp. cases	OR*	95% CI	Exp. cases	OR*	95% CI	Exp. cases	OR**	95% CI
Fungicides	63	0.8	0.6–1.1	28	1.0	0.6–1.7	91	1.0	0.7–1.3
Herbicides	20	1.2	0.7–2.2	8	2.4	0.9–6.7	28	1.4	0.8–2.3
Fumigants	1			1			2		
Insecticides	45	0.8	0.6–1.3	27	1.4	0.8–2.4	72	1.0	0.7–1.4
Molluskicides	5	1.0	0.3–3.2	1	0.5	0–4.8	6	0.9	0.3–2.5
Rodenticides	5	0.5	0.2–1.3	0			5	0.4	0.1–1.2
For animal breeding	4	0.7	0.4–1.1	2	0.9	0.4–1.9	6	0.7	0.5–1.1
Seed treatments	18	0.7	0.4–1.3	2	0.4	0.1–2.0	20	0.7	0.4–1.2

*OR adjusted for age and area.
**OR adjusted for sex, age, and area.

TABLE 3. Exposed cases, OR and 95%CI for MM (ICD IX: 203) by type of treatment

	Men			Women			Men and women		
	Exp. cases	OR*	95% CI	Exp. cases	OR*	95% CI	Exp. cases	OR**	95% CI
Fungicides	32	1	0.6–1.7	17	1.0	0.6–1.9	49	1.0	0.7–1.5
Herbicides	8	1.4	0.6–3.5	3	3.2	0.7–14.7	11	1.6	0.8–3.5
Fumigants	3	3.5	0.7–17.8	0			3	2.8	0.6–12.2
Insecticides	28	1.2	0.7–2.1	11	0.8	0.4–1.8	39	1.1	0.7–1.6
Molluskicide	2	0.9	0.2–4.9	0			2	0.6	0.1–2.8
Rodenticides	5	1.1	0.4–3.4	1	0.6	0.1–6.2	6	1.0	0.4–2.7
For animal breeding	2	0.7	0.3–1.4	3	1.0	0.5–2.3	5	0.8	0.4–1.4
Seed treatments	8	0.7	0.3–1.6	5	1.0	0.3–3.0	13	0.7	0.4–1.5

*OR adjusted for age and area.
**OR adjusted for sex, age, and area.

have found to be at risk of NHL men and women exposed to phenylimides, and women exposed to the nitro derivatives. Considering individual pesticides women exposed to dinocap were at significantly increased risk; men exposed to vinclozolin and both genders exposed to ziram were also at risk but not at significant level. Exposure to different classes of fungicides was also associated with leukemia, and women seemed to be particularly at risk. Significantly increased risk was observed for exposure to nitro derivatives and dinocap among women, nonsignificantly increased risk was observed for exposure to dithiocarbammates. Those exposed to cyclohexanes insecticides were at risk for NHL, and among men, those exposed to some classes of organophosphates compounds, such as phosphonates and phosphoroamidon; furthermore, also use of hydrocarbons derivatives and insecticide oils was associated with NHL.

TABLE 4. Exposed cases, OR, and 95%CI for HDs (ICD IX: 201) by type of treatment

	Men			Women			Men and women		
	Exp. cases	OR*	95% CI	Exp. cases	OR*	95% CI	Exp. cases	OR**	95% CI
Fungicides	14	0.7	0.4–1.3	3	0.3	0.1–1.2	17	0.6	0.3–1.0
Herbicides	5	0.4	0.1–1.3	1	0.5	0.1–4.0	6	0.4	0.2–1.2
Fumigants	0			0			0		
Insecticides	12	0.7	0.3–1.4	4	0.4	0.1–1.5	16	0.6	0.3–1.1
Molluskicide	0			0			0		
Rodenticides	2	0.9	0.2–4.3	0			2	0.7	0.1–2.9
For animal breeding	0	0.8	0.3–1.8	0			0	0.7	0.4–1.5
Seed treatments	3	0.5	0.1–1.6	1	1.0	0.1–8.2	4	0.5	0.2–1.6

*OR adjusted for age and area.
**OR adjusted for sex, age, and area.

The use of some particular classes of insecticide was found to be at risk also for leukemia, namely cyclohexanes and a significantly increased risk was observed among men exposed to insecticide oils. Those, among women, exposed to herbicides of the chemical classes of amides and triazines were at increased risk, and both genders exposed to organotin were at increased risk. When we looked at leukemia and exposure to herbicides, we found a nonsignificant increased risk for use of phenoxy herbicides (OR 1.7; 95% CI 0.7–4.2, 8 exposed cases) and triazines (OR 1.7; 95% CI 0.6–4.71, 6 exposed cases) for men and women together.

Exposure to different classes of pesticides was also associated with a slightly increasing risk of MM. Increased ORs (adjusted for age, sex, and areas) were observed among men and women exposed to dithiocarbammates, fungicides (OR 1.6; 95% CI 0.9–2.8, 21 exposed cases), carbamates (OR = 1.4; 95% CI 0.6–3.2, 8 exposed cases), and tyolophosphates insecticides (OR 1.4; 95% CI 0.7–2.7, 14 exposed cases). Slight increased risk was observed for exposure to diphenylethanes (OR = 1.3; 95% CI 0.6–2.6, 13 exposed cases).

Phenoxy Herbicides and NHL

Principal aim of our study was to study the possible association between herbicides, particularly phenoxy herbicides and NHL. In TABLE 5 results for exposure to phenoxy herbicides are shown. When we restricted the analysis to subjects who never used protective equipments, we found a nonsignificantly increased risk for the use of phenoxy herbicides (OR = 2.4; 95% CI 0.9–7.6), a significantly increased risk for the use of 2,4-D (OR = 4.4; 95% CI 1.1–29.1), and a nonsignificantly increased risk for 4-chloro-2-methylphenoxyacetic acid (MCPA) (OR = 3.4 95% CI 0.8–23.2).

TABLE 5. Risk of NHL and exposure to phenoxy acid herbicides in Italy

	Cases	Controls	OR*	95% CI
Phenoxy herbicides				
Overall	32	28	1.1	0.6–1.8
Probability of use >low and lack of protective equipment	13	6	2.4	0.9–7.6
2,4-D				
Overall	17	18	0.9	0.5–1.8
Probability of use >low and lack of protective equipment	9	3	4.4	1.1–29.1
MCPA				
Overall	18	19	0.9	0.4–1.8
Probability of use >low and lack of protective equipment	7	3	3.4	0.8–23.2

*OR adjusted by gender, age, and area.

Final Remarks

Exposure to pesticides has been associated with cancer development. Farmers and other agricultural populations may experience an excess risk of several cancers as suggested in different cohort studies. One limit of the cohort studies is represented by the crude classification of exposure based in some cases only on the possession of pesticides purchase license, which inevitably resulted in the misclassification of those effectively exposed. Differences in the results among studies may also reflect different distribution of prevalent crops and consequently different exposure, seasonal variations are additive problems.

More detailed information on exposure came from case–control studies in which some chemical classes of pesticides have been associated with different types of cancers.[1,6] Hematolymphopoietic malignancies are the more frequently associated cancer with a variety of agricultural chemical exposures. The association between NHL and pesticides was intensively investigated and studies conducted with this aim may serve as "paradigm" to understand the complexity of the epidemiological studies on cancer and pesticides.

A critical point is the exposure assessment. In fact, the design and interpretation of epidemiological investigations are hampered by the complexity of exposure patterns and the difficulty in documenting past exposure. The major difficulty is in evaluating the risk associated with specific pesticides. Change in usage of specific chemicals over time and variations in work practices are additional problems. Consequently, misclassification of exposure may seriously affect risk estimates.

Differences in study results could be due to the choice of exposure variables. For example, use of crop as risk variable, or the use of broad categories, such as "herbicides" or "fungicides" or "insecticides" that include different classes of chemicals with different properties are roughly indicative of different patterns of exposures. Differences may also be due to the fact that within chemical classes active ingredients with diverse chemical structures and different mutagenic, carcinogenic, or immunotoxic properties may be included. Furthermore, different commercial formulations may have different percentages of active ingredients and contain inert ingredients or solvents.

Results of the Italian multicenter case–control study showed that exposures to nitro derivatives and phenylimides among fungicides, hydrocarbon derivatives, phosporoamidon, and insecticides oils among insecticides, and amines ad triazines herbicides were associated with the pathologies under investigation. Men and women experienced both similar and different risks. Important were the different operations in which women were involved in Italy, particularly the task of reentry or assisting in the treatment. These tasks, too, may entail pesticide contamination, in particular skin contaminations.[27,28] In the past, moreover, a different perception of risk may have been experienced, and consequently attitudes about using protective equipments might have been different.

The mode and types of pesticides to which women were exposed are often different from exposures experienced by men.

Finally, the positive association found between phenoxy herbicides exposures and NHL, when we restricted the analysis to the subjects who never used personal protective equipment, confirms previously reported associations[9,11,12,16,17] and the importance of considering exposure variables, such as the lack of protective equipment. The use of protective equipment, in fact, is one of the most important factors affecting the levels of exposure.[29] In a previous study in the United States a higher risk was observed among farmers who did not regularly use protective equipment when applying pesticides.[11]

The strategy adopted in our study, that is, (*a*) the collection of information on crops grown and specific crop diseases, (*b*) data on the circumstances of exposure and different activities performed by farmers, and (*c*) the involvement of local expert agronomists in order to infer exposure in agriculture, have probably improved the accuracy of pesticide exposure estimates. A merit of our study is also its overall large sample size—in spite of the small number of subjects exposed to phenoxy herbicides—and a good definition of pathology.

In conclusion, our results may contribute to the assessment of the carcinogenicity of pesticides suggesting the role of specific chemical classes, including phenoxy acid herbicides, as risk factors for hematolymphopoietic malignancies.

ACKNOWLEDGMENTS

The authors are highly indebted to L. Bellesini A. Cappelli, G. Caspanella, A. Fiorio, C. Lugaresi, D. Macrelli, M. Minolfi, P. Pasquinelli, and M. Pinna for their interpretation of occupational histories in agriculture in terms of past exposure to pesticides. The work has been carried out with the cooperation of: S. Alberghini Maltoni, S. Barcellini, G. Barni, V. Cacciarini, R. Carlini, M. Casale, G. Castellino, G. Cremaschi, L. Davico, A. Fiorio, R. Gibilisco, L. Guzzo, R. Hirvas, S. Legrotti, L. Migliaretti, R. Monteleone, G. Osella, T. Palma, G. Panizza, C. Picoco, G. Piergiovanni, G. Righetti, E. Scarpi, R. Sguanci, M. Tedeschi, D. Tiberti, G. Tonini, P. Trada, T. Vescio, and M. Zanetta.

The grant was sponsored by the National Cancer Institute, Grant NCI: CA51086

This work was also funded by the European Community (Europe Against Cancer Programme), and by "The Italian Alliance Against Cancer" (Lega Italiana per la Lotta contro i Tumori).

REFERENCES

1. Dich, J. *et al.* 1997. Pesticides and cancer. Cancer Causes Control **8:** 420–443.

2. IARC. 1986. Some Halogenated Hydrocarbons and Pesticides Exposures - IARC Monographs on the Evaluation of Carcinogenic Risks of Chemicals to Humans, Vol. 41, IARC Lyon. www.iarc.it
3. U.S. EPA 2004. www.epa.gov/
4. BLAIR, A. & S.H. ZAHM. 1991. Cancer among farmers. Occup. Med. **3:** 335–354.
5. BURNS, C.J. 2005. Cancer among pesticide manufacturers and applicators. Scand. J. Work Environ. Health **31**(Suppl 1): 9–17.
6. JAGA, K. & C. DHARMANI. 2005. The epidemiology of pesticide exposure and cancer: a review. Rev. Environ. Health **20:** 15–38.
7. HARTGE, P., S. DEVESA & S. FRAUMENI JR. 1994. Hodgkin's and non-Hodgkin's lymphomas. Cancer Surv. **19–20:** 423–453.
8. ZAHM, S.H. & A. Blair. 1992. Pesticides and non-Hodgkin's lymphoma. Cancer Res. **52**(Suppl 19): 5485s–5488s.
9. HARDELL, L. *et al.* 1981. Malignant lymphoma and exposure to chemicals, especially organic solvents, chlorophenols and phenoxy acids: a case-control study. Br. J. Cancer **43:** 169–176.
10. PEARCE, N.E. *et al.* 1987. Non-Hodgkin's lymphoma and farming: an expanded case-control study. Int. J. Cancer **39:** 155–161.
11. HOAR, S.K. *et al.* 1986. Agricultural herbicide use and risk of lymphoma and soft-tissue sarcoma. JAMA **256:** 1141–1147.
12. ZAHM, S.H. *et al.* 1990. A case-control study of non Hodgkin's lymphoma and the herbicide 2,4-dichlorophenoxyacetic acid (2,4-D) in eastern Nebraska. Epidemiology **1:** 349–356.
13. CANTOR, K.P. *et al.* 1992. Pesticides and other agricultural risk factors for non-Hodgkin's lymphoma among men in Iowa and Minnesota. Cancer Res. **52**(Suppl)**:** 2447–2455.
14. WOODS, J.S. *et al.* 1987. Soft tissue sarcoma and non-Hodgkin's lymphoma in relation to phenoxyherbicide and chlorinated phenol exposure in western Washington. J. Natl. Cancer Inst. **78:** 899–910.
15. PERSSON, B. *et al.* 1993. Some occupational exposures as risk factors for malignant lymphomas. Cancer **72:** 1773–1778.
16. HARDELL, L. & M. ERIKSSON. 1999. A case-control study of non-Hodgkin lymphoma and exposure to pesticides. Cancer **85:** 1353–1360.
17. MC DUFFIE, H. *et al.* 2001. Non Hodgkin's lymphoma and specific pesticides exposures in Man: Cross-Canada study of pesticides and health. Cancer Epidemiol. Biomarkers Prev. **10:** 1155–1163.
18. SMITH, J.G. & A.J. CHRISTOPHERS. 1992. Phenoxy herbicides and chlorophenols: a case control study on soft tissue sarcoma and malignant lymphoma. Br. J. Cancer **65:**.442–448.
19. SARACCI, R. *et al.* 1991. Cancer mortality in workers exposed to chlorophenoxy herbicides and chlorophenols. Lancet **338:**1027–1032.
20. KOGEVINAS, M. *et al.* 1995. Soft tissue sarcoma and non-Hodgkin's lymphoma in workers exposed to phenoxy herbicides, chlorophenols, and dioxins: two nested case-control studies. Epidemiology **6:** 396–402.
21. LILIENFELD, D.E. & M.A. GALLO. 1989. 2,4-D, 2,4,5-T, and 2,3,7,8-TCDD: a review. Epidemiol. Rev. **11:** 28–58.
22. NANNI, O. *et al.* 1996. Chronic lymphocytic leukaemias and non-Hodgkin's lymphomas by histological type in farming-animal breeding workers: a population case-control study based on a priori exposure matrices. Occup. Environ. Med. **53:** 652–657.

23. ZHENG, T. et al. 2001. Agricultural exposure to carbamate pesticides and risk of non-Hodgkin lymphoma. J. Occup. Environ. Med. **43:** 641–649.
24. MILIGI, L. et al. 1999. Occupational, environmental, and life-style factors associated with the risk of hematolymphopoietic malignancies in women. Am. J. Ind. Med. **36:** 60–69.
25. SENIORI COSTANTINI, A. et al. 2001. A multicenter case control study in Italy on hematolymphopoietic neoplasms and occupation. Epidemiology **1:** 78–87.
26. MILIGI, L. et al. 2003. Non-Hodgkin's lymphoma, leukemia and exposures in agriculture: results from the Italian multicenter case-control study. Am. J. Ind. Med. **44:** 627–636.
27. APREA, C. et al. 1994. Biological monitoring of exposure to organophosphorus insecticides by assay of urinary alkylphosphates: influence of protective measures during manual operations with treated plants. Int. Arch. Occup. Environ. Health **66:** 333–338.
28. APREA, C. et al. 2002. Evaluation of respiratory and cutaneous doses of chlorothalonil during re-entry in greenhouses. J. Chromatogr. B. Analyt. Technol. Biomed. Life Sci. **778:** 131–145.
29. DOSEMECI, M. et al. 2002A. quantitative approach for estimating exposure to pesticides in the agricultural health study. Ann. Occup. Hyg. **46:** 245–260.

Geographic Model and Biomarker-Derived Measures of Pesticide Exposure and Parkinson's Disease

BEATE RITZ AND SADIE COSTELLO

UCLA School of Public Health, Los Angeles, California 90095-1772, USA

ABSTRACT: For more than two decades, reports have suggested that pesticides and herbicides may be an etiologic factor in idiopathic Parkinson's disease (PD). To date, no clear associations with any specific pesticide have been demonstrated from epidemiological studies perhaps, in part, because methods of reliably estimating exposures are lacking. We tested the validity of a Geographic Information Systems (GIS)-based exposure assessment model that estimates potential environmental exposures at residences from pesticide applications to agricultural crops based on California Pesticide Use Reports (PUR). Using lipid-adjusted dichlorodiphenyldichloroethylene (DDE) serum levels as the "gold standard" for pesticide exposure, we conducted a validation study in a sample taken from an ongoing, population-based case–control study of PD in Central California. Residential, occupational, and other risk factor data were collected for 22 cases and 24 controls from Kern county, California. Environmental GIS–PUR-based organochlorine (OC) estimates were derived for each subject and compared to lipid-adjusted DDE serum levels. Relying on a linear regression model, we predicted log-transformed lipid-adjusted DDE serum levels. GIS–PUR-derived OC measure, body mass index, age, gender, mixing and loading pesticides by hand, and using pesticides in the home, together explained 47% of the DDE serum level variance (adjusted $r^2 = 0.47$). The specificity of using our environmental GIS–PUR-derived OC measures to identify those with high-serum DDE levels was reasonably good (87%). Our environmental GIS–PUR-based approach appears to provide a valid model for assessing residential exposures to agricultural pesticides.

KEYWORDS: pesticides; Geographic Information Systems (GIS); validation; exposure assessment; biomarker

Address for correspondence: Beate Ritz, M.D., Ph.D., Associate Professor of Epidemiology and Environmental Health Sciences, UCLA School of Public Health, Box 951772, 650 Charles E. Young Drive, Los Angeles, CA 90095-1772. Voice: 310-206-7458; fax: 310-206-6039.
e-mail: britz@ucla.edu

INTRODUCTION

Parkinson's disease (PD) is a complex movement disorder and the second most common neurodegenerative disorder affecting the elderly after Alzheimer's disease. PD continues to grow in public health importance due to the aging of populations in Western nations and the considerable personal and societal burden it represents in the form of loss of quality of life and high health/nursing care costs.[1,2] PD is considered to have a multifactorial etiology with environmental exposures likely playing a major role. For more than two decades now, reports have suggested that pesticides and herbicides may cause idiopathic PD. The pesticide rotenone causes PD-like degeneration in an animal model[3]; the herbicide paraquat is structurally similar to the toxin 1-methyl-4-phenyl-1,2,3,6-tetrahydropyridine (MPTP) that induces Parkinsonism in humans,[4] dieldrin has been found in PD brains,[5] and some carbamate fungicides seem to enhance the neurotoxicity of MPTP in animals, raising questions about the biologic interaction of various pesticides.[6,7]

The first epidemiological studies to suggest a link with pesticides were the ecological studies reporting an excess of PD in rural as compared to urban populations.[8–12] Case–control studies investigating pesticide exposures from occupational and non-occupational sources reported two- to six-fold increased risks for PD in exposed subjects.[13–22] However, the conclusion drawn by Checkoway *et al.*[23] in the end 1990s, that "no clear associations with any specific pesticide have been demonstrated from epidemiologic studies," (p. 635) still holds today. Relying on study subjects to recall specific chemical usage for periods in the distant past to evaluate exposures in studies of chronic diseases with long latencies may result in substantial information bias, as has recently been suggested by a German study.[22] As almost all published PD and pesticide case–control studies relied on retrospective self-reports of pesticide use, the validity of their results hinges on whether the exposure assessment suffered from differential recall bias.

An extensive and detailed assessment of occupational exposures to specific pesticides was performed for a large group of licensed pesticide applicators enrolled in the Agricultural Health Study. Results from this study are still forthcoming for PD. Large prospective occupational cohort studies like the Agricultural Health Study[24] are extremely labor-, time-, and cost-intensive, and likely to be rare resources. Pesticides applied from the air or ground may drift from their intended treatment sites, such that there are measurable concentrations of pesticides detected in the air, in plants, and in animals up to several hundred meters from application sites.[25–28] Thus, alternative methods of estimating exposures in rural communities are sorely needed, but accurate exposure assessment may be particularly challenging. Geographic Information System (GIS)-based methods of assessing exposures to pesticides have become popular in recent years and may prove to be an effective solution to this problem. We developed a GIS-based exposure assessment model to estimate pesticide exposures in the residential environment from applications to agricultural crops based on California Pesticide Use Reports (PUR) and land-use maps. We conducted a validation pilot study to examine how well our environmental GIS–PUR model derived exposure estimates predicted biomarkers, specifically dichlorodiphenyldichloroethylene (DDE) serum levels. Although more than 600 unique agricultural pesticides have been reported in the PUR and are available for modeling, only organochlorines (OCs), which provide a biomarker (dichlorodiphenyltrichloroethane [DDT]/DDE serum levels) for longer term exposures, could be used for model validation. OC compounds are stored in adipose tissue, the lipid components of blood, and breast milk. They are resistant to metabolism and have long half-lives; therefore, measurements in humans potentially represent cumulative exposures over many years.[29] In this article, we will describe our environmental GIS–PUR model and the results from the validation study involving biomarkers.

METHODS

Environmental GIS–PUR Model

Employing a geographic model developed by Rull and Ritz,[30] California PUR data and geocoded subject residential histories were linked to obtain estimates of exposures to pesticides in the residential environment based on proximity to agricultural pesticide applications. Briefly, since 1974, agricultural pesticide applications in California are recorded in the PUR system documenting the name of the pesticide's active ingredient, the poundage applied, the crop and acreage of the field, the application method, and the date and location (geographic section). We created a cumulative exposure intensity score for a given residence based on the weighted average of OC applications in a public land use (PLS) section (total lbs. active ingredient ÷ total PLS section acreage) within a 1000-m buffer from a residence between 1974 and 1989. The following OCs were used in the targeted counties and included in our GIS–PUR model: aldrin, benzene hexachloride (BHC), chlordane, dicofol, dieldrin, dienochlor, endosulfan, heptachlor, lindane, methoxychlor, and toxaphene.

Validation Study Population

To validate our GIS–PUR model derived OC pesticide exposure estimates, we employed data from subjects enrolled in a population-based Parkinson's study at UCLA (the PEG Study).[31] We recruited newly diagnosed PD patients for the PEG study from among current residents of Kern, Fresno, and Tulare counties, California, with the help of healthcare providers, mostly neurologists, practicing in this region. We selected 22 patients for whom we had obtained and stored serum samples at random from among all Kern county PEG cases. We also selected 24 controls with serum samples; specifically, controls were recruited from among: (*a*) a random sample of age- and gender-matched Medicare beneficiaries residing in Kern county in the year 2000 and (*b*) residential parcels randomly sampled from Kern county GIS shape files for the years 1998–2000.

OC Serum Analysis Methods

After obtaining subject's informed consent, a blood sample was drawn and stored in a $-20°C$ freezer at UCLA prior to shipment to Pacific Toxicology Laboratory (Chatsworth, California) for analyses of serum OC levels. Serum was tested for 13 OC pesticides and metabolites; however, as only DDE was found above the detection limit in most subjects, we focused in our analyses on this metabolite. Each serum sample was brought to room temperature and thoroughly mixed by vortexing. Two milliliters of serum was transferred to a 16 mL $\times$ 125 mL culture tube. Then, 6 mL of hexane and 20 mL of internal standard was added to the sample. The tube was then tightly capped and rotated at 50 rpm for 2 h. After rotating, 5 mL of the hexane phase was transferred into a graduated centrifuge tube. The volume was then reduced to 1.0 mL under a gentle stream of nitrogen gas after which the centrifuge tube was vortexed briefly before the sample was transferred to an autosampler tube. Analysis was performed on a gas chromatograph with electron capture detection. The column was a 30 m $\times$ 0.25 mm inner diameter DB-35 column with a 0.25-mm film thickness. Calibrators, controls, and blank samples were run with every sample batch.

Questionnaire Data

Each subject provided a detailed demographic and residential history, including dates of residence, and landmarks or cross streets when exact street addresses could not be recalled. We also collected information on residential use of pesticides and whether a subject ever worked on a farm, nursery, or orchard, or as a professional pesticide applicator and used pesticides occupationally. Specifically, information on ever having mixed, loaded, or applied pesticides, and work practices were obtained.

Statistical Methods

All lab results and questionnaire data were entered into a Microsoft Access database (Microsoft Corp., Redmond, WA) by the interviewers. First, we generated descriptive statistics and conducted bivariate analyses. Pearson's correlations of the continuous variables and their relation to lipid-adjusted blood levels of DDE were examined. Using a manual step-wise variable selection technique, linear regression models were built in SAS Software (version 9.1, SAS Institute, Inc., Cary, NC). Variables were kept in the model if they were deemed important for the control of confounding or if their inclusion increased the adjusted r^2. Our final "basic" linear regression model included the log-transformation of lipid-adjusted serum DDE as the outcome, and our environmental GIS–PUR-derived model estimate of OC exposure, age (centered at age 70 years), body mass index (BMI) (centered at 27), the square of BMI, sex, ever mixed and loaded pesticides by hand, and ever used pesticides in the home as predictor variables. The basic regression model was also applied after excluding influential points ($N = 44$) and separately for subjects who had an OC exposure estimate above zero ($N = 26$). Each regression model was tested for overall fit with the F-test ($F =$ the ratio of regression mean square to the residual mean square).

RESULTS

The mean age of the PD cases and matched controls was similar (71.3 and 70.2 years, respectively), but controls were more often male (15 versus 7 male cases) and had a slightly higher BMI, compared to cases (28.8 versus 26.4) (TABLE 1). The mean level of lipid-adjusted DDE in the serum was 1.2 mcg/g lipid in cases and 0.7 mcg/g lipid in controls; cases also had higher model-derived OC measure (38.6 versus 8.1). Only two cases and three controls reported mixing and loading pesticides by hand, while about half of all cases and controls reported using pesticides in their homes.

In TABLE 2, we present the proportion of variance explained (adjusted r^2) for our basic regression model, for the basic regression model after two influential observations were removed, and for a regression model from which we removed all subjects with an environmental GIS–PUR-derived pesticide exposure

TABLE 1. Demographic characteristics of the subjects

	Case ($N = 22$)	Control ($N = 24$)
Age, mean (SD)	71.3 (12.1)	70.2 (10.2)
Male	7 (31.8)	15 (62.5)
BMI, mean (SD)	26.4 (6.7)	28.8 (5.8)
Lipid-adjusted DDE, mean (SD), units are mcg/g lipid	1.2 (0.9)	0.7 (0.8)
OC exposure estimate, mean (SD)	38.3 (72.9)	8.1 (15.6)
Mixed and loaded pesticides by hand	2 (9.1)	3 (12.5)
Used pesticides in the home	12 (54.6)	13 (54.2)

NOTE: Values are number (percentage) of individuals, unless otherwise specified.

TABLE 2. Predictors of the natural log of lipid-adjusted serum DDE levels in three linear regression models

Model type	N	Adjusted r^2
Basic model[†]	46	0.412
Basic model without influential observations[‡]	44	0.470
Basic model for those with GIS–PUR-based OC estimates greater than zero	26	0.485

[†]The basic regression model used the log-transformation of lipid-adjusted serum DDE levels as the outcome and the GIS–PUR model based OC estimate, age, BMI, BMI^2, sex, ever mixed or loaded pesticides by hand, and ever applied pesticides in the home as predictor variables.
[‡]Influential observations had dffits statistic values greater than $2*\text{sqrt}[(k + 1)/(n - k - 1)]$ where k = number of predictor variables and n = sample size.

estimate of zero. The basic regression model with the total study population explains 41% of the variance ($P = 0.002$), whereas the basic regression model with the two influential observations removed explains 47% of the variance ($P < 0.001$). Excluding the 20 subjects for whom our environmental GIS–PUR-derived exposure estimated zero OC exposure between 1974 and 1989, the r^2 increased to 0.49 ($P = 0.005$).

The correlation between our environmental GIS–PUR model derived estimate of OCs and log-transformed lipid-adjusted blood DDE level was estimated to be 0.32 ($P = 0.03$, TABLE 3). The correlation with age was of similar magnitude (Pearson's correlation coefficient 0.35, $P = 0.017$). Age, gender, and mixing and loading pesticides by hand were all important predictors in our linear regression model (all with P value 0.004), as were pesticide use in the home and BMI (P-values 0.007 and 0.045, respectively). Our environmental GIS–PUR exposure estimate alone predicted 6%, mixing and loading of

TABLE 3. Basic regression model results and correlation coefficients ($n = 46$ Kern county PEG study subjects)

	Estimate from linear regression	Standard error	P	Correlation coefficient[†]	P
GIS–PUR-derived OC estimates	0.005	0.003	0.036	0.321	0.030
BMI	−0.06	0.03	0.045	−0.145	0.337
BMI^2	0.004	0.002	0.096	0.061	0.687
Age	0.04	0.01	0.004	0.349	0.017
Female	1.00	0.32	0.004	—	—
Mixed and loaded pesticides by hand	1.45	0.47	0.004	—	—
Used pesticides in the home	0.90	0.31	0.007	—	—

Regression model adjusted $R^2 = 0.412$

[†]Pearson's correlation coefficients of each continuous variable with the log-transformed DDE serum measure.

pesticides by hand predicted 13%, and home pesticide use predicted 11% of the variance of log-transformed DDE blood levels when modeled alone (data not shown). Meat, poultry, seafood, fruit, and vegetable consumption all failed to alter the adjusted r^2 when added to the model.

Using the lipid-adjusted DDE blood test as the "gold standard," the sensitivity of our GIS–PUR estimate of OCs is 38%, whereas the specificity is much higher at 87%.

DISCUSSION

Our environmental GIS–PUR pesticide exposure measures were correlated with lipid-adjusted DDE blood levels and predicted a portion of the variance of the log-transformed serum levels in a linear regression model. Thus, residential exposure to OC pesticides seems to be an important contributor to exposure in this population. Occupational exposures, such as mixing and loading pesticides, and home pesticide use are also components of exposure, but are much less prevalent (only 10% of the subjects reported handling pesticides in this manner). Although we did not take the use of personal protective equipment or type of pesticides handled into account, mixing and handling of pesticides alone seems to be a good indicator of blood levels in this elderly population.

Our GIS–PUR measure reflects potential low-level exposures to pesticides in a residential environment. The unexplained portion of the log-transformed lipid-adjusted DDE serum level variance may include dietary exposures that we were unable to assess adequately. In another small pilot study of 30 slightly younger subjects of lower socio-economic status who resided in Kern county and frequented a neurological clinic, we found that apart from having "ever lived in Mexico" reporting to have "loaded and mixed pesticides" also was a major contributor to DDE serum levels (data not shown). DDT was used widely in Mexico long after use had been banned in the United States.

Although the sensitivity of using our GIS–PUR-derived OC measure to identify subjects with high-serum DDE levels is poor (38%), our specificity is quite good (87%). Thus, we are less likely to classify unexposed subjects as having been exposed, important for a population-based case–control study.

Using DDT/DDE as Gold Standard

Even though DDT has been banned in the United States since 1972, it lasts in the soil of temperate areas for 5–30 years. It may evaporate into the air and significant concentrations of DDT have been found in the atmosphere over treated agricultural plots. DDT adheres strongly to soil, and remains on the soil surface layers; thus, people who work or live around or with contaminated soil might be exposed by accidentally ingesting the soil, having skin contact with the soil, inhaling DDT vapor, or breathing in DDT in dust. Because DDT bioconcentrates in aquatic organisms and bioaccumulates in the food chain, the main source of DDT exposure in the general population is from eating meat, fish, poultry, and dairy products. DDT is stored in fatty tissues and has a prolonged physiological half-life of up to 11–14 years, which makes it one of few candidates for a biomarker of long-term pesticide exposure. DDE, the metabolite of DDT, can likely be found in the serum of everyone reading this article.[32–34]

Limitations

The small sample size restricted the statistical power and precision of our validation study and prevented us from conducting more extensive multivariate analyses. We did not have PUR information on pesticide use before 1972 when DDT was commonly used and prior to its use being banned in the United States. Thus, when we chose DDE as our biomarker; we did so under the assumption that a farmer who previously treated a crop with DDT may most likely have substituted this agent with another OC pesticide still available. We relied on self-report of residential and occupational history and pesticide use. However, while there may have been recall error for exact addresses of past residences, we believe that our method of eliciting such information in combination with the extensive maps for residential parcels employed resulted in very accurate data. While subjects may not recall specific pesticides well, they will recall their work practices underscoring our results for mixing/loading of pesticides as a predictor of exposure.

DDE measures in serum may be less accurate than those derived from adipose tissue, but lipid adjustment is expected to increase reliability.[35] Exposure misclassification of DDE levels would be expected to be nondifferential with respect to our environmental GIS–PUR-derived measures of OC exposure, and therefore associations with DDE would likely be attenuated.

Our GIS PUR model based exposure assessment also had some limitations. We picked a somewhat arbitrary buffer radius for measuring proximity of homes to land on which pesticides have been used (1000 m). Additional factors that could influence residential exposure to agriculturally applied pesticides include wind patterns and pesticide application equipment type (plane, truck, etc.), whether windows of the home are kept open, and the amount of time the occupant spent at home.

CONCLUSION

Our GIS–PUR approach appears to provide a valid model for assessing exposures to agricultural pesticides in the residential environment. The GIS–PUR model derived OC estimates in conjunction with other predictor variables explained almost half of the variance in our model for our lipid-adjusted DDE biomarker. Although our sensitivity is poor, the specificity of our model is good, reducing exposure misclassification bias in case–control settings commonly applied to study rare diseases.

REFERENCES

1. FINDLEY, L. *et al*. 2003. Direct economic impact of Parkinson's disease: a research survey in the United Kingdom. Mov. Disord. **18:** 1139–1145.
2. HAGELL, P. *et al*. 2002. Resource use and costs in a Swedish cohort of patients with Parkinson's disease. Mov. Disord. **17:** 1213–1220.
3. BETARBET, R. *et al*. 2000. Chronic systemic pesticide exposure reproduces features of Parkinson's disease. Nat. Neurosci. **3:** 1301–1306.
4. LANGSTON, J.W. *et al*. 1983. Chronic Parkinsonism in humans due to a product of meperidine-analog synthesis. Science **219:** 979–980.
5. FLEMING, L. *et al*. 1994. Parkinson's disease and brain levels of organochlorine pesticides. Ann. Neurol. **36:** 100–103.
6. DI MONTE, D.A. *et al*. 1986. Comparative studies of paraquat and 1-methyl-4-phenylpyridine (MPP+) cytotoxicity. Biochem. Biophys. Res. Commun. **137:** 303–309.
7. BOCCHETTA, A. & G.U. CORSINI. 1986. Parkinson's disease and pesticides. Lancet **2:** 1163.
8. BURGUERA, J. *et al*. 1992. Mortality from Parkinson's disease in Spain (1980–1985). Distribution by age, sex and geographic areas. Neurologia **7:** 89–93.
9. MORANO, A. *et al*. 1994. Risk-factors for Parkinson's disease: case–control study in the province of Caceres, Spain. Acta Neurol. Scand. **89:** 164–170.
10. BEN-SHLOMO, Y. *et al*. 1993. The epidemiology of Parkinson's disease in the Republic of Ireland: observations from routine data sources. Ir. Med. J.. **86:** 190–194.
11. SVENSON, L. *et al*. 1993. Geographic variations in the prevalence rates of Parkinson's disease in Alberta. Can. J. Neurol. Sci. **20:** 307–311.
12. TANNER, C. *et al*. 1987. Environmental factors in the etiology of Parkinson's disease. Can. J. Neurol. Sci. **14:** 419–423.
13. KOLLER, W. *et al*. 1990. Environmental risk factors in Parkinson's disease. Neurology **40:** 1218–1221.
14. HUBBLE, J. *et al*. 1993. Risk factors for Parkinson's disease. Neurology **43:** 1693–1697.
15. SEMCHUK, K.M. *et al*. 1992. Parkinson's disease and exposure to agricultural work and pesticide chemicals. Neurology **42:** 1328–1335.
16. SEMCHUK, K.M. *et al*. 1993. Parkinson's disease: a test of the multifactorial etiologic hypothesis. Neurology **43:** 1173–1180.
17. STERN, M. *et al*. 1991. The epidemiology of Parkinson's disease—a case–control study of young-onset and old-onset patients. Arch. Neurol. **48:** 903–907.
18. TANNER, C. *et al*. 1989. Environmental factors and Parkinson's disease: a case–control study in China. Neurology **39:** 660–664.
19. HO, S. *et al*. 1989. Epidemilogic study of Parkinson's disease in Hong Kong. Neurology **39:** 1314–1317.
20. BUTTERFIELD, P.G. *et al*. 1993. Environmental antecedents of young-onset Parkinson's disease. Neurology **43:** 1150–1158.

21. HERTZMAN, C. et al. 1994. A case–control study of Parkinson's disease in a horticultural region of British Columbia. Mov. Disord. **9:** 69–75.
22. SEIDLER, A. et al. 1996. Possible environmental, occupational, and other etiologic factors for Parkinson's disease: a case–control study in Germany. Neurology **46:** 1275–1284.
23. CHECKOWAY, H. et al. 1998. Genetic polymorphism in Parkinson's disease. Neurotoxicology **19:** 635–644.
24. ALAVANJA, M.C. et al. 1996. The agricultural health study. Env. Health Perspect. **104:** 362–369.
25. FROST, K.R. & G.W. WARE. 1970. Pesticide drift from aerial and ground applications. Agric. Eng. **51:** 460–467.
26. CURRIER, W.W. et al. 1982. Drift residues of air-applied carbaryl in an orchard environment. J. Econ. Entomol. **75:** 1062–1068.
27. CHESTER, G. & R.J. WARD. 1984. Occupational exposure and drift hazard during aerial application of paraquat to cotton. Arch. Environ. Contam. Toxicol. **13:** 551–563.
28. MACCOLLOM, G.B. et al. 1986. Drift comparisons between aerial and ground orchard application. J. Econ. Entomol. **79:** 459–464.
29. LADEN, F. et al. 1999. Predictors of plasma concentrations of DDE and PCBs in a group of U.S. women. Environ. Health Perspect. **107:** 75–81.
30. RULL, R.P. & B. RITZ. 2003. Historical pesticide exposure in California using pesticide use reports and land-use surveys: an assessment of misclassification error and bias. Environ. Health Perspect. **111:** 1582–1589.
31. KANG, G. et al. 2005. Clinical characteristics in early Parkinson's disease in a Central Californian population-based study. Mov. Disord. **20:** 1133–1142.
32. Agency for Toxic Substances and Disease Registry (ATSDR). 1994. Toxicological profile for 4,4-DDT, 4,4-DDE, and 4,4-DD. Public Health Service, U.S. Department of Health and Human Services. Atlanta, GA.
33. Centers for Disease Control and Prevention. 2005. Third National Report on Human Exposure to Environmental Chemicals. Department of Health and Human Services. National Center for Environmental Health. Division of Laboratory Sciences. Atlanta, GA.
34. TURUSOV, V. et al. 2002. Dichlorodiphenyltrichloroethane (DDT): ubiquity, persistence, and risks [review]. Environ. Health Perspect. **110:** 125–128.
35. PHILLIPS, D.L. et al. 1989. Chlorinated hydrocarbon levels in human serum: effects of fasting and feeding. Arch. Environ. Contam. Toxicol. **18:** 495–500.

The Construction Industry

KNUT RINGEN[a] AND ANDERS ENGLUND[b]

[a]*The Center To Protect Workers Rights, Seattle, Washington 98166, USA*

[b]*Director, Medical and Social Affairs, Swedish Work Environment Authority, 1244 Stockholm, Sweden (Retired) (current address: Trollvagen 2s, 1244 Saltsjobaden, Sweden)*

The construction industry is, in terms of occupational safety and health, a high-risk industry. In virtually all countries, construction workers account for roughly 20–25% of all known work-related mortality and morbidity even though they typically account for 7–9% of all employment. The industry has unique characteristics that challenge traditional models of occupational health and risk management. Among these challenges are temporary employment, temporary worksites where workers employed by many different employers may be working alongside each other, frequently improvised work practices, risks caused by both one's own work, and imposed from work done by other trades. It is an industry with mostly small, and in terms of safety and health, unsophisticated employers. Compounding these problems is a culturally diverse workforce, with about one-third of workers coming from foreign countries, and about one-third working as self-employed independent contractors. Many of the most critical exposures come from end-use of chemicals and the disturbance of hazards in place during renovation, repair, renovation and maintenance of existing structures.[1–3]

Although the construction industry is a very large economic sector, with 8–15% of total Gross National Product (GNP), there has been little safety and health research focused on this industry. Most of the research that exists has been on traumatic injuries, and less so on musculo–skeletal disorders, but there is much less research on exposures and health outcomes for chemical and other chronic disease risks.[4]

In the Conference session on the construction industry, four papers presented different perspectives on the key challenges facing occupational safety and health in the industry.

(*a*) Professor Järvhom from Sweden discussed what is known about the epidemiology of chemical carcinogenesis in the industry.[5]

Address for correspondence: Knut Ringen, 2610 SW 151st Pl, Seattle, WA 98166, Voice: 206-444-9811; fax: 206-444-9832.
  e-mail: knutringen@msn.com

(b) Dr. van Duivenbooden and colleagues from the Netherlands discussed challenges about characterizing risk and implementing preventive measures for solvents.[6]
(c) Dr. Foà and Dr. Campo from Italy presented a paper on a large multinational effort in Europe to determine if asphalt fumes pose a significant risk to construction workers and if so how to prevent such risk.[7]
(d) Dr. Bingham and Dr. Ringen from the United States presented data from interviews with construction workers employed in atomic weapons facilities to illustrate how little information workers have available to them, even though the industry often expects workers to largely protect themselves.[8]

ESTIMATING EXPOSURE VARIANCE

Perhaps the most significant scientific development in construction safety and health over the past decade has been the elucidation of exposure variance. Järvholm, in his paper, mentions this as the key limitation to drawing conclusions about risk from epidemiological studies.[5] For more than a decade, in the United States a large study consortium from the Center To Protect Workers Rights and the National Institute for Occupational Safety and Health has been studying variance.[9] TABLE 1 summarizes the range of exposures measured with personal exposure monitoring devices for identical work tasks, as performed in real world construction environments. As can be seen, the range of measured exposures varies by as much as 50-fold. TABLE 2 shows examples of the high degree of variance in measures of exposures during common construction tasks. These data suggest that a conservative estimate of the upper and lower 95% confidence interval for the geometric mean dose should incorporate a Geometric Standard Deviation (GSD) of approximately 4.25.

TABLE 1. Examples of exposure measurement ranges by task

Work task measured	Exposure range (mg/m^3)	Exposure limit (mg/m^3)
Hot work 1995–1996, total particulate	<1.0200–37.2900	10*
Hot work 1995–1996, manganese	0.0005–1.3105	0.2**
Abrasive blasting (hematite), 2002	0.52–25.66	0.05***
Avon Lake Boilermakers, 2004 manganese during welding	0.006–0.146	0.2**
Abrasive blasting (Coal slag), 2004	20.42–90.11	0.05***
Abrasive blasting (Steel grit), 2004	0.89–57.5	0.05***

*ACGIH[18] Threshold Limit Value (TLV) for insoluble particulate for which an applicable agent-specific TLV is lacking.
**ACGIH[18] TLV for manganese.
***ACGIH[18] TLV and NIOSH[19] Recommended Exposure Limit (REL) for silica.
Source: Reference 5.

TABLE 2. Examples of uncertainty

Investigator	Exposure measured	GSD
Rappaport[20]	Magnesium during welding	4.0–4.5
Woskie[21]	Silica during highway construction	1.5–5.7
Goldberg[22]	Lead during bridge rehab, task specific	1.5–12.7
Goldberg[22]	Lead during bridge rehab, multitask	3.5–5.6

In a recent NIOSH-CPWR workshop on estimating variance in exposure measurements, three multivariate models to predict exposures based on empirical data from exposures to magnesium during welding,[10] asphalt fumes during road paving operations,[11] and radiation in atomic weapons facilities,[12] found that the models were predictive of approximately 40% of the actual range of exposures.

IMPLICATIONS FOR PREVENTION

Given the difficulties of estimating exposure risk *a priori*, it would seem a natural decision, at least if one applies the precautionary principle,[13] that workers in the construction industry should be provided with maximum protection from exposure. This is clearly not the case. Apart from the large range of distribution of exposure measurements shown in TABLE 1, it is especially noteworthy that so much of the ranges for the various work tasks are astoundingly far above the recommended occupational exposure limits.

van Duivenbooden and colleagues, in their paper,[6] point to an important aspect of why the industry performs so poorly in terms of occupational health although paradoxically showing that it does not have to be so. Much of construction is end-use of products, supplies, and materials that have been produced by manufacturers that supply the industry. Manufacturers of supplies and materials that are used in construction increasingly operate in multinational markets, and are reluctant to adopt more stringent precautions in one country than is required by the norm of all nations. Thus, these multinational companies tend to drive safety and health toward a lower common denominator.

The effort by the asphalt industry described by Foà and Campo[7] to perform an international collaborative science-based effort to determine risks and prevention efforts is therefore an encouraging exception to the general trend (this study is also described in Ref. 11).

Yet, in most countries the biggest detriment to prevention comes from a lack of domestic political will. The industry operates according to short-term contracts in a very competitive environment, which tend to drive decisions toward the lowest cost (in the short term) bidder, and immediate cost-saving is usually favored over long-term quality.[14] The construction industry in most countries has always been bifurcated between an established, highly unionized, high-performance, and high-cost sector, and a more informal, less skilled, nonunion, and lower cost sector. This bifurcation is becoming more pronounced as countries enter into free-trade agreements where there are big differences in the level of economic development between countries. Thus, in western Europe, lower-cost labor from the eastern states is increasingly dominating construction employment, as is employment of workers from central and south America in the United States. To the extent it can be measured, the workers in the low-income sector typically have significantly higher injury rates than workers in the higher income sector,[15] and we can infer from this that most likely, they have much greater exposures to chronic health risks as well. To protect themselves from exposures, these workers largely have to rely on themselves.

The paper by Bingham and Ringen[8] points out that this is an unreasonable (and often illegal) approach to safety and health. The workers they interviewed were by the standards of the industry sophisticated and highly skilled, and worked in some of the most advanced (in terms of construction safety and health) construction workplaces in the United States. Yet, even though a signal risk in these workplaces was radiation exposure, less than half of these workers could say if they had worked in areas with potential for radiation exposure, or if they had been adequately monitored for such exposure. If the best of workers are not well enough informed to protect themselves, how can we expect that the lesser skilled workers, who typically also have a poor understanding of the language of the country in which they are working, will have any chance of being adequately protected?

It is the duty of the employer in the United States and under EU Directive 92/57[16] it is also the duty of the client or owner to declare whether a worker has an opportunity for hazardous exposure and then to protect the worker from such exposure and monitor to make sure that exposure does not happen. Instead of this being the rule, it is the exception in most construction workplaces. And that, most likely, is the greatest occupational health risk that workers face in the construction industry.

A decade ago, we wrote that in construction, it is not enough to think about what needs to be done in terms of individual workplaces; we must think industrywide, because that is how construction workers are employed.[17] When we said "industry-wide" we used to think regionally within countries, and then nationwide. Clearly, as we have seen in this Conference session, a nationwide perspective is no longer adequate. With markets for supplies, materials, and labor being increasingly multinational, we have to think in terms of international prevention strategies. There are no shortages of challenges to occupational safety and health professionals in the construction industry, whether they are researchers or practitioners, and these challenges are growing bigger. There is, however, a serious shortage of resources being devoted to occupational safety and health, and also a shortage of high-quality occupational safety and health professionals who choose to work in this industry. That too is a paradox. Work in this industry can be remarkably rewarding, if nothing else because one is bound to see meaningful improvements from one's work.

To end on a positive note, we have seen major improvements in construction safety and health over the past two decades. This is true for both the science base and the practice. All of the papers presented at this Conference session document this progress in one form or another. All of these papers show in one form or another that it is possible for occupational safety and health professionals to make significant change through science-based advocacy. All in all, we think Ramazzini would have been (at least modestly) pleased with the recent progress.

REFERENCES

1. RINGEN, K., A. ENGLUND, L. WELCH, et al., Eds. 1995. Occupational Medicine State-of-the-Art Reviews: Construction Safety and Health. Hanley and Belfus. Philadelphia.
2. RINGEN, K., A. ENGLUND & J. SEEGAL. 2000. Construction workers. In Occupational Health: Recognizing and Preventing Work-Related Disease, 4th ed. Barry S. Levy & David H. Wegman, Eds.: 749–765. Lippincott, williams, and wilkins. Philadelphia.
3. RINGEN, K., J. SEEGAL & J. WEEKS, Eds. 1998. Construction. In Encyclopaedia of Occupational Health and Safety, 4th ed. Jeanne M. Stellman, Ed.: 93.1–93.52, International Labour Office. Geneva.
4. VAN DUIVENBOODEN, C., M.H.W. FRINGS-DRESEN, K. RINGEN. 2005. Construction workers and occupational health care. Scand. J. Work Environ. Health 31(Suppl. 2): 1–116
5. JÄRVHOLM, B. 2006. Carcinogens in the construction trades. Ann. N. Y. Acad. Sci. (this issue)
6. SPEE, T., C. VAN DUIVENBOODEN & J. TERWART. 2006. Exposure to solvents, epoxy resins and other chemicals in the painting and construction trades. Ann. N. Y. Acad. Sci. (this issue)
7. FOÀ, V. & L. CAMPO. 2006. Exposure to PAH in asphalt road pavers by environmental and biological monitoring. Ann. N. Y. Acad. Sci. (this issue)
8. BINGHAM, E. & K. RINGEN. 2006. Radiation exposure monitoring for construction workers in nuclear facilities. Ann. N. Y. Acad. Sci. (this issue)
9. SUSI, P., M. GOLDBERG, P. BARNES & E. STAFFORD. 2000. The use of a task-based exposure assessment model (t-beam) for assessment of metal fume exposures during welding and thermal cutting. Appl. Occup. Environ. Hyg. 15: 26–38.
10. RAPPAPORT, S.J. 2005. Variability of exposures among construction workers. CPWR-NIOSH meeting on Variance in Construction Worker Radiation Exposure Monitoring. Silver Spring, CPWR, July 27.
11. HERRICK, R. 2003. Estimating exposures in the asphalt industry for an international epidemiological cohort study of cancer risk. Am. J. Ind. Med. 43: 3–17.
12. DEMENT, J. 2005. Radiation exposure analyses: Hanford & Savannah river construction workers. CPWR-NIOSH Meeting On Variance In Construction Worker Radiation Exposure Monitoring. Silver Spring, CPWR, July 27.
13. THE PRECAUTIONARY PRINCIPLE: IMPLICATIONS FOR RESEARCH AND POLICY-MAKING. 2004. Statement of the Collegium Ramazzini. Approved by the Fellows, 25 October, 2003. Am. J. Ind. Med. 45: 348–349.

14. RINGEN, K., A. ENGLUND, L. WELCH, *et al*. 1995. Why construction is different. Occup. Med. **10:** 255–259.
15. DONG, X. & J.W. PLATNER. 2004. Occupational fatalities of Hispanic construction workers from 1992 to 2000. Am. J. Ind. Med. **45:** 45–54.
16. EUROPEAN COUNCIL. 1992. Directive on the Implementation of Minimum Safety and Health Requirements at Temporary or Mobile Construction Sites. Brussels: Directive 92/57/EEC.
17. RINGEN, K., A. ENGLUND, L. WELCH, *et al*. 1995. Perspectives on the future. Occup. Med. **10:** 445–452.
18. AMERICAN CONFERENCE OF GOVERNMENTAL INDUSTRIAL HYGIENISTS THRESHOLD LIMIT VALUES. 2005. ACGIH, Cincinnati, OH.
19. NATIONAL INSTITUTE FOR OCCUPATIONAL SAFETY AND HEALTH RECOMMENDED EXPOSURE LIMIT. 1997. NIOSH Pocket Guide to Chemical Hazards. US DHHS (NIOSH) Publication No. 97–140. Washington, DC.
20. RAPPAPORT, M.S., M. WEAVER, D. TAYLOR, *et al*. 1999. Application of mixed models to assess exposures monitored by construction workers during hot processes. Ann. Occup. Hyg. **43:** 457–469.
21. WOSKIE, S.R., A.J. KALIL, D. BELLO & M.A. VIRJI. 2002. Exposures to quartz, diesel, dust and welding fumes in heavy and highway construction. Am. Ind. Hyg. Assoc. J. **63:** 447–457.
22. GOLDBERG, M., S.M. LEVIN, J.T. DOUCETTE, *et al*. 1997. A task-based approach to assessing lead exposure among iron workers engaged in bridge rehabilitiation. Am. J. Ind. Med. **31:** 310–318.

Frequency and Quality of Radiation Monitoring of Construction Workers at Two Gaseous Diffusion Plants

EULA BINGHAM,[a] KNUT RINGEN,[b] JOHN DEMENT,[c] WILFRID CAMERON,[b] WILLIAM McGOWAN,[a] LAURA WELCH,[b] AND PATRICIA QUINN[b]

[a]*University of Cincinnati Medical Center, Cincinnati, Ohio 45267, USA*
[b]*Center to Protect Workers' Rights, Silver Spring, Maryland 20910, USA*
[c]*Duke University Medical Center, Durham, North Carolina 27710, USA*

> ABSTRACT: Construction workers were and are considered temporary workers at many construction sites. Since World War II, large numbers of construction workers were employed at U.S. Department of Energy nuclear weapons sites for periods ranging from a few days to over 30 years. These workers performed tasks during new construction and maintenance, repair, renovation, and demolition of existing facilities. Such tasks may involve emergency situations, and may entail opportunities for significant radiation exposures. This paper provides data from interviews with more than 750 construction workers at two gaseous diffusion plants (GDPs) at Paducah, Kentucky, and Portsmouth, Ohio regarding radiation monitoring practices. The aim was to determine the extent to which workers believed they were monitored during tasks involving potential radiation exposures. The adequacy of monitoring practices is important for two reasons: (*a*) Protecting workers from exposures: Construction workers were employed by sub-contractors, and may frequently been excluded from safety and health programs provided to permanent employees; and (*b*) Supporting claims for compensation: The Energy Employees Occupational Illness Compensation Program Act (EEOICPA) requires dose reconstruction of radiation exposures for most workers who file a claim regarding cancer. The use of monitoring data for radiation to qualify a worker means that there should be valid and complete monitoring during the work time at the various nuclear plants or workers may be unfairly denied compensation. The worker interviews from Paducah and Portsmouth were considered especially useful because these sites were designated as Special Exposure Cohorts (SECs) and the workers did not have to have a dose reconstruction to qualify for compensation for most cancers. Therefore, their responses were less likely to be affected by compensation concerns. Interview questions included asking for

Address for correspondence: Eula Bingham, University of Cincinnati Medical Center, 3223 Eden Avenue, Cincinnati, OH 45267. Voice: 513-558-5728; fax: 513-558-5062.
    e-mail: eula.bingham@uc.edu

Ann. N.Y. Acad. Sci. 1076: 394–404 (2006). © 2006 New York Academy of Sciences.
doi: 10.1196/annals.1371.061

information regarding whether monitoring was performed, how often, and the maintenance (calibration) of monitoring equipment (devices).

KEYWORDS: radiation; construction; workers; gaseous diffusion plant

INTRODUCTION

A pilot medical screening program at the Oak Ridge Reservation for former construction workers was started in 1996 by a Consortium consisting of the University of Cincinnati Medical Center (UC), the Center to Protect Workers' Rights (CPWR), and Duke University Medical Center (DU) and was supported by the U.S. Department of Energy (DOE). This program was one of a series funded by the DOE, frequently referred to as the Former Workers Program (FWP), to determine if workers employed in atomic weapons facilities were at "significant" risk for work-related illnesses. The Oak Ridge Reservation plant sites covered by the Oak Ridge Building Trades Medical Screening Program included X-10 (renamed as the Oak Ridge National Laboratory or ORNL), Y-12, and K-25, a Gaseous Diffusion Plant (GDP). In 2004, this program expanded to include medical screening for construction workers at two other GDPs that were historically administered by Oak Ridge Operations: the Portsmouth GDP in Ohio and the Paducah GDP in Kentucky.

In 2000, the Energy Employees Occupational Illness Compensation Program Act (EEOICPA) was enacted, and under Subpart B, workers can apply to receive compensation for cancers associated with radiation exposure. Workers employed at Amchitka Island and the three GDPs were designated by the Congress as Special Exposure Cohorts (SECs), which means that they are not subject to a dose reconstruction to determine total radiation doses received on the job before being eligible to receive compensation for 24 different types of specified cancers strongly associated with radiation exposure. This decision to create SECs was based on the information that the Congress had when the legislation was passed that radiation dose measurements were inadequate and/or not valid at these sites so that it would not be possible to provide an accurate and fair dose reconstruction for workers.

Ascertainment of cancer risk with retrospective radiation dose reconstruction is based on the assumption that there are accurate and valid records for each worker of his/her exposures over the years they worked at the DOE nuclear weapons production or research sites. Conversely, the SECs are based on the premise that exposure records are not available or accurate. There is a provision in the law that allows additional SECs to be added if a petition challenging the validity of existing radiation dose records is approved.

During the first phases of the Building Trades Medical Screening Program at the Oak Ridge Reservation, we noticed that responses to questions regarding radiation exposures were of concern. Verbal reports indicated that many construction workers were not monitored for radiation. Based on the need for such information when workers filed for compensation under the EEOICPA and the fact that construction workers at Oak Ridge clearly had reported episodes of potential radiation exposure, we considered it important to include specific questions on radiation monitoring in the worker history interview when the medical screening program moved to the GDPs at Portsmouth and Paducah. We developed questions to determine the extent to which construction workers were monitored and included them in the worker history interview. The Portsmouth and Paducah GDP worker cohorts were especially pertinent because they had already been designated as SECs and considered to have enough radiation exposure to cause more than 25 cancers without National Institute Occupational Safety and Health (NIOSH) performing a dose reconstruction, thus workers would have nothing to gain by saying they were not monitored.

METHOD

The medical screening program at the Portsmouth and Paducah GDPs followed closely the process developed for the Oak Ridge Reservation's sites (K-25, X-10, Y-12), the Hanford Nuclear Reservation, and the Savannah River Site. A series of papers have been published describing the program reporting the results of screening construction workers at several sites.[1–5]

The screening programs include the following elements:

1. Workers are enrolled and a determination is made about whether they are eligible or not.
2. Workers who are eligible complete a detailed work history interview, which is usually conducted with a specially trained retired worker as the interviewer, who relies on a structured questionnaire with answers directly entered into computer screens. The interview typically takes about an hour and is used to initially ascertain any risks reported by the worker, so that a tailored occupational medical screening exam can be offered.
3. The medical screening exam is provided by clinics under contract to the program, using specified protocols, and includes a medical history, physical examination, and specified tests. There is a core set of procedures that all workers receive, and then there may be additional tests tailored for each worker based on the work history interview. Instructions for each exam are included in the exam authorization sent to the clinic in advance of each exam.
4. The work history contains questions about construction tasks and materials worked with, which are specific to the worker's trade, but not specific to a particular DOE site. In addition, there are questions about specific site risks. To construct the site-specific questions, a history of the sites indicating major construction and information on processes in various buildings and at different plant locations was compiled and

included dates of the various activities when known together with accidents or incidents available from various reports.

In the work histories for Paducah and Portsmouth we added two sets of questions. The first set of questions was intended to give us information regarding the frequency and quality of radiation monitoring:

(1) When did you begin wearing a radiation badge or dosimeter?
(2) Did you continue to wear a radiation badge or dosimeter after it was first issued?
(3) Was the same badge or dosimeter always assigned to you? (badge had your name or ID number)
(4) How often was the film in the badge changed or the badge replaced with a new badge?
(5) Did you ever wear a radiation badge or dosimeter under a lead apron?

The second set of questions was directed at whether construction workers at these two GDP sites ever had radiation contamination, were ever checked for contamination, or ever had remediation for contamination:

(1) Did you ever have to be decontaminated or scrubbed down?
(2) Why were you scrubbed down?
(3) Aside from a normal physical examination, have you at work ever had:
    (a) Blood drawn?
    (b) Chest X ray performed?
    (c) Urine or feces collected?
    (d) Did your tools or equipment ever have to be decontaminated or replaced when you finished work?

More than 780 construction workers (393 at Portsmouth GDP and 389 at Paducah GDP) were interviewed and then went for medical screening. This article focuses on findings from the two sets of radiation monitoring questions in the revised work history.

RESULTS

The participants at Paducah and Portsmouth are typical of construction workers at other nuclear facilities where we have performed medical screening. Demographic information regarding the participants is presented in TABLE 1. They are older workers with an average age in the early 1960s, and they have on average 20 years of experience in the construction trades. They had several years of experience working in these GDPs. In other words, they were generally highly skilled journeymen who had completed a multiyear apprenticeship in their trade, and then at least a decade of work in their trade.

The results of the interview are presented in FIGURES 1–8 and TABLE 2.
(1) Responses to the question "When did you begin wearing a radiation badge/dosimeter?" are found in FIGURE 1. About 44% of the construction

TABLE 1. Demographics of Participants Completing Screening

		Portsmouth GDP			
Mean Age (SD)	Age Range	Mean Years at Site (SD)	Mean Years in Trade (SD)	Avg Year First Employed	Avg Year Last Employed
62.6 (11.3)	34.0-87.0	4.2 (5.9)	17.5 (16.6)	1973.8	1985.0
		Paducah GDP			
Mean Age (SD)	Age Range	Mean Years at Site (SD)	Mean Years in Trade (SD)	Avg Year First Employed	Avg Year Last Employed
68.1 (12.5)	36.0-(9.0)	3.3 (4.6)	24.5 (16.6)	1966.1	1975.0

Trades: Asbestos Workers, Boilermakers, Carpenters, Cement Finishers, Electrician, Iron Workers, Laborers, Operating Engineers, Painters, Plasterers, Cement Masons, Plumbers, Pipe & Sprinkler Fitters, Sheet Metal Workers, Teamsters.

workers at Paducah GDP had begun wearing a badge or dosimeter either when hired or later on. It is striking that a similar percentage, 42%, of the Paducah cohort reported they had "never" worn a badge or dosimeter. At the Portsmouth GDP facility, 53% of the construction workers examined reported wearing a badge or dosimeter at some time, whereas 32% reported that they had never worn one.

(2) FIGURE 2 provides the results of the question "Did you continue to wear a badge/dosimeter after first issued?" Construction workers at both sites reported that almost half of them did not continue to wear a badge/dosimeter after one was issued.

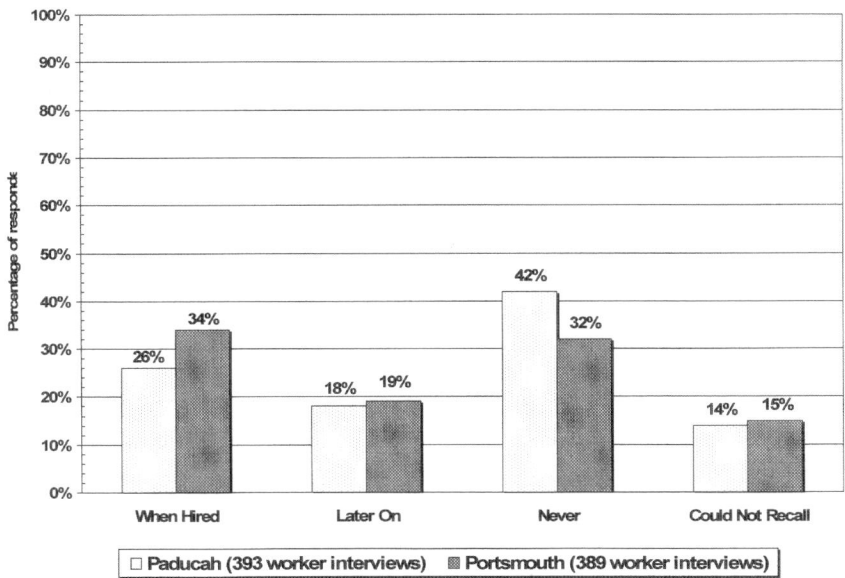

FIGURE 1. When did you begin wearing a radiation badge/dosimeter?

TABLE 2. Why were you scrubbed down?

Example Responses:
1. Got a high reading on hands or body;
2. Exposed to yellow dust
3. While working building the cells in the X-326 building, just in street clothes. We were told that this was clean area. We were worked there for about a month and then we were told that it is a hot area. They sent us to the showers and they gave us new shoes and coveralls. I am concerned about all the radiation I took home to my family.
4. I was running conduit overhead. I was dressed in my street clothes. When I climbed down the ladder, I noticed that I was covered in a greenish-yellow dust. I went to my foreman who told me to go to a health physics. They took me to the showers and told me to shower. I could not get clean even after several showers and they took me to the hospital where I showered several more times in an unknown solution. They had to incinerate my clothes.

(3) In an effort to determine whether data from the badges or dosimeters were kept specifically for each individual, we asked the question in FIGURE 3. It is clear that many construction workers could not recall this information. However, about 23% knew that they did not wear a badge or dosimeter that was specifically assigned to them.
(4) FIGURE 4 reflects the fact that about 25% of the construction workers recalled that there was no change of film in their badges or exchange of old badges for new ones.
(5) FIGURE 5 presents data that indicates construction workers were not asked to place the badge or dosimeter where it would not detect radiation.

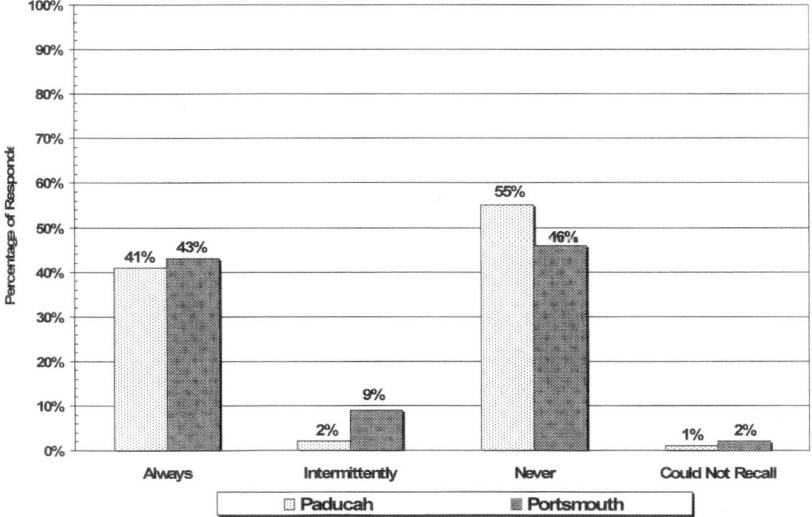

FIGURE 2. Did you continue to wear a badge/dosimeter after first issued?

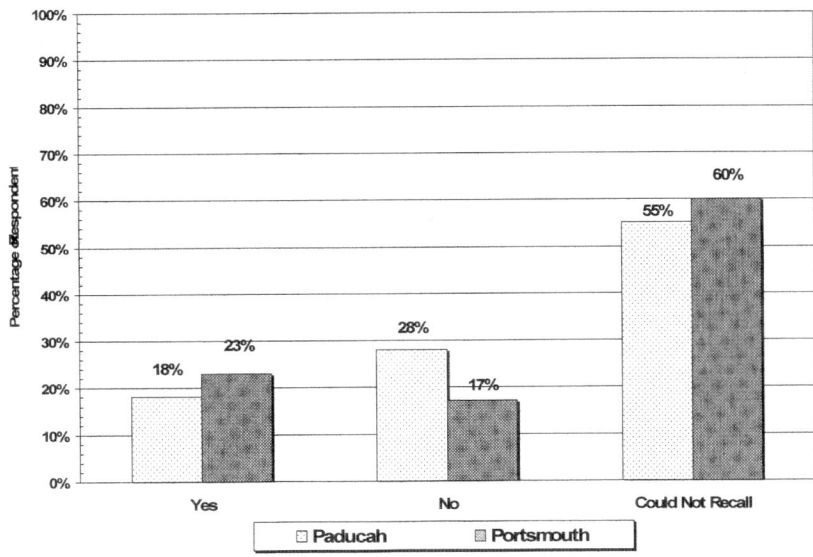

FIGURE 3. Was the same badge/dosimeter (had your name or I.D. number on it) always assigned to you?

(6) FIGURE 6 reports the results of the number of construction workers who underwent decontamination procedures, and in TABLE 2, we have included typical answers from construction workers who had been contaminated.

(7) FIGURE 7 reports the interview results regarding whether blood was drawn, or chest X rays were performed other than for an ordinary physical exam. More interesting is the response to the collection of urine or feces. Almost 20% of the construction workers interviewed indicated that urine/feces were collected and all but one worker indicated that this occurred after 1970. The two facilities opened in the early 1950s and the fact that a substantial number of the construction workers were there before 1970 indicates that there was no recognition of exposure (or no exposure) before 1970, but after that time the DOE began to test workers when contamination was recognized.

(8) Responses to the last question are reported in FIGURE 8. Almost 20% of the construction workers reported that tools or equipment they used had to be decontaminated or replaced.

CONCLUSION

The DOE facilities have some of the most sophisticated occupational safety and health programs in the United States. They use great care in hiring

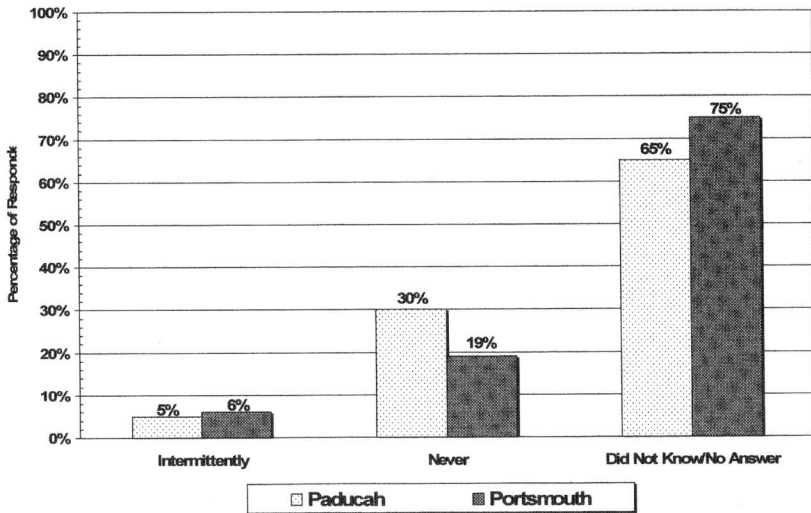

FIGURE 4. How often was the film in the radiation badge changed or the badge exchanged for a new one?

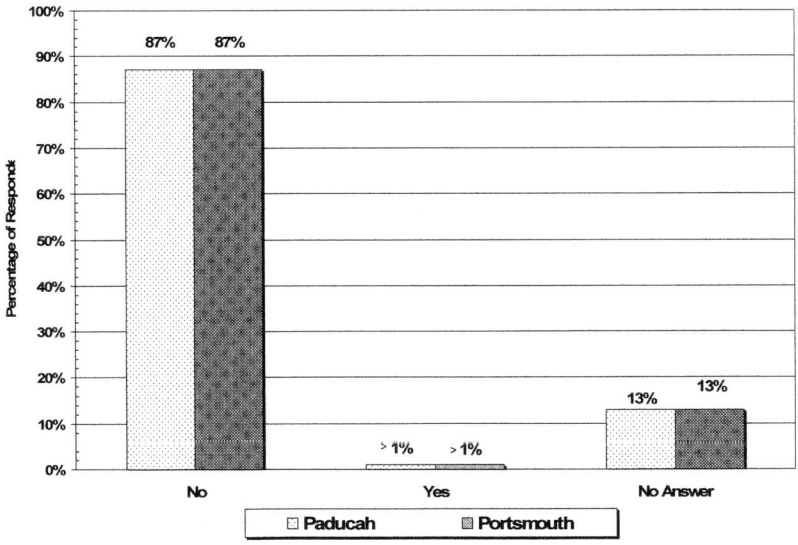

FIGURE 5. Did you ever wear a radiation badge or dosimeter under a lead apron?

construction contractors to come in and perform work, usually requiring that contractors who want to bid for work demonstrate a superior safety and health record. Thus in many ways, the DOE facilities are at the top end of occupational safety and health sophistication.

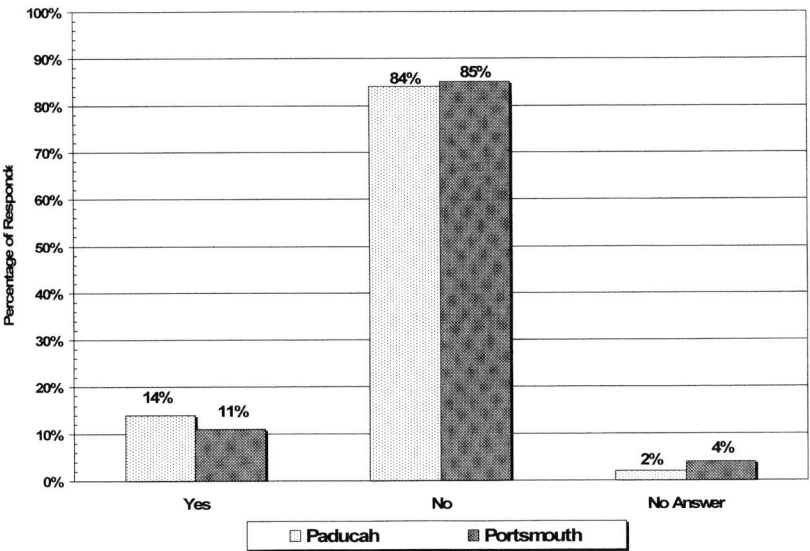

FIGURE 6. Did you ever have to be decontaminated or scrubbed down?

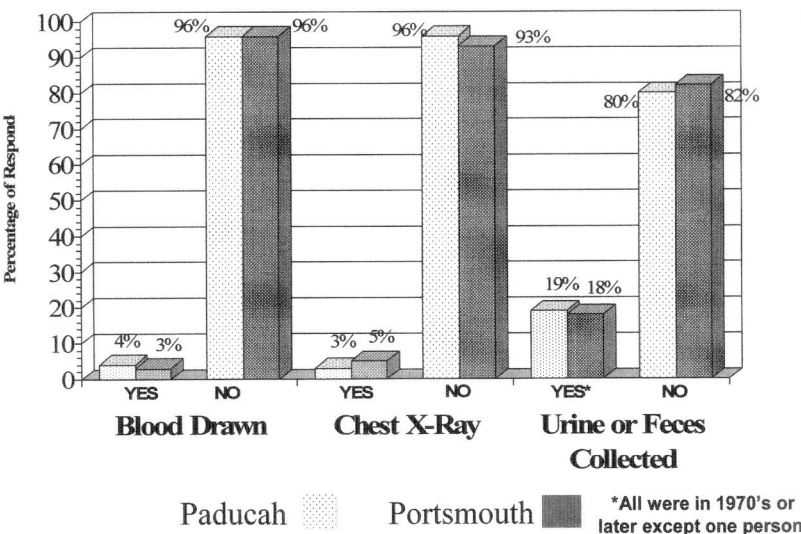

FIGURE 7. Aside from a normal physical, have you at work ever had:

Nevertheless, more than 30% of the construction workers at these two sites reported never wearing a radiation-monitoring badge or dosimeter. Only a small percentage of construction workers knew they had a specific badge or dosimeter and even fewer knew about maintenance or calibration. A very small number of workers reported putting badges or dosimeters under lead

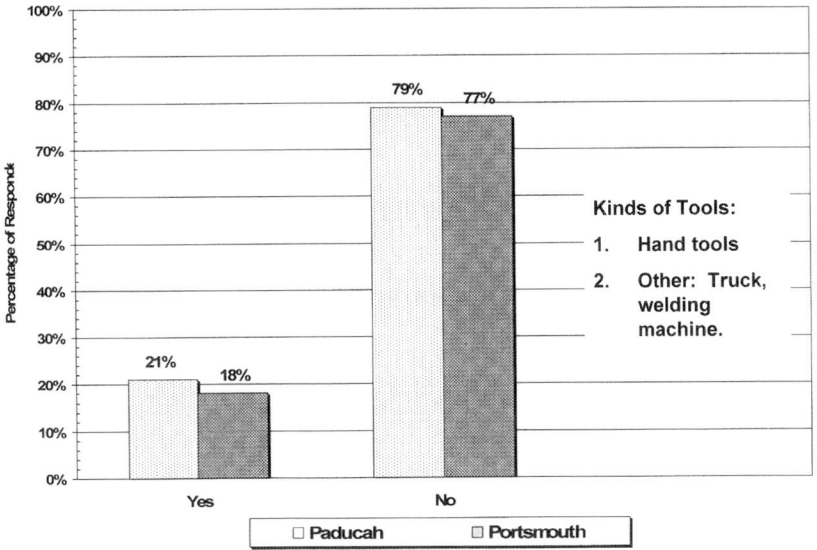

FIGURE 8. Did your tools or equipment have to be decontaminated or replaced when you finished work?

aprons. More than 10% of the workers reported decontamination procedures. About 20% reported contaminated tools/equipment so it can be assumed these construction workers were in "hot" areas.

These are highly skilled and experienced workers. Yet their knowledge of the potential for exposure to one of the most critical risks they might face in these workplaces is clearly very deficient. For anyone with experience in construction safety and health, this is not surprising, but it is alarming. In construction, workers are expected to look after their own safety and health to a much greater degree than in general industry, and these findings clearly demonstrate that this is also the case for radiation risks. At the same time, these findings beg the question, how can one expect workers to look after themselves when the information they receive or have available to them about a critical risk is so limited and deficient? And more importantly, while the reality of the construction industry dictates that workers look after themselves, the law says that it is the employer's responsibility to assure that workers are protected. These findings show clearly, that in these sophisticated workplaces that are owned by the government, where the very best construction employers perform work, the workers were greatly let down in terms of basic occupational health protection.

One can only wonder: if conditions were so deficient in such a sophisticated work environment, how deficient is the level of protection for all those workers who are employed elsewhere in the construction industry?

ACKNOWLEDGMENT

Grant for this study was sponsored by the U.S. Department of Energy through cooperative agreement number DE-FC03-96SF21263.

REFERENCES

1. DEMENT, J. *et al*. 2005. Surveillance of hearing loss among older construction and trade workers at Department of Energy Nuclear Sites. Am. J. Ind. Med. **48:** 348–358.
2. WELCH, L. *et al*. 2004. Screening for beryllium disease among construction trade workers at Department of Energy Nuclear Sites. Am. J. Ind. Med. **46:** 207–218.
3. DEMENT, J. *et al*. 2003. Surveillance of respiratory diseases among construction and trade workers at Department of Energy Nuclear Sites. Am. J. Ind. Med. **43:** 559–573.
4. BINGHAM, E. *et al*. 1998. Exposure profiles of former construction workers. Eur. J. Oncol. **3:** 329–334.
5. RICE, C. *et al*. 1998. A new approach to identifying bystander exposures among construction workers. Eur. J. Oncol. **3:** 335–338.

Evaluation of Exposure to PAHs in Asphalt Workers by Environmental and Biological Monitoring

LAURA CAMPO,[a] MARINA BURATTI,[a] SILVIA FUSTINONI,[a] PIERO E. CIRLA,[a] IRENE MARTINOTTI,[a] OMAR LONGHI,[a] DOMENICO CAVALLO,[b] AND VITO FOÀ[a]

[a]*Department of Occupational and Environmental Health, University of Milan and Fondazione Ospedale Maggiore "Policlinico, Mangiagalli e Regina Elena," 8-20122 Milan, Italy*

[b]*Department of Chemical and Environmental Sciences, University of Insubria at Como, 3-22100 Como, Italy*

ABSTRACT: In the present article we assessed exposure to polycyclic aromatic hydrocarbons (PAHs) in Italian asphalt workers (AW, $n = 100$), exposed to bitumen fumes and diesel exhausts, and in roadside construction workers (CW, $n = 47$), exposed to diesel exhausts, by means of environmental and biological monitoring. 1-Hydroxypyrene (OH-Py) was determined in urine spot samples collected, respectively, after 2 days of vacation (baseline), before, and at the end of the monitored work shift, in the second part of the workweek. Median airborne levels during the work shift of 15 PAHs (both vapor and particulate phases), from naphthalene (NAP) to indeno(1,2,3-cd)pyrene, ranged from below 0.03 to 426 ng/m^3. Median excretion values of OH-Py in baseline, before- and end-shift samples were 228, 402, and 690 ng/L for AW and 260, 304, and 378 ng/L for CW. Lower values were found in nonsmokers compared to smokers (e.g., in AW 565 and 781 versus 252 and 506 ng/L in before-shift and end-shift samples, respectively). In all subjects a weak correlation between personal exposure to the sum of airborne 15 PAHs and OH-Py was observed ($r = 0.30$). The results of this article show that AW experienced a moderate occupational exposure to airborne PAHs, resulting in a significant increase of urinary OH-Py during the workday and the workweek. The contribution of working activities to internal dose was in the same order of magnitude of the contribution of cigarette smoking.

KEYWORDS: polycyclic aromatic hydrocarbons; asphalt workers; biological monitoring; 1-hydroxypyrene; bitumen fumes

Address for correspondence: Laura Campo, University of Milan, Dept. of Occupational and Environmental Health, and Via S. Barnaba, 8-20122 Milano, Italy. Voice: +39-02-50320116; fax: +39-02-503-20111.
e-mail: laura.campo@unimi.it

INTRODUCTION

Workers employed in the road construction and maintenance companies are potentially exposed to bitumen fumes during either production or laying of asphalt. Bitumen is a dark, semisolid material of natural origin or present as byproduct of distillation of petroleum oil. For its binding properties, bitumen is mixed with inert materials like sand, stone, filler, and various additives to form asphalt used in road paving.[1] Bitumen (called asphalt in the United States) contains a large number of different compounds among which polycyclic aromatic hydrocarbons (PAHs), amines, alkanes, cycloalkanes, aromatic, and heterocyclic compounds containing oxygen, sulphur, and nitrogen.

The International Agency for Research on Cancer (IARC) has classified bitumen fumes in group 3 (inadequate evidence that bitumen fumes alone are carcinogenic to humans) and extracts of steam-refined bitumens and air-refined bitumen in group 2B (possibly carcinogenic to humans),[2,3] whereas the American Conference of Governmental Industrial Hygienist (ACGIH) has classified bitumen in group A4 (not classifiable as a human carcinogen) and has recommended a threshold limit value (TLV) of 0.5 mg/m^3 as maximal time-weighted average (TWA) during an 8-h workshift.[4] Nevertheless, a recent European epidemiological study on cancer mortality among asphalt workers (AW) has concluded that workers employed in road paving, asphalt mixing, and other job related to bitumen fumes might experience a small increase in mortality for lung cancer.[5] The reasons of this evidence are not completely clear and confounding factor as tobacco smoking and a possible occupational exposure to coal tar in the past could play a role in determining the observed enhanced risk of cancer.[6]

In addition to being present in trace amounts in bitumen, PAHs are widespread contaminants produced during the incomplete combustion of organic matter arising from different sources (industrial, domestic, natural) and also found in tobacco smoke and charbroiled, or smoked food.[7] For this reason, exposure to PAHs can occur via different pathways, namely respiratory, dermal, and gastrointestinal routes.

IARC has classified some PAHs as probably carcinogenic to humans (class 2A) (namely benzo[a]anthrancene [BaA], benzo[a]pyrene [BaP], and dibenzo[a,h,]anthracene) or possibly carcinogenic to humans (class 2B) (naphthalene [NAP], benzo[k]fluoranthene [BkF], benzo[b]fluoranthene [BbF], and indeno[1,2,3-cd]pyrene [IPY]).[3,8]

To evaluate the total PAHs intake in occupationally exposed subjects, biological monitoring is usually performed. The use of urinary 1-hydroxypyrene (OH-Py), a metabolite of pyrene (PYR), has been proposed as a biological marker of exposure to PAHs because PYR is always present in PAHs mixtures and a good correlation has been found between OH-Py and both airborne PYR and total airborne PAHs.[9–11]

The primary objective of this article was to assess exposure to PAHs by means of environmental and biological monitoring in Italian AW and to evaluate the contribution of occupational exposure and cigarette smoking to the internal dose. For the sake of comparison, a group of roadside construction workers (CW), not exposed to bitumen fumes, was also studied.

MATERIALS AND METHOD

Subjects, Samples Collection, and Working Conditions

The study was conducted in Spring–Summer 2003 in urban, suburban, and rural areas of Milan and Lodi provinces, Italy. Altogether 147 male subjects participated: 100 AW (56 nonsmokers, 44 smokers), exposed to bitumen fumes during laying (91 subjects) or production of concrete asphalt (9 subjects) and 47 CW (17 nonsmokers, 30 smokers), not exposed to bitumen fumes. Both groups were also exposed to diesel exhausts coming from working machines and from the eventually present surrounding autovehicular traffic.

Their average ages were 42 (19–75) and 40 (20–63) years for AW and CW, respectively.

To reduce the possible confounding effect of diet on biological monitoring, the subjects were asked to temporarily abstain from PAH-rich food (i.e., smoked and grilled food, tea, coffee, etc.) the evening before sampling and the day of sampling. Before sampling, each subject was interviewed by an occupational health physician who filled in a questionnaire with information about lifestyle, medical history, and occupational activity. Written informed consent was obtained from each subject.

Personal exposure to airborne PAHs was assessed by personal air samples collected with active samplers worn by the workers in the respiratory zone during the first part of a work shift (typically 7 AM–11 AM). Information was collected about the asphalt type, whereas application temperature and weather conditions (temperature, humidity, and wind strength) were measured.

For biological monitoring, spot urine samples were collected from each worker in three different moments of the same workweek: a baseline sample on Monday morning (around 7 AM), before-shift (around 7 AM), and end-shift (around 5 PM) samples after 2 or more consecutive workdays (same day of air sampling). Specimens collected in the morning were obtained as second urination of the day. Samples were immediately refrigerated, and stored in polyethylene tubes at -20°C till analysis.

Airborne PAHs

Air samplers, used to assess personal exposure to airborne PAHs, consisted in Teflon filters (37-mm diameter, 2 μm porosity, Zefluor filter; Supelco, Milan,

Italy), to collect PAHs on particulate matter, connected in series with a XAD-2 sorbent tube (200 mg, ORBO 42 LG; Supelw), to collect PAHs present in vapor phase. Air was pumped through the samplers at 2 L/min. PAHs were desorbed with acetonitrile and analyzed by high-performance liquid chromatography (HPLC) with fluorimetric detection.[12] A total of 15 priority PAHs, listed by the U.S. Environmental Protection Agency (EPA), were quantified (NAP, acenaphthene [ACE], fluorene [FLE], phenanthrene [PHE], anthracene [ANT], fluoranthene [FLU], PYR, chrysene [CHR], BaA, BkF, BbF, BaP, dibenzo[a,h]anthracene [dBA], benzo[g,h,i]perylene [BPE], and IPY). Analytical limits of detection (LOD), calculated for a 4-h sampling period with an average air volume collection of 0.48 m^3, are reported in TABLE 1.

OH-Py in Urine

Urinary OH-Py was analyzed by HPLC with fluorimetric detection according to a published method.[13] Briefly, urine samples (1 mL) were diluted with 1 mL of acetate buffer (0.2 M, pH 5.0), containing 250 U of β-glucoronidase and 20 U of sulphatase and incubated for 16 h at 37°C. Samples were then purified by extraction with 3 mL of ethyl acetate, the organic phase (2.5 mL) was transferred into a glass vial and taken to dryness at 40°C under a stream of nitrogen. The residue was dissolved with 200 μL of mobile phase (acetonitrile-water-acetic acid 30:69.5:0.5) and 40-μL aliquots of the resulting solution were

TABLE 1. Summary statistics of personal exposure to PAHs (ng/m^3) in study subjects divided in AW ($n = 100$) and CW ($n = 47$) and LOD

	LOD	AW		CW		
		>LOD*	Median (min–max)	>LOD*	Median (min–max)	*P*
NAP	2	100	426 (2–2319)	100	371 (113–877)	NS
ACE	2	59	9 (<2–581)	6	2 (2–4)	<0.01
FLE	0.2	96	33.5 (0.2–284.6)	91	8.6 (0.2–52.6)	<0.01
PHE	1.4	99	51.8 (<1.4–1096.0)	100	13.3 (2.3–39.9)	<0.01
ANT	0.4	55	0.7 (<0.4–97.7)	36	0.4 (<0.4–2.5)	<0.01
FLU	0.4	70	2.8 (<0.4–147.1)	75	1.3 (0.4–4.3)	<0.01
PYR	0.6	100	26.3 (1.2–282.2)	79	1.0 (<0.6–4.9)	<0.01
CHR	0.08	33	0.11 (<0.08–9.83)	94	0.69 (0.11–2.88)	<0.01
BaA	0.08	28	0.10 (<0.08–20.06)	94	0.64 (0.11–4.35)	<0.01
BkF	0.03	74	0.15 (<0.03–18.50)	94	0.27 (0.04–1.12)	0.03
BbF	0.2	69	0.8 (<0.2–26.5)	82	0.9 (0.2–3.1)	NS
BaP	0.03	84	0.33 (<0.03–40.25)	97	0.61 (0.04–2.71)	NS
dBA	0.07	48	0.16 (<0.07–18.61)	58	0.16 (0.08–0.63)	NS
BPE	0.3	69	1.1 (<0.3–13.9)	61	0.8 (0.4–2.4)	0.04
IPY	0.3	12	0.4 (<0.3–6.0)	55	0.6 (0.4–2.3)	<0.01
Σ15 PAHs	—	—	607 (127–2973)	—	405 (157–940)	<0.01
Σ 6 PAHs	—	—	2.7 (0.6–75.4)	—	3.7 (0.9–12.2)	NS

NOTE: *percentage of samples above LOD.
NS = not significant.

injected into chromatograph equipped with a reverse-phase Supelcosil-DP column (50 mm length, 4.6 mm internal diameter, 5 μm particle size; Supelco). The flow of the mobile phase was 4 mL/min. The wavelengths used for quantification were 340 nm for excitation and 390 nm for emission. The LOD of the method is 50 ng/L. Creatinine was determined using Jaffe's colorimetric method.

Statistical Analysis

Statistical analysis was performed using the SPSS 12.0 package for Windows (SPSS Inc., Chicago, IL). For urinary OH-Py statistical analysis, concentrations were expressed both as ng/L and as ng/g creatinine. Data were log transformed to assure normal distribution (Kolmogorov-Smirnov test). To compare groups the Student's *t*-test for independent samples was used, whereas to compare different moments of exposure within groups (baseline versus before-shift and end-shift samples) the Student's *t*-test for paired samples was used. The Pearson's correlations were used to test the associations between variables. A P value of 0.05 was considered statistically significant.

RESULTS

Subjects and Working Conditions

AW and CW were matched for major characteristics, except for the percentage of smokers, higher in CW (64%) than in AW (44%). In both groups average consumption was 20 ± 11 cigarettes per day.

Weather conditions during the field studies were generally good, with absence of rain: median temperature 25 and 14°C, atmospheric pressure 1004 and 1008 mBar, air humidity 45 and 59%, and wind speed 2.4 and 3.1 km/h were measured for AW and CW, respectively.

During road paving, a median asphalt temperature of 130°C (from 120 to 260°C) was measured.

Airborne PAHs

Data on personal exposure to airborne PAHs (as sum of vapor and particle phase) in workers divided according to job title are reported in TABLE 1. Each measured polycyclic compound as well as the total amount of airborne PAHs (sum of all 15 compounds, Σ15 PAHs), and the six high-boiling PAHs classified in group 2A or 2B by IARC (namely, BaA, BaP, dBA, BkF, BbF, and IPY) (Σ6 PAHs) are shown.

In AW, 4 out of 15 PAHs were found to be the most representative: PYR (detected above the LOD in 100% of samples), NAP (100%), PHE (99%), and

FLE (96%). The most abundant compound was NAP (contributing for 77% to the total amount, median value: 426 ng/m^3), followed by PHE (9%, 52 ng/m^3), FLE (6%, 34 ng/m^3), and PYR (5%, 26 ng/m^3). With regard to CW, the most frequently detected compounds were: NAP (100% of samples), PHE (100%), FLE (91%), and the high-boiling CHR (94%), BaA (94%), BkF (94%), and BaP (97%). The most abundant compound was again NAP (contributing for 92% to the total amount, 371 ng/m^3), followed by PHE (3%, 13 ng/m^3) and FLE (2%, 9 ng/m^3); PYR contribution to the total PAHs amount was in this case negligible (<1%, 1 ng/m^3). Median values of low-boiling compounds (NAP to PYR) in AW were higher than those observed in CW, with major differences in PHE (fourfold higher) and PYR (30-fold higher). Among high-boiling PAHs, CHR, BaA, and BkF were somewhat higher in CW than in AW, however, their levels were in any case very low, in the range 0.1–0.7 ng/m^3.

OH-Py in Urine

Summary statistics for urinary OH-Py, expressed in ng/L, are presented in TABLE 2. The results are stratified by job title and smoking habit. In AW significant differences were found between urinary OH-Py in the three moments of sampling: median OH-Py increased during the workday (before shift 402 ng/L versus end shift 690 ng/L, $P < 0.01$) as well as during the workweek. A twofold increase from baseline to before shift (228 ng/L versus 402 ng/L, $P < 0.01$) and a threefold increase from baseline to end shift (228 ng/L versus 690 ng/L, $P < 0.01$) was observed in all subjects as well as in subjects divided according to smoking habit (FIG. 1A). In these subjects baseline, before- and end-shift median creatinine excretion was 1.6, 1.6, and 2.0 g/L. When the statistical analysis was repeated with OH-Py (ng/g creatinine) the same results were obtained.

In CW, considered altogether or grouped according to smoking habits, OH-Py in different urine samples exhibits small variations, with values close to the baseline level throughout the workweek (FIG. 1B). In these subjects baseline, before- and end-shift median creatinine excretion was 1.7, 1.5, and 1.6 g/L. When the comparisons were performed using OH-Py (ng/g creatinine), in all subjects and in smokers, end-shift levels were higher than baseline or before-shift levels; such difference was not observed in nonsmokers.

In both job titles, OH-Py, adjusted and unadjusted for creatinine, was significantly higher in smokers compared to nonsmokers at all sampling times ($P \leq 0.01$).

In baseline samples, collected after a weekend away from work, AW and CW had comparable levels of OH-Py. On the contrary, before-shift and end-shift values were higher in AW than in CW ($P = 0.03$ and $P = 0.01$, respectively). Moreover, higher differences were evidenced when the confounding effect of smoking was excluded: in nonsmokers the amount of OH-Py was threefold higher in AW than in CW, both in before- and end-shift samples ($P < 0.01$).

TABLE 2. Urinary OH-Py at different sampling times in subjects divided in AW ($n = 100$) and CW ($n = 47$) and in smokers and nonsmokers

	Baseline OH-Py (ng/L) median (min–max)			Before-shift OH-Py (ng/L) median (min–max)			End-shift OH-Py (ng/L) median (min–max)		
	All	Nonsmokers*	Smokers*	All	Nonsmokers*	Smokers*	All	Nonsmokers*	Smokers*
AW**	228 (<50–1883)	170 (<50–1758)	289 (<50–1883)	402 (<50–2255)	252 (<50–2114)	565 (<50–2255)	690 (<50–5835)	506 (<50–3799)	781 (123–5835)
CW***	260 (<50–1065)	123 (<50–729)	421 (66–1065)	304 (<50–1568)	94 (<50–978)	402 (86–1568)	378 (59–2000)	153 (59–1297)	489 (108–2000)
P for comparison between jobs	NS	NS	NS	0.03	0.003	NS	0.01	0.01	0.02

NOTE: * = Smokers higher than nonsmokers in both job titles and all different sampling times ($P < 0.01$).
** = baseline lower than before shift ($P < 0.01$), baseline lower than end shift ($P < 0.01$), and before shift lower than end shift ($P < 0.01$) in all subjects and in both smokers and nonsmokers.
*** = No difference between baseline, before shift, and end shift.
NS = not significant.

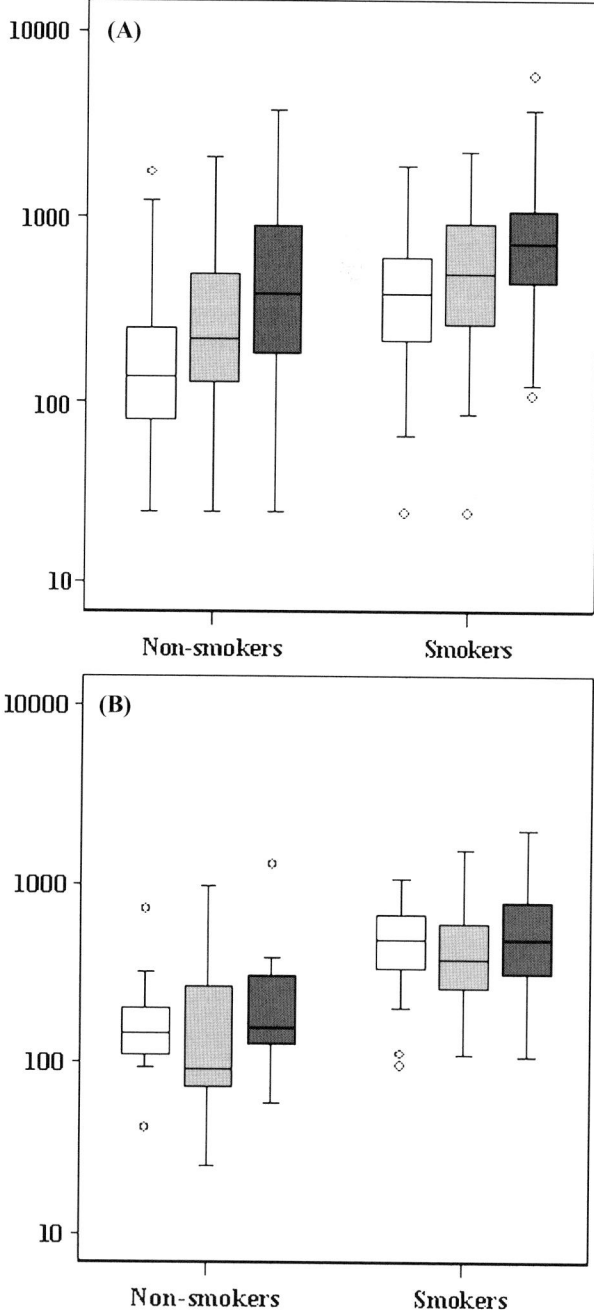

FIGURE 1. Box-plots of baseline (white), before-shift (light gray) and end-shift (dark gray) OH-Py in AW (**A**) and CW (**B**) divided according to smoking habit. ○ = outliers.

Among smokers, OH-Py was higher in AW than in CW only considering end-shift samples ($P < 0.01$).

The comparisons performed using OH-Py (ng/g creatinine) showed that end-shift levels were not different in AW and CW, either considering all subjects or smokers; in nonsmokers higher levels in AW that in CW were confirmed.

Correlations

Pearson's correlation coefficients between selected indices of airborne exposure to PAHs and OH-Py level in the different moments of urine sampling are reported in TABLE 3. Subjects were considered all together or divided according to smoking habit. Selected indices of airborne exposure to PAHs, as pure compounds or mixture, were considered: NAP, for its abundance and toxicology relevance; PHE, representative of the 3-rings PAHs, for its abundance; PYR, representative of the 4-rings PAHs, for its abundance and because urinary OH-Py is a product of its metabolism; BaP, representative of the 5-rings PAHs, for its toxicological relevance; $\Sigma 15$ PAHs, as an index of whole PAHs exposure; and $\Sigma 6$ PAHs, as an index of carcinogenic PAHs exposure.

In all subjects significant correlations were observed between baseline OH-Py and both before- and end-shift OH-Py ($r = 0.48$ and 0.42, respectively) as well as between before- and end-shift OH-Py ($r = 0.79$). These correlations generally improved when only nonsmokers were considered.

Considering airborne PAHs exposure, NAP was weakly correlated with PHE ($r = 0.33$), but not with PYR. PHE and PYR were well correlated to each other ($r = 0.63$). Their correlations with $\Sigma 6$ PAHs was weak and limited to NAP and PHE ($r = 0.25$ and 0.24, respectively). BaP was correlated with $\Sigma 6$ PAHs ($r = 0.71$) and with $\Sigma 15$ PAHs ($r = 0.20$), but not with low-boiling PAHs or with OH-Py. The correlations between NAP, PHE, or PYR and $\Sigma 15$ PAHs are poorly informative, being merely attributable to the major contribution of these compounds to the sum of PAHs.

The correlation between OH-Py and various indices of airborne exposure to PAHs was limited to low-boiling compounds and 15 ΣPAHs, with the best correlation coefficients between end-shift OH-Py and PHE ($0.39 \leq r \leq 0.51$). A weak association was also found between end-shift OH-Py and PYR ($r = 0.27$). Again these correlations always improved when nonsmoking subjects were considered. Lower correlation coefficients were generally observed when similar analysis was performed using OH-Py (ng/g creatinine).

DISCUSSION

In bitumen-related working activities, exposure to PAHs is found to be some orders of magnitude lower than that of coke oven and aluminium industry,

TABLE 3. Correlations, as Pearson's coefficient r, between urinary OH-Py and selected airborne PAHs in all subjects and in subjects divided according to smoking habit

		Before-shift OH-Py	End-shift OH-Py	NAP	PHE	PYR	BaP	Σ15 PAHs	Σ6 PAHs
Baseline OH-Py	All	0.48**	0.42**	0.22*	NS	NS	NS	0.26**	NS
	Smokers	0.44**	NS	NS	NS	NS	NS	NS	NS
	Nonsmokers	0.39**	0.49**	NS	NS	NS	NS	0.39**	NS
Before-shift OH-Py	All	—	0.79**	NS	0.29**	0.22*	NS	0.26**	NS
	Smokers	—	0.70**	NS	NS	NS	NS	NS	NS
	Nonsmokers	—	0.81**	NS	0.44**	0.37**	NS	0.35**	NS
End-shift OH-Py	All	—	—	NS	0.39**	0.27**	NS	0.30**	NS
	Smokers	—	—	NS	0.34**	0.32**	NS	0.26*	NS
	Nonsmokers	—	—	NS	0.51**	0.35**	NS	0.33**	NS
NAP	All	—	—	—	0.33**	NS	0.20*	0.77ᵃ	0.25*
	Smokers	—	—	—	0.49**	NS	NS	0.83ᵃ	0.28*
	Nonsmokers	—	—	—	0.25*	NS	0.28*	0.72ᵃ	NS
PHE	All	—	—	—	—	0.63**	NS	0.64ᵃ	0.24**
	Smokers	—	—	—	—	0.62**	NS	0.67ᵃ	0.26*
	Non-smokers	—	—	—	—	0.64**	NS	0.64ᵃ	NS
PYR	All	—	—	—	—	—	NS	0.48ᵃ	NS
	Smokers	—	—	—	—	—	NS	0.42ᵃ	NS
	Nonsmokers	—	—	—	—	—	NS	0.57ᵃ	NS
BaP	All	—	—	—	—	—	—	0.20*	0.71**
	Smokers	—	—	—	—	—	—	NS	0.69**
	Nonsmokers	—	—	—	—	—	—	0.27*	0.73**
Σ15 PAHs	All	—	—	—	—	—	—	—	0.41**
	Smokers	—	—	—	—	—	—	—	0.45**
	Nonsmokers	—	—	—	—	—	—	—	0.37**

NOTE: * = $P < 0.05$.
** = $P < 0.01$.
ᵃ these correlations are merely attributable to the relevant amount of NAP, PHE, and PYR in the sum of 15 PAHs.
NS = not significant.

where the worst exposure conditions are met.[7] Nevertheless, there is a health concern for road paving workers regarding possible exposure to PAHs, because of possible carcinogenic properties of some of them.[2,3]

The prevalence of low-boiling PAHs (NAP to PYR) observed in the present study is in good accordance with those observed by other authors (see TABLE 4).[14–17] Among these compounds, NAP, recently classified as possible carcinogenic to humans,[8] was the most abundant compound. The high-boiling PAHs (CHR to IPY) were also present in the majority of the samples, their abundance being two to three orders of magnitude lower than the low-boiling ones. The median levels of airborne BaP here reported are in the same range of those found in Italian urban settlements.[13,18–20]

When we compare the results of the present investigation with those of other European studies conducted on AW, we note that the median levels of airborne exposure to PAHs in Italy are similar to those found in Norway and

TABLE 4. Summary of exposure to Σ15 PAHs and PYR (μg/m³) and of urinary end-shift OH-Py (ng/L) in recent published papers on AW and in the present investigation

Country	No. subjects	Σ15 PAHs (μg/m³) mean (min–max)	PYR (μg/m³) mean (min–max)	End-shift OH-Py[a] (ng/L) mean (min–max)	Year, reference
Sweden	28 NS	0.59 (0.02–7.6)[b,c]	0.032[c] (0.004–0.20)	210[c] (50–880)	1999, 21
Hungary	10 NS	0.08	—	400	2001, 24
Norway	320 S/NS	0.02–5.33	0.56	—	2002, 15
Finland	53 S/NS	3.29 (<0.01–2.49)	0.12	1500 (<20–8500)	2002, 16
Finland	26 NS	5.7[c] (0.87–46)	0.08[c] (0.01–1.2)	560[c] (<140–5140)	2003, 14
U.S.A.	20	—	0.2[d] (<0.01–1.7)	1440[d]	2004, 23
Italy	100 S/NS	0.81[d] (0.29–3.1)	0.026[d] (0.001–0.28)	690[d] (50–5835)	This study

NOTE: [a] Original values of OH-Py were transformed in ng/L, assuming a mean creatinine excretion of 1.2 g/L when necessary; [b] the sum of 10 PAHs, among which some methyl-PAHs, was given; [c] = geometrical mean; [d] = median; S = smokers; NS = nonsmokers.

in Sweden,[15,21] but 10-fold inferior to those found in Finland.[14,16] The low levels of airborne PAHs here reported could be explained by many reasons, as for instance, asphalt composition, application technologies and, therefore, asphalt temperature. It has been observed that the temperature of application is one of the most important factors in determining the chemical composition of asphalt fumes and in particular it was observed that the emission of asphalt fumes and volatile compounds was doubled for every 15°C of increase in asphalt temperature.[22] In the present study concrete asphalt was applied, with temperatures in the range 120–170°C, with a median of 130°C. Only in one working situation, with use of mastic asphalt, a temperature of 260°C was measured. On the contrary, in Finnish studies, where mastic or remixed asphalt was applied, higher temperature ranges of 160–250°C or 130–200°C were reported.[14,16] Furthermore, differences in sampling and/or analytical methods could as well contribute to some extent to the observed dissimilarity.

In good accordance with the low levels of airborne PAHs, also OH-Py levels found in the present study are lower than those reported by other authors (TABLE 4).[14,16,21,23,24] The increase in OH-Py excretion in AW during the workday and the workweek, together with the observation that before- and end-shift OH-Py in AW was higher than in CW, suggests that OH-Py is a sensible biomarker, which can be used on a group basis to evidence exposure even at low levels of airborne PAHs such as those found here.

The results of urinary OH-Py reported in TABLE 2 are expressed in ng/L, because creatinine adjustment can introduce biases and reduce precision, due to creatinine variation with age, body mass index, physical activities, and other physiological factors.[25] Moreover, our statistical analysis showed that comparisons between OH-Py at different sampling times and between AW and CW led to similar results using corrected or uncorrected values. The only exception was that of CW, for which end-shift OH-Py (ng/g creatinine) was higher than baseline or before-shift level, and not different from AW. The difference in distribution of smokers in AW and CW may partially explain this observation so that when nonsmokers were considered such discrepancy was not found.

One of the critical issues with the study of a real life situation, such as that examined in the current report, is the selection of time for urine collection. For the determination of OH-Py, the end of the workweek was recommended.[26] Our experience supports this recommendation and suggests that the assessment of low-level PAHs exposure by means of OH-Py may be done using urine samples collected after, at least, two consecutive workdays.

The eventuality that the observed differences among end and before shift in AW could be ascribed to the presence of a circadian cycle in OH-Py excretion is ruled out by the pattern of OH-Py in CW. In fact in these subjects OH-Py does not show any significant variation in different moments of the workweek (see FIG. 1). Moreover, in AW, the increase of OH-Py during the workweek, in accordance with previous studies,[21,23] suggests an accumulation of this compound into the body. This is in good accordance with results from experimental animals and humans showing that, following inhalation or ingestion of PYR, only about 80% of the total OH-Py excretion is completed within 12 h.[27–29] Otherwise, the similarity in baseline OH-Py between AW and CW is in line with a relatively rapid elimination kinetic, and suggests that, after 2 days away from occupational exposure, OH-Py excretion is determined mostly by nonoccupational sources such as diet, smoking, or a general environmental exposure.

Because diet represents the major source of PAHs daily intake in the general population, we asked study subjects to refrain from PAHs-rich food the evening before and the day of the sampling. This allowed us to avoid peaks of excretion due to excessive dietary intake, but of course could not eliminate the background level due to consumption of widespread contaminated food. This background intake of PAHs may be, at least in part, responsible of the excretion of OH-Py observed in baseline samples (about 250 ng/L) both in AW and in CW and may explain the significant correlation found between baseline OH-Py and either before-shift or end-shift OH-Py.

When the difference observed in AW between end- and before-shift OH-Py (about 250 ng/L, irrespectively of smoking), attributable to occupational exposure, was compared to PYR inhaled during the work shift (about 380 ng, estimated from 26 ng/m^3 [median work-shift exposure] $\times$ 1.8 m^3/h [moderate workload ventilation rate] $\times$ 8 h/shift [work-shift duration]), we observed some inconsistencies. Particularly, given the fact that only some percentage of inhaled PYR (<5–20%) is excreted as OH-Py in urine,[29,30] the inhalation exposure alone does not give reason of the appreciable increase in the excretion of OH-Py. A significant contribution to the total PAHs absorbed could be due to dermal exposure. Dermal contamination was proved to be a major route of exposure for coke oven workers,[31] and recent studies support this evidence also for AW.[23,32,33] In particular, the results of a study conducted in the United States suggest that the impact of dermal exposure could be eight times that of inhalation exposure and that the excretion of urinary OH-Py could be increased by a dermal exposure that occurred during the preceding 32 h.[23] Moreover, in a Finnish study, the correlation between PYR contamination at wrists and airborne PYR was closer than that between urinary OH-Py and airborne PYR ($r = 0.69$ versus $r = 0.40$, respectively). Moreover, dermal PYR contamination contributed to a larger extends to explain the variability of OH-Py excretion than did airborne PYR ($R^2 = 47\%$ versus $R^2 = 16\%$, respectively). The same observation was valid for dermal PHE, airborne PHE, and the excretion of urinary phenanthrols.[32] In the light of all these observations, also the weak correlation found in our study between PYR and OH-Py ($r = 0.22$) appears to be largely justified. In order to better determine the total PAHs intake, further work to assess dermal exposure in AW and CW is in progress in our laboratory.

The influence of tobacco smoking on HO-Py is well known.[13,16,34] Also in this study urinary HO-Py was significantly affected by smoking habit with levels approximately two- or threefold higher in smokers than in nonsmokers, both in AW and in CW. The effect of smoking on OH-Py was in the same order of magnitude of the effect of occupational exposure. In fact, in nonsmoking AW, an increase in OH-Py attributable to occupational exposure of approximately 250–300 ng/L was found comparing before-shift and end-shift samples. A similar increase was found, at each sampling time, comparing nonsmoking and smoking CW. Again the smoking effect was evident when different job titles were compared: although, at each time point, OH-Py was always higher in nonsmoking AW than CW, in smoking subjects the same was true only for end-shift samples.

In conclusion, the results of this study show that AW experienced a moderate occupational exposure to airborne PAHs, resulting in a significant increase of urinary OH-Py during the workday and the workweek. The contribution of working activities to internal dose was in the same order of magnitude of the contribution of cigarette smoking.

ACKNOWLEDGMENTS

The project was supported by the Italian Ministry of Education, University and Research (MIUR) as a 2003 COFIN project, and by the Italian Institute for Safety at Work (ISPESL, contract no. B/47/DML/03)

REFERENCES

1. CONCAWE. 1992. Bitumen and bitumen derivatives. Product dossier no. 92/104. Brussels, Belgium.
2. INTERNATIONAL AGENCY FOR RESEARCH ON CANCER. 1985. IARC Monographs on the Evaluation of Carcinogenic Risks to Humans. Polynuclear Aromatic Compounds, Part 4, Bitumen, Coal-tars and Derived Products, Shale-oils and Soots. Vol. 35. IARC, Lyon, France.
3. INTERNATIONAL AGENCY FOR RESEARCH ON CANCER. 1987. IARC Monographs on the Evaluation of Carcinogenic Risks to Humans. Overall Evaluations of Carcinogenicity: an Updating of IARC Monographs Volumes 1–42. (Suppl. 7). IARC, Lyon, France.
4. AMERICAN CONFERENCE OF GOVERNMENTAL INDUSTRIAL HYGIENISTS. 2004. TLVs and BEIs based on the documentation of the threshold limit values for chemical substances and physical agents & biological exposure indices. ACGIH, Cincinnati, OH, U.S.A.
5. BOFFETTA, P., I. BURSTYN, T. PARTANEN, *et al*. 2003. Cancer mortality among European asphalt workers: an international epidemiological study. I. Results of the analysis based on job titles. Am. J. Ind. Med. **43:** 18–27.

6. BOFFETTA, P., I. BURSTYN, T. PARTANEN, *et al.* 2003. Cancer mortality among European asphalt workers: an international epidemiological study. II. Exposure to bitumen fume and other agents. Am. J. Ind. Med. **43:** 28–39.
7. BRANDT, H.C.A. & W.P WATSON. 2003. Monitoring human occupational and environmental exposures to polycyclic aromatic compounds. Ann. Occup. Hyg. **47:** 349–378.
8. INTERNATIONAL AGENCY FOR RESEARCH ON CANCER. 2002. IARC Monographs on the Evaluation of Carcinogenic Risks to Humans. Some Traditional Herbal Medicines, Some Mycotoxins, Naphthalene and Styrene. Vol. 82. IARC, Lyon, France.
9. JONGENEELEN, F.J.R., B.M. ANZION, P.T.J. SCHEEPERS, *et al.* 1988. 1-Hydroxypyrene in urine as a biological indicator of exposure to polycyclic aromatic hydrocarbons in several work environments. Ann. Occup. Hyg. **32:** 35–43.
10. BOUCHARD, M. & C. VIAU. 1999. Urinary 1-hydroxypyrene as a biomarker of exposure to polycyclic aromatic hydrocarbons: biological monitoring strategies and methodology for determining biological exposure indices for various work environments. Biomarkers **4:** 159–187.
11. MCCLEAN, M.D., R.D. RINEHART, L. NGO, *et al.* 2004. Inhalation and dermal exposures among asphalt paving workers. Ann. Occup. Hyg. **48:** 663–671.
12. NATIONAL INSTITUTE FOR OCCUPATIONAL AND SAFETY AND HEALTH. 1985. Polynuclear aromatic hydrocarbons by HPLC: 5506. Manual of Analytical Methods, Edition IV. NIOSH, Cincinnati, OH
13. BURATTI, M., O. PELLEGRINO, G. BRAMBILLA & A. COLOMBI. 2000. Urinary excretion of 1-hydroxypyrene of exposure to polycyclic aromatic hydrocarbons from different sources. Biomarkers **5:** 368–381.
14. VÄÄNÄNEN, V., M. HÄMEILÄ, H. KONTSAS, *et al.* 2003. Air concentrations and urinary metabolites of polycyclic aromatic hydrocarbons among paving and remixing workers. J. Environ. Monit. **5:** 739–746.
15. BURSTYN, I., B. RANDEM, J.E. LIEN, *et al.* 2002. Bitumen, polycyclic aromatic hydrocarbons and vehicle exhaust: exposure levels and controls among Norwegian asphalt workers. Ann. Occup. Hyg. **46:** 79–87.
16. HEIKKILÄ, P., R. RIALA, M. HÄMEILÄ, *et al.* 2002. Occupational exposure to bitumen during road paving. AIHA J. **63:** 156–165.
17. HICKS, J.B. 1995. Asphalt industry cross-sectional exposure assessment study. Appl. Occup. Environ. Hyg. **10:** 840–848.
18. SCIARRA, G. 2003. Valori di riferimento ambientali e biologici degli idrocarburi policiclici aromatici. G. Ital. Med. Lav. Erg. **25:** 83–93.
19. MINOIA, C., S. MAGNAGHI, G. MICOLI, *et al.* 1997. Determination of environmental reference concentration of six PAHs in urban areas (Pavia, Italy). Sci. Total Environ. **198:** 33–41.
20. PASTORELLI, R., M. GUANCI, J. RESTANO, *et al.* 1999. Seasonal effect on airborne pyrene, urinary 1-hydroxypyrene, and benzo(a)pyrene diol epoxide-hemoglobin adducts in the general population. Cancer Epidem. Biomarkers Prev. **8:** 561–565.
21. JÄRVHOLM, B., G. NORDSTRÖM, B. HÖGSTEDT, *et al.* 1999. Exposure to polycyclic aromatic hydrocarbons and genotoxic effects on nonsmoking Swedish road pavement workers. Scand. J. Work Environ. Health **25:** 131–136.
22. BURSTYN, I., H. KROMHOUT & P. BOFFETTA. 2000. Literature review of levels and determinants of exposure to potential carcinogens and other agents in the road construction industry. AIHA J. **61:** 715–726.

23. McCLEAN, M.D., R.D. RINEHART, L. NGO, *et al.* 2004. Urinary 1-hydroxypyrene and polycyclic aromatic hydrocarbon exposure among asphalt paving workers. Ann. Occup. Hyg. **48:** 565–578.
24. SZANISZLÓ, J. & G. UNGVÁRY. 2001. Polycyclic aromatic hydrocarbon exposure and burden of outdoor workers in Budapest. J. Toxicol. Environ. Health Part A **62:** 297–306.
25. IKEDA, M., T. TSUKAHARA, J. MORIGUCHI, *et al.* 2003. Bias induced by the use of creatinine-corrected values in evaluation of beta2-microgloblin levels. Toxicol. Lett. **145:** 197–207.
26. BUCHET, J.P., J.P. GENNART & F. MERCADO-CALDERON. 1992. Evaluation of exposure to polycyclic aromatic hydrocarbons in a coke production and graphite electrode manufacturing plant: assessment of urinary excretion of 1-hydroxypyrene as a biological indicator of exposure. Br. J. Ind. Med. **49:** 761–768.
27. MITCHELL, C.E. & K.W. TU. 1979. Distribution, retention, and elimination of pyrene in rats after inhalation. J. Toxicol. Environ. Health **5:** 1171–1179.
28. VIAU, C., M. BOUCHARD, G. CARRIER, *et al.* 1999. The toxicokinetics of pyrene and its metabolites in rats. Toxicol. Lett. **108:** 201–207.
29. BRZÉZNICKI, S., M. JAKUBOWSKI & B. CZERSKI. 1997. Elimination of 1-hydroxypyrene after human volunteer exposure to polycyclic aromatic hydrocarbons. Int. Arch. Occup. Environ. Health **70:** 257–260.
30. INTERNATIONAL PROGRAMME ON CHEMICAL SAFETY. 1998. Selected non-heterocyclic polycyclic aromatic hydrocarbons. Environmental Health Criteria 202. IPCS, Geneva, World Health Organisation.
31. VAN ROOIJ, J.G.M., M.M BADELIER-BADE & F.J. JONGENEELEN. 1993. Estimation of individual dermal and respiratory uptake of polycyclic aromatic hydrocarbons in 12 coke oven workers. Br. J. Ind. Med. **50:** 623–632.
32. VÄÄNÄNEN, V., M. HÄMEILÄ, P. KALLIOKOSKI, *et al.* 2005. Dermal exposure to polycyclic aromatic hydrocarbons among road pavers. Ann. Occup. Hyg. **49:** 167–178.
33. HICKS, J.B., 1995. Asphalt industry cross-sectional exposure assessment study. Appl. Occup. Environ. Hyg. **10:** 840–848.
34. VAN ROOIJ, J.G.M., M.M.S. VEEGER, M.M. BODELIER-BADE & F.J. JOGENEELEN. 1994. Smoking and dietary intake of polycyclic aromatic hydrocarbons as sources of interindividual variability in the baseline excretion of 1-hydroxypyrene in urine. Int. Arch. Occup. Environ. Health **66:** 55–65.

Carcinogens in the Construction Industry

BENGT JÄRVHOLM

Department of Public Health and Clinical Medicine, Umeå University, SE-901 85 Umeå, Sweden

ABSTRACT: The construction industry is a complex work environment. The work sites are temporary and rapidly changing. Asbestos has been widely used in construction industry, but the risks were primarily detected in specialized trades, such as insulation workers and plumbers. Today, the majority of cases related to asbestos exposure will occur in other occupational groups in the construction industry. In a large cohort of Swedish construction workers, insulators and plumbers constituted 37% of all cases of pleural mesothelioma between 1975 and 1984 while they constituted 21% of the cases between 1998 and 2002. It is estimated that 25–40% of all male cases of pleural mesothelioma in Sweden are caused by asbestos exposure in the construction trades. There are many other known carcinogens occurring in the construction industry, including PAHs, diesel exhausts, silica, asphalt fumes, solvents, etc., but it is difficult to estimate exposures and thus the size of the risk. The risk of cancer is less easy to detect with traditional epidemiological methods in the construction industry than in other industrial sectors. It is not sufficient to rely upon broad epidemiological data to estimate the risk of cancer due chemicals in the construction industry. Thus, a strategy to decrease exposure, e.g., to dust, seems a feasible way to reduce the risk.

KEYWORDS: cancer; construction industry

INTRODUCTION

The workplace for construction workers is often temporary, and employees often work in small groups far from planning departments and production managers. Working conditions may be different compared to traditional industrial workers. The exposure to physical and chemical agents in the construction trades will depend on working techniques and handling and will often be difficult to predict and estimate. In some countries like Sweden, the construction

---

Address for correspondence: Bengt Järvholm, M.D., Ph.D., Professor of Occupational and Environmental Medicine, Department of Public Health and Clinical Medicine, Umeå University, NUS, SE-901 85 Umeå, Sweden. Voice: +46-90-785-2241; fax: +46-90-785-2456.
e-mail: bengt.jarvholm@envmed.umu.se

trades are made up of skilled workers who often have years of training. In other countries, several construction trades are consider "unskilled" and require brief or no training in the safe handling of dangerous agents. Exposure to carcinogens may vary between jobs, between countries, and over time in the construction trades. This makes estimation of risk difficult. To study the total risk of cancer in "construction workers" is of very limited value due to the variable exposure. Studies of Japanese and German construction workers found risk of death from cancer similar to that of the general population (SMR = 0.98, 95% CI = 0.90–1.07 and SMR = 0.89, 95% CI = 0.79–1.00).[1,2] This article will show that some construction workers have a considerable and preventable risk of cancer.

LUNG CANCER

Studies of construction workers usually show a moderately increased risk of lung cancer (relative risk [RR] of about 1.1–1.3) and highly increased risk in some trades.[2–4] The latter are typically studies of insulation workers.[5–7] It is understandable that the risk varies as the exposure to possible carcinogens may vary between jobs, within jobs, and over time just as variable exposure to asbestos may partly explain the variable risk of lung cancer over time. Some activities of exposure to some established or probable lung carcinogens are listed in TABLE 1.

Exposure to silica is common for some workers in the construction industry, e.g., rock workers. No study found convincingly that such exposure causes an increased risk of any type of cancer among construction workers, but this is hard to investigate due to the variable exposure.

Exposure to asphalt fumes has been discussed as a cause of lung cancer. Asphalt contains bitumen, which is a high boiling mineral oil fraction with a complex chemical composition including polycyclic aromatic hydrocarbons

TABLE 1. Examples of exposure to lung carcinogens occurring in the construction trade and classification according to IARC

Carcinogen	Classification according to IARC (ref)	Exposure
Asbestos	I	Insulation
Silica	I	Tunnel work, drilling in rocks
Polycyclic aromatic hydrocarbons	I (varies according to substance/mixture)	Coal tar used by roofers
Hexavalent chromium	I	Welding in stainless steel
Diesel exhausts	IIa	Heavy equipment operations, truck driving especially in tunnels, etc.
Radon	I	Tunnel work

(PAH). Previously, coal tar was a common additive to asphalt; today this is rare. A recent study of asphalt workers from eight countries showed variable risks between countries, but a slight significantly increased risk overall (SMR = 1.17, 95% CI = 1.04–1.30).[8] A possible confounding influence from other factors could not be ruled out, and a case reference study in the cohort is ongoing in some countries. Smoking habits were known among the Swedish bitumen workers, and these workers were found to have a similar risk of lung cancer as the general population (SMR = 0.88, 95% CI = 0.57–1.29).[9] However, some asphalt workers handle mastic asphalt with sometimes a content of coal tar and may also work indoors using this material. Thus, within the group of asphalt workers there may be large differences in exposure to probable carcinogens, such as PAH. Earlier studies from Denmark have indicated an increased risk of lung cancer among mastic asphalt workers.[10]

Tunnel work could mean exposure to high concentrations of radon and diesel exhausts. Because radon has caused lung cancer in miners, it is reasonable to regard such exposure as a risk for construction workers in similar environments.

Exposure to diesel exhaust is also difficult to estimate, and the risk of lung cancer among diesel-exposed construction workers has varied from no increased to a slightly increased risk.[3] A recent study indicated a possible association between exposure to diesel exhaust and multiple myeloma in construction workers.[11]

MESOTHELIOMA

Asbestos is known to cause mesothelioma of the pleura or peritoneum. It has been estimated that 80% of all male cases of mesothelioma can be attributed to occupational exposure to asbestos.[12] Asbestos has long been used in the construction industry, e.g., by insulation workers applying asbestos on pipes and spraying it on walls, steel structures, etc. Plumbers who worked on pipes insulated by asbestos were also exposed. In a Swedish cohort of construction workers who participated in health examinations through Bygghälsan between 1971 and 1993, the occurrence of pleural mesothelioma has been investigated through a linkage with the Swedish Cancer Register. Between 1975 and 1984, there were in total 38 cases of which 14 (37%) occurred among insulation workers and plumbers. The Bygghälsan cohort is estimated to include more than 80% of all construction workers in Sweden during 1971–1993. Between 1998 and 2002, the construction workers in this cohort constituted about 20–25% of all male cases of pleural mesothelioma (FIG. 1). Because not all construction workers were included in the cohort, it is reasonable to estimate that 25–40% of all male cases of pleural mesothelioma had a considerable part of their exposure to asbestos in the construction trades and that this exposure caused their disease. Pleural mesothelioma caused by occupational asbestos exposure is certainly preventable.

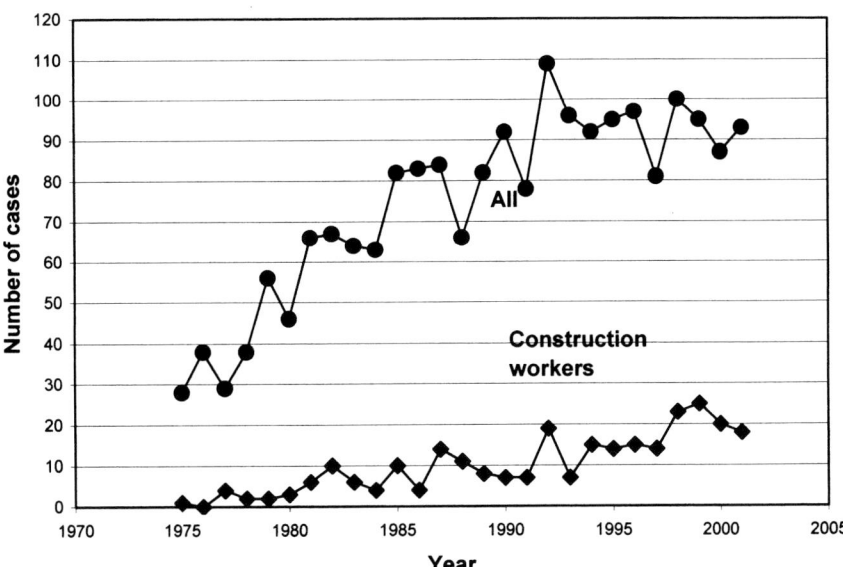

FIGURE 1. Incident cases of pleural mesothelioma among men in Sweden and in a cohort of Swedish construction workers (Bygghälsokohorten) between 1971 and 2002.

The occupational titles of the men in the construction worker cohort that had pleural mesothelioma diagnosed between 1998 and 2002 are described in FIGURE 2. About 20% of the cases with pleural mesothelioma were plumbers and insulation workers, while occupational groups with similar numbers of cases were concrete workers, carpenters, and electricians. The latter groups were probably indirectly exposed to asbestos, i.e., they did not work with asbestos themselves, but worked in areas where other occupational groups handled asbestos.

OTHER TUMORS

Construction workers are a large occupational group in many countries. Studies of associations between occupational titles and cancer based on death certificates or cancer registers often include construction workers. Such linkages have sometimes shown increased risk and sometimes not. A review of all such studies would be very extensive and is beyond the scope of this paper. An increased risk of lip cancer has been found among workers in the construction trades. A possible agent is exposure to UV-radiation possibly in combination with dust.[13] A few recent studies have linked cancer in the upper respiratory or digestive tract to exposure in the construction trades,[14–16] but it is still uncertain if the link is causal. Cement dust or other inorganic dusts have also been suggested as causative agents.

FUTURE EXPOSURE TO CARCINOGENS

Asbestos has been recognized as carcinogen in most western industrialized countries. Many such countries have heavy restrictions or prohibition of asbestos use. However, some countries have not attempted or have been unable to control exposure to asbestos in the construction industry, and it is reasonable to believe that in the future we will have reports of high risk for asbestos-related cancers among construction workers from such countries. Because of the temporary work sites, low control from supervisors, etc., it is very difficult to have very low exposure to asbestos if it is widely used. Indirect or secondary exposure is hard to avoid because workers with secondary exposure may not be aware of such exposure. The Swedish experience (FIG. 2) shows that secondary exposure could be the cause of a majority of the mesothelioma cases among construction workers. Indeed, 25–40% of all cases of mesothelioma in a country may be attributable to indirect exposures.

Even in countries where exposure to asbestos has been brought to a minimum, there are still carcinogens around that will be difficult to eliminate during the next decades (TABLE 1). Apart from those in TABLE 1, there are

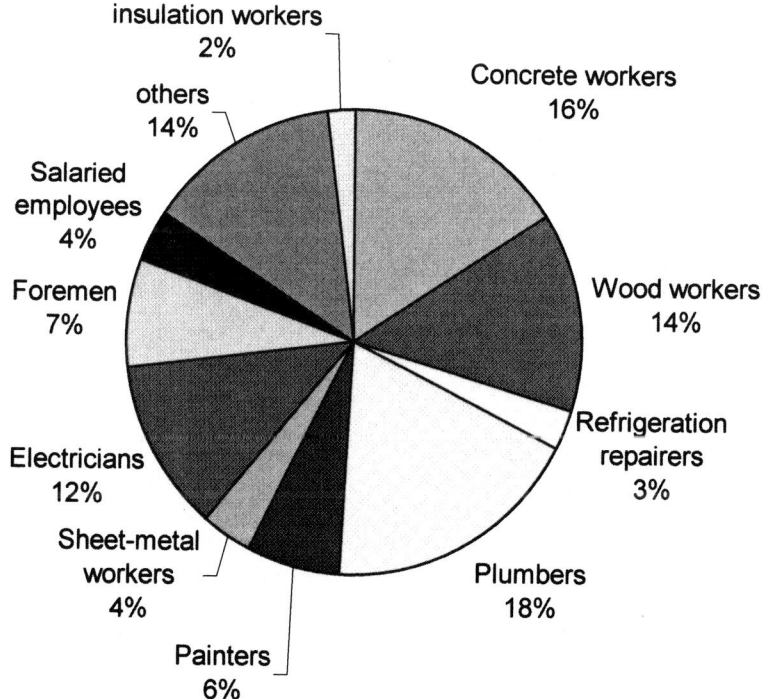

FIGURE 2. Occurrence of pleural mesothelioma ($n = 108$) between 1998 and 2002 according to occupational title among men in the Swedish construction worker cohort.

other suspected or possible carcinogens including solvents, formaldehyde, hard wood, etc. Some animal studies and studies of lung cancer risk in the general environment indicate that biodurable particles may constitute a risk for cancer.[17,18]

Because the application of epidemiology in the construction trades is difficult, mainly due to the difficulties in estimating exposure in this industry, there is less of a science base to build on to prevent or even estimate risk. This does not mean that such studies are not needed or cannot be performed. Well-designed surveillance or cohort studies can measure both overall risk and sometimes detect new risks, especially if the exposure is common. But such studies are expensive and require more resources than most countries are willing to provide, and it is not realistic to expect that industry itself will find ways to finance research. Therefore, government agencies have a more significant duty to protect construction workers than might be the case in some other industries, such as manufacturing.

A strategy to eliminate all possible carcinogenic substances seems less feasible, at least in the next decades, because the political will to do so does not exist. Even with well-established carcinogens, such as asbestos and benzene, progress has been slow in most countries. Therefore, the most practical alternative approach is to decrease exposure in general, which will mean a lower risk. How low the risk will be will depend on the exposure, but even this concept is a hard sell even though it is well established that a low exposure environment generally means a clean work environment as well, which in turn often means a more productive environment. During the last decades, it has been obvious that there is a large interindividual and intraindividual variability of exposure in many branches and jobs.[19] High variability usually means that it is difficult to predict the concentration just by experience or by a short visit at the workplace.

A recent study of the construction industry in the Toronto region showed very high variability and very high exposures in some tasks.[20] In 73 measurements of total dust, concentrations varied from non-detectable to 848 mg/m^3. The highest levels (325, 346, and 848 mg/m^3) were recorded during rock crushing and fireproof mixing. Measurements with a direct-reading instrument showed also very variable concentrations of dust. The highest level (61 mg/m^3, respirable dust) was for a "laborer." The concentration of elemental carbon, a marker of diesel exhausts, varied from 4.9 to 146 µg/m^3 in "laborers." The latter concentration is comparable to exposure in mines where diesel equipment is used in confined spaces. Exposure to Silica was low, and only 4 of 40 samples analyzed for silica showed detectable levels (all four below 0.05 mg/m^3, the current MAC in Sweden). However, a U.S. study found high levels of Silica, often exceeding the MAC in the construction industry.[21]

Measurements in different German industries have shown variable and high concentrations of dust in the construction trades (GESTIS stoffdatenbank, http://www.hvbg.de/d/bia/fac/stoffdb/ Oct 6, 2005). The German findings indicate that the construction trade is the industry with the highest exposure to dust. Thus, workers in the construction trades have a variable and sometimes very high exposure to dust of variable composition. From data from a limited number of countries, it seems that the exposure is higher than in most other industrial branches.

FUTURE RISKS

The construction workers during the next decades will certainly be exposed to carcinogens. The level of the exposure will depend on job task and country and will vary over time. To estimate the risk is difficult. However, good working practice, with low levels of exposure through skillful technical and organizational activities, will keep the risk low. Highly carcinogenic substances should naturally not be introduced and should be eliminated, but it is unlikely that elimination of those carcinogens will occur in some construction trades (TABLE 1). Currently, the levels of total dust are very high in most construction work environments. The knowledge of how to decrease such exposure should be obvious for most trained occupational hygienists. By focusing on the reduction of overall dust concentrations, significant gains will also be made in reducing exposure to carcinogens.

REFERENCES

1. SUN, J. et al. 2002. Mortality among Japanese construction workers in Mie Prefecture. Occup. Environ. Med. **59:** 512–516.
2. ARNDT, V. et al. 2004. All-cause and cause specific mortality in a cohort of 20 000 construction workers; results from a 10 year follow up. Occup. Environ. Med. **61:** 419–425.
3. JÄRVHOLM, B. & D. SILVERMAN. 2003. Lung cancer in heavy equipment operators and truck drivers with diesel exhaust exposure in the construction industry. Occup. Environ. Med. **60:** 516–520.
4. DEMENT, J. et al. 2003. Cancer incidence among union carpenters in New Jersey. J. Occup. Environ. Med. **45:** 1059–1067.
5. JARVHOLM, B. & A. SANDÉN. 1998. Lung cancer and mesothelioma in the pleura and peritoneum among Swedish insulation workers. Occup. Environ. Med. **55:** 766–770.
6. HENDERSON, D.W. et al. 2004. After Helsinki: a multidisciplinary review of the relationship between asbestos exposure and lung cancer, with emphasis on studies published during 1997-2004. Pathology **36:** 517–550.
7. ULVESTAD, B. et al. 2004. Cancer incidence among members of the Norwegian trade union of insulation workers. J. Occup. Environ. Med. **46:** 84–89.
8. BOFETTA, P. et al. 2003. Cancer mortality among European asphalt workers: an international epidemiological study: exposure to bitumen fumes and other agents. Am. J. Ind. Med. **43:** 28–39.

9. BERGDAHL, I. & B. JARVHOLM. 2003. Cancer morbidity in Swedish asphalt workers. Am. J. Ind. Med. **43:** 104–108.
10. HANSEN, E.S. 1991. Mortality of mastic asphalt workers. Scand. J. Work Environ. Health **17:** 20–24.
11. LEE, W.J. *et al.* 2003. Multiple myeloma and diesel and other occupational exposures in Swedish construction workers. Int. J. Cancer **107:** 134–138.
12. JARVHOLM, B. *et al.* 1999. Pleural mesothelioma in Sweden - an analysis of the incidence according to the use of asbestos. Occup. Environ. Med. **56:** 110–113.
13. HAKANSSON, N. *et al.* 2001. Occupational sunlight exposure and cancer incidence among Swedish construction workers. Epidemiology **12:** 552–557.
14. BOFETTA, P. *et al.* 2003. Occupation and larynx and hypopharynx cancer: an international case-control study in France, Italy, Spain and Switzerland. Cancer Causes Control **14:** 203–212.
15. JI, J. & K. HEMMINKI. 2005. Occupation and upper aerodigestive tract cancers: a follow-up study in Sweden. J. Occup. Enivron. Med. **47:** 785–795.
16. JANSSON, C. *et al.* 2005. Occupational exposures and risk of esophageal and gastric cardia cancers among male Swedish construction workers. Cancer Causes Control **16:** 755–764.
17. POPE, C.A. III. *et al.* 2002. Lung cancer, cardiopulmonary mortality, and long-term exposure to fine particulate air pollution. JAMA **287:** 1132–1141.
18. OBERDORSTER, G. 2002. Toxicokinetics and effects of fibrous and nonfibrous particles. Inahl. Toxicol. **14:** 29–56.
19. KROMHOUT, H. *et al.* 1993. A comprehensive evaluation of within- and between-worker components of occupational exposure to chemical agents. Ann. Occup. Hyg. **37:** 253–270.
20. VERMA, D.K. *et al.* 2003. Current chemical exposures among Ontario construction workers. Appl. Occup. Environ. Hyg. **18:** 1031–1047.
21. RAPPAPORT, S. *et al.* 2003. Excessive exposure to silica in the US construction industry. Ann. Occup. Hyg. **47:** 111–122.

Epoxy Resins in the Construction Industry

TON SPEE,[a] COR VAN DUIVENBOODEN,[a] AND JEROEN TERWOERT[b]

[a]*Arbouw, NL-1005 AC Amsterdam, The Netherlands*

[b]*IVAM, University of Amsterdam, NL-1001 ZB Amsterdam, The Netherlands*

ABSTRACT: Epoxy resins are used as coatings, adhesives, and in wood and concrete repair. However, epoxy resins can be highly irritating to the skin and are strong sensitizers. Some hardeners are carcinogenic. Based on the results of earlier Dutch studies, an international project on "best practices,"—Epoxy Code—with epoxy products was started. Partners were from Denmark, Germany, the Netherlands, and the UK. The "Code" deals with substitution, safe working procedures, safer tools, and skin protection. The feasibility of an internationally agreed "ranking system" for the health risks of epoxy products was studied. Such a ranking system should inform the user of the harmfulness of different epoxies and stimulate research on less harmful products by product developers.

KEYWORDS: epoxy resins; construction industry

INTRODUCTION

Workers in the construction industry are end-users of chemicals and, as such, are exposed to a number of hazardous or toxic substances. Examples are the relatively new materials like polyurethane and epoxy resins. Asbestos and tar, which have known carcinogenic properties, have been banned in the Netherlands for almost two decades, but workers are still exposed to these substances during renovation and demolition work. In addition, many seemingly inert materials release toxic substances during work: sand and concrete can hardly be seen as toxic materials but they emit respirable quartz dust during work tasks involving drilling, sanding, and milling.[1,2] Respirable crystalline quartz is carcinogenic.[3] Respirable crystalline quartz, as well as diesel exhaust fumes, are listed on the Dutch list of carcinogenic substances.[4] Construction workers are exposed to diesel exhaust fumes if they are working close to diesel engines and generators.

Address for correspondence: Ton Spee, Arbouw, P.O. Box 8114, NL-1005 AC Amsterdam, The Netherlands, Voice: 020-580-5580; fax: 020-580-5555.
e-mail: spee@arbouw.nl

A new development in the construction industry is the use of secondary materials, mainly in road works. Waste and recycled materials are used as substitutes for primary raw materials. This is good for the environment but less good for the workers. A report of the Ministry of Housing, Spatial Planning, and the Environment from 1991 showed that these materials frequently contain toxic substances like polycyclic aromatic hydrocarbons (PAH) (tar), asbestos, and heavy metals (lead, mercury, cadmium, vanadium).[5] Control measures are necessary to prevent the workers from developing serious health effects from exposures to these risks. This article focuses on one group of toxic substances in the construction industry: epoxies and on control measures to reduce their risks and their effects.

Epoxy resins have unique technical properties that have made them very popular as "problem solvers" in the construction industry. These properties are:

- Excellent adhesion to various substrates, making epoxies good corrosion-protective coatings for metals and strong adhesives;
- Resistance to mechanical wear, making them very suitable as industrial floorings;
- Liquid tight, chemically resistant, and easy to clean, making them excellent as floor and wall coatings as well as joint fillers for tiling in the food industry, professional kitchens, petrol stations, etc.;
- Fast curing, making them popular as concrete repairing agents in flat galleries;
- No shrinkage during curing and easy to sand, making epoxies the most used wood repairing agents.

Usually epoxies are supplied as 2-component products, consisting of a resin component (A) and a hardener (B). Some products consist of three components, the third component being a filler. At curing, the resin and the hardener combine into a tough and solid network. Chemically, epoxies are characterized by the "epoxide group," which is very reactive toward amino compounds in the hardener. Unfortunately, skin proteins are also amino compounds, and that is the reason epoxies are such potent skin sensitizers.

EPOXY ALLERGY

Skin contact with epoxy resins or their hardeners frequently causes allergic contact dermatitis (eczema). The contact dermatitis is generally involves the hands or forearms and sometimes the face. Workers who have acquired an epoxy allergy will be faced with an increasingly stronger skin reaction after each contact with the products. Avoiding every contact is the only option. In practice, this means that the worker must change professions. In the Dutch construction industry, it has been estimated that one out of every five workers who frequently use epoxies (e.g., floor layers) develops an epoxy allergy.[6]

In companies with a low standard of prevention, this number is even higher. Workers find it hard to fully prevent skin contact with epoxies: obvious times of skin contact are during the weighing and mixing of the two components, manual transportation of (open) cans or vessels, direct contact during application, and continued wearing of contaminated or even soaked clothes or shoes.

GOOD PRACTICE AGREEMENT

Since the 1990s, Arbouw, as well as employers' associations and trade unions, observed an increase in the number of epoxy allergies in the Dutch construction industry. For this reason employers' associations and trade unions in the construction industry agreed on "good practice guidelines." Arbouw took the initiative and coordinated the development of these agreements. The agreements were defined in two criteria documents: one covering "epoxy-based concrete repair agents" and the other covering all other applications of epoxies in the construction industry.[7,8] The agreements contained recommendations on product selection, safer work procedures, and skin protection.

In 2000, discussions with the Dutch manufacturers of epoxy products aimed at the development of health risk-based selection criteria for epoxy products. However, most manufacturers of epoxy resins are multinationals, and a number of institutes outside the Netherlands deployed activities for the prevention of epoxy-related skin disease. Therefore, it was decided to set up a European project on this theme.

Epoxy Code

With support of the SME 2003 scheme of the European Agency on Safety and Health at Work in Bilbao, IVAM and Arbouw set up a project known as "Epoxy Code." Partners in this 1-year project were the Health and Safety Executive (HSE; UK), the Bau-Berufsgenossenschaften (BG Bau; Germany), and the Aalborg BST Center (Denmark). The project had two main objectives: to develop an international "Code of Practice" for working with epoxies, and to perform a study of the feasibility of an internationally harmonized "ranking system" for the health risks of epoxies.

Such a ranking system should enable construction companies to select the least hazardous epoxy product for each specific application. A pilot version of a ranking system had been developed in close cooperation with the Dutch manufacturers of epoxy products.

INTERNATIONAL CODE OF PRACTICE FOR EPOXY WORKERS

All partners in the project collected data on "good practices" related to substitution of epoxies by other materials, substitution of epoxies by "less hazardous" epoxies, tools that may reduce skin contact, safer working methods and procedures, skin protection and skin care. Some examples of potential alternatives for epoxy products are:

- *Tiling adhesives*: mineral inorganic adhesives, e.g., cement-based. Many tilers prefer cement-based products because of their high quality.
- *Joint fillers*: silica-based fillers.
- *Metal coatings* for corrosion protection: use uncoated stainless steel or aluminum.
- Decorative epoxy-bound "*gravel/terrazzo floors*": other floor coverings, e.g., parquet, carpets, linoleum.
- *Concrete repair*: cement-based products.

To reduce skin contact, various tools and safer working procedures are available. Epoxy "kits" with a well-defined mixing ratio obviate the need for weighing and prevents residual monomers in the hardened product. Piercable dual packaging allows mixing within the package itself, which reduces direct skin contact. A splash protection shield attached to rollers that are used for applying coatings prevents spattering. Spatulas with relatively long handles, glue spreaders with handles, etc., keep workers from coming into close contact with the product. For the so-called "injection resins" that are used in concrete repair, fully closed systems are available. For the application of cast floors, a trolley can be used. "Good Housekeeping" is also important and includes some simple measures:

- Allow epoxy product on tools (e.g., trowels) to cure and subsequently scrape it off instead of cleaning with volatile solvents;
- Use disposable tools where possible (rollers etc.);
- Close used packages with epoxy resin or hardener immediately.

SKIN PROTECTION AND SKIN CARE

Using protective gloves remains indispensable when working with epoxies. Leather or cotton gloves do not provide any protection against epoxies, nor do latex gloves meant for household use. Use of long-sleeved gloves made of nitrile, neoprene, or butyl rubber over thin cotton inner gloves to absorb moisture is recommended. Use of gloves only once and frequent changing of gloves is also recommended.

CLASSIFICATION OF EPOXY PRODUCTS

Although virtually all epoxy products that are used in the construction industry are very harmful to the skin, but not all epoxy products are equally harmful. Differences are not reflected in the current classification and labeling with R phrases, S phrases, and hazard symbols required by the European Dangerous Preparations Directive.[9] Every epoxy resin is labeled with the R classifications R36 and R38 ("Irritating to the eyes and skin") as well as R43 ("May cause sensitization by skin contact"). Labeling of the hardener component is more variable: most polyamines are classified as "corrosive" (R34) and skin sensitizing (R43); polyaminoamides and epoxy-amino oligomers are only classified as "irritating" to skin and eyes (R36, R38). This EU classification and labeling system is based on the hazard of the product, irrespective of its exposure potential during actual use; differences in "sensitizing potential" are not reflected.

Other existing classification systems, such as "COSHH-essentials"[10] and the Danish "MAL code"[11,12] system, do not reflect the differences in risk that exist between the various epoxy products either. Both systems classify practically all epoxies in the construction industry in same class.

Arbouw and IVAM worked to develop a classification system for epoxies that would complement the current EU classification and labeling system.

SUBSTITUTION TOOL

The major objectives of classifying and ranking epoxies on the basis of their health risks are to clarify the relative risks to the user of various epoxy products with a similar application and to stimulate research in the field of less hazardous products.

When the supplier is willing to specify the class number in the Material Safety Data Sheet (MSDS) and on the label, the information will be easily accessible for the user. It is the intention to specify the class numbers in the Product group Information System of Arbouw (PISA) as well. Former experiences with other product groups (e.g., solvent-based paints and concrete mold release agents) found that such a classification of products resulted in real market shifts toward less hazardous products.

A practicable system must be easily accessible and independently verifiable, contain a limited number of classes that represent a real distinction, and provide the ability to choose among available products, and should, ideally, be agreed upon internationally.

"PENALTY POINTS"

In 2000, discussions with Dutch suppliers of epoxies focused on criteria for classifying epoxies according to their health risks. The discussions resulted in a proposal for a classification based on the following principles:

TABLE 1. Classification criteria

1. Amount of residual monomer epichlorohydrin
2. Toxic or very toxic components (symbol T or T+)
3. Carcinogenic, mutagenic, reprotoxic, or sensitizers for the respiratory tract (based on R phrases)
4. Sensitizing or corrosive hardeners (R43, R34, R35)
5. Amount of reactive diluent in the resin
6. Amount of VOC (volatile organic compounds)
7. Lack of product information
8. Boiling point of the amines in the hardener
9. Boiling point of reactive diluents
10. Molecular weight of the amines in the hardener
11. Molecular weight of reactive diluents
12. Amount of free amine in the hardener

- Products are classified on the basis of their individual components.
- The toxicological characteristics of the components are derived from the R phrases, the physical–chemical properties of the components, and the content of the component in the product (source: MSDS and additional information from supplier) are used.
- Each criterion leads to the attribution of a specified amount of "penalty points" to the product. TABLE 1 presents the criteria that have been developed.
- The total amount of penalty points the product gets, determines the rank class of the product. This resulted in a ranking system with four classes, where class 1 products have the lowest health risk.
- When information is lacking or incomplete, 150 penalty points are added to the score.

Small residues of the monomer epichlorohydrin remain in epoxy resins. Usually, the content is below 1% and does not have to be stated on the MSDS. However, epichlorohydrin is a genotoxic carcinogen. For this reason, European manufacturers of epoxy resins have agreed upon additional voluntary limit values for the monomer content. In the criteria for the classification system, this limit value (max. 10 ppm in the resin) has been adopted. Some examples of other criteria are:

- the molecular weights of the amine compounds in the hardener component and the reactive diluents in the resin, as an indicator of the "sensitizing potential";
- the boiling points of the amine compounds in the hardener component and the reactive diluents in the resin, as an indicator of the inhalation risk;
- the content of reactive diluent in the resin because these are even more potent sensitizers than the epoxy resin itself.

ENCOURAGING TEST RESULT

In order to test the proposed classification system, a test classification was performed with 44 epoxy products in various fields of application (FIG. 1). Six suppliers of epoxy products took part in the test and provided the necessary data. The results were encouraging. Distinguishing among the various products based on their health risks appeared to be possible. This was also the case for products within the same field of application (e.g., water-based coatings). The classification system is supported by the Dutch manufacturers and suppliers of epoxy products. Therefore, the system seems to be able to serve as a user-friendly "substitution tool" for small-and medium-sized enterprises (SMEs) in the construction industry. The user who wishes to choose an epoxy product that has "relatively lower health risks" may use the class numbers for this. However, safe work practices and skin protection (gloves) will always remain crucial.

FOLLOW-UP

The Epoxy Code project was completed in September 2004. The project partners will continue to promote the Code of Practice through brochures, leaflets, and training materials.[8] Training courses will be continued in some countries. In the Netherlands, the classification system for the health risks of epoxies will be introduced in cooperation with the suppliers. This will involve the statement of the class numbers in the MSDS and on the label as well as in the PISA (Product Group Information System) of Arbouw. Arbouw and IVAM and the Dutch epoxy suppliers wish to get international support for the classification system. An internationally harmonized system will make it easier to get the necessary product information from the manufacturers. The aim and principles of a classification system for the health risks of epoxies was understood and supported by the partner organizations in the UK (HSE), Germany (BG Bau), and Denmark (Aalborg BST Center). However, some partners had some doubts regarding how to implement the system. These doubts partly reflect specific national circumstances. Some feared that the manufacturers would not be willing to cooperate in a voluntary system that comes "on top of" the current EU labeling system. It is true that the proposed classification system is a voluntary system, and Dutch suppliers have seen its advantages. The main advantage for these suppliers is the opportunity to distinguish themselves from a number of suppliers that offer cheap and relatively low-standard products. However, it was expected that this advantage would not be enough reason for the suppliers in some of the partner countries to cooperate. Other partners were more optimistic about this point, but still had some questions on the exact implementation of the criteria, and specifically how they should be defined and interpreted. Therefore, discussions on small adaptations of the criteria are ongoing.

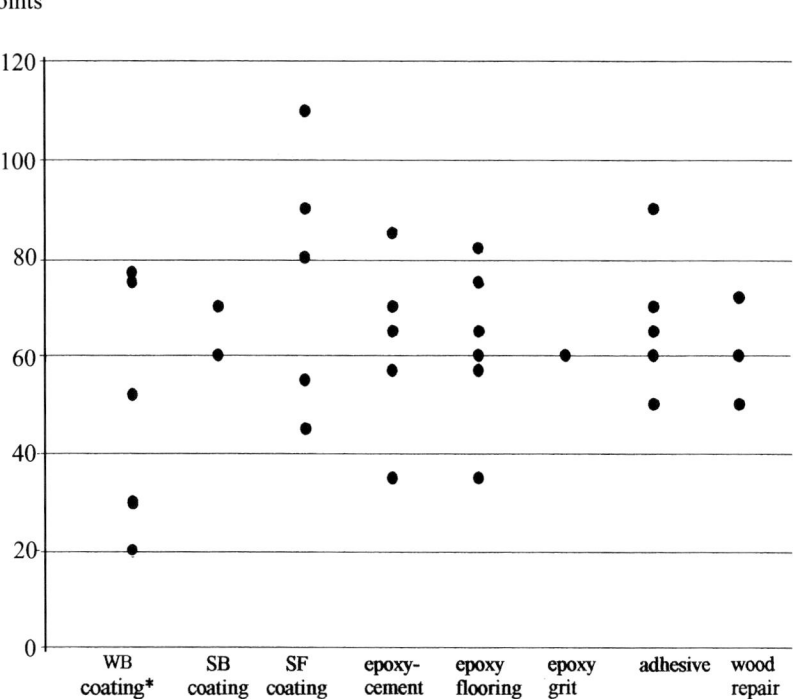

FIGURE 1. Results of the test classification.

CONCLUSIONS

Workers in the construction industry are at risk for exposure to numerous hazardous chemicals. The best way to prevent such exposures is to promote the substitution of less hazardous chemicals where possible. Unfortunately, most chemical product manufacturers do not seem to understand how their products are used in construction tasks, and most small construction employers have limited occupational safety and health expertise. In this article, we show that:

- Developing simple classification systems and product labeling for risk that can be understood by any employer or worker is likely to result in the substitution of lower risk chemicals, and such substitution results in fewer cases of occupational disease, which was the case in the use of solvent-based paints in the Netherlands.

- Even though the development of health hazard-based classification systems makes sense in terms of safety and health, they may not necessarily be adopted voluntarily by all manufacturers, particularly if they operate in international markets. Although the epoxy industry in the Netherlands saw the advantage of adopting a risk classification system, the industries in some of the other countries were less favorable towards such systems so far.
- Therefore, voluntary systems in themselves are not adequate and should be considered as an alternative to having a legal foundation for requiring adherence to the systems, and unless such legal foundations have international application, most workers will continue to be inadequately protected.

ACKNOWLEDGEMENT

The Epoxy Code project has been supported by the European Agency for Safety and Health at Work. Cooperation with Ulrich Goergens (BG Bau), Klaus Kersting (BG Bau), Bob Rajan (HSE), and Christian Pedersen (Aalborg BST Center) is gratefully acknowledged.

REFERENCES

1. LUMENS, M.E.G.L. & T. SPEE. 2001. Determinants of exposure to respirable quartz dust in the construction industry. Ann. Occup. Hyg. **45:** 585–595.
2. TJOE NIJ, E., D. HÖHR, P. BORM, et al. 2004. Variability in quartz exposure in the construction industry: implications for assessing exposure-response relations. J. Occup. Environ. Hyg. **1:** 191–198.
3. STEENLAND, K., A. MANNETJE, P. BOFETTA, et al. 2001. Pooled exposure-response analyses and risk assessment for lung cancer in 10 cohorts of silica-exposed workers: an IARC multicentre study. Cancer Causes Control **12:** 773–784.
4. ARBEIDSINSPECTIE. 2005. Nationale MAC lijst 2005 (National list of Maximum Accepted Concentrations 2005). SDU Uitgevers, Den Haag ISBN 90 12 010699 0.
5. VAN DER WEGEN, G., R. VAN SELST & W. UBACHS. 1991. Definities, toepassingsgebieden en milieuhygiënische aspecten van bouwstoffen (Definitions, application and environemntal aspects of construction materials). Ministry of Housing, Spatial Planning and the Environment, Leidschendam, the Netherlands. Report No 1991/1, ISBN90 346 2590 7 (in Dutch).
6. VAN PUTTEN, P.B., P.J. COENRAADS & J.P. NATER. 1984. Hand dermatoses and contact allergic reactions in construction workers exposed to epoxy resins. Contact Dermatitis **10:** 146–150.

7. ARBOUW. 1999. A-blad Epoxygebonden betonreparatiemiddelen (A-document epoxy based concrete repair agents). Stichting Arbouw, Amsterdam, The Netherlands (in Dutch).
8. ARBOUW. 2004. Werken met epoxyproducten in de afbouw- en onderhoudssector. Stichting Arbouw, Amsterdam, The Netherlands (in Dutch). Available in English: Controlling Skin Diseases When Handling Epoxy resins (Health and Safety Executive, Bootle, U.K.) (Also available in Danish and German).
9. Council Directive 67/548/EEC of 27 June 1967 on the approximation of laws, regulations and administrative provisions relating to the classification, packaging and labelling of dangerous substances.
10. BALSAT, A., J. DE GRAEVE & P. MAIRIAUX. 2003. A structured strategy for assessing chemical risks, suitable for small and medium-sized enterprises. Ann. Occup. Hyg. **47:** 549–556.
11. ARBEJDSTILSYNET. 1993. Executive order on the determination of code-numbers. Order no. 301 (13 May, 1993). Danish Working Environment Authority, DK-2100 København, Denmark.
12. ARBEJDSTILSYNET. 1993. Executive order on work with code-numbered products. Order no. 302 (13 May, 1993). Danish Working Environment Authority, DK-2100 København, Denmark.

Health Effects of Hazardous Waste

STEVE M. DEARWENT,[a] M. MOIZ MUMTAZ,[b] GAIL GODFREY,[a] THOMAS SINKS,[c] AND HENRY FALK[d]

[a]*Division of Health Assessment and Consultation, Agency for Toxic Substances and Disease Registry, Atlanta, Georgia 30333, USA*

[b]*Division of Toxicology and Environmental Medicine, Agency for Toxic Substances and Disease Registry, Atlanta, Georgia, 30333, USA*

[c]*Office of the Director, NCEH/ATSDR, Atlanta, Georgia 30333, USA*

[d]*Coordinating Center for Environmental Health and Injury Prevention, Centers for Disease Control, Atlanta, Georgia 30333, USA*

ABSTRACT: Since 1995, the Agency for Toxic Substances and Disease Registry (ATSDR) has evaluated environmental contaminants and human health risks at nearly 3000 sites. Hazardous substances at these sites include newly emerging problems as well as historically identified threats. ATSDR classifies sites according to the degree of hazard they represent to the public. Less than 1% of the sites investigated are considered urgent public health hazards where chemical or physical hazards are at levels that could cause an immediate threat to life or health. Approximately 20% of sites have a potential for long-term human exposures above acceptable risk levels. At almost 40% of sites, hazardous substances do not represent a public health hazard. Completed exposure pathways for contaminants in air, water, and soil have been reported at approximately 30% of evaluated sites. The most common contaminants of concern at these sites include heavy metals, volatile organic compounds, and polychlorinated biphenyls. This article reviews ATSDR's ongoing work by examining the historic hazard of lead, the contemporary hazard of asbestos, and the emerging issue of perchlorate contamination.

KEYWORDS: ATSDR; hazardous waste; Superfund

INTRODUCTION

The Agency for Toxic Substances and Disease Registry (ATSDR) is tasked with evaluating the public health effects of hazardous substances in the environment. These evaluations include broad public health assessments of hazardous waste sites that analyze numerous contaminants and multiple exposure

Address for correspondence: Steve M. Dearwent, Agency for Toxic Substances and Disease Registry, 1600 Clifton Road, NE, Mailstop E31, Atlanta, GA 30333. Voice: 404-498-0488; fax: 404-498-0420.
e-mail: sdearwent@cdc.gov

pathways, as well as more focused health consultations that address specific substances or single exposure pathways. Other important agency activities include health surveillance and exposure registries, emergency response to hazardous substance releases, and toxicologic assessments for hazardous substances. ATSDR places a strong emphasis on education and training for affected communities and their healthcare providers. The agency encourages public involvement by conducting informational meetings for residents living in or near contaminated areas.

Congress created ATSDR in 1980 as part of the Comprehensive Environmental Response, Compensation, and Liability Act (CERCLA), commonly known as the "Superfund" Act. CERCLA provides ATSDR with the congressional mandate to implement areas of this legislation aimed at protecting the public from exposure to hazardous substances in the environment. As the lead agency within the Public Health Service for implementing the health-related provisions of CERCLA, ATSDR is charged with assessing the presence and nature of health hazards at specific Superfund sites and preventing or reducing exposures that may lead to adverse health outcomes. Furthermore, the agency is responsible for expanding the knowledge base on health effects associated with hazardous substance exposures.

APPROACH TO EVALUATING SITES AND THE RESULTING RECOMMENDATIONS

When conducting a public health assessment, ATSDR initially evaluates contaminant exposure scenarios at a site by gathering and analyzing environmental sampling data. Often ATSDR does not collect the environmental sampling data directly. Instead, the agency reviews information provided by federal and state government agencies, potentially responsible parties, and the public. If a review of environmental sampling data and exposure pathways indicates that residents have come into contact with hazardous substances, a health effects evaluation is warranted.

ATSDR evaluates exposure pathways by examining environmental and human variables that could lead to contact with contaminants of concern. A pathway analysis includes five principal elements: the source of contamination, its transport through an environmental medium, the point of exposure, the route of exposure, and the receptor population.[1] Completed exposure pathways have all five elements present and exposure to a contaminant has occurred at some time. Identification of a completed exposure pathway does not, however, imply the occurrence of an adverse health effect. Exposures may or may not be at levels that result in unacceptable risks. Thus, even if exposure has occurred, human health effects do not necessarily result.

Health effects evaluations follow the determination of a completed exposure pathway. When conducting a health effects evaluation, ATSDR relies on existing scientific information, including the results of medical, toxicologic, and epidemiologic studies to determine the health effects that may result from exposures.[1] Potential for exposure and health effects for sensitive subpopulations, such as children, women of child bearing age, and the elderly, receive special attention. Such sensitive groups may be most likely to exhibit adverse effects associated with hazardous substance exposures.

While conducting public health assessments, ATSDR may recognize the need to better document or prevent exposures or illnesses in a particular community. ATSDR may initiate toxicologic research or an exposure investigation, health study, or health education program. Exposure investigations are conducted to better define site-specific exposures by collecting additional environmental or biologic samples, usually to define a worst-case exposure scenario for potentially affected populations. Health studies are used to assess morbidity or mortality rates in a community and usually require population-based approaches. Medical assessments or analyses of epidemiologic data are used. Health education programs include community- and provider-oriented methods for disseminating information on exposure prevention or mitigation. Toxicologic research is targeted toward developing comprehensive, chemical-specific profiles, and filling critical data gaps needed for the toxicologic assessment of environmental chemicals found at hazardous waste sites.

DESCRIPTIVE STATISTICS FOR SITE EVALUATIONS DURING 1995–2004

Since 1995, ATSDR has assessed environmental contamination and associated human health risks at nearly 3000 sites. Of these, 23.3% (688/2948) were complete public health assessments that evaluated multiple contaminants and exposure pathways. Most of the evaluations (74.8%, 2205/2948) were health consultations that focused on a single contaminant or exposure pathway. The remaining site evaluations were exposure investigations or the issuance of health advisories. ATSDR uses five categories to characterize the overall public health conclusions from site assessments. These five categories define whether site-related exposures pose a hazard, pose no hazard, or cannot be fully evaluated because of a lack of critical information. The categories are used to develop site-specific recommendations and follow-up activities.

When a site assessment determines that a hazard exists, the site is classified as either an *urgent public health hazard* or a *public health hazard*. For sites categorized as urgent public health hazards, residents must be exposed to chemical or physical hazards at levels that can cause immediate threats to life or health. Urgent public health hazards are typically issued because of short-term exposure scenarios and, therefore, often require expeditious follow-up. Data for the 10-year period from 1995 to 2004 show that the frequency of sites categorized as urgent public health hazards has remained relatively

TABLE 1. Frequency of health hazard categories by year of occurrence

Health hazard category	1995	1996	1997	1998	1999	2000	2001	2002	2003	2004
Urgent hazard	1%	0%	0%	1%	0%	0%	0%	0%	0%	0%
Health hazard	38%	32%	19%	17%	18%	16%	17%	28%	31%	26%
Indeterminate hazard	25%	16%	18%	20%	23%	21%	23%	17%	15%	30%
No apparent hazard	28%	46%	51%	54%	47%	51%	46%	46%	48%	37%
No hazard	7%	5%	11%	8%	11%	12%	13%	8%	6%	2%
Other	1%	1%	1%	0%	1%	0%	1%	1%	0%	5%

constant, at less than 1% (TABLE 1). The frequency of sites deemed urgent public health hazards was notably higher in prior years, averaging 4% from 1992 to 1994.[2] A public health hazard is designated for sites where sufficient evidence exists of chronic, site-related exposure to hazardous substances that could result in adverse health effects. The frequency of sites categorized as public health hazards has fluctuated over the last 10 years. Approximately 25% of all site evaluations during the last decade have resulted in this classification (TABLE 1).

When a site assessment determines that no hazard exists, the site is classified as either *no apparent public health hazard* or *no public health hazard*. *No apparent public health hazard* is used to characterize sites at which exposure to site-related chemicals occurs at levels unlikely to cause adverse health effects. Data from 1995 to 2004 indicate that 46% of the sites evaluated by ATSDR during this period were so designated—an increase from 1992 to 1994 when less than 15% of sites fell into this category.[2] *No public health hazard* is used at sites where no human exposure to site-related hazardous substances occurs. The frequency of sites in this category has been declining and averaged 8% annually during 1995–2004 (TABLE 1).

Sites that cannot be fully evaluated because critical information is lacking are categorized as *indeterminate public health hazards*. At these sites, data needed to verify exposure or a hazard associated with the exposure are lacking. In the last decade, 21% of sites evaluated by ATSDR were deemed *indeterminate public health hazards*.

From the nearly 3000 assessments conducted in the last decade, ATSDR identified completed exposure pathways at about 30% (880/2948) of the sites. The most common contaminants at sites that had completed exposure pathways included heavy metals, volatile organic compounds, and polychlorinated biphenyls (TABLE 2). Lead was the most frequent contaminant found in completed exposure pathways from 1995 through 2000, while arsenic was the most prevalent contaminant in completed exposure pathways from 2001 to 2004. At many sites that had completed exposure pathways, people were exposed to site-related contaminants through more than one environmental medium. Routes of exposure via different environmental media are presented in TABLE 3. The most frequent exposure pathway during the 10-year period (1995–2004) was

TABLE 2. Frequently occurring contaminants at sites with completed exposure pathways

Chemical	Sites with completed exposure pathways (%)
Lead	12
Arsenic	10
Trichloroethylene (TCE)	8
Volatile Organic Compounds N.O.S.	7
Tetrachloroethylene (PCE)	6
Cadmium	6
Polychlorinated Biphenyls (PCBs)	6
Chromium	5
Mercury	5
Manganese	5

water (51%), followed by soil (47%), air (30%), and finally, biota (12.5%). People are often exposed to site-related contaminants through more than one environmental medium; therefore, these categories are not mutually exclusive.

CASE STUDIES

The following contaminant-specific case studies are provided to illustrate a few of ATSDR's pertinent experiences in evaluating communities' hazardous substance exposures. These case studies are derived from site evaluations that documented historical or current hazards. They examine the historical hazard of lead, the contemporary hazard of asbestos, and the emerging contaminant perchlorate.

Lead

The history of lead exposure in the United States is well documented. The link between blood lead levels in the U.S. population and the use of lead in consumer products, particularly paint and fuel, is also well documented. Many communities, however, are still contending with adverse lead exposures because of their residential proximity to mining, smelting, and related industries

TABLE 3. Frequency of documented exposures via different environmental media

Environmental medium	1995	1996	1997	1998	1999	2000	2001	2002	2003	2004
Air	34%	27%	25%	25%	28%	27%	20%	30%	37%	32%
Soil	47%	52%	49%	42%	49%	50%	50%	49%	38%	45%
Water	62%	55%	46%	49%	45%	53%	61%	52%	48%	52%
Biota	16%	18%	11%	13%	12%	13%	8%	15%	15%	17%

that rely on this heavy metal. A site in Omaha, Nebraska, illustrates this scenario and the community-wide repercussions.[3]

The Omaha Lead site was proposed for the National Priorities List in 2002 and officially listed in April 2003. A lead refinery operated on the west bank of the Missouri River in downtown Omaha from the early 1870s until 1997. The 23-acre refinery was the primary source for soil lead contamination at the Omaha site, although other small contributors were likely associated with some contamination (e.g., lead-based paints, leaded gasoline, and minor industrial sources). The site encompasses residential properties, childcare facilities, and schools that were contaminated with lead through air emissions from the lead-refining operations. The site covers about 8840 acres and, based on 2000 census data, is home to approximately 86,000 residents. Of these residents, approximately 11% (N~9,700) are children 6 years of age or younger.

An initial exposure pathway evaluation examined soil lead measurements from more than 17,000 locations. Data indicated a high potential for lead exposure of residents living in the surrounding community. Analyses of blood lead surveillance data during 2000–2002 indicated that 9.7% of children 6 years of age and younger had blood lead levels of 10 micrograms per deciliter (μg/dL) compared with 5.5% for the county and 2.0% for the state. According to data from the Center for Disease Control and Prevention's (CDC) National Report on Human Exposure to Environmental Chemicals, only 1.6% of children 1–5 years old sampled over the 4-year period 1999–2002 had blood lead levels greater than or equal to 10 μg/dL.[4] A subsequent review of 9600 records from 2003 found that 484 (6.2%) of the children in the site area had levels of 10 μg/dL or greater and 32.9% (159/484) of these children had blood lead levels between 15 and 44 μg/dL. Five children had blood lead levels exceeding 45 μg/dL.

ATSDR concluded that ongoing lead exposure put children living in or near the Omaha Lead site at risk of lead-related health effects.[3] Because 30% of the children in the area were tested, ATSDR estimated that 1600 children living in or near the site had elevated blood lead levels (> 10 μg/dl) during 2000–2002. The U.S. Environmental Protection Agency (EPA) has continued to remediate lead-contaminated soil from properties, particularly homes that have young children, homes that have children who have elevated blood lead levels, schools, and daycare facilities. ATSDR recommended health education regarding all lead hazards, promoting primary prevention activities, and facilitating annual blood lead testing for all children 6 years and less living in or near the site.

Asbestos

As with lead, asbestos has a long, well-documented history of causing adverse health effects. Although asbestos is typically considered an occupational hazard, in the past few years ATSDR has worked at sites at which community asbestos exposures were a major concern. Some of these exposures were linked to commercial and residential development that disturbed naturally occurring asbestos deposits, and other community exposures were associated with industrial mining and milling of contaminated ore.[5–7] A public health tragedy in Libby, Montana, provides a good example of community-wide asbestos exposures that resulted from the mining and milling of asbestos-contaminated ore.

In 1999, ATSDR received a request from the EPA to evaluate potential asbestos exposure and related human health concerns in Libby, Montana. Vermiculite ore, mined and processed in Libby from the 1920s until 1990, was contaminated with tremolite asbestos. In response, ATSDR conducted several exposure and epidemiologic analyses to evaluate the effect of historical and current asbestos exposures in the Libby community. Epidemiologic projects included a review of historical mortality data and the implementation of a medical testing program to identify and examine current residents for asbestos-related health effects.

ATSDR's analysis of death certificate data evaluated malignant and nonmalignant respiratory diseases, mesothelioma, and digestive cancers for the 20-year period from 1979 through 1998. All mortality data were geographically referenced and analyzed at various geographic scales to focus the investigation on populations specific to the Libby area. Mortality rates were compared with rates for the state of Montana and the U.S. population. For the 20-year period reviewed, standardized mortality rates for asbestosis in Libby were 40–80 times higher than in Montana and the United States, respectively. Mortality from lung cancer was also significantly elevated, with a 20–30% excess over this time period. Mesothelioma mortality was increased but difficult to quantify because reliable statistics on this rare cancer are not available at the state and national levels. Overall, when compared with Montana and U.S. mortality, there was a 20–40% increase in malignant and nonmalignant respiratory deaths in this small Montana community.[6]

The medical testing program implemented in Libby was the largest project of its kind ever undertaken by ATSDR. A total of 7307 residents participated in the initial medical screenings conducted in 2000 and 2001. The project included face-to-face interviews to gather personal history information, a three-view chest X ray to identify changes in the lungs and lung lining, and spirometry for evaluating lung function. Persons eligible for testing included former vermiculite workers from Libby and persons who lived, worked, or played in Libby for at least 6 months before December 31, 1990. A couple of the many significant findings derived from the medical testing program included an increased prevalence of pleural abnormalities (18%, 1186/6668) and an increased risk for pleural abnormalities with increasing age and increasing length of residence in the Libby area.[7]

Other ATSDR activities initiated in response to the problems uncovered in Libby included creation of a tremolite exposure registry, development of a tremolite asbestos toxicologic profile, a case series review for affected residents, a study evaluating the usefulness of computed tomography (CT) scans in assessing asbestos exposures, and assessments of processing facilities receiving contaminated ore from the Libby mine. The agency created the tremolite asbestos registry to track affected persons and ensure that the latest medical recommendations and research findings were communicated to these individuals and their healthcare providers. The registry is also used to educate registrants about asbestos-related disease and its diagnosis and treatment, and will be a useful resource in the future for conducting epidemiologic studies on tremolite asbestos exposure and disease progression. Following the first 2 years of medical screening in Libby, ATSDR conducted a study on the usefulness of CT scans in identifying lung problems associated with asbestos exposure.[8] This study of exposed residents indicated that low-dose, high-resolution CT can be helpful for screening individuals who have indeterminate chest radiographs, particularly in high-risk persons. The case series included a review of medical records, radiographs, and CT scans of patients who have asbestos-related disease to learn more about asbestos-related illness among persons who have only environmental, non-work-related exposure to asbestos.

Asbestos-contaminated vermiculite from Libby was shipped to more than 200 processing facilities in the United States. Once the scope of the public health problems in Libby was identified, ATSDR initiated the National Asbestos Exposure Review to evaluate sites that received and processed the contaminated ore. Evaluations of the first 28 sites, including those that processed the largest quantities of contaminated ore, indicated that former workers were at highest risk.[9] Household contacts and former residents were also at a high, yet unquantifiable, risk.

Perchlorates

ATSDR's experience in assessing perchlorate contamination at sites is somewhat limited. Evaluation of sites with perchlorate contamination is difficult because of the limited number of epidemiologic studies examining the human health risks associated with low-level, chronic exposure in environmental media. The National Academy of Sciences (NAS) has recently published a reference dose for perchlorate (0.0007 mg/kg per day) that EPA subsequently adopted.[10] ATSDR has also adopted a Minimum Risk Level based on the NAS findings.[11] Even so, the concentration of perchlorate that may cause adverse health effects in humans is a subject of continuing scientific debate.[12–15] This uncertainty is reflected in the variability in environmental guidance values currently promulgated by many state agencies.

ATSDR's recently released perchlorate toxicologic profile provides current and complete information on the environmental characteristics and human health effects linked to perchlorate.[11] The agency has partnered with state health departments, primarily in the western United States, to evaluate perchlorate groundwater contamination at a few locations. Because perchlorate is primarily used as an oxidizer in solid rocket fuels, much of this contamination is associated with U.S. Department of Defense (DOD) sites or private companies working on DOD projects. Of the 14 sites where ATSDR has evaluated perchlorate contamination, many (57%, 8/14) are associated with current or former federal facilities. Most (85.7%, 12/14) of these evaluations resulted in classifications of *no apparent public health hazard* or *no public health hazard* although two sites were classified as *public health hazards* because of historical exposures to groundwater contamination that migrated offsite.

CONCLUSION

During ATSDR's 20-year history, the agency has evaluated numerous hazardous waste sites around the country. For many of these sites, the contaminants of concern are thoroughly studied and have well-established environmental guidance values along with supporting epidemiologic data. The challenge at many of these sites is finding the necessary information to fully evaluate exposure pathways and potential adverse health effects associated with site-related exposures. Many site evaluations conducted by the agency provide reassuring information that public health hazards do not exist. For those sites for which completed exposure pathways do exist, the agency maintains consistent approaches in quantifying potential public health effects as well as clearly communicating these findings and recommendations to stakeholders in the community and at other federal, state, or local public health agencies.

Although many sites do not pose public health hazards, the frequency of agency involvement at sites that have valid public health concerns has remained constant. Asbestos-associated morbidity and mortality in Libby, Montana, just surfaced within the last 5 years; and this is arguably the most important site ATSDR has worked on since its inception. Well-known hazards like asbestos and lead along with recently identified and emerging environmental public health threats like perchlorate and perfluorocarbons will require appropriate agency responses for many years to come.

REFERENCES

1. AGENCY FOR TOXIC SUBSTANCES AND DISEASE REGISTRY. 2005. Public Health Assessment Guidance Manual. US Department of Health and Human Services. Atlanta, GA. Available at URL: http://atsdr1.atsdr.cdc.gov:8080/HAC/HAGM/
2. JOHNSON, B.L. & C.T. DEROSA. 1997. The toxicological hazard of superfund hazardous waste sites. Rev. Environ. Health **12:** 235–251.
3. AGENCY FOR TOXIC SUBSTANCES AND DISEASE REGISTRY. 2005. Public Health Assessment entitled Omaha Lead, Omaha, NB. US Department of Health and Human Services. Atlanta, GA.

4. CENTERS FOR DISEASE CONTROL AND PREVENTION, NATIONAL CENTER FOR ENVIRONMENTAL HEALTH. 2005. Third National Report on Human Exposure to Environmental Chemicals. US Department of Health and Human Services. Atlanta, GA.
5. AGENCY FOR TOXIC SUBSTANCES AND DISEASE REGISTRY. 2005. Health Consultation Entitled Asbestos Exposures at Oak Ridge High School. US Department of Health and Human Services. Atlanta, GA.
6. AGENCY FOR TOXIC SUBSTANCES AND DISEASE REGISTRY. 2002. Health Consultation Entitled Mortality in Libby, Montana, 1979–1998. US Department of Health and Human Services. Atlanta, GA.
7. PEIPENS, L.A., M. LEWIN, S. CAMPOLUCCI, et al. 2003. Radiographic abnormalities and exposure to asbestos-contaminated vermiculite in the community of Libby, Montana. Environ. Health Perspect. **111:** 1753–1759.
8. MURAVOV, O.I., W.E. KAYE, M. LEWIN, et al. 2005. The usefulness of computed tomography in detecting asbestos-related pleural abnormalities in people who had indeterminate chest radiographs: the Libby, MT, experience. Int. J. Hyg. Environ. Health **208:** 87–99.
9. ANDERSON, B.A., S.M. DEARWENT, J.T. DURANT, et al. 2005. Exposure pathway evaluations for sites that processed asbestos-contaminated vermiculite. Int. J. Hyg. Environ. Health **208:** 55–65.
10. NATIONAL ACADEMY OF SCIENCE. 2005. Health Implications of Perchlorate Ingestion. National Academies Press. Washington, DC.
11. AGENCY FOR TOXIC SUBSTANCES AND DISEASE REGISTRY. 2005. Toxicological Profile for Perchlorates. US Department of Health and Human Services. Atlanta, GA.
12. GINSBERG, G. & D. RICE. 2005. The NAS perchlorate review: questions remain about the perchlorate RfD. Environ. Health Perspect. **113:** 1117–1119.
13. GIBBS, J.P., A. ENGEL & S.H. LAMM. 2005. The NAS perchlorate review: second-guessing the experts. Environ. Health Perspect. **113:** A730–A732.
14. JOHNSTON, R.B. Jr., R. CORLEY, L. COWAN & R.D. UTIGER. 2005. The NAS perchlorate review: adverse effects? Environ. Health Perspect. **113:** A728–A729.
15. STRAWSON, J., M.L. DOURSON & Q.J. ZHAO. 2005. The NAS perchlorate review: is the RfD acceptable? Environ. Health Perspect. **113:** A728–A729.

Cancer Mortality in an Area of Campania (Italy) Characterized by Multiple Toxic Dumping Sites

PIETRO COMBA,[a] FABRIZIO BIANCHI,[b] LUCIA FAZZO,[a] LUCIA MARTINA,[c] MASSIMO MENEGOZZO,[d] FABRIZIO MINICHILLI,[b] FRANCESCO MITIS,[e] LOREDANA MUSMECI,[a] RENATO PIZZUTI,[c] MICHELE SANTORO,[c] STEFANIA TRINCA,[a] MARCO MARTUZZI,[e] AND "HEALTH IMPACT OF WASTE MANAGEMENT CAMPANIA" WORKING GROUP*

[a]*Department of Environment and Primary Prevention, Istituto Superiore di Sanità, 00161 Rome, Italy*

[b]*Unit of Epidemiology, CNR Institute of Clinical Physiology, 56127 Pisa, Italy*

[c]*Campania Region Health Authority—Regional Epidemiological Observatory, 80100 Naples, Italy*

[d]*Campania Region Environmental Protection Agency, 80100 Naples, Italy*

[e]*WHO European Centre for Environment and Health, Rome Office, 00187 Rome, Italy*

ABSTRACT: Several recent studies have documented that a widespread practice of dumping toxic wastes has taken place for many years in the Provinces of Naples and Caserta. Extensive programs of environmental monitoring are currently ongoing in the area. In this frame, the Department of Civil Defence of the Italian Government has appointed an *ad hoc* study group in order to assess the health status of the population resident in the area of interest. The first investigation performed by the study group has been a geographic study on cancer mortality and occurrence of malformations in 196 municipalities constituting the two Provinces. The study detected an area located in the southeastern part of the Province

---

*Members of the working group are: R. Bertollini, M. Martuzzi, F. Mitis (WHO European Centre for Environment and Health), C. Carboni, P. Comba, L. Cossa, P. De Nardo, L. Fazzo, L. Musmeci, S. Trinca (Istituto Superiore di Sanità), F. Bianchi, N. Linzalone, F. Minichilli, A. Pierini (National Research Council), E. Lorenzo, L. Martina, R. Pizzuti, M. Santoro (Campania Region Health Authority), E. Lionetti, M. Menegozzo (Campania Region Environmental Protection Agency), M. Fusco (Naples Cancer Registry), G. Scarano (Campania Region Birth Defects Registry), S.Menegozzo (Campania Region Mesothelioma Registry), G. Doddi, M. Leonardi, L. Madeo, G. Martini, N. Mazzei, R. Pizzi (Department of Civil Defence), A. Savarese (Legambiente Campania), C. Bove (Local Health Unit Caserta1), A. D'Argenzio (Local Health Unit Caserta2), A. Simonetti (Local Health Unit Naples1), A. Parlato (Local Health Unit Naples2), F. Peluso (Local Health Unit Naples3), R. Palombino (Local Health Unit Naples4), and F. Giugliano (Local Health Unit Naples5).

Address for correspondence: Pietro Comba, Department of Environment and Primary Prevention, Istituto Superiore di Sanità, Viale Regina Elena 299, 00161 Roma, Italy. Voice: 39-06-49902249; fax: 39-06-49387083.

e-mail: comba@iss.it

of Caserta and in the northwestern part of the Province of Naples, where cancer mortality and congenital malformations show significantly increased rates with respect to expected figures derived from the regional population. The area highlighted by the study is, in general terms, overlapping with the area where most illegal dumping of toxic wastes took place. It is now recommended that mortality studies be extended to take into account other health outcomes, to search for correlations with environmental exposures, and consider possible confounding factors.

KEYWORDS: toxic wastes; dumping sites; cancer; malformations

INTRODUCTION

The possible health effects associated with residential proximity to waste disposal sites have been the object of many epidemiological studies over the last 20 years. Several reports of increased risk of different diseases are available, but causal links have not been adequately proven; for a comprehensive review the reader is referred to Vrijheid,[1] who examines 50 papers published from 1980 through 1998. Difficulties of conducting epidemiological studies around waste disposal sites include exposure assessment, because many different chemicals released in the environment may be absorbed by contact, inhalation, and ingestion of food and water.[2] Furthermore, the sites of interest are often located in areas characterized by other sources of environmental pollution, and the resident population is often sparse. Finally, some of the reported increases in risk are small, making their interpretation difficult.

Different study designs have been adopted in epidemiological studies of waste disposal sites. Ecological and geographical studies have reported increases in lung cancer risk in men[3–5] and in women,[5] bladder cancer in both sexes[6,7] and in men only,[3,5] leukemia,[8] childhood leukemia,[9] liver cancer in men,[4] prostate cancer,[4] gastric cancer in both sexes[3] and in men only,[4] uterine cancer,[4] rectum cancer,[3] breast cancer,[3] perinatal mortality,[10] and birth defects.[10,11]

Cohort studies have reported increases of low birth weight[12,13] and birth defects.[14–17] Case–control studies have reported increases in the risk of low birth weight,[18] birth defects,[19–21] bladder cancer, and leukemia in women,[22] and cancer of the pancreas, liver, prostate, and non-Hodgkin's lymphoma.[23]

The issue of possible health effects of waste disposal sites was specifically addressed in Campania, a Region of Southern Italy, where Naples is located. Since the 1980s, thousands of illegal and uncontrolled sites of urban, toxic, and industrial waste disposal, including land filling and unauthorized incineration, have been known to be active in this region (FIGS. 1 and 2). The issue of waste treatment in Campania has been constantly surrounded by controversy and social conflict, periodically making national headlines. Local communities are worried about severe health effects and express frustration over their compromised level of well-being and quality of life. A Commissioner appointed by the

FIGURE 1. Waste disposal site.

FIGURE 2. Treatment of waste that was illegally burnt.

national government and holding special power has held executive authority for waste treatment and disposal policy in the region since 1994. For a review, the reader is referred to the annual report on "Ecomafia."[24] A census of waste dumping sites has been made available by Campania Region's Environmental Protection Agency since 2003.[25]

Possible adverse health effects of the waste cycle in Campania were first investigated with respect to childhood cancer mortality in the Province of Caserta.[26] More recently, two papers published in *Lancet Oncology* have addressed the question of waste and health in Campania. A first report[27] identified a "triangle of death" consisting of three municipalities; subsequently, Bianchi *et al.*[28] revealed a more complex picture and advocated a different and wider approach.

The Department of Civil Defence of the Italian Government requested the World Health Organization (WHO) to design and conduct an epidemiological study on the health impact of the waste cycle in Campania. To this end, a working group comprising WHO, National Research Council, Istituto Superiore di Sanità, Campania Region Health Authority Epidemiologic Observatory, and Campania Region Environmental Protection Agency was appointed, and a network of cooperation has been constructed with local health authorities, cancer and malformation registries, and environmental organizations. The working group was set up to include technical expertise, access to local information, and the ability to establish a dialogue with the numerous stakeholders involved.

The purpose of this article is to present the findings of the first phase of the study, concerning the distribution of cancer mortality and birth defects in the Provinces of Naples and Caserta, the part of Campania most severely affected by illegal waste dumping sites.

MATERIALS AND METHODS

Mortality

The geographical distribution of mortality in the 196 municipalities of the Provinces of Naples and Caserta, with a population of around 4 million people, was analyzed using data from ISTAT (National Bureau of Statistics); mortality records were provided by the Regional Epidemiological Observatory of Campania. A total of 20 causes of death were studied for the period 1994–2001, including all-cause mortality, all-cancer mortality, and a set of cancer causes reported in excess in previous epidemiological studies conducted in the surroundings of waste landfills or incinerators. Availability of such data at the municipality level, relatively small units on average with the exception of large cities, allows the analysis of risk variability by small area and in relation to the spatial occurrence of exposures produced by known sources. The analyses were carried out separately for men and women. At the provincial level, analyses were carried out using standardized mortality rates with national reference and through standardized mortality ratios (SMRs) with regional reference. At the municipality level, analyses were carried out using SMRs, with 95% confidence intervals, and hierarchical Bayesian estimators (BMRs, and 95% uncertainty intervals), calculated as suggested by Besag et al[29] and Mollié.[30] BMRs improve the quality of the risk estimates by removing part of the large random variability due to sparseness of data, and the possible "confounding by location" from unknown, spatially structured determinants.[31] Risk estimates (SMRs and BMRs) were mapped at the municipality level.

Congenital Malformations

This component of the study is based on data recorded by the Campania Region Birth Defects Registry, which include data for the population residing in the Provinces of Naples and Caserta, 1996–2002. For a long time the Campania registry has been part of the EUROCAT & ICBDMS networks of Congenital anomalies.[32,33] Cases have been ascertained at birth and during the postnatal period; fetal deaths after the 20th week and therapeutic abortion prior to week 24 have also been included. The number of observed birth defects in each municipality is contrasted to the corresponding number of expected cases based on regional demographic figures. As in the mortality analysis, both SMRs and BMRs are computed and mapped.

RESULTS

Results show that the cancer mortality profile of the area is characterized by numerous excesses, as summarized in TABLE 1. In the Province of Naples, 8 of 19 (42%) of all the risks that were estimated are in significant excess for men, and 11 of 19 (58%) for women. Excess risks range from 6.1% (mortality for all causes) to 32.9% (cancer of pleura) in men, and from 7.3% (all-cause mortality) to 27.3% (cancer of esophagus) in women. Fewer excesses were found for the Province of Caserta (16% for men, 11% for women) even though some risk estimates are high (cancer of the stomach: 29.3% for men, 18.2% for women). At the municipality level, results indicate the occurrence of increased risks for several causes of cancer death, in 11 municipalities of the southeastern part of the Province of Caserta and 13 adjacent municipalities of the northern part of the Province of Naples. In 19% of the municipalities of the Province of Caserta and in 43% of the municipalities of the Province of Naples all-cause mortality is statistically significantly elevated in men; in 23% of the municipalities of the Province of Caserta and in 47% of municipalities of the Province of Naples all-cause mortality is statistically significantly elevated in women (FIG. 3).

The specific cancer causes frequently in excess in these municipalities (cancer of the stomach, kidney, liver, trachea, bronchus and lung, pleura, bladder)

TABLE 1. Mortality in the provinces of Naples and Caserta, 1994–2001.

	Province of Naples						Province of Caserta			
	Men		Women				Men		Women	
Causes of death	Deaths	SMR	Deaths	SMR			Deaths	SMR	Deaths	SMR
All-cause mortality	95,951	106.1*	91,888	107.3*			27,676	102.5*	24,984	102.9*
All cancers	29,185	108.7*	20,206	109.2*			8,158	102.3*	5,179	98.2
All cancers (0–14)	137	101.9	95	102.6			37	100.3	30	118.3
Cancer of the esophagus	262	98.8	126	127.3*			76	97.4	25	88.6
Cancer of the stomach	1,696	100.3	1,090	98.1			649	129.3*	375	118.2*
Cancer of rectum	655	101.6	584	106.8			191	99.1	158	101.3
Cancer of the liver and biliary ducts	1,910	117.6*	1,572	114.1*			490	102	417	105.5
Cancer of the pancreas	821	103.4	840	108.9*			231	98.5	202	91.5
Cancer of the larynx	728	111.8*	66	126.6			203	105.6	10	67.4
Cancer of the trachea, bronchus, and lung	9,681	114.1*	1,845	126.5*			2,566	102.1	345	82.7
Cancer of the pleura	212	132.9*	109	125.8*			30	64.1	21	85.3
Soft Tissues Sarcoma	107	114.4	87	88.6			26	95.8	24	86.5
Cancer of the breast			3,475	110.7*					871	98.1
Cancer of the testis	34	93.8					7	68.6		
Cancer of the bladder	1,745	110.7*	348	117.5*			445	93.3	81	95.5
Cancer of the kidney	494	105.7	304	120.7*			128	92.7	56	77.6
Cancer of the brain	526	98.6	416	97.3			176	114.2	116	96
Non-Hodgkin lymphomas	682	100	649	109.1*			182	91.4	142	83.6
Leukemias	894	95.9	789	103.7			288	105	214	98.8
Ill-defined causes	2,860	125.4*	1,246	104.8			434	66.7	299	90.2

*lower limit of 95% confidence interval > 100.

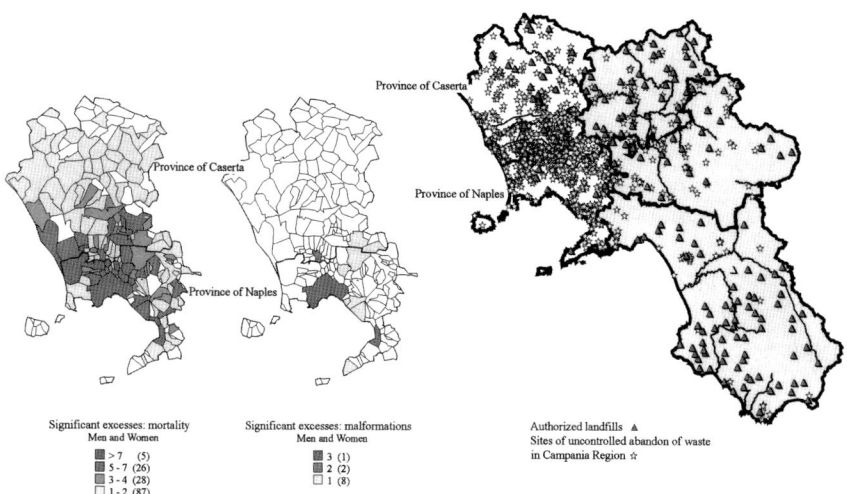

FIGURE 3. Municipalities with significant excesses of mortality and congenital malformations and waste disposal sites in the provinces of Naples and Caserta.

have, as main determinants, risk factors linked to lifestyle and individual circumstances (diet, tobacco smoke, infections) and to occupational exposures. However, in several epidemiological studies they have been observed in relation to residence near waste treatment sites, even though causality remains unconfirmed.

Two sets of municipalities with significant excesses in the total of congenital malformations and on some specific groups were identified. First, the southern part of the Province of Caserta, where there are frequent excesses for congenital malformations overall and for the group of urogenital malformations; second, Naples and its northeastern neighboring municipalities, where there are frequent excesses for the total of congenital malformations, cardiovascular malformations and, to a lesser extent, urogenital malformations (FIG. 3).

DISCUSSION

Consistently with previously published data,[34] residents in the Provinces of Naples and Caserta have cancer mortality significantly raised compared to the Campania region. Increased risks are observed for several cancers; similarly, the occurrence of congenital malformations is significantly in excess. The reliability of the malformation data collected, in terms of accuracy and completeness, is documented by the comparability of the Campania data with other European and International registries.[32,33]

Analysis of mortality data by municipalities entails multiple comparison of observed counts with expected numbers of deaths. Even under the null hypothesis of uniform risk across municipalities, using a 5% threshold for statistical significance would result in approximately 5% of the municipalities (around 10 of them) differing statistically from the null. However, even discounting for this proportion of spurious significance, the statistically significant large risks observed in the municipalities of the two provinces far exceed what can be expected by chance alone. In addition, BMRs remove substantial part of the random variability and are not subject to the inflation of statistically significant risk estimates due to multiple comparison. Thus, the occurrence of excesses is unlikely to be due to chance. On the other hand, the number of expected cases may be systematically overestimated, compared to the value of real interest, that is, the number of cases that would be expected in a situation of no exposure to environmental risk factors linked to waste management. Expected numbers were calculated using, as reference, the regional rates. These are based, in turn, on the data for the whole population of Campania, which includes the exposed areas and whose average exposure, therefore, is nonzero. Hence, it is likely that a comparison between exposed municipalities with unexposed ones, rather than with the regional average, could produce generally higher risk estimates.

Within the two Provinces, there is geographical variability in mortality across municipalities, by cause of cancer death. The distributions of these excesses might provide some indication as to the possible role of environmental exposures, in particular those associated with the cycle of waste treatment. Increased risks in cancer mortality tend to be more frequent in municipalities in the southern part of the Caserta Province and in the northern part of the Naples Province. A similar pattern is also shared by the distribution of risk of congenital malformations (FIG. 3). This similarity may be casual or may be explained, in part, by the presence of one or more risk factors occurring with more intensity in the area between the two Provinces. The area where municipalities at higher risk of mortality and congenital malformations aggregate overlaps with the presence of landfills and sites of uncontrolled disposal of waste (FIG. 3).

These recurring patterns of mortality and congenital malformations that were observed are suggestive, but the underlying complexity should be emphasized; concern is more than justified, even without the support from epidemiological findings, but oversimplifications are likely to result in sensationalism and be of little help in dispelling the social controversy.[28]

The Province of Caserta and some municipalities of the northern part of the Province of Naples show high rates of mortality from gastric cancer, a disease that is decreasing in industrialized societies, but still frequent in rural areas. The most important risk factors of gastric cancer are infection from *Helicobacter pylori*.[35–37] A diet poor in fresh vegetables and fruit, and rich in salt and nitrites adds to gastric cancer risk,[38–42] while occupational exposures do not seem to play an appreciable role.[43] Increases in gastric cancer risk near landfills were reported by Griffith *et al.*[3] and Goldberg *et al.*[4] as previously mentioned.

The southern part of Caserta Province and some municipalities in the northern part of Naples Province show high mortality rates from kidney cancer. Tobacco smoking, diet, and some drugs have been associated with kidney cancer; among occupational risk factors an etiologic role has been suggested, but not confirmed, for combustion products, heavy metals, and solvents.[44–48] An increased risk of renal cancer for the population resident in the neighborhood of landfills has been reported by Goldberg et al.[23] Several municipalities in the southern part of Caserta Province and in the northern part of Naples Province are characterized by high mortality rates from liver cancer. Figures produced by the Naples Cancer Registry indicate a high occurrence of hepatocellular carcinoma in this area, where both B and C type hepatitis show high incidence rates.[49,50] Exposure to hepatotoxic chemicals may have particularly strong adverse effects among subjects who are seropositives for hepatitis virus, as has recently been shown by Mastrangelo et al.[51] with reference to vinyl chloride. An increased risk of liver cancer near landfills has been reported by Goldberg et al.[4,23] Elevated mortality from lung cancer has been described in the southern part of Caserta Province and in the northern part of Naples Province. Tobacco smoking, and several occupational carcinogens (namely arsenic, asbestos, chromates, chloromethylethers, nickel, and polycyclic aromatic hydrocarbons) are established causes of lung cancer; recently the etiologic role of urban air pollution has been documented.[52–55] An increased risk of lung cancer near landfills has been reported by Griffith et al.,[3] Goldberg et al.,[4] and Mitis et al.[5]

Several municipalities of the northwestern part of the Province of Naples have high mortality rates from pleural mesothelioma, which can be attributed to a variety of occupations with exposure to asbestos, which have been described in the area.[56] High mortality from bladder cancer occurs in the south of Caserta Province and in the north of Naples Province. Cigarette smoking is the main cause of this disease, followed by occupational exposure to aromatic amines in textile and rubber industries and by exposure to solvents, paints, leather dust, inks, some metals, polycyclic aromatic hydrocarbons, combustion products, diesel exhaust, and environmental exposure to water chlorination.[57,58] An increased risk of bladder cancer in areas close to landfills has been reported by Budnick et al.,[6] Griffith et al.,[3] Mallin,[7] Lewis Michl et al.,[22] and Mitis et al[5]

The findings of the present study indicate the presence of an area characterized by elevated cancer mortality rates and by the elevated occurrence of birth defects, corresponding with the area where most waste disposal sites are concentrated. In the area under study, in addition, there are many environmental stressors, deriving from intensive agriculture, widespread industrial activities, and a very high population density.

These preliminary findings are consistent with a possible contributory role of waste-related exposures in determining ill health in the area over time. The consistency of the patterns may reflect an effect on cancer mortality, which requires a long latency time (variable by cancer type) to become manifest, while congenital malformations might be affected by concurrent exposures. Given the sizeable excesses observed, the past and present waste disposal activities, legal and illegal, and considering that waste disposal facilities are being built or planned, more detailed investigation is warranted. Besides further epidemiological analysis, it seems advisable to continue and optimize the existing biomonitoring activities. More generally, it is desirable to strengthen the participation of public health officials in indentifying and adopting appropriate policies on waste management in the region. Above all, immediate action is essential to ensure that the most extreme instances of population exposure to environmental contamination due to waste treatment are prevented.

ACKNOWLEDGMENTS

The authors are very grateful to Colin Soskolne for his support and helpful comments to previous draft versions of this article. The study has been commissioned and partly funded by the Department of Civil Defence of the Italian Government.

REFERENCES

1. VRIJHEID, M. 2000. Health effects of residence near hazardous waste landfill sites: a review of epidemiologic literature. Environ. Health Perspect. **108** (Suppl 1): 101–112.
2. JOHNSON, B.L. 1999. Impact of Hazardous Waste on Human Health. US: CRC Press. Boca Raton, Florida.
3. GRIFFITH, J., R.C. DUNCAN, W.B. RIGGAN & A.C. PELLOM. 1989. Cancer mortality in U.S. counties with hazardous waste sites and ground water pollution. Arch. Environ. Health **44:** 69–74.
4. GOLDBERG, M.S., N. AL HOMSI, L. GOULET & H. RIBERDY. 1995. Incidence of cancer among persons living near a municipal solid waste landfill site in Montreal, Quebec. Arch. Environ. Health **50:** 416–424.
5. MITIS, F., M. MARTUZZI, R. BERTOLLINI, et al. 2004. Studio di mortalità nelle vicinanze di due discariche di rifiuti di Torino. *In* Valutazione del rischio sanitario e ambientale nello smaltimento di rifiuti urbani e pericolosi. L. Musmeci, Ed.: 73–85. Istituto Superiore di Sanità. Roma. (Rapporti ISTISAN 04/5).
6. BUDNICK, L.D., J.N. LOGUE, D.C. SOKAL, et al. 1984. Cancer and birth defects near the Drake Superfund site, Pennsylvania. Arch. Environ. Health **39:** 409–413.
7. MALLIN, K. 1990. Investigation of a bladder cancer cluster in northwestern Illinois. Am. J. Epidemiol. **132** (Suppl 1): S96–S106.
8. GREISER, E., I. LOTZ, H. BRAND & H. WEBER. 1991. Increased incidence of leukemias in the vicinity of a previous industrial waste dump in North Rhine-Westfalia, West Germany. Am. J. Epidemiol. **134:** 755.
9. CUTLER, J.J., G.S. PARKER, S. ROSEN, et al. 1986. Childhood leukemia in Woburn, Massachusetts. Public Health Rep. **101:** 201–205.

10. Belli, S., A. Binazzi, P. Comba, et al. 2004. Analisi della mortalità causa-specifica in prossimità di impianti per lo smaltimento di rifiuti solidi urbani. *In* Valutazione del rischio sanitario e ambientale nello smaltimento di rifiuti urbani e pericolosi. L. Musmeci, L., Ed.: 63–72. Roma: Istituto Superiore di Sanità. (Rapporti ISTISAN 04/5).
11. Fielder, H.M.P., S. Monaghan, C. Poon-King & S.R. Palmer. 1997. Report on the health of residents living near the Nant-Y-Gwyddon landfill site using routinely available data. Cardiff. Welsh Combined Centers for Public Health.
12. Vianna, N.J. & A.K. Polan. 1984. Incidence of low birth weight among Love Canal residents. Science **226:** 1217–1219.
13. Berry, M. & F. Bove. 1997. Birth weight reduction associated with residence near a hazardous waste landfill. Environ. Health Perspect. **105:** 856–861.
14. Goldman, L.R., B. Paigen, M.M. Magnant & J.H. Highland. 1985. Low birth weight, prematurity and birth defects in children living near the hazardous waste site, Love Canal. Hazard. Waste Hazard. Mater. **2:** 209–223.
15. Lagakos, S.W., B.J. Wessen & M.L. Zelen. 1986. An analysis of contaminated well water and health effects in Woburn, Massachussets. Public Health Rep. **101:** 201–205.
16. Deane, M., S.H. Swan, J.A. Harris, et al. 1989. Adverse pregnancy outcomes in relation to water contamination, Santa Clara County, California, 1980–1981. Am. J. Epidemiol. **129:** 894–904.
17. Minichilli, F., N. Linzalone, A. Pierini, et al. 2004. Studio epidemiologico sul rischio di malformazioni congenite in prossimità di siti di discarica in due regioni italiane. *In* Valutazione del rischio sanitario e ambientale nello smaltimento di rifiuti urbani e pericolosi. L. Musmeci, Ed.: 86–104: Istituto Superiore di Sanità. Roma. (Rapporti ISTISAN 04/5).
18. Goldberg, M.S., L. Goulet, H. Riberdy & Y. Bonvalot. 1995. Low birth weight and preterm births among infants born to women living near a municipal solid waste landfill site in Montreal, Quebec. Environ. Res. **69:** 37–50.
19. Dolk, H., M. Vrijheid, B. Armstrong, et al. 1998. Risk of congenital anomalies near hazardous-waste landfill sites in Europe: the EUROHAZCON study. Lancet **352:** 423–427.
20. Elliott, P., D. Briggs, S. Morris, et al. 2001. Risk of adverse birth outcomes in populations living near landfill sites. Br. Med. J. **323:** 363–368.
21. Vrijheid, M., H. Dolk, B. Armstrong, et al. 2002. Chromosomal congenital anomalies and residence near hazardous waste landfill sites. Lancet **359:** 320–322.
22. Lewis-Michl, E.L., L.R. Kallenbach, N.S. Geary, et al. 1998. Investigation of cancer incidence and residence near 38 landfills with soil gas migration conditions: New York State, 1980–1989. ATSDR/HS-98-93. Atlanta: Agency for Toxic Substances and Disease Registry.
23. Goldberg, M.S., J. Siemiatycki, R. DeWar, et al. 1999. Risks of developing cancer relative to living near a municipal solid waste landfill site in Montreal, Quebec, Canada. Arch. Environ. Health **54:** 291–296.
24. Legambiente. 2005. Rapporto Ecomafia 2005. Legambiente, Rome.
25. Andrisani, M.G., P. Bianco, R. Belluomo, et al. 2003. Emergenza rifiuti regione Campania, aggiornamento sul censimento dei siti inquinati- litorale Domitio Flegreo ed Agro Aversano. ARPA Campania. Napoli.
26. Trinca, S., P. Comba, A. Felli, et al. 2001.Chilhood mortality in an area of southern Italy with numerous dumping grounds: application of GIS and preliminary

findings. *In* First European Conference Geographic Information Sciences in Public Health. Sheffield, U.K.
27. SENIOR, K. & A. MAZZA. 2004. Italian "Triangle of death" linked to waste crisis. Lancet Oncol. **5:** 525–527.
28. BIANCHI, F., P. COMBA, M. MARTUZZI, *et al.* 2004. Italian "Triangle of death". Lancet Oncol. **5:** 710.
29. BESAG, J., J. YORK & A. MOLLIÉ. 1991. Bayesian image restoration, with two applications in spatial statistics. Ann. Inst. Stat. Math. **43:** 1–59.
30. MOLLIÉ, A. 2000. Bayesian mapping of Hodgkin's disease in France. Spatial epidemiology. *In* Methods and Applications. P. Elliot, J.C. Wakefield, N.G. Best & D.J. Briggs, Eds.: 267–285. Oxford University Press. Oxford.
31. CLAYTON, D., L. BERNARDINELLI & C. MONTOMOLI. 1993. Spatial correlation in ecological analysis. Int. J. Epidemiol. **22:** 1193–1202.
32. EUROCAT Working group, Report 8, Surveillance of Congenital Anomalies in Europe, 1980–1999. EUROCAT (European Surveillance of Congenital Anomalies), 2002. Available at http://www.eurocat.ulster.ac.uk
33. THE INTERNATIONAL CENTRE FOR BIRTH DEFECTS, 2003. ICBDMS (International Clearinghouse for Birth Defects Monitoring Systems). Annual Report 2003 with data for 2001. Available at htpp://www.icbd.org.
34. MARTUZZI, M., F. MITIS, A. BIGGERI, *et al.* 2002. Environment and health status of the population in areas with high risk of environmental crisis in Italy. Epidemiol. Prev. **26** (Suppl 6): 1–53.
35. EUROGAST (THE) STUDY GROUP. 1993. An international association between Helicobacter pylori infection and gastric cancer. Lancet **341:** 1359–1362.
36. FORMAN, D., P. WEBB & J. PARSONNET. 1994. H pylori and gastric cancer. Lancet **343:** 243–244.
37. INTERNATIONAL AGENCY FOR RESEARCH ON CANCER. 1994. Schistosomes, liver flukes and Helicobacter pylori. Lyon: International Agency for Research on Cancer.
38. BUIATTI, E., D. PALLI, A. DECARLI, *et al.* 1989. A case–control study of gastric cancer and diet in Italy. Int. J. Cancer **44:** 611–616.
39. BUIATTI, E., D. PALLI, A. DECARLI, *et al.* 1990. A case–control study of gastric cancer and diet in Italy: II. Association with nutrients. Int. J. Cancer **45:** 896–901.
40. GONZALEZ, C.A., E. RIBOLI, J. BADOSA, *et al.* 1994. Nutritional factors and gastric cancer in Spain. Am. J. Epidemiol. **139:** 466–473.
41. LA VECCHIA, C., M. FERRARONI, B. D'AVANZO, *et al.* 1994. Selected micronutrient intake and the risk of gastric cancer. Cancer Epidemiol. Biomarkers Prev. **3:** 393–398.
42. KELLEY, J.R. & J.M. DUGGAN. 2003. Gastric cancer epidemiology and risk factors. J. Clin. Epidemiol. **56:** 1–9.
43. COCCO, P., D. PALLI, E. BUIATTI, *et al.* 1994. Occupational exposures as risk factors for gastric cancer in Italy. Cancer Causes Control **5:** 241–248.
44. MCLAUGHLIN, J.K., W.J. BLOT, S.S. DEVESA & J.F. FRAUMENI. 1996. Cancer Epidemiology and Prevention. *In* Renal Cancer. D. Schottenfeld & J.F. Fraumeni, Eds.: 1142–1155. Oxford University Press. Oxford. New York.
45. DHOTE, R., N. THIOUNN, B. DEBRE & G. VIDAL-TRECAN. 2004. Risk factors for adult renal cell carcinoma. Urol. Clin. North Am. **31:** 237–247.
46. GUO, J., T. KAUPPINEN, P. KYYRONEN, *et al.* 2004. Risk of esophageal, ovarian, testicular, kidney and bladder cancers and leukemia among Finnish workers exposed to diesel or gasoline engine exhaust. Int. J. Cancer **111:** 286–292.

47. LINDBLAD, P. 2004. Epidemiology of renal cell carcinoma. Scand. J. Surg. **93:** 88–96.
48. ZHANG, Y., K.P. CANTOR, C.F. LYNCH & T. ZHENG. 2004. A population-based case–control study of occupation and renal cell carcinoma risk in Iowa. J. Occup. Environ. Med. **46:** 235–240.
49. UTILI, R., B. GALANTI, G. DA VILLA, *et al*. 1983. Hyperendemicity of viral hepatitis in the Neapolitan area: an epidemiological study. Boll. Ist. Sieroter Milan. **62:** 145–152.
50. DA VILLA, G., F. PICCININO, C. SCOLASTICO, *et al*. 1998. Long-term epidemiological survey of hepatitis B virus infection in a hyperendemic area (Afragola, Southern Italy): results of a pilot vaccination project. Res. Virol. **149:** 263–270.
51. MASTRANGELO, G., U. FEDELI, E. FADDA, *et al*. 2004. Increased risk of hepatocellular carcinoma and liver cirrhosis in vinyl chloride workers: synergistic effect of occupational exposure with alcohol intake. Environ. Health Perspect. **112:** 1188–1192.
52. ALBERG, A.J. & J.M. SAMET. 2003. Epidemiology of lung cancer. Chest **123** (Suppl 1): 21S–49S.
53. HARRISON, R.M., D.J. SMITH & A.J. KIBBLE. 2004. What is responsible for the carcinogenicity of PM2.5? Occup. Environ. Med. **61:** 799–805.
54. POPE, C.A., 3rd, R.T. BURNETT, M.J. THUN, *et al*. 2002. Lung cancer, cardiopulmonary mortality, and long-term exposure to fine particulate air pollution. JAMA **287:** 1132–1141.
55. VINEIS, P., F. FORASTIERE, G. HOEK & M. LIPSETT. 2004. Outdoor air pollution and lung cancer: recent epidemiologic evidence. Int. J. Cancer **111:** 647–652.
56. MENEGOZZO, M., S. TRINCA, F. CAMMINO, *et al*. 2004. Geographical distribution of mortality from malignant pleural neoplasms and of former asbestos-exposed workers in the Campania region. Epidemiol. Prev. **28:** 150–155.
57. PIRASTU, R., I. IAVARONE & P. COMBA. 1996. Bladder cancer: a selected review of the epidemiological literature. Ann. Ist. Super. Sanità. **32:** 3–20.
58. TALASKA, G. 2003. Aromatic amines and human urinary bladder cancer: exposure sources and epidemiology. J. Environ. Sci. Health Part C Environ. Carcinog. Ecotoxicol. Rev. **21:** 29–43.

Hazardous Waste

Recognition of the Problem and Response

HENRIK HARJULA*

Organisation for Economic Co-operation and Development (OECD), 75775 Paris, Cedex 16, France

*The opinions expressed in this article are those of the author and do not necessarily represent the views of the OECD or of the member governments.

ABSTRACT: Two accidents, Seveso 1976 and Bhopal 1984, paved the way toward the international control of hazardous substances and waste, and public dissemination of information on their releases and transfers. The final facility cleanup in Seveso took place in 1982. The remaining dioxin-containing residues were packed in 41 drums. However, the drums disappeared and were found several months later in France. This was the starting point for the development of international control on transboundary movements of waste within the OECD. Between 1983 and 1992 OECD developed five binding Council Acts on the classification and control of transboundary movements of waste. This work also provided the foundation for the adoption of a series of directives and the Shipment Regulation 259/93 within the European Community, aimed at monitoring and controlling transboundary movements of hazardous waste, and of a global instrument in 1989 under UNEP: Basel Convention on the Control of Transboundary Movements of Hazardous Waste and Their Disposal. The Bhopal 1984 and another similar accident shortly thereafter led to the development of a mandatory Toxics Release Inventory (TRI) in the United States in 1986, to inform the public of releases and transfers of potentially hazardous pollutants and waste. OECD work has supported the rapidly increasing implementation of Pollutant Release and Transfer Registers (PRTRs) since 1994. The Basel 1989, Rotterdam 1998, and Stockholm 2001 Conventions all require strict control of the production and management of hazardous chemicals and waste. Yet, production of hazardous chemicals and generation of hazardous waste are still in rise.

KEYWORDS: hazardous waste; shipment of waste; waste controls; PRTRs; hazardous chemicals

Address for correspondence: Henrik Harjula, Organisation for Economic Co-operation and Development, Environment Directorate, 2, rue André-Pascal, 75775 Paris, Cedex 16, France. Voice: +33-1-45-24-98-18; fax: +33-1-44-30-61-79.
 e-mail: henrik.harjula@oecd.org

Origin of the Hazardous Waste Problem

Today, people take Sunday strolls along the paths of Bosco delle Querce, or Seveso Oak Forest Park, roughly 15 km north of Milan. One would not suspect that beneath the lush green carpet lurk the poisonous remains of a chemical disaster of July 10, 1976. A little after noon that Saturday, a valve broke at the Industrie Chimiche Meda Societa Azionaria chemical plant in Meda, releasing a toxic cloud that enshrouded the municipality of Seveso and other communities in the area. Some 3000 kg of chemicals were released into the air, including 2,4,5 trichlorophenol and dioxins.

The first signs of health problems, burn-like skin lesions, appeared on children a few hours after the accident. Beginning in September of that year, chloracne, a severe skin disorder usually associated with dioxin, broke out on some of the people most exposed to the cloud.

The Seveso incident, however, did not end right there. The dioxin-containing residues remaining in the plant were removed from the reaction vessels in the summer of 1982 and packed, together with contaminated auxiliary agents, into 41 drums. With the agreement of the Italian government the contaminated waste was to be removed by a recognized waste disposal firm to a licensed and officially controlled refuse disposal site. The drums were carried abroad, and suddenly disappeared without leaving any trace. More than half a year later, some suspect looking drums were found in France, and identified to be those same Seveso drums.

Seveso was not the only hazardous waste incident at that time. Tons of municipal incinerator ash from Philadelphia was dumped on a rural Haitian beach one night in 1986 by a barge called the Khian Sea. The ship had entered the port with a permit to unload fertilizer. In fact, this cargo contained some of the most toxic chemicals on the planet: dioxins and furans, and heavy metals, such as lead, cadmium, mercury, and arsenic. Nearly one-fourth of the 13,000 tons of waste had been unloaded from the barge before the Haitian government intervened and ordered the ash reloaded onto the barge. But the Khian Sea disappeared under the cover of darkness, leaving approximately 3000 tons of toxic ash on Haiti's beach.

The Khian Sea returned to Philadelphia with the remainder of its deadly cargo. The ship spent the next 2 years vainly seeking a dumping ground for its cargo. It crossed the Atlantic, sailed around the coast of West Africa, through the Mediterranean, down the Suez Canal, and into the Indian Ocean. When it finally pulled into the Singapore harbor it had a new name, the Pelicano, a new owner, and an empty hull.[1] These and many other incidents gave the impetus to active development of control procedures for transboundary movements of hazardous and other waste within the OECD, and beyond.

Reasons for the Hazardous Waste Problem

In early 1980s the OECD Waste Management Policy Group noted that: (*a*) hazardous waste, which crosses frontiers destined for disposal in another country is likely to be waste considered highly hazardous, that is, requiring incineration or physicochemical treatment in the generator country, as well as being restricted from legal sea dumping; (*b*) over 2 million tons of hazardous waste are estimated to cross national frontiers of OECD European countries annually on the way to legal disposal either at sea or ashore. This figure represents 8–10% of all such wastes generated in these countries; (*c*) transport of highly hazardous wastes via ship to certain developing countries from industrialized nations, could be very profitable in economic terms, in the short run. Hence, the occurrence of such "north-to-south" movements of highly hazardous wastes should be considered as a real possibility; (*d*) implications of possible shipments of highly hazardous wastes to developing countries need to be assessed. Legality of the imposition of controls by industrialized nations on such exports may need to be studied; and (*e*) methods providing for monitoring and control of transboundary movements of hazardous waste should be developed.

There are a number of potential stimuli for causing generators of waste to consider export as a means of dealing with hazardous waste. This list includes, but may not be limited to, the following: (*a*) rising costs of disposal in the home country; (*b*) diminishing capacity for treatment and disposal of certain types of wastes in the home country; (*c*) potential future liability for any damages caused by wastes disposed into or onto land in the home country; (*d*) tightening laws, regulations, and policies concerning disposal of certain types of wastes, for example, prescriptive disposal routes, such as incineration being required for liquids containing certain organic constituents; (*e*) tightening laws, regulations, and policies governing onsite disposal operations for wastes performed by a generator on his own premises; (*f*) economic growth that may result in increasing generation of waste; (*g*) existence of foreign disposal facilities which may serve several countries; (*h*) market opportunities for materials that can be recovered from wastes otherwise destined for disposal; and (*i*) existence of an appropriate foreign disposal facility that is closer than a similar domestic facility.

Legal disposal will almost certainly become increasingly costly as a function of time. A generator will normally seek least cost legal disposal for his wastes. If export possibilities are available to legal and less costly facilities, compared to more costly disposal in the home country, then export is a likely choice.

The Response

In the beginning of the 1980s many OECD countries had adopted or were in the process of developing regulatory measures to enable their authorities to monitor the management of hazardous waste from the place of generation to the place of disposal within their jurisdiction. The central goal of such systems was to ensure that discarded harmful substances are managed so as to minimize the possibility for adverse effects. However, it became rapidly clear that, in the case of transboundary movements, such national monitoring systems were not adequate, since countries had generally insufficient knowledge about consignments of waste imported into their territory to exercise proper control.

Control of Transboundary Movements of Hazardous Waste

In the autumn of 1982, the OECD Waste Management Policy Group proposed that guidelines be developed for the export and import of hazardous wastes, bearing in mind the different levels of development of environmental regulations among countries and the different levels of expertise. In February 1984, the OECD Council, the governing body of OECD, decided that member countries shall control the transboundary movements of hazardous wastes and, for this purpose, shall ensure that the competent authorities of the countries concerned are provided with adequate and timely information concerning such movements. Moreover, a comprehensive set of guiding principles concerning such control was recommended to member country governments. The principles included in the OECD Council Decision and Recommendation C(83)180(Final)[2] concern the basic strategies needed to properly monitor and control international movements of potentially hazardous wastes. In order to further explore these issues, OECD convened a Seminar on the Legal and Institutional Aspects of Transboundary Movements of Hazardous Wastes in Paris, June 1984.[3]

Development of the OECD Control Procedure

Between 1983 and 1992 five binding OECD Council Acts and three Council Resolutions relating to transboundary movements of waste were adopted.

The OECD was the first international body to take legal action to monitor and control exports of hazardous wastes to nonmembers. In response to the mandate of OECD Council Resolution C(85)100,[4] a Council Decision-Recommendation C(86)64(Final)[5] on Exports of Hazardous Wastes from the OECD Area was adopted to control exports of hazardous wastes to non-OECD countries as strictly as exports of hazardous wastes to OECD member countries, and not to allow movements of hazardous wastes to nonmember countries to occur without the consent of the appropriate authorities of the importing country and of any nonmember countries of transit, unless the hazardous wastes were directed to adequate treatment and disposal facilities in the importing country.

The Council Decision C(88)90(Final) on Transfrontier Movements of Hazardous Wastes[6] defined the terms *waste*, *disposal*, and *hazardous waste* for the purposes of controlling transboundary movements of waste. The Decision established a "Core List" of wastes to be controlled. In addition to the wastes covered by the "Core List," all other wastes considered to be, or legally defined as hazardous in the country of export or the country of import, were also subject to control under the terms of this Decision. The proposed OECD system for monitoring and controlling transfrontier movements of hazardous waste was compatible with existing international agreements governing transport of dangerous goods.[7] The prior notification information included data concerning which, if any, of the transport of dangerous goods protocols was applicable. Moreover, packaging of the wastes for transport and classification concerning transport hazards had to conform to United Nations recommendations. The proposed OECD hazardous waste notification and shipment documentation was in no way to supplant or interfere with documentation demanded by any existing international agreement concerning transport of dangerous goods. The Decision also established a classification system for wastes subject to a transboundary movement, known as the International Waste Identification Code (IWIC). However, this code is no more required under the Council Decision C(2001)107/FINAL.[8]

Council Resolutions C(89)1(Final) and C(89)112(Final) on the Control of Transfrontier Movements of Hazardous Wastes[9] expressed support to the work initiated under the auspices of the United Nations Environment Programme (UNEP) that aimed at preparing a global convention on transboundary movements of hazardous wastes. The earlier work on the OECD draft international agreement[10] provided the foundation for these global efforts, which finally resulted in the adoption of the Basel Convention on the Control of Transboundary Movements of Hazardous Wastes and Their Disposal in 1989.[11]

Council Decision-Recommendation C(90)178/FINAL on the Reduction of Transfrontier Movements of Wastes[12] was the first OECD Act concluded after the adoption of the Basel Convention. It was also the first Act regulating transfrontier movements of nonhazardous waste, in addition to those of hazardous waste. The Decision called for member countries to reduce to a minimum the exports of all waste for final disposal, in accordance with the environmentally sound and economically efficient management practices. It also encouraged member countries to establish additional and appropriate waste management infrastructure within their own territory and to develop bilateral or regional plans to ensure the environmentally sound management of those wastes, in the case where such infrastructure cannot be established. The Decision recognized the desirability of appropriately controlled international trade in waste materials destined for recovery, and that efficient and environmentally sound management of waste may justify some transboundary movements in order to make use of adequate recovery or disposal facilities in other countries. To that effect the Decision instructed the OECD Environment Committee to develop and implement a program of activities concerning waste destined for recovery operations.

Council Decision C(92)39/FINAL Concerning the Control of Transfrontier Movements of Wastes Destined for Recovery Operations[13] was developed in response to the mandate given by Decision-Recommendation C(90)178/

FINAL and established an intra-OECD mechanism (green, amber, and red tiers) to control transboundary movements of waste destined for recovery operations within the OECD area. It provided a streamlined, simplified, and efficient means of controlling such transboundary movements, compatible with the environmentally sound management of wastes as required by the Basel Convention. The Decision was adopted as an agreement or arrangement pursuant to Article 11 paragraph 2 of the Basel Convention.

Actions of the European Community

The adoption of the OECD actions provided the foundation for European Community (EC) legislation in this area. A Directive 84/631/EEC on the supervision and control within the EC of the transboundary shipment of hazardous waste was adopted in December 1984. This Directive largely transposed the "Principles" included with OECD Council Decision-Recommendation C(83)180(Final) into a legally binding form for the use by the EC Member States.[14]

Directive 84/631/EEC was supplemented in July 1985 by a further Directive 85/469/EEC providing details of precisely how the notification of transfrontier movements of hazardous wastes must be done in the EC. In particular, specific forms and procedures for notification were laid down.[15]

Directive 86/279/EEC set forth conditions governing control of exports of hazardous wastes from the EC area.[16] This Directive closely resembled its earlier OECD counterpart C(86)64(Final). In August 1988, the European Commission proposed two new Directives (88/C295/03 and 88/C295/04) which would define "wastes," "disposal," and "hazardous wastes" within the EC.[17,18] These proposals were developed based on the OECD Council Decision C(88)90(Final) that was adopted by the OECD Council on May 27, 1988.

Actions of the United Nations Environment Programme

In June 1987 the Governing Council of the United Nations Environment Programme decided to proceed with the development of a global convention concerning the control of transboundary movements of hazardous wastes. Similar work in OECD was specifically mentioned by the Governing Council and the results achieved at OECD were to provide a foundation for the UNEP efforts.[19] The UNEP Secretariat adopted a very ambitious schedule, since a convention was to be proposed for adoption by late March, 1989.

On March 22, 1989, the Basel Convention on the Control of Transboundary Movements of Hazardous Wastes and Their Disposal was adopted.[11] This Convention contains the basis for a method of identification, notification, and control of transfrontier movements of hazardous and other wastes. Management of these wastes is also the subject of the Convention since both the generation rates and disposal are taken into account. The goals of the Convention are to minimize the generation of wastes subject to the Convention, as well as their transboundary movements. Disposal of these wastes should be performed so that man and the environment will be protected from "the adverse effects which may result from such wastes."

In keeping with the instructions of the UNEP Governing Council,[19] some portions of the Basel Convention are taken verbatim or are close paraphrases of the OECD draft international agreement.[10] For example, wastes subject to control are essentially those of the OECD Core List of Council Decision C(88)90(Final). The obligations placed upon generators, exporters, importers, and disposers are founded upon the OECD text as well. In addition, there are a number of requirements concerning generation and management of hazardous wastes, provisions for aid and assistance to developing countries, as well as a delineation of illegal traffic in hazardous wastes. OECD governments whole heartedly supported the UNEP efforts. OECD Council Resolutions C(89)1(Final) and C(89)112(Final) clearly reflected this support.[9]

Regional Agreements

Following the adoption of the Basel Convention in March 1989, some regional agreements were negotiated. In December 1989, 68 less industrialized countries from Africa, the Caribbean and the Pacific, together known as the ACP countries, joined with officials from the EC in a treaty prohibiting the international trade in wastes.[20] The agreement, known as the Lome IV Convention, bans all shipment of hazardous or radioactive wastes from EC countries to ACP countries. In addition, the ACP countries agreed not to import any wastes from non-EC countries. The Lome Convention was the first international treaty to ban the trade in radioactive wastes, and the first commitment by EC countries to ban hazardous waste exports.

Under the Basel Convention, two regional agreements were negotiated during 1990s: the Bamako and Waigani Conventions. The Bamako Convention[21] was adopted in January 1991 and entered into force in April 1998. The Convention was convened under the auspices of the Organization of African Unity (OAU) that includes every African nation except South Africa and Morocco. The treaty is called "Bamako Convention on the Ban of the Import into Africa and the Control of Transboundary Movement and Management of Hazardous Wastes within Africa." Its purpose is to: (*a*) protect human health and the environment from dangers posed by hazardous wastes by reducing their generation to a minimum in terms of quantity and/or hazard potential; (*b*) adopt precautionary measures and ensure proper disposal of hazardous waste; and (*c*) prevent "dumping" of hazardous wastes in Africa.

The Waigani Convention[22] was adopted in September 1995 and entered into force in October 2001. The Parties of the Waigani Convention are: Australia, Cook Islands, Federated States of Micronesia, Fiji, Kiribati, New Zealand, Papua New Guinea, Samoa; Solomon Islands, and Tuvalu. The objectives of the Convention are to: (*a*) prohibit the importation of hazardous and radioactive waste into Pacific Developing Parties; (*b*) reduce the transboundary movement of hazardous wastes to a minimum consistent with their environmentally sound management; (*c*) treat and dispose hazardous waste as close as possible to their source of generation in an environmentally sound way; and (*d*) minimize the generation of hazardous wastes.

Harmonization of the International Control Procedures

Following the adoption of the Basel Convention in 1989 and the OECD Council Decision C(92)39/FINAL in 1992, the European Community started to implement the agreed procedures into the EC legislation. Consequently, a Council Regulation 259/93 on the Supervision and Control of Shipments of Waste within, into and out of the EC was adopted in February 1993. The Regulation entered into force in May 1994 and was binding in its entirety and directly applicable in all EC Member States.[23]

The OECD Council Decision C(92)39/FINAL provided a framework for the OECD member countries to control transboundary movements of recoverable wastes within the OECD area in an environmentally sound and economically efficient manner. Compared to the Basel Convention, it gave a simplified and more explicit means of controlling such waste movements. It also facilitated transboundary movements of recoverable wastes between OECD member countries in the case where an OECD country is not a party to the Basel Convention. The United States is the only OECD country that is not a party to the Basel Convention.

The development of the Basel Convention during 1990s, in particular the adoption of two detailed lists of wastes as new Annexes VIII and IX to the Convention in November 1998, gave impetus to revise the OECD Council Decision C(92)39/FINAL, in order to harmonize procedures and requirements and to avoid duplicate activities with the Basel Convention. This revision resulted in the adoption of Council Decision C(2001)107 in June 2001. An addendum to this Decision, C(2001)107/ADD1, which includes the harmonized notification and movement documents and the instructions to complete them, was adopted by the Council in February 2002. Finally, the addendum was incorporated into the Decision as Section C of Appendix 8 and the complete version of the Decision was issued as C(2001)107/FINAL in May 2002.[24] Provisions of the revised OECD Decision have been harmonized with those of the Basel Convention in particular with regard to the classification of wastes subject to control. However, certain procedural elements of the original OECD Decision

C(92)39/FINAL, which do not exist in the Basel Convention, such as time limits for approval process, tacit consent, and preconsent procedures have been retained.

Since the OECD Council Decision is binding to member countries, the European Community is legally required to revise the current Regulation 295/93 in order to implement the OECD Decision C(2001)107/FINAL. This work is currently in progress.[25]

Another Problem: Releases and Transfers of Hazardous Chemicals and Waste

In 1984 a deadly cloud of methyl isocyanate killed in Bhopal, India, over 7000 people within days and another 15,000 in the following years. Around 100,000 people are suffering chronic and debilitating illnesses for which treatment is largely ineffective. In addition, the waste management practices since the opening in 1970 of the facility in question were such that solid and tarry chemical waste was just dumped on the site, causing contamination of large land areas and the ground water. The site was not yet remediated by 2004, although the industrial activity ceased in 1984.[26]

Shortly after the Bhopal incident, there was a serious chemical release at a sister plant in West Virginia, United States. These incidents underscored demands by industrial workers and communities in several U.S. states for information on hazardous materials. Public interest and environmental organizations around the country accelerated demands for information on toxic chemicals being released to water, air or land, or transferred "beyond the fence line" of the facility. Against this background, the Emergency Planning and Community Right-to-Know Act (EPCRA) was enacted in 1986 (http://www.epa.gov/tri/whatis.htm).

Pesticides, in particular obsolete pesticides, provide a particular problem, since their classification is not yet uniform. Most often they are classified as hazardous waste, however, differing practices exist. Accounting for the paucity of data from many regions and individual countries, estimates based on existing inventories and previous experience would indicate that virtually all developing countries and economies in transition hold obsolete pesticides stockpiles. In total it could be estimated that global obsolete pesticides stockpiles in developing countries and economies in transition amount to something in the order of 400,000–500,000 tons.[27]

The condition of obsolete pesticide stockpiles varies from securely contained, well-stored products that can still be used in the field, subject to analysis, to products that have entirely leaked from corroded or otherwise damaged containers into the surrounding environment.

The FAO Obsolete Pesticides Programme has defined six key factors that lead to the accumulation of obsolete pesticides in developing countries.[28] These are: (*a*) product bans; (*b*) inadequate storage and poor stock management; (*c*) unsuitable products or packaging; (*d*) donation or purchase in excess of requirements; (*e*) lack of coordination between donor agencies; and (*f*) commercial interests of private sector and hidden factors.

Response

The primary purpose of the United States Emergency Planning and Community Right-to-Know Act is to inform communities and citizens of chemical hazards in their area. This Act requires businesses to report the locations and quantities of chemicals stored onsite to state and local governments in order to help communities prepare to respond to chemical spills and similar emergencies. It also requires Environmental Protection Agency and the States to annually collect data on releases and transfers of certain toxic chemicals from industrial facilities, and make the data available to the public in the Toxics Release Inventory, TRI (http://www.epa.gov/tri/whatis.htm).

OECD started in 1994 to develop further these inventories and started to call them Pollutant Release and Transfer Registers (PRTRs). The OECD Council Recommendation 1996[29] and the subsequent Guidance Manual 1996[30] provided a catalyst for the development of PRTRs across the OECD countries and elsewhere. Since 1996, the number of OECD countries with operating PRTR systems has more than doubled. By 2005 at least 15 OECD countries had an operational PRTR in place. Many more countries, within the OECD and beyond, have already taken concrete steps toward the establishment of a PRTR.

The Aarhus Convention created a wider framework and process for the potential integration of current national PRTRs. To this end, a protocol on PRTRs was formally adopted in Kiev, Ukraine, May 21 2003.[31] More than 30 states took part in the negotiations and 36 countries and the European Community signed the protocol in Kiev. This would mean that over 30 new countries will establish a PRTR within the next few years.

What is a PRTR?

PRTRs have their origin in Principle 10 of the Rio Declaration on Environment and Development. A PRTR is an environmental database or inventory of potentially harmful chemicals and/or pollutants released to air, water, and soil, and transferred offsite for treatment. According to the OECD Council Recommendation C(96)41/FINAL, as amended by C(2003)87, and the PRTR Protocol, the core elements of a PRTR system are: (*a*) a listing of chemicals, groups of chemicals, and other relevant pollutants that are released to the environment or transferred offsite; (*b*) integrated multimedia reporting of releases and transfers to air, water, and land; (*c*) reporting by source, covering point sources and nonpoint sources, where appropriate; (*d*) periodic reporting, preferably annually; and (*e*) making data available to the public. A PRTR brings together in one place, information about what pollutants are being released, or transferred, where, how much, and by whom.[32]

Controls for Chemicals

Since it has proven impossible to control hazardous waste quality and quantity solely through waste-related activities, it is relevant to pay some attention in this context also to activities that attempt to address future environmental hazards already in the product phase; namely the conventions and agreements toward the control of chemicals.

The Rotterdam Convention on the Prior Informed Consent Procedure (PIC) for Certain Hazardous Chemicals and Pesticides in International Trade was adopted in 1998. Dramatic growth in chemicals production and trade during the past three decades had highlighted the potential risks posed by hazardous chemicals and pesticides. Countries lacking adequate infrastructure to monitor the import and use of such substances were particularly vulnerable. In the 1980s, UNEP and FAO developed voluntary codes of conduct and information exchange systems, culminating in the PIC procedure introduced in 1989. The Rotterdam Convention replaced this arrangement with a mandatory PIC procedure on 24 February 2004 when the Convention entered into force (http://www.pic.int/).

This procedure is not a recommendation to ban the global trade or use of specific chemicals. It is rather an instrument to provide importing parties with the power to make informed decisions on which of these chemicals they want to receive and to exclude those they cannot manage safely. If trade takes place, requirements for labeling and provision of information on potential health and environmental effects will promote the safe use of these chemicals.

The Stockholm Convention on Persistent Organic Pollutants (POPs), was adopted in 2001 in response to the urgent need for global action to protect human health and the environment from the POPs. These are chemicals that are highly toxic, persistent, bioaccumulative, and move long distances in the environment. The Convention seeks the elimination or restriction of production and use of all intentionally produced POPs, that is, industrial chemicals and pesticides. It also seeks the continuing minimization and, where feasible, ultimate elimination of releases of unintentionally produced POPs, such as dioxins and furans. Stockpiles must be managed and disposed of in a safe, efficient, and environmentally sound manner. The Convention imposes certain trade restrictions. The Convention entered into force, thus becoming international law, on May 17, 2004 (http://www.pops.int/).

In February 2001, the European Commission adopted a White Paper setting out the strategy for a future community policy for chemicals. The main objective of the new chemical strategy is to ensure a high level of protection for human health and the environment, while ensuring the efficient functioning of the internal market and stimulating innovation and competitiveness in the chemical industry. It will provide a step-by-step approach to phase out and substitute the most dangerous substances, the ones that cause cancer, accumulate in the human bodies and in the environment, and affect the ability to reproduce. To this end, the European Commission has prepared a proposal for a Regulation of the European Parliament and of the Council concerning the Registration, Evaluation, Authorization and Restriction of Chemicals (REACH) (http://europa.eu.int/comm/environment/chemicals/reach.htm).

Environmental Liability

These days, we are confronted with cases of severe damage to the environment resulting from human acts. The accident with the Erika oil tanker and the mining waste incident near the Doñana nature reserve in the southern Spain, are only two examples of cases where human activities have resulted in substantial damage to the environment, involving the suffering and death of hundreds of thousands of birds and other animals.

So far, many countries have established national environmental liability regimes that cover damage to persons and goods, and they have introduced laws to deal with liability for, and cleanup of, contaminated sites. However, until now, these national regimes have not addressed the issue of liability for damage to nature. This may be one reason why economic actors have focused on their responsibilities to other people's health or property, but have not tended to consider their responsibilities for damage to the wider environment. This has been seen traditionally as a "public good" for which society as a whole should be responsible, rather than something the individual actor who actually caused the damage should bear. The introduction of liability for damage to nature, as specified in the Directive 2004/35/EC on Environmental Liability with Regard to the Prevention and Remedying of Environmental Damage,[33] is expected to bring about a change of attitude that should result in an increased level of prevention and precaution. EC Member States need to implement this directive by April 30, 2007.

Within the context of the Basel Convention, a Protocol on Liability and Compensation was adopted at the Fifth Conference of Parties in December 1999. The Protocol negotiations began in 1993 in response to the concerns of developing countries about their lack of funds and technology for coping with illegal dumping or accidental spills of hazardous waste. The objective of the Protocol is to provide for a comprehensive regime for liability as well as adequate and prompt compensation for damage resulting from the transboundary movement of hazardous wastes and other wastes, including incidents occurring due to illegal traffic in those wastes. The Protocol addresses who is financially responsible in the event of an incident. Each phase of a transboundary movement, from the point at which the wastes are loaded on the means of transport to their export, international transit, import, and final disposal, is considered. The Protocol has not yet entered into force.[34]

Where Do We Stand Today?

OECD Council Decision C(2001)107/FINAL, EC Shipment Regulation 259/93 and the Basel Convention provide a comprehensive international framework for supervision and control of waste and hazardous waste generation and movement within the European Union (EU), OECD, and globally. The effectiveness is still expected to increase, once the EU has completed the harmonization of the Shipment Regulation with the OECD and Basel control mechanisms.

Despite the positive development in controlling hazardous waste, their generation and transboundary movements are in rise. The generation of hazardous waste among the Basel parties increased some 14% from 1993 to 2000.[35] Within the OECD countries the generation of hazardous waste increased some 15% from 1995 to 2001.[36]

The number of global transboundary movements of hazardous and other waste increased from 1000 to 5000 movements from 1993 to 1999. At the same time the total amount of transported waste increased from 2 to 7 million tonnes/annum. Most of the shipments took place within the EU, and were destined for recovery, rather than for final disposal. No increase could be observed in shipments from OECD to non-OECD countries from 1993 to 1999.[35]

Why do we have the increased hazardous waste generation and increased amount of transboundary movements? What are the achievements to date of national legislations and Basel, Rotterdam, and Stockholm Conventions?

Within the OECD area industrial hazardous waste generation increased some 15% between 1995 and 2001. During the same period GDP increased 18% and industrial production 19%. The world chemicals industry has grown from sales of approximately USD 1500 billion in 1998 to USD 1900 in 2003,[37] that is, a 27% increase in 7 years or 3.9% per year. The growth has not been even, for example, in China, output has grown at a rate of 15% per year, while output in the United States has grown only at a rate of 2% per year. Similar rates to the United States have been found in Western Europe and Japan, while India and the Middle East have experienced accelerated growth rates.

According to the North American Commission for Environmental Cooperation, over the 8-year period from 1995 to 2002 total releases and transfers of pollutants decreased by 7% in Canada and United States.[38] Onsite releases decreased by 21%, with a 15% decrease reported by the Canadian NPRI facilities and a 21% decrease by TRI facilities. Offsite releases, transfers to disposal, mainly to landfills, decreased by 14% in NPRI, however, they increased by 49% in the U.S. TRI, with a total North American increase of 38%. Transfers offsite for further management increased in both countries, with NPRI showing a 70% increase and TRI an 18% increase.

The figures provided above clearly indicate that industrial production is in increase, production of chemicals is in increase, and the hazardous waste generation is in increase. However, most alarming signal is that while production of chemicals has increased and direct releases decreased, the amount of chemicals in products is increasing.[39]

The amounts of exports and imports of hazardous waste are increasing, since wastes are more and more becoming tradable goods and source of income. In addition, destruction and disposal of hazardous waste is getting rather expensive along the tightening environmental requirements and, therefore, this waste is most often destined to the cheapest available option for disposal or destruction. Recent information also indicates that illegal transports of hazardous waste are increasing, which will provide a particular challenge for national authorities and relevant international conventions. The reasons for illegal shipments are not totally clear, but could be related to complex control procedures and increasing domestic disposal costs.

The national and international activities during the last 25 years have considerably improved the control and management of hazardous chemicals and waste.[40] However, those activities have failed to reduce the production of hazardous chemicals and generation of hazardous waste.

REFERENCES

1. VALLETTE, J. & H. SPALDING, Eds. 1990. The International Trade in Wastes. A Greenpeace Inventory, 5th ed. Greenpeace. Washington DC.
2. OECD. 1984. Council decision and recommendation on transfrontier movements of hazardous wastes C(83)180(Final). (http://webdomino1.oecd.org/horizontal/oecdacts.nsf/).
3. OECD. 1985. Transfrontier movements of hazardous wastes. Proceedings of a seminar on legal and institutional aspects of transfrontier movements of hazardous wastes. Paris 12–14 June, 1984. OECD. Paris.
4. OECD. 1985. Council resolution on international co-operation concerning transfrontier movements of hazardous wastes C(85)100. OECD. Paris.
5. OECD. 1986. Council decision-recommendation on exports of hazardous wastes from the OECD area C(86)64(Final). OECD. Paris. (http://webdomino1.oecd.org/horizontal/oecdacts.nsf/).
6. OECD. 1988. Council decision on transfrontier movements of hazardous wastes C(88)90(Final). (http://webdomino1.oecd.org/horizontal/oecdacts.nsf/).
7. UNITED NATIONS ECONOMIC AND SOCIAL COUNCIL. 1988. Recommendations on the transport of dangerous goods. Fifth revised edition. ST/SG/AC.10/1/Rev.5, UN-New York.
8. OECD. 2001. Guidance manual for the implementation of the OECD decision C(2001)107/FINAL. ENV/EPOC/WGWPR(2001)6/FINAL. (http://www.oecd.org/env/waste/).

9. OECD. 1989. Council resolution C(89)1(Final) on the control of transfrontier movements of hazardous wastes and council resolution C(89)112(Final) on the control of transfrontier movements of hazardous wastes. OECD. Paris.
10. OECD. 1985. Conference on international cooperation concerning transfrontier movements of hazardous wastes. Basel, Switzerland. 26–27 March, 1985. Background Papers, OECD. Paris.
11. UNEP. 1989. Basel convention on the control of transboundary movements of hazardous wastes and their disposal. Basel Convention Series No 01/03. Geneva. (http://www.basel.int).
12. OECD. 1991. Council decision-recommendation on the reduction of transfrontier movements of wastes C(90)178/FINAL. OECD. Paris.
13. OECD. 1995. Guidance manual on the OECD control system for transfrontier movements of wastes destined for recovery operations, including the Council Decision C(92)39/FINAL Concerning the control of transfrontier movements of wastes destined for recovery operations. OCDE/GD(95)26. OECD. Paris. (http://www.oecd.org/env/waste/).
14. EUROPEAN COMMUNITY. 1984. Council Directive 84/631/EEC on the supervision and control within the European Community of the transfrontier shipment of hazardous waste. Official Journal L 272, 12.10.1985, p.1.
15. EUROPEAN COMMUNITY. 1985. Commission directive 85/469/EEC of 22 July 1985 adapting to technical progress Council Directive 84/631/EEC on the supervision and control within the European community of the transfrontier shipment of hazardous waste. Official Journal L 272, 12.10.1985, p.1.
16. EUROPEAN COMMUNITY. 1986. Council directive 86/279/EEC of 12 June 1986 amending Directive 84/631/EEC on the supervision and control within the European community of the transfrontier shipment of hazardous waste. Official Journal L 181, 4.7.1986, p. 13.
17. EUROPEAN COMMUNITY. 1988. Proposal for a Council Directive 88/C295/03 amending Directive 75/442/EEC on waste. Official Journal C 295, 19.11.1988, p.3.
18. EUROPEAN COMMUNITY. 1988. Proposal for a council directive 88/C295/04 on hazardous waste. Official Journal C 295, 19.11.1988, p.8.
19. UNEP. 1987. Report of the governing council 17 June, 1987. Decision 14/30 on "Environmentally Sound Management of Hazardous Wastes". UNEP. Nairobi.
20. AFRICAN, CARIBBEAN AND PACIFIC GROUP OF STATES, ACP GROUP. 1995. Agreement Amending the Fourth ACP-EC Convention of Lomé Signed in Mauritius on 4 November 1995. ACP-CE 2163/95. (http://www.tralac.org/pdf/lome.pdf).
21. OAU (Organisation of African Unity). 1991. Bamako Convention on the Ban of the Import into Africa and the Control of Transboundary Movement and Management of Hazardous Wastes within Africa. Adopted 30 January 1991, Bamako, Mali. (http://www.ban.org/Library/bamako_treaty.html).
22. PACIFIC ISLANDS FORUM. 1995. Convention to Ban the Importation into Forum Island Countries of Hazardous and Radioactive Wastes and to Control the Transboundary Movement and Management of Hazardous Wastes within the South Pacific Region, Waigani Convention. Adopted 16 September 1995, Waigani. (http://www.greenyearbook.org/agree/haz-sub/waigani.htm).
23. EUROPEAN COMMUNITY. 1993. Council Regulation (EEC) No 259/93 of 1 February 1993 on the supervision and control of shipments of waste within, into and out of the European Community. Official Journal L 30, 6.2.1993, p. 1. (http://europa.eu.int/eur-lex/en/consleg/pdf/1993/en_1993R0259_do_001.pdf).

24. OECD. 2001. Decision of the council concerning the control of transboundary movements of wastes destined for recovery operations. Decision C(2001)107/FINAL as amended by Decision C(2004)20. (http://webdomino1.oecd.org/horizontal/oecdacts.nsf).
25. EUROPEAN COMMISSION. 2003. Proposal for a regulation of the European parliament and of the council on shipments of waste COM(2003) 379 final. Brussels, 30.06.2003. (http://europa.eu.int/eur-lex/en/com/pdf/2003/com2003_0379en01.pdf).
26. AMNESTY INTERNATIONAL. 2004. Clouds of injustice, Bhopal disaster 20 years on. (www.amnesty.org).
27. FAO. 2001. Baseline study on the problem of obsolete pesticide stocks. FAO. Rome. (http://www.fao.org/documents/show_cdr.asp?url_file=////docrep/003/X8639E/x8639e00.htm).
28. FAO. 1995. Prevention of accumulation of obsolete pesticide stocks. Provisional Guidelines. No. 2. FAO. Rome.
29. OECD. 1996. Recommendation of the Council on Implementing Pollutant Release and Transfer Registers C(96)41/Final, as amended by C(2003)87. (http://webdomino1.oecd.org/horizontal/oecdacts.nsf).
30. OECD. 1996. Pollutant Release and Transfer Registers, PRTRs: a tool for environmental policy and sustainable development - guidance manual for governments. OCDE/GD(96)32. OECD. Paris.
31. UNECE. 2003. Protocol on pollutant release and transfer registers under the Aarhus Convention. Adopted on 21 May 2003, Kiev. (http://www.unece.org/env/pp/prtr.htm).
32. WEXLER, P. & H. HARJULA. 2005. Pollutant release and transfer registers, PRTRs. *In* Encyclopedia of Toxicology, 2nd ed. P. Wexler, Ed. Editor-in-Chief Vol. 3: 463–467. Elsevier. Oxford.
33. EUROPEAN COMMUNITY. 2004. Directive 2004/35/EC of the European Parliament and of the Council of 21 April 2004 on environmental liability with regard to the prevention and remedying of environmental damage. Official Journal L 143, 30 April 2004, p. 56. (http://europa.eu.int/comm/environment/liability/index.htm).
34. UNEP. 2003. Basel convention on the control of transboundary movements of hazardous wastes and their disposal and protocol on liability and compensation for damage resulting from transboundary movements of hazardous wastes and their disposal. Basel Convention Series No. 01/03. Basel Convention. Geneva. (http://www.basel.int/pub/protocol.html).
35. UNEP. 2002. Global trends in generation and transboundary movements of hazardous wastes and other wastes. Basel convention series/SBC No 02/14. Basel Convention. Geneva.
36. OECD. 2005. OECD Environmental data compendium 2004. OECD. Paris.
37. AMERICAN CHEMISTRY COUNCIL. 2004. Guide to the Business of Chemistry, August 2004, p. 122.
38. CEC. 2005. Taking Stock 2002. North American Pollutant Releases and Transfers Commission for Environmental Cooperation of North America. Montreal. (http://www.cec.org/takingstock/index.cfm?varlan=english).
39. NET. 2004. Cabinet confidential. Toxic products in the home. National Environmental Trust. Washington DC (http://www.net.org/proactive/newsroom/release.vtml?id=28715).
40. OECD. 2001. OECD Environmental outlook. OECD. Paris.

New Directions in Managing Hazardous Wastes from an Industry Perspective

LYNN D. JOHNSON

EHS Advantage, LLC, Delran, New Jersey 08075, USA

> ABSTRACT: In past decades waste was left to the lowest level of staff to dispose of at the lowest cost. Today, business managers in leading companies consider waste to be part of their responsibility. There is money to be made by eliminating waste problems. Leading companies are moving to the top of the waste disposal hierarchy and eliminating waste for themselves and their customers.
>
> KEYWORDS: waste; hazardous waste; waste disposal hierarchy

Industrial waste disposal has come a long way in the last 50 years. Much has been done to reduce the volume of waste and the impact that its disposal has on the environment. In this article I will discuss a little bit of the history, some current data on changing waste volumes, the driving forces for waste reduction, and the changing industrial management of waste. Finally, I will show you examples of a major change—business management is taking responsibility for managing wastes and is making profit out of it.

In the good old days just 50 years ago, waste was something the front office did not worry about. If it was liquid and more or less miscible in water, it went down the sewer into the river. If it was solid, or too thick to flow, it was dumped on the back lot behind the factory. Responsibility for waste disposal rested near the bottom of the company. If there was enough land at the plant site, the fork lift operator was directed to dump it at the back of the plant. If it had to go offsite, someone in the purchasing department arranged with a waste hauler to carry it away. There was not much concern about where it went because the hauler was responsible for it as soon as he picked it up. The main concern was to keep the cost low.

Progressive companies in those days put their waste in a hole in the ground and covered it with dirt, or sent it offsite to a government licensed "sanitary landfill" where it was put in a hole in the ground with lots of other waste and

---

Address for correspondence: Lynn D. Johnson, 152 Fenwick Court, Delran, NJ 08075. Voice: 856-461-2165; fax: 856-461-8913.
e-mail: LJohnson321@comcast.net

covered with dirt. Raw materials were comparatively cheap and the cost of disposal was almost nothing; no one worried much about the wasted cost. This stuff was waste and waste was something we threw away.

A major contributor to this thinking was the state of analytical techniques— we could not analyze the messy chemical mixture in solid wastes and waste waters and did not know that half of it was good raw material and a quarter was product. Common handbooks of the time were of little help; they listed most organics as simply "insoluble" in water. The first commercial gas chromatograph came on the market when Perkin Elmer introduced its Model 154 Vapor Fractometer in 1955.[1] I toured a big refinery in Baton Rouge, Louisiana as an eighth grade student in 1958. ESSO (now Exxon Mobile) had a huge research lab there. The chemists proudly showed us their new gas chromatograph. They had had it about a year and were delighted to be able to determine common light hydrocarbons at a 10th of a percent concentration in a few minutes work. It was a great aid in helping them understand and solve production problems. However, the thought of measuring organic compounds routinely at part per billion levels was still unthinkable.

A few years later as a young chemical engineer, I persuaded an analytical chemist to use his gas chromatograph on waste water samples. He was reluctant to risk fouling his expensive chromatograph with my junk, but he gave it a try. He was successful and in a few weeks presented me with a short list of results that included products and raw materials he had identified in waste water streams. I turned that into a table showing pounds and dollars lost annually and presented it to the plant production manager. Useless waste suddenly became lost profit; the production manager left my office determined to get some of it back.

In September 1968, Finnegan began commercial introduction of its model 1015 computerized gas chromatograph–mass spectrometer with an integrated data processing system.[2] That combination ultimately led to routine analysis of wastes in the part per billion and lower range—levels where most organics are soluble in water and do have effects on life forms. The early instruments of course concentrated on product problems; it would be well into the 1970s before the new analytical techniques were routinely applied to analysis of wastes. Analytical chemistry has been the handmaiden of environmental control. Without the development of new analytical techniques we could never have made the progress we have made in the last 20 years.

Enough of ancient history. Let's look at progress in the last couple of decades.

Hazardous waste volumes are going down. TABLE 1 is data from one surrogate measure of hazardous waste volume for the United States. This is the Toxic Release Inventory (TRI), a report collected annually from industry since 1988. These data are onsite and offsite disposal and other releases. It includes air emissions as well as waste sent to landfills, incineration, and other waste treatment methods. As you can see, there has been a steady downward trend in spite of growth in the size of the economy.

TABLE 1. TRI, United States[8]

Year	Waste quantity thousands of metric tons
1988	1.43
1993	0.91
1998	0.81
2003	0.59

TABLE 2. Japan waste data[9]

Year	Waste quantity millions of metric tons
1990	104
1995	83
2000	57
2002	50

Waste volumes are changing in Japan also. TABLE 2 shows changes since 1980 in the estimated volume of waste disposed in Japan. There has been relatively little change in municipal waste volume, so the change in this chart largely reflects reductions in industrial waste volume. Waste volume was cut in half between 1990 and 2002 and the goal is to cut it in half again by 2010. These numbers are not directly comparable to the TRI data from the United States as TRI focuses on a specific, though long, list of chemicals and the Japanese data counts everything disposed.

Australia has a similar pollutant reporting register as the United States. It is called the National Pollutant Inventory. It is in its 6th year of reporting and is still having some growing pains, just as the United States did in its first years. The number of reports continues to climb, almost fivefold since the first year of reporting in 1999. The system is beginning to settle down as companies learn what is expected of them and gain experience in measuring and calculating the emissions. TABLE 3 contains the last 3 years of data. I have taken out the major combustion emissions of carbon monoxide, nitrogen oxides, particulate matter, and sulfur dioxides. The result shows about a 2% decline in chemical pollutants in the last 2 years. This is not dramatic, but with the number of filings still rising at 10% per year, this suggests a clear trend of downward emission rates.

TABLE 3. Australia national pollutant inventory[10]

Year	Waste volume thousands of metric tons
2002	385
2003	383
2004	379

TABLE 4. IBM hazardous waste data[11]

Year	Waste quantity thousands of metric tons
1993	60
1998	65
1999	45
2000	25
2001	20

A few companies are measuring and reporting their hazardous waste volume. TABLE 4 is data from the worldwide manufacturing operation of IBM Corporation. It includes hazardous waste sent to recycling, treatment, incineration, and landfill. As you can see, IBM has cut its hazardous waste volume by two-thirds since 1998. In a few minutes I will show you some examples of projects that helped accomplish this.

The conclusion is clear—volumes are going down in spite of industrial production continuing to go up. Let's look at the driving forces behind this trend.

Historically, several forces came to bear on companies in the 1970s, 1980s, and 1990s forcing companies to recognize their waste as a problem and deal with it. Public resistance to new waste disposal sites grew. New laws and regulations were passed. Public opinion about holding the manufacturer responsible for its waste took hold. Open dumping became a crime instead of a nuisance. New liabilities for individuals were put in place. In most of the world today you can certainly be fined for illegal waste disposal and in most places you can be sent to jail for this crime. Stockholders begin to worry about the impact of waste generation and disposal on the value of their holdings, and begin to bring pressure to bear on companies.

Cost was always a force for change. By 1980, every manager recognized that the stuff being thrown away was raw material purchased and every manager wanted to make as much of it into saleable product as possible. It cost money to buy it, it cost money to process and handle it, and it cost money to dispose it of. By the mid 1980s it was also clear to managers that it might cost even more money in the future if it was not properly disposed of.

In recent years the forces are changing. Wastes are no longer considered just a "cost"; more often today they are referred to as "lost profit." This change is more significant than you might think. A "cost" is something a low level employee creates; "profit" is what the main stream of the business is about. "Costs" can be ignored if they are small; "profit" deserves daily attention.

Another new force arriving in the last decade is customers. Customers are looking at where their wastes come from. If it is a contaminant in their raw material supply they are going back to their suppliers for a change. Most companies are starting to think about what happens to their product when their customer processes or uses it. That can lead to actions, such as reformulation to

TABLE 5. Waste disposal hierarchy

- Eliminate generation
- Reduce generation
- Recycle/reclaim
- Burn for energy recovery
- Incinerate
- Biodegradation
- Secure landfill
- Long-term storage

eliminate chlorinated hydrocarbon solvents—and if you are the supplier of the chlorinated solvents you need to pay attention or your sales will dry up. Some customers are starting to look at the total life cycle impact of their product including the waste generated in making the raw materials. That carries them upstream to question supplier's waste generation and disposal practices even though the waste is not their direct responsibility. Customers have become a force for change in waste generation throughout the value chain of a product from basic raw material production to final disposal or recycle by the end consumer.

Takeback laws are an emerging force in the world today. More than 20 countries have some form of law on the books that will require manufacturers to take back the item at the end of its useful life. These laws are not yet widespread and there are some practical difficulties in implementing them. As a few examples take hold and work for items such as computers and automobiles, I think we can expect similar laws to spread throughout the economy and the world in the next few decades.

TABLE 5 is a list we call the "waste disposal hierarchy." It has been around industry for several decades now and it is the commonly used list of preferences in guiding waste reduction work. Eliminating waste is the preferred choice. Destruction is the next best choice. The commonly used destruction methods are incineration for burnable materials and biodegradation for dilute aqueous wastes. Landfill in a well-designed landfill with appropriate liners, leachate collection, cover and long-term financial assurance is usually the last choice.

Thirty years ago the choice was usually incineration for hydrocarbon liquids, biodegradation for waste water, and landfill for everything else. These were at the bottom of the hierarchy and the least expensive options. As I noted earlier, waste disposal was the responsibility of someone low in the plant or low in the purchasing department. All they could do was send it to disposal and they were expected to keep costs down. Waste minimization or elimination required changing the process, which meant the business had to be involved to pay for the engineering and the process modifications. Thirty years ago most business people did not include waste in the area they considered their responsibility.

In the last decade there has been a major change. Today, the business people are involved and taking responsibility for their whole process from raw material to finished product at the ultimate consumer. Understand that when I use the term "business people," I don't mean the production manager at a plant site, or even the plant manager. I mean the people who decide what products a company makes, where it makes them, and what markets it sells to. These are the people responsible for the profits the company makes. Today's business manager understands eliminating waste throughout the value chain eliminates the cost of waste disposal and the potential future costs of waste site cleanup. It also opens profitable new markets by eliminating waste problems for customers.

The common question today is "Why are we making this waste and what can we do to eliminate it." Industry is moving up the waste disposal hierarchy and the business people are driving that change.

One indication of the change in corporate interest in environmental issues can be seen in their advertisements. One advertisement that caught my eye appeared recently in the journal *Scientific American*.[3] It was a picture of a 40 story tall windmill for production of electricity being advertised by General Electric. This ad was a clear, graphic, illustration that environmental issues are being integrated into business thinking by one of the largest companies in the world.

General Electric has been moving in this direction for over a decade but they recently put their strategy together with a name, a corporate purpose, and goals. They call it "Ecomagination," but the underlying driving force is there is profit to be made in products that reduce waste for their customers. To quote their Chief Executive Officer: "Ecomagination is GE's commitment to address challenges such as the need for cleaner, more efficient sources of energy, reduced emissions and abundant sources of clean water. And we plan to make money doing it. Increasingly for business, 'green' is green."[4] Environmental issues are no longer relegated to site staff and the yard operator or purchasing agent who send the waste wherever they can. There is money to be made in solving environmental problems and that means business people will back projects with investment and talent.

GE touts 17 major projects including wind turbines, clean coal power plants, photovoltaic systems for homes, desalination systems for clean water, and high efficiency engines for locomotives, aircraft, ships, and industrial prime movers. They have $10 billion in sales from these products now and expect to double that to $20 billion by 2010. That is small by comparison to their $150 billion/year revenue, but it is big business. "Green is Green."

Herman Miller is another example of the change in the way industry manages waste.[5] Herman Miller is an office furniture manufacturer in Zeeland, Michigan, with $1.3 billion sales. You would not think an office furniture manufacturer would have much in the way of environmental problems but these folks are listening to their customers.

They have adopted a design for the environment philosophy that includes examining the chemicals they use in their products and choosing the safest available. They are concerned with emissions of volatile organics from their products and they test and measure those organics. They choose materials with as much recycle content as possible and they choose materials that can be recycled at the end of the product useful life. Finally, they also design to disassemble the product at the end of its useful life so that it can be recycled. An office chair may last 10 to 20 years, but the buyer of a Herman Miller chair need not send it to a landfill when it wears out.

This company is also actively reducing its own generation of waste. They have a goal of zero landfill, zero air, and water emissions from manufacturing, zero hazardous waste generation, and 100% green energy for power by 2020.

You do not get this kind of action from environmental staff. This company's business managers have taken over responsibility for environmental issues and they are making money from it.

If you are a supplier to Herman Miller you need to think about how to supply them a product that is totally used in their manufacture, because Herman Miller is going to eliminate its waste. As a supplier, you want to be part of the solution, not part of the problem.

Armstrong World Industries makes flooring products, acoustical ceiling tile, and wood cabinets. They have $3 billion sales, 15,000 employees, and 42 plants in 12 countries. They are also in reorganization under U.S. bankruptcy laws because of huge liabilities for asbestos from past operations, so cost cutting is even more vital to their survival than it is to most companies.

Armstrong has an interesting recycle program for ceiling tiles.[6] Acoustical ceiling tiles contain up to 82% recycled fiber content and the fiber in old tiles works as well as new fiber. Armstrong has reclaimed 17 million square feet of ceiling tile since they started the program in 1999. Companies, such as Microsoft, Carolina Power, and the World Bank have taken advantage of the program. If you have a truckload of used tiles, they will send a truck for it. This is not the old "compliance if I have to" mentality. These business people are solving an environmental problem for their customers and providing themselves a new raw material source. It is money in both pockets.

IBM is one more example of industry moving up the waste hierarchy. They reported the following projects and related annual savings in their 2002 annual environmental report.[7]

Eliminate Process Solvent	$5.6 million
Substitute Powder Coating for Liquid Paint	$463,000
Reduce Waste Generation	$3.8 million
Packaging Redesign	$2.8 million

There are three things to note about this list. First, these are big bucks. These projects make a $12 million dollar per year difference in the cost of IBM's production, not a small number even for them. Second, projects of this magnitude do not get done by the fork lift operator, or the purchasing department or even environmental staff. These projects were backed and funded by the individual businesses within IBM who reap the rewards of the savings.

Third, all of them are at the top of the waste disposal hierarchy. They eliminate waste at the source.

The world is far from perfect yet. There is still a lot of waste to deal with. None of the waste volume data cited earlier has reached zero, or even holds much promise of approaching zero in the next two or three decades. But, industry has come a long way. Where waste disposal was once left to the lowest level of staff to dispose of at the lowest cost, today's business managers consider waste disposal to be part of their responsibility. There is money is to be made by eliminating the problem for the company and for its customers. Business management has taken responsibility for wastes. Companies are moving to the top of the waste disposal hierarchy and eliminating waste for themselves and their customers.

REFERENCES

1. ETTRE, L.S. 2005. Fifty years of GC instrumentation. LCGC, February, 2005. http://www.lcgcmag.com/lcgc/article/articleDetail.jsp?id=147842 Last accessed June 28, 2005.
2. FINNIGAN, R.E. 2003. Reflections on the development of laboratory instrumentation, a seminal breakthrough in modern mass spectrometry. Am. Lab. 28–29.
3. SCIENTIFIC AMERICAN, September, 2003, page 7.
4. GE PRESS RELEASE: "GE Launches Ecomagination to Develop Environmental Technologies" May 9, 2005. http://ge.ecomagination.com/@v=06242005_1232@/index.html Last accessed June 27, 2005.
5. HERMAN MILLER WEBSITE: http://www.hermanmiller.com Last accessed May 20, 2005.
6. ARMSTRONG INDUSTRIES WEBSITE: http://www.armstrong.com/commceilingsna/ Last accessed June 28, 2005.
7. IBM 2002. Environment and Well Being Progress Report. Page 36. http://www.ibm.com/ibm/environment/annual2002 Last accessed June 28. 2005.
8. 2003 TRI Public Data Release eReport. US Environmental Protection Agency. May 2005. http://www.epa.gov/tri/tridata/tri03/2003eReport.pdf) Last accessed June 27, 2005.
9. THE FIRST PROGRESS REPORT OF THE FUNDAMENTAL PLAN FOR ESTABLISHING A SOUND MATERIAL-CYCLE SOCIETY. Report from the Central Environment Council, Ministry of Environment, Japan. February 21, 2005. http://www.env.go.jp/en/rep/waste/plan_sound.pdf Last accessed June 27, 2005
10. AUSTRALIA NATIONAL POLLUTANT INVENTORY, years 2002-2004, http://www.npi.gov/au/index.html Last accessed June 27, 2005. This table includes all chemicals except carbon dioxide, sulfur dioxide, nitrogen oxides and particulate matter.
11. IBM. 2002. Environment & well-being progress report. Page 38. http://www.ibm.com/ibm/environment/annual2002 Last accessed June 27, 2005.

Developments in Management and Technology of Waste Reduction and Disposal

PHILIP RUSHBROOK

University of Northampton, School of Environmental Sciences, Northampton NN2 7AL, United Kingdom

ABSTRACT: Scandals and public dangers from the mismanagement and poor disposal of hazardous wastes during the 1960s and 1970s awakened the modern-day environmental movement. Influential publications such as "Silent Spring"[1] and high-profile disposal failures, for example, Love Canal and Lekkerkerk, focused attention on the use of chemicals in everyday life and the potential dangers from inappropriate disposal. This attention has not abated and developments, invariably increasing expectations and tightening requirements, continue to be implemented. Waste, as a surrogate for environmental improvement, is a topic where elected representatives and administrations continually want to do more. This article will chart the recent changes in hazardous waste management emanating from the European Union legislation, now being implemented in Member States across the continent. These developments widen the range of discarded materials regarded as "hazardous," prohibit the use of specific chemicals, prohibit the use of waste management options, shift the emphasis from risk-based treatment and disposal to inclusive lists, and incorporate waste producers into more stringent regulatory regimes. The impact of the changes is also intended to provide renewed impetus for waste reduction. Under an environmental control system where only certainty is tolerated, the opportunities for innovation within the industry and the waste treatment and disposal sector will be explored. A challenging analysis will be offered on the impact of this regulation-led approach to the nature and sustainability of hazardous waste treatment and disposal in the future.

KEYWORDS: solid wastes management; hazardous wastes; regulations; treatment technologies; disposal; Europe

---

Address for correspondence: Philip Rushbrook, Visiting Professor, University of Northampton, School of Environmental Sciences, Park Campus, Northampton, NN2 7AL, United Kingdom. Voice: +44 207 271 1357; fax: +44 207 271 1345.

e-mail: philip.rushbrook@ogc.gsi.gov.uk

DISCOVERY

In 500 BC, the first documented regular collection was in the city-state of Athens. Later in 300 BC, the giant Palace of Knossos on Crete, seat of the Minoan civilization, had the first known waste disposal pit. In spite of these achievements in ancient civilizations, throughout most of recorded history the public has largely ignored waste. Too many other problems could lead to an early death. Being anxious about a pile of broken pots, decaying manure, or rotting vegetables does not seem to have been a major concern among populations for the last 2000 years.

The word "waste" arrived in the English language in the 15th century by way of earlier words used by the Romans, Franks, and Normans, all derived from Latin *vastus*. Four waste-related words "midden," "muck," "filth," and "rubbish" first appeared earlier in the 14th century, while the word for "soap" was not used until a 100 years later. Towns in this period of European history must have been very dirty, stinking places in which to live.

Major health problems aggravated by poor waste management only become obvious when urban population densities increase and large quantities of waste accumulate too close to where people lived. This did not happen in Europe on a countrywide scale until the Industrial Revolution in the 19th century. The development of mass industry at this time, particularly in the fields of the metallurgical, textiles, and inorganic chemicals (chlor-alkali) manufacture, brought about a complementary mass production of chemical wastes. Many of the byproducts from these processes have the potential to be highly polluting and safe handling and disposal physical and chemical characteristics there had to be learnt the hard way. The era of *hazardous* waste had arrived. This waste, like that from homes, was also initially ignored until workplace conditions and local environmental deterioration became intolerable. Reluctantly, factory owners began an organized approach to their waste disposal, usually beginning with back-of-the-factory disposal sites.

Relatively quickly, the simple treating of some hazardous wastes (e.g., by dewatering, thickening, neutralizing) was found to be more economical than direct dumping, particularly when wastes had to be transported to out-of-town sites. Over time, accidents, public criticism, environmental destruction, political pressures, legislation, and court judgments built the framework of mandated regulations and legal principles that underpin hazardous waste management today[2]:

(1) Polluter Pays Principle
(2) Precautionary Principle
(3) Duty of Care
(4) Proximity Principle
(5) Prior Informed Consent.

CHANGES IN WASTE PERCEPTIONS

For most of the last century waste controls and regulations were orientated toward protecting public health from municipal waste. During the 1970s hazardous waste practices became an increasing source of public concern. Scientific studies demonstrated population crashes in grebes due to thin egg syndrome from the uptake of chlorinated pesticides by adult birds, illnesses in homes built over chemical disposal sites (Love Canal, New York)[3] and chemically contaminated land (Lekkerkerk, the Netherlands), and indiscriminate dumping of drummed waste polluting groundwater (e.g., Malkins Bank, U.K.). These, as well as other incidents and the Superfund legislation in the United States, all acted as strong drivers to tighten the control of waste. Much of this tightening was to protect environmental systems rather than simply public health. The well-publicized environmental disasters involving wastes also spawned a new era of intensive waste research, mostly financed through the central government.

The 1970s was the "Age of Enlightenment" when numerous waste disposal and industrial sites were observed and measured. The studies defined the problems and the scientific research and clean up necessary to understand the environmental effects to water and air from wastes and ultimately, better rules to manage them.

The 1980s can be characterized as the "Age of Understanding:" a frenetic 10 years when new discoveries were made on the fate of chemicals in the environment and treatment technology improvements. The 1990s led to an "Age of Regulation" when copious national and international waste legislation was implemented. Environmental concerns from hazardous waste production, handling, and treatment became intensively controlled through zoning and planning rules, design codes and licensing/permitting conditions, environmental monitoring and inspection, detailed record keeping and tracking, tight emissions and effluent discharge standards, strict engineering and performance criteria, diversion and waste reduction targets, materials prohibitions, fiscal liabilities, bonds, and taxes.

The focus has shifted away from research defining risk-based policies to rule-based policies on waste practices where policy makers determine what is acceptable or not acceptable. Inevitably, limits and standards tighten because the technology can be developed to achieve them. High-income countries are now in the territory of diminishing returns, where increasingly resource-intensive controls achieve smaller relative improvements (FIG. 1). The environmental "regulatory ratchet" on waste management is turning ever tighter.

DEFINITION

While there is a huge range of individual chemicals in existence that are or will become wastes, handling their treatment and disposal properly is not

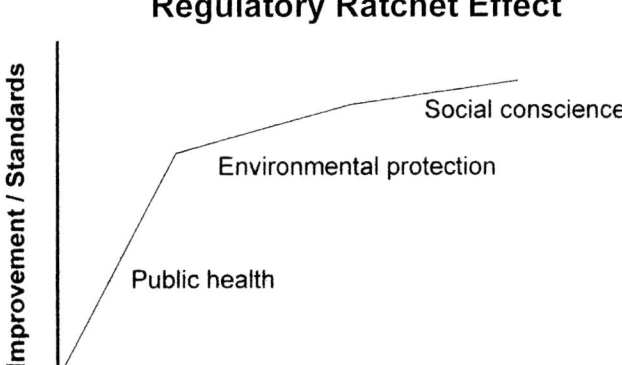

FIGURE 1. Regulatory ratchet effect.

insurmountable. Fortunately, most chemical wastes fall into a small range of categories based on their generic chemical compositions (TABLE 1).

Chemicals are ubiquitous in modern life. Nevertheless, what precisely constitutes a *hazardous waste* has been imponderable because spent chemicals and process byproducts first became recognized as wastes but defining them for regulatory and management purposes is difficult. The first law controlling a hazardous waste is believed to be the Alkali Act of 1863 in the United Kingdom, that led to the creation of the Alkali Inspectorate. The principal difficulty in defining a hazardous waste is that all wastes, for example, household, agricultural, construction, or chemical, can be hazardous to health in some situation or another, so how to define those wastes that have a high degree of hazard? At its most basic, a hazardous waste can be viewed as a discarded product or

TABLE 1. A simplified hazardous waste classification system

Chemical form	Waste group
Inorganic	Spent acids
	Spent alkalis
	Metal plating solutions
	Cyanides
	Inorganic chemicals/spent catalysts
	Paints/resins
	Reactive wastes, e.g., phosphorous compounds, peroxides
Organic	Organic solvents
	Oil/tarry residues
	Halogenated organic compounds
	Nonhalogenated organic compounds
	Infectious/microbiological waste

NOTE: Based on UNEP 1991.[12]

TABLE 2. Example of "hazardous" waste characteristics used in waste definitions

Hazardous characteristics
Explosive
Flammable liquids/solids
Spontaneous combustion
Emit flammable gases in water
Oxidizing
Organic peroxides
Poisonous (acute)
Infectious substance
Corrosive
Liberate toxic gases in air or water
Toxic (delayed or chronic)
Ecotoxic
Yield another material after disposal with a characteristic above

Source: Extract from Annex III of the Basel Convention in UNEP 1989.[4]

material that contains such high concentrations of an industrial chemical or a mixture of chemicals that it needs to be controlled and destined for special treatment or disposal. More recently wastes containing microbiological agents have been added. No classification system is perfect, and a variety of hazardous waste definitions have been produced. The most desirable approach is to relate the chemical characteristics of a waste to its risk of causing ill health. Unfortunately, analysts lack the capability to measure it systematically. One example from the United Kingdom was used until 2000: a hazardous waste was defined as the presence of one or more substances from a specified list that had been shown to demonstrate a flash point below 21°C or that a 5 cm^3 quantity (about the size of a sweet or piece of candy) ingested by a 20-kg child would cause serious injury. This would appear to be a pragmatic definition, but analysts assessing chemical hazard obviously could not test waste mixtures on 20-kg children!

The more widely adopted approach to defining hazardous wastes is the use of inclusive or exclusive lists defined in binding regulations. An inclusive list is the most common. All wastes containing chemical compounds on a specified list and/or all wastes from a list of specified processes are automatically deemed hazardous unless it can be demonstrated by the producer that they fail to possess certain hazardous characteristics (TABLE 2). The Basel Convention (UNEP 1989)[4] hazardous waste classification system is an example of the inclusive approach. Originally used for transfrontier movement of waste, the Convention drew heavily from earlier work undertaken by the OECD (Organisation for Economic Co-operation and Development) in 1985.[5] Over time, it has become the basis for many national definitions and in Europe the starting point for the European Union's recently implemented "European Waste Catalogue."[6]

TABLE 3. Simplified view of wastes and treatment/disposal

	Recovery	Treatment	Solidification	Incineration	Land
Acids, alkalis		X			
Heavy metals, toxic inorganics		X	X		Residues
Reactive compounds		X			
Nontoxic inorganics	X				X
Solvents, oils, resins, paints, organic sludge	X			X	
Organic Compounds	X	X		X	
Halogenated hydrocarbons				X	

TECHNOLOGY CHOICES

Hazardous waste management suffers from a perception among most of the general public, a fear of the unfamiliar. In reality, the common treatment types available for hazardous wastes are well established and have changed little throughout the 20th century. However, the complexity of individual process designs and the safeguards to control emissions and maximize efficiencies have developed continually.

Almost all hazardous waste management can be treated and rendered to a state where its residual potential for harm is currently regarded by regulatory authorities as acceptable via one or other of the five general methods: recovery, treatment, solidification, incineration, and land deposition. The categories of hazardous wastes most suited to each method of treatment are illustrated in TABLE 3. In high-income countries, organic chemical wastes, where recovery and reuse are not possible, are almost universally disposed off by high-temperature incineration. Past distinctions between halogenated and nonhalogenated organics are being largely superseded by legislation requiring all of these wastes to be incinerated in double-chamber facilities at high at temperatures of 1000–1200°C, in an excess of air, with residence times for combustion gases of 2 s or more to ensure efficient destruction. Relatively mild organic wastes such as volatile fatty acids and petroleum sludges are amenable to lagoon aeration and biological digestion, respectively. The separation (cracking) of oil-water emulsions, concentrates the oil fraction for recovery or subsequent disposal and permits separate discharge (after treating biologically) of the aqueous fraction.

Inorganic chemical wastes are not generally suitable for incineration. More commonly, their chemically active components are detoxified by a variety of pretreatment technologies prior to land disposal of the residue by direct deposition in engineered landfills or in a solidified concrete matrix (or similar immobilizing agent). Pretreatment is tailored to the chemical properties of the material being handled. For example, spent inorganic acids and alkalis can be

TABLE 4. A summary of common hazardous waste treatment processes[a]

Physical	Chemical	Biological
25 technologies	11 technologies	7 technologies
Filtration	Neutralization	Activated sludge
Distillation	Oxidation	Aerated lagoon
Flocculation	Reduction	Anaerobic digestion
Dewatering	Precipitation	Composting
Pretreatment of bulk solids		Disposal
3 technologies		11 options
Crushing		Incineration
Dissolution		Solidification
		Land

[a] UNEP (1991) Hazardous waste: policies and strategies.[12]

neutralized and dewatered, heavy metal containing sludges are thickened by dewatering in filter presses, and spent cyanides are oxidized.

A wide variety of processes are available to reduce the hazard posed by a waste stream. The most common processes are listed in TABLE 4 and a full list, including fledgling and noncommercialized processes, is presented in TABLE 5. A more detailed exposition of hazardous waste treatment technologies is presented in Batstone *et al.*[7]

TABLE 5. A detailed list of hazardous waste treatment processes [a]

Physical	Sedimentation	Biological
Air striping	Liquid-liquid extraction	Activated sludge
Suspension freezing	Steam stripping	Aerated lagoon
Carbon adsorption	Ultrafiltration	Anaerobic digestion
Centrifugation	Zone refining	Composting
Dialysis	Chemical	Enzyme treatment
Distillation	Calcination and sintering	Trickling filter
Electrodialysis	Catalysis	Stabilization ponds
Electrophoresis	Chlorinolysis	Pretreatment of bulk solids
Evaporation	Electrolysis	Crushing
Filtration	Hydrolysis	Cryogenics
Flotation	Neutralization	Dissolution
Freeze crystallization	Oxidation	
Freeze drying	Ozonolysis	
Magnetic separation	Photolysis	
Ion exchange	Precipitation	
Liquid ion exchange	Reduction	
Steam distillation		
Resin absorption		
Reverse osmosis		

NOTE: Not all are in wide-scale commercial use.
[a] UNEP (1991) Hazardous waste: policies and strategies.[12]

TABLE 6. Recent EU laws/international agreements

Waste Management Framework Directive
Landfill Directive
Incineration Directives
WEEE (Waste Electrical and Electronic Equipment) Directive
ELV (End of Life Vehicles) Directive
European Waste Catalogue
Restriction of Hazardous Substances Directive
Waste Oils Directive
Groundwater Directive
Packaging Directive
Montreal Protocol—CFCs ban
POPs Convention
Kyoto Protocol—COs emissions
European Carbon Trading Scheme

NOTE: Not an exhaustive list. POP = persistent organic pollutants.

THE PRESENT

Present-day waste operations in Europe are characterized by tightening regulations restricting the secondary use of recovered wastes, the increasing direction of wastes to prescribed types of treatment process, and higher standards for emissions control, facility operation, and waste tracking. An illustrative listing of the growing body of European legislation, which the Member States are required to transpose into their national laws, and various international agreements is given in TABLE 6. A "Duty of Care" is applied to all involved in producing, transporting, and treating hazardous wastes. This is a responsibility to manage the wastes in their possession properly. The highest expectation is on the producer to ensure their waste is finally disposed off in a safe and legal manner at a permitted site.

Within Europe the trend is toward a single definition for hazardous wastes, the "European Waste Catalogue" and a policy of avoidance rather than the production of waste, primarily through using economic measures and legal penalties. Treatment and disposal costs are rising as a response to increasing the standards of operations required by new laws, bans on using less-expensive disposal options (i.e., the recent ending of codisposal in landfills containing municipal wastes and prohibition some years ago of incineration at sea) and, more recently, a substantial reduction in the allowable reuse of some wastes, for example, combustion of waste oils in boilers and the reconditioning of refrigerators with (chlorofluorocarbons) CFC-containing insulation. New hazardous waste definitions in Europe coming into force have designated a further 180 wastes as hazardous and hence requiring specialized disposal, including television cathode ray tubes and computer monitors, fluorescent tubes and mercury lamps, and chemically contaminated soils.[6,8]

In the highly controlled hazardous waste sector, the barriers to entry are very high for new waste operators or new treatment processes. The numbers of operators are declining steeply as the cost of operation rises and producers hesitate to consign and change processes to avoid sending wastes for disposal. In the United Kingdom, the number of waste oil recycling facilities has declined from around 100 to 30,[9] while the number of secure landfills permitted to deposit hazardous wastes has declined from around 200 to 10. In 2004 the U.K. national quantity of recorded hazardous waste decreased from 5 million tonnes by 700,000 tonnes. The reason is not yet known.

An increasing adoption of the precautionary principle by courts and regulators is influencing the siting and operation of treatment plants and use of particular compounds. The Montreal Protocol and Stockholm Convention have brought about the elimination and/or restriction of whole classes of compounds. In the Montreal Protocol CFCs and hydrochlorfluorocarbons (HCFCs) are being phased out to reduce ozone depletion in the upper atmosphere. Whereas in the Stockholm Convention a dozen Persistent Organic Pollutants (POPs) are being eliminated or severely restricted to prevent damage to ecosystems and human health: aldrin, chlordane, dieldrin, endrin, heptachlor, hexachlorobenzene, mirex, toxaphene, polychlorinated biphenyls (PCBs), dichloro diphenyl trichloroethane (DDT) and polychlorinated dibenzodioxins (PCDDs), and polychlorinated dibenzofurans (PCDFs).

Parallel to the tightening of regulations, there are now higher penalties on waste producers and waste operators for breaches of standards or noncompliance. Tolerance of mistakes is decreasing and there is an increasing recourse to criminal litigation to resolve disputes. The traditional consensus-building approach to regulation has been: advice first, enforcement second. How long this can continue is open to question given the strict liabilities now expected from waste producers and operators.

The concentration of hazardous waste treatment and disposal on fewer sites will inevitably increase professionalism at these places and improve training and status for all staff. The use of environmental taxes on landfill disposal on gate fees and producer-run recycling for end-of-life vehicles, scrap electronic and electrical goods and postconsumer packaging will inevitable drive waste reduction and materials substitution within businesses in these sectors. Across Europe different models of hazardous waste provision exist. In some countries (e.g., France) many operations are run by state-controlled companies or franchised to a private operator, sometimes with a degree of geographical exclusivity to protect market share and hence, income. Elsewhere, the hazardous waste market is entirely run by private sector providers in an open market. The ability for one or other market model to survive in the future is still to be determined.

THE FUTURE

The role of all responsible operators and regulators is to reduce as far as practicable health, environmental and social problems arising from wastes.

Wastes, including hazardous wastes, will continue to be the surrogate for general public concern about local environmental concerns. Society and its elected representatives will inevitably introduce further controls, limits, and bans on the use and disposal of chemicals as findings from future epidemiological and environmental research become interpreted by policy officials and the popular media. When and in what circumstances the modern-day penchant of relentless incremental regulation will be satisfied is yet to be reached.

There may well be a further broadening of the precautionary principle where more industrial processes and hazardous waste operations are prevented just in case they cause ill health. One area has been studies seeking to demonstrate a causal link between chemicals in land disposal sites and various genetic abnormalities in infants born nearby.[10] A recent U.K. government-funded academic study undertook an extensive survey into potential health effects from waste facilities and concluded that no identifiable links had been demonstrated.[11] Nevertheless, this has not settled the matter, and the future will be characterized by more studies searching for conclusive evidence and continuing public scepticism.

Society, in general, will move toward a "zero waste" tolerance, in spite of the diminishing returns in environmental improvement this will bring. To achieve zero waste will consume additional resources, energy, materials, finance, and its ultimate impact on sustainability of resource use will need to be better understood. Removing the next y amount of an air pollutant that requires say, $5y$ additional energy and material resources, is not, in the long run, a sustainable way of delivering an environmental balance.

The number of waste management facilities will continue to decline, as the investment cost to remain compliant increases, until perhaps only a handful of regional multitreatment centers remain: similar to the rationalization of provincial cinemas in European countries toward multiscreen sites in large cities. Materials substitution will continue with a preference to adopting demonstrably benign processes. Whether or not this can be achieved in reality is highly uncertain. More waste types will become designated as hazardous and the cost of managing their treatment will continue to rise. The penalties for failure will stiffen.

Even with this scenario for a future of relentless control and protection, it remains questionable if society will ever become at ease with its hazardous waste and the methods used for its elimination.

CONCLUSIONS

Hazardous waste, however defined, will continue to be produced in an increasingly industrialized world. Popular doubts and suspicions about waste management fuel fear and contempt of the unknown. Science can answer these concerns but practitioners in epidemiology have to redouble their efforts to present their activities and findings in an understandable manner.

Each new generation of researchers tackle the prevailing issues with renewed vigor and optimism. With hazardous wastes there is a need for them to do four things[2]:

(1) Find new ways to determine if or which hazardous waste management practices cause elevated health or environmental problems or have the courage to state that no problem exists in the language of the layman.
(2) Achieve actual sustainability in the use of resources to treat and dispose off wastes and the operation of environmental emissions control technologies.
(3) Break out of the diminishing returns being achieved from the prevailing monitoring-and-design-refinements-to-achieve-compliance genre.
(4) Innovate and transfer ideas and methods from other fields to achieve simpler product design, better waste avoidance, and efficient waste processing.

In the future, life inevitably will be more complex and competition globally for everything will increase. Environmental choices may well become more distorted as some resources run out and the dominant societies set the environmental agenda that others follow. In the short term (and perhaps medium term too) to comply with this agenda the sustainability of increasingly sophisticated waste management operations may actually become more elusive. Prolonged concerns over waste generation and resource use may only ease when the global population stabilizes and affluence plateaus: a challenging prospect.

REFERENCES

1. CARSON, R.L. 1962. Silent Spring. Houghton Mifflin. Boston.
2. RUSHBROOK, P.E. & R. ZGHONDI. 2005. Better Healthcare Waste Management. WHO Regional Office for Eastern Mediterranean, Cairo, Egypt. Co-published with World Bank.
3. GLAUBINGER, R.S., P.M. Kohn & R. Remirez. 1979. Love Canal Aftermath: learning from a tragedy. Chem. Eng. **86:** 86–92.
4. UNEP. 1989. Basel Convention on the Control of Transboundary Movements of Hazardous Wastes UNEP/IG, 80/3, 22 March 1989. UNEP, Nairobi, Kenya.
5. OECD (ORGANISATION FOR ECONOMIC CO-OPERATION AND DEVELOPMENT). 1985. Conference on International Cooperation Concerning Transfrontier Movements of Hazardous Wastes. Basel, Switzerland. 26–27. March 1985.
6. CEC (COMMISSION OF THE EUROPEAN COMMUNITY). 2000. European Waste Catalogue Commission Decision 3 May 2000. Decision 2000/532/EC. Official Journal, L 226, 6.9.2000 p.3. Available at http://europa.eu.int/eur-lex/en/consleg/pdf/2000/en_2000D0532_do_001.pdf.
7. BATSTONE, R. et al. Eds. 1989. The Safe Disposal of Hazardous Wastes. Three volumes. World Bank Technical Paper 93. The World Bank, Washington DC.
8. DAY, M. 2005. Special changes to hazardous waste. Wastes Manag. 22–23.
9. HOLMES, I. 2005. 'Is waste oil a burning issue?' Wastes Manag. 18–19.

10. DEPARTMENT FOR ENVIRONMENT, FOOD AND RURAL AFFAIRS (DEFRA). 2004. Review of Environmental and Health Effects of Waste Management. U.K. Government. Defra, London, U.K. Her Majesty's Stationery Office.
11. VRIJHEID, M. 1998. Potential Human Health Effects of Landfill Sites. Report to the Environment Agency, North West Region, U.K. Produced by: Environmental Epidemiology Unit, London School of Hygiene and Tropical Medicine, U.K.
12. UNEP. 1991. Hazardous waste: policies and strategies. A UNEP training manual. UNEP, Paris, France. ISBN 92 807 1311 6.

Systematic Approach to Evaluating Trade-Offs among Fuel Options

The Lessons of MTBE

J. MICHAEL DAVIS[a] AND VALERIE M. THOMAS[b]

[a]*National Center for Environmental Assessment, Office of Research and Development, U.S. Environmental Protection Agency, Research Triangle Park, North Carolina 27711, USA*

[b]*School of Industrial and Systems Engineering, Georgia Institute of Technology, Atlanta, Georgia 30332, USA*

ABSTRACT: The fuel additive methyl tertiary butyl ether (MTBE) has been used in an effort to improve air quality in the United States, but other undesirable effects, particularly the contamination of water resources, were eventually judged to outweigh any air quality benefits it may have offered. The experience with MTBE offers many lessons, including the need to evaluate potential positive and negative environmental impacts associated with fuel choices using a comprehensive approach that combines a product life-cycle perspective with the risk assessment paradigm. Such an approach, referred to as "comprehensive environmental assessment" (CEA), is illustrated here by highlighting some of the issues that might be considered in evaluating reformulated gasoline (RFG) produced with MTBE, ethanol, or no oxygenate.

KEYWORDS: MTBE; methyl tertiary butyl ether; ethanol; oxygenates; alkylates; risk assessment; life-cycle assessment; multimedia assessment; comparative assessment; comprehensive environmental assessment

INTRODUCTION

Methyl tertiary butyl ether (MTBE) is an additive that is generally used to increase the octane rating of gasoline and to add oxygen to petroleum fuels. Its use as an octane booster in the United States dates to the late 1970s, when it served as a substitute for tetraethyl lead in leaded gasoline. As an oxygenate additive, MTBE came into greater use after the 1990 Amendments to the Clean Air Act prescribed the addition of oxygen to unleaded gasoline to help

Address for correspondence: J. Michael Davis, Ph.D., National Center for Environmental Assessment, Mail Drop: B 243-01, U.S. Environmental Protection Agency, Research Triangle Park, NC 27711. Voice: 919-541-4162; fax: 919-685-3331.
  e-mail: Davis.Jmichael@epa.gov

address certain air quality problems. Although the 1990 Amendments did not specifically require the use of MTBE, the additive was generally favored by refiners for various reasons of cost, blending qualities, and other desirable characteristics. However, in recent years MTBE has attracted a considerable amount of unfavorable attention because of other properties associated first with consumer complaints about inhalation of MTBE vapors and later with contamination of water resources.

The experience with MTBE illustrates the phenomenon of unintended environmental consequences, that is, the creation of an unanticipated problem resulting from a technological application intended to address a different environmental issue. In the case of MTBE, the hope was that the addition of oxygen to gasoline would help reduce ambient air levels of carbon monoxide and ozone as well as certain toxic air pollutants such as benzene. Moreover, refiners found other beneficial features in MTBE. It could be produced with excess isobutylene from the gasoline refining process. In addition to increasing the octane number of gasoline, it had other favorable blending characteristics in terms of volatility, driveability parameters, and adding volume to fuel. The last is especially noteworthy, given the close balance between refining capacity and consumer demand for gasoline in the United States. In effect, MTBE extends the fuel supply when used at 15% volume in winter-season oxygenated gasoline or 11% volume in reformulated gasoline (RFG). These and other features of MTBE led to its dominance as an oxygenate additive in the U.S. fuel supply during the 1990s when MTBE accounted for roughly 90% of the fuel oxygenates market, with ethanol making up nearly all of the remaining 10%.[1]

In contrast to the above positives, MTBE had some significant negative features as well. Concerns about health risks gained national attention when the winter oxyfuel program began in the fall of 1992, first in Alaska and then in other locales.[2] Motorists, mechanics, and others in contact with fuel vapors containing MTBE sometimes complained of eye and nose irritation, headache, nausea, and various other irritant effects. A major effort was made by the U.S. Environmental Protection Agency (EPA), the Centers for Disease Control and Prevention, and other organizations to conduct studies, both field (human) and laboratory (human and animal), to investigate these symptom complaints and expand the database on MTBE exposure and health effects.[3] These studies were in addition to a series of toxicity studies in laboratory animals that had been initiated in the late 1980s.[4] Based on the first reports from these animal studies, in 1991 EPA derived an inhalation reference concentration (RfC) of 0.5 mg/m^3 for MTBE, based on renal toxicity effects. An RfC has been defined as an estimate (with uncertainty spanning perhaps an order of magnitude) of an inhalation exposure to the human population (including susceptible subgroups) that is likely to be without an appreciable risk of adverse health effects over a lifetime. After the results of additional, longer term studies became available in the early 1990s, EPA revised the RfC to 3 mg/m^3 for MTBE.[5]

The short-term effects of MTBE reflected in symptom complaints, such as eye and nose irritation or headache, were not specifically addressed by the RfC for chronic inhalation exposure, which was based primarily on chronic kidney effects (although it should be noted that swollen periocular tissue was among the effects discussed in the RfC derivation). Various health risk assessments were conducted in the mid-1990s, and generally they concluded that MTBE probably did not constitute an imminent public health threat.[6–10] However, these assessments also emphasized that more data were needed to better understand the health risks of MTBE, particularly its acute effects. A subsequent study[11] of acute effects in self-described sensitive human volunteers under controlled conditions provided mixed results, and to date many questions and paradoxes about the health issues surrounding MTBE remain unresolved.[12]

While the inhalation health effects of MTBE were being investigated and debated, another problem was starting to attract attention. More than 400,000 confirmed releases of fuel from federally regulated underground fuel storage tanks (UFST) have been reported in the United States since 1988,[13] with possibly half of these leaking tanks containing gasoline blended with MTBE.[1] Although an incident of MTBE-contaminated ground water resulting from a leaking UFST was noted in 1984,[14] it was not until after the mid-1990s that the problem of groundwater contamination started to gain wider recognition as more reports of MTBE contamination emerged.[15–18] Levels of MTBE detected in ground water were generally low, but some reports of concentrations in the hundreds or thousands of micrograms of MTBE per liter of water (μg/L) added to a growing sense that MTBE contamination was a widespread and significant threat to water resources, especially when associated with prominent locales such as Lake Tahoe and Santa Monica, California.[19]

The problem of groundwater contamination by MTBE owed to the fact that MTBE, in contrast to other chemicals in gasoline, was more water soluble, less likely to adhere to soil particles, and more resistant to biodegradation. Consequently, MTBE tended to be at the front of a plume of gasoline leaking from a UFST and persist longer than other notable constituents of gasoline such as benzene, toluene, ethylbenzene, and xylenes (BTEX). These properties made MTBE stand out as a problem distinct from, and even greater than, the long-standing issue of conventional gasoline leakage from UFSTs. Moreover, MTBE contamination was found in both subsurface and surface waters, with the latter typically being associated with the presence of motorized watercraft.

Apart from any potential health concerns associated with such water contamination, a principal issue was the unpleasant odor and taste of MTBE in drinking water that was unacceptable for many people even at relatively low concentrations.[1] In 1999, a "Blue Ribbon Panel" of independent experts appointed by the EPA to evaluate the water quality concerns associated with MTBE came to the conclusion that *"in order to minimize current and future threats to drinking water, the use of MTBE should be reduced substantially ... [or] ... phased out completely."*[20] Citing this recommendation, EPA later issued an "Advance

Notice of Intent To Initiate Rulemaking Under the Toxic Substances Control Act To Eliminate or Limit the Use of MTBE as a Fuel Additive in Gasoline."[21] Section 6 of the Toxic Substances Control Act (TSCA) provides the EPA broad authority to regulate chemical substances to prevent "unreasonable risks to health or the environment." Thus, in the span of just a few years, MTBE had gone from being a major feature of U.S. fuel programs for improving air quality to being viewed as an unreasonable risk to water quality.

LESSONS LEARNED

The experience with MTBE offers lessons that could be useful in averting similar mistakes in the future. For one, a multimedia environmental perspective is crucial. It is unwise to focus on just one part of the environment, such as air, while ignoring the potential for impacts in other areas, such as water. Legislative, organizational, and other factors have contributed to this tendency,[22] but it is clear that the environmental impacts of large-scale applications of technology cannot be evaluated separately in the manner of the fabled blind persons touching different parts of an elephant. In the same vein, it is important to examine such issues from the perspective of the entire life cycle of a product. From "cradle to grave," the product life cycle encompasses various stages in the manufacturing, distribution, use, and disposal of a product and its byproducts. In the case of MTBE, for many years the seeming advantages related to the manufacturing and use of this additive overshadowed the problems of leakage related to storage. Focusing exclusively on one facet of a product's life cycle, such as its use, can lead to unintended consequences just as focusing on a single environmental medium can.

Another lesson from the MTBE experience is the importance of recognizing not only that every technology option has advantages and disadvantages but also that these trade-offs need to be compared between different options. In other words, MTBE (or any other choice of fuel or fuel additive) is not necessarily good or bad *per se*; rather, it is better or worse than any given alternative choice. Indeed, it may be simultaneously better and worse than an alternative, given the various dimensions along which such technological alternatives can be evaluated.

Other "lessons learned" have been well described by Erdal and Goldstein[23] in an extensive account of the history of the MTBE experience and its implications for environmental public policy. The question that arises, however, is how can these lessons be applied to avert future technological debacles? To answer that question, it is instructive to first look back to the late 1980s and early 1990s when the EPA Office of Research and Development started developing a strategy for conducting research that would provide needed information to assess the benefits and risks of various fuels and fuel additives. The result was a document titled "Alternative Fuels Research Strategy,"[24] which included discussion of MTBE and other oxygenates (e.g., ethyl tertiary butyl ether, ETBE), noting:

> Compared to gasoline, the ethers MTBE and ETBE have relatively large aqueous solubilities and would likely leach more rapidly through soil and groundwater. Also, limited data suggest that ethers may be persistent in subsurface environments.

and

> Very little is known about emissions and releases from MTBE and ETBE storage and distribution, making this area an appropriate target for research. Effects on existing equipment and controls... need to be evaluated.

Had these warnings been heeded, it seems likely that at least some of the problems that arose with MTBE could have been reduced or avoided. The more important point, however, is that such cautionary statements could be articulated at all in the face of quite limited information. By understanding how these prescient statements were formed, it should be possible to evaluate the potential environmental impacts of other fuel or technological options.

COMPREHENSIVE ENVIRONMENTAL ASSESSMENT

The Alternative Fuels Research Strategy[24] essentially combined a product life-cycle framework with the risk assessment paradigm. This approach is here referred to as "comprehensive environmental assessment" (CEA) to denote that it is more than either life-cycle assessment (LCA) or risk assessment alone and that it attempts to examine the environmental impacts of technology in a broad, systematic manner. The term comprehensive is relative. In the present context it does not necessarily encompass other dimensions such as economic, political, security, or societal factors, although the CEA approach could be extended to include nonenvironmental considerations.

The general features of CEA are highlighted in FIGURE 1. As listed in Column 1, the life cycle of a product typically comprises several stages, including feedstock production or extraction, manufacturing processes, distribution, storage, use, and disposal of the product and waste byproducts. At any stage of the life cycle, pollutants may enter one or more environmental pathways: air, water, soil, and food web (Column 2). It is important to identify these primary contaminants and, to the extent possible, the transport and transformation processes they undergo. The idea is to characterize the primary as well as secondary or byproduct pollutants associated with the entire life cycle for all relevant environmental media (Column 3).

The existence of a contaminant in the environment does not necessarily mean that humans or other specific organisms are exposed to it. Thus, it is important to go beyond a conventional LCA and apply a key feature of risk assessment, namely exposure assessment. In other words, what is the potential

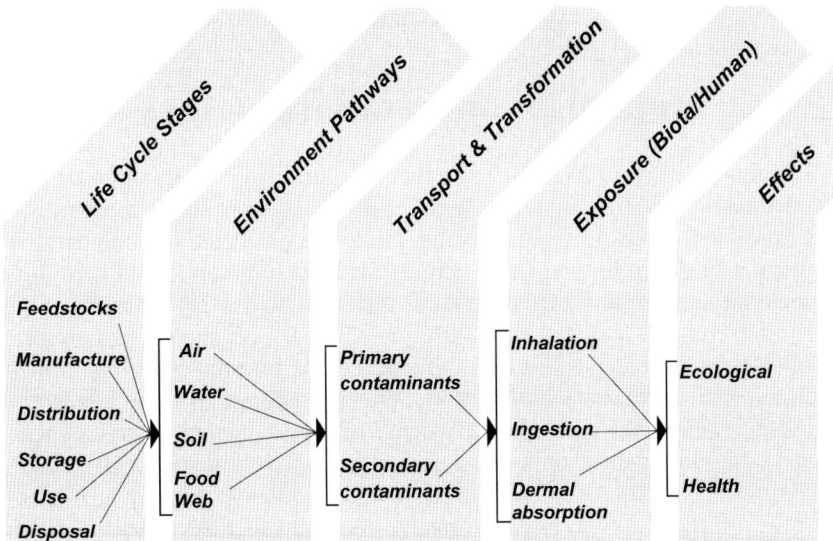

FIGURE 1. Comprehensive environmental assessment.

for humans and other organisms (biota) to come into contact with primary and secondary pollutants via all pertinent routes, for example, inhalation, ingestion, and dermal absorption (Column 4)? Although exposure analysis is much more complicated than simply identifying relevant routes of exposure, the list in Column 4 is intended to reinforce the idea of multiple causal chains of events originating from various sources and leading ultimately to multiple effects. Empirical data on exposure concentrations may not be available for all of the pollutants in question, but it may nevertheless be possible to estimate potential exposure levels using conventional assumptions.[25–27] In particular, microenvironmental and high-end exposure scenarios should be considered, not just "typical" or "average" exposure levels. This scenario-driven approach to exposure assessment is another feature that distinguishes the CEA approach from LCA-based methods of analyzing impacts.[28] Rather than taking a macrolevel view of exposure using nationwide emissions data, for example, the CEA approach focuses especially on the possible upper tails of exposure distributions. Of course, the upper tail of a distribution is by definition limited in size and should be kept in perspective relative to an overall exposure distribution.

Once exposure potential has been characterized, the health and ecological hazards associated with respective contaminants need to be described qualitatively and quantitatively (Column 5). To characterize *risk* quantitatively, the dose–response characteristics of a toxicant must be considered in relation to exposure potential. Some pollutants may pose low risk because the exposure potential is low or the hazard potential is low, or both. In other cases, risk may be relatively high when exposure potential is low but hazard potential is high, or vice versa.

COMPARISON OF OXYGENATE OPTIONS

The above description is little more than an outline of primary topic areas to be covered by the CEA approach. To describe this concept a little more fully, a few examples of questions or issues that might be considered in a comparative evaluation of different oxygenate options for RFG are highlighted here. It is beyond the scope of this article to denote the full extent and complexity of the issues that could be included in actually applying the CEA approach; therefore, it follows that no conclusions or judgments can be inferred from this limited treatment of RFG options. Nevertheless, one may gain some sense of how the CEA approach could be applied, particularly for a comparative assessment. TABLE 1 highlights three options related to the oxygenation of RFG: MTBE, ethanol, or no oxygenate. These oxygenate options are somewhat, but not entirely, arbitrary. As the two historically dominant fuel oxygenates, MTBE and ethanol are clearly relevant. The "no-oxygenate" option has long been proposed especially for use in California and has gained backing in recent legislation passed by the U.S. Congress.[29] Other oxygenates could be added to the list, along with conventional (non-reformulated) gasoline as a basic point of reference for RFG. However, the three options should suffice for illustrative purposes.

Arrayed vertically in the first two columns of TABLE 1 are the headings or topic areas that are listed in FIGURE 1, with the addition of global climate as another topic of significance and particular relevance to fuel formulations. Beginning with Feedstocks, each of the oxygenate options has distinct issues that could be considered. In the case of MTBE, the primary feedstock is methanol, which in North America is produced predominantly from methane. Among the questions that could be asked in relation to the feedstocks for MTBE, one might ask what are the implications of increased or decreased methane extraction for methanol production. Given that methane is an important greenhouse gas, is there likely to be a net increase or decrease (or no change) in methane emissions associated with changes in the amount of MTBE derived from methanol?

The primary feedstock for ethanol in North America is corn. Large scale corn farming operations typically use herbicides such as atrazine, and atrazine is among the most prevalent water pollutants in the United States.[30] Would increased use of ethanol produced from corn result in greater use of atrazine and thereby more contamination of water resources?

Under the no-oxygenate option for RFG, it is important to realize that eliminating oxygenate from gasoline would require other changes in formulation.[31] If nothing else, the loss of volume from either MTBE or ethanol would necessitate a significant increase in other components of gasoline to make up for the net loss in the fuel supply. Given regulatory proscriptions against

TABLE 1. Examples of issues to be included in CEA of RFG oxygenate options

		RFG with:		
		MTBE	Ethanol	No oxygenate
Source characterization	Feedstocks	Methanol: methane	Corn: atrazine	Iso-octane: hydrofluoric acid
	Production	VOCs	Air toxics, odors	VOCs
	Distribution	Small/chronic releases	Large/acute releases	Small/chronic releases
	Storage	Materials compatibility, product containment		
	Use (evaporative & combustion emissions)	Air toxics, NO, CO, ozone, etc.	Acetaldehyde, PAN, alkylates, etc.	Alkylates, toluene, etc.
Environmental quality	Air	Formaldehyde, tert butyl formate	Acetaldehyde, ozone	VOCs, secondary organic aerosols
	Subsurface	MTBE, tert butanol	Increased BTEX	Alkylates
	Surface Water	MTBE	Oxygen depletion	Alkylates
Exposure	Human	Acute/chronic; personal and population exposures; cumulative and aggregate; single chemical and mixtures		
	Biota	Acute/chronic, terrestrial/aquatic		
	Acute			
	Chronic			
Human health effects		Neurobehavioral, respiratory, organoleptic (sensory), etc. Cancer potency, inhalation RfC, RfD		
Ecosystem effects	Terrestrial	Organism, population, community, ecosystem		
	Aquatic (marine & freshwater)			
Global climate	CO_2	Increase? decrease? no net change?		
	methane			
	N_2O			
	CO			
	NO_2			
	VOCs			

PAN = peroxyacetylnitrate.

increases in aromatics, such as benzene or toluene, a likely choice to make up for the loss in volume would be alkylates, such as iso-octane. Typical feedstocks for iso-octane have been sulfuric acid and hydrofluoric acid. One might ask what would be the possible environmental impacts of a significant increase in such feedstocks. For example, would there be an increase in the potential for accidents involving tank cars of hydrofluoric acid? What would be the human health and safety, as well as ecological, ramifications of such accidents?

Second under Source Characterization in TABLE 1, some of the types of emissions associated with Production are denoted. One question for consideration is what are the relative amounts of volatile organic compound (VOC) emissions from production facilities. The importance of this information was underlined by the discovery that several ethanol plants in the U.S. Midwest were emitting toxic air pollutants in much greater amounts than had been initially estimated or allowed under the EPA permitting process.[32] In addition, residents in the vicinity of some ethanol plants have complained about unpleasant odors. Given the significance of odor issues surrounding MTBE, odors associated with alternatives to MTBE would seem to warrant consideration even if no direct health or environmental impacts result from odors *per se*.

Under Distribution it would be useful to evaluate differences in impacts related to small chronic releases versus large acute releases. At a fuel distribution terminal, for example, both types of releases could occur, but their effects could be rather different for different fuel formulations. Also, the probability of different types of releases would likely differ because of differences in the processes used to add oxygenates to blendstock. Whereas MTBE is conventionally blended at the refinery before being transported by pipeline to fuel terminals, ethanol is typically "splash-blended" at the distribution point because its hydrophilic properties make it less suitable for pipeline distribution.

Storage was the critical life-cycle stage associated with the downfall of MTBE. Even a small leak of just 1 gallon of gasoline containing 11% volume MTBE could, under certain conditions, contaminate more than 4 million gallons of water at a concentration of 20 μg/L MTBE. Small leaks of ethanol are not expected to pose comparable threats because ethanol is readily biodegraded.[33] On the other hand, it is also possible that the preferential microbial metabolism of ethanol in gasoline leaking from a UFST could impede the biodegradation of other constituents of gasoline.[34] If so, greater use of ethanol in gasoline could, under some conditions and within certain time frames, result in increased contamination of ground water by chemicals such as benzene, a gasoline constituent that is a known human carcinogen. As for alkylates, some analyses suggest that these components of gasoline may share certain chemical properties with MTBE and consequently may pose similar problems for groundwater contamination from UFST releases.[35]

The Use stage in TABLE 1 refers primarily to the operation of a motor vehicle, including its refueling. Both evaporative and combustion emissions are of importance in this stage. The volatility of a fuel obviously has implications for the type and amount of evaporative emissions from the fuel, but other factors can complicate the picture further. For example, comingling of different formulations of RFG can result in greater evaporative emissions than either formulation by itself.[36] Also, certain components of fuels may degrade the materials in gaskets and seals and significantly increase evaporative emissions from a vehicle's fuel system. When fuels are combusted, the emissions are complex mixtures of compounds (such as benzene) found in the fuel itself as well as byproducts of the combustion process (e.g., aldehydes, 1,3-butadiene). A great deal of research has been devoted to characterizing the chemical profile of combustion emissions from different fuel formulations.[37]

Up to this point, we have been dealing with sources of emissions or releases. Once a pollutant enters the environment, transformation processes may produce new or additional compounds, and transport processes may distribute the original contaminants and/or the secondary byproducts in various environmental media, possibly translocating pollutants far away from the point of release. Ozone is a well-known example of a secondary air pollutant that is produced from primary precursor pollutants that undergo atmospheric transport and transformation processes. Other air pollutants are also relevant here, including secondary organic aerosols, aldehydes, and numerous other compounds in lesser amounts. Characterizing air quality in terms of these primary and secondary pollutants has been primarily attempted through modeling based on emissions data,[38] although environmental measurements of selected air pollutants have been correlated with different fuel oxygenates in some studies.[39,40] Far less research has been devoted to investigating transport and transformation processes in other environmental media (e.g., subsurface, water). Nevertheless, the importance of such processes is clear and merits examination, as illustrated, for example, by increasing reports of tert butanol contamination as a presumed byproduct of the MTBE degradation processes.[41]

The existence of a chemical in the environment may imply some potential for exposure, but exposure does not actually occur until an organism comes into contact with a substance or agent. Exposure is a function of not only the amount or concentration of a compound but its temporal characteristics, for example, the duration and frequency of contact. Thus, exposure may be acute or chronic in duration, and singular or repeated in frequency. For MTBE in water, attention has tended to focus on chronic, repeated, low-level exposure conditions, whereas for ethanol the primary scenario of concern has been acute, single-event, high-concentration exposures such as large releases that might occur with barge or tanker accidents.[42] However, these examples are only a partial list of conditions that need to be considered in a comprehensive evaluation.

Exposure can be even more complex when one considers the possibility of both aggregate and cumulative exposures. Aggregate exposure through multiple environmental pathways and routes might occur if, for example, an individual were to ingest a chemical in drinking water as well as having dermal contact with it while bathing and inhaling it while driving, refueling, or performing other activities. Cumulative exposure to multiple contaminants could involve the primary as well as secondary pollutants related to the respective oxygenate options. Furthermore, such exposures could occur with a mixture of chemicals as well as a variety of single chemicals in succession.

Without exposure, there can be no (direct) effects. Ecological and human health effects, as well as effects on global climate, constitute broad headings of impacts to be considered as part of a CEA. These general headings may be subdivided in various ways, but commonly used divisions include acute and chronic effects under human health, and aquatic and terrestrial effects under ecosystems. Other subdivisions could be equally valid and useful. Global climate change subsumes both ecological and human health effects but is probably best handled as a separate heading in itself. Further subdivisions of human health would include more specific classes of effects, such as carcinogenicity, neurotoxicity, developmental and reproductive toxicity, immunotoxicity, and respiratory toxicity.

Beyond identifying these qualitative types of hazards, such effects may be amenable to quantitative dose-response characterization. In the case of a carcinogen, for example, a cancer potency estimate may be available. For noncancer effects, an oral reference dose (RfD) or an inhalation RfC may exist. If so, and if adequate exposure information is available, it may be possible to characterize health risks quantitatively, at least for some specific chemicals from an array of primary and secondary pollutants. Similarly, for ecological effects, subdivisions of broader headings are possible for characterizing such effects qualitatively and/or quantitatively. Although the present discussion does not attempt to illustrate such effects in any detail, a CEA must ultimately focus on effects, along with exposure information, if the risks associated with different fuel options are to be characterized. Quantitative assessment of specific risks is desirable, but it is important to bear in mind that qualitative assessment may be all that is possible in the face of limited information.

PROCESS AND OUTCOMES

The CEA approach is consistent with recent efforts to integrate LCA and risk assessment methods.[43,44] However, CEA should not be viewed as a form of LCA or as simply an extension of LCA or LCA impact assessment. Despite the familiarity of the concepts of LCA and risk assessment and the fact that various LCAs have been conducted on fuel alternatives, none to date has been as comprehensive as called for in the CEA approach. For example, Row *et al.*[45] compared conventional gasoline with an MTBE blend but limited their consideration of impact categories to air emissions (acid rain precursors, greenhouse gas emissions, ground-level ozone precursors, hazardous air pollutants, and particulate matter); leaks to ground water or other water-borne impacts were not considered. Similarly, Brinkman et al.[46,47] developed an LCA comparing U.S. alternative fuels using the GREET (Greenhouse gases, Regulated Emissions, and Energy used in Transportation) model. Their study went beyond previous uses of the GREET model, which had focused entirely on energy use, and extended the analysis to include greenhouse gas and criteria pollutant emissions, including VOCs, carbon monoxide, nitrogen oxides, particulate matter, and sulfur dioxide, associated both with fuel production and with vehicle operation. However, impacts to water were not included in their analysis. Reinhardt and Jungk[48] summarized an extensive set of LCA-based evaluations of the environmental trade-offs of fossil fuels versus ethanol fuels from different feedstocks in Europe, but their analysis was limited to ecological impacts. A broader effort using LCA methods to compare 22 combinations of conventional and alternative fuels as well as various vehicle technologies (internal combustion, fuel cell) was carried out for the German government by Patyk.[49] This analysis focused on impact categories comprising energy consumption, greenhouse effects, stratospheric ozone depletion, photochemical smog, eutrophication, acidification, and carcinogenicity. The Patyk study did not include MTBE among the fuel options considered, but even if it had, the selection of impact categories would not have captured the groundwater impacts associated with MTBE.

Many other examples of LCAs could be cited, but in general the strength of an LCA is also its limitation, namely the reliance on quantitative data. The importance and value of quantifying inputs and impacts is self-evident, but in instances where quantitative data are lacking, LCA-based evaluations may fail to address key issues. The CEA approach incorporates quantitative LCA information to the extent possible, but CEA is not constrained to considering only those impacts that can be quantified. Recall that when the potential for MTBE contamination of ground water from leaking UFSTs was identified in the Alternative Fuels Research Strategy,[24] quantitative characterization of the risks of such leakage was not possible. At that time (ca. 1988–1992), only a qualitative statement was possible. Nevertheless, if such a qualitative warning had been recognized and heeded, it could have provided a useful guide to risk managers. There is no reason to think that the CEA approach could not continue to provide valuable guidance through a combination of quantitative and qualitative methods.

The CEA approach is based on the idea of looking at the "big picture" in a systematic manner and is consistent with a growing body of research[50,51] that considers various chains and interactions of possible environmental events that could lead to impacts on ecosystems and human health. The actual process of applying the CEA approach therefore requires a broad range of technical expertise, including individuals familiar with product-specific life-cycle stages as well as experts in areas such as environmental transport and transformation, exposure, ecosystem effects, and health effects. Along with stakeholders and other interested parties, such experts are brought together to interact directly so that they can be informed by each other and in so doing achieve a "whole greater than the sum of its parts," that is, a broader perspective than would have been possible by merely assembling their individual points of view. Various formal techniques for collective decision analysis could be used for this purpose.[52–54]

The primary objective of CEA is to identify in as much detail as possible all the issues associated with each option and the sequences of environmental events that might result, without regard to which ramifications would be considered good or bad, desirable or undesirable. The outcome of this process is not a collective judgment about which option is the best (or worst) but an array of potential impacts associated with each option. These potential impacts may be expressed, indeed may only be expressible, in qualitative terms. However, even if every risk identified through the CEA process could be quantified, it would be difficult if not impossible to compare such impacts solely on scientific grounds. How does one weigh human health against ecosystem integrity, or carcinogenicity against developmental toxicity, for example, without invoking a value judgment about which kinds of risks are worse or less acceptable? Nonetheless, to the extent that the array of trade-offs posed by alternative technological options can be articulated by applying the CEA approach, risk managers would be in a better position to make informed decisions. Moreover, even if factors such as social policy, energy independence, or other nonenvironmental considerations outweigh potential environmental impacts in decisions regarding fuel options, the CEA approach can guide risk management efforts to mitigate potential environmental impacts associated with different options. For example, changes in feedstocks, manufacturing processes, or other facets of the life cycle might be made to reduce certain impacts.

Another benefit of the CEA approach is that it can identify data needs and information gaps that could be filled through appropriately directed information collection and research efforts. In some cases, basic efforts directed at data collection and analysis could fill critical information gaps. In particular, environmental monitoring is essential to determine whether excessive releases or emissions are associated with a given choice among fuel options. In addition, environmental monitoring could confirm whether or not expected reductions in pollutant levels result from adopting a particular type of technology. In other instances, targeted research efforts may be necessary to resolve key questions identified by the CEA approach.

Monitoring, research, and other sources of information can be used to refine CEA findings. Consequently, the CEA approach is iterative. It must be repeated periodically as more information becomes available to refine our understanding of previously identified issues and as new issues come to light. Furthermore, CEA is consistent with and supportive of adaptive management approaches.[55]

CONCLUSION

The MTBE experience has provided many important lessons, especially regarding the need to evaluate the potential environmental impacts of fuel and other technological options in a comprehensive, comparative manner. The approach outlined here incorporates a life-cycle perspective in combination with the risk assessment paradigm to provide more useful information to assist risk managers in making decisions about alternative technologies.

ACKNOWLEDGMENTS

Several EPA colleagues provided invaluable intellectual contributions to the creation of the *Alternative Fuels Research Strategy*,[24] especially Dr. Judith A. Graham. The authors thank multiple reviewers for comments and input for this manuscript and previous versions.

REFERENCES

1. U.S. Environmental Protection Agency. 1998. Oxygenates in Water: critical Information and Research Needs. EPA/600/R-98/048. U.S. EPA; Office of Research and Development. Washington, DC. Available at http://www.epa.gov/ncea/pdfs/oxy_h2o.pdf
2. GRAHAM, J. 1995. Overview of the health-related issues and framework for the research. *In* Proceedings of the Conference on MTBE and Other Oxygenates: a Research Update. Falls Church, VA, July 26–28, 1993: p. 10–15. U.S. EPA; Office of Research and Development; report EPA/600/R-95/134. Research Triangle Park, NC. Available at http://www.epa.gov/ncea/pdfs/mtbe/0850-A.pdf
3. U.S. Environmental Protection Agency. 1995. Proceedings of the Conference on MTBE and Other Oxygenates: a Research Update July 26–28, 1993. Falls Church, VA. EPA/600/R-95/134. U.S. EPA; Office of Research and Development. Research Triangle Park, NC. Available at http://www.epa.gov/ncea/pdfs/mtbe/0850-A.pdf
4. DUFFY, J.S., J.A. DEL PUP & J.J. KNEISS. 1992. Toxicological evaluation of methyl tertiary butyl ether (MTBE): testing performed under TSCA consent agreement. J. Soil Contam. **1:** 29–37.
5. IRIS (Integrated Risk Information System) (database). 1993 Printout of reference concentration for chronic inhalation exposure (RfC): methyl tert-butyl ether. Available at http://www.epa.gov/iris.
6. U.S. Environmental Protection Agency. 1993. Assessment of Potential Health Risks of Gasoline Oxygenated With Methyl Tertiary Butyl Ether (MTBE). EPA/600/R-93/206. U.S. Environmental Protection Agency; Office of Research and Development. Washington, DC. Available at http://www.epa.gov/ncea/pdfs/mtbe/gasmtbe.pdf
7. U.S. Environmental Protection Agency. 1994. Health Risk Perspectives on Fuel Oxygenates. EPA 600/R-94/217. U.S. EPA; Office of Research and Development.

Research Triangle Park, NC. Available at http://www.epa.gov/ncea/pdfs/mtbe/oxyrisk.pdf
8. Health Effects Institute. 1996. The potential health effects of oxygenates added to gasoline: a review of the current literature, a special report of the institute's oxygenates evaluation committee. Health Effects Institute. Cambridge, MA. Available at http://www.healtheffects.org/Pubs/oxysum.htm.
9. National Research Council. 1996. Toxicological and Performance Aspects of Oxygenated Motor Vehicle Fuels. National Academy Press. Washington, DC. Available at http://darwin.nap.edu/books/0309055458/html/.
10. Interagency Oxygenated Fuels Assessment Steering Committee. 1997. Interagency assessment of oxygenated fuels. National science and technology council, committee on environment and natural resources and office of science and technology policy. Washington, DC. Available at http://www.epa.gov/oms/regs/fuels/ostpfin.pdf
11. FIEDLER, N., K. KELLY-MCNEIL, S. MOHR, et al. 2000. Controlled human exposure to methyl tertiary butyl ether in gasoline: symptoms, psychophysiologic and neurobehavioral responses of self-reported sensitive persons. Environ. Health Perspect. **108:** 753–763. Available at http://ehp.niehs.nih.gov/members/2000/108p753-763fiedler/108p753.pdf
12. DAVIS, J.M. & W.H. FARLAND. 2001. The paradoxes of MTBE. Toxicol. Sci. **61:** 211–217.
13. ROTHENSTEIN, C. 2004. FY 2004 Semi-Annual Mid-Year Activity Report (memo to EPA UST/LUST Regional Division Directors, Regions 1–10). EPA UST/LUST Regional Division Directors, Regions 1–10. May 13, 2004. Available at http://www.epa.gov/oust/cat/ca_04_12.pdf
14. MCKINNON, R.J. & J.E. DYKSEN. 1984. Removing organics from groundwater through aeration plus GAC. J. Am. Water Works Assoc. **76:** 42–47.
15. SQUILLACE, P.J., J.S. ZOGORSKI, W.G. WILBER & C.V. PRICE. 1996. Preliminary assessment of the occurrence and possible sources of MTBE in groundwater in the United States, 1993–1994. Environ. Sci. Technol. **30:** 1721–1730.
16. HAPPEL, A.M., E.H. BECKENBACK & R.U. HALDEN. 1998. An Evaluation of MTBE Impacts to California Groundwater Resources. UCRL-AR-130897. Lawrence Livermore National Laboratory. Livermore, CA.
17. State of Maine. 1998. The presence of MTBE and other gasoline compounds in maine's drinking water supplies—a preliminary report. Bureau of Health, Department of Human Services; Bureau of Waste Management & Remediation, Department of Environmental Protection; Maine Geological Survey, Department of Conservation. Augusta, ME. Available at http://www.maine.gov/dhs/ehu/wells/MTBE.PDF.
18. BUSCHECK, T.E., D.J. GALLAGHER, D.L. KUEHNE & C.R. ZUSPAN. 1998. Occurrence and behavior of MTBE in groundwater. In Underground Storage Tank Conference: '98 & Beyond. State of California Water Resources Control Board. Sacramento, CA.
19. DOYLE, J. & S. SWARD. 1998. MTBE leaks a ticking bomb: gas additive taints water nationwide. San Francisco Chronicle. December 14, 1998.
20. Blue Ribbon Panel on Oxygenates in Gasoline. 1999. Achieving Clean Air and Clean Water: The Report of the Blue Ribbon Panel on Oxygenates in Gasoline. EPA420-R-99-021. U.S. Environmental Protection Agency. Washington, DC. September 15, 1999. Available at http://www.epa.gov/otaq/consumer/fuels/oxypanel/r99021.pdf

21. U.S. Environmental Protection Agency. 2000. Methyl Tertiary Butyl Ether (MTBE); Advance Notice of Intent to Initiate Rulemaking Under the Toxic Substances Control Act to Eliminate or Limit the Use of MTBE as a Fuel Additive in Gasoline. Fed. Regist. **65:** 16093–16109. Available at http://www.epa.gov/fedrgstr/EPA-TOX/2000/March/Day-24/t7323.htm.
22. FRANKLIN, P.M., C.P. KOSHLAND, D. LUCAS & R.F. SAWYER. 2000. Clearing the air: using scientific information to regulate reformulated fuels. Environ. Sci. Technol. **34:** 3857–3863.
23. GOLDSTEIN, B. & S. ERDAL. 2000. Methyl tert-butyl ether as a gasoline oxygenate: lessons for environmental public policy. Ann. Rev. Energy Environ. **25:** 756–802.
24. U.S. Environmental Protection Agency. 1992. Alternative Fuels Research Strategy (External Review Draft). EPA 600/AP-92-002. Office of Research and Development; U.S. Environmental Protection Agency. Washington, DC. Available at http://www.epa.gov/ncea/pdfs/mtbe/altfuel.pdf
25. U.S. Environmental Protection Agency. 1997. Exposure Factors Handbook. EPA/600/P-95/002Fa. National Center for Environmental Assessment, Office of Research and Development, U.S. EPA. Washington, DC. Available at http://www.epa.gov/ncea/pdfs/efh/front.pdf
26. WILLIAMS, P.R.D., C.A. CUSHING & P.J. SHEEHAN. 2003. Data available for evaluating the risks and benefits of MTBE and ethanol as alternative fuel oxygenates. Risk Anal. **23:** 1085–1115.
27. WILLIAMS, P.R.D., P.K. SCOTT, S.M. HAYS & D.J. PAUSTENBACH. 2000. A screening level assessment of household exposures to MTBE in California drinking water. Soil Sediment Groundwater, March (MTBE Special Issue): pp. 63–69.
28. BARE, J.C., G.A. NORRIS, D.W. PENNINGTON & T. McKONE. 2002. TRACI: the tool for the reduction and assessment of chemical and other environmental impacts. J. Indus. Ecol. **6:** 49–78.
29. U.S. Congress. 2005. Energy Policy Act of 2005. PL 109-58. U.S. Government Printing Office. Washington, DC. Available at http://frwebgate.access.gpo.gov/cgi-bin/getdoc.cgi?dbname=109_cong_bills&docid=f:h6enr.txt.pdf
30. U.S. Environmental Protection Agency. 2006. Consumer Factsheet on: ATRAZINE. U.S. Environmental Protection Agency and Office of Ground Water and Drinking Water. National Primary Drinking Water Regulations. February 21, 2006. Available at http://www.epa.gov/safewater/contaminants/dw_contamfs/atrazine.html.
31. U.S. Department of Energy. 2002. Supply Impacts of an MTBE Ban. Energy Information Agency, U.S. Department of Energy. Available at http://www.eia.doe.gov/oiaf/servicerpt/fuel/mtbe.html.
32. Minnesota Pollution Control Agency. 2002. Air Emissions From Ethanol Plants. Air Quality/General/#1.21/October. Available at http://www.pca.state.mn.us/publications/aq1-21.pdf.
33. POWERS, S.E., D. RICE, B. DOOHER & P.J.J. ALVAREZ. 2001. Will ethanol-blended gasoline affect groundwater quality? Using ethanol instead of MTBE as a gasoline oxygenate could be less harmful to the environment.
34. ALVAREZ, P.J.J. & C.S. HUNT. 2002. The effect of fuel alcohol on monoaromatic hydrocarbon biodegradation and natural attenuation. Rev. Latinoam. Microbiol. **44:** 83–104.
35. GSCHWEND, P. 1999. Alkylates: environmental issues. Presentation for Blue Ribbon Panel to Review the Use of Oxygenates in Gasoline. Washington, DC

May 24, 1999. Available at http://www.epa.gov/otaq/consumer/fuels/oxypanel/blueribb.htm#Washington2.
36. FRENCH, R. & P. MALONE. 2004. Phase equilibria of ethanol fuel blends. Fluid Phase Equilibria **226:** 97–110.
37. Auto/Oil Air Quality Improvement Research Program. 1997. Auto/Oil Air Quality Improvement Research Program: Program Final Report. Coordinating Research Council, Auto/Oil Air Quality Improvement Research Program. Atlanta, GA.
38. HARLEY, R.A., A.G. RUSSELL, G.J. MCRAE, et al. 1993. Photochemical modeling of the Southern California Air Quality Study. Environ. Sci. Technol. **27:** 378–388.
39. SCHIFTER, I., M. VERA, L. DIAZ, et al. 2001. Environmental implications on the oxygenation of gasoline with ethanol in the metropolitan area of Mexico City. Environ. Sci. Technol. **35:** 1893–1901.
40. ANDRADE, M.D., R.Y. YNOUE, R. HARLEY & A. MIGUEL. 2004. Air quality model simulating photochemical formation of pollutants: the Sao Paulo metropolitan area, Brazil. Int. J. Environ. Pollut. **22:** 460–475.
41. DEEB, R.A., K.H. CHU, T. SHIH, et al. 2003. MTBE and other oxygenates: environmental sources, analysis, occurrence, and treatment. Environ. Eng. Sci. **20:** 433–447.
42. RICE, D.W., S.E. POWERS & P.J.J. ALVAREZ. 1999. Health and Environmental Assessment of the Use of Ethanol As a Fuel Oxygenate: Report to the California Environmental Policy Council in Response to Executive Order D-5-99. Vol. 4: Potential Ground Water and Surface Water Impacts. Available at http://www-erd.llnl.gov/ethanol/etohdoc/vol4/chap01.pdf
43. NISHIOKA, Y., J.I. LEVY, G.A. NORRIS, et al. 2002. Integrating risk assessment and life cycle assessment: a case study of insulation. Risk Anal. **22:** 1003–1017.
44. SONNEMAN, G., F. CASTELLS & M. SCHUHMACHER. 2004. Integrated Life-Cycle and Risk Assessment for Industrial Processes. Lewis Publishers. Boca Raton, FL.
45. ROW, J., M. RAYNOLDS, G. WOLOSHYNIUK, et al. 2002. Life-cycle value assessment (LCVA) of fuel supply options for fuel cell vehicles in Canada. Drayton Valley, AB, Canada: The Pembina Institute Available: http://www.pembina.org/pdf/publications/report020610.pdf [25 February, 2004].
46. BRINKMAN, N., M. WANG, T. WEBER & T. DARLINGTON. 2005. Well-to-Wheels Analysis of Advanced Fuel/Vehicle Systems—a North American Study of Energy Use, Greenhouse Gas Emissions, and Criteria Pollutant Emissions. Argonne National Laboratory. Available at http://www.transportation.anl.gov/pdfs/TA/339.pdf
47. U.S. Department of Energy. 2005. The Greenhouse Gases, Regulated Emissions, and Energy Use in Transportation (GREET) Model. U.S. Department of Energy; Transportation Technology R&D Center; Argonne National Lab. Available at http://www.transportation.anl.gov/software/GREET/index.html.
48. REINHARDT, G.A. & N.C. JUNGK. 2000. Which bioethanol is best? An ecological comparison of bioethanol from different crops versus conventional fuel. *In* Proceedings of the 13th International Symposium on Alcohol Fuels (ISAF XIII), Implementing the Transition to a Sustainable Transport System; July 2000.: p. 1–11. ISAF XIII. Stockholm, Sweden.
49. PATYK, A. 2000. Finding the best option for the environment—a comparison of 22 combinations of conventional and alternative fuels and drive systems. *In* Proceedings of the 13th International Symposium on Alcohol Fuels (ISAF XIII),

Implementing the Transition to a Sustainable Transport System; July 2000: p. 1–8. ISAF XIII. Stockholm, Sweden.
50. THOMAS, V., T. THEIS, R. LIFSET, et al. 2006. Industrial ecology: policy potential and research needs. Environ. Eng. Sci. **20:** 1–9.
51. THOMAS, V.M. & T.E. GRAEDEL. 2003. Research issues in sustainable consumption: toward an analytical framework for materials and the environment. Environ. Sci. Technol. **37:** 5383–5388.
52. SHATKIN, J.A. & S. QIAN. 2004. Classification schemes for priority setting and decision making: a selected review of expert judgment, rule-based, and prototype methods. *In* Comparative Risk Assessment and Environmental Decision Making. Linkov, I. & A. Ramadan, Eds.: 213–244. Kluewer. Amsterdam.
53. HERTWICH, E.G. & J.K. HAMMITT. 2001. A decision-analytic framework for impact assessment. Part 1. LCA and decision analysis. Int. J. Life Cycle Assess. **6:** 5–12.
54. HERTWICH, E.G. & J. HAMMIT. 2001. A decision-analytic framework for impact assessment, Part 2: midpoints, endpoints, and criteria for method development. Int. J. Life Cycle Assess. **6:** 265–272.
55. LINKOV, I., F.K. SATTERSTROM, G.A. KIKER, et al. 2005. From optimization to adaptation: shifting paradigms in environmental management and their application to remedial decisions. Integr. Environ. Assess. Manag. **1:** 1–7.

Environmental Impacts and Costs of Energy

ARI RABL AND JOSEPH V. SPADARO

Ecole des Mines, F-75272 Paris, Cedex, France

ABSTRACT: Environmental damage is one of the main justifications for continued efforts to reduce energy consumption and to shift to cleaner sources such as solar energy. In recent years there has been much progress in the analysis of environmental damages, in particular thanks to the ExternE (External Costs of Energy) Project of the European Commission. This article presents a summary of the methodology and key results for the external costs of the major energy technologies. Even though the uncertainties are large, the results provide substantial evidence that the classical air pollutants (particles, No$_x$, and SO$_2$) from fossil fuels impose significant public health costs, comparable to the cost of global warming from CO$_2$ emissions. The total external costs are relatively low for natural gas (in the range of about 0.5–1 eurocents/kWh for most EU countries), but much higher for coal and lignite (in the range of about 2–6 eurocents/kWh for most EU countries). By contrast, the external costs of nuclear, wind, and photovoltaics are very low. The external costs of hydro are extremely variable from site to site, and the ones of biomass depend strongly on the specific technologies used and can be quite large for combustion.

KEYWORDS: environmental damage; external costs; dose–response functions; atmospheric dispersion models; electricity production; impact pathway analysis; life-cycle assessment

INTRODUCTION

Air pollution causes considerable damage to human health, flora and fauna, and materials. The damage costs are externalities to the extent that they are not reflected in the prices of goods. The damage costs, also called external costs, should be taken into account in decisions that affect the emission of pollutants. The external costs can be internalized via taxes, tradable permits, or other environmental regulations.

In recent years there has been much progress in the analysis of environmental damage costs thanks to several major projects to evaluate the ExternE (External

Address for correspondence: Ari Rabl, Ecole des Mines, 60 boulevard St. Michel, F-75272 Paris, Cedex 06, France. Voice: 33-1-4051-9152; fax: 33-1-4634-2491.
 e-mail: ari.rabl@ensmp.fr

Costs of Energy) in Europe[1–3] and also in the United States.[4,5] Of these, the ExternE Project of the European Commission has the widest scope and is the most up-to-date. The present article, by participants of all ExternE Projects since 1992, presents an overview of the methodology and results for electricity production.

The quantification of damage costs has many important applications:

(1) Guidance for environmental regulations, for example, determining the optimal level of the limit for pollutant emission;
(2) Finding the socially optimal level of a pollution tax;
(3) Identify technologies with the lowest societal cost, for example, coal, natural gas, or nuclear for electricity production;
(4) Evaluate the benefits of improving pollution abatement of an existing installation such as a waste incinerator;
(5) Optimize the dispatching of power plants (i.e., which power plants are used at what time);
(6) "Green accounting," that is, including corrections for environmental damage in the traditional accounts of GNP.

DAMAGE COSTS—METHODOLOGY

Here we can only give a very brief discussion of the key features of the methodology used for the analysis of impact pathways, that is, of the chain emission—dispersion—impact—cost. For a more complete presentation we refer to the reports of the ExternE Project or to review papers.[6,7]

Over the years, numerous dispersion models have been developed. Usually separate models are used for the local and the regional domains. In the local domain, up to about 50 km from the source, pollutant deposition and aerosol formation by chemical transformation are relatively insignificant and concentrations are influenced primarily by meteorological parameters, such as wind speed and wind direction. Beyond 50 km, one must account for removal of the pollutant from the air by chemical reactions and deposition, both dry and wet.

For atmospheric dispersion ExternE uses the Gaussian plume model ISC (Industrial Source Complex)[8] at the local scale. Regional concentrations are calculated using the Lagrangian trajectory models of EMEP (European Monitoring and Evaluation Programme) (see for example Ref. 9) and of the Windrose Trajectory Model (WTM); the latter is an adaptation of the Harwell Trajectory Model.[10] ISC and WTM are implemented in the EcoSense software[11] used for the impact calculations of ExternE. EcoSense also contains databases for receptors.

Impacts are quantified using dose–response functions, also known as exposure–effect, exposure–response, or concentration–response functions (CRF) in the case of air pollutants. They relate the pollutant concentration to the resulting impact on a receptor (human health, crop, etc.). Impacts on human health include asthma attacks, hospital admissions, chronic bronchitis, restricted activity days, and mortality. ExternE calculates mortality impacts of air pollution as a reduction in life expectancy, expressed as Years Of Life Lost (YOLL), rather than a number of premature deaths. That is necessary to allow more meaningful comparisons with other causes of death, for instance accidents for which the YOLL per death are much higher than for air pollution (of course such comparisons are not perfect because the affected individuals differ in age and health status).

For health impacts, the CRFs are derived from a survey of epidemiological studies.[2] In view of the available epidemiological evidence, we assume that the CRFs are approximately straight lines, without threshold, for the air pollutants of greatest concern here, especially particulate matter (PM).[12] Of course, epidemiology is very uncertain at low doses or concentrations and the linear model may not be correct; however, there is no clear evidence why other models would be better. Also, if there is a threshold below current concentrations it has no effect on the calculation of incremental damage costs (i.e., for changes relative to current conditions, which is what is reported in external cost studies and here).

For crops and building materials, the CRFs have nonlinear shapes. For agricultural crops there is even the possibility of a small benefit (fertilizer effect) when the background concentrations of SO_2 and NO_x are sufficiently low. For crops, one calculates the losses or gains in yield, and for building materials, the increase in cleaning and repair costs due to air pollution.

Monetization is a convenient method for aggregating health impacts and environmental burdens with different physical units into a single damage indicator. To obtain the damage costs, one multiplies the number of impacts (e.g., cases of asthma attack) by the unit cost per impact, for example,€ per asthma attack.

For health impacts, the unit costs include the cost of treatment and wage and productivity losses, which are market based, as well as nonmarket costs that take into account an individual's Willingness-to-Pay (WTP) to avoid the risk of pain and suffering. If the WTP for a nonmarket good has been determined correctly, it is like a price, consistent with prices paid for market goods. Economists have developed several techniques for valuing nonmarket goods. In recent years, contingent valuation (CV) has become the method of choice; it obtains WTP estimates by asking individuals how much they are willing to pay to achieve a benefit.[13] This method is problematic and the uncertainties are large, but for most environmental nonmarket costs no better alternative is available.

The most important impact is mortality. As ExternE evaluates mortality according to the reduction of life expectancy, one needs the Value Of a Life Year (VOLY). But by contrast to the numerous studies of so-called Value of Statistical Life (VSL) (an unfortunate and often misunderstood name for what is really the "willingness to pay for avoiding the risk of an anonymous premature death"), there have been very few studies until now to determine VOLY. For the 1998 and 2000 reports ExternE had calculated VOLY by assuming that VSL is a discounted sum of annual VOLYs; choosing 3.4 million € for VSL (a weighted mean of European studies), this implied a VOLY of approximately 100,000 €/life year. The current choice for VOLY is 50,000 €/life year, based on a CV by the ExternE team.[3] This study was carried out because at the time there was no CV for VOLY in Europe; it was based on a questionnaire that had already been tested in Canada and it was carried out in France, Italy, and the United Kingdom by teams at Université de Paris 1, FEEM, and the University of Bath.

The unit costs for crops and building materials are based on market prices. As crop damages are relatively small, they are estimated simply on the basis of quantity times constant price, without consideration of induced effects (compensatory producer behavior). The uncertainties of the current monetary values are large (e.g., they are not determined by general equilibrium techniques, and costs of global warming depend on controversial assumptions about VSL in poorer countries and intergenerational discounting).

The damage calculations distinguish the upstream emissions from those during the utilization phase. This point is especially important for primary pollutants such as PM: the damage per kg of pollutant is much higher if the pollutant is emitted at street level in a large city rather than from a tall stack in a rural zone. For secondary pollutants (created by chemical transformation in the atmosphere) this variation with site is much weaker. For upstream emissions we assume typical industrial installations in Europe.

The damage costs quantified here cover global warming for the greenhouse gases (for energy only CO_2, N_2O, and CH_4 are significant), and human health, crop losses, and damage to materials for the other pollutants. While we believe that these are the most important damage categories, it is difficult to be complete in this type of work. Among missing categories we mention reduction of visibility due to air pollution (generally deemed of less concern in the EU than the United States), noise (gas buses are quieter than the diesel buses), road damage (gas buses are heavier than the diesel buses), and impacts caused by the use of oil (oil spills, supply security, etc.).

UWM: A SIMPLE MODEL FOR DAMAGE COST ESTIMATION

A simple and convenient tool for the development of typical values is the "uniform world model" (UWM), first presented by Curtiss and Rabl[14] and further developed, with detailed validation studies, by Spadaro[15] and Spadaro and Rabl.[16] More recently Spadaro and Rabl[17] have extended it to toxic metals and their pathways through the food chain. The UWM is a product of a few factors; it is simple and transparent, showing at a glance the role of the most important parameters of the impact pathway analysis. It is exact for tall stacks in the limit where the distribution of either the sources or the receptors is uniform and the key atmospheric parameters do not vary with location. In practice the agreement with detailed models is usually within a factor of two for stack heights above 50 m. For policy applications one needs typical values, and the UWM is more relevant than a detailed analysis for a specific site.

The UWM for the damage cost D_{uni} of a particular impact due to the inhalation of a primary pollutant is shown in this equation

$$D_{uni} = p \; s_{CR} \; \rho / v_{dep}$$

where

p = cost per case (price) (€/case),
s_{CR} = CRF slope ([cases/yr]/[person ($\mu g/m^3$]),
ρ = average population density (person/m^2) within 500–1000 km of source, and
v_{dep} = deposition velocity of pollutant (dry + wet) (m/sc).

The receptor density is the average over land and water. For secondary pollutants the equation has the same form, but with an effective deposition velocity that includes the transformation rate of the primary into the secondary pollutant. With this model it is easy to transfer to the results from one region to another (assuming that CRF and deposition velocity are the same): simply rescale the result in proportion to the regional average receptor density ρ and the cost per case.

Of course, there is no dependence on site or stack height for globally dispersing pollutants such as CO_2. For As, Hg, Pb, and dioxins the variation with site, for a given ρ, is in the range of about 0.7–1.5, small because noninhalation pathways dominate. Variation with site, for a given ρ is also small for secondary pollutants, a range of about 0.5–2.0 because the formation of the secondary pollutants is slow and occurs mostly far from the source. Variation with stack height is negligible for noninhalation pathways and for secondary aerosols (nitrates and sulfates).

For primary air pollutants the variation with site and stack height is strong and the result of D_{uni} can be improved by using the following correction factors: a range of 0.5–5 for site (higher if near big city), and 0.6–3 for stack conditions (higher for low stacks, up to 15 for ground level emissions in big city).

These correction factors have been derived by evaluating the results of more than a hundred detailed EcoSense calculations. Of course, such rules can only yield rough estimates; site-specific calculations should be carried out when more precise results are needed.

RESULTS FOR COST PER KG OF POLLUTANT

The impacts quantified by ExternE so far are global warming, health, damage to buildings and materials, and loss of agricultural production. Apart from

TABLE 1. Impacts evaluated and key assumptions[3]

Atmospheric dispersion models	
Local range:	Gaussian plume model ISC.
Regional range (Europe):	Harwell Trajectory Model as implemented in ECOSENSE software of ExternE. Ozone impacts based on EMEP model.
Global warming	19 €/t_{CO2eq}
Impacts on health	
Form of dose–response functions	Linearity without threshold.
Chronic mortality	CRF slope f_{CR} = 4.1E-4 YOLL per person per year per μg/m^3 derived from increase in age-specific mortality due to $PM_{2.5}$,[18] by integrating over age distribution.
Acute mortality	For SO_2 and ozone, with 0.75 YOLL per death.
Nitrate and sulfate aerosols	CRF for nitrates same as for PM_{10}. CRF for sulfates same as for $PM_{2.5}$ (slope = 1.7 times slope of PM_{10} functions).
Micropollutants	Cancers due to As, Cd, Cr, Ni, dioxins, benzene, butadiene; IQ decrement due to Pb.
Impacts on plants	Loss of crops due to SO_2 and ozone.
Impacts on buildings and materials	Corrosion and erosion due to SO_2 and soiling due to particles.
Impacts not quantified but potentially significant	Reduced visibility due to air pollution; eutrophication and acidification; disposal of residues from fossil fuels or incineration.
Monetary valuation	Based on WTP to avoid a damage
Valuation of premature death	Proportional to reduction of life expectancy, with VOLY = 50,000 €.
Valuation of cancers	2 M€ per cancer.

global warming due to CO_2, CH_4, and N_2O, more than 95% of the costs are due to health impacts, especially mortality. Morbidity (especially chronic bronchitis, but also asthma, hospital admissions, etc.) accounts for almost 30% of the damage cost of PM, NO_x, and SO_2. The impacts evaluated and the key assumptions are listed in TABLE 1. The resulting damage costs in €/kg of pollutant are shown in FIGURE 1 for typical sources with stack heights above 50 m in Central Europe; for $PM_{2.5}$ ground level sources are also shown.

RESULTS FOR ENERGY PRODUCTION

Once the cost per kg has been determined, multiplication by the emitted quantities of the pollutants yields the cost per activity, for instance per kWh of electricity produced by a power plant. A few results for power plants in France are shown in FIGURE 2.

A complete accounting of the damage costs should involve a life-cycle assessment (LCA), that is, a complete inventory of emissions over the entire chain of processes involved in the activity, such as air pollution from the ships,

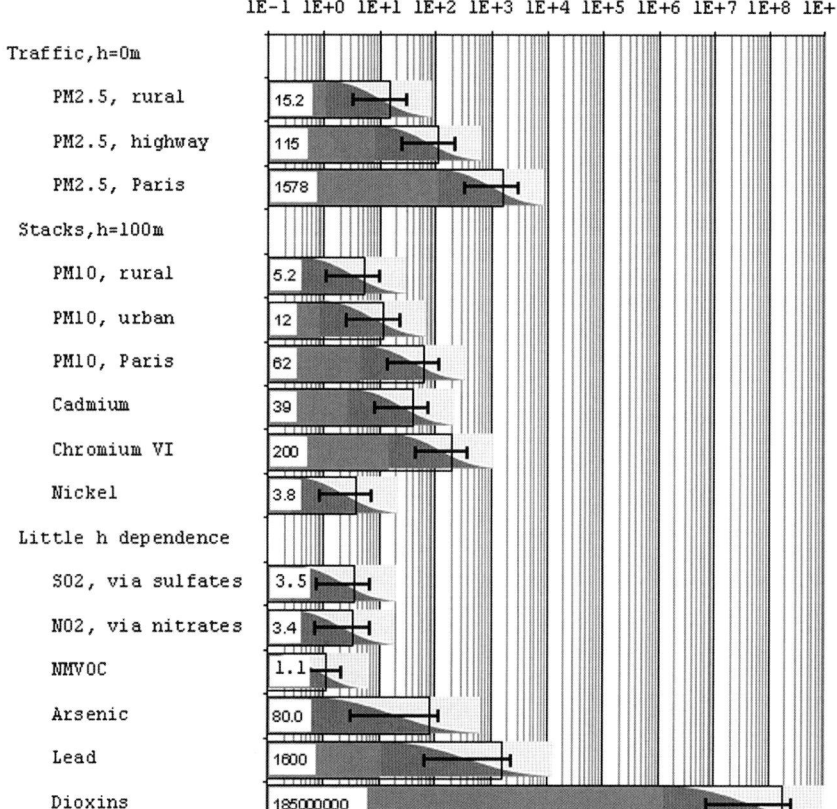

FIGURE 1. Results for damage costs of the most important air pollutants (typical values, for LCA applications in the EU15).[3] The error bars indicate the 68% confidence interval[21]; on the logarithmic scale they are symmetric around the median (= geometric mean of lognormal distribution). The broad hollow bars and the numbers are the mean which is larger than the median. The gray S-shaped curve indicates the probability that the true cost is above a specified value.

trucks, or trains that transport the fuel to the power plant. That has been done for the fuel chain results of ExternE.

Recently updated and more complete results for power production in Europe have been calculated during the ExternE-Pol phase of the ExternE series[19]; they are shown in FIGURES 3 and 4. These figures include the entire fuel cycle, based on detailed LCA inventories provided by the Ecoinvent database, by contrast to FIGURE 2 which includes only the power plants. Also, the methodology for nuclear damage costs in FIGURE 3 is different from the one in FIGURE 2.

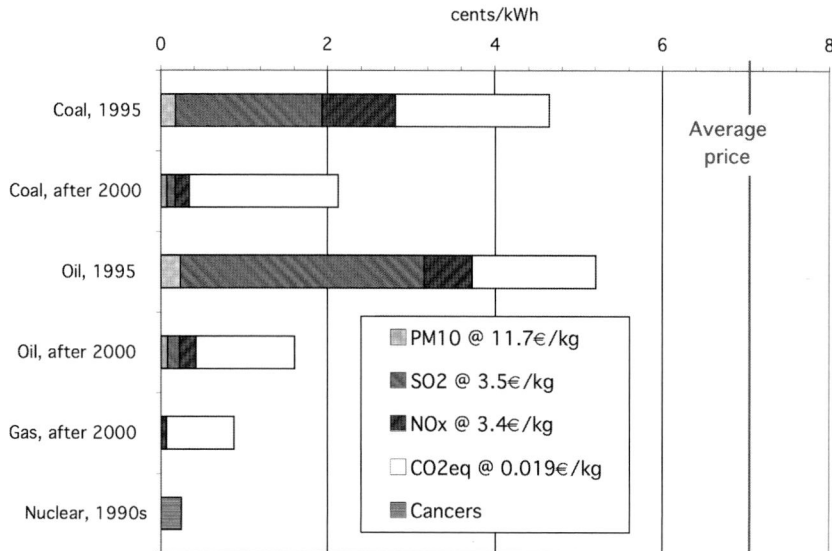

FIGURE 2. Typical damage costs for electric power plants in France, both for plants operating during the mid-90s and for plants that respect the new EU regulations issued in 2000. Costs for nuclear are upper bound (0% effective discount rate); 90% of cancers from nuclear occur only after the first 100 year. Retail price of electricity is about 7 eurocents/kWh, production cost of base load electricity about 3 eurocents/kWh.

For the fossil fuel chains by far the largest portion of the external costs comes from air pollutants emitted by the power plant, the main impact categories being global warming and public health. Air pollutants from upstream and downstream activities make a relatively small contribution (roughly 10% of the greenhouse gases).

The emission of toxic metals is highly uncertain and variable from one source of fuel to another; the numbers shown are not necessarily typical. However, they are in any case so small that their contribution to the total damage cost is negligible.

CONCLUSIONS

A key argument for the necessity of quantifying damage costs has emerged from recent epidemiology. Whereas in the past there was a general belief that it would be sufficient to reduce the concentration of air pollutants below their no-effect thresholds, epidemiological studies have not been able to find evidence for such no-effect thresholds at the population level for the most important air pollutants, and linear dose–response functions appear increasingly plausible (although other forms cannot be ruled out by epidemiological studies

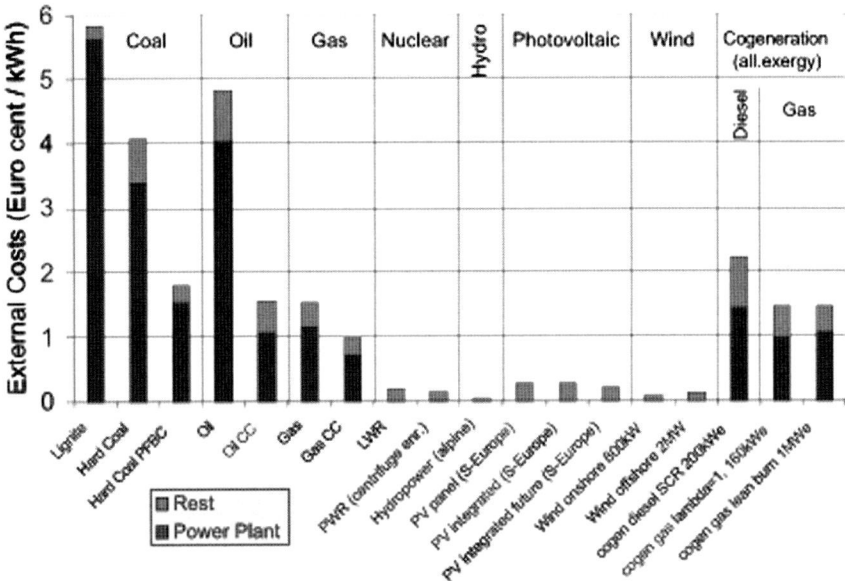

FIGURE 3. Damage costs of current and advanced electricity systems, associated with emissions from the operation of power plant and with the rest of energy chain.[19]

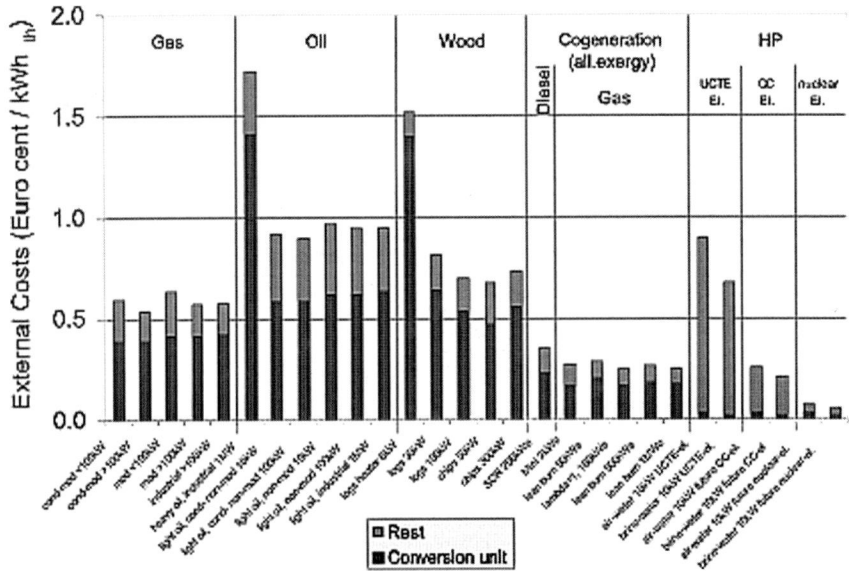

FIGURE 4. Damage costs of heating systems, associated with emissions from the operation of boiler/cogeneration unit and with the rest of energy chain.[19]

because of their large uncertainties at low doses or concentrations). If there are no safe levels of air pollution, we have no natural criterion for deciding how far to reduce the emission of pollutants. As the cost per kg of avoided pollutant increases sharply as emissions are reduced, there is a serious risk of spending too much on the fight against air pollution. This could in fact lead to a deterioration of public health if the money thus spent is not available for more cost-effective measures. A comparison of costs and benefits is needed for rational policy making.

The most relevant damages caused by air pollution can be quantified and monetized using the methodology of ExternE. However, damage cost estimates still show large uncertainties, and the reader may wonder whether it is meaningful to use them as basis for decisions. The first reply is that even a threefold uncertainty is better than infinite uncertainty. Second, in many cases the benefits are either so much larger or so much smaller than the costs that the implication for a decision is clear even in the face of uncertainty. Third, if in the other cases the decisions are made without a significant bias in favor of either costs or benefits, some of the resulting decisions will err on the side of costs, others on the side of benefits. Rabl, Spadaro and van der Zwaan[20] have examined the consequences of such unbiased errors and found a very reassuring result: the extra social cost incurred because of uncertain damage costs is remarkably small, less than 10–20% in most cases even if the damage costs are in error by a factor three according to the uncertainty estimates of Rabl and Spadaro.[21] But without any knowledge of the damage costs, the extra social cost (compared to the minimal social cost that one would incur with perfect knowledge) could be very large.

ACKNOWLEDGMENT

This work has been supported in part by the ExternE project series of the EC DG Research.

REFERENCES

1. EXTERN, E. 1995. ExternE: externalities of Energy. Published by European Commission, Directorate-General XII, Science Research and Development. Office for Official Publications of the European Communities, L-2920 Luxembourg.
2. EXTERN, E. 1998. ExternE: externalities of Energy. Vol. 7: Methodology 1998 Update (EUR 19083); Vol.8: Global Warming (EUR 18836); Vol.9: Fuel Cycles for Emerging and End-Use Technologies, Transport and Waste (EUR 18887); Vol.10: National Implementation (EUR 18528). Published by European Commission, Directorate-General XII, Science Research and Development. Office for Official Publications of the European Communities, L-2920 Luxembourg. Results are also available at http://ExternE.jrc.es/publica.html.

3. EXTERN, E. 2004. New results of ExternE, after the NewExt project. Available at http://www.externe.info
4. ORNL/RFF. 1994. External Costs and Benefits of Fuel Cycles. Prepared by Oak Ridge National Laboratory and Resources for the Future. Russell Lee, Ed.: Oak Ridge National Laboratory, Oak Ridge, TN 37831.
5. ROWE, R.D., C.M. LANG, L.G. CHESTNUT, et al. 1995. The New York Electricity Externality Study. Oceana Publications. Dobbs Ferry. NY.
6. RABL, A., J.V. SPADARO & P.D. MCGAVRAN. 1998. Health risks of air pollution from incinerators: a perspective. Waste Manag. Res. **16**: 365–388.
7. RABL, A. & J.V. SPADARO. 2000. Public health impact of air pollution and implications for the energy system. Annu. Rev. Energy Environ. 25: 601–627.
8. BRODE, R.W. & J.F. WANG, 1992. User's Guide for the Industrial Source Complex (ISC2) Dispersion Models Volumes I–III. EPA–450/4–92–008a, EPA–450/4–92–008b, EPA–450/4–92–008c, U.S. Environmental Protection Agency, Research Triangle Park, NC. p. 27711.
9. SIMPSON, D. 1992. Long period modelling of photochemical oxidants in Europe, Calculations for July 1985. Atmos. Environ. **26A**: 1609–1634.
10. DERWENT, R.G. & K. NODOP. 1986. Long-range transport and deposition of acidic nitrogen species in north-west Europe. Nature **324**: 356–358.
11. KREWITT, W., A. TRUKENMUELLER, P. MAYERHOFER, et al. 1995. ECOSENSE—an integrated tool for environmental impact analysis. In Space and Time in Environmental Information Systems. H. Kremers & W. Pillmann, Eds.: Umwelt-Informatik aktuell, Band 7. Metropolis-Verlag. Marburg.
12. WILSON, R. & J.D. SPENGLER, Eds. 1996. Particles in Our Air: Concentrations and Health Effects. Harvard University Press. Cambridge, MA.
13. MITCHELL, R.C. & R.T. CARSON. 1989. Using Surveys to Value Public Goods: the Contingent Valuation Method. Resources for the Future. Washington, DC.
14. CURTISS, P.S. & A. RABL. 1996. Impacts of air pollution: general relationships and site dependence. Atmos. Environ. **30**: 3331–3347.
15. SPADARO, J.V. 1999. Quantifying the effects of airborne pollution: impact models, sensitivity analyses and applications. Doctoral thesis, Ecole des Mines, 60 boul. St. Michel, Paris, France.
16. SPADARO, J.V. & A. RABL. 2002. Air pollution damage estimates: the cost per kilogram of pollutant. Int. J. Risk Assess. Manag. **3**: 75–98.
17. SPADARO, J.V. & A. RABL. 2004. Pathway analysis for population-total health impacts of toxic metal emissions. Risk Anal. **24**: 1121–1141.
18. POPE, C.A., R.T. BURNETT, M.J. THUN, et al. 2002. Lung cancer, cardiopulmonary mortality, and long-term exposure to fine particulate air pollution. JAMA **287**: 1132–1141.
19. RABL, A., J. SPADARO, P. BICKEL, et al. 2004. ExternE-Pol. Externalities of Energy: extension of accounting framework and policy applications. Final Report contract No. ENG1–CT2002–00609. EC DG Research. Brossels.
20. RABL, A., J.V. SPADARO & B. VAN DER ZWAAN 2005. Uncertainty of pollution damage cost estimates: to what extent does it matter? Environ. Sci. Technol. **39**: 399–408.
21. RABL, A. & J.V. SPADARO. 1999. Environmental damages and costs: an analysis of uncertainties. Environ. Int. **25**: 29–46.

Worldwide Governmental Efforts to Locate and Destroy Chemical Weapons and Weapons Materials

Minimizing Risk in Transport and Destruction

RALF TRAPP

Consultant; formerly Organisation for the Prohibition of Chemical Weapons

ABSTRACT: The article gives an overview on worldwide efforts to eliminate chemical weapons and facilities for their production in the context of the implementation of the 1997 Chemical Weapons Convention (CWC). It highlights the objectives of the Organisation for the Prohibition of Chemical Weapons (OPCW), the international agency set up in The Hague to implement the CWC, and provides an overview of the present status of implementation of the CWC requirements with respect to chemical weapons (CW) destruction under strict international verification. It addresses new requirements that result from an increased threat that terrorists might attempt to acquire or manufacture CW or related materials. The article provides an overview of risks associated with CW and their elimination, from storage or recovery to destruction. It differentiates between CW in stockpile and old/abandoned CW, and gives an overview on the factors and key processes that risk assessment, management, and communication need to address. This discussion is set in the overall context of the CWC that requires the completion of the destruction of all declared CW stockpiles by 2012 at the latest.

KEYWORDS: chemical weapons; destruction; risk; Chemical Weapons Convention

INTRODUCTION

The Chemical Weapons Convention (CWC) defines chemical weapons (CW) as toxic chemicals and their precursors (unless intended for purposes not prohibited and as long as their types and quantities correspond to such purposes), munitions, and devices specifically designed for their dissemination, and equipment specifically designed for the employment of such munitions or devices.[1] The term applies to them separately, but also together. CW utilize toxicity in humans as their primary effect. Toxicity is understood in its broad meaning (lethal, incapacitating, or otherwise harmful), but for the selection of toxic chemicals for CW purposes only acute toxicity is relevant; this is not so, however, with regard to the risk assessments of their storage, transportation, and destruction. CW include inter alia agents stored in bulk, agent-filled munitions, as well as unfilled chemical munitions, devices, and equipment specially designed for CW purposes.

A number of States possess CW stockpiles. Six of them have joined the 1997 CWC and have declared their stockpiles to the Organisation for the Prohibition of Chemical Weapons (OPCW); these States Parties are destroying their CW stockpiles as required under the CWC. There also remain "old and abandoned" chemical weapons (O/ACW) in many countries—CW that have been left behind in the past on former battlefields, test ranges, production or storage sites, or disposal areas. Such O/ACW, also sometimes called "non-stockpile CW," are usually in deteriorating condition and may be leaking. They often constitute a significant environmental and health hazard. In some cases, the fill remains toxic despite severe deterioration of the shell or container surrounding it and despite partial degradation of the agent.

CW pose considerable risks throughout their physical existence, from manufacture through storage to their ultimate disposal. These risks are essentially twofold: there are risks that emanate from the characteristics of the weapons themselves, and there are risks associated with activities to which they are subjected. There are risks if these weapons were to be used in conflict, both with regard to troops on the battlefield and civilians living nearby or downwind. There are also security concerns associated with CW stockpiles, and after September 11, 2001, the possibility has to be considered that terrorists may attempt to manufacture or otherwise acquire CW, or that they might attack CW stockpiles to release toxic chemicals into the environment.

For all these reasons, CW disarmament has long been an aspiration of the international community, aiming at the complete, irreversible, independently verifiable, and safe destruction of all CW and the dismantling of all facilities that have been used in the past to produce them.

THE CHEMICAL WEAPONS CONVENTION

The elimination of the menace of chemical warfare involves a range of policy options for States, from unilateral disarmament, de-proliferation, and the application of nonproliferation policies to the participation in the CWC. Security Council resolution 1540 requires all States take measures to ensure so that non-State actors such as terrorists or criminals cannot gain access to CW and related materials. This resolution, which was adopted under Chapter

VII of the UN Charter and binds all States, complements the disarmament and nonproliferation regime established by the CWC.

The CWC, the first global disarmament regime banning an entire category of weapons of mass destruction under strict international verification, entered into force on April 29, 1997. Today, 173 States participate in the regime.[2] Most of the countries with a history of past manufacture of CW or with chemical industries with the potential of being misused to that end, have now joined the treaty and are taking part in its disarmament and nonproliferation regime; however, some CW–capable States still remain outside the regime. Of particular concern are North Korea and some countries in the Middle East. Twelve States with former CW production capabilities have joined the treaty, declared their facilities, and are destroying them or converting them for peaceful purposes, subject to verification through on-site inspection by the OPCW. Six States Parties have declared CW stockpiles and are destroying them: of the 71.4 thousand metric tones of CW agents declared by them, 11.97 thousand tones (16.8 %) have already been destroyed. To provide assurances that these weapons have in fact been destroyed, and that no new CW are being manufactured by any of the States Parties, the OPCW has conducted more than 2200 on-site inspections at 862 sites located in 72 States Parties since 1997. More than 121,000 inspector days have so been utilized by the OPCW for verification purposes. The focus of that inspection effort has been, and continues to be, on verifying the destruction of CW at declared CW destruction facilities. This verification includes the confirmation of the agent amounts and identity, assurances that no CW have been diverted before their destruction, and confirmation of the completeness of destruction and of the integrity of the destruction process.

An overview on the progress made in CW disarmament is given in FIGURES 1 and 2. The data show that the progress made in CW destruction falls short of the target set in the CWC, which aims at eliminating all CW stockpiles within 10 years with an option of extending that deadline by up to 5 years. Based on the requirements of the CWC, therefore, one should expect an acceleration of the pace of CW destruction in coming years.

THE CWC AND THE CONCEPT OF RISK ASSESSMENT AND MANAGEMENT

The concept of risk is inherent in the CWC at several levels. The risk associated with the transportation of CW from stockpiles to destruction facilities, and the risk associated with their destruction, is only one aspect.

At the top level, there is what the Convention calls the "risk to the object and purpose of the Convention." Activities, facilities, items of equipment, and chemical compounds are addressed with respect to whether and how they might affect the overarching objective of CW disarmament within prescribed time

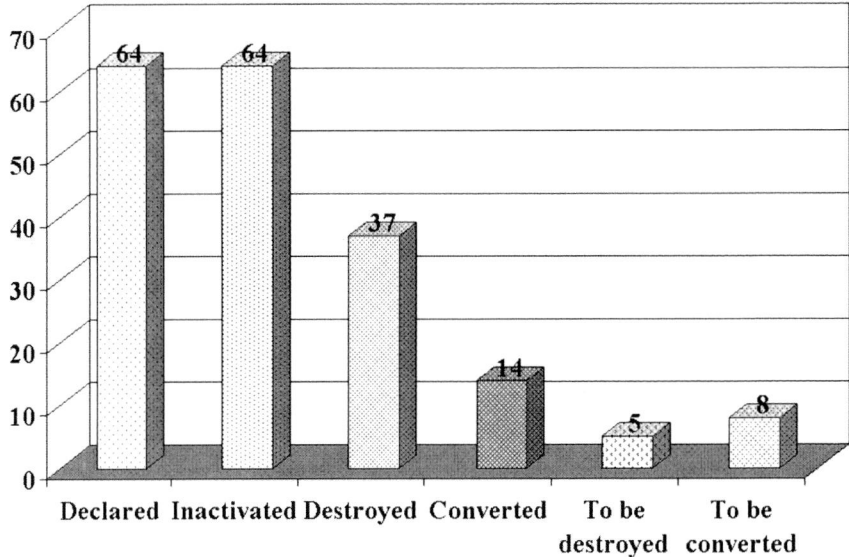

FIGURE 1. Status of destruction or conversion of former CW production facilities as of August 2005.

frames as well as the possibility of the re-emergence of new CW. For example, in the CWC's verification regime, risk is used as a concept to create different levels of assurances for different types of chemicals, equipment, facilities, and activities. The Schedules of Chemicals use risk as a parameter for classifying

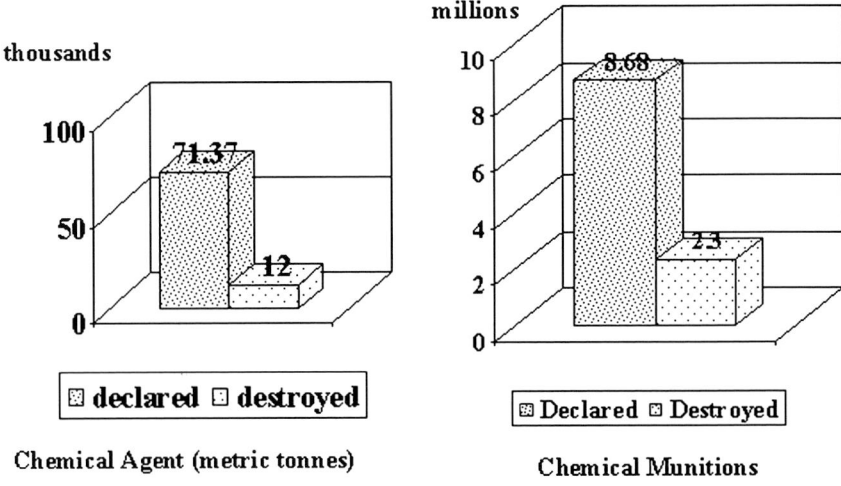

FIGURE 2. Status of destruction of declared CW storkpiles as of August 2005.

toxic chemicals and precursors into three different categories, each with different requirements and verification measures. In this context, risk is associated with the toxic and other properties of the chemicals, and the assessment is based on past experience from weaponization or use of a toxic chemical as a chemical weapon or on established and well-understood structural relationships between these chemicals and certain high-risk CW agents, for example, nerve agents.[3]

The CWC also uses risk in a strict technical sense, both with respect to the operations of the OPCW (in particular the conduct of inspections), and as applied to the timing of destruction obligations of the States Parties. Risk assessment and management are organic parts of the OPCW inspection methodology, to ensure the safety of the inspectors as well as of the people present at and around facilities being inspected, and of the environment. Risk assessment and management during inspections are basic responsibilities of OPCW inspection teams, notwithstanding the responsibilities for safety on-site that the operators of an inspected facility and the State Party being inspected have.[4] But the OPCW's responsibility for risk assessment and management does not extend to the destruction operations themselves. For the assessment, management, and communication of risks associated with the storage, transportation, and destruction of the CW, the possessor States Parties bear sole responsibility.[5] The OPCW can be used as a forum for consultation and cooperation between States Parties on these issues, if they decide so, but the OPCW does not establish or enforce safety standards relate to CW destruction operations. The following remarks therefore do not reflect any official OPCW policy or views.

RISKS ASSOCIATED WITH CW DESTRUCTION

As a first observation, risk assessment of CW destruction needs to address all the different factors and processes involved in a comprehensive and systematic manner, taking account of the characteristics and properties of the weapons as well as the circumstances of the locations where they are and the activities that are being carried out with them. It should also address technology choices at the different points in the chain from storage to completed destruction including waste treatment and disposal. Risk assessment therefore has to start with recognition that storage of CW is not risk-free.

This article does not attempt to provide an actual analysis of the issues related to risk assessment and management of CW and their destruction, which would require a much more extensive treatment. All that can be done in the space available is to highlight some of the issues that need to be assessed and to identify important parameters and processes that any such a risk assessment should consider.

CW STORAGE

When one looks at the risk associated with CW storage, it should be recalled that there is an overarching risk associated with State possession of CW, which was the reason for the international efforts to create a legal framework within which States have agreed to destroy these weapons under international verification within agreed time frames. This, of course, is a matter of State obligations as well as policy. These obligations are binding on the CWC States Parties; they not only bind them on an individual basis, but compliance with them also affects their collective interests as parties of the regime. Risk assessment and management therefore do not happen in a vacuum, and there are constraints on policy choices that flow directly from these legal obligations that States Parties of the CWC have accepted. In a technical sense, the risk associated with the storage of CW relates to the properties and conditions of the weapons (or containers) themselves, the conditions prevailing at the storage facility, and natural processes as well as human activities that may interfere with or affect these weapons and/or conditions.

The OPCW has published the data contained in Table 1, regarding the types and quantities of CW agents declared by the six possessor States Parties.[6] These data form a starting point for the risk assessment. Other factors that also need to be considered include, inter alia, the composition of the agent mixtures (which may contain, for example, solvents, stabilizers, and impurities), the storage form (bulk or weaponized), the design of the containers and weapons, the age of the stockpile, and the level of maintenance that has been applied over time.

CW, containers, and agent mixtures are designed for long-term storage and transport. Weapons may contain a burster charge but are normally not fused. The short-term storage risks are usually assessed to be relatively low, but that assessment no longer holds when long-term storage is assessed. On the one hand, that is the consequence of a stockpile aging, with an increasing likelihood of agent leakage resulting from a variety of processes taking place inside and with or around the weapons. With regard to external processes, longer storage times mean that low-probability events are more likely to occur at some point in time.

The first issue to consider is unintended agent release from leaking weapons or containers, leading to contamination of surfaces and air close to the leaking weapon and, to a lesser degree, downwind hazards depending on the circumstances, the agent type and quantity, and meteorological conditions. Such agent release could be the result of pressure build up inside the weapon due to degradation of the agent mixture, corrosion of the munitions or containers, or damage caused by maintenance and handling of the weapons or container. These risks tend to be local and can be managed with adequate building design and storage configuration, proper maintenance procedures, and the installation of monitoring, protective and remedial (decontamination, medical) systems.

External events can also lead to agent releases at storage locations. Natural events that could destroy storage bunkers and release agent from weapons or containers include earthquakes, flooding, or lightening. External man-made events could range from sabotage or conventional attack on a storage facility, including by terrorists, to accidents such as plane crashes into a storage site.

Many of these risks are reasonably well understood and have been assessed multiple times in the past. What should be noted is that the event probabilities may well be changing as a stockpile ages. Leakage will become a more frequent event and will need to be addressed both in the context of routine maintenance of a stockpile and in terms of the need for containment, and if necessary emergency destruction of leaking items.

Assumptions with regard to certain human risk factors are changing. A risk that can no longer be neglected is that of sabotage or a terrorist attack on a stockpile. If the purpose of such an act would be for terrorists to acquire CW, responses need to be found by way of increasing physical security and protection of the stockpiles, and by way of how the weapons are stored (for example, storing fuses at separate locations). Measures also include intelligence, interception, and other law enforcement measures. At the international level, the First CWC Review Conference (held in April and May 2003) recognized that the protection of CW stockpiles is the responsibility of the possessor States Parties, but that at the same time the OPCW could provide a forum for the exchange of experience between States Parties about effective measures to provide physical security at these facilities.[7] Risk assessments should also address the possibility that after a release of toxic agent, the first responders themselves will be likely targets of follow-on attacks; established response mechanisms might fail or be obstructed as a consequence thereof.

TRANSPORTATION TO CW DESTRUCTION FACILITIES

The first step in the process of CW destruction is the transportation of the weapons from a storage location to a destruction facility. Even if storage and destruction facilities are colocated, some movement of the weapons will be required. These have in fact been designed to be transported. A well-maintained CW stockpile should not pose major transportation hazards, but circumstances under which the weapons and containers will be transported and the likelihood of incidents and accidents during loading, movement, and unloading must be considered.

These transportation risks are determined by the characteristics weapons and containers to be moved (e.g., their resistance to vibration and mechanical shock in the case of external events, their number, size, and agent fill), the characteristics of the packaging and of the transport vehicles (such as load

TABLE 1. Types and amounts of chemical agents declared to the OPCW[1]

IUPAC name for chemical	Common name of chemical	Convention schedule	Quantity declared (MTs)	Quantity destroyed (MTs)
Category 1				
O-isopropyl methylphosphonofluoridate	GB (sarin)	Sch. 1: A (1)	15,048.177	5,504.390
O-pinacolyl methylphosphonofluoridate; (O-(1,2,2-trimethylpropyl)-methylphosphonofluoridate)	GD (soman)	Sch. 1: A (1)	9,174.819	
O-ethyl N,N-dimethyl Phosphoramidocyanidate	GA (tabun)	Sch. 1: A (2)	2.283	0.379
O-ethyl S-2-diisopropylaminoethyl methyl phosphonothiolate	VX	Sch. 1: A (3)	4,032.224	443.239
O-isobutyl-S-[2-(diethylamino) ethyl] methylthiophosphonate	VX	Sch. 1: A (3)	15,557.937	
O-ethyl S-2-(dimethylamino)ethyl methylphosphonothiolate	EA 1699	Sch. 1: A (3)	0.002	
bis (2-chloroethyl) sulfide	Sulfur mustard, mustard gas, H, HD, mustard gas in oil product	Sch. 1: A (4)	13,852.527	1,320.578
Mixture of bis (2-chloroethyl) sulfide and 2–chlorovinyldichloroarsine	Mixture of mustard gas and lewisite	Sch. 1: A (4)	273.259	
Mixture of bis (2-chloroethyl) sulfide and 2–chlorovinyldichloroarsine in 1,2-dichloroethane	Mixture of mustard gas and lewisite in dichloroethane	Sch. 1: A (5) Sch. 1: A (4)	71.392	
Mixture of bis (2-chloroethyl) sulfide and 2–chlorovinyldichloroarsine	Mixture of distilled mustard and lewisite	Sch. 1: A (5) Sch. 1: A (4)	0.400	
		Sch. 1: A (5)		

Continued.

TABLE 1. *Continued.*

IUPAC name for chemical	Common name of chemical	Convention schedule	Quantity declared (MTs)	Quantity destroyed (MTs)
2-chlorovinyldichlorarsine	Lewisite, L	Sch. 1: A (5)	6,745.615	9.223
Methylphosphonyl difluoride	DF	Sch. 1: B (9)	443.967	103.281
O-ethyl O-2-diisopropylaminoethyl methylphosphonite	QL	Sch. 1: B (10)	46.185	0.477
Mixture of 60% bis (2-chloroethyl) sulfide and 40% bis(2–chloroethylthioethyl) ether	HT	Sch. 1: A (4)	3,535.536	0.174
Mixture of 72% isopropyl alcohol and 28% isopropylamine	OPA	Unscheduled	730.545	566.749
Unknown	Unknown		4.645	0.046
	Toxic waste (degraded sulfur mustard)		1.705	1.705
Category 2				
2-chloro-ethane -1-ol	2-chloroethanol	Unscheduled	301.300	301.300
bis(2-hydroxyethyl) sulfide	Thiodiglycol	Sch. 2 : B (13)	51	51.000
Carbonyl dichloride	Phosgene	Sch. 3 : A (1)	10.622	10.622

carried, characteristics of the vehicles, maintenance status), and by external factors, such as the possibility of accidents and the characteristics of the locations through which such movements are to take place. Other risk management measures include route planning and planning of emergency response measures at locations to be transited, escorting procedures, availability of monitoring devices and response systems in case of accident or vehicle failure, and the like.

Today there is a considerable amount of practical experience with moving CW to destruction facilities (both over land and sea), and it appears that the risks are well understood and can be managed. Incidents would be expected to be of local significance, usually involving a small number of items in a well-understood and controllable context.

Risks related to deliberate interference with transport activities (protestors, criminal activity, terrorists) also need to be addressed, and assessment of these risks might lead to a different conclusion.

OLD AND ABANDONED CW

One should expect the situation to be different with regard to O/ACW. These are CW that were left behind at former production, testing, and storage locations, or that were disposed of after previous wars by land-burial or other means, or that were fired at battlefields but failed to explode or otherwise release the agent. As long as these old weapons reside in the ground and remain undisturbed, risk considerations relate to corrosion, deterioration of the weapon or container, agent stability, leakage of agent or degradation products into soil and water, and the subsequent behavior of the leaked chemicals (evaporation, leakage or mass transport into other environmental compartments such as groundwater, chemical, and biological degradation).

The variety of chemical agents that need to be considered in risk assessments concerning O/ACW is considerable larger than the number of agents that are found in stockpiles. Lohs and Stock list a total of 51 different chemical warfare agents that had been stockpiled and/or used between 1914 and 1942, a triple the number of agents currently present in stockpiles.[8]

Risks associated with the recovery of O/ACW are determined by the characteristics and numbers of O/ACW involved, and of the location where they are being recovered (such as vicinity to populated areas, existing infrastructure and response systems, location of destruction or storage facilities to receive the O/ACW, etc.). Once O/ACW have been recovered, they are often stored for a certain period of time before they are being destroyed (either at or near the location of recovery or, sometimes, at a dedicated destruction facility located elsewhere). Intermediate storage of recovered O/ACW can pose considerable risks, given that the munitions or devices may not be stable and the information about their internal state may not be certain, and given that they often contain explosives and may be fused. It is not unusual that a variety of old weapons are stored together, not necessarily all CW but also some conventional, containing high explosives. Accidents in storage can occur and may be difficult to contain given the relative instability of the munitions and the fact that they may contain explosive charges or fills.

The transportation and destruction of these O/ACW pose another set of risks, which is different from stockpile CW given the physical state in which these weapons are. Deterioration makes them vulnerable to mechanical as well as thermal stress. Leaking weapons need to be sealed. O/ACW transportation will require the use of air-and water-tight transport vessels, which contain decontaminants to control any agent leakage, and items may have to be overpacked to control contamination. Also, one needs to account for the potentially large variety of these weapons, in particular in locations related to the past development, production, and testing of CW. Disposal operations may have to deal with a large variety of munitions types, some experimental in nature, and it is critical to ensure that personnel are well-trained and have proper access to design information on such old weapons so they better understand and manage the risks associated with their handling, movement, and disassembly for destruction.

SEA-DUMPED CHEMICAL WEAPONS

A separate issue is the risk associated with sea-dumped CW. The CWC gives States Parties discretion over whether to declare past sea dumping of CW, and also whether to recover and destroy such sea-dumped CW. In practice, States have avoided declaration of past sea dumping and destruction of CW, except countries that have a history of accidental recovery of sea-dumped CW by fisherman or of incidents when such weapons are washed ashore, have programs in place to recover and destroy them under safe conditions. Risk studies have been carried out many times in recent decades, both with regard to assessing the risk of leaving the weapons on the seabed and with regard to the risk involved in recovering them from the seabed. The risks of leaving the weapons where they are relate to those agents that do not rapidly hydrolyze or that degrade into reaction products that continue to have considerable toxicity. Arsenicals and mustard gas are examples. Mustard gas does degrade with time, but has a tendency to coat itself with a layer of polymerization products that form a physical barrier for water. As a consequence, hydrolysis of mustard gas, which is normally very fast, may not occur. Under such conditions, OCW containing mustard gas can remain a serious hazard for decades and can release highly toxic material if containers or munitions are disturbed and opened or if they disintegrate as a result of corrosion. However, the risk is local, and once the agent is distributed into a body of water it will begin to dissolve and hydrolyze. Also, water temperatures are such that at deeper location, the agent will solidify. The majority view seems to continue to be that the risk associated with the recovery of CW from the seabed outweighs the risk of leaving the weapons at their dumping sites, and efforts therefore continue to focus on surveying and monitoring. Over time, the risk of sea-dumped weapons disintegrating and releasing agent or degradation product into the local environment will increase, but so will the risk associated with a potentially much widespread release of agent if recovery measures go wrong. The result of risk assessments is unlikely to change unless technologies for genuine *in situ* destruction would become available.

DESTRUCTION OF CHEMICAL WEAPONS

Destruction of CW involves a number of processes, each of which must be included in the risk assessment. The discussion of risk assessment and management for CW destruction sometimes focuses almost instantly on a comparison of the different destruction technologies at hand. This, however, is much too narrow a perspective. CW destruction begins with the movement of the weapons and containers into an area where the weapons and containers will be removed from their packaging (which will also need to be destroyed as hazardous waste, albeit not necessarily at the destruction facility if they are not contaminated), opened, the agent will be drained from the weapon or container, and the metal parts will be separated from the agent. This separation of the agent from the weapon/container is usually considered the most risky step in the destruction operation. There result three separate material streams into the different aspects of the destruction facility: agent; a mixed stream of energetics, small metal parts and agent residues; and contaminated large metal parts without energetics. The precise steps involved in these processes depend on the types of weapon or container to be destroyed, the agent fill, and the configuration of and equipment used by the destruction facility. Alternatively, in the case of certain bulk storage configurations, the agent can be pumped from the storage tank(s) through pipelines (and measuring devices such as flow meters) into the destruction facility and fed into the destruction unit.

These initial processes, and the associated risks, only indirectly depend on the destruction technology used, but the possibility that agent might be released during these operations is at least as significant a risk factor as potential agent release in the destruction process itself. Containment, monitoring of air concentrations of agent, emergency response systems and back-up systems for critical processes, training and protection of personnel can be used to manage the risks associated with the separation of the agent from the weapons or container and the feeding of agent as well as contaminated metal parts into their respective aspects of the destruction facility.

As for technology choices for the destruction itself, a recent authoritative study has provided an overview on the currently available technologies that have been demonstrated to work for the destruction of agent quantities of 1 kg or more.[9] In addition to the transport of CW and the removal of agents from the weapons, the study discussed four high-temperature destruction methods (incineration, plasma pyrolysis, molten metal technology, and hydrogenolysis), several processes to dispose off arsenicals, and a number of low-temperature processes (hydrolysis of mustard and nerve agents, chemical reactions with amines and other reagents, electrochemical oxidation, and solvated electron technology). The study also addressed effluent treatment, and issues related to old recovered munitions. It did not make particular recommendations with regard to the selection of one or several preferred technologies, but instead provided information on each of them so that countries that need to take decisions on which technology to chose can do so on an informed basis, taking into account the particular national conditions and standards they have to meet.

REFERENCES

1. Convention on the Prohibition of the Development, Production, Stockpiling and Use of Chemical Weapons and on their Destruction (Chemical Weapons Convention), Article II, paragraph 1. Available at www.opcw.org.
2. OPCW. Note by the Technical Secretariat, Office of the Legal Adviser: Status of participation in the Chemical Weapons Conventions as at 1 September 2005; document S/519/2005 dated 5 September 2005.
3. Guidelines for Schedules of Chemicals, Annex on Chemicals of the Chemical Weapons Convention, Section A.
4. OPCW. Decision of the Conference of the States Parties "Procedures Concerning the Implementation of Safety Requirements for Activities of Inspectors and Inspection Assistants, in accordance with Part II, paragraph 43, of the Verification Annex," document C-I/DEC.8 dated 14 May 1997.
5. Chemical Weapons Convention Paragraph 10 of Article IV.
6. Report of the OPCW on the Implementation of the Convention on the Prohibition of the Development, Production, Stockpiling and Use of Chemical Weapons and on their Destruction in 2003, Annex 2 of document C-9/5, dated 30 November 2004.
7. OPCW. Report of the First CWC Review Conference, paragraph 7.93 of document RC-1/5 dated 9 May 2003.
8. LOHS, K. & T. STOCK. 1997. Characteristics of chemical warfare agents and toxic armaments wastes. *In* The Challenge of Old Chemical Munitions and Toxic Armaments Wastes, SIPRI Chemical and Biological Warfare Studies No. 16. T. Stock and K. Lohs, Eds.: 15–34. Stockholm International Peace Research Institute (sipri), Oxford University Press. Oxford and New York.
9. PEARSON, G.S. & R.S. MAGEE. 2002. Critical evaluation of proven chemical weapons destruction technologies (IUAPC Technical Report). Pure Appl. Chem. **74:** 187–316.

Options for the Destruction of Chemical Weapons and Management of the Associated Risks

RON G. MANLEY

Arms Control, Assessment & Analysis, Dorset BH 23 3LH, United Kingdom

ABSTRACT: The destruction of chemical weapons is a hazardous operation. The degree of hazard posed, however, is not uniform and is dependent on the specific chemical agent and the configuration of the weapon or bulk storage vessel in which it is contained. For example, a highly volatile nerve agent in an explosively configured munition, such as a rocket, poses a very different hazard from that of a bulk storage container of viscous mustard gas. Equally the handling of recovered, often highly corroded, World War (WW)I or WWII chemical munitions will pose a very different hazard from that associated with dealing with modern chemical weapons stored under the appropriate conditions. Over the years, a number of technologies have been developed for the destruction of chemical weapons. Each has its advantages and disadvantages. None of them provide a universal solution to the problem. When assessing options for the destruction of these weapons and the management of the associated risks, therefore, it is important to give due consideration and weight to these differences. To ensure that the destruction technology selected takes due account of them and that the resulting overall risk assessment accurately reflects the actual risks involved.

KEYWORDS: chemical weapons; chemical munitions; destruction; neutralization; incineration; risk

INTRODUCTION

For destruction purposes, chemical weapons can be conveniently divided into two categories, namely, stockpiled and nonstockpiled chemical weapons. Stockpiled chemical weapons are those produced since World War (WW)II and retained for military purposes. Chemical weapons recovered from old battlefields, test ranges, burial sites, and sea dumps are classified as nonstockpiled chemical weapons. Although they are all chemical weapons, the hazards

---

Address for correspondence: Ron Manley, Arms Control, Assessment & Analysis, 19 Palmerston Avenue, Christchurch, Dorset BH23 3LH, United Kingdom. Voice: +44-1202-567710; fax: +44-1202-567710.
    e-mail: ron.gmanley@ntlworld.com

posed by stockpiled weapons are significantly different from those posed by nonstockpiled weapons.

Stockpiled Chemical Weapons

Chemical weapons stored in military stockpiles can, generally, be expected to be in relatively good condition, clearly marked to indicate their status and content, and appropriately stored. They may consist of bulk stocks of chemical agents ready for filling into munitions, unfilled chemical munitions, and filled chemical munitions of various types. The filled munitions may or may not be explosively configured.

A national stockpile may contain a single chemical agent or, as in the case of the Russian Federation and the United States, include a range of chemical agents, each chosen to produce a different impact on the battlefield. A list of the principal stockpiled chemical agents and the approximate quantities declared under the Chemical Weapons Convention is contained in TABLE 1.[1]

Mustard gas and Lewisite are primarily vesicants attacking the body's skin and mucous membranes. Nerve agents, such as GB, GD, VX, and VR, on the other hand, are extremely toxic both via inhalation and the percutaneous route. The physical properties and toxicity of a given chemical agent can have a significant impact on the hazards associated with its destruction and the choice of destruction method.

Nonstockpiled Chemical Munitions

Chemical munitions recovered from old battlefields, test-ranges, burial sites, or recovered from the sea are usually heavily corroded and, if not already leaking, are prone to leakage during either transport or destruction. They are frequently fully explosively configured and the status and stability of the individual explosive components will generally be unknown. Often the external

TABLE 1. Stockpiled chemical agents and the quantities declared under the Chemical Weapons Convention (CWC)

Chemical agent	IUPAC name	Agent/tonnes
Lewisite (L)	2-chlorovinyldichlorarsine	6,745
Mustard gas	bis(2-chloroethyl) sulfide	17,387
Mustard gas/ Lewisite mixtures	As above	344
Sarin (GB)	O-isopropyl methylphosphonofluoride	15,048
Soman (GD)	O-pinacolyl methylphosphonofluoride	9,175
VX	O-ethyl S-2-diisopropylaminoethyl methyl phosponothiolate	4,032
Russian VX (VR)	O-isobutyl S-(2-diethylamino ethyl) methyl phosphonothiolate	15,558

markings indicating that it is a chemical weapon and the type of chemical fill will have completely disappeared. This, coupled with the fact that during both WWI and WWII the same munition casing was frequently used for high explosives, chemical agents, chemical smoke, and incendiary compositions, can lead to a significant increase in the risk associated with their destruction.[2] The problem is further complicated by the fact that the range of chemicals weaponized during WWI and WWII was much more extensive than that present in modern stockpiled weapons. In WWI, for example, around 50 different chemical compounds, in many different combinations, were weaponized and deployed in the battlefield.[3] Many of them, such as a number of the arsenic-based chemical agents, were solids rather than the high-boiling-point liquids routinely found in modern, stockpiled, chemical weapons.

Preparation for Destruction

In all of the major destruction facilities for chemical weapons, currently in use throughout the world, the first step in the destruction process is to separate the chemical agent from its container. Where the chemical agent is stored in bulk containers this may be a relatively straightforward process and even where the agent is contained in a nonexplosively configured munition, such as an aerial bomb, it may be possible either to remove the filling plug or punch a hole and drain the weapon. In some situations, however, particularly when the munition is explosively configured, this will be, by far, the most hazardous step in the whole destruction process.

This is especially true when dealing with nonstockpiled munitions, where heavy corrosion and the questionable status of any explosive components present make removal of the chemical agent particularly hazardous and difficult. Disassembly of this class of chemical weapon, therefore, is a slow, highly skilled, process and one that is difficult to fully automate.

Disassembly of stockpiled munitions can also pose difficulties. One example of this is the M55 rocket, one of the chemical weapons in the U.S. stockpile. This weapon consists of: a solid fuel rocket motor and its ignition system; a warhead filled with either the nerve agent GB or VX; a fuse; and an explosive charge to burst the warhead. There is no way to safely disassemble this weapon and the normal destruction process involves chopping the weapon into sections using remotely operated equipment located within a specially designed containment facility. Even where the weapons are not explosively configured, removing the chemical agent will frequently not be a straightforward operation. Some of the chemical weapons in the Russian stockpile, for example, are filled with the nerve agent GD mixed with a polymer to increase the viscosity of the mixture. The resulting gel-like material cannot be readily drained from the munition via its filler plug.

The presence of degradation products may also complicate the process of recovering the chemical agent. Mustard gas, for example, can, over time, polymerise to form a yellow sponge-like material that is very difficult to dissolve and remove from a munition or container. It can also, under certain storage conditions, react with the metal case of the munition to produce hydrogen gas, which may explosively ignite when the munition is subsequently punched or drilled.[4] Even where the munitions are filled with low-viscosity chemical agents in good condition, it will normally not be possible to drain all of the chemical agent from the weapon.

In all situations the key requirement is for appropriately designed containment facilities that will enable this first step in the destruction process to be carried out in a way that will neither put the plant operators at risk nor result in any significant release of chemical agent to the surrounding environment. Designing containment facilities to deal with the destruction of nonexplosively configured chemical weapons, although complex and expensive, does not, in its self, pose major engineering challenges. After all, similar containment facilities were necessary for the production of the chemical weapons in the first place. In addition, this kind of technology is increasingly being used within the commercial chemical industry for the production of certain highly toxic chemicals.

Where the munitions are explosively configured, however, the position is somewhat different. The normal practice when dealing with explosives is to design the facility so that, in the event of an accidental detonation, the explosive force is directed away from the operators and safely vented to the atmosphere. In the case of an explosively configured chemical munition, however, this would also result in the immediate release of the toxic chemical agent to the atmosphere. In situations where such a release is judged to be unacceptable, the facility must be designed to ensure that, in the event of an accidental detonation, the containment will not be breached while at the same time continuing to provide full protection to the plants operators from the effects of the blast. This is clearly a much more complex and difficult problem and one that requires very specialized engineering design.

Over the years, a number of concepts have been put forward for dealing with the chemical weapon as a whole and thus avoiding this complex separation process. One of the proposed technologies, developed by General Atomics in the United States and known as "cryofracture," involves cooling the weapons in liquid nitrogen.[5] At these temperatures, steel becomes brittle and on removal from the liquid nitrogen the weapon can be broken into small pieces, either using a remotely operated press or a hammer mill. The frozen mixture of steel, explosive, and chemical agent fragments can then be fed, directly, to a destruction plant such as, a rotary kiln, high-temperature fluidized bed furnace or, where there are no explosives present, to one of the proven low-temperature chemical destruction processes.

The adoption of "cryofracture," as an alternative to disassembly for the disposal of stockpiled weapons, has been given serious consideration in the United States. It was decided, however, that, although the technology was viable, it was unlikely to offer significant advantages over the existing, well-proven disassembly process, currently in use at U.S. facilities.[6]

A number of the proposals for eliminating the disassembly step, including "cryofracture," are being specifically targeted at dealing with nonstockpiled chemical weapons. In this case, the increased level of risk associated with the disassembly of these weapons is considered sufficient to justify the time and the effort that may be required to fully develop them. Some of these technologies have already undergone development and been used on a limited basis for the disposal of nonstockpile munitions. In the United States, for example, the Explosive Destruction System (EDS) has been developed and successfully tested in the field for this purpose.[7] In this system, the chemical munition is placed inside a specially designed explosive containment chamber and opened with a shaped explosive charge. The contents of the munition are then broken down chemically before being removed from the chamber. A French company has developed a solution that involves the detonation of the weapon, using either its own charge or an externally placed charge, at the bottom of a large pool of aqueous decontaminating solution.[8] Although the practicality of both these techniques have been fully demonstrated, they are designed for dealing with small numbers of recovered chemical weapons, particularly those that are considered to be too hazardous to be dealt with using the standard approach of first removing the chemical agent from the munition.

Disassembly or breakdown of the chemical weapon, in order to recover the bulk of the chemical agent for separate destruction, therefore, remains the first step in the destruction process for the vast majority of chemical weapons. This process will, inevitably, result in some or all of the following waste streams: chemical agent; drained or partially drained containers or munitions; explosive components; contaminated packing materials and other contaminated solid wastes, such as protective suits, etc.; and contaminated waste water and exhaust air from the containment facility. Each of these will require treatment, before they can be safely released to the environment. The selected destruction technology or technologies must be capable of treating these very different waste streams, either together or separately. Irrespective of the technology chosen for the final destruction of the chemical agent, the major part of any chemical weapons destruction facility will be devoted to dealing with this initial disassembly or separation phase and, in almost all situations, this phase will pose the greatest potential hazard to the plant operators and the surrounding environment.

Destruction

Historically, incineration and chemical hydrolysis have been the principal methods used for the destruction of chemical agent. Although alkaline hydrolysis has been used quite extensively, by a number of countries, its principal disadvantage is that it produces large volumes of aqueous waste—between four and five times the volume of chemical agent being destroyed—containing organic salts that require further treatment before they can be released to the environment. In addition the other waste streams from the disassembly process, such as the contaminated metal components, explosives and propellants, etc., still need to be decontaminated or destroyed, usually by some form of thermal process.

For the destruction of their nerve agents and mustard gas, the Russian Federation developed and have used for a number of years, a hybrid process that combines an initial chemical neutralization step with a secondary incineration step.[9] Mustard gas and all the nerve agents, with the exception of VR, are neutralized by treatment with monoethanolamine at around 100°C. In the case of VR, chemical breakdown is achieved using potassium isobutylate. Both these processes convert the chemical agent into a relatively nontoxic waste, which is combustible and, therefore, readily disposed off in the subsequent incineration step. The principal advantage of this approach is that the chemical agent is effectively destroyed in a low temperature batch process before being fed to the incinerator.

Direct incineration, however, accounts for by far the largest amount of chemical agent destroyed and it is estimated that, since WWII, around 80% of the chemical agent destroyed, as opposed to being buried or sea dumped, has been processed using some form of this technology.

Incineration is ideally suited for the destruction of the majority of chemical agents, including all types of nerve agents and mustard gas. It has been extensively demonstrated that it can be used for all of the waste streams arising from the disassembly or breakdown of these chemical weapons, including partially decomposed chemical agent and is favored because it produces only sterile ash, clean metallic waste, and gaseous products. The waste gases can be readily treated in a pollution abatement system to remove any noxious materials prior to their release to the atmosphere. Over the years, it has been demonstrated that appropriately designed and operated incineration plants can consistently achieve very high destruction removal efficiencies (DREs) of better than 99.9999% and meet the most stringent environmental standards.

Despite its proven safety record for the destruction of chemical weapons, incineration has a poor image with the general public. A key argument used against incineration is that, because it is a continuous process, it is not possible to delay the release of the waste products to the environment until after they have been tested and shown to be safe. It is true that, although continuous monitoring of the exhaust gases is carried out, it is not practicable, in the event of something going wrong, to completely eliminate the risk of some material being released during the short period between detection and shut down of the system. Thousands of hours of operation, however, have demonstrated that the risk to the general public from such an incident is extremely low.

Even incineration, however, is not a complete solution to chemical weapon destruction problem. For example, although it has been successfully used for the destruction of small amounts of arsenic-based chemical agents, such as Lewisite and Diphenylchloroarsine, it is not considered to be the ideal solution for the destruction of these materials.[10] The problem is that the arsenic oxides produced as a byproduct of the incineration process are, in themselves, toxic and their complete removal from the exhaust gas stream, prior to release to the atmosphere, is both complex and expensive. Alternative technologies are, therefore, being developed and used for the disposal of the large stocks of Lewisite and Lewisite/mustard gas mixtures, currently either undergoing or awaiting destruction, within the Russian Federation.

Alternative Technologies

In recent years growing opposition from the general public to incineration, particularly within the United States, has led to an extensive search for possible alternative technologies for the destruction of chemical weapons. As a result, a wide range of different technologies have been put forward as possible candidates. Most of these, although theoretically capable of destroying chemical agent, require a great deal of further development before they could be seriously considered as viable alternatives to incineration. Some, such as alkaline hydrolysis and neutralization with organic amines, however, have been shown to be practical alternatives to incineration. Others, such as super critical water oxidation, plasma pyrolysis, hydrogenolysis, electrochemical oxidation, solvated electron technology (SET), etc., have been sufficiently developed to enable their potential as alternatives to incineration to be properly assessed.[11] It should be noted, however, that most of these alternative technologies are primarily targeted at the destruction of the chemical agent, which, although a key component, is only one of the waste streams arising from the chemical weapon destruction process.

Although most of the U.S. stockpile will continue to be destroyed by incineration, a decision has been taken to use hydrolysis as the main destruction process at four of its nine storage sites.[12] The stockpiles at Aberdeen Proving Ground (APG), Maryland and at Newport, Indiana are particularly suited to this alternative approach as they consist solely of bulk chemical agent stored in 1-ton containers. Mustard gas at APG is being destroyed by hydrolysis with hot water. The relatively nontoxic hydrolysate is then shipped to a commercial waste disposal plant for secondary treatment by biodegradation before being released to the environment. The empty 1-ton containers are disassembled, cut in half and then thoroughly decontaminated before being released for disposal as commercial scrap. By mid-2005 more than 95% of the APG stockpile had been safely hydrolyzed using this technique. At Newport, Indiana, VX is being destroyed by hydrolysis with hot, aqueous, sodium hydroxide. Once again, it is planned to use biodegradation as a secondary process to treat the hydrolysate prior to its release to the environment. The destruction facility at Newport recently became operational and around 1% of the stockpile has so far been processed.

Hydrolysis coupled with secondary treatment of the hydrolysate is also the selected solution for the disposal of the filled chemical munitions stored at Pueblo, Colorado and Blue Grass, Kentucky. At both of these sites, however, it will still be necessary to first carry out the complex and hazardous disassembly operation to separate the chemical agent from the munitions and any explosives present. It is not clear, therefore, that the substitution of neutralization for incineration at these sites will lead to any significant reduction in the overall hazard associated with the destruction process.

The chemical weapons stored in the Russian Federation stockpiles are not explosively configured and so they are spared this additional destruction hazard. It is their intention to continue using their established chemical neutralization process for the destruction of their nerve-agent-filled munitions but, due to public pressure, they intend to replace the secondary incineration step with a "bituminization" process. This involves mixing the waste from the first stage with hot bitumen and casting it into blocks for subsequent burial.

None of the currently selected alternative technologies remove the need to first separate the chemical agent from its munition or container and any explosives or propellants present. This remains, particularly in the case of nonstockpile munitions, the most hazardous part of the destruction operation and is the area where the continued application of good design and risk-management techniques will yield the greatest benefits.

CONCLUSION

The destruction of chemical weapons has become a highly emotive issue and there is a danger that this could adversely impact on our judgment of the risks associated with their destruction. Although there is no question that the destruction of these weapons is a hazardous operation, the actual level of risk, as has been illustrated in this article, may vary considerably and will be highly dependent on the specific chemical agent and the precise configuration of the weapon or bulk storage vessel in which it is contained.

Over the years a number of technologies have been developed for the destruction of chemical weapons. Each has its advantages and disadvantages. None of them provide a universal solution to the problem.

When assessing options for the destruction of these weapons and the management of the associated risks it is essential, therefore, to give due consideration and weight to different types of chemicals and weapon configurations involved and to ensure that the destruction technology selected takes due account of them and that the resulting overall risk assessment accurately reflects the actual risks involved.

REFERENCES

1. OPCW. 2003. List of chemical agents declared and destroyed as at 31 December 2003, Annex 2, Annual Report. Available at: www.opcw.org.
2. STOCK, T. & K. LOHS. 1997. Old chemical munitions and warfare agents: detoxification & degradation. *In* The Challenge of Old Chemical Munitions and Toxic Armament Wastes. T. Stock & K. Lohs, Eds.: 35–54. Sipri. Stockholm, Sweden.
3. MANLEY, R.G. 1997. European experience with the disposal of old chemical weapons. *In* Analytical Chemistry Associated with the Destruction of Chemical Weapons. NATO ASI Series, 1. Disarmament Technologies, Vol. 13. M. Heyl & R. McGuire, Eds.: 15–26. Kluwer Academic Publishers. The Netherlands.
4. MANLEY, R.G. 1997. UNSCOM's experience with chemical warfare agents & munitions in Iraq. *In* The Challenge of Old Chemical Munitions and Toxic Armament Wastes. T. Stock & K. Lohs, Eds.: 241–252. Sipri. Stockholm, Sweden.
5. JOHNSON, L. 1998. Cryofracture technology for the disposal of chemical munitions. Presented at the 1st International Chemical Weapons Demilitarisation Conference. CWD 1998. Bournemouth, United Kingdom. 23–25 June 1998.
6. NATIONAL RESEARCH COUNCIL, COMMITTEE ON REVIEW & EVALUATION OF ALTERNATIVE TECHNOLOGIES. 1999. Review and evaluation of alternative technologies for demilitarisation of assembled chemical weapons. 101. National Academy Press, Washington, DC.
7. DIBERARDO, R., B. HAROLDSEN & A. SPEIGHT. 2001. The explosive destruction system: a system to destroy vintage chemical weapons. Presented at the 4th International Chemical Weapons Demilitarisation Conference. CWD 2001. Gifu City, Japan. 22–24 May 2001.
8. GUIR, F. 1997. The technical challenge of dismantling and destroying old & abandoned chemical weapons. *In* The Challenge of Old Chemical Munitions and Toxic Armament Wastes. T. Stock & K. Lohs, Eds.: 156–165. Sipri. Stockholm, Sweden.
9. PETROV, G.S., V.I. KHOLSTOV & V.P. ZOUBRILIN. 1998. Practical actions of Russia on preparation for destruction of stockpiled Lewisite & mustard. *In* Arsenic and Old Mustard: Chemical Problems in the Destruction of Old Arsenical and Mustard Munitions. NATO ASI Series, 1. Disarmament Technologies, Vol 19. J.F. Bunnett & M. Mikolajczyk, Eds.: 79–90. Kluwer Academic Publishers. The Netherlands.
10. MARTENS, H. 1997. The German programme for the disposal of old chemical weapons. *In* The Challenge of Old Chemical Munitions and Toxic Armament Wastes. T. Stock & K. Lohs, Eds.: 166–178. Sipri. Stockholm, Sweden.
11. PEARSON, G.S. & R.S. MAGEE. 2002. Critical evaluation of proven chemical weapon destruction technologies. Pure Appl. Chem. Vol. 74, No. 2.
12. FLAMM, K. 2005. Chemical demilitarisation program management. Presented at the 8th International Chemical Weapons Demilitarisation Conference. CWD 2005. Edinburgh, Scotland, United Kingdom. 12–14 April 2005.

Health and Environmental Threats Associated with the Destruction of Chemical Weapons

JIŘÍ MATOUŠEK

Masaryk University Brno, Faculty of Science, EU Research Centre of Excellence for Environmental Chemistry and Ecotoxicology, CZ-625-00 Brno, Czech Republic

ABSTRACT: Still existing arsenals of chemical weapons (CW) pose not only security threats for possible use in hostilities by state actors or misuse by terrorists but also safety threats to humans and biota due to leakages and possible accidents. The Chemical Weapons Convention (CWC) commits the States Parties (SPs) to destroy CW using technologies taking into consideration human health and environmental protection. It does not allow methods, routinely used up to the 1970s, such as earth burial, open-pit burning, and sea dumping. Long-term health and environmental threats and some accidents that have already occurred in the known localities of the sea-dumped and earth-buried arsenals of Nazi-German armed forces in the Baltic Region and of Imperial Japanese forces in the Far East Region are analyzed according to the impact of major CW and ammunition types (i.e., sulfur mustard—HD, tabun—GA, arsenicals—DA, DC, DM, arsine oil, and chloroacetophenone—CN). Any possible operations and handling with CW envisaged by the CWC as well as their verification are summarized taking into account the health threat they pose. CW and toxic armament waste to be destroyed and applied technologies (both developed and under current use in operational CW destruction facilities [CWDF]) are reviewed as are systems of health safety and environmental protection of the destruction/demilitarization stems from the extraordinary high toxicity of supertoxic lethal agents in man and biota. Problems of currently used Russian and U.S. standards for maximum allowable workplace concentrations and general population limits and possibilities of their determination by available analytical instrumentation are discussed.

KEYWORDS: chemical weapons; chemical weapons convention; verification; destruction; health and environmental threats; workplace safety; outdoor safety; maximum allowable concentrations; SPs; CWC; CW; sea water disposal

Address for correspondence: Jiří Matoušek, Masaryk University Brno, Faculty of Science, EU Research Centre of Excellence for Environmental Chemistry and Ecotoxicology, Kamenice 126/3, CZ-625-00 Brno, Czech Republic, Voice: +420-549492860; fax: +420-549492840.
e-mail: matousek@recetox.muni.cz

Ann. N.Y. Acad. Sci. 1076: 549–558 (2006). © 2006 New York Academy of Sciences.
doi: 10.1196/annals.1371.069

INTRODUCTION

It is generally believed that destruction of chemical weapons (CW) poses extremely high health and environmental threats. Therefore, the build-up of chemical weapons destruction facilities (CWDF) pursuant to the commitments of the Chemical Weapons Convention (CWC) in the States Parties (SPs) possessing chemical arsenals or findings of old and abandoned CW on their territory or under their jurisdiction is sometimes met with strict refusal of local population. There is no doubt that destruction of former CW inventory is a risky undertaking like any other handling with this inventory and toxic armaments waste. But it is evident that this risk is time-limited, and it is generally lower than the time-unlimited security and safety risk associated with stockpiles of CW. The risk of destruction can be very considerably diminished by appropriate technological, organizational, and safety arrangements just in the design of CWDFs taking into account workplace and outdoor safety and environmental protection.

SECURITY AND SAFETY THREATS OF EXISTING CHEMICAL ARSENALS

Arsenals of CW pose immanent security concerns due to the possibility of being used by state actors or misused by terrorists. Existing arsenals of CW of current SPs to the CWC are now envisaged to exist until the year 2012, according to the agreed extension of the time for destruction of 10 years by another 5 years in the case of major possessors of CW, that is, Russia and the United States. Among current 178 SPs, CW were declared also by India, Albania, South Korea, and Libya. There are several possessors of CW among non-SPs, presumably in signatory countries that have not yet ratified (e.g., Israel) and in states that have not yet signed (e.g., neighbors of Israel, such as Egypt, Syria, and also North Korea). That is why the universality of the CWC is needed to rid mankind of the dark heredity of chemical warfare.

CW in stockpiles and also nonstockpiled weapons, such as old, abandoned or even inadequately destroyed CW, pose a considerable safety threat, that is, health and environmental, for frequent leakages and possible accidents during any handling, mainly of already obsolete and commissioned weaponry. These threats are given in the first line (TABLE 1) by the extremely high toxicity of super-toxic lethal agents, both in bulk and in munitions through the most common route of intake, inhalation. It is evident how low concentrations can evoke initial symptoms of intoxication effective to incapacitate personnel and how low concentrations cause death in comparison with some industrial toxic compounds used in the initial phases of chemical warfare in the First World War (WW-I).

TABLE 1. Acute toxicities of the main chemical warfare agents by inhalation (mg. min. m^{-3})

Agent	Incapacitation ICt$_{50}$	Lethality LCt$_{50}$
S-mustard (HD)	200	1,500
Lewisite (L)	300	1,500
Tabun (GA)	100	400
Sarin (GB)	55	100
Soman (GD)	25	70
VX, V-gaz	5	36
Compare some toxic industrial chemicals:		
Phosgene (CG)	1,600	3,200
Cyanogen chloride (CK)	7,000	11,000

RISK ACTIVITIES PREDICTED BY THE CWC

In civilian environment, many people are afraid of emissions from facilities for recycling and destruction of communal and industrial waste (including dangerous waste) that are, as a matter of fact, necessary alternatives to doing nothing and accumulating such waste. Similarly, activities required by the CWC associated with elimination of CW stockpiles obviously pose some risks, but these risks are generally lower compared with security and safety risks of the existence of CW, the possibility of their use and misuse, and existing leakages and possible accidents. Risks predicted by the CWC, associated with eliminating CW arsenals are summarized in TABLE 2.

TOXIC AGENTS TO BE DESTROYED AND PROPOSED/APPLIED TECHNIQUES

Compared with reports of the number of possessors of CW from the late 1980s (in excess of 25), obviously politically influenced in the last decade of the Cold War and East–West confrontation, the good news of entering the new millennium is that the present number is only 6 among the current 178 SPs to the CWC according to their declarations on the CW possession. The representative list of actual toxic agents under consideration is given by remaining stockpiles in the United States and Russian Federation, the only major possessors of CW. Main U.S. agents to be destroyed are GB, VX, HD, L, T, and DM. The situation is similar in Russia, having among G-agents beside GB and huge stocks of GD, not possessing the T-agent (having instead a special type of H for winter conditions), still having the HL-mixture and another type of V-agent with the same molecular weight as VX (the O-isobutyl N,N-diethyl analog of VX), depicted as R-33 or simply V-gaz and a small amount of already obsolete agents

TABLE 2. Short overview of the risk activities predicted by the CWC [1]

Closing, sealing, visiting storage sites.
Checking declarations.
Any operations and on-site inspections at destruction/conversion of former production facilities.
Any handling and on-site inspections of acting small-scale production facilities.
Any handling and on-site inspections at CWDF, including all operations starting with loading cargo containers at storage sites, transport, filling of destruction equipment, controlling the destruction process, checking completeness of destruction, till the safe disposal of nontoxic waste, scrap metal, etc.
On-site inspections on challenge.
Protection against and verification of (alleged) use of CW.
Any handling and on-site inspections at extremely dangerous operations connected with location, excavation, transport, demilitarization, and destruction of old and abandoned chemical weaponry and toxic armaments wastes.

(e.g., CG).[2] Declared stockpiles by three SPs (India, South Korea, and Libya) contain only a partial assortment of the same agents. There is no evidence on other types of CW agents (CWA), presumed in arsenals of the non-SPs. Old and abandoned CW, declared by a couple of SPs stem mainly from the arsenals prepared for the WW-II and exceptionally also from earlier times.

Research on technologies for destruction and disposal of CW, both in bulk and in munitions, started long before the commencement of negotiations on the prohibition of CW because such operations are not only an imperative of the ban and elimination of chemical arsenals but they were connected with upgrading of chemical weaponry (and thus elimination of decommissioned weapons) under increasing awareness of environmental protection, no more allowing such barbaric procedures for destruction/disposal, like ocean dumping, earth burial, and open-pit burning, typical still after the WW-II mainly when destroying German and Japanese CW under Allied supervision. Last known event of this type was the operation CHASE, that is, dumping of the U.S. nerve agent munitions in the Caribic during the 1970s. In addition to the main technological procedure, that is, incineration, commonly used for chemical waste worldwide, there is a relatively long list of technologies in different stages of realization. The CWC leaves the choice of destruction technique to the given SPs, obviously under strict requirements of effectiveness, workplace safety, and environmental protection. The overview of suggested technologies for CW destruction is shown in TABLE 3.

This list of destruction techniques encompasses some curious procedures, like the underground nuclear explosions, probably very effective but hardly acceptable because of posing other risks and inability of checking end point of the destruction process as required by the CWC, or the otherwise attractive proposal by N. Platé, on the capability to destroy two kinds of armaments simultaneously (i.e., CW and missile engines, pursuant to the bilateral U.S.–U.S.S.R. Intermediary Nuclear Force–IMF agreement). There are already attempts to

TABLE 3. Technologies for destruction of CW stockpiles[3,4]

Two-Stage Technology (Soviet)—chemical deactivation + detoxification
Incineration, for example, JACADS (Johnson Atoll Chemical Agent Disposal System—United States)
Underground nuclear explosion—Arzamas-16 according to Academician Trutnev
Disproportionation using liquid propellant missile engine together with elimination of missile engine according to Academician N. A. Plate
Biological methods—biodegradation of organophosphates
Supercritical water oxidation
Wet air oxidation
Molten salt processes
Photochemical processes:
   Gas phase process (photochemical)
   Aqueous phase process
   Ozone/ultraviolet irradiation
   Laser-stimulated photodegradation
   Catalytic ozone oxidation, gamma irradiation
Electrochemical techniques
"Neutralization": hydrolysis, oxidation, etc.
Chemical reprocessing: chlorolysis, catalytic dehydrochlorination
Thermic processes:
   Plasma reactors
   Microwave plasma process
   Infrared thermal process
   Radiofrequency thermal process
Solvated electron technology

---

assess the available and potential alternative techniques taking into consideration technological, economic, and health aspects.[3,4] One review assessment contains alternatives to incineration technologies, dealing with toxicity of end products.[5] Already operational (or under final stage of preparation) technologies include prevailing thermic processes (i.e., both classical incineration and high temperature plasma technique) such as the Russian two-stage technology and supercritical water oxidation.

HEALTH RISKS OF CW DESTRUCTION TECHNOLOGIES

It can be expected that any handling with CW will be associated with high health risks due to extremely high acute toxicities of the agents to be destroyed. According to the profound toxicological studies (obviously only by extrapolation from data accumulated from experimental toxicology in animals), the maximum allowable workplace and outdoor concentrations for principal CWA have been suggested in both the United States and Russia. The values shown in TABLE 4 are based on the lowest observable adverse effect level (LOAEL) and no observable adverse effect level (NOAEL) approach for inhalation of contaminated air. The problem is how to detect and monitor the presence of

TABLE 4. Russian and U.S. standards for maximum allowable concentrations of the main CWA (mg/m^3) in air[3,4]

Chemical agent	Russia Work area limit	GPL^a	United States Work area limit	GPL^a
Sarin (GB)	2×10^{-5}	2×10^{-7}	1×10^{-4}	3×10^{-6}
Soman (GD)	1×10^{-5}	1×10^{-7}	2×10^{-5}	3×10^{-6}
VX, V-gaz	5×10^{-6}	5×10^{-8}	1×10^{-5}	3×10^{-6}
Mustard (HD)	2×10^{-4}	2×10^{-6}	3×10^{-3}	1×10^{-4}
Lewisite (L)	2×10^{-4}	4×10^{-6}	3×10^{-3}	3×10^{-3}

^{a} General Population Limit.

toxic agents in the workplace and ambient atmosphere in such a low concentrations. Comparing the indoor limits with the detection limits of existing devices of chemical reconnaissance based on detection tubes, simple field laboratories and detection kits can lead to a conclusion that such equipment is useless for this purpose. Nevertheless, our critical assessment[6] shares the view of, for example, Chebotarev et al.[7] that such equipment is available everywhere and represents a suitable base for monitoring at the storage and destruction site. The safe handling must be assured by technical measures at the site of the technological equipment by its hermetising and fitting with corresponding barriers and containments at the site and of the crew by its physical protection and last but not least by the telemetric control system, etc. It means that the cases of use of standard military devices, not only mentioned above but also of automatic monitors, principally designated to detect threshold battlefield concentrations will be limited mainly for accidental situations. The air monitoring in the protected zone of personnel will be intermittent sampling of air to reach the average concentration for a longer period of time. For the outdoor limits, the concentrations are still generally 1–2 orders lower than the limits for the workplace environment. This means that it would be still more difficult to determine such concentrations by means of any regular chemical reconnaissance device. On the other hand, it is obvious that the mentioned standards are generally constructed for long-term repeated exposition (8 h a day, 5 days a week, 20 years). It lays in the principle that something like this, in practical terms, is impossible. Therefore, the standards indicated in TABLE 4 are under continuous reconsideration to introduce more practical higher values corresponding to shorter exposition times (in the order of hour or less). The reason for installing modern monitoring devices based on enzymatic inhibition or on any other suitable physicochemical principle like ionization principle or ion mobility or remote (off-site) sensing is to watch both indoor and outdoor atmosphere in the case of incidents or accidents to assure early warning and implementation of previously prepared and planned emergency measures.[6]

Any of the verification exercises anticipated by the CWC is associated with objective analytical methods involving the air monitoring. It is possible to say that the verification carried out as on-site inspection at facilities for CW destruction, including all operations beginning at the storage sites, that is, safe loading to cargo containers, transport to the proper destruction, filling into the destruction equipment, and verifying completeness of the destruction process, belong generally not to physically easy tasks. From the analytical point of view, they are not as complicated as the last indicated task in TABLE 2. It is not necessary to carry out something like systematic analysis of unknown sample because known, that is, declared, compounds can be detected in huge amounts such that for these purposes even simple detection equipment can be used. This procedure is however associated with high danger for inspectors due to possible leakages of the technological equipments. Therefore, air monitoring within the framework of the verification will be at the same time a measure of workplace safety from the point of view of inspectors and of handling personnel.

ENVIRONMENTAL CONCERNS OF CW DESTRUCTION TECHNOLOGIES

All hygienic and toxicological standards dealing with actual CWA, as mentioned above, have been derived from the experimental toxicology because there are not enough data from accidental intoxications of humans. Similarly, with the exception of the frequent toxicological data in experimental animals (suggesting for very high acute toxicity) generally not enough relevant ecotoxicological data are available. There is perhaps only one exception, dealing with the studies on aquatic species, associated with the fate of sea-dumped CW inventories of the Nazi-German and Japanese imperial armed forces after the WW-II and studies on abiotic decompositions of CWA in sea water, frequently used in the 1960s and 1970s in connection with disposal operations. It is possible to find some ecotoxicological studies in aquatic organisms in the connection with the environmental impact of the sea-dumped CW inventories mainly in the Baltic and adjacent areas, such as Skagerrak and Norwegian trench.[8-10]

HEALTH AND ENVIRONMENTAL THREATS POSED BY THE SEA-DUMPED AND EARTH-BURIED CW

The health and environmental threats of the sea-dumped chemical warfare inventories depend on the munitions types, kind of CWA and environmental conditions (prevailing temperature, water composition, character of sea bed, etc.). The release of CWA into seawater is determined by thickness, composition, and construction of metallic casing influencing speed and type of corrosion, possible leakages, and like. Release of CWA proceeds generally very slowly through small holes and leakages, and the released agent is readily mixed with ambient water creating very low concentrations. The risk posed by concrete CWA (and present admixtures) is given beside its toxicity (ecotoxicity) by its solubility, hydrolysis in seawater, reactions with sediments, etc.

Sea-dumped munitions containing phosgene release this compound, which is readily decomposed to hydrochloric acid and carbon dioxide, creating a negligible environmental threat. Tabun (GA) was contained in the munitions sunk mainly in the Norwegian trench and Skagerrak. The German munitions fill contained about 20% of chlorobenzene for stabilizing which enhances otherwise relatively good water solubility of the active agent (120 g/L). Dissolved tabun hydrolyses relatively rapidly (half-life 8.5 h at pH 7 and 20°C), forming another toxicant—hydrogen cyanide. In spite of the high acute toxicity of tabun and hydrogen cyanide, and ecotoxicity of chlorobenzene, neither traces of these compounds nor any damage of marine biota were found, most likely due to the long time of release. Sampling and chemical analyses performed by the Norwegians did not detect any traces of the original fill.[11] Some health and environmental threats are posed by mustard gas (H, HD), especially viscous mustard with addition of a resinous-like substance, due to its freezing point (13.8°C for pure substance) and very limited solubility. Its maximum concentration is about 0.07% at 20°C. At lower temperatures, typical for the bottom layers of seawater (0°C), the solubility decreases to 0.03%. The relatively rapid hydrolysis depends on the dissolution rate. At heterogenous conditions, the hydrolysis rate is only about 0.01 min^{-1} at 0°C. The released mustard gas can therefore remain at the bottom layers in a form of slowly decomposing jelly-like spots on the sites of dumping, producing a long-term threat. Swedes have found sulfur mustard and its toxicologically harmless degradation product, thiodiglycol, near the sunken shipwrecks in Skagerrak in the concentrations of the parts per trillion (ppt) range. Tests to determine the toxic effects in fish have shown that mustard gas concentrations of 10 parts per million (ppm) have a lethal effects on eels but not on flounders.[12] Russians tested toxic effects of low concentrations in fish (*Poecella reticulata*), zooplankton (*Daphnia magna*), and gastropod mollusks (*Lymnaea stagnalis*), and showed no effects in fish, no effects in gastropods, and 67% lethality in zooplankton at 0.33 mg/L (in 1 day) and 33% lethality in zooplankton species at 0.0033 mg/L (in 3 days).[8] Aromatic arsenicals, such as diphenylchloroarsine—DA, diphenylcyanoarsine—DC, adamsite—DM, and the mixed agent—the so-called arsine oil, are generally almost nonsoluble in sea water and will stay at the bottom as solid sediments washed out by sea streams. The main process will be a slow solution and heterogenous hydrolysis. The concentration of arsenic determined in the neighborhood of sunken munitions was by one order higher than its background concentration in the Baltic.[9] Similar behavior is expected in case of chloroacetophenone (CN). The Russian study has found no lethal effects of water over adamsite and CN in quoted aquatic species.[8]

In cases of buried munitions, very slow corrosion and thus slow release of toxic chemicals and local contamination of the soil layer and consequently of ground water represent a long-term health and environmental threat. This is valid mainly for the Munster/Oertze area in Lower Saxony (Germany) as well as for any other generally unknown localities on territories and under jurisdiction of (former) possessors of CW who practiced this method of disposal. It is very surprising how good is the state of, for example, encased adamsite, in excavated munitions, buried in the sandy soil of the mentioned area for many dozens of years. It was one of the fatal mistakes of Professor Fritz Haber ("Father" of the chemical warfare), predicting during the WW-I that: "after some fifty years, there will be no traces of buried chemical munitions."

For both, sunken and earth-buried munitions another danger exists as for any other conventional old and abandoned munitions items, namely the possibility of explosion or sudden release of toxic agent during any manipulations.

CONCLUSIONS

All operations associated with the destruction of CW and its verification, starting with any handling on storage sites, including handling with occasionally found and improperly disposed old and abandoned chemical munitions and toxic armament wastes, to transportation and proper destruction, until the disposal of nontoxic waste and scrap metal, pose health and environmental risks. This is mainly due to possible leakages and releases rather at accidental events (including sabotage and terrorism) that at own destruction operation at CWDF where appropriate system of workplace and outdoor safety is adopted just in their design. Environmental risk might depend on the toxicity of end products and their disposal which should be actually excluded by the design of the technology.

There is no doubt that timely, limited destruction of CW poses less security and safety risk than retention of existing CW stockpiles, old and abandoned CW, and improperly disposed CW.

REFERENCES

1. MATOUŠEK, J. 1994. Workplace safety at the operations connected with the implementation of the Chemical Weapons Convention. Second Moscow International Conference on Chemical Disarmament, MOSCON 94, Moscow.
2. BELETSKAYA, I.P. & S.S. NOVIKOV. 1995. Russian chemical weapons. Herald Russ. Acad. Sci. **65:** 5–10.
3. HART, J. & C. MILLER, Eds. 1998. Chemical Weapon Destruction in Russia: Political, Legal, and Technical Aspects. Oxford University Press. Oxford.
4. PEARSON, G.S. & R.S. MAGEE. 2002. Critical evaluation of proven chemical weapon destruction technologies. Pure Appl.Chem. **74:** 187–316.
5. PICARDI, A., P. JOHNSTON, & R. STRINGER. 1991. Alternative Technologies for the Detoxification of Chemical Weapons. Greenpeace International. Washington, DC.

6. MATOUSEK, J. 1997. Methods and means for air monitoring associated with the destruction of chemical weapons. *In*: Analytical Chemistry Associated with the Destruction of Chemical Weapons. M. Heyl & R. McGuire, Eds.: 181–187. Kluwer Academic Publishers. Dordrecht.
7. CHEBOTAREV, O.V. *et al*. 1994. Express-analysis on the objects of storage and destruction of CW utilising regular means of chemical reconnaissance and chemical control (in Russian). Russ. Chem. J. **38:** 69–73.
8. GORLOV, V.G. *et al*. 1993. Complex analysis of the hazard related to the captured German chemical weapons dumped in the Baltic Sea. National Report of the Russian Federation, Moscow.
9. WIBBERENZ, G. 1992. Gefährdungen durch Giftgas in der Ostsee. PFK, Kiel, Germany.
10. MATOUSEK, J. 2002. Old scrap munitions and toxic armaments wastes in the Central European and Baltic Region. SECOTOX–7th Regional Meeting of the Central and East European Section: Trends and Advances in Environmental Chemistry and Toxicology, Brno, Proceedings, 246–252 Masaryk University Brno.
11. STOCK, T. & K. LOHS, Eds. 1997. The Challenge of Old Munitions and Toxic Armaments Wastes. Oxford University Press. Oxford.
12. NATO/CCMS. 1995. Cross-Border Environmental Problems Emanating from Defence-Related Installations and Activities, Vol. 2. Chemical Contamination. Report No 205. NATO, Brussels.

Results of Long-Term Carcinogenicity Bioassay on Sprague-Dawley Rats Exposed to Aspartame Administered in Feed

FIORELLA BELPOGGI, MORANDO SOFFRITTI, MICHELA PADOVANI, DAVIDE DEGLI ESPOSTI, MICHELINA LAURIOLA, AND FRANCO MINARDI

The end judges everything
—HERODOTUS (480-425 B.C.)
The History

Cesare Maltoni Cancer Research Center, European Foundation of Oncology and Environmental Sciences "B. Ramazzini," 40010 Bentivoglio, Bologna, Italy

ABSTRACT: Aspartame (APM) is one of the most widely used artificial sweeteners in the world. Its ever-growing use in more than 6000 products, such as soft drinks, chewing gum, candy, desserts, etc., has been accompanied by rising consumer concerns regarding its safety, in particular its potential long-term carcinogenic effects. In light of the inadequacy of the carcinogenicity bioassays performed in the 1970s and 1980s, a long-term mega-experiment on APM was undertaken at the Cesare Maltoni Cancer Research Center of the European Ramazzini Foundation on groups of male and female Sprague-Dawley rats (100–150/sex/group), 8 weeks old at the start of the experiment. APM was administered in feed at concentrations of 100,000, 50,000, 10,000, 2,000, 400, 80, or 0 ppm. Treatment lasted until spontaneous death of the animals. The results of the study demonstrate that APM causes: (*a*) an increased incidence of malignant tumor-bearing animals, with a positive significant trend in both sexes, and in particular in females treated at 50,000 ppm ($P \leq 0.01$) when compared to controls; (*b*) an increase in lymphomas–leukemias, with a positive significant trend in both sexes, and in particular in females treated at doses of 100,000 ($P \leq 0.01$), 50,000 ($P \leq 0.01$), 10,000 ($P \leq 0.05$), 2000 ($P \leq 0.05$), and 400 ppm ($P \leq 0.01$); (*c*) a statistically significant increased incidence, with a positive significant trend, of transitional cell carcinomas of the renal pelvis and ureter in females and particularly in those treated at 100,000 ppm ($P \leq 0.05$); and (*d*) an increased incidence of malignant

Address for correspondence: Morando Soffritti, M.D., Cesare Maltoni Cancer Research Center, European Ramazzini Foundation, Castello di Bentivoglio, Via Saliceto, 3, 40010 Bentivoglio, Bologna, Italy. Voice: +39-051-6640460; fax: +39-051-6640223.
e-mail: crcfr@ramazzini.it; www.ramazzini.it

Funding for this research was provided entirely by the European Foundation on Oncology and Environmental Sciences "B. Ramazzini."

schwannomas of the peripheral nerves, with a positive trend in males ($P \leq 0.05$). The results of this mega-experiment indicate that APM, in the tested experimental conditions, is a multipotential carcinogenic agent.

KEYWORDS: aspartame; carcinogenicity; long-term bioassays; rat

INTRODUCTION

The introduction of artificial sweeteners as substitutes for sucrose began during World Wars I and II, when the use of saccharin became prevalent due to its low cost and the wartime shortage of table sugar.[1] In the following years, two additional artificial sweeteners were introduced to the market: cyclamate in the 1950s and aspartame (APM) in 1981. Since the 1970s, the growing obesity problem in industrialized countries, due in part to fast food and soft drink consumption, has lead to an increased demand for reduced-calorie foodstuffs. Given the lucrative market for these so-called "diet" or "light" products, additional new-generation sweeteners have emerged, including acesulfame-K, sucralose, and neotame.[2]

With the expansion of the artificial sweetener market, concerns have arisen among consumers regarding the safety of these sweeteners and their possible long-term health effects. At the center of this debate has been the question of the potential carcinogenic risks associated with artificial sweetener use. Until now, no adequate epidemiological or experimental animal studies have been available.

Most epidemiological studies, aimed to evaluate the relationship between artificial sweetener intake and cancer, have focused on sweetener consumption in general, and not on single compounds.[3] This limitation is attributed to the fact that most consumers use multiple artificial sweeteners, as different sweeteners are often blended together in food products. Moreover, given the fact that wide consumption of artificial sweeteners emerged in the 1980s and 1990s, epidemiological studies are, by definition, limited in terms of exposure to the compounds.

Most long-term carcinogenicity bioassays performed on rodents over the last 30 years have not been adequately designed to assess carcinogenic risk. The sensitivity of these studies in detecting risk has been greatly limited by the following factors: (*a*) the number of animals per sex per group was usually 50 or less; (*b*) the experiments were usually truncated at 104 weeks (or earlier) from the start of the experiment, thus not allowing the tested compound to express its carcinogenic potential; and (*c*) the conduct of the experiments was often inadequate with incomplete or nonsystematic histopathological analysis for all organs and tissues.

Because of the globalization of the industrialized diet and the ever-increasing use of artificial sweeteners among billions of people in both industrialized and developing countries, and because of the inadequacy of experimental data to evaluate the potential carcinogenic effects of artificial sweeteners, the European Ramazzini Foundation (ERF) began an integrated project of mega-experiments to test the carcinogenic potential of artificial sweeteners in the late 1990s. In the framework of this project, a long-term carcinogenicity bioassay on APM was begun in 1997, in which the sweetener was administered in feed to 1800 Sprague-Dawley rats for the life span. This article reports the complete results of this study.

BACKGROUND

APM, the methyl ester of the dipeptide L-α-aspartyl-L-phenylalanine, has a molecular weight of 294.3 and the following structural formula:

$$^{+}_{3}HN-\underset{\underset{\underset{O^{-}}{\overset{\parallel}{C}}}{\overset{|}{\underset{|}{C}H_{2}}}}{\overset{H}{\underset{|}{C}}}-\overset{O}{\overset{\parallel}{C}}-\overset{H}{\underset{|}{N}}-\underset{\underset{\underset{}{\bigcirc}}{\overset{|}{C}H_{2}}}{\overset{H}{\underset{|}{C}}}-\overset{O}{\overset{\parallel}{C}}-OCH_{3}$$

Under particular conditions (extreme pH, high temperature, lengthy storage times), APM may be contaminated by the diketopiperazine cycloaspartylphenylalanine (DKP).[4]

APM was accidentally discovered in the early 1960s, in connection with a research project planned at the Searle & Co. laboratories to find an inhibitor of the gastrointestinal secretory hormone gastrin for the treatment of ulcers.[5] For more than 30 years, APM has been increasingly used as a food additive due to its very strong, sweet taste, estimated to be 200 times that of sucrose.

After saccharin, APM is the second most used artificial sweetener in the world,[6] with an estimated consumption of more than 8000 tons per year in the United States alone.[7] The worldwide production of APM is assumed to be over 16,000 tons per year.[8] More than 6000 products contain APM, including soft drinks, chewing gum, table-top sweeteners, candy, desserts, yogurt, and some pharmaceutical products, such as vitamins and sugar-free cough drops. APM is estimated to be consumed by over 200 million people worldwide.[9]

According to dietary surveys performed in the United States among APM consumers during the period 1984–1992, the average daily intake of APM ranges from 2 to 3 mg/kg of body weight (h.w.) in the general population.[10] These surveys also show that consumption by children and young women range from about 2.5 to 5 mg/kg b.w./day.[10] The Acceptable Daily Intake (ADI) of APM in the United States is 50 mg/kg b.w. and in Europe is 40 mg/kg b.w.[10]

APM is metabolized in rodents, nonhuman primates, and humans in the gastrointestinal tract into three constituents (aspartic acid, phenylalanine, and methanol) which are then absorbed and enter into the systemic circulation.[11] After absorption, these compounds follow the same metabolic path as when ingested through other foods: aspartate and phenylalanine are used as amino acidic building blocks for protein synthesis or transformed, respectively, into alanine plus oxalacetate[12] and tyrosine (and, partially, into phenylethylamine and phenylpyruvate).[13] Methanol is oxidized to formaldehyde and then to formic acid.[14]

APM has been tested for genotoxicity in both *in vivo* and *in vitro* tests. *In vitro*, an assay to measure the induction of unscheduled DNA synthesis in rat hepatocytes was reported to be negative, suggesting the absence of induced DNA damage by APM.[15] APM was also evaluated *in vitro* in a chromosomal aberration test, a sister chromatide exchange (SCE) test, and in a micronuclei test on human lymphocytes.[16] In the chromosomal aberration test, statistically significant increases (2.5-4.2-fold, compared to control values) in the percentage of aberrant cells or in the number of chromosomal aberrations per cell were observed in all doses. No effect of APM was observed in the SCE test. In the micronuclei test, a statistically increased incidence in cells with micronucleus was observed at the highest dose of treatment.

In vivo results of a test for the induction of chromosomal aberration in bone marrow cells of male Swiss mice, after the administration by gavage of a mixture of APM (up to 350 mg/kg) and acesulfame potassium (up to 150 mg/kg), were negative. A dose-related increase in the percentage of cells with chromosomal aberrations was noted with increasing doses of the two sweeteners; however, the increase was not statistically significant.[17] In a peripheral blood micronuclei test conducted on p53 haploinsufficient mice exposed for 9 months to 50,000, 25,000, 12,500, 6250, 3125, or 0 ppm to APM in feed, the results were judged to be positive in females on the basis of a significant trend test and the increased frequency of micronucleated erythrocytes observed in the 50,000 ppm group.[18]

Epidemiological studies to evaluate the relationship between APM intake and the development of cancer in humans are not currently available, with the exception of one study in which an increased incidence of brain tumors in the United States between the 1970s and 1980s was linked to agents/situations of risk of environmental origin, and among them, consumption of APM.[19]

Four long-term experimental bioassays were performed on rodents in the 1970s and early 1980s. Two long-term feeding carcinogenicity bioassays on APM were performed on Sprague-Dawley rats and one on mice by Searle

& Co., the results of which were reviewed by the FDA and summarized in the Federal Register of 1981.[20] To date, the details of these experiments have not been published. A fourth experiment was performed on Wistar rats by Japanese researchers and the results published in 1981, without exhaustive experimental details.[21,22] The results of these four experiments did not show any carcinogenic effects of APM in the tested experimental conditions. The study design, conduct, and results of these experiments were discussed by the ERF in a previous article.[23]

In 2005, a carcinogenicity study on APM was performed by the U.S. National Toxicology Program on genetically altered strains of mice, namely p53 haploinsufficient, Tg AC hemizygous, and Cdkn2a deficient male and female mice which develop, with increased susceptibility and decreased latency, lymphomas or sarcomas, squamous cell papillomas/carcinomas of the forestomach and brain tumors, respectively.[18] Feed containing 50,000, 25,000, 12,500, 6250, 3125, or 0 ppm was administered for 40 weeks to groups of 15 males and 15 females. Although the Technical Report states that in the tested experimental conditions, no evidence of carcinogenic effects was observed, the conclusions of the study also include the following qualification: "because this is a new model, there is uncertainty whether the study possessed sufficient sensitivity to detect a carcinogenic effect."[18]

In light of the inadequacies and uncertainties surrounding the available epidemiological and long-term experimental data on APM, the Cesare Maltoni Cancer Research Center (CMCRC)/ERF decided to perform a life-span megaexperiment which would evaluate the carcinogenic potential of APM when administered in feed to Sprague-Dawley rats.

MATERIALS AND METHODS

The APM used as a food grade material was produced by Nutrasweet and supplied by Giusto Faravelli S.p.A. in Milan, Italy. Its purity was >98%. The impurities included DKP <1.5% and L-phenylalanine <0.5%. The method used to determine its purity was an infrared absorption spectrophotometer assay. APM was added to the standard Corticella pellet diet, used for 30 years at the CMCRC/ERF Laboratory, at concentrations of 100,000, 50,000, 10,000, 2000, 400, 80, or 0 ppm, to simulate an assumed daily intake by humans of 5000, 2500, 500, 100, 20, 4, or 0 mg/kg b.w. The APM daily consumption in mg/kg b.w. for both males and females was calculated considering the average weight of a rat as 400 g for the duration of the experiment and the average consumption of feed as 20 g per day. APM was administered in feed *ad libitum* to Sprague-Dawley rats (100–150/sex/group), 8 weeks old at the start of the experiment. The treatment lasted until natural death. Control animals received the same feed without APM. Upon death, all animals underwent complete necropsy. The general protocols of the experiment, including methods of tumor reporting and

TABLE 1. Long-term carcinogenicity bioassay on ASPARTAME, administered with feed, supplied *ad libitum*, to male and female Sprague-Dawley rats

			Groups																									
	I: 100,000 ppm				II: 50,000 ppm				III: 10,000 ppm				IV: 2,000 ppm				V: 400 ppm				VI: 80 ppm				VII: 0 ppm (control)			
	Male		Female		Male		Female		Male		Female		Male		Female		Male		Female		Male		Female		Male		Female	
Site / Histotype	No.	%	No.	%	No.	%	No.	%	No.	%	No.	%	No.	%	No.	%	No.	%	No.	%	No.	%	No.	%	No.	%	No.	%
Skin																												
Acanthoma	0	—	0	—	0	—	0	—	0	—	0	—	0	—	0	—	0	—	0	—	0	—	1	0.7	0	—	0	—
Dermatofibroma	0	—	0	—	0	—	0	—	0	—	0	—	0	—	0	—	0	—	0	—	1	0.7	0	—	1	0.7	0	—
Subcutaneous tissue																												
Basal cell adenoma	0	—	0	—	1	1.0	0	—	0	—	0	—	0	—	0	—	0	—	0	—	0	—	0	—	0	—	0	—
Fibroma	1	1.0	0	—	1	1.0	1	1.0	1	1.0	1	1.0	1	0.7	0	—	5	3.3	1	0.7	3	2.0	1	0.7	1	0.7	0	—
Fibromyxoma	0	—	0	—	0	—	0	—	0	—	0	—	0	—	0	—	0	—	0	—	0	—	0	—	0	—	0	—
Lipoma & fibrolipoma	0	—	0	—	3	3.0	0	—	2	2.0	1	1.0	10 (13)	6.7	0	—	4	2.7	0	—	4	2.7	0	—	4	2.7	1	0.7
Fibroangioma	1	1.0	0	—	0	—	0	—	0	—	0	—	0	—	0	—	0	—	0	—	0	—	1	0.7	0	—	0	—
Fibrosarcoma	0	—	0	—	0	—	0	—	0	—	0	—	1	0.7	1	0.7	0	—	1	0.7	1	0.7	2	1.3	1	0.7	1	0.7
Liposarcoma	0	—	0	—	0	—	0	—	0	—	0	—	0	—	0	—	0	—	0	—	1	0.7	1	0.7	0	—	1	0.7
Rhabdomyosarcoma	0	—	0	—	0	—	0	—	0	—	0	—	0	—	0	—	0	—	0	—	0	—	0	—	0	—	1	0.7
Interscapular fat pad																												
Fibrolipoosarcoma	0	—	1	1.0	0	—	0	—	0	—	0	—	0	—	0	—	0	—	0	—	0	—	0	—	0	—	0	—
Mammary glands																												
Adenoma & fibroma & fibroadenoma	4	4.0	58 (87)	58.0	7	7.0	59 (88)	59.0	4	4.0	57 (92)	57.0	4	2.7	68 (96)	45.3	9	6.0	89 (134)	59.3	2	1.3	67 (89)	44.7	6	4.0	89 (126)	59.3
Lipoma & fibrolipoma	1	1.0	0	—	1	1.0	1	1.0	1	1.0	2	2.0	1	0.7	0	—	0	—	0	—	0	—	0	—	3	2.0	2	1.3
Adenocarcinoma	1	1.0	7	7.0	18 (19)	18.0	18 (19)	18.0	1	1.0	7 (8)	7.0	1	0.7	12	8.0	2	1.3	16 (18)	10.7	1	0.7	15 (23)	10.0	0	—	8 (9)	5.3
Fibrosarcoma	0	—	0	—	1	1.0	1	1.0	0	—	0	—	0	—	0	—	0	—	0	—	0	—	0	—	0	—	0	—
Liposarcoma	0	—	0	—	0	—	0	—	0	—	0	—	0	—	1	0.7	0	—	1	0.7	0	—	1	0.7	0	—	0	—
Carcinosarcoma	0	—	0	—	0	—	0	—	0	—	0	—	0	—	0	—	0	—	0	—	0	—	1	0.7	0	—	0	—
Zymbal glands																												
Sebaceous carcinoma	0	—	0	—	0	—	0	—	0	—	0	—	0	—	0	—	0	—	0	—	0	—	1	0.7	0	—	0	—
Squamous cell carcinoma	1	1.0	1	1.0	3	3.0	1	1.0	2	2.0	2	2.0	0	—	2	1.3	3	2.0	3	2.0	1	0.7	1	0.7	2	1.3	2	1.3
Ear ducts																												
Acanthoma	0	—	0	—	0	—	1	1.0	0	—	0	—	0	—	0	—	0	—	1	0.7	0	—	1	0.7	0	—	0	—
Squamous cell carcinoma	3	3.0	6	6.0	2	2.0	7	7.0	7	7.0	7	7.0	7	4.7	0	—	3	2.0	2	1.3	3	2.0	4	2.7	5	3.3	9	6.0

Continued

TABLE 1. Continued.

			I: 100,000 ppm				II: 50,000 ppm				III: 10,000 ppm				IV: 2,000 ppm				V: 400 ppm				VI: 80 ppm				VII: 0 ppm (control)			
		Male		Female		Male		Female		Male		Female		Male		Female		Male		Female		Male		Female		Male		Female	
Site Histotype[b]	No.	%	No.	%	No.	%	No.	%	No.	%	No.	%	No.	%	No.	%	No.	%	No.	%	No.	%	No.	%	No.	%	No.	%
Nasal cavities[b]																												
Adenoma	0	—	0	—	1	1.0	0	—	2	2.0	2	2.0	1	0.7	1	0.7	0	—	1	0.7	1	0.7	2	1.3	0	—	0	—
Squamous cell carcinoma	1	1.0	0	—	0	—	0	—	0	—	0	—	0	—	0	—	0	—	0	—	0	—	0	—	1	0.7	0	—
Olfactory neuroblastoma	0	—	1	1.0	0	—	0	—	0	—	0	—	0	—	0	—	0	—	0	—	0	—	0	—	0	—	0	—
Oral cavity & lips																												
Acanthoma	0	—	2	2.0	1	1.0	0	—	1	1.0	1	1.0	1	0.7	0	—	1	0.7	0	—	0	—	1	0.7	0	—	0	—
Squamous cell carcinoma	1	1.0	2	2.0	0	—	1	1.0	1	1.0	1	1.0	0	—	1	0.7	2	1.3	4	2.7	0	—	0	—	0	—	2	1.3
Pharynx																												
Squamous cell carcinoma	0	—	0	—	0	—	0	—	0	—	0	—	0	—	0	—	0	—	0	—	0	—	0	—	1	—	1	0.7
Lung																												
Adenoma	1	1.0	0	—	0	—	2	2.0	0	—	0	—	1	0.7	2	1.3	0	—	0	—	0	—	0	—	0	—	0	—
Fibroma	0	—	0	—	0	—	0	—	0	—	0	—	0	—	0	—	1	0.7	0	—	0	—	0	—	0	—	0	—
Fibroangioma	0	—	0	—	1	1.0	0	—	0	—	0	—	0	—	2	1.3	0	—	0	—	1	0.7	0	—	0	—	0	—
Squamous cell carcinoma	0	—	0	—	0	—	1	1.0	0	—	0	—	0	—	0	—	1	0.7	1	0.7	1	0.7	0	—	0	—	0	—
Hemangiosarcoma	1	1.0	0	—	1	1.0	0	—	0	—	0	—	0	—	0	—	0	—	0	—	0	—	0	—	0	—	0	—
Pleura																												
Mesothelioma	0	—	0	—	0	—	0	—	0	—	0	—	0	—	0	—	0	—	0	—	0	—	0	—	1	0.7	0	—
Stomach - Forestomach																												
Polyp	0	—	0	—	0	—	0	—	0	—	0	—	0	—	0	—	0	—	1	0.7	0	—	0	—	0	—	0	—
Acanthoma	3	3.0	2	2.0	3	3.0	0	—	2	2.0	3	3.0	3	2.0	0	—	2	1.3	1	0.7	1	—	1	0.7	1	0.7	4	2.7
Squamous cell carcinoma	0	—	1	—	1	1.0	0	—	0	—	0	—	0	—	0	—	0	—	0	—	1	0.7	0	—	0	—	0	—
- Glandular																												
Polyp	0	—	0	—	0	—	0	—	0	—	0	—	0	—	0	—	0	—	0	—	1	0.7	0	—	0	—	0	—
Leiomyosarcoma	0	—	0	—	0	—	0	—	0	—	0	—	0	—	1	0.7	0	—	0	—	0	—	0	—	0	—	0	—
Neuroendocrine malignant tumor	0	—	1	1.0	0	—	0	—	0	—	0	—	0	—	0	—	0	—	0	—	0	—	0	—	0	—	0	—
Intestine																												
Adenomatous polyp	0	—	0	—	0	—	2	2.0	1	1.0	1	1.0	1	0.7	1	0.7	0	—	0	—	2	1.3	0	—	1	—	1	0.7
Fibroma	0	—	0	—	0	—	0	—	0	—	0	—	0	—	0	—	0	—	0	—	0	—	0	—	1	0.7	0	—
Leiomyoma	0	—	0	—	0	—	0	—	0	—	0	—	0	—	0	—	2	1.3	0	—	0	—	0	—	0	—	1	0.7
Adenocarcinoma	0	—	0	—	0	—	0	—	0	—	1	1.0	1	0.7	1	0.7	1	0.7	0	—	0	—	0	—	0	—	2	1.3
Leiomyosarcoma	0	—	0	—	1	1.0	0	—	0	—	0	—	0	—	0	—	0	—	0	—	0	—	0	—	1	0.7	0	—

Continued

TABLE 1. Continued.

	Groups														
	I: 100,000 ppm				II: 50,000 ppm				III: 10,000 ppm				IV: 2,000 ppm		
	Male		Female		Male		Female		Male		Female		Male		Female	
Site / Histotype	No.	%	No.	%	No.	%	No.	%	No.	%	No.	%	No.	%	No.	%
Salivary glands																
Adenoma	0	—	0	—	0	—	0	—	0	—	0	—	0	—	0	—
Basal cell carcinoma	0	—	0	—	0	—	0	—	0	—	0	—	0	—	0	—
Liposarcoma & Fibroliposarcoma	0	—	0	—	0	—	0	—	0	—	0	—	1	0.7	0	—
Liver																
Hepatoma	0	—	0	—	1	1.0	1	—	0	—	0	—	1	0.7	0	—
Cholangioma	0	—	4	4.0	2	—	2	2.0	0	—	0	—	2	1.3	0	—
Fibroangioma	0	—	0	—	0	—	0	—	0	—	1	1.0	0	—	0	—
Hepatocarcinoma	1	1.0	0	—	0	—	0	—	0	—	0	—	1	0.7	0	—
Cholangiocarcinoma	0	—	0	—	0	—	0	—	0	—	1	1.0	0	—	0	—
Hemangiosarcoma	0	—	0	—	1	1.0	0	—	0	—	0	—	0	—	0	—
Pancreas																
Exocrine adenoma	0	—	0	—	0	—	1	1.0	0	—	0	—	0	—	0	—
Islet cell adenoma	9	9.0	0	—	6	6.0	1	1.0	7	7.0	0	—	5	3.3	2	1.3
Islet cell carcinoma	0	—	0	—	0	—	0	—	0	—	0	—	0	—	0	—
Kidneys																
Adenoma	0	—	1	1.0	0	—	0	—	0	—	0	—	0	—	1	0.7
Nephroma	0	—	0	—	0	—	0	—	0	—	0	—	0	—	1	0.7
Fibroangioma	0	—	0	—	0	—	0	—	0	—	0	—	0	—	0	—
Tubular cell carcinoma	0	—	0	—	0	—	0	—	0	—	0	—	0	—	0	—
Pelvis[c]																
Papilloma	0	—	3	3.0	0	—	1	1.0	1	1.0	4	4.0	2	—	2	1.3
Transitional cell carcinoma	1	1.0	4	4.0	1	1.0	3	3.0	1	1.0	3 (4)	3.0	1	0.7	3 (4)	2.0
Bladder																
Papilloma	0	—	0	—	1	1.0	0	—	0	—	0	—	0	—	0	—
Transitional cell carcinoma	0	—	0	—	0	—	0	—	2	2.0	0	—	0	—	1	0.7
Prostate																
Adenoma	1	1.0			2	2.0			2	2.0			0	—		
Seminal vesicles																
Rhabdomyosarcoma	0	—			0	—			0	—			1	0.7		
Testes																
Interstitial cell adenoma	1	1.0			3 (5)	3.0			2	2.0			3 (4)	2.0		
Interstitial cell carcinoma	1	1.0			0	—			0	—			0	—		

	Groups												
	V: 400 ppm				VI: 80 ppm				VII: 0 ppm (control)				
	Male		Female		Male		Female		Male		Female		
Site / Histotype	No.	%	No.	%	No.	%	No.	%	No.	%	No.	%	
Salivary glands													
Adenoma	0	—	0	—	0	—	0	—	0	—	1	0.7	
Basal cell carcinoma	1	0.7	0	—	0	—	0	—	0	—	0	—	
Liposarcoma & Fibroliposarcoma	0	—	0	—	0	—	0	—	1	0.7	0	—	
Liver													
Hepatoma	4	2.7	2	—	0	—	1	0.7	0	—	2	1.3	
Cholangioma	4	2.7	2	1.3	0	—	0	—	0	—	3	2.0	
Fibroangioma	0	—	1	—	0	—	0	—	0	—	0	—	
Hepatocarcinoma	0	—	1	0.7	0	—	0	—	0	—	0	—	
Cholangiocarcinoma	0	—	0	—	0	—	1	0.7	0	—	0	—	
Hemangiosarcoma	0	—	0	—	0	—	0	—	0	—	0	—	
Pancreas													
Exocrine adenoma	0	—	1	0.7	0	—	0	—	0	—	0	—	
Islet cell adenoma	10	6.7	1	0.7	3	2.0	4	2.7	7	4.7	1	0.7	
Islet cell carcinoma	0	—	1	0.7	0	—	0	—	0	—	0	—	
Kidneys													
Adenoma	0	—	0	—	0	—	0	—	1	0.7	0	—	
Nephroma	0	—	0	—	0	—	0	—	0	—	0	—	
Fibroangioma	0	—	0	—	1	0.7	0	—	0	—	0	—	
Tubular cell carcinoma	0	—	1	0.7	0	—	0	—	0	—	0	—	
Pelvis[c]													
Papilloma	2	1.3	2	1.3	0	—	3	2.0	0	—	1	0.7	
Transitional cell carcinoma	3	—	3	2.0	0	—	1	0.7	0	—	0	—	
Bladder													
Papilloma	0	—	0	—	1	0.7	0	—	0	—	0	—	
Transitional cell carcinoma	1	0.7	0	—	0	—	0	—	0	—	0	—	
Prostate													
Adenoma	0	—			0	—			1	0.7			
Seminal vesicles													
Rhabdomyosarcoma	1	0.7			0	—			0	—			
Testes													
Interstitial cell adenoma	3 (4)	2.0			5	3.3			2 (4)	1.3			
Interstitial cell carcinoma	0	—			0	—			0	—			

Continued

TABLE 1. Continued.

			Groups																								
	I: 100,000 ppm				II: 50,000 ppm				III: 10,000 ppm				IV: 2,000 ppm				V: 400 ppm				VI: 80 ppm				VII: 0 ppm (control)		
	Male		Female		Male		Female		Male		Female		Male		Female		Male		Female		Male		Female		Male		Female	
Site Histotype	No.	%	No.	%	No.	%	No.	%	No.	%	No.	%	No.	%	No.	%	No.	%	No.	%	No.	%	No.	%	No.	%	No.	%
Ovaries																												
Adenoma & cystadenoma			1	1.0			0	—			1	1.0			8	5.3			2	1.3			4	2.7			5	3.3
Granulosa &/or theca cell tumor			2	2.0			0	—			0	—			2	1.3			2	1.3			2	1.3			1	0.7
Granular cell tumor			0	—			0	—			1	1.0			0	—			0	—			0	—			0	—
Sertoli cell tumor			0	—			0	—			0	—			0	—			0	—			0	—			1	0.7
Fibroma			1	—			0	—			0	—			0	—			0	—			0	—			0	—
Fibroangioma			3(4)	3.0			3	3.0			1	1.0			1	0.7			1	0.7			1	0.7			0	—
Adenocarcinoma			1	1.0			1	1.0			0	—			1	0.7			0	—			0	—			0	—
Malignant Sertoli cell tumor			0	—			0	—			0	—			1	0.7			0	—			0	—			0	—
Uterus																												
Polyp			42	42.0			39	39.0			28	28.0			23	15.3			35	23.3			51	34.0			38	25.3
Adenoma & fibroadenoma			0	—			0	—			0	—			1	0.7			0	—			1	0.7			2	1.3
Fibroma			0	—			2	2.0			0	—			0	—			0	—			0	—			0	—
Fibroangioma			1	1.0			1	1.0			1	1.0			2	1.3			0	—			1	0.7			3	2.0
Leiomyoma			0	—			0	—			0	—			1	0.7			0	—			0	—			0	—
Squamous cell carcinoma			0	—			0	—			2	2.0			2	1.3			1	0.7			0	—			3	2.0
Adenocarcinoma			2	2.0			4	4.0			1	1.0			3	2.0			1	0.7			0	—			0	—
Hemangiosarcoma			0	—			0	—			0	—			1	0.7			0	—			0	—			3	2.0
Leiomyosarcoma			0	—			1	1.0			0	—			0	—			0	—			1	0.7			1	0.7
Sarcoma botryoides			0	—			0	—			0	—			0	—			0	—			1	0.7			0	—
Malignant schwannoma			3	3.0			1	1.0			4	4.0			3	2.0			4	2.7			2	1.3			6	4.0
Vagina																												
Polyp			1	1.0			1	1.0			0	—			0	—			0	—			2	1.3			0	—
Acanthoma			0	—			0	—			0	—			0	—			0	—			0	—			0	—
Fibroma			0	—			0	—			0	—			1	0.7			1	0.7			0	—			0	—
Fibroangioma			0	—			0	—			0	—			1	0.7			0	—			0	—			0	—
Benign schwannoma			0	—			0	—			0	—			1	0.7			1	0.7			0	—			0	—
Hemangiosarcoma			0	—			0	—			0	—			1	0.7			0	—			0	—			0	—
Peritoneum																												
Lipoma & fibrolipoma	0	—	0	—	0	—	0	—	0	—	0	—	0	—	0	—	0	—	0	—	0	—	0	—	1	0.7	0	—
Fibroangioma	0	—	0	—	1	1.0	0	—	0	—	0	—	0	—	0	—	1	0.7	0	—	1	0.7	0	—	1	0.7	0	—
Mesothelioma	0	—	0	—	1	1.0	2	2.0	0	—	0	—	3	2.0	3	2.0	0	—	0	—	0	—	1	0.7	0	—	0	—
Hemangiosarcoma	0	—	0	—	0	—	0	—	3	3.0	0	—	1	0.7	1	0.7	0	—	0	—	0	—	0	—	0	—	0	—
Pituitary gland																												
Adenoma	35	35.0	32	32.0	45	45.0	35	35.0	53	53.0	27	27.0	54	36.0	37	24.7	55	36.7	34	22.7	57	38.0	29	19.3	66	44.0	51	34.0
Adenocarcinoma	2	2.0	1	—	1	1.0	3	3.0	2	2.0	1	1.0	6	4.0	3	2.0	1	0.7	3	2.0	5	3.3	1	0.7	1	0.7	1	—
Thyroid gland																												
Follicular adenoma	0	—	1	—	0	—	1	1.0	1	1.0	0	—	1	0.7	0	—	0	—	0	—	0	—	0	—	0	—	1	0.7
C-cell adenoma	5	5.0	5(6)	5.0	5	5.0	2	2.0	2	2.0	7	7.0	6	4.0	8(9)	5.3	6	4.0	7	4.7	4	2.7	5(6)	3.3	3	2.0	7(8)	4.7
Fibroangioma	0	—	0	—	0	—	0	—	0	—	0	—	0	—	0	—	0	—	0	—	0	—	0	—	0	—	0	—
C-cell carcinoma	0	—	2	2.0	0	—	2	2.0	3	3.0	0	—	2	1.3	2(3)	1.3	2	1.3	3	2.0	1	0.7	2	1.3	1	0.7	2	1.3
Parathyroid gland																												
Adenoma	0	—	0	—	0	—	0	—	0	—	0	—	0	—	1	0.7	0	—	1	0.7	1	0.7	2	1.3	0	—	0	—

Continued.

TABLE 1. Continued.

Site Histotype	I: 100,000 ppm Male No.	%	I: 100,000 ppm Female No.	%	II: 50,000 ppm Male No.	%	II: 50,000 ppm Female No.	%	III: 10,000 ppm Male No.	%	III: 10,000 ppm Female No.	%	IV: 2,000 ppm Male No.	%	IV: 2,000 ppm Female No.	%	V: 400 ppm Male No.	%	V: 400 ppm Female No.	%	VI: 80 ppm Male No.	%	VI: 80 ppm Female No.	%	VII: 0 ppm (control) Male No.	%	VII: 0 ppm (control) Female No.	%
Adrenal glands																											
Cortical adenoma	2	2.0	8	8.0	2	2.0	2 (4)	2.0	5	5.0	7	7.0	4	2.7	11 (13)	7.3	1	0.7	9	6.0	2	1.3	10 (11)	6.7	4	2.7	4	2.7
Pheochromocytoma	19 (22)	19.0	29 (34)	29.0	21 (25)	21.0	24 (32)	24.0	19 (26)	19.0	24 (28)	24.0	15 (18)	10.0	37 (45)	24.7	25 (28)	16.7	22 (28)	14.7	20 (22)	13.3	26 (28)	17.3	18	12.0	36 (49)	24.0
Ganglioneuroma	0	–	0	–	0	–	0	–	0	–	0	–	0	–	0	–	0	–	1	0.7	0	–	0	–	0	–	1	0.7
Cortical adenocarcinoma	0	–	1	1.0	0	–	2	2.0	0	–	2	2.0	0	–	2	1.3	0	–	2	1.3	0	–	2	1.3	1	0.7	3	2.0
Pheochromoblastoma	2	2.0	0	–	4	4.0	5	5.0	1	1.0	1 (2)	1.0	0	–	2	1.3	0	–	2	1.3	0	–	4	2.7	0	–	0	–
Ganglioneuroblastoma	0	–	0	–	0	–	0	–	0	–	0	–	0	–	0	–	0	–	0	–	0	–	0	–	1	0.7	0	–
Central nervous system																												
- Brain																												
Granular cell tumor	0	–	0	–	0	–	0	–	2	2.0	1	1.0	0	–	0	–	0	–	0	–	0	–	0	–	0	–	0	–
Astrocytoma	0	–	1	1.0	0	–	0	–	0	–	0	–	0	–	0	–	0	–	0	–	0	–	0	–	0	–	0	–
Oligodendroglioma	1	1.0	0	–	2	2.0	0	–	0	–	0	–	1	0.7	0	–	0	–	0	–	2	1.3	1	0.7	0	–	0	–
Multiform glioblastoma	0	–	0	–	0	–	1	1.0	0	–	0	–	0	–	1	0.7	0	–	0	–	0	–	0	–	0	–	0	–
Medulloblastoma	0	–	0	–	0	–	0	–	0	–	1	1.0	0	–	0	–	0	–	0	–	0	–	0	–	0	–	0	–
- Meninges																												
Benign meningioma	1	1.0	1	1.0	5	5.0	1	1.0	2	2.0	1	1.0	2	1.3	2	1.3	3	2.0	2	1.3	1	0.7	1	0.7	2	1.3	0	–
Malignant meningioma	0	–	0	–	0	–	0	–	0	–	0	–	1	0.7	0	–	0	–	0	–	0	–	0	–	0	–	0	–
Peripheral nervous system[d]																												
-Cranial nerves																												
Benign ganglioneuroma	0	–	0	–	1	1.0	0	–	0	–	0	–	0	–	0	–	1	0.7	0	–	0	–	0	–	0	–	0	–
Benign schwannoma	0	–	0	–	0	–	0	–	0	–	0	–	0	–	2	1.3	0	–	0	–	0	–	0	–	0	–	0	–
Malignant schwannoma	3	3.0	1	1.0	3	3.0	1	1.0	2	2.0	1	1.0	2	1.3	1	0.7	1	0.7	0	–	1	0.7	1	0.7	1	0.7	0	–
- Other peripheral nerves																												
Malignant schwannoma	1	1.0	1	1.0	0	–	0	–	0	–	0	–	0	–	2	1.3	2	1.3	0	–	0	–	1	0.7	0	–	0	–
Bones																												
- Head																												
Chondroma	0	–	0	–	0	–	0	–	1	1.0	1	1.0	1	0.7	0	–	0	–	0	–	0	–	0	–	0	–	0	–
Osteoma	0	–	2	2.0	0	–	0	–	3	3.0	1	1.0	1	0.7	1	0.7	2	1.3	1	0.7	4	2.7	1	0.7	0	–	0	–
Osteosarcoma	3	3.0	4	4.0	2	2.0	3	3.0	5	5.0	3	3.0	9	6.0	4	2.7	5	3.3	3	2.0	7	4.7	9	6.0	5	3.3	7	4.7

Continued.

TABLE 1. Continued.

														Groups													
		I: 100,000 ppm				II: 50,000 ppm				III: 10,000 ppm				IV: 2,000 ppm				V: 400 ppm				VI: 80 ppm				VII: 0 ppm (control)	
		Male		Female		Male		Female		Male		Female		Male		Female		Male		Female		Male		Female		Male		Female
Site / Histotype	No.	%	No.	%	No.	%	No.	%	No.	%	No.	%	No.	%	No.	%	No.	%	No.	%	No.	%	No.	%	No.	%	No.	%
Others																												
Osteosarcoma	0	–	0	–	0	–	0	–	0	–	0	–	1	0.7	1	0.7	1	0.7	1	–	0	–	0	–	2	1.3	2	1.3
Soft tissues																												
Fibroma	0	–	0	–	0	–	0	–	1	1.0	0	–	0	–	0	–	0	–	0	–	0	–	0	–	0	–	0	–
Lipoma & fibrolipoma	0	–	2	2.0	0	–	1	1.0	0	–	1	1.0	0	–	0	–	0	–	0	–	0	–	0	–	0	–	0	–
Fibroangioma	0	–	1	1.0	0	–	0	–	0	–	1	1.0	0	–	0	–	0	–	0	–	0	–	0	–	1	0.7	1	0.7
Liposarcoma	0	–	0	–	0	–	0	–	0	–	0	–	0	–	0	–	0	–	1	0.7	0	–	0	–	0	–	0	–
Hemangiosarcoma	1	1.0	0	–	1	1.0	0	–	1	1.0	1	1.0	0	–	0	–	0	–	0	–	0	–	0	–	1	0.7	0	–
Heart																												
Myxoma	0	–	2	2.0	0	–	0	–	1	1.0	0	–	1	0.7	0	–	1	0.7	1	0.7	1	0.7	2	1.3	1	0.7	0	–
Benign schwannoma	0	–	1	1.0	0	–	0	–	0	–	0	–	1	0.7	0	–	1	0.7	1	0.7	0	–	0	–	0	–	0	–
Malignant schwannoma	1	1.0	0	–	1	1.0	0	–	1	1.0	1	1.0	0	–	0	–	0	–	0	–	1	0.7	0	–	1	0.7	0	–
Thymus																												
Benign thymoma	1	1.0	2	2.0	0	–	6	6.0	0	–	5	5.0	0	–	6	4.0	0	–	4	2.7	1	0.7	7	4.7	2	1.3	2	1.3
Fibroangioma	0	–	0	–	0	–	0	–	1	1.0	0	–	1	0.7	0	–	0	–	0	–	0	–	0	–	0	–	0	–
Malignant thymoma	0	–	0	–	0	–	0	–	0	–	0	–	0	–	0	–	0	–	0	–	1	0.7	1	0.7	1	0.7	1	0.7
Spleen																												
Fibroma	0	–	0	–	1	1.0	1	1.0	1	1.0	1	1.0	0	–	0	–	1	0.7	2	1.3	0	–	1	0.7	0	–	0	–
Fibroangioma	4	4.0	2	2.0	1	1.0	2	2.0	0	–	0	–	1	0.7	0	–	1	0.7	0	–	0	–	1	0.7	0	–	0	–
Fibrosarcoma	0	–	0	–	1	1.0	0	–	0	–	0	–	0	–	1	0.7	0	–	0	–	0	–	0	–	0	–	0	–
Hemangiosarcoma	2	2.0	0	–	0	–	0	–	1	1.0	0	–	1	0.7	0	–	0	–	0	–	1	0.7	0	–	1	0.7	1	0.7
Lymph nodes																												
Hemangioma & fibroangioma	0	–	0	–	1	1.0	0	–	2	2.0	0	–	0	–	0	–	1	0.7	1	0.7	0	–	1	0.7	0	–	0	–
Hemolymphoreticular tissues	0	–	1	1.0	0	–	0	–	0	–	0	–	0	–	0	–	0	–	0	–	1	0.7	0	–	0	–	0	–
Benign histiocytoma																												
Lymphomas & leukemias[a]	29	29.0	25	25.0	20	20.0	25	25.0	15	15.0	19	19.0	33	22.0	28	18.7	25	16.7	30 (31)	20.0	23	15.3	22	14.7	31	20.7	13	8.7
Unknown																												
Squamous cell carcinoma	0	–	0	–	0	–	0	–	0	–	0	–	0	–	0	–	0	–	0	–	0	–	0	–	0	–	1	0.7
Osteosarcoma	0	–	0	–	0	–	0	–	0	–	1	1.0	0	–	0	–	0	–	1	0.7	1	0.7	0	–	0	–	0	–

Number and percentage of male and female Sprague-Dawley rats bearing various types of benign and Malignant Tumors[a]

[a]Between brackets the numbers of tumors (one animal can bear more than one tumor)
[b]See Table 3
[c]See Table 4
[d]See Table 5
[e]See Table 6

statistical analysis, were described in detail in previous publications.[23,24] The experiment was conducted according to the Italian law regulating the use of animals for scientific purposes.[25]

RESULTS

The study proceeded smoothly without unexpected occurrences. The biophase ended at 151 weeks, with the death of the last animal at the age of 159 weeks. Results of the study are reported in previous publications.[23,24]

Water consumption did not differ among males and females of treated and control groups. A dose-related difference in food consumption was observed in both sexes during the experiment. A slight decrease in body weight was observed in females treated at the highest dose; no substantial differences were observed among treated males, compared to controls. No differences were observed in survival among males or females of the treated groups, compared to controls.

The occurrence of benign and malignant tumors among male and female rats is shown in Table 1. The differences observed among treated and control animals were as follows:

1. an increase in malignant tumor-bearing animals with a significant positive trend in males ($P \leq 0.05$) and in females ($P \leq 0.01$) and a statistically significant difference in females treated at 50,000 ppm ($P \leq 0.01$), compared to controls (Table 2);
2. an increased incidence of hyperplasia of the olfactory epithelium with a significant positive trend in males and females (Table 3). It is noteworthy that among females treated at the highest dose, one case of dysplastic hyperplasia, one adenoma, and one olfactory neuroblastoma were observed. The neuroblastoma invaded the cranium, compressing the forebrain and was positive for chromogranin A immunohistochemical staining;
3. an increase in the incidence of dysplastic hyperplasias, dysplastic papillomas, and carcinomas of the renal pelvis and ureter were observed in females (Table 4). Carcinomas in females occurred with a positive trend ($P \leq 0.05$) and specifically in females exposed at 100,0000 ppm ($P \leq 0.05$), compared with controls. Carcinomas were also observed among males treated at 100,000, 50,000, 10,000, and 2000 ppm. In females, when dysplastic lesions and carcinomas are combined, they show a significant positive trend ($P \leq 0.01$) and a statistically significant increase in those treated at 100,000 ($P \leq 0.01$), 50,000 ($P \leq 0.01$), 10,000 ($P \leq 0.01$), 2000 ($P \leq 0.05$), and 400 ppm ($P \leq 0.05$). An increased incidence of deposits of calcium (mineralization) was observed in females, particularly in those treated at 100,000 ppm (39%), 50,000 ppm (25%), or 10,000 ppm

TABLE 2. Long-term carcinogenicity bioassay on ASPARTAME, administered with feed, supplied *ad libitum*, to male (M) and female (F) Sprague-Dawley rats

Group No.	Concentration (ppm)	Animals		Tumor-bearing animals [a,b,c,d]	
		Sex	No.	No.	%
			MALIGNANT TUMORS		
I	100,000	M	100	43	43.0
		F	100	51	51.0
		M+F	200	94	47.0
II	50,000	M	100	38	38.0
		F	100	58	58.0[##]
		M+F	200	96	48.0
III	10,000	M	100	34	34.0
		F	100	40	40.0
		M+F	200	74	37.0
IV	2,000	M	150	60	40.0
		F	150	67	44.7
		M+F	300	127	42.3
V	400	M	150	48	32.0
		F	150	70	46.7
		M+F	300	118	39.3
VI	80	M	150	44	29.3
		F	150	64	42.7
		M+F	300	108	36.0
VII	0 (control)	M	150	53	35.3*
		F	150	55	36.7**
		M+F	300	108	36.0

[a] The tumors rates are based on the number of animals examined (necropsied).
[b] *P-Values* corresponding to pairwise comparisons between the controls and the dosed group are near the dosed group incidence.
[c] *P-Values* associated with the trend test are near the control incidence.
[d] Bilateral and multiple tumors were plotted as single independent tumors.
*Statistically significant ($P < 0.05$) using Cochran-Armitage trend test.
**Statistically significant ($P < 0.01$) using Cochran-Armitage trend test.
##Statistically significant ($P < 0.01$) using Poly-*k* test ($k = 3$).

(19%), compared with controls (8%). The same effect was not observed among males of the various groups. No difference was observed in the incidence of acute and chronic nephropathies among males and females of all groups. It must be noted that the nephropathy is common in the natural dying process and for this reason, is more frequently observed when animals are allowed to die spontaneously;

4. a dose-related increased incidence in malignant schwannomas of peripheral nerves was observed, with a significant positive trend in males ($P \leq 0.05$), while in females, nine malignancies were observed among treated animals of the different dosage groups and none among controls (Table 5). All lesions, in males and females, diagnosed as malignant schwannoma, were positive for S100 immunohistochemical staining. The occurrence

TABLE 3. Long-term carcinogenicity bioassay on ASPARTAME, administered with feed, supplied *ad libitum*, to male (M) and female (F) Sprague-Dawley rats

PRENEOPLASTIC AND NEOPLASTIC LESIONS OF OLFACTORY EPITHELIUM

Group No.	Concentration (ppm)	Animals		Animals with preneoplastic and neoplastic lesions [a,b,c]					
				Hyperplasia		Adenoma		Olfactory neuroblastoma	
		Sex	No.	No.	%	No.	%	No.	%
I	100,000	M	100	14	14.0[##]	0	–	0	–
		F	100	19[d]	19.0[##]	1	1.0	1	1.0
		M+F	200	33	16.5	1	0.5	1	0.5
II	50,000	M	100	12	12.0[##]	0	–	0	–
		F	100	21	21.0[##]	0	–	0	–
		M+F	200	33	16.5	0	–	0	–
III	10,000	M	100	7	7.0[##]	2	2.0	0	–
		F	100	17	17.0[##]	0	–	0	–
		M+F	200	24	12.0	2	1.0	0	–
IV	2,000	M	150	4	2.7	1	0.7	0	–
		F	150	13	8.7	1	0.7	0	–
		M+F	300	17	5.7	2	0.7	0	–
V	400	M	150	9	6.0[##]	0	–	0	–
		F	150	11	7.3	1	0.7	0	–
		M+F	300	20	6.7	1	0.3	0	–
VI	80	M	150	3	2.0	0	–	0	–
		F	150	5	3.3	2	1.3	0	–
		M+F	300	8	2.7	2	0.7	0	–
VII	0 (control)	M	150	1	0.7*[##]	0	–	0	–
		F	150	6	4.0*[##]	0	–	0	–
		M+F	300	7	2.3	0	–	0	–

[a] The tumors rates are based on the number of animals examined (necropsied).
[b] P-Values corresponding to pairwise comparisons between the controls and the dosed group are near the dosed group incidence.
[c] P-Values associated with the trend test are near the control incidence.
[d] One hyperplasia with atypia.
*Statistically significant ($P < 0.05$) using Cochran-Armitage trend test.
[##] Statistically significant ($P < 0.01$) using Poly-k test ($k = 3$).

of malignant schwannomas mostly involved cranial nerves (72%). The other cases arose from spinal nerve roots. Among three males treated at the highest dose, metastases were observed in the submandibular lymph nodes in two cases, and in the lung and liver in the third case;

5. a dose-related increased incidence in lymphomas–leukemias was observed, with a significant positive trend in males ($P \leq 0.05$) and in females ($P \leq 0.01$). When compared to controls, a statistically significant difference was observed in females treated at doses of 100,000

TABLE 4. Long term carcinogenicity bioassay on ASPARTAME, administered with feed, supplied *ad libitum*, to male (M) and female (F) Sprague-Dawley rats

PRENEOPLASTIC AND NEOPLASTIC LESIONS OF THE TRANSITIONAL CELL EPITHELIUM OF THE RENAL PELVIS AND URETER

Group No.	Concentration (ppm)	Sex	Animals No.	Animals with preneoplastic or neoplastic lesions [a,b,c,d]							
				Dysplastic hyperplasias		Dysplastic papillomas		Carcinomas		Total	
				No.	%	No.	%	No.	%	No.	%
I	100,000	M	100	3	3.0	0	—	1	1.0	4	4.0
		F	100	8	8.0	3	3.0	4	**4.0**[#]	15	**15.0**[##]
		M+F	200	11	5.5	3	1.5	5	2.5	19	9.5
II	50,000	M	100	2	2.0	0	—	1	1.0	3	3.0
		F	100	6	6.1	1	1.0	3	**3.0**	10	**10.1**[##]
		M+F	200	8	4.0	1	0.5	4	2.0	13	6.5
III	10,000	M	100	2	2.0	0	—	1	1.0	3	3.0
		F	100	6	6.0	1	1.0	3[d]	**3.0**	10	**10.0**[##]
		M+F	200	8	4.0	1	0.5	4	2.0	13	6.5
IV	2,000	M	150	4	2.7	0	—	1	0.7	5	3.3
		F	150	6	4.0	1	0.7	3[d]	**2.0**	10	**6.7**[#]
		M+F	300	10	3.3	1	0.3	4	1.3	15	5.0
V	400	M	150	4	2.7	1	0.7	0	—	5	3.4
		F	150	5	3.3	1	0.7	3	**2.0**	9	**6.0**[#]
		M+F	300	9	3.0	2	0.7	3	1.0	14	4.7
VI	80	M	150	3	2.0	0	—	0	—	3	2.0
		F	150	4	2.7	1	0.7	1	**0.7**	6	**4.0**
		M+F	300	7	2.3	1	0.3	1	0.3	9	3.0
VII	0 (control)	M	150	1	0.7	0	—	0	—	1	0.7
		F	150	2	1.3	0	—	0	—*[#]	2	**1.3****[##]
		M+F	300	3	1.0	0	—	0	—	3	1.0

[a] The tumor rates are based on the number of animals examined (necropsied).
[b] p-Values corresponding to pairwise comparisons between the controls and the dosed group are near the dosed group incidence.
[c] p-Values associated with the trend test are near the control incidence.
[d] One animal bears bilateral tumor.
*Statistically significant ($P < 0.05$) using Cochran-Armitage trend test.
**Statistically significant ($P < 0.01$) using Cochran-Armitage trend test.
#Statistically significant ($P < 0.05$) using Poly-k test ($k = 3$).
##Statistically significant ($P < 0.01$) using Poly-k test ($k = 3$).

($P \leq 0.01$), 50,000 ($P \leq 0.01$), 10,000 ($P \leq 0.05$), 2000 ($P \leq 0.05$), and 400 ($P \leq 0.01$) ppm (Table 6). Lymphomas–leukemias are neoplasias arising from hemolymphoreticular tissues and their aggregation is widely used in experimental carcinogenesis. The reason is that both solid and circulating phases are present in many lymphoid neoplasms, and the distinction between them is artificial.[26]

TABLE 5. Long-term carcinogenicity bioassay on ASPARTAME, administered with feed, supplied *ad libitum*, to male (M) and female (F) Sprague-Dawley rats

				MALIGNANT SCHWANNOMAS OF PERIPHERAL NERVES					
				Animals with tumors [a,b,c]					
Group No.	Concentration (ppm)	Animals		Cranial nerves		Other nerves		Total	
		Sex	No.	No.	%	No.	%	No.	%
I	100,000	M	100	3	3.0	1	1.0	4	4.0
		F	100	1	1.0	1	1.0	2	2.0
		M+F	200	4	2.0	2	1.0	6	3.0
II	50,000	M	100	3	3.0	0	–	3	3.0
		F	100	1	1.0	0	–	1	1.0
		M+F	200	4	2.0	0	–	4	2.0
III	10,000	M	100	2	2.0	0	–	2	2.0
		F	100	1	1.0	0	–	1	1.0
		M+F	200	3	1.5	0	–	3	1.5
IV	2,000	M	150	2	1.3	0	–	2	1.3
		F	150	1	0.7	2	1.3	3	2.0
		M+F	300	3	1.0	2	0.7	5	1.7
V	400	M	150	1	0.7	2	1.3	3	2.0
		F	150	0	–	0	–	0	–
		M+F	300	1	0.3	2	0.7	3	1.0
VI	80	M	150	1	0.7	0	–	1	0.7
		F	150	1	0.7	1	0.7	2	1.3
		M+F	300	2	0.7	1	0.3	3	1.0
VII	0 (control)	M	150	1	0.7	0	–	1	0.7*#
		F	150	0	–	0	–	0	–
		M+F	300	1	0.3	0	–	1	0.3

[a] The tumors rates are based on the number of animals examined (necropsied).
[b] P-Values corresponding to pairwise comparisons between the controls and the dosed group are near the dosed group incidence.
[c] P-Values associated with the trend test are near the control incidence.
*Statistically significant ($P < 0.05$) using Cochran-Armitage trend test.
#Statistically significant ($P < 0.05$) using Poly-k test ($k = 3$).

Concerning the incidence of brain malignant tumors, a controversial issue in the experiments performed in the 1970s and early 1980s, 12 malignant tumors (10 gliomas, 1 medulloblastoma, and 1 meningioma) were observed in our study, without dose relationship, in males and females treated with APM, while none were observed in controls.

CONCLUSIONS

In our experimental conditions, APM causes an increased incidence of malignant tumor-bearing animals, with a positive significant trend in both sexes and a significant increase in the incidence of tumors at various sites, including carcinomas of the renal pelvis and ureter in females, malignant schwannomas

TABLE 6. Long-term carcinogenicity bioassay on ASPARTAME, administered with feed, supplied *ad libitum*, to male (M) and female (F) Sprague-Dawley rats

		HEMOLYMPHORETICULAR NEOPLASIAS			
			Animals	Animals with lymphomas–leukemias[a,b,c]	
Group No.	Concentration (ppm)	Sex	No.	No.	%
I	100,000	M	100	29	29.0
		F	100	25	25.0[##]
		M+F	200	54	27.0
II	50,000	M	100	20	20.0
		F	100	25	25.0[##]
		M+F	200	45	22.5
III	10,000	M	100	15	15.0
		F	100	19	19.0[#]
		M+F	200	34	17.0
IV	2,000	M	150	33	22.0
		F	150	28	18.7[#]
		M+F	300	61	20.3
V	400	M	150	25	16.7
		F	150	30	20.0[##]
		M+F	300	55	18.3
VI	80	M	150	23	15.3
		F	150	22	14.7
		M+F	300	45	15.0
VII	0 (control)	M	150	31	20.7*[#]
		F	150	13	8.7**[#]
		M+F	300	44	14.7

[a] The tumors rates are based on the number of animals examined (necropsied).
[b] P-Values corresponding to pairwise comparisons between the controls and the dosed group are near the dosed group incidence.
[c] P-Values associated with the trend test are near the control incidence.
*Statistically significant ($P < 0.05$) using Cochran-Armitage trend test.
**Statistically significant ($P < 0.01$) using Cochran-Armitage trend test.
[#] Statistically significant ($P < 0.05$) using Poly-k test ($k = 3$).
[##] Statistically significant ($P < 0.01$) using Poly-k test ($k = 3$).

of the peripheral nerves in males, and lymphomas–leukemias in females. The carcinogenic effects were shown even at a daily dose of 20 mg/kg b.w., about half the current ADI for humans in Europe and the United States.

The results of our study are not consistent with the data made available by the producers of APM. The interpretation of our results and the explanation for this difference have been extensively discussed in our previous publications.[23,24] The distinctive characteristics of the CMCRC/ERF long-term carcinogenicity bioassays, that is, that they are planned using a large number of animals per sex and per group and that animals are observed until spontaneous death have been, in our opinion, critical. Had we truncated the experiment after just 2 years, we would have most likely not revealed the carcinogenic evidence of APM.

The results of our study demonstrate the necessity of an extensive review of the regulations governing the use of APM as a food additive. The data also call for additional long-term bioassays on another species and a different calendar of exposure to better quantify APM's carcinogenic risk. In our opinion, it is of vital importance to also re-analyze the adequacy of the long-term carcinogenicity bioassays performed on other old- and new-generation artificial sweeteners currently in use worldwide.

Given the ever-increasing use of artificial sweeteners in both industrialized and developing countries, we consider our integrated project on artificial sweeteners to be of the highest priority for the protection of public health, in particular the health of children and pregnant women who are among the most vulnerable populations. In light of this goal, and given what in our view are inadequate data to date on the carcinogenicity of artificial sweeteners, we are conducting additional research, not only on APM, but also on other widely diffused artificial sweeteners and blends used in thousands of foods, beverages, and pharmaceutical products.

ACKNOWLEDGMENTS

We thank the U.S. National Toxicology Program for convening a group of pathologists to provide a second opinion for a set of malignant lesions and for their help in the statistical analysis. We also thank the research staff of the CMCRC/ERF and Kathryn Knowles for her support in the preparation of the manuscripts.

REFERENCES

1. BRIGHT, G. 1999. Low-calorie sweeteners—from molecules to mass markets. World Rev. Nutr. Diet. **85:** 3–9.
2. LINDLEY, M.G. 1999. New developments in low-calorie sweeteners. World Rev. Nutr. Diet. **85:** 44–51.
3. WEIHRAUCH, M.R. & V. DIEHL. 2004. Artificial sweeteners—do they bear a carcinogenic risk? Ann. Oncol. **15:** 1460–1465.
4. BUTCHKO, H.H. et al. 2002. Preclinical safety evaluation of aspartame. Regul. Toxicol. Pharmacol. **35:** S7–S12.
5. MAZUR, R.H. 1984. Discovery of aspartame. In Aspartame Physiology and Biochemistry. L.D. Stegink & L.J. Filer, Jr. Eds.: 3–9. Dekker. New York, NY.
6. FRY, J. 1999. The world market for intense sweeteners. World Rev. Nutr. Diet. **85:** 201–211.
7. U.S. NATIONAL LIBRARY OF MEDICINE. 2006. Hazardous Substances Data Bank. http://toxnet.nlm.nih.gov/ [accessed 10 May 2005].

8. USA FOOD NAVIGATOR. www.foodnavigator-usa.com/news/printNewsBis.asp? id=59163. [accessed 10 May 2005].
9. ASPARTAME INFORMATION CENTER. 2006. http://www.aspartame.org. [accessed 10 May 2005].
10. BUTCHKO, H.H. et al. 2002. Intake of aspartame vs the acceptable daily intake. Regul. Toxicol. Pharmacol. **35:** S13–S16.
11. RANNEY, R.E. et al. 1976. Comparative metabolism of aspartame in experimental animals and humans. J. Toxicol. Environ. Health **2:** 441–451.
12. STEGINK, L.D. 1984. Aspartate and glutamate metabolism. In Aspartame Physiology and Biochemistry. L.D. Stegink & L.J. Filer. Jr., Eds.: 47–76. Dekker. New York, NY.
13. HARPER, A.E. 1984. Phenylalanine metabolism. In Aspartame Physiology and Biochemistry. L.D. Stegink & L.J. Filer. Jr., Eds.: 77–109. Dekker. New York, NY.
14. OPPERMAN, J.A. 1984. Aspartame metabolism in animals. In Aspartame Physiology and Biochemistry. L.D. Stegink & L.J. Filer. Jr., Eds.: 141–159. Dekker. New York, NY.
15. JEFFREY, A.M. & G.M. WILLIAMS. 2000. Lack of DNA-damaging activity of five non-nutritive sweeteners in the rat hepatocyte/DNA repair assay. Food Chem. Toxicol. **38:** 335–338.
16. RENCUZOGULLARI, E. et al. 2004. Genotoxicity of aspartame. Drug Chem. Toxicol. **27:** 257–268.
17. MUKHOPADHYAY, M., A. MUKHERJEE & J. CHAKRABARTI. 2000. In vivo cytogenetic studies on blends of aspartame and acesulfame-K. Food Chem. Toxicol. **38:** 75–77.
18. NATIONAL TOXICOLOGY PROGRAM. 2005. Toxicology studies of aspartame (CAS No. 22839-47-0) in genetically modified (FVB Tg.AC hemizygous) and B6.129-Cdkn2a^{tm1Rdp} (N2) deficient mice and carcinogenicity studies on aspartame in genetically modified B6.129-Trp53tm1Brd(N5) haploinsufficient mice (feed studies). Genetically Modified Model Report NTP GMM1: 5–66.
19. OLNEY, J.W. et al. 1996. Increasing brain tumor rates: is there a link to aspartame? J. Neuropathol. Exp. Neurol. **55:**1115–1123.
20. U.S. FOOD AND DRUG ADMINISTRATION. 1981. Aspartame: Commissioner's final decision. Fed. Regist. **46:** 38285–38308.
21. ISHII, H. 1981. Incidence of brain tumors in rats fed aspartame. Toxicol. Lett. **7:** 433–437.
22. ISHII, H. et al. 1981. Toxicity of aspartame and its diketopiperazine for Wistar rats by dietary administration for 104 weeks. Toxicology **21:** 91–94.
23. SOFFRITTI, M. et al. 2006. First experimental demonstration of the multipotential carcinogenic effects of aspartame administered in the feed to Sprague–Dawley rats. Environ. Health Perspect. **114:** 379–385.
24. SOFFRITTI, M. et al. 2005. Aspartame induces lymphomas and leukaemias in rats. Eur. J. Oncol. **10:** 107–116.
25. DECRETO LEGISLATIVO 116. 1992. Attuazione della direttiva n. 86/609/CEE in materia di protezione degli animali utilizzati a fini sperimentali o ad altri fini scientifici. Supplemento ordinario alla Gazzetta Ufficiale **40:** 5–25.
26. HARRIS, N.L. et al. 2001. WHO Classification of tumors of haematopoietic and lymphoid tissues: introduction. In Tumors of Haematopoietic and Lymphoid Tissues. E.S. Jaffe, et al., Eds.: 12–13. IARC Press. Lyon.

Results of a Long-Term Carcinogenicity Bioassay on Sprague-Dawley Rats Exposed to Sodium Arsenite Administered in Drinking Water

MORANDO SOFFRITTI, FIORELLA BELPOGGI, DAVIDE DEGLI ESPOSTI, AND LUCA LAMBERTINI

Cesare Maltoni Cancer Research Center, European Foundation of Oncology and Environmental Sciences "B. Ramazzini", Bologna, Italy

ABSTRACT: Arsenic (As) is a metal found in nature whose acute and chronic toxic effects have been known for decades. Hundreds of millions of people are at risk of exposure to As and its various chemical forms which can occur in the occupational and general environment in air, water, soil, food, and medicines. Several epidemiological studies have shown that prolonged exposure to As can induce various types of malignant tumors in humans, namely, skin, lung, liver, kidney, and bladder cancers. These effects have been observed particularly in geographic areas where people are exposed to well water with high concentrations of As. While the risks of As at high concentrations are well documented, there is still a great deal of uncertainty regarding the risk of exposure to As at very low levels. This uncertainty is due to the absence of adequate epidemiological data and the insufficiency of experimental data currently available. Given the limited evidence demonstrating the carcinogenic potential of As in animals, a long-term carcinogenicity bioassay on sodium arsenite ($NaAsO_2$) was performed at the Cesare Maltoni Cancer Research Center (CMCRC) of the European Ramazzini Foundation (ERF). $NaAsO_2$ was administered with drinking water at concentrations of 200, 100, 50, or 0 mg/L, for 104 weeks to Sprague-Dawley rats (50/sex/group), 8 weeks old at the start of the study. The animals were monitored until spontaneous death at which time each animal underwent complete necropsy. Histopathological evaluation of all pathological lesions and of all organs and tissues collected was routinely performed on each animal. The results demonstrate that in our experimental conditions $NaAsO_2$ induces sparse benign and malignant tumors among treated rats. The types of tumors observed are infrequent in the strain of Sprague-Dawley rats of the colony used in our laboratory, namely, lung adenomas and carcinomas, kidney adenomas/papillomas and carcinomas, and bladder carcinomas.

Address for correspondence: Morando Soffritti, M.D., Cesare Maltoni Cancer Research Center, European Ramazzini Foundation, Castello di Bentivoglio, Via Saliceto, 3, 40010 Bentivoglio, Bologna, Italy. Voice: +39-051-6640460; fax: +39-051-6640223.
    e-mail: crcfr@ramazzini.it   www.ramazzini.it

Notably, an elevated incidence of these types of oncological lesions is also observed among people living in geographical areas where As is present at higher concentrations in drinking water.

KEYWORDS: sodium arsenite; carcinogenicity; long-term bioassays; rat

INTRODUCTION

Arsenic (As) is a metalloid widely distributed in the earth's crust, found particularly in igneous and sedimentary rocks. It can exist in four valency states: -3, 0, $+3$, and $+5$. Under reducing conditions, the $+3$ valency state as arsenite may be the dominant form; the $+5$ valency state as arsenate is generally the more stable form in oxygenized environments.[1] Arsenic compounds usually occur in trace quantities in rock, soil, water, and air. However, concentrations may be higher in certain areas as a result of weathering and anthropogenic activities including metal mining and smelting, fossil fuel combustion, and pesticide use.[2]

In soils, a global average concentration level of 5 mg/kg was estimated by Koljonen,[3] but concentrations may vary considerably among geographic regions. In some areas of South-West England, near old smelters or mining areas, concentrations of As range from 24 to 161,000 mg/kg.[4]

In natural waters, As is mostly found in inorganic forms as oxyanions of trivalent arsenite or pentavalent arsenate. Background concentrations of As in groundwater range from less than 0.5 to 5000 µg/L.[5] Most high levels of As occur naturally, but cases of As pollution by mining are numerous, albeit localized. In surface waters, levels of dissolved As range from 0.1 to 1.7 µg/L in uncontaminated stream waters,[6] however, in some areas characterized by volcanic activity, As levels may reach 3000 µg/L.[7]

Concentrations in air in remote locations range from less than 1 to 4 ng/m^3; however, in cities, concentrations may reach up to 200 ng/m^3 and may be greater than 1000 ng/m^3 in the vicinity of industrial sources, particularly near nonferrous smelters.[2] It has been estimated that the atmospheric flux of As is about 75,540 tons/year of which 60% is of natural origin and the remaining 40% is derived from anthropogenic sources.[8]

The estimated world production of As (expressed as As trioxide equivalent) was 35,000 tons in 2002, representing a slight decrease compared to the world production at the end of 1990s (about 40,000 tons in 1998 and 1999).[9] As is mostly used for the production of wood preservatives, but also for the production of some agricultural chemicals, pesticides (mainly herbicides), glass, nonferrous alloys, and semiconductors.[9] In the United States, As consumption is predicted to decline drastically because of regulations aimed toward ceasing use of chromated copper arsenate as a wood preservative.[9]

In both humans and rodents, ingested As compounds are quickly absorbed and enter the bloodstream. Inhaled As is not absorbed as readily as ingested

As, even though a study on smelter workers showed an 80% absorption rate of inhaled As.[10] Once absorbed, arsenate is reduced to arsenite in the blood through reactions with glutathione and then transported to the liver. Arsenite is detoxified by methylation in the liver both enzymatically by methyltransferases [to monomethylarsonic acid (MMA) and dimethylarsinic acid (DMA)] and nonenzymatically (to methanearsonic acid).[11] An unusual feature of the metabolism of As is that there are deep interspecies differences in methylation: humans are more sensitive to As toxicity than are several other species because As methylation in humans is believed to be less efficient.[11,12]

In humans, ingestion of high levels of As (>1 g) induces acute effects characterized by severe gastrointestinal damage which may result in shock, multi-organ failure, and death.[2] Subacute exposure affects primarily the respiratory, gastrointestinal, cardiovascular, nervous, and hematopoietic systems.[1] Chronic As toxicity is mainly manifested in the skin in forms, such as hyperpigmentation and hyperkeratosis, with pigmentation affecting trunks and limbs and keratosis affecting hands and feet. Chronic lung disease, peripheral neuropathy, and peripheral vascular disease have been frequently associated with chronic exposure to As.[1]

Few studies have been conducted to evaluate the immunotoxicity of As. From studies on humans, it appears that inorganic As is immunotoxic[13,14] and As-induced apoptosis could be a major mechanism of immunosuppression.[15]

Reproductive and developmental effects have been studied in humans and in experimental animals. Human data, albeit limited, suggest that exposure to high concentrations of As in drinking water during pregnancy may increase the risk of fetal and neonatal mortality.[16,17] The results in rodents show that MMA and DMA have developmental toxicity effects.[1]

In humans, As is a chromosomal mutagen (an agent that induces mutations involving more than one gene). Micronuclei, chromosomal aberration, and aneuploidy have been detected in both peripheral lymphocytes, and urothelial cells of people exposed to elevated levels of As.[1]

In prokaryotic *in vitro* systems, As has been reported to be nonmutagenic.[18–20] In mammalian and human cell *in vitro* systems, As has been shown to be genotoxic. Moore *et al*.[21] reported sodium arsenite and, to a lesser extent sodium arsenate, to induce chromosomal aberrations, micronuclei polyploidy, and endoreduplication in the L5178Y/TK $\pm$ mouse lymphoma assay. Chromosome alteration in Chinese hamster ovary cells are also reported to be induced by arsenite.[22] In human lymphocytes and fibroblasts, inorganic As was reported to induce dose-dependent chromosomal aberrations and DNA-protein cross-links; the effects were observed to be more potent with sodium arsenite than with sodium arsenate.[23,24]

Epidemiological studies on cancer risks in relation to As exposure in drinking water include mostly ecological studies that can provide important information on causal inference due to large exposure contrasts and limited population migration. Several epidemiological studies have shown that prolonged exposure to As in drinking water can induce various types of malignant tumors in humans, namely, tumors of the skin, lung, liver, kidney, and urinary bladder. Many reports several decades ago described skin cancers following ingestion of arsenical medicine, exposure to arsenical pesticides, and As-contaminated water.[25-27] Typical As-associated skin tumors include squamous-cell carcinoma and multiple basal-cell carcinoma.[27-29] Arsenic in drinking water was reported to increase the risk for lung cancer in epidemiological studies which included large population groups and different levels of exposure.[30-32] Exposure to As was also reported to induce liver angiosarcoma and to increase the risk for liver cancer, mainly hepatocarcinoma.[30,31,33] Moreover, a positive correlation between exposure to As in drinking water and cancers of the kidney and urinary bladder was found in many epidemiological studies.[30-32,34]

Various As compounds have been tested for carcinogenicity by oral administration in rats and mice, by intratracheal administration in hamsters, and by transplacental exposure in mice.[1]

In a Fischer rat study, groups of 36 males, 10 weeks of age at the start of the experiment, were administered DMA at concentrations of 0, 12.5, 50, and 200 ppm in drinking water for 104 weeks. A statistically significant increase of the incidence of transitional cell carcinomas of the urinary bladder was observed in the two highest dose groups.[35]

Studies were also conducted on mice to evaluate the potential carcinogenicity of different As compounds. DMA was administered in drinking water at various concentrations to different mice strains, namely: (*a*) groups of 10–14 male A/J mice were given 0, 50, 200, or 400 ppm beginning at 6 weeks of age and lasting for 25 or 50 weeks, respectively. After 50 weeks, an increase (although nonsignificant) in the incidence of lung tumors was observed in treated mice[36]; (*b*) groups of 20–30 K6/ODC transgenic mice were administered 0, 10, or 100 ppm of DMA beginning at 7 weeks of age and lasting for 5 months. One additional group was administered sodium arsenite at 10 ppm. The incidence of squamous skin tumors increased in all treated groups compared to the control[37]; (*c*) groups of 29–30 male $p53^{+/-}$ heterozygous or $p53^{+/+}$ mice were exposed to 0, 50, 200 ppm DMA in drinking water for 80 weeks. In the $p53^{+/-}$ mice, a nonsignificant increase in the incidence of total tumors was observed, while in $p53^{+/+}$ mice a significant increase in the incidence of total tumors was observed, but with no dose-dependence.[38]

In an other study, groups of 25 male and female C3H mice were exposed exclusively during fetal life to 0, 42.5, and 85 ppm of inorganic As. Males and females were then observed for 72 and for 90 weeks, respectively. A dose-related increase in the incidence of hepatocarcinomas and adrenal cortical adenomas was observed in males; in females, a dose-related increase in the incidence of total ovarian tumors (benign and malignant) and lung carcinomas was observed.[39]

The carcinogenicity of As trioxide, calcium arsenate, and As trisulphide was evaluated in groups of 30 male Syrian golden hamsters, treated by intratracheal instillation beginning at 8 weeks of age, once a week for 15 weeks and followed until death (113–121 weeks after the initial instillation). A group of 20 males served as a control. Each compound, containing 0.25 mg As, was suspended in 0.1 mL of saline buffer; the controls received buffer solution alone. A statistically significant increase of the incidence of lung tumors (benign and malignant combined) was observed in animals treated with calcium arsenate.[40]

IARC reviewed the aforementioned studies and concluded that there is sufficient evidence in experimental animals to determine the carcinogenicity of DMA, but there is limited evidence to confirm the carcinogenicity of sodium arsenite, calcium arsenate, and As trioxide.[1]

Given the limited evidence demonstrating the carcinogenic potential of As and the importance of obtaining additional experimental data to better assess the carcinogenic risks among people exposed to As in drinking water, a long-term carcinogenicity bioassay on sodium arsenite ($NaAsO_2$) was performed at the Cesare Maltoni Cancer Research Center (CMCRC) of the European Ramazzini Foundation (ERF). The results of the study are presented in this article.

MATERIALS AND METHODS

The sodium arsenite ($NaAsO_2$) used was produced by Sigma of St. Louis, MO and supplied by Prodotti Gianni of Milan, Italy. Its purity was 98% by titration. $NaAsO_2$ was administered in drinking water at concentrations of 200, 100, 50, or 0 mg/L *ad libitum* for 104 weeks to male (M) and female (F) Sprague-Dawley rats (50/sex/group), 8 weeks old at the start of the experiment. Rats were bred from the colony used at the CMCRC/ERF laboratories for nearly 30 years. Extensive historical data are available on the tumor incidence among untreated rats. Each morning, leftover solution from the previous day was removed and glass drinking bottles were washed and refilled with fresh solution. Control animals received tap water. Animals were fed with the standard Corticella diet (Corticella S.p.A., Bologna, Italy), used for more than 30 years in our laboratories. The experiment was performed according to Good Laboratory Practices using the Standard Operating Procedure (SOP) of the CMCRC/ERF.

After weaning at 4–5 weeks of age, the experimental animals were randomized in order to have no more than one male and one female from each litter in the same group. Animals were then housed in groups of five in makrolon cages (41 cm × 25 cm × 15 cm) with stainless-steel wire tops and a shallow layer of white wood-shavings as bedding. Cages were kept in rooms used exclusively for this experiment at a temperature of 21 ± 2°C and relative humidity of 50–60%. A light/dark cycle of 12 hours was maintained using both natural and artificial light sources.

Mean daily drinking water and feed consumption were measured once weekly per cage for the first 13 weeks, and then every 2 weeks until 111 weeks of age. Body weight was measured individually once weekly for the first 13 weeks and then every 2 weeks until 111 weeks of age. Measurement of body weight continued every 8 weeks until the end of the experiment. The animals were clinically examined for gross changes every 2 weeks for the duration of the study.

The biophase ended at 159 weeks, with the death of the last animal at 167 weeks of age. Upon death, all animals underwent complete necropsy. Histopathology was routinely performed on the following organs and tissues of each animal from each group: skin and subcutaneous tissue, the brain (three sagittal sections), pituitary gland, Zymbal glands, salivary glands, Harderian glands, cranium (five sections, with oral and nasal cavities and external and internal ear ducts), tongue, thyroid, parathyroid, pharynx, larynx, thymus and mediastinal lymph nodes, trachea, lung and mainstem bronchi, heart, diaphragm, liver, spleen, pancreas, kidneys, adrenal glands, esophagus, stomach (fore and glandular), intestine (four levels), urinary bladder, prostate, gonads, interscapular brown fat pad, subcutaneous and mesenteric lymph nodes, and other organs or tissues with pathological lesions. All organs and tissues were preserved in 70% ethyl alcohol, except for bones, which were fixed in 10% formalin and then decalcified with 10% formaldehyde and 20% formic acid in water solution. The normal specimens were trimmed, following SOP. Trimmed specimens were processed as paraffin blocks and 3–5 μm sections of every specimen were obtained. Sections were routinely stained with Hematoxylin-Eosin.

All slides were examined microscopically by the same group of pathologists, following the same criteria of histopathologic evaluation and classification. A senior pathologist reviewed all tumors and all other lesions of oncologic interest. Multiple tumors of different types and sites, of different types in the same site, of the same types in bilateral organs, of the same types in the skin, subcutaneous tissue or mammary glands, or at distant sites of diffuse tissue (i.e., bones and skeletal muscle) were plotted as single/independent tumors. Multiple tumors of the same type in the same tissue and organ, apart from those mentioned above, were plotted only once.

Three statistical tests were used to analyze neoplastic and non-neoplastic lesion incidence data. The χ^2 test and the Fisher's exact test[41] were used to evaluate differences in tumor incidence between treated and control groups. The Cochran-Armitage trend test[42,43] was used to test for linear trends in tumor incidence.

RESULTS

A dose-related lower intake of water containing various levels of $NaAsO_2$ was observed in both male and female rats (FIG. 1). In females, water consumption became similar between the group treated at 50 mg/L and the control after 88 weeks of age. A dose-related lower intake of feed was also observed in both male and female rats (FIG. 2). This difference was less marked between the

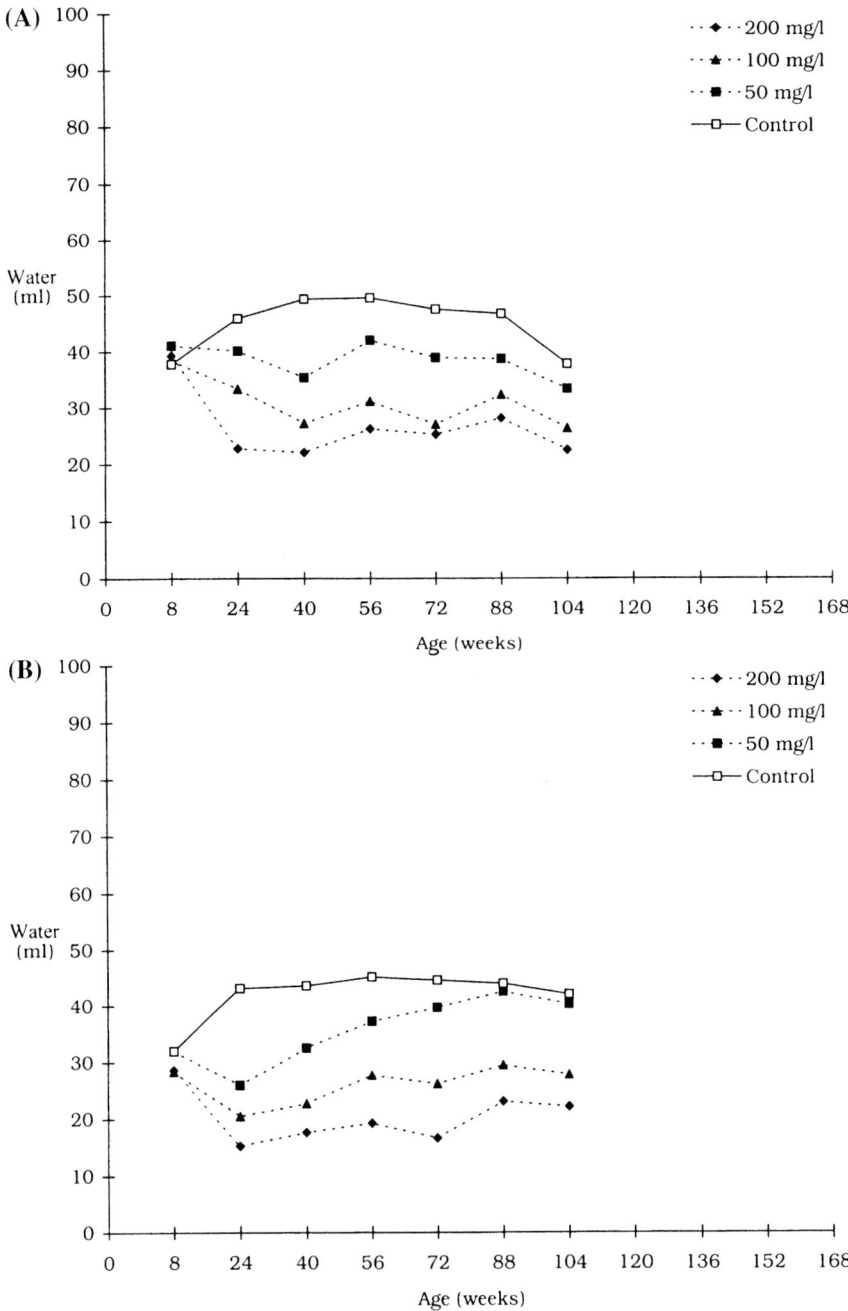

FIGURE 1. Mean daily water consumption in male (**A**) and female (**B**) Sprague-Dawley rats (- - ♦ - -200 mg/L; - - ▲ - - 100 mg/L; - - ■ - - 50 mg/L; □ Control).

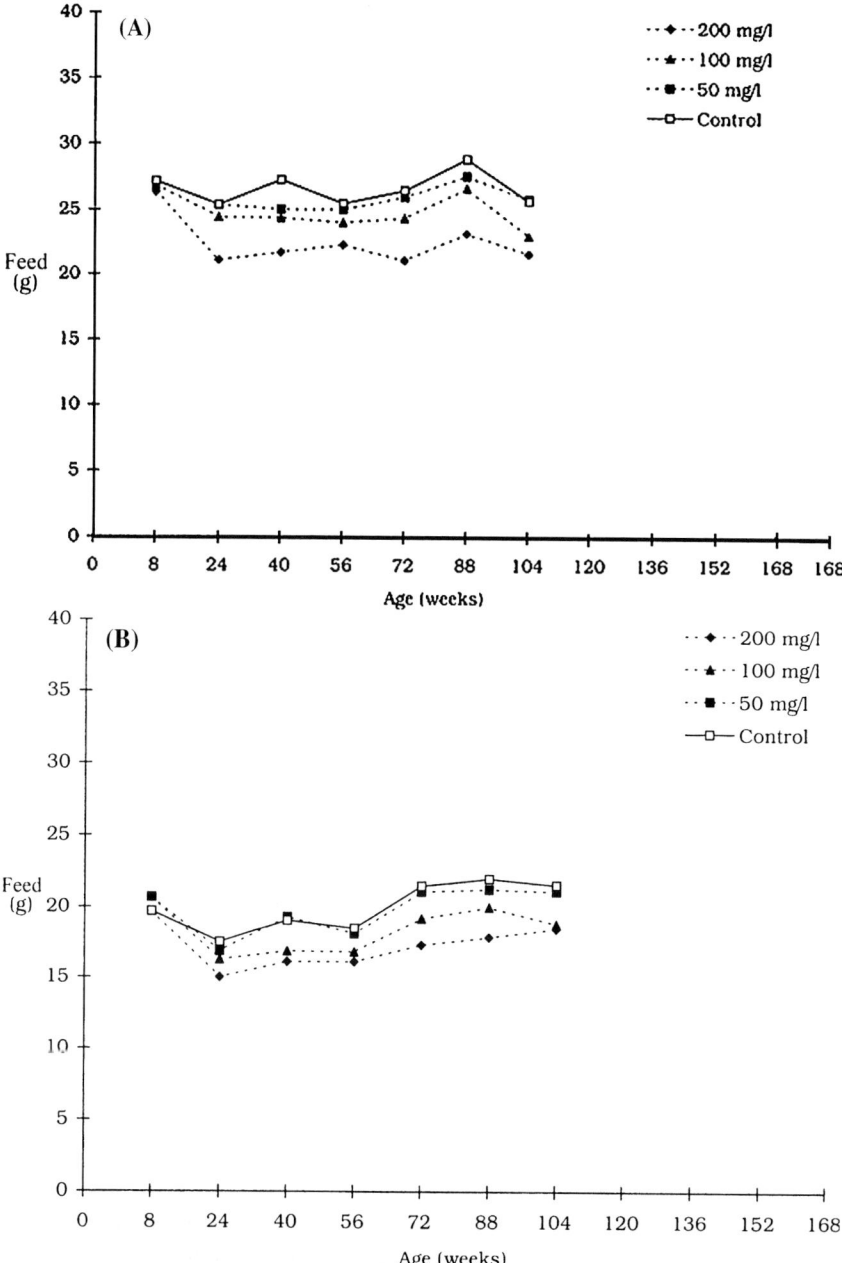

FIGURE 2. Mean daily feed consumption in male (**A**) and female (**B**) Sprague-Dawley rats (- - ♦ - - 200 mg/L; - - ▲ - - 100 mg/L; - - ■ - - 50 mg/L; □ Control).

group treated at 50 mg/L and the control in both males and females. A dose-related difference in mean body weight was observed in males. The difference was more evident in the males treated at 200 mg/L (circa 15% when compared with controls). Differences in mean body weight were also observed in females of the groups treated at 200 and 100 mg/L. Mean body weight was about 20% less in females treated at 200 mg/L compared with control and about 10% less in females treated at 100 mg/L. No treatment-related differences in body weight were observed in females treated at 50 mg/L. Differences in survival rates were observed in both males and females; a slight decrease in the survival rate was observed in males treated at 200 and 100 mg/L, particularly from 40 weeks of age until 88 weeks of age, whereas in females, a decrease in survival rate was observed from 104 weeks of age until the end of the experiment.

Long-term exposure of sodium arsenite administered in drinking water to Sprague-Dawley rats has been shown to induce toxic effects on the kidneys at concentrations as high as 200 mg/L and, to a lesser extent 100 mg/L and 50 mg/L. Nephropathies were characterized by diffuse acute/chronic inflammation, tubular enlargement with deposits of ialin casts and marked fibrosis around glomeruli with distension of Bowman's space.

The main oncologic results of the experiment are reported in TABLE 1 for males and TABLE 2 for females. Among males treated at 100 mg/L, a slightly increased incidence of animals bearing malignant tumors and a statistically significant increased number of total malignant tumors ($P < 0.05$) were observed when compared to controls. Sparse very infrequent benign and malignant tumors were observed in the treated groups, namely, one adenocarcinoma of the lung in a male treated at 200 mg/L; one carcinoma of the kidney and one papilloma of the pelvis in a male treated at 100 mg/L and two papillomas of the renal pelvis in another rat treated at the same dose. Renal pelvis papillomas were also observed in two males treated at 50 mg/L.

Among females treated at 100 mg/L, a slightly increased incidence of animals bearing malignant tumors and an increased number of total malignant tumors were observed when compare to controls. Among the females treated at 200 mg/L, one adenocarcinoma of the lung was observed. The same group also included two animals bearing kidney adenomas, two bearing kidney carcinomas, and one bearing a renal pelvis carcinoma. In the group treated at 100 mg/L, three animals were observed bearing kidney adenomas, one bearing a kidney carcinoma and one bearing a renal pelvis papilloma. One animal bearing a bladder carcinoma was also observed among the females treated at 100 mg/L.

It must be noted that among the untreated Sprague-Dawley rats used in our laboratories over the last 20 years (2265 males and 2274 females), the overall incidence of lung adenomas was 0.2% in males (range: 0–2.0%) and 0.1% in females (range: 0–1.0%), while the overall incidence of lung carcinomas was 0.1% in both males (range: 0–1.0%) and females (range: 0–1.3%). The overall incidence of the kidney adenomas was 0.1% in males (range: 0–1.3%)

TABLE 1. Incidence of the preneoplastic and neoplastic lesions in male Sprague-Dawley rats treated with sodium arsenite in drinking water from 8 weeks of age for 104 weeks and then observed until spontaneous death

Dose ppm (mg/L)	Animals at start	Malignant tumors			Animals bearing neoplastic lesions of the lung[a]				Animals bearing neoplastic lesions of the kidney[a,b]										Animals bearing neoplastic lesions of the bladder			
		Tumor-bearing animals		Total tumors Per 100 animals	Adenomas		Adenocarcinomas		Adenomas		Carcinomas		Pelvis papillomas		Pelvis carcinomas		Total		Papillomas		Carcinomas	
		No.	%		No.	%	No.	%	No.	%	No.	%	No.	%	No.	%	No.	%	No.	%	No.	%
200	50	15	30.0	32.0	0	–	1	2.0	0	–	0	–	0	–	0	–	0	–	0	–	0	–
100	50	23	46.0	56.0*	0	–	0	–	0	–	1	2.0	2(3)	4.0	0	–	2c(4)	4.0	0	–	0	–
50	50	17	34.0	38.0	1	2.1	0	–	0	–	0	–	2	4.0	0	–	2	4.0	0	–	0	–
0	50	17	34.0	38.0	0	–	0	–	0	–	0	–	0	–	0	–	0	–	0	–	0	–

*Statistically significant ($P < 0.05$) using χ^2 test.
[a]The tumor rates are based on the number of animals examined.
[b]Between parentheses the number of tumors (one animal can bear more than one tumor).
[c]One animal bears both kidney carcinoma and papilloma of the renal pelvis.

TABLE 2. Incidence of the preneoplastic and neoplastic lesions in female Sprague-Dawley rats treated with sodium arsenite in drinking water from 8 weeks of age for 104 weeks and then observed until spontaneous death.

Dose ppm (mg/L)	Animals at start	Tumor-bearing animals		Malignant tumors	Total tumors Per 100 animals	Animals bearing neoplastic lesions of the lung[a]				Animals bearing neoplastic lesions of the kidney[a,b]									Animals bearing neoplastic lesions of the bladder			
						Adenomas		Adeno-carcinomas		Adenomas		Carcinomas		Pelvis papillomas		Pelvis carcinomas		Total		Papillomas		Carcinomas	
		No.	%	No.		No.	%	No.	%	No.	%	No.	%	No.	%	No.	%	No.	%	No.	%	No.	%
200	50	19	38.0	27	54.0	0	—	1	2.1	2	4.3	2	4.3	0	—	1	2.2	5	10.0	0	—	0	—
100	50	25	50.0	31	62.0	0	—	0	—	3 (4)	6.0	1	2.0	1	2.0	0	—	5 (6)	10.0	0	—	1	2.0
50	50	17	34.0	21	42.0	0	—	0	—	0	—	0	—	1	2.0	0	—	1	2.0	0	—	0	—
0	50	18	36.0	23	46.0	0	—	0	—	0	—	0	—	1	2.0	0	—	1	2.0	0	—	0	—

[a]The tumor rates are based on the number of animals examined.
[b]Between parentheses the number of tumors (one animal can bear more than one tumor).

and 0.2% in females (range: 0–2.0%), while the overall incidence of kidney carcinomas was 0.2% in males (range: 0–0.3%) and 0.3% in females (range: 0–1.8%). With regard to historical data on the transitional cell epithelium of the renal pelvis and ureter, no papillomas were observed in either males or females, while only one carcinoma was observed in a female (overall incidence: 0.04% and range: 0–1.0%). No carcinomas in the transitional cell epithelium of the bladder were observed in either males or females.

CONCLUSIONS

In our experimental conditions, it has been shown that sodium arsenite administered for 104 weeks in drinking water to male and female Sprague-Dawley rats, 8 weeks old at the start of the experiment and kept under observation until spontaneous death, induces an increased incidence (albeit not statistically significant) of benign and malignant tumors of the lung, kidney, and bladder.

Because these benign and malignant tumors are extremely rare in our extensive historical controls, the observation of adenomas and carcinomas of the lung, of adenomas and carcinomas of the kidney, of papillomas and one carcinoma of the renal pelvis transitional cell epithelium, and of one carcinoma of the bladder transitional cell epithelium among treated rats should not be considered casual.

The biological significance of these results are reinforced if we consider that these tumors are among the same types observed in humans living in geographical areas with an elevated concentration of As in drinking water.

The aforementioned results have shown that Sprague-Dawley rats represent a good animal model to express the carcinogenic potential of As in drinking water. In light of the utility of this model and the results observed, in our opinion, a life-span mega-experiment exposing large groups of male and female Sprague-Dawley rats to As in drinking water at doses much lower than 50 mg/L, starting from embryonic life and lasting until spontaneous death, is urgently required to provide a more adequate scientific basis to the current exposure standards: 50 μg/L in developing countries and 10 μg/L in industrialized countries.[44,45]

REFERENCES

1. International Agency for Research on Cancer. 2004. Monographs on the evaluation of carcinogenic risks to humans. Vol. 84. IARC. Lyon, France.
2. International Programme on Chemical Safety (IPCS). 2001. Arsenic and arsenic compounds, 2nd ed. Environmental Health Criteria 224. WHO. Geneva, Switzerland.
3. KOLJONEN, T. 1992. Geochemical Atlas of Finland. Geological Survey of Finland. Espoo, Finland.
4. FARAGO, M.E. *et al*. 1997. Health aspects of human exposure to high arsenic concentrations in soil in south-west England. *In* Arsenic Exposure and Health

Effects. C.O. Abernathy, R.L. Calderon, W.R. Chappel, Eds.: 210–226. Chapman & Hall. New York.
5. SMEDLEY, P.L. & D.G. KINNIBURGH. 2002. A review of the source, behaviour and distribution of arsenic in natural waters. Appl. Geochem. **17:** 517–568.
6. MATSCHULLAT, J. 2000. Arsenic in the geosphere—A review. Sci. Total Environ. **249:** 297–312.
7. QUEIROLO, F. *et al.* 2000. Total arsenic, lead, cadmium, copper, and zinc in some salt rivers in the northern Andes of Antofagasta, Chile. Sci. Total Environ. **255:** 85–95.
8. CHILVERS, D.C. & P.J. PETERSON. 1987. *In* Lead, mercury, cadmium and arsenic in the environment. T.C. Hutchinson & K.M. Meema, Eds.: 279–303. John Wiley & Sons. Chichester, UK.
9. US Department of Interior – US Geological Survey. Mineral commodity summaries. 2005. US government printing office. Washington, DC.
10. PERSHAGEN, G. & M. VAHTER. 1979. Arsenic. National Swedish Environment Protection Board. Stockholm, Sweden.
11. CHAN, P.C. & J. HUFF. 1997. Arsenic carcinogenesis in animals and humans: mechanistic, experimental, and epidemiological evidence. J. Environ. Sci. Health. **15:** 83–122.
12. VAHTER, M. 1999. Methylation of inorganic arsenic in different mammalian species and population groups. Sci. Progr. **82:** 69–88.
13. OSTROWSKY-WEGMAN, P. *et al.* 1991. Lymphocyte proliferation kinetics and genotoxic findings in a pilot study on individuals chronically exposed to arsenic in Mexico. Mutat. Res. **250:** 477–482.
14. GONSEBATT, M.E. *et al.* 1994. Lymphocyte replicating ability in individuals exposed to arsenic via drinking water. Mutat. Res. **313:** 293–299.
15. HARRISON, M.T. & K.L. MCCOY. 2001. Immunosuppression by arsenic: a comparison of cathepsin L inhibition and apoptosis. Int. Immunopharmacol. **1:** 647–656.
16. ENGEL, R.R. & A.H. SMITH. 1994. Arsenic in drinking water and mortality from vascular disease: an ecological analysis in 30 counties in the United States. Arch. Environ. Health **49:** 418–427.
17. HOPENHAYN-RICH, C. *et al.* 2000. Chronic arsenic exposure and risk of infant mortality in two areas in Chile. Environ. Health Perspect. **108:** 667–673.
18. ROSSMAN, T.G. *et al.* 1980. Absence of arsenite mutagenicity in *E. Coli* and Chinese hamster cells. Environ. Mutagen. **2:** 371–379.
19. ZEIGER, E. *et al.* 1992. Salmonella mutagenicity tests: V. Results from the testing of 311 chemicals. Environ. Mol. Mutagen. **19**(Suppl 21): 2–141.
20. LANTZSCH, H. & T. GEBEL. 1997. Genotoxicity of selected metal compounds in the SOS chromotest. Mutat. Res. **389:** 191–197.
21. MOORE, M.M., K. HARRINGTON-BROCK & C.L. DOERR. 1997. Relative genotoxic potency of arsenic and its methylated metabolites. Mutat. Res. **386:** 279–290.
22. KOCHHAR, T.S. *et al.* 1996. Effect of trivalent and pentavalent arsenic in causing chromosome alterations in cultured Chinese hamster ovary (CHO) cells. Toxicol. Lett. **84:** 37–42.
23. DONG, J.L. & X.M. LUO. 1994. Effects of arsenic on DNA damage and repair in human fetal lung fibroblasts. Mutat. Res. **315:** 11–15.
24. RASMUSSEN, R.E. & D.B. MENZEL. 1997. Variation in arsenic-induced sister chromatid exchange in human lymphocytes and lymphoblastoid cell lines. Mutat. Res. **386:** 299–306.

25. HUTCHINSON, J. 1888. On some examples of arsenic-keratosis of the skin and of arsenic-cancer. Trans. Pathol. Soc. London **39:** 352–393.
26. ROTH, F. 1957. After-effects of chronic arsenism in Moselle wine-makers. Dtsch. Med. Wochenschr. **82:** 211–217.
27. NEUBAUER, O. 1947. Arsenical cancer: A review. Br. J. Cancer **i:** 192–251.
28. NEUMANN, E. & R. SCHWANK. 1960. Multiple malignant and benign epidermal and dermal tumours following arsenic. Acta Derm. Venereol. **40:** 400–409.
29. YEH, S., S.W. HOW & C.S. LIN. 1968. Arsenical cancer of skin. Histologic study with special reference to Bowen's disease. Cancer **21:** 312–339.
30. CHEN, C.J. & C.J. WANG. 1990. Ecological correlation between arsenic level in well water and age-adjusted mortality from malignant neoplasms. Cancer Res. **50:** 5470–5474.
31. TSAI, S.M., T.N. WANG & Y.C. KO. 1999. Mortality for certain diseases in areas with high levels of arsenic in drinking water. Arch. Environ. Health **54:** 186–193.
32. HOPENHAYN-RICH, C., M.L. BIGGS & A.H. SMITH. 1998. Lung and kidney cancer mortality associated with arsenic in drinking water in Cordoba, Argentina. Int. J. Epidemiol. **27:** 561–569.
33. International Agency for Research on Cancer. 1980. Monographs on the evaluation of carcinogenic risks to humans. Vol. 23. IARC. Lyon, France.
34. SMITH, A.H. *et al.* 1998. Marked increase in bladder and lung cancer mortality in a region of northern Chile due to arsenic in drinking water. Am. J. Epidemiol. **147:** 660–669.
35. WEI, M. *et al.* 2002. Carcinogenicity of dimethylarsinic acid in male F344 rats and genetic alterations in induced urinary bladder tumors. Carcinogenesis. **23:** 1387–1397.
36. HAYASHI, H. *et al.* 1998. Dimethylarsinic acid, a main metabolite of inorganic arsenics, has tumorigenicity and progression effects in the pulmonary tumors of A/J mice. Cancer Lett. **125:** 83–88.
37. CHEN, Y. *et al.* 2000. K6/ODC transgenic mice as a sensitive model for carcinogen identification. Toxicol. Lett. **116:** 27–35.
38. SALIM, E.I. *et al.* 2003. Carcinogenicity of dimethylarsinic acid in p53 ± heterozygous knockout and wild type C57BL/6J mice. Carcinogenesis **24:** 335–342.
39. WALKEES, M.P. 2003. Transplacental carcinogenicity of inorganic arsenic in the drinking water: Induction of hepatic, ovarian, pulmonary and adrenal tumors in mice. Toxicol. Appl. Pharmacol. **186:** 7–17.
40. YAMAMOTO, A., A. HISANAGA & L.N. ISHINISHI 1987. Tumorigenicity of inorganic arsenic compounds following intratracheal instillations to the lungs of hamsters. Int. J. Cancer **40:** 220–223.
41. HASEMAN, J.K. 1978. Exact sample sizes with the Fisher-Irwin test for 2×2 tables. Biometrics **34:** 106–109.
42. ARMITAGE, P. 1971. Statistical Methods in Medical Research. John Wiley & Sons. New York.
43. GART, J.J., K.C. CHU & R.E. TARONE. 1979. Statistical issues in interpretation of chronic tests for carcinogenicity. J. Natl. Cancer Inst. **62:** 957–974.
44. World Health Organization. 1998. Guidelines for drinking-water quality, 2nd ed., Addendum Vol. 2. Geneva, Switzerland.
45. World Health Organization. 2004. Guidelines for drinking-water quality, 3rd ed., Vol. 1. Geneva, Switzerland.

Use of Carcinogenicity Bioassays in the *IARC Monographs*

VINCENT JAMES COGLIANO

International Agency for Research on Cancer, 69372 Lyon, France

ABSTRACT: Carcinogenicity bioassays generally provide the best means of assessing the potential for a chemical to be a human carcinogenic hazard. The results of carcinogenicity bioassays are usually the key determinants of *IARC Monograph* evaluations. Along with carcinogenicity bioassays and epidemiological studies, the International Agency for Research on Cancer (IARC) also encourages the consideration of mechanistic data and other relevant data. During 2005 IARC is updating the Preamble to the *IARC Monographs*, which describes the principles and procedures used in developing the *Monographs*, including the criteria that guide the evaluations. Proposed revisions to the Preamble make some changes in the criteria for evaluating carcinogenicity in experimental animals to reflect the greater confidence that can be placed in Good Laboratory Practice (GLP) studies. Other changes will give more specific guidance for the evaluation of mechanistic data. Sections on mechanistic data will be given more prominence in future *Monographs* and will be more closely linked with toxicokinetics. Future *Monographs* will also include a new section on susceptible individuals, populations, and life stages that will often be based on the understanding of mechanisms. In addition, the draft Preamble discusses IARC's procedures for promoting impartial evaluations by avoiding conflicts of interests and ensuring that working groups are free from interference.

KEYWORDS: *IARC Monographs*; bioassays; carcinogenicity; hazard identification; risk assessment; mechanisms of carcinogenesis; conflicts of interests; freedom from interference

INTRODUCTION

The long-term carcinogenicity bioassay has been the mainstay of carcinogen risk assessment, at the International Agency for Research on Cancer (IARC) and elsewhere. Although epidemiological evidence is often preferred when assessing the risks associated with human exposure to potential carcinogens, such evidence is not available for many chemical compounds. Even when epidemiological studies are available, they are often limited by difficulties in characterizing the exposures of the study population. Sometimes quantitative data on the level of exposure to the chemical compound are lacking, and other times there is not enough exposure information for an adequate analysis of whether confounding factors can be ruled out with reasonable confidence as a potential explanation for the observed effects.

As a result of the lack of informative epidemiological studies for many chemical compounds, long-term carcinogenicity bioassays generally provide the best means of assessing the potential for a chemical to be a human carcinogenic hazard. This is because high-quality, long-term bioassays can be conducted on chemical compounds for which there is widespread human exposure or there is some evidence or a suspicion of carcinogenic activity. This is often not the case for epidemiological studies, because for some chemical compounds there are simply no exposed populations with adequate information on confounding exposures or on the chemical compound itself.

One program that uses the results of long-term carcinogenicity bioassays is the *IARC Monographs*. The *Monographs* represent an international expert-consensus approach to carcinogen hazard identification. The objective is to critically review and evaluate the published scientific evidence on carcinogenic hazards to which humans are exposed. Since its beginning in 1971, the *IARC Monographs* have evaluated approximately 900 agents and identified approximately 400 of these as carcinogenic or potentially carcinogenic to humans. Of the 900 evaluations, fewer than 100 have been based on *sufficient evidence of carcinogenicity* in humans. This means that more than 800 (nearly 90%) of all evaluations have been determined by studies in experimental animals. The same is true at other national and international programs that assess carcinogenic hazards.

During 2005 IARC is amending the Preamble to the *IARC Monographs*, which describes the principles and procedures used in developing *Monographs*, including the scientific criteria that guide the evaluations. The objective is to reflect new scientific understanding that has developed since the Preamble was last amended in 1991, and to incorporate procedural changes that have occurred during that time. Some changes are being made to the criteria for evaluating long-term carcinogenicity bioassays in experimental animals. Some of the changes will affect the way that long-term carcinogenicity bioassays are evaluated by *Monograph* working groups.

IARC's Criteria for Evaluating Carcinogenicity Bioassays

The *IARC Monographs* classify evidence of carcinogenicity in experimental animal bioassays by choosing one of the following descriptors.[1]

Sufficient Evidence of Carcinogenicity

The working group considers that a causal relationship has been established between the agent or mixture and an increased incidence of malignant neoplasms or of an appropriate combination of benign and malignant neoplasms in (*a*) two or more species of animals or (*b*) in two or more independent studies in one species carried out at different times or in different laboratories or under different protocols.

Exceptionally, a single study in one species might be considered to provide sufficient evidence of carcinogenicity when malignant neoplasms occur to an unusual degree with regard to incidence, site, type of tumor, or age at onset.

Limited Evidence of Carcinogenicity

The data suggest a carcinogenic effect but are limited for making a definitive evaluation because, for example, (*a*) the evidence of carcinogenicity is restricted to a single experiment; (*b*) there are unresolved questions regarding the adequacy of the design, conduct, or interpretation of the study; or (*c*) the agent or mixture increases the incidence only of benign neoplasms or lesions of uncertain neoplastic potential, or of certain neoplasms which may occur spontaneously in high incidences in certain strains.

Inadequate Evidence of Carcinogenicity

The studies cannot be interpreted as showing either the presence or absence of a carcinogenic effect because of major qualitative or quantitative limitations, or no data on cancer in experimental animals are available.

Evidence Suggesting Lack of Carcinogenicity

Adequate studies involving at least two species are available which show that, within the limits of the tests used, the agent or mixture is not carcinogenic. A conclusion of evidence suggesting lack of carcinogenicity is inevitably limited to the species, tumor sites, and levels of exposure studied.

The first such criteria were formulated in 1982,[2] before current Good Laboratory Practice (GLP) studies were common. Prior to that time, studies in experimental animals were often small studies of limited scope. Study design and conduct were highly variable, consequently, the evaluation criteria emphasized replication by independent investigators.

Today the situation is different. Only a small number of organizations conduct GLP studies for a diverse set of agents (among them the U.S. National

Toxicology Program). Consequently, GLP studies are not likely to be repeated. At the same time, GLP studies are performed according to rigorous protocols that have been widely reviewed and are considered to provide definitive results. As a result, considerable confidence can be placed in findings from GLP studies.[3]

Accordingly, IARC has proposed to update its criteria for reproducibility in the definition of *sufficient evidence*.[4]

Sufficient Evidence of Carcinogenicity

The Working Group considers that a causal relationship has been established between the agent or mixture and an increased incidence of malignant neoplasms or of an appropriate combination of benign and malignant neoplasms in (*a*) two or more species of animals, (*b*) both sexes of a single species in a study conducted under GLP (e.g., a U.S. National Toxicology Program study), or (*c*) in two or more independent studies in one species carried out at different times or in different laboratories or under different protocols.

A single study in one species and sex might be considered to provide sufficient evidence of carcinogenicity when malignant neoplasms occur to an unusual degree with regard to incidence, site, type of tumor, age at onset, or strong findings of tumors at multiple sites.

In this proposed definition of *sufficient evidence*, the word "exceptionally" has been removed from the "single-study" criterion.[4] This last sentence of the definition is meant to give greater weight to bioassay findings that fall outside the range of most studies by virtue of an unusual result. This change responds to a recommendation that the phrase "to an unusual degree" is already sufficiently restrictive in limiting the use of single-study results.[3]

The "single-study" criterion has also been broadened to include "strong findings of tumors at multiple sites." In addition, "age at exposure" has been added to the list of conditions that limit a conclusion of *evidence suggesting lack of carcinogenicity* in animals. The phrase "conditions of exposure" has been added to cover other factors such as exposure route.[3,4]

Consideration of Mechanistic Data to Supplement Bioassay Results

Mechanistic data can support the results of cancer bioassays when they indicate that the mechanisms causing cancer in experimental animals operate in humans, thus increasing concern for human carcinogenicity. Other times the mechanistic data can show that the mechanisms causing cancer in experimental animals do not operate in humans, and this decreases concern for human carcinogenicity. The application of mechanistic data to support or discount bioassay results has sometimes been controversial, as some critics believe that health agencies are too reluctant to use mechanistic data when these data would discount the bioassay results, whereas other critics believe the same agencies have accepted mechanistic hypotheses that are not sufficiently tested.[5]

IARC encourages the consideration of mechanistic data along with data from long-term bioassays. To reflect the growing importance of mechanistic data in carcinogenicity assessments, the discussion of mechanistic data is being expanded in the new Preamble.[4] The new text appears earlier in the section, immediately after the discussion of toxicokinetics. This gives mechanistic data more prominence and provides a closer link between toxicokinetics and mechanisms. Accordingly, the title of the section is being changed to put mechanisms first, "Mechanistic and other relevant data." The new Preamble advises working groups convened to develop future *Monographs* to identify the possible mechanisms of carcinogenesis that might be operating. Mechanisms can be discussed at several levels, from structural changes at the molecular level to effects at the organism level. Future *Monographs* should summarize key data that are consistent or not consistent with the operation of each alternative mechanism and identify data gaps that suggest an inadequately tested mechanistic hypothesis.

Significant data gaps or conflicting data may also suggest the operation of other mechanisms. Failure to consider the potential for involvement of more than one mechanism can lead to premature and false conclusions, because associations observed between one mechanism's markers and tumors cannot rule out the operation of other mechanisms. This is similar to the problem of confounding factors in epidemiology: there may be strong associations between exposure and disease, but if confounding factors are not examined thoroughly, the associations may be spurious.[6] The draft Preamble encourages the working group to "consider whether multiple mechanisms might contribute to tumor development, whether different mechanisms might operate in different dose ranges, whether separate mechanisms might operate in humans and experimental animals, and whether a unique mechanism might operate in a susceptible group. The possible contribution of alternative mechanisms must be considered before concluding that tumors observed in experimental animals are not relevant to humans." It is also necessary to consider that an uneven level of experimental support for different mechanistic hypotheses can reflect disproportionate resources spent on investigating one hypothesis and does not exclude the contribution of other mechanisms.[4]

The conclusion that a mechanism operates in experimental animals is strengthened by findings of consistent results in different experimental systems, by biological plausibility of the mechanism, and by coherence of the overall database. Strong support can be obtained from studies that experimentally challenge the hypothesized mechanism, by demonstrating that suppression of key mechanistic processes leads to suppression of tumor development.

Current or anticipated levels of human exposure are not used to determine whether a mechanism operates in humans. In terms of the risk assessment

paradigm, a conclusion that a mechanism does not operate in humans is a matter of hazard, not exposure or risk. Such a conclusion should be valid in the case of accidental and unanticipated human exposures that are difficult to foresee at present.[4]

Future *Monographs* will also include a new section on susceptible individuals, populations, and life stages. This section builds on the knowledge of toxicokinetics and mechanisms discussed in earlier sections. Susceptibility can arise from polymorphisms of metabolism, from the presence of disease, or from exposure to a carcinogenic agent at a critical period of development, for example, infancy, puberty, or old age. In addition, exposures to other toxic agents can alter the kinetics or dynamics of a carcinogen, either increasing or decreasing the effects from a given exposure. The draft Preamble cites these and other examples of factors that can lead to increased susceptibility.[4]

A New Outline for Presentation of Mechanistic Data

Although the draft Preamble does not prescribe a standard outline for *Monograph* Section 4 (which reviews mechanistic and other relevant data), the order in which topics are discussed suggests the following outline.[3,4]

MECHANISTIC AND OTHER RELEVANT DATA

Toxicokinetic Data (Absorption, Distribution, Metabolism, Excretion)

This section reviews the potential for the agent and its metabolites to be distributed to various organs and tissues.

Mechanistic Data

This section identifies the possible mechanisms of carcinogenesis that might be operating, reviews the data that are consistent or not consistent with each alternative mechanism, and identifies significant data gaps and data that may suggest the operation of other mechanisms.

Susceptible Populations and Life Stages

This section builds on the knowledge of toxicokinetics and mechanisms to identify those who might be more susceptible. This includes, for example, susceptibility that arises from polymorphisms of metabolism, from the presence of disease, from exposure to the agent at a critical period of development (e.g., infancy, puberty, or old age), and from exposure to other agents that can alter the kinetics or dynamics of the agent being evaluated.

Other Forms of Toxicity that are Relevant to Carcinogenicity

This section reviews toxicological effects that are relevant to the evaluation, including developmental and reproductive toxicity. It is not an encyclopedia of chronic toxic effects, but should focus on, for example, toxic effects that confirm distribution and biological effects at the sites of tumor development, or toxicity that alters physiology in a way that could lead to tumor development.

Additional Relevant Data

This section reviews structure-activity relationships, the toxicological implications of physical and chemical properties, and any other data relevant to the evaluation that are not included elsewhere.

Use of Mechanistic Data When There are No Bioassays

At the present time, a classification as *possibly carcinogenic to humans* (Group 2B) requires that there be some positive evidence in either human or experimental animal studies.[1] This requirement was developed at a time when a larger number of bioassays in experimental animals were being conducted. It was also a time when mechanisms of carcinogenesis were not very well understood.

This situation, too, has changed. Scientific understanding of mechanisms of carcinogenesis has advanced to the point where the U.S. National Toxicology Program has been conducting fewer long-term carcinogenicity bioassays and is replacing them with mechanistic studies. As a result, many chemical compounds may be suspected of being potentially carcinogenic even though they have never been tested in a long-term bioassay in experimental animals. Instead, they give positive results in experimental models that investigate the processes by which a cell acquires and accumulates the molecular and cellular modifications that contribute to the development of cancer.

For many years, scientists have been making progress in understanding how to use mechanistic models when assessing potential carcinogenicity. A recent IARC workshop on the use of short- and medium-term tests for carcinogens concluded that, in the absence of carcinogenicity bioassays in experimental animals, strong mechanistic data could be used in an evaluation of potential carcinogenicity.[7] This reflects the increasing ability of mechanistic data to provide an indication of carcinogenic potential. Accordingly, the draft Preamble proposes that an agent can be characterized as *possibly carcinogenic to humans* "solely on the basis of strong evidence from mechanistic and other relevant data."[4] This follows the recommendation of a recent IARC Advisory Group.[3]

IARC's Procedures for Promoting Impartial Evaluations

Good data alone are not sufficient to assure good evaluations. It is also important to have expert and impartial working groups review and evaluate the data. Two criteria guide IARC's selection of working group members: (*a*) knowledge and experience and (*b*) absence of real or apparent conflicts of interests. Consideration is also given to demographic diversity and balance of scientific findings and views.[4,6]

To identify conflicts of interests early, IARC now requires all potential participants to submit a Declaration of Interests[8] before invitations are extended. To assure public confidence that interested parties do not have links to the working group and that special interests cannot influence the meeting, experts with a real or apparent conflict of interests may participate only in a limited capacity. They will not serve as meeting chair or subgroup chair, draft text that pertains to cancer data, or participate in the evaluations. Such experts will be invited only when necessary (someone who has published significant relevant research and comparable knowledge cannot be found among those without conflicts). The declarations are updated and reviewed again at the opening of a meeting, and conflicting interests are disclosed at the meeting and in the published *Monographs*.[4,6]

In addition, Working Groups must be free from interference, before, during, and after a meeting. The draft Preamble states, "It is not acceptable for Observers or third parties to contact participants before a meeting or to lobby them at any time."

To enforce this, IARC posts the list of participants on its web site (http://monographs.iarc.fr) approximately 2 months before each meeting, together with the following statement.[9]

> IARC requests that you do not contact or lobby meeting participants, send them written materials, or offer favors that could appear to be linked to their participation ... IARC will ask participants to report all such contacts and will publicly reveal any attempt to influence the meeting.

In the invitation letters, *Monograph* meeting participants are requested to resist and report attempts at interference. They are also reminded of this again during the meeting. Based on IARC's experience over the past 12 months, these measures appear to be working to increase transparency while, at the same time, maintaining freedom from interference.[9]

ACKNOWLEDGMENT

The speaker gratefully acknowledges contributions of the staff of the *IARC Monographs* program: Robert Baan, Kurt Straif, Yann Grosse, Béatrice Secretan, Fatiha El Ghissassi, Sandrine Egraz, Martine Lézère, Helene

Lorenzen-Augros, and Jane Mitchell. They participated in the development of the revised principles and procedures that are described in this presentation and in the draft Preamble.

In January 2006, after this presentation was given in October 2005, IARC completed the amendment of its guidelines (known as the Preamble). The final wording differs slightly from what is presented here and can be found on the *Monographs* program web site, http://monographs.iarc.fr/.

REFERENCES

1. IARC. 1992. Preamble. IARC Monographs on the Evaluation of Carcinogenic Risks to Humans, Vol. 54. 13–32. IARC, Lyon, France.
2. IARC. 1982. IARC Monographs on the Evaluation of Carcinogenic Risks to Humans: Chemicals, Industrial Processes and Industries Associated with Cancer in Humans. Suppl. 4. 7–24. IARC, Lyon, France.
3. IARC. 2005. Report of the Advisory Group to Recommend Updates to the Preamble to the IARC Monographs. IARC Internal Report No. 05/001. Available at http://monographs.iarc.fr. Accessed 10 July 2006.
4. IARC. 2005. Draft Preamble to the IARC Monographs. Available at http://monographs.iarc.fr. Accessed 10 July 2006.
5. TOMATIS, L. 2002. The IARC monographs program: Changing attitudes towards public health. Int. J. Occup. Environ. Health. **8:** 144–152.
6. COGLIANO, V.J., R.A. BAAN, K. STRAIF, *et al.* 2004. The science and practice of carcinogen identification and evaluation. Environ. Health Perspect. **112:** 1269–1274.
7. MCGREGOR, D.B., J.M. RICE & S. VENITT, Eds. 1999. The Use of Short- and Medium-term Tests for Carcinogens and Data on Genetic Effects in Carcinogenic Hazard Evaluation. IARC Scientific Publications. Lyon, France.
8. WHO. 2005. Declaration of interests for WHO experts. Available at http://monographs.iarc.fr. Accessed 10 July 2006.
9. COGLIANO. V.J., R.A. BAAN, K. STRAIF, *et al.* 2005. Transparency in the IARC Monographs. Lancet Oncol. **6:** 747.

Chemical Hazards in Health Care

High Hazard, High Risk, but Low Protection

MELISSA A. McDIARMID

University of Maryland School of Medicine, Baltimore, Maryland 21201, USA

ABSTRACT: It is counter-intuitive that the healthcare industry, whose mission is the care of the sick, is itself a "high-hazard" industry for the workers it employs. Possessing every hazard class, with chemical agents in the form of pharmaceuticals, sterilants, and germicidals in frequent use, this industry sector consistently demonstrates poor injury and illness statistics, among the highest in the United States, and in the European Union (EU), 34% higher than the average work-related accident rate. In both the United States and the EU, about 10% of all workers are employed in the healthcare sector, and in developing countries as well, forecasts for the increasing need of healthcare workers (HCW) suggests a large population at potential risk of health harm. The explosion of technology growth in the healthcare sector, most obvious in pharmaceutical applications, has not been accompanied by a stepped up safety program in hospitals. Where there is hazard recognition, the remedies are often voluntary, and often poorly enforced. The wrong assumption that this industry would police itself, given its presumed knowledge base, has also been found wanting. The healthcare industry is also a significant waste generator threatening the natural environment with chemical and infectious waste and products of incineration. The ILO has recommended that occupational health goals for industrial nations focus on the hazards of new technology of which pharma and biopharma products are the leaders. This unchecked growth cannot continue without a parallel commitment to the health and safety of workers encountering these "high tech" hazards. Simple strategies to improve the present state include: (*a*) recognizing healthcare as a "high-hazard" employment sector; (*b*) fortifying voluntary safety guidelines to the level of enforceable regulation; (*c*) "potent" inspections; (*d*) treating hazardous pharmaceuticals like the chemical toxicants they are; and (*e*) protecting HCWs at least as well as workers in other high-hazard sectors.

KEYWORDS: healthcare workers; chemical hazards

Address for correspondence: Melissa A. McDiarmid, M.D., M.P.H., Occupational Health Program, School of Medicine, University of Maryland, 405 W. Redwood Street, Second Floor, Baltimore, MD 21201. Voice: 410-706-7464; fax: 410-706-4078.

 e-mail: mmcdiarm@medicine.umaryland.edu

INTRODUCTION

Viewing hospitals and other healthcare activities, such as ambulatory, home, and long-term nursing care, as an industry sector has only recently been considered, even among occupational health professionals. However, the healthcare workplace contains every hazard class and poses daily risks to workers who provide care to the sick—the chief "product" this industry sector delivers. Of interest, Bernardino Ramazzini did not fail to recognize the hazards of healthcare workers in his treatise *The Diseases of Workers*, which included chapters on "healers of inunction," "apothecaries," and "mid wives."[1]

Metrics used to characterize an industry would classify healthcare as a large "service sector" member. It employs about 10% of all workers in both the United States and the European Union (EU), with a forecast for increasing job growth with the aging of the "baby boomer" population who will increasingly need both acute and chronic (nursing) care services.[2] Although both the "product" and mission of the healthcare sector is care of the sick, as an industry, it performs consistently poorly regarding safety, demonstrating among the highest illness and injury statistics in the United States and in the EU, accident rates 34% higher than the average.[3,4]

Hazards

Infectious agents and musculoskeletal injury resulting from manual patient lifting have been the principal focus of employee safety programs to date. However, every hazard class is encountered in healthcare, both those typical of industry (compressed gases, hazardous energy) and the exotic and emerging discoveries of pharmaceutical and biopharma products (TABLE 1). These and other chemical hazards, such as sterilants, germicidals, and anesthetic gases have been more slowly recognized as posing a risk to workers. Of special concern have been the sterilant ethylene oxide, an alkylating agent, and the anticancer drugs, many of which are also genotoxic and pose reproductive risks.[5,6]

TABLE 1. Hazards of the healthcare sector*

- Chemical – glutaraldehyde, anesthetic gases, hazardous (anticancer) drugs, disinfectants, lab reagents
- Physical – ionizing radiation, heat, hazardous energy
- Biologic – infectious agents (TB, HIV, HBV, HCV, SARS)
- Musculoskeletal – patient handling, prolonged standing
- Work Organization – shift work, prolonged hours (double shifts), stress, violence, staffing

*All hazards preparedness.

Evidence of Exposure

Environmental monitoring studies for hospital chemical hazards, especially the anticancer drugs have demonstrated troubling results, even in recent years, showing widespread work area (pharmacies and clinics) contamination.[7,8] Such contamination of work surfaces in clinical areas, albeit at low concentrations, raises questions not only about worker safety, but about patient safety as well. The opportunity for patients to contact hazardous anticancer drugs on surfaces in treatment areas or on drug infusion bags or on other clinical supplies they use, is clearly apparent.

Underrecognition of current healthcare hazards makes recognizing the new and emerging ones, even more challenging. These hazards are the products of complex technological advances and reside largely in pharma and biopharma innovations. They include the new and potentially hazardous therapies rapidly expanding in clinical use. Because the favorable risk–benefit ratio of these drugs for ill patients does not transfer to well workers, caution is warranted as these new agents are introduced in the clinical setting. Not without adverse effects, many biopharmaceuticals, such as monoclonal antibodies, may cause significant organ system failure, toxicity, including reproductive toxicity, and malignancy in treated patients and therefore require safe handling.[9]

Barriers

A principal barrier to achieving an integrated safety management plan in the healthcare sector has been the biased view of the safety and health community itself. Understandably preoccupied with concerns of the so-called "dirty industries," such as manufacturing and construction, safety professionals and even governmental agencies have, until very recently, viewed the healthcare sector, with a "blind eye." Illustrating this, consider the International Labor Organization (ILO) chemical safety card campaign, which intends to make basic chemical information and means of protection easily accessible for workers. Surprisingly, when this database was queried for the 13 group 1 International Agency for Research on Cancer (IARC) carcinogens that are used in anticancer therapy, only two were found in the database (TABLE 2). Viewing pharmaceuticals as chemical hazards is obviously a "blind spot" for even the public health agencies charged with protecting worker safety.

Solution for Change

Acknowledging the realities of the rapid deployment of innovative, yet toxic therapies within a healthcare facility requires that its safety program be poised to anticipate potential problems and to address them comprehensively. Seg-

TABLE 2. Anticancer drugs considered carcinogenic by IARC

Group 1–Human carcinogens
Denotes ILO chemical safety card exists*
  *Arsenic trioxide
  Azathioprine
  Chlorambucil
  Chlornaphazine
  Cyclosporine
  *Cyclophosphamide
  Melphalan
  Myleran
  Semustine
  Tamoxifen
  Thiotepa
  Treosulfam
Combination therapies:
  ECB—Etoposide/Cisplatin/Bleomycin
  MOPP—Mustargen/Oncovin/Procarbazine/Prednisone

menting safety management by focusing only on patients or workers will disenfranchise other potentially exposed populations. Other populations present in the healthcare environment, whose safety also requires assurance include patient's families, visitors, contractors, and volunteers. Understanding a shared responsibility for healthcare safety, the U.S. Occupational Safety and Health Administration (OSHA) began in the mid 1990s, a partnership with the Joint Commission on the Accreditation of Healthcare Organizations (JCAHO), a nonprofit agency that accredits most of the 5000 U.S. hospitals. FIGURE 1 displays the premise of this partnership. It depicts the shared hazards that exist in the healthcare environment that threaten patients, workers, family members, and visitors. This partnership understands that the "environment of care" is an "environment of work" and recommends an integrated safety hazard management approach that protects all who enter the facility.[10]

Other strategies for change also exist (TABLE 3). Strategically exploiting and enlarging existing resources, guidance documents, and standards, even when only generally applicable to the healthcare sector, will help in the short term. For example, seeking to have the ILO chemical safety card program include hazardous pharmaceutical drugs should be pursued. Hazard communication training of workers must be expanded to include healthcare workers. Partnering

TABLE 3. Potential opportunities for improvement of safety in healthcare

(1) International 'patient safety movement'
(2) Chemical safety card program of ILO
(3) Partnering with affected stakeholders, unions, manufacturers, professional organizations
(4) Partnering with environmental movement in healthcare

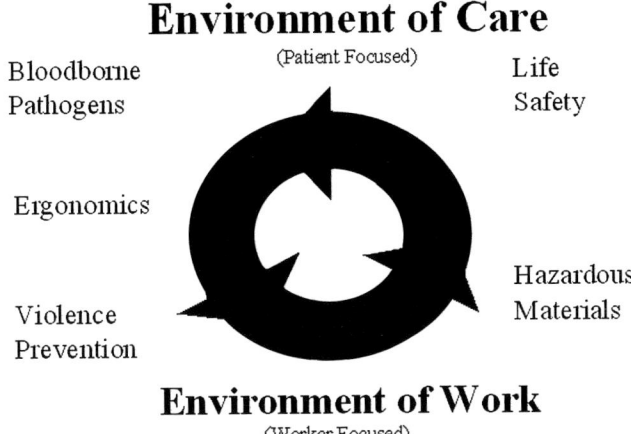

FIGURE 1. Safety management overview for healthcare.

with affected professional organizations, unions, and industry sectors, as the National Institute for Occupational Safety and Health (NIOSH) has done, to develop the Safe Handling Guidelines for Hazardous Drugs,[11] described in a subsequent article, is another useful tool employed to move the recognized standard of professional handling—practice of hazardous drugs to a more protective level.

The International Patient Safety movement is an obvious linkage that can be made to promote the safety of all parties who enter a healthcare organization. The example of "fugitive" pharmaceutical dust contaminating patient care areas and materials does not discriminate between patient and worker exposed and exemplifies why the comprehensive, rather than segmented approach to safety is required.

As well, the environmental movement in healthcare (i.e., healthcare without harm) may be a sympathetic constituency, which could share outreach and hazard awareness activities.

The ILO has recommended that occupational health goals for industrial nations focus on the hazards of new technology of which pharma and biopharma products are the leaders. This unchecked growth cannot continue without a parallel commitment to the health and safety of workers encountering these "high tech" hazards. Improving the present state therefore requires:

1) recognizing healthcare as a "high-hazard" employment sector;
2) fortifying voluntary safety guidelines to the level of enforceable regulation;
3) "potent" inspections;
4) treating hazardous pharmaceuticals like the chemical toxicants they are; and
5) protecting HCWs at least as well as workers in other high-hazard sectors.

REFERENCES

1. RAMAZZINI, B. 1713. Revised with Translation and Notes. 1940. W.C. Wright, Ed.: University of Chicago Press. Chicago, IL.
2. BUREAU OF LABOR STATISTICS, U.S. DEPARTMENT OF LABOR. 2004. Tomorrow's Jobs. Washington, DC.
3. BUREAU OF LABOR STATISTICS, U.S. DEPARTMENT OF LABOR. 2005. Injuries, Illnesses and Fatalities Program. http://www.bls.gov/iif/
4. EUROPEAN AGENCY FOR SAFETY AND HEALTH AT WORK. 2002. Safety and health good practice on line for the healthcare sector. FACTS 29.
5. WEAVER, V.N. *et al*. 1993. Occupational chemical exposure in an academic medical center. J. Occup. Med. **35:** 701–707.
6. BLACK, L.A. & A.C. PRESSON. 1997. Hazardous drugs. *In* The Healthcare Worker, State of the Art Reviews in Occupational Medicine, Vol. 12. M.A. McDiarmid & E.R. Kessler, Eds.:669–685. Hanley & Belfus. Philadelphia, PA.
7. CONNOR, T.H. *et al*. 1999. Surface contamination with antineoplastic agents in six cancer treatment centers in Canada and the United States. Am. J. Health Syst. Pharm. **56:** 1427–1432.
8. RUBINO, F.M. *et al*. 1999. Measurement of surface contamination by certain antineoplastic drugs using high performance liquid chromatography: applications in occupational hygiene investigations in hospital environments. Med. Lav. **90:** 572–583.
9. MANUFACTURERS PRODUCT INFORMATION FOR INFLIXIMAB (Centocor); Rituximab (Genentech); Trastuzumab (Genentech).
10. JOINT COMMISSION ON THE ACCREDITATION OF HEALTHCARE ORGANIZATIONS. OSHA AND THE ENVIRONMENTAL OF CARE COMPATIBILITIES. PLANT, TECHNOLOGY AND SAFETY MANAGEMENT. 1997. Oakbrook Terrace, Illinois.
11. NATIONAL INSTITUTE FOR OCCUPATIONAL SAFETY AND HEALTH. NIOSH, U. S. CENTERS FOR DISEASE CONTROL AND PREVENTION. NIOSH ALERT: PREVENTING OCCUPATIONAL EXPOSURE TO ANTINEOPLASTIC AND OTHER HAZARDOUS DRUGS IN HEALTHCARE SETTINGS. 2004. DHHS Pub. No. 2004-165.

National and International Response to Occupational Hazards in the Healthcare Sector

BRIGITTE FRONEBERG

International Labour Organization, CH-1211 Geneva, Switzerland

ABSTRACT: The health care sector is one of the largest, most rapidly expanding areas of employment and is increasingly in need of qualified staff especially in the area of nursing. The health care sector is complex and comprises a variety of largely different professions; occupational hazards and exposures differ accordingly. Rates of absenteeism, reported work-related ill-health, and early retirement or departure from professions are comparatively high, especially among the nursing staff. While classical health hazards are addressed by international and national regulations, underlying causes of ill-health and departure from the profession, such as psychological stress, violence, pressing time schedules, and poor work organization are less well heeded. Practical guidance and quality information have become increasingly available from national and international Occupational Safety and Health (OSH) institutions and can be easily accessed through the Internet. They will undoubtedly benefit the profession, but difficulties not related to OSH will warrant political solutions. This presentation provides access to relevant international and European Union (EU) legislation and to valuable information resources for health care workers available from the Internet.

KEYWORDS: health care sector; health care worker; occupational safety and health; legislation; good practice; Internet resources

WORK AND HEALTH IN THE HEALTHCARE SECTOR

A healthy and safe workplace is a recurrent topic of discussion in serious talks, strategies, program texts, and legislation. In the minds of occupational health and safety experts, it means a workplace that is free of the risk of diseases and accidents, physical, chemical, and biological exposure, and ergonomic problems. It is also increasingly understood that psychosocial aspects of the workplace and the organization of work, including possibilities for career development and reconciling work–life balance, play an important role in

---

Address for correspondence: Brigitte Froneberg, International Labour Organization, 4, route des Morillons, CH-1211 Geneva 22, Switzerland. Voice: +41-22-799-8373; fax: +41-22-799-6878.
e-mail: froneberg.brigitte@baua.bund.de

ensuring health at work. There is a growing body of evidence from reported good practice and research that high quality workplaces are feasible. However, they have not yet become a reality for a vast majority of workers and are rarely realized in hospital and other healthcare work.

The healthcare sector is rapidly expanding and is one of the largest single employment sectors in Europe and the third largest sector of the U.S. economy. With an estimated 35 million healthcare workers (HCW) worldwide, among them about 18.5 million physicians and nurses, the sector often outranks such established sectors as agriculture, mining, electricity and gas, transport, and communication, particularly in established market economies. Public spending on healthcare is outpacing economic growth in most OECD countries, forcing governments to find new funds or to pass a larger share of the costs on to individuals.[1] Development and diffusion of medical technologies and new drugs are some of the main drivers of rising health expenditures, while the cost of nursing has little part in it.

With the healthcare sector's rapid growth, an increasing shortage in nursing staff becomes visible. The International Council of Nurses relates that most industrialized countries are or will be facing nursing shortages.[2] Estimations from the U.S. Bureau of Labor Statistics in 2001 indicate that in the United States more than 1 million new nurses will be needed by the year 2010.[3] Shortages are expected to rise to 30% by the year 2020. Underlying causes for the increased nursing demand are given by the OECD as follows: economic expansion, population growth, an aging population, technological advances, and higher patient expectations.[4] The apparent nursing shortage and understaffing have been linked to a number of indicators for insufficient care, such as patient mortality, adverse events, accidents, and nosocomial infections.[5]

The growing shortage in adequately trained nursing staff attracts considerable public attention and is addressed by recent research with a view to advise appropriate policy measures.[6,7] The reported causes are manifold—low pay, insufficient appreciation, restricted career options, demanding work schedules, heavy physical workload, and psychological stress—all leading to a general unattractiveness of the nursing professions. At the same time, the importance of nurses' social function is more important than ever. Rates of absenteeism and early retirement among nursing staff and healthcare workers in general are comparatively high; most common medical causes are accidents, musculoskeletal disorders, work-related stress, skin diseases, and infectious diseases.[8]

HCW can be classified on the basis of their qualifications and their actual roles in the provision of health and medical services; they include physicians, nurses, midwives, therapists, social workers, laboratory and X ray technicians, pharmacists, and other personnel, such as cleaners, administrative staff, dietary staff, laundry workers, and maintenance workers. The sector is a labor-intensive industry and covers a highly diversified range of activities. While some risks and hazards are common to the whole sector, others are rather specific to certain categories of HCW or to certain work practices of the industry.

Generally, HCW are exposed to a great variety, combination, and concentration of hazards at the workplace that can be broadly classified as biological, chemical, and physical hazards, ergonomic factors, organizational problems, and psychosocial hazards.

Chemical agents used in hospitals and other health facilities include anesthetic agents, disinfectants, chemical sterilizing agents, drugs, and cytostatic or laboratory reagents. Some of these substances are known irritants or allergens to the skin and respiratory tract, while others, such as ethylene oxide, formaldehyde, hexachlorophene, are known mutagens, teratogens, and human carcinogens.

THE WAY FORWARD: LEGISLATION, GUIDANCE, GOOD PRACTICE, AND INFORMATION MANAGEMENT

Rates of reported work-related ill-health, untimely departure from work, and the apparent low attractiveness of the nursing profession in itself indicate noticeably that the WHO and ILO objectives of occupational health and safety and decent work for all are not yet well-established practices in the average healthcare setting. Still, the body of basic international and national legislation addressing the issue is growing and increasingly implemented into national practice. Promising guidance instruments addressing the complex hazard mix of the healthcare sector are emerging and distributed via widely accessed OSH Internet platforms, accompanied by encouraging good practice reports from many countries and settings.

The following examples provide an overview and access to relevant legal documents and illustrate exemplary resources:

1. International regulation and guidance

    The ILO offers a general framework to promote decent work for all, for example, in the *Decent Work* Agenda.[9] General guidance for safety and health at work is given in instruments, such as the ILO Convention "C155 Occupational Safety and Health Convention, 1981,"[10] the accompanying "Recommendation R164,"[11] and the proposed "Promotional Framework for Occupational Safety and Health."[12] The ILO "Guidelines on Occupational Safety and Health Management Systems (ILO–OSH 2001)"[13] provides an adaptable outline for the integration of occupational safety and health (OSH) into overall company management thus encouraging and promoting a general safety culture at work as a voluntary shared effort.

    Specific hazards of the sector are addressed by the following tripartite standards and interagency instruments:

    - "C170 Chemicals Convention, 1990"[14] and the accompanying "Recommendation R177, 1990,"[15] both representing international efforts to upgrade the national measures and harmonize regulatory standards. Emphasis is laid on the need to establish a coherent national policy of

chemical safety ranging from the classification and labeling of chemicals to the control in all aspects of the use of chemicals. Particular emphasis would thus be placed on roles and responsibility of the competent authority, suppliers, and employers, as well as duties and rights of workers.
- "Globally Harmonized System of Classification and Labeling of Chemicals (GHS),"[16] a common effort by OECD, CETDG, and ILO, designed to cover all chemicals including pure substances and mixtures and to provide for the chemical hazard communication requirements of the work place, transport of dangerous goods, consumers, and the environment.
- "International Chemical Safety Cards (ICSCs),"[17] developed by the International Programme on Chemical Safety (IPCS), a joint activity of UNEP, ILO, and WHO. The ICSCs project is developed in the context of the cooperation between the IPCS and the Commission of the European Communities. An ICSC summarizes essential health and safety information on chemicals for their use at the "shop floor" level by workers and employers in factories, agriculture, construction, and other work places. ICSCs are not legally binding documents, but consist of a series of standard phrases, mainly summarizing health and safety information collected, verified, and peer reviewed by internationally recognized experts, taking into account advice from manufacturers and Poison Control Centers. The ICSCs are available on the World Wide Web in numerous languages.
- "Chemical Control Banding"[18] is a complementary approach to protecting worker's health by focusing resources on exposure controls, also developed by IPCS. A comprehensive "WHO Guideline for HCW" is under preparation and will be available by the end of 2006.

2. EU regulation and guidance

EU instruments are generally in good agreement with ILO regulation, are usually far more specific, and become within specified time frames by transposition into national regulation individual EU member state law. Within the given context of the presentation, special mention is made to the "Council Directive 89/391/EEC of 12 June 1989 on the introduction of measures to encourage improvements in the safety and health of workers at work,"[19] the so-called Occupational Safety and Health Framework Directive and several Daughter Directives, such as the "Council Directive 98/24/EC of 7 April 1998 on the protection of the health and safety of workers from the risks related to chemical agents at work,"[20] the "Commission Directive 2000/39/EC of 8 June 2000 establishing a first list of indicative occupational exposure limit values in implementation of Council Directive 98/24/EC on the protection of the health and safety of workers from the risks related to chemical agents at work,"[21] further the "Directive 2000/54/EC of the European Parliament and of the Council of 18 September 2000 on the protection of workers from risks related to exposure to biological agents at work,"[22] the "Commission Directive of 29 May 1991 on establishing indicative limit values by implementing Council Directive 80/1107/EEC on the protection of workers from the risks related to exposure to chemical, physical and biological agents at work (91/322/EEC),"[23] and the "Council Directive 90/269/EEC of 29 May 1990 on the minimum health and safety requirements for the manual handling of loads where there is a risk particularly of back injury to workers."[24]

The "Council Directive on the protection of workers from the risks related to occupational exposure to carcinogens and mutagens of 1999" is currently under repeal. The Community strategy on health and safety at work 2002–2006 includes a proposal to extend the scope of the directive by including substances detrimental to reproduction; increasing the number of agents covered; reconsidering the appropriateness of the limit values; and by making procedures within the directive simpler and more adaptable to scientific progress.[25] The proposal is currently under consultation with the EU trade union and employer organizations.

Chemical safety and health is further addressed by the "Guidelines of a non-binding nature for implementing certain provisions of Directive 98/24/EC45 of the Council, on the protection of the health and safety of workers from risks related to chemical agents at work."[26] The guidelines cover the following topics: analytical methods for the measurement of indicative occupational exposure limit values; identification, assessment and control of risks arising from the presence of hazardous chemical agents (HCAs) in the workplace; general principles for preventing risks related to HCAs, and specific prevention and protection measures for controlling these risks, medical surveillance, and biological monitoring of workers exposed to lead and its ionic compounds.

3. Guidelines

Comprehensive guidance for the healthcare sector is rare, given the diversity of professions and possible exposures within the sector. Mention is made to the "US NIOSH Guidelines for Protecting the Safety and Health of healthcare Workers"[27] of 1998 and the three-pronged approach of the New Zealand Department of Labour: "Guidelines for the Provision of Facilities and General Safety and Health in the Healthcare Industry,"[28] "Health and Safety Guidelines for Home-Based healthcare Services,"[29] and "Responsible Care™ Manager's Handbook,"[30] thus targeting major facets of the sector in detail and calling on responsible management.

4. Internet resources

Client-oriented information management is crucial in times of increasing fragmentation and time pressure. A wealth of quality-ensured information and guidance for HCW is made available on the Internet by major national OSH institutions, such as NIOSH[31] and HSE[32] and by the extremely well-organized multilingual platform of the European Agency for Safety and Health at Work[33] in Bilbao, which in addition provides easy access to all major national and international OSH institutions worldwide.

Careful analysis of available data and trends, supported by research and public discussion seems thus to lead into the right direction. healthcare sector difficulties will have to be addressed by a mix of measures and cannot be solved by better implementation of OSH legislation and measures alone. Nevertheless, "Erkenntnis ist der erste Schritt zur Besserung" as it states in a German proverb. The healthcare sector may thus still become an area where the investment into good safety and health practice will show the pay-off not only by avoidance of unreasonable cost to employer, society, and employee, but also by building a trustworthy haven for future clients to come, such as the now aging baby-boom generations.

REFERENCES

1. OECD. 2003. Health at a glance: OECD indicators 2003. [cited 2005 Sep 14]. Available from URL: http://www.oecd.org/document/11/0,2340,en_2649_34631_16502667_1_1_1_1,00.html
2. INTERNATIONAL COUNCIL OF NURSES. Workforce forum. [cited 2005 Sep 14]. Available from URL: http://www.icn.ch/forum2002overview.pdf
3. VAN EYCK, K. 2003. Women and international migration in the health sector. Final Report of Public Services International's Participatory Action Research.
4. SIMOENS, S., M. VILLENEUVE & J. HURST. 2005. Tackling nurse shortages in OECD countries. OECD health working papers no. 19, DELSA/ELSA/WD/HEA (2005)1
5. BUCHAN, J. & L. CALMAN. 2004. The Global Shortage of Registered Nurses: an Overview of Issues and Actions. ICN, Geneva.
6. HASSELHORN, H.M., B. MÜLLER & P. TACKENBERG. 2005. Sustaining work ability in the nursing profession – investigation of premature departure from work (Nurses Early Exit Study – NEXT). [cited 2005 Sep 14]. Available from URL: http://www.next.uni-wuppertal.de/
7. BARRON, D. & E. WEST. 2005. Leaving nursing: an event-history analysis of nurses' careers. J. Health Serv. Res. Policy **10:** 150–157.
8. HSE. 2005. Health services. [cited 2005 Sep 14]. Available from URL: http://www.hse.gov.uk/healthservices/
9. ILO. Decent Work. 1999. International Labour Office Geneva. [cited 2005 Sep 14]. Also available from URL: http://www.ilo.org/public/english/standards/relm/ilc/ilc87/rep-i.htm
10. ILO. 1981. C155 Occupational Safety and Health Convention. [cited 2005 Sep 14]. Available from URL: http://www.ilo.org/ilolex/cgi-lex/convde.pl?C155
11. ILO. 1981. R164 Occupational Safety and Health Recommendation. [cited 2005 Sep 14]. Available from URL: http://www.ilo.org/ilolex/cgi-lex/convde.pl?R164
12. ILO. 2005. Promotional framework for occupational safety and health. [cited 2005 Sep 14]. Available from URL: http://www.ilo.org/public/english/protection/safework/promoframe.htm
13. ILO. Guidelines on Occupational Safety and Health Management Systems (ILO-OSH 2001). International Labour Office Geneva. [cited 2005 Sep 14].

Also available from URL: http://www.ilo.org/public/english/protection/safework/managmnt/index.htm
14. ILO. 1990. C170 Chemicals Convention. [cited 2005 Sep 14]. Available from URL: http://www.ilo.org/ilolex/cgi-lex/convde.pl?C170
15. ILO. 1990. R170 Chemicals Recommendation. [cited 2005 Sep 14]. Available from URL: http://www.ilo.org/ilolex/cgi-lex/convde.pl?R177
16. ILO. Globally harmonized system of classification and labelling of chemicals (GHS).[cited 2005 Sep 14]. Available from URL: http://www.ilo.org/public/english/protection/safework/chemsfty/ghs.htm
17. ILO. International Chemical Safety Cards (ICSCs). [cited 2005 Sep 14]. Available from URL: http://www.ilo.org/public/english/protection/safework/cis/products/icsc/index.htm
18. ILO. Chemical Control Banding. [cited 2005 Sep 14]. Available from URL: http://www.ilo.org/public/english/protection/safework/ctrl_bandingindex.htm
19. ECC. Council Directive 89/391/EEC of 12 June 1989 on the introduction of measures to encourage improvements in the safety and health of workers at work. [cited 2005 Sep 14]. Available from URL: http://europa.eu.int/smartapi/cgi/sga_doc?smartapi!celexapi!prod!CELEXnumdoc&lg=en&numdoc=31989-L0391&model=guichett
20. ECC. Council Directive 98/24/EC of 7 April 1998 on the protection of the health and safety of workers from the risks related to chemical agents at work (fourteenth individual Directive within the meaning of Article 16(1) of Directive 89/391/EEC). [cited 2005 Sep 14]. Available from URL: http://europa.eu.int/smartapi/cgi/sga_doc?smartapi!celexapi!prod!CELEXnumdoc&lg=en&numdoc=31998L0024&model=guichett
21. ECC. Commission Directive 2000/39/EC of 8 June 2000 establishing a first list of indicative occupational exposure limit values in implementation of Council Directive 98/24/EC on the protection of the health and safety of workers from the risks related to chemical agents at work. [cited 2005 Sep 14]. Available from URL: http://europa.eu.int/smartapi/cgi/sga_doc?smartapi!celexapi!prod!CELEXnumdoc&lg=en&numdoc=32000L0039&model=guichett
22. ECC. Directive 2000/54/EC of the European Parliament and of the Council of 18 September 2000 on the protection of workers from risks related to exposure to biological agents at work (seventh individual directive within the meaning of Article 16(1) of Directive 89/391/EEC). [cited 2005 Sep 14]. Available from URL: http://europa.eu.int/smartapi/cgi/sga_doc?smartapi!celexapi!prod!CELEXnumdoc&lg=en&numdoc=32000L0054&model=guichett
23. ECC. Commission Directive of 29 May 1991 on establishing indicative limit values by implementing Council Directive 80/1107/EEC on the protection of workers from the risks related to exposure to chemical, physical and biological agents at work (91/322/EEC). [cited 2005 Sep 14]. Available from URL: http://europa.eu.int/smartapi/cgi/sga_doc?smartapi!celexapi!prod!CELEXnumdoc&lg=en&numdoc=31991L0322&model=guichett
24. ECC. Council Directive 90/269/EEC of 29 May 1990 on the minimum health and safety requirements for the manual handling of loads where there is a risk particularly of back injury to workers (fourth individual Directive within the meaning of Article 16 (1) of Directive 89/391/EEC). [cited 2005 Sep 14]. Available from URL: http://europa.eu.int/smartapi/cgi/sga_doc?smartapi!celexapi!prod!CELEXnumdoc&lg=en&numdoc=31990L0269&model=guichett

25. WEILER, A. 2005. Annual review of working conditions in the EU: 2004-2005. European Foundation for the Improvement of Living and Working Conditions Dublin. [cited 2005 Sep 14]. Available from URL: http://www.eurofound.eu.int/publications/htmlfiles/ef05126.htm
26. ECC. 2004. Guidelines of a non-binding nature for implementing certain provisions of Directive 98/24/EC45 of the Council, on the protection of the health and safety of workers from risks related to chemical agents at work. [cited 2005 Sep 14]. Available from URL: http://europa.eu.int/eur-lex/lex/LexUriServ/site/en/com/2004/com2004_0819en01.pdf
27. NIOSH. 1998. Guidelines for Protecting the Safety and Health of healthcare Workers. [cited 2005 Sep 14]. Available from URL: http://www.cdc.gov/niosh/hcwold1.html#injury
28. New Zealand Department of Labour. Guidelines for the Provision of Facilities and General Safety and Health in the Healthcare Industry. [cited 2005 Sep 14]. Available from URL: http://www.osh.dol.govt.nz/order/catalogue/pdf/healthcareg.pdf
29. NEW ZEALAND DEPARTMENT OF LABOUR. Health and Safety Guidelines for Home-Based healthcare Services. cited 2005 Sep 14]. Available from URL: http://www.osh.dol.govt.nz/order/catalogue/ipp/home-healthcare.pdf
30. NEW ZEALAND DEPARTMENT OF LABOUR. Responsible Care[TM] Manager's Handbook. The New Zealand Chemical Industries Council. [cited 2005 Sep 14]. Available from URL: http://www.osh.dol.govt.nz/order/catalogue/responsiblecare.shtml
31. http://www.cdc.gov/niosh/topics/healthcare/
32. http://www.hse.gov.uk/healthservices/information.htm
33. http://agency.osha.eu.int/good_practice/sector/healthcare/

Hazardous Anticancer Drugs in Health Care

Environmental Exposure Assessment

THOMAS H. CONNOR

NIOSH MS C-23, Cincinnati, Ohio 45226, USA

ABSTRACT: Exposure of healthcare workers to anticancer drugs became problematic in the 1970s. Shortly thereafter, studies began documenting exposure of healthcare workers to these drugs. Investigations employing biological markers, such as urine mutagenicity, chromosomal aberrations, sister chromatid exchanges, and micronuclei, demonstrated associations between occupational exposures and elevated marker levels. Other analytical methods emerged to monitor workplaces where drugs were handled. These contemporary studies uncovered widespread contamination of drugs on work surfaces, trace amounts in air samples, and their presence in the urine of workers. Vials containing these drugs are often contaminated with the drug when they are shipped. Most workplace surfaces are contaminated with the drugs being prepared and used in that area. Other anticancer/hazardous drugs would most likely be used in these areas. The interior surfaces of biological safety cabinets and isolators, floors, countertops, carts, storage bins, waste containers, treatment areas, tabletops, chairs, linen, and other items are all potential sources of exposure to anticancer drugs. Patient body fluids contain the drugs and/or metabolites, often more biologically active than the parent compounds. An exposure assessment of areas where anticancer/hazardous drugs are handled must consider every potential source and route of exposure. Data from surface contamination and inhalation studies suggest that dermal exposure is the primary route of exposure. Assessment of exposure is the first step in providing a safe work environment for these workers. However, because of the many drugs to which they are exposed, any assessment can only be an estimation of the overall exposure.

KEYWORDS: antineoplastic drugs; occupational exposure; exposure assessment

---

Address for correspondence: Thomas H. Connor, Ph.D., NIOSH MS C-23, 4676 Columbia Parkway, Cincinnati, OH 45226. Voice: 513-533-8399; fax: 513-533-8138.
  e-mail: tmc6@cdc.gov

Ann. N.Y. Acad. Sci. 1076: 615–623 (2006). © 2006 New York Academy of Sciences.
doi: 10.1196/annals.1371.021

INTRODUCTION

According to the World Health Organization, more that 11 million new cases of cancer are diagnosed every year worldwide.[1] This number is expected to grow to 16 million by the year 2020. However, a diagnosis of cancer is not the dreaded death sentence that it used to be. Cancer patients are enjoying the benefits of the war on cancer. These benefits range from reduced side effects to improved and extended quality of life and in some cases, complete cures. A significant aspect of the treatment regimen for cancer patients is chemotherapy—treatment with drugs designed to kill cancer cells. Such drugs are often called anticancer drugs and have been in clinical use for decades. They are critical in the treatment of cancer and certain noncancer diseases.[2] An interesting historical note that was pivotal in ushering in the modern era of chemotherapy was the observation that exposure to mustard gas, used as a weapon in World War I, resulted in the hospitalization of veterans many years later with bone marrow toxicities. After World War II, the use of mustard gas analogs (nitrogen and sulfur mustards) led to remission in Hodgkin's disease. This initial success provided direction for today's status in which there are approximately 100 different anticancer drugs in use with many more under development. Chemotherapy has indeed opened many new avenues for today's cancer patient and provided hope for a disease that at one time had a very bad prognosis.

These drugs suppress cell proliferation and cause cell death, either directly by binding to DNA, RNA, or proteins in the cell or indirectly by inhibiting production of the same. Typically, these drugs cannot distinguish between normal and cancerous cells. As a purported consequence of this lack of selectivity, secondary malignancies were reported in patients who received anticancer drugs for other, usually solid, primary malignancies. The most commonly seen secondary malignancies were leukemia and bladder cancer reported after a latency period of 1–10 years.[3] While newer generation drugs, such as monoclonal antibodies may target sites other than genetic material and accordingly be more selective in their mechanism of action and as a result "safer," this is currently an exception and not the rule. The secondary malignancies, first observed in the 1970s, served notice of the potential problem. This problem is not just for cancer patients, but for workers in the pharmaceutical industry and members of the healthcare service team. The toxicity and hazards associated with the development of new drugs, the continued formulation of drugs currently employed as anticancer agents, and the preparation and administration of these drugs in the clinical setting all suggest this is an occupational concern. As the number of patients, the use of combinations of drugs, and higher doses of drugs increases, along with the development of more potent drugs, the potential for worker exposure to these hazardous drugs will also most likely continue to increase.

In terms of occupational exposure, a hazardous drug is defined as an agent that presents a danger to healthcare personnel due to its inherent toxicity. These drugs are identified based on one or more of the following characteristics: carcinogenicity; teratogenicity or other developmental toxicity; reproductive toxicity; organ toxicity at low doses; genotoxicity; or structure and toxicity profiles that mimic existing hazardous drugs.[4–6] Hazardous drugs include anticancer and cytotoxic agents, some hormonal agents, immunosuppressants, antiviral medications, monoclonal antibodies, and several other miscellaneous drugs. A list of drugs that require special handling should be posted in every facility where hazardous drug preparation and/or administration take place.

BIOLOGICAL EVIDENCE OF EXPOSURE

Falck and co-workers[7] reported that nurses who worked in environments where anticancer drugs were prepared and administered had higher levels of mutagenic substances in their urine when compared to nonexposed workers. This study suggested that nursing personnel were being occupationally exposed to anticancer drugs, many of which are mutagenic. The results of this study was confirmed in many other efforts examining urine mutagenicity, chromosomal aberrations, sister chromatid exchanges, and other end points in pharmacists and nurses who handle anticancer drugs.[8–10]

In addition to various acute toxic effects resulting from exposure to anticancer drugs,[6] a review of 14 studies described an association between exposure to antineoplastic drugs and adverse reproductive effects, nine of which showed some positive association.[11] The most common reproductive effects found in these studies were increased fetal loss,[12,13] congenital malformations,[14] low birth weight and congenital abnormalities,[15] and infertility.[16]

SOURCES FOR WORKPLACE EXPOSURES

Many of the toxicological end points reported above are nonspecific and only serve as indirect measures of exposure. Such studies only imply causality and do not necessarily validate events in the exposure–disease continuum. Consequently, more direct methods of determining exposure have been developed. These methods include environmental air and surface sampling techniques to assess workplace contamination and analysis of the urine to determine the presence of parent drugs and/or metabolites of hazardous drugs handled by healthcare workers. All workers who come in contact with anticancer and other hazardous drugs have the potential to be exposed to them. These workers include: pharmacy and nursing personnel, physicians, operating room personnel, environmental services workers, workers in research laboratories and animal care facilities, veterinary care workers, shipping and receiving personnel, and waste disposal personnel. Exposure to anticancer drugs in the workplace may result from one or more of the common routes of exposure.

Dermal and inhalation routes are the likely routes of exposure to anticancer drugs in healthcare facilities. Therefore, surface wipe sampling and sampling for airborne drugs have been employed to determine workplace contamination with anticancer drugs. This methodology is similar to methods used in other occupational settings to determine the level and extent of contamination of the workplace and to establish safe working levels for other hazardous substances.

Surface Wipe Sampling: The method employed for determining chemical contamination in the healthcare facility has been the measurement of a number of marker anticancer drugs using wipe samples.[17] Sampling and analytical procedures have been developed for some of the more commonly used anticancer drugs that have been employed as markers of overall surface contamination. The more common drugs sampled include: cyclophosphamide, ifosfamide, 5-fluorouracil, methotrexate, paclitaxel, doxorubicin, and platinum-containing drugs.[17]

Since the early 1990s, studies by a number of researchers have examined environmental contamination of areas where anticancer drugs are prepared and administered in healthcare facilities.[18–31] Using wipe samples, all investigators measured detectable levels of one or more anticancer drugs in various locations, such as surfaces in biological safety cabinets (BSCs), floors, countertops, storage areas, tables and chairs in patient treatment areas, and locations adjacent to drug-handling areas. All of the studies reported some level of contamination with at least one drug, and several reported contamination with all the drugs for which assays were performed.

Several studies have documented that the outer surfaces of anticancer drug vials are often contaminated with the drug contained in the vial.[19,32–38] Various methods have been used to measure the amount of drug on the outer surface of the vials. These include wipe sampling, rinsing, and total emersion of the vials using a suitable solvent. However, because of the nature of the surfaces being sampled, it is difficult to determine the recovery efficiencies with drug vials, which likely results in underreporting of contamination. Once the samples are collected, analytical methods similar to those that have been used for surface wipe sampling have been applied for determination of the external contamination levels.

Studies of surface contamination with anticancer drugs typically employ a collection matrix (e.g., tissue or filter paper wipes) and a solvent system proven to aid recovery of the drugs being studied.[17] Specific strategies have been developed for collecting wipe samples for other chemicals in various industries[39,40] and similar methods have been applied to the sampling of cytotoxic drugs. If a sampling program is established for a healthcare facility, a sampling scheme should be developed which incorporates the areas of interest based on published studies. Such a program would require the appropriate analytical techniques necessary to identify and quantify the drugs that are being measured. Several analytical methods have been employed by researchers and are available in the published literature.[17] These include high-performance liquid chromatography with ultraviolet detection (HPLC-UV), gas chromatography coupled with mass spectrometry or tandem mass spectrometry (GC-MS or GC-MS/MS) or high-performance liquid chromatography-tandem mass spectroscopy (LC-MS/MS). With the use of GC-MS (or GC-MS/MS) for drugs, such as cyclophosphamide and ifosfamide, derivatization is required prior to analysis.[18] Platinum-containing compounds can be analyzed using either voltammetry[29,41] or inductively coupled plasma mass spectrometry (ICP-MS).[22,42]

Measurement of Airborne Anticancer Agents: Several studies have measured airborne concentrations of antineoplastic drugs in healthcare settings.[18,20,28,41,43-48] In most cases, the percentage of air samples containing measurable airborne concentrations of anticancer drugs was low, and the actual concentrations of the drugs, when present, were also low. Most studies have employed glass fiber or paper filters to capture airborne particulates. These results may be attributed to the inefficiency of sampling and analytical techniques used in the past.[48] A solid sorbent material may be more efficient at collecting particulate forms of anticancer drugs. Both particulate and gaseous phases of one antineoplastic drug, cyclophosphamide, have been reported in two studies.[28,48]

Workplace exposure levels have been established for toxic chemicals in many occupational settings. However, no exposure levels have been established for airborne concentrations of anticancer drugs. There are, however, some exposure limits for soluble platinum salts and inorganic arsenic, which would include some of the anticancer drugs, such as platinum-containing compounds and arsenic trioxide.[6,49] Some pharmaceutical manufacturers have developed occupational exposure limits (OELs) that are used to set exposure limits in manufacturing facilities.[50]

While sources of exposure of healthcare providers to anticancer drugs include inhalation, dermal or possibly oral, a major route of exposure is inhalation via droplets, particulates, and vapors. Many procedures can result in aerosol generation (i.e., drug injection into an intravenous [i.v.] line, cleaning of air from the syringe or infusion line, and leakage at the tubing, syringe, i.v. spike, or stopcock connection, clipping used needles and crushing used syringes.[51,52] Drug particles can become airborne after drying of contaminated areas. Vaporization of antineoplastic agents has been recorded with various drugs, such as BCNU, ifosfamide, thiotepa, and cyclophosphamide.[28,53]

Ingestion: Inadvertent ingestion is another problematic issue. When food or beverages are prepared, stored, or consumed in work areas, they may become contaminated with airborne particles of cytotoxic drugs or by dermal contact. Hand-to-mouth exposure is a most likely route since most surfaces in areas where anticancer drugs are handled have demonstrated contamination. Because surface contamination has been reported outside of areas where anticancer drugs are handled,[24] exposure may result in adjacent areas where food or beverages are present.

SIMULATED EXPOSURE STUDIES

The use of fluorescent markers has been employed in some situations to simulate environmental contamination with anticancer drugs. Kromhout and co-workers[51] developed a semiquantitative fluorescent method to evaluate environmental contamination and Spivey and Connor[52] employed a fluorescent marker to demonstrate sources of environmental contamination during simulated drug preparation and administration. Prepared test kits that use a fluorescent marker are available to evaluate worker skills and training during drug preparation and administration.[54]

CONCLUSIONS

As the need for more anticancer drugs increases in the future, the potential for healthcare workers to be exposed to higher levels of more potent drugs becomes apparent. Past and current evidence indicates that workplace settings where anticancer drugs are prepared and administered to patients are contaminated with the drugs that have been used as markers of contamination. Since, the number of drugs that are typically assayed for is a small percentage of the known hazardous drugs, it is easy to speculate that contamination of the workplace with hazardous drugs is an ubiquitous event. Recent studies have documented the presence of these same marker drugs in the urine of healthcare workers, indicating systemic exposure to them. All this evidence highlights the critical need to reduce the risk of exposure to hazardous drugs to workers in the healthcare environment.

REFERENCES

1. WORLD HEALTH ORGANIZATION (WHO) http://www.who.int/en/
2. CHABNER, B.A., C.J. ALLEGRA, G.A. CURT & P. CALABRESI. 1996. Antineoplastic agents. *In* Goodman and Gilman's The Pharmacological Basis of Therapeutics, 9th ed. J.G. HARDMAN & L.E. LIMBIRD, Eds.: 1233–1287. McGraw-Hill. New York.
3. ERLICHMAN, C. & M. MOORE. 1996. Carcinogenesis: a late complication of cancer chemotherapy. *In* Cancer Chemotherapy and Biotherapy: Principles and Practice, 2nd ed. B.A. CHABNER & D.L. LONGO, Eds.: 45–58. Lippincott-Raven. Philadelphia.
4. AMERICAN SOCIETY OF HOSPITAL PHARMACISTS. 1990. ASHP technical assistance bulletin on handling cytotoxic and hazardous drugs. Am. J. Hosp. Pharm. **47:** 1033–1049.
5. OSHA TECHNICAL MANUAL, TED 1-0.15A, Section VI, Chapter 2, Jan 20, 1999 http://www.osha.gov/dts/osta/otm/otm_vi/otm_vi_2.html#2
6. NIOSH ALERT: PREVENTING OCCUPATIONAL EXPOSURES TO ANTINEOPLASTIC AND OTHER HAZARDOUS DRUGS IN HEALTH CARE SETTINGS 2004. U.S. Department of

Health and Human Services, Public Health Service, Centers for Disease Control and Prevention, National Institute.
7. FALCK, K., P. GRÖHN, M. SORSA, et al. 1979. Mutagenicity in urine of nurses handling cytostatic drugs. Lancet **1:** 1250–1251. For Occupational Safety and Health, DHHS (NIOSH) Publication No 2004-165.
8. BAKER, E.S. & T.H. CONNOR. 1996. Monitoring occupational exposure to cancer chemotherapy drugs. Am. J. Health Syst. Pharm. **53:** 2713–2723.
9. SORSA, M. & ANDERSON, D. 1996. Monitoring of occupational exposure to cytostatic anticancer agents. Mutat. Res. **355:** 253–261.
10. SESSINK, P.J.M. & R.P. BOS. 1999. Drugs hazardous to healthcare workers: evaluation of methods for monitoring occupational exposure to cytostatic drugs. Drug Saf. **20:** 347–359.
11. HARRISON, B.R. 2001. Risks of handling cytotoxic drugs. *In* The Chemotherapy Source Book, 3rd ed. M.C. Perry, Ed.: 566–582. Lippincott, Williams and Wilkins. Philadelphia.
12. SELEVAN, S.G., M.-L. LINDBOHM, R.W. HORNUNG & K. HEMMINKI. 1985. A study of occupational exposure to antineoplastic drugs and fetal loss in nurses. N. Engl. J. Med. **313:** 1173–1178.
13. STÜCKER, I., J.-F. CALIIARD, R. COLLIN, et al. 1990. Risk of spontaneous abortion among nurses handling antineoplastic drugs. Scand. J. Work Environ. Health **16:** 102–107.
14. HEMMINKI, K., P. KYYRÖNEN & M.-L. LINDBOHM. 1985. Spontaneous abortions and malformations in the offspring of nurses exposed to anesthetic gases, cytostatic drugs, and other potential hazards in hospitals, based on registered information of outcome. J. Epidemiol. Comm. Health **39:** 141–147.
15. PEELEN, S., N. ROELEVELD, D. HEEDERIK, et al. 1999. Toxic Effects on Reproduction in Hospital Personnel. Reproductie-Toxische Effecten Bij Ziekenhuispersonel. Elsevier. Netherlands.
16. VALANIS, B., W.M. VOLLMER & P. STEELE. 1999. Occupational exposure to antineoplastic agents: self-reported miscarriages and stillbirths among nurses and pharmacists. J. Occup. Environ. Med. **41:** 632–638.
17. TURCI, R., C. SOTTANI, G. SPAGNOLI & C. MINOIA. 2003. Biological and environmental monitoring of hospital personnel exposed to antineoplastic agent: a review of analytical methods. J. Chromatog. B. **789:** 169–209.
18. SESSINK, P.J.M., R.B. ANZION, P.H.H. VAN DER BROEK & R.P. BOS. 1992a. Detection of contamination with antineoplastic agents in a hospital pharmacy department. Pharm. Week Sci. **14:** 16–22.
19. SESSINK, P.J.M., K.A. BOER, A.P. SCHEEFHALS, et al. 1992b. Occupational exposure to antineoplastic agents at several departments in a hospital: environmental contamination and excretion of cyclophosphamide and ifosfamide in urine of exposed workers. Int. Arch. Occup. Environ. Health **64:** 105–112.
20. MCDEVITT, J.J., P.S.J. LEES & M.A. MCDIARMID. 1993. Exposure of hospital pharmacists and nurses to antineoplastic agents. J. Occup. Med. **35:** 57–60.
21. PETHRAN, A., K. HAUFF, H. HESSEL & C.-H. GRIMM. 1998. Biological, cytogenetic, and ambient monitoring of exposure to antineoplastic drugs. J. Oncol. Pharm. Pract. **4:** 57.
22. MINOIA, C., R. TURCI, C. SOTTANI, et al. 1998. Application of high performance liquid chromatography/tandem mass spectrometry in the environmental and biological monitoring of healthcare personnel occupationally exposed to cyclophosphamide and ifosfamide. Rapid Commun. Mass Spectrom. **12:** 1485–1493.

23. RUBINO, F.M., L. FLORIDIA, A.M. PIETROPAOLO, *et al.* 1999. Measurement of surface contamination by certain antineoplastic drugs using high-performance liquid chromatography: applications in occupational hygiene investigations in hospital environments. Med. Lav. **90:** 572–583.
24. CONNOR, T.H., R.W. ANDERSON, P.J. SESSINK, *et al.* 1999. Surface contamination with antineoplastic agents in six cancer treatment centers in Canada and the United States. Am. J. Health Syst. Pharm. **56:** 1427–1432.
25. MICOLI, G., R. TURCI, M. ARPELLINI & C. MINOIA. 2001. Determination of 5-fluorouracil in environmental samples by solid-phase extraction and high-performance liquid chromatography with ultraviolet detection. J. Chromatogr. B. **750:** 25–32.
26. VANDENBROUCKE, J. & H. ROBAYS. 2001. How to protect environment and employees against cytotoxic agents, the UZ Ghent experience. J. Oncol. Pharm. Pract. **6:** 146–152.
27. CONNOR, T.H., R.W. ANDERSON, P.J. SESSINK & S.M. SPIVEY. 2002. Effectiveness of a closed-system device in containing surface contamination with cyclophosphamide and ifosfamide in an i.v. admixture area. Am. J. Health Syst. Pharm. **59:** 68–72.
28. KIFFMEYER, T.K., C. KUBE, S. OPIOLKA, *et al.* 2002. Vapor pressures, evaporation behaviour and airborne concentrations of hazardous drugs: implications for occupational safety. Pharm. J. **268:** 331–337.
29. SCHMAUS, G., R. SCHIERL & S. FUNCK. 2002. Monitoring surface contamination by antineoplastic drugs using gas chromatography-mass spectrometry and voltammetry. Am. J. Health Syst. Pharm. **59:** 956–961.
30. WICK, C., M.H. SLAWSON, J.A. JORGENSON & L.S. TYLER. 2003. Using a closed-system protective device to reduce personnel exposure to antineoplastic agents. Am. J. Health Syst. Pharm. **60:** 2314–2320.
31. ZEEDIJK, M., B. GREIJDANUS, F.B. STEENSTRA & D.R.A. UGES. 2005. Monitoring exposure of cytotoxics on the hospital ward: measuring surface contamination of four different cytostatic drugs from one wipe sample. Eur. J. Hosp. Pharm. Sci. **11:** 18–22.
32. ROS, J.J.W., K.A. SIMONS, J.M. VERZIJL, *et al.* 1997. Practical applications of a validated method of analysis for the detection of traces of cyclophosphamide on injection bottles and at oncological outpatient center. Ziekenhuisfarmacie **13:** 168–171.
33. HEPP, R. & G. GENTSCHEW. 1998. External contamination of commercially available cytotoxic drugs. Krankenhauspharmaxie **19:** 22–27.
34. DELPORTE, J.P., P. CHENOIX & P.H. HUBERT. 1999. Chemical contamination of the primary packaging of 5-fluorouracil RTU solutions commercially available on the Belgian market. Eur. Hosp. Pharm. **5:** 119–121.
35. NYGREN, O., B. GUSTAVSSON, L. STRÖM & A. FRIBERG. 2002. Cisplatin contamination on the outside of drug vials. Ann. Occup. Hyg. **46:** 555–557.
36. FAVIER, B., L. GILLES, C. ARDIET & J.F. LATOUR. 2003. External contamination of vials containing cytotoxic agents supplied by pharmaceutical manufacturers. J. Oncol. Pharm. Pract. **9:** 15–20.
37. MASON, H.J., J. MORTON, S.J. GARFITT, *et al.* 2003. Cytotoxic drug contamination on the outside of vails delivered to a hospital pharmacy. Ann. Occup. Hyg. **47:** 681–685.
38. CONNOR, T.H., P.J.M. SESSINK, B.R. HARRISON, *et al.* 2005. Surface contamination of chemotherapy drug vials and evaluation of new vial-cleaning techniques: results of three studies. Am. J. Health Syst. Pharm. **62:** 475–484.

39. ASTM D666-01 STANDARD PRACTICE FOR FIELD COLLECTION OF ORGANIC COMPOUNDS FROM SURFACE USING WIPE SAMPLING. ASTM International. www.astm.org.
40. OCCUPATIONAL SAFETY AND HEALTH ADMINISTRATION. EVALUATION GUIDELINES FOR SURFACE SAMPLING METHODS. http://www.osha.gov/dts/sltc/methods/surfacesampling/t-006-01-0104-m.html Accessed June 15. 2005.
41. NYGREN, O. & C. LUNDGREN. 1997. Determination of platinum in workroom air and in blood and urine from nursing staff attending patients receiving cisplatin chemotherapy. Int. Arch. Occup. Environ. Health **70:** 209–214.
42. SPEZIA, S., B. BOCCA, G. FORTE, et al. 2005. Comparison of inductively coupled plasma mass spectrometry techniques in the determination of platinum in urine: quadrupole vs. sector field. Rapid Commun. Mass Spectrom. **19:** 1551–1556.
43. KLEINBERG, M.L. & M.J. QUINN. 1981. Airborne drug levels in a laminar-flow hood. Am. J. Hosp. Pharm. **38:** 1301–1303.
44. DE WERK NEAL A., R.A. WADDEN & W.L. CHIOU. 1983. Exposure of hospital workers to airborne antineoplastic agents. Am. J. Hosp. Pharm. **40:** 597–601.
45. MCDIARMID, M.A., T. EGAN, M. FURIO, et al. 1986. Sampling for airborne fluorouracil in a hospital drug preparation area. Am. J. Hosp. Pharm. **43:** 1942–1945.
46. PYY, L., M. SORSA & E. HAKALA. 1988. Ambient monitoring of cyclophosphamide in manufacture and hospitals. Am. Ind. Hyg. Assoc. J. **49:** 314–317.
47. STUART, A., A.D. STEPHENS, L. WELCH & P.H. SUGERBAKER. 2002. Safety monitoring of the coliseum technique for heated intraoperative intraperitoneal chemotherapy with mitomycin C. Annals Surg. Oncol. **9:** 186–191.
48. LARSON, R.R., M.B. KHAZAELI & H.K. DILLON. 2003. A new monitoring method using solid sorbent media for evaluation of airborne cyclophosphamide and other antineoplastic agents. Appl. Occup. Environ. Hyg. **18:** 120–131.
49. AMERICAN CONFERENCE OF GOVERNMENT INDUSTRIAL HYGIENISTS. 2004. Threshold Limit Values for Chemical Substances and Physical Agents Biological Exposure Indices. ACGIH. Cincinnati, OH.
50. SARGENT, E.V., B.D. NAUMANN, D.G. DOLAN, et al. 2002. The importance of human data in the establishment of occupational exposure limits. Hum. Ecol. Risk Assess. **8:** 805–822.
51. KROMHOUT, H., F. HOEK, R. UITTERHOEVE, et al. 2000. Postulating a dermal pathway for exposure to antineoplastic drugs among hospital workers. Applying a conceptual model to the results of three workplace surveys. Ann. Occup. Hyg. **44:** 551–560.
52. SPIVEY, S. & T.H. CONNOR. 2003. Determination of sources of workplace contamination with antineoplastic drugs and comparison of conventional IV drug preparation versus a closed system. Hosp. Pharm. **38:** 135–139.
53. CONNOR, T.H., M. SHULTS & M.P. FRASER. 2000. Determination of the vaporization of solutions of mutagenic antineoplastic agents at 23° and 37° C using a desiccator technique. Mutat. Res. **470:** 85–92.
54. HARRISON, B.R., R.J. GODEFROID & E.A. KAVANAUGH. 1996. Quality-assurance testing of staff pharmacists handling cytotoxic drugs. Am. J. Health Syst. Pharm. **53:** 402–407.

Beyond Managing Healthcare Risks
The Health-Promoting Hospital Initiative in Mexico

CARLOS SANTOS-BURGOA

Director General for Health Promotion, Ministry of Health, 06700 Mexico D.F., Mexico

ABSTRACT: The hospital industry is unique for having within it "customers" exposed to a complex mix of risks. A model is proposed that combines both the risk assessment and the promoting hospital models. This model acts in three stages: exposure elimination and protection, health aptitudes and culture, and hospital population action, and includes specific operations that can be tracked through specific effectiveness factors. Being tested in a small community hospital, there is an opportunity to apply it within the current Mexican Health reform that moves the financial risk from the patient to the provider and thus may support health promotion.

KEYWORDS: health promotion; health promoting hospital; health reform; hospital industry; risk assessment; evidence; exposure; standards

This short communication uses the Collegium Rammazzini's principle[1] to "... assesses present and future potential for injury and disease attributable to the environment and transmits its views to policy-making bodies, authorities, agencies and the public..." The key question is: how can we add value to the people at the hospital setting to prevent disease and promote their health? We need to bridge the knowledge (what we already know) and health promotion (how much health we can produce from such knowledge). We want to approach this question for a unique industry. As any industry, it brings raw materials into a "production process." Entering into the process, this industry includes a highly varied mixture of risks: ergonomic, chemical, physical including ionizing and non-ionizing radiation (EMF, RFF), biologic, psychosocial (importantly stress), as well as new risks derived from the advanced use of nanotechnology and the expandable use of proteomics.

Address for correspondence: Dr. Carlos Santos-Burgoa, M.D., M.P.H., Ph.D., Director General for Health Promotion, Ministry of Health, Guadalajara No. 46, 1er. Piso, Col. Roma, c.p. 06700 Mexico D.F., México. Voice: +52-55 52113139 fax: +52-55 52860207.
e-mail: csantos@salud.gob.mx; c.santosburgoa@ipade.mx

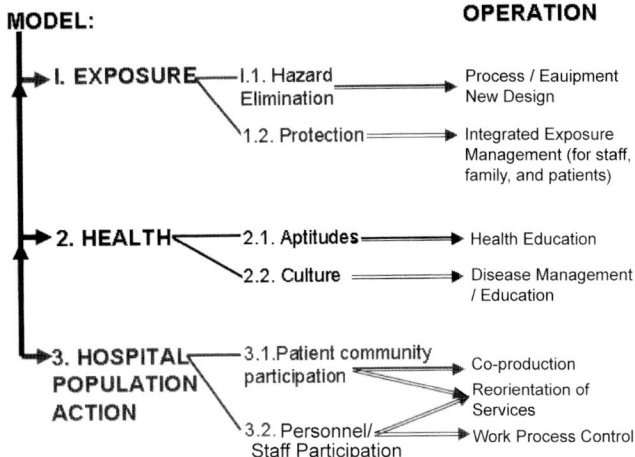

FIGURE 1. Proposed Health Promoting Hospital components.

What makes this industry unique is that it not only includes such a complex mixture of hazards, but it is an industry that includes the "customers" inside its production area. That is, workers (medical doctors, nurses, technicians, support personnel, cleaning, security, etc.) take care within the industry to patients. Family members and/or friends frequently accompany the patients, a usual social pattern in our Latin American countries. The patients have peculiarities that are completely different from the usual healthy worker model under which we usually build our occupational standards: patients spend many days within the environment, are susceptible, and may be immunologically compromised, and have different diseases due to different exposure patterns, i.e., respiratory disorders due to inhalation exposure. In 1 week, a 30-bed hospital will have 22,200 person-hours of exposure. Personal protection is differential, not protecting the most vulnerable, with most of the people inexperienced in its use, and no generalized awareness of such a need. No other hazardous industry has this further complexity.

To address this challenge, two approaches have been taken: the Risk Assessment–Risk Management paradigm,[2] a resource-intensive disease by disease or hazard by hazard approach, or the Health Promoting Hospital approach from the European Network,[3] which has 18 strategies involving patients, staff, and the community, and addresses co-production, environments, disease management, and lifestyles.

In Mexico, we have a context that is particularly conducive to start addressing this challenge. We are going through a profound health reform[4] (2004–2010) that will provide universal health insurance to all citizens. A major Infrastructure Master Plan[5] is being built and consists of 18 healthcare networks, including 10 specialized care hospitals and more than 100 general hospitals.

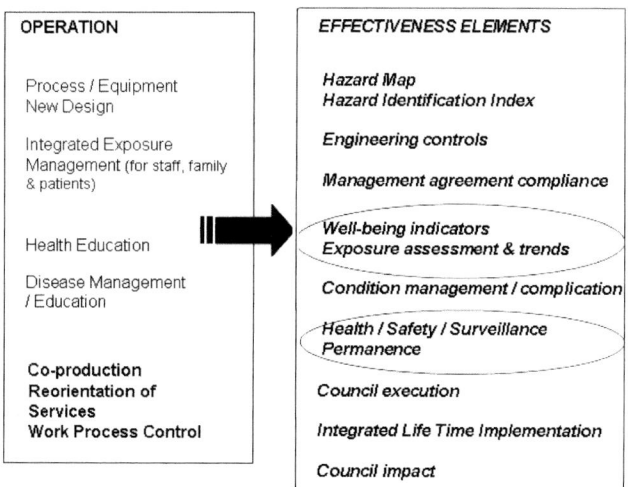

FIGURE 2. Effectiveness elements for follow-up in model implementation.

The financial basis for this reform is built on the principle that each family carries its own public budget. This is a one-time chance to design the medical units, select the equipments and materials, and organize them so that they may be friendlier to workers, patients and families, and the environment. It is also a chance to organize care aimed at transformation of the medical unit from a curative setting to a "health production unit," but how to use this opportunity in a sustainable way, especially when we do not have the local culture to internalize these factors and there is little national expertise to drive the state of the art in medical unit design and technology. We propose a Health Promoting Hospital model (FIG. 1) that addresses the determinants of health, both the positive enhancing determinants, such as environmental, psychosocial, lifestyle, as well as the hazards and risks. The model acts on three stages: on exposure elimination or protection, on health aptitudes and culture, and on action with the hospital population, both staff of patients and their families. It is clearly translated into defined operations that can be translated into specific performance indicators. To develop the model requires an organizational development framework that, in a strategic mapping approach, would require the support of human resources, information systems, and, most importantly, organizational change based on a shared concern and grasp of the opportunity. There are specific critical effectiveness factors that must be addressed for each stage of the model, such as the hazard map, the trends in exposure estimates, and the availability of a consultative council, and so forth (FIG. 2).

We are starting with a new small community hospital in a mountainous region of southern Mexico (Tlapa, Guerrero), and we have developed an easy-to-handle spreadsheet for project control, with specific indices for each activity in the model. All initiatives have opportunities and limitations. The biggest opportunity is the health reform that moves the financial risk burden from the patient to the provider and thus may support health promotion. The reform carries with it a new infrastructure that includes public and private investment, as well as well-developed specifications that could be influenced. The obstacles include resistance to change, more in established hospitals, the "let someone else do it" attitude, the overburden to hospitals, and the built-in hazardous environment with large capital invested. The biggest challenge is to intervene with evidence-based public health and demonstrate its cost-effectiveness to assure its sustainability to meet expectation with reality. We believe this can become an exceptional experience to benchmark future approaches in improving the working and living experiences in this industry.

REFERENCES

1. COLLEGIUM RAMAZZINI. 2004. Purpose Statement. Carpi, Italy.
2. RICCI, P. 2005. Environmental and Health Risk Assessment and Management: Principles and Practices. Springer. Verlag. Netherlands.
3. WHO-EURO. 2004. Standards for Health Promotion in Hospitals. Barcelona, Spain.
4. CONGRESO DE LA UNIÓN. 2004. Ley General de Salud (last published reform). Dirección General de Bibliotecas, Cámara de Diputados, Mexico.
5. SECRETARÍA DE SALUD. 2003. Plan Maestro de Infraestructura. Mexico D.F.

Handling Anticancer Drugs

From Hazard Identification to Risk Management?

MARJA SORSA,[a] MERVI HÄMEILÄ,[b] AND EIJA JÄRVILUOMA[c]

[a] *CA ConsultArt Ltd, Menninkäisentie 8 B, FI-02100 Espoo, Finland*
[b] *Finnish Institute of Occupational Health, F1-00290 Helsinki, Finland*
[c] *Helsinki University Central Hospital (HUCH) Pharmacy, F1-00290 Helsinki, Finland*

> ABSTRACT: The potential health hazards involved with antineoplastic agents have been known for decades. Many anticancer drugs are recognized as carcinogens and genotoxins in experimental assays. Second cancers have been recorded in follow-up studies with treated patients. The first findings on occupational exposures among hospital personnel administering chemotherapy were reported only in 1979. Since that time a magnitude of studies have been published using various chemical and biological exposure measurements. These findings prompted improvements in the handling practices of personnel working with anticancer drugs. In spite of strict guidelines for the safe handling of cancer chemotherapeutic agents and extensive improvements in the handling facilities in hospitals, also recent studies have revealed detectable, even if generally much decreased, amounts of indicator drugs in air and surface wipe samples, also including biological samples of personnel in hospital pharmacies and cancer therapy wards. Follow-up measurements show that application of strict safety precautions in hospitals decreases the biological exposure and/or effect markers to the level of unexposed referents. Open information and constant tutoring of personnel to avoid the hazards when working with anticancer drugs is absolutely necessary with the increasing use of these important drugs.
>
> KEYWORDS: anticancer drugs; biomonitoring; biomarkers; cytostatics; exposure assessment

IDENTIFICATION OF OCCUPATIONAL HAZARDS WITH ANTICANCER DRUGS

The real concern about potential occupational hazards involved at work with antineoplastic agents came quite late, only in 1979 when Falck *et al.*[1] reported

Address for correspondence: Prof. Marja Sorsa, CA ConsultArt Ltd., Menninkäisentie 8 B, FI-02100 Espoo, Finland. Voice: 358-40-50-60-805; fax: 358-9-463-221.
e-mail: marja.sorsa@transmix.fi

of patients and the increasing treatments outside hospital wards. An effective risk management needs to consider several aspects, such as:

1. improvements in handling practices;
2. standardization of analytical methods to allow comparative studies;
3. quantitative risk assessment of new anticancer drugs and multiple exposures;
4. improved psychosocial work atmosphere;
5. prevention of carelessness due to false illusion of constant safety;
6. development of closed system "isolators" and robotics for drug preparation;
7. prevention of unnecessary transport of drugs;
8. preparedness for accidental spillage and clean-up procedures;
9. preventing shipments of contaminated drug vials from companies.

In order to achieve better quantitative risk assessments well-conducted epidemiological studies are needed on persons occupationally exposed to anticancer drugs. Also, reproductive health outcome should be included among the end points studied.

Constant education of personnel and tutoring of new employees are necessary to keep up the motivation of good handling practices and proper occupational safety measures.

REFERENCES

1. Falck, K., P. Gröhn, M. Sorsa, et al. 1979. Mutagenicity in urine of nurses handling cytostatic drugs. Lancet **i:** 1250–1251.
2. Siebert, D. & U. Simon. 1973. Cyclophosphamide. Pilot study of genetically active metabolites in urine of a treated human patient. Mutat. Res. **19:** 65–72.
3. Sorsa, M., K. Hemminki & H. Vainio. 1985. Occupational exposure to anticancer drugs—potential and real hazards. Mutat. Res. **154:** 135–149.
4. Sorsa, M. & D. Anderson. 1996. Monitoring of occupational exposure to cytostatic anticancer agents. Mutat. Res. **355:** 253–261.
5. International Agency for Research on Cancer. 2001. IARC monographs on the evaluation of the carcinogenic risk of chemicals to humans Vol. **76**. Some Antiviral and Antineoplastic Drugs, and Other Pharmaceutical Agents. Lyon, France.
6. Department of Health and Human Services (Niosh). 2004. Occupational hazards in hospitals. Antineoplastic agents. DHHS (NIOSH) Public. **102:** 1–13.
7. Connor, T.H., R.W. Anderson, P.J.M. Sessink, et al. 1999. Surface contamination with antineoplastic agents in six cancer treatment centers in Canada and the United States. Am. J. Health. Syst. Pharm. **56:** 1427–1432.
8. Ziegler, E., H.J. Mason & P.J. Baxter. 2002. Occupational exposure to cytotoxic drugs in two UK oncology wards. Occup. Environ. Med. **59:** 608–612.
9. Hämeilä, M., K. Aaltonen, T. Santonen, et al. 2003. Altistuminen solunsalpaajille apteekki- ja hoitotyössä. Final report. Finnish Institute of Occupational Health, Helsinki, Finland.

10. NYGREN, O., B. GUSTAVSSON, L. STRÖM & A. FRIBERG. 2002. Cisplatin contamination observed on the outside of drug vials. Ann. Occup. Hyg. **46:** 555–557.
11. MASON, H.J., J. MORTON, S.J. GARFITT, *et al*. 2003. Cytotoxic drug contamination on the outside of vials delivered to a hospital pharmacy. Ann. Occup. Hyg. **47:** 681–685.
12. SESSINK, P.J.M., M.C.A. VAN DE KERKHOF, R.B.M. ANZION, *et al*. 1994. Environmental contamination and assessment of exposure to anti-neoplastic agents by determination of cyclophosphamide in urine of exposed pharmacy technicians: is skin absorption an important exposure route? Arch. Env. Health **49:** 165–169.
13. TURCI, R., C. SOTTANI, A. RONCHI & C. MINOIA. 2002. Biological monitoring of hospital personnel occupationally exposed to antineoplastic agents. Toxicol. Lett. **134:** 57–64.
14. PETHRAN, A., R. SCHIERL, K. HAUFF, *et al*. 2003. Uptake of antineoplastic agents in pharmacy and hospital personnel. Part I: monitoring of urine concentrations. Int. Arch. Occup. Environ. Health **76:** 5–10.
15. PILGER, A., I. KÖHLER, H. STETTNER, *et al*. 2000. Long-term monitoring of sister chromatid exchanges and micronucleus frequencies in pharmacy personnel occupationally exposed to cytostatic drugs. Int. Arch. Occup. Environ. Health **73:** 442–448.
16. MALUF, S.W. & B. ERDTMANN. 2000. Follow-up study of the genetic damage in lymphocytes of pharmacists and nurses handling antineoplastic drugs evaluated by cytokinesis-block micronuclei analysis and single cell gel electrophoresis assay. Mutat. Res. **47:** 21–27.
17. ÛNDEGER, Û., N. BASARAN, A. KARS & D. GÜC. 1999. Assessment of DNA damage in nurses handling antineoplastic drugs by the alkaline COMET assay. Mutat. Res. **439:** 277–285.
18. BURGAZ, S., B. KARAHALIL, Z. CANH, *et al*. 2002. Assessment of genotoxic damage in nurses occupationally exposed to antineoplastics by the analysis of chromosomal aberrations. Human Exp. Toxicol. **21:** 129–135.
19. DUTCH EXPERT COMMITTEE ON OCCUPATIONAL STANDARDS. 2005. Health-based calculated occupational cancer risk values. Cyclophosphamide. Cisplatin. Gezondheidsraad. 02, 03 OSH.

from Finland that mutagenicity of concentrated urine samples was increased among oncology unit nurses, as compared with hospital office clerks. An earlier finding[2] of increased mutagenicity in urine samples from patients receiving chemotherapy was confirmed, also in the study of Falck et al.,[1] where the main objective was method development to apply urinary mutagenicity as an unspecific exposure marker for environmental/occupational genotoxins.

Since the first findings, several groups reported both positive and negative results, using the urine mutagenicity assay and other biomonitoring tests, among the personnel admixing and administering anticancer drugs in the hospital environment.[3] Inconsistencies in the results are not necessarily truly contradictory. Together with methodological, individual, and sampling variation, this may reflect differences in the efficiency of preventive actions to avoid exposures (see Ref. 4).

Considering the urine mutagenicity method, it should be stressed that positive results in the assay can only be expected for such anticancer drugs that are active *per se* in the bacterial mutagenicity assays and/or produce active urinary metabolites, taking into account the toxicokinetics of the drugs and the time of urine sampling. With the wide variety of drugs handled daily, and the increasing use of combination chemotherapy and new drugs coming into use (see TABLE 1), it is practically recommendable for occupational safety to handle all antineoplastic drugs as if they would be carcinogens. However, the potency and the proof of evidence available on their carcinogenicity varies considerably. So far, the evidence on cancer risk, based on experimental genotoxicity, animal cancer bioassays, and epidemiology on second cancers is strongest with the group of alkylating cytostatics.[5]

TABLE 1. Some commonly used anticancer drugs

Alkylating agents/monofunctional
   dacarbazine, procarbazine
Alkylating agents/bifunctional
   busulfan, chlorambucil, cyclophosphamide, iphosphamide, melphalan, thiotepa
Antimetabolites
   azathioprine, cytarabin, fluorouracil, mercaptopurine, methotrexate
Enzymes
   asparaginase
Platinum agents
   carboplatin, cisplatin
Podophyllotoxins
   etoposid, tenoposid
Spindle inhibitors
   taxol and derivatives, vinblastine, vincristine
Topoisomerase I inhibitors
   topotekan, irinotekan
Topoisomerase II inhibitors
   amsacrine, mitoxanthrone
Tumor antibiotics
   bleomycin, daunorubicin, mitomycin

EXPOSURE ASSESSMENT STUDIES

Increasing numbers of people are today in contact with anticancer drugs and thus potentially exposed (see TABLE 2). Number of cancer patients is increasing in all countries, as is the use of chemotherapy, also outside the hospital wards in policlinics and at home treatment of cancer patients. "Carry over" of drug aerosols and dust may further expand the potential contacts and exposures.

The hospital personnel, both in pharmacies and in wards may be exposed to anticancer drugs through breathing, skin contact, or even unintentional ingestion. In a recent document by NIOSH,[6] several risky operations have been listed especially risky for possible skin contact: for example, counting tablets from multidose vials, breaking tablets, preparing, handling, administering, and disposing drug solutions, cleaning spills, handling clothing, bed linens, excretes, or vomits of patients under therapy. Also during transport of drugs accidental spillage may occur.

ENVIRONMENTAL MONITORING

Hundreds of reports and publications have been made on hygienic measurements performed in hospitals and pharmacies on specific indicator drugs. There is a clear correlation between positive analytical findings and sloppy handling of the drugs (see Ref. 4). Today, with improved analytical methods using tandem mass spectrometry and specific internal standards, the detection limit has been much lowered. It is, in fact, surprising that in spite of strict handling practices with anticancer drugs, detectable amounts have still been observed in air and surface wipe samples taken from hospital pharmacies and oncology awards even rather recently.[7–9] However, standardizations in the analytical methods are urgently needed to allow comparative studies and interlaboratory calibrations to be made.

Unexpectedly, several recent reports suggest that a potential exposure source in hospitals is in the transport of drugs. Surfaces of the unopened drug vials purchased from the drug companies have been found to be contaminated with antineoplastic drugs.[10,11] This is of special concern to hospital pharmacies

TABLE 2. Groups of persons potentially exposed to anticancer agents in hospitals

* Pharmacy personnel preparing solutions
* Hospital staff in drug administration
* Physicians and nurses in patient care
* Cleaning personnel
* Laundry personnel
* Family members and friends of patients
* Scientists and laboratory personnel
* Drug transport personnel

TABLE 3. Methods to assess exposure and effects of anticancer drugs in hospital work areas

Chemical Analyses	Environmental Samples	Air samples Wipe samples Gloves
	Biological Samples	Drugs Metabolites
Genotox Analyses	Exposure Biomonitoring	Adducts DNA SS breaks U-Mutagens
	Effect Biomonitoring	CA, SCE, MN Point mutation
EPI Studies	Exposed and Referent Cohorts	Symptoms Cancer Reproductive Health effects

receiving large amounts of anticancer drugs; naturally also to the pharmaceutical industry and its packaging and transport personnel.

BIOMONITORING OF OCCUPATIONAL EXPOSURES

The sensitivity of chemical analyses has long been a limiting factor in monitoring occupational exposures from biological samples.[4] Sessink et al.[12] improved the analytical techniques for the detection of cyclophosphamide, 5-fluorouracil, and methothrexate and were able to detect these indicator drugs in the urine samples of pharmacy technicians. More recently, Turci et al.[13] from Italy and Pethran et al.[14] from Germany reported that in spite of strict safety precautions, specific anticancer drugs handled during the work day were identified in urine samples of hospital and pharmacy workers. These results strongly point to the need of constantly keeping up with the safety instructions and improving and standardizing the analytical methods for routine monitoring also to include new pharmaceutical products being introduced to cancer chemotherapy.

Various biological methods have been introduced to measure exposure-related effects, biomarkers, in the cells of potentially exposed persons. These are based on the covalent binding of genotoxic anticancer drugs or their metabolites into proteins or nucleic acids inducing adducts, breaks, and/or somatic mutations in the cells of the exposed persons. The end points measured may be specific for the indicator drug, as some adducts, or specific only to the genotoxicity of the environmental exposure (see TABLE 3). Cytogenetic monitoring provides a tool to detect accumulated unrepaired damage in lymphocytes over years, while adducts in hemoglobin may give a scope reflecting exposures during a few months, that is, the lifetime of erythrocytes.

Both positive and negative results have been reported on exposure-related effects using various biomonitoring methods (see Ref. 4). The improved working conditions have clearly decreased reports on positive biomonitoring results, even in cases when environmental monitoring has detected contamination in the work environment. Accidental exposures may occur in spite of safety precautions, and such cases have been associated with cytogenetic damage, measured as increased sister chromatid exchanges or micronucleated lymphocytes.[15] Also, in countries where traditionally less concern is given to occupational health and safety, biomarkers revealing alkali labile sites[16] and/or single-strand breakage in DNA[17] or increased chromosome aberrations[18] have been detected and reported.

A well-planned biomonitoring study requires careful combination of chemical and biological analyses with a proper environmental exposure survey of the types and amounts of anticancer drugs handled.

RISK MANAGEMENT

The example of occupational handling of antineoplastic drugs in hospitals provides a possibility for true quantitative cancer risk assessment. So far, this has been reached for a few anticancer drugs only. The estimates generally end up in very small lifetime occupational risks in the order of 1:100,000 work life (40 years) risks, as calculated, for example, for cyclophosphamide and cisplatin.[19]

The reality in hospital work is, however, much more complicated. The potential and true exposure situations vary enormously during days of work life. Today, increasing numbers, frequencies, and amounts of different drugs are handled, increasing the complexity of practical risk management. The work needs good collaboration within hospitals, their personnel, and integration of occupational health and safety workers with different expertise from analytical chemistry, industrial hygiene, and toxicology.

Has everything been done? Obviously not, while the new exposure assessments still reveal that the detected hazards are not being completely managed in practice. Successful risk management requires a susceptible soil to flourish. The ingredients derive from an open and supportive work environment. Even if codes of good conduct are available, negligence due to carelessness, too tight work pace, or unconcerned working colleagues may destroy even the best instructions.

Concerning future developments in cancer care, the potential of exposure hazards will certainly be increasing due to expanding number of persons involved. Cancer care and cancer chemotherapy are expanding due to the number

10. NYGREN, O., B. GUSTAVSSON, L. STRÖM & A. FRIBERG. 2002. Cisplatin contamination observed on the outside of drug vials. Ann. Occup. Hyg. **46:** 555–557.
11. MASON, H.J., J. MORTON, S.J. GARFITT, et al. 2003. Cytotoxic drug contamination on the outside of vials delivered to a hospital pharmacy. Ann. Occup. Hyg. **47:** 681–685.
12. SESSINK, P.J.M., M.C.A. VAN DE KERKHOF, R.B.M. ANZION, et al. 1994. Environmental contamination and assessment of exposure to anti-neoplastic agents by determination of cyclophosphamide in urine of exposed pharmacy technicians: is skin absorption an important exposure route? Arch. Env. Health **49:** 165–169.
13. TURCI, R., C. SOTTANI, A. RONCHI & C. MINOIA. 2002. Biological monitoring of hospital personnel occupationally exposed to antineoplastic agents. Toxicol. Lett. **134:** 57–64.
14. PETHRAN, A., R. SCHIERL, K. HAUFF, et al. 2003. Uptake of antineoplastic agents in pharmacy and hospital personnel. Part I: monitoring of urine concentrations. Int. Arch. Occup. Environ. Health **76:** 5–10.
15. PILGER, A., I. KÖHLER, H. STETTNER, et al. 2000. Long-term monitoring of sister chromatid exchanges and micronucleus frequencies in pharmacy personnel occupationally exposed to cytostatic drugs. Int. Arch. Occup. Environ. Health **73:** 442–448.
16. MALUF, S.W. & B. ERDTMANN. 2000. Follow-up study of the genetic damage in lymphocytes of pharmacists and nurses handling antineoplastic drugs evaluated by cytokinesis-block micronuclei analysis and single cell gel electrophoresis assay. Mutat. Res. **47:** 21–27.
17. ÛNDEGER, Û., N. BASARAN, A. KARS & D. GÜC. 1999. Assessment of DNA damage in nurses handling antineoplastic drugs by the alkaline COMET assay. Mutat. Res. **439:** 277–285.
18. BURGAZ, S., B. KARAHALIL, Z. CANH, et al. 2002. Assessment of genotoxic damage in nurses occupationally exposed to antineoplastics by the analysis of chromosomal aberrations. Human Exp. Toxicol. **21:** 129–135.
19. DUTCH EXPERT COMMITTEE ON OCCUPATIONAL STANDARDS. 2005. Health-based calculated occupational cancer risk values. Cyclophosphamide. Cisplatin. Gezondheidsraad. 02, 03 OSH.

of patients and the increasing treatments outside hospital wards. An effective risk management needs to consider several aspects, such as:

1. improvements in handling practices;
2. standardization of analytical methods to allow comparative studies;
3. quantitative risk assessment of new anticancer drugs and multiple exposures;
4. improved psychosocial work atmosphere;
5. prevention of carelessness due to false illusion of constant safety;
6. development of closed system "isolators" and robotics for drug preparation;
7. prevention of unnecessary transport of drugs;
8. preparedness for accidental spillage and clean-up procedures;
9. preventing shipments of contaminated drug vials from companies.

In order to achieve better quantitative risk assessments well-conducted epidemiological studies are needed on persons occupationally exposed to anticancer drugs. Also, reproductive health outcome should be included among the end points studied.

Constant education of personnel and tutoring of new employees are necessary to keep up the motivation of good handling practices and proper occupational safety measures.

REFERENCES

1. FALCK, K., P. GRÖHN, M. SORSA, et al. 1979. Mutagenicity in urine of nurses handling cytostatic drugs. Lancet **i:** 1250–1251.
2. SIEBERT, D. & U. SIMON. 1973. Cyclophosphamide. Pilot study of genetically active metabolites in urine of a treated human patient. Mutat. Res. **19:** 65–72.
3. SORSA, M., K. HEMMINKI & H. VAINIO. 1985. Occupational exposure to anticancer drugs—potential and real hazards. Mutat. Res. **154:** 135–149.
4. SORSA, M. & D. ANDERSON. 1996. Monitoring of occupational exposure to cytostatic anticancer agents. Mutat. Res. **355:** 253–261.
5. INTERNATIONAL AGENCY FOR RESEARCH ON CANCER. 2001. IARC monographs on the evaluation of the carcinogenic risk of chemicals to humans Vol. **76**. Some Antiviral and Antineoplastic Drugs, and Other Pharmaceutical Agents. Lyon, France.
6. DEPARTMENT OF HEALTH AND HUMAN SERVICES (NIOSH). 2004. Occupational hazards in hospitals. Antineoplastic agents. DHHS (NIOSH) Public. **102:** 1–13.
7. CONNOR, T.H., R.W. ANDERSON, P.J.M. SESSINK, et al. 1999. Surface contamination with antineoplastic agents in six cancer treatment centers in Canada and the United States. Am. J. Health. Syst. Pharm. **56:** 1427–1432.
8. ZIEGLER, E., H.J. MASON & P.J. BAXTER. 2002. Occupational exposure to cytotoxic drugs in two UK oncology wards. Occup. Environ. Med. **59:** 608–612.
9. HÄMEILÄ, M., K. AALTONEN, T. SANTONEN, et al. 2003. Altistuminen solunsalpaajille apteekki- ja hoitotyössä. Final report. Finnish Institute of Occupational Health, Helsinki, Finland.

Chemical Safety and Health Conditions among Hungarian Hospital Nurses

ANNA TOMPA,[a] MÁTYÁS JAKAB,[b] ANNA BIRÓ,[b] BALÁZS MAGYAR,[a] ZOLTÁN FODOR,[b] TIBOR KLUPP,[b] AND JENÖ MAJOR[b]

[a]*Department of Public Health, Semmelweis University, 1096 Budapest, Hungary*

[b]*Department of Cytogenetics and Immunology, National Institute of Chemical Safety, 1096 Budapest, Hungary*

ABSTRACT: In the present study genotoxicological and immunotoxicological follow-up investigations were made on 811 donors including 94 unexposed controls and 717 nurses with various working conditions from different hospitals (The Hungarian Nurse Study). The nurses were exposed to different chemicals: cytostatic drugs, anesthetic, and sterilizing gases, such as ethylene oxide (ETO) and formaldehyde. The measured biomarkers were: clinical laboratory routine tests, completed with genotoxicological (chromosome aberrations [CA], sister chromatid exchange [SCE]), and immune-toxicological monitoring (ratio of lymphocyte subpopulations, lymphocyte activation markers, and leukocyte oxidative burst). The highest rate of genotoxicologically affected donors (25.4%) was found in the group of cytostatic drug-exposed nurses. Comparing geno- and immunotoxicological effect markers, we found that among genotoxicologically affected donors the frequency of helper T cell (Th) lymphocytes, the ratio of activated T and B cells increased, whereas the oxidative burst of leukocytes decreased. In hospitals with lack of protective measures increased CA yields were observed compared to those with ISO 9001 quality control or equivalent measures. Anemia, serum glucose level, thyroid dysfunctions, benign, and malignant tumors were more frequent in the exposed groups than in controls. The hygienic standard of the working environment is the basic risk factor for the vulnerability of nurses. On the basis of these results, it is suggested, that the used cytogenetic and immunological biomarkers are appropriate to detect early susceptibility to diseases. The Hungarian Nurse Study proved that the use of safety measures could protect against occupational exposure at work sites handling cytostatic drugs, anesthetic, and sterilizing gases.

KEYWORDS: cytostatics; anesthetics; sterilizing agents; chromosome aberrations; sister chromatid exchange; unscheduled DNA synthesis; HPRT mutations; genotoxicological monitoring; immune toxicology; tumors; health status; risk assessment

Address for correspondence: Prof. Anna Tompa, M.D., Ph.D., Semmelweis University, Department of Public Health, P.O. Box 370, 1445 Budapest, Hungary.
e-mail: tompa.okbi@okk.antsz.hu

INTRODUCTION

Occupational exposure to cytostatic drugs, anesthetic, and sterilizing gases with mutagenic and human carcinogenic capacity is a major hazard of the healthcare personnel. The most effective means of minimalization of work site exposures is the strict quality control (e.g., EU standard ISO 9001).[1]

In Hungary different recommendations have been published in order to reduce the risk. Since 1994, guidelines addressing each step in the hazardous process of handling cytostatic drugs have been developed in Hungary in order to decrease exposure and protect healthcare personnel, patients, and the environment. In 2000, Hungary conformed her legislation to the directives of the European Union (67/548/EEC) by the Public Act No. XXV (2000) on Chemical Safety, and the 26/2000 order of the Ministry of Health on the protection against occupational carcinogenic substances.[1,2] In 2004, a methodological guideline of the National Institute of Pharmacy, Hungary on manufacturing and use of mixed cytostatic infusions was issued.[3] The guidelines in these regulations on the protection against exposure to carcinogenic agents and for the performance of environmental and biological monitoring at workplaces with higher cancer risk were largely based on previous results of the follow-up genotoxicology monitoring of nurses exposed to chemotherapeutic agents. These investigations were carried out by the Department of Genotoxicology of the National Institute of Chemical Safety.

There was no publication of experimental or morbidity data combined with genotoxicological monitoring before our study on the real risk of occupational effects induced by cytostatic, anesthetic, and sterilizing drugs on the nurse's health. Sporadic data without genotoxicological investigations are only available on the odd ratios of the most characteristic symptoms as infertility, disorders in menstruation, hair loss, skin rush, and lightheadedness.[4–6]

An unusual cluster of breast cancer incidence appeared among nurses in the Newborn Unit of a county hospital in Northern Hungary in 1993. A cluster of cancer cases, that is, eight breast cancers, two ovarian carcinomas as well as uterus, colon and brain tumors, and other malignant tumors occurred during the previous 12 years (1981–1993) in the same Newborn Unit of the county hospital in Northern Hungary among the staff using an ethylene oxide (ETO) gas sterilizer.[7] The sterilizer was inappropriately repaired and run, consequently high ETO emissions occurred during sterilization. In addition the staff members of the unit were also exposed to low-level environmental radon. ETO is a well-known human carcinogen.[8] ETO and ionizing radiation can increase both chromosome aberration (CA) and sister chromatid exchange (SCE).[8,9] CA and SCE frequencies were significantly higher in ETO and radon-exposed subjects compared to that of sterilizing units in other hospitals.[7] Formaldehyde, widely used for sterilization, is also clastogenic.

In the last decades, a number of antineoplastic drugs have been introduced to the treatment of cancers. Most of these drugs however have been classified to be carcinogenic to humans according to their mutagenic, clastogenic, and carcinogenic properties.[10–13] Since 1983, several guidelines for the handling of anticancer drugs have been issued with the purpose of decreasing exposure and protecting personnel.[14] Occupational exposure of oncological nurses handling neoplastic drugs may occur in different ways: by inhalation of airborne antineoplastic agents, absorption through skin contact, human excreta handling, ingestion during drug preparation and administration, and during disposal of equipment.[15,16]

In the operation theaters the main genotoxicological hazard for the anesthesiology personnel is the use of Halothane. A possible additional exposure may occur during the use of X ray for imaging.

The genotoxic effects of these agents underline the use of biomonitoring among personnel with potential work site exposure. The evaluation of genotoxic effects of occupational exposures to cytostatics is of primary interest in chemical safety and primary prevention. In biomonitoring studies of effects in human populations exposed to genotoxic agents, cytogenetic methods with human peripheral blood lymphocytes (PBLs) have been extensively used.[17] In lymphocytes of oncology nurses and pharmaceutists increased number of CAs, SCEs, and gene mutations were found.[18–22]

Recognizing the hazardous effect of both cytostatic and gas-sterilizing agents on the health of nurses we have introduced a follow-up multiple end point genotoxicological monitoring since 1992. The aim of our human genotoxicology monitoring is to assess and minimize genotoxic risks and keep exposure as low as possible. In the National Institute of Chemical Safety of Hungary, a multiple end point genotoxicology monitoring system has been developed since the mid 1980s and was used for the risk assessment of different human populations, control subjects, and those occupationally exposed to various genotoxic agents. The monitoring is able to detect and follow-up the alterations in work-related conditions.[23]

The genotoxicological effect of chemical exposures detected in peripheral leukocytes raised the possibility of phenotypic and functional alterations in immune-competent cells. Therefore in 2000 we were among the first to introduce the detection of immune-toxicological biomarkers to the genotoxicological monitoring system. Based on our previous experience, in the present study we have chosen the ratio of lymphocyte subpopulations (T, helper T cell [Th], cytotoxic T cell [Tc], B, and natural killer [NK] cells), Th/Tc ratio, and the activation (receptor for interleukin-2 [IL-2R] expression) of T cells as a means of indicating immunological alteration.[24,26] Donors were considered to have immune alterations if at least two of these parameters showed a considerable difference compared to the normal range of values.

Here we present the results of the so-called Hungarian Nurse Study obtained with cytogenetic methods (CA and SCE) and immunotoxic investigations in the examined groups of nurses preparing and/or handling cytostatic infusions at the bedside, in nurses exposed to gas-sterilizing chemicals, for example, formaldehyde and ETO, and in personnel exposed to different anesthetic gases. We also present the correlation between the observed biomarkers and the disease burden of nurses.

MATERIALS AND METHODS

Donors and Sample Selection

Altogether 500 nurses (700 tests) handling cytostatic drugs, 86 nurses (117 tests) exposed to sterilizing gases, 131 nurses (140 tests) exposed to anesthetic gases collected during the whole study were investigated. In this study 92.1% of the investigated personnel were women. Results were compared to 94 healthy, age-matched controls (all women, working in medical care, but occupationally not exposed to cytostatic drugs, sterilizing, or anesthetic gases). A total of 717 exposed nurses were annually investigated for a period of 13 years, from 1992 to 2004. The nurses were divided into subgroups according to the protective measures used at the working place. The subgroups were: uncontrolled, where no specific protection was used, controlled, where specific protective measures were introduced, and quality controlled by the EU standard ISO 9001 (ISO).

Each donor was personally interviewed by filling in a routine questionnaire including demographic data, exposure and lifestyle (smoking and drinking habits), diseases, exposure to known or suspected mutagens, occupational history including duration of exposure, and the use of protective devices during work. All retrospective medical records were available. Active smoker subjects were considered "Smokers." "Drinkers" consumed less than 80 g pure alcohol regularly (a liter of beer or equivalent) daily. Heavy drinkers were excluded.

Having the informed donors' written permission, blood samples were collected from each donor by venipuncture. The samples were processed both for cytogenetic analysis and for routine clinical check-up including hematology, liver (SGOT, SGPT, and GGT enzymes), and kidney function tests as well as the investigation of risk factors (i.e., serum glucose, serum cholesterol [total high-density lipoprotein (HDL), and low-density lipoprotein (LDL)], triglyceride rate, and urine hyppuric acid concentrations), and for determination of SCN levels as a marker of smoking.

Cytogenetic Analysis

Blood samples were processed for CA and SCE using standard cell culture methods, which were identical in both protocols: 0.8 mL samples of heparinized blood were cultured in duplicates in 10 mL RPMI-1640 medium (Gibco) supplemented with 20% fetal calf serum (Flow) and 0.5% Phytohemagglutinin-P (Difco), without antibiotics; then 5 μg/mL 5-Bromo-2-deoxyuridine (BrdU, Sigma) was added at 22 h of incubation. For CA and

SCE analysis, the cultures were incubated at 37°C, in the presence of 7% CO2, for 50 h and 72 h, respectively. Culture harvest, slide preparation,[26] and staining were made following the standard methods using 5% Giemsa stain (Fluka) for CA, and according to the Fluorescent-Plus-Giemsa method for SCE.[27] All microscope analyses were performed on coded slides by the same (two) observers. Characterization of CA was performed according to Carrano and Natarajan[28] in 100 metaphases per donor in the first mitotic cycle with 46 ± 1 chromosomes. Mitoses containing only achromatic lesions (gaps) and/or aneuploidy (i.e., 46 ± 1 chromosomes per mitosis) were not considered aberrant. A total of 50 of the second divisions per donor were scored for SCE.

Immune Phenotyping of PBLs by Flow Cytometry

Heparinized whole blood was mixed and incubated at room temperature for 20 min with the appropriate amount of fluorescein-isothyocyanate- (FITC), phycoerythrin- (PE), peridinin chlorophyll protein- (PerCP), or allophycocyanin- (APC) labeled monoclonal antibodies against surface antigens. The erythrocytes were removed by lysis with the addition of fluorescent-activated cell sorting (FACS) Lysing solution (Becton Dickinson). After washing with phosphate-buffered saline (PBS), samples were analyzed within 4 h after labeling, or fixed with 2% paraformaldehyde. Four-color analysis was performed on a Becton Dickinson FACSCalibur flow cytometer. The studied antigens were the following: CD3, CD4, CD8, CD14, CD19, CD25, CD45, CD56, and CD71. The monoclonal antibodies were purchased from Becton Dickinson. Cell Quest Software 3.1 was used for analysis.

Measurement of Oxidative Burst Activity of Neutrophil Granulocytes

The measurement of oxidative burst was carried out from heparinized blood samples, using the Burst test (Phagoburst) kit, which uses the conversion of dihydrorhodamine 123 to rhodamine 123 by reactive oxygen intermediates produced upon stimulation of neutrophil granulocytes. Cells were analyzed by a FACSCalibur flow cytometer. Cell Quest Pro Software was used for analysis. The mean fluorescence of rhodamine 123 correlates with oxidation quantity per individual leukocyte.

Statistical Analysis

Statistical analyses were performed by the Student's t-test for CA, SCE; $P < 0.05$ was considered to be significant. The calculation of relative risk (RR) was based on the ratio of the incidence of exposed and nonexposed persons.

RESULTS

In the present study we have examined altogether 811 donors between 1992 and 2004. The subgroups according to working conditions and the number of donors in each subgroup are summarized in TABLE 1. The demographic data (mean age, percentage of smokers and drinkers) of the nurses and the unexposed subjects are presented in TABLE 2. All donors were in reproductive age between 18 and 50 years.

Cytogenetic (CA and SCE) and immunology alterations in the first investigations of the follow-up study are presented in TABLE 3. In the "uncontrolled" subgroup of nurses exposed to cytostatics and anesthetics we observed significantly increased frequency of CAs and SCEs compared to the unexposed. A significant increase in the mean SCE frequency was observed in nurses exposed to anesthetic gases and sterilizing agents, in the "uncontrolled" subgroup compared to the "controlled." However, both CAs and SCEs were significantly decreased in the subgroup "ISO" compared both to "controlled" and "uncontrolled." Similar results in the cumulative mean CA frequencies could be obtained through the whole follow-up (TABLE 4) and in the first investigations of the study.

An increase of immune alterations occurred in all the investigated exposed groups compared to the unexposed (TABLE 3). Immune alterations occurred in nearly a third of the nurses exposed to sterilizing gases under controlled working conditions. Among nurses exposed to anesthetic gases, the group working under controlled conditions was considerably less affected immunologically than the uncontrolled subgroup.

The comparison of gene- and immunotoxicological effect markers among nurses exposed to cytostatics is shown in TABLE 5. The genotoxicologically affected (CA > 4% and SCE > 7.5 per mitosis) donors showed an increase in the frequency of Th lymphocytes, activated (IL-2R positive) T and Th cells, and transferrin receptor positive B cells. However, the frequency of NK cells and the oxidative burst of leukocytes to opsonized *Escherichia coli* stimulus decreased.

The most important biological, lifestyle, and exposure confounding factors, that is, age, smoking, drinking, and ionizing irradiation, were also investigated.

TABLE 1. The number of subjects and investigations in the subgroups of nurses according to protective measures during work

Working environment	Exposure		
	Cytostatic drugs	Sterilizing gases	Anesthetic gases
	Tests/cases		
Uncontrolled	475/339	96/66	50/44
Controlled	211/147	21/20	90/87
ISO	14/14	—	—

TABLE 2. Demographic data of nurses and unexposed controls

Groups	n	Average age (year ± SE)	Smokers[a] %	Drinkers[b] %
Nonexposed (women)	94	39.9 ± 1.4	26.6	36.2
Cytostatic drugs	500			
Uncontrolled	339	34.2 ± 0.6	43.9	38.4
Controlled	147	33.2 ± 0.7	43.5	49.0
ISO	14	28.9 ± 2.1	42.9	85.7
Anesthetic gases	131			
Uncontrolled	44	40.9 ± 1.4	47.7	68.2
Controlled	87	39.6 ± 1.0	33.3	63.2
Sterilizing gases	86			
Uncontrolled	66	42.3 ± 1.2	36.4	39.4
Controlled	20	44.9 ± 2.4	20.0	45.0

[a] active smokers.
[b] less than 80 g pure alcohol regularly.

SCE values in smokers were significantly increased in all investigated subgroups compared to nonsmokers (TABLE 6). In the subgroups exposed to anesthetic and sterilizing gases, SCE was increased in the uncontrolled subgroup of nurses, compared to the controlled subgroup, regardless to smoking as a confounding factor.

Some of the nurses were exposed not only to cytostatics, but also to ionizing radiation (X ray and diagnostic and/or therapeutic radiochemicals). Other subgroups of nurses exposed to anesthetics and sterilizers were also occupationally exposed to X ray and ionizing alpha-radiation (environmental

TABLE 3. Cytogenetic and immune alteration data of nurses (first investigations of the follow-up study)

Groups	n	CA % (mean ± SE)	SCE 1/mitoses (mean ± SE)	Immune alterations %
Nonexposed	94	1.72 ± 0.25	6.44 ± 0.14	6.4
Cytostatic drugs				
Uncontrolled	339	2.32 ± 0.19[a]	6.57 ± 0.06[b]	9.9
Controlled	147	1.77 ± 0.21	6.78 ± 0.09	12.5
ISO	14	0.79 ± 0.39[a]	6.20 ± 0.16[b]	n.d.
Anesthetic gases				
Uncontrolled	44	1.45 ± 0.22	6.67 ± 0.16[b]	19.5
Controlled	87	1.63 ± 0.20	6.24 ± 0.09	10.3
Sterilizing gases				
Uncontrolled	66	5.56 ± 0.68[a]	6.77 ± 0.14[b]	n.d.
Controlled	20	1.33 ± 0.36	6.08 ± 0.19	31.3

[a] significant to the nonexposed, (Student's t-test, $P < 0.05$).
[b] SCE significant to the controlled (Student's t-test, $P < 0.05$).
n.d. = not determined.

TABLE 4. Cumulative structural CA frequencies (excluding gaps) in the exposed groups (all investigations)

Groups	n	CA % (mean ± SE)
Cytostatic drugs		
Uncontrolled	475	2.50 ± 0.17
Controlled	211	2.07 ± 0.20
ISO	14	0.79 ± 0.39[a]
Anesthetic gases		
Uncontrolled	50	1.56 ± 0.23
Controlled	90	1.62 ± 0.20
Sterilizing gases		
Uncontrolled	96	5.46 ± 0.47[a]
Controlled	21	1.26 ± 0.35

[a]Significant in comparison to the controlled (Student's t-test, $P < 0.05$).

radon exposure, in the case of ETO-exposed sterilizers only). Significant increases in mean CA and SCE were observed in donors exposed to cytostatics and sterilizers with radiation, in comparison to the corresponding subgroup without radiation. Ionizing radiation also insignificantly increased the values of cytogenetic end points in case of exposure to anesthetic gases. The effect of ionizing radiation on CA and SCE in the exposed groups (all investigations) is shown in TABLE 7.

The clinical data and the incidence of different pathological conditions in the exposed groups were compared to the unexposed group (TABLE 8). There was a considerable increase in the donor's blood glucose level in the exposed in comparison to the unexposed. Similarly, the incidence of anemia, thyroid alterations, myoma, and other benign tumors in the unexposed group was low, whereas in the exposed groups it was increased.

In TABLE 9 the RRs of different diseases are summarized in the exposed nurses.

TABLE 5. Comparison of gene- and immunotoxicological effect markers (nurses exposed to cytostatics)

Measured parameter	Genotoxically nonaffected	Genotoxically affected
Th (%)	43.79 ± 0.56	46.93 ± 0.90[a]
NK (%)	12.21 ± 0.43	10.67 ± 0.65[a]
Activated T cells (%)[b]	12.70 ± 0.78	16.30 ± 1.10[a]
Activated Th cells (%)[b]	18.53 ± 1.10	23.42 ± 1.10[a]
Activated B cells (%)[c]	40.78 ± 2.37	50.62 ± 3.57[a]
Oxidative burst (MFI)[d]	526.37 ± 32.67	334.74 ± 17.49[a]

[a]Significant changes (Student's t-test, $P < 0.05$).
[b]As measured by IL-2R expression.
[c]As measured by transferrin receptor expression.
[d]Mean fluorescence intensity.

TABLE 6. Comparison of the SCE results in smokers and nonsmokers

Groups	SCE 1/mitoses (mean ± SE)	
	Smokers	Nonsmokers
Cytostatic drugs		
Uncontrolled	6.74 ± 0.08	6.38 ± 0.07
Controlled	6.91 ± 0.11	6.59 ± 0.14
ISO	6.54 ± 0.26	5.85 ± 0.10
Anesthetic gases		
Uncontrolled	6.92 ± 0.24	6.37 ± 0.18
Controlled	6.44 ± 0.17	6.10 ± 0.08
Sterilizing gases		
Uncontrolled	6.96 ± 0.18	6.58 ± 0.20
Controlled	6.55 ± 0.41	5.85 ± 0.18

The RR of thyroid alterations were increased in all exposed groups, whereas the RR of breast and ovary cancer was only increased in nurses exposed to sterilizing gases. The RR of myoma was increased in all exposed groups, especially in the subgroups exposed to sterilizing and anesthetic gases. Although the incidence was increased in all exposed groups the RR of other benign tumors could not be calculated because their incidence was 0 in the unexposed group.

DISCUSSION

In the last two decades, several studies were published about the hazards of handling cytostatic agents by health professionals, predominantly, among

TABLE 7. Ionizing radiation influencing CA and SCE in the exposed groups (all investigations)

Exposure	n	CA % (mean ± SE)	SCE 1/mitoses (mean ± SE)
Cytostatic drugs			
Cytostatics	658	2.28 ± 0.13	6.69 ± 0.04
Cytostatics + ionizing radiation[a]	42	3.21 ± 0.76[c]	7.08 ± 0.18[c]
Anesthetic gases			
Anesthetics	62	1.47 ± 0.20	6.32 ± 0.10
Anesthetics + ionizing radiation[a]	78	1.72 ± 0.22	6.50 ± 0.12
Sterilizing gases			
ETO	57	2.04 ± 0.30	6.31 ± 0.12
ETO + ionizing radiation[b]	60	7.07 ± 0.60 [c]	6.71 ± 0.16[c]

[a] Work site radiation (radioactive isotopes, X ray).
[b] Environmental radiation (Radon).
[c] Significant changes (Student's t-test, $P < 0.05$).

TABLE 8. Burden of chronic noninfectious diseases among hospital nurses

	%			
Diseases	Unexposed	Cytostatic drugs	Anesthetic gases	Sterilizing gases
Increase in blood glucose level	21.3	33.8	30.53	31.4
Anemia	8.5	11.8	16.79	10.47
Metabolic-X syndrome	17.0	20.0	22.14	22.09
Thyroid	1.1	9.2	9.16	8.14
Breast + Ovary cc.	0.94	0.6	0.76	15.12
Myoma	1.1	4.0	11.45	9.30
Other benign tumors	0.0	8.4	8.40	12.79
Allergy	45.7	35.2	45.80	39.5

nurses and pharmacists. It is difficult to provide definitive data on the type and degree of risk for those exposed to chemotherapeutic agents.

A novel and promising approach to risk assessment is the use of immune toxicological methods. Chemical exposure can alter the ratio of lymphocyte subpopulations and may cause changes in the activation of lymphocytes among oil-industry, health-service, and metallurgy workers and thus specific immunological markers can be monitored to assess exposure.[24,25] In the Hungarian Nurse Study we have found an increase in immune alterations in all the exposed groups compared to the unexposed controls. In anesthetic-gas-exposed nurses, the use of protective measures resulted in the decrease of immune alterations. In cytostatic-exposed nurses, the genotoxicologically affected donors showed an activation of lymphocytes, which we consider to be a consequence of chemical exposure.[25] However, the frequency of NK cells and the oxidative burst of leukocytes to opsonized *E. coli* stimulus decreased, as a sign of the suppression of innate immune responses.

In the present study we have carried out a multiple end point genotoxicology monitoring of the nurses exposed to cytostatics since 1992 in order to follow-up the improvement in work-related conditions. Similarly, multiple

TABLE 9. RR of different clinical conditions among hospital nurses

	RR		
Diseases	Cytostatic drugs	Anesthetic gases	Sterilizing gases
Increase in blood glucose level	1.6	1.4	1.5
Anemia	1.4	2.0	1.2
Metabolic-X syndrome	1.2	1.3	1.3
Thyroid alterations	8.4	8.3	7.4
Breast + Ovary cc.	0.6	0.8	16.1
Myoma	3.6	10.4	8.5
Allergy	0.8	1.0	0.9

end point genotoxicology monitoring was carried out among nurses exposed to sterilizing and anesthetic gases. In our study we have assessed the clinical changes and health condition of these groups, besides genotoxicological and immunotoxicological alterations. Some cytostatic and sterilizing drugs are carcinogenic, mutagenic, or teratogenic, as it was well documented earlier in animal experiments[29–32] and some anesthetic gases were clastogenic in human studies.[33,34]

At the uncontrolled workplaces genotoxicological end points indicated occupational exposure to cytostatics[13] and sterilizing gases. Exposure to Halothane also increased CAs (data not shown) although the cytogenetic end points did not indicate exposure to anesthetic gases *in toto*. Anesthetic gases, also harmful to health, often do not cause increased levels of CA, however, SCE frequencies were still increased significantly in an operating room personnel as compared to controls. Exposure to formaldehyde, as a sterilizing agent can cause increased SCE among pathologists.[35]

The confounding effect of smoking, as a known SCE inducer[36] cannot be excluded in all the subgroups as SCE values in smokers were significantly increased in all investigated subgroups compared to nonsmokers. Cigarette smoke, is similar to some environmental and occupational chemical pollutants. The confounding effect of smoking, as a known SCE inducer[37] could not be excluded. SCE values in smokers were significantly increased in all investigated subgroups compared to nonsmokers. Environmental and occupational chemical pollutants, for example, cigarette smoke, styrene, can also alter immune functions.[38,39]

We also proved an additive effect between ionizing irradiation and ETO exposure in uncontrolled conditions.[7,23] However, at all the controlled workplaces the cytogenetic end points remained at the nonexposed control level, and a strict quality control according to ISO 9001 resulted in an even lower CA yield corresponding to the healthy unexposed population.

Accounts have been reported of various acute symptoms experienced by nurses handling chemotherapy, such as hair loss, irregular menstrual cycle, and skin and eye irritation.[4–6] We have also recorded these symptoms during the study. However, without safety regulations, nurses are more susceptible to getting anemia, benign and malignant tumors, and thyroid dysfunctions (cf. TABLE 8). Improvements in working conditions have reduced the symptoms of exposure to these chemicals. Regular clinical check-ups of the donors have shown some differences in laboratory data between the exposed groups. Exposure to cytostatics, anesthetic, and sterilizing gases increase the risk of thyroid alterations. Our investigation found an eight-fold increase in thyroid complaints caused by several pathological changes in the thyroid gland in all groups of the exposed donors. Cancer of the thyroid gland was relatively rare, although the cancer of the breast and/or ovary was high in nurses exposed to ETO, as we published earlier.[7] Sterilizing gases, especially ETO may induce breast and ovary cancer, whereas the risk of myoma is elevated in the case of both anesthetic and sterilizing gases. Other benign tumors were increased in all exposed groups.

These results underline the importance of cancer education in medical staff. The total burden of cancer on the global community and the healthcare professionals is increasing; therefore cancer education must be an integral part of cancer prevention among students and nurses and among staff of oncology departments. Our study confirmed that the handling of chemotherapeutic agents gives a definite risk to healthcare professionals, highlighting that the risk is not restricted to neoplasms but extends to all chronic noninfectious diseases. Adequate education and training seems to be fundamental to provide information about risk reduction. The most important problem is changing the beliefs, attitudes, and behavior of so-called "experienced" practitioners, who were trained before the appearance of guidelines. The increased awareness and the need for universal precautions in hazardous environments, like sterilizing units, or anesthetic gas exposure, are necessary to solve this problem.

Hygienic conditions of the working environment are the basic risk factors for the sensitivity of nurses to chemical agents. Cytogenetic and immunological biomarkers are appropriate to detect early susceptibility to diseases. The Hungarian Nurse Study proved that the use of safety measures could protect against occupational exposure at work sites handling cytostatic drugs, anesthetic, and sterilizing gases.

ACKNOWLEDGMENTS

The authors are thankful to Mrs. Irén Rétháti, Mrs. Anna Herczeg, Ms. Andrea Tóth, Mrs. Ildiko Bárdi, Mrs. Éva Czifra, Mrs. Tünde Szeremlei-Szabóné, Mrs. Andrea Hegedüs, Mrs. Margit Tölyhi, and Mrs. Zsuzsanna Szép-Kis for the excellent technical help, and Dr. Edwin A. König in the contribution of chromosome analysis. This work was supported by the grants NKFP 1/016/2001 and NKFP 1/B-047/2004.

REFERENCES

1. PUBLIC ACT NO. XXV (11.April 2000) on Chemical Safety. 2000. Hungary. Magyar Közlöny **38:** 2058–2071. (In Hungarian)
2. 26/2000 ORDER OF THE MINISTRY OF HEALTH. 2000. Magyar Közlöny **99:** 6179–6275. (In Hungarian).
3. METHODOLOGICAL GUIDELINES OF THE NATIONAL INSTITUTE OF PHARMACY. 2004. Hungary on manufacturing and use of mixed cytostatic infusions. Gyógyszereink **54:** 135–144. (In Hungarian)
4. SHORTRIDGE, L.A., G.K. LEMASTERS, B. VALANIS & V. HERTZBERG. 1995. Menstrual cycles in nurses handling antineoplastic drugs. Cancer Nurs. **18:** 439–444.
5. VALANIS, B., W. VOLLMER, K. LABUHN & A. GLASS. 1997. Occupational exposure to antineoplastic agents and self-reported infertility among nurses and pharmacists. J. Occup. Environ. Med. **39:** 574–580.

6. KRSTEV, S., B. PERUNICIC & A. VIDAKOVIC. 2003. Work practice and some adverse health effects in nurses handling antineoplastic drugs. Med. Lav. **94:** 432–439.
7. TOMPA, A., J. MAJOR & M.G. JAKAB. 1999. Breast cancer cluster influenced by environmental and occupational factors among hospital nurses in Hungary. Pathol. Oncol. Res. **5:** 117–122.
8. IARC. 1988. IARC monographs on the evaluation of the carcinogenic risk of chemicals to humans. Vol. 36. Allyl Compounds, Aldehydes, Epoxides and Peroxides. Ethylene oxide. IARC. Lyon, France. 198–226.
9. IARC 1990. Cancer: causes, occurrence and control. *In* Radiation. L. Tomatis, A. Aitio, N.E. Day, *et al.* Eds.:157–158. IARC Scientific Publications No. 100. WHO, IARC. Lyon, France.
10. SIEBER, S.M. & R.H. ADAMSON. 1975. Toxicity of antineoplastic agents in man: chromosome aberrations, antifertility effects, congenital malformation and carcinogenic potential. Adv. Cancer Res. **22:** 57–155.
11. SORSA, M., K. HEMMINKI & H. VAINIO. 1985. Occupational exposure to anticancer drugs—potential and real hazards. Mutat. Res. **154:** 135–149.
12. SARDAS, S., H. CUHRUK, A.E. KARAKAYA & Y. ATAKURT. 1992. Sister-chromatid exchanges in operating room personnel. Mutat. Res. **279:** 117–120.
13. JAKAB, M.G., J. MAJOR & A. TOMPA. 2001. Follow-up genotoxicological monitoring of nurses handling antineoplastic drugs. J. Toxicol. Environ. Health **62:** 307–318.
14. MAYER, D.K. 1992. Hazards of chemotherapy. Implementing safe handling practices. Cancer **70:** 988–992.
15. NEAL, A.W., R.A. WADDEN & W.L. CHLOU. 1983. Exposure of hospital workers to airborne antineoplastic agents. Am. J. Hosp. Pharm. **40:** 597–601.
16. HIRST, M., D.G. MILLS, S. TSE, *et al.* 1984. Occupational exposure to cyclophosphamide. Lancet **1:** 186–188.
17. CARRANO, A.V. & A.T. NATARAJAN. 1988. Considerations on population monitoring using cytogenetic techniques. ICPEMC Publ. No. 14. Mutat. Res. **204:** 379–406.
18. NORPPA, H., M. SORSA, H. VAINIO, *et al.* 1980. Increased SCE frequencies in lymphocytes of nurses handling cytostatic agents. Scand. J. Work. Environ. Health **6:** 299–301.
19. WAKSVIK, H., O. KLEPP & A. BROGGER. 1981. Chromosome analyses of nurses handling cytostatic agents. Cancer Treat. Rep. **65:** 607–610.
20. NIKULA, E., K. KIVINIITTY, J. LEISTI & P.J. TASKINEN. 1984. Chromosome aberrations in lymphocytes of nurse handling cytostatic agents. Scand. J. Work. Environ. Health **10:** 71–74.
21. CHRYSOSTOMOU, A., R. SESHADRI & A.A. MORLEY. 1984. Mutation frequency in nurses and pharmasists working with cytotoxic drugs. Aust. N. Z. J. Med. **14:** 831–834.
22. ANWAR, W.A., S.I. SALAMA, M.M. EL SERAFY, *et al.* 1994. Chromosomal aberrations and micronucleus frequency in nurses occupationally exposed to cytotoxic drugs. Mutagenesis **9:** 315–317.
23. MAJOR, J., M.G. JAKAB & A. TOMPA. 1996. Genotoxicological investigation of hospital nurses occupationally exposed to ethylene-oxide. I. Chromosome aberrations, sister chromatid exchanges, cell cycle kinetics, and UV-induced DNA synthesis in peripheral blood lymphocytes. Environ. Mol. Mutagen. **27:** 84–92.

24. BIRÓ, A., É. PÁLLINGER, J. MAJOR, *et al.* 2002. Lymphocyte phenotype analysis and chromosome aberration frequency of workers occupationally exposed to styrene, benzene, polycyclic aromatic carbohydrates or mixed solvents. Immunol. Lett. **81:** 133–140.
25. BIRÓ, A., É. PÁLLINGER, A. FALUS & A. TOMPA. 2004. Characterization of chemically exposed groups by immunotoxicological methods. Magyar Onkológia. **48:** 137–139.
26. MOORHEAD, P.S., P.C. NOWELL, W.J. MELLMAN, *et al.* 1960. Chromosome preparations of leukocytes cultured from human peripheral blood. Exp. Cell. Res. **20:** 613–616.
27. PERRY, P. & S. WOLFF. 1974. New giemsa method for the differential staining of sister chromatids. Nature **251:** 156–158.
28. CARRANO, A.V. & A.T. NATARAJAN. 1988. Considerations on population monitoring using cytogenetic techniques. ICPEMC Publ. No.14. Mutat. Res. **204:** 379–406.
29. GALLOWAY, S.M. & S. WOLFF. 1979. The relation between chemically induced sister-chromatid exchanges and chromatid breakage. Mutat. Res. **61:** 297–307.
30. ROSSELLI, F., L. ZACCARO, M. VENTURI & A.M. ROSSI 1990. Persistence of drug-induced chromosome aberrations in peripheral blood lymophocytes of the rat. Mutat. Res. **232:** 107–114.
31. LYNCH, D.W., T.R. LEWIS, W.J. MOORMAN, *et al.* 1984. Carcinogenic and toxicologic effects of inhaled ethylene oxide and propylene oxide in F344 rats. Toxicol. Appl. Pharmacol. **76:** 69–84.
32. LYNCH, D.W., T.R. LEWIS, W.J. MOORMAN, *et al.* 1984. Sister-chromatid exchanges and chromosome aberrations in lymphocytes from monkeys exposed to ethylene oxide and propylene oxide by inhalation. Toxicol. Appl. Pharmacol. **76:** 85–95.
33. JALOSZYNSKI, P., M. KUJAWSKI, M. WASOWICZ, *et al.* 1999. Genotoxicity of inhalation anesthetics halothane and isoflurane in human lymphocytes studied in vitro using the comet assay. Mutat. Res. **439:** 199–206.
34. KARELOVA, K., A. JABLONICKA, J. GAVORA & L. HANO. 1992. Chromosome and sister-chromatid exchange analysis in peripheral lymphocytes, and mutagenicity of urine in anesthesiology personnel. Int. Arch. Occup. Environ. Health **64:** 303–306.
35. SHAHAM, J., R. GURVICH & Z. KAZUFMAN. 2002. Sister chromatid exchange in pathology staff occupationally exposed to formaldehyde. Mutat. Res. **514:** 115–123.
36. PENDZICH, J., G. MOTYKIEWICZ, J. MICHALSKA, *et al.* 1997. Sister chromatid exchanges and high-frequency cells in men environmentally and occupationally exposed to ambient air pollutants: an intergroup comparison with respect to seasonal changes and smoking habit. Mutat. Res. **381:** 163–170.
37. PENDZICH, J., G. MOTYKIEWICZ, J. MICHALSKA, *et al.* 1997. Sister chromatid exchanges and high-frequency cells in men environmentally and occupationally exposed to ambient air pollutants: an intergroup comparison with respect to seasonal changes and smoking habit. Mutat. Res. **381:** 163–170.
38. TANIGAWA, T., S. ARAKI, A. NAKATA & S. SAKURAI. 1998. Increase in the helper inducer (CD4+CD29+)T lymphocytes in smokers. Ind. Health **36:** 78–81.
39. TULINSKA, J., N. DUSINSKA, & E. JAHNOVA. 2000. Changes in cellular immunity among workers occupatoionally exposed to styrene in a plastic lamination plant. Am. J. Ind. Med. **38:** 576–583.

Residual Hazard Assessment Related to Handling of Antineoplastic Drugs

Safety System Evolution and Quality Assurance of Analytical Measurement

ROBERTA TURCI AND CLAUDIO MINOIA

Laboratory for Environmental and Toxicological Testing, Salvatore Maugeri Foundation, 27100 Pavia, Italy

ABSTRACT: Despite improvement of operating procedures and publication of safety guidelines, contamination is still observed in healthcare settings where antineoplastic drugs (ADs) are handled. Even after cleaning work areas, some residual contamination may still be present. Zero percent contamination is not a realistic goal, but the scientific community should set zero contamination as its main goal. The strategies to reach this objective may be traced based on the followings: (*a*) a wider number of drugs should be monitored; (*b*) safety equipment and devices must be available to the workers; (*c*) the likely source of widespread contamination in workplaces is the safety cabinet; (*d*) direct determination of the parent drug or its metabolite in urine is the recommended approach because it provides higher sensitivity and specificity; (*e*) reliable analytical methods are necessary to measure the extent of contamination; and (*f*) analytical methods intended to be applied for routine testing must be assessed through method validation studies. These studies rely on the determination of overall method performance parameters including uncertainty measurement. Our laboratory has developed and validated a number of analytical methods for the determination of several drugs in environmental and biological samples. Surveys were carried out in several hospitals, and there has been progressive, significant decrease in the number of positive samples, mostly due to the improvement of working procedures and safety measures.

KEYWORDS: occupational exposure; antineoplastic drugs; biological monitoring; environmental monitoring; quality assurance

---

Address for correspondence: Roberta Turci, via S. Maugeri 10, Laboratory for Environmental and Toxicological Testing, Salvatore Maugeri Foundation, 27100 Pavia, Italy. Voice: +39-0382-592312; fax: +39-0382-592067.

e-mail: rturci@fsm.it

INTRODUCTION

It has been nearly 30 years since occupational exposure to cytostatic antineoplastic drugs (ADs) has been recognized as a potential health hazard.[1] With the aim of improving operating procedures and workplace safety, guidelines have been published in several countries.[2-6] In Italy, the National Institute for Occupational Health and Prevention, a public research body working under control of the Ministry of Health, established a Working Group on Antineoplastic Drugs Safe Handling in 1997. Two years later, after further revision and update, that work resulted in the Official Italian Guidelines, entitled "Guidelines for Protecting the Safety and Health of Health Care Workers Exposed to Antineoplastic Drugs."[7] The main points of these guidelines are: implementation of handling procedures; use of specialized equipment and adequate personal protective equipment; training and education programs; health surveillance; and exposure assessment. The establishment of a registry of hospital personnel exposed to ADs is also recommended.

The strategy used to prevent or minimize occupational exposure to ADs applies a combination of interventions, such as engineering controls, administrative and work practice controls, and personal protective equipment. Biological and environmental monitoring are essential tools for the collection of adequate exposure data and effective intervention.

Rapid, inexpensive, and reliable analytical methods are needed. Nevertheless, a recent review has highlighted the need for standardization and more stringent quality assurance of the analytical methods used for exposure assessment.[8] Certified reference materials for ADs are not yet available. Hence it is of utmost importance that experienced laboratory personnel work to provide reliable procedures and accurate data in this field.

To comply with the current guidelines, our research activities focused on the development and validation of new analytical methods for the determination of antineoplastic agents in both biological and environmental samples.[9-20] The outcome of our work is presented herein.

METHODS

Validation studies rely on the determination of overall method performance parameters.[21] In TABLE 1, the most important performance characteristics are listed. Method validation enables chemists to demonstrate that a method is "fit for purpose," that is, it is reliable enough so that any decision based on it can be taken with confidence. To judge whether or not a method is suitable for its intended purpose, it is necessary to include the required uncertainty expressed either as a standard combined uncertainty or as expanded uncertainty.[22] The key stages in the uncertainty estimation process are:

TABLE 1. Performance characteristics

- Confirmation of Identity
- Selectivity/Specificity
- Limit of Detection
- Limit of Quantitation
- Working and Linear Ranges

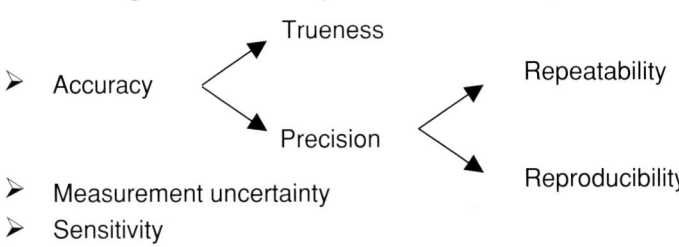

- Accuracy
- Measurement uncertainty
- Sensitivity
- Ruggedness (or Robustness)
- Recovery

---

1. Specification of what is being measured;
2. Indentification sources of uncertainty associated with the method;
3. Quantification of uncertainty components associated with each potential source of uncertainty; and
4. Calculation of total uncertainty following appropriate rules.

The main tools for evaluating the sources of uncertainty are ruggedness, precision, and recovery studies.[22] Once the individual components have been calculated they must be combined to give standard and expanded uncertainties for the method. In case all the individual uncertainty components are proportional to the analyte concentration, they should be converted to relative standard deviations. For example, if a result that is, concentration, is affected by the parameters *repeatability, calibration,* and *sampling,* the combined uncertainty, u(y) is given by:

$$\frac{u_{(y)}}{y} = \sqrt{\left(\frac{u_{(Rep.)}}{(Rep.)}\right)^2 + \left(\frac{u_{(Cal.)}}{(Cal.)}\right)^2 + \left(\frac{u_{(Samp.)}}{(Samp.)}\right)^2}$$

The expanded uncertainty $U(y)$ is calculated from the combined standard uncertainty $u(y)$ using:

$$U(y) = kxu(y),$$

where k is an appropriate coverage factor, which depends on the degree of confidence required. A coverage factor is normally in the range of 2 to 3.

The expanded uncertainty is "an interval which is expected to include a large fraction of the distribution of values reasonably attributable to the measurand."[22] A coverage factor of 2 gives an interval containing about 95% of the distribution of values.

For the last 2 years we have been working to apply this approach to method validation in an attempt to provide more accurate data for exposure assessment.

RESULTS

In TABLE 2, the main results of our work are presented. For the most important antineoplastic agents, the instrumental technique used, precision (CV%), trueness, limit of detection (LOD), limit of quantitation (LOQ), and combined relative uncertainty at a given concentration level, are listed. It is noteworthy that the method for the determination of urinary cyclophosphamide (CP) and ifosfamide (IF)[19] provides significantly improved sensitivity compared with the previous one, which was published in 1998.[10] In addition, use of high-performance instrumentation, like HPLC-MS/MS, enables simultaneous analysis of different drugs with low LODs. The evaluation of uncertainty further increases the confidence that can be placed on the results. For example, the histograms showing the contribution of each individual component to the combined uncertainty calculated for urinary CP and IF are reported in FIGURES 1 and 2.[19] As can be seen, the relative combined uncertainty at the lowest quality control level (25 ng/L) was about 25% for both CP and IF.

DISCUSSION AND CONCLUSIONS

Many efforts have been made during the last decades to reduce exposure levels in healthcare settings where ADs are handled. Based on the findings from surveys carried out in Italian settings, it can be concluded that adherence to guidelines has significantly lowered contamination levels. During the last 5 years, the percentage of positive samples for urinary CP has decreased from 50% to 1%. At the same time, the concentrations measured in biological samples continue to decrease. This means that more sensitive and accurate methods are still needed. For this reason, the main components that can be sources of uncertainty must be studied, and combined standard uncertainty must be calculated.

Calibration (TABLE 2) was the most critical step, and the uncertainty component associated with calibration decreased at higher concentration levels, as expected. This is due to the very low concentration levels studied. However, a method is conventionally considered to be reliable when the total uncertainty is less than 30%. Our results comply with this rule, making the methods at issue suitable for application to exposure assessment.

TABLE 2. Performance parameters

Drug	Instrumental technique	CV%	Trueness	LOD	LOQ	Relative combined uncertainty (conc.)
CP (biol.)[19]	HPLC-MS/MS	<15%	98%–108%	10 ng/L	20 ng/L	<25% (25 ng/L)
CP (environm.)[11]	HPLC-MS/MS	<10%	96% – 120%	2 ng/wipe	2 ng/wipe	15% (30 ng/wipe)
MTX (environm.)[13]	HPLC-MS/MS	<20%	97%– 118%	2 ng/wipe	5 ng/wipe	30% (3 ng/wipe)
5-FU (environm.)[14]	HPLC-UV	<13%	89–106%	1 µg/wipe	1 µg/wipe	<20% (3 µg/wipe)
GCA (biol.)[17]	HPLC-MS/MS	<7%	92.4%	0.05 µg/L	0.2 µg/L	< 15% (2 µg/L)
2dFdU[17]	HPLC-MS/MS	<8%	93.6%	0.3 µg/L	1.0 µg/L	<30% (2 µg/L)
DOXO (biol.)[18]	HPLC-MS/MS	<15%	86%– 99%	0.04 µg/L	0.10 µg/L	<20% (0.2 µg/L)
DAUNO (biol.)[18]	HPLC-MS/MS	<15%	86%– 99%	0.01 µg/L	0.03 µg/L	<10% (0.2 µg/L)
IDA (biol.)[18]	HPLC-MS/MS	<15%	86%– 99%	0.01 µg/L	0.03 µg/L	<10% (0.2 µg/L)
EPI (biol.)[18]	HPLC-MS/MS	<15%	86%– 99%	0.04 µg/L	0.10 µg/L	<20% (0.2 µg/L)
Pt (biol.)[20]	ICP-MS (Q-DRC)	<5%	99.3–101.7%	0.18 ng/L	0.4 ng/L	0.094% (1.5 ng/L)

CP = cyclophosphamide; MTX = methotrexate; 5-FU = 5-fluorouracil; GCA = gemcitabine; 2dFdU = 2′,2′-difluorodeoxyuridine Doxo = doxorubicin; Epi = epirubicin; Dauno = daunorubicin; Ida = idarubicin; Pt = platinum; HPLC-MS/MS = high-performance liquid chromatographic - tandem mass spectrometry; HPLC-UV = high-performance liquid chromatographic - ultraviolet detector; ICP-MS (Q-DRC) = inductively coupled plasma-mass spectrometry (quadrupole with dynamic cell reaction); LOD = limit of detection; LOQ = limit of quantitation.

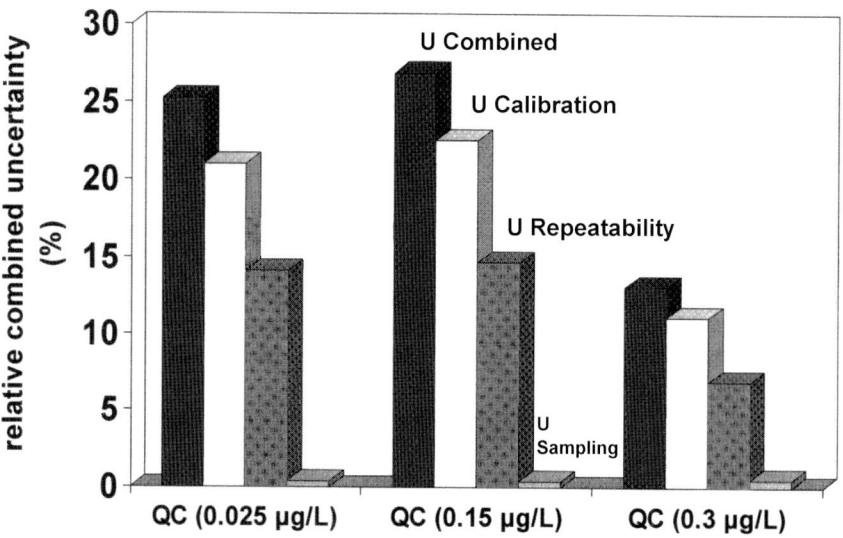

FIGURE 1. Uncertainty components (cyclophosphamide in urine).[19]

After providing reliable analytical results, they must then be interpreted. Researchers should now focus their efforts on understanding reference intervals to ensure a uniform interpretation of the analytical results across countries.

A comprehensive validation study is expensive. Each laboratory should therefore decide what degree of validation is required for a particular analytical

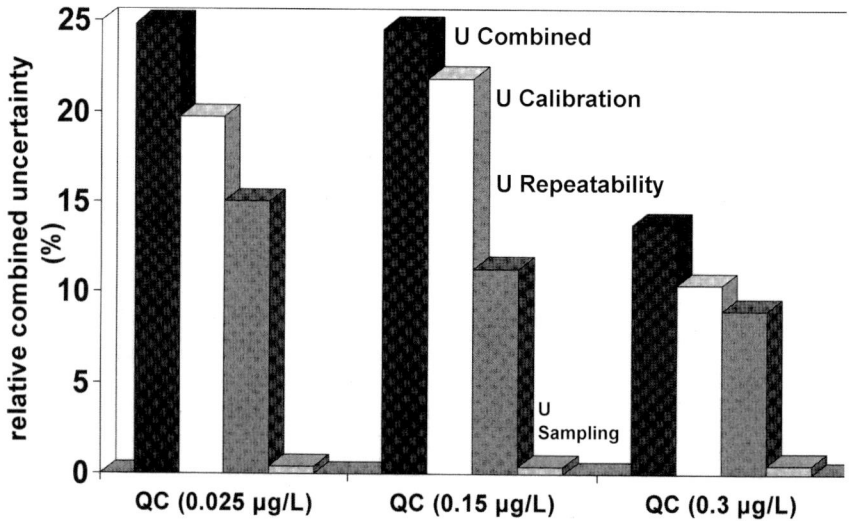

FIGURE 2. Uncertainty components (ifosfamide in urine).

problem, striking a balance between performance and costs. In accordance with guidelines, cost- and time-effective monitoring systems should be established in all countries to control exposure levels in healthcare settings where ADs are handled. Qualified researchers should be in charge of developing appropriate sampling and analytical methods, which must be validated and standardized. These procedures must then be applied to exposure assessment in healthcare settings where surveys should be carried out on a regular basis.

In spite of all the progress made, contamination still occurs. We must remember that some of these drugs are classified by the International Agency for Research on Cancer (IARC) as certain carcinogenic,[23] so there is no safe exposure level above zero. While scientists and physicians are doing their best and aiming at zero contamination, workers must play an active role as well, taking their own share of responsibility.

REFERENCES

1. FALCK, K. et al. 1979. Mutagenicity in urine of nurses handling cytostatic drugs. Lancet **i**: 1250–1251.
2. CLINICAL ONCOLOGICAL SOCIETY OF AUSTRALIA. 1983. Guidelines and recommendations for safe handling of antineoplastic agents. Med. J. Aust. **1**: 426–428.
3. AMERICAN SOCIETY OF HOSPITAL PHARMACISTS. 1990. ASHP technical assistance bulletin on handling cytotoxic and hazardous drugs. Am. J. Hosp. Pharm. **47**: 1033–1049.
4. GUIDELINES FOR THE HANDLING AND DISPOSAL OF HAZARDOUS PHARMACEUTICALS (INCLUDING CYTOTOXIC DRUGS). 1993. Canadian Society of Hospital Pharmacists. Ottawa, Ontario, Canada.
5. OCCUPATIONAL SAFETY AND HEALTH ADMINISTRATION. 2000. OSHA Technical Manual. Hospital Investigations: Health Hazard. Section VI http://www.osha-slc.gov/SLTC/hazardousdrugs/index.html.
6. NATIONAL INSTITUTE FOR OCCUPATIONAL SAFETY AND HEALTH (NIOSH). 2004. Preventing Occupational Exposure to Antineoplastic and other Hazardous Drugs in Healthcare Settings. NIOSH Publication No. 2004-165, September 2004 www.cdc.gov/niosh/docs/2004–165.
7. PROVVEDIMENTO 5 AGOSTO. 1999. Documento di linee-guida per la sicurezza e la salute dei lavoratori esposti a chemioterapici antiblastici in ambiente sanitario, G.U. 236, 7.10.1999.
8. TURCI, R. et al. 2003. Biological and environmental monitoring of hospital personnel exposed to antineoplastic agents: a review of analytical methods. J. Chromatogr. B. Analyt. Technol. Biomed. Life Sci. **789**: 169–209.
9. SOTTANI, C. et al. 1998a. High-performance liquid chromatography tandem mass spectrometry procedure with automated solid phase extraction sample preparation for the quantitative determination of paclitaxel (Taxol®) in human plasma. Rapid Commum. Mass Spectrom. **12**: 251–255.
10. SOTTANI, C. et al. 1998b. Liquid-liquid extraction procedure for trace determination of cyclophosphamide in human urine by high-performance liquid chro-

matography tandem mass spectrometry. Rapid Commum. Mass Spectrom. **12:** 1063–1068.
11. MINOIA, C. *et al.* 1998. Application of high performance liquid chromatography/tandem mass spectrometry in the environmental and biological monitoring of health care personnel occupationally exposed to cyclophosphamide and ifosfamide. Rapid Commum. Mass Spectrom. **12:** 1485–1493.
12. TURCI, R. *et al.* 2000. Determination of methotrexate in human urine at trace levels by solid phase extraction and high-performance liquid chromatogram/tandem mass spectrometry. Rapid Commum. Mass Spectrom. **14:** 173–179.
13. TURCI, R. *et al.* 2000. Determination of methotrexate in environmental samples by solid phase extraction and high performance liquid chromatography: ultraviolet or tandem mass spectrometry detection? Rapid Commum. Mass Spectrom. **14:** 685–691.
14. MICOLI, G. *et al.* 2001. Determination of 5-fluorouracil in environmental samples by solid-phase extraction and high performance liquid chromatography with ultraviolet detection. J. Chromatogr. B. **750:** 25–32.
15. SOTTANI, C. *et al.* 2000. Rapid and sensitive determination of paclitaxel (Taxol®) in environmental samples by high-performance liquid chromatography tandem mass spectrometry. Rapid Commum. Mass Spectrom. **14:** 930–935.
16. TURCI, R. *et al.* 2002. Biological monitoring of hospital personnel occupationally exposed to antineoplastic agents. Toxicol. Lett. **134:** 57–64.
17. SOTTANI, C. *et al.* 2004a. Validated procedure for simultaneous trace level determination of the anti-cancer agent gemcitabine and its metabolite in human urine by high-performance liquid chromatography with tandem mass spectrometry. Rapid Commun. Mass Spectrom. **18:** 1017–1023.
18. SOTTANI, C. *et al.* 2004b. Trace determination of anthracyclines in urine: a new high-performance liquid chromatography/tandem mass spectrometry method for assessing exposure of hospital personnel. Rapid Commun. Mass Spectrom. **18:** 2426–2436.
19. SOTTANI, C. *et al.* 2005. Highly sensitive high-performance liquid chromatography/selective reaction monitoring mass spectrometry method for the determination of cyclophosphamide and ifosfamide in urine of health care workers exposed to antineoplastic agents. Rapid Commun. Mass Spectrom. **19:** 2794–2800.
20. SPEZIA, S. *et al.* 2005. Comparison of inductively coupled plasma mass spectrometry techniques in the determination of platinum in urine: quadrupole vs. sector field. Rapid Commun. Mass Spectrom. **19:** 1551–1556.
21. EURACHEM GUIDE. 1998. The Fitness for Purpose of Analytical Methods. Eurachem, 1st ed. LGC. Teddington, UK. www.eurachem.ul.pt
22. EURACHEM/CITAC GUIDE. 2000. Quantifying Uncertainty in Analytical Measurement, 2nd ed. www.eurachem.ul.pt/
23. INTERNATIONAL AGENCY FOR RESEARCH ON CANCER. 2004. Monographs on the Evaluation of the Carcinogenic Risk of Chemicals to Humans: Pharmaceutical Drugs. IARC. Lyon, France. *www.iarc.fr*

Framing the Future in Light of the Past
Living in a Chemical World

Children today live in an environment that is very different from that encountered by all previous human generations. In today's world there are more than 80,000 synthetic chemicals. Most of these materials were invented in only the past 50 years. They did not exist previously on the planet. Children are especially at risk of exposure to the 3000 synthetic chemicals that are produced in quantities of 1 million kg or more per year. These high-production-volume chemicals are dispersed widely in the environment—in air, food, water, and consumer products in homes, schools, and communities. They are found in measurable quantities in the bodies of most persons in the industrially developed countries. Little is known of their possible toxicity. Only about half of the high-production-volume chemicals have ever been tested for their potential to cause toxicity. Fewer than 20% have been tested for their potential to interfere with development.

Children are uniquely susceptible to chemicals. This great vulnerability reflects the juxtaposition of two phenomena in early life: first, that infants and children have disproportionately greater exposures than adults to environmental chemicals; and second, that children are exquisitely sensitive to these exposures, because they are poorly equipped to metabolize and excrete many toxic compounds, and because they are progressing through the complex, delicate, and easily disrupted stages of early development.

Patterns of illness among children in the industrially developed nations have changed substantially. Infant mortality has declined. Life expectancy has increased. No longer are the infectious diseases the leading causes of illness and death. Today the major diseases confronting children in the developed world are a group of chronic conditions termed the "new pediatric morbidity." These include:

- Asthma, which has doubled in frequency;
- Birth defects, which have become the leading cause of infant death. Certain birth defects, such as hypospadias, have increased sharply in frequency;
- Neurodevelopmental disorders, such as dyslexia, mental retardation, attention deficit/hyperactivity disorder (ADHD) and autism;
- Leukemia and brain cancer in children and testicular cancer in adolescents, all of which have increased in incidence since the 1970s despite declining mortality;

- Obesity and type 2 diabetes.

Evidence is accumulating that toxic chemicals are responsible for at least some of these changing patterns of disease. Well-studied examples include phocomelia in infants exposed prenatally to thalidomide; adenocarcinoma of the vagina in girls exposed prenatally to diethylstilbestrol (DES); asthma and pneumonia caused by smoke and particulate air pollutants; neurodevelopmental toxicity in infants exposed to lead, PCBs, methylmercury and, ethyl alcohol; and small head circumference at birth in infants exposed *in utero* to organophosphate pesticides.

Past discoveries of etiologic associations between toxic environmental exposures and diseases in children have led to successful programs of exposure control and disease prevention. Examples include reductions in the use of alcohol and tobacco during pregnancy, minimization during pregnancy of diagnostic X rays, and removal of lead from gasoline. Sadly, though, the interval between initial recognition of these hazards and their eventual control has typically been far too long. Early warnings have frequently been ignored. The price of delayed action has been widespread disease.

Prevention of disease of environmental origin in children was a major element of the 2005 conference "Framing the Future in Light of the Past: Living in a Chemical World," convened in Bologna by the Collegium Ramazzini. Panelists presented new research findings in children's environmental health and considered new approaches to disease prevention:

- In the talk, "Only one chance to develop a brain: consequences of developmental neurotoxicity," Philippe Grandjean discusses the negative outcomes on the health of children and the broader effects on society that result from the current lack of strong regulatory policies for industrial chemicals.
- Giorgio Tamburlini's discussion, "New developments in children's environmental health in Europe," focuses on the progress made in the development of strong protective policies. Professor Tamburlini stresses the urgent need to support implementation of these policies.
- The health risks associated with children's exposures to toxic chemicals in the developing world are the topics of both Mathuros Ruchirawat's, "Environmental impacts on children's health in Southeast Asia" and Raul Harari's, "Children's health and the environment in Latin America."
- Finally, Richard Jackson's, "The epidemic of child obesity," provides an overview of the serious negative impact on children's health of socioeconomic factors in the modern environment.

All of these presentations work together to illustrate that diseases of environmental origin are a growing problem among children worldwide. Prevention of these diseases is dependent on several factors. First, there must be enhanced research that generates knowledge of etiologic associations. But additionally, prevention demands that this knowledge be translated into action, specifically into child-protective risk assessment, child-protective legislation, expanded right-to-know, enhanced pre-market toxicity testing of chemicals, and wider application of the Precautionary Principle in the face of early warnings of danger. All of these elements are needed to protect the health of children worldwide against environmental threats to health and to ensure the future of human life on earth.

—PHILIP J. LANDRIGAN
Department of Community and Preventive Medicine
Center for Children's Health and the Environment
Collegium Ramazzini, and The Mount Sinai School of Medicine
New York, New York 10029-6574, USA

—JENNY PRONCZUCK DE GARBINO
International Programme on Chemical Safety
World Heath Organization
Geneva 27, Switzerland

—BROOKE NEWMAN
The Mount Sinai School of Medicine
New York, New York 10029-6574, USA

Children's Environment and Health in Latin America

The Ecuadorian Case

RAUL HARARI AND HOMERO HARARI

IFA (Corporation for Development of Production and Work Environment), Casilla de Correo 17-08-8386, Quito, Ecuador

> ABSTRACT: Environmental health problems of children in Latin America and Ecuador are complex due to the close relationship that exists between social and environmental factors. Extended poverty and basic problems, such as the lack of drinking water and sanitation, are common. Infectious diseases are the greatest cause of morbidity and mortality among children. Development in industry and the introduction of chemical substances in agriculture add new risks including pesticide use, heavy metal exposure, and air pollution. Major problems can be divided into (*a*) lack of basic infrastructure, (*b*) poor living conditions, (*c*) specific environmental problems, and (*d*) child labor. Reproductive health disorders are frequent in developing countries like Ecuador. Issues related to children's health should consider new approaches, creative methodologies, and the search for independent predictors to separate environmental from social problems. Only with knowledge of the specific contribution of each factor, can it be possible to develop a strategy for prevention.
>
> KEYWORDS: child labor; child health; malnutrition; poverty; infrastruction

INTRODUCTION

Environmental health problems of children in Latin America are complex due to the close relationship that exists between social and environmental problems. Linked to these are methodological and technical difficulties which do not allow addressing or knowing many of the issues in a timely and specific manner that, despite presenting evidence, cannot be confirmed or associated to causes where all the necessary efforts should be directed in order to generalize conclusions and to prevent or assist them.

Address for correspondence: Raul A. Harari, IFA, Domingo de Brieva N38-107 y Villalengua. Urbanización Granda Centeno., P.O. Box 17-08-8386, Casilla de Correo 17-08-8386 Quito, Ecuador. Voice: 00593-2-2439929; fax: 00593-2-2275662.
    e-mail: ifa@ifa.org.ec

Ann. N.Y. Acad. Sci. 1076: 660–677 (2006). © 2006 New York Academy of Sciences.
doi: 10.1196/annals.1371.082

Within that framework, an attempt will be made to place the main expressions of environmental health suffered by Ecuadorian children by considering it as one of the most affected within the Latin American spectrum, albeit not exclusive of the other regional countries. There is a big difference between concerns in Europe and in Latin American countries.[1]

With the aim of organizing the various related issues, the topic will be presented under the following scheme:

1. Basic problems suffered by our countries and that corresponding directly to deficits in infrastructure.
2. Problems indirectly related to deficits in infrastructure or deriving from general living conditions.
3. Certain specific environmental problems where there is systematized evidence or information.
4. Certain particular social problems related to children's activities, such as child labor or those derived from their working conditions.

THE LATIN AMERICAN AND ECUADORIAN CONTEXTS

Many of the problems with infrastructure are a consequence of an economic model that has been characterized as having very clear priorities geared toward maintaining the relationships of our countries with the world market. The fostering of exports at the cost of the internal market, which has emphasized payment of their foreign debt (Ecuador already owed 16.585 billion dollars in 2003), uses up natural resources, whether agricultural or petroleum, without the benefit of their restoration or adequate transformation, respectively.[2]

Budgetary reductions by the State primarily affect social spending (education, health, and other basic services) and restrict investment. Over 40% of the national budget of Ecuador goes to reducing external debt.

Because of uneven conditions of competition established or maintained with developed countries, there is pressure to lower production costs or force increases in productivity without improving technology. This results in increasing the workday that decreases or maintains the problems of unemployment and forces unfavorable or regressive working conditions upon workers.[3]

The majority of governments in Latin America, even while achieving up to an average of 5% annual economic growth, have not been able to lower their levels of unemployment to less than 7%. Unemployment can reach 20% during times of crisis and lead to vast migrations, such as that of Ecuador and certain Central American countries. This year, Ecuador will grow slightly over 3%.[4]

Salaries form an ever-decreasing portion of national income; attrition rates and school desertion levels are high, and the quality of education is suspect at best. Nonetheless, "dollarization" of the Ecuadorian currency has caused inflation to slow down and has avoided devaluations that hurt persons with fixed incomes, including workers. Although workers have insufficient discretionary income, it is at least more stable than it was previously.[5]

On its side, health is directed to facilitating, if not fostering, private actions, or to promoting self-management, and therefore economic responsibility is passed on to society or is even privatized instead of being assumed by the State. Thus, the Welfare State that existed in Latin America has been replaced by a state of abandonment that is not only economic or budgetary in nature but also political and devoid of control mechanisms.

Society has not been capable of responding to such an onslaught of pressure: social organizations have been dispersed or destroyed, and new ones have not been formed. The labor unions in Latin America have played an important role in grouping forces not only of workers but also of other social sectors, but now face employment policies that have systematically lowered labor stability, created temporary jobs with a large rotation of workers, or have led to outsourcing the work force. As a result, labor unions have been weakened, have disappeared or have not been formed due to a lack of sufficient number of workers or to the lack of stability of workers in their places of employment. This is a result of the so-called flexible work in poor countries.

Rural farmers have been impoverished and/or have sold their lands or they are no longer productive in light of a concentrating market that does not allow them to enter it or compete in costs. In part, they have become the work force of export product companies under conditions of great labor contractual weakness.

The lack of infrastructure and services is, to a great extent, the result of a reduction in fiscal spending for public workplaces. Environmental problems have an unavoidable result.[6]

However, there are also specific problems that derive from unplanned urban sprawling, or from industrialization executed without benefit of a strict prior control.

As part of the implementation and development of the current model, there are specific situations that arise. Such is the case of Child Labor that adds new facets and repercussions to the already debilitated resistance of our children. Within this framework, the results and consequences in Ecuador regarding the environmental health of children is presented below.

BASIC PROBLEMS ASSOCIATED DIRECTLY TO A LACK OF BASIC INFRASTRUCTURE

Official information indicates that the population covered with basic services is still low (TABLE 1).

Although it may appear obvious, it is necessary to remember that the lack of drinking water implies a high risk of contracting infectious and parasitic illnesses. The lack of sewer systems or of a system for the disposal of black

TABLE 1. Indicators of environmental sanitation in Ecuador, 2001

Area	Public services			
	Sewage system	Electrification	Waste collection	Drinking water
Country	48.0	89.7	62.7	85.2
Costa Region	36.1	84.1	64.2	80.2
Sierra Region	61.9	83.3	63.3	90.3
Amazonic Region	34.2	58.4	39.7	No data
Insular Region	30.8	97.3	91.6	No data

SOURCE: SIISE, Version 3.5, Ecuador, 2002.

waters creates the right venue for epidemics and generates endemic areas that, along with a lack of waste collection, complete a framework of difficulties with profound sanitary consequences. These deficiencies affect the rural and urban-marginal areas and specially the poor inhabitants for whom the aforementioned conditions are aggravated, thus comprising periodic cycles of grave problems coupled with vicious circles in certain areas, all of which give rise to problems, such as gastrointestinal pathologies. Among them, the indigenous (25% of total population) and Afro-Ecuadorian populations are the hardest hit.[7]

Within this framework, other problems arise, such as food hygiene, contamination of the social areas intended for housing, recreation, and even education. Many schools have very deficient basic services.

The deficit in electricity and phone service, albeit not as marked as the previous problems, nonetheless comprise a lack of access to a basic service that hinders economic and social development

Housing does not escape this reality, and the problems derived from it are grave, considering the overcrowding or houses in precarious conditions where poor families find shelter in order to overcome their dire situation. Houses in poor conditions comprise 26.5% of the total number of houses in the country. However, the housing deficit is even greater (1.4 million houses increasing by 200,000 per year), in relation to the existing population.[8]

A grave problem is malnutrition. It is estimated that in Ecuador, up to 50% of minors under the age of 6 have varying degrees of malnutrition. However, malnutrition, considering that it is started even while in the womb and thus resulting in low birth weights, affects development of the central nervous system, the immune, digestive, and respiratory systems. The relationship between malnutrition and infectious diseases is well known.[9]

PROBLEMS INDIRECTLY RELATED TO A LACK OF INFRASTRUCTURE: LIVING CONDITIONS

Problems exist, which do not exclusively depend on the lack of infrastructure but rather on the difficulties in making use of them, in maintaining them, or for

TABLE 2. Poverty indicators, Ecuador (poverty and urban indigence, 2003–2004)

Characteristics	Total country
Unsatisfied basic needs (national)	61.30%
Extreme poverty (national)	31.90%
Urban poverty	41.40%
Urban indigence	7.60%

SOURCE: SIISE, Version 3.5, Ecuador, 2002 and INEC 2003–2004.

the execution of actions that compensate for their deficit. Under the concept of living conditions, we will group the following problems that are translated into impacts on health issues (TABLES 2 and 3).

Poverty is the underlying problem that contextualizes the entire landscape of countries, such as Ecuador. The poor, who comprise 61% of the total population do not have their basic needs met, and included among them are the 31% who are in extreme poverty, that is, have an income of less than one dollar per day to cover all of their needs. Despite salaries now maintaining their buying power, salaries barely cover 60% of the staple foods, which comprises a set of minimum supplies for a family with five members.[10]

Unemployment, despite the vast migration of Ecuadorians (approximately 2 million) who have already traveled abroad due to a lack of jobs starting in 2000, is maintained at about 11%, and underemployment (informal jobs) is at 59%.[11]

TABLE 3. Poverty indicators, Ecuador, 1990–2001

Indicator	Years	
	1990	2001
I. Poverty		
Total	74.0%	61.0%
Urban	56.0%	46.0%
Rural	95.0%	86.0%
II. Unemployment		
Total	7.0%	11.0%
Male	5.0%	8.0%
Female	9.0%	14.0%
Subemployment	45.0%	59.0%
III. Education		
Primary matriculation rate	88.9%	90.1%
Secondary matriculation rate	43.1%	44.6%
University matriculation rate	10.9%	11.9%
Illiteracy	11.7%	9.0%
IV. Malnutrition		
Prevalence of chronic malnutrition (height $\times$ age)		10.0%
Prevalence of global malnutrition (weight $\times$ age)		9.0%

SOURCE: Agenda Económica 2004. Indicadores Económicos y Sociales del Ecuador. Serie 3. Colegio de Economistas de Pichincha, Ecuador, 2004.

Education continues to be quantitatively and qualitatively deficient, and only elementary education reaches levels of 90%. Desertion and repetition of school years are quite high.[12]

Ecuador, according to the Human Development Index (established in the Human Development Report published by the United Nations Development Programme, 2002), is located 93rd among 173 countries in average human development at all levels, only ahead of Bolivia among South American countries.[13]

The health profile demonstrates that among the causes of general mortality, significant advances have been made in chronic and degenerative illnesses during the last two decades. General morbidity is maintained, with infectious and contagious pathologies present. These data lead to the general conclusion that there is a transition process among the health–illness profiles in Ecuador. Nonetheless, it is necessary to delve deeper into this analysis, as it is probable that if the population is divided by socio-economic strata, the poorer will show the highest frequency of infectious and contagious diseases than groups with higher incomes. With this, a duality in behavior would be produced in the morbidity–mortality rates, because besides the general trend mentioned regarding infectious and contagious pathologies toward those chronic and degenerative ones, a bifurcation would also be produced in the distribution of these pathologies by socio-economic strata, or at the very least, more pronounced trends in some rather than in the first 10 causes of morbidity of children under a year of age remain virtually unchanged since 1988. Although child mortality has decreased significantly up to 1998, it has begun to rise again starting in that same year. This indicates that there are underlying problems that, despite important strides, have produced peaks and valleys, and also produce certain problems that influence general indicators.[14,15]

There are a series of problems that could at least be controlled, such as certain illnesses of infants that are preventable and for which there are effective vaccines, such as measles, diphtheria, and poliomyelitis. Vaccination coverage is high and with it many of these pathologies have been drastically reduced. However, we can still mention certain zoonoses and transmissible diseases that could be decreased through sanitary education actions with specific measures and programs, and that tend to increase under unfavorable conditions. Among them are human rabies, Chagas disease, malaria, onchocercosis, brucellosis, murine typhus, leishmaniose, and leptospirosis. These pathologies periodically present some peaks which remind us that the underlying causes allowing for their re-emergence still persist.[16]

In this sense, the cholera epidemic that shook Peru and Ecuador in particular during the early 1990s clearly presented the structural weakness that exists in these countries, because an unusual extension of a pathology was caused, which had not emerged since a century ago.[17]

Tuberculosis could at least be limited through preventive programs, control of active patients, and rigorous treatments. Nonetheless, its relationship to the conditions of life and work is undeniable, and is still located among the 15 main causes of mortality for 2004.[18]

Accidents at home and on the roads could be the object of preventive or corrective measures that could help reduce such problems. The indices from these two problems are important, because the deficiencies in public transportation, given the existence of obsolete transport units that are poorly maintained as well as the lack of adequate selection of drivers, and along with secondary roads in poor state, comprise a set of risks that are increased themselves.[19]

Informality has led to the need of producing in the homes, which in turn leads to the introduction of pesticides, chemical substances, equipment, machinery, and processes in the living quarters that imply additional risks to those already present, given the poorly constructed houses.

Food intoxications are not always produced by infectious problems either; rather, the consumption of poorly kept products could be avoided by taking into account the potentially grave consequences. The population, given the constant deterioration of its buying power, must accept forms of intake that not only disregard its protein deficiency but that also do not respect the basic levels of quality.[20]

Other problems endured by children refer to problems in mental health, mistreatment, harassment, rape, and other types of social and family violence.[21]

Health services, despite certain progress in their quantitative offer, often do not respond adequately. A growing process of decentralization provides autonomy to public health centers, although it obligates them to obtain community resources through payment of their services with which they are hindered in their actions, because they treat precisely those who are most impoverished. Within this framework, over 30% of the population does not have access to such medical services and thus resolves its problems in an improvised manner or by resorting to traditional medicine.

In opposition, while we have been able to associate all public, private nongovernmental, religious, and military sectors into a so-called General Health System, the public sector sees its budget constrained, with this being precisely the one called to spearhead this system and set its policies. The health budget comprises barely 5% of the General State Budget. As a result, the populations having sufficient economic resources can obtain health care at private centers, while popular sectors receive costly and deficient health care. Certain private sectors, such as the military or other corporations, receive special and acceptable services, and non-governmental organizations strive to compensate for the deficit in health care. The Ecuadorian Social Security Institute, an institution that only caters to affiliated workers, and in certain cases, to their relatives, represents Social Security in Ecuador. However, this institution has not been foreign to the crisis of the health sector overall, and despite having resources, it has not been able to offer adequate care at all levels. One of the main problems is its coverage that only extends to 9% of the total population and to 18% of the economically active population of Ecuador. As a result, there are a few programs for prevention in environmental and occupational safety and health prevention, well-organized services are scarce, and there does not exist specific services for children exposed to environmental and occupational exposures.[22,23]

Although life expectancy has grown, the situation has still not undergone the necessary changes in regard to maternal health care, because maternal mortality continues to be high despite the important efforts expended in order to reduce it. Arterial hypertension, hemorrhages, deficiencies during labor, and abortions, represent the most frequent causes for maternal mortality.[24,25]

SPECIFIC ENVIRONMENTAL PROBLEMS

Among these, we must include the following:

- contamination of the urban air;
- the effects stemming from petroleum, agricultural, mining, and industrial production;
- deforestation;
- problems with drug production in Colombia, a country bordering Ecuador.

A study performed in Ecuador indicated that when leaded gasoline was being used, effects on human health were present in association with its presence in the students of certain schools in Quito.[26]

These after-effects of urban transportation have become evident also in contamination with carbon monoxide. Students from certain schools in Quito present respiratory symptoms associated to carboxy-hemoglobin, higher in areas of greater traffic than those of certain suburban schools.[27]

Added to this presently is the concern for lead substitutes used in gasoline, because it is now known that there is proof of the presence of benzene in congested urban areas.[28]

Furthermore, in the areas of influence of an oil refinery, levels of benzene in the air were found along with values of t,t-muconic acid that have been found among the neighbors of the said plant.[29]

The problems that stem from oil exploration affect the environment. Children from nearby indigenous populations to areas of oil exploitation present particular health problems that they did not have previously. Certain types of leukemia not previously associated with exposure to hydrocarbons have provided proof of this statement.[30]

Agricultural production, without a doubt, severely affects minors. During family production or self-consumption, this is very noticeable because they use pesticides without having enough information on them, by applying inadequate techniques without the use of personal protection equipment; these are stored in their homes and it is quite common for children to work in these parcels and be exposed due to fumigation or other work performed in the family vegetable cultivation plot. In a flower-growing area of Cuenca in Ecuador, there were three groups of minors under the age of 12, which presented levels of diminished AChE, especially in open-air sites surrounding the plantations of summer flowers.[31]

Moreover, the students of a school neighboring a flower plantation presented effects associated to exposure to pesticides stemming from that plantation.[32]

Minors living near to banana plantations are practically fumigated by the chemicals sprayed from small planes in neighboring areas.[33]

The industry, generally located in urban areas, affects neighboring populations. This affectation is produced due to its closeness, and ranges from smoke and gases, to dust, soot, and other chemical contaminants.[34]

Through non-treated water, chemical products are brought to urban rivers that cross densely populated areas.

Industrial waste only recently has been the target of public control; however, due to the short time these controls have been in place, we cannot yet expect major results.[35]

Mining not only affects certain areas, such as those of Nambija, Ponce Enríquez, Zaruma, Portovelo, and San Gerardo, but also extends its influence through water contaminated with mercury used in the amalgam of artisan gold panning. Children are at risk from these contaminants through their panning for gold which they perform by themselves or next to their parents, or even through the consumption of the contaminated fish they eat.[36]

Deforestation seriously affects provinces, such as that of Esmeraldas, because it implies desertification processes that increase the presence of dust, decrease the flows of rivers, and remove productive and nutritional sources farther away. Recent long periods of drought on the Coast, as well as frosts in the Highlands, have produced devastating effects on agricultural production, a percentage of which is dedicated to self-sustenance by the very farmers, as a reflection of climate change that also impinges on Ecuador.[37]

A current problem is that of fumigations applied throughout the Ecuadorian–Colombian border with the purpose of destroying coca plantations. Aerial fumigation used along with glyphosate, solvents, and other toxic products causes dermatological, respiratory, and digestive problems in children of fumigated areas that are now seen more frequently than prior to such fumigations. There are insufficient comprehensive studies that allow understanding this problem, but it is evident that the problems endured chronically by these populations due to lack of services will generate complaints, lawsuits, and evidence that must be further studied.[38]

Regarding respiratory illnesses, the environmental causes can be underlying such problems and to a lesser or greater extent, the social causes. There are controversies regarding the causes and factors associated to the emergence of asthma. A study performed at the city of Esmeraldas, where there is an oil refinery, demonstrated that under conditions of stable operation of the said refinery, social factors appeared as the ones affecting the emergence of asthma and asthma-like manifestations the most.[39] Children attending schools near to the refinery were more exposed to heavy metals, such as nickel.[40]

Another study close to the three refineries in Ecuador is not conclusive regarding the presence of asthma-like respiratory problems in neighboring populations, although it is in regard to the presence of malnutrition, stunting, and certain chronic respiratory pathologies, such as bronchitis.[41]

A growing concern is the presence of Persistent Organic Pollutants. In Ecuador, a study performed recently found that certain organic-chlorinated pesticides, whose importation was banned in 1985, are still present as residues in the soil. Electrical transformers containing polychlorinated biphenyls (PCBs) have been out of control since their introduction into Ecuador, and we know of only a mere portion of the destination of approximately 150,000 such transformers existing in past decades. We do not know with certainty all of the sources of dioxins and furans or their environmental concentrations in even sites of origin known as landfills or polluting industries.[42–44] The rate of accumulated incidence of cancer in children aged under 14 years has grown between 1994 and 2002, especially of hematopoietic origin; the lymphomas and cancer of the brain have increased their frequency in Quito, Guayaquil, and Cuenca. Although there are no studies on these matters, their presence is starting to form a part of the epidemiological map of children's health in Ecuador (TABLES 4 and 5).[45]

A recent study of children of mothers who worked in flower plantations during their pregnancies, as compared to children whose mothers were not in contact with pesticides during their mothers' pregnancies, showed that there are significant differences in the presence of visual-motor problems and blood pressure among the first group with regard to the second group, independent of the possible effects of other factors, such as malnutrition.[46]

TABLE 4. Average annual incidence of tumors per 100,000 by site and age group (Quito residents, 1985–1992 and 2000–2002, males)

	1985–1992			2000–2002		
	Age group (years)			Age group (years)		
Site	0–4	5–9	10–14	0–4	5–9	10–14
Bones, joints, and articular cartilage of limbs	0.2	0.2	0.8	1.0	2.0	11.0
Hematopoietic, reticular, endothelial system	5.4	6.5	4.5	57.0	35.0	32.0
Brain	1.1	2.0	0.8	16.0	12.0	9.0
Lymph nodes	0.9	3.3	2.8	12.0	20.0	9.0
All sites	11.9	14.5	10.7	133.0	87.0	76.0

SOURCE: Registro Nacional de Tumores, SOLCA, Quito, Ecuador, 2004.

CHILD LABOR

One of the main environmental problems affecting the health of children is child labor. Within the working environment shared or not with adults, in formal or informal production, in the city or in rural areas, minors under the age of 18 work in all kinds of activities, including those considered dangerous. It has been calculated that over 800,000 minors are working in Ecuador, with their participation largely in agriculture being one of the highest with 23%.[47]

Minors are exposed to pesticides, heavy metals, chemical substances of all kinds, ergonomic problems, mental and physical overloads, and insecurity coupled to grave accidents.[48,49]

Every economic sector can show varying facets of this situation: for example, in agriculture, it is possible that ancestral practices, removal from centers of study, and poverty, among other factors, may promote conditions for family agricultural practices surrounding child labor. Nevertheless, in formal agriculture, child labor can also be found: this is the case with flower plantations, banana plantations, and other plantations producing strawberries, broccoli, and other produce items. One of the grave problems is that the minors work close to their parents in agricultural activities by helping to fumigate without any type of personal protective equipment, and the next day they return to their schools to study, which undoubtedly affects their studies and performance.[50]

In the construction industry, we could say that there is a combined situation: the contractors who are hired by construction developers include minors either because of their family relationship or to cut costs. The problem arises because of the high informality in employment practices: minors represent a permanent source of employment and perform strenuous efforts, which added to a prolonged work shift, dwindles their energy for studying at night.[51]

Production of roof tiles also is a source of exposure to lead with health effects in children.[52]

Some children (987 between the ages of 5 and 7; 2116 between the ages of 8 and 11; and 6065 between the ages of 12 and 14) work producing cinder blocks and bricks between evening and 6 am, when some go to school and others return to their homes.[53]

Finally, in industry, under the guise of being "apprentices," minors are made to work as adults in their midst without any protection and are paid much less.[52]

It is also important to point out what happens at the level of formal agricultural production. The case of banana plantations is evident. A report of a study that was done by IFA (Corporation for development of Production and Work Environment) with the support of the International Labor Organization reveals that over 50,000 minors had allegedly been working in those plantations prior to the public outcry that ensued and prior to programs applied to reduce them, although they did not altogether eliminate them. Some of them (10%) work in fumigation and the majority are exposed to aerial fumigations.[54]

TABLE 5. Average annual incidence of tumors per 100,000 by site and age group (Quito residents, 1985–1992 and 2000–2002, females)

| | 1985–1992 | | | 2000–2002 | | |
| | Age group (years) | | | Age group (years) | | |
Site	0–4	5–9	10–14	0–4	5–9	10–14
Bones, joints, and articular cartilage of limbs	0.2	0.2	0.8	0.0	1.5	2.5
Hematopoietic, reticular, endothelial system	4.3	3.9	4.4	8.1	7.0	6.0
Brain	1.0	1.3	0.6	2.5	0.5	1.0
All sites	8.7	9.5	10.5	21.6	14.0	17.0

SOURCE: Registro Nacional de Tumores, SOLCA, Quito, Ecuador, 2004.

Within the flower industry, there is talk of percentages fluctuating between 5% and 20% of working minors under the age of 18. The presence of personal, family, economic, and cultural changes has been identified among them, some of which have not always been positive. The impact on their health has been detected through a study performed indicating the high presence of cephalea, migraines, tremors, and decrease of the acetyl-cholinesterase, among other disorders.[55]

Child labor is also performed in garbage dumps that have no measures of control or safety and expose minors to a myriad of biological, chemical, and ergonomic risks. No specific control programs exist, which address the health of minors or the use of child labor.[56]

REPRODUCTIVE HEALTH IN ECUADOR

As an intermediate link between environmental factors and the health of children, it is important to consider the state of reproductive health, whether due to the occupation by their parents or to the conditions of pregnancy, and occupational and environmental risks of abortions, congenital malformations, or premature deliveries.[57] In Ecuador, maternal mortality is as high as 76.4 maternal deaths per 100,000 live births and is difficult to lower. Congenital malformations of the heart (Codes Q20–Q24 of the CIE 10) occur in about 9.5 cases per 100,000 live births. The data on abortions are not segregated by cause, but account for between 8% and 14% of the total number of maternal deaths. Equally high is the number of abortions presented by the Ecuadorian Social Security Institute, and, if we were to consider that of the women treated, there are also workers, a study of this population could potentially separate the occupational causes from other causes.[58] For example, female workers at flower plantations have twice the relative risk of abortions compared with females in a non-exposed reference group.[59] Premature delivery is frequent in Ecuador, and it would appear to be related to both living conditions and working conditions.[60] Also DDT was found in breast milk among females in two rural areas.[61]

UNDERDEVELOPMENT AND THE ENVIRONMENT: OVERLAPPING, SUM, OR INDEPENDENCE BETWEEN THEM?

Latin American physicians and environmentalists addressed social causes of certain problems. Social distribution by class, strata, ethnic group, etc., and certain priority problems were denounced. However, neither the State nor the society could react based on them. The drastic changes suffered by Latin America during the past few years, in particular the process of globalization, exacerbated these difficulties and also put forth the technical and methodological limitations in order to understand these phenomena. The complexity of the problems and the urgency regarding these increased the need and availability in order to address them in a more comprehensive manner.[62]

We require a methodological aperture in order to incorporate the social variables into the environmental issues in Latin America. However, we also require an integration of the environmental variables with the social variables in order to produce a set that in its results may ponder the impacts in a better way, so that we may establish the necessary and more precise measures of prevention and care.[63,64]

TOWARD A NEW PERSPECTIVE OF THE ENVIRONMENTAL HEALTH OF CHILDREN IN LATIN AMERICA

In Latin America in general, and in Ecuador in particular, a combination of components influence the environmental health of children. The complexity of the social structure which presents multiple elements that do not favor the well-being of people cause many health problems to be related to basic problems of infrastructure, such as nutrition, housing, education, and health care. However, although the environmental problems comprise a portion of the manner in which society has developed and generally accompany its misalignments and contradictions, it is necessary to strengthen the effort to analyze its specific weight and its own influence. This is why the analysis of risks attributable to the environmental health of children represents the harsh Latin American reality and requires being addressed in a particular manner.

One issue that has been widely considered in the social medical literature is malnutrition. It is known that the majority of the causes for malnutrition can be found in poverty and inequality. Recent studies present proof that, besides the problems of growth and development which originate from malnutrition, there are also a series of environmental risks that produce specific and independent effects, which are aggregated to deterioration in health. Although health and poverty are intimately linked, it is still necessary to disarticulate the influence that other components of their reality may be producing, and have a more precise knowledge of the environmental impact within the framework of health in poverty, which should also be considered.

Poverty and environmental problems do not necessarily co-exist permanently: in the working environment, workers are exposed to toxic substances that can produce adverse health effects even in the absence of poverty. This can apply especially to the use of child labor. Despite its being a social problem, child labor is the result of a combination of elements that are originated in addition to the manner of production implemented by the majority of countries in Latin America: these are linked to state-run and corporate strategies for use of the labor force, and are circumscribed within employment regulations and policies, social security, and above all, to official decisions regarding protection of labor rights. A State that leans toward deregulation, one that limits or hinders union implementation of workers, and one that does not promote collective hiring, paves the way for informal employment or the so-called "black labor" as part of the precarious jobs it generates. Within this framework, child labor easily fits into all of its worst manifestations, and it has but few limits in order to be reduced or overcome.[65]

Therefore, in Ecuador, and probably in other countries as well, it is very important to develop new strategies for research and a wider frame of reference, which connect the study of social and environmental problems and, at the same time, allow for their specific impacts to be understood, which many times remain hidden or masked into the weave of social problems.

It is not enough either to quantitatively compare the problems between developed and developing countries, or to consider them as part of a fatal or identical sequence between the two. Nor does it have to do with an epidemiology of poverty from the strictly economic standpoint, which could be "reductionistic." Rather, it is necessary to widen the frontiers of epidemiology in order to address new and old problems under better technical conditions than in previous decades, and regarding problems that are or are not the same, or are not presented in the same manner in different countries. It is necessary to make an effort to further the search for independent predictors of the probable causes, or social or environmental factors, or both, which influence the occupational and environmental health of the inhabitants of our countries. Fortunately, there are certain strides that have been made regarding these issues, which help to generate a useful theoretical and technical framework of reference in order to understand these problems in an integral way.[66–71]

REFERENCES

1. TAMBURLINI, G. et al. 2002. Children's health and environment: a review of evidence. Environmental Issue Report No. 29. World Health Organization, Regional Office for Europe. European Environment Agency. EEA. Copenhagen.
2. VÁSQUEZ, S.L. & N.G. SALTOS. 2005. Ecuador: su realidad. 2004–2005. Edición actualizada. Fundación "José Peralta". Décimotercera Edición. Quito, Ecuador.

3. HARARI, R. 2004. La economía de exportación y la salud: los casos de petróleo, banano y flores. En "Efectos sociales de la globalización: Petróleo, Banano y Flores". Tanya Korovkin, Compiladora. CEDIME-Abya-Yala. Primera Edición. Ecuador.
4. Banco Central del Ecuador. 2005.
5. VICUÑA, L. 2005. Ecuador Siglo XXI. Política Económica. Escuela Superior Politécnica del litoral-Instituto de Ciencias Humanísticas Económicas. Guayaquil, Ecuador.
6. PAULSON, S. 1998. Desigualdad Social y degradación ambiental en América latina. Embajada Real de los Países Bajos. Programa Bosques, Árboles Comunidades Rurales. FAO. Ediciones Abya-Yala. Ecuador.
7. Sistema Integrado de Indicadores Sociales del Ecuador. 2002. SIISE. Versión 3.5. Ecuador.
8. Instituto Nacional de Estadísticas y Censos. INEC. 2001. VI Censo de Población y Vivienda. Ecuador.
9. CÓRDOVA, P.A. 2004. Seminario de Evaluación de la Realidad Económica y Social del Ecuador. Colegio de Economistas de Pichincha. Primera Edición. Ecuador.
10. CÓRDOVA, P.A. 2004. Seminario de Evaluación de la Realidad Económica y Social del Ecuador. Colegio de Economistas de Pichincha. Primera Edición. Ecuador.
11. Sistema Integrado de Indicadores Sociales del Ecuador. SIISE. 2002. Versión 3.5. Ecuador.
12. Sistema Integrado de Indicadores Sociales del Ecuador. SIISE. 2002. Versión 3.5. Ecuador.
13. Programa de Naciones Unidas para el Desarrollo (PNUD). 2002. Informe Sobre Desarrollo Humano. Ediciones Mundi-Prensa. New York, USA.
14. Instituto Nacional de Estadísticas y Censos (INEC). 1998. Anuario de Estadísticas Hospitalarias. Camas y Egresos Hospitalarios. Ecuador.
15. Instituto Nacional de Estadísticas y Censos (INEC). 1998. Anuario de Estadísticas Vitales. Nacimientos y Defunciones. Ecuador.
16. JARRÍN, A.E. 2004. Estadísticas de las Enfermedades 1990–2003. Dirección Nacional de Epidemiología. Ministerio de Salud Pública. Ecuador.
17. CIFUENTES, M. & J.H. SOLA. 1992. El cólera: una respuesta desde la Comunidad Andina. Centro Andino de Acción Popular (CAAP) y Comité de Emergencia en lucha contra el cólera (CELCO). Quito, Ecuador.
18. Ministerio de Salud Pública-Instituto Nacional de Estadísticas y Censos-Objetivos Desarrollo del Milenio-Secretaria nacional de Planificación y Desarrollo-Secretaría Técnica del Frente Social-Organización Panamericana de la Salud-Fondo de Naciones Unidas para Población-UNICEF-PNUD-UNIFEM. Indicadores Básicos de Salud. Ecuador. 2005.
19. Ministerio de Salud Pública-Instituto Nacional de Estadísticas y Censos-Objetivos Desarrollo del Milenio-Secretaria nacional de Planificación y Desarrollo-Secretaría Técnica del Frente Social-Organización Panamericana de la Salud-Fondo de Naciones Unidas para Población-UNICEF-PNUD-UNIFEM. Indicadores Básicos de Salud. Ecuador. 2005.
20. Ministerio de Salud Pública-Instituto Nacional de Estadísticas y Censos-Objetivos Desarrollo del Milenio-Secretaria nacional de Planificación y Desarrollo-Secretaría Técnica del Frente Social-Organización Panamericana de la Salud-Fondo de Naciones Unidas para Población-UNICEF-PNUD-UNIFEM. Indicadores Básicos de Salud. Ecuador. 2005.
21. UNICEF. 1998. La situación de los niños en el mundo.

22. CONASA-MSP-MODERSA-OPS-UNFPA. 2005. Marco general de la reforma estructural de la Salud en el Ecuador. Quito, Ecuador.
23. ZUMÁRRAGA, P.R. 2001. Historia de un atraco (De la Seguridad Social a la Inseguridad Total). Casa de la Cultura Ecuatoriana Benjamín Carrión. Ecuador.
24. MSP. Mapa del Sistema de Salud. 2002. Estadísticas y Registros de Salud. Ecuador.
25. HERMIDA, C. 2005. Fora, policies and systems for maternal mortality in Ecuador. Global Forum Update on Research for Health. Volume 2. p.86.
26. BOSSANO, F. & J. OVIEDO. 1996. Contaminación por plomo en Quito. Revista Ciudad Alternativa, pp. 85–89. Quito, Ecuador.
27. ESTRELLA, B. *et al*. 2005. Acute respiratory diseases and carboxyhemoglobin status in school children of Quito, Ecuador. Environ. Health Perspect. **113:** 607–611.
28. Universidad Central del Ecuador-Escuela de Ingeniería Química-PETROECUADOR. Estudio DE la "Calidad del aire de la Ciudad de Quito". Departamento de Petróleos, Energía y Contaminación. Quito, Ecuador. 2003.
29. LÓPEZ, R. *et al*. 2005. Condiciones de salud y alteraciones cromosómicas en áreas aledañas a la Refinería de Petróleo de la Ciudad de Esmeraldas. Informe de Trabajo. Ministerio de Salud Pública-Proceso de Ciencia y Tecnología-IFA. Ecuador.
30. SAN SEBASTIÁN, A.-K. & M. SAN SEBASTIÁN. 2002. Incidence of childhood leukemia and oil exploitation in the Amazon Basin of Ecuador. Int. J. Epidemiol. **31:** 1021–1027.
31. HARARI, R. *et al*. 2004. Gonzalo Albuja, Homero Harari. Ambiente y salud en la floricultura: tres estudios de caso. En P. Comba y Harari. R. Compiladores: El ambiente y la salud: la Epidemiología Ambiental. IFA-Istituto Superiore di Sanitá-Abya-Yala. Ecuador.
32. HARARI, R. *et al*. 2005. Exposición ambiental a plaguicidas provenientes de la una florícola y salud de los estudiantes de un colegio vecino. Informe Preliminar de Trabajo. Ecuador.
33. HARARI, R. & H. HARARI. 2005. Producción bananera y exposición a plaguicidas en la Costa Ecuatoriana. IFA-FENACLE-FONDO AGIL. Ecuador.
34. HARARI, R. *et al*. 1995. Paul Segovia y William Vargas. Mapa Territorial de Riesgos: una técnica participativa. IFA-ILDIS. Ecuador.
35. Desarrollo y Autogestión. 2002. Estudio Nacional de Línea de Base del Trabajo Infantil en basurales del Ecuador. Programa IPEC-OIT. Ecuador.
36. ALLEN, S., S.A. COUNTER, *et al*. 2005. Mercury levels in urine and hair of children in an Andean gold-mining settlement. Int. J. Occup. Environ. Health **11:** 132–137.
37. Ministerio del Ambiente. 2005. Ecuador.
38. Tribunal Constitucional del Ecuador. Ecuador exigirá a Colombia el cese de fumigaciones en la Frontera. 2005. Comité Interinstitucional sobre las Fumigaciones. Ecuador.
39. HARARI, R., F. Forastiere, A. Vaca Rodríguez, *et al*. 2004. Pobreza y factores de riesgo para el asma y las sibilancias entre niños afroecuatorianos. En P. Comba & R. Harari, Compiladores: El ambiente y la salud: la Epidemiología Ambiental. IFA-Istituto Superiore di Sanitá-Abya-Yala. Ecuador.
40. HARARI, R. & F. Forastiere. 2004. Emisiones de una refinería de petróleo y exposición a níquel en su entorno: un estudio de caso. En Pietro Comba y Raúl Harari Compiladores: El ambiente y la salud: la Epidemiología Ambiental. IFA-Istituto Superiore di Sanitá-Abya-Yala. Ecuador.

41. HARARI, R., G. ALBUJA, L. ALVARADO, H. HARARI,. 2003. Exposición a emisiones de tres refinerías de petróleo y salud de los estudiantes de áreas vecinas. Ecuador.
42. Ministerio del Ambiente-ESPOL-ICQ. 2004. Invenytarios de Plaguicidas COP's en Ecuador.
43. Ministerio del Ambiente-COALDES. 2003. Inventario de Bifenilos Policlorados PCB's. Ecuador.
44. Ministerio del Ambiente-ESPOL-ICQ. 2004. Inventario preliminar de Dioxinas y Furanos en el Ecuador. Ecuador.
45. FABIAN CORRAL, F.C. *et al.* (Eds.). 2004. National Cancer Registry (NCR). SOLCA-Quito. Agreement SOLCA-MSMP-INEC. Cancer epidemiology in Quito and other Ecuadorian regions. Ecuador.
46. GRANDJEAN, P. *et al.* 2006. Pesticide exposure and stunting as independent predictors of neurobehavioral deficits in Ecuadorian school children. Pediatrics **117:** 546–556.
47. Ministerio de Trabajo y Empleo-Desarrollo y Autogestión-Organización Internacional del Trabajo. 2005. Diario El Comercio, A9. Lunes 21 de Diciembre de 2005. Ecuador.
48. HARARI, R., F. Forastiere & O. Axelson. 1997. Unacceptable "occupational" exposure to toxic agents among children in Ecuador. Am. J. Ind. Med. **32:** 185–189.
49. HARARI, R. *et al.* 1999. Health conditions in Ecuadorian working Children. Eur. J. Oncol. **4:** 627–629.
50. HARARI, R. *et al.* 2005. Exposición ambiental a plaguicidas provenientes de la una florícola y salud de los estudiantes de un colegio vecino. Informe Preliminar de Trabajo. Ecuador.
51. HARARI, R. *et al.* 2002. Informe de Trabajo sobre Trabajo Infantil en el sector de la construcción en Quito, Ecuador. IFA-OIT. Ecuador.
52. HARARI, R. & M.R. CULLEN, 1995. Childhood lead intoxication associated with manufacture of roof tiles and ceramics in the Ecuadorian Andes. Arch. Environ. Health **50:** 393.
53. PEÑAFIEL, C. 2005. Trabajo infantil en las bloqueras. Diario El Comercio D10, Redacción Ambato, Domingo 20 de Noviembre del 2005. Ecuador.
54. HARARI, R. *et al.* 2002. Línea de Base para trabajo infantil en banano y Erradicación de las Peores Formas de Trabajo infantil en Ecuador. IFA-OIT. Ecuador.
55. INNFA. 2001. Trabajo Infantil en floricultoras de las zonas de Cayambe y Cotopaxi. Ecuador.
56. Centro de Desarrollo y Autogestión-IPC-OIT. 2002. Estudio Nacional de Línea de Base del Trabajo Infantil en Basurales en Ecuador. DYA-IPEC-OIT. Ecuador.
57. Ministerio de Salud Pública-Instituto Nacional de Estadísticas y Censos-Objetivos Desarrollo del Milenio-Secretaria nacional de Planificación y Desarrollo-Secretaría Técnica del frente Social-Organización Panamericana de la Salud-Fondo de Naciones Unidas para Población-UNICEF-PNUD-UNIFEM. Indicadores Básicos de Salud. Ecuador. 2005.
58. Ministerio de Salud Pública-Instituto Nacional de Estadísticas y Censos-Objetivos Desarrollo del Milenio-Secretaria nacional de Planificación y Desarrollo-Secretaría Técnica del frente Social-Organización Panamericana de la Salud-Fondo de Naciones Unidas para Población-UNICEF-PNUD-UNIFEM. Indicadores Básicos de Salud. Ecuador. 2005.
59. HARARI, R. & V. DÁVALOS. 1998. Aborto y ocupación materna. Informe de Investigación. IFA-PPM-IESS. Ecuador. 1998.

60. FREIRE, R. & R. HARARI. 1994. Parto prematuro y ocupación materna. Informe de Investigación. Ecuador.
61. BOLAÑOS, M. 1985. Determinación de DDT en leche materna. SESA. Ecuador.
62. BARRETO, M.L. 2004. The globalization of epidemiology: critical thoughts from Latin América. Int. J. Epidemiol. **33:** 1132–1137.
63. PEARCE, N. 2004. The globalization of epidemiology: introductory remarks. Int. J. Epidemiol. **33:** 1137–1131.
64. LOEWENSON, R. 2004. Epidemiology in the era of globalization: skills transfer of new skills? Int. J. Epidemiol. **33:** 1144–1150.
65. HARARI, R. 2001. Trabajo Infantil y salud. INNFA-IFA. Ecuador.
66. LEVITSKY, D. & B. STRUPP. 1995. Malnutrition and the brain: changing concepts, changing concerns. J. Nutr. **125:** 2212S–2220S.
67. ALKON, A. 2003. Developmental and contextual influences on autonomic reactivity in young children. Dev. Psychobiol. **42:** 64–78.
68. FENSKE, R.A. *et al.* 2002. Children's exposure to chlorpyrifos and parathion in an agricultural community in Central Washington State. Environ. Health Perspect. 110.
69. CURL, C.L. *et al.* 2003. Organophosphorus pesticide exposure of urban and suburban preschool children with organic and conventional diet. Environ. Health Perspect. **111:** 377–382.
70. SLOTKIN, T.A. *et al.* 2002. Functional alteration in CNS catecholamine systems in adolescence and adulthood after neonatal chlorpyrifos exposure. Dev. Brain Res.**133:** 163–173.
71. LARREA, C. & I. KAWACHI. 2005. Does economic inequality affects children malnutrition? The case of Ecuador. Soc. Sci. Med. **60:** 165–178.

Environmental Impacts on Children's Health in Southeast Asia

Genotoxic Compounds in Urban Air

MATHUROS RUCHIRAWAT,[a,b] PANIDA NAVASUMRIT,[a] DAAM SETTACHAN,[a] AND HERMAN AUTRUP[c]

[a]*Laboratory of Environmental Toxicology, Chulabhorn Research Institute, Bangkok 10210, Thailand*

[b]*Department of Pharmacology, Faculty of Science, Mahidol University, Bangkok 10400, Thailand*

[c]*Department of Environmental and Occupational Medicine, University of Aarhus, DK-8000 Aarhus, Denmark*

ABSTRACT: Air pollution is a serious problem in many countries in Southeast Asia, particularly in major metropolises with high levels of traffic congestion generating significant amounts of genotoxic substances. The contribution of such environmental exposure to children's illnesses, such as respiratory diseases and cancer, is a public health concern. Innercity children may have higher levels of exposure to genotoxic substances in the air than those living in rural areas. This study was conducted in Bangkok, where ambient levels of polycyclic aromatic hydrocarbons (PAHs) and benzene are relatively high. Bangkok school children were exposed to total PAHs at about sixfold higher levels than those in rural areas, with levels of urinary 1-hydroxypyrene (1-OHP) also being significantly higher. PAH–DNA adduct levels in lymphocytes were fivefold higher in Bangkok children. Benzene exposure in Bangkok school children was more than twofold higher than the levels measured in children from the rural areas. This is in agreement with the biomarkers of internal dose, that is, blood benzene and urinary *trans, trans*-muconic acid (t,t-MA) levels. The potential health risks from exposure to PAHs and benzene were assessed through the use of DNA damage and DNA repair capacity as markers of early biological effect. DNA strand breaks were significantly higher in Bangkok school children, while DNA repair capacity was significantly lower. It appears that children in major cities in developing countries may have an increased health risk for the development of certain diseases, such as cancer due to exposure to genotoxic substances in their environment.

Address for correspondence: Mathuros Ruchirawat, Laboratory of Environmental Toxicology, Chulabhorn Research Institute, Vipavadee Rangsit Highway, Lak Si, Donmuang, Bangkok 10210, Thailand. Voice: +66-2-574-0615; fax: +66-2-574-0616.

e-mail: mathuros@tubtim.cri.or.th

KEYWORDS: air pollution; Bangkok; biomarker; children; DNA damage

INTRODUCTION

Many countries in Southeast Asia have undergone rapid economic development and industrialization, which, in some way, has resulted in negative impacts on the environment and health of the population. The main environmental and health issues in countries of the region are: contamination of surface and ground water by chemicals, water supply, and sanitation; indoor and outdoor air pollution; industrial pollution; pesticides; solid and hazardous waste management; and changes to the ecosystem, including climate change.[1] In addition to the classical health problems of the developing world, such as infectious diseases, emerging environmental threats to health are toxic chemicals in the environment as a result of industrialization and urbanization. Urban air pollution has been identified by the WHO as one of the most significant priority areas for concern in the Asia and Pacific region.[2]

With the Southeast Asian and Western Pacific regions together making up approximately 50% of the world's population of children, the issue of children's environmental health certainly is a significant one in this region.[3] Children are among the most susceptible members of the population. Reasons for this greater susceptibility are both behavioral and physiological in nature, and include the fact that there is a significant amount of time after exposure at childhood for the possible manifestation of resultant health effects later in life.[4] Clearly, the association between environmental exposure to toxic chemicals and illnesses in children is an urgent public health concern. There is a need for a better understanding of the mechanisms underlying the potentially greater risks in children compared to adults, as well as the potentially greater susceptibility as a result of gene–environment interactions.[5,6] With this knowledge, it will be possible to protect children's health against these environmental threats.

Genotoxic Carcinogens in Urban Air

Countries in Southeast Asia share many common environmental problems, such as air pollution in major cities with a large population and high levels of traffic congestion. This is a problem that is not evenly distributed within the various different nations, but rather one that is concentrated in the various different capital cities, including Bangkok, Thailand. This is because these cities tend to be significantly more urbanized than most other cities.

Most of the studies that have been conducted in Asian developing countries with regards to air pollution have concentrated on particulate matter (PM_{10}), SO_2, NO_2, and CO, and their effects on the respiratory and cardiovascular systems as well as on mortality rates.[7] However, carcinogenic compounds in urban air pollution also pose a significant threat to health and deserve serious attention. Polycyclic aromatic hydrocarbons (PAHs) and benzene are major genotoxic carcinogens found in urban air pollution from motor vehicle emissions. Pollutants, such as these are believed to be associated with many types of health problems, including cancer.[8,9]

Studies have been conducted in our laboratory to try to establish a link between exposure to these compounds and potential health effects in several high-risk and vulnerable groups of the population, such as children, by measuring ambient exposure levels, as well as biomarkers of exposure and early effects.[10–12]

PAHs

PAHs are a class of compounds that are generally byproducts of incomplete combustion of fossil fuels. Some members of this class of compounds, for example, benzo[a]pyrene, have been classified as probable human carcinogens by the International Agency for Research on Cancer (IARC). Exposure to PAHs occurs through inhalation of contaminated indoor or ambient air, or consumption of contaminated food or water. Exposure may be monitored through environmental samples and the use of biomarkers of exposure.[13–16]

PAH exposure is a problem not confined specifically to Asia, yet it can be observed that it is relatively pronounced in Bangkok when compared to many other cities in the same as well as in other parts of the world.[11,17–21] One of the assumptions in the studies conducted in our laboratory is that since PAH's are traffic-related air pollutants and are normally particle-associated, their concentrations should be highest near the source, that is, on the roads, and that the people most at risk are those who spend significant portions of their day directly at or very near these sources (TABLE 1). A previous study conducted in Thai traffic police yielded similar exposure levels directly at the roadside.[12] From the data presented in TABLE 1, it can be observed that while PAH levels are high directly on the road, they drop by approximately 50%

TABLE 1. Ambient air levels of total PAHs at various locations in Bangkok

Sampling locations	Number of sites	Total PAHs (ng/m^3)
On the road	9	65.43 ± 5.74
		56.64 (46.60–83.88)
At the roadside	7	36.23 ± 8.37
		32.18 (17.90–83.04)
500 m from the roadside	3	2.42 ± 0.40
		2.56 (1.67–3.04)

Values are expressed as mean ± SE on the first line and median (min–max) on the second line.

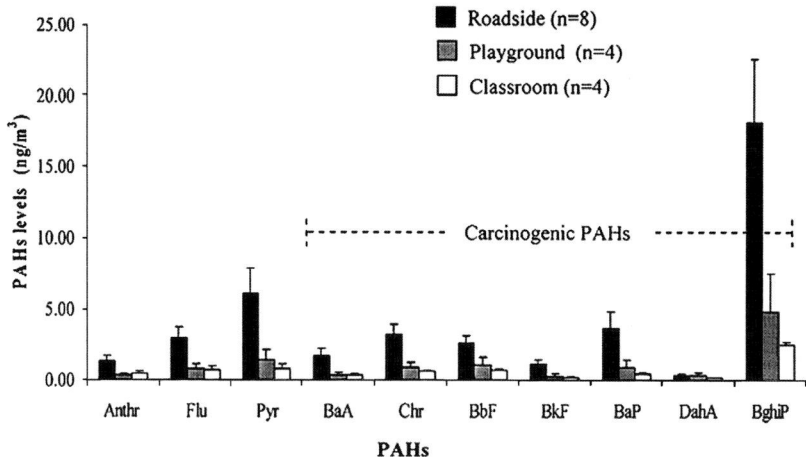

FIGURE 1. Pattern of PAHs at different sampling sites in Bangkok.

at the roadside and by more than 90% even at a relatively short distance of 500 meters away behind the street canyon. Also, the fact that the predominant species of PAH found is benzo[g,h,i]pyrelene indicates that the major source of these PAHs is combustion of petroleum (FIG. 1). Benzo[a]pyrene (B[a]P) accounts for about 10% of the total carcinogenic PAHs.

Many schools are situated in the vicinity of main roads, thus school children are regularly exposed to high levels of traffic-related air pollution (TABLE 2). The data were collected from 10- to 12-year-old school children (boys) from four schools in Bangkok and two schools in Chonburi, a city located approximately 100 km outside of Bangkok and which is primarily rural in nature. Clearly, the potential for exposure in Bangkok school children is much greater than that in rural school children, with Bangkok school children potentially exposed to 40-fold higher levels than Chonburi school children are exposed to at the roadsides in front of school, greater than 10-fold higher levels than Chonburi school children are exposed to on the playgrounds, and 8- to 9-fold higher levels than Chonburi school children are exposed to in the classrooms (all differences significant at $P < 0.001$).

To assess the health risks to which these children are exposed, it is important to quantify actual exposures and the resultant changes that are effected (TABLE 3). Individual exposures, as collected through personal air pumps collecting air particulate onto glass fiber filters and analyzed by high-performance liquid chromatography (HPLC), is approximately six times greater in Bangkok school children. When levels of carcinogenic PAHs are converted to B[a]P equivalents, Bangkok children are exposed to nearly 10-fold greater levels than Chonburi children. The observation that the morning levels of urinary

TABLE 2. Ambient air levels of total PAHs at roadsides and within school areas

Sampling location	Ambient PAH levels (ng/m^3)	
	Chonburi	Bangkok
At the roadside (in front of school)	1.17 ± 0.40 0.60(0.31–2.91) $n = 7$	41.14 ± 10.17*** 34.59 (12.01–99.95) $n = 8$
On the playground	0.91 ± 0.36 0.72 (0.28–1.92) $n = 4$	11.31 ± 5.19*** 8.54 (2.64–25.54) $n = 4$
In the classroom	0.84 ± 0.49 0.39 (0.28–2.31) $n = 4$	7.03 ± 1.03*** 6.83 (5.06–9.39) $n = 4$

Values are expressed as mean ± SE on the first line and median (min–max) on the second line.
***Significant differences from Chonburi at $P < 0.01$.

1-hydroxypyrene (1-OHP) were also significantly higher in Bangkok school children ($P < 0.01$) would seem to indicate that significant exposures were taking place outside the school as well, most likely as children traveled to and from school to their homes. Afternoon levels were not significantly different from morning levels, even in the Bangkok school children. This confirms our previous finding that exposure to total PAHs at these levels (∼6 ng/m^3) is not enough to cause elevated levels of urinary 1-OHP in children.[11] One-hydroxypyrene is a metabolite of pyrene, which is normally associated with

TABLE 3. PAHs exposure levels and biomarkers of exposure in school children

	School locations	
	Chonburi	Bangkok
Individual exposure		
Total PAHs (ng/m^3)	1.06 ± 0.17 0.59 (0.25–3.67) $n = 32$	6.33 ± 0.35*** 6.05 (2.35–13.91) $n = 43$
B[a]P equivalent (ng/m^3)	0.19 ± 0.12 0.12 (0.03–0.44) $n = 32$	1.83 ± 0.28* 1.72 (1.28–2.59) $n = 42$
Urinary 1-OHP		
Morning (μmol/mol creatinine)	0.15 ± 0.02 0.12 (0.03–0.47) $n = 29$	0.20 ± 0.01** 0.16 (0.08–0.43) $n = 42$
Afternoon (μmol/mol creatinine)	0.13 ± 0.01 0.10 (0.04–0.30) $n = 27$	0.24 ± 0.03*** 0.19 (0.05–1.48) $n = 43$

Values are expressed as mean ± SE on the first line and median (min–max) on the second line.
*,**,*** Significant differences from Chonburi at $P < 0.05$, 0.01, and 0.001, respectively.

the volatile fraction of PAHs. There is evidence that 1-OHP may correlate better with relatively higher occupational exposures rather than environmental exposures.[15,16]

PAH–DNA adducts represent both biomarkers of exposure and biomarkers of biologically effective dose since PAHs are metabolized to electrophilic intermediates, which bind to cellular macromolecules, such as DNA and proteins to form adducts. The formation of DNA adducts is believed to be the first step in the initiation of carcinogenesis. PAH–DNA adducts also reflect accumulated exposure over a longer period of time than 1-OHP.

The potential impact on children's health can be observed by the significantly greater number of PAH–DNA adducts observed in Bangkok children when compared to rural school children, with preliminary findings suggesting more than fivefold higher levels (unpublished data). However, we are aware that the greater number of PAH–DNA adducts in Bangkok school children may not be from traffic-related sources alone. Alternative sources are cooking (especially grilling), consumption of contaminated food, passive smoking, and other nontraffic-related activities associated with combustion. Detailed lifestyle questionnaires were used to identify potential exposure to PAHs through other sources. It was found that the pattern of lifestyle and food consumption in both groups was similar, and that the contribution of PAHs exposure from other lifestyle-related sources and food, if any, should be the same for both groups.

Benzene

Benzene is present in the environment from both natural and anthropogenic sources, including the burning of gasoline and cigarette smoking. It is classified as a known human carcinogen.[9] Exposure to benzene is primarily through inhalation of contaminated air, and may be monitored through measuring blood benzene levels or metabolites, such as urinary t,t-MA and S-phenylmercapturic acid (SPMA), which are commonly used biomarkers of exposure.[22,23]

As stated previously, benzene is also a traffic-related air pollutant. Relatively high levels of benzene have been reported in many cities, including Cotonou, Mexico City, and Bangkok, while many European cities have reported lower levels.[24–29] As is the case with PAHs, benzene levels are also higher nearer the roads, and levels drop by 60% 500 m away from the main roads (TABLE 4). Children in schools in the vicinity of main roads are exposed to this pollutant on a daily basis, with Bangkok school children being potentially exposed to approximately three- to fourfold higher levels than rural school children are exposed to within the school compounds (TABLE 5). Individual exposure levels, collected by diffusive badges attached near the breathing zone of the student subjects and analyzed by gas chromatography/mass spectrometry, are twofold higher in Bangkok school children than in rural school children ($P < 0.01$). Measured levels of blood benzene and

TABLE 4. Ambient levels of benzene at various locations in Bangkok

Sampling locations	Number of sites	Levels of benzene (ppb)
At the roadside	7	33.71 ± 6.90
		28.32 (15.49–65.70)
500 m away from the roadside	3	12.39 ± 1.83
		11.21 (9.09–21.32)

Values are expressed as mean ± SE on the first line and median (min–max) on the second line.

urinary t,t-MA, biomarkers of benzene exposure, mirror differential exposure levels in city and rural school children (differences significant at $P < 0.01$; TABLE 6). As with PAH exposures, it is likely that benzene exposures are also occurring outside of school, and this is reflected in the significantly higher levels of urinary t,t-MA in the mornings before school commences ($P < 0.01$). Afternoon levels were also significantly higher than morning levels in Bangkok school children, indicating a significant exposure during the school day for these children ($P < 0.05$).

Genotoxic/Carcinogenic Heavy Metals

A different class of pollutants that may also be traffic related is the heavy metals, particularly arsenic (As), lead (Pb), nickel (Ni), and chromium (Cr). Blood levels in Bangkok school children were not significantly higher than in rural school children for any of the metals studied. With Pb being phased out of gasoline in Thailand since 1995, it was expected that this would not be a significant problem, and indeed blood Pb levels measured in these school children are considered to be normal (TABLE 7). Interestingly, levels in city school children were significantly lower than those in rural school children ($P < 0.001$), suggesting nontraffic-related sources as the main sources of exposure.

TABLE 5. Ambient levels of benzene at the roadside and in schools

School area	Levels of benzene (ppb)	
	Chonburi	Bangkok
Roadside	4.49 ± 0.59	17.75 ± 2.23***
	4.20 (2.20–7.40)	14.85 (6.60–41.80)
	$n = 10$	$n = 20$
Playground	2.77 ± 0.48	8.91 ± 1.09**
	2.20 (1.80–4.40)	9.05 (4.50–14.00)
	$n = 6$	$n = 8$
Classroom	2.60 ± 0.78	6.93 ± 0.54*
	2.20 (1.50–4.10)	6.50 (6.20–8.50)
	$n = 3$	$n = 4$

Values are expressed as mean ± SE on the first line and median (min–max) on the second line.
*,**,*** Significant differences from Chonburi at $P < 0.05, 0.01, 0.001$, respectively.

TABLE 6. Benzene exposure levels and biomarkers of exposure in school children

	School location	
	Chonburi	Bangkok
Individual exposures (ppb)	2.54 ± 0.23	5.50 ± 0.40**
	2.20 (1.20–5.40)	4.60 (2.40–12.30)
	$n = 30$	$n = 41$
Blood benzene (ppb)	46.23 ± 4.32	77.97 ± 11.67**
	47.24 (7.03–92.76)	65.63 (18.81–470.75)
	$n = 30$	$n = 41$
Urinary MA (mg/g creatinine)		
Morning	0.04 ± 0.04	0.07 ± 0.01**
	0.03 (0.01–0.11)	0.06 (0.02–0.36)
	$n = 28$	$n = 35$
Afternoon	0.06 ± 0.01	0.17 ± 0.03**,§
	0.05 (0.01–0.18)	0.13 (0.01–0.66)
	$n = 28$	$n = 35$

Values are expressed as mean ± SE on the first line and median (min–max) on the second line.
** Significant differences from Chonburi at $P < 0.01$.
§ Significant difference from the corresponding morning at $P < 0.05$.

Of the six heavy metals measured, Ni was higher than normal levels as specified by the United States Department of Health and Human Services' Agency for Toxic Substances and Disease Registry (ATSDR), while manganese (Mn) was marginally higher.

POTENTIAL HEALTH RISK FROM EXPOSURE TO GENOTOXIC SUBSTANCES

The potential health risk from exposure of children to genotoxic substances was assessed through the measurement of levels of PAH–DNA adducts using the P^{32}-postlabeling technique, DNA strand breaks through the Comet assay, and DNA repair capacity through the cytogenetic challenge assay.[30–32] The results from these assays provide information on possible early biological effects that may be indicative of health risks.

PAH–DNA adduct levels are significantly higher in Bangkok school children when compared to rural school children (unpublished data), reflecting the significantly higher exposure levels observed in Bangkok school children. DNA damage and repair capacity is also significantly different in city and rural school children (TABLE 8). DNA strand breaks are significantly higher, while DNA repair capacity, which is inversely correlated with the number of dicentric chromosomes or chromosomal breaks (deletions) per metaphase, is lower in Bangkok school children. Taken together, this would seem to indicate that Bangkok school children were at an elevated risk for development of diseases, such as cancer.

TABLE 7. Heavy metal levels in blood of school children

Group	Heavy metal (ppb)					
	Pb	As	Cd	Cr	Mn	Ni
Bangkok $n = 44$	62.86 ± 3.50 67.09*** (25.21–112.32)	6.70 ± 0.61 8.05 (2.64–25.00)	0.44 ± 0.05 0.44 (nd-1.29)	13.74 ± 1.01 25.23 (3.27–27.82)	17.12 ± 1.01 18.69 (9.74–26.80)	3.72 ± 0.77 2.70 (nd-12.07)
Chonburi $n = 25$	100.70 ± 6.35 92.88 (49.23–185.08)	8.30 ± 0.68 9.19 (3.81–16.03)	0.49 ± 0.03 0.37 (nd-0.84)	12.66 ± 0.85 24.90 (4.76–22.00)	16.49 ± 0.86 18.61 (8.06–28.69)	3.15 ± 0.52 2.28 (nd-9.68)

Values are expressed as mean ± SE, median, and range on the first, second, and third lines of each group, respectively.
*** Represents a significant difference from Chonburi school children at $P < 0.001$.

TABLE 8. DNA damage and repair capacity in school children

	School locations	
	Chonburi	Bangkok
DNA strand breaks		
Olive tail moment (μm)	0.13 ± 0.01	$0.22 \pm 0.01^{***}$
	0.11 (0.00–0.28)	0.23 (0.03–0.45)
	$n = 28$	$n = 41$
DNA repair capacity		
Dicentric/metaphase	0.19 ± 0.01	$0.25 \pm 0.02^{*}$
	0.19 (0.12–0.28)	0.24 (0.14–0.42)
	$n = 18$	$n = 21$
Deletion/metaphase	0.27 ± 0.01	0.31 ± 0.02
	0.27 (0.20–0.34)	0.28 (0.18–0.54)
	$n = 18$	$n = 21$

Values are expressed as mean ± SE on the first line and median (min–max) on the second line.
*,*** Significant differences from Chonburi at $P < 0.001$.

CONCLUSIONS

It is obvious that children living in capital cities of developing countries, such as Bangkok, are exposed to higher levels of genotoxic air pollutants, such as PAHs and benzene than children in more rural areas. The level of exposure to PAHs and benzene in these children is considered relatively low when compared to those through occupational exposures, or to those who spend a considerable amount of time at the roadside. Nevertheless, it can also be observed that Bangkok school children have higher levels of DNA adducts and DNA damage, as well as lower DNA repair capacity than children in rural schools. This would seem to indicate an increased risk for health effects, particularly since these exposures are still ongoing.

The WHO Air Quality Guidelines for Europe specify a unit risk of 9×10^{-5} per ng/m^3 of B[a]P.[33] This equates to one additional cancer case per 100,000 exposed individuals as a consequence of a lifetime exposure to 0.1 ng/m^3. Taking into account the B[a]P equivalents calculated in Bangkok school children of 1.83 ng/m^3, the lifetime cancer risk associated with this exposure level based on WHO Air Quality Guidelines is approximately 18 additional cancer cases per exposed population of 100,000. As for benzene, the lifetime leukemia risk has been estimated at 2 additional deaths per population of 100,000 at a benzene exposure level of 1 ppb.[34,35] By extrapolation, at the mean level of individual exposure to benzene in Bangkok school children of 5.5 ppb, the lifetime risk is calculated to be at about 10 additional cases in an exposed population of 100,000. The level of acceptable risk is generally considered to be 1×10^{-6}, or one additional cancer case in a population of 1 million, so the cancer risk to Bangkok school children from exposure to either of the compounds is above what is considered to be generally acceptable.

It should be kept in mind that there are no "safe" exposure levels for carcinogenic compounds, such as benzene and certain PAHs, for example benzo[a]pyrene. At the same time, it cannot be concluded that exposure to PAHs and benzene are solely responsible for the observed effects on DNA damage and DNA repair capacity. Concurrent exposure to other compounds, for example, 1,3-butadiene from traffic congestion, and chromium[36] cannot be ruled out, even though the health risk posed by chromium in Bangkok is minor due to the relatively low exposure level. In addition to gaining a better understanding of the risks involved with exposure of children to polluted urban air, monitoring exposure in children through measurement of potential exposure to contaminated air as well as through biomarkers of exposure and early effect will allow for the timely initiation of preventive or corrective measures such that the health risks from these exposures may be minimized.

REFERENCES

1. WORLD HEALTH ORGANIZATION. 2004. Report of the WHO/UNEP/ADB High-Level Meeting on Health and Environment in ASEAN and East Asian Countries, Manila, Philippines. December 2004. Report Series No.: RS/2004/GE/ZO(PHL).
2. WORLD HEALTH ORGANIZATION. 2002. The World Health Report 2002: Reducing Risks, Promoting Healthy Life. WHO, Geneva, Switzerland.
3. UNITED NATIONS. 2003. World Population Prospects. The 2004 Revision Database. New York: United Nations. Available: http://esa.un.org/unpp/index.asp?panel=1 [accessed 17 August 2005].
4. NATIONAL ACADEMY OF SCIENCES. 1993. Pesticides in the Diet of Infants and Children. National Academy Press, Washington, DC.
5. SUK, W.A., K. MURRAY & M.D. AVAKIAN. 2003. Environmental hazards to children's health in the modern world. Mut. Res. **544:** 235–242.
6. SUK, W.A. *et al.* 2003. Environmental threats to children's health in Southeast Asia and the Western Pacific. Environ. Health Persp. **111:** 1340–1347.
7. HEI INTERNATIONAL SCIENTIFIC OVERSIGHT COMMITTEE. 2004. Executive Summary. Health Effects of Outdoor Air Pollution in Developing Countries of Asia: a Literature Review. Special Report 15. Health Effects Institute, Boston, MA.
8. INTERNATIONAL AGENCY FOR RESEARCH IN CANCER. 1985. IARC Monograph on evaluation of the carcinogenic risk of chemicals to humans: polynuclear aromatic hydrocarbons, Part 4. IARC Monographs No. 35.
9. INTERNATIONAL AGENCY FOR RESEARCH ON CANCER. 1987. IARC Monographs on evaluation of the carcinogenic risk to humans. Overall evaluations of carcinogenicity: an updating of IARC monographs. Vol. 1–42.
10. NAVASUMRIT, P. *et al.* 2005. Environmental and occupational exposure to benzene in Thailand. Chem. Biol. Interact. **153/154:** 75–83.
11. RUCHIRAWAT, M. *et al.* 2005. Measurement of genotoxic air pollutant exposures in street vendors and school children in and near Bangkok. Toxicol. Appl. Pharmacol. **206:** 207–214.

12. RUCHIRAWAT, M. et al. 2002. Exposure to genotoxins present in ambient air in Bangkok, Thailand—particle associated polycyclic aromatic hydrocarbons and biomarkers. Sci. Total Environ. **287:** 121–132.
13. GRIMMER, G. et al. 1991. Excretion of hydroxy derivatives of PAH of the masses 178, 202, 228 and 252 in urine of coke and road workers. Int. J. Environ. Anal. Chem. **43:** 177–186.
14. JONGENEELEN, F.J. et al. 1987. Determination of hydroxylated metabolites of polycyclic aromatic hydrocarbons in urine. J. Chromatogr. **413:** 227–232.
15. ØVREBØ, S. et al. 1995. Biological monitoring of polycyclic aromatic hydrocarbon exposure in a highly polluted area of Poland. Environ. Health Perspect. **103:** 838–843.
16. PASTORELLI, R. et al. 1999. Seasonal effect on airborne pyrene, urinary 1-hydroxypyrene, and benzo[a]pyrene diol epoxide-hemoglobin adducts in the general population. Cancer Epi. Biomarkers Prev. **8:** 561–565.
17. KYRTOPOULOS, S.A. et al. 2001. Biomarkers of genotoxicity of urban air pollution. Overview and descriptive data from a molecular epidemiology study of populations exposed to moderate-to-low levels of polycyclic aromatic hydrocarbons: the AULIS project. Mutat. Res. **496:** 207–228.
18. VYSKOCIL, A. et al. 2000. Assessment of multipathway exposure of small children to PAH. Environ. Toxicol. Pharmacol. **8:** 111–118.
19. PANTHER, B.C. et al. 1999. A comparison of air particulate matter and associated polycyclic hydrocarbons in some tropical and temperate urban environments. Atmos. Environ. **33:** 4087–4099.
20. CARICCHIA, A.M. et al. 1999. Polycyclic aromatic hydrocarbons in the urban atmospheric particulate matter in the city of Naples (Italy). Atmos. Environ. **33:** 3731–3738.
21. MOTYKIEWICZ, G. et al. 1998. Molecular epidemiology study in women from upper Silesia, Poland. Toxicol. Lett. **96/97:** 195–202.
22. DUCOS, P. et al. 1990. Improvement in HPLC analysis of urinary trans,trans-muconic acid, a promising substitute for phenol in the assessment of benzene exposure. Int. Arch. Occup. Environ. Health **62:** 529–534.
23. BOOGAARD, P.J. & N.J. VAN SITTERT. 1995. Biological monitoring of exposure to benzene: a comparison between s-phenylmercapturic acid, trans,trans-muconic acid, and phenol. Occup. Environ. Med. **52:** 611–620.
24. BAUMBACH, G. et al. 1995. Air pollution in a large tropical city with a high traffic density: results of measurements in Lagos, Nigeria. Sci. Total Environ. **169:** 25–31.
25. AYI FANOU, L. et al. 2006. Survey of air pollution in Cotonou, Benin—air monitoring and biomarkers. Sci. Total Environ. **358:** 85–96.
26. ORTIZ, E. et al. 2002. Personal exposure to benzene, toluene and xylene in different microenvironment at the Mexico city metropolitan zone. Sci. Total Environ. **287:** 241–248.
27. MAITRE, A. et al. 2002. Exposure to carcinogenic air pollutants among policemen working close to traffic in an urban area. Scand. J. Work Environ. Health **28:** 402–410.
28. BONO, R. et al. 2003. Ambient air levels and occupational exposure to benzene, toluene, and xylenes in north-western Italy. J. Toxicol. Environ. Health **66:** 519–531.

29. SØRENSON, M. *et al.* 2003. Urban benzene exposure and oxidative DNA damage: influence of genetic polymorphisms in metabolism genes. Sci. Total Environ. **309:** 69–80.
30. AUTRUP, H. *et al.* 1999. Biomarkers for exposure to ambient air pollution—comparison of carcinogen-adduct levels with other exposure markers and markers of oxidative stress. Environ. Health Perspect. **107:** 233–238.
31. MØLLER, P. *et al.* 2000. The Comet assay as a rapid test in biomonitoring occupational exposure to DNA-damaging agents and effect of confounding factors. Cancer Epidemiol. Biomarkers Prev. **9:** 1005–1015.
32. AU, W. 2003. Mutagen sensitivity assays in population studies. Mutat. Res. **544:** 273–277.
33. WHO. 2000. Air Quality Guidance for Europe. Second edition. World Health Organization.WHO Regional Publications. European Series no. 91, Copenhagen.
34. CRUMP, K.S. 1994. Risk of benzene-induced leukemia: a sensitivity analysis of the Pliofilm cohort with additional follow-up and new exposure estimates. J. Toxicol. Environ. Health **42:** 219–242.
35. USEPA. 1988. Carcinogenic Effects of Benzene: An Update. US Environmental Protection Agency.Washington, DC.
36. HARRISON, R.M. *et al.* 2004. What is responsible for the carcinogenicity of PM2.5? Occup. Environ. Med. **61:** 799–805.

New Developments in Children's Environmental Health in Europe

GIORGIO TAMBURLINI

Institute of Child Health IRCCS Burlo Garofolo, 34137 Trieste, Italy

ABSTRACT: Important developments have taken place in Europe regarding children's environmental health (CEH) over the last few years. In 1999 the Third Ministerial Conference on Environment and Health identified CEH as a priority area and started a process of scientific review and policy development that culminated at the Fourth Ministerial Conference held in Budapest in June 2004 with the adoption of the Children's Environment and Health Action Plan for Europe (CEHAPE). The rationale of the CEHAPE is based on a thorough review of the scientific evidence on CEH and on a study that quantified for the first time the burden of disease related to the main environmental exposures of children and adolescents in Europe. The Action Plan suggests actions and policies to achieve the four main priority goals: clean air, safe water, chemical and physical agents, and injuries. Over the same period, the European Commission (EC) has strengthened its focus on environment and health issues, has supported research on CEH, and has developed a proposal for a new EU regulatory framework for chemicals that has clear implications for children and for the reproductive period. The proposed new system, called REACH (Registration, Evaluation, and Authorization of Chemicals), currently under examination by the European Parliament, aims at reducing risks to human health and improvement of environmental quality through the better and earlier identification of the properties of chemical substances. The EC also adopted policies and action plans that are very relevant to children, such as the EU European Environment and Health Strategy, referred to as the SCALE initiative (Science, Children, Awareness, Legislation, Evaluation), and the 2004–2010 Environment and Health Action Plan.

KEYWORDS: children's environmental health; European region; action plan

INTRODUCTION

Driven primarily by advances in research leading to a clearer understanding of the links between environmental exposures and child health, important

Address for correspondence: Giorgio Tamburlini, Direzione Scientifica, Istituto per l'Infanzia Burlo Garofolo, Via dell'Istria 65\1, 34137 Trieste, Italy. Voice: +39-0403785478, +39-0403785419; fax: +39-0403785210.
  e-mail: tamburli@burlo.trieste.it

developments have taken place in Europe with respect to children's environmental health (CEH) over the last decade. The process had its cornerstone in the Declaration on CEH signed at the Third Ministerial Conference on Environment and Health, held in London in 1999,[1] and culminated in the Children's Environment and Health Action Plan for Europe (CEHAPE) approved by the representatives of the 52 countries belonging to the WHO European Region at the Fourth Ministerial Conference on Environment and Health held in Budapest in June 2004.[2] Over the same period, the European Commission (EC) has strengthened its focus on environment and health issues by supporting research and by developing policies and action plans that are very relevant to children, such as REACH (Registration, Evaluation, and Authorization of Chemicals), the EU European Environment and Health Strategy, and the 2004–2010 Environment and Health Action Plan. This article provides an overview of these recent developments focusing particularly on the contribution of science to the political process and the increasing attention to CEH and the key contribution of science

The Environment and Health Process in Europe, initiated in 1989 and led by the European office of WHO, has significantly contributed to include environment and health in the political agenda through several mechanisms such as the multisectoral European Environment and Health Committee (EEHC)—a coalition of Member States, intergovernmental organizations, and civil society organizations working together—and five yearly ministerial conferences to agree on priorities and commitments. At the Third Ministerial Conference on Environment and Health held in London in 1999, the European ministers committed themselves to developing policies and actions to achieve safe environments in which children could reach the highest attainable level of health. The Declaration highlighted the urgency of taking action to protect children from environmental risk factors, identified priority areas for action, and started a process that led the EEHC to propose in 2001 "The Future of our Children" as the overall theme for the Fourth Ministerial Conference to be held in Budapest.

The developments in CEH in Europe have been strongly influenced by the evidence produced by the scientific community and voiced by the civil society, in accordance to the principles of international agreements and conventions, such as the International Convention on the Rights of the Child.[3] Research on CEH in Europe, which initially lagged well behind the United States, was stimulated by the pioneering work on the effects of lead in the 1970s[4,5] and got boosted by the impressive amount of valuable research carried out in the United States thereafter.[6] Research acivity rapidly developed over the last decade, with studies, that focused mainly on the effects of outdoor air pollution[7,8] and neurotoxicants[9,10] and contributed to draw the attention on the issue brought the issue of CEH to the attention of NGOs and of the broader public. A CEH-focused NGO, the International Research and Information Network on Children's Health, Environment and Safety (INCHES)[11] was established in 1999 and several other NGOs, including the European Public Health Alliance

TABLE 1. Estimates of deaths and DALYs attributable to selected environmental factors as a proportion of deaths and DALYs from all causes among children and adolescents in the European region by age group[14,15]

Environmental factor	% of deaths from all causes			% of DALYs from all causes		
	0–4	5–14	15–19	0–4	5–14	15–19
Outdoor air pollution	6.4%	—	—	—	—	—
Household solid fuel use	4.6%	—	—	3.1%	—	—
Water, sanitation, and hygiene	9.6%	0.8%	—	7.9%	1.0%	—
Lead	—	—	—	4.4%	—	—
Injuries	6.0%	41.2%	59.9%	7.3%	29.8%	27.1%
All	26.5%	42.1%	59.9%	22.7%	30.8%	27.1%
Absolute number of deaths and DALYs from all causes	216,194	37,784	77,952	10,997,728	5,604,270	8,564,440

NOTE: The results presented in the table refer to the analysis based on the estimated exposure levels and counterfactual scenarios (details can be found in the original paper) and refer to the year 2001.
DALY: disability adjusted life years.

(EPHA)[12] started focusing on CEH issues. The monograph jointly published in 2002 by WHO and the European Environment Agency (EEA), by offering a thorough review of the available scientific evidence,[13] contributed to a wider understanding of the issues related to children's health and environment and drew the attention of policy makers at the level of the EC as well as at country level. Furthermore, in preparation of the Budapest Conference, a study was carried with support by WHO to assess the environmental burden of disease among children and adolescent in the European Region of WHO, which includes 52 Member States, encompassing all 25 EU countries, the countries of the Community of Independent States including the Central Asian Republics, countries of south-east Europe, and Turkey. The study estimated, for the first time, the burden of disease in children and adolescents in Europe related to injuries and to exposure to some of the main environmental hazards, such as lack of clean water and sanitation, indoor and outdoor air pollution and lead, with a subregional breakdown, thus providing crucial information for the identification of priorities (TABLE 1).[14,15]

THE DEVELOPMENT OF THE CEHAPE

The idea of a children's environment and health action plan for Europe was born at the Third Ministerial Conference on Environment and Health in 1999. The need for such a Plan moves from the recognition that "in many countries of the WHO European Region children's health has significantly improved in recent years, but not all children enjoy these improved conditions and environmental risk factors are unequally distributed within and between countries, due to different economic standards and public policies," that "all developing organisms, especially during embryonic and fetal periods and the early years of life, are often particularly susceptible and may be more exposed than adults to many environmental risk factors" that "despite differences in sensitivity and exposure to many toxic agents, safety standards for chemicals and maximum doses of exposure are still based mostly on criteria used for adults" and therefore that "because of children's special situation, special attention needs to be paid to them to protect their health today, to prevent, for example, respiratory infections or injuries, but also to avoid adverse health effects, such as cancer or cardiovascular disorders, in later life, and to avoid intergenerational effects in the future, such as birth defects."[16] With all these considerations in mind, it became obvious that regionwide political commitment to action was needed. Preparation for the Conference was coordinated by WHO European Office and by the EEHC. It involved eight large preconference meetings with very active involvement of all 52 Member States of the WHO European Region, as well as NGOs, representatives of industry and Trade Unions. The overall goal of this planning process was to define the rationale, structure, and objectives of CEHAPE. The CEHAPE was then approved at the Budapest Conference at the highest political level, thus affirming the commitments of all Member States to the mitigation of environmental threats to children's health and setting the scene for action in each nation. CEHAPE is aimed at ensuring the reduction and, where possible, elimination of children's exposure to environmental risk factors and focuses on four regional priority goals: ensure safe water and adequate sanitation; ensure protection from injuries and adequate physical activity; ensure clean outdoor and indoor air; and aim at chemical-free environments.

Policy tools to support this process were developed for the Budapest Conference and include a book providing evidence on issues in CEH and guidance and tools showing countries how to transform the overall CEHAPE framework into national action plans suited to each country's circumstances, priorities, and resources.[17] A collection of successful experiences implemented in Member States for protecting children from environmental risks[18] is also included. Part I of the book provides the scientific evidence on children's susceptibility to environmental risk factors, with an overview of environmental risk factors and their effects on health. Part II is the core of the publication and includes tables of child-specific actions, that is, concrete ways in which a Member State can work to reduce children's exposure to environmental risk factors, selecting the combination of actions that best suits their own national priorities. Part III focuses on the tools required to ensure implementation of national CEHAPs, such as a set of child-specific indicators, to ensure that Member States will be able to monitor the implementation of the national CEHAPs, and gives arguments for dealing with uncertainty by applying a precautionary approach.

ACTION BY THE EC: THE DEVELOPMENT OF REACH

The EC has substantially strengthened its policy focus on CEH. Not only has the EC supported its Member States' plans and activities for implementation of environmental health plans and for carrying out research under the 5th and 6th Research Programme Frameworks, but has developed comprehensive and far-reaching policies such as REACH.[19] REACH was proposed by the EC in October 2003 and presents a new trans-European regulatory framework for chemicals, replacing over 40 existing directives and regulations (TABLE 2). REACH requires that safety and toxicity information be made publicly available on all chemicals produced or imported in Europe in volumes above 1 ton/year per manufacturer/importer. Under REACH, the chemical industry must be able to demonstrate that the chemical can be used safely, and how. All actors in the supply chain will also be obliged to ensure the safety of the chemical substances they handle. Innovation of safer substances will

TABLE 2. EU regulations on chemicals. Comparison between the present system and the new proposed system (REACH)[19]

Present system	REACH
There are gaps in our knowledge about many of the chemicals on the European market.	REACH will provide safety information about chemicals produced or imported in volumes higher than 1 ton/year per manufacturer/importer.
The "burden of proof" is on the authorities: they need to prove that the use of a chemical substance is unsafe before they may impose restrictions.	The "burden of proof" will be on industry. It has to be able to demonstrate that the chemical can be used safely, and how. All actors in the supply chain will be obliged to ensure the safety of the chemical substances they handle.
Notification requirements for "new substances" start at a production level of 10 kg. Already at this level, one animal test is needed. At 1 ton, a series of tests including other animal tests have to be undertaken.	Registration will be required when production/import reaches 1 ton. As far as possible, animal testing will be minimized.
It is relatively costly to introduce a new substance on the market. This encourages the continued use of "existing," untested chemicals and inhibits innovation.	Innovation of safer substances will be encouraged under REACH through: more exemptions for research and development; lower registration costs for new substances; and the need to consider substitute substances for decisions on authorization and restrictions.
Public authorities are obliged to perform comprehensive risk assessments that are slow and cumbersome.	Industry will be responsible for assessing the safety of identified uses, prior to production and marketing. Authorities will be able to focus on issues of serious concern.

be encouraged by providing exemptions for research and development; lower registration costs for new substances; and the need to consider substitute substances for decisions on authorization and restrictions. All substances of very high concern will be subject to authorization that will be granted only if the producer or importer can show that risks from the use in question can be adequately controlled, or that the socio-economic benefits of the use of the substance outweigh the risks. In the latter case, the authorization decision will take into account substitution plans showing, for example, that the industry is researching substitutes. Third parties will also be able to provide information to the EEA about possible substitute substances or technologies. Substances that will be subject to authorization include those that are carcinogenic, mutagenic, or toxic to reproduction; persistent, bioaccumulative and toxic, as well as substances identified as having serious and irreversible effects to humans and the environment, that is, certain endocrine-disrupting substances. The latter will be identified on a case-by-case basis and subject to authorization.

It is envisaged that the European chemicals industry will benefit from a single EU regulatory system, decision making with set deadlines, and a high-quality image for their products. Downstream users of chemicals will get relevant information on the safe use of each chemical substance they buy. They will have closer contacts with their suppliers, will be able to ensure better protection of their workers, and their products will be safer for consumers and the environment. If fully adopted, REACH will hasten the end of the vast ongoing toxicologic experiment in which chemicals are being tested on children worldwide instead of in the laboratory. The benefits of the REACH system are twofold: risks to human health will be reduced and environmental quality will be improved through the better and earlier identification of the properties of chemical substances. The benefits will come gradually as more and more substances are phased into the new system. A feasibility study examining potential impacts of REACH was produced in June 2003.[20] The study examined four chemicals and found that, under REACH, the toxic properties of chemicals will be assessed more quickly, and hazards will be more likely identified before significant damage has occurred. The anticipated benefits to environment and human health are expected to be significant. The Commission's Impact Assessment developed an illustrative scenario which put the health benefits in the order of magnitude of 50 billion over a 30-year period.[19] The expected environmental benefits have not been expressed in monetary terms.

THE EUROPEAN ENVIRONMENT AND HEALTH STRATEGY AND ACTION PLAN

The EC launched an Environment and Health Strategy in June 2003, referred to as the SCALE initiative (Science, Children, Awareness, Legislation, Evaluation), proposing an integrated approach involving closer cooperation between the health, environment, and research areas.[20] Its added value is the development of a system integrating information on the state of the environment, the ecosystem, and human health. The Strategy's ultimate goal is to develop an environment and health "cause-effect framework" that will provide the necessary information for the development of EC policy, dealing with sources and impact pathways of health stressors. The overall environmental impact on human health will be made more efficient by taking into account effects such as cocktail effects, combined exposure, and cumulative effects. The Strategy set out a long-term approach that will be implemented in cycles, gradually expanding its coverage as our knowledge base improves.

The European Environment and Health Strategy and Action Plan represent the Commission's contribution to the Fourth Ministerial Conference on Environment and Health. It has been developed to be consistent with the Budapest Ministerial Declaration and the CEHAPE. In addition to improving well-being, the aim of the Action Plan is to maximize the potential economic benefits, because spending on remedial actions and lost productivity often outweighs costs of prevention. The Plan explicitly builds on the evidence provided by recent studies including estimates on premature death due to air pollution,[8,21] costs of chronic obstructive pulmonary disease,[22] and on the environmental burden of disease study.[15,16] The Action Plan for the period 2004–2010 is designed to give the EU the scientifically grounded information needed to help all 25 EU Member States reduce the adverse health impacts of certain environmental factors and is designed to fit with existing actions at regional, national, European, and international level, and particularly the pan-European Environment and Health process and the CEHAPE. The Action Plan has been developed in close cooperation with experts from the Member States, and representatives of the main stakeholders. It has three main themes: improving the information chain to understand the links between sources of pollution and health effects; filling the knowledge gap by strengthening research, and addressing the emerging issues on environment and health; and reviewing policies and improving communication (TABLE 3).

The first step is to assess the contribution that environmental factors make to health problems and completing the information chain from the pollution sources through different pathways to the human health effects. Improvements in environment and health monitoring are required to ensure that both policies are properly coordinated. In addition, the EU Framework Programme for Research will reinforce the scientific research efforts to analyze and improve our knowledge on the causal links between environmental factors and human health. Only when sufficiently clear evidence is available, also in line with the precautionary principle,[23] appropriate policy options can be developed in order to review and if necessary revise existing policy responses and develop new ones. Biomonitoring is considered key to elucidating children's exposure in Europe to a wide range of environmental factors, comparing spatial and temporal trends in exposure to these factors in different areas, identifying

TABLE 3. European Environment and Health Strategy and Action Plan 2004–2010[24]

IMPROVE THE INFORMATION CHAIN by developing integrated environment and health information to understand the links between sources of pollutants and health effects.	Action 1: Develop environmental health indicators Action. Action 2: Develop integrated monitoring of the environment, including food, to allow the determination of relevant human exposure. Action 3: Develop a coherent approach to biomonitoring in Europe. Action 4: Enhance coordination and joint activities on environment and health.
FILL THE KNOWLEDGE GAP by strengthening research on environment and health and identifying emerging issues.	Action 5: Integrate and strengthen European environment and health research. Action 6: Target research on diseases, disorders, and exposures. Action 7: Develop methodological systems to analyze interactions between environment and health. Action 8: Ensure that potential hazards on environment and health are identified and addressed.
RESPONSE: REVIEW POLICIES AND IMPROVE COMMUNICATION by developing Awareness Raising, Risk Communication, Training & Education, and by reviewing and adjusting risk reduction policy to give citizens the information they need to make better health choices, and to make sure that professionals in each field are alert to environment and health interactions.	Action 9: Develop public health activities and networking on environmental health determinants through the public health program. Action 10: Promote training of professionals and improve organizational capacity in environment and health. Action 11: Coordinate ongoing risk reduction measures and focus on the priority diseases. Action 12: Improve indoor air quality. Action 13: Follow developments regarding electromagnetic fields

disproportionately exposed populations, and clarifying relations between low exposures to environmental factors and health effects and also health effects that occur at a low incidence, by increasing the power of studies. A preliminary overview of existing environment and health biomonitoring[25] showed that, although efforts are not equally distributed among European countries due to differences in resources and public interest, a substantial amount of biomarker data on children are collected and that resources are increasingly devoted to these efforts. Many of the biomonitoring activities integrate data on environmental factors with health data and address both children and their parents through markers of exposure, effect, and susceptibility. Similar aspects are addressed in nearly all EU countries: for example, exposures to heavy metals, PCBs (polychlorinated biphenyls), and dioxins. End points such as asthma, allergy and neuro-developmental disorders, as well as exposures to genotoxic agents are covered, and multidisciplinary is typical of many activities. In many cases establishment of biobanks (archives of biological materials/samples) is part of the activity allowing for later follow-up. The establishment of periodically updated biomonitoring and reporting systems may provide authorities with a more comprehensive view on the actual exposure at the population level and guide them in the development of regulatory strategies for disease prevention and in the evaluation of the impact of environmental measures on children's health. Working groups have been established, that, in addition to developing a general methodology, made many substance-specific recommendations, to be taken into account in the Commission initiatives targeted on these substances, for example, the Dioxin & PCB Strategy (COM[2001] 593), the Endocrine Disrupter Strategy (COM[1999] 706), and the forthcoming Mercury Strategy.

During this initial period the Action Plan gives priority on gaining a better understanding of the links between environmental factors and respiratory diseases, neuro-developmental disorders, cancer, and endocrine-disrupting effects. For these multicausal diseases and conditions, the Action Plan will set up targeted research actions to improve and refine knowledge of the relevant causal links, and at the same time, health monitoring will be improved to obtain a better picture of disease occurrence across the Community. The other key information aspect is to monitor exposure through the environment, including food, to the factors most linked to the occurrence of these diseases. In order to develop a coherent framework for integrated exposure monitoring, three pilot projects were carried out on substances for which data collection and monitoring is already in place (dioxins, PCBs, heavy metals, and endocrine disrupters). The concerns of children are integrated throughout the Action Plan. A number of major child health issues will be covered in the monitoring, as will exposure to the environmental stressors to which children are particularly sensitive. The proposals in the Action Plan on indoor air pollution are a case in point, as the scientific evidence shows that the health impacts of, for instance, Environmental Tobacco Smoke (ETS) are particularly evident for children.

CONCLUSIONS: FUTURE CHALLENGES AND THE ROLE OF SCIENCE

Now that the main principles, policies, and plans have been developed, the challenges remain in their full implementation. Now that the main principles, policies, and plans have been developed, the challenges remain in their full implementation. Primarily, in the capacity of linking together the various strategic components and in increasing the collaboration among countries, particularly between the EU and the remaining countries of the European region, and among research institutions.

The main challenges can be easily identified along three main directions: expanding the knowledge basis, finalizing the regulatory work, and developing and implementing action plans at country level. So far, scientists and epidemiologists have focused much more on causal links (association studies) than on the assessment of the impact of policies and interventions. Strengthening policy and intervention research and getting started with Europe-wide coordination of biomonitoring activities are both necessary to expand the knowledge basis. Integration of local, regional, and national initiatives in a wider European perspective needs, above all, sound and harmonized study design, reliable tools for sampling and analysis with consistent/comparable protocols, common to each participating country and/or geographical area within a country. Coordination of biomonitoring activities through Europe may allow for a better integration of information by bringing together available knowledge and actively promoting exchange of experiences between teams and countries and enable a more effective use of resources by shared development of tools and strategies. Research projects may be grafted on such a surveillance framework, allowing reduction in costs by using already existing infrastructure. Long-term cohort studies linking biological information with environmental exposures and lifestyles, such as the U.S. Children's Study,[26] are an ideal setting for providing new insights both on causation links and impact of environmental policies, allowing for comparisons in exposures and outcomes and nested intervention studies. They do exist in many European countries and are being started in many others. A Europe-wide project should harmonize as much as possible current efforts in this direction. REACH is an innovative and ambitious proposal that needs now to be adopted and implemented. The proposal is currently under examination by the European Parliament. The challenge resides in finalizing a regulatory framework with a right balance between different needs, such as those of the industry and the citizens particularly in the new Member States, acknowledging that these needs may be easier to be conciliated in a long-term perspective. It is expected that the final decision on REACH will be reached by the European Parliament and Council in autumn 2006. The Commission expects entry into force of the Regulation for spring 2007. Thereafter it will take about a year for the REACH Agency to be operational. Accordingly the operational requirements of REACH are expected to start from 2008 onward.[19]

The Health and Environment ministries in the European Member States should honor the commitments taken at Budapest by supporting the implementation of CEHAPE. This work is being supported by the WHO European Office through training of health care providers and guidance on advocacy, information, and communication. The EU should continue to invest in this area, despite recent cuts in the budget, and advocacy work will have to be strengthened at country level to maintain the momentum. NGOs as well as leading scientists will have a key role to stimulate, support, and monitor the CEHAPE policy development process. It is crucial that the EC in the implementation of the Budapest conclusions will continue to cooperate actively with WHO on all aspects of the environment and health interaction.

Science played a key role in these important developments in CEH. Scientists now have an increased responsibility in assisting authorities at supranational and national level in meeting all these challenges and bringing further the process, in the best interest of today's children as well as of next generations and in the perspective of a sustainable and more equitable development.

ACKNOWLEDGMENTS

I would like to thank the Special Programme on Environment and Health of the WHO European Office, and particularly its Director Roberto Bertollini, for giving me the chance of getting involved in these exciting developments over the last 6 years and again Roberto Bertollini, Lucianne Licari, and Leda Nemer for sharing with me the most important steps of the process.

REFERENCES

1. THIRD MINISTERIAL CONFERENCE ON ENVIRONMENT AND HEALTH, LONDON, June 1999. Available at http://www.euro.who.int/Document/E69046.pdf. Accessed on February 6th, 2006.
2. FOURTH MINISTERIAL CONFERENCE ON ENVIRONMENT AND HEALTH HELD IN BUDAPEST, June 2004. Available at http://www.euro.who.int/budapest2004. Accessed on February 6th, 2006.
3. CONVENTION ON THE RIGHTS OF THE CHILD. Available at http://www.unicef.org/crc/. Accessed on February 6th, 2006.
4. LANDRIGAN, P.J. et al. 1975. Neuropsychological dysfunction in children with chronic low-level lead absorption. Lancet **1:** 708–712.
5. NEEDLEMAN, H.L. et al. 1979. Deficits in psychologic and classroom performance of children with elevated dentine lead levels. N. Engl. J. Med. **300:** 689–695.
6. LANDRIGAN, P. & G. TAMBURLINI. 2005. Children's health and the environment: a transatlantic dialogue. Environ. Health. Perspect. **113:** A646–A647.
7. VAN DER ZEE, S. et al. 1999. Acute effects of urban air pollution on respiratory health of children with and without chronic respiratory symptoms. Occup. Environ. Med. **56:** 802–812.
8. KUNZLI, N. et al. 2000. Public-health impact of outdoor and traffic-related air pollution: a European assessment. Lancet **356:** 795–801.
9. GRANDJEAN, P. et al. 1999. Methylmercury exposure biomarkers as indicators of neurotoxicity in 7-year-old children. Am. J. Epidemiol. **150:** 301–305.
10. WALKOWIAK, J. et al. 2001. Environmental exposure to polychlorinated biphenyls and quality of the home environment: effects on psychodevelopment in early childhood. Lancet **358:** 1602–1607.
11. INCHES (INFORMATION NETWORK ON CHILDREN'S HEALTH, ENVIRONMENT AND SAFETY). Available at http://www.inchesnetwork.net/. Accessed on February 6th, 2006.

12. EPHA (European Public Health Alliance). Available at http://www.epha.org/. Accessed on February 6th, 2006.
13. Tamburlini, G. et al. 2002. Children's health and environment: a review of evidence. Environmental issue report no. 29. WHO European office and European Environment Agency, Copenhagen, 2002.
14. Valent, F. et al. 2002. Burden of disease and injuries attributable to selected environmental factors among Europe's children and adolescents. WHO. Geneva.
15. Valent, F. et al. 2004. Burden of disease attributable to selected environmental factors and injury among children and adolescents in Europe. Lancet **363:** 2032–2039.
16. CEHAPE (Children's Environment and Health Action Plan for Europe). Available at http://www.euro.who.int/childhealthenv. Accessed on February 6th, 2006.
17. Licari, L. et al. 2005. Children's health and environment: developing national action plans. WHO European Office, Copenhagen. Available at http://www.euro.who.int/eprise/main/WHO/InformationSources/Publications/Catalogue/20050812_1. Accessed February 6th, 2006.
18. Nemer, L., K. von Hoff, Eds. Children's health and environment. Case studies summary book. WHO European Office, Copenhagen, 2004. Available at http://www.euro.who.int/childhealthenv/Policy/20040921_1. Accessed February 6th, 2006.
19. REACH (Registration, Evaluation, and Authorisation of Chemicals). Available at http://www.europa.eu.int/comm/environment/chemicals/reach.htm. Accessed February 6th, 2006.
20. The Impact of the New Chemicals Policy on Health and the Environment—Final Report, June 2003. Available at http://europa.eu.int/comm/environment/chemicals/pdf/envhlthimpact.pdf. Accessed on February 6th, 2006.
21. Nerrièr, E. & C. Boudet. 2004. Impact sanitaire de la pollution atmospherique urbaine. French Agency for Environmental Health Safety. Available at http://www.afsse.fr/documents/Rapport_1.pdf. Accessed on February 6th, 2006.
22. Loddenkemper, R., G.J. Gibson & Y. Sibille. 2003. European Lung White Book. European Respiratory Society (ERS) and the European Lung Foundation (ELF), Bruxelles.
23. The Precautionary Principle: protecting public health, the environment and the future of our children. Background document. WHO European Office, Copenhagen, 2004. Available at http://www.euro.who.int/InformationSources/Publications/Catalogue/20041119_1. Accessed on February 6th, 2006.
24. The European Environment and Health Strategy and Action Plan. 2004–2010. Available at http://europa.eu.int/comm/environment/health/pdf/com2004416.pdf. Accessed on February 6th, 2006.
25. Draft Baseline Report on "Biomonitoring of Children" in the framework of the European Environment and Health Strategy (COM[2003]338 final). Available at http://www.env-health.org/IMG/doc/BR_Biomonitoring_ES. Accessed on February 6th, 2006.
26. The National Children's Study. Available at http://nationalchildrensstudy.gov/. Accessed on February 6th, 2006.

Toxicogenomics—A New Systems Toxicology Approach to Understanding of Gene–Environment Interactions

KENNETH OLDEN

National Institutes of Health, U.S. Department of Health and Human Services, Research Triangle Park, North Carolina 27709, USA

> ABSTRACT: Toxicogenomics is a new interdisciplinary area of research being developed to monitor the expression of multiple genes, proteins, and metabolites simultaneously. It combines new technologies in genomics, proteomics, and metabolomics with traditional tools of pathology and toxicology to study biological response to drugs and other environmental xenobiotics. The biological response to environmental exposure is so complex and involves so many interactive factors that the use of a systems biology analytical approach is required. In my opinion, the development of the field of toxicogenomics will provide powerful and relatively inexpensive tools to identify biomarkers and to relate exposure and biological events during disease progression.
>
> KEYWORDS: gene–gene interactions; gene–environment interactions; environmental response machinery; susceptibility; genomics; proteomics; metabolomics; toxicogenomics; biomarkers

The scale and complexity of biomedical research has changed dramatically over the past 25–30 years. In large measure, this has been driven by knowledge and technologies developed from investments in the various genome projects. Now that the gene discovery phase of the human genome project is complete[1,2] and so few genes responsible for the major diseases were discovered, we realize that the complex human phenotype (e.g., chronic diseases) cannot be explained by the action of single genes and their multiple protein products but rather by gene–gene and gene–environment interactions. Such studies are consistent with several large twin cohort studies that show that only a small fraction of several major diseases can be attributed to genetics alone.[3–6] This has led to a paradigm shift with respect to the etiology of complex diseases and the potential role of environmental xenobiotics; and the myth that diseases are caused by

---

Address for correspondence: Kenneth Olden, Ph.D., Sc.D., Cell Adhesion and Metastasis Section, Laboratory of Molecular Carcinogenesis, National Institute of Environmental Health Sciences, National Institutes of Health, U.S. Department of Health and Human Services, Research Triangle Park, NC 27709. Voice: 919-541-0367; fax: 919-541-0039.
e-mail: olden@niehs.nih.gov

"bad genes" has been "blown away" with an avalanche of experimental data, including haplotype mapping studies.[7,8]

The practice of linking genes, the environment, and behavior in studies of health and disease is relatively recent. So, the major challenge facing biomedical researchers is to determine how genes and environmental agents interact to influence health and disease. The lack of knowledge that we have about the earliest events in disease development is largely due to the multifactorial nature of disease risk. This information gap is largely the consequence of the lack of appreciation of the fact that most diseases arise from the complex interactions between genes and the environment as a function of the age or stage of development of the individual. The relationship between genes and the environment is best captured by the quotation: "genetics loads the gun, but the environment pulls the trigger." A loaded gun by itself causes no harm. It is only when the trigger is pulled that the potential harm is released or initiated. Similarly, one can inherit a predisposition for a devastating disease, yet never develop the disease unless exposed to the environmental trigger(s). That is, most chronic diseases result from an unfavorable combination of genetic variations and environmental exposures. Whether an environmental exposure causes illness or not is dependent on the efficiency of the so-called "environmental response machinery" (i.e., the complex of metabolic pathways that can modulate response to environmental perturbations) that one has inherited. Thus, elucidating the causes of most chronic diseases will require an understanding of both the genetic and environmental contribution to their etiology. Unfortunately, the exploration of the relationship between genes and the environment has been hampered in the past by the limited knowledge of the human genome, and by the inclination of scientists to study disease development using the exposure to a single environmental agent. Rarely in the past were interactions between multiple genes or between genes and environmental agents considered in studies on the causes of human illness.

Traditionally, toxicologists and environmental health scientists have relied on pathology, biochemistry, genetics, and cell biology approaches to study one or a few genes or proteins in a single pathway. These approaches are too costly, time-consuming, and woefully inadequate for analyzing more than a few events simultaneously. Furthermore, they typically do not address issues related to susceptibility, low-dose exposures, and possible interactions between drugs and environmental xenobiotics due to co-exposure. Over the years, there has been a tendency to modify existing test systems to meet ever-increasing requirements because investigators have not been very receptive of innovation in toxicity testing. These practices have increased the burden on both government and industry in terms of costs, animal use, and time required for safety assessment.

Historically, the effort to understand the interplay between the environment and human health has been considered primarily the domain of toxicology. However, the field of toxicology has recently begun the process of reinventing itself in light of the rapid technological and conceptual change in molecular biology and genomics to become an interdisciplinary science. The grand goal of toxicology in the postgenome era is to characterize the entire set of genes and proteins that are affected when humans are exposed to drugs or environmental xenobiotics. What has long been needed to achieve this objective is a systems biology approach for monitoring dynamic changes in activity and quantity of molecular constituents of cells and tissues in animals with known genetic backgrounds. This is especially true for the assessment of drugs and environmental xenobiotics that exert their effects through multiple mechanisms, depending on dose, timing, and duration of exposure, and cell or tissue type. Using "omic" technologies (i.e., genomics, proteomics, metabolomics, and bioinformatics), environmental health scientists can conduct large-scale studies of the effects of toxicants on gene expression at the mRNA and protein levels, while simultaneously monitoring metabolite profiles in body fluids and tissues to gain insight into the activity state of all relevant genes and gene products. By integrating information obtained from these various experimental approaches, one can develop biomarkers of toxicity to predict specific adverse events long before they can be detected using current diagnostic approaches. Furthermore, toxicogenomics data generated in multiple species will enhance the understanding of complex interactions between genes, the environment and human health, and can improve our ability to use laboratory animals to predict effects of chemicals in people.

Once biological indicators are discovered that represent responses to specific environmental stressors, public health, or clinical intervention strategies can be designed. Genes and their protein products can "tell stories" that existing diagnostic and screening tests cannot.

The talks presented in this session will highlight the state of this area of research from the vantage point of advancement in technologies and development of useful databases.

REFERENCES

1. VENTER, J.C., M.D. ADAMS, E.W. MYERS, et al. 2001. The sequence of the human genome. Science **291**: 1304–1351.
2. LANDER, E.S., L.M. LINTON, B. BIRREN, et al. 2001. Initial sequencing and analysis of the human genome. Nature **409**: 860–921.
3. POWELL, J.J., J. VANDE MATER & M.E. GERSCHWIN. 1999. Evidence for the role of environmental agents in the initiation and progression of autoimmune conditions. Environ. Health Perspect. **107**: 667–672.
4. VERKASALA, P.K., J. KOPRIA, M. KASKENUVUA, et al. 1999. Genetic predisposition, environmental and cancer incidence: National Twin Study in Finland. Int. J. Cancer **83**: 743–749.
5. TANNER, C.M., R. OHTMAN, S.M. GOLDMAN, et al. 1999. Parkinson's disease in twins: an etiologic study. JAMA **281**: 341–346.
6. LICHTENSTEIN, P., V.H. NIELS, P.K. VERKASALA, et al. 2000. Environmental and heritable factors in the causation of cancer: analysis of cohorts of twins from Sweden. N. Eng. J. Med. **343**: 78–85.

7. HEIN, D.W.. 2002. Molecular genetics and function of NAT1 and NAT2: role in aromatic amine metabolism and carcinogenesis. Mutat. Res. **506-507:** 65–77.
8. OLDEN, K. 2007. Gene-gene and gene-environment interactions.: In Handbook on Genomic Medicine. H. Willard & G. Ginsburg, Eds. Elsevier Inc. New York, New York. In press.

Toxicoproteomics in Liver Injury and Inflammation

B. ALEX MERRICK

National Institute of Environmental Health Sciences, Research Triangle Park, Durham, North Carolina 27709, USA

ABSTRACT: Toxicoproteomics, in applying proteomics to toxicology, seeks to identify critical proteins and pathways in biological systems responding to adverse chemical exposures and environmental stressors using global protein expression technologies. Toxicoproteomics is being exploited for the discovery of new biomarkers and toxicity signatures in target organs, such as liver, in major biological processes, such as inflammation, in mapping serum, plasma, and other biofluid proteomes, and in parallel proteomic and transcriptomic studies. The new field of toxicoproteomics is uniquely positioned toward discovery of new biomarkers and signatures of tissue injury and a better understanding of protein expression responses during toxicity and environmental disease.

KEYWORDS: proteomics; toxicoproteomics; biomarkers; tissue injury; liver; environmental toxicity

INTRODUCTION

A broad definition of toxicogenomics (FIG. 1) includes the disciplines of transcriptomics, proteomics, and metabonomics.[1] The technologies used to measure these three strata of molecular expression each convey different types of information. The number of genes in cells and tissues expressed at any one time can be most comprehensively measured by whole genome queries using DNA microarrays. Estimates of the human genome are about 25,000 genes.[2] The idea that activation of a single gene transcribing one mRNA (transcript) and giving rise to one translation product (protein) is clearly an oversimplified model when considering the complexity of mammalian systems. Toxicogenomic researchers are interested in changes in those portions of the genome that respond to toxicant exposure by gene activation of proteins (protein profiling) or mRNAs (transcript profiling) in affected tissues at critical times in the toxicity process. A critical component of toxicogenomics is the relatively

---

Address for correspondence: B. Alex Merrick, National Institutes of Environmental Health Sciences, 111 TW Alexander Dr, P.O. Box 12233, Research Triangle Park, Durham, NC 27709. Voice: 919-541-1531; fax: 919-541-4704.
  e-mail: merrick@niehs.nih.gov

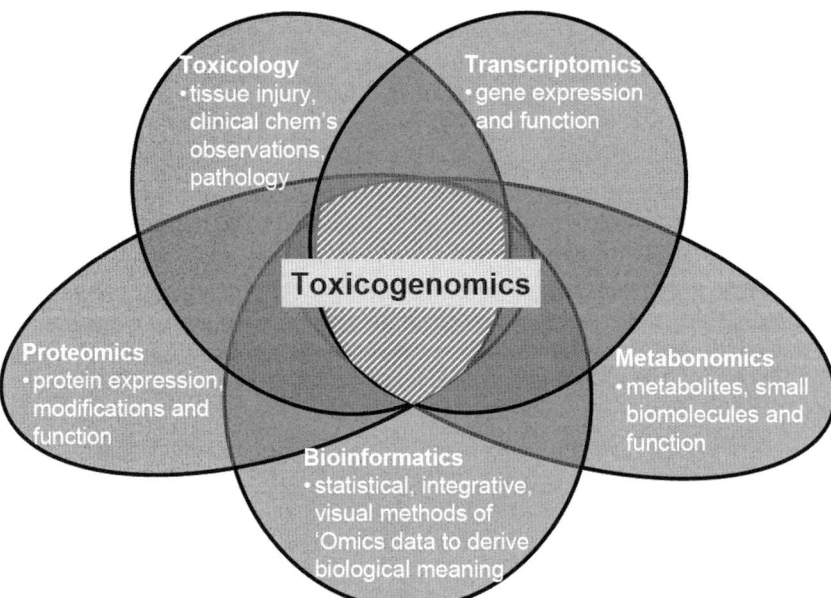

FIGURE 1. Toxicogenomics is the study of a response of a genome to environmental stressors and toxicants. Changes in gene expression are broadly defined as alterations in transcripts, proteins, or metabolites that make up the respective toxicogenomic subdisciplines of transcriptomics, proteomics, and metabonomics. Application of these omics technologies will enable linkage of traditional toxicology observations and pathology with specific genes and proteins (biomarkers) or groupings of gene expression products (signatures).

new discipline of bioinformatics that uses advanced computing techniques and biostatistics to manage, analyze, and integrate large, complex data sets from gene expression experiments.

Toxicoproteomics is a relatively new discipline that applies global protein measurement technologies (proteomics) to toxicology and seeks to discover critical proteins and pathways responding to adverse chemical exposures and environmental stressors. A major goal for toxicoproteomics is to translate identified protein changes into new biomarkers and signatures of chemical toxicity that can provide much greater definition than current indicators. New biomarkers will be useful for classifying toxicants in experimental animals for health risk assessment and eventually for use in biomonitoring the human population for environmental contaminants. "Biomarkers" in biochemical and molecular toxicology are often thought of as singular measures of a protein, enzyme activity, gene expression product (i.e., PCR product), or a small molecule (i.e., metabolite) that are associated with health, disease, and toxicity. "Toxicity signatures" represent more complex data outputs typical of omics data. In the context of gene expression studies, a toxicity signature is a critical set of up- and downregulated transcripts or proteins in a biological sample that predicts an adverse outcome.[3] Desirable features of biomarkers and signatures are their (*a*) association with specific tissue injury to mechanistically similar injurious agents and environmental stressors; (*b*) reflection of adverse effects in a dose- and time-related manner; and (*c*) incorporation of the genetic diversity of surrogate organisms that serve as models for human toxic response and disease.

Use of bioinformatics in toxicoproteomics is crucial in accomplishing these goals. Large amounts of data generated from toxicoproteomics experiments involving protein occurrence, abundance, identity, sequence, structure, properties, and interactions need to be stored for analysis. Currently, common standards have not yet been established for all forms of proteomics data. Researchers envision that open access to proteomic databases will spur continued development of robust data analysis algorithms, statistical treatment, visualization tools, and eventual integration with other toxicogenomics data.[4,5]

Many toxicoproteomic studies have focused upon the liver as a primary target organ for studying chemical-induced injury and serum for derivation of new biomarkers and signatures for the temporal process of injury and repair. Since inflammation is a frequent accompaniment to organ injury, toxicoproteomic studies are being designed to derive signatures of inflammation as a complex set of overlying biological processes that occur during liver damage. This article will focus upon use of toxicoproteomics as part of the toxicogenomics field's efforts to advance understanding of liver injury and inflammation.

Proteomic Platforms for Toxicoproteomic Studies

The complexity of a "proteome" (total protein expression of a specific cell, organ, tissue, or biofluid) presents numerous challenges for comprehensively describing the many attributes of protein expression during a single analysis.[6] Such attributes include protein sequence identity, quantity, posttranslational modifications (PTM), interactions, structure, and function. Some of these challenges in proteomic analysis include defining the identities and quantities of an entire proteome in a particular spatial location (i.e., serum, liver, mitochondria), the existence of multiple protein forms and complexes, the evolving structural and functional annotations of the human and rodent proteomes, and integration of proteomics data with transcriptomics or other expression data. Primary aims of proteomic analysis are to achieve maximal proteome depth, high-throughput, quantitative protein measurement, and timely analysis through use of discovery-oriented platforms. Proteomic platforms represent combinations of technologies to describe protein attributes by the separation, quantitation, and identification of all proteins in a biological sample. The following proteomic platforms represent some of the primary technologies being used for separating and identifying proteins during toxicoproteomic studies (see Ref. 6).

Two-Dimensional Gel Electrophoresis and Mass Spectrometry

Two-dimensional gel (2D gel) electrophoresis systems have been combined with mass spectrometry in an established and adaptable platform since 1975 that is the most commonly used proteomic platform to separate and comparatively quantitate protein samples.[7] Current state-of-the-art 2D gels use immobilized pH gradient (IPG) gels to separate proteins first by charge and then subsequently resolve by mass using SDS gel electrophoresis for effective separation of complex protein samples. Fluorescent staining is often the most sensitive means of protein detection (μg–mg). After electronic alignment (registration) of stained proteins in 2D gels by image analysis, intensities of identical protein spots are compared among treatment groups and a ratio (fold change) is calculated for each protein using specialized software. The combination of 2D gel separation of proteins with a ready means of protein identification by mass spectrometry forms a versatile and discovery-oriented platform for use in toxicoproteomic studies.[8] A downside to this platform is limited coverage of a proteome that can be separated and visualized on 2D gels by even the most sensitive fluorescent stains.

Multidimensional LC–MS/MS

In this proteomic platform, liquid chromatography (LC) is used to separate protein digests (ng–μg peptides) by charge (strong anion exchange) and hydrophobicity (C18) immediately prior to entry into tandem mass spectrometers (MS/MS) for protein identification.[9] The platform has also been called "shotgun proteomics" since entire protein lysates are trypsin digested into thousands of peptide fragments without any fractionation prior to separation and identification. Advantages of this newer platform are the potential for detection and identification of low abundance proteins that may not be observed in gel-based protein separation methods, although the LC–MS/MS platform is only semiquantitative. Use of isotopic labeling methods does permit quantitation by this platform but at reduced sample throughput and lengthy analysis time.

SELDI-TOF Mass Spectrometry

Retentate chromatography–mass spectrometry (RC–MS) is a high-throughput proteomic platform that creates a laser-based mass spectrum from a chemically absorptive surface. The principle of this approach is the adsorptive retention (pg–ng protein) of a subset of sample proteins on a thin, chromatographic support (i.e., hydrophobic, normal phase, weak cation exchange, strong anion exchange, or immobilized metal affinity supports). The absorptive surfaces are placed on thin metal chips that can be inserted into a MALDI-type mass spectrometer. The laser rapidly desorbs proteins from each sample on a metal chip to create a mass spectrum profile. RC–MS can be performed upon any protein sample but thus far this platform has found greatest use in the analysis of serum and plasma for disease biomarker discovery.[10] The lead commercial platform of RC–MS proteomic platform is the SELDI-TOF-MS instrument or, surface-enhanced laser desorption ionization time-of-flight.[11] Analysis of samples is relatively rapid (100/day) and only a few μL of sample are necessary. A downside is that protein identification of peaks is not readily achieved without additional analysis.

Antibody Arrays

Protein arrays (any mass parallel array of proteins, peptides, capture ligands, or adsorptive surfaces for protein analysis) represent a promising new proteomic tool that closely emulates the design for parallel analysis of DNA microarray technology.[12] Many different types of capture molecules can be arrayed but the most prevalent are antibody arrays that directly separate proteins from each other by affinity binding to specific protein targets. Generally, commercial antibody array platforms have widely varying sensitivities (pg–μg peptides) that fall into three classes based on targeted proteins: cytokine/chemokine arrays, cellular function protein arrays, and cell signaling arrays. Not all proteins for any given cell type or biofluid (i.e., serum/plasma) are presently represented on available antibody arrays, which precludes this platform as a tool for new protein discovery. However, antibody arrays do provide mass parallel analysis (similar to DNA microarrays) as a rapid screen for protein alterations that may be relevant to tissue injury or disease.

Liver Injury

The liver is the primary site of xenobiotic metabolism and is a major organ for biotransformation and elimination of foreign substances from the body. The biotransformation and metabolic capabilities of the liver also render it particularly susceptible to injury from environmental contaminants, natural products, and viral/bacterial agents. The phenotypes of liver injury are wide ranging and include necrosis, hepatitis, cholangitis, cholestatsis, steatosis, fibrosis, cirrhosis, hepatomegaly, and cancer. Chronic exposure to environmental agents may produce one or more combinations of phenotypes to liver injury. The differing cellular types of the liver may also be targets for injury and disease. While the liver primarily consists of hepatic parenchymal cells containing biotransformation enzymes, several nonparenchymal cells are also important for liver function and may increase in numbers in some toxic and pathologic conditions to significantly alter hepatic transcript and protein profiles. In this regard, some of the more important nonparenchymal cells include stellate cells, pit cells (NK cells), Kupffer cells, biliary cells, and endothelial cells.[13] Specific cell separation techniques for these specialized cells involve centrifugal elutriation, immunomagnetic beads, density gradient centrifugation, and others which have been recently reviewed.[14] Contributions of these specialized cells in their resting and active states to the overall protein profile of the liver may reveal protein profiles that might include cytokines, chemokine, and angiogenic factors that contribute to liver organ toxicity. Furthermore, the liver is a highly vascular organ with arterial circulation to deliver oxygenated blood and nutrients and a highly specialized venovasculature (hepatic sinusoidal endothelium). Contribution of cytokines and chemokines by circulating and infiltrating leukocytes is an important component to the overall liver protein expression profile in liver injury and inflammation.

Toxicoproteomic Analysis of Liver and Blood

The rationale for collecting liver and blood for toxicoproteomic analysis is that blood and the injured organ are partners in injury, recovery, and repair. Blood is the common biofluid that contacts all organs of the body. Proteomic analysis of serum or plasma proteomes (the soluble portion of blood), can uniquely reveal signs of specific organ toxicity or pathology from the peptides and proteins passively leaked or actively secreted during dysfunction or injury. In addition, blood also carries the nutrients, intermediary metabolites, amino acids, peptides, proteins, and enzymes needed by organs for health and recovery during organ toxicity.

FIGURE 2 shows a strategy for toxicoproteomic study of liver during acute injury. Transcriptomic analysis would also be carried out in liver and whole blood[15] (containing leukocytes) to determine changes in mRNA. Representative liver toxicants (i.e., bromobenzene, dichlorobenzenes, acetaminophen, monocrotaline, lipopolysaccharide [LPS], others) are being studied at the NIEHS National Center for Toxicogenomics as model agents for producing hepatic necrosis and inflammation. A general approach for studying acute liver injury is shown for model hepatotoxicants, such as acetaminophen.[16,17] Male rats are treated with vehicle, low dose (noninjury) and high dose (injury causing) of chemical toxicants and then rats are euthanized at 6 h, 24 h, and 48 h. Metabolic activation often produces electrophiles that consume GSH or engage other conjugative systems and once overwhelmed may ultimately lead to necrosis. The 6, 24, and 48 h time points would be "phenotypically anchored"[18] to histopathology and conventional serum chemistries (see ALT/AST time course in box on upper right). For acetaminophen and many other agents, these times represent preinjury, injury, and recovery time periods, respectively.

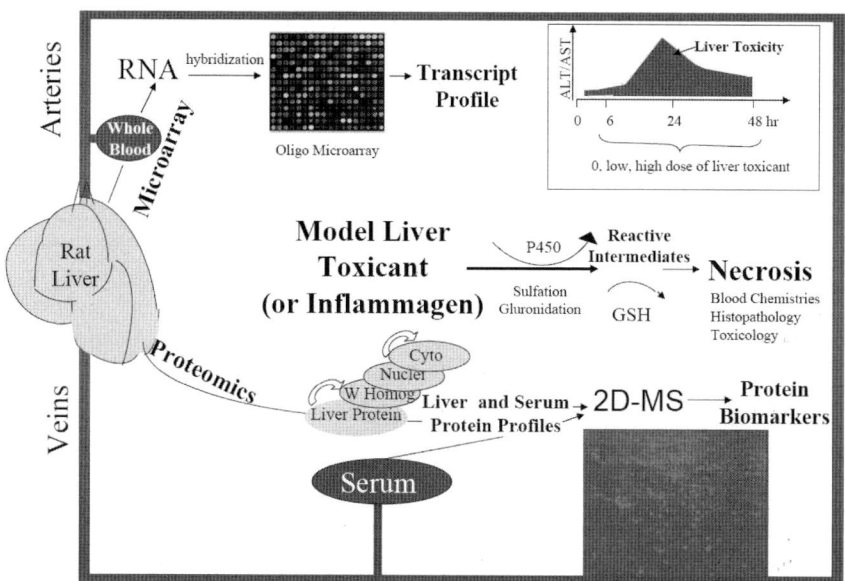

FIGURE 2. Integrative strategy for biomarker discovery during liver injury using toxicogenomic and toxicoproteomic analysis of liver and serum/blood in rats. Liver from each rat is used for both proteomic or transcriptomic analysis. Protein is isolated from serum and liver for differential protein expression (protein profiling) in a whole homogenate (W homog) or in subcellular fractions, such as nuclei and cytosol (Cyto). Rats are sacrificed after various times of toxicant treatment (upper right inset). Similarly, RNA is isolated from whole blood as well as liver for differential transcript expression (transcript profiling) using DNA microarrays. The upper inset shows a typical experiment for study of protein or gene expression during acute liver injury using a vehicle control as a "0" dose, nontoxic or "low" dose, and a liver-injury producing or "toxic" dose. Animals would be euthanized at 6 h as a "pretoxic" time, at 24 h as a "toxic" time and at 48 h as a "recovery" time period in which liver toxicity can be initially described by histopathology and serum chemistries, such as ALT and AST. Adjustments in time of euthanasia can be made to accommodate different toxicants. Model hepatotoxicants (inflammagens, such as LPS) can be profiled for various phenotypes of liver injury, such as necrosis and inflammation. Data from protein and transcript profiles of liver and serum/blood are integrated during the course of toxicity for biomarker discovery reflecting liver injury processes.

The choice of proteomic platform for toxicoproteomic studies depends on the amount, availability, and complexity of the biological sample. The 2D-MS proteomic platform involves the 2D gel electrophoresis for separation of proteins. This is followed by mass spectrometry to identify proteins from serum and liver or subcellular fractions to determine differentially expressed proteins. Additional proteomic platforms, such as liquid chromatography combined with tandem mass spectrometry (i.e., LC–MS/MS), can also be used to increase the number of identified altered proteins. In addition to analyzing whole liver homogenates, profiling of subcellular fractions allows for enrichment of subcellular structures and enhancement of differential expression of toxicant-affected proteins for biomarker discovery as previously described.[19]

Integration of Proteomics and Transcriptomics

The formulation of a strategy to integrate transcript, protein, and toxicology data is predicated on the idea that combined protein and transcript expression will bring more information to bear on environmental toxicology problems in identifying affected, biochemical and regulatory pathways that can lead to biomarker discovery.[20] The integration of toxicogenomic and toxicoproteomic data is often *first* approached by analysis of the concordance of altered gene changes from DNA microarray analysis of liver with the profile of altered proteins. A *second* level approach is to compare affected pathways that contain common altered gene transcripts and proteins representing pathway activation or repression. A *third* level integration could be the comparison of posttranslationally altered proteins after toxicant exposure with specific gene transcript changes in kinases, proteases, phosphorylases, conjugating enzymes, and others that target specific proteins for modification. A *fourth* method might be subcellular proteomic analysis to link affected transcripts for specific structural, enzymatic, or functions of organelles. Profiling of enriched fractions of nuclei, mitochondria, endoplasmic reticulum, plasma membranes, or cytosol can give insight into the subcellular site or subcellular structures of liver injury or upon affected protein trafficking or signaling pathways. A *final* level integration could occur by kinetically profiling protein changes in the serum proteome over the course of pretoxic, toxic, and recovery stages of chemical exposure. Such changes in the serum proteome represent the dynamic nature between target organ injury and blood during time of repair and recovery. Proteomics may eventually link protein expression patterns of multiple organ systems involved in chemical exposure that may contribute to a better understanding of mechanisms of tissue injury and produce useful biomarkers of environmental disease.

Strategy for Biomarker Development in Acute Liver Injury

The liver is a strategic organ between incoming absorbed materials from the environment and the rest of the body. Its many roles in nutrient processing, synthetic abilities, biotransformation, and phagocytic clearance can also make the liver susceptible to a variety of metabolic, toxic, microbial, circulatory, and carcinogenic challenges. These stressors may result in a variety of disorders that include acute hepatic necrosis with accompanying involvement of the immune system. Inflammation occurs in a variety of liver pathologies as a histopathological cellular feature or as measurable cytokines and chemokines. It is suspected that such immunomodulatory changes, so often associated with chemically induced hepatotoxicity, may be responsible for hepatocellular injury while simultaneously be involved in subsequent regenerative processes that allow liver to recover from injury.[21] This incomplete understanding of immune system activities in tissue injury and regeneration presents a challenge in toxicogenomics in assigning gene expression changes as potential biomarkers/signatures for either injury or repair. The discernment of inflammatory signatures from toxicoproteomic studies by study of model inflammagens, like LPS, will be helpful in dissecting injurious and reparative protein signatures from liver and blood during hepatotoxicity. For example, results of DNA microarray analysis of RNA extracted from whole blood of rats were recently reported to demonstrate discriminating toxicity signatures during acute inflammation after treatment with LPS in rats.[22] Parallel toxicoproteomic analysis of serum and liver from model hepatotoxicants, like acetaminophen in rats, have been performed to derive more descriptive biomarkers of organ injury, repair, and recovery.[16,17] NIEHS is conducting toxicogenomic and toxicoproteomic analysis of whole blood/serum and liver in rats for an ambitious compendium of model hepatotoxicants and inflammagens. The aim of these studies is to derive sets of descriptive biomarkers that robustly describe the time course of acute liver injury, repair, and recovery and provide insight into the molecular mechanisms that underlie each of these events.

Biomonitoring assesses human exposures to natural and synthetic chemicals by measuring biological changes in an individual's tissues and fluids for use in epidemiology and public health. While the application of toxicoproteomics as a tool for discovery of new proteins as potential biomarkers and signatures from xenobiotic exposure is well under way in experimental toxicology and drug development,[3] it is just in its early stages as a tool for human biomonitoring studies. Translation of proteins into reproducible biomarkers with the accompanying specimen standards and population reference values will take time since basic data standards and bioinformatic practices in proteomics are still under development.[4] Even so, toxicoproteomics has contributed to biomonitoring advances for some occupational exposures as recently documented for benzene.[23]

SUMMARY

Toxicoproteomics is part of a comprehensive gene expression strategy for biomarker discovery and involves protein profiling using global protein separation and identification technologies. A rationale for discovery of new biomarkers during liver injury is described using gene expression studies that include differential protein and gene expression. Both liver and serum/blood are included in protein and gene profiling studies. Liver injury involves many biological processes, such as necrosis and inflammation. In this article, an initial strategy has been proposed to examine protein and gene expression profiles of several hepatotoxicants and inflammagens with the goal of determining their individual contributions to acute liver injury and recovery. It is hoped that these and similar studies will play a role in biomarker and signature development and will have eventual use in biomonitoring strategies.

ACKNOWLEDGMENT

This research was supported by the Intramural Research program of the NIH, National Institute of Environmental Health Sciences

REFERENCES

1. WATERS, M.D., K. OLDEN & R.W. TENNANT. 2003. Toxicogenomic approach for assessing toxicant-related disease. Mutat. Res. **544:** 415–424.
2. PENNISI, E. 2003. Human genome. A low number wins the GeneSweep Pool. Science **300:** 1484.
3. MERRICK, B.A. & M.E. BRUNO. 2004. Genomic and proteomic profiling for biomarkers and signature profiles of toxicity. Curr. Opin. Mol. Ther. **6:** 600–607.
4. KREMER, A., R. SCHNEIDER & G.C. TERSTAPPEN. 2005. A bioinformatics perspective on proteomics: data storage, analysis, and integration. Biosci. Rep. **25:** 95–106.
5. BLUEGGEL, M., D. CHAMRAD & H.E. MEYER. 2004. Bioinformatics in proteomics. Curr. Pharm. Biotechnol. **5:** 79–88.
6. WETMORE, B.A. & B.A. MERRICK. 2004. Toxicoproteomics: proteomics applied to toxicology and pathology. Toxicol. Pathol. **32:** 619–642.
7. RIGHETTI, P.G., A. CASTAGNA, F. ANTONUCCI, et al. 2004. Critical survey of quantitative proteomics in two-dimensional electrophoretic approaches. J. Chromatogr. A. **1051:** 3–17.
8. YATES, J.R. 2004. Mass spectral analysis in proteomics. Annu. Rev. Biophys. Biomol. Struct. **33:** 297–316.
9. MACDONALD, N., S. CHEVALIER, R. TONGE, et al. 2001. Quantitative proteomic analysis of mouse liver response to the peroxisome proliferator diethylhexylphthalate (DEHP). Arch. Toxicol. **75:** 415–424.
10. PETRICOIN, E., J. WULFKUHLE, V. ESPINA, et al. 2004. Clinical proteomics: revolutionizing disease detection and patient tailoring therapy. J. Proteome. Res. **3:** 209–217.
11. ISSAQ, H.J., T.P. CONRADS, D.A. PRIETO, et al. 2003. SELDI-TOF MS for diagnostic proteomics. Anal. Chem. **75:** 148A–155A.
12. CUTLER, P. 2003. Protein arrays: the current state-of-the-art. Proteomics **3:** 3–18.
13. MERRICK, B.A. & J.H. MADENSPACHER. 2005. Complementary gene and protein expression studies and integrative approaches in toxicogenomics. Toxicol. Appl. Pharmacol. **207:** 189–194.

14. KMIEC, Z. 2001. Cooperation of liver cells in health and disease. Adv. Anat. Embryol. Cell. Biol. **161:** III–XIII, 1–151.
15. IIDA, M., C.H. ANNA, J. HARTIS, *et al.* 2003. Changes in global gene and protein expression during early mouse liver carcinogenesis induced by non-genotoxic model carcinogens oxazepam and Wyeth-14,643. Carcinogenesis **24:** 757–770.
16. HEINLOTH, A.N., R.D. IRWIN, G.A. BOORMAN, *et al.* 2004. Gene expression profiling of rat livers reveals indicators of potential adverse effects. Toxicol. Sci. **80:** 193–202.
17. FOSTEL, J., D. CHOI, C. ZWICKL, *et al.* 2005. Chemical effects in biological systems—data dictionary (CEBS-DD): A compendium of terms for the capture and integration of biological study design description, conventional phenotypes and omics data. Toxicol. Sci. **88:** 585–601.
18. PAULES, R. 2003. Phenotypic anchoring: linking cause and effect. Environ. Health Perspect. **111:** A338–A339.
19. BRUNO, M.E., C.H. BORCHERS, J.M. DIAL, *et al.* 2002. Effects of TCDD upon IkappaB and IKK subunits localized in microsomes by proteomics. Arch. Biochem. Biophys. **406:** 153–164.
20. MERRICK, B.A. & K.B. TOMER. 2003. Toxicoproteomics: a parallel approach to identifying biomarkers. Environ. Health Perspect. **111:** A578–A579.
21. MARKIEWSKI, M.M., R.A. DEANGELIS & J.D. LAMBRIS. 2006. Liver inflammation and regeneration: two distinct biological phenomena or parallel pathophysiologic processes? Mol. Immunol. **43:** 45–56.
22. FANNIN, R.D., J.T. AUMAN, M.E. BRUNO, *et al.* 2005. Differential gene expression profiling in whole blood during acute systemic inflammation in lipopolysaccharide-treated rats. Physiol. Genom. **21:** 92–104.
23. VERMEULEN, R., Q. LAN, L. ZHANG, *et al.* 2005. Decreased levels of CXC-chemokines in serum of benzene-exposed workers identified by array-based proteomics. Proc. Natl. Acad. Sci. USA **102:** 17041–17046.

Gene Expression Alterations in Immune System Pathways following Exposure to Immunosuppressive Chemicals

RACHEL M. PATTERSON AND DORI R. GERMOLEC

National Institutes of Environmental Health Sciences, Research Triangle Park, North Carolina 27719, USA

ABSTRACT: Exposure to environmental agents can affect a number of adverse immunological outcomes, including changes in the incidence of infectious disease. Diethylstilbestrol (DES), dexamethasone (DEX), cyclophosphamide, and 2,3,7,8-tetrachlorodibenzo-p-dioxin (TCDD) are immunosuppressive chemicals that can induce similar pathophysiological end points in the thymus; however, the mechanism of toxicity is different for each compound. We examined differential gene expression in the spleen and thymus following chemical exposure and correlated these changes with alterations in functional immune end points and our knowledge of the known mechanisms of action. RNA from the spleen and thymus has been analyzed using Illumina Sentrix arrays and BeadStudio software. Preliminary data suggest that DES induced the greatest number of gene changes in the spleen, while DEX induced the most changes in the thymus. In both spleen and thymus, genomic analysis revealed gene expression changes that were common to multiple chemicals and that may be associated with xenobiotic-induced immune system perturbations, including alterations in genes associated with apoptosis, antigen processing and presentation, and response to biotic stimulus. This was particularly evident in the thymus, where there were many similarities in the expression profiles, as well as gene alterations unique to a single compound. In contrast, expression profiles in spleen were more distinct. The category of genes most profoundly affected by all four chemicals was response to biotic stimulus: there were both clusters of genes modulated by multiple chemicals and genes altered by a single chemical. The distinct gene profiles may specifically relate to cellular targets and mechanism of action.

KEYWORDS: gene expression; spleen; thymus; immunosuppression; microarray; genomics

---

Address for correspondence: Dori R. Germolec, 111 TW Alexander Dr, P.O. Box 12233, Research Triangle Park, NC 27719. Voice: 919-541-3230; fax: 919-541-0870.
e-mail: germolec@niehs.nih.gov

INTRODUCTION

The ability to evaluate the expression of thousands of genes and establish patterns of gene expression associated with specific biomolecular events has the potential to become one of the most powerful tools available to biomedical scientists. The use of this technology as a screening tool to identify potential immunotoxicants would be of considerable significance as alterations in immune function can lead to increased incidence of hypersensitivity disorders, autoimmune or infectious diseases, or neoplasias. Recent advances in genomics-based identification of gene families and gene polymorphisms associated with immune disorders have answered basic questions regarding immune mechanisms and individual susceptibility and moved forward our understanding of disease processes. Building on the success of basic research, gene array technology holds great promise to identify the genes associated with individual variation in responses to exogenous factors including infectious agents, therapeutic drugs, and environmental chemicals.[1]

Large-scale gene expression analysis may be accomplished in a number of ways. Full-genome arrays and specialized microarray panels that contain specific sets of DNA sequences may be directly or indirectly related to the immune system. Genome-wide, large-scale DNA-based arrays may be useful for generalized toxicological screening, but the management and interpretation of changes relevant to the immune system can be difficult. "Data overload" may be minimized by identifying and exploiting the patterns of gene expression in subpopulations of T and B lymphocytes, macrophages, and dendritic cells and may present a fairly complete picture of differentiation and proliferation that would (*a*) allow for the monitoring of the progress of the immune response, and (*b*) help identify defects at the cellular and molecular level.[2] Considerable progress has been made in the mapping of immune responses to specific pathogens and tumors such that changes in gene expression may be used to predict resistance to disease.[3–5] Analytic approaches have shown that the clustering of genes based on similar expression patterns in immune versus all tissues can been used to establish associations between specific genes and processes such as antigen processing and chemokine-mediated responses.[6]

While a number of laboratories are using genomics as a tool to investigate the underlying mechanisms of immunotoxicity,[7–10] to date there have been few attempts to evaluate the utility of genomic analyses as a screening tool for immunotoxicants. The present article compares functional end points and gene expression changes in the spleen and thymus from B6C3F1 mice following *in vivo* treatment with well-characterized immunosuppressive agents including diethylstilbestrol (DES), dexamethasone (DEX), cyclophosphamide (CPS), and 2,3,7,8-tetrachlorodibenzo-*p*-dioxin (TCDD). The B6C3F1 mouse is routinely used in National Toxicology Program carcinogenesis and toxicity studies, and the biology, immune parameters, and response to TCDD, DES, DEX, and CPS have been studied extensively in this strain. The four agents

TABLE 1. Treatment regimen for evaluation of differential gene expression induced by immunosuppressive chemicals

Chemical	DES	DEX	CPS	TCDD
High dose	8.0 mg/kg/day	5.0 mg/kg/day	50.0 mg/kg/day	3.0 μg/kg/day
Low dose	0.8 mg/kg/day	0.5 mg/kg/day	5.0 mg/kg/day	0.3 μg/kg/day
Vehicle	Corn oil	Ethanol/saline	Phosphate buffered saline	Corn Oil
Route of exposure	Subcutaneous injection	Intraperitoneal injection	Intraperitoneal injection	Gavage

Female B6C3F1 mice (4 per group) were treated as described above for 5 consecutive days.

examined induce similar pathophysiological changes in the thymus and modify resistance to challenge with syngeneic tumors or infectious agents; however, their putative mechanisms of action and cellular targets differ. Preliminary results suggest that while a large number of differentially expressed genes are associated with an individual chemical, xenobiotic-induced immune system perturbation may also share common patterns of altered gene expression.

MATERIALS AND METHODS

Animals and Treatment

Female B6C3F1 mice were obtained from Charles River (Raleigh, NC) and Taconic (Germantown, NY) (CPS) at 4–8 weeks of age and maintained on a 12-h light/dark cycle at 20–22°C. The animals were fed Harlan Teklad Laborarory Diet (NIH07) and were provided food and water *ad libitum*. At 11–12 weeks of age, the mice were treated with one of four experimental chemicals or a matched vehicle daily for 5 days as described in TABLE 1. The four chemicals used in this study were: DES, DEX, CPS (Sigma-Aldrich Corp., St. Louis, MO), and TCDD (Research Triangle Institute, Research Triangle Park, NC). Mice were monitored daily for acute toxicity. On the sixth day the mice were euthanized by CO_2 asphyxiation, the spleen and thymus were excised, and body and organ weights were measured. All experiments were conducted in accordance with the National Institutes of Health Animal Research Advisory Committee guidelines.

RNA Isolation and Amplification

Total RNA was extracted via the Qiagen RNeasy kit (Qiagen, Valencia, CA) using the standard protocol, DNase digestion, and clean-up procedures, and quantified spectrophotometrically. For quality control, RNA purity and integrity were evaluated by denaturing gel electrophoresis, OD 260/280 ratio, and analysis on an Agilent 2100 Bioanalyzer (Agilent Technologies, Palo Alto,

CA). Total RNA was amplified and purified using the Ambion Illumina® RNA amplification kit (Ambion, Austin, TX) according to the manufacturer's instructions. Briefly, 400 ng of total RNA was reverse transcribed to cDNA using a T7 oligo(dT) primer. Second-strand cDNA was synthesized, *in vitro* transcribed, and labeled with biotin-16-UTP. The cRNA was quantified using the Molecular Probes RiboGreen® RNA Quantitation Kit (Molecular Probes, Eugene, OR).

Microarray

The labeled cRNA samples were hybridized to a custom Illumina® Sentrix Array Matrix (SAM; Illumina Inc., San Diego, CA) for 16–18 h at 55°C, following the manufacturer's instructions. The SAM contains 96 identical oligonucleotide arrays, consisting of 710 genes, 698 user-selected and housekeeping genes, and 12 negative control sequences. There were two 50-mer probes representing each gene. The SAM was washed, blocked with casein in phosphate buffered saline (PBS), incubated with streptavidin-Cy3, dried, and scanned on the Illumina BeadArray Reader GX.

Data Analysis

Microarray data were analyzed using Illumina BeadStudio software and DAVID (Database for Annotation, Visualization and Discovery).[11] Intensity data were normalized using a rank invariant algorithm. Differential expression of each gene, relative to the respective control was evaluated using an error model that calculates a P value as a function of intensity differential and biological, technical, and nonspecific variation. A differential expression score (DIFF) of ± 20 corresponds to a P value of 0.01. Only genes that were determined to be differentially expressed at $P \leq 0.01$ by both probes for a given gene were considered significantly altered and included in further analysis. Hierarchical clustering was done at $P \leq 0.05$ using a Pearson correlation. Gene ontology was obtained using DAVID software and focused on the immune-relevant categories shown in TABLE 2.

RESULTS

General Parameters of Toxicity

Consistent with published literature, all four chemicals induced thymic atrophy, as evidenced by decreased organ weight and cellularity compared to

TABLE 2. Gene ontology (GO) categories represented in the Illumina® Sentrix Gene Array Matrix

GO Category	Number of genes in array	Examples of specific genes in GO category* (GenBank symbol)
Immune response	144	Tnfrsf5, Cd7, H2-M3, Ccl24, Rag2
Antigen processing/presentation	18	Fcer1g, Fcgr3, H2-Ob, Ha-Aa, B2m, H2-Ea, H2-Dma, H2-K, H2-Ab1
Apoptosis	74	Bcl2, Casp1, Casp3, Traf1, Ltbr, Bok, Apoe, Bnip3, Tcf7, Gzma, Birc3, Bak1, Myc, Dedd, Casp 8, Casp 9
Cytokine metabolism/production	13	Irf1, Cebpb, Map2k3,
Humoral immune response	14	H2-Bf, C3, C1qa, Cr2, Cd79b, Igj
Immune cell activation	21	Egr1, Rag1, Was, Fcgr2b, Il7
Inflammatory response	45	Cd14, Il1b, Cxcl13, Tnfrsf1a, Mif
Innate Immune Response	45	Il16, Ccl5, Cxcl9, Ccl19,Ccl8, Ccl25
Lymphocyte Activation/proliferation	18	Cd8a, Icos1, Cxcr4, Blr1, Cdkn1a,
Response to biotic stimulus	192	Tnfrsf7,Tnfrsf6, Tnfrsf13b, Ly6d, Icam1, Ii, Ltb, Blr1, Relb, Lat, Irf1, Il18, IL15, IL7, H2-Dmb1, H2-Bf, H2-D4, Cd2, Cd4, Cd5, Cd8b,Cd22, Cd28,

In GO classification, individual genes may appear in multiple GO categories. For simplicity, in this table genes are represented in one GO category only.

control (data not shown). In the spleen, CPS and DEX exposure resulted in reduced organ weight and cellularity, while DES exposure induced splenomegaly. Exposure to TCDD did not affect spleen size.

Gene Expression Categorization—Thymus

Of the approximately 700 genes represented in the array, 254 genes were differentially expressed in thymus by one or more of the four chemicals at $P \leq 0.01$, as compared to their respective control. Hierarchical clustering of all genes induced or suppressed by any chemical showed that the gene expression profiles resulting from exposure to DES, DEX, or CPS clustered close to each other (FIG. 1 A), indicating that they modulate many of the same genes. The DIFF scores also suggested that a large number of genes were altered by multiple chemicals. Expression of 94 (37%) genes was significantly changed by one chemical, 57 (22%) by two chemicals, 86 (34%) by three chemicals, and 17 (7%) by all four chemicals under study. The groups of genes modulated by three or four chemicals were designated "common" genes; genes modified by only one chemical were designated "unique" to that chemical. In total, DES altered 145 genes; DEX altered 214 genes; CPS altered 98 genes; and TCDD altered 73 genes. The expression profiles of DES and CPS showed a high

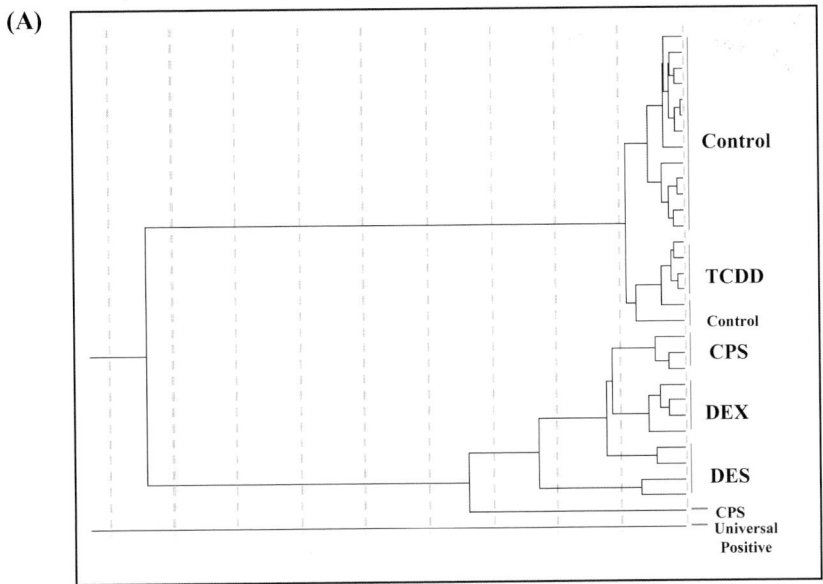

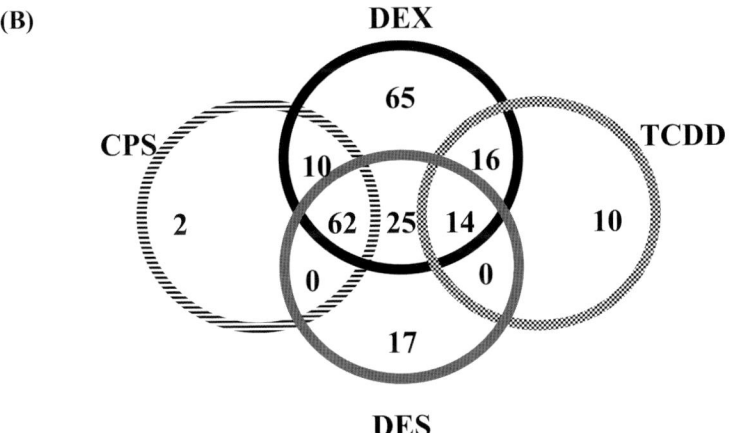

FIGURE 1. Differential gene expression in mouse thymus induced by immunosuppressive chemicals. (**A**). This dendrogram depicts the hierarchical clustering of genes significantly modulated by high-dose DEX, DES, CPS, and/or TCDD (338 genes) at $P \leq 0.05$ using Pearson correlation. Greater distances between nodes, such as the connection between the group containing TCDD and control samples and the group containing DES, DEX, and CPS samples, indicate greater differences in gene expression patterns; nodes that are in close proximity indicate similar expression patterns. (**B**). Venn diagram showing interrelationships of gene expression profiles for each chemical. Values represent the total number of genes modulated by chemical treatment in thymus at a significance level of $P \leq 0.01$. In addition to the genes represented in the figure, all four chemicals modified expression of 17 genes.

degree of homology with DEX, 97% of CPS-modulated genes, and 81% of DES-modulated genes were similarly affected by DEX. The overlap in gene expression patterns between chemicals is shown in FIGURE 1 B. Notably, DEX, DES, and CPS similarly modified 62 genes, 60% of the group of "common" genes. DEX had the most profound effect of all chemicals, inducing both the largest number of overall gene expression changes and the largest number of changes unique to a single chemical. In addition to the genes represented in FIGURE 1, all four chemicals modified the expression of 17 genes.

Gene Expression Categorization—Spleen

At $P \leq 0.01$, 259 genes were differentially expressed in spleen by one or more of the four chemicals, compared to the respective control. Hierarchical clustering demonstrated that DES induced an expression profile distinctly different from the other chemicals (FIG. 2 A). Expression of 80 (69%) genes was significantly modulated by one chemical, 59 (23%) by two chemicals, 18 (7%) by three chemicals, and 2 (1%) by all four chemicals. In total, DES altered 186 genes, DEX altered 97 genes, CPS altered 64 genes, and TCDD altered 12 genes. The overlap in gene expression patterns between chemicals is shown in FIGURE 2 B. In addition to the genes represented in FIGURE 2 B, all four chemicals modulated two genes. Consistent with the clustering model, DES had the most profound effect in the spleen, inducing both the largest number of overall gene expression changes and the largest number of changes unique to a single chemical. The CPS profile was most similar to DEX; 59% of CPS-modulated genes were also affected by DEX. The DEX profile included groups of unique genes (31%), genes similarly altered by CPS (33%) and DES (33%), and a group of genes modified by DES but in the opposite direction (14%). As in the thymus, most of the "common" genes (85%) were modulated by DES, DEX, and CPS.

DISCUSSION

The ability to detect the induction of inflammation and to recognize defects in self-recognition, host responses to foreign antigens, and immune surveillance at the molecular level could serve as an important tool for evaluating the potential of chemicals to cause adverse immune responses. Gene arrays are already being used in immunotoxicology research to help identify biochemical pathways that are altered following chemical exposure. In this study, differentially expressed genes were categorized using the Gene Ontology (GO) Consortium biological function designations. Not surprisingly, genes involved in apoptotic pathways were modulated by all four chemicals in thymus and by DEX, DES, and CPS in spleen in a pattern that mirrored the changes in organ

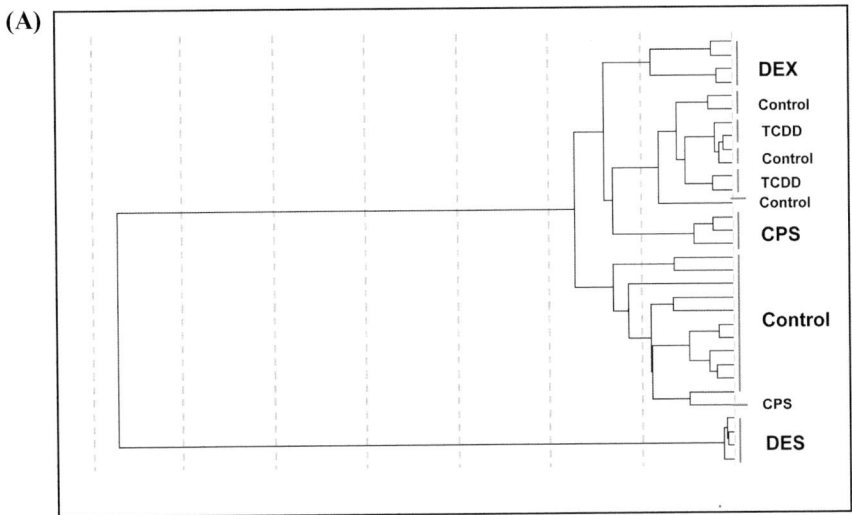

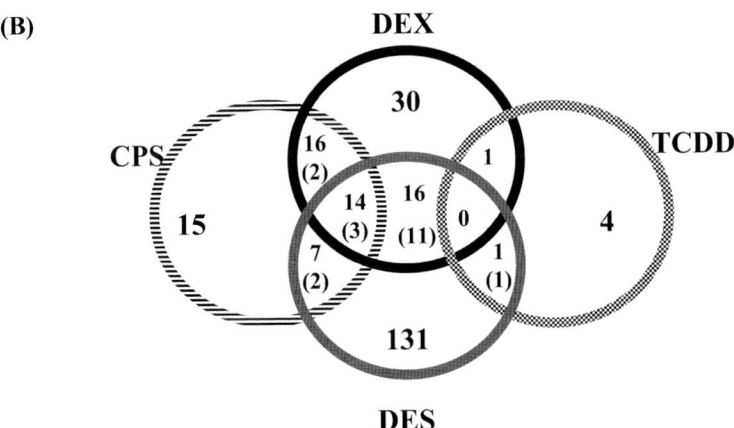

FIGURE 2. Differential gene expression in mouse spleen induced by immunosuppressive chemicals. (**A**) This dendrogram depicts the hierarchical clustering of genes significantly modulated by high-dose DEX, DES, CPS, and/or TCDD (347 genes) at $P \leq 0.05$ using a Pearson correlation. Greater distances between nodes, such as the connection between the DES samples and the group containing TCDD, DEX, and CPS samples, indicate greater differences in gene expression patterns; nodes that are in close proximity indicate similar expression patterns. (**B**) Venn diagram showing interrelationships of gene expression profiles for each chemical. Values not in parentheses represent the total number of genes modulated by chemical treatment in spleen. In the case of two or more chemicals, all compounds similarly affect the genes. Values in parentheses represent genes altered by two or more chemicals but in opposing directions. All differential expression was measured at a significance level of $P \leq 0.01$. In addition to the genes represented in the figure, all four chemicals modified the expression of two genes.

weight and cellularity. Antigen presentation/processing genes were induced by all four chemicals in thymus, but suppressed by DES and DEX, and unaffected by CPS and TCDD, in spleen. Consistent with published reports[12–15] of T cell toxicity by the chemicals in this study, CPS, DEX and DES, but not TCDD, altered CD3, CD4, and CD8 in thymus. Much mechanistic work has been performed to understand the immunomodulatory properties of TCDD, but the exact mechanism of TCDD-induced immunotoxicity remains controversial. Dose, exposure duration, and age at exposure are all factors that may affect the cellular targets and immune deficits induced by this chemical.

Host resistance assays have been used in immunotoxicology testing to evaluate responses to infectious agents and/or neoplastic cells, and to relate functional alterations from primary screens with disease end points that may be more relevant to what is experienced in the human population.[16–18] If predictive sets of genes or gene families can be associated with these end points, our ability to evaluate species-specific responses and extrapolate rodent-derived data to humans will be increased. The GO category most profoundly affected by all four chemicals was response to biotic stimulus, which includes genes important in nonspecific inflammatory responses, as well as, responses to bacteria, virus, protozoa, parasites, and oxidative stress. In thymus, DEX, DES, CPS, and TCDD altered common groups of genes associated with response to biotic stimulus, and DEX and DES also altered unique gene clusters in that GO. In the spleen, the only GO affected by TCDD was response to biotic stimulus. DEX, CPS, and DES all modulated both unique and common responses to biotic stimulus genes.

There were many similarities in expression profiles in thymus, especially when considering the changes in gene expression induced by DES, DEX, and CPS. However, there were also distinct features characterizing each chemical profile. In contrast, the expression profiles in the spleen were more distinct than similar, although several common gene clusters were identified. While there is still much work to be done with regard to the generation of data that might allow the mapping of hypothetical gene expression patterns identifying adverse effects on the immune system, increased use of DNA-based technologies has the potential to facilitate the evaluation of chemical-induced toxicity in immune cells and tissues. Studies such as the present one suggest that transcriptional profiling technology can be used to discover common pathways involved in immune system dysfunction and to discriminate distinct mechanisms of action from chemicals that have similar biologic effects.

ACKNOWLEDGMENTS

This research was supported by the Intramural Research program of the NIH, National Institute of Environmental Health Sciences.

REFERENCES

1. VAN DER POUW KRAAN, T.C.M.T., P.V. KASPERKOVITZ, N. VERBEET, et al. 2004. Genomics and the immune system. Clin. Immunol. **111:** 175–185.
2. GELIEBTER, J., A. MITTELMAN & R.K. TIWARI. 2003. Molecular phenotyping of the immune system by microarray analysis. Cancer Invest. **21:** 293–303.
3. MOCELLIN, S., E. WANT, M. PANELLI, et al. 2004. DNA array-based gene profiling in tumor immunology. Clin. Cancer Res. **10:** 4597–4606.
4. PULENDRAN, B. 2005. Variegation of the immune response with dendritic cells and pathogen recognition receptors. J. Immunol. **174:** 2457–2465.
5. SETTE, A., W. FLERI, B. PETERS, et al. 2005. A roadmap for the immunogenomics of category A-C pathogens. Immunity **22:** 155–161.
6. HUTTON, J.J., A.G. JEGGA, S. KONG, et al. 2004. Microarray and comparative genomics-based identification of genes and gene regulatory regions of the mouse immune system. BMC Genomics **5:** 82–97.
7. FISHER, M.T., M. NAGARKATTI & P.S. NAGARKATTI. 2004. Combined screening of thymocytes using apoptosis-specific cDNA array and promoter analysis yields novel gene targets mediating TCDD-induced toxicity. Toxicol. Sci. **78:** 116–124.
8. KINSER, S., Q. JIA, M. LI, et al. 2004. Gene expression profiling in spleens of deoxynivalenol-exposed mice: immediate early genes as primary targets. J. Toxicol. Environ. Health Part A **67:** 1423–1441.
9. LUYENDYK, J.P., W.B. MATTES, L.D. BURGOON, et al. 2004. Gene expression analysis points to hemostasis in livers of rats cotreated with lipopolysaccharide and ranitidine. Toxicol. Sci. **80:** 203–213.
10. PRUETT, S.B., C. SCHWAB, Q. ZHENG, et al. 2004. Suppression of innate immunity by acute ethanol administration: a global perspective and a new mechanism beginning with inhibition of signaling through toll-like receptor 3. J. Immunol. **173:** 2715–2724.
11. DENNIS Jr, G., B.T. SHERMAN, D.A. HOSACK, et al. 2003. DAVID: database for annotation, visualization, and integrated discovery. Genome Biol. **4:** P3.
12. MIYAUCHI, A., C. HIRAMINE, S. TANAKA, et al. 1990. Differential effects of a single dose of cyclophosphamide on T cell subsets of the thymus and spleen in mice: flow cytofluorometry analysis. Tohoku J. Exp. Med. **162:** 147–167.
13. BERKI, T., L. PALINKAS, F. BOLDIZSAR, et al. 2002. Glucocorticoid (GC) sensitivity and GC receptor expression differ in thymocyte subpopulations. Int. Immunol. **14:** 463–469.
14. CALEMINE, J.B., R.M. GOGAL Jr., A. LENGI, et al. 2002. Immunomodulation by diethylstilbestrol is dose and gender related: effects on thymocyte apoptosis and mitogen-induced proliferation. Toxicology **178:** 101–118.
15. KERKVLIET, N.I. & J.A. BRAUNER. 1990. Flow cytometric analysis of lymphocyte subpopulations in the spleen and thymus of mice exposed to an acute immunosuppressive dose of 2,3,7,8-tetrachlorodibenzo-p-dioxin (TCDD). Environ. Res. **52:** 146–154.
16. LUSTER, M.I., C. PORTIER, D.G. PAIT, et al. 1993. Risk assessment in immunotoxicology. II. Relationships between immune and host resistance tests. Fundam. Appl. Toxicol. **21:** 71–82.
17. SELGRADE, M.K. 1999. Use of immunotoxicity data in health risk assessments: uncertainties and research to improve the process. Toxicology **133:** 59–72.
18. GERMOLEC, D.R. 2004. Sensitivity and predictivity in immunotoxicity testing: immune end points and disease resistance. Toxicol. Lett. **149:** 109–114.

Transcriptional Profiling and Functional Genomics Reveal a Role for AHR Transcription Factor in Nephrogenesis

KENNETH S. RAMOS

Department of Biochemistry and Molecular Biology, and The Center for Genetics and Molecular Medicine, University of Louisville, Kentucky 40292, USA

ABSTRACT: Transcriptional profiling and functional genomics experiments using E11.5 metanephros organ cultures from $Ahr^{-/-}$ and $Ahr^{+/+}$ have shown that aryl hydrocarbon receptor (AHR) transcription factor is involved in the regulation of mesenchymal-to-epithelial transition (MET) during nephrogenesis. This response is mediated by alterations in the post-transcriptional control of Wilms' tumor suppressor ($Wt1$) gene and Wt1 splicing. In this article, biologically relevant gene predictor sets of the nephrogenic response were calculated for target genes of interest. The predictability of the gene set for each target was quantified by the coefficient of determination which provided a good criterion for identification of predictor sets that define the complex gene–gene interactions co-regulated by Ahr and Wt1. A subset of the signature genes was found to be co-regulated by Ahr and Wt1 and was responsible for shifts in renal cell transdifferentiation.

KEYWORDS: development; functional genomics; aryl hydrocarbon receptor; wilm's tumor suppressor gene; nephrogenesis

INTRODUCTION

The maintenance of cellular homeostasis is essential for the preservation of health at all levels of organization in living organisms. Chronic human diseases often arise as a result of environmental interactions that alter the normal function of genes and give rise to altered phenotypes that can no longer support cellular functions. The study of cellular phenotype control mechanisms and the genetic and environmental factors that bring about a shift from normal to disease phenotypes has dominated the field of molecular biology. Until recently, such investigations have been limited to analyses of single genes and,

---

Address for correspondence: Dr. Kenneth S. Ramos, Department of Biochemistry and Molecular Biology, School of Medicine, University of Louisville Health Sciences Center, Louisville, KY 40292. Voice: 502-852-5217; fax: 502-852-1114.
  e-mail: kenneth.ramos@louisville.edu

therefore, our understanding of the genetic complexity of polygenic disorders is limited.

With sequencing of the human genome nearing completion, a new wave of scientific investigations is upon us that focus on technological applications to help decipher the meanings encrypted on the DNA code, and the genetic responses of the human body to the foods we consume, the drugs we use, and the constant changes in our surrounding environment. The level of molecular sophistication is such that studies of a single individual and its unique response profiles are now possible to define global patterns of gene expression in response to changes in the environment, or genetic polymorphisms that define the inherent susceptibility (or not) of an individual to a drug, a chemical in the environment, or a product used in the home. The study of gene responses has paved the way for a complete understanding of the genetic fingerprints that define similarities and differences in response to any one the variables listed above. Thus, transcriptional profiling has emerged as a powerful tool to conduct genome-wide analyses of host interactions with the environment.[1] Global measurements of gene expression have been made possible by technological advances that allow the simultaneous analysis of thousands of transcripts in a biological sample. Currently, the dominant technology utilized for global measurements of gene expression is DNA microarrays, a technology that involves the use of a solid matrix as a platform on which thousands of surface-bound DNAs are hybridized against a pool of sample RNA to measure gene expression. Such analyses allow simultaneous comparisons of normal and disease tissues, as well as control versus treated samples.

The work summarized in this article focuses on the use of transcriptional profiling to study developmental regulatory programs involved in nephrogenesis and the role of aryl hydrocarbon receptor (AHR) and Wilms' tumor suppressor gene (Wt1) in this process. The aryl hydrocarbon receptor gene (Ahr) has been identified as a liver cytosolic protein that mediates transcriptional activation of cytochrome *P4501a1* gene in response to planar aromatic hydrocarbons.[2] In elegant studies by several independent groups, it has been established that the Ahr locus encodes a PAS (Per-Arnt-Sim) homology domain transcription factor that binds with high affinity to hydrocarbon ligands and interacts with DNA via a basic helix-loop-helix DNA-binding domain.[3–5] These studies prompted intensive worldwide investigations spanning over 20 years detailing the molecular mechanisms responsible for the nuclear functions of the AHR protein and its role in mediating the biochemical and toxic effects of planar aromatic hydrocarbon ligands of the receptor.[6–8] It was not until the development of Ahr knockout mice that the environmental health science research community began to appreciate the role of Ahr as a developmental transcription factor involved in organogenesis.[9] More recently, this laboratory published evidence implicating AHR in the regulation of kidney formation (i.e., nephrogenesis) in mice.[10] Kidney formation in mammals proceeds through a series of carefully integrated cell–cell and cell–matrix interactions that involve specific induction of metanephric blastema by the ureteric bud and epithelialization. Ligands of the Ahr have been shown to disrupt this carefully orchestrated process. The molecular mechanisms involved in this deleterious response are detailed here and discussed within the framework of environmental systems biology.

RESULTS

Exposure of mice to benzo(a)pyrene (BaP), a polycyclic aromatic hydrocarbon generated as a combustion byproduct and identified as a probable human carcinogen, has been associated with several diseases, such as atherosclerosis and cancer.[11] The biochemical and toxic actions of BaP are closely linked to the AHR since toxicity is antagonized following blockade of AHR or AHR-related functions using genetic or pharmacological approaches.[12] AHR participates in transcriptional and post-transcriptional regulation of cytochrome *P450* genes involved in cellular metabolism of BaP to reactive chemical intermediates.[13] These intermediates in turn interact with cellular macromolecules and cause structural and functional deficits that disrupt cellular function. BaP binds AHR to activate AHR signaling in mammalian cells and alter patterns of gene expression.

Consecutive challenge of metanephric kidneys in short-term organ culture with a BaP concentration well within the range of the levels encountered by humans in the environment[14] is associated with inhibition of nephrogenesis.[10] This response involves the AHR since it can be antagonized by α-nathoflavone, an antagonist of AHR signaling and inhibitor of cytochrome P450s. The inhibition of nephrogenesis is associated with inhibition of metanephric cell differentiation and alterations in the relative abundance of Wt1 splice variants (FIG. 1).

The *Wt1* gene has been identified as a master switch in the regulation of kidney developmental programs. *Wt1* gene encodes a Cis2-His2 zinc finger DNA-binding protein that mediates transcriptional repression/activation by binding DNA sequences containing the 5'-GCGGGGGCG-3' as well as (TCC)n repeats. Wt1 functions as a transcriptional repressor or a transcriptional activator depending upon cellular context, interacting protein partners, and/or stage of differentiation.[15] Many of the genes involved in nephrogenesis, including insulin-like growth factor, frizzle-like protein, and E-cadherin, have been identified as downstream transcriptional targets of Wt1. The function of *Wt1* gene is regulated at the transcriptional, post-transcriptional, and post-translational levels.[16] At the post-transcriptional level, Wt1 undergoes differential splicing on two different exons to give rise to four major variants expressed in the kidney and involved in transcriptional control as well as RNA processing. One alternative splicing event results in the inclusion/exclusion of exon 5, a sequence that encodes for a stretch of 17 amino acids within the N-terminus of the four zinc fingers. Another significant splicing event involves

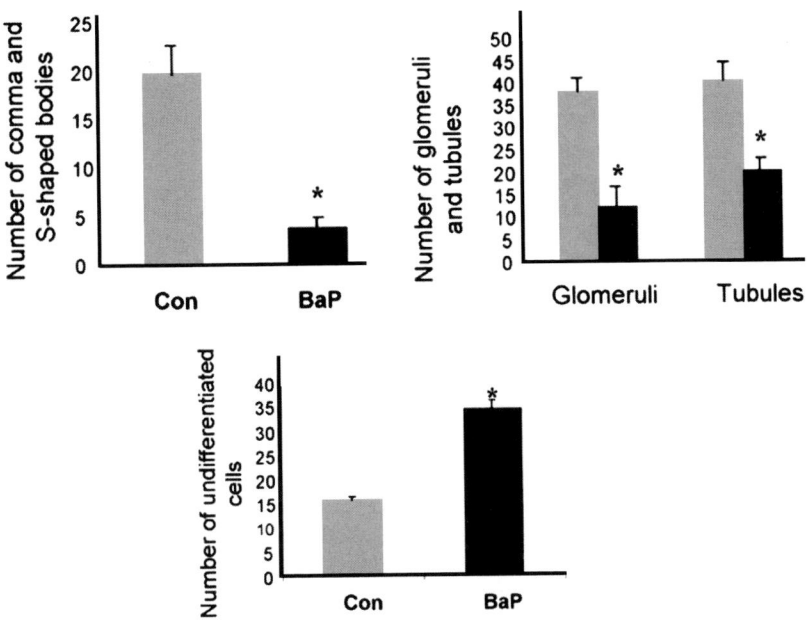

FIGURE 1. Deficits in murine metanephric differentiation following challenged with 3 μM benzo(a)pyrene. Kidneys were harvested and fixed *in situ* for microscopic evaluation. Histograms shown are the results of morphometric analysis of control and BaP-treated cultures. Data are presented as the mean ± SD and representative of four different experiments. (Modified from Falahatpisheh & Ramos, 2003[10])

a splice site in exon 9 resulting in the insertion of lysine–threonine–serine between zinc fingers 3 and 4 in the DNA-binding domain. These four splice variants are all expressed in the kidney in temporal, spatial, and evolutionarily stable ratios. The essential role of Wt1 in kidney development was established by the phenotype of Wt1 null mice, which fail to develop metanephric kidneys and die *in utero*.

To study the influence of AHR activation by BaP on Wt1 signaling, metanephric kidney cultures were isolated from mice of varying Ahr genotype and treated with BaP under experimental conditions known to interfere with nephrogenesis.[10] Total RNA was isolated from $Ahr^{+/+}/Ahr^{-/+}$ and $Ahr^{-/-}$ mice and processed for global measurements of gene expression using the murine genome Affymetrix GeneChip MG·U74Av2 high density oligonucleotide array. This platform contained all sequences (∼6000) in the Mouse UniGene database that had been functionally characterized at the time the experiment was completed. In addition, ∼6000 EST clusters are also represented on this single array. In parallel studies, cultures were processed for measurements of cellular metabolite levels using NMR-coupled MS. The fluorescence intensities obtained from microarray experiments were processed for background correction, normalization, and statistical analysis as previously described.[17] The data set was also processed for K-means clustering and calculation of predictor networks.[18] Discrete networks of biological activity regulated by BaP during nephrogenesis included glial derived neurotrophic factor, small frizzle-like protein, insulin-like growth factor receptor, and syndecan (Falahatpisheh *et al.*, unpublished observations). These genes have previously been identified as genes involved in nephrogenesis.[19] The surprising finding was that modulation of gene expression by BaP was not regulated in an AHR-dependent manner (Falahatpisheh *et al.*, unpublished observations). Novel targets of hydrocarbon interference in the developing kidney included sry, oncostatin M, pinin, GATA1, Sox-18, and Nrf2. BaP-induced deficits in nephrogenesis could be reproduced in kidneys of AHR knockout mice suggesting that AHR is an essential protein in renal development. Metabolic profiling experiments established that hydrocarbon treatment interfered with phospholipids signaling cascades and taurine, a sulfur-containing amino acid (Ramos *et al.*, unpublished observations). This independent set of observations was integrated in a systems biology paradigm to begin to define biological networks of regulatory activity. Biological networks were constructed based on the expression of predictors that are correlated under diverse conditions and that are predictive of the behavior of the system as a whole. The connectivity properties of the gene co-expression network were defined using functional genomics approaches, such as silencing RNA methodology. Boolean idealizations were systematically evaluated using the coefficient of determination (CoD) algorithm developed by Kim *et al.*[20] The CoD methodology detects multivariate nonlinear influences on gene expression. This algorithm identifies associations between the expression patterns of individual genes by determining whether the transcriptional levels of a small gene set predict the associated transcriptional state of another gene. A key goal in networks analysis is the development of analytical tools to delineate how individual gene actions are integrated into complex biological systems. Some of the genes predicted to be part of the AHR gene regulatory network included lymphocyte antigen 6 complex, locus e, insulin-like growth factor, binding protein 4, and tumor necrosis factor receptor.[21] The complexity of this regulatory network was visualized using a linkage diagram (FIG. 2). This diagram was built using the ranks for each predictor gene and each target gene. The inclusion of any singular gene does not represent a causal relationship between predictor and target, but rather that gene a given gene in combination with two other genes was a good predictor of a specific target.

CONCLUSIONS

Evidence was presented that relevant components of the biological response to AHR ligands could be unraveled using modern molecular technologies. AHR was shown to play an essential role in kidney development, and this response involves complex transcriptional and post-transcriptional interactions with

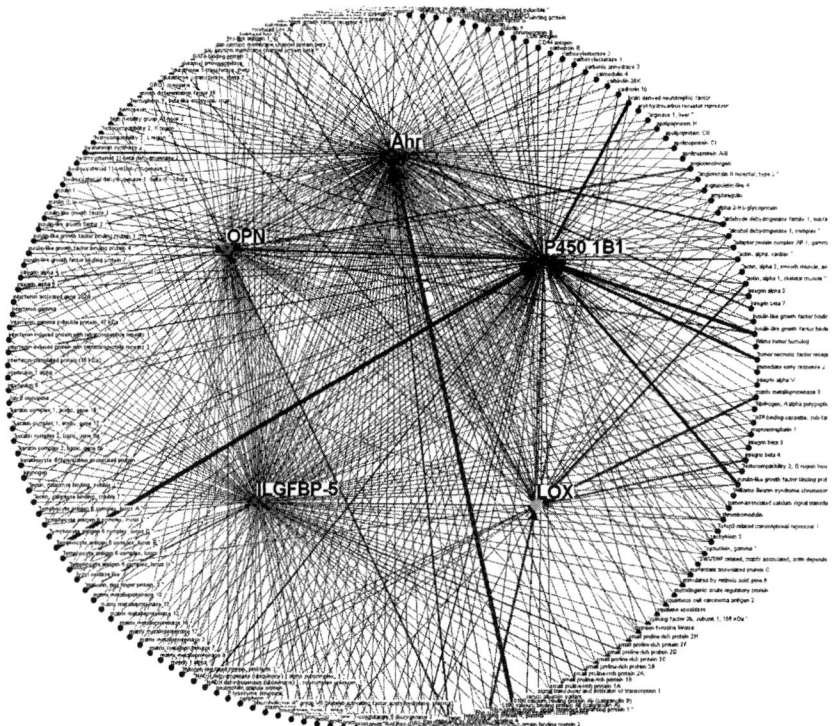

FIGURE 2. Gene–gene interaction networks regulated by ligands of the aryl hydrocarbon receptor. This linkage diagram was constructed using all three gene combinations for each pre-determined target meeting established cutoff parameters as described.[21] The thickness of the line denotes the selection frequency for individual predictor relationships. (Reprinted from Johnson et al. 2004 with permission)

Wt1 and downstream targets of this gene. On the basis of these findings, we conclude that toxicogenomics technologies have emerged as powerful tools to evaluate the molecular response of biological systems to environmental contaminants. Putative targets for developmental interference in mammals can be identified by transcriptional profiling and confirmed using functional genomics approaches. The study of global transcriptional profiles coupled to intervention strategies using genetically modified model systems facilitates detailed measurements of dose- and time-dependent molecular effects and for resolution of biological regulatory networks of activity.

ACKNOWLEDGMENTS

The contributions of Teresa Fan, Hadi Falahatpisheh, Charlie Johnson, Andrew Lane, and Adrian Nanez to the published and unpublished work presented in this communication are gratefully acknowledged. The work was supported in part by grants from the National Institute of Environmental Health Sciences and the National Heart Lung and Blood Institute.

REFERENCES

1. RAMOS, K.S. 2003. EHP toxicogenomics: a publication forum for the post-genomic era. EHP Toxicogenomics **111**: A13.
2. SCHMIDT, J.V. & C.A. BRADFIELD. 1996. Ah receptor signaling pathways. Ann. Rev. Cell Dev. Biol. **12**: 55–89.
3. WALKER, M.K. et al. 2000. Molecular characterization and developmental expression of the aryl hydrocarbon receptor from the chick embryo. Comp. Biochem. Physiol. C. Toxicol. Pharmacol. **126**: 305–319.
4. THOMAS, R.S. et al. 2002. Sequence variation and phylogenetic history of the mouse Ahr gene. Pharmacogenetics **12**: 151–163.
5. CARVER, L.A. & C.A. BRADFIELD. 1997. Ligand-dependent interaction of the aryl hydrocarbon receptor with a novel immunophilin homolog in vivo. J. Biol. Chem. **272**: 11452–11456.
6. LAHVIS, G.P. et al. 2005. The aryl hydrocarbon receptor is required for developmental closure of the ductus venosus in the neonatal mouse. Mol. Pharmacol. **67**: 714–720.
7. KERZEE, J.K. & K.S. RAMOS. 2001. Constitutive and inducible expression of Cyp1a1 and Cyp1b1 in vascular smooth muscle cells: role of the Ahr bHLH/PAS transcription factor. Circ. Res. **89**: 573–582.
8. WALISSER, J.A. et al. 2004. Gestational exposure of Ahr and Arnt hypomorphs to dioxin rescues vascular development. Proc. Natl. Acad. Sci. USA **101**: 16677–16682.
9. BUNGER, M.K. et al. 2003. Resistance to 2,3,7,8-tetrachlorodibenzo-p-dioxin toxicity and abnormal liver development in mice carrying a mutation in the nuclear localization sequence of the aryl hydrocarbon receptor. J. Biol. Chem. **16**: 17767–17774.
10. FALAHATPISHEH, H.M. & K.S. RAMOS. 2003. Ligand activated Ahr signaling leads to disruption of nephrogenesis and altered Wilms' tumor suppressor gene splicing. Oncogene **14**: 2160–2171.
11. NEBERT, D.W. et al. 2000. Role of the aromatic hydrocarbon receptor and [Ah] gene battery in the oxidative stress response, cell cycle control, and apoptosis. Biochem. Pharmacol. **59**: 65–85.
12. KERZEE, J.K. & K.S. RAMOS. 2000. Activation of c-Ha-ras by benzo(a)pyrene in vascular smooth muscle cells involves redox stress and aryl hydrocarbon receptor. Mol. Pharmacol. **58**: 152–158.
13. MILLER, K.P. & K.S. RAMOS. 2001. Impact of metabolism on the biological effects of benzo(a)pyrene and related hydrocarbons. Drug Metab. Rev. **33**: 1–35.
14. Toxicological profile for polycyclic aromatic hydrocarbons. US EPA. 1990. Publication no. PB 91-181537.
15. DRUMMOND, I.A. et al. 1992. Repression of the insulin-like growth factor II gene by the wilms' tumor suppressor WT1. Science **257**: 674–678.
16. DISCENZA, M.T. et al. 2004. Activation of the WT1 tumor suppressor gene promoter by Pea3. FEBS Lett. **560**: 183–189.

17. JOHNSON, C.D. et al. 2003. Genomic profiles and predictive biological networks in oxidant-induced atherogenesis. Physiol. Genomics **13:** 263–275.
18. BRUN, M. et al. 2005. Clustering: revealing intrinsic dependencies in microarray data. *In* Genomic Signal Processing and Statistics. EURASIP Book Series on Signal Processing and Communications, Ed.: 129–162. Hindawi Publishing Corporation.
19. DAVIES, J.A. et al. 2003. Development of an siRNA-based method for repressing specific genes in renal organ culture and its use to show that the Wt1 tumour suppressor is required for nephron diffentiation. Hum. Mol. Genet. **13:** 235–246.
20. KIM, S., E.R. DOUGHERTY, M.L. BITTNER, et al. 2000a. Multivariate measurement of gene expression relationships. Genomics **67:** 201–209.
21. JOHNSON, C.D. et al. 2004. Unraveling gene-gene interactions regulated by ligands of the aryl hydrocarbon receptor. EHP Toxicogenomics **112:** 403–412.

Results of Long-Term Carcinogenicity Bioassays on Coca-Cola Administered to Sprague-Dawley Rats

FIORELLA BELPOGGI, MORANDO SOFFRITTI, EVA TIBALDI,
LAURA FALCIONI, LUCIANO BUA, AND FRANCESCA TRABUCCO

Cesare Maltoni Cancer Research Center, European Foundation for Oncology and Environmental Sciences "B. Ramazzini," 40010 Bentivoglio, Bologna, Italy

ABSTRACT: Coca-Cola was invented in May 1886 in Atlanta, Georgia by a pharmacist who, by accident or design, mixed carbonated water with the syrup of sugar, phosphoric acid, caffeine, and other natural flavors to create what is known as "the world's favorite soft drink." Coca-Cola is currently sold in more than 200 countries and in early 2000, the company sold its 10 billionth unit case of Coca-Cola branded products. Given the worldwide consumption of Coca-Cola, a project of experimental bioassays to study its long-term effects when administered as substitute for drinking water on male and female Sprague-Dawley rats was planned and executed. The objective of the project was to study whether and how long-term consumption of Coca-Cola affects the basic tumorigram of test animals. The bioassays were performed on rats beginning at different ages, namely: (*a*) on males and females exposed since embryonic life or from 7 weeks of age; and (*b*) on males and females exposed from 30, 39, or 55 weeks of age. Overall, the project included 1999 rats. During the biophase, data were collected on fluid and feed consumption, body weight, and survival. Animals were kept under observation until spontaneous death and underwent complete necropsy. The results indicate: (*a*) an increase in body weight in all treated animals; (*b*) a statistically significant increase of the incidence in females, both breeders and offspring, bearing malignant mammary tumors; (*c*) a statistically significant increase in the incidence of exocrine ademonas of the pancreas in both male and female breeders and offspring; and (*d*) an increased incidence, albeit not statistically significant, of pancreatic islet cell carcinomas in females, a malignant tumor which occurs very rarely in our historical controls. On the basis of the results of this study, excessive consumption

Address for correspondence: Morando Soffritti, M.D., Cesare Maltoni Cancer Research Center, European Ramazzini Foundation, Castello di Bentivoglio, 40010 Bentivoglio (BO), Italy. Voice: +39-051-6640460; fax: +39-051-6640223.
e-mail: crcfr@ramazzini.it; www.ramazzini.it

Funding for this research was provided entirely by the European Foundation on Oncology and Environmental Sciences "B. Ramazzini."

of regular soft-drinks should be generally discouraged, in particular for children and adolescents.

KEYWORDS: Coca-Cola; carcinogenicity; long-term bioassay; rat

INTRODUCTION

Cancer is one the principal public health problems faced by industrialized countries, both because of its epidemiological dimension and because of the growing environmental influences behind its increase to today's epidemic proportions. Diet is often cited as an important factor in the occurrence of cancer; however, influence of diet on health cannot be easily isolated or quantified given the varied composition of the average diet over a lifetime and the complex mixture of substances present in food. While the overall diet is difficult to analyze, specific nutrients and food components have been implicated both in causing cancer and as protective factors in the carcinogenic process.

There are several sources of food components believed to induce cancer. One is the presence of naturally occurring substances, such as aflatoxin found in groundnuts and corn. Another potential source of risk are substances which are added directly to food in order to improve its quality, taste, or stability.

It has long been suggested that total calorie intake may have an important effect on the incidence of cancer. Data by Tannenbaum in the 1940s and 1950s demonstrated that underfed mice developed fewer spontaneous mammary tumors, primary lung adenomas, and chemically induced tumors.[1,2] In addition, human obesity has been identified as a risk factor for some types of cancer.[3,4]

Long-term carcinogenicity bioassays on rodents, which reproduce exposure situations experienced by humans as much as possible, is in our opinion, a good approach to study the influence of dietary products on the incidence of spontaneous tumors in a controlled environment.[5,6] In an attempt to evaluate the interaction between one such product and tumor incidence in rodents, an experimental carcinogenicity bioassay was performed at the Cesare Maltoni Cancer Research Center (CMCRC) of the European Ramazzini Foundation (ERF). The product tested is a beverage mixture which is both caloric and widely consumed in most of the world—Coca-Cola.

Coca-Cola was invented in Atlanta, Georgia, on May 8, 1886 by pharmacist John Pemberton. The regular cola beverage contains water, sugar (about 10% by weight), carbon dioxide, caramel coloring, acidulates, flavoring substances, and caffeine, all of which are approved for use around the world. Today Coca-Cola, recognized as the world's number one soft drink brand, is currently sold in more than 200 countries. In 2001, the company sold its 10 billionth unit case of Coca-Cola branded products.[7]

The experimental project described in this article was designed to evaluate the long-term effects of Coca-Cola on the spontaneous development of tumors when administered as a substitute for drinking water to Sprague-Dawley rats for the life span.

Since it is known that the age of the animals at the start of the experiment may affect the possible modulating effects of the test compounds, the project includes experiments performed on groups of male and female breeder rats (30, 39, and 55 weeks old at the start of the experiment), all the offspring of each litter descending from the aforementioned breeders (exposed since intrauterine life), and one group exposed beginning at 7 weeks of age.

MATERIALS AND METHODS

The Coca-Cola used was supplied by an Italian reseller, in 1-L glass bottles delivered every 2–3 weeks. During the experiments, the bottles were stored at a room temperature of $22° \pm 3°C$. Before administering the Coca-Cola to the rodents, the CO_2 was first eliminated by mechanical shaking for 60 minutes.

The male (M) and female (F) Sprague-Dawley rats used in this experiment were bred from the colony used at the CMCRC/ERF laboratories for nearly 30 years. Extensive historical data on more than 15,000 rodents are available on the tumor incidence among untreated rats.

At 4 weeks of age, the breeders were identified by ear punch, separated by sex, and assigned to experimental groups so as to have no more than one male and one female from each litter in the same group. Rats were housed 5 per cage until the time of mating, at which time parents were placed in breeding cages (one male with one female). After 7 days, males were re-housed 5 per cage and females were housed individually. After 8 weeks, the females were also re-housed 5 per cage.

All available pups of all litters from treated and untreated dams were included in the various groups of offspring. After weaning (at 4–5 weeks), offspring were identified by ear punch, weighed, separated by sex, and assigned sequentially, litter by litter, to the exposed and control group, respectively. The two groups of 7-week-old rats were randomized in order to have no more than one male and one female from each litter in the same group.

Animals were housed in makrolon cages (41 cm × 25 cm × 15 cm) with stainless-steel wire tops and a shallow layer of white wood shavings as bedding. Cages were kept in rooms used exclusively for this experiment at a temperature of $21° \pm 2°C$ and relative humidity of 50–60%. A light–dark cycle of 12 hours was maintained using both natural and artificial light sources.

Animals were given the standard Corticella pellet diet (Corticella S.p.A., Bologna, Italy), analyzed for nutritional components and possible contaminants. Every 24 hours, Coca-Cola or drinking water (control) were disposed of and the bottles were cleaned and refilled.

The animals were checked three times a day from Monday to Saturday, and twice on Sundays and holidays for clinical symptoms and behavior.

Coca-Cola was administered *ad libitum* instead of drinking water to groups of breeders (55 – 110 per sex per group) 30, 39, or 55 weeks old at the start of the experiment, to their offspring exposed since intrauterine life (24 – 110 per sex per group), and to a group of 80 males and 80 females, 7 weeks old at the start of the experiment. The administration of Coca-Cola lasted until spontaneous death. Equivalent groups of breeders and offspring administered drinking water served as controls. The entire project involved 1999 rodents.

Mean daily drinking water and feed consumption were measured on the day prior to the start of the breeders' treatment, after 7 days of treatment before mating and then, post-weaning, once weekly for the first 13 weeks from the start of the experiment, then every 2 weeks until 104 weeks. Individual animal weight of breeders and of offspring after weaning was measured once weekly for the first 13 weeks, every 2 weeks until 104 weeks, and then every 8 weeks until the end of the experiments. In order to detect and register all gross lesions, the animals were examined every week for the first 13 weeks, and then every 2 weeks until the end of the experiment.

Upon death, all animals underwent complete necropsy. Histopathology was routinely performed on all macroscopically observed pathological lesions (with a margin of surrounding normal tissue) and on skin and subcutaneous tissue, the brain, pituitary gland, Zymbal glands, salivary glands, Harderian glands, cranium (with oral and nasal cavities and external and internal ear ducts, 5 levels), tongue, thyroid and parathyroid, pharynx, larynx, thymus and mediastinal lymph nodes, trachea, lung and mainstem bronchi, heart, diaphragm, liver, spleen, pancreas, kidneys and adrenal glands, esophagus, stomach (fore and glandular), intestine (4 levels), bladder, prostate, uterus, gonads, interscapular fat pad, subcutaneous and mesenteric lymph nodes, and any other organ or tissue with pathological lesions.

All organs and tissues were preserved in 70% ethyl alcohol, except for the bones which were fixed in 10% formalin and then decalcified with 10% formaldehyde and 20% formic acid in water solution. The normal specimens were trimmed, following Standard Operating Procedures (SOP) at the CMCRC/ERF laboratories. Trimmed specimens were processed as paraffin blocks, and 3–5 μm sections of every specimen were obtained. Sections were routinely stained with hematoxylin-eosin. Specific stainings were performed when needed. All slides were examined microscopically by the same group of pathologists; a senior pathologist reviewed all tumors and any other lesion of oncological interest.

Statistical analysis was performed using the χ^2 test in order to evaluate the significance in tumor incidence differences between treated and control groups.

RESULTS

In this article, we present the results of data aggregated from all breeders and all offspring. Offspring include both rodents whose treatment began during embryonic life and those rodents whose treatment began at 7 weeks of age.

A treatment-related difference in fluid consumption was observed, in both sexes of breeders and offspring, with treated animals consuming more than two times as much fluid compared to controls. An opposite treatment-related difference in food consumption was observed in both sexes of breeders and offspring during the experiment, with treated animals consuming about 40% less feed compared to controls.

An increase in body weight was observed, in both breeders and offspring, among both sexes in treated groups compared to the controls.

No differences in survival were observed among treated males and females breeders compared to the controls. A slight decrease in survival was observed in female offspring compared to the controls.

The occurrence of benign and malignant tumors among male and female rats is shown in TABLE 1. The differences observed among treated and control animals were as follows:

1. a slight increase in the incidence of malignant tumor-bearing animals was observed in female breeders compared to controls (TABLE 2);
2. a statistically significant increase in females bearing malignant mammary tumors and total malignant mammary tumors, both among breeders ($P < 0.01$) and offspring ($P < 0.01$), compared to the controls (TABLE 3);
3. a statistically significant increased incidence in animals bearing adenomas of the exocrine component of the pancreas was observed among male ($P < 0.01$) and female ($P < 0.05$) breeders and male ($P < 0.01$) and female ($P < 0.01$) offspring compared to the controls. No exocrine carcinomas were observed (TABLE 4). Concerning the incidence of islet cell adenomas, no differences were observed in treated breeders or offspring of either sex compared to the controls. An increased incidence of islet cell carcinomas was observed in treated female breeders and treated female offspring compared to the controls. Only one case of carcinoma was observed in treated male offspring. It must be noted that in our historical controls, we observed only 1 (0.04%) islet cell carcinoma out of 2274 untreated females and 5 (0.2%) islet cell carcinomas out of 2265 males.

CONCLUSIONS

Although Coca-Cola is one of the most consumed soft-drinks worldwide, to our knowledge this study represents the first time that a long-term

TABLE 1. Long-term carcinogenicity bioassays on Coca-Cola, administered instead of drinking water, supplied *ad libitum*, to male and female Sprague-Dawley rats

Number and percentage of male and female sprague-dawley rats bearing various types of benign and malignant tumors[a]

Site / Histotype	I: Coca-Cola (Breeders) Male No.	%	Female No.	%	II: Drinking water (Breeders) Male No.	%	Female No.	%	III: Coca-Cola (Offspring) Male No.	%	Female No.	%	IV: Drinking water (Offspring) Male No.	%	Female No.	%
Skin																
Acanthoma	0	—	0	—	3	1.3	0	—	1	0.4	0	—	0	—	0	—
Epithelioma	1	0.4	0	—	0	—	0	—	0	—	0	—	0	—	0	—
Dermatofibroma	4	1.7	0	—	3	1.3	1	0.4	2(3)	0.8	0	—	1	0.3	1	0.4
Squamous cell carcinoma	2	0.9	1	0.4	1	0.4	1	0.4	0	—	0	—	1	0.3	1	0.4
Basal cell carcinoma	1	0.4	0	—	0	—	1	0.4	0	—	0	—	1	0.3	0	—
Sebaceous adenocarcinoma	0	—	1	0.4	0	—	1	0.4	0	—	1	0.4	0	—	0	—
Subcutaneous tissue																
Fibroma	6	2.6	0	—	5	2.1	0	—	4	1.6	0	—	2	0.7	0	—
Lipoma & fibrolipoma	0	—	0	—	0	—	0	—	1	0.4	1	0.4	2	0.7	1	0.4
Liposarcoma	0	—	1	0.4	0	—	0	—	1	0.4	1	0.4	2	0.7	0	—
Hemangiopericytosarcoma	0	—	0	—	1	0.4	0	—	0	—	0	—	0	—	0	—
Rhabdomyosarcoma	0	—	0	—	0	—	0	—	0	—	0	—	0	—	1(2)	0.4
Interscapular fat pad																
Fibroangioma	0	—	0	—	0	—	0	—	0	—	0	—	1	0.3	0	—
Mammary glands[b]																
Fibroma & fibroadenoma	11(13)	4.7	83(120)	35.3	10(12)	4.3	83(122)	35.3	16	6.4	118(180)	48.8	18(21)	6.2	124(179)	44.8
Lipoma & fibrolipoma	3	1.3	3	1.3	0	—	1	0.4	5(6)	2.0	3	1.2	2	0.7	0	—
Fibroangioma	0	—	0	—	1	0.4	0	—	0	—	0	—	0	—	0	—
Adenocarcinoma	4	1.7	30(41)	12.8	0	—	14	6.0	0	—	43(54)	17.8	2	0.7	24(28)	8.7
Fibrosarcoma	0	—	5	2.1	0	—	1	0.4	1	0.4	4	1.7	1	0.3	2	0.7
Liposarcoma	2	0.9	2	0.9	1	0.4	0	—	2	0.8	0	—	0	—	1	0.4
Leiomyosarcoma	0	—	0	—	0	—	0	—	0	—	0	—	0	—	1	0.4
Hemangiosarcoma	0	—	0	—	1	0.4	0	—	0	—	0	—	0	—	0	—
Carcinosarcoma	0	—	0	—	0	—	0	—	0	—	0	—	0	—	1	0.4

Continued.

TABLE 1. Continued.

Number and percentage of male and female sprague-dawley rats bearing various types of benign and malignant tumors[a]

Site Histotype	I: Coca-Cola (Breeders)				II: Drinking water (Breeders)				III: Coca-Cola (Offspring)				IV: Drinking water (Offspring)			
	Male		Female		Male		Female		Male		Female		Male		Female	
	No.	%	No.	%	No.	%	No.	%	No.	%	No.	%	No.	%	No.	%
Zymbal glands																
Sebaceous adenoma	0	—	0	—	0	—	0	—	0	—	0	—	0	—	2	0.7
Carcinoma	7	3.0	7	3.0	7	3.0	3(4)	1.3	5	2.0	3	1.2	7	2.4	8	2.9
Ear ducts																
Carcinoma	13	5.5	18	7.7	14	6.0	15	6.4	9	3.6	6	2.5	20(22)	6.9	16(17)	5.8
Nasal cavities																
Carcinoma	3	1.3	3	1.3	5	2.1	2	0.9	2	0.8	0	—	0	—	1	0.4
Olfactory neuroblastoma	0	—	1	0.4	2	0.9	3	1.3	4	1.6	1	0.4	0	—	0	—
Oral cavity & lips																
Acanthoma	0	—	0	—	0	—	2	0.9	0	—	0	—	1	0.3	0	—
Carcinoma	1	0.4	2	0.9	6	2.6	6	2.6	4	1.6	4	1.7	8	2.7	10	3.6
Malignant ameloblastoma	0	—	0	—	0	—	1	0.4	0	—	0	—	0	—	0	—
Pharynx																
Acanthoma	0	—	0	—	0	—	0	—	0	—	1	0.4	0	—	0	—
Carcinoma	1	0.4	0	—	1	0.4	1	0.4	0	—	1	0.4	1	0.3	0	—
Larynx																
Carcinoma	0	—	0	—	0	—	0	—	0	—	0	—	0	—	1	0.4
Lung																
Adenoma	2	0.9	1	0.4	0	—	0	—	1	0.4	0	—	2	0.7	1	0.4
Fibroangioma	0	—	0	—	0	—	0	—	0	—	0	—	0	—	1	0.4
Adenocarcinoma	0	—	0	—	1	0.4	1	0.4	0	—	0	—	1	0.3	0	—
Leiomyosarcoma	0	—	0	—	0	—	0	—	0	—	0	—	0	—	1	0.4
Squamous cell carcinoma	0	—	0	—	0	—	0	—	0	—	0	—	0	—	1	0.4

Continued.

TABLE 1. Continued.

Number and percentage of male and female sprague-dawley rats bearing various types of benign and malignant tumors[a]

Site / Histotype	I: Coca-Cola (Breeders)				II: Drinking water (Breeders)				III: Coca-Cola (Offspring)				IV: Drinking water (Offspring)			
	Male		Female		Male		Female		Male		Female		Male		Female	
	No.	%	No.	%	No.	%	No.	%	No.	%	No.	%	No.	%	No.	%
Stomach																
- Forestomach																
Acanthoma	3	1.3	8	3.4	1	0.4	2	0.9	8	3.2	8	3.3	5	1.7	5	1.8
Leiomyoma	0	–	0	–	0	–	0	–	1	0.4	0	–	0	–	0	–
Carcinoma	0	–	0	–	1	0.4	0	–	0	–	0	–	1	0.3	0	–
Leiomyosarcoma	0	–	0	–	1	0.4	0	–	0	–	0	–	0	–	0	–
- Glandular stomach																
Adenocarcinoma	0	–	0	–	1	0.4	1	0.4	0	–	1	0.4	0	–	0	–
Intestine																
Adenoma	1	0.4	0	–	0	–	0	–	0	–	0	–	1	0.3	0	–
Leiomyoma	0	–	0	–	0	–	0	–	0	–	1	0.4	0	–	1	0.4
Adenocarcinoma	0	–	1	0.4	1	0.4	0	–	1	0.4	0	–	0	–	0	–
Leiomyosarcoma	0	–	0	–	0	–	1	0.4	0	–	0	–	0	–	0	–
Salivary glands																
Adenocarcinoma	1	0.4	0	–	0	–	0	–	0	–	0	–	0	–	0	–
Liver																
Cholangioma	0	–	0	–	0	–	0	–	0	–	0	–	1	0.3	1	0.4
Hepatocarcinoma	2	0.9	2	0.9	3	1.3	0	–	1	0.4	0	–	11	3.8	4	1.4
Cholangiocarcinoma	0	–	0	–	0	–	0	–	0	–	1	0.4	0	–	0	–
Hemangiosarcoma	0	–	0	–	0	–	0	–	1	0.4	0	–	1	0.3	2	0.7
Pancreas[c]																
Exocrine adenoma	18	7.7	10	4.3	1	0.4	1	0.4	23	9.2	16	6.6	1	0.3	1	0.4
Islet cell adenoma	12	5.1	3	1.3	13	5.5	5	2.1	14	5.6	8	3.3	19	6.5	7	2.5
Islet cell carcinoma	0	–	2	0.9	0	–	0	–	1	0.4	3	1.2	0	–	0	–

Continued.

TABLE 1. Continued.

Number and percentage of male and female sprague-dawley rats bearing various types of benign and malignant tumors[a]

Site / Histotype	I: Coca-Cola (Breeders) Male No.	%	I: Coca-Cola (Breeders) Female No.	%	II: Drinking water (Breeders) Male No.	%	II: Drinking water (Breeders) Female No.	%	III: Coca-Cola (Offspring) Male No.	%	III: Coca-Cola (Offspring) Female No.	%	IV: Drinking water (Offspring) Male No.	%	IV: Drinking water (Offspring) Female No.	%
Kidneys																
Adenoma	2	0.9	0	—	0	—	0	—	2	0.8	1	0.4	1	0.3	0	—
Lipoma & fibrolipoma	1	0.4	0	—	0	—	0	—	1	0.4	1	0.4	0	—	0	—
Fibroangioma	0	—	0	—	0	—	0	—	0	—	0	—	1	0.3	0	—
Adenocarcinoma	0	—	0	—	0	—	1	0.4	0	—	2	0.8	0	—	1	0.4
Liposarcoma	0	—	0	—	0	—	0	—	0	—	0	—	1	0.3	0	—
Nephroblastoma	0	—	1 (2)	0.4	0	—	0	—	0	—	0	—	0	—	1	0.4
Pelvis & ureteres																
Transitional cell carcinoma	0	—	0	—	0	—	0	—	0	—	0	—	0	—	1	0.4
Bladder																
Papilloma	0	—	0	—	1	0.4	0	—	0	—	0	—	0	—	0	—
Transitional cell carcinoma	0	—	1	0.4	0	—	0	—	0	—	1	0.4	0	—	0	—
Seminal vesicles																
Adenocarcinoma	1	0.4			0	—			2	0.8			0	—		
Prostate																
Adenoma	2	0.9			0	—			1	0.4			1	0.3		
Adenocarcinoma	2	0.9			1	0.4			1	0.4			1	0.3		
Testes																
Interstitial cell adenoma	10 (12)	4.3			19 (23)	8.1			11 (14)	4.4			20 (22)	6.9		
Ovaries																
Cystadenoma			0	—			4	1.7			1	0.4			0	—
Granulosa cell tumor			4 (5)	1.7			5 (7)	2.1			9 (13)	3.7			10 (15)	3.6
Leiomyoma			0	—			0	—			0	—			1	0.4
Sertoli cell tumor			0	—			1 (2)	0.4			4 (6)	1.7			3 (4)	1.1
Adenocarcinoma			0	—			0	—			1	0.4			0	—
Arrhenoblastoma			0	—			0	—			0	—			1	0.4

Continued.

TABLE 1. Continued.

Number and percentage of male and female sprague-dawley rats bearing
Various types of benign and malignant tumors[a]

Site Histotype	I: Coca-Cola (Breeders)				II: Drinking water (Breeders)				III: Coca-Cola (Offspring)				IV: Drinking water (Offspring)			
	Male		Female		Male		Female		Male		Female		Male		Female	
	No.	%	No.	%	No.	%	No.	%	No.	%	No.	%	No.	%	No.	%
Uterus																
Polyp			18	7.7			18	7.7			18	7.4			25	9.0
Leiomyoma			2	0.9			3	1.3			1	0.4			2	0.7
Fibroangioma			0	—			1	0.4			0	—			2	0.7
Granular cell tumor (Abrikosoff's tumor)			0	—			0	—			0	—			1	0.4
Adenocarcinoma			7	3.0			9	3.8			13	5.4			17	6.1
Squamous cell carcinoma			3	1.3			7	3.0			3	1.2			1	0.4
Liposarcoma			1	0.4			0	—			0	—			0	—
Hemangiosarcoma			1	0.4			0	—			0	—			0	—
Malignant schwannoma			0	—			0	—			1	0.4			1	0.4
Sarcoma			1	0.4			2	0.9			0	—			0	—
Uterus & vagina																
Malignant schwannoma			0	—			1	0.4			1	0.4			2	0.7
Sarcoma			1	0.4			0	—			1	0.4			3	1.1
Vagina																
Sarcoma			0	—			1	0.4			0	—			0	—
Peritoneum																
Lipoma	0	—	0	—	0	—	1	0.4	1	0.4	0	—	0	—	0	—
Liposarcoma	0	—	0	—	0	—	0	—	1	0.4	1	0.4	0	—	0	—
Hemangiosarcoma	0	—	0	—	0	—	0	—	0	—	1	0.4	0	—	0	—
Mesothelioma	1	0.4	0	—	0	—	0	—	0	—	0	—	1	0.3	2	0.7
Pituitary gland																
Adenoma	44	18.7	89	37.9	35	14.9	88	37.4	70	28.1	132	54.5	59	20.3	128	46.2
Adenocarcinoma	0	—	1	0.4	0	—	0	—	0	—	0	—	0	—	1	0.4

Continued.

TABLE 1. Continued.

Number and percentage of male and female sprague-dawley rats bearing various types of benign and malignant tumors[a]

Site Histotype	I: Coca-Cola (Breeders)				II: Drinking water (Breeders)				III: Coca-Cola (Offspring)				IV: Drinking water (Offspring)			
	Male		Female		Male		Female		Male		Female		Male		Female	
	No.	%	No.	%	No.	%	No.	%	No.	%	No.	%	No.	%	No.	%
Thyroid gland																
Follicular adenoma	2	0.9	0	–	1	0.4	0	–	0	–	0	–	0	–	1	0.4
C-cell adenoma	8	3.4	5	2.1	6	2.6	5	2.1	18	7.2	8	3.3	21	7.2	13	4.7
Follicular carcinoma	0	–	1	0.4	0	–	1	0.4	0	–	1	0.4	0	–	2	0.7
C-cell carcinoma	3	1.3	2	0.9	2	0.9	0	–	3	1.2	4	1.7	3	1.0	1	0.4
Parathyroid gland																
Adenoma	2	0.9	1	0.4	0	–	0	–	1	0.4	0	–	0	–	2	0.7
Adrenal glands																
Cortical adenoma	0	–	4	1.7	1	0.4	12 (13)	5.1	1	0.4	10	4.1	1	0.3	19 (20)	6.9
Pheochromocytoma	65 (84)	27.7	52 (68)	22.1	82 (120)	34.9	62 (88)	26.4	81 (115)	32.5	36 (49)	14.9	130 (190)	44.7	87 (111)	31.4
Cortical adenocarcinoma	0	–	4	1.7	0	–	3 (4)	1.3	0	–	2	0.8	1	0.3	5	1.8
Pheochromoblastoma	15 (24)	6.4	7 (9)	3.0	6 (9)	2.6	2 (3)	0.9	12 (16)	4.8	6 (8)	2.5	9 (11)	3.1	9 (10)	3.2
Central nervous system																
- Brain																
Astrocytoma	1	0.4	0	–	0	–	0	–	0	–	0	–	0	–	0	–
Oligodendroglioma	2	0.9	1	0.4	7	3.0	1	0.4	6	2.4	1	0.4	5	1.7	2	0.7
Multiform glioblastoma	0	–	0	–	0	–	0	–	1	0.4	0	–	0	–	0	–
- Meninges																
Benign meningioma	0	–	0	–	1	0.4	0	–	0	–	0	–	2	0.7	0	–
Malignant meningioma	1	0.4	0	–	0	–	0	–	0	–	0	–	2	0.7	0	–
Peripheral nervous system																
-Major peripheral nerves																
Benign schwannoma	0	–	0	–	1	0.4	0	–	0	–	1	0.4	1	0.3	0	–
Malignant schwannoma	2	0.9	0	–	1	0.4	0	–	1	0.4	1	0.4	0	–	1	0.4
-Ganglia & paraganglia																
Ganglioneuroma	0	–	0	–	0	–	0	–	0	–	0	–	1	0.3	1	0.4
Pheochromocytoma	0	–	0	–	0	–	1	0.4	1	0.4	0	–	1	0.3	0	–

Continued.

TABLE 1. Continued.

Number and percentage of male and female sprague-dawley rats bearing various types of benign and malignant tumors[a]

Site Histotype	I: Coca-Cola (Breeders)				II: Drinking water (Breeders)				III: Coca-Cola (Offspring)				IV: Drinking water (Offspring)			
	Male		Female		Male		Female		Male		Female		Male		Female	
	No.	%	No.	%	No.	%	No.	%	No.	%	No.	%	No.	%	No.	%
Bones - Head																
Osteosarcoma	11	4.7	7	3.0	2	0.9	6	2.6	7	2.8	1	0.4	12	4.1	6	2.2
Bones - Other																
Chordoma	0	–	0	–	0	–	1	0.4	0	–	1	0.4	0	–	0	–
Osteosarcoma	0	–	2	0.9	1	0.4	0	–	1	0.4	3	1.2	3	1.0	1	0.4
Soft tissues																
Lipoma	1	0.4	0	–	0	–	0	–	0	–	0	–	0	–	1	0.4
Fibrosarcoma	0	–	0	–	0	–	0	–	1	0.4	0	–	0	–	0	–
Liposarcoma	2	0.9	2	0.9	0	–	0	–	0	–	0	–	0	–	0	–
Heart																
Myxoma	0	–	0	–	0	–	0	–	0	–	0	–	1	0.3	0	–
Malignant schwannoma	0	–	0	–	1	0.4	1	0.4	1	0.4	1	0.4	0	–	3	1.1
Thymus																
Benign thymoma	1	0.4	1	0.4	0	–	0	–	0	–	1	0.4	0	–	0	–
Malignant thymoma	1	0.4	0	–	0	–	1	0.4	1	0.4	4	1.7	2	0.7	2	0.7
Spleen																
Fibroma	0	–	0	–	1	0.4	0	–	0	–	0	–	0	–	0	–
Leiomyoma	0	–	0	–	0	–	0	–	1	0.4	0	–	0	–	0	–
Fibroangioma	2	0.9	0	–	4	1.7	0	–	2	0.8	0	–	1	0.3	1	0.4
Hemangiosarcoma	1	0.4	0	–	0	–	0	–	2	0.8	0	–	0	–	1	0.4
Lymph nodes																
Fibroangioma	4	1.7	0	–	1	0.4	1	0.4	2	0.8	2	0.8	4	1.4	2	0.7
Hemolymphoreticular tissues																
Lymphomas & leukemias	51	21.7	31	13.2	62 (64)	26.4	39	16.6	49	19.7	40	16.5	49 (52)	16.8	42	15.2

[a] Number in parenthesis is the number of tumors.
[b] See Table 3.
[c] See Table 4.

TABLE 2. Long-term carcinogenicity bioassays in total malignant tumors on COCA-COLA, administered instead of drinking water, supplied *ad libitum*, to male (M) and female (F) Sprague-Dawley rats

Group No.	Treatment	Animals Age	Sex	No.	Malignant tumors Tumor-bearing animals No.	%	Tumors No.	Per 100 animals
I	Coca-Cola	30, 39, 55 weeks (breeders)	M	235	107	45.5	140	59.6
			F	235	111	47.2	164	69.8
			M+F	470	218	46.4	304	64.7
II	Drinking water (control)	30, 39, 55 weeks (breeders)	M	235	108	46.0	136	57.9
			F	235	98	41.7	131	55.7
			M+F	470	206	43.8	267	56.8
III	Coca-Cola	embryo or 7 weeks (offspring)	M	249	98	39.4	126	50.6
			F	242	118	48.8	174	71.9
			M+F	491	216	44.0	300	61.1
IV	Drinking water (control)	embryo or 7 weeks (offspring)	M	291	122	41.9	153	52.6
			F	277	133	48.0	188	67.9
			M+F	568	255	44.9	341	60.0

TABLE 3. Long-term carcinogenicity bioassays for mammary malignant tumors on COCA-COLA, administered instead of drinking water, supplied *ad libitum*, to male (M) and female (F) Sprague-Dawley rats

Group No.	Treatment	Animals				Malignant tumors			
		Age	Sex	No.		Tumor-bearing animal		Tumor	
						No.	%	No.	Per 100 animals
I	Coca-Cola	30, 39, 55 weeks (breeders)	M	235		6	2.6	6	2.6
			F	235		37	15.7**	48	24.0**
			M+F	470		43	9.1	54	11.5
II	Drinking water (control)	30, 39, 55 weeks (breeders)	M	235		2	0.9	2	0.9
			F	235		15	6.4	15	6.4
			M+F	470		17	3.6	17	3.6
III	Coca-Cola	embryo or 7 weeks (offspring)	M	249		3	1.2	3	1.2
			F	242		47	19.4**	58	24.0**
			M+F	491		50	10.2	61	12.4
IV	Drinking water (control)	embryo or 7 weeks (offspring)	M	291		3	1.0	3	1.0
			F	277		29	10.5	33	11.9
			M+F	568		32	5.6	36	6.3

**Statistically significant ($P < 0.01$) using χ^2 test.

TABLE 4. Long-term carcinogenicity bioassays for tumors of the pancreas on COCA-COLA, administered instead of drinking water, supplied *ad libitum*, to male (M) and female (F) Sprague-Dawley rats

Group No.	Treatment	Animals Age	Sex	No.	Exocrine Adenomas No.	%	Exocrine Carcinomas No.	%	Islet cell Adenomas No.	%	Islet cell Carcinomas No.	%
I	Coca-Cola	30, 39, 55 weeks (breeders)	M	235	18	7.7**	0	—	12	5.1	0	—
			F	235	10	4.3*	0	—	3	1.3	2	0.9
			M+F	470	28	6.0	0	—	15	3.2	2	0.4
II	Drinking water (control)	30, 39, 55 weeks (breeders)	M	235	1	0.4	0	—	13	5.5	0	—
			F	235	1	0.4	0	—	5	2.1	0	—
			M+F	470	2	0.4	0	—	18	3.8	0	—
III	Coca-Cola	embryo or 7 weeks (offspring)	M	249	23	9.2**	0	—	14	5.6	1	0.4
			F	242	16	6.6**	0	—	8	3.3	3	1.2
			M+F	491	39	7.9	0	—	22	4.5	4	0.8
IV	Drinking water (control)	embryo or 7 weeks (offspring)	M	291	1	0.3	0	—	19	6.5	0	—
			F	277	1	0.4	0	—	7	2.5	0	—
			M+F	568	2	0.4	0	—	26	4.6	0	—

* Statistically significant ($P < 0.05$) using χ^2 test.
** Statistically significant ($P < 0.01$) using χ^2 test.

carcinogenesis bioassay has been conducted to evaluate the beverage mixture for its potential carcinogenic effects.

The results of the study conducted at the CMCRC/ERF have shown the following:

1. an increase of fluid consumption in all animals, both breeders and offspring, administered Coca-Cola as a substitute for drinking water and a general decrease in food consumption;
2. an increase in body weight in all treated animals.

When compared to controls, animals treated with Coca-Cola also demonstrated the following oncological effects:

1. a statistically significant increase of the incidence in females, both breeders and offspring, bearing malignant mammary tumors. In our opinion, this observation confirms the correlation between the increase in body weight and an increased risk of mammary cancer;
2. a statistically significant increase in the incidence of exocrine adenomas of the pancreas in both male and female breeders and offspring;
3. an increased incidence, albeit not statistically significant, of pancreatic islet cell carcinomas in females. Because of the rarity of pancreatic iselt cell carcinomas in our historical controls, the biological significance of the increased incidence of this malignant tumor cannot be underestimated.

Although humans do not consume this beverage under the same conditions designed in our experiment, the results nevertheless confirm that an exaggerated ingestion of high caloric beverages, such as regular soft-drinks, can lead to a marked increase in body weight which in turn presents an increased risk for developing cancer.

ACKNOWLEDGMENTS

We thank the research staff of the CMCRC/ERF and Kathryn Knowles for her support in the preparation of the manuscripts.

REFERENCES

1. TANNENBAUM, A. 1940. The initiation and growth of tumors. I: effects of underfeeding. Am. J. Cancer **38**: 335–350.
2. TANNENBAUM, A. & H. SILVERSTONE. 1953. Nutrition in relation to cancer. Adv. Cancer Res. **1**: 451–501.
3. CALLE, E. *et al*. 2003. Overweight, obesity, and mortality from cancer in a prospectively studied cohort of U.S. Adults. N. Engl. J. Med **348**: 1625–1638.

4. RAPP, K. et al. 2005. Obesity and incidence of cancer: a large cohort study of over 145 000 adults in Austria. Br. J. Cancer **93:** 1062–1106.
5. SOFFRITTI, M. et al. 1999. Mega-experiments to identify and assess diffuse carcinogenic risks. Ann. N. Y. Acad. Sci. **895:** 43–55.
6. SOFFRITTI, M. et al. 1996. Results of experimental bioassays on the chemopreventive effects of vitamin A and N-(4-hydroxyphenyl)-retinamide (HPR) on mammary cancer. In The Scientific Bases of Cancer Chemoprevention. C. Maltoni, M. Soffritti & W. Davis, Eds.: International Congress Series 1120 : 241–248. Elsevier. Amsterdam, The Netherlands.
7. THE COCA-COLA COMPANY ANNUAL REPORT. 2005. http://www.cocacola.com [accessed 16 May 2006]

Occupational Kidney Cancer

Exposure to Industrial Solvents

NACHMAN BRAUTBAR, MICHAEL P. WU, ELION GABEL, AND LEE REGEV

University of Southern California School of Medicine, Los Angeles, California 90048, USA

ABSTRACT: We report seven cases of renal cell carcinoma in workers diagnosed with occupational exposure via skin contact and inhalation to industrial solvents containing benzene. The clinical significance of these cases are: (*a*) all seven patients diagnosed with kidney cancer were seen by private physicians who missed addressing occupational history to industrial solvents; (*b*) emphasize the importance of taking an in-depth history including occupational history in any patient presented to the clinician, especially like in these cases, kidney cancer; and (*c*) demonstrate the importance of educating workers. We believe that there exist more patients with renal cancer whose diagnosis has been rendered "idiopathic" due to the lack of detailed occupational, environmental, personal, and family history.

KEYWORDS: renal cell carcinoma; kidney cancer; case reports; benzene; industrial; organic solvents; occupation

INTRODUCTION

Kidney cancer is the general term for malignancies of the kidney. Malignancy of the kidney comprises a heterogeneous class of tumors arising from different cell types within the nephroma. Histopathologic classification of kidney tumors include hypernephroma (renal cell carcinoma), also called nonpapillary cancer, and comprises 75–80% of kidney cancer, papillary renal cell cancer which comprises 10–15%, chromophobe renal cell carcinoma, collecting duct carcinoma, and other unclassified renal cell carcinoma.[1] In the United States, the incidence of kidney cancer is 17 cases per 100,000 person-years and reaches the maximum at the age of 65 years.[2] Kidney cancer is attributed to both genetic and environmental factors. Studies have linked the genetic predisposition of kidney cancer to several conditions, including mutations that lead to Von

---

Address for correspondence: Nachman Brautbar, 6200 Wilshire Boulevard, Suite 1000, Los Angeles, CA 90048. Voice: 323-634-6500; fax: 323-634-6501.
e-mail: brautbar@aol.com

Ann. N.Y. Acad. Sci. 1076: 753–764 (2006). © 2006 New York Academy of Sciences.
doi: 10.1196/annals.1371.012

Hippel–Lindau syndrome, hereditary papillary renal carcinoma, and tuberous sclerosis Type I and II.[3] Cigarette smoking is a risk factor for kidney cancer. The International Agency for Research on Cancer (IARC) has found sufficient evidence for the carcinogenic role of tobacco smoking in renal pelvis carcinoma.[4] Subsequent epidemiological studies have demonstrated cigarette smoking as a risk factor for kidney cancer. The percentage of kidney cancer attributed to cigarette smoking is between 21% and 30% among men and 9% and 24% among women.[5–7] Diet has been suggested as a risk factor for kidney cancer, however, there is no agreement in the scientific literature.[8] Alcohol is a suspected risk factor, however, it remains only as a possibility. Epidemiological studies associate kidney cancer with excess body weight, overweight, and obesity as measured by the body mass index (BMI).[1] In the United States, obesity attributes 21% of kidney cancer risk.[9] Patients with diabetes also have an increased risk of kidney cancer, mainly those with long-term and uncontrolled diabetes mellitus.[10] Occupational studies have also attempted to address the risk of exposure to asbestos, hydrocarbon, lead, cadmium, and chlorinated industrial solvents.[1]

Many epidemiological studies have investigated the association between solvent exposure and kidney cancer. Case–control studies and cohort studies have been described in the peer-reviewed, scientific literature. TABLE 1 describes selected case–control studies of kidney cancer and exposure to industrial solvents.[11–24] Statistically significant odds ratio (OR) range from 1.2 to 6.4, with some studies controlling for risk factors as age, sex, cigarette smoking, and weight. TABLE 2 summarizes cohort studies that describe exposure to solvents and increased risk of kidney cancer.[25–30]

PATIENTS AND METHODS

We describe here seven patients who were evaluated based on physical examination and review of records, who have developed kidney cancer, and have a history of exposure to industrial solvents. Exposure dose were determined either by job description, material safety data sheets, court records, and/or history of exposure. Although the employers and industry never provided actual levels, based on the collected information, exposures were determined to be substantial. Potential confounding factors are addressed in the "Discussion" section.

RESULTS

The patients studied here have a common denominator of substantial exposure to industrial solvents from 5.5 years to 44 years (average 35 years). TABLE 3 summarizes the seven patients' age, occupation, exposure history and

TABLE 1. Case–control studies of increased risk of kidney cancer and exposure to industrial solvents

Study	Study design	Control group	Exposure	Result	95% CI	Controlled variables
Ref. 11	Death certificate	Noncancer deaths	Petroleum or petrochemicals	OR = 2.27	1.01–5.43	—
Ref. 12	Interviews	Other cancer sites	Aviation gasoline	OR = 2.6	1.2–5.8*	Age, smoking, socioeconomic status
			Jet fuel	OR = 2.5	1.1–5.4	Age, smoking, socioeconomic status
Ref. 13	Occupational histories	Matched for hospital and from population	Petroleum refining and distribution	OR = 4.3†	1.7–10.9*	Age, smoking, weight
Ref. 14	Occupational histories	Other cancer sites	Truck drivers	OR = 3.1†	1.1–8.5	Age, smoking, alcohol
Ref. 15	Interviews	Matched for hospital, sex, and age	Chemical, petrochemical, plastics industries; gasoline and petroleum	OR = 4.0	1.6–9.8	Sex, smoking
Ref. 16	Interviews	General population	Moderate levels of hydrocarbon	OR = 2.7†	1.2–6.5	Weight, education
Ref. 17	Questionnaires	Matched on year of birth, gender, and survival status	Paper and pulp; printing and publishing	OR = 2.20†	1.02–4.72	Obesity, smoking coffee consumption
			Chemicals	OR = 4.19†	1.09–16.1	Obesity, smoking coffee consumption
			Gasoline	OR = 1.72	1.03–2.87	—
Ref. 18	Questionnaires	Matched for expected age	Solvents	OR = 1.54	1.11–2.14	Age, sex, education
			Cutting oils or mists	OR = 1.99	1.23–3.23	Age, sex, education
			Petroleum products	OR = 1.96	1.28–2.99	Age, sex, education
Ref. 19	Interviews	Central Population Register	Gasoline	OR = 2.1†	1.1–4.1	Age, BMI, smoking
			Solvents	OR = 6.4‡	1.8–23	Age, BMI, smoking
			Cutting oils	OR = 2.1†	1.0–4.3	Age, BMI, smoking

Continued.

TABLE 1. Case–control studies of increased risk of kidney cancer and exposure to industrial solvents (*Continued*)

Study	Study design	Control group	Exposure	Result	95% CI	Controlled variables
Ref. 20	Occupational histories	Other cancer sites	Firefighters	OR = 3.51	2.09–5.92	—
			Painters	OR = 1.59	1.00–2.43	—
Ref. 21	Interviews	General population	Dry-cleaning solvents	OR = 1.4†	1.1–1.7	Age, smoking status, BMI, education
			Gasoline	OR = 1.6†	1.2–2.0	Age, smoking status, BMI, education
			Petroleum products	OR = 1.6†	1.3–2.1	Age, smoking status, BMI, education
Ref. 22	Interviews	Matched by region, sex, age	Chemicals (very long duration)	OR = 3.1†	1.2–7.9	Age, smoking
			Organic solvents	OR = 1.6†	1.1–2.3	Age, smoking
			Solvents	OR = 2.1‡	1.0–4.4	Age, smoking
			Paints	OR = 1.3†	1.0–1.6	Age, smoking
			Color films	OR = 2.9†	1.1–8.3	Age, smoking
			Cutting fluids	OR = 1.3†	1.0–1.8	Age, smoking
			Tar, pitch, mineral oil	OR = 2.1‡	1.0–4.5	Age, smoking
				OR = 1.2† (medium exposure); OR = 1.2† (high exposure)	1.0–1.5 (medium exposure); 1.0–1.6 (high exposure)	
				OR = 1.4† (substantial exposure)	1.0–1.9 (substantial exposure)	
				OR = 1.4‡ (medium exposure); OR = 1.3‡ (high exposure)	1.0–1.9 (medium exposure); 1.0–1.8 (high exposure)	

Continued.

TABLE 1. Case–control studies of increased risk of kidney cancer and exposure to industrial solvents (*Continued*)

Study	Study design	Control group	Exposure	Result	95% CI	Controlled variables
Ref. 23	Questionnaires	General population	Mineral oil	OR = 1.3[†]	1.0–1.8	Age, smoking
			Polycyclic aromatic hydrocarbons	OR = 1.3[†]	1.0–1.6	Age, smoking
			Benzene	OR = 1.8[†]	1.3–2.6	Age, education, BMI, smoking, alcohol, meat intake
			Mineral, cutting or lubricating oil	OR = 1.5[†]	1.2–1.8	Age, education, BMI, smoking, alcohol, meat intake
Ref. 24	Telephone interviews	General population	Automotive dealers and service stations	OR = 2.2[†]	1.0–4.9	Age, BMI, smoking, meat intake, fruit intake, hypertension, family history
			Automotive repair shops	OR = 4.0[†]	1.3–12.2	Age, BMI, smoking, meat intake, fruit intake, hypertension, family history
			Mechanics and repairers	OR = 2.0[†]	1.1–3.5	Age, BMI, smoking, meat intake, fruit intake, hypertension, family history
			Automobile mechanics	OR = 2.9[†]	1.5–5.8	Age, BMI, smoking, meat intake, fruit intake, hypertension, family history
			Garage and service station	OR = 3.1[†]	1.0–9.1	Age, BMI, smoking, meat intake, fruit intake, hypertension, family history

[†] OR reported for males only.
[‡] OR reported for females only.
* 90% CI.
CI: confidence interval.

TABLE 2. Cohort studies of increased risk of kidney cancer and exposure to industrial solvents

Study	Study design	Population	Result	95% CI	Controlled variables
Ref. 25	Death certificates	Newspaper pressmen	SMR = 303[†,a]	—	—
Ref. 26	Vital statistics and occupational history	Oil refinery workers, only in maintenance department	SMR = 1,750[†]	293–5781	—
		Oil refinery workers, ever in maintenance department	SMR = 1,112[†]	186–3675	—
Ref. 27	National cancer registries	Service station workers	SIR = 1.3[†]	1.0–1.7	—
Ref. 28	Neste Group workers, Finnish Cancer Registry	Oil and chemicals company workers	SIR = 1.97	1.29–2.88	—
Ref. 29	Swedish Cancer-Environment Registry	Manufacturing industry	SIR = 1.06[†,b]	—	Age, region
Ref. 30	Refinery and chemical plant sites	Baton Rouge workers	SMR = 192	113–303	—

[†]For males only.
[a]$P < 0.05$, [b]$P < 0.01$.
SIR: standardized incidence ratio.
SMR: standardized mortality ratio.

TABLE 3. Summary of patients reported here with occupational exposure to industrial, nonchlorinated solvents

Patient ID	Age (years)	Occupation	Exposure	Duration (years)	Diagnosis (age in years)	Confounding factors
1	60 (deceased)	Automobile mechanic	Industrial solvent	40	Renal cell cancer (35)	Smoked for 2 years; obesity
2	74	Airline mechanic	Industrial solvent	33	Renal cell cancer (67)	Smoked for 35 years; lifetime use of analgesics
3	45	Route serviceman,	Organic solvent	5.5	Renal cell cancer (40)	Smoked for 15 years
4	41	Automobile mechanic	Organic solvent	20	Renal cell cancer (35)	Smoked for 20 years
5	51 (deceased)	Mechanic	Industrial solvent	30	Renal cell cancer (48)	Smoked for 10 years; asbestos; obesity
6	58	Offset stripper	Organic, press and printing solvents	44	Renal cell cancer (50)	Hypertension; overweight
7	65	Machine operator	Industrial, organic solvents and petrochemicals	41	Renal cell cancer (63)	Smoked for 30 years; asbestos; pesticides; chlorinated solvents

duration, diagnosis, and confounding factors. No patient had a family history of renal cancer.

DISCUSSION

Confounding Factors

Six of our patients were smokers. Smoking is a risk factor for kidney cancer, however, regression analysis by the epidemiological study of Hu *et al.* found that the risk of kidney cancer remained elevated after accounting for smoking (OR = 1.8, 95% confidence interval (CI) = 1.3–2.6 for exposure to benzene).[24] Therefore, exposure to industrial solvents containing benzene should be considered as a risk aside from smoking. This finding of smoking and kidney cancer is not surprising because cigarette smoking contributes to the cumulative dose of benzene.[4] Indeed, Hu *et al.* demonstrated the causation role of benzene in kidney cancer. In all likelihood, cigarette smoking is probably an additive factor upon exposure to solvents containing benzene. Three of our patients had increased BMI, a risk factor for kidney cancer. Several epidemiological studies have found significantly increased risk for kidney cancer from exposure to industrial solvents, after regression analysis and controlling for BMI and smoking.[13,17,21,23,24] Therefore, exposure to industrial solvents is considered a distinctive risk factor for kidney cancer, even when accounting for smoking history and BMI.

Although the mechanism is not clear, the common denominator in exposure to industrial solvents is the contents of benzene. Industrial solvents (produced from crude oil) cannot be produced without benzene contamination.[31] Benzene has been shown to be a substantial risk factor for kidney cancer.[23] Because the common denominator in the industrial solvent exposure (nonchlorinated) is also a benzene content, it is reasonable to conclude that benzene, which is a known human carcinogen, is acting as a carcinogenic agent either by itself or in combination with solvent mixture in the causation of kidney cancer.

Benzene in Industrial Solvents

For the last several decades, most organic solvents have been refined from crude oil by a process commonly known as fractional distillation. Benzene is a natural constituent of crude oil. When crude oil is refined to produce refined petroleum distillates, the benzene in the crude oil is not destroyed, but rather ends up as a residual constituent of petroleum distillates. The concentration of benzene in the distillates is largely a function of the quality of the fractional distillation process and other processing methods such as catalytic cracking (which usually increases the benzene content) and hydrotreatment (which usually decreases the benzene content). The highest benzene concentration in

TABLE 4. Duration of exposure to industrial solvents and increased risk of kidney cancer

Study	Exposure	Exposure duration (years)	Result	95% CI
Ref. 16	Moderate levels of hydrocarbon	16–30	OR = 2.4	1.0–6.2
Ref. 19	Hydrocarbons	<20	OR = 2.1[†]	1.2–3.6
Ref. 26	Petrochemicals	5–15	SMR = 606	101–2002
Ref. 28	Petrochemicals	≥5	SIR = 2.87	1.61–4.73
Ref. 30	Petrochemicals	20–29	SMR = 330*	—

[†]OR reported for males only.
*Authors report statistical significant results.
SIR: standardized incidence ratio.
SMR: standardized mortality ratio.

solvents has been reported in hexane-based solvents.[32] Following adoption of Occupational Safety and Health Administration's (OSHA) regulation limiting benzene in petroleum hydrocarbon mixtures to no more than 0.1%, the petroleum industry began upgrading fractional distillation and other refining processes.[33] During the 1980s and 1990s, the benzene content of organic solvents decreased rather appreciably, although benzene remains a contaminant of most petroleum solvents today.

Duration of Exposure

Duration of exposure in our cases is described anywhere from 5.5 to 44 years. The scientific literature has addressed the duration of exposure to industrial solvents and the risk of kidney cancer.[16,19,26,28,30] TABLE 4 summarizes these studies, with a range in duration of exposure to industrial solvents from 5 to 30 years. Therefore, the range of exposure described in our patients studied here is within the range described by other studies addressing industrial solvents and increased risk of kidney cancer.

Other Exposures

Patients 5 and 7 have histories of exposure to asbestos. Asbestos exposure has been described as a risk factor for kidney cancer, therefore it is probable that asbestos exposure in patients 5 and 7 also played a substantial role.[21,34–36] Whether the concomitant exposure to asbestos and solvents is additive or synergistic remains unknown for kidney cancer. Patient 7 has been exposed to pesticides. The exposure to pesticides was deemed possible, nevertheless is included in the exposure history. To the best of our knowledge, exposure to pesticides has not been shown to increase risk for kidney cancer. Patient 7 has also been exposed to chlorinated solvents. Some chlorinated solvents have been shown to be a causative risk factor in kidney cancer, therefore it is probable that exposure to chlorinated solvents (if occurred at substantial levels) was also a contributing factor in the causation of kidney cancer in patient 7.[21,37]

CONCLUSION

Based on the reviewed scientific literature and description of our patients, we recommend to occupational physicians for their patients mandatory detailed occupational history, social history, body weight, family history, or other causes, and include exposure to industrial solvents in causation analysis of kidney cancer.

ACKNOWLEDGMENTS

The authors thank Jill Tremblay for her efforts in transcribing this manuscript. Funding for this manuscript was solely provided by Nachman Brautbar, Inc.

DISCLOSURE

Some of the comments in this manuscript have been presented in civil court. Nachman Brautbar, M.D. serves from time to time as an expert in litigation related to toxic exposures. He has served as a retained expert in cases of solvent exposure and kidney cancer. This manuscript was not authored for litigation process. Funding for this manuscript was solely provided by Nachman Brautbar, M.D., Inc. The opinions expressed in this manuscript do not necessarily reflect that of any organizations the authors are affiliated with.

REFERENCES

1. LINDBLAD, P. & H. ADAMI. 2002. Kidney cancer. *In* Textbook of Cancer Epidemiology. H. Adami, D. Hunter & D. Trichopoulos, Eds.: 467–485. Oxford University Press, Inc. New York, NY.
2. FERLAY, J., F. BRAY, P. PISANI, *et al*. 2001. GLOBOCAN 2000: Cancer Incidence, Mortality and Prevalence Worldwide. International Agency for Research on Cancer. Lyon, France.
3. GNARRA, J.R. 1998. Von Hippel-Lindau gene mutations in human and rodent renal tumors—association with clear cell phenotype. J. Natl. Cancer Inst. **90:** 1685–1687.

4. INTERNATIONAL AGENCY FOR RESEARCH ON CANCER. 1986. Tobacco smoking. IARC Monographs on the evaluation of the carcinogenic risk of chemicals to humans. IARC **38:** 35–394.
5. MCLAUGHLIN, J.K., J.S. MANDEL, W.J. BLOT, et al. 1984. A population-based case-control study of renal cell carcinoma. J. Natl. Cancer Inst. **72:** 275–284.
6. MCLAUGHLIN, J.K., P. LINDBLAD, A. MELLEMGAARD, et al. International renal-cell cancer study. I. Tobacco use. Int. J. Cancer **60:** 194–198.
7. YUAN, J.M., J.E. CASTELAO, M. GAGO-DOMINGUEZ, et al. 1998. Tobacco use in relation to renal cell carcinoma. Cancer Epidemiol. Biomarkers Prev. **7:** 429–433.
8. WOLK, A., P. LINDBLAD & H.O. ADAMI. 1996. Nutrition and renal cell cancer. Cancer Causes Control **7:** 5–18.
9. BENICHOU, J., W.H. CHOW, J.K. MCLAUGHLIN, et al. 1998. Population attributable risk of renal cell cancer in Minnesota. Am. J. Epidemiol. **148:** 424–430.
10. LINDBLAD, P., W.H. CHOW, J. CHAN, et al. The role of diabetes mellitus in the aetiology of renal cell cancer. Diabetologia **42:** 107–112.
11. GOTTLIEB, M.S. & J.K. CARR. 1981. Mortality studies on lung, pancreas, esophageal, and other cancers in Louisiana. *In* Banbury Report 9: Quantification of Occupational Cancer. R. Peto & M. Schneiderman, Eds.: 195–204. Cold Spring Harbor Laboratory. Cold Spring Harbor, NY.
12. SIEMIATYCKI, J., R. DEWAR, L. NADON, et al. 1987. Associations between several sites of cancer and twelve petroleum-derived liquids. Results from a case-referent study in Montreal. Scand. J. Work Environ. Health **13:** 493–504.
13. ASAL, N.R., J.R. GEYER, D.R. RISSER, et al. 1988. Risk factors in renal cell carcinoma. II. Medical history, occupation, multivariate analysis, and conclusions. Cancer Detect. Prev. **13:** 263–279.
14. BROWNSON, R.C. 1988. A case-control study of renal cell carcinoma in relation to occupation, smoking, and alcohol consumption. Arch. Environ. Health **43:** 238–241.
15. JENSEN, O.M., J.B. KNUDSEN, J.K. MCLAUGHLIN, et al. 1988. The Copenhagen case-control study of renal pelvis and ureter cancer: role of smoking and occupational exposures. Int. J. Cancer **41:** 557–561.
16. KADAMANI, S., N.R. ASAL & R.Y. NELSON. 1989. Occupational hydrocarbon exposure and risk of renal cell carcinoma. Am. J. Ind. Med. **15:** 131–141.
17. PARTANEN, T., P. HEIKKILA & S. HERNBERG, et al. 1991. Renal cell cancer and occupational exposure to chemical agents. Scand. J. Work Environ. Health. **17:** 231–239.
18. MCCREDIE, M. & J.H. STEWART. 1993. Risk factors for kidney cancer in New South Wales. IV. Occupation. Br. J. Ind. Med. **50:** 349–354.
19. MELLEMGAARD, A., G. ENGHOLM, J.K. MCLAUGHLIN, et al. 1994. Occupational risk factors for renal-cell carcinoma in Denmark. Scand. J. Work Environ. Health **20:** 160–165.
20. DELAHUNT, B., P.B. BETHWAITE & J.N. NACEY. 1995. Occupational risk for renal cell carcinoma. A case-control study based on the New Zealand Cancer Registry. Br. J. Urol. **75:** 578–582.
21. MANDEL, J.S., J.K. MCLAUGHLIN, B. SCHLEHOFER, et al. 1995. International renal-cell cancer study. IV. Occupation. Int. J. Cancer. **61:** 601–605.
22. PESCH, B., J. HAERTING, U. RANFT, A. KLIMPEL, et al. 2000. Occupational risk factors for renal cell carcinoma: agent-specific results from a case-control study

in Germany. MURC Study Group. Multicenter urothelial and renal cancer study. Int. J. Epidemiol. **29:** 1014–1024.
23. HU, J., Y. MAO & K. WHITE. 2002. Renal cell carcinoma and occupational exposure to chemicals in Canada. Occup. Med. (Lond.) **52:** 157–164.
24. ZHANG, Y., K.P. CANTOR, C.F. LYNCH, *et al*. 2004. A population-based case-control study of occupation and renal cell carcinoma risk in Iowa. J. Occup. Environ. Med. **46:** 235–240.
25. PAGANINI-HILL, A., E. GLAZER, B.E. HENDERSON, *et al*. 1980. Cause-specific mortality among newspaper web pressmen. J. Occup. Med. **22:** 542–544.
26. BERTAZZI, P.A., A.C. PESATORI, C. ZOCCHETTI, *et al*. 1989. Mortality study of cancer risk among oil refinery workers. Int. Arch. Occup. Environ. Health. **61:** 261–270.
27. LYNGE, E., A. ANDERSEN, R. NILSSON, *et al*. 1997. Risk of cancer and exposure to gasoline vapors. Am. J. Epidemiol. **145:** 449–458.
28. PUKKALA, E. Cancer incidence among Finnish oil refinery workers, 1971–1994. J. Occup. Environ. Med. **40:** 675–679.
29. MCLAUGHLIN, J.K., H.S. MALKER, B.J. STONE, *et al*. 1987. Occupational risks for renal cancer in Sweden. Br. J. Ind. Med. **44:** 119–123.
30. SHALLENBERGER, L.G., J.F. ACQUAVELLA & D. DONALESKI. 1992. An updated mortality study of workers in three major United States refineries and chemical plants. Br. J. Ind. Med. **49:** 345–354.
31. MEHLMAN, M.A. 2004. Benzene: a haematopoietic and multi-organ carcinogen at any level above zero. Eur. J. Oncol. **9:** 15–36.
32. SHELL OIL COMPANY. 1977. Benzene content of Shell hydrocarbon solvents. Houston, TX. July.
33. OCCUPATIONAL SAFETY AND HEALTH ADMINISTRATION. 1978. Benzene. Final rules. Standard number: 1910.1028. Federal Register. **43:** 27962–27971.
34. SELIKOFF, I.J., E.C. HAMMOND & H. SEIDMAN. 1979. Mortality experience of insulation workers in the United States and Canada. Ann. N. Y. Acad. Sci. **330:** 91–116.
35. ENTERLINE, P.E., J. HARTLEY & V. HENDERSON. 1987. Asbestos and cancer: a cohort followed up to death. Br. J. Ind. Med. **44:** 396–401.
36. MACLURE, M. 1987. Asbestos and renal adenocarcinoma: a case-control study. Environ. Res. **42:** 353–361.
37. BRUNING, T., B. PESCH, B. WIESENHUTTER, *et al*. 2003. Renal cell cancer risk and occupational exposure to trichloroethylene: results of a consecutive case-control study in Arnsberg, Germany. Am. J. Ind. Med. **43:** 274–285.

Occupation and Breast Cancer
A Canadian Case–Control Study

JAMES T. BROPHY,[a,b,c] MARGARET M. KEITH,[a,b,c] KEVIN M. GOREY,[c] ISAAC LUGINAAH,[d] ETHAN LAUKKANEN,[e] DEBORAH HELLYER,[a] ABRAHAM REINHARTZ,[a] ANDREW WATTERSON,[b] HAKAM ABU-ZAHRA,[f] ELEANOR MATICKA-TYNDALE,[c] KENNETH SCHNEIDER,[f] MATTHIAS BECK,[g] AND MICHAEL GILBERTSON[b]

[a]*Occupational Health Clinics for Ontario Workers (OHCOW), Toronto, Ontario, Canada M6A 3B6*

[b]*University of Stirling, Stirling UK FK9 4LA*

[c]*University of Windsor, Windsor, Ontario, Canada N9B 3P4*

[d]*University of Western Ontario, London, Ontario, Canada N6A 3K7*

[e]*Prince Edward Island Cancer Treatment Centre, Charlottetown, Prince Edward Island, Canada CIA 8T5*

[f]*Windsor Regional Cancer Centre, Windsor, Ontario, Canada N8W ZX3*

[g]*University of York, York, UK YO10 5DD*

ABSTRACT: A local collaborative process was launched in Windsor, Ontario, Canada to explore the role of occupation as a risk factor for cancer. An initial hypothesis-generating study found an increased risk for breast cancer among women aged 55 years or younger who had ever worked in farming. On the basis of this result, a 2-year case–control study was undertaken to evaluate the lifetime occupational histories of women with breast cancer. The results indicate that women with breast cancer were nearly three times more likely to have worked in agriculture when compared to the controls (OR = 2.80 [95% CI, 1.6–4.8]). The risk for those who worked in agriculture and subsequently worked in automotive-related manufacturing was further elevated (OR = 4.0 [95% CI, 1.7–9.9]). The risk for those employed in agriculture and subsequently employed in health care was also elevated (OR = 2.3 [95% CI, 1.1–4.6]). Farming tended to be among the earlier jobs worked, often during adolescence. While this article has limitations including the small sample size and the lack of information regarding specific exposures, it does provide evidence of a possible association between farming and breast cancer. The findings indicate the need for further study to determine which aspects of farming may be of biological importance and

Address for correspondence: James T. Brophy, Ph.D., 171 Kendall, Point Edward, Ontario, Canada N7V 4G6. Voice: 519-337-4627; fax: 519-337-9442.
e-mail: jbrophy@ohcow.on.ca

to better understand the significance of timing of exposure in terms of cancer risk.

KEYWORDS: breast cancer; occupation; environment; farming; Canada

INTRODUCTION

The lifetime risk for breast cancer among Canadian women is approximately 1 in 9. Over the past 30 years, there has been a 25% increase. The majority of cases cannot be explained by the currently known or suspected risk factors. Family history of breast cancer, particularly with respect to having two or more relatives with breast cancer and mutation of the *BRCA1* and *BRCA2* gene, can explain less than 10% of breast cancer cases.[1] Factors that increase cumulative estrogen load have been found to increase risk. There is evidence of an association with diet, alcohol use, body mass index, reproductive history, age, physical activity, and socioeconomic status.[2] The recent increase in incidence may be linked to the combination of identified risk factors and those requiring further study, such as occupational and environmental exposures.[3] Increasing evidence suggests that synthetic chemicals, particularly those that mimic estrogen (xenoestrogens), may increase risk by acting as endocrine disruptors.[4] Such exogenous chemicals include organochlorine pesticides, polycyclic aromatic hydrocarbons, organic solvents, and plastics.[5-8] Animal bioassays have identified over 200 chemical substances that trigger breast cancer.[9] Another factor that has implications for research into the possible role of occupational and environmental exposures is the multistage developmental process that characterizes cancer. Toxic insults, either singular or in combination, may influence the initiation, promotion, and progression of carcinogenesis. Both dose and timing of exposure may be important in terms of risk. It has been suggested that there are critical moments in breast development when the emerging cells may be more susceptible to tumor initiation and progression.[1,4] There may be particular vulnerability during periods of morphological and biochemical change, that is, beginning during gestation and continuing through puberty to time of first pregnancy and possibly throughout the reproductive years.[9] There may also be a combined impact from exposure to carcinogens and hormonally active substances. It has been suggested that genotoxic agents, in conjunction with estrogen, can affect cell repair mechanisms thereby allowing damaged cells to reproduce.[1]

There is a significant gap in our understanding of work-related exposures and breast cancer risk.[10] Many substances shown to induce breast cancer in experimental mammals exist in high concentrations in occupational settings.[11] In spite of the continuing increase in the incidence of cancer in Canada and the existence of carcinogens in occupational environments, there remain no registries or systematic methods to record the occupational histories of cancer patients in general, nor breast cancer patients in particular.

The failure to document lifetime occupational histories and corresponding workplace exposures results in an underestimation of occupationally related cancers and a corresponding lack of substantive prevention-related activity.[12]

METHODS

A hypothesis-generating study, entitled Computerized Recording of Occupations Made Easy (CROME), was conducted between 1995 and 1999 through a collaborative effort of the Windsor Regional Cancer Centre (WRCC), the Ontario Occupational Disease Panel (ODP), and the Occupational Health Clinics for Ontario Workers (OHCOW). The study area, Windsor-Essex, Ontario, Canada, has extensive manufacturing and agricultural activity. The CROME study gathered the occupational histories of 299 breast cancer cases, which were then compared to 237 women with cancers other than breast or ovary. It found an elevated risk for breast cancer among women 55 years of age or younger who had ever worked in farming (OR = 9.05 [95% CI, 1.06–77.43]).[13] There were a number of important limitations to the study including: small sample size; the use of hospital-based controls; and the failure to adequately capture data regarding potential confounders beyond those of age, socioeconomic status, and body mass index. Moreover, detailed occupational descriptions were absent.

On the basis of the results from CROME, a second population-based case–control study, entitled Lifetime Occupational Histories Record (LOHR), was undertaken in 2000 in the same geographical study area to further explore possible associations between breast cancer and occupation in general and farming in particular. LOHR had several improvements over the previous study: it used randomly selected community controls rather than hospital controls; it captured more detailed occupational descriptions; and it collected data for a broader range of potential confounders. Over a 2.5-year period, all female patients treated at the Windsor Regional Cancer Centre (WRCC) with histologically confirmed new incident primary breast cancer were invited to participate. None of the cases had participated in the previous study. The medical records department screened breast cancer patients to confirm pathology and date of diagnosis. A letter was mailed to each eligible patient outlining the study and was followed up by a telephone call. Five hundred sixty-four eligible breast cancer patients participated in the study. Three patients declined participation, resulting in a 99% plus response rate. Community controls were chosen at random using city directory software[14] and were recruited by letter and a scripted follow-up telephone call to improve response rate among the less literate. The information provided to potential controls about the research did not specify a particular focus on occupational or environmental risk factors for breast cancer; the research was simply referred to as a "Risk History

Study." The controls were approximately matched by age and by geographical area. Five hundred ninety-nine eligible community controls participated out of 1146 contacted representing a response rate of 52.2%. All subjects, signed informed consents and each was offered a $20 stipend as compensation for their time.

A comprehensive lifetime history questionnaire was administered to each subject by a trained interviewer. The questionnaire gathered data regarding height and weight (body mass index), marital status, income, education, age of menarche, menstrual history, pregnancy and breast-feeding history, menopausal status, hormone use, family breast cancer history, residential history by three-digit postal code, hobbies, and complete occupational history including age at the start and end of each job. The questionnaire also included questions about a range of occupational exposures: asbestos, man-made mineral fibers, dusts, second-hand tobacco smoke, engine exhaust, other smoke or particulate, metal-working fluids, solvents, paints, strippers, and pesticides. Agricultural workers were also asked about chemical exposures. Subjects' recall regarding specific agents proved to be limited and much of the exposure information was deemed unreliable or was missing. As a result, data regarding exposure to specific agents was not included in the analysis. Jobs were categorized by coders, who were blind as to the case–control status of the subject data, using National Occupational Classification (NOC)[15] codes and the North American Industrial Classification System (NAICS).[16] The NOC codes, which provided more specificity than the NAICS, were included in the analysis. Similar or related occupations were grouped together to provide adequate statistical power.

FINDINGS

Included in the LOHR analysis were data from 564 female breast cancer cases and 599 female controls. The statistical program SPSS Version 10 was used to conduct a three-step multivariate analysis to test the hypothesis of a possible association between breast cancer risk and occupation. Logistic regression analysis was used to calculate odds ratios and their 95% confidence intervals.[17] In the initial step, cases and controls who had ever been employed in agriculture were compared while controlling for duration of employment using five ordinal variables: none, 0.5–5 years, 6–10 years, 11–20 years, and 21 or more years. Due to small sample size duration did not reach statistical significance for any of the specific periods.

In the next step, the odds ratios for the independent variables (e.g., farming) indicate the effect of each variable, after adjusting for covariates, on the probability of developing breast cancer. The following ordinal covariates were included in the model: age at diagnosis (mean age of cases = 60.33, mean

TABLE 1. Descriptive profile of 564 female breast cancer cases and 599 female community controls

	Breast cancer cases		Community controls	
	Yes %	No %	Yes %	No %
Ever pregnant[1]	495 (43%)	65 (5.6%)	527 (45.8%)	64 (5.6)
Ever use hormone replacement	261 (22.4%)	303 (26.1%)	263 (22.6%)	336 (28.9%)
Ever smoke tobacco	249 (21.4%)	315 (27.1%)	273 (23.5%)	326 (28%)
Ever breast feed	378 (32.5%)	186 (16%)	370 (31.8%)	229 (19.7%)
Ever used oral contraceptives[2]	289 (25%)	273 (23.6%)	339 (29.3%)	256 (22.1%)
Mother ever had cancer[3]	125 (10.9%)	431 (37.5%)	146 (12.7%)	447 (38.9%)
Ever reside on a farm or live within a mile of a farm[4]	253 (22.2%)	305 (26.7%)	247 (21.6%)	337 (29.5%)

[1] Missing cases = 12 (1%).
[2] Missing cases = 6 (.5%).
[3] Missing cases = 14 (1.2%).
[4] Missing cases = 21 (1.8%).

age of controls = 58.64); education level; annual household income; body mass index; number of pregnancies; years of oral contraceptive use; months of breast feeding; years of cigarette smoking; alcohol use and marital status. This step also included the following dichotomous covariates (TABLE 1): ever pregnant; ever used hormones; ever smoked tobacco; ever breast feed; ever used oral contraceptives; mother ever had cancer; and ever reside on a farm or live within a mile of a farm. Number of years of residence in Essex County was included as a continuous variable within the model.

The final step in the conditional logistic model included the major occupational groups: automotive-related manufacturing; clerical; communications; dry cleaning; education or library; petrochemical; finance or insurance; food processing; food service; hair dressing; manufacturing or engineering managers; office professionals; skilled sales; health care; janitorial; other manufacturing; plastics; printing, painting, or construction; retail; social service; textile; transportation or security; animal care; sports or arts; pest control; postal; mining or logging; landscaping; home care; and unemployed outside the home. In the final step of the model all the occupations interacted with age, but only four occupations remained within the model (agriculture, retail, and the interactions of agriculture with automotive-related manufacturing and of agriculture with health care).

At this stage there were 1026 subjects (cases = 506; controls = 520) with 137 missing (11.8%). As shown in TABLE 2, the results indicate that women with breast cancer were nearly three times more likely to have worked in agriculture ($n = 154$) when compared to the controls ($n = 133$) (OR= 2.80 [95% CI, 1.6–4.8]). Although the individual contribution of automotive-related

TABLE 2. Logistic regression-estimated odds ratios (OR) of women ever employed in agriculture, automotive-related manufacturing, or health care

	Odds ratio (OR)	95.0% C.I.	
		Lower	Upper
Ever worked in agriculture	2.8[a]	1.6	4.8
Worked in agriculture and then in automotive-related manufacturing	4.1[b]	1.7	9.9
Worked in agriculture and then in health care	2.3[c]	1.1	4.6
Worked in automotive-related manufacturing (but never agriculture)	0.76	0.59	1.10
Worked in health care (but never agriculture)	0.85	0.62	1.17
Ever worked in retail	1.0	1.0	1.05
Age	1.0	1.0	1.02

[a]Sig. = 0.0002
[b]Sig. = 0.002
[c]Sig. = 0.02

manufacturing alone was not significant (OR = 0.76 [95% CI, 0.59–1.10]), the risk for those who worked in agriculture and subsequently worked in automotive-related manufacturing was further elevated (OR = 4.0 [95% CI, 1.7–9.9]). The individual contribution of health care alone was not significant (OR = 0.85 [95% CI, 0.62–1.17]), however, the risk for those employed in agriculture and subsequently employed in health care was elevated (OR = 2.3 [95% CI, 1.1–4.6]). Agricultural jobs tended to be among the first worked, often during adolescence.

There is a modest body of literature regarding breast cancer risk in agriculture, health care, and the automotive industry.

FARMING OCCUPATIONS AND BREAST CANCER RISK

While some studies of farming populations have shown an elevated risk for breast cancer, as well as other cancers,[18–20] several large cohort studies found no elevated risk for breast cancer.[21–23] Female farmers and laborers have not been as extensively studied as their male counterparts. While most studies did not indicate specific exposures, it is plausible that agricultural chemicals may play a role. There is evidence of an association between breast cancer and some pesticides, such as dichlorodiphenyltrichloroethane (DDT), its metabolite dichlorodiphenyldichloroethylene (DDE), polychlorinated biphenyls (PCBs), hexachlorobenzene, hexachlorocyclohexane, heptachlor epoxide, and triazine herbicides; others are under review.[24–26] A large number of pesticides are also hormonally active.[27] The herbicide, atrazine, for example, is one of the most widely used agricultural chemicals. The triazine pesticides are considered endocrine disruptors and are suspected human carcinogens.[28] Some agricultural chemicals, such as organochlorine pesticides, are persistent and bioaccumulate in the adipose tissue.[29] A case–control study that controlled for both traditional breast cancer risk factors as well as exposures among women engaged in farming, found that women who reported being present in the fields during or shortly after pesticide application had an increased risk of developing breast cancer (OR= 1.8 [95% CI, 1.1–2.8]).[30] Among those who reported using pesticides without protective clothing, an increased risk of breast cancer was identified (OR = 2.0 [95% CI, 1.0–4.3]); while women with protective clothing did not have an elevated breast cancer risk (OR = 0.8 [95% CI, 0.4–1.8]). The researchers concluded that, while farming may not present an elevated risk *per se*, farming women who were not adequately protected from exposure to pesticides might have an elevated risk.

A Canadian study found that, among the combined pre- and postmenopausal group, there was an increased breast cancer risk among women who had ever been employed in fruit and vegetable farming (OR = 3.11, 90% [CI 1.24–7.81]).[31]

A recent study examining the breast cancer risk of Hispanic agricultural workers in California associated three specific pesticide exposures—chlordane, malathion, and 2,4-dichlorophenoxyacetic acid (2,4-D)—with elevated breast cancer risk.[32]

HEALTHCARE OCCUPATIONS AND BREAST CANCER RISK

A number of known or suspected carcinogens are present in the healthcare setting. Nurses and other healthcare workers are potentially exposed to ionizing radiation, antineoplastic drugs, anesthetic waste gases, and viruses possibly associated with cancer risk.[33] A number of hormonally active chemicals are, or have been, used in medicine and laboratory work.[34] These include: nonylphenol (used in detergents and plastics); ethylene oxide (a sterilant); bisphenol A (used in polycarbonate plastics); butyl benzyl phthalate; and polychlorinated biphenyls (PCBs). These substances have been shown to display estrogenic activity in human breast cell bioassays.[35] Studies of shift work involving nurses have found statistically significant increases in breast cancer (OR = 1.6 (95% CI, 1.0–2.5)) and a relative risk (RR = 1.36 (95% CI, 1.04–1.78)), respectively.[36,37] These elevations occurred among women who worked night shifts over long periods. It is hypothesized that melatonin is disrupted, thereby affecting estrogen levels. Breast cancer risk among nurses and other healthcare workers was examined in administrative, cohort, and case-controls studies.[38] Of 10 administrative and cohort studies, 8 found a positive association with breast cancer[39–46] while 2 did not.[47,48] Among six case–control studies[31,49–53] there were mixed results. Elevated risk is generally noted in several studies that examined breast cancer among registered hospital nurses but these studies shared methodological limitations; most did not control for known or suspected risk factors. Some of the findings varied depending on which comparison group was used or whether menopausal status was examined. None controlled for specific exposures and all nurses were grouped into one occupational category assuming that this broad title would be an appropriate surrogate for their exposures. In one case–control study an elevated breast cancer risk was revealed only when the study population was separated into occupational subgroups.[53] None of the existing studies assess timing of exposure.

AUTOMOTIVE-RELATED MANUFACTURING OCCUPATIONS AND BREAST CANCER RISK

A few reports have addressed women autoworkers and risk of breast cancer. A cohort study published in 1994 found no association between female breast cancer and automotive manufacturing.[54] A recent study, however, found a weak association (OR = 1.18 [95% CI, 1.02–1.35]) with soluble metal working fluid (MWF) exposure.[55] There is wide use of chlorinated solvents in the automotive industry and growing evidence that these chemicals may increase breast cancer risk, possibly through endocrine disruption.[9,56] Organic solvents have produced mammary tumors in animal studies. Organic solvents have been detected in breast milk, subjecting the ducts to constant exposure. Interestingly, the majority of breast tumors reside in the ductular system.[57]

DISCUSSION

The LOHR study primarily found and then tested associations with specific occupations. It provides evidence of an association between farming and breast cancer risk as well as an interactive effect between occupational farming exposures and subsequent exposures in other occupational environments. It might be hypothesized that agents or conditions present in agricultural settings initiate the breast cancer process at a vulnerable period (adolescence) and that subsequent exposures to agents or conditions (e.g., shift work) in automotive-related industry, health care, or other industries may act as promotors.

The LOHR study had some limitations. While it attempted to gather information through the interview process about exposures, it was not able to accurately identify specific causative agents. Unfortunately, many of the patients and community controls were not aware of or could not reliably recall their exposures. It is possible that the actual breast cancer risk for some of the women in the LOHR cohort, that is, those who had exposure to specific pesticides, is even higher because the aggregation of the unexposed with the exposed in the analysis may have diluted the findings. Such nondifferential misclassification decreases the probability of detecting associations and tends to underestimate the actual risks.[58] Another limitation of the LOHR study was the small sample size, which necessitated the grouping of occupational categories thereby increasing the risk of misclassification. There was a significant percentage difference in the response rate of cases (99%) versus controls (52%) raising the issue of possible recruitment or selection bias. Steps had been taken, however, to minimize the risk of such bias. Selection of controls was made independent of the exposures of interest. Information provided about the research did not specify a particular focus on occupational or environmental risk factors (the independent variable) for breast cancer (the dependant variable) in order to reduce the likelihood that those interested in specific exposures would respond.

The results of the LOHR study call for research to determine which aspects of farming may be of biological importance. The further development of our understanding regarding breast cancer risk and farming is an important public health concern given the prevalence of potential pesticide exposure and disease in rural communities. Moreover, the interaction between early and subsequent exposures requires further study and consideration. A clearer understanding is needed regarding the effects of farming exposures during the early periods of life when breast tissue is most vulnerable.

A third study, entitled, Lifetime Histories Breast Cancer Research (LH-BCR)[59,60] was initiated in 2004 to evaluate more specific exposures among agricultural and other workers. Open-ended job description questions will provide comprehensive data for expert exposure assessment.[61-63] Estrogen and progesterone receptor status of the tumor will be obtained from pathology reports and included in the analysis. A larger sample size—1000 cases and 1000 controls—will provide added statistical power.[64] Such occupational and environmental breast cancer research may ultimately serve to inform the formulation of breast cancer prevention and early detection strategies. This may be accomplished through the identification of current work practices or exposures that can be modified to minimize breast cancer risk; and through the identification of specific populations at potentially higher risk for breast cancer from past exposures who can be then encouraged to pursue more diligent early detection efforts. It may also encourage previously exposed women to avoid further potentially harmful exposures. The results may also serve to shape public health and regulatory policy regarding prevention strategies.

CONCLUSION

The LOHR study indicates that women who have a history of work in agriculture have an elevated risk for breast cancer. The risk for those who worked in agriculture and were subsequently employed in health care or the automotive industry is further elevated. While occupational categories in this study serve as surrogates for exposure, it is plausible that exposure to agricultural chemicals is a causative factor. Because many women who worked in farming began during adolescence, it is plausible that the timing of exposure is of significance in terms of risk.

ACKNOWLEDGMENTS

The research was sponsored by the Occupational Health Clinics for Ontario Workers (OHCOW), Windsor Regional Cancer Centre (WRCC), and University of Windsor. Funding was provided for the Lifetime Occupational Histories Record (LOHR) study by: the Workplace Safety and Insurance Board Research Advisory Council; Windsor-Essex County Cancer Foundation; Green Shield Foundation; and Canadian Auto Workers (CAW) union locals. Support was provided by: Nicole Mahler, Robert Park, Jeff Desjarlais, Kathy Mayville, Mary Cook, Janet Davis, Julie Durocher, Jeremy Garman, Michael Lax, Rory O'Neill, Eileen Senn, Gregory Siwinski, Ann Sovan, and Peter Infante. Funding was provided for the current Lifetime Histories Breast Cancer Research (LHBCR) by the Canadian Breast Cancer Foundation and the Breast Cancer Society of Canada.

REFERENCES

1. DAVIS, D.L., D. AXELROD, L. BAILEY, et al. 1998. Rethinking breast cancer risk and the environment: the case for the precautionary principle. Environ. Health Perspect. **106:** 523–529.
2. CANADIAN BREAST CANCER INITIATIVE. 2001. Summary report: review of lifestyle and environmental risk factors for breast cancer. Health Canada (Cat. No. H39-586/ 2001E).
3. DAVIS, D.L., M. PONGSIRI & M. WOLFF. 1997. Recent developments on the avoidable causes of breast cancer. Ann. N. Y. Acad. Sci. **837:** 513–523.
4. BIRNBAUM, L.S. & S.E. FENTON. 2003. Cancer and developmental exposure to endocrine disrupters. Environ. Health Perspect. **111:** 389–394.
5. BRUCKER-DAVIS, F., K. THAYER & T. COLBORN. 2001. Significant effects of mild endogenous hormonal changes in humans: considerations for low-dose testing. Environ. Health Perspect. **109**(Suppl. 1): 21–26.
6. BACCARELLI, A., A.C. PESATORI & P.A. BERTAZZI. 2000. Occupational and environmental agents as endocrine disrupters: experimental and human evidence. J. Endocrinol. Invest. **23:** 771–781.
7. DEGEN, G.H. & H.M. BOLT. 2000. Endocrine disrupters: update on xenoestrogens. Int. Arch. Occup. Environ. Health **73:** 433–441.
8. KENNEDY, S. 2000. Endocrine disrupters: overview and a pathologist's perspective. Toxicol. Pathol. **28:** 418–419.
9. BRODY, J.G. & R.A. RUDEL. 2003. Environmental pollutants and breast cancer. Environ. Health Perspect. **111:** 1007–1019.
10. GOLDBERG, M.S. & F. LABRECHE. 1996. Occupational risk factors for female breast cancer: a review. Occup. Environ. Med. **53:** 145–156.

11. EPSTEIN, S., D. STEINMAN & S. LEVERT. 1997. Hazards in the workplaces. *In* the Breast Cancer Prevention Program. S. Epstein, D. Steinman & S. Levert, Eds.: 273–296. MacMillan, USA.
12. BROPHY, J. 2004. Occupational histories of occupational cancer patients in a Canadian treatment centre and the generated hypothesis regarding breast cancer and farming (letter). Int. J. Occup. Environ. Health **10**: 116–118.
13. BROPHY, J., M.M. KEITH, K.M. GOREY, *et al.* 2002. Occupational histories of cancer patients in a Canadian cancer treatment centre and the generated hypothesis regarding breast cancer and farming. Int. J. Occup. Environ. Health **8**: 346–353.
14. POLK CITY DIRECTORY. 1999. Infotyme Software for Leamington, Windsor, Windsor Suburban, Ontario. Multi-Dimensional Intelligence, R.L. Polk, Southfield, Michigan.
15. HUMAN RESOURCES DEVELOPMENT CANADA. 1992. National Occupational Classification. Ottawa, Canada.
16. STATISTICS CANADA. 1998. North American Industrial Classification System. Ottawa, Canada.
17. CHECKOWAY, H., N. PEARCE & D. CRAWFORD-BROWN. 1989. Research Methods in Occupational Epidemiology. Oxford Press. New York.
18. MCDUFFIE, H. 1994. Women at work: agriculture and pesticides. J. Occup. Med. **36**: 1240–1246.
19. MCDUFFIE, H. 2005. Host factors and genetic susceptibility: a paradigm of the conundrum of pesticide exposure and cancer associations. Rev. Environ. Health **20**: 77–100.
20. DAVIS, D.L., A. BLAIR & D. HOEL. 1992. Agricultural exposures and cancer trends in developed countries. Environ. Health Perspect. **100**: 39–44.
21. COOGAN, P.F., R.W. CLAPP & P.A. NEWCOMB, *et al.* 1996. Variation in female breast cancer risk by occupation. Am. J. Ind. Med. **30**: 430–437.
22. CANTOR, K.P., P.A. STEWART & L.A. BRINTON, *et al.* 1995. Occupational exposures and female breast cancer mortality in the United States. J. Occup. Environ. Med. **37**: 336–348.
23. MORTON, W.E. 1995. Major differences in breast cancer risks among occupations. J. Occup. Environ. Med. **37**: 328–335.
24. CLAPP, R.W., G.K. HOWE & M.M. JACOBS. 2005. Environmental and occupational causes of cancer: a review of recent scientific literature. Lowell Center for Sustainable Production, University of Massachusetts Lowell. www.sustainableproduction.org
25. INTERNATIONAL AGENCY FOR RESEARCH ON CANCER (IARC). 2005. Monographs on the evaluation of carcinogenic risks to humans. http://monographs.iarc.fr/
26. WATTERSON, A. 1995. Environmental and Occupational Carcinogens and Breast Cancer: Public Health Concerns and Public Health Failures. DeMontfort University, Leicester, UK.
27. JANSSENS, J.P., E.V. HACKE, H. GEYS, *et al.* 2001. Pesticides and mortality from hormone-dependent cancers. Eur. J. Cancer Prev. **10**: 459–467.
28. DICH, J., S.H. ZAHM & A. HANBERG, *et al.* 1997. Pesticides and cancer. Cancer Causes Control **8**: 420–443.
29. ARONSON, K., A. MILLER, C. WOLLCOTT, *et al.* 2000. Breast adipose tissue concentrations of polychlorinated biphenyls and other organochlorines and breast cancer risk. Cancer Epidemiol. Biol. Prevent. **9**: 55–63.

30. DUELL, E.J., R.C. MILLIKAN, D.A. SAVITZ, *et al.* 2000. A population-based case-control study of farming and breast cancer in North Carolina. Epidemiology **11**: 523–531.
31. BAND, P.R., N.D. LE, R. FANG, *et al.* 2000. Identification of occupational cancer risks in British Columbia. A population-based case-control study of 995 incident breast cancer cases by menopausal status, controlling for confounding factors. J. Occup. Environ. Med. **42**: 284–310.
32. MILLS, P.K. & R. YANG. 2005. Breast cancer risk in hispanic agricultural workers in California. Int. J. Occup. Environ. Health **11**: 123–131.
33. PEIPENS, L.A., C. BURNETT & T. ALTERMAN, *et al.* 1997. Mortality pattern among female nurses: a 27-State Study, 1984-1990. Am. J. Pub. Health **87**: 1539–1543.
34. DEBRUIN, L.S. & P.D. JOSEPHY. 2002. Perspectives on the chemical etiology of breast cancer. Environ. Health Perspect. **110**(Suppl 1): 119–128.
35. SOTO, A.M., T.M. LIN, H. JUSTICIA, *et al.* 1992. An 'in culture" bioassay to assess the estrogencity of xenobiotics. *In* T. Colborn, Ed.: 295–309. Chemically Induced Alterations in Sexual Development: The Wildlife/Human Connection. Princeton Scientific Publishing. Princeton, NJ.
36. DAVIS, S., D. MIRICK & R. STEVENS. 2001. Night shift work, light at night, and risk of breast cancer. J. Natl. Cancer Inst. **93**: 1557–1562.
37. SCHERNHAMMER, E.S., F. LADEN, F.E. SPEIZER, *et al.* 2001. Rotating night shifts and risk of breast cancer in women participating in the nurses' health study. J. Natl. Cancer Inst. 2001; **93**: 1563–1568.
38. LIE, J.-A.S. & K. KJAERHEIM. 2003. Cancer risk among female nurses: a literature review. Eur. J. Cancer Prev. **6**: 517–526.
39. PETRALIA, S.A., M. DOSEMECI, E.E. ADAMS, *et al.* 1999. Cancer mortality among women employed in health care occupations in 24 U.S. States, 1984–1993. Am. J. Ind. Med. **36**: 159–165.
40. RUBIN, C.H., C.A. BURNETT, W.E. HALPERIN, *et al.* 1993. Occupation as a risk identifier for breast cancer. Am. J. Pub. Health **83**: 1311–1315.
41. PEIPINS, L.A., C. BURNETT, T. ALTERMAN, *et al.* 1997. Mortality patterns among female nurses: a 27-State Study, 1984 through 1990. Am. J. Pub. Health **87**: 1539–1543.
42. RIX, B.A. & E. LYNGE. 1996. Cancer incidence in Danish health care workers. Scand. J. Soc. Med. **24**: 114–120.
43. SANKILA, R., S. KARJALAINEN, E. LAARA, *et al.* 1990. Cancer risk among health care personnel in Finland. Scand. J. Work Environ. Health **16**: 252–257.
44. MORTON, W.E. 1995. Major differences in breast cancer risks among occupations. J. Occup. Environ. Med. **37**: 328–335.
45. PETRALIA, S.A., W.H. CHOW, J. MCLAUGHLIN, *et al.* 1998. Occupational risk factors for breast cancer among women in Shanghai. Am. J. Ind. Med. **34**: 477–483.
46. POLLAN, M. & P. GUSTAVSSON. 1999. High-risk occupations for breast cancer in the Swedish female working population. Am. J. Pub. Health **89**: 875–881.
47. CALLE, E., T. MURPHY, C. RODRIGUEZ, *et al.* 1998. Occupation and breast cancer mortality in a prospective cohort of US women. Am. J. Epidemiol. **148**: 191–197.
48. GUNNARSDOTTIR, H. & V. RAFNSSON. 1995. Cancer incidence among icelandic nurses. J. Occup. Environ. Med. **37**: 307–312.
49. COOGAN, P.F., R.W. CLAPP, P.A. NEWCOMB, *et al.* 1996. Variation in female breast cancer risk by occupation. Am. J. Ind. Med. **30**: 430–437.

50. HABEL, L.A., J.L. STANFORD, T.L. VAUGHAN, et al. 1995. Occupation and breast cancer risk in middle-aged women. J. Occup. Environ. Med. **37:** 349–356.
51. PETRALIA, S.A., M. DOSEMECI, et al. 1999. Cancer mortality among women employed in health care occupations in 24 U.S. States, 1984–1993. Am. J. Ind. Med. **36:** 159–165.
52. ZHENG, T., T.R. HOLFORD, M.S. TAYLOR, et al. 2002. A case-control study of occupation and breast-cancer risk in Connecticut. J. Cancer Epidemiol. Prev. **7:** 3–11.
53. GUNNARSDOTTIR, H.K., T. ASPELUND, T. KARLSSON, et al. 1997. Occupational risk factors for breast cancer among nurses. Int. J. Occup. Environ. Health **3:** 254–258.
54. DELZELL, E., C. BEALL & M. MANCALUSO. 1994. Cancer mortality among women employed in motor vehicle manufacturing. J. Occup. Med. **36:** 1251–1259.
55. THOMPSON, D., D. KRIEBEL, M.M. QUINN, et al. 2005. Occupational exposure to metalworking fluids and risk of breast cancer among female autoworkers. Am. J. Ind. Med. **47:** 153–160.
56. HANSEN, J. 1999. Breast cancer risk among relatively young women employed in solvent-using industries. Am. J. Ind. Med. **36:** 43–47.
57. LABRECHE, F. & M. GOLDBERG. 1997. Exposure to organic solvents and breast cancer in women: a hypothesis. Am. J. Ind. Med. **32:** 1–14.
58. BLAIR, A., A. LINOS, P.A. STEWART, et al. 1993. Evaluation of risks for non-Hodgkin's lymphoma by occupation and industry exposures from a case-control study. Am. J. Ind. Med. **23:** 301–312.
59. BROPHY, J., M. KEITH, I. LUGINAAH, et al. 2003. Occupational histories of breast cancer patients. 3 year research grant from Canadian Breast Cancer Research Foundation—Ontario Chapter.
60. BROPHY, J., M. KEITH, I. LUGINAAH, et al. 2003. Occupational histories of breast cancer patients. 2 year research grant from Breast Cancer Society of Canada.
61. SIEMIATYCKI, J. 1991. Risk Factors for Cancer in the Workplace. CRC Press. Boca Raton.
62. AHRENS, W. & P. STEWART. 2003. Retrospective exposure assessment. *In* Exposure Assessment in Occupational and Environmental Epidemiology. M. Nieuwenhuijsen: Ed.: Oxford University Press. Oxford.
63. JOFFE, M. 1992. Validity of exposure data derived from a structured questionnaire. Am. J. Epidemiol. **135:** 564–570.
64. SIEMIATYCKI, J. 1995. Future etiologic research on occupational cancer. Environ. Health Perspect. **103**(Suppl 8): 209–215.

Adverse Health Effects of Fluoro-Edenitic Fibers

Epidemiological Evidence and Public Health Priorities

CATERINA BRUNO, PIETRO COMBA, AND AMERIGO ZONA

Department of Environment and Primary Prevention, Istituto Superiore di Sanità, I-00161 Rome, Italy

ABSTRACT: Subsequent to the detection of a cluster of mesothelioma cases in the Sicilian town of Biancavilla, located at the slopes of Etna volcano, *ad hoc* epidemiological studies and environmental monitoring suggested an etiological role of an asbestiform fiber present in a stone quarry. The fiber was shown to constitute a new mineral species named fluoro-edenite. Fluoro-edenitic fibers were found in the materials extracted from the quarry and used in the local building industry, as well as in soils. Besides the risk of mesothelioma, residents in Biancavilla showed a significantly increased mortality from chronic obstructive pulmonary disease, which was particularly evident among women. In the light of these findings, Biancavilla was defined a site of national interest for environmental reclamation. The first preventive action involved termination of quarrying activity, covering with asphalt of roads previously paved with local soil materials, and removal of sources of dust in the urban area. Concurrent to the implementation of environmental cleanup, some specific "second generation" studies are now being designed and performed, namely morbidity surveys based on hospital discharge cards, monitoring of fibers in sputum and health surveillance in selected population groups. In this frame, special emphasis is given to the issue of communication, both to the general public and to target groups like family doctors, teachers, and media professionals. This experience could represent a useful basis for the elaboration of a strategy to approach similar environmental issues.

KEYWORDS: environmental exposure; fluoro-edenite; mesothelioma; chronic obstructive pulmonary disease; Biancavilla

A descriptive epidemiological study on mortality for malignant pleural neoplasms in Italy, identified some geographic clusters of cases.[1] Among these

Address for correspondence: Caterina Bruno, M.D., Department of Environment and Primary Prevention, Istituto Superiore di Sanità, Viale Regina Elena 299, I-00161 Rome, Italy. Voice: +39-06-49902461; fax: +39-06-49902999.
e-mail: caterina.bruno@iss.it

areas, a town located in a volcanic area in eastern Sicily, Biancavilla, was considered of special interest, because no industrial activities related to asbestos had occurred in the area. This observation thus prompted the first epidemiological investigation.[2] Using data provided by the Italian National Institute for Statistics (ISTAT), cause-specific mortality among residents in Biancavilla, 1980–1993 was examined, taking into consideration causes with known causal association with asbestos (malignant neoplasms of the pleura, peritoneum, and lung), and causes with suspected relation to the same exposure (malignant neoplasms of the ovary). For each cause, observed mortality was compared to the expected value, based on cause-sex-calendar year-specific mortality rates of Sicily's population, with the aid of epidemiological databank of the Italian National Board for Energy, New Technology and Environment (ENEA). Standardized mortality ratios (SMRs) and their 95% confidence interval (CIs) in accordance with Poisson distribution, were calculated. In the study period, mortality from malignant pleural neoplasms in Biancavilla showed a significant increase (observed [obs.] 9 cases, expected [exp.] 2.31, SMR 390, 95% CI 178–740): the increase was relatively stronger among women (obs. 5, exp. 0.74, SMR 676, 95% CI 219–1577), in patients 65 years old or less (obs. 5, exp. 0.89, SMR 562, 95% CI 182–1311), and in the period 1988–1993 (obs. 5, exp. 1.15, SMR 435, 95% CI 141–1015). The evidence of the mesothelioma cluster in Biancavilla, based on mortality figures, is shown in TABLE 1.

Seventeen cases of pleural mesothelioma were identified by means of active search between 1980 and 1997, 7 females and 10 males: hospital medical documentation relative to cases identified was collected and thoroughly examined. Patients or their next of kin were interviewed about previous fiber exposure: 9 of them had no evidence of occupational exposure; 2 had possible professional exposure, which could not be excluded for 5 other cases. For 1 case occupational exposure could not be evaluated, as the subject had been living in Biancavilla only since 1996. Exposure categories were attributed according to the Italian National Registry of Mesotheliomas guidelines (ReNaM).[3] An annual incidence rate around 7.0×10^4 was estimated. An environmental survey suggested that the stone quarry located in Monte Calvario, northeast of Biancavilla, could represent the cause of the asbestos exposure

TABLE 1. Mortality from pleural neoplasm in Biancavilla, data from pleural neoplasm mortality national surveillance (100 observed = expected)

Publication year (reference)	Observational period	Observed deaths	SMR	95% CI
1996[1]	1988–1992	4	417	-
2000[20]	1988–1994	8	579	249–1,141
2002[21]	1988–1997	12	580	299–1,013

for Biancavilla's inhabitants. The quarry was mined for more than 50 years, with a peak around 1960s and 1970s. The survey revealed the occurrence of a fibrous, yellow and brown tremolitic amphibole; large quantities of acicular and asbestiform Ca-amphiboles were present in the materials extracted from the quarry, extensively used in the local building industry.[4] Samples collected at the front of the quarry, were examined with scanning electron microscopy (SEM) and with energy dispersive spectroscopy X ray microanalysis. X ray investigation showed mineralogical phases between actinolite and tremolite [$Ca_2(Mg,Fe)_5Si_8O_{22}(OH,F)_2$]. The fibrous habit of the amphibole phases was evident under SEM at $>10,000\times$.

To assess the extent to which the amphibole fibrous phases were used in the local building industry, an investigation was performed to evaluate in which substance materials from Monte Calvario were used, the percentage of dwellings in which the materials were used, period of building construction or restructuring, type of materials and their destinations.

Two sample sets (25 and 38 specimens, respectively) from walls, plaster, and concrete, located inside and outside of the buildings were obtained. For the second set, building dates were known. After a proper treatment in laboratory, samples were examined under the SEM ($2500\times$) for their morphology, and with an energy-dispersion spectrometry system for X ray microanalysis to investigate the samples chemical composition. Seventy-two percent of the first sample set showed the same types of amphibole phases detected in the Monte Calvario quarry. In 71% of the second sample set, fibrous phases were identified. Fiber concentrations in building materials range from a few 1000 up to more than 4×10^4 fibers/mg (dwellings built around 1960s).

The autopsy samples of lung tissue of an 86-year-old woman who died from pleural mesothelioma, were processed and examined for the mineral fibers, which were the same tremolite–actinolite fibrous amphibole found in the quarries.

The amphibole was finally identified as fluoro-edenite.[5,6] Information on chemical and structural data for amphibole along the edenite-fluoro-edenite joins are reported in the scientific literature, but are mostly referred to synthetic samples. Edenitic compositions are quite rare in amphiboles, probably for structural instability.[7] The association between pleural mesothelioma and the mineral fibers in the building materials suggested in the first epidemiological investigation,[2] was then evaluated as causally attributable to fluoro-edenitic fibers.[8]

The preliminary findings of a pilot study using sputum as an exposure indicator for fluoro-edenite in Biancavilla citizens, showed in 6 out of 12 subjects at least one positive sample for the presence of this mineral fiber (concentration range 0.05–10 f/gr).[9]

Cases of environmental exposure to natural asbestiform fibers have already been described, mostly tremolite, in the international scientific literature.[10] Because of the similarities between tremolite and fluoro-edenite, a fibrogenic activity for fluoro-edenite was hypothesized. To investigate the association between mortality from chronic obstructive pulmonary disease (COPD) and exposure to fluoro-edenite, 36 municipalities located in the volcanic area of mount Etna were studied.[11] An ecological regression model was applied with mortality from COPD as the dependent variable, mortality from mesothelioma as a proxy for exposure to fluoro-edenite, and lung cancer mortality, an urban–rural index, a deprivation index, and an aging index as the predictors of COPD mortality. A significant association was found between COPD mortality and pleural neoplasm mortality among the women in this study. The results suggest an etiological role for fluoro-edenite in nonmalignant respiratory diseases. Studies on *in vitro* and *in vivo* properties of the fiber were performed.[12–14]

An epidemiological study on health conditions of populations living in high environmental risk areas and in sites of national interest for environmental reclamation in Sicily, based on mortality (period 1995–2000) and hospitalization data (period 2001–2003),[15] showed some increases of the occurrence of certain diseases as shown in TABLE 2. These results confirm the observation for the increase of malignant pleural neoplasms; the excess for chronic pulmonary diseases might be due to a possible fibrogenic property of fluoro-edenite. Acute respiratory diseases could be determined by other irritant materials often present in volcanic areas. No explanation has been found for cardiovascular diseases increase, which deserves further investigation.

Public health recommendations were elaborated, in order to decrease exposure to fluoro-edenite, and consequently the risk of new cases of fiber-related diseases: termination of quarrying activity, removal of sources of dust downtown Biancavilla, and asphalting roads previously paved with local soil materials. These recommendations were successfully implemented when Biancavilla was included among the sites of national interest for environmental reclamation (for a detailed discussion, see Ref. 16). In addition to enforcement of stringent environmental regulation, guidelines for modifying individual behavior were provided, with emphasis on quitting cigarette consumption, in order to reduce lung cancer risk, and on avoiding do-it-yourself activity (i.e., decorating or repairing home, or making things for home on your own), which

TABLE 2. Mortality and morbidity increases for specific disease, in Biancavilla (modified from Ref. 15)

Disease (ICD-9 Code)	Mortality		Morbidity	
	Male	Female	Male	Female
Malignant pleural neoplasms (163)	X	X		X
Cardiovascular diseases (390–459)	X	X	X	X
Cardiovascular ischemic diseases (410–414)		X		X
Respiratory diseases (460–519)	X	X	X	X
Acute respiratory diseases (460–466, 480–487)	X		X	X
Chronic pulmonary diseases (490–496)	X	X		X

could imply fiber dispersion, unless properly done.[17] Communication to the general population, and to target groups as well, is of special importance in order to convey proper information, answer specific questions, discuss comprehensible fears, on a mutually respectful relationship basis.[18] In fact, citizens were informed about the preventive nature of the recommendations, even before the fiber was identified as fluoro-edenite.[19]

In conclusion, the town of Biancavilla is experiencing an epidemic of pleural mesothelioma as a consequence of the presence of fluoro-edenitic fibers, and an increase in nonmalignant respiratory diseases, which deserves further investigations.

A major environmental reclamation program is currently being developed, in order to achieve significant reduction in fiber exposure levels. In the meanwhile, continuing environmental monitoring, epidemiological surveillance, and health education campaigns are warranted.

REFERENCES

1. DI PAOLA, M., M. MASTRANTONIO, M. CARBONI, *et al.* 1996. Mortality from malignant pleural neoplasms in Italy in the years 1988–1992. Rome, Italy: Rapporti ISTISAN 96/40. Italian.
2. PAOLETTI, L., D. BATISTI, C. BRUNO, *et al.* 2000. Unusually high incidence of malignant pleural mesothelioma in a town of Eastern Sicily: an epidemiological and environmental study. Arch. Environ. Health **55:** 392–398.
3. CHELLINI, E., E. MERLER, C. BRUNO, *et al.* 1996. National Register of Asbestos-Related Mesothelioma. Fogli d'informazione ISPESL: 19–106. Italian.
4. GIANFAGNA, A., L. PAOLETTI & P. VENTURA. 1997. Fibrous amphibole phases in the volcanic products of Monte Calvario (eastern Sicily). Eur. J. Mineral. **18**(Suppl): 117–119. Italian.
5. GIANFAGNA, A. & R. OBERTI. 2001. Fluoro-edenite from Biancavilla (Catania, Sicily, Italy): cristal chemistry of a new amphibole end-member. Am. Mineral. **86:** 1489–1493.
6. GIANFAGNA, A., P. BALLIRANO, F. BELLATRECCIA, *et al.* 2003. Characterization of amphibole fibres linked to mesothelioma in the area of Biancavilla, Eastern Sicily, Italy. Mineral. Mag. **67:** 1221–1229.
7. RAUDSEPP, M., A.C. TURNOCK & F.C. HAWTHORNE. 1991. Amphibole synthesis at low pressure: what grows and what doesn't. Eur. J. Mineral. **3:** 983–1004.
8. COMBA, P., A. GIANFAGNA & L. PAOLETTI. 2003. The pleural mesothelioma cases in Biancavilla are related to the new fluoro-edenite fibrous amphibole. Arch. Environ. Health **58:** 229–232.
9. PUTZU, M.G., C. BRUNO, A. ZONA, *et al.* 2006. Fluoro-edenitic fibres in the sputum of subjects from Biancavilla (Sicily). Environ. Health **16:** 15(1): 20 [Epub ahead of print].
10. PASETTO, R., P. COMBA & A. MARCONI. 2005. Mesothelioma associated with environmental exposures. Med. Lav. **96:** 330–337.
11. BIGGERI, A., R. PASETTO, S. BELLI, *et al.* 2004. Mortality from chronic obstructive pulmonary disease and pleural mesothelioma in an area contaminated by natural fiber (fluoro-edenite). Scand. J. Work Environ. Health **30:** 249–252.

12. TRAVAGLIONE, S., B. BRUNI, L. FALZANO, et al. 2003. Effects of the new-identified amphibole fluoro-edenite in lung epithelial cells. Toxicol In Vitro **17:** 547–552.
13. TRAVAGLIONE, S., B.M. BRUNI, L. FALZANO, et al. 2006. Multinucleation and pro-inflammatory cytokine release promoted by fibrous fluoro-edenite in lung epithelial A459 cells.Toxicol. In Vitro **20(6):** 841–850.
14. SOFFRITTI, M., F. MINARDI, L. BUA, et al. 2004. First experimental evidence of peritoneal and pleural mesotheliomas induced by fluoro-edenite fibres present in Etnean volcanic material from Biancavilla (Sicily, Italy). Eur. J. Oncol. **9:** 169–175.
15. FANO, V., A. CERNIGLIARO, S. SCONDOTTO, et al. 2005. Health conditions of populations living in high environmental risk areas and in sites of national interest for environmental reclamation in Sicily—Analysis of mortality (period 1995–2000) and hospitalization data (period 2001–2003). Regione Siciliana – Assessorato Sanità – Dipartimento Osservatorio Epidemiologico. http://www.epicentro.iss.it/focus/amb_sal/Rapporto_sicilia.pdf [Visited on 05/09/05]. Italian.
16. CORI, L., M. COCCHI & P. COMBA. Epidemiologic studies in the sites of national interest for environmental reclamation. Rome, Italy: Rapporti ISTISAN 05/1. Italian.
17. COMBA, P., C. BRUNO & R. PASETTO. 2003. Indications of public health in areas naturally polluted with asbestiform fibers. G. Ital. Med. Lav. Erg. **25:** 405–407. Italian.
18. LAMBERT, T.W., L.C. SOSKOLNE, V. BERGUM, et al. 2003. Ethical perspectives for public and environmental health: fostering autonomy and the right to know. Environ. Health Perspect. **111:** 133–137.
19. MANNA, P. & P. COMBA. 2001. Communicating with health authorities and the public about asbestos risk in Biancavilla (CT). Epidemiol. Prev. **25:** 28–30. Italian.
20. DI PAOLA, M., M. MASTRANTONIO, M. CARBONI, et al. 2000. Asbestos exposure and mortality from malignant pleural neoplasms in Italy (1988–1994). Rome, Italy: Rapporti ISTISAN 00/9. Italian.
21. MASTRANTONIO, M., S. BELLI, A. BINAZZI, et al. 2002. Mortality from malignant pleural neoplasms in Italian municipalities. Rome, Italy: Rapporti ISTISAN 02/12. Italian.

Glycol Ethers: A Ubiquitous Family of Toxic Chemicals

A Plea for REACH Regulation

ANDRÉ CICOLELLA

Health Risk Assessment Unit, National Institute of Risks and Industrial Environment, 60550 Verneuil-en-Halatte, France

ABSTRACT: Glycol ethers (GE) are chemicals used since the 1930s as solvents in paints, inks, varnishes, and cleaning agents, mainly in water-based products, cosmetics, and drugs. World production approximates 1 million tons. Nineteen GE are produced or imported each year; over 1000 tons in European Union (EU) have been classified as high production volume chemicals (HPVCs). First animal data were published in 1971 and 1979 showing severe reprotoxicity for some GE. Two alerts were launched in the United States in 1982 and 1983, but the first partial GE regulation only occurred in 1993 in the EU. Although these chemicals may expose a very large population, basic toxicity data, more especially carcinogenicity, are still lacking (3/32 GE). However, experimental data were sufficient to lead developmental toxicity risk assessment since the early 1980s. Risk indices over 1000 have been calculated for consumers and workers exposed to reprotoxic GE in domestic and industrial activities. The first ban was decided in 1999 in France, but was only for drugs and cosmetics. Not surprisingly, since the late 1980s, human studies have found results similar to those in animal data: spontaneous abortions, malformations, testicular toxicity, and hematotoxicity. Despite this highly coherent set of data, and although substitution products are available, reprotoxic GE have been and still remain widely used in the world. The case of GE shows the failure of the present system based on *a posteriori* risk assessment. This pleads for the change of paradigm through the European REACH regulation based on the "No data, no market" principle. Ethics in REACH management should also be considered.

KEYWORDS: glycol ethers; precautionary principle; REACH

Address for correspondence: André Cicolella, Head of the Health Risk Assessment Unit, National Institute of Risks and Industrial Environment, INERIS BP N 2, 60550 Verneuil-en-Halatte, France. Voice: 33-344-556-202; fax: 33-344-556-899.
e-mail: andre.cicolella@ineris.fr

INTRODUCTION

Since the first directive issued in 1967, the European Union (EU) has played a major role in chemical risk management. Twenty-nine following adaptations have classified more than 3000 substances in 15 hazard classes, over 100,000 listed in 1981. Risk assessment methodology has been specified in 1993–1994 regulations to assess in priority high production volume chemicals (HPVCs). With 141 chemicals listed, 59 final reports, 28 recommendations out of 2747 HPVCs, this system has obviously failed. In 2001, the EU Commission decided to change the paradigm in "The white paper on a strategy for a future Chemicals Policy." The Registration Evaluation and Authorization of Chemicals (REACH) program was proposed with the principle: "No data, no market." The chemical industry has severely criticized the process as unrealistic, bureaucratic, and costly. Now the process is entering in the final step, it seems necessary to point out the inability of the old paradigm to protect adequately health of European people and the necessity to shift toward a new one. The case of glycol ethers (GE) illustrates the two main reasons of this failure: lack of basic data, even for the more severe effects such as CMR (carcinogenic, mutagenic, reprotoxic), and postponed implementation of risk management, even once scientific evidence is obvious.

THE GE CASE

Generality

GE are a family of chemicals developed in the 1930s by Union Carbide Company under the brand name of Cellosolve®.[1–3] From the 1960s, they were considered as nontoxic because of their weak acute toxicity and because they have been widely used in industrial, domestic, cosmetic uses, and even as a drug excipient (EGEE, a GE now classified as teratogen, was used in an anti-acne drug till 1995). Main uses were in water-based products (paintings, varnishes, inks, cleaning agents), but also include nonsolvent applications: paintings, brake fluids

GE are made up with two subgroups: ethylene GE (E series) and propylene GE (P series). In each subgroup, ethers can be esterified into ethers-esters, the most common being acetates. GE acetates are rapidly transformed in the corresponding ethers when introduced in the body and are consequently considered as toxicologically equivalent. Both subgroups have the same technological properties to mix water, polymer resins, and organic solvents together but are quite different from a toxic point of view owing to metabolic differences: E series are generally much more toxic than P series, which can be considered in return as good substitutes. In 1988, world production was 700,000 tons (81% E series, 19% P series), with 19 GE classified as HPVCs. GE are readily absorbed following inhalation or oral administration. Besides inhalation, skin exposure to the liquid and inhalation of the vapor may be significant methods of exposure. Millions of workers and a much larger number of consumers have thus been exposed to GE for several decades.

CMR Risk Assessment

For long, all GE have been considered as very weak toxicants. The first scientific articles dealing with reprotoxicity only came in the 1970s and those with genotoxicity/carcinogenicity in the late 1990s. Not all GE were tested, the poorest data being for carcinogenicity. In the early 1980s, animal data were sufficient to perform a risk assessment based on developmental toxicity for EGME, EGMEA, EGEE, and EGEEA, for workers and consumers (TABLE 1). Safety factors classically recommended are 100 (10 for extrapolation from animal to humans, 10 for human susceptibility). For teratogenic compounds, the World Health Organization and the California EPA recommend an additional safety factor of 10. The critical period chosen, i.e., the time to give adverse effects, was the day, according to EPA guidelines on developmental toxicity. Risk index (RI) was calculated for two scenarios (maximal and minimal) as follows:

$$RI = DD \text{ (estimated daily dose)} / RfD_D \text{ (reference dose for developmental effects)}.$$

Very high figures in the range 1000–10000 have been found for popular consumer goods, such as water-based paints (0.9% EGME), window cleaner (4% EGME + 15% EGEE), and parquet floor varnish (32% EGEE) (FIG. 1). It was obvious therefore that pregnant consumers or workers exposed in these conditions have long been at risk, even when using products with GE respecting the EU 0.5% concentration limit in consumer products.

TABLE 1. Main GE experimental data, adapted from INSERM[3]

	EGME (EGMEA)	EGEE (EGEEA)	EGBE	2PG1ME	Data availability (All GE in 1989)
Male reprotoxicity	+++	+++	−	−	22/32
Female reprotoxicity	++	++	+	−	8/32
Developmental toxicity	+++	+++	++	−	30/32
First data	1979	1971	1979	1972	
Genotoxicity (aldehyde metabolite)	+++	+++	+++	−	24/32
First data	1996	1996	1996	1996	
Carcinogenicity	?	?	++	−	1/32 3/32 (2002)
First data	No data	No data	1998	1999	

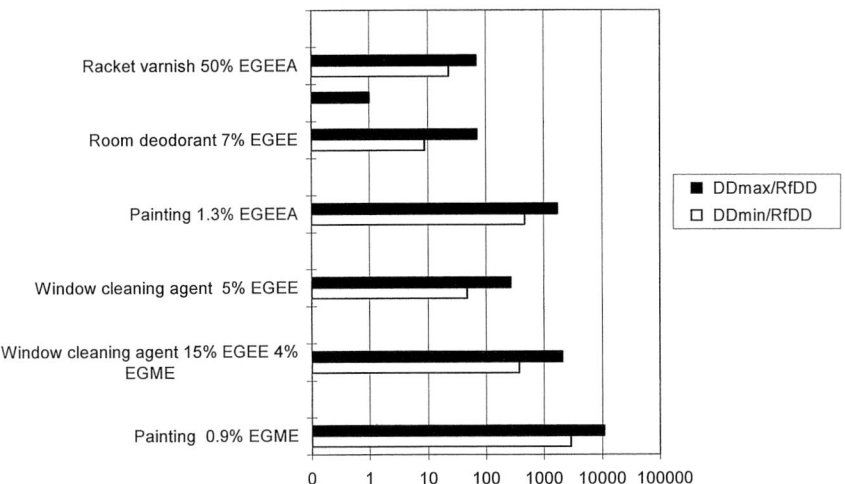

FIGURE 1. Developmental toxicity risk index for consumers adapted from Cicolella.[4]

Human Data

Not surprisingly, because of similar metabolism between humans and animals, the same toxic effects have been seen among workers (shipyards, semiconductor industry, and foundry). Cancer data remain scarce. Only one study has suggested EGME involvement in testicular cancer among aeromechanics.

CMR Risk Management

Ten years after the first publication, a first alert was launched in May of 1982 by the State of California (Department of Health Services and Department of Industrial Relations) for four GE (EGME, EGMEA, EGEE, and EGEEA) stating: *"Glycol ethers have damaged the reproductive systems of test animals, raising the possibility that they may cause similar effects in humans."* [5] US National Institute for Occupational Safety and Health (NIOSH) published in May, 1983 a Current Intelligence Bulletin no. 39: *"NIOSH recommends that EGME and EGEE be regarded in the workplace as having the potential to cause adverse reproductive effects in male and female workers.... Of particular concern are those studies in which exposure of pregnant animals to concentrations of EGME and EGEE at or below their respective Occupational Safety and Health Administration (OSHA) permissible exposure limits (PEL's) led to increased incidences of embryonic death, teratogenesis, or growth retardation. Exposure of male animals resulted in testicular atrophy and sterility."* [6] This alert was based on 13 studies for EGME and 12 studies for EGEE. Others are quoted for structurally related GE.

TABLE 2. EGME and EGEE occupational limit values (2005)

GE	NOEL	ACGIH, France	OSHA	NIOSH (OSHA Project 1993)
EGME	10 ppm	5 ppm ACGIH project 2005 (0.1 ppm)	25 ppm	0.1 ppm
EGEE	50 ppm	5 ppm	200 ppm	0.5 ppm

However, no regulation was issued before 1992, when Swedish National Chemicals Inspectorate (KEMI) recommended phasing out the same four GE. In October 1993, EU classified them as class 2 reprotoxic. Consequently, their use in consumer goods was automatically limited to 0.5%. For occupational use, consequences of this classification were special labeling with signs and risk phrases on tins (*R60: May hamper fertility; R61: Risk for Child Health during pregnancy*) and obligation of substitution, however only if technically feasible. If not, occupational limit values (OLV) had to be adhered. These OLV are very close to the no observed effects level (NOEL) for developmental toxicity in animals (American Conference of Governmental Industrial Hygienists, France) or even much higher (US Occupational Safety and Health Administration) and do not offer adequate protection to pregnant workers (TABLE 2).

On this basis, French Drug Agency decided a first ban in 1999 for EGME, EGMEA, EGEE, and EGEEA, which was extended to other reprotoxic GE once they were classified by EU (DEGME, EGDME, DEGDME, and TEGDME). In 2000, the French Consumers Safety Commission (FCSC) recommended a ban on reprotoxic E series and for all GE not adequately evaluated. In 2002, the French Higher Council of Public Health (CSHPF) issued a report on risk due to exposure to 0.5% in domestic products and concluded similarly that 0.5% level recommended by EU in domestic products was inadequate to protect consumers' health. A plan especially devoted to GE was then added in the French National Environmental Health Plan. EU has decided on no ban yet. Overall use of the most reprotoxic GE has dramatically decreased in the decade (3% of the 1993 use). However, one part of the shift was partly toward EGBE, which may not be totally safe because this substance has been proven to have "some evidence" of carcinogenicity according to the National Toxicology Program evaluation.

DISCUSSION

GE have been widely used in industrial and domestic products for more than 70 years. Toxicological data have not been available during 40 years. Although reprotoxicity was clearly assessed in the 1980s for the most used GE, it took several years to get regulations and much more time to get decreasing use. Today, however, even the most toxic GE still remain on the world market, exposing population to high reprotoxic risk. As far as carcinogenicity is concerned, lack of data continues to hamper risk assessment; only 3 GE out of 32 have been tested for chronic toxicity.

The case of GE pleads for a change of paradigm in chemical risk management, based on precautionary principle as foreseen in REACH regulation. Because of their set of severe toxicity data, neither industry could try to put on the market, nor European public agencies could accept, firstly so severe reprotoxic chemicals, secondly HPVCs without chronic toxicity data. REACH regulation represents a chance to change dramatically chemical risk management system and better protect public health. However, this change would not be complete if the question of ethics in risk assessment and scientific practices was not put on the EU agenda. Organizing international cooperation to produce scientific data on GE toxicity and publish them has encountered unfair opposition[7] and illustrates the quotation by Michael Davis, the former U.S. assistant secretary of energy for environment, safety, and health under Clinton Administration, on industry strategy to postpone regulation as expressed by a former tobacco industry leader: "Doubt is our product since it is the best means for competing with the 'body of fact' that exists in the mind of the general public."[8] EU has also a leading role to play on that topic in relationship with UNESCO and other international agencies.

REFERENCES

1. HARDIN, B., Ed. 1984. Proceedings of the 1st health risks of glycol ethers international symposium. Environ. Health Perspect. **57:** 1–332.
2. CICOLELLA, A., B. HARDIN & G. JOHANSON (Eds.). 1996. Proceedings of the 2nd health risks of glycol ethers international symposium. Occup. Hyg. **2:** 1–456.
3. INSERM. 1999. Glycol Ethers: What Health Risks? In French.
4. CICOLELLA, A. 1999. Consumers risk assessment. *In* INSERM Glycol Ethers: What Health Risks? 309–322. In French.
5. State of California, Hazard Evaluation System and Information Service, Hazard Alert #2; May 1982.
6. National Institute of Occupational Safety and Health Current Intelligence Bulletin 39 Glycol ethers, 2-methoxyethanol and 2-ethoxyethanol; May 2, 1983.
7. BALTER, M. 1994. Toxic tiff spreads beyond France. Science **264:** 898.
8. DAVIS, M. 2005. Doubt is their product. Sci. Am. **292:** 73–79.

Controlling Exposure to Chemicals
A Simple Guide

ALASTAIR HAY

Molecular Epidemiology Unit, LIGHT Laboratories, Clarendon Way, School of Medicine, University of Leeds, Leeds LS2 9JT, United Kingdom

ABSTRACT: Controlling exposure to chemicals in the workplace has been made easier by the use of a guide published by the U.K. Health and Safety Executive (HSE). Known as COSHH (Control of Substances Hazardous to Health Regulations) Essentials, the guide is a simple five-step procedure to devise appropriate control strategies to reduce exposures to various substances under different conditions. U.K. health and safety law requires risk assessments prior to use of hazardous substances and installation of appropriate control strategies before work commences. A 1996 survey of 1500 safety managers and trade union safety representatives revealed that the majority had little understanding of occupational safety limits for chemicals. Small- and medium-sized companies had little understanding of limits, and most could not develop control strategies. A new approach was required. COSHH Essentials is it. Developed over 3 years by a working group of hygienists and toxicologists representing HSE, industry, trade unions, and independent experts, the guide is now available in both paper-based and internet versions. It applies a hazard banding approach validated by data for 111 substances that have well-founded U.K. occupational exposure limits. New users select an appropriate hazard band for chemicals based on risk phrases. Details about dustiness for powders or volatility for liquids are inserted, and the guide allocates substances to one of four exposure bands linked, in turn, to specific control strategies. Now accessible through the HSE web site, COSHH Essentials will offer control strategies for both single chemicals and whole processes. To date over 300,000 risk assessments have been carried out using the internet version of COSHH Essentials.

KEYWORDS: COSHH Essentials; chemical controls; chemical exposure; risk assessment; small companies

---

Address for correspondence: Alastair Hay, Molecular Epidemiology Unit, LIGHT Laboratories, Clarendon Way, School of Medicine, University of Leeds, Leeds LS2 9JT, UK. Voice: 00-44-113-343-5682, 00-44-113-343-6602; fax: 00-44-113-343-6503.
  e-mail: a.w.m.hay@leeds.ac.uk

INTRODUCTION

Health and Safety law in the United Kingdom has a long pedigree. The more significant statutes are the 1974 Health and Safety at Work Act[1] and the Control of Substances Hazardous to Health Regulations, colloquially known as COSHH.[2] Further regulations and modifications to COSHH emphasize the responsibility of the management to have procedures in places of work to control exposure to chemicals and to ensure the safety of the workforce.[3,4] For those chemicals in more widespread use, which present a greater risk to the workforce, there were some 650 statutory occupational exposure limits. These were known as Occupational Exposure Standards (OES) for chemicals where evidence of a threshold effect existed and Maximum Exposure Limits (MELs). Substances with MELs included those with no identifiable threshold and included substances such as carcinogens and reproductive toxins, or those where a threshold effect could be identified but the costs of achieving it were considered excessive.[5]

Concern that some chemicals did not fit either category of limit led to a revision in the limits-setting system. A new single Workplace Exposure Limit was introduced in February 2005 to replace the former two-tier limit system.[6] With the number of chemicals available for use estimated to be in excess of 100,000, it is evident that the 600–700 exposure limits will only provide guidance on appropriate controls for a small percentage of chemical substances used at work.

U.K. Health and Safety law requires management to carry out a risk assessment for all procedures at work. This has presented a major problem for small- and medium-sized companies. Large organizations have far more expertise upon which to draw on and are able to perform risk assessments more easily. Often employing their own occupational hygienists they are able to devise in-house control limits and control procedures to minimize exposures. Whereas small- to medium-sized enterprises (SMEs) have been unclear about what controls to institute and have found risk assessments both difficult and expensive.

Industry's perception and use of occupational exposure limits was assessed in a survey commissioned by the Health and Safety Executive (HSE) in 1996. Telephone interviews were carried out with 1000 randomly selected users of chemicals. A total of 400 companies had some use of chemicals (all user group) and 600 used chemicals daily (heavy users). In addition 150 interviews were conducted with trade union health and safety representatives.[7,8] Interviews covered basic information on the chemicals used, sources of information, risk reduction measures, and understanding of safety legislation and limits.

Results indicated that most users relied heavily on information provided by suppliers or on personal experience. Far fewer obtained information from organizations such as trade associations or the HSE. Awareness of the COSHH legislation was limited to 45% of the "all user" group and 53% of the "heavy user" group. Awareness of the occupational exposure limits was only 19% of the all user group and 32% of the heavy user group. Results from trade union safety representatives indicated that they were better informed about legislation and exposure limits than the safety officers or management in the firms surveyed.[9]

Results of the survey presented significant problems for the HSE. For the first time, the organization had asked those whom it advised and policed how they were dealing with health and safety legislation. The survey indicated that managers were ill informed. A completely new approach was required that would take into account both the limited knowledge base about hygiene in the workplace and the lack of understanding of how to perform risk assessments. The HSE recognized that it was no longer sufficient simply to reiterate that managers needed to carry out proper risk assessments. The law had required this for years but the telephone survey indicated that the vast majority had little idea about what to do.[8] The survey did show that managers were concerned about their workforce and were more likely than not to have some control procedures in place. Most of the guidance managers relied on came from chemical suppliers. Controls in place relied on information printed on a label or in a material safety data sheet. It was apparent that this situation was not likely to change much in the future.

After considerable discussion and consultation, the HSE concluded that the primary objective had to be a workable procedure to help those using the chemicals. The procedure had to enable them to work out how to handle chemicals safely while controlling exposure to most substances being used and not just those for which there were occupational exposure limits.

Development of COSHH Essentials

In 1996, the Advisory Committee on Toxic Substances (ACTS) in the HSE established a working group comprised of occupational hygienists, toxicologists, and epidemiologists from the HSE, individuals nominated by industry and trade unions, and independent scientists. After meetings and consultations with user groups over the next two and half years, the Advisory Group devised a procedure that it considered both acceptable and easy to use. The COSHH Essentials guide was the outcome of this process.[9]

To deal with the wide range of chemical substances in use, COSHH Essentials adopted a generic approach to risk assessment. Well-recognized principles were the foundation: risk is a combination of hazard and exposure. Even though a substance may be particularly hazardous, if there is no exposure there is no risk. The procedure adopted was based on the following flow diagram:

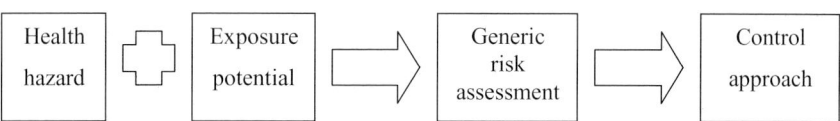

Because most interviewed in the HSE industry survey indicated that they used information from suppliers to help devise control procedures for chemicals, the working group decided to construct a guide that used this information.[9] Details about the hazardous properties of chemicals can be found in material safety data sheets. Schemes devised in the United Kingdom included those by the Association of the British Pharmaceutical Industries,[10] the Royal Society of Chemistry,[11] and the Chemical Industries Association.[12] These schemes were reviewed, but all were too sector specific, hence the need for something more generic.

Hazardous chemicals sold in the European Union (EU) are given Risk phrases (R phrases) summarizing their hazardous properties. In the United Kingdom the EU classification system is implemented by the Chemicals Hazard Information and Packaging for Supply Regulations (CHIP).[13]

The assignment of R phrases to substances and preparations is based on well-defined criteria agreed across the EU. The classification system addresses all relevant toxicological end points and offers an internationally recognized and accepted means for identifying toxicological hazards.[14]

COSHH Essentials allocates solids and liquids to one of five inhalation toxicological hazard bands labeled A–E and one for skin and eye exposure. Each of the inhalation hazard bands represents a different target airborne exposure range covering a log concentration range. Allocation of each R phrase into one of the five inhalation hazard bands takes account of three key factors:

(a) evidence that the toxicological end point had an identifiable dose threshold,
(b) the seriousness of resultant health effects, for example, cancer or developmental effects are more serious than mild eye irritation, and
(c) relative exposure levels at which toxic effects occur, or potency. Thus a more stringent control would be advocated for a substance causing severe kidney damage at exposures an order of magnitude lower than another substance causing the same health effects.

Two of the five bands, the target airborne concentrations they represent, and the matching R phrases are shown in TABLE 1.[9] For certain R phrases, such as R48, referring to a danger of serious damage to health by prolonged exposure, some adjustment between bands was necessary to accommodate conversion of vapor exposures in parts per million (ppm) to a weight measure in milligrams per cubic meter (mg/m^3).[9]

To evaluate the effectiveness of the scheme and the appropriate allocation of R phrases to hazard bands, 111 substances with health-based occupational exposure limits were chosen. The substances were then matched to one of the five hazard bands based on the R phrases for each. In situations where R phrases might allocate a substance to two hazard bands the more stringent, that is the one with the lower target airborne concentration was selected. For 98% of the 111 substances evaluated, COSHH Essentials led to selection of

TABLE 1. Example of allocation of R phrases to hazard groups, showing airborne concentration ranges for two hazard bands

Hazard group	Target airborne concentration range	R phrases
A	>1–10 mg/m^3 dust >50–500 ppm vapor	R36, R38 all substances that do not have R phrases in groups B–E
B	>0.1–1 mg/m^3 dust >5–50 ppm vapor	R20/21/22, R40/20/21/22

NOTE: Adapted from Reference 17.

a control strategy that provided a standard of control equivalent to or greater than that required to comply with the occupational limit.[9]

Control Procedures

Although many factors could be considered for precise degrees of control, the scheme would not work if it were too complicated. In practice, there were only four categories of control based on the degree of containment. These were: general ventilation, engineering containment, industrial closed systems, and special controls. Other factors also come into play such as the degree of training, supervision of staff, and selection of appropriate personal protective equipment. The greater the degree of containment the more important these additional factors are. All are integral to a proper control strategy.[15]

General ventilation does not require any special engineering. With small- or medium-scale use of low-hazard materials, natural ventilation provided by doors and windows was considered sufficient. Occasionally additional forced ventilation might be necessary. This is control approach 1.

Engineering containment refers to local exhaust ventilation ranging from single-point extracts close to the source of emission to some form of ventilated partial enclosure [control approach 2].

Industrial closed systems are self-evident. Small breaches of containment will occur for the collection of quality control samples, for example [control approach 3].

Specialist advice is required for very hazardous substances, or use of very large quantities of chemicals and risks need to be assessed on a case-by-case basis.[15]

Exposure Potential

Many factors can lead to the generation of dust or vapor in air. The inherent physical properties of a material and how it is handled determine airborne

TABLE 2. Definitions of exposure predictor bands for the first two of the four categories for solids

	Solids
Exposure predictor band	Description
EPS1	Gram quantities of medium/low dusty material
EPS2	Gram quantities of high dusty material and kilogram/ton quantities of low dusty material

NOTE: EPS: Exposure Predictor: Solids. Adapted from Reference 17.

concentrations. Most hazardous substances encountered in the workplace are present either as solids, which generate dust exposures, or liquids, which create either vapor or aerosol exposures.

For a solid, dustiness is the important physical property. For simplicity, COSHH Essentials divides dustiness into three categories: *High* for fine light powders like cement dust; *medium* for crystalline granular materials like soap powder or sugar; and *low* for pellets or nonfriable material such as polyvinyl chloride (PVC) pellets or waxes.

For liquids, volatility is the most important factor. COSHH Essentials divides liquids into *low, medium,* or *high* volatility based on their boiling point and the operating temperature of the process.[15] The scale of an operation is probably the most important factor determining exposure although packaging, transport, and use are also important issues. An individual weighing a few grams of flour in a kitchen is unlikely to produce high concentrations of flour dust in the air; an employee emptying a 50 kg sack in a bakery may well do so.[15] COSHH Essentials takes account of quantity in a simple way. Measures for solids are grams, kilograms, or tons. The scale of use for liquids is milliliters, liters, or cubic meters.

A model was developed which combined the effects of physical properties and scale of use. This combination lead to identification of what became known as exposure predictor bands, a measure of the quantity of material likely to become airborne. Four exposure predictor bands covered the range of likely situations. The lowest exposure predictor band for solids consists of gram quantities of low or medium dusty materials. The next band would be gram quantities of very dusty material, or kilogram or ton amounts of low dusty material (TABLE 2). The highest band is for ton quantities of medium or very dusty material, known as Exposure Predictor: Solids 4, or EPS4. A similar categorization was determined for liquids.[15]

It was apparent that ton quantities of very dusty or highly volatile material could not be included because this combination would generate air concentrations higher than those permitted by law.[2] Specialist advice is required in these circumstances. However, the scheme does accommodate ton quantities of less

TABLE 3. Relating exposure predictor bands to control approach for solids: predicted dust in air exposure ranges (mg/m^3)

Control approach	Exposure predictor band mg/m^3			
	EPS1	EPS2	EPS3	EPS4
1	0.01–0.1	0.1–1	1–10	>10
2	0.001–0.01	0.01–0.1	0.1–1	1–10
3	<0.001	0.001–0.01	0.01–0.1	0.1–1

NOTE: 1. general ventilation; 2. engineering containment; 3. industrial closed systems EPS: Exposer Predictor: Solids.
Adapted from Reference 17.

dusty or less volatile materials. COSHH Essentials allocates control strategies that will not result in air concentrations of solids exceeding 10 mg/m^3 or vapors or aerosols exceeding 500 ppm.

To advise on the appropriate control strategy, the scheme uses three degrees of containment namely general ventilation, engineering containment, and industrial closed systems. When exposure cannot be contained within reasonable bounds by these measures, specialist advice is recommended. These control strategies were coupled with exposure predictor bands to assess what airborne concentrations would be likely. For example, if general ventilation was available for handling gram quantities of low/medium dusty material, airborne concentrations were predicted to range from 0.01–0.1 mg/m^3. For very dusty material in this type of environment, airborne dust would likely be in the 0.1–1 mg/m^3 range (TABLE 3). This approach was applied for the three control options using each of the four exposure predictor bands.[15] Moving from one exposure predictor band to the one above will increase predicted air concentration ranges by a factor of 10. In contrast, adopting the next most stringent control option going from general ventilation to engineering containment will reduce predicted air concentrations by a factor of 10.

When developing the scheme, it was necessary to link the five inhalational hazard bands with control options. This was done by considering the target airborne concentrations for substances in each hazard band using a particular control option and exposure predictor band. For example, solids allocated to hazard band A would need to be controlled within a target airborne range of 1–10 mg/m^3. With general ventilation as the control option, the scheme predicts that with *low* dusty material used in gram quantities, exposures would be predicted to range from 0.01–0.1 mg/m^3. If it were kilogram or even ton quantities of the same material, concentrations are predicted to range from 0.1–1.0 mg/m^3 (TABLE 4). Both categories of use are less than or equal to the target airborne concentration for hazard band A dusts (TABLE 1), which is 1–10 mg/m^3. Thus, hazard band A solids, if not very dusty, can be used with only general ventilation, even in ton quantities with the knowledge that exposures will be less than any likely occupational exposure limit.

TABLE 4. Matching exposure predictor bands and control approach to hazard group

Control approach	Solids			
	Exposure predictor band mg/m^3			
	EPS1	EPS2	EPS3	EPS4
1	0.01–0.1 A, B, C	0.1–1 A, B	1–10 A	>10
2	0.001–0.01 D	0.01–0.1 C	0.1–1 B	1–10 A

NOTE: A, B, C, and D refer to hazard bands. Table truncated version of original as example of matching exercise. EPS: Exposure Predictor: Solids.
Adapted from Reference 17.

The mapping exercise was repeated for each of the hazard bands to ensure that target airborne concentrations and exposures predicted from application of particular control options, scale of use, and physical properties all matched. This resulted in a table with hazard bands matched with exposure predictor bands, a truncated version of which is shown in TABLE 4.

Subsequent rearrangement of the table allows it to be used when the hazard band of the substance is known, and this will lead to a control option seen in TABLE 5. For example, exposures to a solid in hazard band A, if not very dusty and used in gram quantities (EPS1) can be controlled by general ventilation. A similar iteration of the scheme was carried out for liquids.[15]

In the initial paper-based version of COSHH Essentials, users are asked to refer to a safety data sheet to find the risk phrases for the substance.[16] These are then matched with the appropriate hazard band and the more stringent chosen if the phrases fall between two bands, that is, band C in preference to band B. Scale of use is then required, in grams if a solid, or milliliters if a liquid. Following this, users describe the physical state or degree of dustiness if a solid or refer to a graph of boiling point and operating temperature to assess volatility of a liquid. Users then are directed to the best control option.

TABLE 5. Moving from hazard though exposure prediction to control options

Hazard group	Solids			
	Exposure predictor band			
	EPS1	EPS2	EPS3	EPS4
A	Control approach 1	Control approach 1	Control approach 1	Control approach 2
B	Control approach 1	Control approach 1	Control approach 2	Control approach 3

NOTE: Table truncated version of original. EPS: Exposure Predictor: Solid.
Adapted from Reference 17.

Illustrated texts with helpful tips called "control guidance sheets" are available for many specified work activities and show how the control options work in practice. Originally only available in hard copy the sheets can now be downloaded from COSHH Essentials on the HSE web site at www.coshh-essentials.org.uk. The web site is heavily used with hundreds of thousands of risk assessments performed since its installation. This successful project to help small businesses control chemicals is an example of successful joint U.K. government/industry/trade union and independent expert cooperation. International versions of the scheme are also now available.

REFERENCES

1. H.S.E. 1992a. Guide to the Health and Safety at Work etc. at 1974, 5th ed. L1. HSE Books. ISBN 0 7176 0441. Her Majesty's Stationary Office (HMSO). Norwich, UK.
2. H.S.E. 1997. Control of Substances Hazardous to Health Regulations 1994. Approved Codes of Practice. HSE Books. ISBN 0 7176 1308 9. Her Majesty's Stationary Office (HMSO). Norwich, UK.
3. H.S.E. 1992. Management of Health and Safety at Work. Management of Health and Safety at Work Regulations 1992. Approved Codes of Practice. L21. HSE Books. ISBN 0 7176 0412 8. Her Majesty's Stationary Office (HMSO). Norwich, UK.
4. H.S.E. 1997. Successful Health and Safety Management, 2nd edition. HSE Books. ISBN 0 7176 1276 07. Her Majesty's Stationary office (HMSO). Norwich, UK.
5. H.S.E. 1998. EH 40/98. Occupational Exposure Limits. HSE Books. ISBN 0 7176 1474 3. Her Majesty's Stationary office (HMSO). Norwich, UK.
6. H.S.E. 2005. EH 40/2005. Workplace Exposure Limits. HSE Books. ISBN 0 7176 2977 5. Her Majesty's Stationary office (HMSO). Norwich, UK.
7. Research International 1997. Industries' Perception and Use of Occupational Exposure Limits. HSE Contract Research Report no. 144. HSE Books. ISBN 0 7176 1407 7.
8. Topping, M.D., C.R. Williams & J.M. Devine. 1998. Industries' perception and use of occupational exposure limits. Ann. Occup. Hyg. **42:** 357–366.
9. Brooke, I.M. 1998. A UK scheme to help small firms control health risks from chemicals: toxicological considerations. Ann. Occup. Hyg. **42:** 377–390.
10. Association of the British Pharmaceutical Industry. 1995. Guidance on Setting In-House Occupational Exposure Limits for Airborne Therapeutic Substances and Their Intermediates. Association of the British Pharmaceutical Industry. London. Her Majesty's Stationary office (HMSO). Norwich, UK.
11. Royal Society of Chemistry. 1996. COSHH in laboratories, 2nd ed. Cambridge, United Kingdom. ISBN 0 85404 427 02.
12. Chemical Industries Association 1997. Control of Substances Hazardous to Health—guidance on allocating occupational exposure band Regulation 7. Chemical Industries Association. London. ISBN 1 85897 048 2.
13. H.S.E. 1997. Chemical Hazard Information and Packaging for Supply. Amendment. Regulations 1997. Approved Guide to the Classification and Labelling of Substances and Preparations Dangerous for Supply, 3rd ed. HSE Books. ISBN 0 7176 1366 6.

14. GARDENER, R.J. & P.J. ALDERSHAW. 1991. Development of pragmatic exposure-control concentrations based on packaging regulation risk phrases. Ann. Occup. Hyg. **35:** 51–59.
15. MAIDMENT, S.C. 1998. Occupational hygiene considerations in the development of the structured approach to select chemical control strategies. Ann. Occup. Hyg. **42:** 391–400.
16. H.S.E. 1999. COSHH Essentials Easy Steps to Control Chemicals. HSE Books. ISBN 0 7176 2421 8. Her Majesty's Stationary Office (HMSO). Norwich, UK.
17. H.S.E. 1999. The technical basis for COSHH Essentials: easy steps to control chemicals. HSE Books. ISBN 0 7176 2434 x. Her Majesty's Stationary Office (HMSO). Norwich, UK.

Progress of Epidemiological and Molecular Epidemiological Studies on Benzene in China

GUILAN LI AND SONGNIAN YIN

Institute of Occupational Health and Poison Control, China CDC, and International Benzene Research Team of the China CDC and NCI, USA

ABSTRACT: Benzene is an organic solvent that has been used in industry for about 100 years throughout the world. Since 1973, a series of toxicological and molecular epidemiological studies on benzene were conducted by researchers at the Chinese Academy of Preventive Medicine (CAPM) (1973–1986) and subsequently by a collaboration between the CAPM and the National Cancer Institute (NCI) in the United States that began in 1986, which was joined by investigators from the University of California at Berkeley, the University of North Carolina at Chapel Hill, and New York University. The findings demonstrated that the risk of leukemia and lymphoma among benzene-exposed workers was significantly increased, with elevated risks for leukemia present not only at higher exposure but also among workers exposed to under 10 ppm. Therefore, the benzene permissible level was decreased to 1.8 ppm (6 mg/m^3) and benzene-induced leukemia is treated as an occupational cancer in China. The benzene permissible level is 1.0 in the United States and in several other developed countries and it has been suggested to be decreased to 0.5 ppm (ACGIH). A number of potential biomarkers are related to benzene exposure and poisoning. Some of these are benzene oxide–protein adducts, chromosome aberration of lymphocytes, and GPA mutations in erythrocytes, a decrease in B cell and CD4$^-$T cell counts in peripheral blood, and altered expression of CXCL16, ZNF331, JUN, and PF4 in lymphocytes. Variation in multiple benzene metabolizing genes may be associated with risk of benzene hematotoxicity, including CYP2E1, MPO, NQO1, and GSTT1.

KEYWORDS: benzene; epidemiology; poisoning; hematotoxicity; genomics; molecular aspects; individual susceptibility; adducts

INTRODUCTION

Benzene is an organic solvent and has been used in industry for about 100 years throughout the world. High benzene exposure has been controlled in

Address for correspondence: Guilan Li, Vice Office Director, National Chemical Assessment Center, National Institute of Occupational Health and Poison Control, 29 Nan Wei Road, Beijing 100050, China.
e-mail: guilanli@263.net.cn

developed countries, but low benzene exposure, including industrial and environmental contamination and its risk, still exists in developed and developing countries.

There were reports of some individual occupational health investigations and of BP in some factories in China in the 1950s–1970s. Since 1978, when the reform and open policy was issued in China, the progress of epidemiological and molecular epidemiological study of workers exposed to benzene in China can be probably divided into three phases.

RESULTS

First Phase

To understand the risk of BP and prevention, the Ministry of Health of China organized a nationwide investigation of workers exposed to benzene and four other chemicals. We found that 500,000 workers were exposed to benzene and benzene mixtures. The geometric average benzene concentration was 18.3 mg/m^3 (5.5 ppm) in 19,969 factories, and the prevalence of chronic BP was 0.5%. In addition, nine cases of aplastic anemia and nine cases of leukemia were found in benzene-exposed workers.[1]

Based on these results, a retrospective cohort study of benzene and leukemia was conducted by the Chinese Academy of Preventive Medicine (CAPM) and cooperating institutions in 12 cities. The period of follow-up was 1972–1981. The mortality of leukemia was 14 of 105 among 28,460 benzene-exposed workers and 2/10^5 among 28,257 non-benzene-exposed workers (Standard Mortality Rate [SMR] = 5.47). Lymphosarcoma, lung cancer, and liver cancer were significantly higher in the benzene cohort than in the control cohort. This was the first report that benzene might be a multiple carcinogen in humans.[2] Based on these findings, prevention measures were enhanced, and the health standard of benzene was decreased to 40 mg/m^3. Benzene-related leukemia was treated as an occupational cancer in China.[3] These findings were cited in a 1987 IARC monograph[4] and by IPCS, WHO in 1993.[5] The mechanism of the action of benzene on blood and hematopoietic function includes benzene–DNA and protein adducts; benzene metabolism patterns of workers exposed to benzene, toluene, and benzene mixtures; diagnostic criteria for BP and use of Chinese herb extracts "XUEZASISHEN" for the treatment of chronic BP were reported at Annual Meeting of the Council of Fellows, Collegium Ramazzini in 1989.[6]

Second Phase

To verify the dose–response relationship between benzene exposure and leukemia and to carry out molecular epidemiology studies to further understand benzene's mechanism of action, CAPM collaborated with National Cancer Institute (NCI) and the NIH in the United States to do an expanded cohort study among exposed and unexposed workers between 1972 and 1987. A total of 75,000 benzene-exposed workers and 35,000 unexposed workers were investigated in the same 12 cities.[7] Historical benzene measurements were collected and benzene exposure estimates were made for job title, work units, and factories from 1950 through 1987. The benzene exposure level was 25–33 ppm in 1950–1974, 11–15 ppm in 1975–1984, and 8 ppm in 1985–1988.[8] There were 38 cases of leukemia in the benzene cohort and 9 cases of leukemia in the nonbenzene cohort. Relative risk (RR) was 2.5. Acute nonlymphocytic leukemia (ANLL) and acute nonlymphocytic leukemia/myelodysplastic syndrome (ANLL/MDS) each showed patterns of increasing risk with increasing average exposure. Risk for non-Hodgkin's lymphoma (NHL) increased with increasing duration (RR = 3.3) among those exposed for 5–9 years and 4.2 for those exposed for 10 or more years.[9-12] This study provided evidence that benzene may cause hematological neoplasm and related disorders at average exposures of less than 10 ppm and cumulative exposure of less than 40 ppm-years.[13]

After the completion of the expanded cohort study, we continued our collaboration of doing molecular epidemiology studies among benzene-exposed workers and BP in Shanghai and Tianjin, working with investigators at NCI and several universities in the United States including University of California (UC) at Berkeley, University of North Carolina (UNC) at Chapel Hill, and New York University (NYU). The main results follow.

1. **Validation of benzene metabolites as a biomarker**

    It is well known that benzene metabolites include phenol, hydroquinone, tt-muconic acid, and S-PMA. They are all related to benzene exposure in the middle or higher level, but are not good markers for lower-level exposure. Qu et al.[14] reported that urine tt-MA and S-PMA are sensitive biomarkers for exposure levels of 0.1–1.0 ppm in shoe makers in Tianjin. Waidyanatha et al.[15] determined urine benzene and metabolites of benzene-exposed workers with gas chromatography mass spectroscopy (GC-MS) and found that urine benzene is a specific biomarker for exposure of less than 1.0 ppm.

2. **Benzene DNA, Alb adducts**

    Benzene DNA adducts[16] have been found in animals exposed to benzene 10 years ago. Recently Yeowell-O'Connell et al.[17] reported that benzene oxide–albumin adduct (BO-Alb) and 1.4-benzoquinone albumin adducts (BQ-Alb) are 2.4 times higher in benzene-exposed workers ($n = 160$) than non-benzene-exposed workers ($n = 102$). This relationship is linear at lower benzene exposures but not at high exposures.

3. **Chromosome aberrations in lymphocytes and glycoprotein A (GPA) mutation of erythrocytes in peripheral blood**

    Since the metaphase preparation method was established, many epidemiological studies focused on benzene exposure and chromosomal aberrations. In most studies using the nonbanding staining method, elevated chromosome aberration levels in peripheral lymphocytes were detected in BP patients and also in nondiseased workers exposed to benzene. Our studies using G-banding indicated that aberrations on the long arms of chromosomes 5 and 7 were increased in BP patients, and structural aberrations were most common on the long arms of chromosome 2 and 10 in nondiseased workers exposed to benzene.[18]

In 1992, in collaboration with the NCI, UC Berkeley, and UNC, we conducted a molecular epidemiological study in Shanghai, China. Forty-three workers exposed to benzene (median = 31 ppm, 8-h time-weighted average) and 44 matched controls were sampled. Specific chromosome aberrations were detected using FISH. The results showed:

1. Benzene exposure was associated with increases in the rates of monosomy 5 and 7 but not monosomy 1 and with increases in trisomy and tetrasomy frequencies of all three chromosomes. Long-arm deletion of chromosomes 5 and 7 was increased in a dose-dependent fashion up to 3.5-fold in the exposed workers.[19]
2. High benzene exposure (>31 ppm, $n = 22$) increased the hyperdiploid frequency of chromosome 9; trisomy 9 was the major form of benzene-induced hyperdiploidy. The level of hyperploidy 9 in exposed workers correlated with their urinary phenol level, a measure of internal benzene dose.[20]
3. Benzene exposure was associated with significant increases in hyperdiploidy of chromosomes 8 and 21. Translocations between chromosomes 8 and 21 were increased up to 15-fold in highly exposed workers. In one highly exposed individual, these translocations were reciprocal and were detectable by reverse transcriptase-polymerase chain reaction (PCR).[21]

   Loss and long (q)-arm deletion of chromosomes 5 and 7 are two of the most common cytogenetic changes in therapy- and chemical-related leukemia. Numerical and structural aberrations in chromosomes 8 and 21 are commonly observed in AML. These data indicate a potential role for aberrations in chromosomes 5, 7, 8, and 21 in benzene-induced leukemogenesis, and these chromosome aberrations may be useful biomarkers of early biological effect for benzene exposure.

   In 2000–2001, in collaboration with the NCI, UC Berkeley, and UNC, we conducted another molecular epidemiological study in Tianjin, China. To determine if selective effects of benzene can occur Zhang et al.[22] employed three-color painting on an 8-square slide to screen numerical changes in all 24 human chromosomes (Octo-Chrome FISH) in a pilot study of 11 subjects (6 exposed to >5 ppm benzene and 5 age- and sex-matched controls). Selective effects were observed on monosomy of chromosomes 5, 6, 7, and 10, and were also observed on trisomy induction with chromosomes 8, 9, 17, 21, and 22. These results suggest that benzene has the capability of producing selective effects on certain chromosomes.[22] These selective effects are under further study in a larger population. Effects of low-level benzene exposure on chromosome aberrations and the dose–response relationship are also in process.

Rothman et al.[23] used the GPA gene loss mutation assay to evaluate 24 workers exposed to benzene and 23 matched control workers in Shanghai. The GPA assay identifies stem cell or precursor erythroid cell mutations expressed in peripheral erythrocytes of MN-heterozygous subjects, distinguishing the NN and Nφ mutation variants. A significant increase in NN GPA variant cell frequency (Vf) was found in benzene-exposed workers as compared with unexposed control workers (Vf 13.9/7.4 per 10^6 cells). In contrast, no significant difference existed between these two groups for the Nφ. The lifetime cumulative occupational exposure to benzene was also associated with NN Vf ($P < 0.01$) but not with Nφ Vf ($P < 0.31$). These findings suggested that NN mutations occur in long-lived bone marrow stem cells, and variants result from loss of the GPA M allele and duplication of the N allele.

4. **Genetic polymorphism and susceptibility to benzene hematotoxicity:**

To evaluate the impact of interindividual variation in activating enzymes (CYP2E1) and detoxifying enzymes (NQOI) of benzene, in 1997 Rothman et al.[24] reported the BP cases ($n = 50$) and control ($n = 50$) study in Shanghai. Subjects with both a rapid fe6-OH and two copies of the $NQOI^{609}C \rightarrow T$ mutation had a 7.6-fold increased risk of BP compared to subjects with a low fe6-OH who carried one or two wild-type NQOI allele, but the CYP2E1 Rsal/Pst1 polymorphism did not influence BP risk.

To further investigate the role of the polymorphisms of benzene-metabolizing enzymes in human susceptibility to BP, Chen et al.[25] analyzed BP cases ($n = 100$) and matched control workers ($n = 90$) with the same job title. PCR and PCR-RFLP were used for genotyping. The results showed that single gene mutation of $NQO1^{609}C \rightarrow T(T/T)$ was increased 2.82-fold in BP compared with those carrying heterozygons (C/T) and wild type (C/C). The subjects with GSTTI null genotype had a 1.91-fold increased risk of BP compared with those carrying GSTT1 non-null genotype. There was evidence that individual with genetic variants in several genes (NQO1, GSTT1, and GSTM1) had a substantially increased risk of BP.

Third Phase: To Identify the Risk of Low Benzene Exposure and Molecular Mechanism of Carcinogenesis

In early reports, benzene exposure was about 100–1000 ppm before 1940 in Europe and America. After WWII, benzene exposure was gradually decreased to 65–259 ppm in 1946; 22–96 ppm in 1949–1957; 18–69 ppm in 1964–1969; and 3–52 ppm after 1970 in a rubber manufacture factory in the United States.[26] In recent years, the benzene health standard has decreased to about 1 ppm in many developed countries. Further studies are needed to determine the safety or risk in low benzene exposure in China, the United States, and other countries.

1. The retrospective and prospective cohort study on the original cohort of 110,000 workers

   We followed up subjects in our cohort study through 1999 and then designed a "case–cohort control study" of hematopoietic malignancies and related disorders (HLD) and lung cancer (LC) in benzene-exposed workers, with a subcohort from the cohort identified to be used as a comparison group. Historical benzene measurements have been collected and are being used to estimate benzene exposure for the study subjects. The main goal is to study the relationship between benzene exposure and the incidence of HLD and mortality from LC. The study is ongoing.

2. Molecular mechanism of benzene toxicity and carcinogenesis.

   Benzene's effects on the blood and bone marrow include leukopenia, pancytopenia, aplastic anemia, and myelodysplastic syndrome (MDS), and leukemia, as established by many animal and epidemiological studies. However, the mechanism of benzene-induced hematotoxicity and leukemogenesis is still unclear. In order to identify the risk and mechanism at the molecular level among workers with low benzene exposure, we performed studies on benzene-exposed and unexposed workers in Tianjin.

   A. The first report from the project was a detailed exposure assessment for each shoe making factory [27.]

      The result showed benzene concentration <1.0 ppm for 109 workers, <10 ppm for 110 workers, >10 ppm for only 31 workers in the exposed group, and <0.04 ppm in the unexposed group of 140 workers.

   B. We then reported on the hematotoxicity in workers exposed to low-level of benzene in *Science* (Dec. 2004).

      Workers were categorized based upon exposure in the month prior to phlebotomy. The result showed benzene concentration <1.0 ppm for 109 workers, <10 ppm for 110 workers, and >10 ppm for 31 workers in the exposed group; the benzene concentration was <0.04 ppm in unexposed group of 140 workers.[28]

      Lan et al.[28] found that all types of WBC and platelets were significantly decreased in 109 workers exposed to <1 ppm benzene compared to controls; further, lymphocyte subset analysis showed that $CD4^+$ T cells, the $CD4^+/CD8^+$ ratio, and B cells were also significantly decreased. Tests for the linear trend using benzene air level as a continuous variable were significant for platelets and each WBC type measured except monocytes and CD8$^+$ T cells. Because benzene affected nearly all blood cell types, toxicity to hematopoietic progenitor cells was suspected. We used peripheral blood from 29 benzene-exposed subjects and 24 matched controls to culture CFU-GM (granulocyte-macrophage), CFU-E (erythroid), and CFU-GEMM (granulocyte erythroid macrophage, megakaryocyte) colonies. Highly significant dose-dependent decreases in colony formation for these progenitor cells were observed.[28] Also, the progenitor cells are more sensitive than mature cells to the hematotoxic effect of benzene.

A total of four single nucleotide polymorphisms (SNPs) in the CYP2E1, MPO, and NQO1 genes were examined for their effects on benzene-induced WBC toxicity. Two genotypes significantly influenced WBC counts in benzene-exposed workers, MPO 463 GG (rs2333227) ($P = 0.04$) and NQO1-465 CT (rs4986998) ($P = 0.014$). In exposed subjects who carry either one ($n = 191$) or both of the "at risk" genotypes ($n = 11$), there was a strong gene-dosage effect (P trend $= 0.004$), which was also present among those exposed to <1 ppm benzene ($P = 0.003$).[28]

C. We hypothesized that genetic variation in cytokines and cellular adhesion molecule genes may modify the relationship between benzene exposure and hematotoxicity. One or more SNPs in each of 18 candidate genes were studied for their association with hematotoxicity in 250 workers exposed to benzene and the 140 unexposed controls that were studied. Lan *et al.* showed that SNPs in several genes (e.g., ILA, IL-10, CSF3, VCAM1) were associated with a highly significant decrease in WBC counts and with several specific WBC subtypes.[29]

D. Microarray analysis of gene expression in benzene-exposed workers.

The new "Omic" technologies include genomics, transcriptomics (gene expression profiling), proteomics, and metabolomics, which can be used to develop novel biomarkers of exposure, susceptibility, expression, and response to benzene. We have applied microarrays to the study of global gene expression in the peripheral blood mononuclear cells of benzene workers ($n = 6$) and matched controls ($n = 6$). Three recently developed software programs—LIMMA, EASE, and HOPACH—were used to analyze the array data.

The expression of 19 known cytokine genes was significantly different between the exposed and control subjects. Six genes were selected for conformation by real-time PCR and of those CXCL16 (chemokine), ZNF331 (zinc finger protein), JUN (oncogene), and PF4 (platelet factor 4) were most significantly affected by benzene exposure. Thus, microarray analysis along with real-time

PCR conformation showed that altered expression of CXCL16, ZNF331, JUN, and PF4 are potential biomarkers of benzene exposure.[30]

CONCLUSION

Since 1973, the Benzene Research Group in the Chinese Academy of Medical Science (1973–1982) and the Chinese Academy of Preventive Medicine (1983–2002), which was renamed the China CDC in 2002, have conducted a series studies on benzene toxicity and occupational epidemiology, then collaborated with NCI (1986–2006) to do the expanded retrospective cohort, prospective cohort study, and molecular epidemiology studies on benzene and leukemia and other cancers, along with investigators from UC Berkeley, UNC, and NYU. It is the largest and longest international collaboration to study the benzene toxicity and carcinogenicity in the world.

The findings are of benefit to the occupational benzene-exposed workers in China, the United States, and other countries. The benzene permissible level had been decreased to 1.8 ppm (6 mg/m^3) from 12 ppm (40 mg/m^3), and leukemia among benzene-exposed workers was treated as occupational cancer in China. The benzene permissible exposure level has been considered and a decrease to 0.5 ppm has been suggested in the United States. Many developed countries have also decreased the benzene permissible exposure level to about 1.0 ppm. A number of biomarkers were found as potential biomarkers related with benzene exposure and poisoning including benzene–oxide protein adducts, chromosome aberrations of lymphocyte, GPA mutations in erythrocytes in peripheral blood, and altered expression of CXCL16, ZNF331, JUN, and PF4. Also, variation in multiple genes involved in benzene metabolism and in the control of hematopoiesis may be important risk factors for BP. Ongoing work will continue to study risk of adverse health effects at lower levels of occupational exposure to benzene and ultimately at environmental exposure levels. Also, it is anticipated that molecular epidemiology studies will continue to produce new insights into the genetic susceptibility and mechanism of benzene-induced BP and malignancies.

In China, benzene exposure levels have gradually decreased as we have shown in the benzene cohort study.[31] However, benzene exposure is still higher in some new private factories established after the 1980s. Therefore, we need to do more to control benzene exposure in accordance with the law for occupational disease prevention and control in China.

REFERENCES

1. YIN, S-N., G. LI, *et al.* 1987. Occupational exposure to benzene in China. Br. J. Ind. Med. **44:** 192–195.

2. YIN, S-N., G. LI, et al. 1987. Leukaemia in benzene workers: a retrospective cohort study. Br. J. Ind. Med. **44:** 124–128.
3. The List of Occupational Disease. 2002. The Ministry of Health and Ministry of Labour Protection Issued No:108: 1–2.
4. IARC. 1987. Monographs on the evaluation of benzene carcinogenic risks to human. [Suppl 7] : 120–122.
5. MCLONNELL, E.I. 1993. Benzene Environmental Health Criteria 150. WHO. Geneva.
6. YIN, S-N., G. LI. 1989. Toxicological and epidemiological study on benzene. Reported in Annual meeting of the Council of Fellows Collegium Ramazzini, Italy. October 1989.
7. YIN, S-N., M.S. LINET, et al. 1994. Cohort study among workers exposed to benzene in China: I. General methods and resources. Am. J. Ind. Med. **26:** 383–400.
8. DOSEMECI, M., G. LI, et al. 1994. Cohort study among workers exposed to benzene in China. II. Exposure assessment. Am. J. Ind. Med. **26:** 401–411.
9. YIN, S-N., R.B. HAYES, et al. 1996. A cohort study of cancer among benzene-exposed workers in China: Overall results. Am. J. Ind. Med. **29:** 227–235.
10. HAYES, R.B., S-N. YIN, et al. 1998. Benzene and the dose-related incidence of hematological neoplasms in China. J. Natl. Cancer Inst. **89:** 123–129.
11. LINET, M.S., S-N. YIN, et al. 1996. Clinical features of hematopoietic malignancies and related disorders among benzene-exposed workers in China. Environ. Health Persp. **104:** 1353–1364.
12. LI, G-L., M.S. LINET, et al. 1994. Gender differences in hematopoietic and lymphoproliferative disorders and other cancer risks by major occupational group among workers exposed to benzene in China. Occup. Med. **38:** 875–881.
13. HAYES, R.B., S-N. YIN, et al. 2001. Benzene and lymphohematopoietic malignancies in humans. Am. J. Ind. Med. **40:** 117–126.
14. QU, Q.M., A. ASSIEH, et al. 2000. Validation of biomarkers in humans exposed to benzene: urine metabolites. Am. J. Ind. Med. **37:** 522–531.
15. WAIDYANATHA, S., N. ROTHMAN, et al. 2001. Urinary benzene as a biomarker of exposure among occupationally exposed and unexposed subjects. Carcinogenesis **2:** 279–286.
16. LI, G., C. WANG, et al. 1996. Tissue distribution of DNA adducts and their persistence in blood of mice exposed to benzene. Environ. Health Persp. **104:** 1337–1338.
17. YEOWELL-O'CONNELL, K., N. ROTHMAN, et al. 2001. Protein adducts of 1,4-benzoquinone and benzene oxide among smokers and nonsmokers exposed to benzene in China. Cancer Epidemiol. Biomarkers. Prev. **10:** 831–838.
18. GAO, Y., C.H. XIN, et al. 2002. Lymphocyte chromosome aberration analysis of BP. J. China Health Inspection **9:** 81–84.
19. ZHANG, L., N. ROTHMAN, et al. 1998. Increased aneusomy and long arm deletion of chromosomes 5 and 7 in the lymphocytes of Chinese workers exposed to benzene. Carcinogenesis **19:** 1955–1961.
20. ZHANG, L., N. ROTHMAN, et al. 1996. Interphase cytogenetics of workers exposed to benzene. Environ. Health Persp. **104** (Supplement 6):1325–1329.
21. SMITH, M.T., L. ZHANG, Y. WANG, et al. 1998. Increased translocations and aneusomy in chromosomes 8 and 21 among workers exposed to benzene. Cancer Res. **58:** 2176–2181.

22. ZHANG, L., Q. LAN, et al. 2005. Use of OctoChrome fluorescence *in situ* hybridization to detect specific aneuploidy among all 24 chromosomes in benzene-exposed workers. Chemico-Biol. Interact. **153-154:** 117–122.
23. ROTHMAN, N., R. HAAS, et al. 1995. Benzene induces gene-duplicating but not gene-inactivating mutations at the glycophorin A locus in exposed humans. Proc. Natl. Acad. Sci. USA. **92:** 4069–4073.
24. ROTHMAN, N., M.T. SMITH, et al. 1997. Benzene poisoning, a risk factor for hematological malignancy, is associated with the NQOI 609C→T mutation and rapid fractional excretion of chlorzoxazone. Cancer Res. **57:** 2839–2842.
25. CHEN, Y., G. LI, et al. 2005. Genetic polymorphisms of NQO1, GSTT1. GSTM1 and susceptibility to chronic BP. Chin. J. Ind. Hyg. Occup. Dis. **23:** 1–5.
26. KIPEN, H.M., R.P. CODY, et al. 1988. Hematologic effects of benzene: A thirty-five year longitudinal study of rubber workers. Toxicol. Ind. Health **4:** 411–430.
27. VERMEULEN, R., G. LI, et al. 2004. Detailed exposure assessment for a molecular epidemiology study of benzene in two shoe factories in China. Ann. Occup. Hyg. **48:** 105–116.
28. LAN, Q.Z., L. LUOPING, et al. 2004. Hematotoxicity in workers exposed to low levels of benzene. Science **306:** 1774–1776.
29. LAN, Q., L. ZHANG, et al. 2005. Polymorphisms in cytokine and cellular adhesion molecule genes and susceptibility to hematoxicity among workers exposed to benzene. Cancer Res. **65:** 9574–9581.
30. FORREST, M.S., Q. LAN, et al. 2005. Discovery of novel biomarkers by microarray analysis of peripheral blood mononuclear cell gene expression in benzene-exposed workers. Environ. Health Perspect. **113:** 801–807.
31. LI, G., P. CHAN, et al. 2002. Review of benzene health standard. The Ministry of Health. Occupational Exposure Limit for Hazardous Agents in the Workplace. GBZ2-2002.

Carcinogen Exposure and Epigenetic Silencing in Bladder Cancer

CARMEN J. MARSIT,[a] MARGARET R. KARAGAS,[b] ALAN SCHNED,[c] AND KARL T. KELSEY[a]

[a]*Department of Genetics and Complex Diseases, Harvard School of Public Health, Boston, Massachusetts 02115, USA*

[b]*Department of Community and Family Medicine, Dartmouth Medical School, Lebanon, New Hampshire 03756, USA*

[c]*Department of Pathology, Dartmouth Medical School, Lebanon, New Hampshire 03756, USA*

> ABSTRACT: Tobacco smoking, certain occupational exposures, and exposure to inorganic arsenic in drinking water have been associated with the occurrence of bladder cancer. However, in these tumors the exposure-associated pattern of somatic alterations in genes in the causal pathway for disease has been poorly characterized. Animal and *in vitro* studies have suggested that arsenic, tobacco carcinogens, and other exposures may act through epigenetic mechanisms. We, therefore, examined, in a population-based study of human bladder cancer ($n = 351$), the relationship between epigenetic silencing of the tumor-suppressor genes, $p16^{INK4A}$, *RASSF1A*, *PRSS3*, and the four *SFRP* genes and exposure to both tobacco and arsenic in bladder cancer. Promotor methylation silencing of each of these genes occurred in approximately 30–50% of bladder cancers. Epigenetic silencing of *RASSF1A* and *PRSS3* and any of the *SFRP* genes were each significantly associated with advanced tumor stage ($P < 0.001$, $P < 0.04$, and $P < 0.005$, respectively). Arsenic exposure, measured as toenail arsenic, was associated with *RASSF1A* ($P < 0.02$) and *PRSS3* ($P < 0.1$) but not $p16^{INK4A}$ or *SFRP* promotor methylation, in models adjusted for stage and other risk factors. Cigarette smoking was associated with a greater than twofold increased risk of promotor methylation of the $p16^{INK4A}$ gene, with greater risk seen in patients with exposures more recent to disease diagnosis, and smoking was also significantly associated with any *SFRP* gene methylation ($P < 0.01$). These results from human bladder tumors, add to the body of animal and *in vitro* evidence that suggests bladder carcinogens play a crucial role in the induction of important epigenetic alterations.

Address for correspondence: Carmen J. Marsit, Department of Genetics and Complex Diseases, Harvard School of Public Health, 665 Huntington Avenue, Building I, Room 611, Boston, MA 02115. Voice: 617-432-4677; fax: 617-432-0107.
e-mail: cmarsit@hsph.harvard.edu

KEYWORDS: bladder cancer; DNA methylation; epigenetic; arsenic; tobacco smoke

INTRODUCTION

Epidemiologic studies have consistently suggested that the contribution of environmental and occupational carcinogen exposures to the occurrence of sporadic human cancer is considerable.[1,2] For example, tobacco smoking is known to cause numerous malignancies, being linked to greater than 90% of lung cancers,[3] while the overwhelming majority of cutaneous malignant melanoma is similarly linked to ultraviolet (UV) exposure.[4] This observational data, while critically important, cannot address the nature of the interaction between any carcinogen and the target cell on which it confers a selective growth advantage. Knowledge of the mechanisms responsible for clonal selection is crucial to extending the suggested causality described by epidemiologic observations to true classification of an exposure as carcinogenic. This understanding will enhance our ability to prevent cancer and to predict risk for novel agents introduced into the environment.

The process of clonal selection is the most widely accepted hallmark of cancer development and is considered the driving force of tumor progression.[5] Consequently, studies that seek a deeper understanding of how exposures to carcinogens act in clonal selection must apply tools that probe the cell, comparing the nature of clonally related molecular alterations with patterns of carcinogen exposure. Approaching this problem, it is important to recognize that there are multiple intercellular targets of any carcinogen. The nature of any lesion induced in clones selected by these exposures (either directly or indirectly, or both) will, by definition, be heritable and distinct from normally regulated tissue. In an effort to more precisely define and understand the complex influence of carcinogen exposure upon individual tumors, scientists have turned to *in vitro* and animal model systems, as well as observational studies of tumors from affected individuals to better define their biological effects.

Work to date has defined a paradigm (one that has proven extremely useful), which states that carcinogens produce heritable genetic alterations, most often characterized as mutations that arise in a structured, step-wise fashion, which is variable in different cells and tissues.[6] Applying this theory to epidemiologic studies has led to further refinements of this paradigm and have led some to assert that carcinogens leave genetic footprints in tumors themselves, in that mutations of important genes (such as *TP53*) harbor specific alterations associated with specific exposures.[7-9] While this paradigm has served admirably, the early work of Tennant *et al.*[10] began to demonstrate that it falls far short of one of its goals; finding *in vitro* models capable of predicting which agents are human carcinogens. Although genetic toxicologic studies have proved useful for broadly classifying carcinogens, they have been unable to fully define the mode of action of all carcinogens and also failed in allowing extrapolation of the data to human cancer risk. Indeed, this paradigm has also yielded only limited insight into how genetic variation in the population affects cancer susceptibility.[11] Further complicating the attempts to broaden this paradigm and encompass a holistic understanding of environmental carcinogenesis was the realization that there are many carcinogens that are not genotoxic. The mechanism of action of potent carcinogens, such as 2,3,7,8-tetrachlorodibenzo-p-dioxin, phenobarbitol, and metals, such as cadmium and arsenic, and hormonal agents, such as diethylstilbestrol remain elusive, but the ability of these agents to cause cancer is beyond dispute.[12–16]

CARCINOGENS AS SELECTIVE AGENTS

Recent advances in understanding of carcinogenesis have suggested that, in addition to or possibly even more important than, traditional genotoxicity, carcinogens may act through growth-promoting mechanisms by mimicking receptor agonists or by altering cell signaling and gene expression pathways to induce clonal selection that is responsible for tumor progression. As noted above, tumors arise from single cells that are selected as a result of alterations of the cellular programming. Through successive selection events, clones expand and become increasingly somatically altered compared to their cell of origin. It is in this process that carcinogens may exert their greatest affect. A prominent example is found in lung adenocarcinomas, a well-defined carcinogen-related disease. In adenocarcinomas arising in smokers, there is a significantly elevated prevalence of oncogenically mutant *KRAS*.[17,18] Strikingly, oncogenic mutation of the *EGFR* gene occurs in a mutually exclusive fashion (exclusive of *KRAS* mutation[19]) in adenocarcinomas predominantly arising in nonsmokers. Together, results suggest that these proteins operate in the same pathway whose inactivation is necessary for lung adenocarcinoma development but the somatic inactivation of this pathway is strongly dependent upon the type of carcinogen exposure and the specific selective pressures that each exposure exerts on the altered cells. Numerous examples of genetic alterations and their associations with particular exposures have been defined and thoroughly reviewed.[7,20,21]

Here, we assert that similar differential selective pressures attributable to carcinogen exposure may, in addition to inducing genotoxic damage that is growth promoting for individual altered cells, also affect the epigenetic state of individual cells. Epigenetic alterations are now considered an alternative method for fulfilling Knudson's two-hit model of inactivation of a tumor-suppressor gene.[22] Recent work has begun to define the human epigenome, and has noted considerable variation in epigenetic profiles between individuals, but the source of this variation remains unclear.[23] Our work, therefore, has focused on assessing the epigenetic alterations present in human cancer and examining the relationships that these cellular somatic alterations have with carcinogen exposures important to human cancer. We will examine, as a model, bladder cancer, a known exposure-related disease, in an attempt to articulate the importance of epigenetic alterations in the carcinogenic modes of action of both genotoxic and nongenotoxic human carcinogens.

EPIGENETIC ALTERATIONS

The study of the contribution of epigenetic alterations to cancer biology is now a vast field, and numerous in depth reviews have been written on this topic (examples include Jones and Laird,[22] Herman and Baylin[24]). In tumor biology, the simplest epigenetic alteration to study, and thus the most thoroughly examined, is DNA methylation, particularly CpG island hypermethylation. CpG dinucleotides, which are underrepresented in the genome, occur at greater than expected frequencies in gene promotor region as islands.[25] The methylation occurs as the addition of a methyl group to the 5'-carbon of the cytosine in the CpG dinucleotides, and in the context of multiple methylated cytosines within the CpG island, can lead to transcriptional silencing of the downstream gene.[26] The presence of the methyl group alone is not sufficient for transcriptional silencing, but this alteration can lead to the binding of a variety of proteins with specific affinity for this mark, such as MeCP2, which in turn recruit repressor complexes and histone deacetylases, thereby leading to alterations in the chromatin conformation in the promotor region, and thus silencing of transcription.[27–30] Although epigenetic silencing of tumor-suppressor genes is considered an aberrant mark of cancer, moderately to highly repetitive, noncoding sequences of the genome, in contrast, are normally methylated, and hypermethylation is considered an important part of normal development, cellular differentiation, and X-chromosome inactivation.[31–35]

DNA methylation has been identified and characterized in most common tumor types, although the specific tumor-suppressor genes targeted and the prevalence of methylation of any individual gene varies greatly across the tumor spectrum.[36] The difference in the prevalence of methylation of specific genes as well as the now well-characterized observation that tumors exhibit global hypomethylation with only specific gene hypermethylation,[37] strongly suggests that hypermethylation does not occur as a whole genomic event, but instead may be targeted, with the individual gene targets differing by tissue type and even tumor histology.[38–40]

TOBACCO SMOKE AND DNA METHYLATION

Epigenetic alterations, particularly DNA methylation, have been thoroughly examined in human lung cancer[41] and, because tobacco is the major etiological contributor to lung cancer, studies have begun to examine the relationship between this exposure and DNA promotor methylation. For example, methylation silencing of the $p16^{INK4A}$ gene occurs more often in squamous cell lung cancer, and has been associated, in these tumors, with an increasing duration of tobacco smoking.[38,40] Even in adenocarcinomas of the lung, methylation of the $p16^{INK4A}$ promotor occurs more often in smokers[42] and has been associated with a higher degree of background anthracocis.[43] The *RASSF1A* gene, located in the 3p21 *LUCA* region, which is highly susceptible to deletion in lung and breast cancer, has been associated with an earlier age at starting smoking in lung cancer patients,[44,45] suggesting that still growing adolescent lungs may be particularly susceptible to tobacco-related epigenetic events.[44,46] Interestingly, in mouse models of lung adenocarcinoma induced by nitrosamine 4-(methylnitrosamino)-1-(3-pyridyl)-1-butanone (NNK) exposure, tumors exhibit $p16^{INK4A}$ deletion, but also exhibit hypermethylation of *DAPK* and *RARB*.[47–50] NNK exposure has also been linked to $p16^{INK4A}$ methylation in rat liver adenomas and hepatocellular carcinomas.[51] Important to note, though, is that tobacco exposure does not lead to global hypermethylation of tumor-suppressor genes, suggesting instead a targeted mechanism for the carcinogen's role in epigenetic alteration. For example, *DAPK* methylation, which is associated with advanced stage lung cancer, was not associated with tobacco smoke exposure.[52] promotor methylation of *MGMT* occurs more commonly among never smokers than ever smokers in lung adenocarcinomas,[53] and *FANCF* methylation occurs more often in patients with a shorter duration of tobacco use.[39] Methylation of the *ESR1* gene (encoding estrogen receptor-α) also occurs at a lower prevalence in NNK-induced lung tumors in mice and rats and in human lung cancer in smokers compared to nonsmokers.[54] Together, these results suggest that for some genes, susceptibilities or exposures other than tobacco smoke may be important in the targeting of these loci for epigenetic alteration.

ARSENIC AS A CARCINOGEN

The mode of arsenic's carcinogenicity is unclear, with only some speculation about arsenite-generated free radicals and reactive oxygen species leading to genotoxic damage.[55] *In vitro* exposures to inorganic arsenic species have demonstrated dose-dependent increases in promotor region hypermethylation of CpG sites, although not to those resulting in altered gene expression, as well as to the occurrence of genomewide hypomethylation.[56–58] In mouse models of methyl- or folate-deficient diets, arsenic exposure through water supply led to hypomethylation in hepatic-derived DNA,[59] as well as to increases in chromosomal aberrations in blood lymphocytes.[60] These are thought to result from the depletion of S-adenosyl methionine (SAM, the universal methyl donor) due to the metabolism of inorganic arsenic to its methylated forms. The genomic hypomethylation observed in these murine models is similar to the characteristic aberrant methylation pattern seen in human solid tumors.[37] Together these studies suggest that inorganic arsenic exposures may produce cellular environments that select for or allow epigenetic alteration to occur.

Bladder Cancer

Bladder cancer is the most common urologic malignancy. Over 60,000 new cases of bladder cancer will be diagnosed in the United States in 2005, and more than 13,000 deaths will result from this disease.[61] Bladder cancer occurs almost three times as often in males than females, and occurs predominantly as transitional cell carcinoma.[62] Bladder cancer patients usually present with early stage disease, with approximately 70% of patients presenting with superficial bladder tumors, compared to the other 30% that present at a higher histologic grade and often involve invasion of the lamina propria or muscular layers of the bladder wall.[63] Even with the majority of tumors presenting at low clinical stage, recurrences, progression, and even death remain common, and efforts to reduce these outcomes are issues of critical clinical importance.

Tobacco carcinogen exposure through active smoking is the main established risk factor for bladder cancer; however, the attributable risk for tobacco exposure is far less than for lung cancer, and the etiology of bladder cancer remains unclear.[62,64] Other risk factors for bladder cancer include occupational exposures (particularly aromatic amine and polycyclic aromatic hydrocarbon exposures)[62,65] as well as drinking water arsenic exposure.[16,66–68] There is also evidence that use of certain hair dyes, exposure to chlorination byproducts, individual fluid intake, and dietary factors may play a role.[69,70]

Numerous somatic alterations have been described in bladder cancer, most notably mutation or functional inactivation of the tumor suppressor *TP53*.[71,72] Other tumor suppressors are subject to epigenetic silencing, through promotor hypermethylation and in bladder cancer, this silencing has been described for genes, such as *p16^{INK4A}* and *RASSF1A*.[73,74] *RASSF1A* promotor hypermethylation, in particular, has been shown to be associated with invasive bladder cancer, the more deadly form of the disease.[74] These important clinical correlates, as well as the potentially powerful sensitivity and specificity of the approaches used in detection of these alterations have made these good candidates for clinical detection of cancer even in urine specimens.[75,76]

EXPOSURES AND EPIGENETIC ALTERATIONS IN BLADDER CANCER

To examine the relationship between DNA promotor hypermethylation and exposures important to bladder carcinogenesis, we have used the resources of a population-based case–control study of bladder cancer in New Hampshire, USA.[16] Using this approach limits the biases encountered in the more widely used hospital-based case series studies, as our case population presents with a wider range of disease types, and allows the results to be more generalizable to the population as a whole. We also make use of exquisite measures of exposure, through well-validated questionnaires (for tobacco exposure information) and toenail arsenic levels (as measures of cumulative arsenic exposure). We strongly believe that only with this type of well-executed epidemiologic design can the appropriate relationships between exposure and somatic alterations, including DNA promotor methylation, be examined.

We have examined the relationship between epigenetic silencing of the tumor-suppressor genes, $p16^{INK4A}$, *RASSF1A, PRSS3*, and the four soluble Frizzled receptor proteins (*SFRPs*), and exposure to tobacco and arsenic in bladder cancer. promotor methylation of each of these genes occurred in approximately 30% of bladder cancers. Cigarette smoking was associated with an approximately two-fold increase in the odds of methylation of the $p16^{INK4A}$ gene, but was not associated with the duration or intensity of smoking. The greatest risks for methylation of this gene, in fact, were seen in those with the most recent tobacco smoke exposure prior to diagnosis (i.e., current smokers and former smokers with <10 years since quitting).[77] Similarly, smoking was also associated with silencing of the *SFRP* genes.[78] Together, these results suggest that continued exposure during the development of malignancy may be driving or selecting this alteration. It is of interest that there was no relationship with duration or intensity of smoking, suggesting that even the lightest smokers may have the particular exposure necessary to drive or select for these specific alterations, a finding different than that seen in lung cancer for $p16^{INK4A}$ promotor methylation,[38] but which may be due to the difference in the routes of exposure between the target tissues.

Interestingly, arsenic exposure at or above the 95th percentile in this population, measured as toenail arsenic, was significantly associated with *RASSF1A* ($P < 0.02$) and *PRSS3* ($P < 0.05$) methylation, in models controlled for confounders.[77] These results provide perhaps the first evidence from human tumors that arsenic may directly or indirectly induce targeted gene silencing. Further, these data suggest that arsenic exerts its carcinogenicity through epigenetic alterations.

From a clinical standpoint, it is also of interest that promotor methylation of the *SFRPs* was significantly associated with invasive tumor stage and patient survival.[78] This result further supports the hypothesis that understanding the relationship between exposures and somatic alteration may be critical in determining patient prognosis.

CONCLUSIONS

It is well established that epigenetic alterations, and particularly transcriptional silencing related to gene promotor hypermethylation are critical and causal events in the development of human cancer. Less well understood, though, is the targeting and/or selection of these events, and how environmental exposures critical to carcinogenesis may play a role. Our work, using the tools of both molecular biology and epidemiologic research provides profound evidence that these somatic alterations may be driven by the exposures causal to disease, and suggest that additional work using similar principles be undertaken to better elucidate the biological mechanisms underlying the environmental agents responsible for human disease.

ACKNOWLEDGMENTS

The authors acknowledge the work and discussion on this work from Hadi Danaee, Mei Liu, and Heather H. Nelson of the Harvard School of Public Health, and Angeline Andrew of the Dartmouth Medical School. This work is supported by the NIEHS Superfund Center Grant 00002, NCI grant R01 CA100679, and NIEHS toxicology and environmental health sciences training grant T32 ES007155.

REFERENCES

1. SCHOTTENFELD, D. & J.F. FRAUMENI. 1996. Cancer Epidemiology and Prevention. Oxford University Press. New York.
2. DOLL, R. & R. PETO. 1981. The causes of cancer: quantitative estimates of avoidable risks of cancer in the United States today. J. Natl. Cancer Inst. **66:** 1191–1308.
3. IARC, I. A. F. R. O. C. 1986. IARC monographs on the evaluation of carcinogenic risks to humans. Tob. Smok. **38:** 163–189.
4. ARMSTRONG, B.K. & D.R. ENGLISH. 1996. Cutaneous malignant melanoma. In Cancer Epidemiology and Prevention. D. Schottenfeld & J.F. Fraumeni, Eds.: 1282–1312. Oxford University Press. New York.
5. CALABRESE, P., S. TAVARE & D. SHIBATA. 2004. Pretumor progression: clonal evolution of human stem cell populations. Am. J. Pathol. **164:** 1337–1346.
6. VOGELSTEIN, B., E.R. FEARON, S.R. HAMILTON, et al. 1988. Genetic alterations during colorectal-tumor development. N. Engl. J. Med. **319:** 525–532.
7. OLIVIER, M., S.P. HUSSAIN, C. CARON DE FROMENTEL, et al. 2004. TP53 mutation spectra and load: a tool for generating hypotheses on the etiology of cancer. IARC Sci. Publ. **157:** 247–270.

8. STOWERS, S.J., R.R. MARONPOT, S.H. REYNOLDS, *et al*. 1987. The role of oncogenes in chemical carcinogenesis. Environ. Health Perspect. **75:** 81–86.
9. VINEIS, P. & P.W. BRANDT-RAUF. 1993. Mechanisms of carcinogenesis: chemical exposure and molecular changes. Eur. J. Cancer **29A:** 1344–1347.
10. TENNANT, R.W., B.H. MARGOLIN, M.D. SHELBY, *et al*. 1987. Prediction of chemical carcinogenicity in rodents from in vitro genetic toxicity assays. Science **236:** 933–941.
11. OLIN, S.S., D.A. NEUMANN, J.A. FORAN, *et al*. 1997. Topics in cancer risk assessment. Environ. Health Perspect. **105**(Suppl 1): 117–126.
12. WILLIAMS, G.M. & J. WHYSNER. 1996. Epigenetic carcinogens: evaluation and risk assessment. Exp. Toxicol. Pathol. **48:** 189–195.
13. STEENLAND, K., P. BERTAZZI, A. BACCARELLI, *et al*. 2004. Dioxin revisited: developments since the 1997 IARC classification of dioxin as a human carcinogen. Environ. Health Perspect. **112:** 1265–1268.
14. WAALKES, M.P. 2003. Cadmium carcinogenesis. Mutat. Res. **533:** 107–120.
15. KARAGAS, M.R., T.A. STUKEL, J.S. MORRIS, *et al*. 2001. Skin cancer risk in relation to toenail arsenic concentrations in a US population-based case-control study. Am. J. Epidemiol. **153:** 559–565.
16. KARAGAS, M.R., T.D. TOSTESON, J.S. MORRIS, *et al*. 2004. Incidence of transitional cell carcinoma of the bladder and arsenic exposure in New Hampshire. Cancer Causes Control **15:** 465–472.
17. MASCAUX, C., N. IANNINO, B. MARTIN, *et al*. 2005. The role of RAS oncogene in survival of patients with lung cancer: a systematic review of the literature with meta-analysis. Br. J. Cancer **92:** 131–139.
18. AHRENDT, S.A., P.A. DECKER, E.A. ALAWI, *et al*. 2001. Cigarette smoking is strongly associated with mutation of the K-ras gene in patients with primary adenocarcinoma of the lung. Cancer **92:** 1525–1530.
19. SHIGEMATSU, H., L. LIN, T. TAKAHASHI, *et al*. 2005. Clinical and biological features associated with epidermal growth factor receptor gene mutations in lung cancers. J. Natl. Cancer Inst. **97:** 339–346.
20. WIENCKE, J.K. 2002. DNA adduct burden and tobacco carcinogenesis. Oncogene **21:** 7376–7391.
21. POIRIER, M.C. 2004. Chemical-induced DNA damage and human cancer risk. Nat. Rev. Cancer. **4:** 630–637.
22. JONES, P.A. & P.W. LAIRD. 1999. Cancer epigenetics comes of age. Nat. Genet. **21:** 163–167.
23. RAKYAN, V.K., T. HILDMANN, K.L. NOVIK, *et al*. 2004. DNA methylation profiling of the human major histocompatibility complex: a pilot study for the human epigenome project. PLoS Biol. **2:** 2170–2182.
24. HERMAN, J.G. & S.B. BAYLIN. 2003. Gene silencing in cancer in association with promotor hypermethylation. N. Engl. J. Med. **349:** 2042–2054.
25. BIRD, A. 1987. CpG islands as gene markers in the vertebrate nucleus. Trends Genet. **3:** 342–347.
26. BIRD, A. 1992. The essentials of DNA methylation. Cell **70:** 5–8.
27. FUKS, F., P.J. HURD, D. WOLF, *et al*. 2003. The methyl-CpG-binding protein MeCP2 links DNA methylation to histone methylation. J. Biol. Chem. **278:** 4035–4040.
28. JONES, P.L., G.J. VEENSTRA, P.A. WADE, *et al*. 1998. Methylated DNA and MeCP2 recruit histone deacetylase to repress transcription. Nat. Genet. **19:** 187–191.
29. NAN, X., F.J. CAMPOY & A. BIRD. 1997. MeCP2 is a transcriptional repressor with abundant binding sites in genomic chromatin. Cell **88:** 471–481.

30. NAN, X., H.H. NG, C.A. JOHNSON, et al. 1998. Transcriptional repression by the methyl-CpG-binding protein MeCP2 involves a histone deacetylase complex. Nature **393:** 386–389.
31. TURKER, M.S. 1999. The establishment and maintenance of DNA methylation patterns in mouse somatic cells. Semin. Cancer Biol. **9:** 329–337.
32. RIGGS, A.D. 1975. X inactivation, differentiation, and DNA methylation. Cytogenet. Cell. Genet. **14:** 9–25.
33. HOLLIDAY, R. & J.E. PUGH. 1975. DNA modification mechanisms and gene activity during development. Science **187:** 226–232.
34. MOHANDAS, T., R.S. SPARKES & L.J. SHAPIRO. 1981. Reactivation of an inactive human X chromosome: evidence for X inactivation by DNA methylation. Science **211:** 393–396.
35. KIM, K.M. & D. SHIBATA. 2002. Methylation reveals a niche: stem cell succession in human colon crypts. Oncogene **21:** 5441–5449.
36. ESTELLER, M., P.G. CORN, S.B. BAYLIN, et al. 2001. A gene hypermethylation profile of human cancer. Cancer Res. **61:** 3225–3229.
37. EHRLICH, M. 2002. DNA methylation in cancer: too much, but also too little. Oncogene **21:** 5400–5413.
38. KIM, D.H., H.H. NELSON, J.K. WIENCKE, et al. 2001. p16(INK4a) and histology-specific methylation of CpG islands by exposure to tobacco smoke in non-small cell lung cancer. Cancer Res. **61:** 3419–3424.
39. MARSIT, C.J., M. LIU, H.H. NELSON, et al. 2004. Inactivation of the Fanconi anemia/BRCA pathway in lung and oral cancers: implications for treatment and survival. Oncogene **23:** 1000–1004.
40. TOYOOKA, S., R. MARUYAMA, K.O. TOYOOKA, et al. 2003. Smoke exposure, histologic type and geography-related differences in the methylation profiles of non-small cell lung cancer. Int. J. Cancer **103:** 153–160.
41. MINNA, J.D., J.A. ROTH & A.F. GAZDAR. 2002. Focus on lung cancer. Cancer Cell **1:** 49–52.
42. DIVINE, K.K., L.C. PULLING, P.G. MARRON-TERADA, et al. 2005. Multiplicity of abnormal promotor methylation in lung adenocarcinomas from smokers and never smokers. Int. J. Cancer **114:** 400–405.
43. HOU, M., Y. MORISHITA, T. ILJIMA, et al. 1999. DNA methylation and expression of p16(INK4A) gene in pulmonary adenocarcinoma and anthracosis in background lung. Int. J. Cancer **84:** 609–613.
44. MARSIT, C.J., D.H. KIM, M. LIU, et al. 2005. Hypermethylation of RASSF1A and BLU tumor suppressor genes in non-small cell lung cancer: implications for tobacco smoking during adolescence. Int. J. Cancer **114:** 219–223.
45. KIM, D.H., J.S. KIM, Y.I. JI, et al. 2003. Hypermethylation of RASSF1A promotor is associated with the age at starting smoking and a poor prognosis in primary non-small cell lung cancer. Cancer Res. **63:** 3743–3746.
46. WIENCKE, J.K. & K.T. KELSEY. 2002. Teen smoking, field cancerization, and a "critical period" hypothesis for lung cancer susceptibility. Environ. Health Perspect. **110:** 555–558.
47. BELINSKY, S.A., D.S. SWAFFORD, S.K. MIDDLETON, et al. 1997. Deletion and differential expression of p16INK4a in mouse lung tumors. Carcinogenesis **18:** 115–120.
48. HERZOG, C.R., E.V. SOLOFF, A.L. MCDONIELS, et al. 1996. Homozygous codeletion and differential decreased expression of p15INK4b, p16INK4a-alpha and p16INK4a-beta in mouse lung tumor cells. Oncogene **13:** 1885–1891.

49. VUILLEMENOT, B.R., L.C. PULLING, W.A. PALMISANO, *et al.* 2004. Carcinogen exposure differentially modulates RAR-beta promotor hypermethylation, an early and frequent event in mouse lung carcinogenesis. Carcinogenesis **25:** 623–629.
50. PULLING, L.C., B.R. VUILLEMENOT, J.A. HUTT, *et al.* 2004. Aberrant promotor hypermethylation of the death-associated protein kinase gene is early and frequent in murine lung tumors induced by cigarette smoke and tobacco carcinogens. Cancer Res. **64:** 3844–3848.
51. PULLING, L.C., D.M. KLINGE & S.A. BELINSKY. 2001. p16INK4a and beta-catenin alterations in rat liver tumors induced by NNK. Carcinogenesis **22:** 461–466.
52. KIM, D.H., H.H. NELSON, J.K. WIENCKE, *et al.* 2001. promotor methylation of DAP-kinase: association with advanced stage in non-small cell lung cancer. Oncogene **20:** 1765–1770.
53. PULLING, L.C., K.K. DIVINE, D.M. KLINGE, *et al.* 2003. promotor hypermethylation of the O6-methylguanine-DNA methyltransferase gene: more common in lung adenocarcinomas from never-smokers than smokers and associated with tumor progression. Cancer Res. **63:** 4842–4848.
54. ISSA, J.P., S.B. BAYLIN & S.A. BELINSKY. 1996. Methylation of the estrogen receptor CpG island in lung tumors is related to the specific type of carcinogen exposure. Cancer Res. **56:** 3655–3658.
55. ROSSMAN, T.G. 2003. Mechanism of arsenic carcinogenesis: an integrated approach. Mutat. Res. **533:** 37–65.
56. ZHONG, C.X. & M.J. MASS. 2001. Both hypomethylation and hypermethylation of DNA associated with arsenite exposure in cultures of human cells identified by methylation-sensitive arbitrarily-primed PCR. Toxicol. Lett. **122:** 223–234.
57. MASS, M.J. & L. WANG. 1997. Arsenic alters cytosine methylation patterns of the promotor of the tumor suppressor gene p53 in human lung cells: a model for a mechanism of carcinogenesis. Mutat. Res. **386:** 263–277.
58. ZHAO, C.Q., M.R. YOUNG, B.A. DIWAN, *et al.* 1997. Association of arsenic-induced malignant transformation with DNA hypomethylation and aberrant gene expression. Proc. Natl. Acad. Sci. USA **94:** 10907–10912.
59. OKOJI, R.S., R.C. YU, R.R. MARONPOT, *et al.* 2002. Sodium arsenite administration via drinking water increases genome-wide and Ha-ras DNA hypomethylation in methyl-deficient C57BL/6J mice. Carcinogenesis **23:** 777–785.
60. MCDORMAN, E.W., B.W. COLLINS & J.W. ALLEN. 2002. Dietary folate deficiency enhances induction of micronuclei by arsenic in mice. Environ. Mol. Mutagen **40:** 71–77.
61. JEMAL, A., T. MURRAY, E. WARD, *et al.* 2005. Cancer statistics, 2005. CA Cancer J. Clin. **55:** 10–30.
62. SILVERMAN, D., A. MORRISON & S. DEVESA. 1996. Bladder cancer. *In* Cancer Epidemiology and Prevention. D. Schottenfeld, & J.J. Fraumeni, Eds.: 1156–1179. Oxford University Press. New York.
63. PASHOS, C.L., M.F. BOTTEMAN, B.L. LASKIN, *et al.* 2002. Bladder cancer: epidemiology, diagnosis, and management. Cancer Pract. **10:** 311–322.
64. KOGEVINAS, M. & D. TRICHOPOULOS. 2002. Urinary bladder cancer. *In* Textbook of Cancer Epidemiology. H. Adami, D.J. Hunter & D. Trichopoulos, Eds.: 446–466. Oxford University Press. London.
65. COLT, J.S., D. BARIS, P. STEWART, *et al.* 2004. Occupation and bladder cancer risk in a population-based case-control study in New Hampshire. Cancer Causes Control **15:** 759–769.

66. BATES, M.N., A.H. SMITH & K.P. CANTOR. 1995. Case-control study of bladder cancer and arsenic in drinking water. Am. J. Epidemiol. **141:** 523–530.
67. HOPENHAYN-RICH, C., M.L. BIGGS, A. FUCHS, et al. 1996. Bladder cancer mortality associated with arsenic in drinking water in Argentina. Epidemiology **7:** 117–124.
68. NATIONAL RESEARCH COUNCIL. 1999. Arsenic in Drinking Water. National Academy Press. Washington, DC.
69. KING, W.D. & L.D. MARRETT. 1996. Case-control study of bladder cancer and chlorination by-products in treated water (Ontario, Canada). Cancer Causes Control **7:** 596–604.
70. WILKENS, L.R., M.M. KADIR, L.N. KOLONEL, et al. 1996. Risk factors for lower urinary tract cancer: the role of total fluid consumption, nitrites and nitrosamines, and selected foods. Cancer Epidemiol. Biomarkers Prev. **5:** 161–166.
71. SIDRANSKY, D., A. VON ESCHENBACH, Y.C. TSAI, et al. 1991. Identification of p53 gene mutations in bladder cancers and urine samples. Science **252:** 706–709.
72. KELSEY, K.T., T. HIRAO, A. SCHNED, et al. 2004. A population-based study of immunohistochemical detection of p53 alteration in bladder cancer. Br. J. Cancer **90:** 1572–1576.
73. CHAN, M.W., L.W. CHAN, N.L. TANG, et al. 2003. Frequent hypermethylation of promotor region of RASSF1A in tumor tissues and voided urine of urinary bladder cancer patients. Int. J. Cancer **104:** 611–616.
74. MARUYAMA, R., S. TOYOOKA, K.O. TOYOOKA, et al. 2001. Aberrant promotor methylation profile of bladder cancer and its relationship to clinicopathological features. Cancer Res. **61:** 8659–8663.
75. FRIEDRICH, M.G., D.J. WEISENBERGER, J.C. CHENG, et al. 2004. Detection of methylated apoptosis-associated genes in urine sediments of bladder cancer patients. Clin. Cancer Res. **10:** 7457–7465.
76. CHAN, M.W., L.W. CHAN, N.L. TANG, et al. 2002. Hypermethylation of multiple genes in tumor tissues and voided urine in urinary bladder cancer patients. Clin. Cancer Res. **8:** 464–470.
77. MARSIT, C.J., M.R. KARAGAS, H. DANAEE, et al. 2006. Carcinogen exposure and gene promotor hypermethylation in bladder cancer. Carcinogenesis **27:** 112–116.
78. MARSIT, C.J., M.R. KARAGAS, A. ANDREW, et al. 2005. Epigenetic inactivation of SFRP genes and TP53 alteration act jointly as markers of invasive bladder cancer. Cancer Res. **65:** 7081–7085.

Causal Relationship from Exposure to Chemicals in Oil Refining and Chemical Industries and Malignant Melanoma

MYRON A. MEHLMAN

Department of Community Medicine, The Mount Sinai Medical Center, New York, New York, 10029 USA

Department of Environmental and Community Medicine, UMDNJ—Robert Wood Johnson Medical School, New Jersey, USA

Department of Preventive Medicine and Community Health, University of Texas Medical Branch at Galveston, Galveston, Texas, USA

> ABSTRACT: Malignant melanoma has been thought to be related mainly to exposure to the sun or radiation. A review of the scientific literature reveals many significant correlations between benzene and benzene-containing solvents in the workplace and the occurrence of malignant melanoma, particularly in sites that have never been exposed to sunlight. A comparison of positive correlations between such exposure and malignant melanoma by independent investigators and negative findings by investigators with industry affiliations reveals that this difference, at least in part, may account for the discrepant findings. Based on independent studies, it is reasonable to conclude that malignant melanoma is causally related to employment-related chemical exposures in the petroleum refining industry.
>
> KEYWORDS: benzene; malignant melanoma; industry studies; independent investigators; petroleum industry; chemical carcinogenesis; cutaneous malignancies

INTRODUCTION

Malignant melanoma is a cancer of melanocytes, cells in the epidermal layer of the skin that produce melanin, the pigmentation in the skin. Malignant melanoma occurs when changes in the genetic material of normal melanocytes results in loss of their normal uniform structure and the appearance of abnormal division and multiplication locally and metastases to regional lymph nodes and then to remote sites that can include nearly every organ of the body. The

---

Address for correspondence: Myron A. Mehlman, 7 Bouvant Drive, Princeton, NJ 08540. Voice: 609-683-4750; fax: 609-683-0438.
e-mail: mehlman@patmedia.net

diagnosis of malignant melanoma may well be underreported when the cause of death is reported as cancer of the organ to which malignant melanoma has metastasized, for example, brain, lung, etc.

It is widely accepted that exposure to sunlight, other sources of UV rays, radiation, and the like can induce changes in normal melanocytes that lead to malignant melanoma. However, malignant melanoma occurs in skin areas never exposed to sunlight and in persons who have had no known risk factors except for employment in petrochemical refineries[1–15] (TABLE 1) and in various groups of petrochemical workers[2,7,13–23] (TABLE 2). Malignant melanoma and similar exposures have been demonstrated in experimental animals.[24]

In studies conducted by industry or industry consultants,[25–31] essentially no statistically significant increase of malignant melanoma was found in workers in petrochemical plants. Many reports have pointed out the flaws and biases in these industry-conducted studies.

The 1995 report of Wong and Raabe[25] of the Mobil Corporation concluded that benzene exposure causes only acute myelocytic leukemia (AML) and that a range of 400–500 ppm-years of benzene exposure is the threshold for leukemia. As reviewed by Infante,[32] "It is difficult to reconcile Wong's opinion in light of

TABLE 1. Studies showing increased risk for development of malignant melanoma at specific petrochemical and refinery sites

Author	Place of employment	Increased risk (95% CI)
Alderson and Rushton, 1982	Refinery coded "B"	5.0*
	Refinery coded "H"	6.7*
AMOCO, 1985	Indiana refinery	1.92*
	Sugar Creek refinery	5.24*
	Texas City refinery	2.63*
Bahn et al., 1976	Mobil Paulsboro refinery	3.0*
Bahn, 1976	Mobil Paulsboro refinery	5.0*
Blot et al., 1977	Refinery workers	SIR = 1.1*
Gun et al., 2004	Australian refineries: males	1.54* (1.3–1.81)
	females	2.74* (1.42–4.8)
Health Watch, 2000	Australian refinery workers	1.52* (1.16–1.97)
	Australian terminal workers	1.58* (1.22–2.02)
	Airport workers	2.57* (1.23–4.72)
Magnani et al., 1987	Refinery workers	4.81* (1.19–15.0)
Monarrez-Espino et al., 2002	Heavy equipment operators, freight handlers	2.8*
Rushton and Alderson, 1980	Refinery workers $P < 0.01$	4.81** (1.19–15.0)
Scottenfeld et al., 1981	Petroleum industry workers, mid-Atlantic	2.78*
Thomas et al., 1982	Three Texas refineries	3.02*

NOTE: CI = 95% confidence interval. *$P < 0.05$, **$P < 0.01$.

TABLE 2. Increased risk of malignant melanoma by occupation in oil refinery workers

Source	Workers/location	Increased risk
Ref. 2	Hired prior to 1945	2.9*
	15 years of exposure	5.2*
	Hourly workers	2.6*
	Refinery products, routine exposure	2.7*
	Heavy oils, routine exposure	2.7*
	Aromatics (benzene), routine exposure	2.6*
Ref. 7	Airport workers	2.6*
Ref. 13	Western Electric workers—PCB	4.0*
Ref. 14	PCB-exposed workers	3.6*
Ref. 15	PCB-exposed workers	4.0*
Ref. 16	PCB-exposed workers	2.6*
Ref. 17	Barge & boatmen workers	5.0*
	Printers	6.7**
	Vehicle drivers	*
Refs. 18 and 19	Trichloroethylene exposed	*
	Paint and varnish exposed	*
Ref. 20	Chemists	2.4*
Ref. 21	Welders	3.0*
	Asbestos exposure	2.4*
	Chemists	5.9*
Ref. 16	PCB exposure	2.6*
Ref. 22	Elevated risk in chemical industry	*
Ref. 23	Lithographers	3.4*

* = Significant at $P < 0.05$.

his own study results[33] reported in the *British Journal of Industrial Medicine* in 1987." Infante[32] went on to say that, "The data in the Dow study, reported by Ott et al.,[34] Bond et al.,[35] and Wong[33] clearly contradict this statement."

The mortality study of Thorpe[29] of 38,000 petroleum workers potentially exposed to benzene has often been cited as showing no evidence of significant excess leukemia and exposure to benzene. This interpretation ignores the comparison of exposed workers with control workers from the same facilities who were not exposed to benzene; in fact, the exposed group had a twofold risk for leukemia relative to the nonexposed group.[29]

The choice of an appropriate comparison group is an important issue. A number of negative studies were based on comparisons with the general population and failed to detect an increased risk for lympho-hematopoietic cancer or leukemia. Teitelbaum et al.,[36] in a letter to the editor of *Blood*, questioned the validity of scientific evidence in a paper written by Bergsagel et al.[30] on the lack of relationship between benzene and multiple myeloma stating that, "Clearly, the current paper was litigation driven." Goldstein and Shalot[37] wrote, "The evidence cited by Bergsagel et al.[30] does not at all support their conclusion that there is no causal relationship between exposure to benzene or benzene-containing solvents and multiple myeloma... The data they present in their Tables 3 and 4 is akin to a fishing expedition in waters known to be sterile."

The industry-sponsored review paper of Paustenbach et al.[31] was critically analyzed by Utterback and Rinsky[38] who emphasized the following:

> Paustenbach et al.[31] have apparently overlooked important information in the literature related to the use and testing of control ventilation in the rubber hydrochloride [RH] plants even though they extensively cite other information from the very same page of that source" [p. 665] ... and "used selected information, sometimes improperly cited, to adjust previously reported benzene exposure estimates for the RH worker cohort.

Savitz and Moure[38] identified some deficiencies and biases in studies done by oil companies or their consultants that included (a) poorly defined exposures and inclusion of many nonexposed workers producing dilution of the sample and erroneous conclusions; (b) failure to examine the incidence of cancer at appropriate times following exposure; and (c) failure to consider confounding influences.

In a case related to a benzene-induced carcinoma, the U.S. District Court for the Eastern Division[39] in considering Mobil Oil Corporation's motion for dismissal based on a report by Gerald Raabe cited flaws in Raabe's report, as follows:

> In this case, Defendant expert, Gerald Raabe, relied on two studies of mortality of employees at the Beaumont Refinery that he did for the Defendant, Mobil Oil Corporation.
>
> In his report to the Court, Raabe asserted that there was an 18% <u>decrease in mortality</u> among Mobil employees and that there was no excess risk of death from lung cancer. (Emphasis added)

In his affidavit, Dr. John Dement, cites several methodological flaws that undermine the [Raabe] study's validity. These flaws are quoted in Judge Cobb's Memorandum Opinion[39] and include

(a) The claim of a 'healthy worker effect' is called 'rather astonishing' and significantly masks deaths by workers with significant asbestos exposure.
(b) The failure to compare a properly selected exposure group to a properly selected control group.
(c) The second study included administrative personnel and therefore diluted the study population of exposed workers.

The Court denied Mobil's request for summary judgment saying

> ... This Court concludes that Mobil's study suffers from too many methodological flaws to be conclusively valid. As such, the Raabe study cannot be considered as conclusive proof of the absence of excessive risk at the Beaumont refinery and genuine issues of material fact remain to be resolved at trial.
>
> THEREFORE, the Defendant's Motion for Summary Judgment is Denied pursuant to the Order entered on this date.

In contrast to studies conducted by industry or industry consultants, low-dose effects of carcinogens have been confirmed by a scientific peer review conducted by the U.S. National Toxicology Program (NTP). NTP's conclusion speaks of the discrepancy between industry and independent study results:

(a) At least in the preliminary report, the panel states that the positive results have been obtained independently by scientists at academic institutions.
(b) The negative results have been carried out in industry labs and/or funded by industry.
(c) This pattern is reminiscent of the state of the scientific research on neurocognitive impacts on lead in the mid-1970s and generations of studies of tobacco health impacts.
(d) Independent scientists found impacts. Industry-funded scientists found no effects. Those debates have now been fully resolved: there are effects.

DISCUSSION/CONCLUSION

Based on the studies showing a significant association between malignant melanoma and exposure to petroleum hydrocarbons and solvents in the workplace, we can conclude that exposure to chemicals such as polycyclic aromatic hydrocarbons (PAHs), polychlorinated biphenyls (PCBs), benzene, aromatic hydrocarbons, and heavy oils in the refining industry are causally related to malignant melanoma in workers exposed to these substances and compounds.

The differences in interpretation of findings between industry and industry-sponsored research and independent, usually university-based researchers are becoming well known, and the bias in favor of industry by paid consultants has been noted by many individual investigators in the field. These differences in conclusions have prompted policy statements from the NTP as described above, and such findings are leading to more complete disclosure in scientific journals and more careful scrutiny of negative results in industry studies. Biases in science based on profit motivation of industry does not only harm workers in the field but also the population in general by not warning about exposure to toxic and cancer-causing substances with attendant increases in cancer risk for all.

REFERENCES

1. ALDERSON, M. & R. RUSHTON. 1982. Mortality patterns in eight U.K. oil refineries. Ann. N. Y. Acad. Sci. **381:** 139–145.
2. AMOCO. 1985. Submission to U. S. Environmental Protection Agency.
3. BAHN, A.K. et al. 1976. Melanoma after exposure to PCBs. N. Engl. J. Med. **295:** 450.
4. BAHN, A.K. 1976. Report on Paulsboro, N.J. Mobil Oil Plant Study, April 26, 1976.

5. BLOT, W.J. *et al*. 1977. Cancer mortality in U.S. counties with petroleum industries. Science **198:** 51–53.
6. GUN, R.T. *et al*. 2004. Update of a prospective study of mortality and cancer incidence in the Australian petroleum industry. Occup. Environ. Med. **61:** 150–156.
7. HEALTH WATCH - ELEVENTH REPORT. 2000. The Australian Institute of Petroleum Health Surveillance Program. The University of Adelaide Department of Public Health, Adelaide University, South Australia. pp 1–75.
8. MAGNANI, C. *et al*. 1987. Occupation and five cancers: a case-control study using death certificates. Br. J. Ind. Med. **44:** 769–776.
9. MONARREZ-ESPINO, J. *et al*. 2002. Occupation as a risk factor for uveal melanoma in Germany. Scand. J. Work Environ. Health **28:** 270–277.
10. RUSHTON, L. & M.R. ALDERSON. 1980. The influence of occupation on health—some results from a study in the U.K. oil industry. Carcinogenesis **1:** 739–743.
11. SCHOTTENFELD, D. *et al*. 1981. A prospective study of morbidity and mortality in petroleum industry employees in the United States—a preliminary report. Banbury Report No. 9. Quantification of Occupational Cancer, pp. 247–265. Cold Spring Harbor Laboratory. Cold Spring Harbor. New York.
12. THOMAS, T.L. *et al*. 1982. Mortality patterns among workers in three Texas oil refineries. J. Occup. Med. **24:** 135–141.
13. MAZZUCKELLI, L.F. & P.A. SHUTTE. 1993. Notification of workers about an excess of malignant melanoma: a case study. Am. J. Ind. Med. **23:** 85–91.
14. SINKS, T. *et al*. 1991. Westinghouse electric. Health Hazard Evaluations Report HETA 89-116-2094.
15. SINKS, T. *et al*. 1992. Mortality among workers exposed to polychlorinated biphenyls. Am. J. Epidemiol. **136:** 389–398.
16. LOOMIS, D. *et al*. 1997. Cancer mortality among electric utility workers exposed to polychlorinated biphenyls. Occup. Environ. Evidence **54:** 720–728.
17. ADELSTEIN, A.M. 1972. Occupational mortality: cancer. Ann. Occup. Hyg. **15:** 53–57.
18. FRITSCHI, L. & J. SIEMIATYCKI. 1996a. Lymphoma, myeloma and occupation: results of a case-control study. Int. J. Cancer **67:** 498–503.
19. FRITSCHI, L. & J. SIEMIATYCKI. 1996b. Melanoma and occupation: results of a case-control study. Occup. Environ. Med. **53:** 168–173.
20. HOAR, S.K. & S. PELL. 1981. A retrospective cohort study of mortality and cancer incidence among chemists. J. Occup. Med. **23:** 485–494.
21. HOLLY, E.A. *et al*. 1996. Intraocular melanoma linked to occupations and chemical exposure. Epidemiology **7:** 55–61.
22. NELEMANS, P.J. *et al*. 1992. Nonsolar factors in melanoma risk. Clin. Dermatol. **10:** 51–63.>
23. NIELSEN, H. *et al*. 1996. Malignant melanoma among lithographers. Scand. J. Work Environ. Health **22:** 108–111.
24. HUFF, J.E. *et al*. 1989. Multiple site carcinogenicity of benzene in Fisher 344 rats and B6C3F1 mice. Environ. Health Perspect. **82:** 125–163.
25. WONG, O. & G.K. RAABE. 1995. Cell-type specific leukemia analyses in a combined cohort of more than 208,000 petroleum workers in the United States and the United Kingdom, 1937–1989. Reg. Toxicol. Pharmacol. **21:** 307–312.
26. MORGAN, R.W. & O. WONG. 1984. An epidemiologic analysis of the mortality experience of Mobil Oil Corporation employees at the Beaumont Texas, Refinery, Submitted to Mobil Oil Corp.

27. RAABE, G.K. *et al.* 1994. An updated mortality study of workers at the Beaumont Texas, Refinery, Dated December 5, 1994.MOB10229–MOB10281.
28. WONG, O. *et al.* 1992. A mortality study of marketing and marine distribution workers with potential exposure to gasoline in the petroleum industry. American Petroleum Institute, Washington, DC.
29. THORPE, J.J. 1974. Epidemiologic survey of leukemia in persons potentially exposed to benzene. J. Occup. Med. **16:** 375–382.
30. BERGSAGEL, D.W. *et al.* 1999. Benzene and multiple myeloma: appraisal of the scientific evidence. Blood **94:** 1174–1182.
31. PAUSTENBACH, D.J. *et al.* 1992. Reevaluation of benzene exposure for the Pliofilm (rubberworker) Cohort, (1936–1976). J. Toxicol. Environ. Health **36:** 177–231.
32. INFANTE, P.F. 1995. Benzene and leukemia: cell types, latency and amount of exposure associated with leukemia. *In* Update on Benzene. M. Imbriani, *et al*, Ed. Advances Occup. Med. Rehabil. Fondazione Salvatore Maugeri Edizioni, Pavia, Italy. **1:** 107–120.
33. WONG, O. 1987. An industry wide mortality study of chemical workers occupationally exposed to benzene. II. Dose response analysis. Br. J. Ind. Med. **44:** 382–395.
34. OTT, M.G. *et al.* 1978. Mortality among individuals exposed to benzene. Arch. Environ. Health **33:** 3–10.
35. BOND, G.G. *et al.* 1986. An update of mortality among workers exposed to benzene. Br. J. Ind. Med. **43:** 685–691.
36. TEITELBAUM, D.T. *et al.* 2000. Benzene and multiple myeloma: appraisal of the scientific evidence. Blood **95:** 2995–2996.
37. GOLDSTEIN, B.E. & S.L. SHALOT. 2000. The causal relation between benzene exposure and multiple myeloma. Letter to the editor. Blood **95:** 1512–1513.
38. UTTERBACK, D.F. & R.A. RINSKY. 1995. Benzene exposure assessment in rubber hydrochloride workers: a critical evaluation of previous estimates. Am. J. Ind. Med. **27:** 661–676.
39. SAVITZ, D.A. & R. MOURE. 1984. Cancer risk among oil refinery workers. A review of epidemiologic studies. J. Occup. Med. **26:** 662–670.
40. UNITED STATES DISTRICT COURT FOR THE EASTERN DISTRICT OF TEXAS, Beaumont Division. Ruth Cowen et al., Plaintiffs, vs. Mobil Oil Corporation, Defendant. Memorandum Opinion signed 17 October 1995 by Howell Cobb, United States District Judge.
41. NATIONAL TOXICOLOGY PROGRAM. (NTP). 2000. Summary points from the National Toxicology Program's Endocrine disruptors low-dose peer review, October 10–12, 2000. Available at http://www.ourstolenfuture.org/NewScience/lowdose/2000-10ntppanelreport.htm. Accessed September 9, 2005.

A Regional Approach to Assess the Impact of Living in a Chemical World

CHRISTOPHER T. DE ROSA, HERALINE E. HICKS, ANNETTE E. ASHIZAWA, HANA R. POHL, AND M. MOIZ MUMTAZ

Division of Toxicology and Environmental Medicine, Agency for Toxic Substances and Disease Registry, CDC, Atlanta, Georgia, 30333, USA

ABSTRACT: In the United States, some 80,000 commercial and industrial chemicals are now in use of which over 30,000 are produced or used in the Great Lakes region. Thus, the environmental quality within the Great Lakes basin has been compromised particularly with respect to persistent toxic substances (PTS). Information derived from wildlife studies, prospective epidemiological and toxicological studies, databases, demographics, and Geographical Information Systems (GIS) demonstrate significant public health implications. Studies of human populations indicate: (*a*) elevated body burden levels of PTSs, (*b*) decrease in gestational age, (*c*) low birth weight (LBW), (*d*) greater risk of male children with birth defects (OR = 3.01), (*e*) developmental and neurological deficits, (*f*) increased risk of infertility, (*g*) changes in sex ratio, and (*h*) fluctuations in thyroid hormones. These findings have been identified in vulnerable populations, such as the developing fetus, children, minorities, and men and women of reproductive age who are more susceptible because of their physiologic sensitivity and/or elevated exposure to toxic chemicals. Typically such health effects are assessed on a chemical specific basis; however, most human populations are exposed to hazardous chemicals as mixtures in air, water, soil, and biota. In this article we present an assessment of the potential for joint toxic action of these substances in combinations in which they are typically found. These evaluations represent an integration of all available scientific evidence in accordance with the "NAS paradigm" for risk assessment. In aggregate, our evaluations have demonstrated a need for community-based frameworks and computational techniques to track patterns of environmentally related exposures and associated health effects.

KEYWORDS: Great Lakes; human health; vulnerable populations; chemical mixtures; persistent toxic substances; syndromic surveillance; community-based research

Address for correspondence: Christopher T. De Rosa, Division of Toxicology and Environmental Medicine, Agency for Toxic Substances and Disease Registry, 1600 Clifton Road N.E., Mail Stop F32, Atlanta, GA 30333. Voice: 770-488-3301; fax: 770-488-7015.
e-mail: CYD0@cdc.gov

INTRODUCTION

The U.S. Great Lakes basin is the largest system of fresh surface water on earth, comprising roughly 18% of the world supply. For over 200 years, the Great Lakes basin has been used as a resource for industry, agriculture, shipping, and recreation. By the early 1960s, the environmental quality of the Great Lakes had deteriorated. Eutrophication, overfishing, and the widespread presence of toxic substances all contributed to the decline. The physical nature of the basin and the long retention time of chemicals in the lakes combine to make this huge freshwater resource a repository for chemicals and their byproducts. Less then 1% of the lakes total volume flows out of the St. Lawrence River each year, allowing toxic chemicals to accumulate in the lakes and the sediments. In the United States, some 80,000 commercial and industrial chemicals are now in use of which over 30,000 are produced or used in the Great Lakes region. Researchers have identified almost 400 contaminants in the water, sediment, and biota in quantifiable amounts. Approximately 10% of the U.S. population lives in the region.[1]

In 1985, 11 of the most persistent and widespread toxic substances were identified as "critical Great Lakes pollutants" by the International Joint Commission (IJC). The critical pollutants are: polychlorinated biphenyls (PCBs), dichlorodiphenyl trichloroethane (DDT), dieldrin, toxaphene, mirex, methylmercury, benzo[a]pyrene (a member of a class of substances known as polycyclic aromatic hydrocarbons [PAHs]), hexachlorobenzene (HCB), furans, dioxins, and alkylated lead.[2] Eight of these persistent toxic substances (PTSs) tend to bioaccumulate in organisms, biomagnify in the food chain, and persist at elevated levels in some areas of the ecosystem of the Great Lakes. Because of the persistence and ubiquitous presence of these chemicals in the environment, toxic effects in wildlife have been documented and results from epidemiological investigations, toxicological studies, databases, demographics, and Geographical Information Systems (GIS) demonstrate significant public health implications (TABLE 1).[3–12] These findings were identified in vulnerable populations by using the traditional elements of disease prevention that include:

- surveillance for patterns of morbidity in susceptible populations by virtue of elevated exposure and/or physiologic sensitivity;
- evaluation of the factors underlying the patterns of morbidity and mortality observed at the population level;
- interventions or control strategies that are strategically targeted to susceptible populations including health education and risk communication, so that individuals can take steps to reduce their exposure and that of their families;
- infrastructure development at the state and local levels to implement such a model of disease prevention; and

TABLE 1. Great Lakes research: identified associations between PTSs exposure and adverse health effects in vulnerable populations

Behavioral	Developmental	Endocrine	Neurological	Reproductive
Inability to respond to negative stimuli; greater number of abnormal reflexes; less mature autonomic responses; less attention to visual and auditory stimuli in newborns.[3]	Increase risk for birth defects in males.[4] LBW associated with elevated maternal PCB levels.[5] Reduction in birth weight due to in utero exposure to DDE.[6] Changes in sex ratio if father had elevated levels of PCBs.[4] Changes in sex ratio if the mother had elevated levels of PCBs.[7]	Decreased levels of thyroxine in men and women and decreased levels of sex-hormone-binding globulinbound testosterone in men.[4] Decreased levels of free thyroxine and total thyroxine and increased levels of thyrotropin in children.[9]	Immature nervous and autonomic responses.[3] Poor performance on the Fagan Test of Intelligence at 6 and 12 months.[4] Negative associations between prenatal methylmercury exposure and McCarthy performance were found in children with higher levels of prenatal PCB exposure at 38 months.[4]	Conception rate and the incidence of a live birth are lower in women who are high fish consumers.[4] Reduction in menstrual cycle length.[4] In utero DDE exposure reduced age at menarche by 1 year.[11] In utero exposure to PCBs results in decrease in gestational age and LBW.[12]
Response inhibition in preschool children exposed prenatally to PCBs.[8]	Suboptimal development of the nervous system–splenium.[8]		Low IQ scores, 2 years behind in reading comprehension, poor short- and long-term memory, and difficulty paying attention.[10] Lower scores on several measures of memory and learning.[4]	

LBW = low birth weight.

- impact assessment to ensure that the interventions undertaken actually serve to improve health status of susceptible populations.

There is evidence that populations are exposed to environmental chemicals, especially chemical mixtures. In order to fully assess the health status of these populations, we must consider the fact that these populations were exposed to mixtures of hazardous chemicals in air, water, soil, and biota. Therefore, in this article we assess the potential for joint toxic action of these chemical substances in combinations in which they are typically found and describe the need for community-based public health research to identify disease outcomes.

TARGETING THE MIXTURES OF CONCERN

We evaluated the health effects of the chlorinated dibenzo-p-dioxin's (CDDs), hexachlorobenzene, p,p'-dichlorodiphenyldichloroethylene (DDE), methylmercury, and PCBs (TABLE 2). Each of these contaminants, apart from causing a primary (critical) effect at the lowest exposure concentration, also have the potential to cause secondary multiple health effects at higher tissue doses. Several health effects, such as neurological and neurodevelopmental, have been well documented both in animals as well as through human epidemiological studies. Also, there is evidence that all five of these chemicals can act on the developing nervous system.[13,14] Because these pollutants have common target organs and similar health effects, there is a possibility of more than one contaminant being present at the same time in a target organ increasing the potential for interactions. Our agency has developed a guidance that includes a simplified strategy to conduct a comprehensive evaluation of the joint toxicity and assessment of potential health hazards from chemical mixtures.[15] This guidance is based on exposure-based screening assessment but also recognizes the prominent roles of community health concerns, community-specific health

TABLE 2. Health effects of representative Great Lakes pollutants

Effects	TCDD	HCB	DDE	Hg	PCB
Liver damage	H, a	H, a	H, a	H, a	
Immunosuppression	a	a	H, a	H, a	a
Thyroid hormones	H, a	H, a		H	H, a
Female repro function	a	H, a	H	H, a	H
Male repro function	H, a	a	a		H, a
Neurological	H, a	H, a	H, a	H, a	H, a
Neurodevelopmental	a	H, a		H, a	H, a
Reprodevelopmental	a		a		a
Other developmental	a	H,a	a	a	a
Cancer	H,a	a	a	a	a

H = Humans; a = Animals.

outcome data, and scientific/biomedical judgment in the determination of the public health implications of exposures to chemical mixtures at hazardous waste sites.

Epidemiological studies have demonstrated an association between exposure to biopersistent chemicals in human breast milk or fish and mild neurodevelopmental deficits. These findings are not directly useful for conducting screening level exposure-based assessments of hazards specific to a community or for scenarios involving exposure to specific mixtures of CDDs, hexachlorobenzene, p,p'-DDE, methylmercury, and PCBs. The component-based approaches recommended in the Agency for Toxic Substances and Disease Registry (ATSDR) mixtures guidance viz., the Hazard Index (HI), and the target-organ toxicity dose (TTD)[16] are useful for the purpose of establishing causal effect relationships and allowing assessments of the possibility of altered neurological development. However, both these approaches are based on potency weighted response or dose additivity and hence do not factor in the role of chemical interactions in the overall joint toxicity. The binary weight-of-evidence (WOE) method[17] was developed to modify the HI to account for interactions by using the available information among binary combinations of mixture components. This method predicts possible interactions in pairs of chemicals based on evaluations of toxicological significance and mechanistic understanding of the interactions. Only a limited amount of evidence is available on the existence of greater-than-additive or less-than-additive interactions between a few pairs of the chemicals of concern: (*a*) hexachlorobenzene potentiation of tetrachlorodibenzo-p-dioxin (TCDD) and reduction of body and thymus weights (a greater-than-additive interaction); (*b*) PCB antagonism of TCDD immunotoxicity (less-than-additive interaction); (*c*) PCB antagonism of TCDD developmental toxicity (less-than-additive interaction); and (*d*) synergism between PCBs and methylmercury in disrupting regulation of brain levels of dopamine that may influence neurological function and development (greater-than-additive interaction). WOE analyses of these data, however, indicate that scientific evidence for these interactions is limited and is inadequate to fully characterize the possible modes of joint action on these toxicity targets. For the remaining pairs, additive joint action at shared targets of toxicity is either supported by data (for a few pairs) or is recommended as a public health protective assumption due to lack of interaction data, conflicting interaction data, and/or lack of mechanistic understanding to reliably project potential nonadditive interactions.[13,14]

This approach can be used to conduct exposure-based assessments of possible noncancer or cancer health hazards from oral exposures to mixtures of 2,3,7,8-TCDD, hexachlorobenzene, p,p'-DDE, methylmercury, and PCBs. Component-based approaches are recommended, because direct experimental toxicological data are not available to characterize health hazards (and dose–response relationships) from exposure to such mixtures. Alternatively, physiologically based pharmacokinetics (PBPK/PD) models could be used for such assessment but they have not yet been developed to predict appropriate target tissue/organ doses of these components.

Joint toxicity assessments and application of these computational methods will be very useful in areas that have high levels of chemical contamination in well-defined geographic areas where vulnerable populations live, such as the areas of concern (AOCs) in the Great Lakes.

AOCs IN THE GREAT LAKES

The Great Lakes AOCs are geographic areas that are particularly contaminated to the extent that the contamination impairs beneficial uses of the area (e.g., contamination of fish and drinking water). Industrial and agricultural discharges to the Great Lakes waterways and land have resulted in the degradation of these areas. A large number of hazardous waste sites, 118, are located within the AOCs and around the Great Lakes resulting from industrial and agricultural uses. Of these 53 are in the Lake Michigan area, 40 in the Lake Erie region, and 15 in the Lake Huron area. The Lake Superior and Ontario have the fewest at 5 waste sites each.

The ATSDR public health assessments, conducted in the AOCs, have revealed 15,000 instances of contaminants exceeding the health-based screening values in a variety of media. Of the 115 hazardous waste sites in the AOCs investigated by ATSDR, ~50% of the sites had completed exposure pathways. The toxic substances detected in these areas were located and in a form that facilitates exposure to the residents of these areas, and ~80% of the sites had multiple chemicals of concern, that is, two or more chemicals. In addition to the public health assessments, ATSDR used demographic data from the U.S. census for identifying at-risk populations. AOC county health outcome data were used for examining patterns of health outcomes potentially associated with area toxic substances. The public health assessments and the EPA Toxic Release Inventory and the EPA National Pollutant Discharge Elimination System data were also used to demonstrate other sources of toxic substances. Excess incidence of low birth weight (LBW) was observed at sites with PCB contamination. Great Lakes research has shown that LBW has been associated with prenatal exposure to PCBs from maternal consumption of contaminated Great Lakes sport fish. This evaluation of the U.S. AOCs has identified potentially vulnerable communities that may require more in-depth investigation.[18]

DETERMINING DISEASE OUTCOME BY COMMUNITY-BASED INVOLVEMENT

Public health interventions have played a significant role in the Great Lakes basin in terms of controlling the overt symptoms of toxicity and reducing body burdens of pollutants in human populations. The research findings of the ATSDR Great Lakes research program have led to a number of success stories using community-based strategies. For example, in one study involving American Indians, 97% of the men knew about fish advisories against consuming local fish prior to the study; however, 80% of the men ate local fish during the last 2 years before participating in the ATSDR study. To encourage the study population to become aware of the hazards of consuming contaminated fish, ATSDR used various risk communication strategies and involved tribal members in the planning and implementation of the study. This was a key element in structuring the study, recruiting participants, and strengthening the research design.

Over 1000 health advisories have been issued for this area and they have been effective in bringing down the levels of individual pollutants significantly in vulnerable populations on both sides (United States and Canada) of this unique body of fresh water.[19] However, a decade after the implementation of Superfund, and despite Congressional efforts to redirect the program, substantial public health concerns remain, and critical information on the distribution of exposures and health effects associated with hazardous waste sites is still lacking.[20] Unfortunately, at present, the same conclusion still applies.[21,22] Nevertheless, we have come a long way from the first report of the IJC that stated in 1918 "situation along the frontier is generally chaotic, everywhere perilous, and, in some cases, disgraceful."

Similarly, in 2001 the Pew Environmental Health Commission concluded that there is a growing disparity between the national public health infrastructure and the ability to monitor the levels of these contaminants in the environment and to assess their potential impacts on the general health status of the U.S. population. Additionally, the Pew Commission concluded that the ATSDR Great Lakes Human Health Effects Research Program, established in 1992, is yielding "compelling data concerning exposure to chemical contaminants and health consequences associated with these exposures."[23] These "comments" by the Pew Commission "underscore" a compelling need for and the value of further development of community-based public health capacity for purposes of tracking chemical exposures and their potential human health impact. This capacity must include the ability to identify and monitor "hot spots" in the human population with respect to elevated exposures to toxic substances, as well as potential clusters of environmentally related diseases. What is needed is a group of signs and symptoms that collectively indicate or characterize a disease or health outcome, that is, "syndromic surveillance" to identify diseases that may have an environmental link. This would allow for a unique intellectual capacity for designing a program to meet surveillance challenges at a local level and have data sets in exposure, toxicology, and epidemiology addressing a wide spectrum of public health issues. Importantly, short-term benefits include the establishment of a community-based framework that will address the recommendations of the NRC and Pew reports, and one can build upon existing programs at the national, state, and local levels. The long-term benefits of this approach would (*a*) allow for the development of short- and long-term interventions, (*b*) provide a framework for a rigorous, technical evaluation of underlying factors that account for morbidity and mortality, (*c*) provide a basis for community-based interventions, and (*d*) facilitate long-term assessments of intervention effectiveness and thus restore the foundation of public health.

The community-based research described earlier resulted in a significant reduction in consumption of contaminated fish and reduction in elevated PCB levels. During the first year of the study, the men reduced their consumption rate from an average of 98 meals per year to 28, and even lower during the second year. The serum PCB levels in men were significantly related to the number of fish meals consumed per year. Therefore, a reduction in consumption led to lower PCB serum levels. These lower PCB levels in men now represented levels that were similar to the general U.S. population. A similar trend was also found in the women of this group.

The experiences demonstrated by researchers and policy makers from the Great Lakes states illustrate the efficacy of directing resources to community-based research to effect change in public health status. These experiences also demonstrate the power and effectiveness of tracking chemical exposures and translating scientific information into public health service on a local level and serve as a model for other ecosystems with public health practice.

The ultimate objective of these efforts is to empower communities by providing them the means to make informed decisions on personally relevant environmental public health issues. In order to help communities make informed decisions, we must use all available scientific knowledge in the areas of exposure, epidemiology, toxicology, chemical mixtures, computational models, and methodologies in parallel with syndromic surveillance to provide answers to their questions pertaining to their health.

ACKNOWLEDGMENT

The authors want to thank Ms. Olga Dawkins for her technical support in the preparation of this manuscript.

REFERENCES

1. U.S. ENVIRONMENTAL PROTECTION AGENCY AND GOVERNMENT OF CANADA. 1995. The Great Lakes: An Environmental Atlas and Resource Book, 3rd ed. U.S. Environmental Protection Agency, Washington, DC USEPA 905-B-001.
2. INTERNATIONAL JOINT COMMISSION (IJC). 1983. An inventory of chemicals substances identified in the Great Lakes ecosystems. Vols. 1-6. International Joint Commission, December 31, Windsor, Ontario, Canada.

3. LONKY, E., J. REIHMAN, T. DARVILL, et al. 1996. Neonatal behavioral assessment scale performance in humans influenced by maternal consumption of environmentally contaminated Lake Ontario fish. J. Great Lakes Res. **22:** 198–221.
4. HICKS, H.E., A. ASHIZAWA & C.T. DE ROSA. 2005. Assessing the health status of vulnerable populations from exposure to persistent toxic substances in the U.S. Great Lakes. Proceedings of the 25th International Symposium on Halogenated Environmental Organic Pollutants and POPs 2389–2390.
5. KARMAUS, W. & X. ZHU. 2004. Maternal concentration of polychlorinated biphenyls and dichlorodiphenyl dichlorethylene and birth weight in Michigan fish eaters: a cohort study. Environ. Health **3:** 1–17.
6. WEISSKOPF, M.G., H.A. ANDERSON, L.P. HANRAHAN, et al. 2005. Maternal exposure to Great Lakes sport-caught fish and dichlorodiphenyl dichloroethylene, but not polychlorinated biphenyls, is associated with reduced birth weight. Environ. Res. **97:** 149–162.
7. WEISSKOPF, M.G., H.A. ANDERSON, L.P. HANRAHAN & THE GREAT LAKES CONSORTIUM. 2003. Decreased sex ratio following maternal exposure to polychlorinated biphenyls from contaminated Great Lakes sport-caught fish: a retrospective cohort study. Environ. Health **2:** 1–14.
8. STEWART, P.W., J. REHIMAN, E.L. LONKY, et al. 2003. Prenatal PCB exposure, the corpus callosum, and response inhibition. Environ. Health Perspect. **111:** 1670–1677.
9. SCHELL, L.M., M.V. GALLO, A.P. DECAPRIO, et al. 2004. Thyroid function in relation to burden of PCBs, p, p'-DDE, HCB, mirex and lead among Akwesasne Mohawk youth: a preliminary study. Environ. Toxicol. Pharmacol. **18:** 91–99.
10. JACOBSON, J.L. & S.W. JACOBSON. 1996. Intellectual impairment in children exposed to polychlorinated biphenyls in utero. N. Engl. J. Med. **335:** 783–789.
11. VASILIU, O., J. MUTTINENI & W. KARMAUS. 2004. In utero exposure to organochlorines and age at menarche. Hum. Reprod. **19:** 1–7.
12. TAYLOR, P.R., J.M. STELMA & C.E. LAWRENCE. 1989. The relation of polychlorinated biphenyls to birth weight and gestational age in the offspring of occupationally exposed mothers. Am. J. Epidemiol. **129:** 395–406.
13. AGENCY FOR TOXIC SUBSTANCES AND DISEASE REGISTRY. 2004a. Interaction profile for persistent chemicals found in breast milk. U.S. Department of Health and Human Services, Agency for Toxic Substances and Disease Registry. Atlanta, Georgia. (http://www.atsdr.cdc.gov/interactionprofiles/ip03.html) Accessed October 2005.
14. Agency for Toxic Substances and Disease Registry. 2004b. Interaction profile for persistent chemicals found in fish. U.S. Department of Health and Human Services, Agency for Toxic Substances and Disease Registry. Atlanta, Georgia. (http://www.atsdr.cdc.gov/interactionprofiles/ip01.html) Accessed October 2005.
15. AGENCY FOR TOXIC SUBSTANCES AND DISEASE REGISTRY. 2001. Guidance manual for the assessment of joint toxic action of chemical mixtures. U.S. Department of Health and Human Services, Agency for Toxic Substances and Disease Registry, Atlanta, Georgia (http://www.atsdr.cdc.gov/interactionprofiles/ipga.html) Accessed October 12, 2005.
16. MUMTAZ, M.M., K.A. POIRIER & J.T. COLMAN 1997. Risk assessment of chemical mixtures: fine-tuning the hazard index approach. J. Clean Technol. Occup. Med. **6:** 189–204.

17. MUMTAZ, M.M. & P.R. DURKIN 1992. A weight of evidence scheme for assessing interactions in chemical mixtures. Toxicol. Ind. Health **8:** 377–406.
18. AGENCY FOR TOXIC SUBSTANCES AND DISEASE REGISTRY. Public Health Implications of Hazardous Substances in the U.S. Areas of Concern. U.S. Department of Health and Human Services, Agency for Toxic Substances and Disease Registry, Atlanta, Georgia In press.
19. U.S.ENVIRONMENTAL PROTECTION AGENCY. 2005. EPA Fact Sheet 2004 National Listing of Fish Advisories. Office Water, EPA-823-F-05-004; September 2005.
20. NATIONAL RESEARCH COUNCIL. 1991. Environmental Epidemiology: Public Health and Hazardous Wastes. Vol.1. National Research Council.Washington DC
21. GARRETT, L. 2000. The Collapse of Global Public Health in "Betrayal of Trust." Hyperion Publishers. New York.
22. DE ROSA, C.T. 2003. Guest Editorial—Restoring the foundation: tracking chemical exposures and human health. Environ. Health Perspect. **111:** A374–375.
23. PEW ENVIRONMENTAL HEALTH COMMISSION. 2001. Strengthening Our Public Defense Against Environmental Threats: Transition Report to the New Administration. Johns Hopkins School of Public Health. Baltimore, MD: Available: http://pewenvirohealth.jhsph.edu/html/reports/Transition˙Technical.pdf.

Low-Dose Risk, Hormesis, Analogical and Logical Thinking

GIOVANNI A. ZAPPONI[a] AND IDA MARCELLO[b]

[a]*Department of Technology and Health, Istituto Superiore di Sanità, 00161 Rome, Italy*

[b]*Department of Environment and Prevention, Istituto Superiore di Sanità, 00161 Rome, Italy*

ABSTRACT: The hormesis theory proposes the low-dose beneficial and high-dose detrimental pattern, existing for specific conditions, as a "general default assumption" for toxicology and carcinogenicity. Crump and Kitchin and Drane underline that in a *post hoc* retrospective scientific literature searching for hormetic dose–response patterns, the consideration of the whole available relevant studies is necessary and, for statistical testing purposes, for instance at a 0.05 standard level, a P value obtained from $1 - (1 - P)^n = 0.05$ (i.e., $P = 0.0005$ for 100 examined cases) should be used (otherwise, by definition, 5 "positive" results are expected by chance over 100 cases). The hypothesis, based on some experimental data on rodents, by Calabrese and Baldwin, of an hormetic effect of 2,3,7,8-TCDD at the 1–10 ng/kgbw/day dose, of Na-saccharine in the $\leq 1\%$ of diet exposure range, of Cadmium Chloride in the 0–5 μmol/kg dose range, single injection, and of neutrons in the 0- to 2-rad dose range, are not confirmed, and, rather, are contradicted, when the whole relevant data presented by international and national agencies are considered. As far as the radiation risk is in particular concerned, a recently published epidemiological study on more than 400,000 nuclear plant workers, co-ordinated by the IARC has indicated a small, but significant risk, at the current exposure limits, and possibly below them. Therefore, the hormesis theory-based criticism of current radiation protection criteria, assumed to be excessively conservative, is not justified. Also not justified is the assumption that "by dismissing hormesis, regulatory agencies such as U.S. EPA deny the public the opportunity for optimal health and avoidance of diseases;" rather, the contrary is here considered true. Analogical considerations are not necessarily logical ones and the single result should be considered in its whole context.

KEYWORDS: hormesis; risk assessment; health protection

Address for correspondence: Dr. Giovanni Zapponi, Alfredo Zapponi, Senior Scientist, Head, Technology and Health Department, Istituto Superiore di Sanità (Italian National Health Institute), 00161, Rome, Italy. Voice: +3906 4990 2915; fax: +3906 4938 7075.
e-mail: giovanni.zapponi@iss.it

INTRODUCTION

Assuming that low doses positively stimulate the mechanisms of biological defense, the hormesis theory supporters propose, as a general principle, the extension also to toxicology and carcinogenicity of the biphasic dose–response trends (low-dose beneficial and high-dose detrimental effects) existing for specific categories of chemicals and conditions (e.g., substances essential for life). Many papers have been published in the last decade, dealing with the presence of hormetic effects in a very large amount of biological processes, mostly concerning effects other than toxicology and carcinogenesis. As far as the assessment of the hormetic principle is concerned, Calabrese has stated that "the hormetic response may be at best assessed within the context of a dose–time response since the low-dose stimulatory response often represents a modest overcompensation response following an initial disruption (i.e., toxicity) in the homeostasis. The initial response displays a dose–dependent toxicity. As the time progresses, a compensatory response occurs at low doses eventually leading to the low-dose stimulation characteristic of hormesis. The nature of the overcompensation response appears to result from biological compensatory processes that allocate resources slightly in excess of those ensuring a return to homeostasis. This "extra" allocation of resources, i.e., adaptive response, leads to the hormetic stimulation.[1] On the other hand, the incorporation of hormesis within the context of risk assessment has been repeatedly proposed as a "default" assumption, based on the result of a pertinent literature review, than on a case-by-case experimental demonstration, because this latter "would likewise create an overwhelming financial burden and would surely be the death knell of any practical significance for hormesis in the regulatory world" (p. 446).[2] However, as underlined by Crump, this assumption is unlikely to be accepted without a suitable and convincing evidence that hormesis really is an universal phenomenon.[3] And, as logic teaches, an universal assumption is intrinsically at risk: in principle, it will be put in a critical position by few cases, or even only one, that clearly demonstrate the opposite.

Hereafter, the discussion is limited to the hormesis hypotheses based on carcinogenic experiments that have been used for risk assessment, also taking into account that the current risk evaluation and assessment rules have been defined as inappropriate by hormesis supporters ("by dismissing hormesis, regulatory agencies such as U.S. EPA deny the public the opportunity for optimal health and avoidance of diseases" (p. 1).[4] The following analysis mainly deals with some experimental carcinogenesis studies, which have been used for demonstrating or suggesting hormetic effects. It is worthwhile noticing that some important problems have been underlined in the procedure adopted for demonstrating hormesis,[3,5] that are worthy of attention. Among them, a possible selection bias in the construction of hormesis dose–response databases and a correlated limitation of data completeness and difficulties arising in the statistical significance of evaluations based on a *"post hoc"* literature searching, rather than on experimental studies data specifically—*"propter hoc"*—designed for this purpose. Moreover, some intrinsic limitations in the data selection [e.g., exclusion dose–response relationships whose control and no observed adverse effect level (NOAEL) incidences are null or close to zero, for which responses below the control are not identifiable] are important.

SOME GENERAL CONSIDERATIONS

Post Hoc and Propter Hoc Evaluations, Statistical Testing, and Completeness

A point raised by Crump and by Kitchin and Drane is that in order to avoid false positives in a *post hoc* literature searching for hormesis, it is necessary to take into account the extension of the examined data set.[3,5] For instance, if 100 dose–response curves eligible for hormesis evaluation are found out in the relevant scientific literature, and if the 5% level is used for statistical testing, it is immediate, by definition of statistical significance, that five positive cases are expected simply due to chance. Therefore, in order to control false positive cases of this type, in the examination of n studies or dose–response relationships, the statistical testing for a 0.05 level should consider a P value obtained from the formula $1 - (1 - P)^n = 0.05$, and not simply 0.05, as underlined by Crump.[3] In the case of 100, 50, 20, and 10 dose–response curves, the P value will, respectively, result equal to 0.0005, 0.001, 0.0026, and 0.005. Clearly, it is not easy to reach these significance levels.

As far as the completeness need in risk evaluation is concerned, it is worthwhile underlining that the International and National Agencies, responsible for the carcinogenicity and toxicity evaluation, classification, and risk evaluation, examine all the relevant available and suitable epidemiological, experimental, *in vitro* studies and data, and the other related information. The final conclusion results from a long, attentive, global evaluation. This is immediately evident, for instance, from the monographs of the International Agency for Research on Cancer (IARC), the World Health Organization documents, the European Union Committee documents, and, at a national level, for instance, the U.S. National Toxicology Program (NTP), the U.S. NCI and U.S. Environmental Protection Agency (EPA) documents, as well as of the ones of a very large number of other institutions. Taking into account this context, it seems somewhat strange that the assumption of a low-dose beneficial effect of a carcinogenic agent could be only based on few data or even on a single dose–response curve, without a reasonably extended and comparative evaluation of all the other available relevant data. This also reminding that, as already mentioned, hormesis theory supporters have heavily criticized regulatory agencies, as the ones above mentioned, because by dismissing the hormesis, they "deny the public the opportunity for optimal health and avoidance of diseases" (p. 1).[4]

Incidences in Experimental Control and NOAEL Groups and Hormesis Generalization

The dose–response curves considered eligible for hormesis are typically characterized by a sensibly high level of control and NOAEL incidences.[3] This is because if the control incidence is null or very close to zero, obviously no response decrease below it will be observable and detectable at a significant level. Evidently, such limitation implies the *a priori* omission of a very large proportion of the available dose–response relationships (a control response considerably low or even null is commonly observed in many studies). Moreover, on the other hand, well-known aspects of "statistical power" clearly indicate that studies with null or very low control response are the most suitable for detecting, at a significant level, a possible small response increase at lower doses (e.g., 5/100 versus 0/100 is statistically significant, while 10/100 versus 5/100 is not, even if the response increase—5%—is the same). These studies cannot be *a priori* excluded in the evaluation. In particular, it seems reasonable to assume that in the dose range for which an hormetic effect is hypothesized based on a specific dose–response relationship, null or almost null responses should be observed in the other available dose–response relationships relevant to the examined effect, whose control incidence is zero or close to zero (and therefore, have been omitted from the analysis). If this does not result and a response increase is observed, in particular at a significant level, the hormesis hypothesis is contradicted in this dose range. This verification is important and easy.

Moreover, because the control incidence is a fundamental parameter in the hormesis assessment, the reference to the historical control, whenever available, is clearly appropriate in order to reduce the uncertainty of the estimate of this parameter, together with a comparative analysis of control incidences in the two genders or of the control incidence observed in other analogous and relevant available studies. This makes part of the completeness requirements above discussed.

In general, the presence of a remarkably high control incidence may in principle be attributed to chance, or to a background exposure and risk, or to a specific high susceptibility, or a combination of these causes. In the case of first hypothesis, a statistical comparative analysis may provide suitable information.

If the background exposure, by itself, is supposed to be sufficient to induce some response increase, some important points emerge. First, in a well-conducted experimental study, an identical background exposure will be present in all the experimental groups, including the ones for which an hormetic effect could be hypothesized. In this case, a contemporary "combined exposure" will exist for treated groups (background and treatment doses, both able or potentially able to induce the same adverse effect under study). This condition is different, for instance, from the case of a previous positively "stimulating exposure" followed by a potentially hazardous exposure. For instance, according to the "low-dose induced adaptation" paradigm, the potentially beneficial exposure should take place before the adverse exposure rather than together with it. Last, in the case of a high control response attributable to background exposure, some level of homeostasis disruption should have taken place also for this experimental group, according to hormesis theory. These aspects, not sufficiently discussed, could also be considered within a "multiple exposure" paradigm.

As an alternative hypothesis, a specific high susceptibility of the exposed animals could be considered the cause of remarkably high control responses. Whenever an hormetic effect is assumed at low doses in such conditions, and this result is proposed to be extrapolated to human population, difficult problems arise. According to a common definition, the response of a susceptible individual takes place at lower exposures and/or in early time periods in comparison to normal subjects, and/or his response will be higher than the normal one at the same exposure level.[6,7] For instance, it is worthwhile reminding that the American Conference of Governmental Industrial Hygienists (ACGIH, various years), when commenting its proposals of threshold limit values, commonly underlines that these exposure limits could not sufficiently protect particularly susceptible subjects. Based on the available epidemiological data, evaluations and regulations concerning susceptibility, as well as on common sense and ethical principles, the hypothesis that a small potentially carcinogenic exposure could in some way be beneficial for cancer susceptible individuals appears to be neither reasonable nor acceptable.

A BRIEF DISCUSSION OF SOME PRESENTED EXAMPLES OF HORMESIS EFFECT IN CARCINOGENESIS BASED ON EXPERIMENTAL STUDIES

Hereafter, some experimental data, presented to support the hormesis theory are examined, also in the light of the above-mentioned points. In most cases, the reported statistical evaluation was the one effected by the authors of the examined experiments. The few other simple statistical evaluations carried in this study were the current ones (Fisher's exact test for small frequencies, Mantel–Haentzel test for trend, χ^2 test when appropriate[8,9]). In the tables hereafter presented, when the statistical significance level is followed by the "N" letter, the difference is an incidence decrease; otherwise it is an increase. Last, in order to prevent possible selection biases, the following analysis has been effected considering the whole relevant comprehensive data and evaluations reported by well-known international and national reviews (e.g., IARC, U.S. NTP, and U.S. NCI).

2,3,7,8-TCDD Carcinogenesis

Calabrese and Baldwin underline an hormetic effect[10] suggested for 2,3,7,8-TCDD by the Kociba *et al.* study.[11] Moreover, in order to confirm this hypothesis, they also present a table reporting the tumor rates, in terms of number of tumors per 100 animals, that, for all tumors, indicates rates, respectively, of 80, 98, and 120 tumors per 100 animals at the 0.001, 0.01, and 0.1 μg/kg body weight/day (i.e., 1, 10, and 100 ng/kg b.w./day) dose levels for male rats, in comparison with a control tumor rate of 162.4 tumors per 100 animals. For female rats, 192, 204, and 244.9 tumor rate per 100 animals is reported for the same dose levels, in comparison with a control tumor rate of 267.4 tumors per 100 animals. Similar differences are reported for most of single tumors rates for 100 animals. The statistical significance of these differences is not presented. Provided that the Kociba *et al*. study[11] has been considered important for risk assessment purposes, it is hereafter discussed together with the NTP study,[12] also used for this scope. In order to avoid a selection bias, the following analysis include all the relevant dose–response obtained by the Kociba *et al*. study[11] and its successive revisions reported by the NTP study,[12] and largely discussed and evaluated by the IARC.[13] All these dose–response relationships include the dose range for which Calabrese and Baldwin suggest the presence of an hormetic effect (TABLES 1–3). As a whole, a statistically significant increasing trend has been found in 17 out of 21 of the above-mentioned dose–response relationships. The tumor incidence is significantly increased in almost all the highest dose groups, in a large part of the intermediate dose groups, and in one of the lowest dose group. In 20 out of 21 cases, the trend is significant and/or the higher dose–response is significantly increased in comparison with the control.

If reference is made to the significance levels of differences among control and treatment groups, reported by NTP12, immediately emerges that:

- For the lowest dose (1.4 ng/kg b.w./day or 1.0 ng/kg b.w./day), the only statistically significant result ($P < 0.05$) is an increase of thyroid follicular adenoma in male rats (NTP, 5/48 versus 1/69, $P = 0.042$).[12]
- At the $P < 0.10$ significance level, response increases have been found at the lowest dose (1.4 ng/kg b.w./day or 1.0 ng/kg b.w./day) for thyroid adenoma in female mice and for adrenal cortex in male rats (respectively, 3/50 versus 0/69, $P = 0.07$ and 9/50 versus 6/72, $P = 0.09$), and a response decrease has been found for lung adenoma in male mice (2/48 versus 10/71, $P = 0.069$).
- As a whole, these data clearly suggest the hypothesis of a response increase at 1.4 ng/kg b.w./day, rather than a decrease, and that this dose could be in principle assumed as a lowest effect dose.
- In particular for the Kociba *et al*. study, on which has been essentially based the hormesis hypothesis for 2,3,7,8-TCDD, the tumor incidence data indicate a very scarce significance level range (P: 0.29–0.69) for the responses lower than the one of control.
- The whole above-presented data clearly point out a consistent evidence of a tumor increase at the two higher doses.

TABLE 1. Kociba et al. (1978)[11] study in Sprague-Dawley rats of 2,3,7,8-TCDD and successive revisions[12,13]

Tumor category	Control	2,3,7,8TCCD dose (ng/kg b.w./day)		
		1	10	100
Kociba et al., 1978				
Hyperplastic nodule (trend: $P < 0.0001$)	8/86 (9.3%)	3/50 (6%) $P = 0.37$ (N)	18/50 (36%) $P < 0.001$	23/49 (47%) $P < 0.001$
Hepatocellular carcinoma (trend: $P < 0.0001$)	1/86 (1.2%)	0/50 (0%) $P = 0.63$ (N)	2/50 (4%) $P = 0.3$	11/49 (22%) $P < 0.001$
Neoplastic nodule, hepatocellular	16/86 (19%)	8/50 (16%)	27/50 (54%)	33/47 (70%)
Hyperplastic nodule, hepatocellular carcinoma (trend: $P < 0.001$)	9/86 (10%)	3/50 (6%) $P = 0.29$ (N)	18/50 (36%) $P < 0.001$	34/48 (71%) $P < 0.001$
Squire, 1980, revision of the Kociba et al. study				
Carcinoma (trend: $P < 0.0001$)[12]	16/86 (19%)	8/50 (16%) $P = 0.44$N	27/50 (54%) $P < 0.001$	33/47 (70%) $P < 0.001$
Goodman and Sauer, 1992, revision of Kociba et al. study				
Hepatocellular adenoma (trend: $P < 0.0001$)	2/86 (2.3%)	1/50 (2%) $P = 0.69$ (NS)	9/50 (18%) $P < 0.01$	14/45 (31%) $P < 0.001$
Hepatocellular carcinoma (trend: $P < 0.01$)	0/86 (0%)	0/50 (0%)	0/50 (0%)	4/45 (9%) $P = 0.01$
Hepatocellular adenoma, hepatocellular carcinoma (trend: $P < 0.0001$)	2/86 (2.3%)	1/50 (2%) $P = 0.69$ N	9/50 (18%) $P < 0.01$	18/45 (40%) $P < 0.001$

N = decrease of response in comparison with control.

TABLE 2. NTP (1982) study in Osborne-Mendel rats of 2,3,7,8-TCDD[12]

Tumor category	Control	2,3,7,8-TCCD dose (ng/kg b.w./day)		
		1.4	7.1	71
Male rats thyroid follicular cell adenoma trend: $P < 0.01$	1/69 (1.4%)	5/48 (10%) $P = 0.042$	6/50 (12%) $P = 0.021$	10/50 (20%) $P = 0.001$
liver neoplastic nodule trend. $P = 0.005$	0/74 (0%)	0/50 (0%)	0/50 (0%)	3/50 (6%) $P = 0.6$
Adrenal cortex adenoma trend: $P = 0.26$	6/72 (8.3%)	9/50 (18%) $P = 0.09$	12/49 (24%) $P = 0.015$	9/49 (18%) $P = 0.09$
Female rats liver neoplastic nodule trend. $P < 0.001$	5/75 (6.6%)	1/49 (2%) $P = 0.19$ N	3/50 (6%) $P = 0.6$	12/49 (24%) $P = 0.039$
Adrenal cortex adenoma or carcinoma (trend: $P < 0.014$)	11/73 (15%)	9/49 (18%) $P = 0.4$	5/49 (10%) $P = 0.6$ N	14/46 (30%) $P = 0.039$
Subcutaneous fibrosarcoma trend: N	0/75 (0%)	2/50 (4%) $P = 0.16$	3/50 (6%) $P = 0.06$	4/49 (8%) $P = 0.023$

N = decrease of response in comparison with control.

TABLE 3. NTP (1982) study in B6C3F$_1$ mice on 2,3,7,8-TCDD[12]

Tumor category	Control	2,3,7,8-TCCD dose (ng/kg b.w./day)		
		1.4	7.1	71
Male mice				
Liver carcinoma	8/73 (11%)	9/49 (18%)	8/49 (16%)	17/50 (34%)
Trend: $P = 0.002$		$P = 0.19$	$P = 0.28$	$P = 0.002$
Liver adenoma	7/73 (9.6%)	3/49 (6.1%)	5/49 (10%)	10/50 (20%)
Trend: $P = 0.024$		$P = 0.37$ N	$P = 0.6$	$P = 0.09$
Lung adenoma or carcinoma	10/71 (14%)	2/48 (4.2%)	4/48 (8.3%)	13/50 (26%)
(Trend: $P = 0.004$)		$P = 0.069$ N	$P = 0.26$	$P = 0.08$

Tumor category	Control	5.7	28.6	286
Female mice				
Subcutaneous fibrosarcoma	1/74 (1.3%)	1/50 (2%)	1/48 (2.1%)	5/47 (11%)
Trend: $P = 0.007$		$P = 0.6$	$P = 0.6$	$P = 0.32$
Liver carcinoma	1/73 (1.4%)	2/50 (4%)	2/48 (4.2%)	6/47 (13%)
Trend: $P = 0.008$		$P = 0.4$	$P = 0.4$	$P = 0.014$
Liver adenoma	2/73 (2.7%)	4/50 (8%)	4/48 (4.2%)	5/47 (10.6%)
Trend: $P = 0.11$		$P = 0.2$	$P = 0.2$	$P = 0.8$
Thyroid adenoma	0/69 (0%)	3/50 (6%)	1/47 (2.1%)	5/46 (11%)
Trend: $P < 0.01$		$P = 0.07$	$P = 0.4$	$P = 0.009$
Hematopoietic all lymphomas	18/74 (24%)	11/50 (22%)	13/48 (27%)	20/47 (42%)
(Trend: $P = 0.011$)		$P = 0.46$ N	$P = 0.4$	$P = 0.029$

N = decrease of response in comparison with control.

- This contrasts with the hypothesis of the hormetic effect in this whole dose range, which has been based on the tumor rate for 100 animals (number of tumors per 100 animals, parameter not used for current risk assessment procedures).

Cadmium Carcinogenesis

The hypothesis of an hormetic effect of cadmium chloride is also presented, based on the incidence data of testicular tumors in rats,[2] reported by Waalkes et al.,[14] in a study on cadmium carcinogenesis in male rats treated with cadmium injections. The dose–response relationships presented in the study are summarized in TABLE 4. According to Calabrese and Baldwin, "the treatments demonstrated a striking U-shaped dose–response relationship with lowest dose having only 1 of 30 rats (3.3%) with testicular cancer" (p. 10).[10] However, in the same paper, Waalkes et al. also discuss in detail the neoplastic and preneoplastic lesions of the prostate, underlining that "a strong positive trend also occurred between dosage levels of cadmium given as a single injection of 0, (pooled control), 1.0 µmol Cd/kg (group 2), and 2.5 µmol Cd/kg (group 3) and both tumor incidence and number of tumorous foci/prostate" (p. 4657).[14] For the neoplastic lesions of prostate, they also report a statistically significant response increase at the 2.5 µmol Cd/kg dose, and an incidence about two-fold higher than the one of control for the lowest dose (1 µmol Cd/kg). The incidence increase of prostatic tumors in the low dose range resulted coherent and correlated with a parallel increase of the number of tumorous foci per prostate (FIG. 1). A significant incidence decrease of pancreatic tumors for the highest dose (40 µmol/kg b.w.) is reported by Waalkes et al.[14] Moreover, Waalkes,[15] in a recent comprehensive review of cadmium-related risks, underlines that the single injection treatment in rats induces a prostate tumor incidence increase below the threshold for significant cadmium-induced testicular toxicity (that is in the dose range for which the hormetic effect has been hypothesized). Last, also the IARC (Monograph Volume 58) reports the significantly increased incidence of prostatic tumors at the 2.5 µmol Cd/kg dose and a significant dose–response trend in the 0–2.5 µmol Cd/kg dose range.[15,16] It seems fairly strange that these results have not been mentioned in the Cadmium hormesis discussion. In conclusion:

- The data produced by the Waalkes et al. experiment indicate a complex pattern of carcinogenic responses, which needs to be considered as a whole.[14]
- As widely discussed in Waalkes et al. study (selected by Calabrese and Baldwin for presenting a possible hormesis "striking" example in the case for testicular tumors), their experimental data also indicate a significantly increasing trend of prostatic tumors in the same dose range where the

TABLE 4. Waalkes et al. (1988) study in male Wistar rats on cadmium chloride for injection[14]

Tumor category	Control	Cadmium chloride, single injection, μmol/kg b.w.					
		1.0	2.5	5.0	10.0	20.0	40.0
Testicular tumors trend: $P < 0.001$	8/45 (18%)	1/30 (3.3%) $P = 0.058S$	3/29 (10%) $P = 0.27S$	3/30 (10%) $P = 0.28N$	4/30 (13%) $P = 0.45N$	21/29 (72%) $P < 0.001$	24/29 (83%) $P < 0.001$
Prostatic tumors trend. n.s. first 3 doses (trend: $P < 0.05$)*	5/44 (11%)	6/27 (22%)	8/26 (31%)	4/28 (14%)	4/23 (17%)	4/26 (15%)	3/29 (10%) $P = 0.046$
Tumors at the injection site (trend: $P < 0.01$)	2/45 (4.4%)	1/30 (3.3)	0/29 (0%)	1/30 (3.3%)	2/30 (6.7%)	1/29 (3.4%)	14/30 (47%) $P < 0.001$
Pancreatic tumors (trend: n.s.)	27/45 (60%)	16/30 (53%)	17/29 (59%)	19/30 (63%)	19/28 (68%)	11/20 (39%)	6/30 (20%) $P = 0.001S$

*As reported by IARC, 1994.
N = decrease of response in comparison with control.

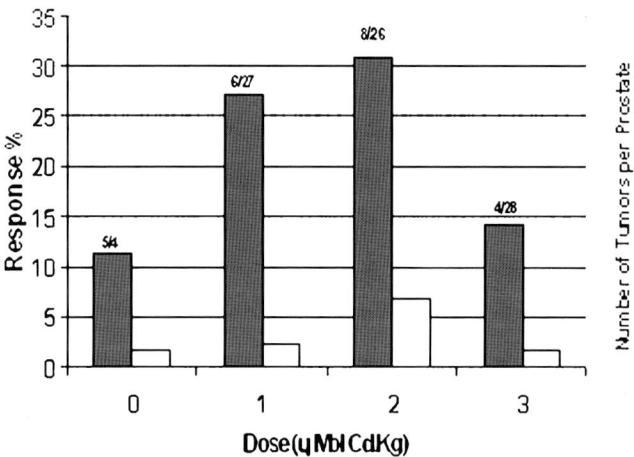

FIGURE 1. Prostatic cancer incidence and number of tumors for prostate (From Ref. 14).

testicular tumor incidences were lower (but not at statistically significant level) than the control response (hormesis hypothesized interval).[14,15]
- The whole above data do not support the hypothesis of a risk reduction for low doses, but rather, suggest a possible risk increase if the prostate tumors are also considered. In fact, the only significant result ($P < 0.05$) reported in the low dose range is an increased incidence of prostatic cancer at the 2.5 μmol/kg b.w. treatment dose.

Na-Saccharin Carcinogenesis

Calabrese and Baldwin underline an hormetic effect suggested by the dose–response relationship for the all tumor incidence in Sprague-Dawley rats (Wisconsin Alumni Research Foundation—WARF Study) and an analogous effect for urinary bladder hyperplasia,[10] based on a study by Food and Drug Administration of 1974 (discussed in a NCI review by Reuber[17]). The considerations presented hereafter are based on the data obtained in these two studies and on the comprehensive review on Saccharin toxicity carried out by the NCI in the same period.[17] The data concerning the total tumor incidence in the WARF study mentioned by Calabrese and Baldwin are reported in TABLE 5, while in TABLE 6 the data by the FDA are reported (not mentioned by Calabrese and Baldwin, even if this study has been considered for bladder hyperplasia). It is worthwhile noticing that the WARF and FDA studies had an analogous experimental design and used the same animals.[17]

The WARF study reports total tumor incidences in female rats at the 0.05% and 0.5% saccharin exposures, that are sensibly lower than the control, even if

TABLE 5. WARF (1974) study in Sprague-Dawley rats on Na-saccharin in the diet[*]

Saccharin dose (% in diet)	Total tumors		
	Male rats	Female rats	Both sexes
0	3/20 (15%)	12/20 (60%)	15/40 (37%)
0.05	2/20 (10%)	6/20 (30%)	8/40 (20%)
		($P = 0.055$N)	($P = 0.068$N)
0.5	2/20 (10%)	9/20 (45%)	11/40 (27%)
5.0	14/20 (70%)	18/20 (90%)	32/40 (80%)
	($P < 0.001$)	($P = 0.032$)	($P < 0.001$)
Trend	$P < 0.005$	$P < 0.01$	$P < 0.005$

[*](from Ref. 10).
N = decrease of response in comparison with control.

at a nonsignificant level (e.g., 6/20 for 0.05% dose versus 12/20 for control, $P = 0.055$), and, in the same dose range, incidences somewhat lower than the control for male rats (2/20 versus 3/20, not significant). The difference among the control response in male and female rats is highly significant (3/20 versus 12/20, $P < 0.01$) (TABLE 5).[17]

On the contrary, the dose relationships by FDA (malignant tumors at all sites) indicate that the control response is never higher than the responses of the treated male and female rats. An evident monotonic increase exist also in the 0–0.1% saccharin dose range for the two genders and, in particular, for the cumulated incidences of the two genders (TABLE 6). The control response reported by the FDA for female rats is much lower (1/27) than the one reported in the WARF study (12/20) (at a highly significant level, 1/27 versus 12/20, $P < 0.001$).[17] Based on the above data, it seems reasonable to assume that the hormesis hypothesis has been essentially founded on a single remarkably high

TABLE 6. FDA (1973) study in Sprague-Dawley rats on Na-saccharine in the diet[*]

Saccharin dose (% in diet)	Malignant tumors at all sites		
	Male rats	Female rats	Both sexes
0	2/29 (7%)	1/27 (4%)	3/56 (5.4%)
0.01	2/28 (7%)	3/30 (10%)	5/58 (8.6%)
0.1	5/29 (14%)	3/32 (9%)	8/61 (13.1%)
1.0	3/28 (17%)	5/32 (16%)	8/60 (13.3%)
		($P = 0.14$)	($P = 0.12$)
5.0	4/24 (17%)	7/29 (24%)	11/53 (20.8%)
		($P = 0.033$)	($P = 0.016$)
7.5	8/26 (27%)	9/32 (28%)	17/58 (29.3%)
	($P = 0.025$)	($P = 0.013$)	($P < 0.001$)
Trend	$P < 0.01$	$P < 0.005$	$P < 0.005$

[*](from Ref. 17).

TABLE 7. FDA (1973) study in Sprague-Dawley rats on Na-saccharine in the diet for bladder hyperplasia*

Saccharin dose (% in diet)	Male rats	Female rats	Both sexes
0	10/73 (14%)	3/85 (4%)	13/158 (8%)
0.01	6/71 (8%)	0/81 (0%)	6/152 (4%)
0.1	4/81 (5%)	0/81 (0%)	4/162 (2.5%)
	($P = 0.053$ NS)		($P < 0.025$ NS)
1.0	4/76 (5%)	3/90 (3%)	7/166 (4%)
5.0	6/64 (9%)	5/88 (6%)	11/152 (7%)
7.5	19/62 (31%)	10/76 (13%)	29/138 (21%)
	($P = 0.089$)	($P < 0.025$)	($P < 0.005$)
Trend	$P < 0.005$	$P < 0.005$	$P < 0.005$

*(from Ref. 10).
N = decrease in response in comparison with control.

control response observed in only one gender (female rats) in the WARF study, but not in male rats in the same study, and not in both genders the FDA study.

The dose-response relationships of urinary bladder hyperplasia resulting from the FDA study are reported in TABLE 7, while the ones of WARF study concerning the lesions of urinary bladder are reported in TABLE 8. In the FDA study, the control incidence is remarkably high in male rats (10/73) and the one of female rats (3/85) is considerably lower (significant difference, 10/73 versus 3/85, $P < 0.05$). When male and female rats are pooled together, the response at the 0.1% of saccharin in the diet is significantly lower than the control (TABLE 7). Also in this case, the hormesis hypothesis is essentially due to the high control incidence resulting in only one gender (male rats). On the contrary, in the WARF study, the hyperplasia incidence was practically absent (only one case in males at each of the two highest doses), so that the whole incidences of bladder hyperplasia and carcinoma are jointly discussed (TABLE 8).[17] In this case, the control responses are null for both male and female rats. Practically no effect is observed in female rats. As a whole, no substantial effect is observed below the 5% dose. In another dose–response relationship, considering only the 0%, 5%, and 7.5% of saccharine in the diet, an increased incidence is found at the highest dose.

Last, effecting "some rough searching" for response increases at low dose in the whole FDA data, as reported by Reuber,[17] significant increases result at the 0.01% dose for lymphosarcomas (male and female rat pooled, 8/14 versus 0/20, i.e., 40% versus 0%, $P < 0.001$) and for mammary carcinoma only in male rats (14/25 versus 6/29, i.e., 56% versus 21%, $P < 0.01$). For both these tumors, these increases were higher than the ones observed at the highest doses, and the trend was not significant, so that these results, presented here only as an example, appeared questionable and have not been used as reference in regulating saccharine exposure; these considerations are in agreement with the methodological approach of the present analysis. In conclusion:

TABLE 8. WARF (1974) study in Sprague-Dawley rats on Na-saccharin in the diet for lesions of urinary bladder (carcinomas plus sporadic hyperplasia)*

Saccharin dose (% in diet)	Male rats	Female rats	Both sexes
0	0/16 (0%)	0/17 (0%)	0/33 (0%)
0.05	0/16 (0%)	1/17 (6%)	1/33 (3%)
0.5	1/15 (7%)	0/15 (0%)	1/30 (3%)
5.0	8/16 (50%)	0/20 (0%)	8/36 (22%)
	($P < 0.01$)		($P < 0.01$)
Trend	$P < 0.01$		$P < 0.01$

*(from Ref. 17).

- If the FDA and WARF studies concerning all tumors are compared, the hypothesis of an hormetic effect, based on the WARF study, is contradicted by the FDA study, indicating a clear monotonic trend and the absence of responses below the control.
- Similar considerations hold for the hypothesis of an hormetic effect for bladder hyperplasia, based on the FDA study, but contradicted by the WARF study, for which an increased hyperplasia incidence exists only at the highest dose, while it is practically absent at low doses. In this latter case, the control incidence is null for both genders, for both hyperplasia and carcinomas of urinary bladder.
- The high control incidence, observed in the two cases for only one gender (female rats for all tumors, male rats for bladder lesions) are the substantial basis for the hormesis hypothesis and result only in one of the two studies.
- One of these two studies has been selected for indicating an hormetic effect for totals tumors and the other for indicating an hormetic effects for hyperplasia, but without any joint evaluation and comparison of the whole data of the two studies.

Neutron-Induced Carcinogenesis

Calabrese and Baldwin discuss three experiments by Broerse *et al.*, proposed as an hormesis example.[10] The incidences of mammary tumors in Sprague-Dawley and Wag/Rij female rats are presented only as percentages (as in the original papers), so that an appropriate statistical evaluation is not possible. In particular, 4 responses lower than the control are reported at the 2 Rads exposure level (12% versus 29%, 15% versus 30%, 15% versus 27%, and 13.3% versus 18.9%), and 2 at the 0.05 Rads exposure level (20% versus 27% and 5.7% versus 18.9%).

The IARC (Monograph Volume No. 75, Table 11) has analyzed in detail the mammary tumors in female rats and mice after exposure to neutrons, and has presented a comprehensive summary of 14 different experimental studies,[18] also including 6 studies by Broerse *et al.*, as jointly reported by the authors in successive publication.[19] The exposure is reported in Gray (Gy). The Broerse *et al.* data presented by the IARC include two cases in which the low-dose responses are lower than the control (Sprague-Dawley rats, 15% at 0.02 Gy, versus 30%, control; and WAG/Rij rats, 20% at 0.05 Gy versus 27%, control), while for other four studies of the same authors no responses lower than the control resulted and the trends were monotonically increasing. In particular, a remarkable response increase emerge even at the lowest dose employed in several experiments (i.e., Sprague-Dawley rats, 40% at 0.05 Gy versus 30%, control; BN/Bi rats, 11% at 0.05 Gy versus 8%, control; WAG/Rij 35% at 0.12 Gy versus 27%, control; BN/Bi rats, 22% at 0.15 Gy versus 8%, control). Even if the data considered by Calabrese and Baldwin and by the IARC are only in part the same, the IARC has based its evaluation on a comprehensive publication by Broerse *et al.*,[19] in which their experiments were reviewed (published after the ones mentioned by Calabrese and Baldwin). It is important to underline, in any case, that the IARC summary has been predisposed for scopes fully independent from an hormesis screening, so that it may be considered a neutral and reliable comprehensive data source. Last, as far as Broerse *et al.* studies are concerned, it is worthwhile mentioning that the authors specify that "In general, linear dose–response curves have been observed for mammary tumourigenesis in three rat strains for both X-rays and fast neutrons" (p. 1)[19] which is obviously different from a U–shaped low-dose trend.

The IARC summary also reports a study showing a significant response increase at low doses [Vogel and Zaldivar, 1972, Sprague-Dawley rats, 21/27 [78%] at 0.05 Gy versus 43/89 (48%), control, $P < 0.05$ and 29/34 (85%) at 0.10–0.12 Gy versus 43/89 (48%), control, $P < 0.005$], and a study indicating a substantially linear and significant dose–response trend together with a response significant increase at high doses [Shellabarger, 1976, Sprague-Dawley rats, 20/167 (12%) control, 28/182 (15%) at 0.01 Gy, 16/89 (18%) at 0.04 Gy, 21/68 (31%) at 0.16 Gy, $P < 0.005$ and 26/45 (58%) at 0.54 Gy, $P < 0.005$].[17] The dose range, reported by Calabrese and Baldwin for an hormetic effect,[10] is included in the dose range examined in this latter study. The remaining five reported studies, whose treatment doses are, however, higher than the ones above discussed (over the hypothesized hormetic effect range) do not indicate responses lower than the control, except for one case for which there is a small, insignificant decrease [Montour *et al.*, 1977, 0/30 (0%) at 0.25 Gy versus 2/31 (6.5%), control, $P = 0.25$].[18]

As a whole, these data do not support an hormetic effect at low neutron doses and, rather, suggest a non-negligible risk at these exposures.

The recently discussed aspects of low-dose radiation risk are very complex and include various adverse effects together with some beneficial effects (e.g., the low-dose induced effects in nonirradiated cells, such as nontargeted induced mutations, enhanced cell growth, genomic instability, neoplastic transformation, apoptosis and removal of damaged cells, adaptation, and others, mostly considered within the "bystander effect" paradigm). Research is in progress on these topics, as well as their whole joint evaluation. The complexity of this topic implies that more than a simple data review of some animal experiments is necessary for appropriate and scientifically founded conclusions.

However, it is worthwhile underlining that recent epidemiological results confirm the presence of non-negligible risks also at very low doses of radiation. In particular, the retrospective cohort study of cancer risk of ionizing radiation in 15 countries (407,391 workers), coordinated by the IARC, indicates that "there is a small excess risk of cancer, even at the low doses and dose rates typically received by nuclear workers in this study."[20] The results of the study include that "The excess relative risk for cancers other than leukemia was 0.97 per Sv, 95% confidence interval 0.14 to 1.97," that "The excess relative risk for leukaemia, excluding chronic lymphocyitic leukaemia was 1.93 per Sv (<0 to 8.47) and that, "On the basis of these estimates, 1–2% of death from cancer among the workers in this cohort may be attributable to radiation" (p. 1).[20] When discussing these results in the light of the current recommendations of the International Commission on Radiological Protection (ICRP) that limits occupational doses to 100 mSv (no more than 50 mSv per year) and 1 mSv per year for the general population, it is observed that "less than 5% of workers in this study received cumulative doses of the order of 100 mSv over their entire career," mostly in the early years when the protection criteria were less stringent than today.

The conclusions of this study are of main importance for the topics of the present paper. First, the criteria on which current risk assessment and protection are based (including a linear and linear quadratic trend for low doses) clearly appear to have not been excessively strict in this case, and, presumably, even not sufficiently strict. Second, these results have been obtained from an extremely large and highly representative sample of human beings (more than 400,000 workers), with a very high statistical power, allowing the identification of effects otherwise not evident. Finally, these data should attentively taken into account when discussing the "excessive" conservativeness level of current ionizing radiation risk assessment and management procedures.

SOME CONCLUSIONS

A brief review of a sample of the experimental results for which an hormetic effect in carcinogenesis has been hypothesized has been here carried out, with the aim of verifying such hypotheses in the light of the other available and relevant experimental (and in one case, also epidemiological) results. In order to prevent a selection bias, the data available from general relevant reviews, effected for other scopes than hormesis searching, were consulted (e.g., by IARC, WHO, U.S. NTP, U.S. NCI, and in general reviews). In one case (cadmium carcinogenesis), the analysis was essentially limited to the single study that has been selected for showing an hormetic effect, because this effect resulted not compatible with other data and with the evaluations presented in the same study.

This examined sample of experimental results is evidently very limited; the discussed cases, however, that mostly deal with important risk agents (except saccharine), appear informative and useful for pointing out some methodological problems, already underlined by other authors.

In general, the hormesis hypotheses were not confirmed, and, rather, they were contradicted by the joint consideration of other relevant data. Therefore, the present study largely confirms the observations and criteria proposed by Crump and by Kitchin and Drane concerning the *post hoc* literature searching for hormesis, and, in particular, the need of investigating where and when hormesis does not result and of verifying its prevalence, as well as the need of statistical criteria appropriate for preventing false positives in *post hoc* analyses, and, last, the need of further studies specifically designed for investigating hormesis possible mechanisms (*propter hoc*).[3,5]

Because of the frequent lack of statistical significance of the proposed hormetic effects above examined, the statistical criterion reported by Crump was not used[3]; it is clear, however, that its application would have provided further support to the absence of a really convincing hormesis evidence in the above examples.

As a last comment, it is important to compare the criteria adopted for hormesis identification with the extremely complex, complete, and detailed work carried out by international and national agencies and institution for the classification of carcinogens and the regulation of carcinogen exposure and risk. It is, for instance, sufficient to give a look to the IARC monographs for immediately ascertaining how many relevant experimental and epidemiological studies and data are regularly examined, and pointing out the thoroughly investigation of both positive and negative results effected by multidisciplinary expert commissions. The extension of these procedures to hormesis investigation seems highly appropriate and necessary; in particular in *post hoc* literature searching; certainly this will not imply the "overwhelming financial burden" for hormesis verification, feared by Calabrese and Cook,[4] but only some more work, aimed at examining all the relevant data.

Finally, based on the above discussion, the statement that "by dismissing hormesis, regulatory agencies such as U.S. EPA deny the public the opportunity for optimal health and avoidance of diseases" (p. 446)[2] seems unacceptable. Rather, the opposite appears more appropriate.

REFERENCES

1. CALABRESE, E.J. 2005. Paradigm lost, paradigm found. The re-emergence of hormesis as a fundamental dose response model in the toxicological sciences. Environ. Pollut. **138:** 378–411.

2. CALABRESE, E.J. & L.A. BALDWIN. 1998. Hormesis as a default parameter in RfD derivation. Hum. Exp. Toxicol. **17:** 444–447.
3. CRUMP, K. 2001. Evaluating the evidence for hormesis: a statistical perspective. Crit. Rev. Toxicol. **31:** 669–679.
4. CALABRESE, E.J. & R.R. COOK. 2005. Hormesis: how it could affect the risk assessment process. Hum. Exp. Toxicol. **24:** 265–270.
5. KITCHIN, K.T. & J.W. DRANE. 2005. A critique of the use of hormesis in risk assessment. Hum. Exp. Toxicol. **24:** 249–253.
6. FINKEL, A.M. 1995. A quantitative estimate of the variations in human susceptibility to cancer and its implications for risk management. *In* Low-Dose Extrapolation of Cancer Risks. Issues and Perspectives. S. OLIN *et al.*, Eds.: 297–328. ILSI (International Life Science Institute) Press. Washington, DC.
7. BERWICK, M. & P. VINEIS. 2000. Markers of DNA repair and susceptibility to cancer in humans: an epidemiological review. J. Natl. Cancer Inst. **92:** 874–897.
8. ARMITAGE, P. & G. BERRY. 1994. Statistical Methods in Medical Research. Blackwell Scientific Publication Ltd. Oxford.
9. HASEMAN, J.K. 1984. Statistical issues in the design, analysis and interpretation of animal carcinogenicity studies. Env. Health Perspect. **58:** 385–392.
10. CALABRESE, E.J. & L.A. BALDWIN. 1998. Can the concept of hormesis be generalized to carcinogenesis? Regul. Toxicol. Pharmacol. **28:** 230–241.
11. KOCIBA, R.J., D.G. KEYES, J.E. BEYER, *et al.* 1978. Results of a 2-year chronic toxicity and oncogenicity study of 2,3,7,8-Tertrachlorodibenzo-p-dioxin in rats. Toxicol. Appl. Pharmacol. **46:** 279–303.
12. NTP (National Toxicology Program). 1982. Bioassay of dibenzo-p-dioxin for possible carcinogenicity (gavage study). Tech. Rept. Sr. No. 201. Research Triangle Park, NC: US DHHS,PHS.
13. IARC (International Agency for Research on Cancer) Monograph Volume 69. 1997. Polychlorinated dibenzo-*para*-dioxins and polichlorinated dibenzofurans. IARC. Lyon.
14. WAALKES, M.P., S. REHEM, C.W. RIGGS, *et al.* 1988. Cadmium carcinogenesis in male Wistar (Crl.(WI) BR) rats: dose–response analysis of tumor induction in the prostate and testes and at the injection site. Cancer Res. **48:** 4656–4663.
15. WAALKES, M.P. 2003. Cadmium carcinogenesis. Mutat. Res. **533:** 107–120.
16. IARC (International Agency for Research on Cancer) Monograph Volume **58**. 1994. Berillium, cadmium, mercury, and exposures in the glass manufacturing industry. IARC. Lyon.
17. REUBER, M.D. 1978. Carcinogenicity of saccharin. Environ. Health Perspect. **25:** 173–200.
18. IARC (International Agency for Research on Cancer) Monograph Volume **75**. Ionizing radiation, Part 1: X- and Gamma (γ) radiation, and neutrons. IARC. Lyon.
19. BROERSE, J.J., L.A. HENNEN & H.A. SOLLEVELD. 1986. Actuarial analysis of the hazard for mammary carcinogenesis in different rat strains after X- and neutron irradiation. Rev. Leuk. Res. **10:** 749–754.
20. CARDIS, E., M. VRIJHEID, M. BLETTNER & E. GILBERT *et al.* 2005. Risk of cancer after low doses of ionising radiation: retrospective cohort study in 15 countries. (on line) BMJ doi:10.1136/bmj. 38499.599861. EO (published 29 June 2005).

Occupational Injury and Illness Meet the Labor Market

Lessons from Labor Economics about Lost Earnings

LESLIE I. BODEN

Department of Environmental Health, Boston University School of Public Health, Boston, Massachusetts 02118, USA

ABSTRACT: Recent labor economics studies in the United States and Canada have demonstrated that occupational injuries and illnesses often lead to substantial lost earnings for workers and their families. Other studies have shown substantial long-term lost earnings attributable to large-scale layoffs, where no health impairment has taken place. This article uses evidence from these and other studies of apparently different situations to draw inferences about how managers' actions and public policy choices can affect the costs of occupational injuries and illnesses. Although primary prevention remains the policy of choice, reduction in the impact of workplace injuries and illnesses can decrease the costs of these events and can provide substantial benefits. This article proposes two hypotheses and discusses the evidence for each: (*a*) Loss of the job held at the onset of illness or injury increases time off work and exacerbates workers' lost earnings. (*b*) Workers' losses may be substantially reduced by policies that encourage employers to rehire people recovering from or disabled by workplace injuries and illnesses.

KEYWORDS: disability; injuries; occupational diseases; cost of illness; displaced workers

INTRODUCTION

Every year, millions of people incur occupational injuries and illnesses. Many are minor and result in little or no time lost from work. For these injuries, no lost earnings result. For other injuries, workers may need more substantial medical treatment, may not be able to return to work immediately, or may have work limitations that make them less productive. Because of this reduced

Address for correspondence: Leslie I. Boden, Ph.D., Department of Environmental Health, Boston University School of Public Health, 715 Albany St. TE-221, Boston, MA 02118. Voice: 617-638-4635; fax: 617-638-4857.
  e-mail: lboden@bu.edu

productivity, their postinjury wages may be lower than they would have been, they may have to work fewer hours, and they may incur more nonwork spells. These consequences of workplace injuries and illnesses can last a few days or a lifetime.

In large part, lost earnings are directly linked to the health consequences of injury or illness. But evidence suggests that workers with occupational injuries and illnesses may continue to incur lost earnings even if they have returned to their preinjury health status. In addition, among those who have long-term health limitations, labor market factors may affect the amount of earnings lost. In others, the medical aspects of workplace injuries may play a minor role in determining economic outcomes for workers and their families.

In this article, I will bring together some ideas from labor economics and data from studies by labor economists to shed light on how occupational injuries and illnesses affect workers' earnings. I will first present estimates of the extent of lost earnings caused by occupational injuries and illnesses. I will then briefly describe how economists think about the factors determining the size of these losses. In particular, I will focus on how job loss can increase the magnitude of the impact of health limitations on earnings. Finally, I will describe some policies that could improve economic outcomes for workers with occupational injuries or illnesses.

Lost Earnings Caused by Occupational Injuries and Illnesses

Research that uses modern statistical methods to estimate the labor market impacts of occupational injuries and illnesses is in its infancy. The first of these studies was published in 1998,[1] and a small number have been completed since. A consistent story emerges from these studies: many injured workers suffer substantial lost earnings.

Lost earnings are actual earnings minus what would have been earned if the injury or illness had not occurred. FIGURES 1 and 2, adapted from Reville,[2] display a conceptual model of lost earnings. FIGURE 1 shows earnings increases over time prior to injury or illness. If the worker remains in good health, earnings will continue to rise, as shown by the dashed line. However, after the onset of injury or of illness, work stops during recovery and earnings are zero until the worker begins to work again. At this point, if wages return to the original earnings path (indicated by the dashed line) the worker has incurred a temporary total disability. The shaded area measures lost earnings.

Some workers never return to the original earnings path. In this case, they have permanent disabilities, as is shown in FIGURE 2, where the shaded area of lost earnings continues indefinitely. FIGURE 2 does not tell us whether workers with long-term lost earnings have long-term health problems, but only whether they have returned to the original earnings path.

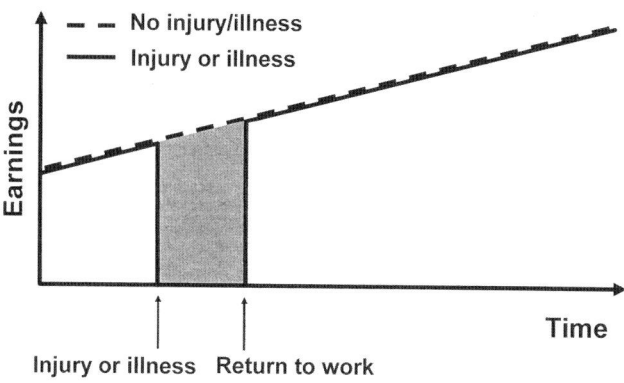

FIGURE 1. A conceptual model of temporary injury-related losses.

If we could observe what the earnings of these workers would have been in the absence of injury or illness, then we could simply subtract them from actual earnings to determine how much was lost. However, a worker is either injured or uninjured at a moment in time. If we observe earnings after onset of an injury or illness (hereafter referred to as injury), we cannot know with certainty what their earnings would have been had they not been injured. We must find a way to estimate counterfactual earnings from another source.

Recent studies have estimated lost earnings of injured workers by identifying workers who are similar to the injured workers in all other observable respects but who were not injured. These recent studies use two methods to estimate uninjured earnings: matching and regression. The matching approach uses a comparison group of uninjured workers and matches each injured worker to one or more uninjured workers with similar relevant characteristics in the immediate preinjury period. Lost earnings are then the difference between the mean

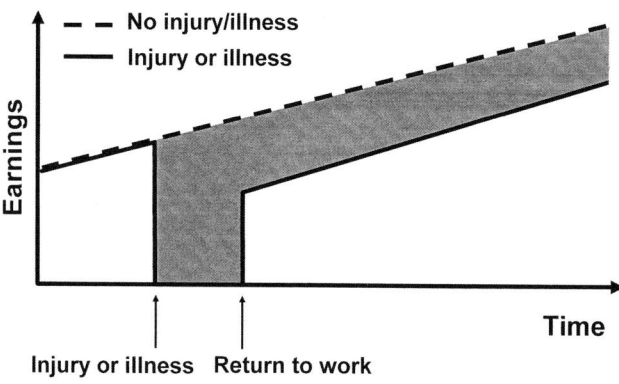

FIGURE 2. A conceptual model of permanent injury-related losses.

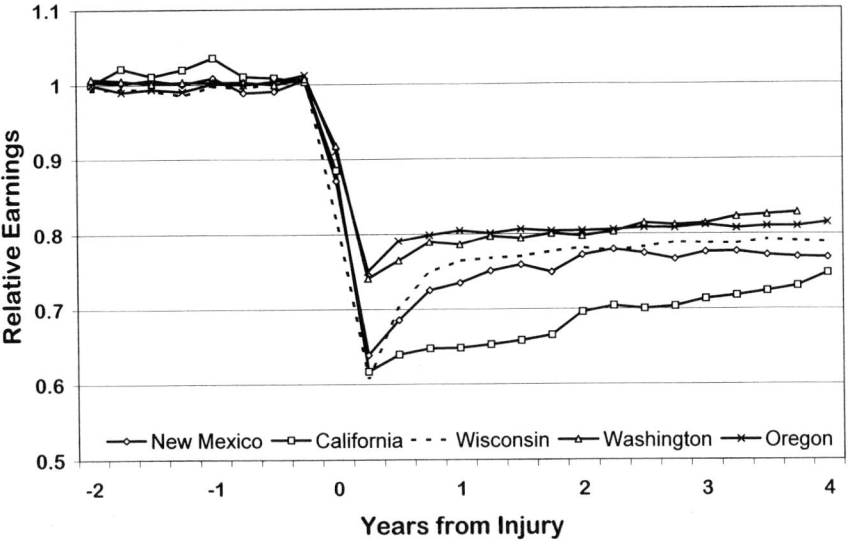

FIGURE 3. Earnings of workers with occupational injuries and illnesses relative to matched uninjured workers: five U.S. states. (From Reville et al.[3])

earnings of injured workers and comparable uninjured workers. Similarly, the regression approach compares regression-adjusted mean earnings between injured and uninjured workers to determine the extent of lost earnings.

Using the matching approach, the most comprehensive study of permanently disabling workplace injuries and illnesses to date[3] estimated lost earnings over a 10-year period. Different states define permanent disability in somewhat different ways, but permanently disabling injuries generally involve a functional impairment from which the worker does not fully recover and that leads to a long-term loss in earnings. Injuries are categorized as temporarily disabling if the worker fully recovers from the occupational injury or illness. In the five U.S. states studied, between 18% and 41% of workers with lost-time injuries are categorized as permanently disabled. The study found that these injured workers incur a substantial initial decline in earnings relative to the comparison group, followed by a period of recovery (FIG. 3). Despite some recovery of earnings, losses remain substantial even 4 years after injury. Average losses for workers in individual states range from 16% to 25% of earnings.

An earlier study of all injuries and illnesses involving at least 1 week of lost work[4] used a regression approach and found 10-year losses of about 5% of earnings. This figure includes many workers with short-term disabilities. However, for the 35% of injured workers losing at least 8 weeks work or receiving permanent disability benefits, lost earnings are much higher—about

18% (author's calculations). A recent study of permanently disabled workers in Ontario estimated even greater losses.[5] On average, Ontario workers with permanent partial disability (PPD) benefits have losses averaging about 40% of preinjury earnings. Like similar workers in the United States, substantial losses persist through the follow-up period of up to 10 years. The higher proportionate losses in Ontario may reflect the fact that only about 10% of workers with lost-time injuries in Ontario receive PPD benefits. Because a much larger proportion of U.S. injured workers obtain PPD benefits, recipients with less severe injuries might not qualify for these benefits if they were injured in Ontario. Ontario PPD cases would tend to be more severe on average, leading to higher average losses.

These North American studies only include a small number of people with chronic occupational illnesses, because these conditions rarely enter the workers' compensation systems in the United States and Canada. However, there are some qualitative similarities between people with chronic occupational diseases and injured workers with permanently disabling conditions. First, many chronic occupational diseases cause long-term functional impairment. Second, workers with chronic occupational diseases tend to be older and have greater job tenure than the average injured worker. The same is true of workers receiving permanent disability benefits. Of course, there are substantial differences as well. As a consequence, we do not know whether the lost earnings of workers with chronic occupational illnesses would be similar to those described in this article.

Factors Influencing Injured Workers' Earnings

Labor markets consist of workers who supply labor, employers who demand labor, and institutions (including laws and regulations) that provide the framework within which wages are determined and jobs and workers are matched. Economists think of wages as largely reflecting workers' productivity. A simplified model of employer behavior leaves out some important considerations, but in many situations is a useful device. It begins with the assumption that employers compete to hire workers and are willing to pay higher wages to more productive workers. Productivity can be affected by education, training, talent, skills, health, motivation, and the match between workers and their jobs. The first six of these are aspects of what economists call human capital.

Anything that diminishes human capital or interferes with its use typically reduces earnings. Workplace injuries can do this in several ways. First, and most obviously, they can impair health and cause functional limitations that directly affect ability to work. These limitations can take many forms. Poor health can cause workers to be off work, can limit their ability to work fulltime, can cause them to be less effective while performing work tasks, can keep them from doing some tasks, and so on. In addition, poor health can distract from the ability to think clearly, focus on job demands, and communicate with others.[6] Poor health can also make it more difficult and costly to maintain and improve skills and can decrease the number of jobs available to apply these skills. The higher cost and lower value of skills can lead employers to invest less to acquire such workers.

The injury and events following the injury can affect workers' employment and earnings over and above the direct impacts on health and thereby on productivity. When injured workers lose much time from work during recovery, employers may incur additional costs by hiring temporary workers or paying overtime. In this case, employers may choose to hire permanent replacements, causing the injured workers to lose their jobs. Reville et al.[3] look at New Mexico workers with PPD cases who have returned to work. These injured workers are 10% less likely to be working for the at-injury employer in the years after injury than matched uninjured workers. Similarly, Galizzi and Boden[7] find that the first postinjury job was for a new employer for 17% of workers with lost-time injuries in Wisconsin. In Oregon, 90% of workers with short-term injuries return to the employer, compared to only 75% of those with PPD benefits (author's calculations).

Factors other than reduced productivity can affect employers' willingness to hire and retain workers and the wages offered to them. A workplace injury may be seen, often incorrectly, as an indication that a worker is "injury-prone." Alternatively, the employer may be suspicious that people filing workers' compensation claims are malingering. Thus, some employers will treat a workplace injury as an indicator of a problem employee. To the extent that this happens, it will reduce the willingness of the at-injury employer to retain the injured worker and will reduce wage offers by other potential employers, limiting future employment and earnings.

To gain insight into the impact of job loss on employment and earnings, we turn to studies of workers who lost their jobs as a result of plant closings or layoffs. As the next section describes, job loss unrelated to poor health or disability can lead to substantial lost earnings. When job loss follows occupational injury or illness, the economic consequences are likely to be magnified.

Impact of Job Loss on Earnings

In research done over the last 30 years, economists have learned a great deal about how job loss affects earnings. Much of this research has focused on displaced workers in the United States. By definition, displaced workers have lost their jobs because of large-scale layoffs or plant closings. They have not been fired, which might indicate that they were less productive than other workers; and they have not quit, which might occur because they had found better jobs. There is no reason to think that displaced workers are less healthy than nondisplaced workers with similar nonhealth characteristics although this issue has not been studied. Even so, displaced workers represent the best opportunity to study the pure effect of job loss on earnings. A plant closing is, for these purposes, a "natural experiment."

The impact of displacement on earnings is the difference between actual postdisplacement earnings and what the same workers would have earned had they not been displaced. The methods used to estimate counterfactual earnings are parallel to those used to measure earnings lost as a consequence of workplace injuries: researchers have typically used the earnings of nondisplaced workers with similar characteristics to estimate counterfactual earnings. Disparities in observed worker characteristics between displaced and nondisplaced workers are handled either by matching or regression. In the United States, most displaced workers have a substantial period of nonemployment after displacement. During this time, they engage in job search, which can take many months. In some cases, they look for a while and then stop looking, but in most cases they find another job. In the United States, the Displaced Worker Survey (DWS) gathers data from workers who have been displaced during the 3 years prior to interview. Depending on the interview year, between 60% and 75% of respondents are employed at the interview date.[8] Studies of displaced workers in France and Germany suggest that displaced workers are less likely to experience nonemployment than their U.S. counterparts, possibly because the United States has less stringent requirements to notify workers of impending plant closures or mass layoffs. Of those who have a period of nonemployment, employment rates after 12 months are only 55% in France and 60% in Germany.[9]

A study of long-tenure Pennsylvania workers in the 1980s found that displacement-related losses were long lasting. Six years after displacement, workers were earning an average of about 25% of predisplacement earnings.[10] The time profile of postdisplacement losses is very similar to those of injured workers with PPD benefits, as can be seen by comparing FIGURES 3 and 4. Another study[11] found similar long-term losses—between 17% and 25%—among workers displaced from jobs in California's durable goods manufacturing industries during the early 1990s. Farber[8] examined the losses of displaced workers who had full-time employment both before displacement and when they were interviewed. Overall, despite being fully employed, their lost earnings averaged 17% at interview—an average of 2 years after displacement.

There have been a few non-U.S. studies of displacement. A study of postwar West German workers displaced in 1988–1996[12] found lower losses than in the United States. In the year of displacement, losses averaged 13.5% of predisplacement earnings but shrank to 6.5% after 2 years. A study of Swedish displaced workers[13] looked only at employment, not lost earnings, and found an employment decline relative to matched controls of 4%–6% that continued for the full 13 years of follow-up. This is comparable to findings about employment in the United States, where, except for workers over 55 years of age, most

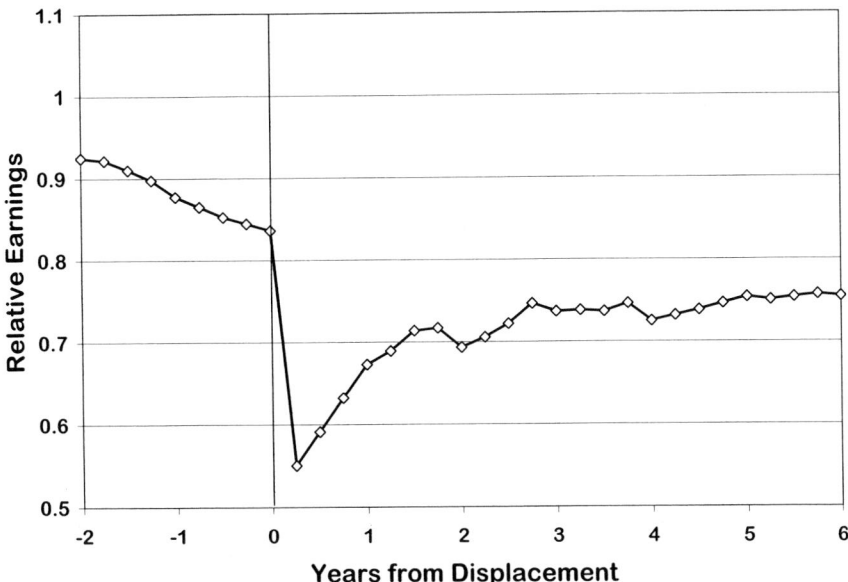

FIGURE 4. Earnings of displaced workers relative to matched non-displaced workers. (Adapted from Jacobson et al.[10])

long-term losses derive from lower postdisplacement wages rather than continuing employment declines.[14–16] A study of Slovenian workers displaced in the early 1990s found a much larger impact in this transition economy than in Western Europe or the United States.[17] Two years after displacement, only one-third of the displaced Slovenian workers had jobs. Those who found jobs faced an average reduction in earnings of 70% compared to otherwise similar nondisplaced workers. The higher costs of displacement may have been related to less-developed labor market and a fairly high unemployment rate.

The overall message from these studies is clear: losing a job puts many workers at risk of a substantial period of nonemployment and, after that, a long-term risk of reduced earnings. What happens when this job loss occurs after a substantial period off work because of an occupational injury or illness?

Displaced Workers: Uninjured and Injured

Workers who lose their jobs because of mass layoffs or injuries have some important features in common. First, to find another job, workers just engage in job search. It may take many months before an acceptable job is found. Second, a portion of the skills and knowledge built up at the old job may not add to productivity at the new job, so the portion of wages associated with that skills and knowledge is lost. Also, if the time off work is long, the worker is not acquiring new skills and may see some old valuable skills and work habits decline. All of these factors will tend to reduce earnings.

Injured workers who lose their at-injury jobs may face problems not encountered by displaced workers. Workers experiencing a long recovery period may not be able to engage in a comprehensive job search, so the period off work will be greater than for otherwise comparable displaced workers. To the extent that their injuries have lasting consequences, they suffer productivity losses over and above the loss of specific human capital incurred because of job loss. We would also expect a longer period between job loss and employment for injured workers because, unlike many displaced workers, injured workers do not receive prior notice of the impending event.

There is limited evidence about the impact of job loss on injured workers. This research is particularly difficult because job loss and lost earnings are both affected by the severity and duration of injury-related health conditions. Because the data typically used by economists to estimate lost earnings lack good measures of injury severity, researchers cannot tell whether job loss caused subsequent reductions in earnings or whether injury severity caused both the job loss and earnings reductions. Galizzi and Boden[7] estimated the impact of job loss on injured workers, but data lacked direct measures of injury severity. To deal with the bias induced by unobserved severity, they used an instrumental variables approach to determine the impact of job loss on duration off work after injury. They found that loss of the pre-employment job dramatically increased durations off work. Reville et al.[18] used a different approach. To control for severity, they stratified injured workers with PPD benefits by their medical disability assessments, which were summarized as percentage disability ratings. They then compared lost earnings of workers who were employed with the at-injury employer 1 year post injury with all those in the same medical disability rating. They found substantially smaller losses—for most groups 30%–40% less—among those who were back to work with the at-injury employer (FIG. 5).

Impact of Workplace Accommodations

Given the value of maintaining employment continuity, a logical question is: What can be done to increase the proportion of injured workers who return to the at-injury job? There is evidence that better medical care and tighter coordination between medical providers and employers reduces time off work and may thus increase the probability of return to the at-injury job.[19,20] Another effective policy is employer accommodation for disabled workers. Burkhauser et al.[21] studied the impact of employer accommodations on job tenure after the onset of a disabling health condition. They estimate that offering job accommodations increases postdisability job duration for the average disabled worker from 2.6 to 7.5 years. Using data from the 1978 Survey of Disability

FIGURE 5. Three-year lost earnings of injured workers by disability category and return to the at-injury employer, California. (Adapted from Reville et al.[18])

and Work and the 1992 Health and Retirement Study (HRS), Burkhauser and Butler[22] find that employer accommodations for disabled men significantly increase the time between the onset of disability and application for Social Security Disability Insurance (SSDI). Based on the HRS, employer accommodations would reduce the number of SSDI applications in the 5 years after onset of disability by about 30%. Krause et al.,[23] in a review of studies of the effectiveness of workplace accommodations for injured workers, concludes that modified work programs tend to reduce lost workdays by about half. While appropriate accommodations allow some injured workers to return to the at-injury job without undue risk, not all injured workers can or should return to the at-injury job. Even with accommodations, job demands or risk at that job may be excessive.

DISCUSSION

This article, based largely on reviewing studies by labor economists, has shown that injury-related lost earnings are substantial. It has also presented evidence that lost earnings are exacerbated when workers do not return to their at-injury jobs. Finally, it has indicated that workplace accommodations are effective in increasing the probability that workers will return to work for the at-injury employer.

There are a variety of reasons that some injured workers are offered accommodations and others are not. In some employment situations, accommodations are infeasible or impractical either because of the nature of the job or the skills and motivation of the injured worker. In others, employer policies offer limited or no support for measures to re-employ workers with health limitations. Work accommodations may or may not be supported by laws and regulations. Despite the passage of the Americans with Disabilities Act, U.S. law tends to offer few incentives for the rehiring of workers by the at-injury employer.[24] Some U.S. states have put programs into place that provide financial incentives for providing job accommodations for injured workers,[25] but these states are in the minority.

Lost earnings and the return-to-work of injured workers have largely been studied in the North American context. Studies of displaced workers in other countries appearing in a recent book[26] cover only a year or two after injury but still provide interesting insights. These studies show little or no wage decline (relative to nondisplaced workers) among displaced workers in France and Germany. Larger declines are documented in the United States, Canada, and the UK. The important difference between these two groups is probably not language, but a combination of a narrower wage distribution and stronger employment protection laws in France and Germany. For countries with less wage disparity, greater employment protections, and more comprehensive social insurance programs, concerns over job loss may be of less importance. However, many countries have less well-developed labor market institutions and limited social insurance programs. For them, the North American findings presented here may well be applicable.

Finally, although lost earnings are the focus of this article, they are not the only important consequence of occupational injuries and illnesses. Injuries and illnesses may also cause a substantial decline in the quality of life of workers and their families,[27,28] may lead to the breakup of marriages,[29] and may have other important economic and noneconomic impacts that are outside the scope of this article.

REFERENCES

1. PETERSON, M.A., R.T. REVILLE, R.K. STERN, *et al.* 1998. Compensating Permanent Workplace Injuries: a Study of California's System. RAND. Santa Monica, CA.
2. REVILLE, R.T. 1999. The impact of a permanently disabling workplace injury on labor force participation and earnings. *In* The Creation and Analysis of Linked Employer-Employee Data, Contributions to Economic Analysis. J.C. Haltiwanger, *et al.* Eds.: 147–174. Elsevier Science. Amsterdam.

3. REVILLE, R.T., L.I. BODEN, J. BIDDLE, et al. 2001. New Mexico Workers' Compensation Permanent Partial Disability and Return-to-Work: an Evaluation. RAND. Santa Monica, CA.
4. BODEN, L.I. & M. GALIZZI. 1999. Economic consequences of workplace injuries and illnesses: lost earnings and benefit adequacy. Am. J. Ind. Med. **36:** 487–503.
5. TOMPA, E., C. MUSTARD, S. SINCLAIR, et al. 2003. Post-accident earnings and benefits adequacy and equity of Ontario workers sustaining permanent impairments from workplace accidents. Working paper #210. Institute for Work and Health. Toronto, ON.
6. LERNER, D., B.C. AMICK III, W.H. ROGERS, et al. 2001. The work limitations questionnaire. Med. Care. **39:** 72–85.
7. GALIZZI, M. & L.I. BODEN. 2003. The return to work of injured workers: new evidence from matched unemployment insurance and workers' compensation data. Labour Econ. **10:** 311–337.
8. FARBER, H. 2003. Job loss in the United States, 1981–2001. NBER Working Paper #9707. National Bureau of Economic Research. Cambridge, MA.
9. KUHN, P.J. 2002. Summary and synthesis. *In* Losing Work, Moving On. P.J. Kuhn Ed.: 1–104 W.E. Upjohn. Kalamazoo, MI.
10. JACOBSON, L.S., R.J. LALONDE & D.G. SULLIVAN. 1993. Earnings losses of displaced workers. Amer. Econ. Rev. **83:** 685–709.
11. SCHOENI, R. & M. DARDIA. 1996. Wage losses of displaced workers in the 1990s. RAND labor and population program working paper series 96–14. RAND. Santa Monica, CA.
12. COUCH, K.A. 2001. Earnings losses and unemployment of displaced workers in Germany. Industrial Labor Rel. Rev. **54:** 559–572.
13. ELIASON, M. & D. STORRIE. 2003. The echo of job displacement. William Davidson Institute Working Paper 618. University of Michigan Business School. Ann Arbor, MI.
14. RUHM, C. 1991. Are workers permanently scarred by job displacement? Amer. Econ. Rev. **81:** 319–324.
15. KLETZER, L.G. & R.W. FAIRLIE. 2003. The long-term costs of job displacement for young adult workers. Industrial Labor Rel. Rev. **56:** 682–698.
16. CHAN, S. & A.H. STEVENS. 2001. Job loss and employment patterns of older workers. J. Labor Econ. **19:** 484–521.
17. ORAZEM, P., M. VODOPIVEC & R. WU. 2004. Worker displacement during the transition: experience from Slovenia. IZA Discussion Paper No. 1297. Bonn.
18. REVILLE, R., T. SEABURY, S.A. NEUHAUSER, et al. 2005. An Evaluation of California's Permanent Disability Rating System. RAND. Santa Monica, CA.
19. KRAUSE, N., J.W. FRANK, L.K. DASINGER, et al. 2001. Determinants of duration of disability and return-to-work after work-related injury and illness: challenges for future research. Am. J. Ind. Med. **40:** 464–484.
20. HABECK, R.V., A.H. HUNT & B. VAN TOL. 1998. Workplace factors associated with preventing and managing work disability. Rehab. Counsel. Bull. **42:** 98–143.
21. BURKHAUSER, R.V., J. BUTLER & Y. KIM. 1995. The importance of employer accommodation on the job duration of workers with disabilities: a hazard model approach. Labour Econ. **2:** 109–130.
22. BURKHAUSER, R.V. & J.S. BUTLER. 1999. The importance of accommodation on the timing of disability insurance applications. J. Hum. Res. **34:** 589–611.
23. KRAUSE, N., L.K. DASINGER, F. NEUHAUSER. 1998. Modified work and return to work: a review of the literature. J. Occup. Rehab. **8:** 113–139.

24. RABINOWITZ, R. Ed. 2002. Occupational Safety and Health Law, 2nd ed. BNA Washington D.C.
25. GALIZZI, M. & L.I. BODEN. 1996. What Are the Most Important Factors Shaping Return to Work? Evidence from Wisconsin. WCRI Cambridge, MA.
26. KUHN, P.J. Ed. 2002. Losing Work, Moving On. W.E. Upjohn. Kalamazoo, MI.
27. STRUNIN, L. & L.I. BODEN. 2003. Family consequences of chronic back pain. Soc. Sci. Med. **58:** 1385–1393.
28. BODEN, L.I. 2005. Running on empty: families, time, and disabling conditions. Am. J. Public Health. **95:** 1894–1897.
29. DEMBE, A.E. 2005. The effect of occupational injuries and illnesses on families. *In* Work, Family, Health and Well-Being. S.M. Bianchi, L.M. Casper & R.B. King, Eds.: 397–411. Lawrence Erlbaum. Mahwah, NJ.

The Economic Costs of Health Service Treatments for Asbestos-Related Mesothelioma Deaths

ANDREW WATTERSON,[a,b] TOMMY GORMAN,[a,c] CARI MALCOLM,[a,b] MAVIS ROBINSON,[d] AND MATTHIAS BECK[e]

[a]*Occupational and Environmental Health Research Group, University of Stirling, Stirling, FK9 4LA Scotland, United Kingdom*

[b]*Public Health Research Group, Department of Nursing and Midwifery, University of Stirling, Stirling, FK9 4LA Scotland, United Kingdom*

[c]*West Dunbartonshire Council: Welfare Rights Representation Unit, West Dunbartonshire, G82 3PU Scotland, United Kingdom*

[d]*Nurse advisor formerly with the Macmillan Mesothelioma Information Project, LS1 3EB Leeds, England, United Kingdom*

[e]*Professor of Public Sector Management, Department of Management Studies, York University, YO10 5DD York, England, United Kingdom*

ABSTRACT: This article explores the complex and neglected picture of occupational and environmental disease healthcare costs specifically relating to asbestos. Diagnosed mesothelioma cases in Scotland in one calendar year were used to investigate the subject in greater depth. Data from UK sources on asbestos disease types recorded in 2000 and their disease treatment costs were obtained. Acute care economic costs of these diseases are estimated. One hundred and twenty diagnosed, recorded, and treated cases of asbestos-related diseases occurred in 2000 in Scotland. Mesothelioma accounted for 100 cases and directly cost Scottish National Health Service hospitals an estimated £942,038. The estimated UK figure in 2000 was at least £16,014,646 because official figures for diagnosed and recorded deaths from mesothelioma are running at over 1700 a year with rises predicted for 2010 of 2000 deaths. By 2003, 50,000 people in the UK had died from diagnosed and recorded mesothelioma since records began. Earlier disease treatment costs would have been significantly lower than those in 2000 but, at 2000 prices, cost to the UK was roughly £471,019,000 in acute hospital expenditure. Figures for primary care costs, including caregiver costs, are incomplete or unknown. These disease costs are substantial and have some international generalizability.

Address for correspondence: Professor Andrew Watterson, Occupational and Environmental Health Research Group, Room 3T11, RG Bomont Building, University of Stirling, Stirling, FK9 4LA, Scotland, United Kingdom. Voice: 01786466283; fax: 01786466344.

e-mail: aew1@stir.ac.uk

Ann. N.Y. Acad. Sci. 1076: 871–881 (2006). © 2006 New York Academy of Sciences.
doi: 10.1196/annals.1371.042

Treatment patterns and costs vary greatly. Many lung cancer cases due to asbestos exposure occur globally for each mesothelioma case. Hence figures provided in this article are certain to be gross underestimates of the total health service and personal economic costs of asbestos illness and treatment in Scotland.

KEYWORDS: asbestos; mesothelioma; hospital economic costs; health care costs

INTRODUCTION

The article is not concerned with the narrow health economics argument that those who die relatively quickly from a terminal illness may save the health service money. Such arguments are of limited value in their own terms because mesothelioma treatment and care costs are significant and will continue to rise over several decades. Older not younger people are the largest group affected by mesothelioma, which often has a latency period of four decades or more. Hence there will be few savings on health costs through younger people dying earlier, and such arguments are morally indefensible. The article is concerned with social justice issues surrounding those who suffer asbestos-related environmental diseases, the costs of such diseases to society, who does and who should pay those costs.[1,2] In 1993 two Harvard researchers felt able to state categorically that until the precautions and studies for asbestos substitutes were adequate: "there is no known substitute for chrysotile asbestos that, if properly applied, is *known* to be safe"[3] (p. 206). Yet the "proper" use of a carcinogen cannot be guaranteed in industrial, commercial, and domestic sites and the exposures and economic costs of asbestos-related diseases are likely to continue and, in some parts of the world, will increase significantly in line with usage or imports of material.[4]

Asbestos-related diseases have been described in the scientific literature for many centuries and include mesothelioma of the pleural cavity and lung cancer. The precise categorization and causation of such diseases has taken longer and there is some debate now about whether asbestos has been conclusively linked to laryngeal and pharyngeal cancer, colon cancer, and uterine cancer.[5] Despite these continuing areas of uncertainty, the economic costs of mesothelioma and asbestosis to society have not been fully calculated. In contrast, insurance companies since early in the 20th century have indicated the potentially prohibitive but rarely fully realized financial costs of worker compensation due to these very diseases, with one company statistician recommending that asbestos workers should not be insured in 1918 because the hazards they faced were too high.[6] Economic theory would suggest that, where the costs of a health hazard are not fully known and where an industry is allowed to externalize these costs, it will operate on an appropriately large scale. Specifically such an industry will tend to expand full costs of the hazards they impose on the workforce.[7]

French researchers have argued that the "no threshold" level for asbestos has led to "very expensive clean up operations that may have been counterproductive"[8] (p. 49). Nowhere do they document either the basis for the economic costing of this assertion or the economic and human costs of recorded asbestos diseases. Few studies from any countries have touched on the hospital or health service economic costs of this disease.[9] Others have concentrated on the costs of preventing asbestos-related disease.[10,11] In the United States in 1987, one small study attempted to work out medical costs faced by asbestos victims using research from 1975 on an average respiratory cancer case and adjusting upwards at 1984 prices. The figure per case was calculated at $18,834, and the estimate assumed no medical advances that prolonged life, cured asbestos-related diseases, or enabled physicians to separate asbestos-caused lung cancers from lung cancers due to other causes.[12] The sum computed for asbestos lung cancer treatment costs based on Selikoff's and others figures was over $3 billion in 2000 rising to just under $5 billion in 2015.

The human costs of asbestos-related diseases have been enormous in terms of physical and emotional pain of those with these diseases and the impacts on their families and caregivers. Secondary to this must be consideration of the economic and social consequences of such occupationally caused or occupationally related diseases. If such costs are adequately worked out, and if those responsible for generating those costs in society are fully penalized for the consequences of their activity along the lines of the "polluter pays principle," this may help to drive the prevention agenda. The UK Health and Safety Executive (HSE) estimated in the 1990s that around the equivalent of one year's growth in the UK economy at that time was lost due to occupational accidents and ill-health.[13] Other studies within Europe and in Australia have revealed a similar picture.[14,15] Following these reports, the HSE argued persuasively that "good health was good business" because occupational diseases cost companies money due to lost production, replacement labor and training costs, compensation claims, and the like.[16] In 2002, economist Joseph Stiglitz estimated that the "asbestos litigation crisis" has cost the American economy tens of thousands of jobs and reduced pensions for employees at bankrupt firms by 25% on average. Some 61 companies have gone bankrupt as a direct result of asbestos liabilities.[17] Hence the workforces exposed to asbestos may face quadruple jeopardy: from disease, from job losses; from reduced pensions; and from no or low compensation due to company liquidation or selling off of assets. In the UK: " one in every hundred men born in 1940s will die of malignant pleural mesothelioma which is almost exclusively a consequence of exposure to asbestos with a lag time that is rarely less than 25 years and often more than 50 years from first exposure"[18] (p. 237). In 2003, there were over 1800 mesothelioma deaths of men and women, and these figures will increase. The care and treatment implications for cardiothoracic surgeons of these figures, with regard to asbestos exposures that continued up to 1980, are considerable.[18] Direct occupational exposures to asbestos continued after

1980 and will continue for many maintenance and support workers in buildings containing asbestos as well as for others working in such buildings and being incidentally exposed for example, health workers in hospitals, teachers, cleaners.

Evidence indicates that in the United States asbestos companies were aware from the 1940s onwards that asbestos killed their workers.[19] Some companies also recognized that they would save money, presumably in terms of pensions, sick pay, and compensation, if they permitted asbestos workers to "work until they dropped dead"[19] (p. 581). In the UK, the funds set aside by government and companies to compensate and help support those workers with asbestos-related disease were minute when compared to the companies' profits. Turner and Newall spent just over £57,000 on compensating registered asbestos victims in addition to £15,690 on worker medical examinations between 1931 and 1948 versus £15 million profits in the same period although they did sometimes cover sanatorium costs on a no prejudice basis.[20] In the UK, campaigners found an unsatisfactory economic and legal positions existed for those with asbestos-related diseases.[1] Writing in 1995, one campaign group noted, "At the Asbestos Victims Support group we offer victims emotional and practical support. Asbestos victims do not want charity or sympathy. They have been denied the right to a happy and healthy retirement. They have been disabled by a material they were told was safe until quite recently by the government Factory Inspectors. That same material is now killing them. Victims often feel a great deal of bitterness. They find it extremely difficult to gain a state social security pension. Legal aid changes have made it increasingly difficult to obtain civil compensation without risking life savings or family homes"[21] (p. 109).

METHODS

The study has drawn on national statistics provided by the Information and Statistics Division in Scotland (ISD), the Office of National Statistics, UK (ONS), the General Registry Office in Scotland (GRO [S]), and HSE publications, costing provided by Scottish Government departments and additional information from those involved with treating and caring for asbestos victims to estimate the current incidence of mesothelioma, the health service resources consumed while treating the disease, and associated economic costs.

GRO(S) death records were consulted to identify our sample, that being all deaths in Scotland during the year 2000 where mesothelioma was the primary or contributory cause of death. GRO death records for 2000 used the tenth version of the *International classification of diseases and related health problems, tenth revision* (ICD10) to code mesothelioma (C45.0–C45.9) (Table 1).

The Scottish Morbidity Record (SMR) is an official database produced by ISD Scotland where information regarding all admissions to Scottish hospitals (day case and inpatient) including patient demographics and clinical details

TABLE 1. ICD10 codes for mesothelioma

Code	Condition
C45.0	Mesothelioma of pleura
C45.1	Mesothelioma of peritoneum
C45.2	Mesothelioma of pericardium
C45.7	Mesothelioma of other sites
C45.9	Mesothelioma, unspecified

relating to the patient's hospital stay is registered. Additional information stored on the SMR database includes the medical specialty under which the individual received care (defined as the division of medicine covering a specific area of clinical activity), type of admission (i.e., booked, transfer, emergency), length of stay (days), diagnosis, and any operations performed. All SMR records for our study sample were extracted from the Historic SMR1 (General/Acute Inpatient and Day Case records 1981–March 1997) and the COPPISH SMR01 (General/Acute Inpatient and Day Case records April 1997 onwards) by a statistician at ISD Scotland. A unique link number was allocated to all records relating to the same patient so that it would be possible to identify individual cases. Each individual's SMR records were then assessed to determine when they were first diagnosed with mesothelioma and all related treatments received concerning the disease from time of diagnosis until death. Time of diagnosis was defined as: (a) the first hospital admission date in the SMR database where the primary or secondary medical condition managed/investigated during the patient's stay was mesothelioma; and/or (b) the date the cancer (malignant mesothelioma) became formally known to the National Health Service (NHS) and was added to the cancer records.

An estimate of the NHS costs of treating mesothelioma in Scotland was then performed. ISD Scotland collects annual data on the cost of providing health care in Scotland and publishes the results in the "Scottish Health Services Cost Book." Statisticians at ISD Scotland provided figures on the specific costs per day for hospital inpatient and day case care for various medical specialties from the Scottish Health Services Cost Book, year ended 31st March 2001 (NHS Scotland) (Table 2). These costs are inclusive of all direct costs associated with hospital admissions including medical, nursing, pharmacy and professions allied to medicine (PAM) staff, drugs, equipment, supplies, and laboratory costs. Other allocated costs such as administration, catering, linen and laundry, portering, heating, cleaning and property maintenance are also included. The total healthcare costs for each subject were then calculated using these figures.

The focus is deliberately on acute hospital costs because only sparse data exist for primary care treatments.

Ethical approval was gained for the study from the relevant University Ethics Committee.

TABLE 2. Day case and inpatient costs by medical specialty

	Cost book specialty	Cost per admission
Day case	General surgery	£351.24
	Cardiothoracic surgery	£757.94
	Medical	£310.00
	Respiratory medicine	£266.77
	Radiotherapy	£421.58
Inpatient	General surgery	£324.51
	Cardiothoracic surgery	£588.82
	Medical	£221.72
	Respiratory medicine	£187.36
	Radiotherapy	£378.27
	Geriatric assessment	£129.17
	Intensive care unit	£1,279.22
	Coronary care unit	£529.02
	General practice	£160.99

RESULTS

A total of 100 mesothelioma deaths (91 male and 9 female) were identified in Scotland during the year 2000. Mesothelioma of the pleura accounted for 60% of the deaths while 30% were classified as mesothelioma, unspecified (Table 3). However, 27 of the 30 cases with unspecified mesothelioma as the main cause of death had been previously diagnosed and/or treated for pleural mesothelioma. Coding of these deaths as unspecified mesothelioma therefore suggests an underestimation of the true number of deaths from pleural mesothelioma.

While the cases were distributed widely across Scotland geographical areas where significant numbers would have worked in the shipyard industry, including Clydebank, had higher numbers of recorded cases. Specifically, the highest number of deaths occurred in Greater Glasgow and accounted for 22% of the total mesothelioma mortality in Scotland during the year 2000.[22]

Occupation and socioeconomic status were recorded for 97 of the 100 cases and classed as "not stated" in the remaining three cases. However, because occupational work histories are not available, these do not provide as useful a data set as would be hoped for. People may have changed jobs and managers may

TABLE 3. Mesothelioma deaths in Scotland during the year 2000 ($n = 100$)

Cause of death	Males (n)	Females (n)	Total (n)
Mesothelioma of pleura	53	7	60
Mesothelioma of peritoneum	4	0	4
Mesothelioma of other sites	5	1	6
Mesothelioma unspecified	29	1	30

TABLE 4. Sample occupations of those dying from mesothelioma in 2000 in Scotland (GRO codes)

Occupation	Numbers
Carpenters and joiners	12
Cleaners, domestics	2
Computer analyst, programmers	1
Construction and related operatives	4
Medical practitioners	1
Metal working production and maintenance fitters	12
Nurses	1
Postal workers, mail sorters	3
Primary and nursery teaching professionals	1
Rail transport inspectors, supervisors, guards	2
University professionals	1

have worked in shipbuilding and engineering or other occupations where exposure could have occurred. What is incontrovertible is that shipyard workers and other engineering workers do of course still provide the majority of mesothelioma cases. The highest incidence of mesothelioma deaths was observed in carpenters and joiners and metal working production and maintenance fitters (Table 4). Analysis of socioeconomic status of the sample revealed that the largest proportion, 55 of the 97 (57%) were classed as IIIM (skilled manual) (FIG. 1).

Time of diagnosis, defined as the date the malignant mesothelioma was formally made known to the NHS and entered in the cancer records was available for 46 of the 100 cases. Time of diagnosis for the remaining 54 cases was estimated from the date of the first hospital admission where the main medical condition managed or investigated during the patient's stay was mesothelioma. Seventy-three cases were diagnosed as mesothelioma of the pleura. Nineteen cases were diagnosed as other types of mesothelioma (other sites, unspecified and peritoneum) and eight cases were diagnosed as other respiratory illnesses.

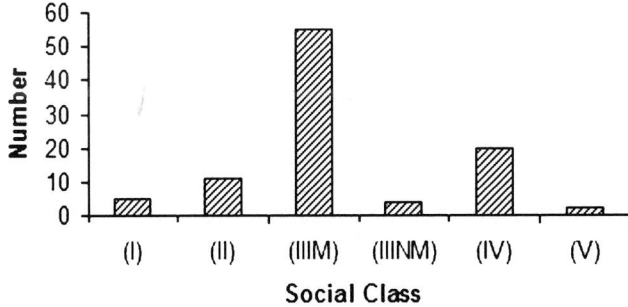

FIGURE 1. Socioeconomic breakdown of the study sample ($n = 97$).

TABLE 5. Total number of days of hospital treatment and total costs of hospital treatment for the 100 cases from diagnosis until death

	Number of days of treatment	Cost
Day cases	103	£ 35,073
Inpatients	3,285	£ 906,965
Total	3,388	£ 942,038

The mean (SD) age at diagnosis was 68.5 (9.2) years for males and 72.8 (10.3) years for females.

HOSPITAL COSTS OF MESOTHELIOMA

The total estimated cost of hospital care for the 100 individuals in Scotland who died in 2000 from the asbestos-related disease, mesothelioma, came to £942,038 of which £35,073 relates to day cases and £906,965 relates to inpatient treatment. The total number of days of hospital treatment from diagnosis until death for the 100 individuals was 3388 (Table 5).

DISCUSSION

The Primary Care Sector

Data for primary care treatments are not readily available. Health Boards do not have data readily to hand on the total costs of treating asbestos-related diseases in acute, primary care and palliative settings. Other costs including primary medical care or general practitioner (GP) costs, practice nurses, and pharmaceutical costs are very high but are not included in our costing. The polypharmacy regime required for an individual mesothelioma patient can exceed £500 per month depending on the patient needs/program.

Palliative Care

Those working in palliative care indicate that in patient bed days for such patients is around £223-18 and home care visits cost is around £64.

UK and Great Britain Wide Figures for Economic NHS Mesothelioma Treatment

The following provide some crude estimates of that cost. In 2000, there were 100 diagnosed, recorded, and treated cases of asbestos-related mesothelioma in

Scotland. Their hospital and NHS day hospital costs totaled £942,038. By 2003, 50,000 people had died just from diagnosed and recorded mesothelioma.[23,24] Earlier disease treatment costs would have been significantly lower than those in 2000 but in terms of figures at 2000 prices, this totals roughly £471,019,000 in acute hospital costs. The official figures for UK and Great Britain diagnosed and recorded deaths from mesothelioma are running at over 1700 a year (2003) with rises predicted for the year 2010 of 2000 deaths. These figures are almost certainly underestimates particularly because most existing projections assume that inflation in the NHS is likely to run between 4% and 9% over the next years.[25] As lung cancer deaths due to asbestos kill at least twice the number of people who die from mesothelioma and asbestosis deaths, the economic costs just to UK hospitals of asbestos are enormous.

Additionally the costs of treating other asbestos-related diseases such as cancer of the pharynx and larynx, cancer of the colon, and cancer of the kidney for which industrial disease compensation has been granted to asbestos-exposed workers in Canada, will not be recorded. This is because the means available to identify such causes have been limited or nonexistent. Hence the economic costs of treating those made ill by asbestos in Scotland are grossly under-recorded here. Nevertheless, the picture is very grim. With mesothelioma and asbestosis alone, the costs of NHS treatment are enormous and will continue to be so over coming decades.[26] Those paying for the NHS therefore cover the costs of employers who have profited from selling and using asbestos. Victims and their families, along with the public, will pay a disproportionate amount of such treatment costs.

The numbers of people with asbestos-related diseases within Scotland are difficult to estimate. Even where there is little doubt about the etiology of diseases, such as asbestosis or mesothelioma of the pleural cavity caused by asbestos exposure, data in the past have been patchy. This has been partly due to the disease recording process, in some instances diagnosis and in some instances disease classification or a combination of all three. ICD did not specifically identify asbestos-caused mesothelioma cases. It is not possible therefore to assess accurately the number of people with the two asbestos-related diseases in question and calculate the economic costs to the NHS of their treatment. The economic costs are accordingly seriously underestimated.

Much asbestos disease research has rightly focused on the human costs of asbestos; how to prevent or reduce asbestos exposures in the future, how to gain financial compensation and effective support and care for those with asbestos-related diseases. Germany collects information about asbestos treatment costs through its insurance schemes and the United States has similar cost data available from completed legal cases. However, the global environmental justice case on asbestos should include charging those companies for the full economic costs of treating the diseases they caused. This will be another small step for justice for the victims and the public as a whole. It may also serve to deter, through heavy economic sanctions, those in the future who think they may kill employees and people in communities exposed to their hazardous materials and processes almost with impunity.

ACKNOWLEDGMENTS

This study emerged from the Convention of Scottish Local Authorities (COSLA) working group on asbestos. The authors alone are solely responsible for the opinions and analysis offered in this article. However, we would like to thank Ana Rodriguez and her colleagues at ISD, Scotland who provided some of the data upon which the analysis is based and who proved most helpful in dealing with our enquiries. We also acknowledge Dr. Clark Mullen's contributions to this article.

CONFLICT OF INTEREST

There were no conflicts of interest in this study. The corresponding author had full access to all the data in the study and had final responsibility for the decision to submit for publication.

REFERENCES

1. GORMAN, T. Ed. 2000. Clydebank: Asbestos the Unwanted Legacy. Clydebank Asbestos Group, Glasgow.
2. GORMAN, T. & A. WATTERSON. 2003. Confronting the continuing problem of asbestos in Scotland: report of a Scottish Public Sector Initiative for the 21st Century. New Solutions. J. Environ. Occup. Health Policy **14:** 77–98.
3. AGOSTINO, R. & R. WILSON. 1993. Asbestos: the hazard, the risk and public policy. *In* Phantom Risk. K.R. Forster, *et al*. Eds.: 183–207. MIT Press. Cambridge, Massachusetts.
4. SHUKER, L., P. HARRISON, S. POOLE, Eds. 1997. Fibrous Materials in the Environment: A Review of Asbestos and Man-Made Mineral Fibers. Medical Research Council, Institute of Health. Leicester, UK.
5. KIELKOWSKI, D., G. NELSON & D. REES. 2000. Risk of mesothelioma from exposure to crocidolite asbestos: a 1995 update of a South African mortality study. Occup. Environ. Med. **57:** 563–567.
6. RAMPTON, S. & J. STAUBER. 2001. Trust Us, We're the Experts. Penguin Putnam. New York.
7. BECK, M. & C. WOOLFSON. 2001. An economic analysis of the proposed reform of the law on involuntary manslaughter. J. Inst. Occup. Health Saf. **5:** 63–74.
8. FOURNIER, E. & M. EFTHYMIOU. 1997. Problems with very low dose risk evaluation: the case of Asbestos. *In* What Risk? R. Bates, Ed.: 49–69. Butterworth-Heinemann. Oxford.

9. BERGERET, A. & G. TERRASSON DE FOUGERES. 1999. Impact social du epistage et de la surveillance medicale chez les personnes exposees a l'amiante. Rev. Mal. Respir. **16:** 1327–1331.
10. LE GALES, C. & A. OUDI. 1984. Contribution methodologique a la determination de valeurs limites d'exposition professionelle a l'aminate. Relation exposition-risque et criteres economiques. Rev. Epidemiol. Sante Publique. **32:** 113–121.
11. MOLL, K.D. & D.P. TIHANSKY. 1977. Risk benefit analysis for industrial and social needs. Am. Ind. Hyg. Assoc. J. **38:** 153–161.
12. SISKIND, F.B. 1987. The cost of compensating asbestos victims under the Occupational Disease Compensation Act of 1983. Risk Anal. **7:** 59–69.
13. HEALTH AND SAFETY EXECUTIVE. 1993. The Costs of Accidents at Work. HMSO. London.
14. WORKSAFE, AUSTRALIA. 1994. The Cost of Work-related Injury and Disease. Australian Government Publishing Service. Canberra.
15. BEATSON, P. & P. LUNDE-JENSEN. 1996. Costs and benefits of preventive measures. Euro ECHO **2:** 1–4.
16. HEALTH AND SAFETY EXECUTIVE. 1999. Good Health is Good Business. HSE, London.
17. www.asbestossolution.org
18. TREASURE, T., D. WALLER, S. SWIFT, et al. 2004. Radical surgery for mesothelioma. Br. Med. J. **328:** 237–238.
19. CASTLEMAN, B. 1999. Asbestos: medical and Legal Aspects, 4th ed. Aspen Law and Business. New Jersey.
20. TWEEDALE, G. 2000. Magic Mineral to Killer Dust: Turner and Newall and the Asbestos Hazard. Oxford University Press. Oxford.
21. LONDON HAZARDS CENTRE. 1995. The Asbestos Hazards Handbook. London Hazards Centre Trust. London.
22. GORMAN, T., R. JOHNSTON, A. MCIVOR & A. WATTERSON. 2004. Asbestos in Scotland. Int. J. Occup. Environ. Health **10:** 183–192.
23. WHITE, C. 2003. Annual deaths from mesothelioma in Britain to reach 2000 by 2010. Br. Med. J. **326(7404):** 1417.
24. BONN, D. 1999. Asbestos—the legacy lives on. Lancet **353:** 1336.
25. WANLESS, D. 2002. Securing our future: taking a long-term view. Final report. Her Majesty's Treasury. London.
26. THE MEDICAL LETTER. 2004. Pemetrexed (Alimata) for mesothelioma. Med. Lett. **46:** 31–32.

Valuing the Adult Health Effects of Air Pollution in Chinese Cities

ROBERT W. MEAD AND VICTOR BRAJER

California State University, Fullerton, California 92834, USA

ABSTRACT: China's ongoing economic growth is accompanied by a large amount of air pollution that exacts significant health and economic costs on its people. Following up on some earlier work focusing upon general mortality and child-specific health effects, this article uses a larger data set, covering more than 90 Chinese cities, along with a set of China-based epidemiological functions, to estimate some of the adult health benefits of reducing urban air pollution. Projecting future air pollution based upon current conditions, it calculates the averted mortality and morbidity effects that would result from the cleanup of particulates, sulfur dioxide, and nitrogen dioxide. The inclusion of nitrogen dioxide in our analysis is particularly important because it is a growing problem and has not been included in most of the more widely known studies that examine Chinese air pollution. Finally, the economic valuation of these pollution-related health effects is developed, using a number of recent, China-based valuation studies.

KEYWORDS: urban air pollution; health effects; economic valuation; China

INTRODUCTION

With a significant percentage of China's cities failing to meet national pollution standards and an even larger number of cities falling short of the more stringent World Health Organization (WHO) standards, the existence of an urban air pollution problem in China is without question. However, determining and valuing the subsequent health consequences of this pollution has been difficult. Limited data availability, few China-based health studies, and rare Chinese valuation studies have forced most previous studies to either make a number of simplifying assumptions or impose limitations in their analysis. Unfortunately, many of these assumptions and limitations generate problems for the subsequent analysis. Limited city-level pollution data have caused previous studies to either project one city across all of China,[1] which

Address for correspondence: Robert W. Mead, Department of Economics, California State University–Fullerton, Fullerton, CA 92834. Voice: 714-278-4479; fax: 714-278-3097.
e-mail: rmead@fullerton.edu

TABLE 1. Descriptive statistics of Chinese pollution and population levels

	TSP ($\mu g/m^3$)	NO$_2$ ($\mu g/m^3$)	SO$_2$ ($\mu g/m^3$)	Population
95 city average	264.84	35.44	52.67	2,861,767
(deviation)	(157.77)	(13.07)	(43.83)	(3,098,796)
95 city median	219	34	40	1,407,124
Highest level	1055	71	244	13,375,900
(city)	(Ge'ermu)	(Beijing)	(Shijiazhuang)	(Shanghai)
Lowest level	51.67	11	2	103,279
(city)	(Beihai)	(Beihai)	(Lhasa)	(Ge'ermu)
China standard	200	80	60	
International standard	90[a]	40	50	

[a]Because the WHO has shifted the focus to particulates smaller in size (PM$_{10}$ or PM$_{2.5}$), this particular standard is now dated.

ignores the variation in pollution levels between cities (TABLE 1) or focus on a single city,[2–6] which fails to explore the overall problem. Another assumption has been the cross-country application of epidemiological studies from other countries to China, yet cultural, social, and chemical variations between countries challenge the validity of this approach.[7–9] Other valuation efforts have used scaling United States or other developed country values to China using ratios of per capita income. This approach assumes per capita income level is either the sole determinant of relative valuations or sufficiently captures all factors, but as discussed elsewhere,[10–12] the evidence for such assumptions is limited and the parameters used are quite varied.

To overcome these limitations and better quantify the health consequences of urban air pollution in China, this article takes on four tasks. First, it uses a relatively recent collection of air pollution levels for 95 Chinese cities. Second, all health functions used are based upon Chinese epidemiological studies. Third, valuing the derived health effects is based either directly or indirectly upon Chinese valuation studies or Chinese cost of illness values. Finally, this study considers nitrogen dioxide (NO$_2$) in addition to sulfur dioxide (SO$_2$) and total suspended particulates (TSP), the two historical components to China's pollution, in its analysis.

DATA

Most of the 2001 pollution levels for the 95 cities are from the Chinese Environmental Yearbook[13] and the remaining values are from Mead and Brajer.[12] We also use Mead and Brajer's population figures for 47 cities and obtain the rest from the China Population Yearbook.[14] The adult population is then derived using the age demographics also found in the China Population Yearbook.

Cities studied include all provincial capitals and all cities with populations larger than 4 million people, as well as 16 of 22 cities with populations between

2 and 4 million. In addition, 37 smaller cities with populations of under a million people are included. The analyzed population exceeds 270 million, which includes over half of China's urban population and over 21% of the national total.

Finally, we use health data from several sources to generate our health functions. In most cases, prevalence rates are reported in the studies we use to obtain health functions, but for mortality, we derive our own baseline rate. The urban death rate stopped appearing in the annual yearbooks[15] after 1999, so we derive a 2002 urban death rate by applying the 1999 city death rate ratio to the 2002 national death rate. Finally, since deaths from the top 10 major diseases account for 92% of total urban deaths, 92% of our derived city death rate is used as the baseline mortality figure.

SCENARIO CONSTRUCTION

We next construct three scenarios, requiring several assumptions about future pollution and population levels. For population, we assume a conservative annual urban increase of 0.7875% based on United Nations projections reported by Heilig.[16]

For each pollutant, we follow the lead of Mead and Brajer and construct three scenarios. The first scenario, or the "Business as Usual" (BAU) case, constitutes a baseline based upon current pollution levels. For SO_2 and TSP, we project the BAU scenario as keeping the 2001 levels constant. For NO_2, we incorporate China's growing automobile fleet and project a BAU annual growth of 7.6%. The two additional scenarios then look at the benefits of successful pollution abatement efforts. First, we examine expected health improvements resulting from cleanup efforts that lower pollution levels to the Chinese clean air standards: 200 $\mu g/m^3$ for TSP, 60 $\mu g/m^3$ for SO_2, and 80 $\mu g/m^3$ for NO_2. Second, because the WHO publishes even more stringent standards, we also calculate the averted morbidity benefits of meeting the WHO standards: 90 $\mu g/m^3$ for TSP, 50 $\mu g/m^3$ for SO_2, and 40 $\mu g/m^3$ for NO_2. For both TSP and SO_2, we project linear declines from their 2001 levels to the targeted standards. For NO_2, the projected cleanup begins when the BAU NO_2 levels exceed the targeted standards.

DEVELOPING HEALTH EFFECT FUNCTIONS

Calculating expected changes in pollution-related health outcomes involves the derivation of China-specific, concentration-response health functions. Here, we rely on the basic natural exponential form as developed in the U.S. EPA Retrospective Analysis[17] on the health benefits of the Clean Air Act. Specifically, we use the following functional form in all of our health effect calculations: $\Delta C = C(e^{b\Delta P} - 1)$, where C is the number of baseline cases (of a particular health effect), ΔP is the change in ambient pollutant concentration, and b is an exponential "slope" risk factor. This b is derived from the odds ratio (OR) (or relative risk factor), which relates changes in pollution levels to the increased odds of developing various health conditions as typically reported in the health literature (or can be easily computed). More specifically, we derive the b-values in the following manner: $b = [\ln(\text{odds ratio})]/(\Delta P)$, where again ΔP represents the change in ambient pollution concentration.

For this study, we derive our b-values exclusively from China-based studies (discussed in the sections below) examining the health impacts of pollution. Using studies conducted in China is important because different countries can generate starkly different health equations.[10–12]

Morbidity—TSP-Induced Cases of Cough, Phlegm, Wheeze, PCP, Asthma, and Acute Bronchitis

To derive concentration-response functions for self-reported asthma, cough, wheeze, phlegm, persistent cough and phlegm (PCP), and acute bronchitis, we draw from Zhang *et al.*'s study,[18] exposure to ambient air pollution and respiratory health effects in four districts of three large Chinese cities (Guangzhou, Wuhan, and Lanzhou). Using logistic regression models, the authors find positive, and statistically significant, associations between the above-mentioned symptoms and outdoor levels of particulate matter, which they translate into prevalence ORs. We then use these values to develop exponential slope factors (our b-values) for Wuhan, Lanzhou, and a Wuhan–Lanzhou average (TABLE 2). Next, we apply these values, as well as the reported baseline prevalence rates, to our 95 cities based on their TSP levels. More specifically, for those cities with TSP levels below 456 $\mu g/m^3$ we use the Wuhan-generated numbers; where TSP is between 456 and 674 $\mu g/m3$ we apply the average of the Wuhan and Lanzhou values; and where TSP is over 674 $\mu g/m^3$ we use the Lanzhou-generated numbers.

TABLE 2. Exponential slope factors (b-values) by symptom

Symptom	Wuhan value	Lanzhou value	Wuhan–Lanzhou average value
Cough	0.00694	0.00179	0.00436
Wheeze	0.01724	0.00369	0.01046
Phlegm	0.00419	0.00081	0.00250
PCP	0.01261	0.00299	0.00780
Asthma	0.01030	0.00042	0.00536
Acute bronchitis	0.00997	0.00284	0.00641

Morbidity—SO_2 and TSP-Induced Hospital Outpatient and Emergency Room Visits

For these health outcomes, we turn to Xu et al.[19] who find support for the existence of significant associations between both SO_2 and TSP and outpatient visits and between SO_2 and emergency room (ER) visits. Using linear regression, they estimate SO_2-related slope coefficients of 41.5 for outpatient visits and 6.8 for ER visits. We combine these with the mean number of each type of visit to calculate relative risk factors for outpatient visits (RR = 1.0226) and ER visits (RR = 1.03736). Resultant SO_2 exponential slope factors, b-values, are 0.0002235 and 0.0003668 for outpatient and ER visits, respectively. For TSP, similar calculations result in a b-value of 0.0001143 for the outpatient visit case. We consider this TSP value to be the base value that we apply to most of the cities in our sample. For those cities with very high annual TSP levels (over 456 $\mu g/m^3$), we adjust this b-value downward, using the ratios obtained in the Zhang et al. study described previously.

Morbidity—TSP-Induced Chronic Bronchitis

Ma and Hong,[20] in a Shanghai study, use logistic regression to determine an odds ratio of 1.29 for chronic bronchitis with each increase of 100 $\mu g/m^3$ in particulate levels. From this ratio, we develop an exponential b-value of 0.002546 for this health endpoint. As before, lower b-values are developed for the higher TSP cities as well.

Mortality—SO_2 and TSP

To develop TSP and SO_2 mortality health equations, we use a study by Xu et al.,[21] which regresses the number of daily deaths against pollution and weather variables, along with indicators for Sundays and the previous day's mortality, to calculate relative risk factors for the two pollutants. For SO_2, their estimated RR factor equals 1.0188; for TSP, the corresponding RR factor is equal to 1.013. These relative risk factors lead to b-value estimates of 0.000186 for SO_2 and 0.000129 for TSP.

Mortality—NO_2

For NO_2 mortality, our analysis relies on two recent Shanghai studies,[22,23] which generate NO_2 results adjusted for the effects of PM_{10} and SO_2, that allow for the derivation of relative risk factors for nonaccident mortality (RR = 1.012 and RR = 1.008 for the two studies, respectively). These factors are then converted, via the exponential concentration-response function, and averaged, yielding a final b-value of 0.0009949.

DEVELOPING ECONOMIC VALUES

There are two methods commonly used to value health outcomes: willingness to pay (WTP) and either cost of illness (COI) for or human capital accounting for mortality. Theoretically, WTP values are preferred because they are based upon individual preferences and behavior, more fully capturing the value of lost leisure time, inconvenience, and discomfort; or for mortality, they more completely capture the overall value of life by assessing the value groups of individuals place on reducing the probability of dying earlier than would otherwise be expected. They are, however, harder to obtain because they require surveying or observing individual behaviors or preferences. COI and human capital values, which either measure direct medical costs or lost days of work or lost future income, are often easier to find or estimate, but may omit intangible costs captured in WTP calculations. As a result, the COI approach generally produces a lower bound for a particular symptom. Accordingly, we attempt to use Chinese WTP calculations but rely on COI values as necessary. We also note that in the case of mortality the value of reduced annual risk of death, or of averted death, is a more accurate term for what is being measured, but the expression commonly used is *value of a statistical life* (VSL).

Morbidity–Colds, Phlegm, and Chronic Bronchitis

Our valuation efforts here make use of a study by Zhou and Hammitt[24] employing the contingent valuation (CV), or survey, method to estimate values for a number of morbidity effects (along with VSL) in three diverse locations—Bejing, Anqing, and the rural areas near Anqing. We therefore value colds at $6, and estimate the number of cases by considering a "cold" to be the overlap of three distinct symptoms—cough, wheeze, and phlegm. The excess number of symptom-days (cough and wheeze for most cities) are then valued at a fixed percentage of the cold figure, using relative values established in a recent U.S. valuation study.[25] TABLES 3 and 4 include these dollar estimates.

Morbidity–PCP

Days of PCP are defined in the Zhang *et al.* study as periods of illness lasting at least 3 months of the year. Merely multiplying the sum of daily cough and phlegm values by 90 would probably overstate the WTP to reduce 90 days of these health effects, since a person's WTP for avoiding a symptom typically declines as the number of days of poor health increase. To compensate for this likely decline, we follow Brajer, Hall, and Rowe[26] and use a WTP adjustment function: $WTP_n = WTP_1 \times N^{-0.5}$, where WTP_n is WTP to reduce N days, WTP_1 is WTP to reduce the first day, and N is the number of days the symptom

TABLE 3. Number of averted health incidents and valuation—China standard

Symptom	Number of cases	Unit value	Total cost
Morbidity			
Colds	85,411,238	$6.00	$512,467,428
Extra coughs	184,300,438	$1.77	$326,211,775
Extra wheeze	256,341,875	$2.78	$712,630,413
Extra phlegm	0	$1.77	$0
PCP	66,436,798	$56.92	$3,781,582,567
Acute bronchitis	141,097,390	$21.35	$3,012,429,292
Chronic bronchitis	3,658,664	$861.00	3,150,109,837
Asthma	38,659,621	$11.29	$436,467,124
ER visits	1,340,487	$27.16	$36,412,989
Outpatient visits	21,594,611	$13.48	$206,228,541
Mortality			
TSP	51,816	$411,000	21,296,376,000
SO_2	18,953	$411,000	7,789,683,000
NO_2	68,180	$411,000	28,021,980,000
Totals			
Morbidity	983,141,560		$12,259,406,788
Mortality	138,924		$57,108,039,000
Grand total			$69,367,445,788

was reduced by reaching the pollution standard (here, of course, $N = 90$). This adjustment produces an economic value of $56.92 for an avoided episode of PCP.

TABLE 4. Number of averted health incidents and valuation—WHO standard

Symptom	Number of cases	Unit value	Total cost
Morbidity			
Colds	247,898,534	$6.00	$1,487,391,204
Extra coughs	593,981,933	$1.77	$1,051,348,021
Extra wheeze	1,527,777,241	$2.78	$4,247,220,730
Extra phlegm	7,365,902	$1.77	$13,037,647
PCP	252,076,862	$56.92	$14,348,215,028
Acute bronchitis	450,171,507	$21.35	$9,611,161,674
Chronic bronchitis	9,141,106	$861.00	$7,870,492,266
Asthma	143,671,916	$11.29	$1,622,055,932
ER visits	1,642,919	$27.16	$44,621,680
Outpatient visits	47,818,309	$13.48	$644,590,805
Mortality			
TSP	130,485	$411,000	$9,548,763,000
SO_2	23,233	$411,000	$137,499,288,000
NO_2	333,548	$411,000	$53,629,335,000
Totals			
Morbidity	3,281,546,229		$40,940,134,988
Mortality	333,548		$200,677,326,000
Grand total			$241,617,460,988

Morbidity–Acute Bronchitis

To value cases of acute bronchitis, we use Yang and Xu's CV survey[27] of Tianjin residents, which studied WTP to prevent respiratory illness. Their result generate a WTP of 176.6 Yuan (or $21.35).

Morbidity—Asthma

Though it is chronic in nature, we avoid imparting upward bias in valuation by treating asthma cases as a single attack requiring treatment. Lacking direct figures specific to asthma, a study[28] on the health costs of smoking in China provides an early 1990s cost for a single treatment of a case of chronic obstructive pulmonary disease. Updating this value to current terms using the Chinese price index for medicine and treatment gives a value of $11.29.

Morbidity—ER Visits and Outpatient Visits

Valuation of the two hospital-related health effects (outpatient visits and ER visits) comes from an international COI study[29] looking at an early intervention for asthma that reports a Chinese cost of 224.6 Yuan for ER visits and 111.5 Yuan for physician visits. Converting into U.S. dollars (at a rate of 8.27 Yuan/$) gives a dollar value of $27.16 for ER visits and $13.48 for outpatient visits.

Mortality

In contrast to the United States, relatively few studies have taken place in countries with significantly lower incomes. Fortunately, we again make use of the Zhou and Hammitt study employing the CV, or survey, method to estimate VSL. Because this valuation study is China based, it provides a starting point from which to develop VSL estimates for averted pollution mortality. Still, even though the study was conducted completely in China, we hesitate to rely solely upon the resulting values. By their own admission, the authors concede that some form of hypothetical bias (or respondent bias) may have existed in the study. When respondents were offered risk reductions of different sizes as part of the CV questioning, there was no significant difference in their WTP responses.

This study uses a hybrid approach to generate a plausible Chinese VSL, averaging a purchasing power parity (PPP)-based conversion of a U.S.-based value of $6.2 million[30] with those reported by Zhou and Hammitt. This process results in an estimate of $411,000. It should be noted that these values are not being ascribed to the life of any individual but to reducing the annual probability of death by a small amount.

RESULTS

Applying the Chinese epidemiological functions to our projected levels of pollution to each city and totaling, the projected potential of successful cleanup activities over the 10-year period 2003–2012 is substantial. Under the China scenario, we estimate nearly a billion morbidity instances would be avoided, including over 85 million colds or multisymptom effects and over 140 million cases of bronchitis, with a total valuation of over $12 billion (TABLE 3). In addition, over 138,000 deaths (totaling more than $57 billion) would also be averted. The total valuation is over $69 billion of which 17% is from averted morbidity.

In the WHO scenario, the results are even more substantial (TABLE 4). Under this scenario we project over 3 billion averted morbidity instances including more than 247 million colds or multisymptom effects, 250 million cases of persistent cough with phlegm, 450 million cases of acute bronchitis, and 143 million cases of asthma. In this scenario, the total morbidity valuation exceeds $40 billion. The number of averted deaths is over 333,000 and valued at over $200 billion. Total valuation here surpasses $240 billion with morbidity accounting for just over one-sixth of the total.

To put these numbers in context, the China scenario suggests that every single adult in our sample would avoid just over 3.5 morbidity symptoms over the 10-year period as a result of cleanup. Under the WHO scenario, the number more than doubles to 8.25 morbidity instances avoided. For an economic perspective, we note that 2001 per capita expenditures for public health were just under $48 of which nearly $29 was by individuals themselves.[15] Over our entire set of cities, the average annual per capita costs in the China scenario are $4.50 for morbidity, $21 for mortality, and $25.50 combined. In this scenario, the morbidity costs alone are nearly 10% of the total per capita health expenditure and the combined costs are nearly equivalent to the individual annual expenditures. While some of the cleaner cities do not incur any costs under this scenario, in some of the dirtier, such as Beijing or Shijiazhuang, the annual per capita cost is over $90, easily exceeding the total per capita health expenditure. Under the WHO scenario, the results are even more dramatic. In this scenario, the average annual per capita costs are just over $15 for morbidity, $73.81 for mortality, and $88.00 combined. Now the morbidity values are a third of the total per capita health expenditure with the mortality and combined totals exceeding the expenditures. Moreover, the annual per capita costs in the dirtier cities now exceed $200.

SUMMARY AND CONCLUSIONS

This article explores the potential economic benefits to China's urban areas should extensive air pollution cleanup efforts be undertaken by projecting two cleanup scenarios, estimating averted adult morbidity and mortality, and then assigning dollar values to the averted cases. In doing so, we deviate from previous studies in several ways. First, we use contemporary pollution levels from a wider spectrum of different cities in China than found in previous work. Second, we rely exclusively on China-based health studies. Third, we use China-based numbers in valuing the health effects. Last, we make an attempt to include some effects of NO_2 pollution, which is becoming increasingly problematic as China's automobile fleet grows.

We close with several observations. First, we find that the limited numbers of both health effects and valuation studies indicate a need for more of these studies, especially since China is undergoing rapid economic growth and social transformation. Second, our focus here on health effects in adults omits a number of other air pollution consequences or cleanup benefits, such as health effects in children,[11] as well as nonhealth benefits. Third, these benefits and valuations are based solely upon a 10-year projection subject to multiple uncertainties. Cleanup efforts involving long-term projects or permanent changes to lower pollution fuels will produce additional benefits for China's urban areas well beyond 2012. Finally, we note that the values estimated here must be viewed as just that, *estimates* of the health benefits resulting from improved air quality. Recent advances in health research have narrowed the bounds of uncertainty and there is some emerging consensus on key economic values, but there are still many questions about where, within a range of probable effects or dollar measures, the "real" values lie.

REFERENCES

1. WORLD BANK. 1997. Clear water, blue skies: China's Environment in the New Century. World Bank. Washington DC.
2. PENG, C. *et al.* 2002. Urban air quality and health in China. Urban Studies **12**: 2283–2299.
3. BRAJER, V. & R.W. MEAD. 2003. Blue skies in Beijing? Looking at the Olympic effect. J. Environ. Dev. **12**: 239–263.
4. KAN, H. & B. CHEN. 2004. Particulate air pollution in urban areas of Shanghai, China: health-based economic assessment. Sci. Total Environ. **322**: 71–79.
5. KAN, H. *et al.* 2004. An evaluation of public health impact of ambient air pollution under various energy scenarios in Shanghai, China. Atmosph. Environ. **38**: 95–102.
6. LI, J. *et al.* 2004. Quantifying the human health benefits of curbing air pollution in Shanghai. J. Environ. Manage. **70**: 49–62.
7. ALBERINI, A. & A. KRUPNICK. 1997. Air pollution and acute respiratory illness: evidence from Taiwan and Los Angeles. Am. J. Agric. Econ. **79**: 1620–1624.
8. CROPPER, M.L. *et al.* 1997. The health benefits of air pollution control in Delhi. Am. J. Agric. Econ. **79**: 1625–1629.
9. MURRAY, F. *et al.* 2001. Assessing health effects of air pollution in developing countries. Water Air Soil Poll. **130**: 1799–1804.

10. BRAJER, V. & R.W. MEAD. 2004. Valuing air pollution mortality in China's cities. Urban Studies **41:** 1567–1585.
11. MEAD, R. & V. BRAJER. 2005. Protecting China's children valuing the health impacts of reduced air pollution in urban China. Environ. Dev. Econ. **10:** 745–768.
12. MEAD, R. & V. BRAJER. 2005. Rise of the automobiles: the costs of increased NO_2 pollution in China's changing urban environment. J. Contem. China **15:** 349–367.
13. ENVIRONMENTAL YEARBOOK. 2002. China Environmental Yearbook 2002 (*Zhongguo huangjing nianjian* 2002). China Environmental Yearbook Press. Beijing.
14. POPULATION YEARBOOK. 2003. China Population Statistics Yearbook (*Zhongguo renkou tongji nianjian*). China Statistics Press. Beijing.
15. STATISTICAL YEARBOOK. (various years). China Statistical Yearbook (*Zhongguo tongji nianjian*). China Statistics Press. Beijing.
16. HEILIG, G.K. 1999. Can China feed itself? A system for evaluating policy options. Obtained from IIASA, International Institute for Applied Systems Analysis website at http://www.iiasa.ac.at/Research/LUC/ChinaFood/data/pop/pop_7.htm
17. U.S. ENVIRONMENTAL PROTECTION AGENCY. 1997. The benefits and costs of the clean air act, 1970 to 1990 (Report). U.S. EPA. Washington DC.
18. ZHANG, J. *et al.* 1999. Effects of air pollution on respiratory health of adults in three Chinese cities. Arch. Environ. Health **54:** 373–381.
19. XU, X. *et al.* 1995. Air pollution and unscheduled hospital outpatient and emergency room visits. Environ. Health Persp. **103:** 286–289.
20. MA, H.B. & C.J. HONG. 1992. Impact of air particulates on chronic respiratory diseases. Chinese J. Pub. Health **11:** 229–232.
21. XU, Z. *et al.* 2000. Air pollution and daily mortality in Shenyang, China. Arch. Environ. Health **55:** 115–120.
22. KAN, H. & B. CHEN. 2003. Air pollution and daily mortality in Shanghai: a time-series study. Arch. Environ. Health **58:** 360–367.
23. KAN, H. & B. CHEN. 2003. A case-crossover analysis of air pollution and daily mortality in Shanghai. J. Occup. Health **45:** 119–124.
24. ZHOU, Y. & J. HAMMITT. 2003. The economic value of air-pollution-related health risks in China: a contingent valuation study. Harvard School of Public Health, Mimeo.
25. LURMANN, F.W. *et al.* 1999. Assessment of the health benefits of improving air quality in Houston, Texas. Final report prepared for city of Houston. November. Sonoma Technology, Inc. Petaluma, CA.
26. BRAJER, V. *et al.* 1991. The value of cleaner air: an integrated approach. Contemp. Policy Iss. **IX:** 81–91.
27. YANG, Z. & L. XU. 2004. Valuing health effects from the industrial air pollution in rural Tianjin, China. J. Environ. Sci. **16:** 157–160.
28. JIN, S-G. et al. 1995. An evaluation of smoking-induced health costs in China (1988–1989). Biomed. Environ. Sci. **8:** 342–349.
29. BUXTON, M.J. et al. 2004. Country-specific cost-effectiveness of early intervention with budesonide in mild asthma. Eur. Respir. J. **24:** 568–574.
30. DOCKINS, C. *et al.* 2004. Value of statistical life analysis and environmental policy: a white paper for presentation to science advisory board—environmental economics advisory committee. Final report (National Center for Environmental Economics, U.S. EPA. 2004).

Chinese Workers and Labor Conditions from State Industry to Globalized Factories

How to Stop the Race to the Bottom

MARINA THORBORG

Department of Intellectual History, Soedertoern Hoegskola, SE-141 89 Huddinge, Sweden

ABSTRACT: This article discusses administrative obstacles in China that hinder the full integration of the rural population into the mainstream of development during a period of rapid industrialization. The Chinese household registration only for urban residents with its golden contents of cradle-to-grave security has become a formidable stumbling block that perpetuates the status of rural migrants as second-class citizens in their own country. Rural migrant workers are excluded from certain types of jobs and are not eligible for many benefits that urbanites have, such as health, education, and unemployment protection. These workers must also pay a number of fees and work for lower minimum wages than the local residents. With a precarious legal existence in urban areas, they are easy prey to unscrupulous officials and employers. Because they are not allowed to form independent trade unions, their best option is to vote with their feet and leave the firms with the worst conditions; this is exactly what they did from 2004. Given this situation, the debate on Corporate Social Responsibility (CSR) took a new turn with not only non-governmental organizations (NGOs) pushing it but with a wider range of employers and, of late, Chinese officials promoting their version of CSR. In the campaign to promote minimum labor standards, the norms set down in the Social Accountability 8000 were included in the CSR, recognizing the right to free collective bargaining and free trade unions but were excluded in the Chinese version even though the World Trade Organization (WTO) agreements recognized these rights.

KEYWORDS: China; migrant labor; work conditions; household registration system; wages; trade unions; CSR; labor standards; labor laws

Chinese society today is in the middle of five major transitions: (*a*) from plan to market; (*b*) from rural to urban; (*c*) from communist, with a party fully

Address for correspondence: Marina Thorborg, Department of Intellectual History, Soedertoern Hoegskola, SE-141 89 Huddinge, Sweden. Voice: +46-8-608-4112; fax: +46-8-608-42-05.
  e-mail: marina.thorborg@sh.se

above the law, to a socialist, one that is ruled by law and the party, but not always above the law; (*d*) from centralization to decentralization and some recentralization; and (*e*) from an unusually egalitarian society for its stage of economic development to one of the most inegalitarian, surpassing both India and Russia.[1]

Until the reforms of 1978, China was predominantly a rural society, where 80% of the population was preoccupied with agriculture and related work, and 20% was urban and derived its main income from the state sector, with a low-wage policy combined with a system of extensive social coverage.[2] Today those with an urban household registration, "hu-kou," have doubled and now make up 40% of the population, with an urban working population encompassing 120 million people.[3]

Since China began its reforms in 1978 and eased its restrictions on moving people, it currently counts more than 120 million rural residents with temporary work permits toiling in urban areas and making up roughly half of the workforce in urban areas.[4] In urban China, the rural migrant population dominates manual labor, but enjoys neither the wages nor the work benefits of the urban population[5] (TABLE 1). Rural migrants make up 68% of the workers in processing and manufacturing, 80–90% in construction, and 52% in the service.[6]

Hence in the urban scene the changes have involved not only a flow of capital and labor from state to collective to private, but also from an urban state-employed labor force with "cradle-to-grave" benefits commonly referred to as the "iron rice bowl," to a migrant, less skilled, rural labor force mostly working in the collective, private, often foreign-invested sector[7] (TABLE 1).

Outside established urban areas, a new phenomenon appeared in the late 1970s: the special economic zone (SEZ). After China began moving toward a market economy in 1978 to accelerate economic development, a number of zones were set up. The earliest and most far-reaching are Shenzhen, just north of Hong Kong, and Zhuhai, north of Macao. Since 1985 these two zones have been enmeshed in what was officially designated the greater Pearl River Delta, where export production in world market factories has been the main development strategy. Rural migrant workers, mainly women, make up the majority of employees in this delta. This region has become a melting pot of foreign investment, technology, and know-how mixed with "round-tripping"[8] Chinese investment combined with low-cost labor, with sweatshops of a bygone European era and state-of-the-art factories.[9] Although the Pearl River Delta is minuscule compared to the rest of China, what happens in this industrially advanced area (types of industries, management, Chinese-foreign cooperation, relationships between workers and management, implementation of the World Trade Organization (WTO) rules, development of cooperation between activists and workers) has implications for the rest of the country and has served as a model on several, earlier occasions.[10] With regard to low wages, excessive overtime, and bad working conditions, China has until now had the worst record, even outdoing such underdeveloped countries as Vietnam and

TABLE 1. Comparison of rural migrants with urban resident employees, work, and wages

	Rural migrants	Urban resident employees
Type of work, tabulated as a percentage of all workers 16–60 years old*		
Self-employed[a]	50+	10+
In public units[a]		70–
In nonpublic units[a]	30	10
Of that; in management, or professional or technical work[a]	Very few	8–
In construction in %[b]	30	
In Guangdong Province working in physical, dirty, and dangerous work[c]	66	
Having signed a contract in %[a]	29	53
Wages		
Average hourly pay in yuan[a]	4.05	5.70
Average annual wage in 2004[d]	780	1,345
How much did it cost to find a job?		
Average cost to highest in yuan[a]	80–10,000	56
Guarantee cash, deposits paid to new job		
Average fee to highest in yuan[a]	66–5,000	57
Defaulting or pocketing workers' pay in % of units[a]	12–24[b]	9
Highest amount defaulted in yuan[e]	45,000	
Being defaulted by 1,000 yuan or more by company in %[e]	72.5	
Joint-ventures, private, or foreign-invested firms defaulting, pocketing, or refusing to pay in %[f]	64.4	
Defaulted pay in time of up to[f]	1 year	

NOTE: *In Type of Work the proportion of rural migrant to all rural migrants is compared to the equivalent proportion among urbanites; hence roughly 30% of all migrants work in construction, but of all workers in construction, 80–90% are rural migrants (see first page of paper), while under Occupational Disease migrants are counted as a proportion of all urban and rural laborers.

Sources:

[a]China Urban Labor Survey in five Chinese cities by the Institute of Population and Economics, Chinese Academy of Social Sciences in Cai Fang & Wang Meiyan, "The Marginalisation of Migrant Workers in China" in Chinese Cross Currents, April 1, 2004.

[b]Xinhua, China News agency, May 26, 2005, according to ACFTU, in CSR, 1:22, p. 5, Frost S., "China View."

[c]Information Times, January 26, 2005, according to Wu Suisheng, Representative of the NPC of Guangdong, in CSR, 1:5, p. 7, Frost S., "China View."

[d]Xinhua, China New Agency, October 10, 2005, quoting Wuyun Qimuge, Vice-director of the Standing Committee of the NPC, National People's Congress in CSR Asia Weekly, 1:44, p.7.

[e]Li Qiang's "Budeyi de feifa shengcun," (Forced illegal existence) in Gaige Neikan, (Internal References for Reform), No. 2, 2003, same as footnote *a* of this table.

[f]Xinhuanet.com, January 24, 2002, same as footnote *a* of this table.

Cambodia. Normally three main explanations for this state of affairs have been given:

(*a*) China's inexhaustible supply of labor; (*b*) the "hu-kou" system, the household registration system; and (*c*) the lack of freedom to organize independent trade unions.

However four more reasons can be forwarded:

(*a*) The decentralization and deregulation of wages; (*b*) local governments turning a blind eye to labor exploitation; (*c*) an exploitative administrative system for migrant labor; and (d) low agricultural procurement prices and the agricultural policy of the government.

The first three reasons mainly refer to China's rural population of almost 800 million people (60% of 1.3 billon) and to "hu-kou," a system originally set up in 1958 to issue local food ration coupons to only those with urban registration, initiated as part of a planned economy, and is now likened to the South African pass system. In the Chinese party state, only the trade union under the absolute authority of the Communist Party has legally been tolerated and often seems to work as a tool of those at higher levels to control and direct workers. The four additional reasons require more detailed evaluation.

The Decentralization and Deregulation of Wages

In the early 1990s China instituted a system of minimum wages as a way of protecting workers particularly in the new industries funded by foreign investment. The minimum wage was to be about 40–60% of the average local wage and be at least sufficient to guarantee a minimum level of survival. However, through decentralization, the setting of wages was to be done by the local authorities, which has resulted in over 100 different minimum wage levels today. After more than a decade with this system, a general trend has been discovered: the more globalized an area is, the lower the minimum wage is in relation to the local wage. In the beginning the first export production zone of Shenzhen in the Pearl River Delta received the highest minimum wages in China due to their general high cost of living. In 1993 the local minimum wage was 40% of the general average wage, while in 1999 it was barely 24%. This means that if a worker with an urban "hu-kou" receives 100 yuan, a rural migrant is paid 24 yuan for doing exactly the same job. Moreover, all the clean, well-paying, easy jobs go to local residents while outsiders, the rural migrants, perform the dirty, noisy, dangerous, and low-paying jobs[11] (TABLE 1). The Shenzhen minimum wage remained the same or even declined during the 1990s. Even the decrease in the capital of Beijing was less steep, ranging from 37% to 27% of the local average wage during roughly the same time. Both Guangzhou and Shenzhen had the lowest minimum wages in relation to the local wages until the policy changed in July 2005. In addition many are working illegally and working long overtime hours (usually unpaid) making their wages even lower.[12]

A survey conducted by the Guangdong Province Department of Labor and Social Security discovered the following three things: (*a*) 85% of approximately 26 million migrants workers work about 10–14 h daily; (*b*) nearly half of them have no day off; and (*c*) most are not paid for overtime (TABLE 1). A survey of the garment industry in the Pearl River Delta released in 2004 revealed that

TABLE 2. Comparison of rural migrants with urban resident employees, old age, unemployment, and medical insurance

	Rural migrants	Urban resident employees
Provision of old age social security		
On the average in %[a]	7	69
Range on the average in %[a]	2–12	63–71
Pearl River Delta in %[g]	4	
Beijing, Nanjing, Wuhan, Xian, Tianjin, Changchun in %[h]	14	88
Unemployment insurance [i,j]	Not eligible	All urban employees
Medical insurance		
Medical insurance in %[a]	8	68
Range of medical insurance in %[a]	4–10	56–72
Getting medical expenses covered or subsidized by employer in %[a]	4	53
"Never got a penny for medical expenses" in % of all workers[e]	93	
If you are entitled to medical expenses? Did you get reimbursed?[a]	60.8	63.4
If you are entitled to medical expenses what are you most likely to get paid for partly or fully?[g]		
On-the-job-injury in %	72	
Serious diseases in %	33	
Common illness in %	23	
Pregnancy, maternity leave in %	18	

NOTE: Sources:
For [a,e] see TABLE 1.
[g] Migrant Workers' Research Group, "The Conditions of Migrants Workers in the Pearl River Delta," in Zhongguo shehui kexue, (China Social Sciences) No. 4, 1995.
[h] Survey by Department of Sociology of Beijing University, Institute of Population and Labor Economics, Chinese Academy of Social Sciences and Australian University and Bates College, USA, 2000.
[i] Wang Fengyu, Li Lulu et al., "Zhongguo chengshi laodongli liudong," (Mobility of labor force in Chinese cities), Beijing Press, 2001.
[j] Asia Monitor Resource Centre Ltd, "At what Price? Workers in China," Hong Kong, 1997.

(*a*) workers' wages had declined during the previous 3 years; (*b*) 15–50% of the workers in the eight factories investigated did not receive the local, legal minimum wage; and (*c*) neither were they paid for sick leave, maternity leave, or overtime[13] (TABLE 2).

This investigation found that the 155 workers interviewed between April 2003 and September 2003 worked on average 308 h monthly compared to the 36 h legally allowed for overtime in addition to 168 regular hours. Although overtime had increased, wages fell and labor costs usually made up only 1–5% of the total retail prices. Some workers were even paid by their employers to lie to independent inspectors sent by overseas suppliers in order to hide their illegal overtime and below-minimum wages.[14] In one factory, workers were lined up the night before the "human rights monitors" arrived and told to recite and remember "correct" answers.[15]

Another survey found that in the footwear industry 11 h was the usual length of the workday with no overtime payment.[16] The first cases of death of migrant workers because of overwork—*karoshi* as it is called in Japanese—have been reported.[17]

In some cases, wages are withheld from workers. In one survey, up to two-thirds of all migrant workers were affected (TABLE 1); another survey stated, "The illegal retention of workers' wages for between one and three months exists in 80% of foreign financed firms."[18] The Chinese press abounds with reports of employers hiring either security guards or local bullies to attack migrant workers who try to secure their back pay.[19] When state enterprises began to lay off workers in the 1990s, the privileged urban population demanded that more types of work go to them rather than to the migrant rural workers.[20] With the safety net of the "iron rice bowl" gone and only one child allowed for the urban population—with this child increasingly seen as a security for old age—the urban authorities felt pressure to consider their own people first in order to ensure social stability.[21] In Shanghai, for example, jobs were classified into three types: category A, open to migrants; category B, under special conditions open to migrants; and category C, closed to rural migrants. The last category included such work as security guards, taxi drivers, kindergarten teachers, telephone operators, front-desk clerks at high-class hotels, and work in insurance and finance. If employers hired cheap migrant workers, they had to pay a fee of 50 yuan monthly for each rural migrant employed, and half of these fees, taken from the toil of migrants, were used for an unemployment fund only open to workers with an urban "hu-kou!"[22] In this way, rural migrants were effectively excluded from well-paying jobs.

Local Governments Turned a Blind Eye to Labor Exploitation

Coastal areas were most developed, paid the highest wages, and had the highest cost of living compared to inland regions. This meant that local-level authorities in coastal areas were afraid of increasing the already low minimum wages for rural migrants because this could mean losing manufacturing jobs to the inland regions. Because of this over the years the income gap between rural migrant labor and the local population has grown. The urban population can also increase their wages by shifting places of work, while after the first job shift, rural migrants could hardly improve their job status.[23]

When in trouble, workers would say, "The company will not help us and the authorities do not protect us. We must have a trade union that acts on our behalf."[24] Hence local governments in the SEZs did not intervene when workers' rights were abused, nor did they, in case of open conflict, side with the workers.

An Exploitative Administrative System of Migrant Labor

Looking at the cost for a migrant, a rural woman worker about to find employment in the Pearl River Delta in 2003 explained why the local government was using velvet gloves with employers. Before this woman could even start out from her home village, she needed to have: (*a*) a border region pass that cost 120 yuan; (*b*) an unmarried status certificate that cost 60 yuan and only valid for 1 year; (*c*) a personal identity card that cost 80 yuan; and (*d*) a planned birth certificate to show that she is not born out of quota, costing 45 yuan and only valid for 1 year. This is a cost of 305 yuan even before she starts out from her village. If she obtained a job in the delta, she would normally have to pay (*a*) a 300-yuan deposit to her employer that was forfeited if she left her job early; (*b*) a temporary residence permit that cost 300 yuan; and (*c*) a work permit for 40 yuan. Hence, even before beginning work, this woman has to spend the equivalent of 640 yuan, which for a beginner would be about two full month's wages. For obvious reasons our woman migrant worker would be desperate to accept any job and working conditions available, meaning that the would-be employer could lower wages and increase unpaid overtime even more. The minimum wage that year in the delta was 547 yuan, but this sum was not always paid, especially to new arrivals. It has been calculated that the 3–4 million migrant workers, the overwhelming majority female, in the Pearl River Delta Region were, on average, paying 600 yuan every year for certificates. Local authorities collect certificate fees of between 1,800 to 2,400 million yuan annually from migrants, without any equivalent outlays. These migrants have no right to free schools for their children, to health care, or to any social benefits. These were supposed to be provided by their home villages[25] (TABLE 2).

In addition, a migrant worker must have all the permits with her at all times or she can be fined by the police or physically abused, put into a detention camp, abused again, and then sent back to her home village after being stripped of all her belongings. Therefore, in the Delta there is a booming trade in false permits, which, if found, carry an even heavier fine or worse treatment. After much public outcry this situation is now changing, particularly in towns and smaller cities, with the most rapid changes occurring in Southern China (TABLE 2).

Difficulties that are faced by the rural migrants are exemplified by the fact that rural migrants make up only half of the workforce in urban China but suffer 90% of all deaths due to unsafe working conditions. Data from Guangdong Province where two-thirds of all migrants live in substandard housing and work in dirty, dangerous environments, show that they are victims of 80% of all industrial deaths and injuries (TABLE 3). In the Pearl River Delta and in Guangzhou in the local, Taiwanese- and Hong Kong-funded factories, where many rural migrants work, accident rates were highest. The lowest rates were recorded in other foreign-funded enterprises with more mod-

TABLE 3. Comparison of rural migrants with urban resident employees, occupational disease, and housing

	Rural migrants	Urban resident employees
Occupational disease and death in 2004		
Occupational disease patients, in %[d]	50+	50−
All industrial accident deaths due to unsafe work conditions in % of all workers[d]	90	10
In Guangdong Province of all industrial deaths and injury in % of all workers[c]	80	20
Quality of housing[i]:		
Sharing accommodation, in %[i]	49	
Sharing accommodation with three or more people, in %[i]	24	
Having tap water in house in %[i]	90	
Having gas in house in %[i]	60	
Having heating in house in %[i]	39	
Beijing, Wuxi, Zhuhai, living under extremely poor conditions in %[i]	70	

Sources: See TABLES 1 and 2.

ern conditions, whereas a low but still double accident rate was found in state-owned factories, explained by a slower work pace and frequent overstaffing (TABLE 4).

Low Agricultural Procurement Prices and Government Agricultural Policy

Our last crucial factor to discuss is why regardless of low pay and abusive work conditions, migrants still kept coming in accelerated numbers to work in the Pearl River Delta, driven by worse situations in their home villages.

TABLE 4. Industrial injuries in different types of factories in the Pearl River Delta and in Guangzhou, 2001-2004, in %

In European-, American-, Japanese-, and Korean-funded factories[k]	1.7
In collective factories[k]	1.9
In state-owned factories[k]	3.5
In Hong Kong- and Taiwan-funded factories[k]	37.5
In local factories[k]	44.9
Not defined[k]	11.0

Source:
[k]Xie Zexian, Professor of Guangdong Business College, survey results from interviewing 582 injured workers in 39 hospitals in the Pearl River Delta and in Guangzhou during the last 3 years reported in Guangzhou Daily, September 12, 2005, according to Stephen Frost, "China View," CSR 1:38, p. 4–5, 2005.

In Guangdong province the 12 million migrant workers in 2002, had by January 2005 swollen to 31 million compared to 79 million permanent registered residents.[26]

The Chinese government releases Document No. 1 on Chinese farming every year. In the 2004 document, agricultural prices increased leading to pay raises of almost 7% for the peasants who, due to a good harvest, managed to get 25% more for selling their produce. Altogether, rural income growth was estimated at 10.6%, while cost of living rose by 3.6% in the cities during 2004.[27]

A NEW SITUATION

For the first time, in the spring of 2004 the Pearl River Delta experienced a labor deficit. According to one estimate, manufacturing jobs were short 1 million workers. Another report from the department of Labor and Social Security calculated a shortfall of 2 million workers in just this region. The enterprises experiencing the worst deficits of labor were those with owners or investors from Hong Kong, Macao, and Taiwan already pinpointed by investigators as being the worst culprits in treatment of labor.[28] The factories producing toys, low-cost electronic products, shoes, clothing, and plastics were worst hit; these were the types of factories known for the worst working conditions. This was documented in an official report by the National Ministry of Labor and Social Security in September 2004.[29]

According to Li Qiang, executive director of the Hong Kong–based China Labor Watch, China's manufacturing wealth was based on the "enormous mental and physical strength of young workers from rural areas."[30] A substantial amount of research has documented the sweatshop conditions of the migrant workforce in global factories and their subcontractors in the delta region during the last 20 years of rapid growth.[31]

Because of these conditions, Guangdong Province has been experiencing a dramatic increase in labor disputes—on a national level doubling every year since 1994—with an army of 31 million migrant workers asking for better work conditions. As independent trade unions are not allowed, only disputes or voting with the feet are solutions.[32]

In a labor shortage situation, the chance increased that some labor demands could be met through litigation, some fee requirements be relaxed, and pressure could be put to better pay and working conditions. In the summer of 2004, cities in Southern China were ordered by the provincial governments to shorten overtime and raise minimum wage rates after years of no pay increases.[33] Bad conditions kept migrant workers away from manufacturing work, and this situation continued in 2005. For example, a factory in the center of the Pearl River Delta in Dongguan moved further inland to a county with lower wages after only 20 of its former 200 workers returned to work after the Spring Festival of 2005.[34] Normally in Guangdong, during the Qing Ming Festival in late spring, workers do not get the day off, but after extensive reporting in the media on shortage of labor in early 2005 some employers gave workers the day off; workers in other firms just left and were not afraid of either loosing their job or getting reduced wages.[35] According to officials, as a direct result of the difficulty in recruiting workers, the Shenzhen authorities decided to increase the minimum wages as of July 1, 2005 from 610 to 690 yuan for the SEZ and from 480 to 580 yuan for the Outer Zone. The new standards will be the highest in China, surpassing Shanghai as the largest increases in the preceding year.[36]

Not surprisingly, the migrant labor shortage was most serious in labor-intensive factories with poor working conditions and low wages. Initially Southern China needed unskilled, especially female, workers and fewer skilled workers. The reasons given for this need, and for the bad conditions and low wages were (*a*) that increased demand for labor rising by some 10% every year in Shenzhen due to rapid economic development, (*b*) that cost and benefits were not equal (wages only increased by 68 yuan over the past 10 years, while food prices were rising continually, the increase amounting to that sum every other year); (*c*) that poor working conditions were exacerbated by poor enforcement of workers' rights; and (*d*) that rising prices of raw materials constrained wage increases.[37]

In order to alleviate the shortage of workers and attract migrants, for example, the Federation of Trade Unions in Jingjiang, with a 50% migrant membership, promised that rural migrants would be treated as urban workers, implying that they would get some support for their children's education; have the right to complain and to suggest solutions, vote and be elected; get one or two free training sessions provided by the trade union; and get some subsidies for medical fees.[38]

Simultaneously, reports on labor shortages in other areas began to appear. Workers were needed not only in light industry but also in the accident-prone mining business in inland regions. A report said that there was basically no shortage of labor, but that wages and labor conditions were so poor that people were leaving or refusing to take up positions. It was reported that when 2000 workers recently left the state mines, the province of Henan allocated 45 million yuan to train peasants from remote and poor areas for work there.[39]

In the year following the first reported labor shortage, the Agricultural Document No. 1 of 2005 stated that 70% of the recently augmented resources for the fields of health, education, and culture were going to be earmarked for rural areas by the central government. The latest Organization for Economic Cooperation and Development (OECD) review of Chinese agricultural policies recommended that China actively contribute to erasing the great and growing economic income disparity between rural and urban areas that was due to largely limited mobility of capital and labor, stressing that this division was aggravated by differences in access to health care, social services, and education: that is, that the "hu-kou" system should be abolished.[40]

LABOR LAWS AND NEW STANDARDS

Since 1949, there has been only one official discourse on the treatment of workers in China expressed in the labor laws, the last one promulgated in 1995, and in the Constitution of China. Although the law of 1995 was regarded as satisfactory, it remained weak on various labor standards and on methods of implementation. In general, China lacks the capacity and basic infrastructure to enforce and monitor this new labor law. This means long legal processing that is hampered because cases need to be filed within 60 days of a violation. Legal education here is seen as critical.[41]

However, both the 1993 Regulation of Handling Labor Disputes and the 1995 Labor Law gave migrant workers and employees the legal right to complain about wages, termination of contracts, occupational health and safety, and fringe benefits to their local Labor Dispute Arbitration Committee. Therefore according to the China Labor and Social Security Handbook, arbitrated labor disputes and collective (more than 30 people) labor disputes increased 15 and 18 times, respectively, and the number of employees involved in disputes from 1993 to 2002 increased by 18 times.[42]

Introducing internationally agreed-upon standards such as the 1998 SA8000 Social Accountability (SA) stressing freedom of association and collective bargaining—to the Chinese light labor-intensive industry where a majority of rural migrants work could as well be viewed as part of (*a*) a process of globalization; (*b*) China's quest for WTO membership; (*c*) a response by Transnational Corporations (TNCs) to human rights campaigns; (*d*) a way for western trade unions to stop wage "dumping;" and (*e*) an attempt by western politicians to appease their electorate by not "loosing" jobs to China.

Allowing the introduction of "Codes of Conduct" in the late 1990s can be seen as another step in the Chinese process of globalization and of conforming to international standards to enable the rapid expansion of Chinese exports.[43] Antisweatshop campaigns from different international and national organizations, nongovernmental organizations (NGOs), foremost among them consumer organizations in western countries, had created public opinion demanding humane living conditions for workers in the supply chain to western products and brand names. These movements and organizations have forced TNCs into proposing "codes of conduct," thereby promoting a "moral economy," institutionalizing "international best practices" to appease consumer organizations, western trade unions, and others. On the international level such initiatives for social responsibility as the Global Compact from 2003 for particularly transnational, corporate participants for promoting 10 principles—in the areas of labor standards, environment, human rights, transparency, and anticorruption—in partnership with the United Nations added importance to local campaigns.[44] In line with what is called Corporate Social Responsibility (CSR) has been the latest development of this type, where four aspects were outlined by Fan Baojin, Chairman of the China Charity Association: economic, legal, moral, and philanthropic responsibility.[45]

Meanwhile western trade unions have voiced concerns about and opposed wage "dumping" and "employee hostile" trade unions in countries where neither freedom of association nor collective bargaining is allowed. In response many TNCs have promoted a more "employer-friendly" type of labor union for companies involved in export production in China. Recent research shows that working conditions in subcontracting firms to TNCs with some type of trade union and "rules of conduct," however shallowly implemented, were better than in those without.[46]

According to these reports a new discussion regarding in the Rules of Conduct seemed to be brewing among the TNCs while some layers of discrimination still continued unabated. Research was aimed at appraising the fact that the Chinese subcontractor firms often failed to "empower" the workers and encourage their initiatives. Other researchers, however, found that Chinese working women fared better under globalized factory regimes than in Chinese state industry laden with a corporatist ideology where seemingly seniority mixed with "old boy network," and paternalism made for an unhealthy cocktail particularly for young women workers.[47] When university graduates were asked to choose the 50 best companies to work for, 32 chose foreign companies.[48]

When the new standard of SA8000 was introduced from the outside, it was questioned by both trade union activists and by Chinese officials, but for different reasons. Research supported by the labor activist group concluded that rather than empowering workers, the labor codes of transnational corporations and their implementation in Chinese subcontractor firms served chiefly to deemphasize the "labor rights" discourse from workers and incorporate it into the workings of transnational capital thereby denying legitimacy to workers' own models of organization and resistance where workers' activism and own initiative was not encouraged by the employers or by the Chinese state.[49] While labor activists thought that labor's own organizing activities were not encouraged, official representatives of China expressed fears that they would be and voiced apprehension about foreign interference. The official view expressed by Ji Mingbo, former secretary of the secretariat of the All-China Federation of Trade Unions (ACFTU), in criticizing the CSR and particularly SA8000, was that, "the collective bargaining and freedom of association clauses in SA8000 were not consistent with either the laws of China or the real situation in the country. SA8000 may provide hostile western forces with the ability to butt into the trade union movement in China."[50] At the first CSR conference at the government level in October 2004, Wang Maolin, member of the Standing Committee of the National People's Congress (NPC), vice director of its Law Committee, president of the China Institute of Multinational Companies, and chairman of the Chinese Association of Productivity Science stated: "Other countries are doing research on and promoting CSR. We have to be aware of and prevent people in the West (xifang) from using CSR issues to influence its development in China in the wrong direction."⁵¹

However, Chen Yuanqiao, a senior engineer at the China National Institute of Standardization, could see some features inherent in CSR being in line with national policies and goals, such as encouragement of environmental protection, work safety, and improvement of working conditions while other features such as freedom of association and right to collective bargaining were inconsistent with the legal and political system of China.⁵²

A totally different view was held by Bao Yujun, President of the China Private Economy Research Association, who warned that demands on CSR should not push enterprises into behaving like the failed state firms in mistaking social for corporate responsibility.⁵³

To counter this situation the Chinese National Textile and Apparel Council (CNTAC) released their own, new Chinese quality control system called CSC9000T in June 2005. It was launched as a management system, applying Simon Zadek's five stages of organizational learning: (*a*) defensive, (*b*) compliance, (*c*) managerial, (*d*) strategic, and (*e*) civic. Most Chinese firms were said to be on the first, or perhaps second stage of learning, while few companies had attained the third stage according to Lucy Lu, Deputy Director of the Office for Promoting Social Responsibility at CNTAC, saying in an official letter, "Rather than go beyond the requirements of Chinese law, we equip companies with the right tools to operate within the legal framework."⁵⁴

In SA8000 the right of workers to freedom of association and collective bargaining was clearly expressed, but in China only the official ACFTU under the direct authority of the Communist Party was allowed.⁵⁵

Further promotion of adherence to international labor standards came from Hong Kong in June of 2005 when an NGO named Students and Scholars Against Corporate Misbehaviour (SACOM) was founded by labor activists and students to actively monitor and campaign against violations of workers' rights, and to lobby for dignity, welfare, health, and safety. SACOM had, in collaboration with the U.S. National Labor Committee (NLC) and labor organizations in Southern China, initiated campaigns and released reports about working conditions in toy, shoe, and apparel firms.⁵⁶ A novel method was to name and shame "blood and sweat factories" (as it is called in Chinese "xiehan gongchang") on the Internet. This is exactly what the Guangdong Department of Labor and Social Security did when it posted the names of 20 such companies and their violations of the labor law on their web site and in a respected Chinese newspaper.⁵⁷ Such exposure was the first of its kind in China.

A totally different way to look at how to stop the race to the bottom is to look at what pushed people to migrate from rural areas. Except the usual "push" arguments, the "extreme neglect of rural areas" has been forwarded as one explanation by specifying further, "the combination of severe poverty, lack of political rights, poor or non-existing facilities, and a fiscal system managed by predatory and irresponsible local officials are the real reasons why people feel compelled to migrate."[58] Other research both confirms and modifies this point by showing that those without political clout migrate far away to other provinces and coastal zones, while those with connections stay closer in order to use their political capital implying there is no level playing field for outsiders.[59]

To alleviate this situation, different solutions have been suggested. One has been to drain the urban "hu-kou" of its "golden content" by gradually erasing the artificial barriers against rural migrants by (*a*) allowing migrants into more types of jobs; (*b*) ending the practice of preferential treatment to urbanites, for example, free schooling, medical treatment, and heavily subsidized housing; and (*c*) eliminating fines for hiring rural migrants, fines which go to provide benefits for "hu-kou" holders. Hence the "hu-kou" system is being both questioned and gradually weakened while rural migrant labor is becoming increasingly aware of their legal rights and their own might. The system of labor standards is in the process of change greatly helped by workers voting with their feet and avoiding the worst work places.

Looking at the latest trends might give some basis for optimism and some clue to future developments. Both the Chinese state and foreign investors are interested in developing high technology industries, services, and environmental protection. Research and development centers are also high on the agenda.[60,61] This means a more skilled and educated workforce is needed. Higher skills would be rewarded not only by a demand for more skilled workers, but also through more training of the existing unskilled workforce where rural migrants make up a majority.

CONCLUSIONS

When workers in rural areas choose to remain where they are rather than migrate back to previously higher paying urban areas, it is clear that worsening labor conditions in urban areas have reached the bottom. This is exemplified by the present shortages of labor in urban sweatshop enterprises. Not surprisingly, the industries in Taiwan first affected by workers leaving are now the same ones suffering in China, namely textiles, garments, and toys. Enterprises with investors from Hong Kong, Taiwan, and Macao committing the worst abuses were also those losing their workers first. Hence in a situation where the formation of trade unions was illegal, workers both voted with their feet and started complaining more after becoming increasingly aware of their newly accorded labor rights and of recent legislative changes. The existence of the "hu-kou" system had made matters worse for rural migrant labor who were treated like second-class citizens in urban areas. This system is now being questioned and is undergoing slow reform, particularly in the fast-growing industrial centers and industrial production zones of Southern China.

Simultaneously, while international labor standards were being introduced in China, a debate ensued on how to reconcile these new required standards with what Chinese authorities perceived were Chinese conditions. For this reason the SA8000 was supposed to be supplanted by, or at least become a supplement to, the Chinese CSC9000T introduced in June of 2005.

Perhaps the "race toward the bottom" is being halted by a combination of outside forces and China's own introduction of a system entitling workers to more legal rights, by gaining membership in the WTO, and by different types of labor standards introduced by TNCs and enforced, however weakly, through their supply chain.

Three changes that would be decisive in stopping the "race to the bottom" are the total abolishment of the "hu-kou" system, the right of workers to collective bargaining, and freedom of association. Progress is being made by loosening up the "hu-kou" system, by giving workers more rights in the 1993 Regulation of Handling Labor Disputes, and by the 1995 Labor Law, which gives migrant workers and employees the legal right to complain to their local level Labor Dispute Arbitration Committee about wages, termination of contracts, occupational health and safety, and fringe benefits.

We hope a turning point has been reached in China that will have a positive impact on both workers and corporations in the rest of the world.

REFERENCES

1. RISKIN, C., Z. RENWEI, L. SHI, Eds. China's Retreat From Equality Income Distribution and Economic Transition. M. E. Sharpe, 2001 and World Development Report 2006: Equity and Development, (WDR –2006), Part I, Inequity within and across countries, p. 25 ff.
2. THORBORG, M. Chinese Employment Policy in 1949–1978 With Special Emphasis on Women in Rural Production, in Chinese Economy Post-Mao, Joint Economy Committee, Congress of the United States, Nov. 1978, p.535 ff, 210 A-4 pages, U.S: Government Printing Office, Washington D.C., Women in Non-Agricultural Production in Post-Revolutionary China, (Dissertation), 230 pp. Uppsala University, 1980.
3. FEI-LING, W. 2005. Organizing through Division and Exclusion China's "hu-kou" system. Stanford University Press.
4. Two types of very different kinds of "hu-kou"s exist, one category called permanent migrant registration – including to 90% already highly educated city and urban residents 2/3 of whom were transferred through their work – and the other one called temporary migrant registration – including over 75% rural migrants and 75 % of them only having received the lowest two levels of education and 87 % of then being engaged in factory jobs or business – according to census statistics from Shenzhen 1990, in Zai Liang and Yiu Por Chen, "Migration, Gender and Returns to Education in Shenzhen, China," in eds Garcia BRIGIDA, Anker RICHARD, Pinelli ANTONELLA in Women in the Labor Market in Changing Economies: Demographic Issues, Oxford University Press, 2003. In this paper we discuss temporary migrant registration.

5. YANG Yunyan and Chen JINYONG, "Transitional Labor Market Segmentation and Competition," in Social Sciences in China, Nr 4, 2000
6. CSR, Asia Weekly: on www.crs.asiacom/index.php/archives/2005/01/29/ (hereafter CSR), 1:37, p.4, China Radio Net, 2005.09.09.and the figure 90 % in construction from Xinhua, China News agency, 2005.05.26., according to ACFTU, in CSR, 1:22, p. 5, Frost Stephen, "China View."
7. THORBORG, M. *Industrialization in East- and Southeast Asia, Trends Concerning Women in the Work Force* in Southeast Asia: Contemporary Perspectives, ed.s HEIKKILÆ-HORN, MARJA-LEENA and SEPPÆNEN JOUKO, HELSINKI, 1990, pp.119–132 and Book Review, *Lin Yi-min, Between Politics and Market Forms Competition and Institutional Change in Post-Mao China* for China Perspectives, No. 41, French Centre for Research on Contemporary China, Hong Kong, pp. 59-62, May-June 2002 and Lee Ching Kwan, "Three Patterns of Working Class Transition in China," in ed.s Rocca Jean-Loius and Mengin Francoise, "Politics in China Moving Frontiers," Palgrave, 2002.
8. "Round-tripping" refers to inland Chinese capital taking a detour via Hong Kong in order to enjoy the same economic benefits as foreign invested capital in Chinese SEZs.
9. THORBORG, M. *The development of special economic zones in Asia, with reference to the People's Republic of China*" in Proceedings of the 31st International Congress of Human Sciences in Asia and North Africa, 1983, Vol. 2, p.756 and 2002 op.cit.
10. SUNG, et al. "The Fifth Dragon (The emergence of the Pearl River Delta)," Addison Wesley, Singapore, 2004,Yeh *et al.* eds, "Building a Competetive Pearl River Delta Region Cooperation, Coordination, and Planning," The University of Hong Kong, 2002, Cheng Joseph Y. S., ed., "The Guangdong Development Model and Its Challenges," 1998, and ed.,"Guangdong in the Twenty-first Century: Stagnation or Second Take-off?" 2000, both City University of Hong Kong Press, and "Guangdong Preparing for the WTO Challenge," The Chinese University Press, 2003 Cheng J., and Macpherson S., ed.s, "Development in Southern China A Report on The Pearl RiverDelta Region including The Special Economic Zones", Longman Asia ltd, Hong Kong, 1995, Lin George C.S., "Red Capitalism in South China Growth and Development of the Pearl River Delta," UBC Press, 1997 and "An Emerging Global City Region Economic and Social Integration Between Hong Kong and the Pearl River Delta" in ed. So Alvin, "China's Development Miracle" M.E. Sharpe, 2003, Wu Weiping, "Pioneering Economic Reform in China's Special Economic Zones," Ashgate, UK, 1999.
11. Asia Monitor Resource Center Ltd, "At what Price? Workers in China," Hong Kong 1997.
12. Oxfam, "Turning the Garment Industry Inside Out Purchasing Practises and Workers' Lives," Oxfam, Hong Kong Briefing Paper, April 2004.
13. CSR, 1:44, p. 6, Frost S., "China View."
14. CSR, 1:14, p. 8, Frost S., "China View."
15. Contemporary News Express, 2005.04.28, according CSR, 1:18, p.10 in Frost S., "China View."
16. YEUNG, G. 2001. "Foreign Direct Investment and Investment Environment in Dongguan Municipality in Southern China," Journal of Contemporary China.
17. CSR, 1:14, p.8.
18. SOLINGER. 2002. "Labor in Limbo: Pushed by the Plan toward the Mirage of the Market" in eds ROCCA Jean-Loius and Mengin FRANCOISE, "Politics in China Moving Frontiers," Palgrave.

19. THORBORG, M. 2005. Where have all the Young Girls Gone, Chinese Fatal Daughter Discrimination in a Comparative Perspective, in China Perspectives, Hong Kong, No.86 in French and No. 56 in English, January-February, 2005.
20. FEN, W. & S. ANAN. 2003. "Double Jeopardy? Female Rural Migrant Laborers in Urban China, The Case of Shanghai" in Garcia Brigida *et al.* in Women in the Labor Market in Changing Economies: Demographic Issues, Oxford University Press.
21. QIANG, L. 2000. "Occupational Mobility of Rural Workers in Chinese Cities", in Social Sciences in China, Nr 4.
22. "17,000 workers strike at Uniden's Shenzhen plant." 2005. http://www.csr-asia.com/index.php/archives /2005/04/23/17000-workers-strike-at-unidens-shenzhen-plant/ "Migration as the Second Best Option Local Power and Off-farm Employment", in China Quarterly, Nr 181, March and Dagens Arbete, (Work of the day), Member magazine of the Metal Workers' Union of Sweden, Dokument, (Document) "Kinas arbetare har fått nog", (China's workers have had enough) Nr 9.
23. LINDSTRÖM, S. 2004. "Gränslösa kläder," (Limitless clothes), Atlas and,"Högt pris i Kina för billiga varor i Väst," (High price in China for cheap products in the West), in Dagens Nyheter, DN, (Daily News) January 19,2005.
24. CSR Asia on www.csr.asiacom/index.php/archives/2005/01/29/ p.1.
25. LI JULIETTE. 2005. "The Pearl River Delta migrant labor shortage" CSR 1: 9, p.2, and Fong Mei,"A Chinese Puzzle" in Asian Business News in The Wall Street Journal 2004.08.16, online,., and "Labor shortage in Pearl River Delta" in www.amrc.org.hk/5210.htm.
26. PETTERSON, T. "Kineser ratar låglönejobb", (Chinese skip low wage jobs) DN, Daily News, Sweden, 2004.12.27.
27. FONG MEI, 2004.08.16.
28. ANDORS, "Women and Work in Shenzhen," Bulletin of Concerned Asian Scholars," Vol.**20:3**, 1988, Fan C. Cindy, "Migration and Gender in China" in Lau Chung-ming and Shen Jianfa eds., China Review, Chinese University Press, Hong Kong, 2000, HRC Human Rights in China, "Institutionalized Exclusion: The tenous legal status of internal migrants in China's major cities," a report November 6, 2002, p.8, Shi Li, "Migration of Rural Labor and Income Growth and Distribution in Rural Areas of China," in Social Sciences in China, Autumn 2000, Solinger Dorothy J., "Citizenship Issues in China's Internal Migration: Comparisons with Germany and Japan," in Political Science Quarterly, Vol.114, No.3, 1999, pp. 455-78 and 2002, MacLaren Anne E., "Chinese Women – Living and Working," RoutledgeCurzon 2004, Ngai Pun, "Made in China Women Factory Workers in a Global Workplace," Hong Kong University Press, 2005.
29. LU YING, 2003. "Protection of Women Workers' Rights," (in Chinese), Women and Gender Study Centre, Sun Yat-Sen Univesrity, Guangzhou.
30. FONG MEI, 2004.08.16.
31. RMRB, Ren-min ri-bao, People's Daily, 2005.04.20.
32. Southern Morning News, 2005.04.05, in CSR, 1:15, p.9,
33. JINGBAO, 2005.05.31. and Southern Metro Daily, 2005.05.31., in CSR, 1:22, p 4.
34. CSR, 1:24, p 7,
35. Workers' Daily, 2005.04.05, in CSR, 1:15, p.8,
36. CSR, 1:37, p.3,
37. OECD,"OECD Review of agricultural Policies China," http://www.oecd.org/infobycountry/0,2981,en_2649_201185_1_70342_1_1_1,00.html
38. OXFAM, 2004.

39. CHAN, J. Wai-ling, "The end of the MFA and the rising tide of labor disputes in China" in CSR, 1:11, p. 6,
40. DE TRENK, Charles *et al*. "Red Chips and the Globalization of China's Enterprises", Asia 2000 Ltd, Hong Kong, 1998.
41. KRØLDRUP, Lars, "Social ansvarighed under lup" (Social responsibility being scrutinised) in Udvikling, (Development) October, Nr 7, 2005, monthly magazine from Danida, Danish Development Agency,
42. CSR, 1:44, p. 8.
43. CHAN, Anita, Labor Relations in Foreign-Funded Ventures, Chinese Trade Unions and the Prospects for Collective Bargaining," in ed. O'LEARY G., Adjusting to Capitalism: Chinese Workers and the State, 1998, and "China's Workers Under Assault: The Exploitation of Labor in a Globalizing Economy," both Armonk, N.Y 2001, Sargeson Sally, "Reworking China's proletariat," N.Y., St. Martin's Press,1999, Tomba Luigi, "Paradoxes of Labor Reform Chinese Labor Theory and Practise From Socialism To Market," University of Hawaii Press,2002 and Ngai 2005.
44. YI, C., S. DÉMURGER, M. FOURNIER, 2004. "Salary Differentials according to Sex in Urban China," in *China Perspectives,* No. 54, July, August 2004, Ping Ping, "Gender Strategy in the Management of State-owned Enterprises and Women Workers' Dependence on the Enterprises," in Social Sciences in China, No.1, 2000.
45. Sina Finance, 2004.07.08, in CSR, 1:30, p. 11.
46. NGAI, 2005.
47. CSR, 1:37, p.5,FROST, S.
48. CSR, 1:42, p.1ff Frost S. and Ho Brian, "GoTone Nanchang CSR Forum."
49. CSR, 1:39, p 1-3, Frost S. and Ho Brian, "The year of CSR in China."
50. CSR, 1:45 p.11, Frost S., "China View."
51. CSR, 1:24, p. 4, "CSC9000T: An update from China" by Frost, S. in response to Lu, L. [Lett. to CSR].
52. SAI, Social Accountability International, "Overview of SA8000" in http://www.sa-intl.og/index.cfm?fuseaction=Page.viewPage&pageId=473 2005.11.15.
53. SACOM http://www.sacom.org.hk in CSR,1:34, p.6,Frost, S., "Disney's supply chain in China."
54. China Business News, 2005.09.23, in CSR, 1:39 p.6,FROST, S., "China View" and CSR, 1:40 p.8, and 1:44, p. 6, Frost, S., "China View" and the Nanfang Weekend. 2005.10.27.
55. HRC, 2002, p.8.
56. GUANG, L. & S. LU SHENG. 2005. "Migration as the Second Best Option Local Power and Off-farm Employment", in China Quarterly, Nr 181.

Applying Cost Analyses to Drive Policy That Protects Children

Mercury as a Case Study

LEONARDO TRASANDE,[a,b] CLYDE SCHECHTER,[c] KARLA A. HAYNES,[a] AND PHILIP J. LANDRIGAN[a,b]

[a]*Center for Children's Health and the Environment, Department of Community and Preventive Medicine, Mount Sinai School of Medicine, New York, New York 10029, USA*

[b]*Department of Pediatrics, Mount Sinai School of Medicine, New York, New York 10029, USA*

[c]*Department of Family Medicine, Albert Einstein College of Medicine, Bronx, New York 10461, USA*

ABSTRACT: Exposure in prenatal life to methylmercury (MeHg) has become the topic of intense debate in the United States after the Environmental Protection Agency (EPA) announced a proposal in 2004 to reverse strict controls on emissions of mercury from coal-fired power plants that had been in effect for the preceding 15 years. This proposal failed to incorporate any consideration of the health impacts on children that would result from increased mercury emissions. We assessed the impact on children's health of industrial mercury emissions and found that between 316,588 and 637,233 babies are born with mercury-related losses of cognitive function ranging from 0.2 to 5.13 points. We calculated that decreased economic productivity resulting from diminished intelligence over a lifetime results in an aggregate economic cost in each annual birth cohort of $8.7 billion annually (range: $0.7–$13.9 billion, 2000 dollars). $1.3 billion (range: $51 million–$2.0 billion) of this cost is attributable to mercury emitted from American coal-fired power plants. Downward shifts in intellectual quotient (IQ) are also associated with 1566 (range: 115–2675) excess cases of mental retardation (MR defined as IQ < 70) annually. This number accounts for 3.2% (range: 0.2–5.4%) of MR cases in the United States. If the lifetime excess cost of a case of MR (excluding individual productivity losses) is $1,248,648 in 2000 dollars, then the cost of these excess cases of MR is $2.0 billion annually (range: $143 million–$3.3 billion). Preliminary data suggest that more stringent mercury policy options would prevent thousands of cases of MR and billions of dollars over the next 25 years.

Address for correspondence: Leonardo Trasande, M.D., M.P.P., Center for Children's Health and the Environment, Department of Community and Preventive Medicine, Mount Sinai School of Medicine, One Gustave L. Levy Place, Box 1057, New York, NY 10029. Voice: 212-241-8029; fax: 212-996-0407.
  e-mail: leo.trasande@mssm.edu

KEYWORDS: methylmercury; mercury; cord blood; lost economic productivity; mental retardation; power plants; electrical generation facilities; environmentally attributable fraction

INTRODUCTION

Methylmercury (MeHg) is a developmental neurotoxicant.[1] Maternal exposure results principally from consumption of seafood contaminated by mercury released from anthropogenic (70%) and natural (30%) sources.[2] Coal-fired electricity generating plants are an important source of environmental mercury and in 1999 accounted for 41% of all anthropogenic mercury emissions in the United States.[3,4]

The toxicity of MeHg to the developing brain was first recognized in the 1950s in Minamata, Japan, where consumption of fish with high concentrations of MeHg by pregnant women resulted in at least 30 cases of cerebral palsy in children; exposed women were affected minimally if at all.[5] A similar episode followed in 1972 in Iraq when the use of a MeHg fungicide led to poisoning in thousands of people[6]; again, infants and children were most profoundly affected.[7,8] The vulnerability of the developing brain to MeHg reflects the ability of lipophilic MeHg to cross the placenta and concentrate in the central nervous system.[9] Moreover, the blood–brain barrier is not fully developed until after the first year of life, and MeHg can cross this incomplete barrier.[10]

Three recent, large-scale prospective epidemiologic studies have examined children who experienced MeHg exposures *in utero* at concentrations relevant to current U.S. exposure levels.[11–20] An assessment of these three prospective studies by the National Academy of Sciences (NAS)[21] concluded that there is strong evidence for the fetal neurotoxicity of MeHg, even at low concentrations of exposure. Moreover, the NAS opined that the most credible of the three prospective epidemiologic studies was the Faroe Islands investigation. In recommending a procedure for setting a reference dose for a MeHg standard, the NAS chose to use a linear model to represent the relationship between mercury exposure and neurodevelopmental outcomes, and based this model on the Faroe Islands data. The NAS found that the cord blood MeHg concentration was the most sensitive biomarker of exposure *in utero* and correlated best with neurobehavioral outcomes. The NAS was not deterred by the apparently negative findings of the Seychelles Islands study, which it noted was based on a smaller cohort than the Faroe Islands investigation and had only 50% statistical power to detect the effects observed in the Faroes.[21]

CHANGES IN MERCURY POLICY IN THE UNITED STATES

Throughout the 1990s the Environmental Protection Agency (EPA) made steady progress in reducing industrial mercury emissions. However, in January 2003, EPA announced a proposal to reverse strict controls on emissions of mercury from coal-fired power plants. This proposed "Clear Skies Act" would slow recent progress in controlling mercury emission rates from electric generation facilities and would allow these releases to remain as high as 26 tons per year through 2010.[22] By contrast, existing protections under the Clean Air Act will limit mercury emissions from coal-fired power plants to as low as 5 tons/year by 2008.[23] After legislative momentum for this proposal faded, the U.S. EPA proposed an almost identical Clean Air Mercury Rule that again failed to examine impacts on health. The U.S. EPA issued a final rule on March 15, 2005.[24] On June 24, 2005, EPA decided to initiate a reconsideration process for parts of the Rule in order to ensure ample opportunity for public comment.[25]

The U.S. EPA's technical analyses in support of "Clear Skies" failed to incorporate or quantify consideration of the health impacts resulting from increased mercury emissions.[26] Researchers at the Harvard School of Public Health and the Northeast States for Coordinated Air Use Management (NESCAUM) have estimated that the Clear Skies Act will save $75 million–$4.9 billion in economic productivity resulting from decreased cognitive toxicity and in healthcare costs resulting from reduced cardiovascular mortality.[27] However, this analysis failed to compare this saving with the greater savings that previous U.S. law (the Clean Air Act) would have provided.

In an effort to inform policy makers more fully about mercury policy choices, we have executed a series of analyses to describe the current health and economic consequences of mercury pollution. In this article, we outline our previous work, and we describe how we are using these analyses to estimate the future health and economic consequences of the Clean Air Mercury Rule, compared with other policy options under consideration. This approach is a model for similar assessments of the implications of environmental policy choices. Comparing the health and economic implications of policy choices can help government officials recognize the real and significant benefits that result from child-protective environmental policy.

METHYLMERCURY-INDUCED LOSS IN COGNITION AND ECONOMIC PRODUCTIVITY

To assess the disease burden and the costs due to MeHg exposure, we used an environmentally attributable fraction (EAF) model. The EAF approach was developed by the Institute of Medicine (IOM) to assess the "fractional contribution" of the environment to the causation of illness in the United States,[28] and it has been used to assess the costs of environmental and occupational disease.[29,30] It was used recently to estimate the environmentally attributable costs of lead poisoning, asthma, pediatric cancer, and neurodevelopmental disabilities in American children.[31] The EAF is defined by Smith et al.[32] as "the percentage of a particular disease category that would be eliminated if environmental risk factors were reduced to their lowest feasible concentrations." The EAF is a composite value and is the product of the prevalence of a risk factor multiplied by the relative risk of disease associated with that risk factor. Its calculation is useful in developing strategies for resource allocation and prioritization in public health. The general model developed by the IOM and used in the present analysis is the following:

$$Costs = disease\,rate \times EAF \times population\,size \times cost\,per\,case$$

"Cost per case" refers to discounted lifetime expenditures attributable to a particular disease, including direct costs of healthcare, costs of rehabilitation, and lost productivity. "Disease rate" and "population size" refer, respectively, to the incidence or prevalence of a disease and the size of the population at risk.

In applying the EAF model, we first reviewed the adverse effects of MeHg exposure. We then estimated the costs of those effects and subsequently applied a further fraction to parse out the cost of anthropogenic MeHg exposure resulting from emissions of American electrical generation facilities.

We decided to apply a no adverse effect level of 5.8 μg/L, based upon the epidemiologic evidence.[14–16] The Faroes study also found that effects on delayed brain stem auditory responses occurred at much lower exposure concentrations.[18] In its report, NAS concluded that the likelihood of subnormal scores on neurodevelopmental tests following *in utero* exposure to MeHg increased as cord blood concentrations increased from levels as low as 5 μg/L to the benchmark dose level (BMDL) of 58 μg/L.[21]

Recent data suggest that the cord blood mercury concentration may on average be 70% higher than the maternal blood mercury concentration,[33] and a recent analysis suggests that a modification of the EPA reference dose for MeHg be made to reflect a cord blood/maternal blood ratio that is greater than 1.[34] If the developmental effects of mercury exposure do, in fact, begin at 5.8 μg/L in cord blood, as suggested by the Faroes[14] and New Zealand[15,16] data and by the NAS report,[21] then effects would occur in children born to women of child-bearing age with blood mercury concentrations $\geq 3.41 (=5.8/1.7)$ μg/L.

To compute intellectual quotient (IQ) decrements in infants that have resulted from these elevated maternal mercury exposures, we used published data from the 1999–2000 National Health and Nutrition Examination Survey (NHANES) on percentages of women of child-bearing age with mercury concentrations ≥ 3.5, 4.84, 5.8, 7.13, and 15.0 μg/L.[35] We assumed conservatively that all mercury concentrations within each segment of the distribution were at the lower bound of the range, and that the probability of giving birth to a child did not correlate with maternal blood mercury level concentrations. In our base case analysis, we assumed that children born to women with mercury concentrations 3.5–4.84 μg/L suffer no loss in cognition, and that successive portions of the birth cohort experience loss of cognition associated with cord blood levels of 8.2, 9.9, 12.1, and 25.5 μg/L, respectively.

To assess the implications on our findings of a range of various possible ratios between maternal and cord blood mercury concentrations, we conducted a sensitivity analysis. In this analysis, we set as a lower bound for our estimate the costs to children with estimated cord blood concentrations ≥5.8 μg/L (assuming a cord/maternal blood ratio of 1), and assumed no IQ impact <4.84 μg/L (assuming a cord/maternal blood ratio of 1.19). This estimate assumed no loss of cognition to children born to women with mercury concentration <5.8 μg/L, and assumed that subsequent portions of the birth cohort experienced cord blood mercury concentrations of 5.8, 7.13, and 15 μg/L, respectively. In this scenario, we estimated no IQ decrement for children with blood mercury concentrations below 4.84 μg/L. We estimated decrements resulting from an incremental increase in blood mercury concentration from 4.84 μg/L to 5.8 μg/L in the population born with cord blood mercury levels between 5.8 μg/L and 7.13 μg/L, and also estimated decrements resulting from increases from 4.84 μg/L to 7.13 μg/L and 4.84 μg/L to 15 μg/L in the percentages of the population between 7.13 μg/L and 15 μg/L, and >15 μg/L, respectively. The results that follow from this calculation are expressed in our analysis as a lower bound for the true burden of mental retardation (MR) resulting from MeHg toxicity to the developing brain.

The Faroes study found that a doubling of mercury concentration was associated with adverse impacts on neurodevelopmental tests ranging from 5.69% to 15.93% of a standard deviation (SD).[36] Assuming that IQ is normally distributed with a SD of 15 points, a doubling of mercury concentration would be associated with a decrement ranging from 0.85 to 2.4 IQ points. The Faroes researchers used a structural equation analysis to estimate MeHg's impact on verbal and motor function at 7 years of age, and found an association between a doubling of blood mercury and loss of 9.74% of a SD on motor function and of 10.45% of a SD on verbal function.[11] This analysis suggests that a doubling in mercury concentration produces a decrement of approximately 10% of a SD, or 1.5 IQ points.

In the New Zealand study, the average WISC-R full-scale IQ for the study population ($n = 237$) was 93. In the group with maternal hair mercury above 6 μg/g (about fourfold higher than in the study population, $n = 61$) the average was 90.[16] This finding provides further support for our use of a loss of 1.5 IQ points for each doubling in our base case analysis. Confounders, such as polychlorinated biphenyls were found not to cause significant confounding.[11,20] In our sensitivity analysis, we therefore chose to set outer bounds of 0.85–2.4 IQ points per doubling, as described by the Faroes researchers.[36] In applying the EAF methodology, we assume that the relationship between cord blood mercury and IQ is relatively linear over the range of exposures studied (>5.8 μg/L).

In our sensitivity analysis, we used the same linear dose–response model that the National Research Council used to set a reference dose for mercury exposure even though the logarithmic model is a better fit for Faroes data (p = 0.06).[21] In our initial analysis, we applied a 0.59–1.24 IQ point decrement (average = 0.93) per microgram/L increase in cord blood mercury concentration, when the Faroes researchers actually found that, for those children whose mothers had hair mercury concentrations <10 μg/g, a 1 μg/L increase of cord blood mercury concentration was associated with adverse impacts on neurodevelopmental tests ranging from 0.395% to 0.833% of a SD, or 0.059–0.124 IQ points (average = 0.093 IQ points).[36] We also varied the cord/maternal blood mercury ratio from 1 to 1.7 in calculating IQ impact from the linear model as part of our sensitivity analysis.

In our revised sensitivity analysis, as an upper bound to our cost estimate, we calculated that children born to women with mercury concentrations below 4.84 μg/L suffer no loss in cognition, and that children born to women with concentrations of 4.84–5.8 μg/L, 5.8–7.13 μg/L, 7.13–15.0 μg/L, and >15.0 μg/L experience losses of cognition of 1.21, 1.84, 2.55, and 5.13 IQ points, respectively. The lower bound estimate assumed that children born to women with mercury concentrations below 5.8 μg/L suffer no loss in cognition, and that children born to women with concentrations of 5.8–7.13 μg/L, 7.13–15.0 μg/L, and >15.0 μg/L experience losses of cognition of 0.06, 0.14, and 0.60 IQ points, respectively.

In attributing mercury emissions to sources, we applied a 70% factor to convert the health and economic burden of MR resulting from MeHg exposure to the cost attributable to anthropogenic MeHg exposure. We next parsed out the proportion of anthropogenic MeHg in fish that arises from American sources, and then isolated the subset of that proportion that is emitted by coal-fired electrical generating plants. In our base case analysis, we applied a 36% factor (the weighted average of American sources of mercury content in fish) to specify the burden of anthropogenic MeHg exposure attributable to American sources. In our sensitivity analysis, we varied the factor used to convert the economic cost of anthropogenic MeHg exposure to the economic cost attributable to American sources from 18% (incorporating industry modeling of mercury deposition) to 36% (using federal data on mercury deposition). We also applied an additional fraction of 41% in our analysis to convert the burden of MR attributable to all American emissions to the burden attributable to American electric power generation facilities.[37] In applying an economic value to each lost IQ point, we applied estimates of lifetime economic productivity for children born in the United States in 2000 from Max et al.,[38] and applied a 1.931% decrement in lifetime earnings in boys, and a 3.255% decrement in lifetime earning in girls, ????.[39]

We found that between 316,588 and 637,233 babies are born each year in the United States with cord blood mercury levels >5.8 μg/L. These infants suffer mercury-related losses of cognitive function ranging from 0.2 to 5.13 IQ points. We calculated that this loss of cognitive function results in an aggregate economic cost in each annual birth cohort of $8.7 billion annually (range: $0.7–

$13.9 billion, 2000 dollars). Of this cost, $1.3 billion (range: $51 million–$2.0 billion) is attributable to mercury emitted from American coal-fired power plants.

METHYLMERCURY-ASSOCIATED MENTAL RETARDATION AND ASSOCIATED SOCIAL COSTS

MR is typically defined on the basis of an IQ below 70.[40] Because MeHg exposure shifts the distribution of IQ in an exposed population downward, the number of children with an IQ score below 70 is increased. We therefore applied our previous model of MeHg-associated loss in cognition and summed the number of excess MR cases among children born with cord blood mercury >5.8 μg/L. We assumed that IQ is normally distributed with a SD of 15.

To obtain percentages of MR attributable to mercury pollution sources, we divided the number of cases of MR attributable to each source by 49,030, our estimate of the number of mentally retarded children in the 2000 U.S. birth cohort, on the basis of a 1.2% prevalence rate.[41,42] To estimate the costs of MR due to each pollution source, we relied upon previously published estimates and applied a 3% discount rate to obtain present value in 2000; this yielded a cost per case estimate of $1,248,648, including direct medical costs.[43] This represents a potential overestimate because the Honeycutt et al.[43] estimate overrepresents the per case cost of MR rather than the excess per case costs associated with MR. Grosse et al. have estimated the excess per case costs to be $244,000 in 2000 dollars but this estimate does not include excess caregiving costs.[42] We apply a range of per case costs, from $244,000 to $1,248,648 for purposes of this article.

Our findings were that downward shifts in IQ resulting from prenatal exposure to MeHg of anthropogenic origin are associated with 1566 excess cases of MR annually, or 3.2% of MR cases in the United States. The costs of caring for these children amount to $2.0 billion/year. After incorporating uncertainties in the relationship of IQ loss with increases in blood mercury levels and applying a range for the true cord/maternal mercury ratio, we estimate that between 115 and 2675 excess cases of MR, or 0.2–5.4% of MR cases in the United States are associated with MeHg toxicity. Applying the sensitivity analysis, we estimate that the true cost of caring for children with MeHg-associated MR ranges between $28 million and $3.3 billion.[44]

After applying an additional fraction of 41% in our analysis to convert the burden of MR attributable to all American emissions to the burden attributable to American electric power generation facilities, we find that mercury from American power plants accounts for 231 cases of MR/year (range: 9–394), 0.5% (range: 0.02–0.8%) of all MR cases in the United States. The total annual costs of MR in children damaged *in utero* by mercury from U.S. power plants ranges between $2 and $43 million[44]

IMPLICATIONS FOR MERCURY POLICY OPTIONS IN AMERICA

Data on the costs of pediatric environmental disease have proven extremely useful in the direct comparison with the costs of other categories of illness, both in the setting of priorities and in allocation of resources.[45–50] However, during the policy debate about the Clear Skies Initiative, policy makers could not compare the health and economic implications of the two leading alternative proposals that were presented by U.S. Senators Jeffords and Carper.[51,52] The environmental implications of these alternative proposals were presented alongside the EPA Mercury Utility Rule in May 2004 by the U.S. Department of Energy,[51] yet EPA did not estimate the health benefits of the Clean Air Mercury Rule or compare the health consequences of more aggressive reductions in mercury emissions. In its analysis, EPA defended its approach to preventing prenatal methylmercury toxicity by arguing that more aggressive mercury emissions standards would cost industry an additional $1.3–2.1 billion (1999 dollars). The technical support documents also state that aggregate economic benefits of reduced IQ decrements from eliminating power plant-attributable mercury exposure in 2020 could not be higher than $210 million (1999 dollars).[53] Geyer and Hahn corroborate EPA's analysis, and estimate that the cost of more stringent mercury regulation as proscribed by the Clean Air Act vastly outweighs the economics benefit of the Clean Air Mercury Rule ($15.4–23.2 billion cost for $60–140 million benefit, 2004 dollars).[54]

The EPA's Clean Air Mercury Rule has been widely criticized as dangerous to the public's health, and the American Academy of Pediatrics, the American Public Health Association and a number of leading medical and public health organizations have joined 13 states in a lawsuit that would force the EPA to implement more stringent marcury emissions standards.[55] In their efforts to block the EPA's Clean Air Mercury Rule, they cite the evidence we have provided in the peer-reviewed literature of the health and economic consequences of MeHa toxicity. Rice and Hammitt have also confirmed taht efforts to reduce exposure to this potent neurotoxicant can be economically sensible. They estimated the economic benfits of the Clean Air Mercury Rule to range from $75 million-$4.9 billion (2000 dollars), but were unable to obtain the necessary data on mercury deposition to quantify the economic consequnces of more stringent regulation.[27]

We have also preliminarily compared the implications of the Clean Air Mercury Rule with alternative proposals that more aggressively reduce emissions. We first applied the Department of Energy estimates of mercury emissions,[51] and assumed that mercury content of American fish will be reduced in proportion to reductions in American emissions, while mercury content of imported fish will be reduced in proportion to reductions in international emissions of the Carper and Jeffords proposals[51,52] and the EPA. We assumed that international emissions will remain otherwise unchanged, and that international mercury emissions will be reduced by the amount reduced by the United States. We then calculated a weighted average of the reduction in mercury content of fish, using most recent data on the relative proportion of imported and American fish consumed in the United States.[56] In the same way as the Harvard/NESCAUM researchers did,[27] we then applied reductions in blood mercury levels among these highly exposed subgroups in proportion to reductions in mercury content of fish to estimate future mercury levels among the 15.7% most highly exposed women of child-bearing age over the years 2005–2025.

Our findings, which are to be presented on completion as part of a more rigorous cost-benefit analysis, are that, while the Clean Air Mercury Rule will likely prevent some 1475 cases of MR and save approximately $4.1 billion in lost economic productivity, special education, healthcare and other costs, more stringent reductions could prevent an additional 4450 cases of MR and save $13.1 billion. This additional cost will be borne by the American public, while the savings resulting from less stringent mercury regulation will be passed to the power plant industry. This preliminary data suggest that the Clean Air Mercury Rule should be reconsidered in light of new knowledge of the public health implications of mercury pollution. Increasing knowledge of the health and economic benefits of more aggressive reductions in mercury pollution has prompted the departments of environmental protection in muliple states to propose more stringent emissions on mercury emissions from coal-fired power plants. The EPA's Clean Air Mercury Rule relies upon companies buying and selling opportunities to emit mercury in the free market, and thus the result of large states, such as Pennsylvania, opting out of EPA's Clean Air Mercury Rule would render it meaningless. In an effort to undermine state efforts to limit the prenatal toxicity of MeHa, EPA has endeavored to discredit our peer-reviewed academic work by publishing a note on its Web site that makes a number of questionable assertions about our work to date,[57] rather than participate in an open forum so that members of the public can weigh the science in their own minds, and determine what interventions are justified to protect fetal brain development. We have invited the authors of this note to submit letters to the editor of the peer-reviewed journals where our findings were published, but they have thus far chosen not to do so. We hope that this and other manuscripts facilitate an open discussion about the correct approach that the federal government should take in its stewardship of the public health, and especially the health of our nation's children.

ACKNOWLEDGMENTS

We are grateful to Drs. Philippe Grandjean, Esben Budtz-Jorgensen, Danielle Laraque, Paul Leigh, and Judith Palfrey for their advice as well as to Drs. Dorothy P. Rice and Wendy Max for their assistance with lifetime earnings estimates.

This research was supported by a grant from the National Institute of Environmental Health Sciences (NIEHS P42 ES07384-07S1), and by the Jenifer Altman Foundation, Physicians for Social Responsibility, and the Rena Shulsky Foundation.

REFERENCES

1. GOLDMAN, L.R. & M.W. SHANNON, FOR THE AMERICAN ACADEMY OF PEDIATRICS COMMITTEE ON ENVIRONMENTAL HEALTH. 2001. Technical report: mercury in the environment: implications for pediatricians. Pediatrics **108:** 197–205.
2. UNEP. 2002. Global Mercury Assessment Report. New York. United Nations Environmental Programme. Available: http://www.chem.unep.ch/mercury/Report/GMA-report-TOC.htm [accessed 17 May 2004].
3. U.S. EPA. 2003. National Emissions Inventories for Hazardous Air Pollutants, 1999. Version 3, July 2003. Washington, DC.U.S. Environmental Protection Agency, Technology Transfer Network, Clearinghouse for Inventories and Emissions Factors. Available: http://www.epa.gov/ttn/chief [accessed 18 May 2004].
4. U.S. EPA. 2003. National Toxics Inventory, 1990, Version 0302, October 2003. U.S. Environmental Protection Agency, Technology Transfer Network, Clearinghouse for Inventories and Emissions Factors. Washington, DC: U.S. Environmental Protection Agency. Available: http://www.epa.gov/ttn/chief [accessed 18 May 2004].
5. HARADA, Y. 1968. Congenital (or fetal) Minamata disease. *In* Minamata Disease (Study Group of Minamata Disease), Eds.: 93–118. Kumamato, Kumamato University. Japan.
6. BAKIR, F., S.F. DAMLUJI, L. AMIN-ZAKI, *et al*. 1973. Methylmercury poisoning in Iraq. Science **181:** 230–241.
7. AMIN-ZAKI, L., S. ELHASSANI, M.A. MAJEED, *et al*. 1974. Intrauterine methylmercury poisoning in Iraq. Pediatrics **54:** 587–595.
8. AMIN-ZAKI, L. 1979. Prenatal methylmercury poisoning. Clinical observations over five years. Am. J. Dis. Child **133:** 172–177.
9. CAMPBELL, D., M. GONZALES & J.B. SULLIVAN. JR. 1992. Mercury. *In* Hazardous Materials Toxicology—Clinical Principles of Environmental Health. J.B. SullivanJR & G.R. Krieger, Eds. : 824–833. Williams and Wilkins. Baltimore, MD,
10. RODIER, P.M. 1995. Developing brain as a target of toxicity. Environ. Health Perspect. **103**(Suppl 6): S73–S76.
11. BUDTZ-JORGENSEN, E., N. KEIDING, P. GRANDJEAN & P. WEIHE. 2002. Estimation of health effects of prenatal methylmercury exposure using structural equation models. Environ. Health **1:** 2.
12. GRANDJEAN, P., E. BUDTZ-JORGENSEN, R.F. WHITE, *et al*. 1999. Methylmercury exposure biomarkers as indicators of neurotoxicity in children age 7 years. Am. J. Epidemiol. **150:** 301–305.
13. GRANDJEAN, P., K. MURATA, E. BUDTZ-JORGENSEN & P. WEIHE. 2004. Cardiac autonomic activity in methylmercury neurotoxicity: 14-year follow-up of a Faroese birth cohort. J. Pediatr. **144:** 169–176.
14. GRANDJEAN, P., P. WEIHE, R.F. WHITE, *et al*. 1997. Cognitive deficit in 7-year-old children with prenatal exposure to methylmercury. Neurotoxicol. Teratol. **19:** 417–428.

15. KJELLSTROM, T., P. KENNEDY, S. WALLIS & C. MANTELL. 1986. Physical and Mental Development of Children with Prenatal Exposure to Mercury from Fish National Swedish Environmental Protection Board. Stage I: Preliminary Tests at Age 4. Report 3080. Solna, Sweden.
16. KJELLSTROM, T., P. KENNEDY, S. WALLIS, et al. 1989. Physical and Mental Development of Children with Prenatal Exposure to Mercury from Fish. National Swedish Environmental Protection Board.Stage II: Interviews and Psychological Tests at Age 6. Report 3642. Solna, Sweden.
17. LANDRIGAN, P.J. & L. GOLDMAN. 2003. Prenatal methylmercury exposure in the Seychelles [letter]. Lancet 362: 666.
18. MURATA, K., P. WEIHE, E. BUDTZ-JORGENSEN, et al. 2004. Delayed brainstem auditory evoked potential latencies in 14-year-old children exposed to methylmercury. J. Pediatr. **144**: 177–183.
19. MYERS, G.J., P.W. DAVIDSON, C. COX, et al. 2003. Prenatal methylmercury exposure from the ocean fish consumption in the Seychelles child development study. Lancet **361**: 1686–1692.
20. STEUERWALD, U., P. WEIHE, P.J. JORGENSEN, et al. 2000. Maternal seafood diet, methylmercury exposure and neonatal neurological function. J. Pediatr. **136**: 599–605.
21. National Research Council. 2000. Toxicological Effects of Methylmercury. National Academy Press. Washington, DC.
22. U.S. EPA. 2004A. EPA Proposes Options for Significantly Reducing Mercury Emissions from Electric Utilities U.S. Environmental Protection Agency. Washington, DC. Available: http://www.epa.gov/air/mercuryrule/hg_factsheet1_29_04.pdf [accessed May 7, 2004].
23. U.S. EPA. 2004b. Clean Air Act. U.S. Environmental Protection Agency. Washington, DC. Available: http://www.epa.gov/oar/oaq_caa.html [accessed 17 May 2004].
24. U.S. EPA. 2005a. Clean Air Mercury Rule: Basic Information. U.S. Environmental Protection Agency. Washington, DC. Available: http://www.epa.gov/air/mercuryrule/basic.htm [accessed 1 April 2005].
25. U.S. EPA. 2005b. Controlling Power Plant Emissions: Decision Process and Chronology. U.S. Environmental Protection Agency. Washington, DC. Available: http://epa.gov/mercury/control_emissions/decision.htm [accessed 1 August 2005].
26. U.S. EPA. 2004c. U.S. Environmental Protection Agency. Section B: Human and Environmental Benefits. 2002 Technical Support Package for Clear Skies. Washington, DC. Available: http://www.epa.gov/clearskies/tech_sectionb.pdf [accessed 17 May 2004].
27. NORTHEAST STATES FOR COORDINATED AIR USE MANAGEMENT. 2005. Economic Valuation of Human Health Benefits of Controlling Mercury Emissions from U.S. Coal-Fired Power Plants. 2005. Boston, MA. Available at http://bronze.nescaum.org/airtopics/mercury/rpt050315mercuryhealth.pdf [Accessed 26 August 2005.]
28. OM (INSTITUTE OF MEDICINE). 1981. Costs of Environment-Related Health Effects: A Plan for Continuing Study. National Academy Press. Washington, DC.
29. FAHS, M.C., S.B. MARKOWITZ, E. FISCHER, et al. 1989. Health costs of occupational disease in New York State. Am. J. Ind. Med. **16**: 437–449.
30. LEIGH, J.P., S. MARKOWITZ, M. FAHS, et al. 1997. Costs of occupational injuries and illnesses. Arch. Intern. Med. **157**: 1557–1568.

31. LANDRIGAN, P.J., C.B. SCHECHTER, J.M. LIPTON, et al. 2002. Environmental pollutants and disease in American children: estimates of morbidity, mortality, and costs for lead poisoning, asthma, cancer, and developmental disabilities. Environ. Health Perspect. **110:** 721–728.
32. SMITH, K.R., C.F. CORVALIN & T. KJELLSTROM. 1999. How much global ill health is attributable to environmental factors? Epidemiology **10:** 573–584.
33. STERN, A.H. 2005. A revised probabilistic estimate of the maternal methyl mercury intake dose corresponding to a measured cord blood mercury concentration. Environ. Health Perspect. **113:** 155–163.
34. STERN, A.H. & A.E. SMITH. 2003. An assessment of the cord blood-maternal blood methylmercury ratio: implications for risk assessment. Environ. Health Perspect. **111:** 1465–1470.
35. MAHAFFEY, K.R., R.P. CLICKNER & C.C. BODUROW. 2004. Blood organic mercury and dietary mercury intake: National Health and Examination Survey, 1999 and 2000. Environ. Health Perspect. **112:** 562–570.
36. JORGENSEN, E.B., F. DEBES, P. WEIHE & P. GRANDJEAN. 2004. Adverse Mercury Effects in 7 Year-Old Children as Expressed as Loss in "IQ." University of Southern Denmark. Odense. Available: http://www.chef-project.dk/PDF/iq04louise5.pdf [accessed 15 May 2004].
37. TRASANDE, L., C. SCHECHTER & P.J. LANDRIGAN. 2005. Public health and economic consequences of environmental methylmercury toxicity to the developing brain. Environ. Health Persp. **113:** 590–596.
38. MAX, W., D.P. RICE, H.-Y. SUNG & M. MICHEL. 2002. Valuing Human Life: Estimating the Present Value of Lifetime Earnings, 2000. San Francisco: Institute for health & Aging.
39. SALKEVER, D.S.. 1995. Updated estimates of earnings benefits from reduced exposure of children to environmental lead. Environ. Res. **70:** 1–6.
40. AMERICAN ASSOCIATION ON MENTAL RETARDATION. 2002. Mental retardation: definition, Classification, and Systems of Supports, 10th ed. American Association on Mental Retardation.Washington, DC.
41. US. CENTERS FOR DISEASE CONTROL AND PREVENTION, NATIONAL VITAL STATISTICS SYSTEM. 2004a. Infant mortality statistics from the 2000 period linked birth/infant death data set. National Vital Statistics Report **50:** 1–27. Available at http://www.cdc.gov/nchs/data/nvsr/nvsr50/nvsr50_12. pdf [Accessed 1 November 2004.]
42. CDC. ECONOMIC COSTS ASSOCIATED WITH MENTAL RETARDATION, CEREBRAL PALSY, HEARING LOSS, AND VISION IMPAIRMENT—UNITED STATES. 2003. *MMWR* 2004; **53:** 57–59.
43. HONEYCUTT, A., L. DUNLAP, H. CHEN & G. AL HOMSI. 2000. The Cost of Development Disabilities. Task Order No. 0621-09. Revised Final Report. Research Triangle Park, NC: Research Triangle Institute. Strike 40.
44. TRASANDE, L., C. SCHECHTER, K.A. HAYNES & P.J. LANDRIGAN. 2006. Mental retardation and prenatal methylmercury toxicity. Am. J. Ind. Med. **49:** 153–158.
45. FAHS, M.C., J. MANDELLBLATT, C. SCHECHTER & C. MULLER. 1992. The cost effectiveness of cervical cancer screening for the elderly. Ann. Intern. Med. **17:** 520–527.
46. EXECUTIVE OFFICE OF THE PRESIDENT. 1997. Office of Science and Technology Policy. Investing in Our Future: a National Research Initiative for America's Children for the 21st Century. The White House.Washington, DC.

47. DE KONING, H.J., B.M. VAN INEVELD, G.J. VAN OORTMARSSEN, et al. 1991. Breast cancer screening and cost-effectiveness: policy alternatives, quality of life considerations and the possible impact of uncertain factors. Int. J. Cancer **49:** 531–537.
48. CARLSON, J.E. 1995. Environmental policy and children's health. Future Child **5:** 34–52.
49. TAYLOR, W.R. & P.W. NEWACHECK. 1992. Impact of childhood asthma on health. Pediatrics **90:** 657–662.
50. ARROW, K.J., M.L. CROPPER, G.C. EADS, et al. 1996. Is there a role for benefit-cost analysis in environmental, health, and safety regulation? Science **272:** 221–222.
51. U.S. GOVERNMENT PRINTING OFFICE. 2005a. S.150, 109th Congress, Clean Power Act of 2005. Available at http://thomas.loc.gov/cgi-bin/bdquery/z?d109:s.00150: [Accessed 26 August 2005.]
52. U.S. GOVERNMENT PRINTING OFFICE. 2005b. S. 843, 108th Congress, Clean Air Planning Act of 2003. http://thomas.loc.gov/cgi-bin/bdquery/z?d108:s.00843: [Accessed 26 August 2005.]
53. US EPA. Cost and Energy Impacts–Technical Support Document, Clean Air Mercury Rule. Available at: www.epa.gov/ttn/atw/utility/cost-TSD-112.pdf [Accessed 1 May 2006].
54. GAYER T., & R.W. HAHN. Costs and Benefits of Regulating Mercury. AEI-Brookings Joint Center Policy Matters 05-32, November 2005. Available at http://www.aei-brookings.org/policy/page/php? id = 235 [Accessed 1 May 2006.]
55. KUEHN, B.M. Medical Groups Sue EPA Over Mercury Rule. JAMA 2005; **294:** 415–46.
56. National Marine Fisheries Service, National Oceanic and Atmospheric Administration. 2004. Fisheries of the United States - 2002. Available at http://www.st.nmfs.gov/st1/fus/current/2002-fus.pdf (Accessed 24 May 2004).
57. EPA, National Center for Environmental Economics. A Note on Trasande et al., "Public Health and Economic Consequences of Methylmercury Toxicity to the Developing Brain." Authored by Charles Griffiths, Al McGartland and Maggie Miller. Available at yosemite.epa.gov/ee/epa/eed.nsf/WPNumberNew/2006-02?OpenDocument[Accessed 1 July 2006].

ABSTRACTS OF POSTER PRESENTATIONS

Spectrum of organophosphorus poisoning in Lok Nayak Hospital

S.K. Agrarwal, V. Sahni, A.K. Kapoor, India
dr_skag@yahoo.com

Aim: Demography, clinical manifestations, outcomes of organophosphorus poisoning. Methodology: Medical records (June 2004–May 2005). Results: 67 patients, mostly (39) 15–30 years. Suicide–most common motive (41). Malathion–most common agent (21). Baygon (12), Diazinon (10), Fenthion (8), Propoxur (6), Carbaryl (2). All had muscarinic symptoms–pulmonary edema, diarrhea, miosis, incontinence. Forty-two had nicotinic symptoms. Intermediary syndrome in 16. Delayed neuropathy in 2. Supportive treatment + atropine ± Pralidoxime (Fenthion, Diazinon). Eight expired, rest recovered. Conclusions: OP poisoning—young adults, suicide—most common motive. Intermediary syndrome in 25%, delayed polyneuropathy in 3–4%.

The legal debate over cancer data and the public right-to-know in a petrochemical region

B.L. Allen, USA
ballen@vt.edu

A current lawsuit and set of appeals over the control and dissemination of cancer data by a U.S. government-mandated cancer collection agency has illuminated ambiguities in the rules and regulations governing public access to cancer data. The dispute began when a group of citizens and researchers wanted access to the Louisiana Tumor Registry's data by zip code as they had anecdotal evidence of elevated cancer rates in communities adjacent to petrochemical plants. Analysis of the legal documents reveals each side's scientific argument in support of their case, but behind the rhetoric are possibly political and economic motivations.

Effects of maternal 4-tert-octylphenol exposure on reproductive tract of male rats at adulthood

M. Aydogan, N. Barlas, Turkey
barlas@hacettepe.edu.tr

In the present study, effects of octylphenol (OP), on male rats were investigated. Rats were treated with OP maternally at doses of 0, vehicle (corn oil), 100, and 250 mg/kg/day. After birth, male rats were allowed to grow until adulthood and then testes and reproductive organs were investigated. At the end of the study, treatment with OP-induced atrophy of seminiferous tubules and prostate glands. Although there were no differences in sperm counts among treatment groups, it has been observed that abnormal sperm percentages in treatment groups increased considerably.

Assessment of DNA damage in workers occupationally exposed to hazardous chemicals

N. Başaran, U. Ündeğer, M. Shubair, A. Kars, A.F. Zorlu, Turkey
nbasaran@hacettepe.edu.tr

Antineoplastic drugs and pesticides are some of the toxic chemicals that may pose potential hazards to occupationally exposed workers. DNA damage in the peripheral lymphocytes of 30 professional oncology nurses and 33 pesticide sprayers was evaluated by the alkaline single-cell electrophoresis "comet" assay. The results were compared to that of controls with comparable age, sex, and smoking habits. The DNA damage observed in the lymphocytes of nurses and pesticide workers was significantly higher than the controls ($P < 0.001$). The observed DNA damage was found to be significantly lower ($P < 0.001$) both in nurses and workers applying the necessary individual safety protections.

Urinary 6-sulfatoxymelatonin concentration and exposure to extremely low frequency electromagnetic fields: a methodological approach

M. Benedetti, P. Comba, M. Nordio, Italy
marta.benedetti@iss.it

Exposure to extremely low-frequency (ELF) magnetic and electric fields has been associated with a number of adverse health effects. The mechanism by which exposure to ELF magnetic fields may cause such effects is by reducing or suppressing the normal nocturnal rise in melatonin. We have recently developed guidelines for assessing the health status of subjects resident close to sources of ELF magnetic fields. Measurement of 6-sufatoxymelatonin, the main melatonin metabolite in daytime and nighttime urine samples is part of this protocol, together with assessment of immune function, heart rate variability, neuropsychological, and psychiatric disorders.

In utero exposure to phthalates and organophosphate pesticides: a prospective Israeli study

T. Berman, U. Wormser, Y. Amitai, E. Richter, Israel
Tr75@pob.huji.ac.il

Recent studies have shown that low-level *in utero* exposure to organophosphate pesticides and phthalates may increase risks for adverse birth effects. We will measure phthalate and organophosphate exposure in pregnant women in Israel, identify primary exposure sources, and determine whether *in utero* exposure to these contaminants is associated with adverse birth effects. We will analyze urine samples collected during late pregnancy for organophosphate and phthalate metabolites using gas chromatography and interview women on their diet, occupation, and use of household pesticides and phthalate-containing products. We will obtain anthropometric measurements (height, weight, head circumference) and genital measures at birth.

High birth prevalence of congenital anomalies in two petrochemical areas of Sicily, Italy

F. Bianchi, S. Bianca, Italy
Fabrizio.bianchi@ifc.cnr.it

Two epidemiological studies on malformed newborns over the period 1990–2002 were recently performed in two petrochemical areas of Sicily (Italy). Data on total and specific congenital anomalies were significantly higher compared with other reference rates. Significant excesses concerned defects of male external genitalia, cardiac septa and great vessels, upper limb reduction. The observed birth prevalence of hypospadias (2.5 times higher than the reference data in both areas) is among the most elevated ever reported in literature. Results reinforce the hypothesis of a causal role of risk factors present in these areas on the etiology of malformations. A case–control study is in progress.

Multi-system disease and various cancers in nonsmoking teachers: Is chalk dust an occupational risk?

E. Bitchatchi, O. Levy, E.D. Richter, E. Fireman, R. Baruch, Israel
elir@cc.huji.ac.il

We report 76 self-referred nonsmoking teachers (age range 40–59 years) with prolonged exposure to chalk dust (SiO_2, As, Cd, Cr, Ni, Pb, Mn, Li, Al, and radioactivity) with multisystem disease ($n = 30$), lung diseases other than cancer ($n = 33$), various autoimmune disorders ($n = 14$), renal disease ($n = 4$), and various cancers ($n = 35$). Air sampling studies were inconclusive. We suggest investigating the hypothesis that chalk exposures may account for these diseases.

Medical monitoring for early detection of latent disease

N. Brautbar, B.S. Levy, USA
brautbar@aol.com

In occupational and environmental health, medical monitoring consists of examinations, tests, and procedures designed for early detection of latent disease. For persons with specific hazardous exposures to toxic or carcinogenic chemicals, medical monitoring is mandated by governmental agencies or recommended by health professionals, including sometimes by medical experts in civil action lawsuits. Courts in several states in the United States have accepted the principle of medical monitoring for those at increased risk of latent disease. We will present considerations in developing a medical monitoring protocol and provide examples of medical monitoring programs.

Precautionary worker health and safety for emerging technologies

K. Burns, A. Hawes, A. Michaels, J. Sass, USA
kmb@sciencecorps.org

The precautionary principle is a health-protective approach that should incorporate stages of research and development of new materials and processes. This is necessary to minimize introduction of hazardous chemicals in the workplace and environment. Impediments to precautionary actions include emphasis in research on established protocols, lack of relevant information and training, lack of incentives for the use of least hazardous chemicals, and minimal incentives to devise safe production methods or produce nontoxic products. Policy and educational strategies to address these limitations will be discussed.

Toward a coherent approach to human biomonitoring in Europe: Action 3 of the EU Environment & Health Action Plan 2004–2010

L. Casteleyn, B. Van Tongelen, Belgium
ludwine.casteleyn@lin.vlaanderen.be

In accordance with the Environment and Health Strategy adopted by the European Commission and Action 3 of the EU Action Plan, several member states confirmed their interest in developing a coherent approach to human biomonitoring. Strongly supported by European Parliament and the European Economic and Social Committee, the ultimate aim is to support environmental and public health policy by better data comparability and accessibility and more effective use of resources through shared development of scientific tools and appropriate strategies. The European Commission, supported by a multidisciplinary working group of member states' representatives is preparing a pilot project for the end of 2006.

Array-based differential methylation hybridization pattern of sodium arsenite-treated keratinocytes

B.C. Christensen, C.J. Marsit, K.T. Kelsey, USA
bchriste@hsph.harvard.edu

Aberrant methylation of tumor suppressor CpG islands is known to be common in cancer. The mechanism responsible for inducing gene silencing remains unclear. Epidemiologic data have confirmed that the human carcinogen arsenic is associated with epigenetic silencing in tobacco-related solid tumors. To investigate the role of exposure to arsenic plays in inducing these aberrations, we assayed global methylation, using array-based differential methylation hybridization in keratinocytes treated with 10 ppb sodium arsenite and compared treated cells to untreated keratinocytes. Highly reproducible and specific global decreases in methylation were seen in >5500 CpG loci while similar increases in methylation were seen in 166 CpG loci. These results suggest that arsenic plays a role in epigenetic alterations common in human cancer.

Presence of asbestos-containing materials in enterprises in Belgium

H. De Raeve, G. Moens, J. Van Bouwel, Belgium
hilde.deraeve@idewe.be

All employers are obliged to make an inventory of all asbestos-containing materials (ACM) in buildings. From October 29, 2002 till July 8, 2004, IBEVE evaluated 402 buildings resulting in 2053 locations with possible ACM. In 26.5% no asbestos was present. In the remaining locations the asbestos found was mainly chrysotile (90.4%). Of these 1840 locations with asbestos, the surface was damaged in 43.2%. The risk for normal use was nonexistent in 83.5%, small in 11.7%, considerable in 1.1%, and unacceptable in 0.1%. The risk during maintenance was nonexistent in 26.6%; small in 51.5%; considerable in 15.5%; and unacceptable in 0.3%.

Living with chemicals: RAGE in the community

J.A. Dix, B. Hillery, USA
hilleryb@oldwestbury.edu

Citizens living on or near toxic sites are stakeholders who must have a voice in identification, mitigation, and remediation of the toxins; otherwise, they become unwitting participants in future epidemiological studies. The Residents Action Group of Endicott NY (RAGE) was formed in October 2002 and soon trichloroethene and other volatile organic compounds (VOCs) were found in residents' homes. VOCs entered the homes via vapor intrusion from a decades-old toxic plume from a nearby IBM manufacturing plant. Sub-slab ventilation systems were subsequently installed in over 450 homes. We chronicle RAGES's role in the mitigation and remediation of the toxic plume.

Comparison of patella lead, blood lead, and tibia lead as predictors of neurobehavioral test scores in South Korean lead workers

Winner of the Irving J. Selikoff and Cesare Maltoni student poster competition
C.D. Dorsey, B.K. Lee, K.I. Bolla, V.M. Weaver, S.S. Lee, G.S. Lee, A.C. Todd, B.S. Schwartz, USA
chimes@jhsph.edu

Few studies have examined trabecular bone lead, an estimate of the bioavailable bone lead and its relation to neurobehavioral test scores. We performed a cross-sectional analysis of the relations of patella, blood, and tibia lead with measures of neurobehavioral and peripheral nervous system function in 652 lead workers. Patella lead was found to be significantly associated with poorer performance on 4 of 19 tests. Blood lead predicted poorer performance on 6 and tibia lead on 5 tests. We conclude that in this study, measurement of patella lead was of less utility in predicting cognitive effects in exposed workers.

Cause-specific mortality study of a population exposed to 50 Hz magnetic fields in a municipal district of Rome

L. Fazzo, M. Grignoli, I. Iavarone, A. Polichetti, M. De Santis, V. Fano, F. Forestiere, S. Palange, R. Pasetto, N. Vanacore, P. Comba, Italy
comba@iss.it

This work was conducted under contract with ISS – APAT. A cohort of 357 subjects resident near 60 kV power lines was studied. Exposure to a 50 Hz magnetic field was evaluated both by modeling and by measures with EMDEX LITE apparatus. Mortality was studied from 1980 through 2003. All-cause mortality complied with expected figures (SMR 0.99, 40 observed). A significant increase of all cancer mortality appeared after 30 years duration of residence (SMR 2.09, 8 observed). A significant increase of pancreatic cancer was observed in the sub-cohort with the highest levels of magnetic fields (SMR 17.56, 3 observed). Morbidity is now being studied in the same cohort.

What is the role of oxidative stress in the mechanism of particulate matter toxicity?

J.R. Froines, A.K. Cho, A. Nel, USA
jfroines@ucla.edu

A major development in understanding the pulmonary and cardiovascular effects from airborne particles has been the elucidation of their ability to induce oxidative stress and proinflammatory effects in the respiratory and cardiovascular tissues. Reactive chemical species in PM initiate oxidative stress responses. These chemical species can be organic or inorganic and act via redox and electrophilic reactions. Assays have been developed to measure this activity. PM has been shown to induce oxidative stress by increasing antioxidant enzymes, decreasing glutathione, and disrupting mitochondrial function. The results provide a roadmap for the identification of the most important characteristics of PM responsible for health effects and form a scientific basis for regulatory intervention.

How many excess deaths? Reanalysis of the updated mortality study in a petrochemical plant producing vinyl chloride (VC) and polyvinyl chloride (PVC) located in Porto Marghera (Venice)

V. Gennaro, M. Ceppi, F. Montanaro, Italy
Valerio.gennaro@istge.it

The mortality of 1658 males (248 deaths) was reanalyzed by Poisson regression, adjusting for main variables. We calculated the relative risks (RR) and the 95% CI for PVC baggers, PVC compound, autoclave, and other blue-collar workers. Technicians and clerks ($n = 202$), were used as a reference. Overall we found a 55% statistically significant increased mortality for all causes of death (RR $= 1.55$; $n = 229$ deaths). Increased mortality (ns) for all tumors (RR $= 1.42$; 118 cases) and other causes were detected also in each specific subgroup of workers. We practically confirm the number of deaths in excess (81) of those previously reported.

Physico-chemical properties influencing biological response to carbon nanotubes

G. Greco, I. Fenoglio, M. Tomatis, J. Muller, D. Lison, B. Fubini, Italy
m.tomatis@unito.it

Carbon nanotubes (CNT) are a newly discovered form of crystalline carbon of widespread use. The toxicity of CNT is still being debated. In the present article we investigated two physico-chemical properties that are known to modulate the toxicity of inhaled particles, the degree of surface hydrophilicity, and the generation of free radicals. We found that the different hydrophobic character of ground and unground nanotubes is in agreement with the observed responses *in vitro* and *in vivo*. We also report the unexpected finding that instead of releasing free radicals, as most toxic particulates, nanotubes act as quenchers.

Determination of radon indoor concentration in some areas of Ospedale Maggiore (hospital) of Bologna

G. Guidarelli, P. Dovesi, D. Tovoli, E. Melecchi, D. Breveglieri, Italy
Guiseppe.guidarelli@ausl.bo.it

Italian national regulations require the monitoring of the concentration of indoor radon in those workplaces prone to a significant presence of gas, setting the action level for remediation activities at an air concentration of 500 Bq/m^3. During 12 months, 15 underground areas of the Ospedale Maggiore of Bologna were monitored using CR39 track dosimeters. The annual mean concentration in every place resulted well below the action level (41 Bq/m^3) and within the interval of expected concentration in plain areas of Emilia Romagna, with a maximum value of 228 Bq/m^3 in a single 4-month period.

Putting the precautionary principles to work at work

A. Hawes, K. Burns, A. Michaels, J. Sass, CA, USA
ahawes@alexanderlaw.com

Workplace standards for carcinogens and reproductive toxins are much weaker than environmental standards. Nonprotective PELs, inadequate controls, poor hazard communication, and few incentives to use safe alternatives put workers at unacceptable risk. Semiconductor manufacturing uses many notorious toxics. IBM's Mortality File reflects disturbing patterns of cancer mortality. Many "clean room" workers have had children with serious birth defects or suffered excess pregnancy loss. California Assembly Bill 815 closes the gap between workplace and environmental "PELs." AB 816 accesses chemical manufacturers' customer lists, facilitating education and enforcement. Sound economic analyses show how significant savings can be made from such proactive health policies.

Identification of work-related asbestos disease in a Canadian community

M. Keith, J. Brophy, Canada
margkeith@yahoo.com

Sarnia is a major center for the Canadian petrochemical industry. A record number of workers have been diagnosed with asbestos-related disease. Government reports noted some of the highest levels asbestos dust ever measured. Conventional occupational health and safety approaches had failed the asbestos-exposed workers and their families. To address this problem, innovative research methods were employed. Health and exposure data were gathered using body and hazard mapping techniques. These worker-centered participator research methods facilitated broad community and national awareness, the establishment of a workers' clinic, and the eventual awarding of over $20 million in related workers' compensation benefits.

Molecular characterization of skin tumors induced by PUVA therapy

L. Lambertini, K. Surin, T.T. Ton, N. Clayton, J.K. Dunnick, Y. Kim, H.L. Hong, T.R. Devereaux, R.C. Sills, USA
lambertini@ramazzini.it

Nonmelanoma skin cancer, including squamous cell carcinoma (SCC), is the most common malignant neoplasm in Caucasians. SCC incidence occupation, amounting to 42.7 cases per 100,000 in the United States with an increasing trend. Moreover, an increase in the relative risk of SCC, ranging from 10 to 20, has been detected in patients who have been administered the widely used PUVA phototherapy to treat chronic skin diseases. Skin hyperplasias and SCCs in PUVA-treated hairless mice have been used to investigate sensitivity and specificity of p53 and PCNA protein expression assessment together with p53 gene mutations evaluation as possible biomarkers of PUVA-induced SCC.

The use of induced sputum (IS) in the evaluation of occupational and environmental lung diseases

Y. Lerman, E.M. Fireman, Israel
ylerman@post.tau.ac.il

We present the usefulness of IS in occupational lung diseases assessment. BAL (bronchoalveolar lavage) and IS yielded similar particles size distribution (PSD) and similar chemical composition of particles in silica and hard metal workers. IS from exposed workers contains a higher percentage of larger particles than nonexposed workers, which were positively correlated with FEV1/FVC. IS from NY firefighters demonstrated inflammation, PSD, and particle composition that was different from controls and consistent with WTC dust exposure. IS is a safe noninvasive technique in the field of research and diagnosis of occupational lung diseases.

The Alien Tort Claims Act: a remedy for chemical genocide

R. Metzger, USA
rmetzger@toxictorts.com

Companies that sell banned chemicals to Third World countries have escaped liability under the "forum non conveniens" doctrine. Such cases may not be dismissed if they are brought under The Alien Tort Claims Act (ATCA), which creates a "civil action by an alien for a ... violation of the law of nations ...." The Convention for the Prevention and Punishment of the Crime of Genocide defines genocide broadly to include causing serious bodily harm or preventing births of members of a national or ethnic group. Predicating an ATCA case on genocide may provide a remedy for foreign nationals against companies whose toxic chemicals harm them.

Fighting fire with fire: using economic arguments in support of worker protection

A. Michaels, A. Hawes, K. Burns, USA
Amichael@email.sjsu.edu

Strong economic pressures exist to use hazardous chemicals. These economic pressures for short-term profits may be countered with persuasive economic arguments phrased in terms of long-term gains. The costs of implementing engineering controls for a commonly used carcinogenic solvent are more than 100 times cheaper than the costs of doing nothing, that is the cost of healthcare for exposed workers and the cost of cleaning up environmental contamination. Such arguments were used in support of California Assembly Bill 815, which was created to close a policy gap that allowed workplace exposure to carcinogens to occur, despite Proposition 65 protections.

Mesotheliomatogenic effect of fluoro-edenite, a new calcium amphibole: final results

F. Minardi, F. Belpoggi, L. Lambertini, D. Degli Esposti, M. Soffritti, Italy
crcfr@ramazzini.it

Following the detection of an increased mortality for malignant pleural mesotheliomas (4 observed versus 0.9 expected) in the town of Biancavilla (about 20,000 inhabitants), near the Etna volcano (Sicily, Italy) in the period 1990–1993, a new fibrous calcium amphibole, fluoro-edenite, was identified as a possible cause. In order to prove the causal relationship, groups of male and female Sprague–Dawley rats were treated by a single intraperitoneal or intrapleural injection at a dose of 25 mg of fibrous or prismatic fluoro-edenite in 1 cc of H_2O. The results indicate that fibrous fluoro-edenite induces peritoneal and, to a much lesser extent, pleural mesothelioma. No mesotheliomas were observed in the animals treated with prismatic fluoro-edenite.

First-hand experience of early childhood chemical exposure and disability

E.M.T. O'Nan, USA
pace@mcdowell.main.nc.us

Individuals severely injured by chemicals during childhood are often disabled for life. However, the many challenges that children with chemical disabilities must overcome are frequently overlooked. These include being denied an education; treatment by physicians who believe chemical injuries to be psychological in origin; and having their families torn apart. Educational and medical records, interviews, and published independent studies showing health and social ramifications, demonstrate these unheeded and especially detrimental effects. It is imperative that these neglected consequences compel adequate precautionary provisions, preventing predictable and exponential increases in chemical injuries and ensuring correct treatment and invaluable disability assistance for the chemically disabled.

Nefarious obstructions to scientific goals

E.M.T. O'Nan, USA
pace@mcdowell.main.nc.us

Nefarious obstructions to scientific goals include both dishonorable and illegal activities. Current laws, studies, and news reports reveal fraudulent concealment of hazardous waste, inappropriate influence by private corporate powers and disinformation from corporate nonprofit front groups and public relations firms. These miscreant activities prevent and discourage science that is needed to avoid chemical injuries through reduced exposures, regulation of dangerous products, and assessment of liability. Full disclosure will allow objective scientific inquiry without corruption by conflicted interests. Scientists may then advocate precautionary principles that will protect health and environment and allow long-denied assistance and treatment for the chemically injured.

Review of occupational exposures to single chemicals or homogeneous groups

B. Papaleo, L. Caporossi, M. De Rosa, A. Pera, Italy
b.papaleo@ispesl.it

We draw a general picture of studies to date on specific occupational exposures to single chemicals (bisphenal A and styrene) or homogeneous groups (pesticides, metals, phthalates). Although the exposure occurs in different ways, the toxic mechanisms of action vary widely and it is hard to establish precisely the conditions of occupational exposure, significant correlations are nevertheless evident between the potential dose and its effects. Investigations to date have focused on the effects on the reproductive system, in males in particular. The effects on other endocrine organs—particularly the thyroid gland—and on the immune and neurological systems are required.

A multi-criteria and fuzzy logic–based approach for the relative assessment of chemical substances hazards

A.N. Paralikas, A.I. Lygeros, Greece
aparal@tee.gr

A new approach for the assessment of the relative hazard chemicals is introduced, treating the issue as a multi-criteria decision-making problem, with often more than one parameter or criteria being taken into account. Based on this approach, a number of hazard indices have been developed, employing a multi-criteria technique. These include the Analytic Hierarchy Process, Fuzzy

Logic, the Substance Fire Hazard Index, the Substance Toxicity Hazard Index, and the Consequences Index, with emphasis given to the major accident hazards of the substances. The challenges and limitations of using the multi-criteria approach in developing relative ranking indices are also discussed.

Damage during chronic intoxication with 2,3,7,8-tetrachloro-p-dibenzodioxin (TCDD)

D. Pelclova, Z. Fenclova, J. Preiss, J. Skrha, M. Pranzy, J. Spacil, Z. Duska, P. Urban, E. Lukas, Czech Republic
Daniela.pelclova@LF1.cuni.cz

In the years between 1965 and 1968, 80 workers in a herbicide production plant became ill with chloracne. They belong to the most severely exposed groups of workers in the world (estimated TCDD average of approximately 5,000 pg/g plasma fat). Among 15 remaining subjects with the highest TCDD levels, five fell ill several decades later with myocardial infarction, one with brain stroke. In 2004, all still had residues of chloracne, retinal angiosclerosis, and chronic conjunctivitis, 87% with hyperlipoproteinemia and neurasthenic syndrome, 60% with hypertension, 20% with porphyrinuria, and 13% with organic psychosyndrome.

Exposures and respiratory health effects from indoor air pollution on women in Mysore, India

P. Rechkemmer, G. Ramachandran, P. Pai, A. Maynard, USA
pandrese@jhsph.edu

A detailed questionnaire and spirometry test was administered to 80 women in Mysore, India in order to assess health effects of cooking fuels. A subset of 30 women was selected for 24-hour gravimetric PM2.5 personal and indoor exposure monitoring. Results revealed that kerosene users had almost twice the personal and indoor exposures to PM2.5 as LPG users. No statistically significant differences were found in the lung function values between kerosene and LPG users. Socioeconomic status, age, and season were significant predictors of cooking fuel choice. However, both kerosene and LPG users spent about the same amount on cooking fuel each month.

Malthusian pressures, environmental impacts, and genocide

E.D. Richter, T. Berman, R. Moses, Hadassah Jerusalem, Israel
tb75@pob.huji.ac.il

Timelines of genocides in Bosnia and Kosovo, (>200,000 dead), Rwanda, (1,000,000 dead), and Darfur (>300,000 dead) indicate that (1) Malthusian-type conflicts over land and water can trigger genocidal scenarios; (2) ethnic and racial conflicts can trigger genocidal scenarios without Malthusian-type conflicts; (3) such conflicts themselves can produce Malthusian scenarios by creating zero-sum or lose–lose situations; and (4) delays in addressing "upstream determinants" of genocide adversely impact on public health, carrying capacity, and sustainability. But "upstream determinants" cannot be addressed without stopping genocide. International early warning and response systems are needed. WHO/ECEH/(Rome 1999 Global Ecological Integrity and Sustainable Development http://www.euro.who.int/document/gch/ecorep)

Genetic consequences of environmental factors—Effects on fathers (ex. Hg vapors +X-ray): investigation of the second generation

I.V. Santoskly, Russia
sanotskly@ixv.comcor.ru

Methods: Male Wistar rats exposed by Hg in concentrations of 15; 4; 1; 0.02 mg/m^{-3}. Half of animals – irradiated (300 r). Female rats – intact. Mortality of control posterity 25–30%. Results: Hg (on "fathers") protected 1PG from infection (0.02 mg/m^3 – absolute protection). Effect on 2PG–mirror reflection of 1 PG (however 0.02 mg/m^3–total protection). Irradiation of 1PG – "fathers" (with Hg) decrease of 1PG – mortality depending on Hg concentration (however 0.02 mg/m^3–absolute protection). 2PG sensitivity had no change. Hg microinfluence gave no protection.

U.S. regulation of atrazine: taking care of business

J.B. Sass, A. Colangelo, USA
jsass@nrdc.org

We provide an overview of U.S. pesticide regulations and then address the evaluation of the herbicide atrazine. Atrazine disrupts hormone function in test animals and may cause cancer. Atrazine is a ubiquitous water contaminant and was recently banned in the European Union. However, the United States continues to allow unfettered atrazine use. By reviewing governmental memos and emails we demonstrate significant political influence on the U.S. atrazine assessment and suggest that unlawful negotiations between the government and the manufacturer of atrazine, Syngenta, may have influenced not only the regulation of atrazine, but also the characterization of its toxicology.

Hematological and cytogenetic biomarkers in industrial workers exposed to cytotoxic drugs in relation to exposure and hygiene conditions

J. Shaham, Y. Lerman, Israel
yshaham@bezequint.net

Our study was conducted to investigate hematological (WBC) and cytogenetic (SCE) biomarkers in relation to exposure to cytotoxic drugs and hygiene conditions. The study population included 174 industrial workers. Adjusted mean levels of WBC, lymphocytes, monocytes, and neutrophiles were lower and SCE was higher among the exposed group compared to the unexposed group. Among the exposed group, significant elevation of the adjusted mean WBC and neutrophiles and significant reduction of the adjusted mean of SCE were found after improvement of hygiene conditions. We suggest that CBC and SCE can be used as biomarkers of exposure and in the surveillance for hygiene changes.

Carbon disulfide-induced changes in cytoskeleton protein content of rat central nerve tissues

F. Song, S. Yu, X. Zhao, K. Xie, Jinan, China
Fysong3707@163.com

To investigate the mechanism of carbon disulfide–induced neuropathy, cerebrums and spinal cords of carbon disufide–intoxicated rats and their age-matched controls were Triton extracted and centrifuged at a high speed to yield a pellet fraction of NF polymer and a corresponding supernatant fraction. The contents of five cytoskeletal proteins in both fractions were then determined by immunoblotting. Results show that in both fractions of spinal cord and cerebrum, the contents of the neurofilament subunits decreased significantly and furthermore, a trend of increasing microtubule proteins was found. These findings suggested that changes of cytoskeleton protein might involve the development of carbon disulfide neurotoxicity.

Toward an integrated approach to children and adult consumers' exposure to chemicals

Steenhout, Belgium
asteen@ulb.ac.be

Consumers are exposed to numerous substances through various routes both indoors and outdoors. The article addresses harmonization and nomenclature issues encountered in preparing a database(*) on lifestyle, time-budget, consumption, age- and gender-related determinants, etc. in European populations including sensitive and vulnerable groups. However, harmonization is not integration. The question is how to keep the number of variables manageable without loss of addressing complexity, sensitivity, and precautionary issues to consumers? The article develops criteria for exposure scenarios, examining how and how many scenarios can be generic given the variety of situations, variability, uncertainty, and the context of an integrated, causal, environment, and health approach. *Acknowledgment to DEFIC-LRI grant.

Effects of low-frequency electro-magnetic fields on circadian rhythms of some blood parameters in Sprague–Dawley rats

A.C. Stelletta, L. Contalbrigo, D. Esposti, D. Falcioni, M. Lauriola, M. Padovani, G. Piccione, M. Morgante, C. Guiliani, F. Belpoggi, Italy
Laura.contalbrigo@unipd.it

Since low-frequency electro-magnetic fields are supposed to influence the ionic membrane exchange, they may also produce some metabolic changes in the normal activity of organism cells. In the framework of a mega-experiment aimed to evaluate the long-term effects, in particular carcinogenic effects, a satellite experiment was performed to determine the interactions between SEMF-50 Hz and circadian rhythms of some blood parameters of energetic metabolism (glycemia, total cholesterol, and triglycerides) in *Rattus norvegicus*. The results of this study highlight that the circadian rhythms of these energetic metabolism parameters are inverted in rats exposed to 50 Hz magnetic sinusoid fields with an intensity of 1000 uT and 100 uT.

Job histories: occupational exposure to carcinogens in patients with cancer, and compensation in France

Thebaud-Mony, France
thebaud@vjf.inserm.fr

In 2 years (2002–2004), the job histories of 250 new patients of three hospitals (Seine-Saint-Denis, France) suffering lung cancer or mesothelioma have been reconstituted, of which 82% of men and 60% of women have been exposed to occupational carcinogens, especially in construction and metal workers. Because of the restrictive rules of the French compensation system, 45% of the exposed patients were considered not eligible for compensation. The discussion emphasizes the gap between the rules of compensation for occupational cancer and the reality of patients' job histories.

Organic solvents: occupational exposure of mothers as a risk factor for congenital malformation (CM) in children

G.I. Tikhonova, T.A. Tkacheva, Russia
tkacheva@ixv.comcor.ru

A case-control study was carried out during 8 follow-up years. Cases: 550 newborns with CM who died during the perinatal period. 1778 controls were randomly selected out of healthy newborns born at the same time. Parents were interviewed (demographic and social data, lifestyle, smoking, medical history of the family). After stratification analysis, odds ratio (OR) for

CM was 2.15 (95% CI 1.73–2.69) when one of the parents was occupationally exposed. After organic solvents exposure of mothers, OR = 3.51 (17 "case group", 16 -"control") (95% CI 1.8–6.9). Pathological heredity and low socioeconomic status exacerbate the negative impact of occupational exposure.

Integration of quality of life with survival for comparative health risk assessment

J. Wang, Taiwan
jdwang@ntu.edu.tw

By adjusting survival function with mean quality of life (QOL) at every time point t and then summing up throughout life span, we come up with healthy life expectancy with a common unit of quality-adjusted life year (QALY). When the utility of QOL is replaced by psychometry, QALY becomes score-year. Both can be used to assess the potential health impacts of different policies in priority setting. Three examples in occupational and environmental health are demonstrated: enforcement of helmet law in Taipei City, the contamination of underground water by chlorinated hydrocarbons from an electronic factory, and offspring of female lead workers.

Medical management guidelines for adults exposed to lead

R.P. Wedeen, USA
wedeen@umdnj.edu

The AOEC convened an expert panel to produce management guidelines for adults exposed to lead. Controversy centered on setting an upper limit for long-term exposure. Based on evidence of adverse effects on blood pressure, the kidneys, neurological performance, and the fetus, a blood lead of 10 mg/dL is indicated as a safe upper limit for long-term exposure. Some physicians on the panel appeared to be influenced by outdated OSHA regulations that guide their practice. Five years of deliberation by the panel demonstrate how the lead industry's influence delays effective public health policy by exploiting the inherent uncertainty of scientific evidence.

Ecojustice and the rights of the first generation and the future

L. Westra, Canada
lwestra@interlog.com

Children are exposed to violent attacks on their physical integrity and their normal function, even long before they are born. They are also far more vulnerable than adults to environmental assaults and exposures because of their different physiology and their specific health needs. Children should be considered the first generation when future generations' rights are protected. Thus, environmental justice from both the moral and legal point of view, does not encompass only North/South issues in its present synchronic aspect as justice among peoples, but it also has even stronger implications from the diachronic point of view as the human race itself appears to be at stake.

Index of Contributors

Abu-Zahra, H., 765–777
Alavanja, M.C.R., 343–354
Alex Merrick, B., 707–717
Ashizawa, A.E., 829–838
Atkinson, M.A.L., 281–291
Autrup, H., 678–690

Balbus, J.M., 331–342
Beck, M., 765–777, 871–881
Belpoggi, F., xvii–xviii, 559–577, 578–591, 736–752
Benvenuti, A., 366–377
Bermejo, J.L., 137–148
Bhuva, U.B., 292–308
Bianchi, F., 449–461
Bingham, E., xvii–xviii, 394–404
Biró, A., 635–648
Boden, L.I., 858–870
Brajer, V., 882–892
Brautbar, N., 753–764
Brophy, J.T., 765–777
Bruno, C., 778–783
Bua, L., 736–752
Buratti, M., 405–420

Cameron, W., 394–404
Campo, L., 405–420
Cao, D-Z., 129–136
Cavallo, D., 405–420
Cicolella, A., 784–789
Cirla, P.E., 405–420
Cogliano, V.J., 592–600
Comba, P., 449–461, 778–783
Connor, T.H., 615–623
Costantini, A.S., 366–377
Costello, S., 378–387

Davis, J.M., 498–515
De Garbino, J.P., 657–659
De Rosa, C.T., 829–838
Dearwent, S.M., 439–448
Degli Esposti, D., 578–591
Dement, J., 394–404
Denison, R.A., 331–342

Dodson, R.F., 281–291
Draper, G., 318–330

Englund, A., 388–393
Esposti, D.D., 559–577

Falcioni, L., 736–752
Falk, H., 439–448
Fazzo, L., 449–461
Fenske, R.A., 355–365
Florini, K., 331–342
Foà, V., 405–420
Fodor, Z., 635–648
Försti, A., 137–148
Froneberg, B., 607–614
Funez, A., 355–365
Fustinoni, S., 405–420

Gabel, E., 753–764
Georgopoulos, P., 54–79
Germolec, D.R., 718–727
Gibbons, C., 80–89
Gilbertson, M., 765–777
Glass, D.C., 80–89
Godfrey, G., 439–448
Gorey, K.M., 765–777
Gorman, T., 871–881
Gray, C.N., 80–89
Guan, J-R., 129–136

Hämeilä, M., 628–634
Hannan, L.M., 29–53
Harari, H., 660–677
Harari, R., 660–677
Harjula, H., 462–477
Hay, A., 790–799
Haynes, K.A., 911–923
Hellyer, D., 765–777
Hemminki, K., 137–148
Hicks, H.E., 829–838
Hoel, D.G., 309–317
Hoffmann, B., 253–265
Hoppin, J.A., 343–354

Infante, P.F., 90–109
Irish, R.S., 355–365

Jakab, M., 635–648
Järvholm, B., 421–428
Järviluoma, E., 628–634
Jöckel, K.-H., 253–265
Jemal, A., 29–53
Jin, X-P., 129–136
Johnson, L.D., 478–485
Jolley, D.J., 80–89
Joshi, T.K., 292–308

Karagas, M.R., 810–821
Katoch, P., 292–308
Keith, M.M., 765–777
Kelsey, K.T., 810–821
Klupp, T., 635–648
Kroll, M., 318–330

Lambertini, L., 578–591
Landrigan, P.J., xvii–xviii, 657–659, 911–923
Laukkanen, E., 765–777
Lauriola, M., 559–577
Li, G., 800–809
Lioy, P.J., 54–79
London, S.J., 343–354
Longhi, O., 405–420
Lu, C., 355–365
Luginaah, I., 765–777
Lynch, C.F., 343–354

Magyar, B., 635–648
Major, J., 635–648
Malcolm, C., 871–881
Manley, R.G., 540–548
Mantovani, A., 239–252
Marcello, I., 839–857
Marsit, C.J., 810–821
Martina, L., 449–461
Martinotti, I., 405–420
Martuzzi, M., 449–461
Maticka-Tyndale, E., 765–777
Matoušek, J., 549–558
McDiarmid, M.A., 601–607
McGowan, W., 394–404

Mead, R.W., 882–892
Mehlman, M.A., xvii–xviii, 110–119, 120–128, 822–828
Menegozzo, M., 449–461
Michaels, D., 149–162
Miligi, L., 366–377
Miller, R.L., 15–28
Minardi, F., 559–577
Minichilli, F., 449–461
Minoia, C., 649–656
Mitis, F., 449–461
Mumtaz, M.M., 439–448, 829–838
Musmeci, L., 449–461

Navasumrit, P., 678–690
Newman, B., 657–659

Olden, K., 703–706

Padovani, M., 559–577
Patterson, R.M., 718–727
Perera, F., 15–28
Pizzuti, R., 449–461
Pohl, H.R., 829–838
Pott, F., 266–280

Quinn, P., 394–404

Rabl, A., 516–526
Ramos, K.S., 728–735
Rauh, V., 15–28
Regev, L., 753–764
Reinhartz, A., 765–777
Ringen, K., 388–393, 394–404
Ritz, B., 378–387
Robinson, M., 871–881
Rodríguez, T., 355–365
Roller, M., 266–280
Ruchirawat, M., 678–690
Rudén, C., 191–206
Ruder, A.M., 207–227
Rushbrook, P., 486–497

Sandler, D.P., 343–354
Santoro, M., 449–461
Santos-Burgoa, C., 624–627
Schechter, C., 911–923